ELECTRO-OPTICS HANDBOOK

Ronald W. Waynant Editor

Food and Drug Administration
Rockville, Maryland

Marwood N. Ediger Editor

Food and Drug Administration
Rockville, Maryland

McGRAW-HILL, INC.

New York San Francisco Washington, D.C. Auckland Bogotá
Caracas Lisbon London Madrid Mexico City Milan
Montreal New Delhi San Juan Singapore
Sydney Tokyo Toronto

Library of Congress Cataloging-in-Publication Data

Electro-optics handbook / Ronald W. Waynant, editor, Marwood N.
  Ediger, editor.
      p.   cm.   —   (Optical and electro-optical engineering series)
      Includes index.
      ISBN 0-07-068663-7
      1. Electrooptical devices—Handbooks, manuals, etc.   I. Waynant,
  Ronald W.   II. Ediger, Marwood N., date.   III. Series.
  TA1750.E44   1993
  621.36—dc20
                                                  93-14691
                                                  CIP

1 2 3 4 5 6 7 8 9 0   DOC/DOC   9 9 8 7 6 5 4 3

ISBN 0-07-068663-7

*The sponsoring editor for this book was Daniel A. Gonneau, the editing
supervisor was Paul R. Sobel, and the production supervisor was Donald
F. Schmidt. It was set in Times Roman by Techna Type.*

Printed and bound by R. R. Donnelley & Sons Company.

This book is printed on acid-free paper.

To our wives and families who tolerated this project and to our colleagues and friends with whom we have enjoyed this field

CONTENTS

Chapter 16. Visible Detectors *Suzanne C. Stotlar*   16.1

Chapter 17. Infrared Detectors *Suzanne C. Stotlar*   17.1

Chapter 18. Imaging Detectors *Frederick A. Rosell*   18.1

Chapter 19. Holography *Tung H. Jeong*   19.1

Chapter 24. Lasers in Medicine   *Ashley J. Welch and M. J. C. van Gemert*   24.1

Chapter 25. Material Processing Applications of Lasers
James T. Luxon   25.1

Chapter 26. Optical Integrated Circuits   *Hiroshi Nishihara, Masamitsu Haruna,*
and Toshiaki Suhara   26.1

Chapter 27. Optoelectronic Integrated Circuits   *Osamu Wada*   27.1

CONTRIBUTORS

Georg F. Albrecht, *Lawrence Livermore National Laboratory, Livermore, California* (CHAP. 5)

George R. Carruthers, *E. O. Hulburt Center for Space Research, Naval Research Laboratory, Washington, D.C.* (CHAP. 15)

Y. L. Chen, *Department of Electrical Engineering, University of Maryland, College Park, Maryland* (CHAP. 22)

James J. Coleman, *Microelectronics Laboratory, University of Illinois, Urbana, Illinois* (CHAP. 6)

Charles M. Davis, *Centerville, Virginia* (CHAP. 21)

J. G. Eden, *Department of Electrical Engineering, University of Illinois, Champaign, Illinois* (CHAP. 20)

Marwood N. Ediger, *Food and Drug Administration, Rockville, Maryland* (CHAP. 1)

T. J. Harris, *Applied Physics Laboratory, Johns Hopkins University, Laurel, Maryland* (CHAP. 11)

Mashamita Haruna, *Department of Electronic Engineering, Osaka University, Osaka, Japan* (CHAP. 26)

P. T. Ho, *Joint Program for Advanced Electronic Materials, Department of Electrical Engineering, University of Maryland, College Park, Maryland* (CHAPS. 9, 22)

Michael Ivanco, *Atomic Energy of Canada Limited, Chalk River Laboratories, Chalk River, Ontario* (CHAP. 7)

Tung H. Jeong, *Chairman, Department of Physics, Lake Forest College, Lake Forest, Illinois* (CHAP. 19)

S. B. Kim, *Department of Chemistry, California Institute of Technology, Pasadena, California* (CHAP. 20)

Chi H. Lee, *Joint Program for Advanced Electronic Materials, Department of Electrical Engineering, University of Maryland, College Park, Maryland* (CHAP. 9)

James T. Luxon, *Associate Dean, Graduate Studies, Extension Services and Research, GMI Engineering and Management Institute, Flint, Michigan* (CHAP. 25)

Sharon Miller, *Food and Drug Administration, Rockville, Maryland* (CHAP. 2)

Hiroshi Nishihara, *Department of Electronic Engineering, Osaka University, Osaka, Japan* (CHAP. 26)

John A. Pasour, *Mission Research Corporation, Newington, Virginia* (CHAP. 8)

Stephen A. Payne, *Lawrence Livermore National Laboratory, Livermore, California* (CHAP. 5)

Martin Peckerar, *Nonelectronic Processing Facility, Naval Research Laboratory, Washington, D.C.* (CHAP. 22)

Jack C. Rife, *Condensed Matter and Radiation Sciences Division, Naval Research Laboratory, Washington, D.C.* (CHAP. 10)

Paul A. Rochefort, *Atomic Energy of Canada Limited, Chalk River Laboratories, Chalk River, Ontario* (CHAP. 7)

G. Rodriguez, *Everitt Laboratory, University of Illinois, Urbana, Illinois* (CHAP. 20)

Frederick A. Rosell, *Westinghouse Electric Corporation, Defense and Space Center, Baltimore, Maryland* (CHAP. 18)

Roland Sauerbrey, *Department of Electrical and Computer Engineering and Rice Quantum Institute, Rice University, Houston, Texas* (CHAP. 3)

William T. Silfvast, *Center for Research in Electro-Optics and Lasers, Orlando, Florida* (CHAP. 4)

Edward J. Sharp, *Department of the Army, U.S. Army Research Laboratory, Fort Belvoir, Virginia* (CHAP. 13)

David H. Sliney, *Department of the Army, U.S. Army Environmental Hygiene Agency, Edgewood, Maryland* (CHAP. 23)

Suzanne C. Stotlar, *Yorba Linda, California* (CHAPS. 16, 17)

Toshiaki Suhara, *Department of Electronic Engineering, Osaka University, Osaka, Japan* (CHAP. 26)

M. E. Thomas, *Applied Physics Laboratory, Johns Hopkins University, Laurel, Maryland* (CHAP. 11)

W. J. Tropf, *Applied Physics Laboratory, Johns Hopkins University, Laurel, Maryland* (CHAP. 11)

Carlton M. Truesdale, *Corning Industries, Corning, New York* (CHAP. 12)

M. J. C. van Gemert, *College of Engineering, The University of Texas at Austin, Austin, Texas* (CHAP. 24)

Osamu Wada, *Deputy Manager, Fujitsu Laboratories, Limited, Optical Semiconductor Devices Laboratories, Atsugi Kanagawa, Japan* (CHAP. 27)

Ronald W. Waynant, *Food and Drug Administration, Rockville, Maryland* (CHAP. 1)

Ashley J. Welch, *College of Engineering, The University of Texas at Austin, Austin, Texas* (CHAP. 24)

Gary L. Wood, *Director, Center for Night Vision and Electro-Optics, Department of the Army, U.S. Army Research Laboratory, Fort Belvoir, Virginia* (CHAPS. 13, 14)

Li Yan, *Department of Electrical Engineering, University of Maryland, Baltimore, Maryland* (CHAP. 9)

Clarence J. Zarobila, *Optical Technologies, Incorporated, Herndon, Virginia* (CHAP. 21)

PREFACE

Our concept for a new handbook on electro-optics integrates sources, materials, detectors and on-going applications. The field of electro-optics now encompasses both incoherent optical sources and lasers that operate from the millimeter wavelength region to the x-ray region. In this handbook we provide coverage of the most important laser sources in this wavelength range. Having chosen a broad range of wavelengths from our sources, we then define the properties of the materials through which these sources might travel. From there we consider the detectors that might be used to observe them. When all the components have been covered, we consider the applications for which electro-optical systems can be used.

The applications for electro-optics systems is growing at a phenomenal rate and will most likely do so for the next fifty years or more. Applications range from the astronomical to the microscopic. Laser systems can track the moon and detect small quantities of atmospheric pollution. Laser beams can trap and suspend tiny bacteria and help measure their mechanical properties. They can be used to clip sections of DNA. The applications that we have included in this handbook are only the beginning of applications for this field.

This handbook is intended as a reference book. It can be used as a starting place to learn more about sources, materials, detectors and their use and applications. Most chapters have a considerable list of references to original research articles, or else refer to books that contain such lists of references. Liberal use is made of tables of data and illustrations that clarify the text. The authors are all experts in their fields.

We make no statement that this handbook is complete although it was our goal to work toward complete coverage of this field. It is a dynamic field continually advancing and changing. We hope to follow these changes and to strive for further completeness in future editions. We believe electro-optics will be part of a new field with new ways of transferring knowledge. We hope to use these new fields to find additional ways to present data and knowledge that will be even more comprehensive.

We are indebted to Daniel Gonneau of McGraw-Hill for suggesting this project and then providing the encouragement and motivation to see it through. As editors we are grateful to the authors who made great sacrifices to complete their contributions and who made our job quite pleasant. We hope that references are made to the authors and their sections because it is with these authors that the knowledge presented here really resides. We would be remiss not to mention Paul Sobel for his help and encouragement during the finishing stages of this book and to thank Eve Protic for her help during the many stages of production.

Ronald W. Waynant
Marwood N. Ediger

ACRONYMS

2DEG	two dimensional electron gas	CBE	chemical beam epitaxy
2PA	two photon absorption	CCD	charge coupled device
III-V	Group III, group V of periodic table	CDRH	Center for Devices and Radiological Health (of FDA)
3HG	third harmonic generation	Ch	choroid
AEL	accessible emission limit	CID	charge-injection device
AM	amplitude modulation	CIE	Commission International de l'Eclairage
ANSI	American National Standards Institute	COD	catastrophic optical damage
AO	acousto-optic	CPM	colliding-pulse mode-locked
APD	avalanche photodiode		
APDs	avalanche photodiodes	cw	continuous wave
APM	additive pulse mode locking	D*	detectivity
		DBR	distributed Bragg reflector
AR	anti reflection	DCG	dichromated gelatin
ASE	amplified spontaneous emission	DFB	distributed feedback
		DFDL	distributed feedback dye lasers
BEFWM	Brillouin enhanced four wave mixing	DIN	Deutsche Institüt für Normung
BH	buried heterostructure	DM	depth of modulation
BLIP	background-limited infrared performance	DODCI	diethyloxadicarbon-cyanine iodide
C/S	coupler/splitter	DOES	double heterostructure optoelectronic switches
CAD	computer-aided design		
CAIBE	chemically assisted ion-beam etching	DoF	depth of focus
		DUT	device under test
CARS	coherent anti-stokes Raman spectroscopy	EA	electron affinity
		EB	electron beam

EBCCD	electron bombarded charge coupled device		GRO	Gamma Ray Observatory
EBS	electron bombardment silicon		GSMBE	gas source molecular beam epitaxy
ECL	emitter-coupled logic		GVD	group velocity dispersion
EKE	electronic Kerr effect		GVDC	group velocity dispersion compensation
EL	exposure limits			
EMI/ESD	electromagnetic impulse/electrostatic discharge		HAZ	heat-affected zone
			HbO	oxyhemoglobin (blood)
EO	electro-optic		HBT	heterojunction bipolar transistors
ESA	excited state absorption			
FAFAD	fast axial flow with axial discharge		HEAO	High Energy Astronomy Observatory
			HEMTs	high-electron-mobility transistors
FDA	Food and Drug Administration		HID	high intensity discharge
FEL	free electron laser			
FELs	free electron lasers		HOE	holographic optical element
FET	field-effect transistors			
FFT	fast Fourier transform		HpD	hematoporphyrin derivative
FGC	focusing grating coupler		HR	high reflection
FHD	flame hydrosis deposition		HR	high resistivity
			HUD	head-up display
FID	free-induction decay		IC	integrated circuit
FM	frequency modulation		ICI	International Commission on Illumination
FOV	field of view			
FTFTD	fast transverse flow with transverse discharge		IDT	interdigital transducer
			IEC	International Electrotechnical Commission
FTP	Fourier transform plane			
FWHM	full width half-maximum		ILD	injection laser diodes
			IO	image orthicon
G-R	generation-recombination		IODPU	integrated optic disk pickup
GRIN-SCH	graded index waveguide separate confinement heterostructure		IOSA	integrated optic spectrum analyzer
			IPC	imaging proportional counter

ir	infrared	MO	magneto-optic
JFETs	junction FETs	MOCVD	metal organic chemical vapor deposition
KTP	KTiOPO$_4$, potassium tellurium phosphate	MOPA	master oscillator power amplifier
LANs	local area networks	MOS	metal-oxide-semiconductor
LAVA	laser assisted vascular anastomosis	MOVPE	metal organic vapor phase epitaxy
LDV	laser Doppler velocimeter	MPE	maximum permissible exposure
LED	light emitting diode	MPI	multiphoton ionization
lidar	light detection and ranging	MQW	multiple quantum well
LIF	laser induced fluorescence	MSM	metal semiconductor metal
LIS	laser isotope separation	MTBF	mean time between failure
LiTaO$_3$	lithium tantalate	MTF	modulation transfer function
LLLTV	low light level television	NA	numerical aperture
LLNL	Lawrence Livermore National Laboratory	NALM	nonlinear amplifying loop mirror
LM	light microscopy	NEP	noise-equivalent power
LPE	liquid phase epitaxy	NHZ	nominal hazard zone
LSI	large scale integration	NIST	National Institute of Standards and Technology
LSO	laser safety officer	nm	nanometers = 10^{-9} meters
LTE	local thermal equilibrium	NO	nitric oxide
LURE	Laboratoire pour l'Utilisation du Rayonment Electromagnetic	NO	non-linear optic
μm	micrometers (microns) = 10^{-6} meters	NOHA	nominal ocular hazard area
MAMA	multianode microchannel array	NRL	Naval Research Laboratory
MBE	molecular beam epitaxy	OD	optical density
MCP	microchannel plate	OEIC	optoelectronic integrated circuits
MES	metal semiconductor	OIC	optical integrated circuit
ml	mode-locked	OKE	orientational Kerr effect
MMIC	monolithic microwave integrated circuit		

OODR	optical-optical double resonance	RIBE	reactive-ion-beam etching
OPD	optical path difference	RIE	reactive ion etching
OPO	optical parametric oscillator	RIKES CARS	Raman-Induced Kerr-effect spectroscopy
OSSE	Oriented Scintillation Spectrometer Experiment	RIMS	resonance ionization mass spectroscopy
PAC	photoactive compounds	RPM	resonant passive mode-locking
PC	photoconductive	SAFAD	slow axial flow with axial discharge
PDT	photodynamic therapy	SAW	surface acoustic waves
PE	pigment epithelium	SBN	strontium barium nitrate
PES	photoelectron spectroscopy	SBS	stimulated Brillouin scattering
PFL	pulse forming line	SEBIR	secondary electron bombardment-induced response
PGC	phase-generated carrier		
PLL	phase-locked-loop	SEC	secondary electron conduction
PM	polarization maintaining	SEED	self-electro-optic effect devices
PMMA	polymethyl methacrylade	SELFOC	self-focusing
PMT	photomultiplier tube	SEVA	slowly varying envelope approximation
PPCM	passive phase conjugate mirror	SHG	second harmonic generation
PVF	polyvinyl fluoride		
PWS	port wine stains	SI	semi-insulating
PZT	lead zirconate	SIT	silicon intensified tube
PZT	piezoelectric transducer	SLB	super lattice buffer
QE	quantum-effect	SLD	superluminescent diodes
QW	quantum-well		
RC	resistance-capacitance	SLM	single longitudinal mode
RE	rare earth	SMF	spectral matching factor
REC	rare earth cobalt	SNR_D	signal to noise ratio of a display
REMPT	resonantly enhanced multiphoton ionization	SNR_{DT}	signal to noise ratio of a display at threshold
RGH	rare gas halide		

SNR_{VO}	signal to noise ratio of video (for white noise)	TO	thermo-optic
		TPF	two-photon fluorescence
SPM	self-phase modulation		
		TVL	threshold limit values
SQW	single-quantum-well	TVL/PH	television lines/picture height
SRS	stimulated Raman scattering		
		uv	ultraviolet
TCDD	tetra chlorodibenzo-*p* dioxin	VCO	voltage-controlled oscillator
TCE	trichloroethane	VCSEL	vertical cavity surface emitting lasers
TEA	transversely excited atmospheric		
		VLSI	very large scale integration
TEA	trienthylamine		
TEM	transmission electron microscopy	VLSIs	very large scale integrated circuits
TGFBS	twin-grating focusing beam splitter	VSTEP	vertical to surface transmission electrophotonic
TGS	triglycerine sulfide		
TGSe	triglycerine selanate	vuv	vacuum ultraviolet
THG	third harmonic generation	WDM	wavelength division multiplexing
TIR	total internal reflection	XPM	cross phase modulation
TMAE	tetraKis-(dimethylamino) ethylene	YAG	yttrium aluminum garnet
TMAH	trimethylaluminum hydride	YLF	$LiYF_4$, lithium yittrium fluoride

CONTENTS IN BRIEF

CHAPTER 1
INTRODUCTION
TO ELECTRO-OPTICS

Ronald W. Waynant and Marwood N. Ediger

1.1 INTRODUCTION

The field of electro-optics has become increasingly more important in the last 20 years as its prodigies and applications have found their way into most facets of science, industry, and domestic use. This near-revolution, which essentially started with the advent of the laser, has been the result of extensive parallel and often symbiotic development of sources, materials, and microelectronics. The combination of these technologies has enabled a great variety of compact devices with ever greater intelligence and performance. If source development was instrumental in initiating the field, materials and detectors were the binding elements. Vast improvements in optical materials have made fiber optics feasible and the availability of high-quality, affordable fibers has, in turn, made optical circuits and a variety of optical sensors possible. Refinement and development of new materials have resulted in an astonishing variety of devices to modulate, polarize, frequency-shift, and otherwise control coherent radiation. In turn, detectors have achieved greater performance and smaller size and cost.

This handbook attempts to cover a broad spectral bandwidth—from x-rays to far infrared. A primary motivation in extending the short-wavelength limit of the source spectrum, and the handbook's coverage of it, is the demand for higher resolving powers in materials and device fabrication applications as well as medical and biological imaging. Figure 1.1 depicts the size of objects of interest in the biological, materials science, and electronics worlds, and the wavelength necessary to resolve them as prescribed by the Rayleigh criterion. The infrared boundaries of the spectrum are also continually being strained by sources, materials, and detectors in the development of a variety of applications such as imaging, optical diagnostics, and spectroscopy.

Each chapter of this handbook falls into one of four categories: sources, materials, and their properties (e.g., nonlinear optics), detectors, and applications. In the remainder of this chapter we present some simple overlying principles of each category and a topical map to aid the reader in finding the desired information.

1.1

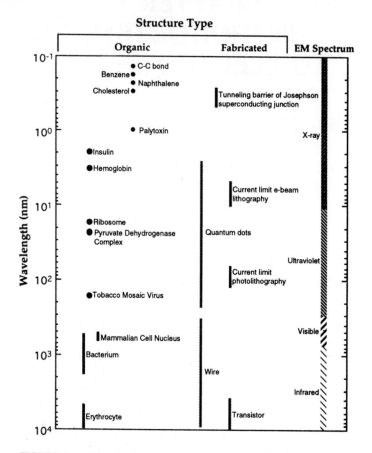

FIGURE 1.1   Relation of object size and resolving wavelength.

1.2 TYPES OF LIGHT SOURCES

Chapter 2 takes a detailed look at incoherent sources, and Chaps. 3 through 8 are devoted to the numerous laser sources grouped in part by media and in part by wavelength. Ultrashort pulse lasers and techniques are covered in Chap. 9.

Although the activity in the field of electro-optics has often been mirrored by events in laser development, incoherent sources still have an important role. Lasers are much newer and more space is devoted to them in the chapters to follow; however, the inescapable fact is that lamps currently have a much greater effect on our everyday lives than do lasers. With hundreds of millions of plasma discharge lamps and billions of incandescent light bulbs in constant use on a worldwide basis, power expenditure on lighting alone approaches the Terawatt level. Even the 22 percent or lower efficiency of most lamps still exceeds that of most lasers.

Arc lamps are characterized by high currents (several amperes) and high pressures (atmospheres) with ballast resistors used to prevent complete runaway. The lamps can be exceedingly bright. Examples include high-pressure (3 to 10 atmosphere) mercury vapor arc lamps, high-pressure metal halide lamps, high-pressure xenon arc lamps, high-pressure sodium arc lamps, as well as xenon flash lamps and rf excited lamps. They are used where high brightness is required for such purposes as movie projection, solar simulation, large-area illumination, and other special-purpose illumination.

Lower-pressure discharges (Torr) are used to excite atomic gases such as mercury vapor, hydrogen, cesium, the rare gases, and other elements. The best-known example of these low-pressure lamps is the fluorescent lamp. The low-pressure discharge gives rise to emissions characteristic of the gas in the tube. Mercury is especially valuable, since a mercury discharge gives about 90 percent of its emission in the mid-ultraviolet at 253.7 nm. This mid-ultraviolet emission is capable of exciting a thin phosphor coating on the inside of the glass tube. The phosphor subsequently fluoresces rather uniformly over the visible spectrum, thereby giving off "white" light. The entire process is quite efficient compared with incandescent bulbs. An essentially similar energy transfer process produces compact fluorescent tubes, germicidal ultraviolet lamps, low-pressure sodium lamps, neon signs, glow lamps, and hollow-cathode lamps.

There is still work to do to understand and improve lamps. Because of the great usage for fundamental necessities of life, improvements such as greater efficiency, lower emission of ultraviolet (uv) and infrared (ir), and longer life can be of great benefit. The current understanding of nonequilibrium plasmas, near local thermal equilibrium (LTE), and LTE plasmas can be found in several references.[1,2] An improved understanding of the mechanisms of these plasmas is the key to producing better light sources.

Lasers are of such importance to modern electro-optics that six chapters have been devoted to them. They are categorized both according to the spectral region in which they emit and according to the type of material used to obtain lasing. This categorization seems to suit the majority of lasers rather well. In Chap. 3 x-ray, vacuum-ultraviolet (vuv), and uv lasers are covered. Most of the lasers in this spectral region are gaseous (atom, ion, or plasma), but occasionally a solid medium is available and more are expected in the future. Chapter 4 considers visible lasers including dye lasers, except solid-state lasers, which have become important enough to warrant both Chap. 5 on conventional solid-state lasers and Chap. 6 on solid state semiconductor lasers. The lasers in these two chapters fall over parts of the visible and infrared. The remainder of the infrared belongs largely to gas lasers and is covered in Chap. 7. Figure 1.2 gives an overview of where the various generic types of lasers fall on the wavelength scale. Specific lasers, most of which have been commercialized or otherwise have noteworthy characteristics, are denoted in detail in Fig. 1.3. Further information on specific lasers can be found in several places in the open literature.[3,4]

Chapter 8 covers free electron lasers (FELs) which operate by magnetically perturbing an accelerated electron beam and which have vast tunability. To date these lasers have operated primarily in the infrared, but they are anticipated to operate tunably in the visible in the near future and eventually may provide ultraviolet and x-ray beams.

Many lasers have yielded to a variety of techniques that produced incredibly short pulsewidths—some only a few femtoseconds wide—and these lasers will be used in a wide variety of electro-optics, physics, chemistry, and biology experiments which will yield new information, new insight, and further progress and products. The techniques for producing ultrashort pulses are given in Chap. 9. Applications of these lasers will grow rapidly as soon as the production of the ultrashort lasers themselves becomes solidly commercialized.

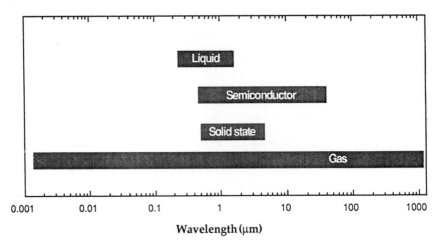

FIGURE 1.2   Location of generic lasers on the wavelength scale.

It is interesting to reflect on the reasons that so many lasers occur in the visible and near infrared. It is primarily a matter of materials, pumping sources, and the basic physics of lasers themselves. Because the human eye responds to radiation in the 400 to 700-nm region, considerable development of materials which transmit in the visible has taken place. Infrared instruments, especially military instruments,

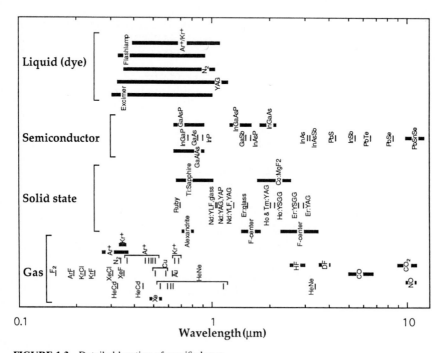

FIGURE 1.3   Detailed location of specific lasers.

have also encouraged development of infrared materials. Most optical sources, lamps, arcs, and flashlamps (and now diode lasers) emit most easily in the infrared as well. In addition, the small signal gain of a laser is directly proportional to the square of the wavelength. Related factors increase the dependence of gain on wavelength to the third or fourth power. For all these reasons, it is much harder to make uv, vuv, or x-ray lasers than it is to make infrared lasers.

1.3 MATERIALS

Materials that are nonabsorbing over a broad bandwidth are critical to source (and detector) development. We first consider the linear optical properties of materials— the responses that are proportional to the incident electric field. Optical materials are covered in two chapters that are roughly divided by wavelength. Material properties in the ultraviolet and shorter wavelengths are dealt with in Chap. 10, while Chap. 11 contains information about visible and infrared optical materials. The special material properties and techniques of optical fibers are covered in Chap. 12. Perhaps the most crucial linear optical specification of a material is its transmission bandwidth, since this determines its suitability for use as a window, filter substrate, or fiber. Figure 1.4 gives a quick survey of the transmission bandwidth of some common optical materials.

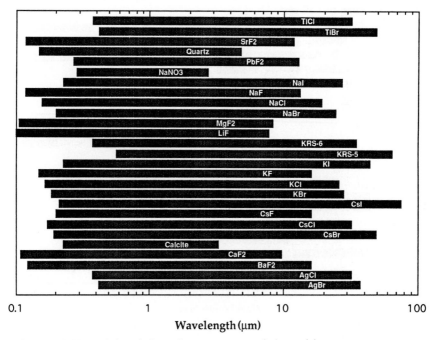

FIGURE 1.4 Transmission windows of some common optical materials.

Optical fields can also induce polarizations in materials that depend upon second- and higher-order powers of the field intensity. These nonlinear material responses lead to a variety of elastic and inelastic interactions between the media and the optical field. Nonlinear interactions of both categories including harmonic generation, four-wave mixing, and stimulated scattering are described in Chap. 13 while phase conjugation is treated in Chap. 14.

1.4 DETECTORS

Detection of optical radiation is often a crucial aspect of many applications in the field of electro-optics. Like lasers and materials, the selection and performance of detectors continue to grow at a remarkable rate. While spectral sensitivity is far from the only meaningful specification regarding detectors, it does provide a convenient reference point for assessing a detector's suitability for use with a source or in an application. Consequently, excepting imaging detectors, the succeeding three chapters describe detectors grouped by spectral response. Detectors for use with wavelengths in the ultraviolet and shorter are presented in Chap. 15. Chapters 16 and 17 undertake the discussion of the myriad of detectors available in the visible and infrared region. Figure 1.5 surveys the spectral coverage of numerous detector

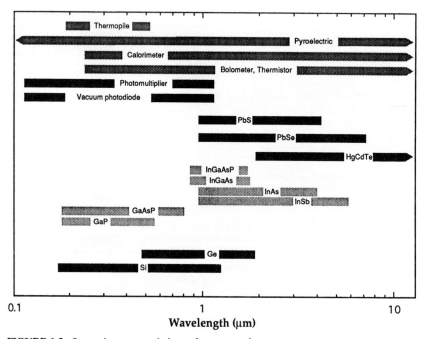

FIGURE 1.5 Spectral response windows of numerous detector types.

types, subgrouped by operational mode including photoemissive, quantum, and thermal devices. Finally, imaging detectors are described in Chap. 18.

1.5 CURRENT APPLICATIONS

Two-thirds of our space has been devoted to the principles of generation of light, transmission of light through optical materials, and detection of light. The remainder of the handbook is devoted to specific applications which extend the techniques and devices to broad areas capable of generating new knowledge.

Holography is the subject of Chap. 19. Holography has a tremendous number of applications, primarily in the area of nondestructive testing. Holographic interferometric and double-exposure techniques can determine small movements in surfaces and thereby detect faults in materials or structures. These techniques are extremely valuable, but holography also has value in creating unique art forms. This chapter sets forth basic principles while also giving new techniques for practical implementation of holography.

Laser spectroscopy is presented in Chap. 20. This topic has many direct and ongoing uses in basic research in physics, chemistry, and biology and in applications such as remote sensing, combustion diagnostics, and medical diagnostics and imaging.

Fiber-optic sensors, covered in Chap. 21, is an emerging field that shows great potential. The ability to position these sensors discretely in perhaps hazardous or inaccessible locations is but a part of their allure. Already fiber sensors are used in numerous applications including surveillance, temperature, pressure, and displacement measurements, and a variety of medical probes.

The principles of lithography for optoelectronics are presented in Chap. 22. Lithography involves many of the topics covered in this handbook including vuv and x-ray sources and optics, holography, material properties, and laser chemistry of resists.

The important subject of laser safety is covered in Chap. 23. While laser safety in the research lab will continue to be an essential concern, as more lasers continue to appear in industrial and consumer settings laser safety issues will become even more diverse and acute.

Chapter 24 presents a broad view of lasers in medicine. Medical applications both in practice and in development include a range of uses such as diagnostics, surgery, laser-induced activation of pharmaceuticals (photodynamic therapy), and imaging. This chapter discusses the current usage of lasers in medicine and surgery, the underlying physics pertinent to those applications, and an array of information regarding tissue optics.

Applications of lasers in material processing are described in Chap. 25. Processes such as hardening, alloying, and cladding where the laser has unique attributes are presented. The advantages and disadvantages of the use of lasers in lieu of traditional tools for welding, cutting, drilling, and marking are detailed. Also, the use of lasers for microelectronic applications is surveyed.

The principles of optical integrated circuits and optoelectronic integrated circuits are covered in Chaps. 26 and 27. Combining an array of electro-optic devices in miniature form involves integrating diode lasers, detectors, materials, and fiber optics. The topics and devices described in these last two chapters are undoubtedly critical to the future of computing, communication, and a continuing development of smaller but smarter devices that will further improve our quality of life.

1.6 REFERENCES

1. J. T. Dakin, "Nonequilibrium Lighting Plasmas," *IEEE Transactions on Plasma Science*, vol. 19, pp. 991–1002, 1991.
2. J. F. Waymouth, "LTE and Near-LTE Lighting Plasmas," *IEEE Transactions on Plasma Science*, vol. 19, pp. 1003–1012, 1991.
3. R. Beck, W. Englisch, and K. Gurs, *Table of Laser Lines in Gases and Vapors*, Springer-Verlag, 2nd ed. New York, 1978.
4. R. Waynant and M. Ediger, *Selected Papers on UV, VUV and X-Ray Lasers*, vol. MS71, SPIE Milestone Book Series, 1993.

CHAPTER 2
NONCOHERENT SOURCES

Sharon Miller

2.1 INTRODUCTION

In this chapter, the fundamentals of noncoherent or "nonlaser" optical radiation are discussed. Noncoherent radiation consists of electromagnetic waves whose amplitude and phase fluctuate randomly in space and time. The most familiar source of noncoherent optical radiation, and the first to be studied by experimental scientists, is the sun. The concepts and terms necessary for the characterization of both the sun and artificial sources are provided. In addition, the practical aspects of proper measurements and the problems associated with this task are discussed.

2.2 DEFINITION OF TERMS

The wavelength region for optical radiation spans from approximately 100 nanometers (nm) to 1000 micrometers (μm). The optical radiation spectrum can be broken up into three basic regions:

Ultraviolet (uv)	100 nm–400 nm
Visible	400 nm–760 nm
Infrared (ir)	760 nm–1000 μm

There is some controversy over the exact cutoff between the uv and visible region (380 to 400) and also between the visible and ir region (760 to 800). In the field of physics, the uv and ir regions are further divided into the "near" (300 to 400 nm), "mid" (200 to 300 nm), and "far" (30 to 200 nm) categories in the uv, and the "near" (760 to 4000 nm), "mid" (4 to 14 μm), and "far" (14 to 1000 μm) categories in the ir. In photobiology or photomedicine, it is more common to use biologically meaningful divisions, or the A,B, and C categories as defined by the CIE.[1] They are: UVC (100 to 280 nm), UVB (280 to 315 nm), and UVA (315 to 400 nm) for the uv region, and IRA (760 to 1400 nm), IRB (1400 to 3000 nm), and IRC (3 μm to 1 mm) for the ir region.

In order to discuss the characteristics of noncoherent optical radiation, it is first necessary to define certain terms and quantities. Tables 2.1 and 2.2 list several of the commonly used SI units[2] of optical radiation for radiometry and photometry.

TABLE 2.1 Radiometric Units

Quantity	Symbol	Units	Definition*
Radiant energy	Q	J (joule)	
Radiant exposure (dose)†	H	J/m^2	dQ/dA
Radiant fluence	F	J/m^2	dQ/dA
Radiant energy density	W	J/m^3	dQ/dV
Radiant power, or flux	Φ	W (watt)	dQ/dt
Radiant intensity	I	W/sr	$d\Phi/d\Omega$
Radiant flux density, or irradiance (dose rate)†	E	W/m^2	$d\Phi/dA$
Radiant exitance	M	W/m^2	$d\Phi/dA$
Radiance	L	$W/m^2{*}sr$	$d^2\Phi/d\Omega\,da$

*dA = element of directed surface area, $da = dA \cos\theta$ = element of spherical surface area, and θ = angle between normal to element of the source and the direction of observation.
†Common terminology in photobiology.

TABLE 2.2 Photometric Units

Quantity	Symbol	Units	Definition
Luminous flux	Φ_v	lm(lumen)	
Luminous exitance	M_v	lm/m^2 (lux)	$d\Phi_v/dA$
Illuminance or (luminous density)	E_v	lm/m^2	$d\Phi_v/dA$
Luminous intensity or (candlepower)	I_v	lm/sr or cd (candela)	$d\Phi_v/dr$
Luminance	L_v	$lm/(m^2{*}sr)$	$d^2\Phi_v/dr\,da$
CGS units:*			
Illuminance	E_v	lm/ft^2 (footcandle)	
Luminance	L_v	$(1/\pi)cd/ft^2$	

*These units still appear in some texts, although the SI units are now the prefered system.

Although most of the quantities in Tables 2.1 and 2.2 are listed in terms of the square meter (m^2), the preferred unit of area for optical radiation is cm^2 because this more closely approximates the sensitive area of most detectors. The nm is the preferred unit of wavelength in the ultraviolet to mid-infrared portion of the electromagnetic spectrum. Thus, when discussing "spectral" quantities, i.e., "per unit wavelength," all terms in Tables 2.1 and 2.2 would be modified by the suffix "per nm" and subscript λ. The spectral quantity may then be integrated over the wavelength region of interest to obtain total flux, intensity, etc., in a specified wavelength band. If one is talking about wavelengths in the infrared region greater than 1000 nm, the convention is to use μm or microns, instead of nm.

2.2.1 Solid Angle

Shown in Fig. 2.1, the solid angle Ω is defined as the area of some "irradiated" surface da_s, divided by the distance from the source r squared.

$$d\Omega = \frac{da_s}{r^2} \qquad (2.1)$$

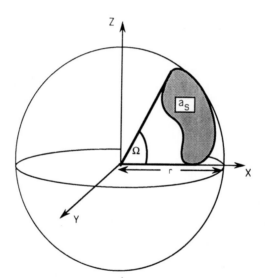

FIGURE 2.1 Pictorial demonstration of "solid angle" in a spherical coordinate system. The solid angle is defined by the area intercepted by the irregular cone on a sphere of radius r.

This distance r equals the radius of a sphere which is centered at the vertex of the solid angle. Thus the area da_s is the area of the intercepted spherical surface. In spherical coordinates,

$$d\Omega = \sin \theta \, d\theta \, d\phi \qquad (2.2)$$

Note that this surface need not be regular. As a basic example, a "free-standing" point source would irradiate an entire spherical surface subtending a solid angle of 4π steradians (sr):

$$\Omega = \int_0^{2\pi} d\phi \int_0^{\pi} \sin \theta \, d\theta = 4\pi \qquad (2.3)$$

A steradian can be envisioned as the three-dimensional equivalent of the unit for linear angle, the radian. The linear angular subtense of a source α is defined as the maximum linear dimension of the source D divided by the distance of observation:

$$\alpha = \frac{D}{r} \tag{2.4}$$

2.2.2 Radiant Intensity and Radiance

In the previous section, the concept of solid angle was introduced. This quantity is an important geometrical concept for the discussion of radiant intensity I and radiance L. The formal definition of intensity in the field of radiometry is different from the definition in the field of physical optics (W/m^2). In the field of radiometry, radiant intensity is defined as the flux of radiation in a given angular direction (W/sr). The radiant intensity in any solid angle is the flux, or power, emitted within that angle divided by the size of that solid angle, in sr. The *radiance* can be thought of as the radiant intensity of a source divided by the projected image area of this source, where the projected area lies in the plane normal to the direction of propagation.

The significant characteristic of radiance is its *property of invariance* through a lossless optical system. It can be shown[3] that, within an isotropic medium, the value of L in the direction of any ray has the same value at all points along that ray, neglecting losses by absorption, scattering, or reflections. This is known as the *radiance theorem.*[4] Knowledge of the radiance of a source enables one to determine the radiant power flowing through any surface if the cross-sectional area of the surface and its solid angle are known. This is especially useful in complex optical systems that may contain multiple aperture and field stops. An aperture stop is defined as any element, be it the edge of a lens or an open diaphragm, that forms a boundary which limits the amount of light that passes through the optical system. The amount of light is limited by the reduction in the number of rays (from an object) reaching the final image plane. The field stop governs the size of the final image and thus determines the FOV (field of view) of the optical system.

The invariance of radiance can be easily visualized with the aid of Fig. 2.2. This configuration could be used as a simple radiometer with a detector of sensitive surface area A_2 and input aperture of area A_1. Assuming X is significantly larger than the diameter of the detector, the detector will subtend the same solid angle at all points of the input aperture. If we place an "extended" (i.e., of finite dimension) source in front of this instrument, such that the radiation field overfills the input aperture, then

$$P = LA\Omega = \frac{LA_1A_2}{X^2} \tag{2.5}$$

where P = total radiant power incident on the detector. (This neglects losses due to absorption, reflection, or scattering.) This equation will hold regardless of the distance between the source and the input aperture, as long as the beam overfills the input aperture. If the distance from the source is increased further, the beam no longer overfills the input aperture. This results in a reduction of the cross-sectional area on the detector and, therefore, a reduction in the power measured by the detector. The quantity $A\Omega$ is solely dependent upon the geometry of the optical system and is known as the throughput, or the "entendue" of the system.

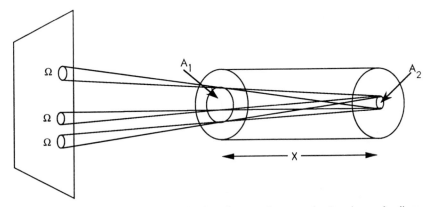

FIGURE 2.2 Schematic diagram of a simple radiometer demonstrating invariance of radiance. As long as the optical radiation from the source overfills the input aperture of area A_1, the power per unit area at the detector surface A_2 will not change.

2.2.3 Photometric Units vs. Radiometric Units

Radiometric quantities are applicable across the entire electromagnetic spectrum. There is another analogous system of units that is applicable only in the visible portion of the spectrum (from 380 to 780 nm) as defined by the CIE. These are used in the field of photometry (the measurement of visible light) and are listed in Table 2.2. In radiometry, the primary unit of radiation transfer, or radiant flux, is the watt (W). In photometry, the corresponding unit is the lumen (lm), which signifies the *visual response* produced by a light source with a given output power (W).

There is no direct method of converting from luminous flux (lumens) to radiant flux, unless the exact spectral distribution is known and is limited to the visible portion of the spectrum. Ordinarily, one would apply the following equations:

$$\Phi_{v\lambda} = K(\lambda)\Phi_{e\lambda}{}^{\dagger} \tag{2.6}$$

$K(\lambda)$ is defined as the spectral luminous efficacy, which is the ratio of any photometric unit to its radiometric equivalent and has units of lumens per watt. If normalized to its peak value K_{max}, it is referred to as spectral luminous efficiency $V(\lambda)$. The term "efficiency" as used here refers to the relative ability of the light to stimulate the visual response in the eye. Thus,

$$V(\lambda) = \frac{K(\lambda)}{K_{max}} \tag{2.7}$$

$V(\lambda)$ is defined as the photopic (daylight-adapted sensitivity of the human eye) spectral luminous efficiency and $V'(\lambda)$ is the scotopic (night-adapted) spectral luminous efficiency. Plots of both $V(\lambda)$ and $V'(\lambda)$ are shown in Fig. 2.3.

For photopic weighting,

$$K_{max} = 673 \text{ lm/W}$$

---

†The subscripts v and e are used to differentiate between photometric and radiometric quantities.

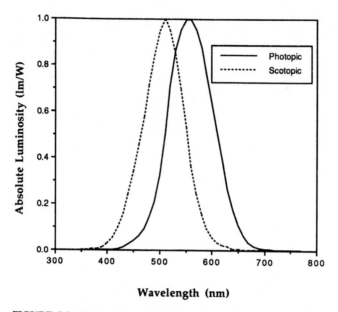

Wavelength (nm)

FIGURE 2.3 Spectral plot of the photopic $V(\lambda)$ and scotopic $V'(\lambda)$ functions. The photopic function defines the spectral daylight-adapted sensitivity of the eye, while the scotopic function defines the spectral night-adapted sensitivity of the eye.

and for scotopic,

$$K_{max} = 1725 \text{ lm/W}$$

Therefore, Eqs. (2.6) and (2.7) and the appropriate above constants can be applied to calculate the equivalent photometric quantities from their radiometric counterparts.

The current standard unit in photometry is defined as the candela, the luminous intensity of monochromatic radiation at 555 nm whose radiant flux is equal to 1/683 W. The human eye is most sensitive at this wavelength. Between 1948 and 1979, the unit of luminous intensity was defined as one-sixtieth of the luminous intensity of 1 cm^2 from a blackbody at the freezing point of platinum. This is consistent with the current definition, which lies within the error limits that resulted from the uncertainty in the freezing point of platinum.[5] Previous to 1948, the standard was based on the output of a group of carbon-filament vacuum lamps. The predictability of this standard was limited by the fact that the lamp intensities were highly dependent on their physical construction or manufacture.

2.3 CHARACTERISTICS

2.3.1 Point Sources

A frequently encountered concept in radiometry is that of a point source. A point source radiates uniformly in all directions (i.e., it is isotropic); the waves emanating from it can be considered to be spatially coherent, and its radiative transfer obeys the inverse square law. The inverse square law simply states that either the irradiance or the intensity of a source falls off in a manner proportional to the square of the distance from the source. Although the ideal point source does not exist, it is acceptable to talk about sources whose dimensions are very small in relation to their distance of observation. A star is a good example, its distance being so great that it subtends an extremely small angle as seen from the earth. In general, if the maximum source dimension D is less than one-tenth of the distance r from the source, then assuming inverse square law behavior will result in an error of less than 1 percent. If D is less than one-twentieth of the distance r, the error will be less than 0.1 percent. However, if one is dealing with a highly collimated light source, the distance necessary to achieve inverse square law behavior will be much greater, owing to the effects of the collimating optics.[6]

2.3.2 Lambertian Sources

Lambertain sources are defined as sources which have a constant radiance L for all viewing angles. A blackbody is, by definition, Lambertian. A good approximation to a Lambertian surface is a white piece of paper. The radiance, or apparent brightness, does not change as a function of viewing angle. An example of an extremely non-Lambertian source would be a bank of fluorescent lamps. When viewed normal to their surface, the dark areas between the bulbs would be clearly visible. However, if one were to view them at grazing incidence, they would appear to be a solid sheet of light. Therefore, the brightness, or radiance, would be different depending on the viewing angle. If we look at the relationships between radiance, intensity, and solid angle from Table 2.1:

$$L = \frac{d^2\phi}{da \; d\Omega} \tag{2.8}$$

$$I = \frac{d\phi}{d\Omega} \tag{2.9}$$

As indicated in Fig. 2.4, $da = dA \cos \theta$, which is the projected area. Thus, from Eq. (2.1):

$$d\Omega = \frac{dA \cos \theta}{r^2} \tag{2.10}$$

Substituting into Eq. (2.9),

$$d\phi = \frac{I \; dA \; \cos \theta}{r^2} \tag{2.11}$$

Therefore, the irradiance E at dA is given by

$$E = \frac{d\phi}{dA} = \frac{I \cos \theta}{r^2} \tag{2.12}$$

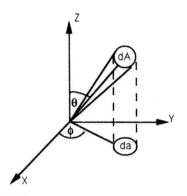

FIGURE 2.4 Spherical coordinate system demonstrating projected area, where $da = dA \cos \theta$.

This equation demonstrates Lambert's cosine law, which states that the irradiance decreases as the observation angle increases. Integrating both sides of Eq. (2.8) and substituting for $d\phi$, we obtain

$$I = \int L \cos \theta \, dA \qquad (2.13)$$

Since the radiance of a Lambertian source is independent of angle θ, Eq. (2.13) can be easily integrated to yield

$$I = LA \cos \theta \qquad (2.14)$$

The product LA will be a fixed quantity; therefore, I is solely dependent on the viewing angle of the source. This decrease in intensity as the viewing angle increases from $0°$ (or from the normal to the surface) is entirely due to the decreased projected area of the source.

2.3.3 Extended Sources

The term "extended source" is commonly used to describe most real sources, i.e., sources of a finite size (as opposed to the ideal point source) that are not necessarily Lambertian. When evaluating the irradiance, or illuminance, from general extended sources, it is often useful to speak of the radiant or luminous exitance M of the source. In lighting engineering, the exitance is the starting point for determining the illuminance of numerous different geometrical configurations. In fact, illumination engineers use tables of "configuration factors" to calculate light levels based on the geometry of their task. Referring to Fig. 2.5, the exitance M can be calculated from a knowledge of the radiance L of the source by the following series of equations:

$$d\Phi = \frac{L(\theta,\phi)dA_1 \cos \theta_1 dA_2 \cos \theta_2}{d^2} \qquad (2.15)$$

where

$$dA_2 = R \, d\phi \, R \sin \theta \, d\theta \qquad (2.16)$$

$$d\Phi = L(\theta,\phi)dA_1 \sin \theta \cos \theta \, d\theta \qquad (2.17)$$

From our definition of exitance in Tables 2.1 and 2.2,

$$dM = \frac{d\Phi}{dA} \qquad (2.18)$$

Therefore, with $dA_1 = dA$

$$M = \int_0^{2\pi} d\phi \int_0^{\pi/2} L(\theta,\phi)\sin \theta \cos \theta \, d\theta \qquad (2.19)$$

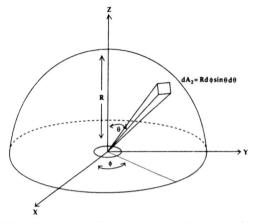

FIGURE 2.5 Spherical coordinate system as an aid to the calculation of the exitance M of a source of area dA_1 at a distance of R from the source.

If L is Lambertian, it will be independent of viewing angle θ and

$$M = 2\pi L \left.\frac{\sin^2\theta}{2}\right|_0^{\pi/2} = \pi L \qquad (2.20)$$

Once the exitance M is known, the illuminance or irradiance can be determined for any source configuration. The general expression is

$$E_2 = M_1 * C_{2-1} \qquad (2.21)$$

where E_2 = the illuminance or irradiance at a point P_2, M_1 is the exitance of A_1, and C_{2-1} is the configuration factor from surface 1 to point 2. The configuration factor is determined by an area integral which describes the relationship between the area of the source and the point P_2. Assume we want to determine the irradiance at the point P_2 of the disk Lambertian source in Fig. 2.6. The formal derivation can be found in Boyd.[7] The final solution is

$$E = \frac{d\Phi}{dA} = \pi L \sin^2 \theta \qquad (2.22)$$

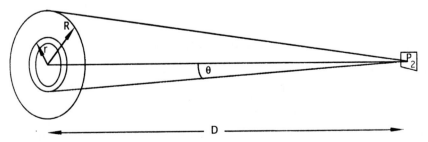

FIGURE 2.6 Figure demonstrating the calculation of irradiance at a point P_2 from a disk Lambertian source.

where $\sin^2\theta = r^2/(r^2 + D^2)$, leading to

$$E = \frac{\pi L r^2}{r^2 + D^2} \qquad (2.23)$$

So, the configuration factor in this instance is

$$C_{2-1} = \frac{r^2}{r^2 + D^2} \qquad (2.24)$$

It is instructional to examine Eq. (2.23) for four cases corresponding to different values for D.

1. $D \gg r$ $\qquad\qquad E = \dfrac{\pi L r^2}{D^2}$

This case obeys the inverse square law since the irradiance is dependent on the fixed quantity $\pi L r^2$ and falls off as a function of the square of the distance from the source.

2. $D = r$ $\qquad\qquad E = \dfrac{\pi L}{2}$

3. $D = 2r$ $\qquad\qquad E = \dfrac{\pi L}{5}$

These last two cases illustrate that, at fewer than 10 source diameters from a finite, extended source, the irradiance falls off more slowly than would be predicted by the inverse square law. In fact, within the distance of one source diameter (2 * radius), the falloff displays an almost linear behavior. Also, consider the case where

4. $r \gg D$ $\qquad\qquad E = \pi L$

This results because $r^2 + D^2 \approx r^2$ as long as D is sufficiently smaller than r as to be insignificant. This demonstrates that the irradiance remains almost constant for a large, planar sheet of light until one is at least half a radius away from it. An excellent reference for configuration factors is the "Catalog of Selected Configuration Factors."[8]

2.4 MEASUREMENTS AND CALIBRATION

Accurate measurements of noncoherent optical radiation are very difficult and can only be accomplished by accounting for all the potential sources of error. For instance, reflectors and apertures will affect the spatial distribution and therefore the intensity variation with distance. Most sources do not produce uniform illumination over an irradiated surface, and this must be considered when choosing the location of the detector. It is important to know the source characteristics and geometry before attempting to perform high-accuracy measurements.

2.4.1 Instrumentation

The power (or flux) from a noncoherent light source can be measured with a variety of instruments, depending on the application or the desired degree of precision. To determine total light output, one could use a photometer or radiometer. Photometers measure visible light in lumens, while radiometers are capable of measuring the entire spectrum of optical radiation, in watts. These instruments measure total light incident upon a detector (either a solid-state or photomultiplier tube) and are therefore highly dependent on the spectral sensitivity of the particular detector. Filters placed in front of the detector can significantly alter the instrument's spectral response, as is demonstrated by Fig. 2.7.

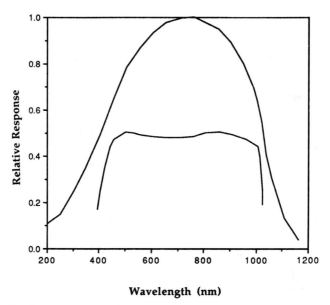

FIGURE 2.7 The spectral response of a bare silicon detector, and of a silicon detector modified by a radiometric filter. *(Adapted from Ref 14 with permission.)*

A photometer's detector is filtered so that its response closely matches that of the human eye. A typical portable photometer uses a photovoltaic or photoconductive cell, connected to a meter which is calibrated directly in lux (lumens/m^2), or footcandles. Photovoltaic detectors operate by generating a voltage as the result of the absorption of a photon. These types of detectors have poor linearity of response at high levels of incident illumination, which must be compensated for by external circuitry. Photoconductive detectors are constructed of materials whose resistance changes with photon absorption. (For a detailed discussion of solid-state detector characteristics, see Chaps. 15–18.)

For laboratory, or low-light-level applications, a photomultiplier tube (PMT) is more likely to be used. The photomultiplier tube is a photoemissive detector as opposed to a photovoltaic or photoconductive detector. The PMT operates via the photoelectric effect. These detectors produce current when light is absorbed by a photoemissive surface which is then amplified in sequential stages. Photomultipliers (PMTs) are extremely sensitive and are capable of rapid response times but require a high voltage (500 to 5000 V) for operation. The disadvantages of these detectors are their fragility—they are extremely sensitive to mechanical shock, interference from electromagnetic fields, moisture, and temperature. Some are recommended for use at temperatures down to 45°C below ambient (≈ -20°C). Thus, in addition to the high-voltage supply, the PMT may require an external cooling device. For these reasons, a PMT-based instrument is not well suited to field use. PMTs also tend to produce high dark currents (signal when no light is incident on the tube) which must be subtracted from the intended signal. Figure 2.8 shows the spectral response characteristics of several commercially available PMTs.

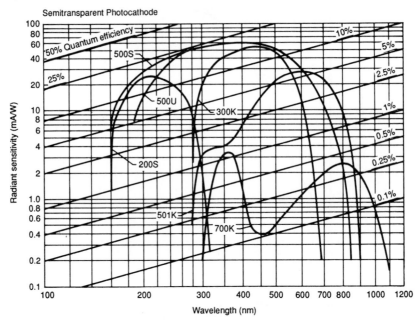

FIGURE 2.8 The spectral response characteristics of several commercially available photo-multiplier tubes (PMTs). *(Modified with permission from Hamamatsu Corp. Jan/89 (Rev), T-7000, p. 77.)*

The spectral sensitivity of a radiometer will depend on what type of semiconductor (for non-PMT instruments) and filters are used in the detector. Currently, most of these types of detectors are classified generically as "photodiodes" which produce current in proportion to the incident illuminance. Ideally, a radiometer will have a detector that is "spectrally flat" over the wavelength region of interest. Silicon detectors (most sensitive in the visible and near-ir region) can be doped to increase the sensitivity in the ultraviolet region. Another technique which is sometimes employed in broadband uv meters is to place fluorescent phosphors in front of the photodetector. The uv radiation impinges on the phosphor surface and is then transformed into visible radiation to which the semiconductor is more sensitive. The most pervasive obstacle in making accurate measurements of uv radiation with unsophisticated instrumentation is the fact that most sources which emit uv also emit visible, and even ir, radiation in much greater intensities. A uv-transmitting, but visible-blocking filter is normally inserted above the phosphor to limit the transmission of the undesired visible wavelengths from the source to the detector. The spectral transmittance of a commercially available filter which exhibits this type of behavior is shown in Fig. 2.9. Unfortunately, these filters' transmittance usually starts to increase again in the near-ir. When it is desired to measure the ir radiation from a source, detectors made of materials other than silicon [e.g., germanium (Ge) and lead sulfide (PbS)] are often chosen. The spectral sensitivity of several commercially available detectors is shown in Fig. 2.10.

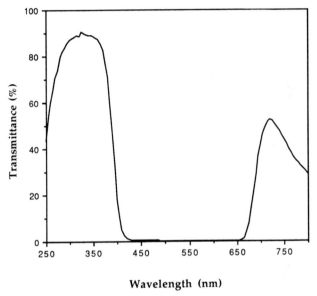

Wavelength (nm)

FIGURE 2.9   The spectral transmittance of a commercially available (Hoya Optics U-330) uv-passing, visible blocking glass filter.

The broadband radiometers and photometers are fairly inexpensive, rugged, and easy to use. The drawback of these instruments is that they are generally calibrated to a particular type of light source and therefore will have limited accuracy when measuring light sources that have a different spectral output. For example, a detector calibrated to measure the visible light from an incandescent lamp will produce misleading results if used to measure the visible light output from a fluorescent lamp. This is because, as will be discussed in Sec. 2.5.2, the incandescent lamp produces more output in the infrared region, and less in the ultraviolet region, than does the fluorescent lamp. This "out-of-band" radiation will interfere with the detection of the intended signal. An additional, obvious drawback of the broadband instruments is that they cannot provide information about the spectral content of a light source unless they are used with narrow bandpass filters. The use of such filters is accompanied by an additional set of

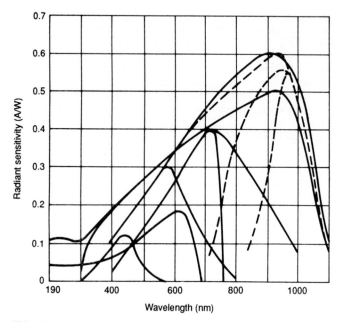

FIGURE 2.10 The spectral response characteristics of several commercially available solid-state detectors. *(Modified with permission from Hamamatsu Corp. Jan/89, CR-4000, p. 1.)*

problems like high signal attenuation (typical transmittance is 15 to 25 percent), high sensitivity to incident angle, and changing transmittance characteristics upon exposure. High levels of uv and ir radiation are especially damaging to both plastic and glass filters.[9,10]

Spectroradiometers. A more accurate method of making power measurements of a light source is to use a spectroradiometer. The major component of the spectroradiometer is the monochromator. Many different configurations have been used over the years. A schematic of a single-grating monochromator with Ebert-Fastie geometry is shown in Fig. 2.11*a* and one with Czerny-Turner geometry is shown in Fig. 2.11*b*. Through use of a prism, or grating, the monochromator separates the broadband light source into its different spectral components and the power per unit wavelength band can be determined. When "white," or broad-spectrum, light is incident on a prism, the shorter wavelengths (uv) are refracted at larger angles than the longer wavelengths (ir). This is because the index of refraction of the glass or quartz is higher at shorter wavelengths and, therefore, these rays travel more slowly through the prism. A narrow slit placed in the path of the refracted light can select a narrow band of wavelengths and thus will produce approximately monochromatic radiation from the previously broadband incident radiation. A mechanism is usually incorporated in the monochromator to automate the rotation of the prism, or grating.

A grating serves the same purpose as a prism but works by the mechanism of diffraction, instead of refraction. Most monochromators contain reflection gratings,

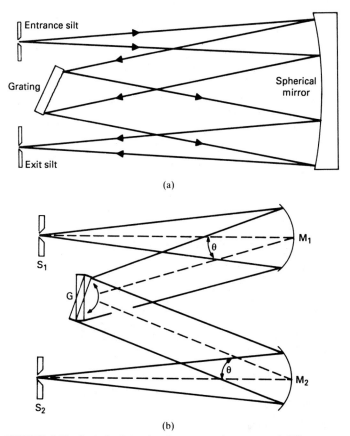

FIGURE 2.11 Optical schematic of two monochromators of different geometries: (a) the Ebert-Fastie geometry monochromator, and (b) the Czerny-Turner geometry monochromator.

as opposed to transmission gratings. Conventionally ruled reflection gratings are usually constructed of thin films of aluminum which have parallel grooves ruled onto them (Fig. 2.12). These master gratings are used to make numerous replica gratings by pressing the masters onto aluminum coated glass substrates. Holographic gratings are manufactured by illuminating a photosensitive layer with a pattern of interference fringes. When light is incident on the grating, its angle of reflection will depend on the spacing of the grooves, the angle of incidence, and the wavelength of the light. This behavior is illustrated through use of the grating equation:

$$m\lambda = d(\sin \alpha \pm \sin \beta) \qquad (2.25)$$

where α is the angle between the incident light and the normal to the grating, d is the grating spacing, and β is the angle of diffraction. The integer m specifies the order of the interference maxima. When $m = 0$, $\alpha = \beta$ and the grating acts as a mirror. The variation of angle with λ produces the desired result of separating the spectral components. The light rays reflected from each groove will recombine at

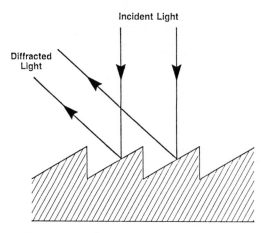

FIGURE 2.12 Diagram showing how collimated light is refracted from a diffraction grating.

a given point in space with different phases owing to the difference in optical path length. Incident light of a given wavelength will be reflected from all the grooves but will be in phase only at certain angles. Concave gratings are often used in monochromators instead of plane gratings as they avoid the need for additional focusing optics.

The *efficiency* with which a grating reflects a particular wavelength is determined by the "blazing," or changing the angle of the groove face. At the blaze wavelength, the angle of reflection from the groove face will equal the angle of diffraction. This results in a higher percentage of the incident light at a particular wavelength being available for detection. The *angular dispersion* of a prism, or grating, is defined as

$$\mathcal{D} = \frac{m}{d \cos \theta_m} \tag{2.26}$$

where d is the grating separation and θ corresponds to the angle of diffraction of the mth order. Thus the angular separation between different wavelengths will increase as the order m increases. The resolving power of a grating is defined as

$$\mathcal{R} = \frac{\lambda}{(\Delta \lambda)_{min}} \tag{2.27}$$

where $\Delta \lambda$ is the minimum resolvable wavelength difference. This is a function of total grating width dN (where N = number of grooves), the angle of incidence, and wavelength λ.

Most gratings today are ruled by a holographic process, as opposed to diamond tool ruling. This approach can reduce stray light to 1 to 10 percent of that observed with conventionally ruled gratings. The grating specifications and the monochromator focal length, along with the width of the slits installed in the monochromator, will determine the bandpass of the system. A range of bandpasses can be found in commercially available units; usually from 20 nm down to 1 nm. Some instruments have variable slit widths, while others are fixed. Resolution also can be increased by lengthening the monochromator. Lengths from 1–10 meters are common in laboratories doing spectral identification where accuracy to 0.001 Å and beyond are needed.

The reduction of stray light is highly desirable in a spectroradiometer system. Stray light includes radiation from sources other than that which is being measured but also includes radiation from the source itself at unwanted wavelengths. It can arise from light leaks in the monochromator housing or from reflections or fluorescence from components inside the monochromator. Stray light can be a significant source of error, especially when performing measurements in the uv region. The potential "swamping" of the detector with visible light (to which it is usually more sensitive) may require the use of a double monochromator. These suffer from loss of throughput but have excellent stray-light rejection. Example specifications for a single monochromator from manufacturer A vs. a double monochromator from manufacturer B are listed in Table 2.3.

TABLE 2.3 Specifications for a Single-Grating Monochromator vs. a Double-Grating Monochromator

	Single	Double
Effective aperture	$f/3.5$	$f/3.5$
Line density, l/mm	1200	1200
Linear dispersion, nm/mm	8	4
Resolution, nm	1.0	0.5
Stray light	10^{-5}	2×10^{-9}

An additional consideration is the elimination of harmonic frequencies (also a form of stray light) from interfering with the fundamental signal. The way this is handled is through the use of order-sorting filters. These are "low-pass" filters (with respect to frequency) which are specified by their "cutoff" wavelength, or wavelength below which their spectral transmittance is close to zero. For example, one manufacturer inserts cutoff filters near the entrance aperture at wavelengths of 400, 600, 1100, and 2000 nm. Most uv-visible spectroradiometers will have only one order-sorting filter installed with a cutoff wavelength in the region between 500 and 600 nm.

Spectrophotometers.   Spectrophotometers differ from spectroradiometers in several ways. First, instead of being used to measure the output of an unknown light source, they usually contain their own light source which is used as a "probe." Their applications range from precise chemical analysis of compounds to evaluation of the physical properties of materials. An important application for these instruments is color analysis and standardization. They are also used to measure the spectral transmittance or reflectance of materials. A schematic diagram of a dual-beam spectrophotometer is shown in Fig. 2.13. This configuration allows for fast measurements which are not susceptible to errors from time-varying fluctuations in the intensity or wavelength distribution of the excitation source.

The main component of the spectrophotometer is the monochromator, just as it was for the spectroradiometer. Depending on the detector and other optics selected, currently available spectrophotometers have the capability to scan from 180 nm up to 10 μm, though different gratings will be necessary for different wavelength regions. One limitation in the use of spectrophotometers is their susceptibility to alterations in the optical path of the sample vs. the reference beam.

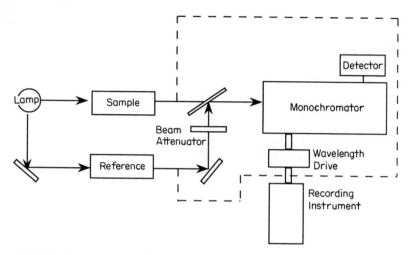

FIGURE 2.13 Schematic diagram of a dual-beam spectrophotometer.

This can arise from the incorrect use of vessels, like cuvettes, which may exhibit significant absorption at particular wavelengths. There can also be problems from focusing effects of nonplanar samples. A curved sample, or vessel, can cause the light beam to either converge or diverge, and thus the measured power per unit area will be different from that of the reference beam. If it is desired to measure the transmittance of a curved sample, it would be preferable to use a spectroradiometer.[11] Then, one can measure the energy from the light source that gets through the sample at a particular wavelength and divide by the inherent energy of the light source at that wavelength. In this way, both the material absorption effects and the refraction effects of the sample are evaluated.

2.4.2 Calibrations

The most common source of error in performing measurements of noncoherent sources is due to calibration of an instrument with a source that is very different spectrally or geometrically from the source being measured. Ideally, one should calibrate with a source of similar intensity level to that being measured or at least at levels within the dynamic range of the instrument. However, if the instrument is fairly linear over several decades of intensity, this is not so critical. It is very important to calibrate the radiometer in the exact same configuration in which it will be used. If this is not possible, the differences should be accounted for with appropriate correction factors. Broadband detectors can be calibrated so that they will yield accurate results for a particular type of light source but they will give meaningless results when used to measure a light source with a different output spectrum. For example, broadband detectors are available from various manufacturers which have been calibrated with a line source in the uv. If this detector is used to measure a source which emits a continuous spectrum, especially if it is shifted from the wavelength of the line source, the results may exhibit more than 50 percent error. The beauty of spectroradiometers is that once they are calibrated, they can be used on light sources with virtually any spectral distribution. The accepted method of calibration is to use a standard of spectral irradiance, or standard lamp, which is calibrated by the National Institute of Standards and Technology (NIST).[12] The standard lamps in use currently are incandescent. They are made with a tungsten filament and are filled with halogen gas. To increase the proportion of ultraviolet radiation they emit, and because of its ability to withstand higher temperatures, they have a quartz envelope instead of glass. They should be operated from a precision-regulated direct-current power supply. The FEL type lamps (the designation of the latest NIST generation of lamp) require a drive current of approximately 8 A, which should be attained gradually via "ramp" circuitry provided by the power supply. These lamps have sufficient output to be used as calibration sources from 250 to 2400 nm. NIST quotes an accuracy of about 5 percent in the uv and 2 percent in the visible and ir. For the wavelength region below 250 nm, the deuterium arc lamp is recommended, as it has significantly higher output in the 200- to 350-nm region than does the tungsten-halogen, with quoted accuracies in this region of 5 to 7.5 percent.

The system response of the spectroradiometer is determined by measuring the output of the standard lamp and comparing that to the known values for spectral irradiance provided by NIST. These are usually given at 5-nm intervals in the uv region and at selected wavelengths (usually mercury lines) in the visible and ir regions. They can be interpolated by the user to obtain values at intermediate wavelengths. A logarithmic interpolation algorithm is preferred over trapezoidal since the output spectrum of the standard lamp is parabolic in shape, similar to a typical blackbody curve. The standard lamp must be measured as closely as possible to the way in which it was calibrated at NIST. It must be aligned properly to the input optics of the spectroradiometer and measured at exactly the same distance (typically 50 cm). Any deviation from this geometry can introduce significant errors to the final result.

The system response is normally stored in the spectroradiometer's computer so that it can be applied to all future measurements. A typical plot of the system response (or calibration curve) for a double-grating monochromator system with PMT detector is shown in Fig. 2.14. Note that it increases dramatically in the uv and near-ir. This is mostly due to the decreased sensitivity of the PMT in these regions but is also affected by the blaze wavelength of the gratings. The system response can fluctuate daily, especially when the detector is a PMT, as they are highly sensitive to temperature and humidity changes. Therefore, if high-precision measurements are required, a system of this sort should be calibrated at least once on the day it is being used. In fact, those who are in the business of making outdoor solar measurements may have to calibrate approximately every 30 min, yielding one calibration run for every solar measurement. This will minimize errors which are due to the temperature sensitivity of the detector.

Spectroradiometers are very delicate instruments and should be carefully maintained. To ensure that the calibrations of the NIST standard lamps are meaningful, several factors have to be taken into consideration. The spectroradiometer should be periodically checked for linearity of response and wavelength accuracy. Linearity can be verified through use of a light source which has small enough dimensions to approximate a point source. By increasing the distance from the input aperture of the spectroradiometer, one can verify whether the signal output decreases by the expected $1/r^2$ factor, i.e., obeys the inverse square law. If not, adjustments to the amplification circuitry may be required. Wavelength alignment accuracy is normally checked by using a low-pressure mercury-arc source, such as the Pen-

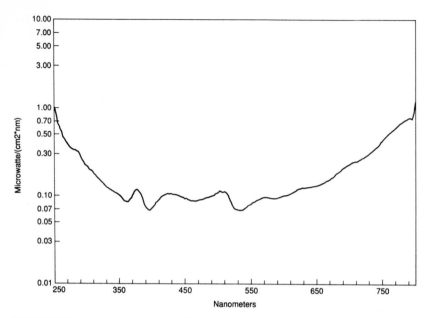

FIGURE 2.14 Spectral plot of a typical system response (or calibration factors curve) from a double-grating spectroradiometer system equipped with a PMT detector.

ray lamp.[13] This lamp emits many narrowband distinct peaks at the characteristic mercury lines. A plot of the spectral output of such a lamp is shown in Fig. 2.15. Depending on the characteristics of the spectroradiometer system, the wavelength alignment accuracy can be adjusted to well below 0.1 nm. Unfortunately, this accuracy is not maintained across a wide spectral region, so it is best to maximize

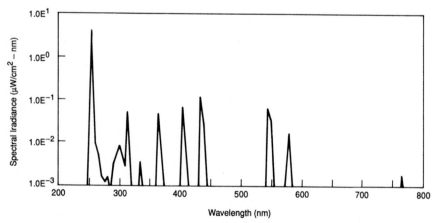

FIGURE 2.15 A logarithmic-scale plot of the spectral output of a Pen-ray® (Ultra-violet Products, Inc.) low-pressure Hg vapor lamp.

the alignment accuracy in the region where precision measurements are most important.

2.4.3 Input Optics

The choice of input optics for the spectroradiometer, or radiometer, will have a great influence on the degree of difficulty of making measurements. Noncoherent sources are notorious for being spatially nonuniform in their output. It is mainly for this reason that input optics such as cosine diffusers and integrating spheres are used. Both will present a uniform, nonpolarized beam of light to the detector. Unfortunately, they will also reduce the intensity of the signal by as much as a factor of 100. The cosine diffuser has a curved surface to intercept more light at large angles of incidence. This design produces a flux at the detector that closely follows the cosine response. The integrating sphere is known as the standard diffuser. A schematic diagram of how the light rays are reflected by the integrating sphere is shown in Fig. 2.16. It basically consists of a hollow metal sphere that is coated with either a pressed powder or a machined reflective material. Two common coatings are barium sulfate ($BaSO_4$) and a form of Teflon called Halon. These materials have fairly flat reflectance characteristics over a broad spectral range. A plot of the reflectance of the two materials is shown in Fig. 2.17. Halon is currently the material of choice (in the 200- to 2400-nm region) as it can be machined, instead of packed, and is less prone to degradation by moisture absorption.

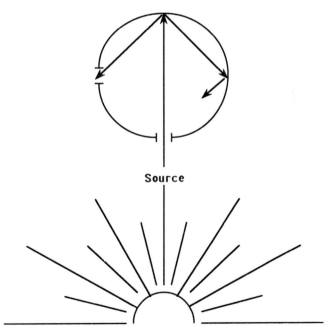

FIGURE 2.16 Diagram demonstrating the manner in which light is reflected inside an integrating sphere.

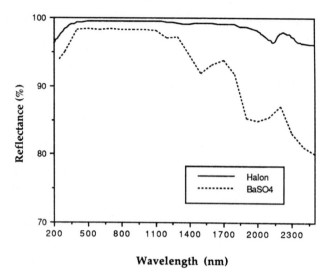

FIGURE 2.17 Spectral reflectance of two coatings typically used in integrating spheres, BaSO₄ and Halon.

2.5 SOURCES OF NONCOHERENT OPTICAL RADIATION

2.5.1 Solar Emission

Human beings have been fascinated with the sun and the behavior of its electromagnetic radiation output for ages. It has proved to be one of the most interesting sources of light because its output is so dynamic. In actuality, its inherent output may not be highly variable but may only appear so to an earth observer. The changes in the amount of radiant intensity that we receive on earth are largely due to the changes in orientation between the sun and the earth. The zenith angle is the angle between the sun's present position and its position at noon on a given day. The larger the zenith angle, the longer the path length through which the solar radiation must travel. This longer path through the earth's atmosphere produces much more scattering of the shorter wavelengths, and the ultraviolet radiation is therefore reduced by a larger proportion at large zenith angles than is the visible and ir radiation. For example, the ultraviolet spectral irradiance at 300 nm decreases by a factor of 10 from noon to 4:00 pm, while total solar irradiance may only decrease by 20 percent.[14] An additional consequence of this phenomenon is the spectral shift in the sun's irradiance with time of day. In the evening, when the sun is at a lower elevation, the ultraviolet and blue wavelengths have been mostly scattered out, leaving only the orange and red wavelengths to penetrate the longer path through the atmosphere.

Wavelengths less than 290 nm are highly absorbed by the atmosphere so, fortunately, a negligible amount of this highly carcinogenic radiation reaches the earth. The ozone layer in the upper atmosphere is primarily responsible for this absorption, and its apparent recent depletion is the cause of a renewed interest in accurate spectral measurements of the solar output. The atmospheric spectral transmittance curves for varying optical air masses are shown in Fig. 2.18 for wavelengths between

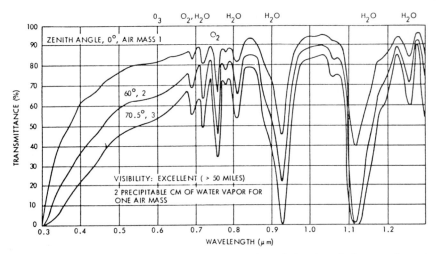

FIGURE 2.18 Plots of the spectral transmittance of solar radiation through the atmosphere at three different optical air masses.

300 and 1300 nm. Other factors which affect the solar spectral irradiance are amount and location of cloud cover, geographical location (including altitude), and latitude. In the winter, the earth is closer to the sun than in the summer, so, for the same zenith angle, the earth actually receives more radiation. However, the sun achieves larger zenith angles in the winter, i.e., it appears to be at a lower elevation in the sky, so the daily uv dose is far less than in the summer. Figure 2.19 shows how the solar irradiance varies with location and time of day. Notice how rapidly the spectrum changes in the 290–350-nm region. This is the region with which scientists who are evaluating the ozone layer depletion are most concerned. The fact that the irradiance changes so dramatically here makes accurate spectral measurements extremely difficult. Stray-light rejection becomes even more important. Therefore, almost all of the international laboratories that are monitoring the solar UVB levels use a double-grating spectroradiometer. This requires the use of a highly sensitive detector, owing to the low throughput of a double-grating monochromator.

The amount of radiation received from the sun can be separated into a direct and a diffuse component. The direct component is the energy within a narrow solid angle which radiates from the solar disk itself. The diffuse component is a result of the scattering of the atmosphere and can be modeled by a hemispherical, uniform source of radiation. On a clear day, about one-fifth of the total irradiance received at the earth's surface is due to this diffuse component. The diffuse term also contains radiation resulting from ground reflectance. A surface covered with fresh snow has the highest reflectance—89 percent, as compared to a grassy surface which may reflect only 17 percent of the total solar energy. Several references on characterizing solar ultraviolet radiation can be found in the work of Green et al.[15]

2.5.2 Artificial Sources of Noncoherent Radiation

There are basically two types of man-made sources of noncoherent optical radiation: the incandescent and the gas discharge lamp. The two most common gas discharge

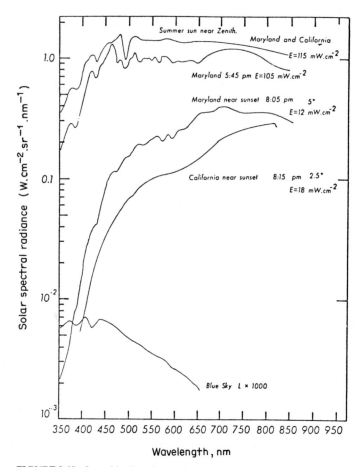

FIGURE 2.19 Logarithmic-scale plot demonstrating the variation of measured solar output (radiance) for different locations and times of day. *(Reprinted with permission from Ref 14.)*

lamps are the glow discharge (e.g., neon and fluorescent lights) and the arc lamp. The incandescent lamp has a continuous-shaped output spectrum that can be modeled fairly accurately as a blackbody, whereas the gas discharge lamp spectrum may have narrow peaks superimposed on an underlying continuum that does not follow the characteristic form of a blackbody spectrum. Figure 2.20 compares the spectrum of a 100-W incandescent lamp and a 40-W fluorescent lamp.

Ballast or Input Power Requirements. The ballast provides the proper current and voltage to start and operate the lamp. For arc lamps, the most important function of the ballast is its current-limiting capability. All arc lamps (except high-intensity carbon arcs) have negative resistance characteristics, which means that the arc discharge will draw an unlimited amount of current almost instantaneously if operated from a nonregulated power supply. Therefore, a ballast must be used

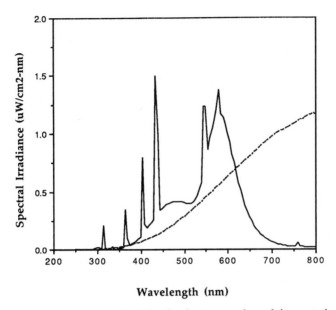

FIGURE 2.20 Linear-scale plot showing a comparison of the spectral irradiance of a 100-W incandescent dotted and 40-W fluorescent lamp solid.

to prevent the lamp from "self-destructing." This can be achieved through the use of an inductive or capacitive reactance in series with the lamp. For mercury and metal halide lamp ballasts, the net reactance is capacitive, while for high-pressure sodium ballasts, it is inductive. An autotransformer can limit the effects of line voltage fluctuation and provide any necessary high voltages. The manufacturer's specifications should be consulted to determine the proper ballast specifications for a given lamp. In most cases, ballasts are not interchangeable between the different high-intensity discharge lamps.

Ballasts are also required for fluorescent lamps and are available in three basic types: rapid-start, instant-start, and preheat. This refers to the cathode heating circuitry and can affect warmup time and efficiency. Attempting to operate a rapid-start lamp from an instant-start ballast, and vice versa, will result in a marked reduction of lamp life. In the case of preheat vs. instant-start ballasts, the different start-up circuitry requirements will prevent the lamp from lighting. Therefore, it is best to operate each lamp from the prescribed ballast.

Incandescent Sources. Although the light from both incandescent and fluorescent lamps is produced by the same mechanism, i.e., the transition of electrons from a higher, excited state to a lower, or ground, state, the method of excitation is different. The incandescent lamp produces light because its metal filament is heated as electric current flows through it. As it reaches temperatures near 600°C, visible radiation and heat are produced. At much higher filament temperatures, 2000°C and above, ultraviolet radiation is produced, as well.

For high-illuminance applications, especially where compact size is desirable, tungsten-halogen lamps are used. Tungsten is the material of choice for the filaments of these lamps because it has a high melting point and can therefore be operated at high temperatures for relatively long periods of time. Tungsten-halogen lamps operate via the regenerative "halogen cycle." As the tungsten filament is heated to sufficient temperatures, some of the tungsten begins to evaporate. In an ordinary tungsten lamp, the evaporated tungsten would be deposited on the inner surface of the lamp envelope, causing bulb blackening and consequently a reduction in light output. The function of the halogen (either iodine or bromine) is to chemically combine with the evaporated tungsten so that it can be redeposited on the filament and thus maintain a longer lifetime. However, if the lamp is not operated long enough for the halogen to reach the required temperature, black deposits of tungsten will adhere to the inside of the quartz or silica envelope as they would in an ordinary tungsten lamp. Tungsten may be used at temperatures up to 3500°C, but rated lifetime decreases as the filament temperature increases.

The total output power of an incandescent lamp is proportional to the filament surface area and goes up as the fourth power of the filament temperature (from the Stefan-Boltzmann law). Tungsten filaments at high temperatures emit significant amounts of uv radiation near 300 nm. This radiation would be absorbed by the glass envelope in conventional incandescent lamps but is transmitted through quartz (Fig. 2.21).

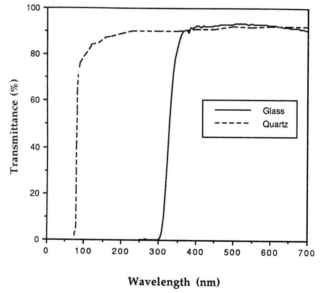

FIGURE 2.21   Spectral transmittance of common glass vs. quartz.

Gas Discharge Lamps   *Fluorescent Lamps.*   By far the most commonly used gas discharge lamp is the fluorescent lamp. The light is generated by a low-pressure (near vacuum condition) mercury arc. This arc strikes when a sufficient voltage potential is established between the cathodes of the lamp. When the inert gas (usually argon, krypton, or neon) inside the tube is ionized, the mercury atoms become excited by collisions with ions and free electrons emitted from the heated cathode, and ultraviolet radiation is produced. The majority of the uv radiation produced is at the characteristic 253.7-nm line of mercury.

The fluorescent lamp tube is coated with a phosphor which converts the uv radiation to visible light. One exception to this is a special-purpose gas discharge lamp, the "germicidal" lamp which has no phosphor and therefore emits the 253.7-nm line almost exclusively. A linear plot of this lamp's output spectrum is shown in Fig. 2.22a. Although the majority of this lamp's output is at the highly bactericidal and mutagenic 253.7-nm wavelength, other uv and visible mercury lines are produced as well.

The phosphors used in fluorescent lamps have been chosen for their maximum sensitivity near the 253.7-nm mercury line. Depending on the chemical composition of the phosphate, different colors of light are produced. For instance, "black light" (actually near-uv radiation) is produced from a mixture of barium silicate. The output of this type of fluorescent lamp is shown in Fig. 2.22b. Notice that this output spectrum contains a significant amount of uv. Many "tints" of white light are available, as well as colored light such as pink, blue, orange, red, or green. The output of several commonly used "white" fluorescent lamps is plotted in Fig. 2.22c to e. The majority of the output arises from the continuum of visible light. However, some uv light produced by the excited mercury atoms still gets through the glass envelope, and these characteristic mercury lines are present in all fluorescent lamp spectra.

Fluorescent lamps have high luminous efficacy, compared with incandescent lamps, with values ranging from 75 to 80 lumens per watt. They are available in wattage designations ranging from as low as 4 W for special-purpose black lights to 96 W for special-purpose white lights and sunlamps. The most common is the 40-W general-purpose fluorescent lamp, available in several designations shown in Fig. 2.22. The wattage designation indicates input power, not the amount of output

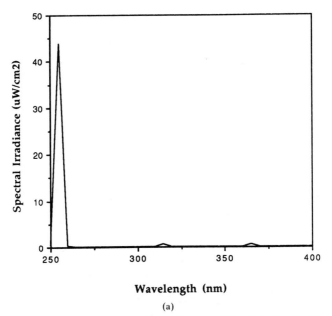

(a)

FIGURE 2.22 Linear-scale plots of the spectral irradiance for the following lamps: (a) germicidal, (b) black light, (c) daylight fluorescent, (d) cool white fluorescent, and (e) warm white fluorescent at a distance of 1 m.

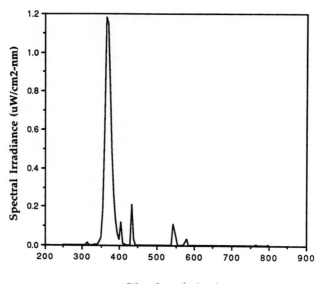

(b)

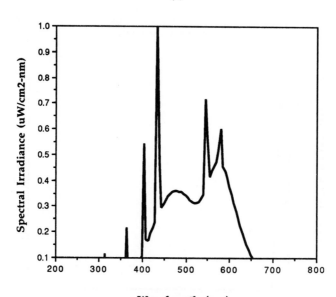

(c)

FIGURE 2.22 (*Continued*)   Linear-scale plots of the spectral irradiance for the following lamps: (*a*) germicidal, (*b*) black light, (*c*) daylight fluorescent, (*d*) cool white fluorescent, and (*e*) warm white fluorescent at a distance of 1 m.

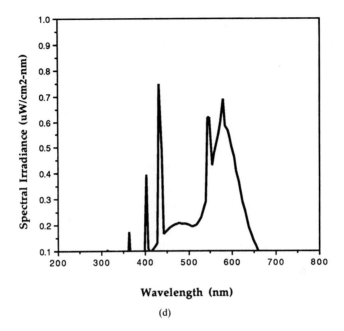

(d)

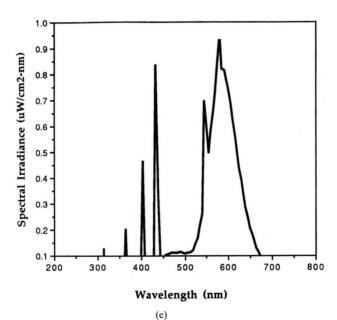

(e)

FIGURE 2.22 (*Continued*)   Linear-scale plots of the spectral irradiance for the following lamps: (*a*) germicidal, (*b*) black light, (*c*) daylight fluorescent, (*d*) cool white fluorescent, and (*e*) warm white fluorescent at a distance of 1 m.

optical radiation. Even though fluorescent lamps are considered relatively efficient, in general, less than a third of the input power is emitted as visible radiation. Slightly more than 50 percent of the input wattage delivered to the lamp is spent on conversion to the 253.7-nm wavelength. About half of this energy is converted to visible radiation by the phosphor. The color rendition of the phosphor, i.e., whether its output is higher in blue or red content, will further determine the lumens per watt that the lamp produces, since the human eye is most sensitive at 550 nm, or green light. The remainder of the input wattage is converted into heat, which is either absorbed through the walls of the glass tube or the electrodes, or transmitted to the surrounding air.

High-Intensity Discharge Lamps. The second most common arc lamp is the high-intensity discharge (HID) lamp. These lamps operate at much higher pressures and temperatures than fluorescent lamps and thus can produce higher output. Actually, the term "high pressure" is relative here as the internal pressure of these lamps is only about two to four times normal atmospheric pressure. The effect of this higher pressure is to shift a larger portion of the output into the longer-wavelength region. At extremely high pressures, more of the output shows up in the continuum, which lessens the appearance of line spectra. There are basically three types of HID lamps: the mercury vapor, the metal halide, and the high-pressure sodium. All three of these lamps are constructed with an inner envelope which contains the discharge. The outer envelope protects the inner tube from drafts and temperature fluctuations and filters out the UVC and UVB radiation. These lamps are available with or without phosphor coatings on the outer envelope. The function of the phosphor is to increase visible light output, affect color rendition, and diffuse the light from the glowing arc.

The mercury vapor lamp operates on a principle similar to that of the low-pressure fluorescent lamps. They are available in 40-W to 1000-W powers, with the 400-W model being the most commonly used (Fig. 2.23). Argon is used as the starting gas which facilitates the vaporization of the liquid mercury. The electrodes are larger and constructed to withstand much greater temperatures than their fluorescent lamp counterparts. The typical output from a mercury vapor lamp has a blue-green appearance, which is why phosphors are normally used to shift the output toward the red end of the spectrum.

The metal halide lamp (sometimes referred to as a multivapor lamp) differs from the mercury vapor lamp only in the inner tube constituents (Fig. 2.24). In addition to mercury and argon, these lamps contain metals such as thallium, indium, scandium, and dysprosium[16] in the form of halide salts. They are available in 400 W, 1000 W, and 1500 W. Most metal halide lamps require a higher open-circuit voltage to start than do their mercury vapor counterparts. After the main arc is struck, the temperature rises to the point where the iodides vaporize and separate into elementary iodine and the three additive metals. The metals form a multilayer vapor sheath around the mercury arc; in the order of blue (indium), green (thallium), and yellow (sodium).

High-pressure sodium lamps are well known for their slightly yellow color (Fig. 2.25) and high efficiency. For comparison, Fig. 2.26 shows the relative luminous efficacy (lumens per watt) of incandescent, fluorescent, and high-intensity discharge lamps, with the high-pressure sodium lamp having the highest luminous efficacy. The construction of these lamps is quite different from other high-intensity arc lamps, the main difference being that the inner tube is made of an aluminum oxide ceramic, which can withstand temperatures up to 1300°C. This ceramic material was chosen for its ability to transmit a high percentage of visible light and its translucence. It is also resistant to the corrosive effects of hot sodium.

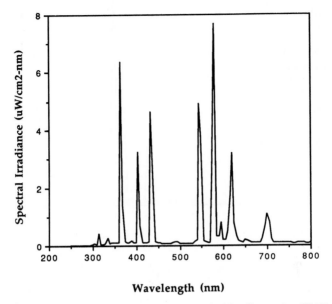

FIGURE 2.23 Linear-scale plot of the spectral irradiance of a 400-W Hg vapor lamp at a distance of 2 m.

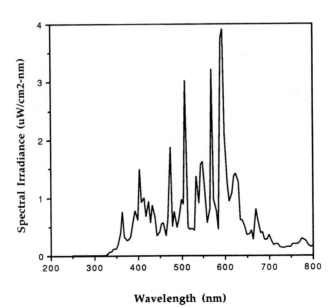

FIGURE 2.24 Linear-scale plot of the spectral irradiance of a 400-W metal halide (multivapor) lamp at a distance of 2 m.

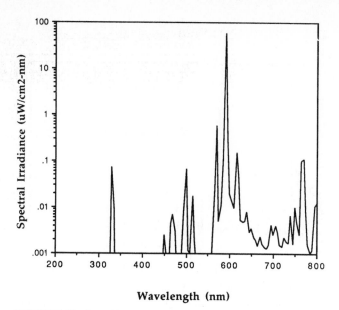

Wavelength (nm)

FIGURE 2.25 Logarithmic-scale plot of the spectral irradiance of a 400-W high-pressure sodium lamp at a distance of 2 m.

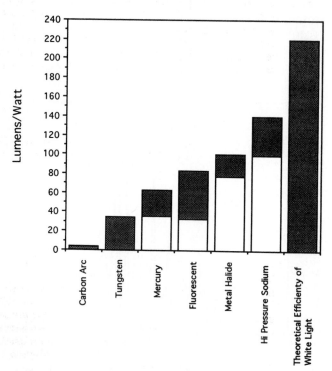

FIGURE 2.26 Relative luminous efficacy of carbon arc, incandescent, fluorescent, and HID lamps. The shaded area indicates the possible range of efficacy which depends on the geometric design and operating conditions of the lamp.

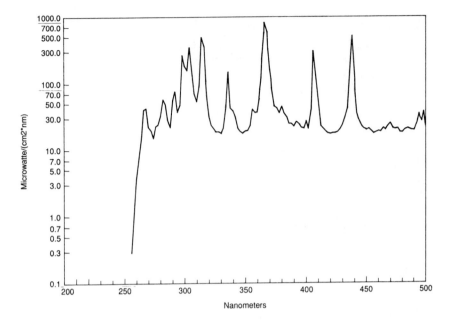

FIGURE 2.27 Logarithmic-scale plot of the spectral irradiance of (*a*) a 1000-W Hg-xenon lamp and (*b*) a 1000-W compact arc xenon lamp.

Other less common sources include compact xenon or mercury-xenon arcs, low-pressure sodium, and microwave-driven lamps. Figure 2.27 shows the spectra of a 1000-W mercury-xenon arc lamp. These lamps can incorporate mercury, mercury-xenon, or xenon gas and have an arc length ranging from 0.3 to 12 mm. The compact arc sources operate at much higher pressures than the original HID lamps. These lamps have the highest radiance of any continuously operating noncoherent source. For example, a 100-W compact arc mercury lamp emits five times the radiance of a 1000-W xenon arc lamp. The primary applications for these lamps are in the motion picture industry, theatrical lighting, and graphic arts. A relatively new uv curing source used in the graphic arts industry is a slender, electrodeless lamp which is excited with microwave radiation. These lamps, from Fusion Systems Corp. (Rockville, Md.), are available with several different types of spectral output (Fig. 2.28*a* to *c*).

Infrared Sources.   Infrared radiation is produced very effectively by almost any incandescent filament lamp. More than 50 percent of the input wattage is radiated as infrared energy in the 770- to 5000-nm region. In general-purpose lighting, it is an undesirable by-product of the total radiant output. As the color temperature of a blackbody increases, the peak of its spectral output shifts away from the longer-wavelength near-infrared region toward the visible region. Heat lamps which emit radiation in the mid- and far-infrared regions are available in powers up to 5000 W. The power distributions of various infrared lamps and heaters are shown in Fig. 2.29.

A highly versatile and reproducible source of infrared radiation is a blackbody cavity. Virtually any shape of cavity can be used, but the most popular are the

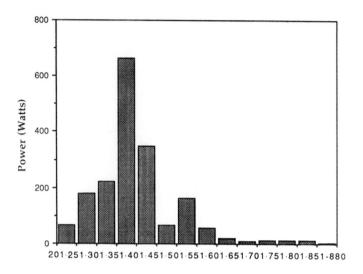

Wavelength (nm)

(a)

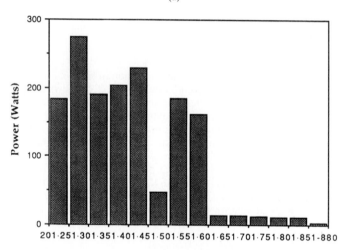

Wavelength (nm)

(b)

FIGURE 2.28 Histogram representation of the spectral irradiance of three microwave-driven lamps available from Fusion Systems, Inc. (Rockville, MD). (*a*) D bulb, (*b*) H bulb, and (*c*) V bulb.

cones and cylinders. A cross-sectional diagram of a blackbody with cone-shaped cavity is shown in Fig. 2.30. Blackbodies can be obtained for temperatures which range from that of liquid nitrogen to approximately 3000°C. For a list of the different types of blackbodies and their characteristics, refer to *The Infrared Handbook*, Table 2.5.[17]

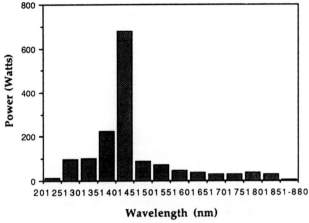

FIGURE 2.28 (*Continued*)   Histogram representation of the spectral irradiance of three microwave-driven lamps available from Fusion Systems, Inc. (Rockville, MD). (*a*) D bulb, (*b*) H bulb, and (*c*) V bulb.

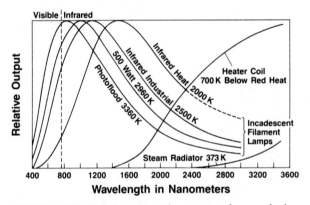

FIGURE   2.29 Relative   spectral   output   of   several   ir sources.   *(Reprinted with permission from Ref 14.)*

Other nondischarge sources include the Nernst Glower, the Globar, and the gas mantle. All of these produce much higher levels of radiation in the 2- to 15-μm region than does a blackbody. The Nernst Glower and the Globar consist of slender, cylindrical rods, which are heated through metallic electrodes at the rod ends. Water cooling is required for the electrodes of the Globar, making it less convenient to use than the Nernst Glower. The gas mantle is found in high-intensity gasoline lamps. It has high visible and far-infrared output, but low near-infrared output. The relative output for the Nernst Glower, Globar, and the gas mantle compared with that of a blackbody is shown in Fig. 2.31.

The first commercially practical electric light source was the carbon arc, developed near the close of the nineteenth century. There are three types of carbon arcs: low-intensity, flame, and high-intensity. The low- and high-intensity arcs are

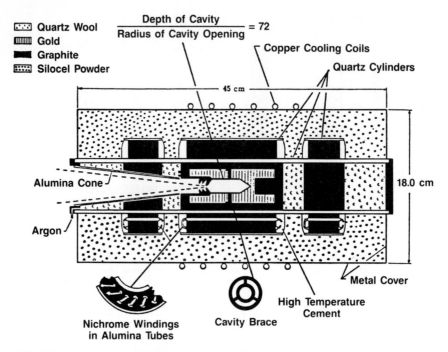

FIGURE 2.30 Cross-sectional diagram of a blackbody cavity showing the various components. *(Reprinted courtesy of the National Institute of Standards and Technology, from NBS Monograph 41,* Theory and Method of Optical Pyrometry, *by H. J. Kostkowski and R. D. Lee, 1962.)*

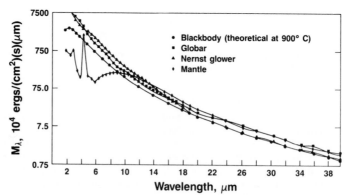

FIGURE 2.31 Spectral plot showing relative output of the Nernst Glower, Globar, gas mantle compared to a 900°C blackbody. *(Reprinted with permission from Infrared Physics, vol. 8, W. Y. Ramsey and J. C. Alishouse "A Comparison of Infrared Sources," 1968, Pergamon Press, Ltd.)*

usually operated on direct current. The flame type adapts to either direct or alternating current. The carbon arc is an open arc which increases in brightness along its length as the distance from the core center increases. The choice of material for the core determines the output spectra. Typical materials are iron for the ultraviolet, rare earths of the cerium group for white light, calcium compounds for yellow, and strontium for red.[16] The spectral output of three different carbon arcs in Fig. 2.32 demonstrates the high intensity available from these sources, and the variability of output for the low-intensity, flame, and high-intensity carbon arc.

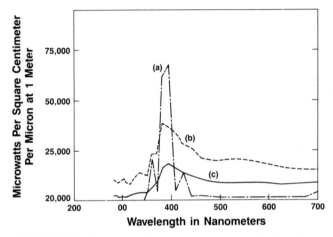

FIGURE 2.32   Spectral plot showing relative output of the low-intensity, flame, and high-intensity carbon arcs.   *(Reprinted from* IES Lighting Handbook *with permission.)*

2.6 REFERENCES

1. Commission International de l'Eclairage (International Commission on Illumination), International Lighting Vocabulary, *Pub. CIE* 17, 1970.

2. "SI Units, Conversion Factors and Abbreviations" (Revised), 1988, *Photochem. Photobiol.* vol. 47, p. 1.

3. Fred E. Nicodemus, "Radiance," *Am. J. Phys.*, vol. 31, no. 50, pp. 368–377, 1963.

4. Robert W. Boyd, *Radiometry and the Detection of Optical Radiation*, Wiley, New York, 1983, p. 75.

5. Ibid, p. 97.

6. David Sliney and Myron Wolbarsht, *Safety with Lasers and Other Optical Sources*, Plenum Press, New York, 1982, p. 766.

7. Boyd, op. cit., p. 23.

8. Robert Siegel and John R. Howell, *Thermal Radiation Heat Transfer*, McGraw-Hill, New York, 1972, Appendix C.

9. Sharon A. Miller, Robert H. James, Stephen Sykes, and Janusz Beer, "Photoaging Effects on the Transmittance of Plastic Filters," *Photochem. and Photobiol.*, 55, pp. 625–628, 1992.

10. J. H. Mackey, H. L. Smith, and A. Halperin, "Optical Studies in X-irradiated High Purity Sodium Silicate Glasses," *J. Phys. Chem. Solids*, vol. 27, p. 1759, 1966.

11. Sharon A. Miller and Robert H. James, "Variables Associated with UV Transmittance Measurements of Intraocular Lenses," *Am. J. Ophthalmol*, vol. 106, pp. 256–260, 1988.

12. James H. Walker, Robert D. Saunders, John K. Jackson, and Donald A. McSparron, "Spectral Irradiance Calibrations," *NBS Special Pub.* 250-20, 1987.

13. L. R. Koller, *Ultraviolet Radiation*, 2d ed., Wiley, New York, 1965.

14. David H. Sliney and Myron Wolbarsht, *Safety with Lasers and Other Optical Sources*, Plenum Press, New York, 1982, p. 188.

15. Alex E. S. Green, K. R. Cross, and L. A. Smith, "Improved Analytic Characterization of Ultraviolet Skylight," *Photochem. Photobiol.* vol. 31, p. 59, 1980.

16. John E. Kaufman and Jack F. Christensen (eds.), *IES Lighting Handbook*, 1984, p. 8–41.

17. William L. Wolfe and George J. Zissis (eds.), *The Infrared Handbook*, Office of Naval Research, Department of the Navy, Washington, D.C., 1978, p. 2–17.

CHAPTER 3
ULTRAVIOLET, VACUUM-ULTRAVIOLET, AND X-RAY LASERS

Roland Sauerbrey

3.1 LASERS IN THE ELECTROMAGNETIC SPECTRUM

The term *laser* is an acronym for a radiation source based on light amplification by stimulated emission of radiation. The foundation for lasers was established with the first formulation of a quantum theory of light by Planck and Einstein.[1] Later, the technological development of lasers was stimulated by the invention of various microwave devices, which led in the 1950s to microwave amplifiers based on stimulated emission of radiation (*masers*). The quest for the development of devices based on this principle but operating at higher frequencies or shorter wavelengths started then and continues up to the present time. It led in 1960 to the construction and operation of the first laser that emitted visible radiation by Maiman.[2] Today, lasers deliver radiation over large portions of the electromagnetic spectrum, ranging from the far infrared to the soft x-ray region (Fig. 3.1).

In principle, most lasers consist of three parts[3] (Fig. 3.2): (1) a pump source, (2) an active medium, and (3) a resonator. Almost every source of energy can, in some ways, be used to pump lasers. Those currently employed for lasers range from a small current in a submicrometer-sized semiconductor heterojunction to nuclear reactors. Of particular importance for short-wavelength lasers are gas discharges and plasma sources. It is the purpose of the pumping process to establish population inversion in the active medium. Population inversion describes a condition of the active medium where the density of states at a higher energy is larger than the density of states at a lower energy.

Three scattering processes characterize the interaction of light and matter: (1) absorption, (2) stimulated emission, and (3) spontaneous emission (Fig. 3.3). In the case of spontaneous emission, a higher excited state decays spontaneously to a state of lower energy, and a photon is emitted. In an absorption event a photon (light quantum) promotes the active medium from a state of lower energy to a higher-energy state. Stimulated emission is the inverse process where photons stimulate an excited state to decay to a state of lower energy, emitting an additional

LASERS IN THE ELECTROMAGNETIC SPECTRUM

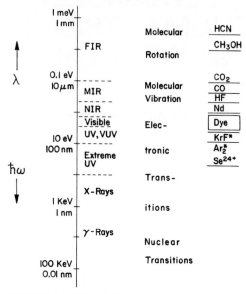

FIGURE 3.1 The left-hand side shows an approximate energy and wavelength scale for the radiation. (The exact energy wavelength conversion is 1 eV $\cong$ 1.2316 μm.) The second column gives the most frequently used name for the respective part of the electromagnetic spectrum. The third column denotes the most common mechanism that produces radiation in any part of the spectrum, and the right-hand column gives some examples of typical lasers.

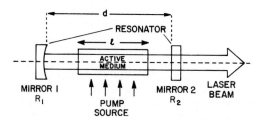

FIGURE 3.2 Typical parts of a laser. The resonator mirrors have the reflectivities R_1 and R_2 and the transmittivities T_1 and T_2. For most dielectric mirrors in the uv and near-vuv in $T_{1,2} \approx 1 - R_{1,2}$. For shorter wavelengths mirror absorption is often important. The length of the active medium is l, and the separation of the resonator mirrors is d.

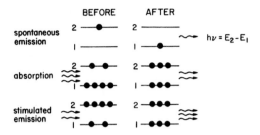

FIGURE 3.3 Fundamental processes for the interaction of radiation and matter. E_1 and E_2 are the energies of levels 1 and 2, which can be in principle any pair of levels in an atom, molecule, or solid. For laser applications mostly level pairs with allowed radiative transitions between them are important.

photon. It is important that in all these three processes energy *and* momentum are conserved. In a stimulated-emission process the energy and the momentum of the newly generated photons equals that of the stimulating photons. In other words, stimulated light is emitted in the same direction and with the same wavelength as the stimulating light.

The number of transitions per volume and time, called the transition rate is, for all three processes, proportional to the density of initial states (for example, state 1 for absorption). In a medium with population inversion, more stimulated (and spontaneous) emission events than absorption events will occur in any given time interval. Consequently, more photons are generated per time than annihilated and the inverted active medium can amplify either the spontaneously generated photons or light inserted in the medium from the outside. In the first case we speak of an amplified spontaneous emission (ASE) device. The second form is called an optical amplifier.

If the amplification in one pass through the active medium is too small to be useful, the active medium can be artificially lengthened by an optical resonator. In its simplest form it consists of two highly reflecting mirrors with a small transmission that enclose the active medium (Fig. 3.2). Spontaneously emitted photons in the direction of the resonator axis are amplified in the active medium, and most of the light is directed back into the active medium by the mirrors, where it is further amplified. When the losses for the light per round trip are smaller than the gain per round trip, the light intensity inside the resonator will grow. The laser has reached threshold when this condition is fulfilled. The light intensity grows until the density of upper laser states (energy E_2, Fig. 3.3) that are produced per time by the pumping process approximately equals the density of photons generated per time by stimulated emission. The laser intensity will then stabilize at this level which is dictated by the pumping process. The laser output has reached saturation.

Figure 3.1 shows the electromagnetic spectrum from the far-infrared (fir) to the gamma-ray region. The middle part indicates the physical process dominating the production of radiation in any particular part of the spectrum. These processes dictate the choice of the active medium for any laser. For example, ultraviolet (uv) radiation is predominantly produced by electronic transitions. Consequently, all ultraviolet lasers utilize electronic transitions in gases, solids, or liquids, as an active medium. The column on the right (Fig. 3.1) shows examples of particular lasers.

Today only the visible part of the spectrum and adjacent parts of the near infrared (nir) and the near-ultraviolet can be continuously covered by laser radiation. These regions can be expanded when nonlinear optical devices such as frequency multipliers or parametric oscillators are coupled with lasers. In the mid- and far-infrared only isolated laser lines are available. The same holds for most of the ultraviolet and the vacuum ultraviolet (vuv). In the extreme ultraviolet and soft x-ray region, amplification of radiation has been observed for many examples, but saturated lasers are only at the beginning of their development.

This chapter summarizes the present status of short-wavelength laser development. The cutoff toward longer wavelengths is placed somewhat arbitrarily at 400 nm, the approximate short-wavelength boundary of the visible spectrum. The emphasis of this article will be on presently available, practical laser sources that emit at wavelengths shorter than 400 nm. The development of short-wavelength lasers, particularly from the vuv to the soft x-ray region, is a field of active research, and the main ideas to develop new shorter-wavelength lasers will be summarized.

3.2   PRINCIPLES OF SHORT-WAVELENGTH LASER OPERATION

In order to discuss the issues important for short-wavelength laser action, we consider the three essential parts of a laser: the active medium, the optical resonator, and the pumping process.

Numerous media are capable of producing uv, vuv, and x-ray photons. For laser action, suitable transitions in these media have to be inverted by an appropriate pumping mechanism.[4] The requirements these media have to fulfill include: (1) transparency or sufficiently small absorption for the laser radiation; (2) the capability of producing sufficiently high gain, which means, in most cases, an allowed optical transition, as well as a relatively narrow linewidth, yielding a high cross section for stimulated emission; and (3) the active medium has to be invertible, at least in the transient regime, meaning that it has to have a suitable energy-level structure and favorable radiative and nonradiative transition rates.

These requirements restrict the available gain media largely to gases and plasmas with few exceptions of uv/vuv lasers in liquids and solids. This may, however, change for potential hard x-ray or gamma-ray lasers, which are discussed at the end of this article.

Optical materials, as well as reflective coatings having reflectivities exceeding 99 percent, are now widely available in the wavelength range between 400 and 200 nm. In the vuv range between 120 and 200 nm, most optical elements such as mirrors of varying reflectivity and transmission, interference filters, prisms, lenses, etc., are available commercially from a few specialized suppliers. The quality approaches that of longer-wavelength optics and is improving continuously, as materials purification and thin film technologies progress. In particular, for high-power uv and vuv lasers, the lifetime of laser optics is, however, still well below optics lifetimes for visible or infrared lasers.

Optics in the xuv and soft x-ray region have for a long time been limited by the nonavailability of highly reflective or highly transparent materials for this wavelength region. Although this is still to some extent the case, there have been some important advancements in recent years. No solid window materials exist below the cutoff wavelength of LiF at 105 nm. Normal incidence reflectivities of metal films are limited to values ≤40 percent below about 100 nm. These materials problems impose severe restrictions on the possibilities of constructing optical cavities for xuv and soft x-ray lasers. There has, however, been some recent progress in the production of highly reflecting mirrors. Using alternating layers of tungsten and carbon matched to the wavelength, several groups were able to generate narrowband, highly reflective (>40 percent) interference mirrors in the 0.5- to 20-nm range.[5] Sophisticated molecular beam epitaxy (MBE) techniques were employed for their production. Although the damage thresholds for these mirrors are still very low, they are beginning to play a major role in the further development of soft x-ray lasers.

The requirements on the pumping process become increasingly demanding for shorter and shorter laser wavelengths. This trend can be observed with existing and relatively mature gas laser systems. For the CO_2 laser at 10 μm, a conventional glow discharge is sufficient to produce strong cw laser output. However, for excimer lasers in the near- or mid-uv around 0.3 μm, only pulsed laser operation is possible, even though the active medium would allow for cw operation. This is due to the high pumping requirements which are characteristic for short-wavelength lasers and which are briefly discussed.

The small-signal gain coefficient g_0 for a homogeneously broadened transition is given by the product of the population inversion n and the cross section σ for stimulated emission: $g_0 = n \cdot \sigma$. The stimulated-emission cross section depends on the radiative transition probability between the upper and lower levels A_{21}, the transition wavelength, and the lineshape function $g(v)$.

$$\sigma(v) = \frac{\lambda_{21}^2 A_{21}}{8\pi} g(v) \tag{3.1}$$

For atoms and ions in gaseous media, the lineshape function is, in general, a Voigt profile:

$$g(v) = \left(\frac{M}{2\pi kT}\right)^{1/2} \int_{-\infty}^{+\infty} \frac{\Delta v_H/2\pi}{[v - v_0 + v_0(v_z/c)]^2 + (\Delta v_H/2)^2}$$
$$\times \exp\left(-\frac{Mv_z^2}{2kT}\right) dv_z \tag{3.2}$$

Here M is the mass of the radiating species, Δv_H the homogeneous linewidth, T the absolute temperature, v_0 the center frequency of the homogeneous transition, v_z the velocity of the radiating species in the direction of the observer (in a laser usually along the resonator or amplifier axis). Obviously, this lineshape function constitutes the statistical average over Boltzmann distributed ensemble of homogeneous emitters with linewidth Δv_H, each radiating at the Doppler shifted frequency $v = v_0(1 + v_z/c)$. Fortunately, in many cases either the velocity distribution or the homogeneous linewidth dominates in width. If the homogeneous linewidth is much larger than the Doppler shift due to the thermal velocity

$$\Delta v_H \gg v_0 \frac{v_{th}}{c} \tag{3.3}$$

where $v_{th} = \sqrt{3kT/M}$, the lineshape function is approximated by a Lorentzian of width Δv_H (FWHM).

$$g(v) = \frac{\Delta v_H/2\pi}{(v - v_0)^2 + (\Delta v_H/2)^2} \tag{3.4}$$

If $\Delta \nu_H \ll \nu_0(\nu_{th}/C)$ the lineshape function is a Gaussian with the full width.

$$\Delta \nu_D = \left(\frac{8kT \ln 2}{Mc^2} \right)^{1/2} \nu_0 \tag{3.5}$$

$\Delta \nu_D$ is called the Doppler linewidth. For electronic transitions in molecular gases, the lineshape may be determined by the internal structure of the molecule, as will be discussed in connection with nitrogen and excimer lasers. Extensive discussions on lineshapes in gases and plasmas can be found in the literature.[6,7]

In weakly ionized plasmas, the most common medium for uv and vuv lasers, the homogeneous linewidth is frequently determined by foreign gas broadening. Since the natural linewidth can usually be safety neglected, the homogeneous linewidth is given by

$$\Delta \nu_H \approx N \sigma_{opt} \nu_{th} \tag{3.6}$$

where N is the gas density and ν_{th} the thermal velocity of the atoms. σ_{opt} is the optical cross section which can be obtained from a Weisskopf theory.[7] The optical cross section is usually about one order of magnitude larger than the gas kinetic cross sections.

In order to estimate the gain in a laser, it is sufficient to consider the cross section for stimulated emission at the line center. For a homogeneously broadened line, we obtain

$$\sigma = \frac{\lambda^2 A_{21}}{4\pi^2 \Delta \nu} \tag{3.7}$$

For a laser to be above threshold, the small signal gain has to exceed the losses. This classical threshold condition, however, is strictly valid only for quasi-cw lasers in sufficiently good resonators and with relatively low gain.[3,8] For many short-wavelength lasers, at least one of these conditions is not fulfilled. High-gain systems show amplified spontaneous emission (ASE). The threshold behavior of an ASE laser is characterized by a gradual transition from a lamplike behavior to a saturated laser, and no sharp threshold is observed.[9,10] An approximate threshold criterion is given by $g \cdot l > 1$, where g is the small signal gain and l is the length of the amplifying medium. If the laser is not quasi-cw—i.e., the gain duration is on the order of or shorter than the photon lifetime in the resonator—again no sharp threshold is observed. The approximate threshold criterion is now that the temporally integrated gain has to exceed a constant value:

$$c \int g(t)dt > K \tag{3.8}$$

where K is a constant on the order of 1.[11] The losses may be characterized by a "loss time" τ_R, which might be the photon lifetime τ_{ph} in a resonator for conventional laser systems, the transit time through the medium for an amplified spontaneous emission (ASE) device, or the gain lifetime τ_g if $\tau_g < \tau_{ph}$. With c being the speed of light, we have as a laser threshold criterion:

$$g_0 = n\sigma \geq \frac{1}{c\tau_R} + \alpha \tag{3.9}$$

where α is the medium absorption coefficient. This yields a minimum or critical inversion density n_c and threshold:

$$n_c = \frac{1}{\sigma c \tau_R} + \frac{\alpha}{\sigma} \qquad (3.10)$$

For a four-level laser scheme given in Fig. 3.4, the minimum pumping power density P can be estimated. For each laser photon, the pump mechanism has to produce an atom or molecule of the active medium in state 2. The pumping energy per atom E_p has to exceed the laser transition energy $h\nu$, yielding $E_p = kh\nu$, with $k > 1$ and h being Planck's constant. Assuming that the upper laser level is only depopulated by radiative processes to the lower laser level (branching ratio equal to 1), the minimum pumping power density is given by

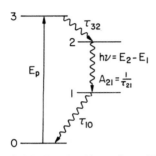

$$P \geq n_c E_p A_{21} \qquad (3.11)$$

With the expression for n_c, assuming no medium absorption ($\alpha = 0$), the minimum pumping power density is

FIGURE 3.4 Four-level laser scheme. The active medium is pumped from its ground state 0 to a highly excited state 3 which requires the pump energy E_p. From there it relaxes to the upper laser state 2, without populating state 1. Consequently, a population inversion between states 2 and 1 is built up that leads to optical gain on the $2 \rightarrow 1$ transition. Photons of energy $b\upsilon$ are guaranteed. The lower laser state 1 relaxes to the ground state.

$$P_{\min} = \frac{E_p A_{21}}{\sigma c \tau_R} \qquad (3.12)$$

When we insert the expressions for σ (8) and E_p in Eq. (3.12), the minimum power density scales like

$$P_{\min} \sim \frac{1}{\eta_Q} \frac{\nu^3 \Delta\nu}{\tau_R} \qquad (3.13)$$

Besides requiring a good quantum efficiency $\eta_Q = k^{-1}$ and low losses (τ_R small), pumping power considerations strongly favor low transition energies and narrow linewidth. Depending on the line broadening mechanism, the increase of P with transition frequency can become even more severe, i.e., for Doppler broadening $\Delta\nu \sim \nu$ which leads to

$$P \sim \nu^4 \qquad (3.14)$$

In general, the minimum pumping power density scales with a high power of the transition frequency:

$$P \sim \nu^\alpha \qquad (\alpha \geq 3) \qquad (3.15)$$

In the light of these simple estimates, it appears unavoidable to invest large power densities in order to achieve lasing in the vacuum ultraviolet (vuv) or soft x-ray region. Assuming as typical values for a uv/vuv laser $E_p \approx 25$ eV, $A_{21} = 10^9$ s^{-1}, $\sigma = 10^{-16}$ cm^2, $c = 3 \cdot 10^{10}$ cm/s, and $\tau_R = 10^{-8}$ s, we obtain $P_{\min} \approx 100$ kW/cm^3 from Eq. (3.12). Since the volumes of the active media of most short-wavelength lasers exceed 1 cm^3, pumping powers well in excess of 100 kW are usually required.

Therefore, most short-wavelength lasers are pulsed lasers because these pumping powers are difficult to sustain for a variety of technical reasons. An exception are the uv atomic transitions, i.e., in the rare gases where the cross sections for stimulated emission are considerably larger ($\sigma \approx 10^{-13}$ to 10^{-14} cm^2) and, consequently, cw laser action can be achieved.

Numerous pump sources are available for short-wavelength lasers. The most common ones are either continuous or pulsed gas discharges for uv/vuv lasers or laser produced plasmas for soft x-ray lasers. The pumping processes will be discussed in detail with each specific short-wavelength laser system.

Of general importance for many short-wavelength systems are questions concerning power extraction and efficiency. First, consider a laser medium of length l with time-independent small-signal gain g_0 in a stable cavity with total output coupling T (Fig. 3.2). If both mirrors have transmissions T_1 and T_2 the total output coupling is $T = T_1 + T_2$. Other losses in the cavity that may be due to absorption or dielectric reflection on optical components are characterized by a loss coefficient α. The total gain or loss per round trip are characterized by $G = 2g_0d$ or $L = 2\alpha d$, respectively. The continuous-wave (cw) output intensity is then given by

$$I_{\text{out}} = T\frac{I_s}{2}\left(\frac{G}{L + T} - 1\right) \tag{3.16}$$

where $I_s = h\nu/\sigma\tau_{\text{eff}}$ is the saturation intensity. Here $h\nu$ is the photon energy and τ_{eff} is the effective lifetime of the upper state that is determined by including both radiative and nonradiative decay of the upper laser state. The saturation intensity characterized the saturated gain g which for an intracavity intensity I is given by

$$g = \frac{g_0}{1 + I/I_s} \tag{3.17}$$

For many laser systems the small-signal gain is proportional to the pumping density P. This can be obtained from our analysis of the four-level laser system shown in Fig. 3.4. If, for example, τ_{32} and τ_{10} are fast compared with all other time constants in the system, the population in state 2 n_2 is equal to the inversion density n. The temporal development of $n = n_2$ is given by

$$\frac{dn}{dt} = R - \sigma n I - \frac{n}{\tau_2} \tag{3.18}$$

where τ_2 is the total lifetime of level 2. In the absence of stimulated emission ($\sigma I \ll 1/\tau_2$), we obtain in equilibrium ($dn/dt = 0$):

$$n = R\tau_2 = \frac{P\tau_2}{h\nu} \tag{3.19}$$

If stimulated emission is important ($\sigma I \geq 1/\tau_2$), the stimulated emission rate σI will depopulate the upper laser state,

$$n = \frac{R\tau_2}{1 + \sigma c\tau_2 I} \tag{3.20}$$

and the inversion density is reduced. Since $g = \sigma n$, Eq. (3.19) is equivalent to Eq. (3.17) for $I_s = 1/\sigma c\tau_2$ and it shows that g is proportional to the pumping rate R.

Consequently, we obtain a linear increase of laser output power with increasing pumping density once the laser has reached the threshold determined by

$$G \geq L + T \tag{3.21}$$

Eq. (3.20) is, of course, equivalent to (3.9). The optimum mirror transmission for a cw laser is given by

$$T_{opt} = \sqrt{G}(\sqrt{G} - \sqrt{L}) \tag{3.22}$$

Most short-wavelength lasers are pulsed lasers. If the gain lifetime exceeds the photon lifetime in the resonator, the conditions for output intensity I_{out} and the optimum transmission I_{opt} given in Eqs. (3.16) and (3.21) are good approximations. In the opposite case, however, the behavior of pulsed lasers can be quite different from that of cw lasers, which is discussed in Ref. 11. As an example, the optimum transmission for a cw and a pulsed laser with identical resonator parameters as a function of the net gain $g_0 - \alpha$ is shown in Fig. 3.5.

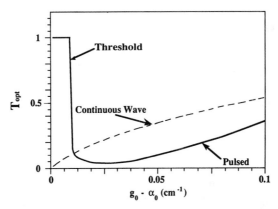

FIGURE 3.5 Optimum transmission as a function of net gain for a pulsed and a cw laser. The solid curve for a pulsed laser was calculated for a gain lifetime of 10 ns, a resonator length of 15 cm, and a constant loss coefficient of $\alpha_0 = 5 \cdot 10^{-3}$ cm^{-1}. The resonator parameters for the cw laser are identical.[11]

Numerous definitions for laser efficiencies can be found in the laser literature. Of particular importance for short-wavelength lasers are those given in Table 3.1.

The quantum efficiency η_Q gives the maximum total efficiency η that can be obtained for a given laser system, when all other processes have an efficiency of 1. The extraction efficiency η_{ext} describes the efficiency a laser system can obtain for a given pumping process. η_{ext} can be maximized by optimizing the resonator parameters. It gives the fraction of excited states that can *not* be extracted in the form of laser photons from the laser medium $(1 - \eta_{ext})$ because they are lost to other processes such as absorption in the medium. The intrinsic efficiency η_{int} characterizes the overall efficiency of the active medium but disregards the efficiency for the production of the pump energy. Of ultimate importance for a well-engineered laser system is its total efficiency η. There are very few laser systems for

TABLE 3.1 Laser Efficiencies

Efficiency	Definition
Quantum efficiency η_Q	$\eta_Q = \dfrac{\text{laser photon energy}}{\text{energy to populate one upper laser state}}$
Extraction efficiency: η_{ext}	$\eta_{ext} = \dfrac{\text{number of extracted laser photons (per time)}}{\text{number of excited states produced (per time)}}$
Intrinsic efficiency: η_{int}	$\eta_{int} = \dfrac{\text{laser energy (power) extracted}}{\text{pump energy (power) deposited}}$
Laser efficiency: η	$\eta = \dfrac{\text{laser energy (power) extracted}}{\text{total energy (power) inserted in laser system}}$

which η exceeds 10 percent, none of them operating in the uv or at shorter wavelengths. Excimer lasers are the most efficient short-wavelength laser sources with $\eta \approx 0.02$ for commercial systems.

In order to optimize the extraction efficiency of a laser system, we consider again the four-level laser system (Fig. 3.4). The total number of decays from the upper level per time is n_2/τ_2 and the stimulated-emission rate is $\sigma I n_2 = gI$. If there is absorption in the system, the absorption rate αI has to be subtracted from the stimulated-emission rate in order to obtain extracted photon density per time. Consequently, the extraction efficiency is[12]

$$\eta_{ext} = \frac{I(g - \alpha)}{n_2/\tau_2} \tag{3.23}$$

Inserting (3.17) yields

$$\eta_{ext} = \frac{I}{I_s}\left(\frac{1}{1 + I/I_s} - \frac{\alpha}{g_0}\right) \tag{3.24}$$

This shows that the extraction efficiency is maximized by adjusting the laser flux to the absorption coefficient α and its bleaching behavior. If the absorption is not bleachable—i.e., α is independent of I—the maximum extraction efficiency is obtained easily from Eq. (3.23):

$$\eta_{ext,max} = 1 - 2\left(\frac{\alpha}{g_0}\right)^{1/2} + \frac{\alpha}{g_0} \tag{3.25}$$

It can be seen from this formula that the maximum extraction efficiency is a very sensitive function of g_0/α. For $\eta_{ext,max} \geq 0.5$ we require $g_0/\alpha \geq 12$. Consequently, even small residual nonsaturable absorptions in a laser medium severely restrict the efficiency for the extraction of excited states in the form of laser photons.

It is not difficult to see the physical origin of this behavior of the extraction efficiency. Consider an amplifier with a nonsaturable absorption (Fig. 3.6). The effective net gain is

$$g_{eff} = g - \alpha = \frac{g_0}{1 + I/I_s} - \alpha \tag{3.26}$$

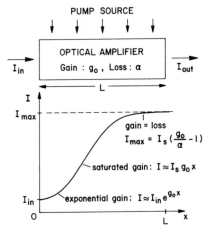

FIGURE 3.6 Saturation behavior of a homogeneously broadened amplifier with nonsaturable loss.

When the small signal gain saturates to a level where $g = \alpha$, the net gain of the amplifier vanishes and no further amplification takes place. The maximum intensity from an amplifier is therefore given for $g_{eff} = 0$ which yields

$$I_{max} = I_s \left(\frac{g_0}{\alpha} - 1\right) \tag{3.27}$$

When I approaches I_{max} in an amplifier, the net amplification equals the nonsaturable absorption and the photons gained are all lost inside the medium to absorption. These excited states, therefore, do not contribute to the output intensity, and the extraction efficiency decreases dramatically. Using Eqs. (3.25) and (3.26), we can easily estimate the maximum useful length of an amplifier. If we integrate the radiation transport equation for an amplifier:

$$\frac{dI}{dx} = g_{eff} I \tag{3.28}$$

in the range $I_s < I < I_{max}$—i.e., for the saturated amplifier case—we obtain approximately

$$I(x) \approx I_s g_0 x \tag{3.29}$$

Setting $I(L_{max}) = I_{max}$ yields

$$L_{max} \approx \frac{1}{g_0}\left(\frac{g_0}{\alpha} - 1\right) \tag{3.30}$$

As an example, consider a discharge pumped excimer amplifier (i.e., KrF) with $g_0/\alpha \approx 10$, $g_0 = 0.1$ cm^{-1}, and a saturation intensity of $I_s \approx 1$ MW cm^{-2}. Using (3.26) we obtain $I_{max} \approx 9$ MW cm^{-2}, which corresponds to a pulse duration of 20 ns, a beam area of 3 cm^2, and an output energy of $E_{out} \approx 540$ mJ. The maximum useful length is $L_{max} \approx 90$ cm. These are typical values for commercial excimer laser amplifiers.

Finally, we note that the cross section for stimulated emission σ, which is the fundamental atomic constant for each laser system, enters the basic laser parameters in two different complementary ways. First, the small-signal gain is proportional to the cross section for stimulated emission. In order to achieve the critical threshold inversion density with minimum pumping requirements, it is desirable to have a system with a large σ. Second, in the presence of losses the maximum output power is proportional to the saturation intensity which decreases with increasing σ and consequently limits the laser output. Laser systems based on atomic transitions have typically narrower linewidth and atomic systems, but molecular lasers are, in most cases, scalable to higher output powers than atomic transition lasers.

3.3 *ULTRAVIOLET AND VACUUM*
ULTRAVIOLET LASERS

Ultraviolet (uv) and vacuum ultraviolet (vuv) lasers are distinct from shorter-wavelength lasers mainly through different techniques for spectroscopy in the respective wavelength ranges. For ultraviolet lasers good optical materials are available and the radiation can be transported in air at least as long as the laser power is not too high. For radiation with wavelengths shorter than 185 nm, the air becomes absorbing mainly owing to absorption in the Schumann-Runge bands of oxygen, and laser radiation in this wavelength range has to be propagated in vacuum. Optical window materials with good transmission are available down to 105 nm, the cutoff wavelength of LiF. Transmission curves for various optical window materials are shown in Fig. 3.7. UV lasers in this chapter are those where wavelength is between 400 and ~185 nm. VUV lasers are in the wavelength range between 185 and 105 nm. Lasers with wavelength somewhat shorter than 105 nm are also sometimes called extreme ultraviolet or xuv lasers. There is no clear wavelength distinction between x-ray and xuv lasers. In this chapter existing uv and vuv lasers are briefly reviewed. The emphasis is on the underlying physics, in particular the spectroscopy and the pumping techniques for these lasers. Commercial systems, if available, will be mentioned briefly.

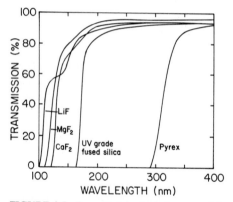

FIGURE 3.7 Transmission of window materials frequently used for uv and vuv lasers. Dielectric losses are included in the transmission.

3.3.1 Gas Lasers

The pumping mechanisms for most gas lasers are electrical gas discharges. We do not treat this important subfield of lasers physics in great detail, but rather quote the essential results. An excellent short review can be found in Verdeyen's textbook on *Laser Electronics*.[3] A more complete text on discharge physics is Ref. 13, and the gas kinetic processes relevant for gas lasers are summarized in Ref. 14.

Ion Lasers.   *Spectroscopy.*   Consider an atom or ion with a ground state that consists of a filled or partially filled electron shell. When one electron is excited to a higher energy level, its binding energy depends strongly on its angular momentum. S-electrons have the lowest energy from the ground state (i.e., the highest binding energy) because they have a high probability to be close to the nucleus. Therefore, these s-electrons are attracted by a less shielded nuclear potential than electrons with angular momentum that are on an average farther away from the nucleus and move in a potential that is more effectively shielded by the remaining electrons in the lower-lying shells. The energy separation between atomic states that have the same principal quantum number but different angular momentum is expressed by the quantum defect $\Delta(l)$.

$$E_{n,l} = -\frac{E_{ion}}{[n - \Delta(l)]^2} \tag{3.31}$$

$E_{n,l}$ is the binding energy of a singly excited electronic state of an atom or ion with ionization energy E_{ion} with principal quantum number n. The quantum defect $\Delta(l)$ is a function of the angular momentum l of the electron and decreases with increasing l. Since the ionization energy of an ion increases strongly with increasing ionization stage, the energy separation between states of the same principal quantum number and different l values can vary widely depending on the ionization stage. For example, for neonlike systems the energy difference between the $3s$ and $3p$ states is on the order of about 2 eV for neutral neon but about 60 eV for neonlike Se^{24+}. This neonlike family of atomic states plays an important role in laser physics. The He-Ne laser uses these states in neutral neon. The rare gas ion lasers make extensive use of these states, and even x-ray lasers are based on these schemes. In this section we describe ultraviolet transitions in rare gas ion lasers. Laser transitions in the ultraviolet can also be seen in metal vapors, but not many metal uv laser lines are commercially available. In order to have sufficiently large energy separations in the uv, at least doubly ionized rare gas ions are used in uv rare gas ion lasers. An energy-level diagram of ArIII is shown in Fig. 3.8. Most of the laser transitions take place between the $4p$ and $4s$ levels, as discussed before. Some $4p \rightarrow 3d$ transitions are also lasing.

Laser Pumping.   The quantum efficiency of the 288.5-nm ArIII laser line is given by (Fig. 3.9)

$$\eta_Q = \frac{E(4p'^3P_1 \longrightarrow 4s'^3D_2^0)}{E(ArI) + E(ArII) + E(4p'^3P_1)} \tag{3.32}$$

with $E(4p'^3P_1 \rightarrow 4s'^3D_2^0) = 4.27$ eV, $E(ArI) = 15.75$ eV, $E(ArII) = 27.6$ eV, $E(4p'^3P_1) = 28.7$ eV, yielding $\eta_Q = 0.059$. These low quantum efficiencies are characteristic of rare gas ion lasers. Since the lower laser level depopulates sufficiently rapidly by radiation, rare gas ion lasers can be operated as cw lasers.

Rare gas ion lasers are pumped by electrical discharges. A schematic diagram of an ion-laser tube is shown in Fig. 3.9. In case of an ArIII laser an argon atom is first doubly ionized by two subsequent electron impact ionization collisions. A third collision with an electron promotes the ArIII ion into the $4p$ state. Since the cross section for electron impact excitation in these systems is typically larger for the $3p \rightarrow 4p$ transitions than for the $3p \rightarrow 4$ transitions, population inversion between the $4p$ and $4s$ states is achieved. The uv ion lasers are pumped by a process that involves at least three consecutive electron collisions. In order to have a sufficient rate for such a sequence of collisions, the gas discharges to pump such lasers are operated at high current densities. Typical for uv ion lasers are currents on the order of 50 to 100 A and a voltage of 300 V across the discharge, corresponding to a total power dissipation on the order of 15 to 30 kW. Ion-laser plasma

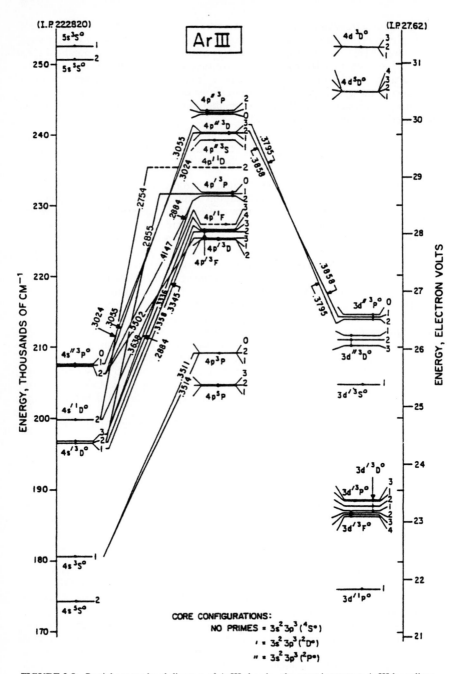

FIGURE 3.8   Partial energy-level diagram of ArIII showing the most important ArIII laser lines.

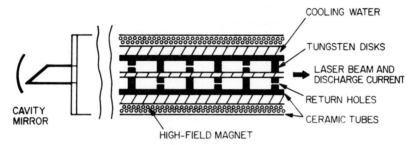

FIGURE 3.9 Schematic diagram of a high-power water-cooled ion-laser tube.

tubes are usually enclosed in a longitudinal magnetic field in order to avoid frequent collisions of ions and electrons with the wall. This measure reduces the adverse effects of the plasma on the tube wall and enhances the pumping density. The high pumping power causes considerable stress on the plasma tube and requires efficient cooling of such systems. Because of the multistep excitation process, the laser output power is usually very sensitive to the discharge current. Typical single-line laser powers from a Xe-ion uv laser are shown in Fig. 3.10.

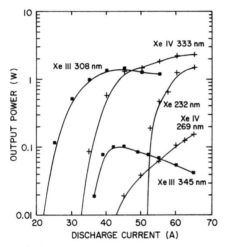

FIGURE 3.10 Output power of uv XeIII and XeIV lines as a function of discharge current in a modern high-power ion laser.

Laser Parameters.   Today commercial systems are available with single-line cw output in the uv well in excess of 1 W (Ref. 15). Some of the shortest wavelengths available are a line at 219 nm in krypton and the Xe 232-nm line (Fig. 3.10). These direct laser lines now have shorter wavelengths than the output that can be obtained from frequency doubling visible ion-laser lines. A total of several hundred uv laser lines have been observed in the rare gases.[16] Some of the more important ones are summarized in Table 3.2.

TABLE 3.2 Output Power of Selected UV Rare Gas Ion-Laser Lines

Gas	λ, nm	Discharge current, A	Output power, mW
Neon	322–338	65	2300
Argon	275	80	94
	288	80	94
	300–308	60	475
	300–308	80	1020
	333–364	80	4300
	438	60	345
Krypton	337–356	60	1080
	337–356	75	1230
Xenon	282	55	230
	308	55	420

The Nitrogen Laser. Spectroscopy. Contrary to the ion lasers treated in the previous section which operate on atomic transitions, the nitrogen (N_2) laser is a molecular gas laser. The N_2 laser was realized first in 1963.[17] More than 440 laser emission lines are known in neutral N_2. Additional laser lines can be found in atomic nitrogen and N_2^+.[18]

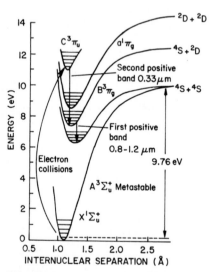

FIGURE 3.11 Partial energy diagram of N_2. Discharge electrons preferentially pump the C state and inversion to the B state is created. This leads to lasing on several vibronic lines of the $C \rightarrow B$ transition. In discharge pumped N_2 lasers the transition at 337 nm in the most intense laser line.

The most important laser transitions in this molecule are the vibronic transitions in the ultraviolet between the electronic $C\,^3\Pi_u$ and $B\,^3\Pi_g$ states of neutral N_2. Besides the $C \rightarrow B$ transitions, laser emission is observed from the $B \rightarrow A$ transition in the near infrared. The transitions important for uv laser action in N_2 are shown in the partial energy diagram given in Fig. 3.11. The potential parameters and radiative lifetimes of the relevant transitions are given in Table 3.3.

The best known, and for discharge pumping the most intense, emission of the N_2-laser is at 337.1 nm, corresponding to the $C(^3\Pi_u, v' = 0) \rightarrow B(^3\Pi_g, v'' = 0)$ transition. Here v' and v'' denote the vibrational levels of the upper C and lower B states, respectively. A more detailed spectral analysis shows that the laser transition takes place on many rotational transitions which leads to a relatively large linewidth of $\Delta\lambda \approx 0.1$ nm for this transition. The Franck-Condon factors to v'' for the $C(v' = 0) \rightarrow B(v'')$ transitions are given in Table 3.4.[19]

TABLE 3.3 Data on N_2

State	T, cm^{-1}	ω_e, cm^{-1}	$\omega_e x_e$, cm^{-1}	r_e, Å	τ
$X^1\Sigma_g^+$	0	2359.61	14.456	1.094	∞
$A^2\Sigma_u^+$	50,206.0	1460.37	13.891	1.293	seconds
$B^3\Pi_g$	59,626.3	1734.11	14.47	1.2123	10 μs
$C^3\Pi_u$	89,147.3	2035.1	17.08	1.148	40 ns

TABLE 3.4 Selected Wavelengths and Franck-Condon Factors for the N_2 (C → B) Transitions

Transition	0-0	0-1	0-2	0-3
Wavelength/nm	337	358	380	406
$f_{0v''}$	0.45	0.33	0.15	0.05

The cross section for stimulated emission on the 0-0 transition at 337.1 nm can now be estimated according to (3.7)

$$\sigma_{00} \approx \frac{\lambda_4 A_{CB} f_{00}}{8\pi c \Delta \lambda} \approx 2 \cdot 10^{-15} \text{ cm}^2 \tag{3.33}$$

In order to have a small-signal gain of $g_0 \approx 1$ cm^{-1}, a population inversion of $n = N(C, v' = 0) - N(B, v'' = 0)$ of $n \approx 5 \cdot 10^{14}$ cm^{-3} is required. Since the radiative lifetime in the B state is considerably longer than in the C state, the N_2 laser bottlenecks. This means that in pure N_2 gas gain exists only until the lower laser state, which is populated mainly by stimulated emission from the upper laser state, has reached the same population density as the upper laser state. This effect represents the main limitation for efficient energy extraction from the molecular nitrogen laser.

Pumping. Pulsed gas discharges are the most frequently used pumping method for uv nitrogen lasers. Their success is rooted in the relative magnitude of the cross sections for electron impact excitation of the B state and the C state from the nitrogen ground state $N_2(X^1\Sigma_g^+, v = 0)$ (Fig. 3.11). Since the electron impact excitation cross section for the process $N_2(X) + e^- \rightarrow N_2(B) + e^-$ is smaller than for $N_2(X) + e^- \rightarrow N_2(C) + e^-$ in the vicinity of their energy thresholds the high-energy tail of the electron energy distribution in a glow discharge will preferentially pump the $N_2(C)$ state and population inversion between the C and the B state can be achieved. Since the C state decays into the B state within about 40 ns while the B state lives for ~10 μs (Table 3.3) the N_2 laser operates only as a pulsed laser in the transient regime and the discharge has to be faster than the 40-ns lifetime of the upper state.

The major challenge for pumping N_2 lasers is therefore the development of fast discharges. For high-power lasers usually striplines with low impedance,often in connection with Blumlein pulse formers, are used.[20] A typical setup is shown in

Fig. 3.12. The discharge between the metal electrodes excites the laser medium transverse to the laser axis. A typical voltage between the metal plates is 20 kV. When the switch, typically a hydrogen thyratron, fires and connects the two metal plates, the left electrode is pulled to zero voltage, the high voltage is suddenly applied across the electrodes, and a fast discharge between the electrodes commences. More detailed analysis of this circuit actually reveals that the voltage pulse-propagation times are dependent on the geometry of the striplines and the dielectric constant of the medium between the metal plates. For optimum energy delivery to the discharge, the wave impedance of the stripline should be on the order of the discharge resistance of typically 0.1 Ω. The wave impedance of a stripline is given by $Z = 377 \times (s/w)(\epsilon_r)^{-1/2}\Omega$, where s represents the gap and w the width.

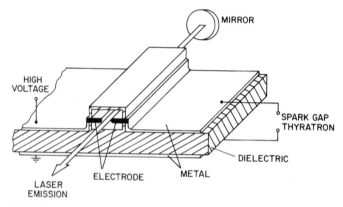

FIGURE 3.12   Schematic of a nitrogen laser with a Blumlein pulse-forming line.

Since the gap and the width are determined by the voltage and the laser length, a high dielectric constant is necessary to lower the stripline impedance. Solid dielectrics are used frequently, but it was demonstrated recently that water gives good results and has the additional advantage of being self-healing in event of a high-voltage breakdown.[21] This work also demonstrates that pseudo-spark switches can be used successfully for fast pulsed power laser pumping.

Typical gas pressures in a N_2 laser range from several 10 torr to more than 1 atm. The laser pulse duration tends to become shorter with increasing gas pressure. While most commercial systems have typical pulse durations between 3 and 10 ns recently also N_2 lasers operating above atmospheric pressure with subnanosecond pulsewidth became available.

Laser Parameters.   Nitrogen lasers are capable of very large gains in excess of $g_0 \approx 1$ cm^{-1}. Despite respectable output powers in the range of 100 kW to 10 MW, the energy output is rarely above 10 mJ, which is due to the short pulse duration caused by bottlenecking in the lower state. Because of its high gain, the nitrogen laser operates superradiant. One mirror is used simply to direct the output in one direction and reduces the beam divergence somewhat, which is usually between 5 and 10 mrad. Typical data for a commercial N_2 laser are given in Table 3.5.[22] Other pumping methods have also been used to operate the N_2 laser. High-power relativistic electron-beam pumping of Ar/N_2 mixtures gives higher intrinsic laser effi-

TABLE 3.5 Typical Nitrogen Laser Parameters

Wavelength	337 nm
Pulse duration	0.5–10 ns
Pulse energy	$\leq$10 mJ
Repetition frequency	$\leq$100 Hz
Beam dimensions	20 × 5 mm
Beam divergence	~10 mrad
Bandwidth	0.1 nm
Efficiency	0.1%

ciencies. Using this method, N_2 lasers can be operated on all four vibronic transitions listed in Table 3.4.[23] Transitions in N_2^+ that give laser emission at 391 and 428 nm can also be excited by electron-beam or discharge pumping of He/N_2 mixtures.

Commercial nitrogen lasers are frequently used to pump dye lasers. Although excimer lasers give higher output powers and shorter wavelength, nitrogen lasers are still used owing to their cost-effectiveness. The short pulse duration of the N_2 laser is sometimes of advantage for studies requiring temporal resolution of ~1 ns or below in laser plasma interaction or laser-induced fluorescence.

Excimer Lasers.   Rare gas halide excimer lasers are the most widely used uv lasers today. Their active medium constitutes a subclass of a wider class of molecules called excimer or exciplex molecules. Their common characteristic is an unbound or repulsive electronic ground state while an electronically excited state forms the lowest bound state of the molecule. A generic potential scheme of such molecules that also includes major formation pathways is shown in Fig. 3.13. After formation of the lowest bound state, this state can decay radiatively into the repulsive ground state. Since the lower laser state decays on the time scale of the vibrational period of a molecule ($\tau_1 \approx 10^{-13}$ s), the stimulated-emission rate is limited by $\sigma I \approx \tau_1^{-1}$, which yields $I \approx 10^{28}$ cm^{-2}s^{-1} corresponding to about 10^{10} W cm^{-2} in the deep uv. For smaller intensities the lower state lifetime and the lower state can always be considered unpopulated. In this sense excimer lasers are ideal lasers because any population in the upper state corresponds to an inversion and the inversion density always equals the upper laser state population density. Figure 3.14 gives an overview of most known excimer emissions of small, neutral molecules that are suitable for laser applications. These molecules cover the visible, uv, and vuv range of the spectrum. Several good review articles on excimers and excimer lasers have been published in the 1980s.[24–26] In this brief summary we focus on the rare gas and rare gas halide excimers.

Rare Gas Excimers

1. *Spectroscopy.*   The atomic ground state of Ne, Ar, Kr, and Xe has the electronic configuration $s^2p^6(^1S_0)$, and the first excited configuration s^2p^5s describes four states. Russel-Saunders notation is often used to describe these states ($^1P_1$ and $^3P_{0,1,2}$), although the terms *singlet* and *triplet* are not very meaningful for the heavier gases. The structure of the excited noble gas atoms is very similar to that of the ground-state alkali metals. A single s electron orbits a core of unit

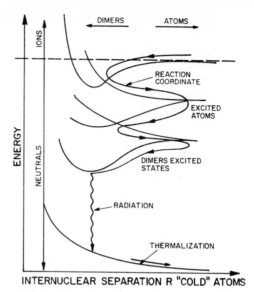

FIGURE 3.13 Energy-level scheme and typical formation pathways of a generic excimer state. Ion recombination in a high-pressure discharge or electron-beam excited gas mixture leads to formation of an excimer state which is the lowest bound state in the molecule. Because of the fast kinetics in the high-pressure gas mixture, most of the initial ion population is converted into excimer molecules. Since the lowest potential curve is repulsive, population inversion between the excimer state and the lower state is achieved and intense laser radiation can be generated.

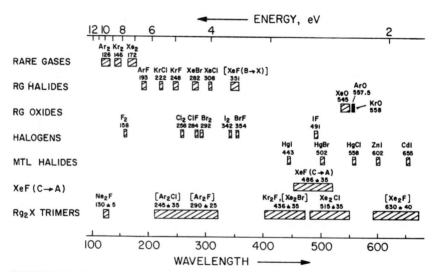

FIGURE 3.14 Overview of small excimer molecules with emission ranging from the vuv to the visible.

positive charge. Ionic bonds may be formed between an excited noble gas atom and electronegative atoms or groups to form excimers. However, in the case of two neutral noble gas atoms, ionic or covalent bonding is not possible, but the positive molecular ion is strongly bound (~1 eV) and gives rise to a set of bound Rydberg states for the excimer. Calculations of the potential curves for Xe_2^*, Kr_2^*, and Ar_2^* have been made. The curves for Ar_2^* are shown in Fig. 3.15 and are typical of the heavier noble gases. These indicate the existence of strongly bound ($\epsilon_b \sim 1$ eV) $^1\Sigma_u^+$ (O_u^+) and $^3\Sigma_u^+$ (O_u^-, $1u$) states which correlate with the atomic $^3P_1$ and $^3P_2$ states, respectively, and rather weakly bound $^1\Sigma_g^+$ (O_g^+) and $^3\Sigma_g^+$ (O_g^-, $1g$) states which also correlate with $^3P_1$ and $^3P_2$. At relatively low pressures, two-body collision processes populate vibrational excimer levels close to the dissociation limit and give rise to broadening to the long-wavelength side of the $^3P_1$ atomic resonance line. This is referred to as the "first continuum." At higher pressures ($\gtrsim 100$ torr), excimers are formed by three-body collisions

$$Ar^*(^3P_{1,2}) + Ar + Ar \longrightarrow Ar_2^*(^{1,3}\Sigma_u^+) + Ar \qquad (3.34)$$

By absorbing energy, the third body enables lower vibrational levels to be populated and emission occurs over a broad continuum (second continuum) centered at 126 nm for Ar_2^*. At even larger wavelengths a third continuum is observed (for argon centered at ~190 nm) that has recently been assigned to the ionic excimer transition $Ar^{2+} + 2Ar \rightarrow Ar^{2+}Ar + Ar$; $Ar^{2+} + A \rightarrow Ar^+ + Ar^+ + h\nu$.[27]

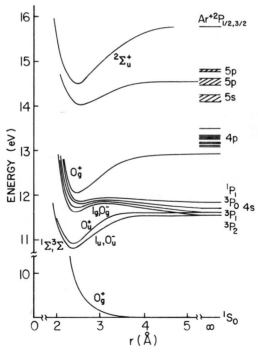

FIGURE 3.15 Partial potential-energy diagram of the argon excimer.

TRANSVERSE EXCITATION

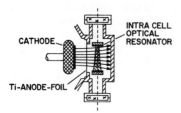

COAXIAL EXCITATION

FIGURE 3.16   Transverse and coaxial excitation geometries used in electron-beam pumped excimer lasers.

2. *Laser Pumping.*   Rare gas excimer lasers are almost exclusively pumped by high-current (1 to 10 kA), high-voltage (0.25 to 2.5 MeV) electron beams. Typical arrangements are shown in Fig. 3.16. The electron-beam source consists of a high-voltage generator such as a Marx bank or pulse transformer, a pulse-forming line to produce ideally square pulses of 10 to 100-ns duration, and a vacuum diode. Electron emission is from a cold cathode which is constructed from graphite, carbon felt, or sharp blades to enhance the local electric fields and produce more efficient and uniform emission. The anode consists of a foil, usually of titanium, aluminum, stainless steel, or aluminized dielectric, which is sufficiently thin (≤ 50 µm) so as to allow efficient penetration by electrons with energies of 200 keV or greater. The maximum current density (J) supplied by a diode is limited by space-charge effects and is given by the Child-Langmuir law

$$J = \frac{2.3 \times 10^3}{d^2} V^{3/2} \text{A cm}^{-2} \tag{3.35}$$

where d is the anode cathode separation (cm) and V is the applied voltage (MV). The pulse duration which may be obtained from a cold-cathode diode is limited by diode closure which is produced by expansion of a plasma from the cathode with a velocity of $\sim 2 \times 10^6$ cm s^{-1} and effectively decreases the anode-cathode separation and therefore the impedance of the diode. However, high-current pulses of up to 1 µs duration may be generated in this way.

When a transverse diode geometry is used, the gas is contained in a cell and is pumped in a direction transverse to the optical axis of the laser. Scattering of electrons by the metallic foil which separates the diode vacuum from the laser gas causes the energy to be deposited nonuniformly in the gas, the greatest excitation density occurring close to the entrance foil. More uniform deposition and therefore more efficient use of the pumping energy may be obtained using a coaxial geometry where the gas is contained in a thin-walled metal tube which acts as the anode and is concentric with a cylindrical cathode. The electrons are accelerated radially inward and excite the gas contained within the anode. Formation of the upper laser state in electron-beam pumped rare gases is quite involved.[24-26] The main kinetic steps are briefly summarized for the example of the Ar$_2^*$ excimer. The electron beam (e_{fast}) generates primarily argon ions and secondary electrons:

$$\text{Ar} + e_{\text{fast}} \longrightarrow \text{A}^+ + e_{\text{fast}} + e_s \tag{3.36}$$

The secondary electrons are cooled rapidly by collisions, and an electron temperature of a few electronvolts is reached (typically $kT_e \approx 1$ to 3 eV). Thus, at high pressures (several atmospheres) the atomic ions undergo rapid three-body collision processes to form molecular ions:

$$Ar^+ + 2Ar \longrightarrow Ar_2^+ + Ar \qquad (3.37)$$

The molecular ions recombine rapidly with secondary electrons to form $Ar^*(4p)$ states that relax rapidly to $Ar^*(4s)$ states

$$Ar_2^+ + e_s^- \longrightarrow Ar^* + Ar \qquad (3.38)$$

This dissociative recombination process is most efficient for slow electrons and therefore constitutes the main heating process for the electrons to stabilize the electron temperature. Excited argon atoms undergo rapid three-body quenching to the argon excimer molecule:

$$Ar^* + 2Ar \longrightarrow Ar_2^*(^{1,3}\Sigma_u) + Ar \qquad (3.39)$$

Both the singlet and the triplet states are formed in this process. The triplet state has a radiative lifetime of 3 μs, while the somewhat higher-lying ($\Delta E \approx 1000$ cm^{-1}) singlet state has a lifetime of 4 ns. The lasing state is consequently the $Ar_2^*(^1\Sigma_u)$ state. In order to obtain sufficient population in this state, the gas pressure has to be high enough to ensure rapid vibrational relaxation and efficient triplet-singlet mixing in the atomic and molecular states. Therefore, the Ar_2^* laser at 126 nm operates typically above 30 atm, while for the Xe_2^* laser at least 10 atm is required. Early work on rare gas excimers is summarized in Refs. 24 and 25.

More recently the Ar_2^* excimer laser, which is particularly attractive owing to its short wavelength of 126 nm, has been operated as a tunable, electron-beam-pumped vuv laser.[28] Wavelength tuning over a 3-nm width from 124.5 to 127.5 nm with a linewidth of 0.3 nm was reported from a coaxially pumped Ar_2^* laser. Frequency tuning was obtained using a MgF$_2$ prism in the cavity. The output pulses had typically a 10-ns pulse duration and a peak power of 2 MW, corresponding to a pulse energy of 20 mJ. Because of their substantial output powers and short wavelength, the rare gas excimers are very interesting vuv lasers. Electron-beam excitation is, however, at the present time the only workable pumping technique, which limits repetition rate and reliability of such lasers. Efforts to pump the Kr_2^* transition in a discharge[29] or the Ar_2^* excimer in a nozzle discharge[30] have not yet led to laser emission.

Rare Gas Halide Excimers. The spectra of rare gas halide excimers were first observed by Golde and Thrush[31] and Velazco and Setser,[32] in 1974. The first rare gas-halide (RGH) excimer laser (XeBr) was reported by Searles and Hart[33] in 1975. Shortly thereafter, lasing from XeF was obtained by Brau and Ewing.[34] Initially, these lasers were pumped by intense electron beams. Subsequently, other rare gas halide lasers were reported. Today, commercial systems employ volume-uniform avalanche discharges with x-ray, uv, or corona preionization. Besides electron-beams excitation, electron-beam-controlled discharges and proton beams have been used in experimental devices. The main transitions of the rare gas halides are shown in Fig. 3.14. They cover the spectrum from the near uv to the vuv. In addition to the diatomic rare gas halides, triatomic rare gas halide excimers[35] can provide tunable coherent photon sources in the visible to the uv region of the spectrum

(Fig. 3.14). Furthermore a four-atomic rare gas halide excimer, Ar_3F at (430 ± 50) nm, was reported in 1986.[36] Rare gas halide lasers have been reviewed frequently. This summary follows the treatments by Brau,[12] Hutchinson,[25] and Obara.[26]

1. *Spectroscopy.* The electronic configuration of an excited rare gas atom is very similar to that of an alkali metal, i.e., a single s electron orbiting a core of unit positive charge, and results in a strong similarity between the ionization potentials and polarizabilities of the metastable states ($^3P_{0,2}$) of Ne, Ar, Kr, and Xe and the ground states of Na, K, Rb, and Cs, respectively. In particular, the excited noble gases form very strong ionic bonds by charge transfer to electronegative atoms such as the halogens forming excimers which radiate in the ultraviolet and vacuum ultraviolet. By taking advantage of the similarity between noble gas halides and alkali halides, the emission wavelengths of many molecules can be predicted.

Some understanding of why some excimers are formed and radiate with high efficiencies while others radiate less strongly or not at all can be gained by considering the mechanisms by which noble gas halides are formed. Excited noble gas atoms (A^*) have relatively low ionization potentials (4 to 5 eV) and can interact with electronegative molecules (RX) acting as halogen donors by a charge transfer or "harpooning" mechanism, e.g.,

$$A^* + RX \longrightarrow A^* + RX^- \tag{3.40}$$

where X is a halogen atom. As shown in Fig. 3.17 for the example of KrF, this charge transfer may take place at relatively large atom-molecule separations (0.5 to 1 nm) where the covalent (A^*, RX) and ionic (A^+, RX^-) potentials

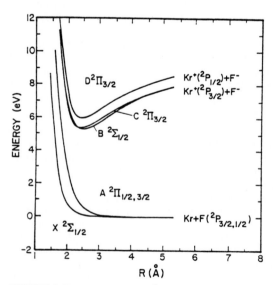

FIGURE 3.17 Potential-energy diagram of KrF.

curves cross. The donor ion RX^- may then dissociate in the field of the noble gas ion to form the ionic excimer $(A^+X^-)^*$ in a vibrationally excited state, e.g.,

$$A^+ + RX^- \longrightarrow (A + RX^-) \longrightarrow (A + X^-)^* + R \qquad (3.41)$$

For the reaction of the excited noble gas atom and the donor molecule to lead to the formation of an excimer, the dissociation energy $D(A - X)^*$ of the excimer of A^* and X must be greater than the dissociation energy $D(R - X)$ of the donor molecule; i.e., the reaction $A^* + RX \rightarrow AX^* + R$ must be exothermic. The excimer is therefore formed in a range of vibrationally excited states. The potential diagram of ArBr is shown in Fig. 3.18. If, as in the case of ArBr, the covalent potentials $A + X^*$ cross the ionic excimer potentials close to their minima, the probability of predissociation $(A^+X^-)^* \rightarrow A + X^*$ is very high, and for this reason no ArBr excimer emission is observed. In ArCl, potential crossings occur at much higher energies, and although predissociation leading to emission from atomic chlorine does occur when the excimer is formed from Ar and Cl_2, excimer emission is observed. The relation between ArCl formation and Cl_2^* formation is shown schematically in Fig. 3.19. The ionic B states correlate with the separated ion pair $A^+(^2P) + X^-(^1S)$. The smaller the halogen ion, the smaller will be the equilibrium length of the ionic $A - X$ bond and so the greater will be the ionic dissociation energy. Since all the halogen atoms have similar electron affinities ($\sim$3 eV), the smallest ions, the fluorides, will have the lowest B states. Hence the $B - X$ emission wavelengths of the

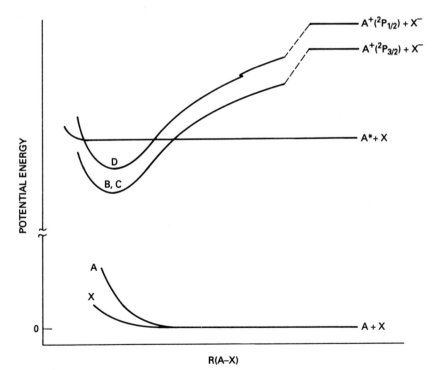

FIGURE 3.18 Potential-energy diagram of ArBr.

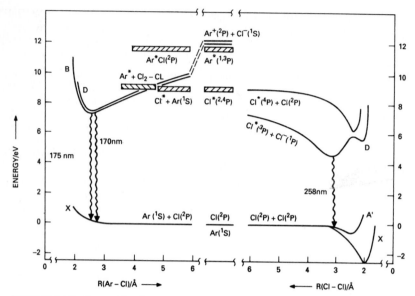

FIGURE 3.19 Competition between ArCl* formation and Cl_2^* formation. Predissociation of the ArCl* excimer in high vibrational levels leads to the formation of $Cl_2^{**}(^4P)$ which forms Cl_2^*.

halides of a given noble gas decrease monotonically with increasing atomic number of the halogen. The ionization potentials of the noble gases increase with decreasing atomic number, and therefore $B - X$ emission wavelengths of a given halide of different noble gases decrease monotonically with decreasing atomic number of the noble gas. The lighter noble gases, He and Ne, would form excimers with ionic potential minima well above the covalent potentials correlating with $A + X^*$. Under these circumstances the formation of stable excimers is unlikely, and the only neon-halide excimer emission to be observed is from NeF. Figure 3.20 shows the emission wavelengths of rare gas halides which corroborate this simple picture.

	Ne	Ar	Kr	Xe	
	108 F	193 L	248 L	351 L	F
	P	175 F	222 L	308 L	Cl
	P	161 F	206 F	282 L	Br
	P	P	185 F	253 F	I

FIGURE 3.20 Emission wavelength of rare gas halide molecules. L: Lasing has been observed. F: Only fluorescence but no lasing. P: Excimer predissociates and no fluorescence is observed.

The interaction of a noble gas atom and a halogen atom gives rise to two states designated by $1^2\Sigma^+$ and $1^2\Pi$ depending on the orientation of the singly occupied $2p$ halogen orbital. However, owing to spin-orbit splitting in the ground state of the halogen atoms, the 2Π state is split, giving rise to $X(1/2)$, $A(1/2)$, and $A(3/2)$ states. Similarly, the charge transfer interaction between the excited noble gas atom and halogen atom gives rise to two ionic states, $2^2\Sigma^+$ and $2^2\Pi$, which in the absence of spin-orbit coupling would be almost degenerate. However, spin-orbit effects are very pronounced in the noble gas ions where, for example, the $Kr^+(^2P)$ ground state is split ($^2P_{1/2}$, $^2P_{3/2}$) by 0.666 eV. Thus the ionic $^2\Pi$ state which correlates with the ^2P ion ground state is also split into $\Omega = 1/2$ and $\Omega = 3/2$ components. The higher-lying $\Omega = 1/2$ state is referred to as the $D(1/2)$ state and the lower $\Omega = 3/2$ state, which is referred to as the $C(3/2)$ state, is almost degenerate with the $B(1/2)$ state which is derived from the $^2\Sigma^+$ state. The normal designation of the spin-orbit corrected potentials is, to label them, A, B, C, etc., in order of increasing energy. Since the B and C states are nearly degenerate, their designation in terms of the axial angular momentum Ω may be reversed.

Radiative transitions may occur on the $D(1/2) \rightarrow X(1/2)$, $B(1/2) \rightarrow X(1/2)$, and $C(3/2) \rightarrow A(3/2)$ transitions. The typical excimer emissions are shown for XeF in Fig. 3.21. In all the noble gas halides, the A states are strongly repulsive, whereas the $X(1/2)$ state is at most only weakly repulsive at the internuclear separations at which transitions from the upper manifolds take place. The $B - X$ bands all show pronounced structure which is expected for the fairly flat potentials of the lower states. However, the $X(1/2)$ states of XeF and XeCl are bound by 1065 and 255 cm^{-1}, respectively, so that for these molecules, transitions terminating in these states show normal bound-bound vibrational structure. At high pressures, the emission bandwidths for $B \rightarrow X$ transitions are typically 2 nm. Weaker bands due to $C \rightarrow A$ emission are observed

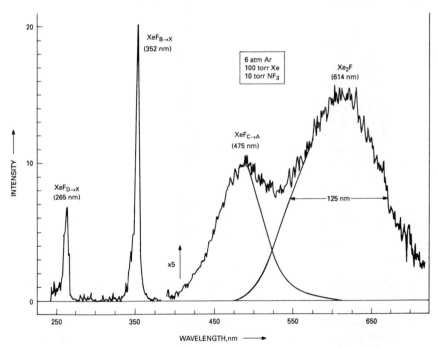

FIGURE 3.21   Emission from an electron-beam-excited Ar, Xe, NF$_3$ mixture. In addition to the XeF($D \rightarrow X$), ($B \rightarrow X$), and ($C \rightarrow A$) transitions, the emission of the triatomic species Xe$_2$ is shown.

at longer wavelengths in each molecule. Because they terminate on the purely repulsive A state, these $C \rightarrow A$ transitions have much greater bandwidth (~ 70 nm for XeF) than the $B \rightarrow X$ emission. $D \rightarrow X$ emission has been observed for most of the noble gas halides. Since the D and B potentials are approximately parallel and separated by the atomic $^2P_{1/2} - {}^2P_{3/2}$ splitting, the $D - X$ bands are similar to the $B - X$ bands, but blue-shifted by an energy comparable to the ionic spin-orbit splitting. However, there is evidence that the D state is strongly quenched, and at high pressures the emission is very weak.

The $B - X$ and $C - A$ transitions are of most importance as laser transitions, and an accurate determination of radiative lifetimes is therefore important. The experimentally measured lifetimes of the $B - X$ transition in KrF and XeF are 9 ns and 16 ns, respectively, and compare reasonably well with the calculated values of 6.5 and 12 ns. The calculated lifetimes for the $C - A$ transitions are much longer (~ 120 ns) and in the case of XeF agree reasonably well with the measured lifetime of 100 ns.

The stimulated-emission cross sections can be calculated from a knowledge of the emission wavelength, bandwidth, and radiative decay times. The cross sections can be calculated from a knowledge of the emission wavelength, bandwidth, and radiative decay times. The cross sections for the $B - X$ bands mostly lie within the range 2 to 5×10^{-16} cm^2, whereas the $C - A$ transitions with longer lifetimes and broader bandwidths have much smaller cross sections, e.g., $\sigma[\mathrm{XeF}(C - A)] = 9 \times 10^{-18}$ cm^2.

2. *Laser Kinetics.* The kinetic processes leading to the formation of the upper laser state in a rare gas halide laser are very complicated because a multitude of neutral and ionic reactions, such as two-body and three-body collisions, superelastic processes, and absorption, all take place simultaneously. The kinetic processes for the different rare gas halide lasers are, in principle, quite similar. The kinetic processes for the discharge-pumped XeCl laser are discussed here as an example.

A typical gas mixture for a self-sustained discharge-pumped XeCl laser is a 3-atm mixture of Xe/HCl/Ne. In the discharge ionization and electron attachment lead to the formation of Xe$^+$ and Cl$^-$. The dominant production reactions for XeCl(B) are Xe$^+$ + Cl$^-$ + $M \rightarrow$ XeCl(B) + M (ion recombination reaction) and NeXe$^+$ + Cl$^-$ $\rightarrow$ XeCl(B) + Ne. A small contribution comes from Xe* + HCl(v) $\rightarrow$ XeCl(B) + H and Xe$_2^+$ + Cl$^-$ $\rightarrow$ XeCl(B) + Xe. Over 23 percent of the electrical energy deposited into the discharge can be utilized to form XeCl(B).

About 65 percent of the formed XeCl(B) contributes via stimulated emission to the intracavity laser flux, but 30 percent of the XeCl(B) is collisionally quenched. At high excitation rates of ~ 3 MW/cm^3, collisions of the XeCl(B) with the discharge electrons, called superelastic collisions, are important. Spontaneous emission is negligible under typical laser conditions.

Only a fraction of the excited states produced can be extracted because the RGH laser mixture contains many absorbers at the laser wavelength and consequently has an extraction efficiency of less than 1. The main absorbers appear to be Cl$^-$ and Xe$_2^+$. The photon extraction efficiency, defined as the ratio of the extracted laser energy to the intracavity laser energy, is typically in excess of $\eta_{\mathrm{ext}} = 70$ percent [Table 3.1, Eq. (3.24)].

If the mixing ratio of Xe/HCl/Ne is varied, the electron energy distribution in the discharge plasma changes. As a result, formation of precursors Xe$^+$, Xe*, Ne$^+$, and Ne* is affected. If helium is used as a diluent gas in place of Ne, the electron temperature changes, resulting in less effective pathways for the XeCl(B) formation.

3. *Laser Pumping.* Discharge technology is well suited to pump high-repetition-rate rare gas halide lasers. They can operate at laser output energies ranging from several millijoules per pulse to more than 1 joule at repetition rates up to several kilohertz. A pumping rate on the order of 1 GW per liter of discharge volume is necessary to produce rare gas halide laser radiation efficiently [Eq. (3.12)]. The discharge resistance in a rare gas halide laser is typically around 0.2 Ω. Therefore, a typical voltage of 20 kV gives a discharge current as high as 100 kA. It is difficult to switch such high currents directly, i.e., by thyratron switches. Consequently, in a typical laser a primary low power and long pulse is produced in a primary circuit, and subsequently this pulse is compressed in the secondary circuit into the secondary high power and short pulse, which can efficiently pump the rare gas halide laser. Discharge pumping circuits developed so far are mainly classified into capacitor transfer circuit, pulse-forming-line (PFL) circuit, magnetic pulse compressor circuit, and spiker sustainer circuit, which are shown schematically in Fig. 3.22.

a. *Capacitor Transfer Circuit.* The capacitor transfer circuit is widely used in relatively small-scale high-repetition-rate commercial rare gas halide lasers. Resonant charge transfer ($C_1 = C_2$ in Fig. 3.22) is frequently employed, because the charge transfer efficiency from C_1 to C_2 is maximized. Figure 3.23 shows a set of typical values for this type of excitation circuit used to pump rare gas halide lasers. If a 4-atm mixture of Xe/HCl/Ne = 1.3/0.1/ 98.6 (percent) is assumed, then the peak value of the primary current I_1 is less than 5 kA, which is within the current ratings of thyratrons. The peak value of the secondary current I_2 increases up to about 18 kA. This increase is due to the fact that L_2 is much smaller than L_1, as shown in Fig. 3.22. The corresponding excitation rate is ~1.6 MW/cm^3, which gives a specific laser energy of 3 J/liter.

Using this type of excitation circuit, a maximum laser efficiency of nearly 3 percent for both XeCl and KrF lasers can be obtained with output energies of around 300 mJ. For ArF lasers the efficiencies are typically between 1 and 2 percent.

b. *Pulse-Forming-Line (PFL) Circuit.* The PFL circuit uses a low-impedance (typically less than 1 Ω) PFL consisting of solid or liquid dielectric materials in place of capacitor C_2 in the capacitor transfer circuit in Fig. 3.22. Coaxial, parallel-plate, and Blumlein pulse-forming lines were used for RGH laser excitation.

The advantage of this circuit is that it makes it possible to inject a quasi-rectangular waveform pulse into a discharge load. The pulse duration and output impedance are simply selected by changing the length and geometry of the PFL, respectively. This pumping system is well suited for high-energy XeCl or KrF lasers with output energies in excess of several joules per pulse.

c. *Magnetic Pulse Compressor (MPC) Circuit.* A magnetic switch consists of a magnetic core made of ferromagnetic material and is fundamentally different from a gas-discharge switch such as a thyratron. A three-stage magnetic pulse compressor is schematically shown in Fig. 3.22. The capacitor C_1 is charged through the transformer from the storage capacitor. Initially, the current through the first magnetic switch is small owing to its large inductance, and charge builds up on the capacitor C_1. When the current through the magnetic

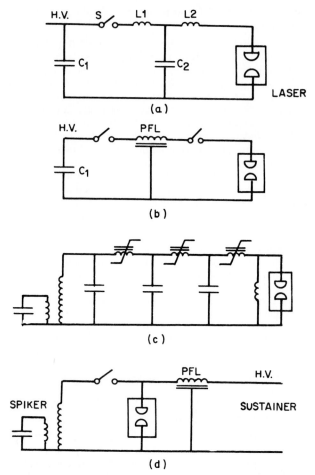

FIGURE 3.22 Schemes of rare gas halide excimer laser excitation circuits. (*a*) Capacitor transfer circuit; (*b*) pulse-forming line; (*c*) magnetic pulse compression; (*d*) spiker sustainer circuit.

switch increases, the inductor saturates and the current rise accelerates, leading to a faster charging of capacitor C_2. If these magnetic switches are connected in series, the primary pulse is successively compressed only by decreasing the saturated inductance of the magnetic coils. The magnetic switch can be used as a long-life or solid-state switch, because it experiences no erosion and can act reliably at a high repetition frequency. It is also used in connection with thyratrons to limit the currents through the thyratrons and thus increase their lifetime.

d. Spiker Sustainer Circuit. The spiker sustainer circuit is an advanced excitation circuit for rare gas halide lasers. The low-impedance pulse-forming line sees initially the discharge load being an open load so that the voltage pulse is reflected owing to the impedance mismatch. To eliminate this unfavorable

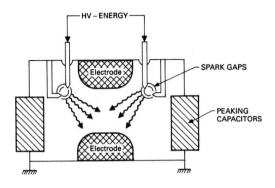

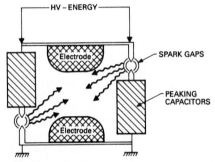

FIGURE 3.23 Preionization techniques. UV and corona preionization are used in small high-repetition-rate lasers, while x-ray preionization is employed for large-volume high-energy excimer amplifiers.

voltage reflection, a high-voltage high-impedance pulser initially breaks down the laser gas mixture and then a second power supply maintains the discharge plasma, for efficient pumping of the excimer laser. The former is called a spiker, the latter a sustainer. This scheme is more complicated than the PFL circuit alone, but higher overall electrical efficiency is achieved, because of better impedance matching.

e. *Commercial Systems.* To initiate a volumetrically uniform avalanche discharge in the 2- to 4-atm rare gas–halogen mixture, preionization of the high-pressure mixture prior to the initiation of the main discharge is indispensable. Spatial uniformity of the preionization in the rare gas–halogen mixture, especially perpendicular to the discharge electric field, is the most important issue. Typically, preionization electron densities are on the order 10^6 to $10^{12}/cm^3$.

A variety of preionization technologies developed so far are shown in Fig. 3.23. The simplest and most convenient preionization technology is uv photopreionization using a photoelectron emission process. The uv photons are generated by the use of a pin-arc discharge or a dielectric surface discharge, both of which are induced in the laser gas mixture, and uv excimer laser beams. UV preionization via pin-spark discharge is widely used in commercially available excimer lasers. X-ray preionization technology was successfully applied to high-pressure excimer lasers, and has been used preferably in large-scale excimer laser devices. The effect of the preionization electron density is significant, and it has been shown experimentally that the excimer laser energy and laser pulsewidth increase with increasing initial electron density.

High-repetition-rate operation of rare gas halide excimer lasers is desirable for high average-power generation. Thyratrons have been commonly used as a switching element. The performance of the switch is one of the repetition-rate-limiting issues. The allowable rise rate of the current for a thyratron is less than 10^{11} A/s, while that of a gas-insulated spark gap reaches 10^{13} A/s.

Because of rapid progress in both the pulsed power technology involved in the modulator and laser gas purification, operations at repetition rates of up to 2 kHz and an average laser power of up to 0.5 kW have been demonstrated separately.

For attaining the nearly endless lifetime of a RGH laser exciter, an all-solid-state circuit is the state of the art and appears to be promising. At present, commercially available high-power semiconductor switches have been developed, but their specifications in terms of hold-off voltage, peak current, and current-rise rate do not yet fulfill the switching requirements for efficient rare gas halide laser excitation.

In addition to the high-repetition-rate exciter, gas purification and aerodynamic technologies such as a fast gas-circulation system and an acoustic damper at repetition rates exceeding the multikilohertz range are required to realize long-life high-repetition-rate operation of the RGH lasers. Lifetimes in excess of 10^9 shots have been demonstrated to date.

A typical commercial excimer laser system using a capacitor transfer system is shown in Fig. 3.24. The thyratron switches the energy stored in the storage capacitor to the peaking capacitor which are connected by a low-inductance connection to the discharge electrodes. A glow discharge in the excimer laser gas mixture is initiated, and high gain on the $B \rightarrow X$ transition is obtained. The rear mirror is highly reflecting. Because of the high gain, the laser operates in the ASE mode and an output coupler is often not necessary. Commercial excimer lasers are capable of operating on all $B \rightarrow X$ transitions that have been demonstrated (Fig. 3.25). Most commonly used are the transitions of ArF (193 nm), KrF (248 nm), XeF (351 nm), and XeCl (308 nm). Table 3.6 summarizes the typical parameters for an advanced

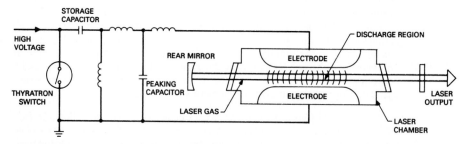

FIGURE 3.24 Excimer laser system using capacitor transfer circuit. The energy is switched by a thyratron from the storage capacitors to the peaking capacitor. A low-inductance connection with the discharge electrodes initiates a glow discharge in the laser chamber. The back mirror of the resonator is usually highly reflecting, while the output coupler is often left out.

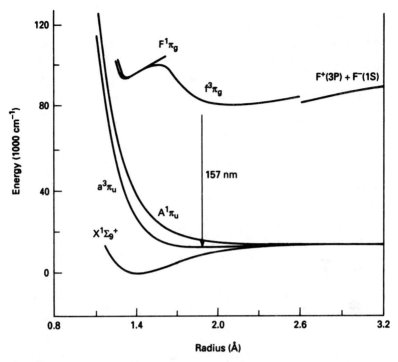

FIGURE 3.25 Partial energy diagram of the F_2 molecule.

TABLE 3.6 Typical Parameters of Industrial Excimer Laser Systems

Laser medium	KrF	XeCl
Wavelength, nm	248	308
Max. stab. pulse energy, mJ	500	500
Max. repetition rate, Hz	250	300
Max. stab. average power, W	125	150
Pulse duration, ns FWHM	26	30
Pulse-to-pulse fluctuations, ($\pm\%$) typ.	6	6
Beam dimensions, mm^2, typ. ($v \times h$)	(28.0 ± 2) $\times$ (8.0 ± 1.0)	
Divergence, mrad, typ. ($v \times h$)	<4.5 $\times$ 1.5	
Angular pointing stability, mrad, typ. ($v \times h$)	0.45 $\times$ 0.15	
Scheduled gas-exchange interval, h (10^6 pulses), typ. Dynamic operation	8 (7.2)	20 (21.6)
Static operation, days, typ.	2	5
Scheduled window-cleaning interval, gas fills (10^6 pulses) typ.	1 (7.2)	1 (21.6)

industrial excimer system. Similar parameters are obtained for research lasers. Since the beam divergence and the linewidth are relatively large in excimer lasers oscillator amplifier versions are available that reduce the beam divergence to less than 50 μrad, while the other laser parameters remain about the same as in Table 3.6.

Excimer lasers are the most widely used ultraviolet lasers today. Their main scientific application is the pumping of dye lasers. In industry, excimer lasers are beginning to be employed for materials processing (mainly laser ablation), and advanced line narrowed systems are being developed for applications in submicrometer lithography.

The Halogens and Interhalogens.   Although the known laser transitions in the homonuclear halogens and heteronuclear halogens (interhalogens) are, in fact, bound-bound electronic molecular transitions, they behave in many respects so similar to the rare gas halide excimers that they are frequently classified as excimer lasers (Fig. 3.14).

Lasing in the ultraviolet and the visible has been observed in all homonuclear and interhalogen molecules. The laser wavelengths are summarized in Table 3.7. The interhalogen laser transitions are discussed in Ref. 37. These lasers are capable of emitting energies in the millijoule range in conventional excimer laser discharges. Owing to the complicated chemistry in providing the halogen donors, this class of lasers has not found many practical applications. Of particular interest among the homonuclear halogen lasers is the fluorine (F_2) laser which emits at 158 nm. This vuv wavelength is the shortest laser wavelength available in a commercial laser system. We therefore briefly discuss the spectroscopy and kinetics of the F_2 laser which is in many ways similar to that of the other homonuclear halogen lasers.

A partial potential energy diagram of F_2 is shown in Fig. 3.25. The upper laser state, the $D'(^1\Sigma_u^+)$ state, correlates to the ion states F^+ and F^-. As in a rare gas halide molecule the electronic transition to the covalent $A_1(^1\Sigma_y^+)$ state is a charge transfer transition which leads to a sizable transition moment. The fluorescence spectrum of the F_2-laser transition is shown in Fig. 3.26. It reflects the complicated vibrational-rotational structure of a bound-bound transition. The observed laser transitions are marked by arrows.

It was first theoretically predicted that unusually high buffer gas pressures of about 8 atm would lead to an efficient operation of the discharge pumped F_2 laser in a helium-fluorine mixture.[38] The main kinetic steps for the formation of the upper state proceed as follows: $He^*(1s2s^3S_1)$ is formed efficiently by the discharge electrons through electron impact excitation or recombination of helium ions. Collisions with F_2 molecules lead to the formation of excited atomic fluorine $F^*(^4P)$:

$$He^* + F \longrightarrow He + F^* + F \tag{3.42}$$

TABLE 3.7 Laser Wavelength of the Halogen and Interhalogen Molecules

	F	Cl	Br	I
F	158	284	354	490
Cl	—	258	314	431
Br	—	—	292	386
I	—	—	—	342

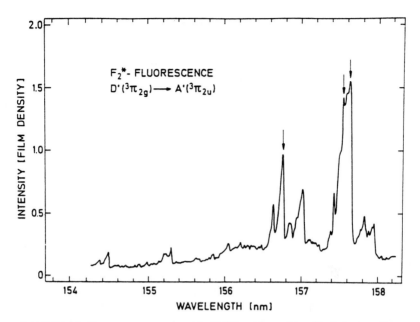

FIGURE 3.26   Fluorescence spectrum of the $F_2(D' - A')$ transition. The lasing transitions are marked by arrows.

Collisions of the excited fluorine atom with F_2 molecules form the upper laser state:

$$F^* + F_2 \longrightarrow F_2(D') + F \qquad (3.43)$$

Since the lower laser state is weakly bound, collisions with neutral helium help to depopulate this state, leading to more efficient laser operation. There are therefore two major reasons for the unusually high pressure regimes for efficient F_2 laser operations. High excitation rates in the discharge are required for sufficient population inversion which is favored by high pressure, and high pressures give sufficient collision rates for effective depopulation of the lower level. F_2 lasers can be operated in pulsed discharges such as those used for rare gas halide lasers (Fig. 3.24). For efficient operation, several modifications are, however, necessary.[39] The high-pressure operation requires a gas container that is capable of withstanding pressures on the order of 10 atm. Also, the electrodes need to be modified for F_2 lasers. The vuv wavelength in combination with high pressures makes the laser output very sensitive to impurities. Cryogenic gas purification is therefore necessary for reasonable gas lifetimes. In Table 3.8 the typical operation conditions for a commercial F_2/ArF laser are given. The main applications for F_2 lasers are still being developed. They are beginning to be used for the pumping of other gas and solid-state lasers (see *Other UV and VUV Gas Lasers* and Sect. 3.2), vuv photochemistry, and materials processing.

Other UV and VUV Gas Lasers.   It is evident from Fig. 3.1 and the previous chapters that there are still only very few deep uv and vuv lasers available for application, and the available wavelengths are severely restricted. Even when frequency shifting by nonlinear optical techniques is employed, it is difficult to generate

TABLE 3.8 Typical Parameters for a
Commercial F_2 (ArF) Laser System

Laser medium	F_2	ArF
Wavelength, nm	158	193
Max. pulse energy, mJ	60	100
Max. repetition rate, Hz	50	50
Max. average power, W	3	5
Pulse-to-pulse stability, $\pm\%$	8	10
Pulse duration, ns	18	10
Gas lifetime, 10^5 shots	5	8
Beam dimensions, mm^2	8 × 13	
Beam divergence, mrad	1 × 3	
Timing jitter, ns	2	

tunable coherent radiation below 200 nm, and the spectral power of the available tunable sources in the vuv is considerably smaller than for longer wavelengths. This establishes a need for direct laser sources at wavelengths below about 200 nm for a variety of initially scientific applications in spectroscopy, photochemistry, imaging, or materials processing. In this discussion several new ideas to develop useful deep uv and vuv lasers are summarized.

The Fluorine Laser Pumped Nitrogen Oxide Laser. The fluorine laser is a fixed-wavelength laser in the vuv at 158 nm. It is appealing to use this laser to pump other possibly tunable sources in the deep uv and vuv. One such system was recently realized by Hooker and Webb.[40] The $X^2\Pi$ ground state of NO is excited by an F_2 laser at 158 nm to the $B'^2\Delta$ excited state. Since the rotational-vibrational transitions of the NO molecule are not in exact resonance with the F_2 laser line, a magnetic field is employed to Zeeman-shift the NO transitions into resonance with the F_2 laser. This measure provides a strong enough excitation of the $B'^2\Delta$ upper laser level to generate gain on transitions to higher-lying vibrational levels of the NO electronic ground state. Strong laser oscillation on a single rotational line of the B'—X(3—10) transition of NO at 218.11 nm was observed. Many more lines between 160 and 250 nm in NO can be excited by the same scheme.

VUV Atomic Laser Transitions Pumped by Laser-Produced Plasmas. Optical pumping was the method employed for realizing the first laser and is still today one of the most frequently used techniques for laser pumping. If a laser is used in a one-photon optical pumping process, only longer-wavelength transitions can be excited. Incoherent, bright optical pump sources are therefore desirable to push deeper into the vuv. It was realized in the 1970s that laser-produced plasmas provide a bright source of soft x-rays, and schemes were developed to use such sources for the pumping of atoms or ions. Most of these schemes rely on the fact that the photoionization cross section for inner-shell ionization is large. In this way core-excited atoms can be generated that then undergo further Auger- and Super Koster-Kronig decays that lead to the population of the upper laser level.

As an example the XeIII laser at 109 nm, first demonstrated in 1986,[41] is discussed. A simplified energy-level scheme is shown in Fig. 3.27. Soft x-rays from a laser-produced plasma generate a hole in the Xe $4d$ shell. The resulting ion state undergoes rapid Auger decay into both the upper and the lower laser state with a

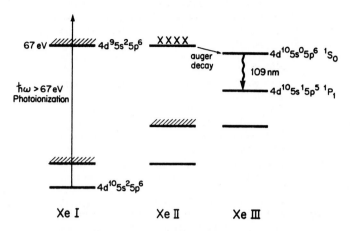

FIGURE 3.27 Partial energy-level diagram of Xe showing the levels relevant to photoionization and to Auger pumping of XeIII.

comparable branching ratio. Inversion results from the higher degeneracy of the lower level. An advanced traveling-wave excitation scheme for this laser is shown in Fig. 3.28.[42] A 500-mJ, 0.5-ns pulse of a Nd:glass laser (1064 nm) is focused by a cylindrical lens under oblique incidence onto a threaded target. The laser has an intensity of $5 \cdot 10^{10}$ W cm^{-2} on target and generates numerous microplasmas along the threaded rod. The soft x-ray spectrum emitted from those plasmas can be approximated by a blackbody spectrum with a temperature on the order of 10 to 20 eV. The soft x-rays pump the Xe gas surrounding the target, leading to inversion in XeIII and emission of 109-nm laser radiation along the target. Owing to the short pulse duration (0.5 ns) of the pump laser, a traveling-wave effect is realized, and the xenon laser emits preferentially in the direction of the traveling excitation pulse. Laser lengths of up to 25 cm have been realized, and the Xe laser output was saturated. Total energies of up to 10 µJ at 109 nm were extracted. The laser

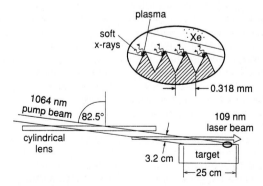

FIGURE 3.28 Pumping geometry for a saturated 109-nm XeIII laser.

shows good spatial coherence and is focusable to intensities in excess of 10^9 W cm^{-2}. Since this laser has short vuv wavelength and is relatively easy to operate at repetition rates of up to 10 Hz, it is presently being developed as a source for vuv imaging applications. Other similar atomic transition lasers have also been demonstrated in ZnIII at 130 nm[43] and neutral Cs at 97 nm.[44] While the ZnIII laser is quite similar to the XeIII laser, the neutral Cs laser is excited by the electrons emitted from the laser-produced plasma.

The Molecular Hydrogen Laser. The molecular hydrogen laser has two emission bands in the vuv: the Lyman band in the vicinity of 160 nm and the Werner band at about 115 nm. Numerous rotational and vibrational lines in both bands have been observed as laser lines with a variety of different pumping techniques. Since the radiative lifetime of the upper laser states are in the 1-ns range (i.e., 0.6 ns for $C^1\Pi_u$, $v' = 1$ state the upper laser state for a Werner band laser), the pump sources have to be sufficiently fast to compete against spontaneous decay. In the 1970s fast discharges and relativistic electron beams were used to pump this laser.[45,46] Recently, intense lasers and photoelectrons generated by the soft x-rays from a laser-produced plasma were employed.[47] With an experimental setup similar to the one shown in Fig. 3.28, saturated output on the 116-nm Werner band line was obtained. As in the case of the XeIII laser, the requirements for pumping this laser are relatively modest. A 580-mJ, 200-ps Nd:glass laser can be used to produce saturated output from the H_2 laser. The H_2 laser usually emits on several lines simultaneously. Since the lower laser states are the higher vibrational states of the electronic ground state of H_2, this laser bottlenecks and the output energy on each individual line rarely exceeds 1 μJ.

Ionic Excimers. At this time excimer lasers, in particular the rare gas halide lasers, clearly are the most useful lasers in the ultraviolet. Therefore, it appears interesting to extend the excimer concept to shorter wavelengths. As outlined in this chapter, rare gas halides are ionically bound states of a rare gas atom Rg^+ and a halogen ion X^-. Similar electronic states should occur in a molecule that is formed by a doubly charged alkali atom A^{2+} and a halogen ion X^-. Such molecules were first postulated in 1985.[48] This idea of isoelectronic scaling of ionically bound states can be extended to other systems. For example, the rare gas alkali ions of the form Rg^+A can be considered isoelectronic to the rare gas excimers (see pages 3.19–3.21). In recent years many of such new molecular states were observed spectroscopically. These alkali halide ionic excimer transitions in $Cs^{2+}F^-$ (185 nm), $Rb^{2+}F^-$ (130 nm), and $Cs^{2+}Cl^-$ (208 nm) can be efficiently excited by photoionization pumping with soft x-rays from laser-produced plasmas.[49] Transitions from the rare gas alkali ions ranging from 60 to 200 nm have been observed by ion-beam excitation of rare gas alkali mixtures.[50] Although the fluorescence efficiencies of ionic excimer molecules appear to be high enough and their cross sections for stimulated emission are comparable with those of the rare gas halide and rare gas excimers, no ionic excimer lasers have been demonstrated to date. Further development of laser pumping techniques is, however, expected to lead to a successful excitation of ionic excimer lasers at many vuv wavelengths.

3.3.2 Ultraviolet Lasers in Solids and Liquids

Solid-state lasers and lasers in liquids (in particular dye lasers) are discussed in Chaps. 4 and 5. Here we concentrate on the few aspects that are important for operation of these lasers at wavelengths below 400 nm.

Dye Lasers in the Ultraviolet. With the development of rare gas halide excimer laser pump sources, the short-wavelength limit for dye lasers could be extended to ~320 nm. Now tunable sources in the ultraviolet between 400 and ~320 nm are commercially available. Dyes are organic molecules diluted in liquids (solvent) that are excited from their electronic ground state (S_0) to the first singlet state S_1 by the pump source, which is for ultraviolet dyes usually a XeCl excimer laser (308 nm) or a nitrogen laser (337 nm). The electronic states formed by the II-electrons of these molecules consist of so many rotational-vibrational substates that they are broadened over an energy range of typically 0.2 to 0.5 eV. In practice, these states behave similar to bands in a solid in the sense that light emission can occur from any substate of the excited S_1 state to any substate of the S_0 ground state. Since the energy bandwidth of these states is considerably larger than thermal energies at room temperature, and since the relaxation within each electronic state proceeds on a subpicosecond time scale, any pump source that populates a set of sublevels in the upper S_1 state will cause rapid population of the lowest sublevels of the S_1 state. Owing to rapid relaxation within the S_0 ground state, the higher-lying sublevels of S_0 will have population densities well below the lowest S_1 levels, and a population inversion between the upper and lower state which will be obtained. The broad excitation bands S_1 and S_0 lead to a wide tuning range for dye lasers. The physics and technology of dye lasers have been reviewed frequently.[51,52] In Table 3.9 the properties of the most frequently used uv-laser dyes are summarized.[53]

Since the emission band of dyes is always Stokes-shifted with respect to their absorption band, the pump wavelength has to be shorter than the wavelength of the dye laser radiation. Since efficient shorter-wavelength sources than XeCl at 308 nm are available (for example, KrF at 248 nm or ArF at 193 nm), the short-wavelength end of dye lasers is obviously not limited by the pump source. Also, energy solvents are transmissive to wavelengths of about 200 mm which would allow for dye laser action below ~320 nm. The main reasons for the present short-wavelength limit of dye lasers are the increasing tendency for dyes to decompose and form chemically active radicals with decreasing pump wavelength,[52] as well as a generally decreasing transition probability from the lowest S_1 states to the S_0 states with increasing transition energy.

All uv dye lasers discussed so far are pulsed lasers. CW dye lasers in the ultraviolet are limited by the availability of strong cw pump sources in the uv, as well as dye lifetime problems. The shortest wavelength available for commercial cw dye lasers is just below 400 nm, with polyphenyl pumped by the uv lines of Ar$^+$ and

TABLE 3.9 Properties of UV Laser Dyes

Dye	Tuning range, nm	Efficiency, %	Pump laser
BM-terphenyl	312–352	4	KrF
P-terphenyl	332–350	8	XeCl
DMQ	346–377	8	XeCl
QUI	368–402	11	XeCl, N$_2$
BiBuQ	367–404	10	XeCl, N$_2$
rBBQ	386–420	7	XeCl, N$_2$

Kr^+ ion lasers. Ultraviolet dye lasers find primarily research applications, for example, for laser-induced fluorescence or nonlinear optics.

Ultraviolet and Vacuum Ultraviolet Solid-State Lasers. Although solid-state lasers are very important for the generation of uv-vuv coherent light by the methods of nonlinear optics, there are few solid-state lasers that emit directly in this spectral region. The problems that are associated with uv operation of solid-state lasers are similar to those for dye lasers. Solid laser materials with favorable optical and mechanical properties[54] for uv laser operation and uv-vuv pump wavelength are rare, and their long-term stability under typical operating conditions for solid-state lasers decreases with decreasing pump and operating wavelength. Owing to the availability of new, convenient short-wavelength pump sources such as the F_2 laser at 158 nm, however, there has recently been substantial progress toward new uv-vuv solid-state lasers. In the 1970s a class of host materials based on rare earth fluorides in combination with Nd^{3+}, Er^{3+}, and Tu^{3+} ions was identified as promising uv-vuv solid-state laser candidates.[55] One of these materials, Nd^{3+} doped LaF_3, was successfully pumped by fluorescence from electron-beam-excited Kr_2^* at 146 nm and showed laser emission at 172 nm.[56] With an F_2-laser pump source at 158 nm, output energies in excess of 3 mJ were obtained from Nd^{3+}: LaF_3 at 172 nm.[57]

3.4 X-RAY LASERS AND GAMMA-RAY LASERS

3.4.1 X-Ray Lasers

Fundamentals of X-Ray Lasers. The physical concepts leading to soft x-ray laser action in laser-produced plasmas have been known for many years but detailed solutions to the practical and theoretical problems of demonstrating soft x-ray or xuv lasers have been found only recently with the use of multiterawatt laser systems. This chapter summarizes the most important aspect of soft x-ray lasers, following recent reviews of the field.[58-60] Shown in Fig. 3.29 is the absolute spectral brightness of xuv and x-ray sources. Even spontaneous x-ray emission from laser-produced plasmas exceeds that of synchrotron sources considerably. XUV and x-ray lasers offer unsurpassed spectral brightness, which could alter new applications in areas such as materials processing or medical imaging.

The only difference between an xuv laser and a more usual visible light laser is that the laser transition is an xuv frequency. Such transitions occur in high-temperature plasmas of highly charged ions and electrons. The reason can be seen in Bohr's elementary model of a hydrogenlike ion illustrated in Fig. 3.30. The positive nucleus, with one orbiting electron, has a system of energy levels in which the transition energies are

$$h\nu = Z^2 h\nu_H \tag{3.44}$$

where Z is the charge number of the nucleus and $h\nu_H$ is the transition energy of the corresponding transition in a hydrogen atom ($Z = 1$).

For example, the Balmer α transition illustrated in Fig. 3.30 emits xuv photons ($h\nu \sim 68$ eV) when $Z = 6$. The hydrogenlike ion is produced by removing five electrons from a carbon atom, leaving a positive nucleus with a single electron,

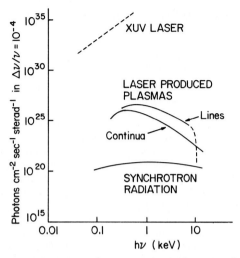

FIGURE 3.29 Comparison of the absolute spectral brightness of xuv and x-ray sources. Synchrotron data are from the SERC synchrotron radiation source (SRS), and laser-produced plasma data are compiled for 10^{12} W Nd glass.

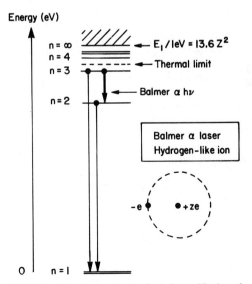

FIGURE 3.30 Energy levels of a hydrogenlike ion of nuclear charge $+Ze$ showing the location of the thermal limit when population inversion is obtained between $n = 2$ and $n = 3$.

which is isoelectronic to a hydrogen atom. Such isoelectronic scaling to higher Z shifts the emitted frequency toward the x-ray region.

Production of highly charged ions in plasmas is described by an equation due to Saha and requires high temperatures such that the energy kT in collisions between electrons and ions is an approximately constant fraction f of the ionization energy $Z^2 E_H$

$$kT = fZ^2 E_H \tag{3.45}$$

Here $E_H = 13.6$ eV is the ionization energy of hydrogen.

A major problem to implement xuv lasers is to find methods of creating a population inversion when nature tends to create thermal equilibrium in which the well-known Boltzman relations apply, giving no amplification at any temperature. Thermal equilibrium is a consequence of thermodynamic detailed balance in which each transition-inducing process is balanced by its inverse process. For example,

$$A + e^- \rightleftharpoons A^+ + 2e^- \tag{3.46}$$

Here collisional ionizations are balanced with three-body recombination. Collisional processes are always balanced by their inverse, and therefore in the limit of high plasma density where collisions are dominant, thermal-equilibrium-level populations prevail. Spontaneous radiative transitions are most rapid across large energy gaps ($A \propto v^2 \propto Z^4$), whereas for collision-induced transitions (of rate $K_{21} n_i n_e$, where n_e and n_i are the number densities of electrons and ions) the opposite is true, with the rate coefficient K_{21} scaling as

$$K_{21} \propto v^1 kT^{-1/2} \propto v^{-3/2} \propto Z^{-3} \tag{3.47}$$

The result is that we can identify in a plasma of lower density a "thermal limit," as illustrated in Fig. 3.30. Above the limit where energy levels are closely spaced collisions dominate, and below the limit radiative rates are greater than collisional rates. Radiative processes are not in detailed balance unless spontaneously emitted photons are reabsorbed by the plasma. This will not occur unless the plasma volume is large. Thus levels below the thermal limit need not be in thermal equilibrium, and with suitable tricks, population inversion can be created—for example, between levels of principal quantum number $n = 3$ and $n = 2$ in Fig. 3.30.

In isoelectronic scaling of laser mechanisms to shorter wavelengths, it is generally necessary to preserve the balance between radiative and collisional processes, so that the thermal limit remains at the same principal quantum number. From the preceding discussion, it follows that the temperature $kT \propto Z^2$ and the particle number density $N \propto Z^7$ require plasmas of progressively higher temperature and density for lasers of higher frequency.

A basic characteristic of any plasma laser is the scaling with frequency v of fluorescent intensity and therefore of pumping intensity per unit of gain coefficient. The emission from laser-produced plasmas under conditions ideal for soft x-ray laser production is Doppler-broadened. Consequently, the linewidth scales as $\Delta v \sim v\sqrt{kT}$ when T is the plasma temperature. Higher x-ray laser transition energies require higher plasma temperatures, and plasma temperature has to scale as $kT \sim v$. This gives $\Delta v \sim v^{3/2}$ and with Eq. (3.13) for the necessary pumping power:

$$P \propto v^{4.5} \tag{3.48}$$

With laser-produced plasmas giving the most intense laboratory xuv brightness, as shown in Fig. 3.29, it is not surprising to find them as the first xuv laser media.

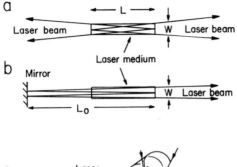

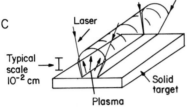

FIGURE 3.31   Schematic single-pass (*a*) and double-pass (*b*) ASE lasers. (*c*) Schematic model of plasma production by a laser beam focused onto a line.

Experiments have mostly involved laser beams concentrated into line foci to create elongated cylindrical plasmas of small diameter from fibers, thin foils, and solid targets as shown in Fig. 3.31. The amplification of spontaneous emission in a single transit along the plasma column gives an exponentially increasing intensity with increasing length which can be detected for small exponents gl, in the axial-transverse intensity ratio $[\exp(gl) - 1]/gl$ and for large values of gl in obvious exponentiation with length. For $gl \geq 8$ the ASE has a narrow beam angle and for $gl > 15$ saturation of the ASE laser occurs with stimulated emission becoming more probable than spontaneous emission.

The characteristics of saturated single-transit ASE in a laser include high power density and narrow bandwidth. The spectral brightness is extreme and, as indicated in Fig. 3.29, exceeds by many orders of magnitude that of any other source.

Collisional Excitation.   A major breakthrough in xuv lasers was achieved in 1985 using the world's most powerful laser (NOVA in the United States) with an ingenious design of target to obtain laser gain on $3p - 3s$ transitions of Ne-like Se^{24+}.[61] The mechanism is illustrated in Fig. 3.32 and relies on strong $3s - 2p$ resonance emission to depopulate the $3s$ level. The $3p$ levels are populated both by collisional excitation from $2p$ to $3p$, by dielectronic recombination, and by cascading from higher levels, and have no allowed radiative decay to $2p$. The temperature is as high as possible without ionization of the Ne-like ions to maximize collisional excitation, and the density as high as possible without collisional thermalization of the $3s$ and $3p$ levels (that is, the local thermal equilibrium (LTE) limit is between $n = 3$ and $n = 4$).

The best conditions are therefore $kT = 1000$ eV and $n_e \sim 5 \times 10^{20}$ cm^{-3}, and the plasma must be small enough in lateral dimension for resonance emission to escape without reabsorption. Both requirements have been achieved with a thin film target (750 Å of Se on 1500 Å of polymer) irradiated in a 200-μm-wide line

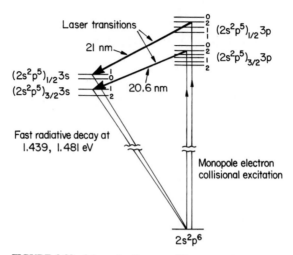

FIGURE 3.32 Schematic diagram of the energy levels and inversion-producing mechanism of the Se^{24+}Ne-like $3p - 3s$ laser.

focus with 10^{12} W cm^{-1} in a 500-ps pulse of wavelength 0.53 μm. A gain coefficient of 4 cm^{-1} on $J = 2$ to 1 transitions at 206 and 209 Å has been observed with a maximum gain x length $gl \sim 14$, giving a factor of 10^6 single-transit amplification! The output power is 10^6 W in a 200-ps pulse and the beam divergence 200 mrad. Refraction in the plasma begins to be a problem at the long (5 cm) length, with the beam being deflected out of the region of gain.

Isoelectronic scaling to Mo^{32+} has been demonstrated irradiating a similar thin-foil target with 2×10^{12} W cm^{-1}, and laser amplification with gl up to 7 has been observed on transitions in the range 106 to 140 Å.

Further progress to shorter wavelength has been achieved with an analogous scheme using $4d$-$4p$ transitions in Ni-like ions. The $3d^{10}$ closed shell of the Ni-like ion is very stable, and the excitation energy of the $4d$ level is significantly smaller relative to the $4d$-$4p$ transition energy than in the analogous Ne-like scheme. Successful isoelectronic extrapolation to Yb^{42+} at 51 Å has been achieved. The scheme has some prospects for laser action in the water window below 44 Å. Using very high pump power, gain has been observed in W^{46+}.

Recombination X-Ray Lasers. A quite different class of xuv lasers is based on transient production of population inversion in a rapidly recombining laser-produced plasma. The initial state is a fully ionized plasma of bare nuclei and free electrons at high density produced by laser irradiation of a solid target. The hot surface layer of plasma explodes and cools adiabatically, as illustrated in Fig. 3.33. When it reaches the density at which the LTE limit is at about $n = 3$ and the levels above $n = 3$ are in Saha-Boltzmann equilibrium with the cool free electrons, population inversion is produced. The equilibrium above $n = 2$ is very rapid, and there is no time lag relative to the cooling process. The $n = 2$ level has a lower LTE population because of depopulation by the Lyman-α transition in inner and outer regions of the plasma and enhances the escape of resonance radiation. The plasma radius must still be small for this to occur. The ground state $n = 1$ is

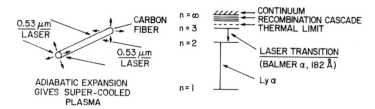

FIGURE 3.33   Schematic diagram of the fiber-target H-like recombination laser.

populated well below the equilibrium level because of the rather slow processes of recombination to it. Population inversion and gain is thus established for the 3-2 Balmer-α transition.

This scheme has been studied in detail for C^{5+} by numerical and analytical modeling, leading to the conclusion that optimum conditions can be obtained with a laser-irradiated carbon fiber target in which the initial plasma is at 10^{21} cm^{-3} and 200 eV, obtained by 70-ps 0.53-μm pulse irradiation of a 7-μm-diameter fiber at 0.3 TW cm^{-1}, with 10 percent absorption of the incident power. The plasma expands to a density of 2×10^{19} cm^{-3} at 30 eV temperature in less than 1 ns when significant gain ($\sim$10 cm^{-1}) is predicted.

Gain of 4 cm^{-1} for length up to 1 cm giving 50 × single-transit amplification has been recorded for hydrogen like C^{5+} Hα at 182 Å with a carbon fiber irradiated as specified earlier. Isoelectronic scaling of the fiber scheme has also been investigated, and modeling has shown a pump-power requirement scaling as Z^4 with gain produced at higher final density scaling as Z^7.

A dramatically higher gain has been seen for C^{5+} Hα laser action in a magnetically confined laser-produced plasma with $gl \sim 6$. The cylindrical plasma is produced by a CO_2 laser pulse of 4×10^9 W and 75-ns duration of 10.6-μm wavelength, focused to a spot on a solid carbon target. An axial field of 9 tesla confines the plasma flow to a narrow cylinder almost 1 cm long in which gain is produced in a cool boundary layer. The mechanism is not well understood and not readily scaled to shorter wavelengths, but the gl/P is almost 1000 times better than the other schemes illustrated in Fig. 3.34, and the long pulse duration favors the future use of a resonator.

The efficiency advantage of H-like recombination arises from the more favorable ratio of laser transition energy to ionization energy of the laser ion, and the situation is slightly more favorable again for recombination from the next lower stable ion configuration (He-like) to Li-like, as shown in Fig. 3.34.

This fact has been exploited in experiments with Al^{10+} in which gain was initially observed on the 103/106 Å 5f-3d transitions using a 6×10^9 W, 1-μm, 20-ns pump laser to irradiate a solid Al target. The gl values were low ($\sim$2) and the gain coefficient was $\leq$2 cm^{-1}, but the gl/P was very high.

Recently, interesting proposals have been discussed to use ultrashort-pulse, ultrahigh-brightness laser sources to pump x-ray lasers.[62] Such lasers could produce initially very cold plasmas by field ionization which would lead to rapid recombination and substantial gain on recombination laser transitions.

Future prospects for xuv laser research are exciting, with expectation of laser action in the "water window" soon, which will open up biological applications. Resonators have been developed using new xuv multilayer optics, and the first demonstration of laser-produced xuv holograms has recently been made.[62]

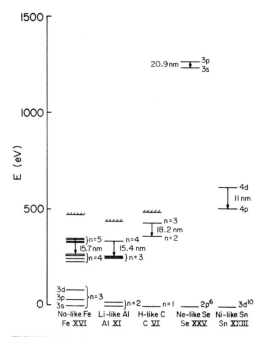

FIGURE 3.34 Comparison of the laser transition energy and excitation energy of the laser upper state for different xuv laser schemes discussed in the text.

3.4.2 Gamma-Ray Lasers

X-ray lasers can probably not be scaled to photon energies above about 10 keV. Already, shortly after the invention of the visible laser, proposals were made to use gamma-ray-emitting electromagnetic transitions in nuclei for shortest-wavelength lenses. This field has been summarized recently.[60] In the more than 30 years since, these proposed gamma-ray lasers (sometimes also called grasers) have not developed significantly beyond the proposal stage. Many very challenging problems have to be overcome to construct gamma-ray lasers. It is generally accepted, for example, that a gamma-ray laser would utilize a Mössbauer line to minimize recoil broadening. At the same time using relations similar to Eqs. (3.11) to (3.15), the specific pump power can be estimated to be on the order of 10^{16} W/cm^3. The solid would rapidly disintegrate at such powers, and the Mössbauer effect would become ineffective. Although proposals have been discussed to overcome this problem, the orders of magnitude for typical gamma-ray laser parameters such as pumping power, photon energy, or laser output power are so unusual that rapid progress toward the construction of gamma-ray lasers appears unlikely at this time.

3.5 REFERENCES

1. A. Einstein, "Zur Quantentheorie der Strahlung," *Physik Zeitsehr*, vol. 18, pp. 121–128, 1917.

2. T. H. Maiman, "Stimulated Optical Radiation in Ruby," *Nature*, vol. 187, pp. 493–494, 1960.

3. Many good textbooks cover the fundamentals of lasers, including:
   A. E. Siegman, *Lasers*, University Science Books, Mill Valley, Calif., 1986.
   P. W. Milonni and J. H. Eberly, *Lasers*, Wiley, New York, 1988.
   A. Yariv, *Quantum Electronics*, 3d ed., Wiley, New York, 1988.
   J. T. Verdeyen, *Laser Electronics*, 2d ed., Prentice-Hall, Englewood Cliffs, N.J., 1989.
   O. Svelto, *Principles of Lasers*, 3d ed., Plenum Press, New York, 1989.

4. Recently also lasers without inversion have been proposed but not yet realized: S. E. Harris, J. E. Field, and A. Imamoglu, "Nonlinear Optical Processes Using Electromagnetically Induced Transparency," *Phys. Rev. Lett.*, vol. 64, pp. 1107–1110, 1990.

5. A. G. Michette, *Optical Systems for Soft X-Rays*, Plenum Press, New York, 1986, p. 336.

6. H. R. Griem, *Spectral Line Broadening by Plasmas*, Academic Press, New York, 1974, p. 408.

7. A. Corney, *Atomic and Laser Spectroscopy*, Clarendon Press, Oxford, 1977, p. 763.

8. A. L. Schawlow and G. H. Townes, "Infrared and Optical Masers," *Phys. Rev.*, vol. 112, pp. 1940–1949, 1958.

9. A. Yariv and R. C. C. Leite, "Super Radiant Narrowing in Fluorescence Radiation of Inverted Populations," *J. Appl. Phys.*, vol. 34, p. 3410, 1963.

10. L. Allen and G. I. Peters, *Phys. Rev.*, vol. A8, p. 2031, 1973.

11. R. Sauerbrey and Z. Ball, "Threshold Behavior and Optimum Transmission for a Pulsed Laser," *Optics Communications*, vol. 95, pp. 153–164, 1992.

12. Ch. A. Brau, "Rare Gas Halogen Excimer," in C. K. Rhodes (ed.), "Excimer Lasers," *Topics in Applied Physics*, vol. 30, pp. 87–137, 1984.

13. A. von Engel, *Ionized Gas*, 2d ed., London, 1965, p. 325.

14. "Applied Atomic Collision Physics," *Series on Pure and Applied Physics* 43, vol. 3; series editors: H. S. W. Massey, E. W. McDaniel, and B. Bederson; volume editors: E. W. McDaniel and W. L. Nighan.

15. A. B. Petersen, "Enhanced CW Ion Laser Operation in the Range 270 nm < λ < 380 nm," *Proceedings of SPIE*, vol. 737, pp. 106–110, 1987.

16. A. B. Petersen, in M. Weber (ed.), *CRC Handbook of Laser Science and Technology*, *Supplement 1*: "Lasers," CRC Press, Boca Raton, Fla., 1990.

17. H. G. Heard, "Ultraviolet Gas Laser at Room Temperature," *Nature*, vol. 200, p. 667, 1963.

18. R. Beck, W. English, and K. Gürs, "Tables of Laser Lines in Gases and Vapors," *Springer Series in Optical Sciences*, vol. 2, Berlin, 1976, p. 130.

19. R. N. Zare, E. P. Lanon, and R. A. Berg, "Frenck-Conden Factors for Electronic Band Systems of Molecular Nitrogen," *J. Molec. Spectroscopy*, vol. 15, pp. 117–1390, 1965.

20. W. A. Fitzsimmons, L. W. Anderson, C. E. Riedhauser, and J. M. Urtilek, "Experimental and Theoretical Investigation of the Nitrogen Laser," *IEEE J. Quan. Electron.*, vol. QE-12, pp. 624–633, 1976.

21. J. Metzner and H. Langhoff, "A High Power N_2 Laser Using Water Filled Strip Lines," *Appl. Phys.*, vol. B54, pp. 100–101, 1992.

22. F. K. Kneubühl and M. W. Sigrist, *Laser*, Teubner, Stuttgart, 1989.

23. R. Sauerbrey and H. Langhoff, "Lasing in an E-Beam Pumped Ar-N_2 Mixture at 406 nm," *Appl. Phys.*, vol. 22, pp. 399–402, 1980.

24. Ch. K. Rhodes (ed.), "Excimer Lasers," *Topics in Applied Physics*, vol. 30, Springer Verlag, Berlin, 1984, p. 194.

25. M. H. R. Hutchinson, "Excimers and Excimers Lasers," *Appl. Phys.*, vol. 21, pp. 95–114, 1980.

26. M. Obara, "Rare Gas Halide Lasers," *Encyclopedia of Physical Science and Technology*, vol. 7, Academic Press, San Diego, 1987.

27. H. Langhoff, "The Origin of the Third Continua Emitted by Excited Rare Gases," *Opt. Commun.*, vol. 68, pp. 31–34, 1988.

28. Y. Uehara, W. Sasaki, S. Kasai, S. Saito, E. Fujiwara, Y. Kato, C. Yamanaka, M. Yamanaka, K. Tsuchida, and J. Fujita, "Tunable Oscillation of a High-Power Argon Excimer Laser," *Opt. Lett.*, vol. 10, p. 487, 1985.

29. T. Sakurai, N. Goto, and C. E. Webb, "Kr_2^* Excimer Emission from Multi-Atmosphere Discharges in Kr, Kr-He, and Ne Mixtures," *J. Phys. D, Appl. Phys.*, vol. 22, pp. 709–713, 1987.

30. T. Ephtimiopoulos, B. P. Stoicheff, and R. I. Thompson, "Efficient Population Inversion in Excimer States by Supersonic Expansion of Discharge Plasmas," *Opt. Lett.*, vol. 14, pp. 624–626, 1989.

31. M. F. Golde and B. A. Thrush, "Vacuum UV Emission From Reactions of Metastable Inert Gas Atoms: Chemiluminescence of ArO and ArCl," *Chem. Phys. Lett.*, vol. 29, pp. 486–490, 1974.

32. J. E. Velazco and D. W. Setser, "Bound-Free Emission Spectra of Diatomic Xenon Halides," *J. Chem. Phys.*, vol. 62, pp. 1990–1991, 1975.

33. S. K. Searles and G. A. Hart, "Stimulated Emission at 281.8 nm from XeBr," *Appl. Phys. Lett.*, vol. 27, pp. 243–245, 1975.

34. C. A. Brau and J. J. Ewing, "354 nm Laser Action on XeF," *Appl. Phys. Lett.*, vol. 27, pp. 435–437, 1975.

35. D. L. Huestis, G. Marowsky, and F. K. Tittel, "Triatomic-Rare-Gas-Halide Excimers in Topics in Applied Physics," *Excimer Lasers*, vol. 30, pp. 181–215, 1984.

36. R. Sauerbrey, Y. Zhu, F. K. Tittel, and W. L. Wilson, "Optical Emission and Kinetic Reactions of a Four-Atomic Rare Gas Halide Exciplex: Ar_3F," *J. Chem. Phys.*, vol. 85, pp. 1299–1302, 1986.

37. M. Digelmann, K. Hohla, F. Rehentrost, and K. L. Kompa, "Diatomic Interhalogen Laser Molecules: Fluorescence Spectroscopy and Reaction Kinetics," *J. Chem. Phys.*, vol. 76, pp. 1233–1247, 1982.

38. M. Ohwa and M. Obara, "Theoretical Evaluation of Diacharge Pumped F_2 Lasers," *Appl. Phys. Lett.*, vol. 51, p. 958, 1987.

39. K. Yamada, K. Miyazaki, T. Hasama, and T. Sato, "High Power Discharge Pumped F_2 Laser," *Appl. Phys. Lett.*, vol. 54, pp. 597–599, 1989.

40. S. M. Hooker and C. E. Webb, "F_2 Pumped NO: Laser Oscillation at 218 nm and Prospects for New Transitions in the 160–250 nm Region," *IEEE J. Quan. Electron.*, vol. QE-26, pp. 1529–1535, 1990.

41. H. C. Kapteyn, R. W. Lee, and R. W. Falcone, "Observation of a Short-Wavelength Laser Pumped by Auger Decay," *Appl. Phys. Lett.*, vol. 57, pp. 2939–2942, 1986.

42. M. H. Shaw, S. J. Benerofe, J. F. Young, and S. E. Harris, "2 Hz 109 nm Mirrorless Laser," *J. Opt. Soc. Am.*, vol. B8, pp. 114–116, 1991.

43. D. J. Walker, C. P. Barty, G. Y. Yin, J. F. Young, and S. E. Harris, "Observation of Super Costa Kronig-Pumped Gain in ZnIII," *Opt. Lett.*, vol. 12, pp. 894–896, 1987.

44. C. P. Barty, D. A. King, G-Y Yin, K. H. Hahn, J. E. Field, J. F. Young, and S. E. Harris, "12.8 eV Laser in Neutral Cesium," *Phys. Rev. Lett.*, vol. 61, pp. 2201–2204, 1988.

45. R. W. Waynant, "Observation of Gain by Stimulation Emission in the Werner Band of Molecular Hydrogen," *Phys. Rev. Lett.*, vol. 28, pp. 533–535, 1972.

46. R. T. Hodgson and R. W. Dreyfus, "Vacuum-UV Laser Action Observed in H_2 Werner Bands 1161 Å–1240 Å," *Phys. Rev. Lett.*, vol. 28, pp. 536–539, 1972.

47. S. J. Benerofe, G. Y. Yin, C. P. Barty, J. F. Young, and S. E. Harris, "116 nm H_2 Laser Pumped by a Traveling-Wave, Photoionization Electron Source," *Phys. Rev. Lett.*, vol. 66, pp. 3136–3139, 1991.

48. R. Sauerbrey and H. Langhoff, "Excimer Ions as Possible Candidates for VUV and XUV Lasers," *IEEE J. Quan. Electron.*, vol. QE-21, pp. 179–181, 1985.

49. S. Kubodera, P. J. Wisoff, and R. Sauerbrey, "Spectroscopy and Kinetics of Ionic Alkali Halide Excimers Excited by a Laser-Produced Plasma," *J. Opt. Soc. Am.*, vol. B9, pp. 10–21, 1992.

50. K. Petkau, J. W. Hammer, G. Herre, M. Mantel, and H. Langhoff, "Vacuum Ultraviolet Emission Spectra of the Helium and Near Alkali Ions," *J. Chem. Phys.*, vol. 94, pp. 7769–7774, 1991.

51. F. P. Schäfer (ed.), "Dye Lasers," *Topics in Applied Physics*, vol. 1, 3d ed., Springer Verlag, Berlin, 1989.

52. F. J. Duarte and L. W. Hillman (eds.), *Dye Laser Principles*, Academic Press, San Diego, 1990.

53. N. Brackmann, *Lambdachrome Laser Dyes*, Lambda Physik, Göttingen, 1986.

54. W. Koechner, *Solid State Laser Engineering*, 3d ed., Springer Series in Optical Sciences, vol. 1, Berlin, 1992.

55. K. H. Yang and J. A. DeLuca, "VUV Fluorescence and Nd^{3+}, Er^{3+}, and Tin^{3+}-Doped Trifluorides and Tunable Coherent Sources from 1650 Å and 2600 Å," *Appl. Phys. Lett.*, vol. 29, pp. 499–501, 1985.

56. R. W. Waynant and P. H. Klein, "Vacuum Ultraviolet Laser Emission from Nd^{3+}: LaF_3," *Appl. Phys. Lett.*, vol. 29, pp. 499–501, 1985.

57. M. A. Dubinskii, A. C. Cefalas, E. Sarantopoulou, S. M. Spyrou, and C. A. Nicolaides, "Efficient LaF_3:Nd^{3+} Based Vacuum-Ultraviolet Laser at 172 nm," *J. Opt. Soc. Am.*, vol. B9, pp. 1148–1150, 1992.

58. M. H. Key, "XUV Lasers," *J. Modern Opt.*, vol. 35, pp. 575–585, 1988.

59. M. H. Key, "Laboratory Production of X-Ray Lasers," *Nature*, vol. 316, pp. 314–318, 1985.

60. R. C. Elton, *X-Ray Lasers*, Academic Press, San Diego, 1990.

61. D. L. Matthews et al., *Phys Rev. Lett.*, vol. 54, pp. 110–113, 1985.

62. E. E. Fill (ed.), *X-Ray Lasers 1992*, Institute of Physics Conference Series 125, IOP Publishing, Bristol, 1992.

CHAPTER 4
VISIBLE LASERS

William T. Silfvast

4.1 INTRODUCTION

This chapter is a summary of visible lasers, with the primary emphasis upon lasers that are commercially available. Although the first laser ever discovered was a visible solid-state laser, operating in a ruby crystal at 694 nm (just barely in the visible spectral region), most visible lasers that have been discovered have occurred in gaseous media (including metal vapors and high-density gaseous plasmas) or in liquid media (organic dyes). For the visible lasers that occur in gaseous media, the output beam generally consists of one or more discrete wavelengths, each having a relatively narrow spectral bandwidth ($\Delta\lambda/\lambda < 0.0001$) corresponding to the width of the atomic or molecular transition associated with that wavelength. For organic dye lasers, the spectral distribution or bandwidth of the gain occurs over a broad wavelength range (of the order of up to 60 to 70 nm with $\Delta\lambda/\lambda = 0.05$ to 0.1). Consequently, the narrow tunable spectral output that is available from such a laser has to be generated by frequency-selective components associated with the laser cavity.

The other two types of visible lasers are solid-state and semiconductor lasers, which are more commonly identified with infrared laser output. However, there are several specific laser materials in these two laser categories that provide laser output that extends into the visible spectral region. These lasers include specially doped solid-state crystal and glass lasers, and "bandgap engineered" semiconductor lasers.

This chapter first reviews visible gaseous lasers, including atomic, ionic, and molecular (excimer) lasers. Second, it deals with dye lasers and summarizes the range of gain media and optical properties of those lasers. Third, it addresses the few solid-state lasers that operate in the visible spectrum. Finally, it concludes with two new types of visible semiconductor lasers. One has been extended from the infrared into the 60 to 70 nm wavelength region of the visible spectrum. The other involves semiconductor laser output in the blue-green spectral region.

Before describing specific laser systems, it is appropriate to define the spectral range of "visible" lasers and to summarize the organization of each of the topics. The visible region of the electromagnetic spectrum is generally accepted as ranging from 400 nm (4000 Å or 0.4 μm) to 700 nm (7000 Å or 0.7 μm). The peak of the visual response of the human eye at normal light levels occurs approximately halfway between these values (550 nm), whereas the peak of the eye's low-light-level response occurs at 500 nm. While the human eye sensitivity drops approximately equally toward either end of the visible range, the imaging quality of the eye is not as great in the blue as in the red (because of chromatic aberrations of the various optical elements of the eye). Thus, when working with blue lasers, one generally finds it difficult to visually determine the image quality, the beam mode quality, or the exact focal position when the beam is focused, without viewing through a supplemental optical element.

The first part of each laser section summarizes some of the significant historical aspects of the laser. Second, a summary of the macroscopic properties is given, including wavelengths, power output, gain media dimensions, gain media composition, excitation power requirements, and other external parameters such as cooling and magnetic field requirements. The third section includes a summary of microscopic properties including spectroscopic notations of energy levels, gain or stimulated emission cross sections, gain linewidth, gain, excitation, and decay mechanisms. The fourth section summarizes some of the commercial lasers of that category and their properties, and the fifth section mentions applications.

4.2 VISIBLE LASERS IN GASEOUS MEDIA

Visible lasers in gaseous media occur primarily in atomic gaseous species. These include atoms and ions of materials that are normally in a gaseous state at room temperature, such as the noble gases of neon, argon, krypton, and xenon, as well as vapors of atomic species that are normally in a liquid or solid state at room temperature. Historically, lasers in ionic species of noble gases have been known as "ion lasers," whereas lasers that have been developed in vaporized species, even though they might occur in ionized atomic species of those vapors, have been referred to as "metal vapor lasers." This is probably not the most logical categorization, but it is one that has evolved and is still used.

Well over 100 visible laser transitions have been discovered over the years in gaseous media. Laser output on these transitions has occurred in over 30 different elements and molecules. Most of these lasers have been observed in ionic species of various gases and vapors using pulsed excitation. Most are not very efficient and therefore not amenable to commercial development and are therefore not well known. Also, considerably more effort has been put into the commercial development of normally gaseous laser media as opposed to vaporized laser media, partially because gaseous media are easier to use than vapors, and partially because many of the early gas lasers had more useful properties than some of the earlier vapor lasers. The factors that drive commercial development include suitability of the laser wavelengths, laser efficiency, power output capabilities, continuous or cw operation, and high repetition rate for pulsed lasers.

4.2.1 Lasers in Atomic Gases

Helium-Neon Laser. Laser action in a mixture of helium and neon gases led to the first gas laser; however, the visible laser transitions in that gas mixture were not among those first discovered. The well-known red He-Ne transition at 632.8 nm was discovered in 1962,[1] nearly two years after the first laser. This laser has probably been used more than any other laser over the years, primarily because of its relatively low cost and compactness, high beam quality, low operating power requirements, and, perhaps most importantly, long operating life. More recently, a green He-Ne laser transition at 543.5 nm has been developed that has found a market niche because of its relatively low cost and its lower power consumption than other visible green lasers.

The He-Ne lasers mentioned above operate with a continuous output by applying a voltage across a narrow-bore glass tube containing a mixture of helium and neon gases with electrodes (anode and cathode) mounted inside the tube as shown in Fig. 4.1. The optimum discharge current is of the order of 10 to 20 mA. The gas mixture is typically optimized at a total pressure of 1 Torr with neon comprising approximately 15 percent of that total. Typical dimensions of the gain medium include a bore size of 1 to 2 mm, and a bore length of 10 to 20 cm, although much longer bore lengths have been successfully constructed. The voltage drop across the gain region is of the order of 60 V/cm, and the tube has a small pin-type anode and a large canister-type aluminum cathode enclosed within a gas reservoir that is either surrounding the bore region or attached adjacent to the bore region. The large aluminum cathode serves to reduce the current density on the cathode, thereby reducing the possibility of gaseous contamination due to cathode sputtering. The large gas reservoir associated with the cathode serves to minimize the effect of helium diffusion through the glass, which would gradually change the pressure ratio of helium to neon in the tube owing to the higher leakage rate of helium. Cleanliness of the He and Ne gases and of the tube during assembly and processing is essential for efficient laser output, because of the sensitivity of the laser excitation process to the presence of impurities in the discharge.

Both the helium and neon gases play an important role in the operation of the He-Ne laser. The laser transitions occur between energy levels in neutral atomic Ne as shown in Fig. 4.2. The visible transitions occur from the 3s spectroscopically defined levels to the 2p levels. Both the red (632.8 nm) and green (543.5 nm) transitions have the same upper laser level and thus both transitions compete for available gain. The red transition has the highest gain with a stimulated-emission cross section of approximately $3.0 \times 10^{-13} \text{cm}^2$ while the green transition cross section is only $1.6 \times 10^{-14} \text{cm}^2$. Thus the only way the green transition can be made to lase is to suppress the gain at 632.8 nm by using laser mirrors that have low reflectivity in the red and *very high* reflectivity at 543.5 nm. This has been successfully achieved in recent years, making the green He-Ne laser a useful addition to the list of available visible lasers. Other properties of the 632.8-nm transition include an emission linewidth of 2 GHz and a gain coefficient of 0.001 to 0.002/cm. The steady-state (cw) population inversion is provided by the rapid decay of the lower laser level (2p) to the ground state. In addition to these two transitions, 10 other visible laser transitions, ranging in wavelength from 540.0 to 640.2 nm, have been reported to occur in neutral neon gas under a variety of excitation conditions but are not generally available commercially.

Helium plays the role of the excitation species for the He-Ne laser as indicated in Fig. 4.2. Energetic electrons within the gas discharge, accelerated by the electric field (produced by the applied voltage), populate the neutral helium metastable levels. These levels then transfer most of their energy to neon via atom-atom collisions, rather than losing it by direct radiative or collisional decay to the He ground state. The 2^1S He metastable level is the pumping source for the two well-known neon transitions. The upper laser levels of those Ne transitions are energetically aligned with the 2^1S level of helium, which leads to a high probability of direct transfer of energy.

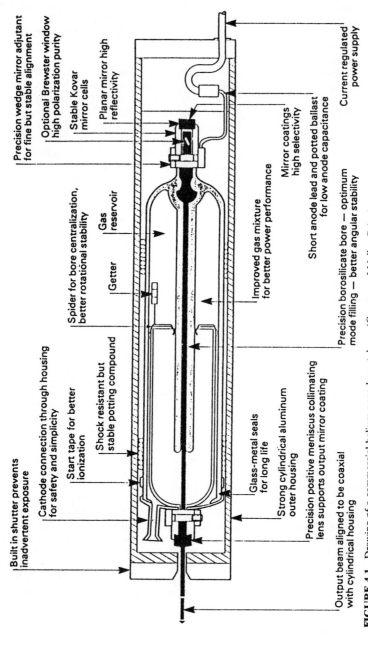

FIGURE 4.1 Drawing of a commercial helium-neon laser tube. *(Courtesy of Melles Griot.)*

Built in shutter prevents inadvertent exposure

Cathode connection through housing for safety and simplicity

Start tape for better ionization

Shock resistant but stable potting compound

Output beam aligned to be coaxial with cylindrical housing

Glass-metal seals for long life

Strong cylindrical aluminum outer housing

Precision positive meniscus collimating lens supports output mirror coating

Spider for bore centralization, better rotational stability

Getter

Gas reservoir

Improved gas mixture for better power performance

Precision borosilicate bore — optimum mode filling — better angular stability

Short anode lead and potted ballast for low anode capacitance

Mirror coatings high selectivity

Current regulated power supply

Planar mirror high reflectivity

Stable Kovar mirror cells

Optional Brewster window high polarization purity

Precision wedge mirror adjutant for fine but stable alignment

4.4

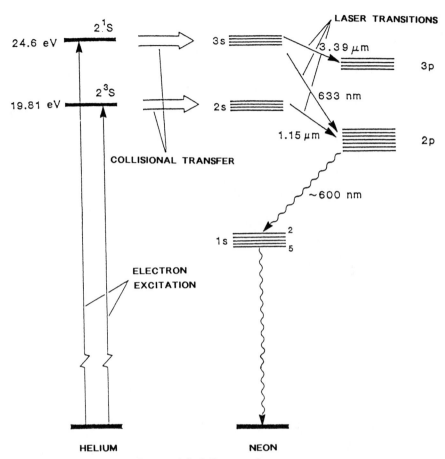

FIGURE 4.2 Energy-level diagram of the helium-neon laser system.

Commercial He-Ne lasers range in size from 10 to 100 cm in length, with the average being approximately 25 to 30 cm. The 25 to 30-cm-length lasers typically have power outputs of 0.5 to 5 mW in a TEM_{00} mode. The lasers require input powers of the order of 10 W to operate and therefore need no special cooling requirements. The laser noise is less than 0.5 percent rms. He-Ne lasers operate for lifetimes of up to 50,000 h and have become the "workhorse" of the laser industry over the years (when high power is not required). These lasers are used for a variety of applications, including surveying, construction, supermarket checkout scanners, printers, and alignment tools or reference beams.

Noble Gas Ion Lasers. Noble gas ion lasers have become known in the laser industry as "ion lasers." This is a historical labeling in that the only other "ion" laser that is available commercially is the He-Cd laser, which has been termed a "metal vapor" laser. Therefore, what are known as ion lasers consist mainly of lasers involving argon, krypton, and xenon gases. The Ar^+ and Kr^+ lasers are available commercially primarily as cw lasers, although a few laser wavelengths are available in a pulsed operating mode. The lasers generally operate at much higher current densities than the He-Ne laser and consequently require more power input but also produce more power output. They are known for their high cw output powers of up to 100 W and higher. The high-power versions require water cooling and also use a magnetic field to confine the plasma in the center of the gain region. A lower-power version operates with air cooling and no magnetic field.

Argon Ion Laser.   The argon ion laser was the second ion laser to be discovered.[2] It was first observed as a pulsed laser and later operated in a continuous mode (cw). Approximately 25 visible transitions in Ar^+ and Ar^{2+}, ranging in wavelength from 408.9 to 686.1 nm, have produced laser output in plasma discharges at various laboratories. It is not uncommon to have a 30 to 100-W laser with the output divided among several visible transitions. The argon laser linewidth in a typical resonator configuration containing several longitudinal modes, is of the order of 2.5 GHz.

Argon ion lasers are generally produced in a high-temperature plasma tube having a bore diameter of the order of 1 mm and lengths up to 50 cm. A diagram of a commercial argon ion laser tube is shown in Fig. 4.3. Because the required discharge current is of the order of 30 to 40 A with a voltage drop of 300 V, the power consumption is extremely high (9 to 12 kW).

The two most efficient and high-powered transitions in argon are the 488.0-nm and 514.5-nm transitions. They tend to produce nearly equal power outputs when broadband mirrors are used to obtain maximum total power output. Other transitions, when operating simultaneously with those two transitions, produce powers that are at most one-third of the power at those wavelengths. The visible lasers occurring in Ar^+ involve transitions from the many levels of the electronic configuration $3p^43p$ to levels of the configuration $3p^43s$ as shown in Fig. 4.4. A large number of transitions occur between these two levels, all of which have gain when the Ar^+ gain medium is operated under the conditions for optimum laser action. The stimulated-emission cross sections for the 488.0- and 514.5-nm transitions are $5 \times 10^{-12} cm^2$ and $4 \times 10^{-12} cm^2$, respectively, and the gain is of the order of 0.5 percent/cm for both transitions.

The excitation mechanism for the argon laser has been shown to be different for pulsed laser operation than for cw operation. For pulsed operation, excitation occurs primarily via electron collisions with ground-state neutral argon atoms directly producing Ar^+ ions in the upper laser level (a single-step process). The cw laser is excited by a two-step process in which first the Ar^+ ground-state ions are produced by electron col-

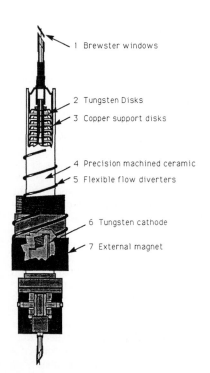

1 Brewster windows

2 Tungsten Disks
3 Copper support disks

4 Precision machined ceramic
5 Flexible flow diverters

6 Tungsten cathode

7 External magnet

FIGURE 4.3   Drawing of a commercial argon ion laser tube.

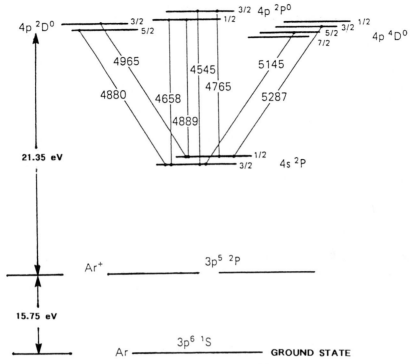

FIGURE 4.4 Energy-level diagram of the Ar+ laser transitions.

lisions with ground-state Ar atoms and then the laser levels are populated in a second electron excitation process involving collisions with the ion ground state. This two-step process leads to a laser power output that is proportional to the square of the discharge current. It is therefore highly desirable to operate this laser at high currents. However, there is a maximum discharge current above which the electrons begin to detrimentally deplete the upper laser level population before stimulated emission can extract the energy. In either pulsed or cw operation, under the conditions required for optimum laser output, the lower laser level has a very rapid decay rate to the Ar+ ground state which provides the necessary depletion of the lower laser level required for a population inversion.

The upper laser levels of the Ar+ laser transitions are approximately 40 eV above the Ar+ ground state and therefore require extremely high electron energies to provide the necessary excitation. Such electron energies can only be provided by operating at very low Ar gas pressures, of the order of 0.1 Torr. As mentioned previously, such low pressures and high electron temperatures lead to very high plasma tube temperatures and hence the need for supplemental external water cooling of the bore. In addition, a magnetic field is generally provided to prevent the electrons from prematurely escaping from the gain region and colliding with the discharge walls, thereby producing additional unwanted heat.

Commercial Ar+ lasers are generally produced in three sizes. These include (1) high-power, large-frame and (2) medium-power, small-frame water-cooled lasers, as well as (3) low-power air-cooled lasers.

The large-frame lasers provide output powers of up to 30 W or more multiline and a single-line power at 514.5 nm of nearly 10 W or more. Such lasers require input powers of the order of 60 kW and cooling-water flow rates in the range of 5 gal/min at a pressure of 60 lb/in^2 gauge. The lasers are approximately 2 m long and have a separate power supply. One commercial version of this laser produces powers of up to 100 W cw.

The small-frame lasers provide cw output powers of approximately 5 W multiline and single-line powers up to 2 W at 514.5 nm. They require input powers of approximately 8 kW and water cooling rates of 2 gal/min at 25 lb/in^2 gauge. The lasers are approximately 1 m long with a separate power supply.

The third category is the small air-cooled Ar$^+$ laser producing 10 mW TEM$_{00}$ mode at 488.0 nm. The beam amplitude noise is less than 2 percent peak to peak and less than 0.1 percent rms at lower frequencies. The laser typically stabilizes in less than 15 min after turn-on. Input powers are of the order of 1 kW. The lasers are compact, with the laser head dimensions of the order of 35 cm long by 15 cm square.

Visible argon lasers are used primarily for phototherapy of the eye, pumping dye lasers, laser printing and cell cytometry, etc.

Krypton Ion Laser.   Krypton ion lasers are closely related to argon ion lasers in that the laser transitions are spectroscopically similar and consequently the means of producing laser output is similar. In fact, krypton ion lasers are generally made by just replacing the argon gas with krypton gas in the same discharge tube. The optimum operating gas pressure is slightly different, and since the wavelengths are different, the laser mirrors have different reflectivity values. Other than that, the lasers are nearly identical.

The principal reason for offering krypton ion lasers commercially is that they provide different laser wavelengths from those of the argon ion laser. In krypton, the laser wavelengths range from 406.7 to 676.4 nm, with the dominant outputs occurring at 406.7, 413.1, 530.9, 568.2, 647.1, and 676.4 nm, and the strongest transition occurring at 647.1 nm. Krypton lasers offer a much broader range of laser wavelengths in the visible spectrum with transitions over most of the color spectrum with the exception of orange.

Some manufacturers provide a laser with a mixture of argon and krypton gases which essentially provides both the strong blue and green transitions of the argon ion laser and the strong red transitions of the krypton ion laser to provide three primary color components. Such lasers are used for color display.

Since krypton lasers can be interchanged with argon ion lasers, the dimensions of the laser head and power supply are similar to those described for argon lasers, with the exception that krypton ion lasers are not available in the compact, air-cooled version of the argon ion laser. This is because the gains are lower for the krypton lasers, making it more difficult to extract useful power in a smaller version of the laser.

4.2.2 Metal Vapor Lasers

Approximately 75 to 80 visible laser transitions have been reported in vapors of over 30 elements. These range from the first ion laser discovered, which was in mercury vapor where oscillation occurs in the red portion of the spectrum at 615.0 nm, to the helium-cadmium laser which oscillates at 441.6 nm at the blue end of the visible spectrum. Other well-known metal vapor lasers include the pulsed copper laser at 510.5 and 578.2 nm and the gold vapor laser at 627.8 nm. Pulsed visible lasers in strontium ions at 407.7, 416.2, and 430.5 nm have also been developed. A laser with perhaps one of the broadest spectral output ranges is the helium-selenium laser producing over 35 laser transitions in the visible spectrum ranging from 446.7 to 653.4 nm.

Helium-Cadmium Laser. The cw He-Cd laser, operating at 441.6 and 325.0 nm, is probably the best known and most widely used metal vapor laser. The blue transition[3] is the shortest-wavelength visible laser available commercially, and the 325.0-nm uv laser was the shortest-wavelength cw uv laser for many years (until a number of transitions in the 200 to 300-nm region were developed in Ar^{2+} and Ar^{3+} by operating those lasers at very high currents and by improving the quality of the cavity mirrors). A number of other laser transitions have also been developed in the helium-cadmium discharge. The most significant of these are the transitions at 537.8 and 533.7 nm in the green and at 636.0 nm in the red. These transitions operate most effectively under different plasma conditions than the blue and uv lasers and generally are optimized with a different type of discharge tube, known as a hollow cathode tube. When all of these blue, green, and red transitions are made to lase simultaneously, the resultant output is a white light laser which for many years was investigated for use in color copying systems.

The blue and uv lasers are operated primarily in a positive column type of dc discharge at currents in the range of 60 to 70 mA in a 1 to 2-mm-bore glass discharge tube. The operation of these lasers is more like that of the He-Ne laser than the noble gas ion lasers. A relatively low current and relatively high gas pressure are required and, therefore, in most lasers no supplemental cooling requirements are necessary. Some models do, however, use a small cooling fan. Typical bore lengths are of the order of 25 to 30 cm with a voltage drop from anode to cathode of over 1000 V. The laser tube is filled with helium gas at a pressure of approximately 5 to 7 Torr and the Cd is fed into the discharge from a side reservoir by a process known as cataphoresis. In this process Cd is vaporized at the anode end of the tube and migrates toward the cathode, where it recondenses in an unheated region of the glass tube. The cataphoresis effect results from the relatively high degree of ionization of Cd in the discharge (primarily Cd^+) and the force the electric field within the plasma exerts upon those ions, thereby pulling them toward the cathode. A few grams of Cd are sufficient to provide up to 5000 h of continuous operation of the laser.

The single-mode blue laser power is typically 15 to 20 mW for the 25 to 30-cm-long discharge. Each laser tube is fitted with a He reservoir in order to replace the helium that is gradually lost, both by diffusion through the tube walls and also by being captured as the Cd condenses in the unheated cathode regions. The He pressure is electronically monitored and adjusted for optimum laser output.

Higher-power versions of this laser are developed by connecting two of the smaller tubes together to share a common cathode, effectively doubling the gain length and thereby significantly increasing the available power. Such tubes offer cw powers of the order of 75 to 100 mW at 441.6 nm.

The blue and uv transitions in Cd are somewhat unique spectroscopically, when compared with other laser transitions, because they involve an inner shell electronic configuration. The upper laser level configuration is $4d^95s^2$ whereas the lower laser level is $4d^{10}5p$. Consequently when electronic transitions occur between these levels, two electrons must change their momentum state, with one 5s electron jumping to fill the 4d shell and the other changing to 5p. Such a transition is therefore less likely to occur than a more typical visible electronic transition that requires only one electron change. These specific Cd^+ transitions have a relatively long radiative lifetime (~700 ns) and a relatively low stimulated-emission cross section of 4 × 10^{-14}cm². The gain coefficient at 441.6 nm is typically 0.0003 cm, which is significantly higher than that of the He-Ne laser.

The unique electronic configuration for these transitions also leads to a large isotope spectral shift at 441.6 nm for different Cd isotopes. Naturally occurring metallic Cd is composed of several isotopes, and the isotope shift between adjacent even isotopes is of the order of the Doppler width (1.1 GHz). The most abundant isotopes are Cd 110 (12 percent), Cd 111 (13 percent), Cd 112 (24 percent), Cd 113 (12 percent), Cd 114 (29 percent), and Cd 116 (8 percent). Consequently, since most lasers typically use naturally occurring Cd, the spectral output of the laser is significantly broader than just the Doppler width of a single isotope. Often only the two strongest isotopes, Cd 112 and Cd 114, are involved in the laser output, which provides a combined bandwidth of 2.6 GHz for the gain medium.

The unique electronic configuration also leads to unique excitation mechanisms, as indicated in Fig. 4.5. Several have been identified for these laser transitions. The first mechanism identified is a process known as Penning ionization in which highly excited (usually metastable) helium atoms transfer their energy to Cd, in a way similar to the operation of the He-Ne laser. However, in the case of Cd, ions are produced by the process, instead of excited neutral atoms as in the case of neon, owing to the much lower ionization potential of Cd than of Ne. In Cd, the

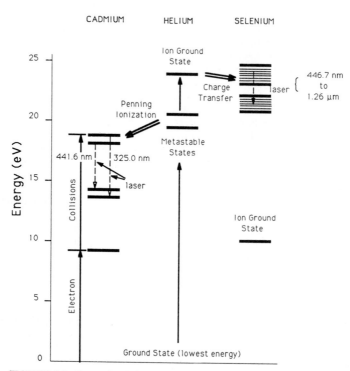

FIGURE 4.5 Energy-level diagram for the helium-cadmium and helium-selenium lasers.

ejected electron in the ionization process takes up the excess energy mismatch between the initially excited He atom and the resulting Cd^+ ionized level. This collisional transfer process is particularly enhanced in Cd since only a single d electron has to be removed from the Cd neutral ground state ($4d^{10}5s^2$) during the collision to result in the population of the Cd^+ upper laser level ($4d^95s^2$).

The second mechanism is a two-step electron collisional excitation where the first electron collision produces Cd^+ ion ground states and the second electron collision produces the Cd^+ upper laser level (similar to the Ar^+ excitation). The two-step electron ionization and the Penning process are the major excitation mechanisms for the conventional He-Cd laser. There is still some question as to which process dominates under various conditions of operation.

The third mechanism, more recently identified, is the photoionization of the Cd atom using very-short-wavelength photons (soft x-rays). For example, when Cd vapor at a pressure of the order of 1 Torr is irradiated with photons of wavelengths ranging from 10 to 70 nm, those photons will directly excite only Cd^+ ions in the upper laser level ($4d^95s^2$). This process was first shown to occur by pumping Cd vapor with short-wavelength laser-produced plasma sources, but it also operates in the He-Cd discharge plasma tube because of the presence of significant He emission in the 30 to 60-nm spectral region during operation of the He-Cd laser. The mechanism is believed to play only a minor role in the operation of the positive column He-Cd laser owing to the relatively low density of Cd for optimum laser output, which allows most of the soft-x-ray emission to escape from the discharge before being absorbed by Cd atoms.

The other transitions in Cd^+, primarily at 537.8, 533.7, and 636.0 nm, are more effectively excited by charge transfer from He^+ ions than by the other processes described for the blue laser. They operate more effectively in a hollow-cathode type of discharge in which a larger number of He^+ ions are produced and are available for excitation of Cd. Several versions of the hollow-cathode type of discharge have been investigated over the years, but there is currently only one design available commercially that produces a "white light" output of 15 to 30 mW and low-noise operation ($<1/2$ percent rms).

Commercial blue lasers are available at power levels ranging from 20 to 100 mW and in sizes ranging from 50 to 200 cm in length with a separate, relatively compact power supply. The lasers operate with noise levels of less than 2 percent rms and typically operate for lifetimes of the order of 3000 to 6000 h. Warmup time to full power is typically 15 min.

Applications for these lasers include printing, microchip inspection, flow cytometry, lithography, and fluorescence analysis.

Helium-Selenium Laser.   The helium-selenium laser[4] operates in a configuration similar to that of the helium-cadmium laser. Cataphoresis is used to distribute the selenium vapor in the laser bore region, and helium at a pressure of a few torr provides excitation via charge transfer from the He^+ ion. This laser has operated on 25 wavelengths simultaneously in the visible spectral region at discharge currents of several hundred mA. The upper laser levels are energetically closely aligned with the He^+ ion ground-state energy and allow the laser to operate via the charge transfer mechanism, where the He^+ ground-state energy is transfered directly to the upper laser levels of Se^+ as indicated in Fig. 4.5. This laser was available commercially for a short time but is no longer on the market. Problems associated with the control of the Se vapor within the laser tube significantly restricted the useful lifetime of the laser.

Mercury Ion Laser. One of the earliest visible lasers was the mercury ion laser operating with a pulsed output at 635.0 nm in the red and 567.7 nm in the green.[5] It was available commercially in the early days of lasers but has not been available recently. The red transition has exceptionally high gain when operated in a hollow-cathode type of discharge. There was never a significant market for a pulsed red laser with peak powers of the order of hundreds of watts since the average power was quite low. The red He-Ne laser satisfied most of the low-power red laser needs, and the much higher power ruby laser was more effective in producing pulsed red laser output. The green Hg laser transition is not as efficient as the red laser and the demand for it in the early days was low since it was competing with the more reliable pulsed and cw Ar ion lasers.

Other Visible Ion Lasers. As discussed earlier, a large number of laser transitions in a variety of vapors have been discovered over the years but have been of no practical significance. Further development of these transitions will probably not take place owing to the rapid advances of more reliable and efficient solid-state lasers in the visible spectral region.

High-Gain Pulsed Metal Vapor Lasers. A separate category of metal vapor lasers, first discovered in 1965, is the high-gain pulsed self-terminating lasers. They require rapid excitation of the heated metal vapors contained within a discharge tube filled with a low-pressure buffer gas, producing pulsed laser output typically lasting 10 to 50 ns. The high gain (as much as 6 dB/cm) can produce an observable laser beam emerging from the gain medium in only a single pass through the amplifier (no mirrors required for feedback). The first laser of this type occurred in lead vapor with the laser oscillating at 722.9 and 406.2 nm. Subsequent to this a number of other lasers in this category were discovered in neutral manganese at 5 wavelengths in the green (from 542.0 to 553.7 nm) and 7 wavelengths in the 1.3-μm range, copper at 510.5 and 578.2 nm, gold at 627.8 nm, and in singly ionized calcium at 370.6 and 373.7 nm, and strontium at 416.2 and 430.5 nm. All the lasers have similar energy-level arrangements that provide the appropriate conditions necessary for the production of high gain.

The general energy-level arrangement for these lasers includes the ground state of the atom or ion, the resonance state (first excited state), and an intermediate level or a set of levels that lie between the resonance level and the ground state but have the same parity as the ground state. The radiative coupling from the resonance state to the lower-lying intermediate states is reasonably high since they are of opposite parity. This energy-level arrangement provides a strong electron collisional excitation from the ground state to the first excited state, but a much lower electron excitation from the ground state to the intermediate states. Thus, when an electric current is rapidly pulsed through the discharge containing the metal vapor and a buffer gas, the resonance levels (upper laser levels) are more efficiently populated than the intermediate levels. This leads to a high-gain inversion between the resonance level and the intermediate levels and a large laser output on transitions between those levels. Careful studies of these lasers have indicated that, while the above description is approximately correct, the lower laser levels (intermediate levels) are not as empty during the laser pulse as first believed. In fact they increase significantly beyond the population of the upper laser level very shortly after the laser pulse terminates. These lasers are relatively efficient and can be operated at very high repetition rates (up to 20 kHz). Consequently they produce very high average power outputs (up to hundreds of watts for some laser systems).

The best-known laser of this category is the copper vapor laser operating in the green at 510.5 nm and yellow (578.2 nm). The other significant visible laser, developed primarily for medical applications, is the gold laser at 627.8 nm.

Copper Vapor Laser. The copper vapor laser[6] has become a useful laser because of its wavelengths, its efficiency, and its high average power output. It has produced average powers as high as 350 W at an efficiency of nearly 1 percent in large noncommercial laser systems designed for laser-assisted isotope enrichment processes. The enclosure for these large lasers is of the order of 3 m long by 120 cm wide by 60 cm high including the power supply. The discharge tube diameter is approximately 8 cm and the laser operates at a partial pressure of neon gas of 40 torr and a partial pressure of Cu vapor of approximately 1 torr. These large laser tubes require a pulsed voltage of 20,000 V and a peak current of 1000 A to produce a pulsed energy of 100 mJ. However, these lasers are not the normal laboratory type of laser. Smaller commercial versions, of the order of 2 to 3 m in length and 20 cm^2 in cross section with up to a 6 cm beam diameter, produce energies of the order of 10 mJ/pulse, average powers of up to 100 W, and repetition rates up to 30 kHz. The lasers have to be operated at temperatures near 1500°C to provide the necessary Cu vapor pressure. Thus the discharge tubes must be made of high-temperature ceramic materials. The Cu is loaded into the bore region in chunks or braided strands of pure Cu metal, and the laser typically operates for 300 to 500 h or more before the Cu must be replaced. Impurities are not as much of a problem in this class of lasers as in the case of ion or He-Ne lasers and, therefore, vacuum requirements are not as stringent while processing the discharge tube. The typical output of the laser is highly multimode owing to the high gain and the large bore diameter.

The relevant Cu levels are the 3d^{10}4s ground state, the 3d^{10}4p upper laser levels (resonance levels), and the 3d^{9}4s^2 intermediate levels which are metastable to radiative decay to the ground state since they are of the same parity (see Fig. 4.6). Calculations indicate that the electron excitation cross sections are significantly higher (factor of 5) from the ground state to the upper laser levels than to the lower laser levels. The gain on the copper transitions is of the order of 0.05 to 0.1/cm. The laser linewidth is approximately 2.3 GHz and the stimulated-emission cross section is 8 × 10^{-14}cm^2 for the green transition. The lower laser level decays very slowly, especially for larger-bore-diameter tubes that limit depopulation of the levels by wall collisions. Consequently at high repetition rates the residual lower laser level population reduces the inversion and must be taken into account when considering the high repetition rate mode of operation.

These lasers are available commercially in industrial versions that require only periodic servicing and are designed for reliable performance over extended periods of time (500 operating hours before Cu is reloaded). With an unstable resonator cavity, the laser beam has a divergence of less than 0.6 mrad with a top hat beam profile. In addition to isotope enrichment (pumping of high-power tunable dye lasers), these lasers are used for pumping of Ti:sapphire lasers, ultrashort pulse amplification, micromachining and materials processing, uv light generation (frequency doubling), high-speed photography, holography, and projection television.

Gold Vapor Laser. Just as in the case of the similarity of the argon and krypton ion lasers, the gold vapor laser[6] is also very similar to the copper vapor laser (see Fig. 4.6). The two primary differences are the wavelength (gold lases at 627.8 nm in the orange) and the operating temperature (gold operates at a temperature approximately 150°C higher). Otherwise the operation of the lasers is similar. The higher operating temperature for gold requires more power to keep the gold vapor at the appropriate temperature, but the discharge tubes designed for copper gen-

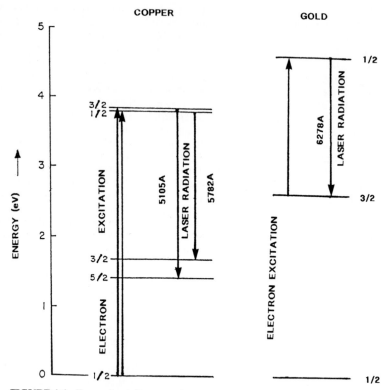

FIGURE 4.6 Energy-level diagrams of the copper vapor and gold vapor visible lasers.

erally also operate at the required higher temperatures for gold. The power output for a gold laser is typically one-fourth that of a copper laser since the gain for gold is lower than that for copper.

The gold laser is used primarily for photodynamic therapy (PDT) involved in cancer treatment. This treatment involves the ingestion of a photosensitive material that accumulates primarily at a cancerous tumor site in the body. The photosensitive material absorbs light primarily in the orange part of the spectrum. Thus when the tumor region is irradiated with the gold laser, or another laser with a wavelength in the orange spectral region, the energy from the laser is primarily absorbed by the tumor, thereby destroying it without damaging surrounding tissue. The copper vapor laser is also used for this treatment by using the copper vapor laser to pump a dye, which is then tuned to the appropriate wavelength for PDT.

4.2.3 Excimer Lasers

Commercially available excimer lasers operate primarily in the ultraviolet and vacuum ultraviolet spectral regions. A few lasers, however, operate in various parts of the visible spectrum on transitions in several diatomic and triatomic excimer molecules.[7] The shortest-wavelength visible system occurs in the Kr_2F excimer molecule and is centered at 436 nm with a bandwidth ranging from 400 to 470 nm. The XeF laser peaks at 486 nm with a bandwidth from 450 to 510 nm, the Xe_2Cl laser peaks at 518 nm and ranges from 480 to 550 nm, and the Xe_2F laser peaks at 620 nm with a bandwidth from 590 to 670 nm. These lasers yield high-power pulses because the stimulated-emission cross sections of the laser transitions are relatively low, thereby allowing large population densities (leading to large energy densities) to accumulate in the upper laser levels before stimulated emission occurs. In addition, the wide spectral widths of many of these laser transitions offer tunability and also the prospect of mode locking for the production of ultrashort laser pulses.

There are also a few rare-gas oxide excimer lasers in the green region of the spectrum and several metal halide lasers in the blue, green, and orange parts of the spectrum. All these lasers have much narrower spectral outputs than the rare-gas halogen excimer lasers (of the order of 10 nm). The only laser of this group that has been extensively investigated is the mercury-bromide laser with a maximum laser output occurring at 505 nm.

Excimer molecules are particularly effective as laser species because their lower laser levels are generally unstable (dissociative) and therefore provide no significant lower laser level population to reduce or quench the gain, as indicated in Fig. 4.7.

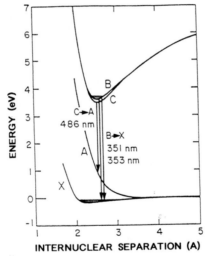

FIGURE 4.7 Energy-level diagram of the xenon-fluoride excimer laser.

The upper level is a strongly bound molecular level that is typically formed by a noble gas ion combining with a halogen atom in an electron-excited discharge. This produces a state known as an excimer excited state. The term "excimer" is coined from the term "excited state dimer" meaning a molecule that is in existence only in its excited state form and is therefore inherently short-lived because of rapid radiative decay (1 to 10 ns) of the excited state to the unstable ground state. Owing to the short radiative lifetime of the upper laser level, extremely high excitation rates are required. Typical excitation occurs in either a fast electron-beam-pumped discharge or a very fast low-inductance transverse-excited pulsed dc discharge.

The use of the reactive halogen species requires specially designed discharge chambers primarily made from glass or quartz, and stainless-steel materials which are resistive to halogen corrosion. The lasers operate most effectively at high pressures (up to 6 atm). They require special safety features in the design that are associated with structural requirements of high-pressure chambers and the toxic nature of halogen gases.

The XeF laser is one of the most important lasers in this category. This laser was first discovered in Ar-Xe-NF_3 mixtures excited by short (2 ns) electron-beam pulses with laser output occurring over the wavelength range of 450 to 510 nm. It was also later observed to lase by photolytic excitation of XeF_2 using 172-nm pump radiation from the fluorescence of Xe_2. The optimum pressure for electron-beam excitation is approximately 8 torr NF_3, 16 Torr Xe, and 6 atm Ar. Electron-beam pulses of the order of 15 kA at an energy of 1 MeV and 8 ns duration have also been used for excitation. With this arrangement, a current density of the order of 100 A/cm^2 is achieved with an energy density of 2 J/cm^2. A 50-μm-thick foil is used as a window to allow the high-energy electrons to enter the high-pressure gain region of the laser. With this arrangement, an output energy in excess of 6 J per pulse was achieved.

This laser transition occurs on the C-A transition of the XeF molecule and has a highly repulsive lower state. The stimulated-emission cross section is of the order of 1×10^{-17} cm^2 and the saturation energy density is approximately 50 mJ/cm^2, which is significantly larger than that of most excimer laser transitions. The small signal gain is of the order of 0.05 cm-1.

This laser is not available commercially. It has been investigated primarily as a possible laser for undersea communications and also as a laser for ultrashort pulse generation, owing to its exceptionally wide gain bandwidth. A mode-locked version of this laser has produced 275-mJ pulses of 250 fs duration in a 2.5 times diffraction limited beam.

4.3 VISIBLE LASERS IN LIQUID MEDIA—ORGANIC DYE LASERS

Lasers in liquid gain media were first demonstrated in liquid solutions of organic dye molecules approximately 6 years after the discovery of the first laser. Dye lasers were developed to take advantage of the broad emission spectrum (radiating over a large wavelength range) typically available from such organic dyes. Such broad emission spectra had the potential to produce gain over a wide range of wavelengths, and thus to make possible very broadly tunable lasers. These laser gain materials evolved as mixtures of highly absorbing and radiating organic dyes, diluted with solvents such as water or alcohol. The dyes are organic compounds that strongly absorb in specific wavelength regions in the visible and ultraviolet. Such dyes had originally been developed for coloring various materials including paints, fabrics, etc., but were thought to be good laser prospects owing to their efficient absorption and emission characteristics. The dyes are mixed with solvents to dilute them to the proper concentration, determined by the desired absorption depth of the pumping radiation as it penetrates the gain medium. Such an arrangement efficiently uses the pump energy to excite the laser levels.

Dye lasers cover the wavelength range from 320 nm to 1.5 μm, but most dye lasers occur in the visible spectrum.[8] They typically lase over a wavelength range or bandwidth that is 5 to 10 percent of their maximum emission wavelength. Therefore, to cover the entire wavelength range mentioned above requires 10 to 20 different dyes inserted consecutively into the laser cavity [they could not be used simultaneously owing to the strong absorption characteristics of each dye at a wavelength range just shorter than its emission (gain) wavelength range].

The highly absorbing, wavelength-sensitive dyes are mixed with water, alcohol, or other solvents that are transparent to both the pumping radiation and the emitted radiation, at a dye concentration that is typically a 10^{-4} molar solution. The concentration is largely determined by both the size of the gain medium and the distance over which the pump radiation must be absorbed.

Dye lasers are produced by pumping the liquid medium with an external light source, either a flashlamp or another laser. Originally, most dye lasers were excited by flashlamps because the high optical absorption and broad spectral pumping band of the dye strongly overlaps the spectral output of flashlamps. In recent years laser pumping has also become a very useful pumping technique. Short-pulse, narrow-emission-spectrum lasers are used to pump very-high-gain, small-diameter dye gain media. Laser pumping is most effective for providing short-duration pump pulses, which allow extraction of the energy before undesirable absorbing species (triplet states and other unwanted impurities) develop in the gain media after pumping begins. Rapid flowing of the dye through the gain medium can also help overcome the detrimental effects of triplet absorption.

The energy levels associated with dye lasers are shown in Fig. 4.8. The singlet manifold (S) consists of a ground state S_0 and an excited state S_1, both of which have a broad spread of energies as shown in Fig. 4.8. Also shown is the triplet manifold (T) with triplet levels T_1 and T_2 which lie well above S_0. T_1 also lies below S_1, and thus energy can be transferred from S_1 to T_1 after S_1 is excited by pump radiation from S_0. Population excited to the high-lying portion of the S_1 band rapidly decays (10^{-13} s) to the lowest-lying levels of S_1, where it remains temporarily until it is extracted by stimulated emission. Absorption from S_0 to S_1 and from T_1 to T_2 is highly efficient as are emissions from S_1 to S_0 and from T_2 to T_1. When S_1 is excited by pump radiation from S_0, it can decay by radiating back to S_0 in a time of the order of 10^{-9} s, or it can be converted to T_1 by slower (10^{-6} s) nonradiative processes. Radiative decay from S_1 to S_0 initiates laser action, whereas nonradiative conversion from S_1 to T_1, with subsequent absorption to T_2, leads to loss in the gain medium. The emission and absorption spectra in the S manifold are shown in Fig. 4.9 along with the profile of the gain for a typical dye (Rhodamine 6G). A decrease in gain at the shorter-wavelength side of the emission spectrum is due to the absorption overlap (involving transitions from S_0 to S_1) with the emission spectrum from S_1 to S_0. The pumping process can be seen in Fig. 4.8 to encompass a range of wavelengths due to the broad nature of the energy levels.

The above-described energy-level arrangement is typical for most dye lasers except for variations in the separations between levels, which affects the wavelength and bandwidth of the emission and absorption, and also for variations of the emission and absorption cross sections. A typical stimulated-emission cross section is of the order of 10^{-16} cm^2. Dye lasers can operate with efficiencies of several percent and are some of the most efficient lasers available. Their principal drawback is the deterioration of the dye solution which requires rapid circulation through the gain medium and frequent replacement.

Because of the broad spectral output and high gain, dye lasers typically operate multimode unless frequency-selective cavity elements are used to restrict the high-

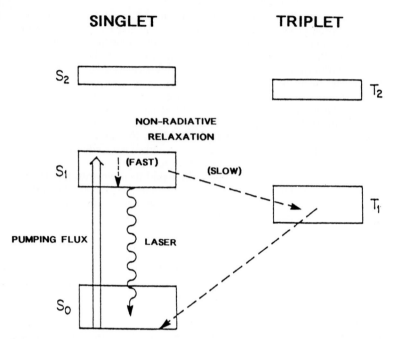

FIGURE 4.8 Energy levels of a typical dye laser molecule.

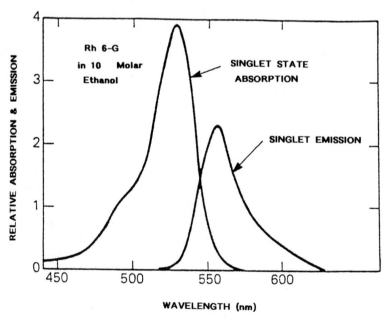

FIGURE 4.9 Typical absorption and emission spectrum for an organic dye laser molecule.

order modes. The simplest frequency-selective tuning element is either a prism and mirror combination or a diffraction grating, located at one end of the laser cavity, with a broadband mirror at the other end as shown in Fig. 4.10. Rotation of the prism or grating provides the wavelength selectivity. Other frequency-selective elements can also be added to the cavity to provide additional frequency narrowing of the laser output for high-resolution spectroscopic applications.

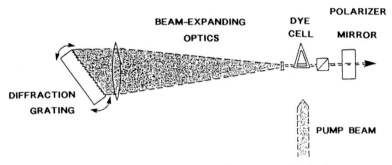

FIGURE 4.10 Tuning arrangement for a narrow-frequency tunable dye laser.

A number of pumping configurations have evolved for organic dye lasers. Early dye lasers used elliptical pumping cavity geometries in which the pumping lamp was at one focus of an elongated reflecting cavity having an elliptical cross section, and the dye medium was at the other focus. Rapidly flowing the dye through the gain medium provided effective cooling of the medium and also rapid removal of the unwanted absorbing species and excess heat. Eventually it was discovered that intense pumping in a very thin, rapidly flowing dye solution (jet stream) produced sufficient gain that gain lengths of the order of less than a millimeter could provide high-power dye laser output. Such pumping arrangements led to complex multi-mirror laser cavities (Fig. 4.11) that are particularly effective for mode locking and the production of ultrashort laser pulses. Synchronous pumping of such mode-locked lasers was obtained by first mode locking a cw visible laser, such as an Ar^+ laser, and then pumping a dye with the short Ar^+ pulses, in exact synchronization with the dye laser pulses. Such synchronization requires precise timing and therefore exact matching of the arrival time of the Ar^+ pump laser pulses with the round-trip transit time of the dye laser pulses in their separate laser cavity.

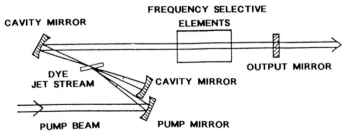

FIGURE 4.11 Multiple-mirror folded laser cavity for a dye laser.

Commercial dye lasers are available that produce cw output of up to 1 Watt over a wavelength range from 0.22 to 1.0μm, pumped by a 5 to 7-W argon ion laser. Pulsed dye lasers are available that can produce tunable laser output of up to 100 J per pulse in a duration of several μs (flashlamp-pumped), as well as laser pulses with a duration of 100 fs and shorter and energies up to many millijoules (powers up to 10 GW and higher).

Dye lasers have been used primarily for a wide range of spectroscopic investigations where wavelength tunability and narrow linewidth are required. They have also been used to generate the shortest light pulses ever produced (6 fs) because of the broad spectral gain bandwidth that is used in conjunction with mode-locking techniques. A more recent application in the medical field involves a high-energy flashlamp-pumped dye laser which is used to irradiate the skin and underlying tissues to remove unwanted tissue ranging from wavelength-sensitive cancerous tissue (PDT) to birthmarks and other undesirable skin blemishes.

4.4 VISIBLE LASERS IN SOLID MATERIALS

The category of lasers in which the gain medium is produced in a solid material has evolved in two directions over the years. The first is the area of crystalline and glass lasers. These lasers involve the use of dielectric materials (hosts) in which specific laser ions (various atoms in their ionic form) are embedded within the hosts at various concentrations that depend upon the desired properties of the amplifier. The second is the area of semiconductor materials in which a semiconductor is produced with special properties that enhance its radiative characteristics and thus lead to laser output. Although both of these types of lasers are produced in solid materials, the trend in the laser community has been to refer to the first type as solid-state lasers and the second type as semiconductor lasers. Presently, there are very few visible lasers in either category. The few that exist will be summarized below.

4.4.1 Visible Solid-State Lasers in Crystalline Hosts

Although the primary spectral region where solid-state lasers emit is in the near infrared, there are a few solid-state lasers that lase in the visible spectral region.[9] The best known of these is the ruby laser, the very first laser. The ruby laser will be described in some detail below. Other visible solid-state lasers have existed only as laboratory systems, since they have not had the necessary properties that would make them commercially attractive.

Ruby Laser. The ruby laser consists of a synthetic aluminum oxide (sapphire) host material in which chromium ions are doped at a relative concentration of the order of 5 percent. The aluminum oxide serves as a matrix which holds the chromium ions (Cr^{3+}). When the chromium ions are located within the aluminum oxide "cages," they are isolated from the fields associated with the sapphire crystal, and the ions behave more atomic-like (narrow emission spectra) than solid-like (broad spectra). In that sense, the Cr^{3+} ions have very long radiative lifetimes (3 ms) and relatively narrow emission linewidths (5 Å) when compared with the typical radiating properties of most solids. Such a long lifetime allows for energy storage by allowing the upper laser level to accumulate population over long periods of time.

The narrow linewidth produces high gain by concentrating the emission from the upper laser level population within a narrow frequency range or gain bandwidth.

In ruby, the laser output occurs just barely in the visible spectrum on two transitions at 694.3 and 692.9 nm, with the longer wavelength transition being significantly stronger. Ruby typically operates in a pulsed mode in which flashlamps are used to pump the ions into the upper laser level. The ruby laser is known as a three-level laser as shown in Fig. 4.12. In such an arrangement, the lower laser level is the ground-state level or source, from which ions are pumped into the upper laser level through a higher-lying intermediate level. Therefore, in order to obtain a population inversion (taking into account the degeneracies of the upper and lower laser levels), nearly one-half of the ions in the ground state must be pumped out of that state before an inversion occurs. This rather difficult task is possible owing to the long lifetime of the upper laser level.

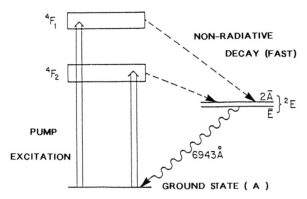

FIGURE 4.12 Energy-level diagram of the ruby laser.

The ruby laser is excited by optical pumping with flashlamps. The wavelength regions over which effective pumping can occur are two 100 nm-wide bands in the blue and green spectral regions. Thus a typical visible-output flashlamp with a blackbody spectral output temperature of the order of 8000 K conveniently matches the absorption spectrum required to excite the upper laser level. The light is absorbed into a band of levels located above the upper laser level, from which rapid decay ($\sim$1 ns) occurs to the upper laser level where it is stored for laser output. The absorption cross section at the peak wavelengths is of the order of 10^{-20}cm^2 whereas the stimulated-emission cross section is 2.5×10^{-20}cm^2.

Ruby lasers are typically produced using polished cylindrical ruby laser rods, of the order of 5 to 10 mm in diameter and 10 cm in length. The rods are mounted within a highly reflecting cylindrical laser cavity having either a single or double elliptical cross section. The laser rod is located at one focus of the ellipse and the flashlamp at the other focus (or foci in the case of the double lamps and a double ellipse).

Laser output can be as high as 100 J in a duration of a few milliseconds for a peak power of nearly 100 kW at a repetition rate of a pulse every several seconds. The low repetition rate is limited by the cooling time required for the crystal. Other ruby lasers operate at pulse repetition rates of up to 5 Hz. The ruby laser, having been the first laser discovered, has had the advantage of extensive development efforts, and yet other solid-state lasers such as the Nd:YAG laser have produced more desirable properties that have reduced the demand for ruby lasers in recent years.

Other Solid-State Lasers.   A number of other solid-state laser materials have been made to lase in the visible spectral region; however, only one of these has been developed to the point of being commercially available. That laser is the Ti:Al_2O_3 laser which has its primary output in the near infrared with a tuning range of from 660 to 1100 nm (1.1 μm). Commercial lasers operating both cw and pulsed are available at wavelengths as short as 680 nm. This laser crystal consists of titanium ions doped in a sapphire (Al_2O_3) host. It is a relatively recent addition to the commercial laser market and offers both wide tunability and short pulse operation under conditions of mode locking. The reader should see the discussions in Chap. 5 for a more detailed description of this laser.

The other solid-state lasers that have been made to lase in the visible spectral region are only laboratory devices that have not shown sufficient efficiency or ease of fabrication to be of commercial significance. Many of them operate only at low temperatures. Nevertheless, ions of tamarium, praseodymium, thulium, erbium, holmium, samarium, and europium have all exhibited visible laser output when grown in host crystals such as YLF. The reader is referred to Ref. 9 for a further discussion of these materials.

4.4.2 Visible Semiconductor Diode Lasers

Semiconductor diode lasers have many characteristics that make them one of the most attractive types of laser. They are extremely small and efficient, have long operating lifetimes, and in many cases are extremely inexpensive. They are also capable of high powers of up to several watts cw from a very small laser. Unfortunately most semiconductor lasers operate in the near-infrared spectral region owing to the inherent nature of their energy levels. However, in recent years significant progress has been made in engineering special semiconductor materials that have provided energy levels in the visible spectral region that can lead to population inversions and gain.

Semiconductor laser materials are produced by doping semiconductors with other materials having either an excess (*n*-type) or deficiency (*p*-type) of electrons when compared with the pure semiconductor materials. Such doping involves implanting a small amount of the *p* and *n* materials (10^{-4} concentration) in the semiconductor in narrow spatial regions located adjacent to each other. Such a spatial region is referred to as a junction. When an electric field is applied across this junction, the excess electrons in the *n*-type material are pulled into the *p*-type material, creating an overpopulation of electrons or a population inversion in the medium. These electrons then recombine with vacancies (holes) of that material and produce recombination radiation. This radiation leads to the generation of laser output if the geometry of the semiconductor is suitably designed. Because the threshold currents are thousands of amperes per square centimeter or higher, only very small gain regions can be excited within the semiconductor volume to avoid excessive heat which would destroy the population inversion.

The approach to making visible semiconductor lasers has been to use special mixtures of materials that lead to large energy gaps consistent with visible emission. Since orderly arrays of atoms (crystals) are required for semiconductor laser operation, obtaining "lattice matching" of these new materials with existing semiconductor substrates has been a difficult problem, owing to the physical size differences of the various atoms associated with different materials. A procedure known as bandgap engineering has evolved relatively recently to produce special crystals using these new materials. Another problem in producing visible diode emission has been in obtaining the proper doping concentration of some materials.

In spite of these difficulties, a significant amount of progress has been made in the last few years in producing visible semiconductor lasers. For example, red and orange diode lasers have been made in InGaP/InGaAlP strained layer quantum well materials at wavelengths as short as 0.6 μm when cooled to very low temperatures. Also, lasers have been made in AlGaAs lattice matched materials at wavelengths as short as 0.68 μm. Another recent breakthrough is the production of blue-green lasers in ZnCdSe quantum well materials embedded in ZnSe, using a ZnSeS waveguide. These blue-green lasers operate at 0.46 μm at liquid nitrogen temperatures and shift to 0.49 μm at 0°C and are projected to lase at 0.52 μm at room temperature.

Only the AlGaAs lasers operating near 0.7 μm are commercially available at this time. High-volume applications of these lasers include inexpensive pen-style laser pointers, bar code scanners, interrupt sensing devices, and gunsights. Reliable performance, small size, and low cost (a few dollars) make these lasers viable alternatives to other visible lasers such as the He-Ne lasers.

The use of such lasers in optically demanding systems is much more limited because of the relatively low beam quality and unstable temporal behavior. The poor beam quality, including an astigmatic output, leads to larger than diffraction-limited beam wavefronts and non-diffraction-limited spot size at focus. Thus applications in such areas as information storage and retrieval, optical probing, atomic spectroscopy, and metrology will have to await further development of these lasers.

4.5 REFERENCES

1. A. D. White and J. D. Rigden, "Continuous Gas Maser Operations in the Visible," *Proc. IRE*, vol. 50, p. 1697, 1962.

2. W. B. Bridges, "Laser Oscillation in Singly Ionized Argon in the Visible Spectrum," *Appl. Phys. Lett.*, vol. 4, pp. 128–130, 1964. Erratum, *Appl. Phys. Lett.*, vol. 5, p. 39, 1965.

3. W. T. Silfvast, "Efficient CW Laser Oscillation of 4416 Å in CdII," *Appl. Phys. Lett.*, vol. 13, pp. 169–171, 1968.

4. W. T. Silfvast and M. B. Klein, "CW Laser Action on 24 Visible Wavelengths in SeII," *Appl. Phys. Lett.*, vol. 17, pp. 400–403, 1970.

5. W. E. Bell, "Visible Laser Transitions in Hg+," *Appl. Phys. Lett.*, vol. 4, pp. 34–35, 1964.

6. W. T. Walter, N. Solimene, M. Piltch, and G. Gould, "Efficient Pulsed Gas Discharge Lasers," *IEEE J. Quantum Electron.*, vol. QE-2, pp. 474–479, 1966.

7. F. K. Tittel, Gerd Marowsky, W. L. Wilson, Jr., and M. C. Smayling, "Electron Beam Pumped Broad-Band Diatomic and Triatomic Excimer Lasers," *IEEE J. Quantum Electron.*, vol. QE-17, pp. 2268–2281, 1981.

8. F. J. Duarte and L. W. Hillman (eds.), *Dye Laser Principles*, Academic Press, New York, 1990.

9. M. J. Weber (ed.), *Handbook of Laser Science and Technology*, vol. I, *Lasers and Masers*, CRC Press, Boca Raton, Fla., 1982.

CHAPTER 5
LASERS, SOLID-STATE

Georg F. Albrecht and Stephen A. Payne

5.1 INTRODUCTION

A solid-state laser is a device in which the active medium is based on a solid material. This material can be either an insulator or a semiconductor; semiconductor lasers are covered in Chap. 7. Solid-state lasers based on insulators include both materials doped with, or stoichiometric in, the laser ions, and materials which contain intrinsic defect laser species, known as F centers.

The physics and engineering of solid-state lasers are both mature fields, and areas burgeoning with new activity. While many concepts and laser designs have been established, each year continues to bring remarkable discoveries that open new avenues of research. This chapter is intended to provide a brief accounting of the known physics of solid-state laser sources and also to convey a sense of the enormity of the field and the likelihood that many new laser materials and architectures will be discovered during the next decade.

The principle of laser action was first reported in 1960 by Maiman. This first system was a solid-state laser; a ruby crystal served as the active element and it was pumped with a flashlamp. With this report of laser action, the main concepts upon which solid-state lasers are based became established (see Fig. 5.1). The idea of optically pumping the laser rod was realized, as was the use of an impurity-doped solid as the laser medium. Last, the concept of a laser resonator, as adapted from the work of Townes and Schawlow,[1] was experimentally demonstrated. Much of the chapter that follows is essentially a tutorial-style exposition of the extensive technical progress which has occurred in each of these three areas. Optical pumping has evolved considerably, by way of optimization of flashlamps, and through the additional use of laser-pumping techniques. The number of impurity-doped solids that have now been lased stands at over two hundred. Optical resonators have also become remarkably sophisticated in terms of the manipulation of the spatial, temporal, and spectral properties of the laser beam.

This chapter is separated into two parts: (1) solid-state laser devices, and (2) solid-state laser materials. The section on device technology provides an understanding of the issues that are important to laser design and also gives numerous "rules of thumb" with which the performance of a system may be assessed. The section on solid-state materials describes the types of laser media that are available and discusses the parameters that are utilized to characterize them. In addition,

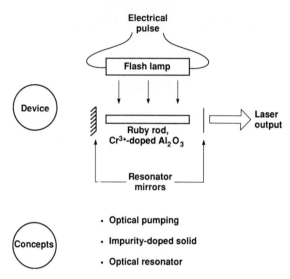

FIGURE 5.1   Schematic description of the ruby system for which laser action was first reported, and a listing of the fundamental concepts that were introduced.

two tabular listings contain the thermal, mechanical, and laser properties of the most significant and useful laser materials.

5.2 SOLID-STATE LASER DEVICES

Two essential ingredients comprise a laser: the active medium in which the inversion is created (the laser material), and the configuration of mirrors and optical switches which surrounds the active medium to enable useful extraction of energy in a specified manner. Many different laser materials exist and a similarly large variety of oscillator and amplifier configurations which combine to provide an enormous variety of pulse and energy formats.[2] In the following section, we first give an overview of pumping mechanisms, useful arrangements of mirrors and switches, and the different possible modes of operation. After that, we briefly describe simple scaling rules for solid-state laser oscillators. Then, some issues surrounding average power operation, beam quality, and some simple amplifier designs will be considered. The frequency conversion of the laser output is described elsewhere (see Chaps. 13 and 14).

Solid-state lasers come in a wide variety of types and sizes, and the following examples provide only an introduction to the systems that are possible. At one end of the power scale are the minilasers, which can be about the size of a sugar cube and deliver a few microwatts of power. Small lasers are used for memory repair in integrated circuits and for a multitude of alignment tasks. As the power of the laser increases, applications such as ranging and wind velocity measurements become accessible. Somewhat larger lasers enable activities like marking, medical, and military applications. Some of the most powerful systems are designed for cutting, drilling, and welding at high rates of material throughput. A modern automotive production line now includes many robot-controlled solid-state lasers. The very largest lasers are quite unique and serve special research purposes. At the top end of the list are the fusion drivers which, as the name suggests, generate enough energy and power to initiate the same process of thermonuclear fusion which powers the stars. Various other chapters in this book are devoted to the description of industrial, medical, and scientific applications.

Regardless of the size of the laser, one always has to start out with an active medium which is pumped by either flashlamps or another laser. This creates an inversion which is then extracted by amplification of an external signal or by spontaneous emission which is generated by the medium itself. This process takes place in a resonator cavity which consists of two or more mirrors and which may contain optical switches as well. Such an assembly is referred to as an oscillator. One significant difference between solid-state lasers and other types of lasers is that they have the capability to store energy. The storage arises from the fact that the typical fluorescent lifetime of an inversion is very much longer (several microseconds to milliseconds) than the time it takes for the lasing process to extract the inversion (microseconds to nanoseconds). Hence, as the active medium is excited by the pump, the inversion can be accumulated over time (it can be stored) and, by use of an optical switch, be extracted at a chosen instant. To get increased amounts of power or energy one can, up to a point, make larger oscillators. Eventually, however, one has to build external amplifiers to further increase the output of a system.

5.2.1 Methods of Pumping

Flashlamp Pumping. There are three methods with which an inversion can be created in a solid-state laser. The cheapest and most common method utilizes flashlamps (Fig. 5.2). A flashlamp essentially consists of a fused silica tube of

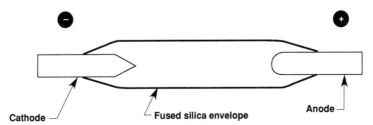

FIGURE 5.2 The basic elements of a flashlamp. The envelope is typically made from fused silica, the electrodes from a tungsten alloy. The shape of the cathode helps to increase lamp life.

suitable diameter and length with an electrode at each end. Once triggered with a short, high-voltage spike, a plasma discharge occurs between the electrodes which converts the supplied electrical power with high efficiency to power radiated as light in the infrared, visible, and ultraviolet. Some of this light is absorbed by the active ion with which the host medium is doped, and by virtue of its energy levels and decay dynamics, an inversion is created.

Pulsed lasers operated in a storage-type mode use different kinds of lamps than continuous-wave (cw) or quasi-cw systems, where the pulse duration is of order of, or long compared to the fluorescent lifetime of the laser ion. The lamp used for a cw Nd:YAG laser is filled with several atmospheres of Kr gas. In this case the plasma mostly radiates in several pronounced lines around 800 nm, which is a wavelength readily absorbed by the neodymium ion. Lifetimes of cw arc lamps are typically measured in hundreds of hours and are limited by sputter deposition of electrode material on the inside of the lamp envelope, which prevents plasma radiation from leaving the lamp and shortens its effective arc length. If one wants to pump a pulsed laser, the gas of choice tends to be Xe. In that case the electrical pulse applied to the lamp is of the order of or shorter than the lifetime of the upper laser level. The energy delivered to the lamp electrodes is typically a few hundred times the energy in a single output pulse, depending on the efficiency of the laser. The plasma radiation pumping the laser is approximately described with a black-body spectrum which extends from the ultraviolet to the infrared spectral regions, whereas the active ion tends to absorb only over a narrow spectral region. As a result, only a small fraction of the lamp light is actually absorbed by the active medium (around 10 percent). Part of this absorbed fraction will end up as inversion in the upper lasing level from where it will either be removed by stimulated emission (lasing) or decay by spontaneous emission. The balance of the energy eventually has to be removed as heat, which has important consequences for average power operation. Nevertheless, the peak power densities possible with such pulsed Xe flashlamps are sufficiently intense that high-power pulsed lasers can be constructed in this way.

The capabilities of pulsed flashlamps are broadly characterized by an explosion fraction which is the ratio of energy delivered to the lamp electrodes to the maximum possible deliverable energy at which the lamp literally explodes. Critically damped operation at low explosion fractions of a few percent generally means lower efficiency, but exponentially longer life. Operation at high explosion fraction (>70 percent) conversely means a dense plasma, good graybodylike radiative efficiencies, but a short lifetime. Well-designed systems operate somewhere in between and make explicit use of this parameter to optimize the overall characteristics of the system. Sales brochures by flashlamp manufacturers or Ref. 2 provide good tutorials on this subject.

A large part of the overall laser efficiency is given by how well flashlamp light is transported to and absorbed by the active medium. Additional constraints are given by the desire to achieve adequate uniformity of energy deposition across the lasing aperture. It is generally observed that the fewer reflections a ray has to make on its way from the plasma to the active medium, the better the efficiency of the pump cavity will be. This leads to "close-coupled" designs where the lamp is brought as close as possible to the laser rod. Additional improvements come from surrounding the lamp and rod with a medium of high refractive index so that a larger solid angle of rays leaving the plasma is refracted toward the rod. Finally, to provide pumping uniformity, the highly reflecting surfaces are made from a diffusing material, often barium sulfate. Such close-coupled pump cavities are employed in many commercial systems and are generally more efficient than those with elaborately designed reflective surfaces. The latter type is a better choice if the precise spatial deposition of pump light in the active medium is of overriding importance. Experience shows that one obtains either good efficiency or a well-controlled energy deposition, but not both.

Flashlamp pumped solid-state lasers are clearly the mainstay of the industry and, with the exception of CO_2 welding lasers, make up all of the higher-power

FIGURE 5.3 A 250-W Nd:YAG laser used in the semiconductor industry. *(Courtesy of Quantronix Corp.)* The complete unit includes a beam delivery system and the means to observe the workpiece through a microscope.

systems. Figure 5.3 shows a typical cw lamp pumped system with applications in the semiconductor processing industry.

Diode and Laser Pumping. The pumping of small solid-state lasers (<1 W) with laser diodes has matured into commercial units within the last few years (Fig. 5.4). Diode pumping allows for more efficient solid-state lasers, because the diode lasers themselves (see Chap. 6) efficiently convert electrical power to radiated power, and the diode laser power is matched to a specific absorption line of the active medium of the solid-state laser. Although, at present, diode lasers are far more expensive than flashlamp pumping, diode laser pumping is the method of choice where efficiency is at a premium, as is the case for most military applications. Diode pumped lasers are likely to become cheaper in the future through increased volume of diode production. Since laser diodes are limited by the peak power they can generate, the total energy output necessarily decreases with a shorter pump pulse duration. The single-shot intensity of a two-dimensional diode array is on the order of 1 kW/cm². As a consequence, laser ions with a long fluorescent lifetime are easier to diode pump efficiently than ions with a short fluorescent lifetime. Presently, the most common device involves the use of AlGaAs diode lasers to pump the Nd^{3+} absorption band near 810 nm in various hosts. Current efforts are aiming for diode pumped solid-state lasers in the several hundred watt range, which requires diode pump arrays capable of producing several hundred W/cm² of *average* optical power and a cooling scheme capable of removing several kW/cm² of waste heat from the back plane of the diode array (Fig. 5.5). As of this writing up to 1000 W of average output power from a Nd^{3+}:YAG laser have been achieved with this method.

Other interesting developments utilize diode lasers based on InGaAs to efficiently pump ions such as Yb^{3+}. Small, single-diode laboratory prototypes have been demonstrated. The next step is to build higher-power versions using diode arrays and architectures which are best suited to the characteristics of a particular ion-host combination. This work is in flux at the present time, but a suitable introduction to the subject is presented in Ref. 3.

It is also possible (and the diode pumping mentioned above is a case in point), to pump a laser with another laser. With the exception of diode pumping, this is

FIGURE 5.4 A 5-W commercial diode pumped Nd^3:YLF laser. *(Courtesy of Spectra Physics.)* The Nd:YLF crystal is located underneath the cooled aluminum housing; the pen points at the diode bar assembly which pumps the YLF crystal. Note the hand for scale.

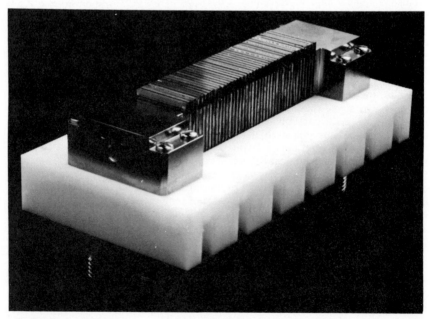

FIGURE 5.5 A diode pump array developed at LLNL. It is capable of 8.5-kW peak power at 25 percent duty cycle. A Nd^{3+}:YAG slab laser using two of these arrays produced 1 KW of average output power at 1.06 μm.

mostly done for scientific applications. Since more than one laser is involved, the individual efficiencies multiply to a fairly low overall system efficiency. On the other hand, laser pumping enables pump fluences and pump pulse formats entirely out of the reach of flashlamps. It is the availability of high pump fluence which makes laser pumping such a useful research tool, since it is often required when lasing newly discovered materials, which can at first have rather high loss levels.

A more sophisticated example of laser pumping is known as "synchronously pumped mode locking." In this case, the pump laser emits a stream of mode-locked pulses, which in turn pump an active medium in a resonator cavity which has the same round trip time as the interpulse spacing of the pump laser (see chapter 9). The pumped laser can have a dye, Ti:sapphire, or other material as its active medium and can, additionally, incorporate pulse shortening and wavelength tuning components of its own. The described technique is a well-established method of producing ultrashort, stable pulses from the laser which is being pumped.

A less complex example would be the pumping of Ti:sapphire with frequency-doubled, Q-switched Nd:YAG lasers. This technique is employed because the upper-level lifetime of Ti:sapphire (3.2 μs) is too short to easily apply conventional flashlamp pumping techniques without approaching the lamp explosion limit discussed above. The wavelength of frequency-doubled Nd:YAG ideally matches the absorption band of Ti:sapphire and has a suitably short pulsewidth. Other reasons for laser pumping are that an upper-state laser level can be exactly matched by the laser which does the pumping, thereby minimizing the deposited waste heat. This sometimes allows operation of lasers at room temperature which have to be operated cryogenically when flashlamp pumped. The ultimate utilization of the high pump fluence available with laser pumping is a technique which goes by the name of "bleach pumping."[4,5] A quantity known as the saturation fluence F_{sat} is given by

$$F_{sat} = \frac{h\nu}{\sigma_{abs}} \tag{5.1}$$

and is an important characteristic parameter for various aspects of laser behavior in the bleach pumping regime. In Eq. (5.1), h is Planck's constant, ν is the frequency of the light, and σ_{abs} is, in this example, the absorption cross section of the active ion. This last item is the enabling factor in bleach-pumped solid-state lasers. As a specific example, an alexandrite laser pulse may be used to pump the 745-nm absorption line of Nd^{3+} in a Y_2SiO_5 host crystal with a sufficient fluence to put nearly all of the Nd^{3+} ions present in the crystal into the upper laser level. With essentially no Nd^{3+} ions left in the ground state, the active medium becomes transparent to further pump radiation (hence the name bleach pumping). This method of pumping makes it possible to achieve efficient laser action on transitions for which the active ion would otherwise not lase because of ground-state absorption. Although a new development, the number of useful wavelengths for solid-state lasers may be greatly extended by this method. Outputs of 0.5 J at several Hertz repetition rates have been demonstrated for the 911-nm transition of $Nd^{3+}:Y_2SiO_5$; it could not be lased at such output levels with conventional flashlamp pumping methods.

Laser pumping has also led to the development of an entirely new class of lasers known as fiber lasers. Here, the dopant ions are incorporated directly into a fiber, which is then end-pumped by a laser source and aligned between two resonator mirrors. The details of fiber fabrication are described in Chap. 12. Many new lasing materials have been developed using this technique since much lower doping levels are needed by virtue of the long path lengths which are possible and the very low passive loss levels. A particularly useful recent development is the erbium fiber amplifier, pumped by an InGaAs diode. This device serves as an optical repeater in fiber optic communications technology.

If one attempts to model the pumping of a laser, the case of bleach pumping a laser is one of the easiest to describe, since the pump laser generally has a linewidth much narrower than the absorption profile. For diode pumping, the linewidths of pump laser and absorption line are comparable in crystals and one readily observes how the diode line center is quickly burned out whereas the wings propagate into the medium for an appreciable distance. Modeling flashlamp pumping is a very difficult and complex task which is best undertaken by Monte Carlo type ray tracing. There appears to be no shortcut to modeling all wavelengths and all directions of emission from the lamp, and following the ray as it encounters all the different surfaces in the pump cavity several times, each time splitting in a reflected and a refracted ray according to Fresnel's laws. This procedure very rapidly requires a fast computer and a lot of time. Models which try to "lump" some of the effects into some global parameter usually end up disagreeing with the measurements in some major way.

5.2.2 Architectures and the Use of Optical Switches

The aligned mirrors that surround the active medium and permit repeated passes of laser light through the inversion form the resonator cavity. In addition to the active medium one can also place a variety of optical switches inside the cavity, such as Pockels cells, acousto-optic switches, and saturable dyes. Depending on the arrangement of the resonator and the optical switches, many different laser architectures become possible. Figure 5.6 shows four different architectures for a simple cw or free-running laser. Figure 5.6a is a simple stable resonator cavity with the output being emitted through a partially transparent mirror at one end. Figure 5.6b shows a trivial variant which emits from both ends and can, for example, be used to produce holograms because the two outputs are phase-coherent.

Yet another related variant is sketched in Fig. 5.6c, where the output emerges from a polarizer and the output coupling is varied by rotating a waveplate. Both mirrors are 100 percent reflective in this case. Figure 5.6d shows a traveling-wave rather than a standing-wave version, since the light is not reflected back on itself. In conjunction with a device called a Faraday rotator, a properly oriented half waveplate, and a polarizer, one direction of travel can be suppressed in favor of the other direction. Such unidirectional ring lasers largely eliminate the phenomenon called spatial hold burning which therefore makes single longitudinal mode operation easier to obtain. Ring lasers can couple out power via schemes a,b, and c. Yet another set of variations is obtained by incorporating an interferometer (for precise wavelength control) in place of one of the mirrors. Such architectures also play important roles in ultrashort pulse generation and frequency conversion.

Given this brief discussion, it becomes clear that, when used in conjunction with optical switches timed in various ways, frequency converters, and other nonlinear elements, the multitude of possible architectures is enormous. The field is further enriched by the possibility of coupling different lasers, in order to inject a single pulse from a mode-locked oscillator into a regenerative amplifier (as will be discussed below), or to achieve mutual phase coherence across individual amplifier apertures. Since each approach, however, has its own merits and drawbacks, a judicious, well-thought-out choice of system architecture remains one of the keys

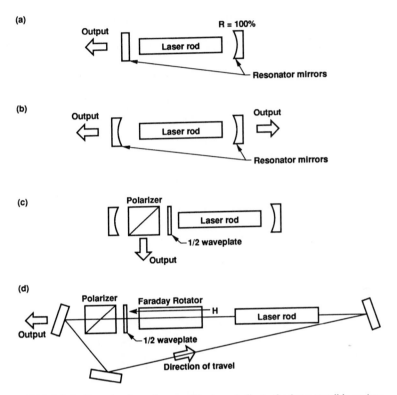

FIGURE 5.6 Four simple cw laser architectures indicate the large possible variety of different arrangements.

of successful laser design. The resulting availability of different pulse formats, in conjunction with variable output energies and wavelengths from different materials, is the origin of the great versatility of solid-state lasers. Ultrashort pulse generation is described in more detail elsewhere in this book and will therefore not be treated here.

In addition to Pockels cells and acousto-optics devices there is, however, one more optical switch which is commonly used, namely, the saturable absorber.[2] In this method of switching, the optical beam enters a cell which contains a material that absorbs at the wavelength of the incident light pulse. If the beam has fluence several times greater than the saturation fluence for the absorption transition as described in Eq. (5.1), the beginning of the pulse will bleach through the cell, making it transparent for the rest of the pulse. Hence such a saturable absorber essentially serves as an intensity threshold filter. The cell can contain a solution of an appropriate dye, or it can be a crystal containing suitable F centers. Switching light by this method provides a simple and inexpensive means of Q switching and mode locking in the laboratory. Compared with Pockels cells and acousto-optic devices, however, the statistics of the bleaching process results in less precise timing control. Furthermore, the poor chemical stability of the dyes renders saturable absorbers unsuitable for lasers, which need to be maintenance-free.

Aside from the architecture surrounding the active medium in the form of a specially configured resonator, the notion of "architecture" in solid-state lasers also can be applied to the shape of the active medium itself. Several examples of shapes used for active media are depicted in Fig. 5.7. The most familiar shape, determined by tradition more than anything else, is that of a rod. For some average

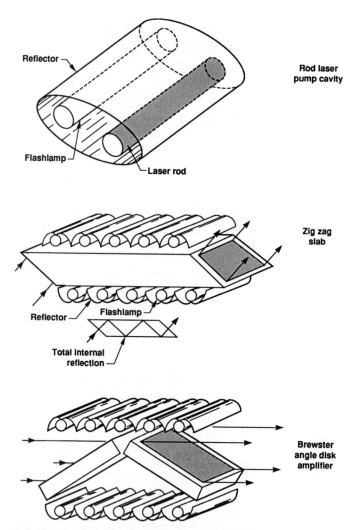

FIGURE 5.7 Three different shapes of active media in solid-state lasers. The round rod is found in almost all commercial lasers and is typically 1/4 in in diameter and 2 to 4 in long. Small zigzag-path slabs (about $2 \times 15 \times 0.5$ cm^3) have just begun to enter the commercial market in Europe. The Brewster angle disk amplifier configuration is a very common amplifier component in large scientific laser systems and has been constructed in sizes up to almost 1/2 m in diameter.

power systems, however, it is advantageous to use a rectangular slab instead. We discuss this case in more detail below. Large amplifiers (as in fusion-class lasers) use the active medium in the shape of disks. These are face-pumped by arrays of flashlamps and oriented at alternate Brewster's angles with respect to the extracting beam.[6] The widest variety, however, is found in diode pumped miniature lasers, and one of the most interesting variants is the monolithic ring resonator.[7] Here, the mirrors of a ring resonator are directly fabricated on the laser material itself. Moreover, by taking the light path out of the two-dimensional plane into three dimensions, the mirror reflections can be made to serve as half-wave plates. Applying a magnetic field through the active medium then ensures unidirectional operation of the ring. Hence the wave plate and the Faraday rotator are all embodied in the architecture of the active medium itself. This highly integrated architecture contributes to the remarkable stability of the output characteristics of this laser. Extremely narrow linewidths have been achieved using carefully studied derivatives of this design (of order tens of Hz for short periods).

5.2.3 Modes of Operation

Some of the different architectures described in the previous section are most advantageously used in modes of operation which correspond to different pulse formats in the output. In this way, pulse durations from femtoseconds to cw can be created, although some regimes are more easily accessed than others. Of equal importance is the difference between three-level and four-level laser systems, as they will be described in more detail in the section on materials. Since a three-level system must have at least half of all ions inverted before any net gain exists, these systems require correspondingly more pumping power. The following discussion is kept general enough to apply to both cases.[2]

CW and Free-Running Operation. The simplest mode of operating a laser is with no switches in the cavity, so that the resonator contains only the active medium. For a continuous pump, the laser operates in a cw mode. Since this problem does not depend on time, the basic equations describing parameters like output power, extraction efficiency, and intracavity intensity do have well-developed approximate analytical solutions which are treated in most textbooks. If no transverse or longitudinal mode control is implemented, the output will fluctuate because of the complex ways in which the longitudinal and transverse modes beat and couple to each other via the active medium. It has, in fact, been found that the new field of chaotic dynamics describes these phenomena correctly. Other output fluctuations will originate from the power supply. A laser with full mode control will operate only on a single longitudinal and transverse mode. After stabilizing the power supply, such lasers can have output fluctuations of less than 1 percent. The most powerful industrial solid-state lasers are cw lasers that operate without mode control and achieve output powers above 1 kW for numerous material working applications. The characteristics of such lasers are closely related to the control of heat flow inside the active medium, which will be discussed in the section on average power issues.

The term "free-running" is generally used to describe a cw laser which runs for times on the order of, or longer than, the storage time of the laser medium, which is typically on the order of 1 to 1000 μs. Figure 5.8*a* shows the pump pulse, the gain and loss in the active medium, and the output energy. As the active medium is pumped sufficiently to exceed threshold, lasing begins with a few output spikes,

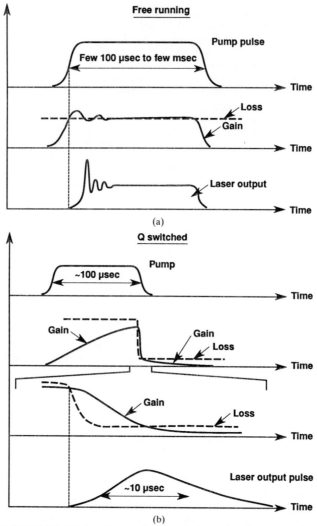

FIGURE 5.8 Timing diagrams for three basic modes of operation. (*a*) Free-running, pulsed oscillator: Depending on how fast the pump pulse causes the gain to rise above the loss initially, the gain and laser output will react with overshoots which will damp out. The loss level includes that of the active medium and the transmission of the resonator mirror, through which the laser light exits the resonator cavity. In cw and quasi-cw lasers the steady-state gain equals the loss, and the excess pump power is converted to laser output. (*b*) Q-switched oscillator: Initially, the Pockels cell causes a high loss in the cavity so that lasing is inhibited and the pump power is integrated as inversion in the upper laser level. When the pump pulse is over, the Pockels cell loss is switched off, lowering the loss in the resonator cavity to a value corresponding again to the transmission of the mirror through which the laser light leaves the cavity. The laser pulse will build up rapidly (note the much expanded time scale) and reach the peak at the time where the gain is equal to the loss.

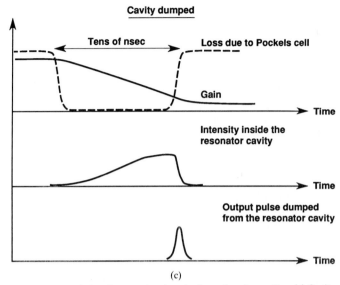

FIGURE 5.8 Timing diagrams for three basic modes of operation. (*c*) Cavity dumped case: The storage of inversion is identical to the Q-switched case. After switching the Pockels cell to enable lasing, the cavity loss now is very small since, for a cavity dumped architecture, the reflectivity of both resonator mirrors is 100 percent. Once the intensity in the cavity has reached the maximum value, the Pockels cell is again switched and ejects the intracavity intensity from the laser in a pulse whose duration is equal to the round-trip time of the resonator cavity.

after which it settles down to the cw output level. The output of such lasers is often temporally shaped by tailoring the current pulse to the flashlamps so that the laser output has the optimum effect on the material to be drilled or welded.

Q *Switching and Cavity Dumping.* This mode of operation requires the addition of an optical switch, such as a Pockels cell and a polarizer, into the resonator. In the "off" state the light is prevented from circulating in the cavity by the Pockels cell, which rejects it from the polarizer surface before it can complete one round trip. Contrast ratios of 1000:1 are easily achieved in properly aligned Pockels cell switches. Since the active medium is now prohibited from lasing, the medium builds up a far higher population inversion density, or gain, than would be possible without the switch. Because of this accumulation of inversion through storage, these devices are called "storage lasers."

Figure 5.8*b* is a plot of the timing sequence of the events, including the pump pulse from the flashlamp, the impact of the Pockels cell voltage on the transmission of the resonator cavity, and the laser output. After having stored the inversion to the maximum level allowed by the pump pulse, the Pockels cell is then turned to the "on" state, and light is now allowed to circulate in a cavity with initially very high gain (The quality factor Q of the resonator cavity has been switched, hence the name Q switching.) This circulation and pulse buildup goes on until the inversion has been depleted to the point where the gain in the active medium is equal to the loss in the cavity, at which instant the output pulse reaches its peak intensity value. From then on the loss exceeds the amplification and the pulse decays with time, although it still continues to remove residual inversion from the active medium. The duration of the output pulse depends on the cavity length, the reflectivity of the output coupler, and how far the system is pumped above threshold. Given a reasonably designed system, pulse durations equal to a few cavity lengths are typical. The basic process is described by a set of coupled differential equations not amenable to analytical solutions. The case of the *optimized* Q-switched oscillator, however, has quite recently been solved analytically in a paper by Degnan.[8] Although not explicitly spelled out in the paper, one interesting practical conclusion that can be derived from it is that all practical, optimized Q-switched oscillators operate about a factor three above threshold.

There are various limits to the Q-switching process which apply to amplifiers as well (as we will see later). Given the high contrast of the Pockels cell–polarizer combination, it is not possible to "outpump" this round trip cavity loss. Spontaneously emitted light from within the inverted medium, however, can find reflective surfaces other than the resonator mirrors, such as the surface of a mounting fixture or surfaces of the active medium itself, through which the path of a light ray can eventually close on itself and experience net round trip gain. In high-performance systems, such parasitic oscillations must be judiciously controlled. They can actually destroy the optical elements of a laser by ablating material from some metal surface and depositing it on optical surfaces. Damage to the laser notwithstanding, the gain ultimately available for lasing can be clamped by parasitic oscillations at a value well below that originally anticipated, greatly reducing the envisioned output characteristics of the device. It should be clear that once such parasitic oscillations have started (reached their own lasing threshold), no amount of extra pumping will increase the gain in the laser further. All the extra pump power will only increase the output power of the parasitic mode.

Finally, there is "amplified spontaneous emission" (ASE) which is due to the fact that, since spontaneously emitted photons travel some distance through a medium with gain, they will be amplified and therefore reduce the available net gain at the time of Q switching. It should be noted that ASE and parasitics are often loosely considered as describing the same phenomenon, but this is not correct. Gain reduction due to ASE is a natural limit in a well-designed laser system and is as unavoidable as fluorescence decay. Avoiding parasitics, however, is an issue of laser design which can be controlled by proper incorporation of absorbers such as edge claddings and by eliminating unwanted feedback through the active medium.

Typical commercial Q-switched oscillators produce output pulses with energies on the order 100 mJ and pulse durations of tens of nanoseconds. There are, however, some specialized laboratory lasers with output energies of more than 10 J per pulse, using Nd^{3+} doped glass.

A mode of operation closely related to Q switching is cavity dumping. The essential architecture is the same as that for Q switching, although the demands on the timing controls are more severe. Following Fig. 5.8c, the first phase of cavity dumped operation is similar to that of Q switching in that energy is stored in the inversion. Since both of the resonator mirrors are 100 percent reflective, the amplified light remains trapped within the cavity. As the peak intracavity intensity is reached, the Pockels cell rapidly switches the cavity transmission off again. This ejects the light circulating in the cavity by reflection off the polarizer in a pulse whose duration is equal to two cavity passes. This technique is used to produce pulses of a few nanoseconds duration, since the pulsewidth now depends on the length of the resonator cavity and not on the amount of inversion stored before switching. The peak power output is significantly limited in this method since the light intensity circulating inside the resonator can become large enough to destroy optical components of the laser. Also, the fall time of the Pockels cell voltage should be shorter than the round trip time of the laser. For short resonator cavities, this can be a technically challenging requirement.

An architecture very closely related to the cavity dumped oscillator is the regenerative amplifier. The principal difference is that the laser oscillation in a regenerative amplifier does not build up from spontaneous emission but is instead initiated by a signal externally injected into the resonator as the Pockels cell is switched to transmission. This injected signal is then trapped in the cavity and amplified until it has reached maximum intensity, at which point it is ejected (dumped) from the cavity. The primary practical challenge to this approach is the degree of synchronism required between the arrival of the injected pulse and the opening of the Pockels cell in the regenerative amplifier. It clearly has to be better than a cavity round trip time. The special light-pulse-activated semiconductor switches developed for this purpose represent an elegant solution at the laboratory level.

Mode Locking. By inserting an acoustooptic mode locker into the cavity it is possible to produce very short pulses in an oscillator (see Chap. 9). With a mode locker in the cavity and the transverse modes suitably constrained, the oscillator can be operated quasi-cw to produce a steady stream of short pulses. It is also possible to add a Pockels cell and a polarizer and produce an output which has the pulse envelope of a Q-switched pulse, but which is composed of a picket fence–like train of individual short pulses from the mode-locking process. External to the cavity one can then pick out a single one of those pulses by placing an additional Pockels cell between two polarizers and applying a short high-voltage pulse at just the instant when the desired pulse is at the Pockels cell. Such an arrangement is commonly called a "single-pulse switchout." This is a standard way to produce individual pulses with durations from 100 ps to 1 ns and energies on the order of 100 μJ, which are sometimes injected into a regenerative amplifier for further amplification. Figure 5.9 shows such a Q-switched mode-locked pulse train and a switched-out individual pulse.

Several other methods of mode locking were previously discussed in the context of saturable absorbers and synchronous pumping.

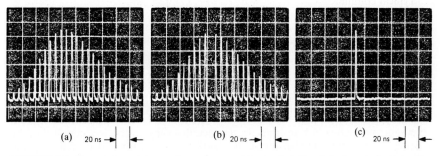

(a) 20 ns →| |←   (b) 20 ns →| |←   (c) 20 ns →| |←

FIGURE 5.9 From left to right the figure shows a (a) complete Q-switched mode-locked pulse train, (b) the same pulse train with the central pulse removed by a Pockels cell switch-out system, and (c) the single picosecond pulse which was switched out from the train. *(From Ref. 9, courtesy of IEEE.)*

Operation on Secondary Transitions. As will be discussed in the section on different host materials, a single lasing ion generally offers more than one lasing transition. Staying with the example of Nd^{3+}, it is possible to lase many of the individual Stark transitions from the $^4F_{3/2}$ metastable level to the $^4I_{9/2}$, $^4I_{11/2}$, and $^4I_{13/2}$ manifolds. This gives rise to numerous transitions grouped around 0.9, 1.05, and 1.3 μm. If we lase into the ground-state manifold, we encounter a three- rather than four-level system. In fact, the transition between a three-level and a four-level system is a gradual one, and depends on temperature, since the degree of "three-levelness" is determined by the thermal population in the lower laser level. Other ions, such as Pr^{3+}, offer an almost bewildering variety of transitions in the same crystal. The Pr^{3+} system exhibits individual laser transitions ranging from 0.45 to 3.6 μm. The efficiency and the ease of obtaining threshold for these different transitions in practice depends not only on the cross section of the individual transition but also on how readily the upper level can be pumped, on the doping, and on its absorption cross section. Moreover, the laser architecture must be constructed to select lasing on one transition and not another. That means that all unwanted transitions have to be successfully suppressed by inserting a wavelength-sensitive element into the resonator cavity. Generally, the most important issues confronting the lasing of two transitions independently are their respective gain in the pumped medium and how far they are separated in wavelength. To appreciate the difficulty of effective gain suppression, consider that, at the least, the competing transition must be kept at threshold in the same oscillator at which the weaker transition is supposed to lase at a certain desired output. Characterizing the weak transition with a desired small signal gain coefficient g_w, the competing strong transition with g_s, and the ratio of cross sections as σ_s/σ_w, the threshold condition (net gain = 1 after one round trip through the resonator) at the strong transition wavelength is given by

$$(g_w l)\frac{\sigma_s}{\sigma_w} = -\ln(T_s) - \ln(\sqrt{R_s}) \tag{5.2}$$

where T_s is the single-pass transmission and R_s is the reflectivity at the strong transition wavelength. Here it is assumed that the same upper laser level is involved in both transitions. From (5.2) one gets

$$T_s = \frac{1}{\sqrt{R_s}} \exp\left[-g_w l \frac{\sigma_s}{\sigma_w}\right] \tag{5.3}$$

Assume one desires to pump the weak transition with a small signal gain coefficient g_w (or store $g_w l$ saturation parameters across the aperture). The mirror reflectivity has some optimized value at the weak transition wavelength and is measured to have a residual reflectivity R_s at the strong transition wavelength. Then $1 - T_s$ is the least single-pass cavity loss one must provide to keep the strong transition at threshold.

As an example, consider a cross-section ratio of $\sigma_s/\sigma_w = 4$, $g_w l = 1$, and $R_s = 0.2$. This results in a single-pass loss requirement of at least 96 percent to suppress to competing high-gain transition. In practice this suppression is accomplished by placing wavelength-selective elements inside the resonator and/or tailoring the reflectivity of the outcoupler to be extremely low at the undesired wavelength. In some anisotropic crystals, transitions can be quite easily selected by inserting a polarizer in the resonator since different wavelength transitions emit in different directions of polarization, as is the case for the 1.053- and 1.047-μm transitions of

Nd:YLF. Since the number of saturation parameters stored at the strong transition wavelength is obviously given by

$$g_s l = g_w l \frac{\sigma_s}{\sigma_w} \tag{5.4}$$

it is clear that the maximum gain which is achievable on the weak transition for a single-gain element will always be determined by parasitics and ASE limitations characterizing the competing strong transition. Hence the power from weaker transitions of the same active ion is weaker not only because the transition itself has a smaller cross section but also because of the measures required to suppress the stronger competing transitions. The same considerations hold true for lasers with broadband transitions which are operated well off the gain peak, such as a Ti:sapphire laser operating at wavelengths larger than 0.9 μm.

5.2.4 Oscillator Scaling Rules of Thumb

Instead of attempting a terse and necessarily very incomplete overview of the theory for different modes of operation, we shall qualitatively describe how the energy flows through a laser from the pump (flashlamp, diodes, or other source) to the active medium and into the output beam, and in the process develop some practically useful rules of thumb. We begin with the more involved case of a pulsed laser operating in storage mode.

Since the transport of pump light to the active medium can be an extremely complicated problem to model, let us begin at the point where part of the pump light absorbed by the active medium decays into heat (more about that later), while the other part accumulates as inversion. The amount of inversion which will have been stored after the end of the pump pulse depends on the ratio of pump pulse duration to fluorescent lifetime. While the pump produces inversion, fluorescence decay depletes it. Hence only for pulses short compared with the fluorescent lifetime will the active medium truly integrate the inversion as accumulated energy delivered by the pumping power. For longer pulse duration, the "fluorescent loss" experienced during pumping causes a significant reduction in the accumulated inversion. The fraction q remaining in the upper state for a square pump pulse of duration T_p and an emission lifetime of time τ_{em} is given by

$$q = \frac{1 - \exp(-T_p/\tau_{em})}{T_p/\tau_{em}} \tag{5.5}$$

For $T_p = \tau_{em}$ the fraction of inversion remaining in the upper laser level is 63 percent. It is clear that longer pump times, although they lead to slightly higher gains, will lessen the efficiency of the laser because much of the pumping energy goes into fluorescence decay rather than stored inversion. Remember, also, that despite the fluorescent loss, it is not possible to pump beyond a parasitic lasing threshold, which then becomes the limiting factor. As the pumping time goes toward infinity, the inversion attained is that for cw pumping, which is a steady-state equilibrium between the pumping rate and the rate of fluorescence decay.

Having followed the buildup of inversion, imagine now that the pumped active medium is located outside of a laser resonator. To find the level of inversion we send a signal through it and measure its amplification factor G after a single pass, taking care that the probe signal is small enough to leave the inversion essentially unchanged. Next, consider the passive losses the light being amplified will experience as it circulates once inside the oscillator cavity. The optical elements, including the lasing material itself, will at least have some scattering losses, in addition to possible losses due to residual surface reflections. Note that reflectivities of resonator mirrors are *not* included in this consideration. The total single-pass transmission of all these lossy elements will be defined as T. Using these two quantities G and T, we then define one of the key parameters in laser design, the gain to loss ratio Γ where

$$\Gamma = -\frac{\ln G}{\ln T} \tag{5.6}$$

Clearly, for fixed cavity losses, additional pumping or a longer active element will increase Γ. The equations describing the extraction efficiency as a function of gain and loss are different, depending on the mode of operation (cw, Q-switched, regenerative amplifier). However, if one plots the behavior of a given extraction efficiency vs. Γ (Fig. 5.10), the generic behavior is described by a nearly linear rise from $\Gamma = 1$ to $\Gamma \approx 5$, a gradual decrease in slope from $\Gamma \approx 5$ to 10, and for higher values of Γ, an asymptotically gradual approach toward unity. For Γ between 5 and 10, one finds typical extraction efficiencies of approximately 50 percent. Figure 5.10 also shows that, for a given gain to loss ratio Γ, optimized cw operation gives the lowest, and Q switching gives the highest, extraction efficiency. Then again, a cw laser often has fewer components in the resonator, and as a result, a higher gain to loss ratio is easier to achieve. Various cases for cavity dumping produce intermediate results, although the ideal cavity dumped resonator approaches the extraction efficiency of Q-switched operation. The practical implication is that for gain to loss ratios near or below $\Gamma \approx 8$ the design should strive to increase Γ before addressing other problems. The laser will be inefficient, no matter what the output reflectivity, and moreover the output becomes sensitive to variations in the input power. Conversely, once a gain to loss ratio of about 20 has been reached, increasing it means a marginal payback in laser performance at best and efficiency will only

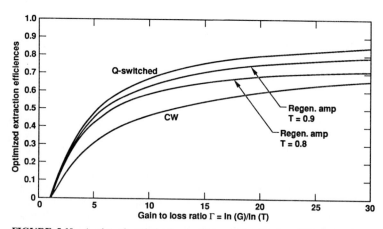

FIGURE 5.10 A plot of optimized extraction efficiencies for different modes of operation (cw, Q-switched, and cavity dumped). For low gain to loss ratios even an optimized extraction efficiency is low; past gain to loss ratios of 15 or 20, the extraction efficiency rises only slowly, regardless of the mode of operation.

be reduced marginally if the gain to loss ratio is dropped to, say, 17. Another important rule of thumb, mentioned earlier, is that an optimized Q-switched laser will always operate about a factor of three above threshold.

For storage-mode lasers, the parasitic oscillations discussed above will eventually limit the achievable gain. It is not possible to give a generically valid rule regarding the gain at which parasitics become important, since their onset can depend very sensitively on the details of the design. For single-pass values of ln G larger than 3, however, it is always valid to question the explicit mechanism by which a design will suppress parasitics. Note that for a typical single-pass resonator transmission of $T = 0.9$, a ln G value of 3 implies a gain to loss ratio of around 30. The three rules discussed in this section provide a simple test any Q-switched laser design should pass, no matter how sophisticated its layout or its intended application may otherwise be.

In cw oscillators, in contrast to storage-mode lasers (Q-switched, regenerative amplifiers, and cavity dumped systems), the gain in the active medium, when placed between two aligned resonator mirrors, never exceeds the gain at threshold. Once the pumping is strong enough for lasing to start, extra pump power is converted only to increased light output, not to more gain in the active medium. For this reason cw lasers do not have a problem with parasitics, since the only gain which ever materializes in a cw laser oscillator is the relatively low threshold gain.

The optimum cw laser extraction efficiency Φ_{ext} which is possible for given values of gain and loss is again determined by the gain to loss ratio as it was defined by Eq. (5.6) and given by

$$\Phi_{ext} = (1 - 1/\sqrt{\Gamma})^2 \tag{5.7}$$

and occurs at an output coupler reflectivity:

$$R_{opt} = T^{2(\sqrt{\Gamma}-1)} \tag{5.8}$$

Given the condition for the gain G at threshold:

$$\ln G_{th} = -1/2 \cdot \ln R_{opt} - \ln T \tag{5.9}$$

it is then easy to show that an optimized cw laser always operates a factor $\sqrt{\Gamma}$ above threshold, regardless of its intracavity loss per pass T. (The equivalent treatment for optimized Q-switched lasers shows that they operate a factor ln Γ above threshold. See above for the discussion of Q switching.)

The existence of an optimum follows from considering the case of strong and weak feedback of the photon flux inside the cavity. If the mirror reflectivity is too high, the intracavity flux is predominantly dissipated by intracavity losses rather than used as output. If the reflectivity is too low there is not enough feedback to generate optimum levels of intracavity flux for inversion extraction. There are some special cases, such as intracavity frequency conversion and mode-locking applications, where it can be advantageous to explicitly increase the intracavity power by increasing the mirror reflectivities. Aside from those, however, typical optimum reflectivities for well-designed Q-switched lasers are around or below 50 percent; those for cw lasers are around or below 85 percent. Diode pumped cw minilasers generally achieve large gain to loss ratios within the active volume and therefore can have significantly lower optimized reflectivities.

Having discussed the extraction of the inversion, let us now turn to the second key parameter, which is a parameter characterizing the laser material and is somewhat different if the system is run in a storage mode or cw. For pulsed lasers, it is the extraction saturation fluence E_{sat} which is given by

$$E_{sat} = \frac{h\nu}{\sigma_{st}} \quad J/cm^2 \tag{5.10}$$

For cw lasers, it is the saturation intensity I_{sat} given by

$$I_{sat} = \frac{E_{sat}}{\tau_{em}} \quad W/cm^2 \tag{5.11}$$

where h is Planck's constant, ν is the photon frequency, σ_{st} is the stimulated emission cross section, and τ_{em} is the emission lifetime.

The saturation fluence defined in (5.1) (but now containing the stimulated-emission cross section σ_{st}) and the gain to loss ratio defined in (5.6) are related by

$$E_{sto} = \ln G \cdot E_{sat} \tag{5.12}$$

where E_{sto} are the joules per unit area stored across the aperture of the active medium through which lasing occurs. With the area of the lasing aperture and the small signal gain known, the total joules stored in the rod can now be calculated.

Conversely, for a desired amount of stored energy, the aperture of the laser can be determined (once we include the fill factor to be discussed shortly). For cw lasers the equivalent statements are made by replacing E_{sat} with I_{sat} and the word "energy" with "power." It helps to understand that it is *not* the gain coefficient or the energy density or the total energy stored that determines the extraction efficiency or the amplification behavior. It is the number of saturation fluences stored at a given gain to loss ratio.

Typical values for E_{sat} and I_{sat} are 0.7 J/cm² and 2.5 kW/cm² for Nd^{3+}:YAG and 5 J/cm² and 15 kW/cm² for Nd^{3+}:glass, respectively. Let us apply some of these numbers to typical lasers. Consider a nominal 100-μm-diameter Nd:glass fiber. Assume we operate it at an output of one I_{sat} at 50 percent extraction efficiency. This corresponds to an $\ln G = 2$ inside the cavity [see Eq. (5.12)] (The intracavity losses, of course, have to be low enough that the resulting gain to loss ratio Γ supports the postulated extraction efficiency of 50 percent.) The product of the fiber area and I_{sat} gives an output of 7.8×10^{-5} cm² $\times$ 15 kW/cm² = 1.2 W. In contrast, the lower I_{sat} of Nd:YAG permits the use of 3-mm-diameter rod, and under the same conditions, we obtain about 150 W output. Similarly, if we want an amplifier design at the 10-J level, YAG will not be able to store this energy without falling victim to parasitics. Glass, however, is a suitable material. The purpose of the saturation parameter, then, is to enable a selection of laser materials and apertures which are commensurate with the desired output characteristics of the laser. Additional material parameters come into play once average power issues are considered. Nevertheless, before the laser can operate properly at a high repetition rate, it must work in a single-shot mode. The proper material selection is as crucial to successful laser design as the proper pumping and extraction conditions.

An important issue which contributes to the overall observed extraction of energy and power is the fill factor, or the fractional utilization of the lasing aperture. This subject is intimately connected to the subject of resonator eigenmodes, which are properly described in the context of diffraction theory.[2] Unfortunately, diffraction theory does not lend itself well to a back-of-the-envelope description. The basic mechanisms involved in forming resonator eigenmodes are these: the slightly focusing resonator mirrors attempt to confine the light rays bouncing back and forth between them, whereas diffraction tries to spread them out. If one imposes the condition on the circulating light wave that its amplitude and phase repeat themselves after one round trip, one derives an eigenvalue problem for the electrical field whose solutions, in rectangular coordinates, are called "Hermite-Gaussian polynomials." In suitably operated resonators the transverse intensity distribution of pure Hermite-Gaussian polynomials are visible. For a typical Q-switched laser, however, the time-integrated output from the lasing aperture is a random, dynamically changing, superposition of these eigenmodes. Provided the pumping is reasonably uniform, the output then appears more or less uniform across the aperture as well, but with random, high spatial frequencies superimposed on it.

To remain within the framework of establishing simple rules of thumb, a critical parameter for estimating the fill factor is ω_0, the $1/e^2$ width of the intensity distribution of the lowest-order mode, which has the shape of a Gaussian bell curve. This parameter depends on the spacing and the curvature of the resonator mirrors, and is, for mirrors of equal curvature, given by

$$\omega_0{}^2 = \frac{\lambda}{\pi[L(2R - L)]^{1/2}} \tag{5.13}$$

where λ is the wavelength and R is the radius of curvature of the mirrors separated by length L. As the fourth-root dependence of ω_0 on the resonator characteristics suggests, almost all other curvatures and spacings (barring extremes) will have much the same value of ω_0 of a few hundred micrometers for visible wavelengths.

The following considerations are related to the fill factor. The highest-order Hermite-Gaussian mode which can fill the rod aperture is the one where the rod radius r allows for decay of the radial intensity profile which is of order ω_0. Since the remaining fraction of extracted lasing aperture Φ_{fill} goes roughly as

$$\Phi_{fill} = \frac{(r - \omega_0)^2}{r^2} \tag{5.14}$$

it is obvious that for a typical 1/4-in laser rod ($r = 3200\ \mu$m) and an ω_0 of 300 μm, the area fill factor Φ_{fill} is of order 80 percent. This is a lower limit, since there might well be other effects like irises or edge bevels which constitute additional aperture limitations. But as (5.14) suggests, particularly in smaller rods, the effect is quite significant. Note that the 80 percent mode fill factor, together with the 60 percent inversion that remains after the fluorescent loss is accounted for, form roughly a factor of two reduction in laser output.

In this section we have treated only stable resonators with uniform reflectivity across the aperture. There are other options, of which the most interesting is probably the super-Gaussian reflectivity unstable resonator.[10] It is an elegant way to provide excellent output intensity profiles in high-gain systems.

5.2.5 Beam Quality[2]

The key characteristics of the output which often determine the actual usefulness of the entire laser system, other than the energy, efficiency, wavelength, and temporal format, is how well the output beam can be propagated and focused. For a uniform active medium, it is the resonator which gives the laser beam its propagation properties, or beam quality. It is properly described by the phase and intensity distribution across the aperture at, say, the output mirror of the laser. A commonly used notion to describe imperfect beam quality is "n times diffraction limited." Its basic meaning is how many times bigger the focal spot of an imperfect beam would be when compared with a spot formed by a beam with a perfectly uniform phase and intensity distribution across an aperture D. One arrives at a similar number when comparing the full-beam divergence angle θ of an imperfect beam with that of a "perfect" beam, given by

$$\theta = \frac{2.4 \lambda}{D} \tag{5.15}$$

if it emanates from a uniformly filled, circular aperture. This jargon, however, has little utility if the beam is very nonuniform and broken up into different hot spots.

The closest one can approach a "diffraction-limited" beam is to generate the lowest-order Gaussian mode (designated TEM_{00}) in a stable resonator. As the aperture of the beam increases, higher-order modes will begin to participate and the beam divergence increases roughly in proportion to the aperture diameter. In an oscillator, it is not straightforward to maintain a smooth, uniphase intensity distribution over an aperture much larger than the area occupied by the lowest-order Gaussian mode. The use of the above-mentioned super-Gaussian unstable resonator is one possible way to achieve this condition. A more conservative approach is to start with a small-diameter Gaussian beam in an oscillator and then expand the beam through amplifiers of increasing apertures. In practice, the *efficient* creation of large-aperture, high-power beams with uniform phase and intensity distribution across this aperture is one of the most difficult tasks in laser design.

5.2.6 Average Power Issues[2,11]

Heat removal from the active medium is a key design issue in optimizing cw lasers and pulsed lasers operating at a significant repetition rate. This heat has its ultimate origin in the properties of the laser ion. Consider, for example, the Nd^{3+} ion whose spectroscopy and energy levels are discussed in Sec. 5.3. The lasing is induced from the $^4F_{3/2}$ to the $^4I_{11/2}$ manifolds. If the laser under consideration is pumped with flashlamps, most of the higher-lying levels are excited by flashlamp radiation, and all of these higher-lying states decay toward the $^4F_{3/2}$ manifold by phonon emission, which manifests itself as an increased temperature of the host lattice. Similarly, after lasing, the lower level again decays by phonon emission into the ground state, generating more waste heat. Although the near-blackbody spectrum used in flashlamp pumping will generate far more heat per inverted ion than pumping a single absorption feature with a diode laser, this simply translates heat-removal issues to regimes of higher output power.

There are several effects through which heat flow in an active medium manifests itself. Since heat is deposited in the volume but removed from the surface, thermal gradients invariably occur. Moreover, these thermal gradients lead to local stresses. Of particular concern to average power solid-state operation are tensile stresses at the cooled surfaces, since it is these stresses which ultimately cause catastrophic failure of the material. However, even before actual breakage, gradients in the refractive index across the beam profile greatly influence laser design considerations.

If the active medium has the shape of a rod, the refractive index will have a roughly parabolic distribution vs. radius. The rod then acts like a thick lens which focuses the beam passing through it, with a focal length f_{rod} given by

$$f_{rod} = [n_0 \sqrt{U} \cdot \tan(L_{rod} \sqrt{U})]^{-1} \tag{5.16}$$

There L_{rod} is the rod length and f_{rod} is measured from the rod end to the focal spot. The refractive index profile in the rod is, like the temperature profile, parabolic in shape and given by

$$n(r) = n_0\left(1 - \frac{U}{2}r^2\right) \qquad (5.17)$$

Hence the parameter U describes the refractive index curvature, and n_0 the refractive index at rod center. Solving 5.17 for U and considering a 3-mm-diameter YAG rod with a center to edge temperature difference of 10 K, one obtains $U \sim 1.8 \times 10^{-3}/cm^2$. To some degree thermal focusing can be accommodated by the design of the resonator cavity, but as the pump power or the rod length are increased, the focus may become located inside the rod itself. Focusing the radiation inside the rod will cause beam distortion and possibly damage, and obviously cannot be corrected for with additional optical elements in the cavity. Aside from the thermal focusing, the radially oriented stresses in the rod depolarize linearly polarized light to such a degree that techniques like Q switching or cavity dumping with Pockels cells become unacceptably inefficient. One of the major limitations to increasing the average power of a rod laser by increasing rod diameter is that an increase in lasing aperture is necessarily accompanied by an increase in the distance that heat must travel in order to be removed by the coolant. Nevertheless, multi-kW welding lasers have been built by spacing the focusing rods to create a waveguide effect which thereby confines the light between the resonator mirrors.

If the active medium has the shape of a slab (Fig. 5.7), these refractive index gradients are oriented such that their net effect on the beam passing through the active medium is reduced to acceptable levels. Laser slabs which utilize such gradient averaging beam paths are generally known as zigzag lasers. In this geometry, the rodlike distortions are now confined to the top and bottom end of the slab, with a central section of the rectangular aperture being relatively free of distortions other than gradients through the slab thickness only. These, in turn, are sampled by the zigzagging path in such a way that beam distortions are, to first order, canceled. This scheme indeed works in the midsection of the slab. The beam, however, also has to enter and leave the slab, and a careful analysis reveals that it is the residual beam distortions picked up at these ends which are so difficult to minimize and which prevent the zigzag laser from emitting a completely distortion-free beam. Nevertheless, zigzag lasers do have their niche, since rods provide no gradient averaging beam propagation at all and scale toward larger average powers only with increasingly cumbersome architectures. The drawback of zigzag lasers is increased system complexity and, for smaller systems, reduced system efficiency.

In analogy to the treatment of laser performance given above, it is also possible to formulate rules of thumb for the heat flow issues as they pertain to average power operation. To retain the simplest possible picture we will treat a slab only, although the essential parameter scaling for rods is obviously the same, apart from different geometrical form factors. The first quantity of importance is the specific heat parameter χ. It is defined as the ratio of waste heat generated per unit of stored inversion. Hence, if the laser is pumped by a short flashlamp pulse so that $0.1\ J/cm^3$ is stored in the inversion just before extraction, and if this pumping deposited $0.2\ J/cm^3$ of heat, χ is equal to 2. Considering the Nd ion as an example, χ in pulsed flashlamp pumped lasers is between 2 and 3.5, while in cw arc lamp pumped systems χ is around 1. For diode pumped systems it is found to be about 0.4.

The next important parameter is the thermal shock R_T, defined as

$$R_T = \frac{(1 - v)\,\kappa\,S_T}{\alpha E} \tag{5.18}$$

where v is Poisson's ratio, κ is the thermal conductivity, α is the coefficient of thermal expansion, E is Young's modulus of elasticity, and S_T is the thermal tensile stress at the slab surface. R_T is a useful figure of merit for thermomechanical stresses. The critical thermal shock R_{Tc} is reached when the tensile stress at the active medium surface reaches the fracture limit. Typical values of R_{Tc} are of order 5 W/cm for crystals and around 1 W/cm for glasses. It is important to realize that R_{Tc} is a statistical quantity. The exact value of R_{Tc} at which a given sample of material actually fractures is given by the intrinsic resistance of the material to resist crack propagation, the size of a flaw (microcrack) which is present somewhere on the material surface owing to polishing or handling, and the likelihood of having a given size flaw on the sample. Since the flaw size is most likely not a known quantity for a given laser slab, it is not surprising that some slabs withstand high-power loading, whereas other seemingly identical slabs break at low-power loading. The likelihood of failure is generally found to be acceptably low if one operates at a thermal shock R_T at or less than 20 percent of the R_{Tc} values measured from material samples.

Since rod barrel surfaces can be etched to remove microcracks (rod surfaces do not sample the extraction beam as slabs do) and also have a comparatively smaller surface area, rods can exhibit a higher value of R_{Tc} and therefore break less often as a result of average power operation.

The connection between the energy stored in the laser as inversion E_{sto} (J), and the surface heat flux Q (W/cm²) removed from it is given by

$$E_{sto} \times f = Q \cdot A \tag{5.19}$$

where A is the cooled area and f is the repetition rate. Furthermore, in considering how a surface stress is created by the presence of the surface heat flux Q and how the stress, in turn, is connected to the operating thermal shock R_T, one finds for a slab of thickness t:

$$Q = \frac{6\,R_T}{t} \tag{5.20}$$

By combining (5.19) and (5.20) in a slab geometry, one obtains one of the important scaling laws for average power slab lasers:

$$E_{sto} = 6\,\frac{R_T}{\chi \cdot f} \cdot \frac{A}{t} \tag{5.21}$$

This expression vividly shows how a given stress, resulting from material properties (R_T), spectroscopy (χ), and geometry (A/t), conspires to limit the stored energy for a given repetition rate. The equivalent expression for a rod of diameter d is

$$E_{sto} = 4\sqrt{2}\,\frac{R_T}{\chi \cdot f} \cdot \frac{A}{d} \tag{5.22}$$

Following this derivation, the equivalent expression for a cw laser is easily obtained. Although much more is involved in describing average power laser behavior, Eqs. (5.21) and (5.22) are important because they allow back-of-the-envelope checks of a given design.

5.2.7 Solid-State Laser Amplifiers

Thus far we have primarily described laser oscillators. To achieve higher energies, amplification of the pulse from the oscillator is required.[2] The regenerative amplifier was mentioned in the section on cavity dumping. In comparison to a single-pass amplifier, the regenerative amplifier is a far more complex setup, although it provides, simultaneously, for efficient extraction of energy from the medium with simultaneous high amplification, and is favored for short pulse amplification in the laboratory. One would, of course, like to make the most efficient use of the inversion stored in such an amplifier, while avoiding the potentially destructive consequences of propagating a high peak power pulse through the amplification medium. Consider a pulse making a single pass through an amplifier in which an inversion is stored. The key parameter describing the amplification process is again the saturation fluence (in *emission*) introduced in Eq. (5.10). If the input fluence to the amplifier remains small compared with the saturation fluence of the transition (called the small signal gain regime), the amplification G proceeds in an exponential manner according to

$$G = \exp(N_i \sigma_{st}\, x) \tag{5.23}$$

where N_i is the inversion density, σ_{st} is the stimulated-emission cross section, and x is the path length in the amplifier. In this regime the amplification factor G is high, but the extraction efficiency of the inversion is consequently low, since saturation of the inversion does not occur. The other extreme is that the input fluence is much larger than the saturation fluence. In this case extraction of the inversion will approach 100 percent, and the energy stored in the amplifier is simply added to the energy in the input beam. For most practical applications, the best place to operate is in between the two extremes, where there is a desirable degree of signal amplification at an acceptable extraction efficiency of the energy originally stored in the amplifier. The basic equation which describes this amplification process is the Frantz-Nodvik equation, given for a four-level system and rectangular pulses by

$$\frac{E_{out}}{E_{sat}} = \ln\left\{1 + \left[\exp\left(\frac{E_{in}}{E_{sat}}\right) - 1\right] \cdot \exp\left(\frac{E_{sto}}{E_{sat}}\right)\right\} \tag{5.24}$$

Note that the saturation parameter E_{sat} enters as the normalization constant. Equation (5.24) is the complete description of an amplifier without loss, valid for all ranges of saturation. It is noteworthy that the amplifier length does not enter.

Two limits are easily seen. For $E_{in}/E_{sat} \ll 1$, (5.24) reduces to

$$\frac{E_{out}}{E_{sat}} = \ln\left\{1 + \left[\exp\left(\frac{E_{sto}}{E_{sat}}\right)\right]\right\} \tag{5.25}$$

which, for E_{sto}/E_{sat} small enough can be roughly approximated by

$$\frac{E_{out}}{E_{sat}} = \exp\left(\frac{E_{sto}}{E_{sat}}\right) = G \tag{5.26}$$

where G is, as above, the small signal amplification factor. Note that one recovers simple exponential amplification only if the injected signal as well as the amplification is small, since the signal grows as it travels through the amplifier and may start to saturate toward the end section of the amplifier.

Similarly, if $\exp(E_{in}/E_{sat})$ is $\gg 1$, (5.24) reduces to

$$\frac{E_{out}}{E_{sat}} = \frac{E_{in}}{E_{sat}} + \frac{E_{sto}}{E_{sat}} \tag{5.27}$$

which means that for heavy saturation the entire stored energy adds to the input signal.

A variety of effects limit the energy which may be generated in a large-scale amplifier chain.[12] First, the amplifier medium cannot be made arbitrarily large since it will lose its storage capability to depletion by ASE, even if the parasitics are successfully controlled. Second, as the peak power in the amplified pulse increases its intensity, it becomes large enough that the electric field of the amplified light wave itself locally modifies the refractive index of the medium through which it travels. The material property which describes the susceptibility to this effect is the nonlinear refractive index n_2. The effect is proportional to the square of the local electric field amplitude of the light wave and adds to the regular, linear refractive index which is observed at low intensities:

$$n = n_0 + n_2 \cdot \langle E^2 \rangle \tag{5.28}$$

The extent to which nonlinear index effects become noticeable depends on the magnitude of n_2 (a material parameter), the electric field strength in the laser pulse, and the length over which the interaction occurs (e.g., the number of passes through an amplifier rod, or the length of an optical fiber). The additional optical path length accumulated due to the nonlinear refractive index after traveling through a medium of length L with a pulse of electric field strength E is given by

$$\delta z = n_2 \langle E^2 \rangle L \tag{5.29}$$

Hence, when the intensity of the pulse is large enough to effect a retardation δz equal to about one wavelength, i.e.,

$$\langle E^2 \rangle = \frac{\lambda}{n_2 * L} \tag{5.30}$$

the nonlinear index effects will noticeably affect beam propagation. Equation (5.30) may be viewed as another simple but useful rule of thumb.

As the phase retardation grows, the first nonlinear index effect to be encountered is known as "self-phase modulation." The phase Φ of a sinusoidally varying electromagnetic wave in a medium with refractive index n is described by

$$2\pi \frac{vt - x}{n\lambda} = \Phi \tag{5.31}$$

The equation shows that a change in the refractive index n (like the addition of a self-induced nonlinear change) alters the phase of a wave, and therefore also the phase relationship between the waves which form a short, phase-coherent pulse. At first this effect contributes to a nearly linear chirp along the pulse, meaning that the beginning of the pulse has, say, a shorter wavelength and the tail a longer wavelength than the peak of the pulse. If such a chirped pulse is refracted off a grating whose dimension is larger than the pulse duration times the speed of light, the longer-wavelength pulse tail can be made to catch up with the shorter-wavelength beginning of the pulse. These effects are at the root of the pulse-shortening techniques so widely used in ultrashort pulse generation, much of which was originally developed using dye lasers.

If the effect of the nonlinear index is allowed to proceed, the tail of the pulse, "seeing" a lower intensity than the peak of the pulse, will eventually catch up with the peak, the pulse will distort and shorten, and the local peak intensity will increase further until the peak power will eventually exceed the bulk damage threshold of the material. Since there is inevitably a transverse intensity modulation across the lasing aperture, one observes, as the consequence of an unchecked nonlinear index effect, a breakup of the beam into little filaments which leave bubblelike tracks throughout the amplifier. This effect has been named "small-scale self-focusing" and has to be judiciously avoided in the design of high peak power amplifiers. Note that one need not construct a fusion-class laser to encounter this effect since it is the watts per cm^2 and the length of medium traversed, not the absolute watts, which causes the effect. Amplification of short mode-locked pulses in the laboratory, particularly in short pulse regenerative amplifiers, is often limited by nonlinear refractive index effects.

One recently developed method circumvents nonlinear refractive index effects in an amplifier by first stretching a short pulse for the purpose of amplification and then recompressing it afterward. This so-called chirped pulse amplification technique[13] has generated great interest as a method to achieve Terawatts of peak power in smaller laboratory systems[14] and has opened up an entirely new regime of laser plasma interaction phenomena.

Amplifier surfaces and other optics are coated with dielectric films to increase or decrease their reflectivities. These coatings can be destroyed by a light pulse which has too high a peak power, often at power levels below that at which self-focusing sets in. The resistance of such coatings to this effect is described by a "damage threshold" which characterizes the fluence the coating can withstand at a given pulse duration. The architecture of an amplifier chain is determined by the

FIGURE 5.11 The Nova laser at Lawrence Livermore National Laboratory is the largest laser currently in existence. Note the beautiful young woman for scale.

requirement to obtain at each stage a sufficient degree of amplification at an acceptable extraction, while avoiding damage to coatings and the active medium itself. As the ultimate example of a laser which can only be built if all of these types of processes are well understood, Fig. 5.11 shows the Nova laser at the Lawrence Livermore National Laboratory. It has delivered pulses with 120,000 J of energy in durations of several nanoseconds.

5.3 SOLID-STATE LASER MATERIALS

5.3.1 Laser-Active Centers

Solid-state lasers are based on a wide variety of insulating materials. All of these materials are conceptually similar, however, in that a laser-active center is incorporated into the solid referred to as the host. The host is an ionic solid (e.g., MgO) and the extrinsic laser center usually carries a positive charge (e.g., Ni^{2+}). As a simple illustration, a two-dimensional view of the $MgO:Ni^{2+}$ system is pictured in Fig. 5.12. The small fraction of Ni^{2+} ions present at the Mg^{2+} sites gives the MgO crystal a green hue. A crystal also may be colored by intrinsic defects, as illustrated on the right-hand side of Fig. 5.12 for NaF. We can imagine here that a fluorine ion is first removed from the perfect NaF lattice, and then the electron is returned to the crystal. The electron becomes bound at the vacancy, since the rest of the lattice creates an effective positive charge centered at the empty chlorine site. This defect species is known as an F center, in which the "F" stands for Farbe, or color. While the F center itself is not laser-active, numerous related defects composed of F centers do comprise an important class of laser materials (e.g., two nearby vacancies sharing an electron or F_2^+).

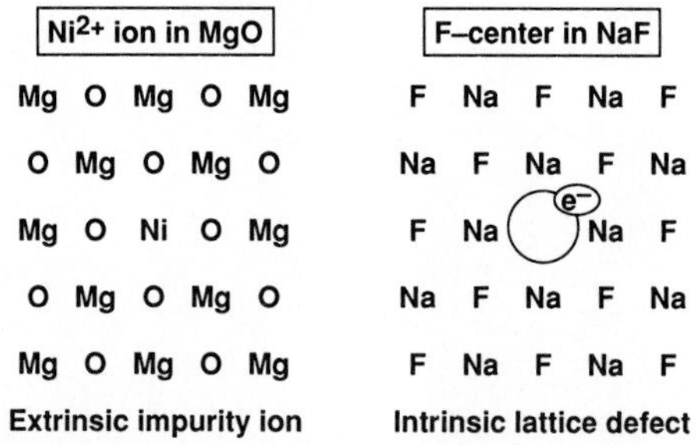

Ni²⁺ ion in MgO					F–center in NaF				
Mg	O	Mg	O	Mg	F	Na	F	Na	F
O	Mg	O	Mg	O	Na	F	Na	F	Na
Mg	O	Ni	O	Mg	F	Na	e⁻	Na	F
O	Mg	O	Mg	O	Na	F	Na	F	Na
Mg	O	Mg	O	Mg	F	Na	F	Na	F
Extrinsic impurity ion					**Intrinsic lattice defect**				

FIGURE 5.12 Example of an impurity-doped host $MgO:Ni^{2+}$, and the F center in NaF.

On the basis of this introduction, and a cursory glance at the periodic table of the elements, it may seem that the number of potential laser media is virtually limitless. The known laser centers are indicated in Table 5.1, along with a very abbreviated list of host materials.[15,16] Among the laser centers are 6 transition metals, 13 trivalent and 3 divalent rare earths, the actinide U^{3+}, and 7 types of intrinsic-defect species. While 30 distinct centers are listed in Table 5.1, it should

TABLE 5.1 Laser Ions and Abbreviated Listing of Host Materials

Laser ions and defects

Transition metal ions: Ti^{3+}, V^{2+}, Cr^{3+}, Cr^{4+}, Co^{2+}, Ni^{2+}

Trivalent rare earth ions: Ce^{3+}, Pr^{3+}, Nd^{3+}, Pm^{3+}, Sm^{3+}, Eu^{3+}, Gd^{3+}, Tb^{3+}, Dy^{3+}, Ho^{3+}, Er^{3+}, Tm^{3+}, Yb^{3+}

Divalent rare earth ions: Sm^{2+}, Dy^{2+}, Tm^{2+}

Actinide ion: U^{3+}

Color centers: F^+, $F_A(II)$, $F_B(II)$, F_2^+, $(F_2^+)_A$, $(F_2^+)^*$, $Tl°(1)$

Examples of laser hosts

Oxide crystals: MgO, Al_2O_3, $BeAl_2O_3$, $YAlO_3$, $CaWO_4$, YVO_4, $Y_3Al_5O_{12}$, $Gd_3Sc_2Ga_3O_{12}$, $LiNdP_4O_{12}$

Fluoride crystals: NaF, MgF_2, CaF_2, LaF_3, $LiYF_4$, $LiCaAlF_6$, $LiSrAlF_6$

Other halide crystals: KCl, KI

Glasses: ZrF_4-BaF_2-LaF_3-AlF_3, SiO_2-Li_2O-CaO-Al_2O_3, P_2O_5-Al_2O_3-K_2O-BaO, fused SiO_2

be emphasized that they lase with varying degrees of proficiency, and each possesses both advantages and disadvantages. For example, Ti^{3+} has a wide tuning range but lacks the ability to effectively store energy; Pm^{3+} is predicted to lase quite efficiently but is radioactive. Color center lasers provide tunable radiation in the infrared but usually require cryogenic cooling.

Each of the laser host materials also possesses both good and bad characteristics. Only certain impurity ions are compatible with a particular host, since the size and charge of the substitutional host metal ion must be similar to that of the impurity. For example, Ni^{2+} can be incorporated into the Mg^{2+} sites of MgF_2 and MgO, and Nd^{3+} into the Y^{3+} sites of $LiYF_4$, Y_2SiO_5, $YAlO_3$, and $Y_3Al_5O_{12}$. In addition to the oxide and halide crystals of Table 5.1, several glassy media are listed, including fluoride-, silicate-, and phosphate-based systems; fused silica primarily serves as the matrix for rare-earth doped fiber lasers. The color center lasers are predominantly associated with alkali halide crystals that have been either irradiated or chemically treated to generate the desired defect centers.

The nature of the energy levels and dynamics of the active center determines the character and effectiveness of the laser. A generic representation of the energy levels and energy flow appears in Fig. 5.13. The laser crystal or glass initially absorbs the light from the pump source (step 1). The absorption of the medium is characterized by the absorption coefficient (with units of cm^{-1})

$$\alpha(\lambda) = N_{gnd}\,\sigma_{abs}(\lambda) \qquad (5.32)$$

where the ground-state number density (cm^{-3}) and the absorption cross section (cm^2) are noted. The fraction of light absorbed at λ in path length l is

$$f = 1 - \exp[-\alpha(\lambda) \cdot l] \qquad (5.33)$$

The energy then commonly relaxes to the emitting excited state (step 2). This level is metastable, in that it often possesses an emission lifetime τ_{em} that is long enough

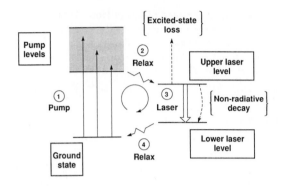

FIGURE 5.13 Generic representation of the energy levels and energy flow (steps 1 to 4) for an idealized laser center. The dashed arrows illustrate two fundamental loss mechanisms.

to store energy [see Eq. (5.5)]. The emission lifetime is, in part, determined by the natural radiative lifetime of the state τ_{rad} but may also be affected by nonradiative processes that lead to a shortening of the excited state lifetime. The competition between the radiative and nonradiative (nr) decay rates can be expressed as

$$\tau_{em}^{-1} = \tau_{rad}^{-1} + k_{nr} \tag{5.34}$$

k_{nr} can, for example, take the form of a thermally mediated process:

$$k_{nr} = A_{nr} \exp\left(-\frac{E_{nr}}{kT}\right) \tag{5.35}$$

where E_{nr} is the activation energy, A_{nr} is a constant, and k is Boltzmann's constant. Nonradiative decay leads to additional heating of the laser medium, as discussed in Sec. 5.2.6.

Gain can potentially be present during step 3, as the center undergoes the transition between the upper and lower laser levels. The stimulated-emission cross section is one of the crucial parameters involved in any laser design since it appears in the expression for the saturation parameters [see σ_{st} in Eq. (5.10)] and may be calculated from the related quantity[17]

$$\sigma_{em} = \frac{\lambda^2 \, g(v)}{8\pi n^2 \, \tau_{rad}} \tag{5.36}$$

where $g(v)$ is the emission lineshape function having units of seconds, and n is the refractive index. One of the subtle aspects of determining the emission cross section involves knowing an accurate value of the radiative lifetime, since nonradiative decay can lead to a shortening of the emission lifetime under some circumstances such that τ_{em} is not equal to τ_{rad}.

Excited state absorption (ESA) is another loss mechanism that reduces the gain of the laser medium. The excited state encounters pump-induced loss due to the excitation to higher-lying states. The effective stimulated-emission cross section is given by the difference of the emission and ESA cross sections:

$$\sigma_{st} = \sigma_{em} - \sigma_{ESA} \tag{5.37}$$

In fact, the reason that many materials luminesce efficiently but do not lase is the result of the presence of pump-induced ESA or loss, rather than gain.

Step 4 of Fig. 5.13 shows that the lower laser level relaxes back to the ground state. The advantage of the four-level scheme is that the lower laser level is initially unoccupied and therefore cannot introduce ground-state absorption loss into the system. A system which does possess a small amount of ground-state absorption loss will have a higher threshold of oscillation. For example, the power threshold of a laser-pumped laser with pump and cavity waists of ω_p and ω_c is given by[17]

$$P_{\text{th}} = \frac{\pi(\omega_p^2 + \omega_c^2)h\upsilon_p(T_{\text{tot}} + 2N_{\text{gnd}}\sigma_{\text{abs}}l)}{4(\sigma_{\text{em}} - \sigma_{\text{ESA}})\tau_{\text{em}}} \tag{5.38}$$

where the ground-state absorption loss $N_{\text{gnd}}\sigma_{\text{abs}}l$ has been added to the sum of the output coupler transmission and passive loss T_{tot} and we take $T_{\text{tot}} \ll 1$. The lower-laser-level absorption can arise from either the equilibrium thermal population or a bottleneck effect, in which the relaxation rate of step 4 is slow compared with the stimulated-emission rate. For the case where the lower laser level is precisely the same as the ground state, the system is three-level in nature and requires considerably more energy for the oscillator to reach threshold and lase, since an inversion is achieved only when at least half of the ground-state population has been pumped to the upper level (for equal degeneracies of the ground and excited states). It is also apparent from Eq. (5.38) that ESA serves to increase the threshold as well.

5.3.2 Host Materials

The host materials that are utilized in laser systems must exhibit adequate optical, mechanical, and thermal properties. In addition, the material must be able to sustain a precise optical polish, be cast or grown within certain economic and time con-straints, and afford the laser centers the spectroscopic properties appropriate for good laser performance. As a result of the numerous and diverse requirements, not many materials turn out to be useful in practical circumstances. The nature of glasses and crystals is discussed below, and the important physical properties are outlined in detail.

Most laser glasses fall into one of several categories: silicates, phosphates, and fluorides. These glass systems may also be mixed, yielding fluorophosphates, sili-cophosphates, etc. In all cases, the glass is considered to consist of two major components: the network former and the modifiers.[18] The network is a covalently bonded three-dimensional system, whereas the modifiers are ionically bonded and tend to disrupt the network structure. The silicate glasses provide a simple de-scription of the interplay between the network and modifiers. First, consider crys-talline quartz, or SiO_2, as illustrated in Fig. 5.14a. Here, every oxygen bridges between two silicons. Fused silica in Fig. 5.14b is similar, although it is glassy, meaning that the highly ordered nature of the system has been eliminated. If modifiers such as Na_2O are added, some of the oxygens become "nonbridging" (Fig. 5.14c). There are several favorable features afforded to the glass by the modifier ions: (1) the melt acquires a much lower viscosity and may be easily poured and cast, and (2) the glass is able to dissolve rare earth ions much more effectively, compared with fused silica. A similar situation exists for other types of glasses, as well. For example, the P_2O_5 in phosphate glasses forms the network, and alkali and alkaline earth oxide compounds are added as modifiers. For the case of fluoride glasses, ZrF_4, ThF_4, or BeF_2 may serve as the network former.

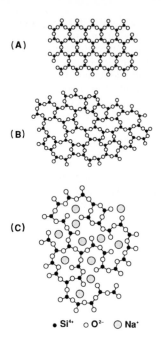

(A)

(B)

(C)

• Si⁴⁺ ○ O²⁻ ○ Na⁺

FIGURE 5.14 Two-dimensional view of the structure of SiO_4 tetrahedron in (a) quartz crystal, (b) fused silica, and (c) sodium silicate glass. (*Adapted from Ref. 18.*)

The growth of most crystals turns out to be considerably more difficult than melting and casting glassy materials. Crystals provide an important advantage, however, since a precisely defined site is available to the laser ion, rather than the broad distribution of sites that characterize a glass. Crystals often have more favorable thermal and mechanical properties as well. For example, the thermal conductivity tends to be much higher, and oxide crystals tend to be very strong mechanically compared with glasses. As a result, it is often worthwhile to generate the crystalline media.

Crystals may be grown many different ways,[19] two examples of which are shown in Fig. 5.15. The Bridgman method typically involves slowly lowering a crucible through a zone in which the temperature abruptly drops from above to below the melting point of the crystal. A seed crystal is sometimes placed at the bottom of the crucible to initiate the growth. Also shown is the Czochralski method, in which a seed is dipped into the melt and then slowly raised as it is rotated. Most crystals are grown at the rate of 0.1 to 10 mm/h. It is important to emphasize that there are many other methods of crystal growth that have not been discussed here (solution and flux growth, flame fusion, gradient freeze, top seeded solution, etc.). All methods are based on the concept of slowly enlarging the seed crystal.

The mechanical, thermal, and optical properties of the hosts dramatically impact the laser design considerations. The numerical values of the important physical quantities are listed in Table 5.2 for representative oxide, fluoride, and chloride crystals, as well as several glasses. The first six columns are related to an assessment of the propensity for the material to break while under thermally induced stress, as described by a quantity related to the thermal shock parameter[20]

$$R'_T = \frac{\kappa\,(1 - v)\,K_{1c}}{\alpha E} \tag{5.39}$$

where κ is the thermal conductivity, v is Poisson's ratio, K_{1c} is the fracture toughness, α is the expansion coefficient, and E is Young's modulus. If R'_T is divided by the square root of half the flaw size that initiates fracture and several design-dependent quantities, the thermal parameter utilized in the previous section is obtained (R_T); see Eq. (5.18). Since the flaw size is highly dependent on the polishing procedure and is generally unknown in any case, R'_T is the most useful embodiment of a figure of merit with which to compare the materials in Table 5.2, whereas R_T is of practical interest for the laser designer.

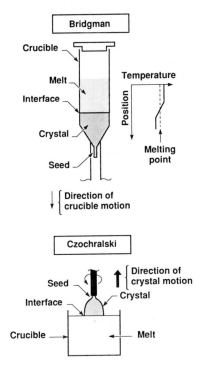

FIGURE 5.15 Schematic drawing of two common crystal growth methods.

The thermophysical properties[20-31] vary substantially among the host media of Table 5.2. E is largest for oxides since their highly ionic nature renders them difficult to deform. The K_{1c} values, which are a measure of the ability of the material to resist crack formation, tend to be highest for oxides, intermediate for the oxide-based glasses, and smallest for fluorides. A similar grouping of values prevails for α, with the notable exception that fused silica (SiO_2) possesses an anomalously low value. The thermal conductivities κ have a large distribution of magnitudes among the crystals, although they tend to be much lower for the glasses. This effect may be understood from the mechanism of thermal conduction, which is inhibited by the enhanced scattering of phonons by the disordered nature of glasses.

The R'_T thermal shock quantities vary by nearly two orders of magnitude in Table 5.2. It should be noted that oxides are better than fluorides and glasses, although perhaps by not as large a margin as many laser designers probably assume to be the case. This is in part true because E appears in the denominator of Eq. (5.39), since the more deformable solids are able to relieve the stresses due to nonuniform pumping and cooling in the laser-active medium. Interestingly, we also see that the KCl crystal fares nearly as well based on the definition of R'_T as do the glassy media.

The change of the refractive index with temperature dn/dT appears in the next column and is seen to exhibit both positive and negative values. The dn/dT parameter occurs in the thermal lens formulation describing a cooled laser rod in a pump chamber. For simplicity, we consider here the case of a block of material of length l that is uniformly heated and unstressed; the optical pathlength change is

$$\Delta p = \left[\alpha(n-1) + \frac{dn}{dT} \right] L \, \Delta T \qquad (5.40)$$

It is apparent that the negative values of dn/dT tend to cancel the positive contribution due to the thermal expansion. (As it turns out, the negative dn/dT also cancels the lensing contribution due to the stress-optic effect present in realistic laser configurations.) From Table 5.2 it is clear that most fluorides and chlorides, and a few glasses, are characterized by negative dn/dT parameters and, as a result, tend to have substantially less thermal lensing distortion than most other media. Actual thermal lensing measurements on flashlamp-pumped laser rods[29,32-34] have confirmed that fluoride laser materials such as $LiYF_4$ and $LiCaAlF_6$ exhibit roughly an order of magnitude less thermal lensing than the typical oxides $Y_3Al_5O_{12}$,

TABLE 5.2 Summary of the Mechanical, Thermal, and Optical Properties of Laser Hosts

Material	E,[a] GPa	v[b]	K_{1c},[c] MPa $\sqrt{m}$	α,[d] 10^{-7}/K	κ,[e] W/(m·K)	R_T',[f] W/$\sqrt{m}$	dn/dT,[g] 10^{-6}/K	n_2,[h] 10^{-13} esu	References
Al_2O_3 (oxides)	405	0.25	2.2	48, 53	28	22	+11.7, +12.8	1.2	20–22
$BeAl_2O_4$	446	0.3	2.6	44, 68, 69	23	14	+9.4, +8.3	1.5	20, 21, 23
$Y_3Al_5O_{12}$	282	0.28	1.4	67	10	4.6	+8.9	2.7	20–22
$Gd_3Sc_2Ga_3O_{12}$	210	0.28	1.2	75	6.0	3.3	+10.1	5.5	20–22
Y_2SiO_5	~110	0.3	0.5	74, 74, 52	4.5	1.9	+9.1, +5.7, +6.7	3.7	24–26
MgF_2 (fluorides)	138	0.27	0.9	131, 88	21	7.6	+0.9, +0.3	0.3	20–22
CaF_2	110	0.3	0.33	260	9.7	0.8	−11.5	0.4	20–22
LiF	91	0.3	0.36	140	11	2.2	−16.7	0.3	20–22
$LiYF_4$	75	0.33	0.27	130, 80	5.8, 7.2	1.1	−2.0, −4.3	0.66	20, 27, 28
$LiCaAlF_6$	96	0.3	0.31	220, 36	4.6, 5.1	0.53	−4.2, −4.6	0.2	29
KCl (chloride)	39	0.3	0.14	400	9.2	0.58	−36.2	2.0	20–22
ED2 (silicate glasses)	92	0.24	0.83	80	1.36	1.2	+2.9	1.41	20, 28, 30
Fused SiO_2	72	0.17	0.75	5.5	1.38	22	+11.8	0.9	20–22
LG-750 (phosphate glasses)	52	0.26	0.45	114	0.62	0.35	−5.1	1.08	20, 30, 31
APG-1	71	0.24	0.62	76	0.83	0.72	+1.2	1.13	31

[a]Young's modulus.
[b]Poisson's ratio.
[c]Fracture toughness.
[d]Expansion coefficient.
[e]Thermal conductivity.
[f]Intrinsic thermal stress resistance.
[g]Refractive index change with temperature.
[h]Nonlinear refractive index.

$BeAl_2O_4$, and $Gd_3Sc_2Ga_3O_{12}$. See Eqs. (5.16) and (5.17) for further discussion of thermal lensing.

The last quantity listed in Table 5.2 is the nonlinear refractive index n_2, which is responsible for modifying the spectral and spatial character of high-intensity pulses. The important observation that fluorides and light cations give the lowest n_2 values is readily apparent. Oxides containing heavy cations, on the other hand, give the largest n_2 values. Since n_2 is traditionally listed in esu, yet the beam irradiance is given in W/cm^2, it is useful to establish the equivalent of the nonlinear index with respect to irradiance in MKSA units. From the conversion rules from esu to MKSA units, we find

$$n_I \text{ (MKSA)} = \frac{4\pi\varepsilon_0}{\varepsilon_0 c} \cdot n_2(\text{esu}) \tag{5.41}$$

n_I (MKSA) is a more intuitive quantity than n_2 (esu).

5.3.3 Rare Earth and Actinide Lasers

As noted in Table 5.1, all thirteen of the trivalent rare earth (RE) ions have been demonstrated to lase. In passing from Ce^{3+} to Yb^{3+} the 4f shell becomes filled with electrons, from $4f^1$ to $4f^{13}$. It is the electronic states that arise from the $4f^n$ shell that are responsible for nearly all of the RE laser transitions. In this section, several examples of laser ions will be discussed.

The energy levels of Pr^{3+} are qualitatively sketched in Fig. 5.16. This discussion of the Pr^{3+} ion is intended to give a brief sketch of how the numerous transitions come about from the $4f^n$ core of RE ions.[36] The $4f^2$ orbital occupancy gives rise to seven electronic states, as a result of the electron-electron repulsion and the exchange (spin) interactions between the 4f electrons. The spin S appears as the degeneracy $2S + 1$ at the upper left of the orbital L designation, where S, P, D, F, G, H, I corresponds to $L = 0, 1, 2, 3, 4, 5, 6$, respectively. Only certain spin

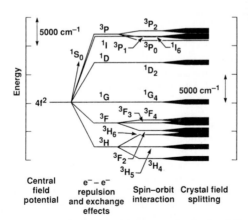

FIGURE 5.16 Origin of the energy levels of a Pr^{3+}-doped solid, showing the intraionic effects of electron repulsion, exchange, and spin-orbit splitting, and the crystal-field splitting due to the host medium.

and orbital combinations may exist in order to assure the antisymmetry of the overall wavefunction. If the interaction between the electron spin and orbital motion is then "turned on," the additional splitting that occurs is described by the J-state subscripts. The overall state designation is $^{2S+1}L_J$, where $J = L - S$, $L - S + 1, \ldots, L + S$. Up till now, only the properties of the RE ion in free space have been considered. The host medium that surrounds Pr^{3+} is responsible for the crystal field interactions, which lift the $2J + 1$ degeneracy of the spin-orbit state and result in a series of lines or in a narrow band of states, as shown in Fig. 5.16. The crystalline field also serves the critical function of inducing electric dipole-based transition strength into the forbidden 4f-4f transitions of the free-ion.

The Yb^{3+} ion is the simplest example of a RE laser ion. The electronic configuration of $4f^{13}$ is only one electron short of being a completely filled shell, $4f^{14}$, and by symmetry may be described as a single $4f_h^1$ hole. This configuration produces only two electronic states: the $^2F_{7/2}$ ground state and the $^2F_{5/2}$ excited state. The absorption and emission spectra[37] of Yb^{3+} in $Y_3Al_5O_{12}$ (YAG) are displayed in Fig. 5.17. The various lines apparent in the figure are due to the crystal field split levels of the $^2F_{7/2}$ and $^2F_{5/2}$ states. The prominent emission line at 1.03 μm is the operating wavelength of the laser, while some of the shorter-wavelength lines are appropriate for laser pumping. The presence of ground-state absorption at the 1.03-μm laser line stipulates that a minimum of 5 percent of the Yb^{3+} ions must be inverted before the system may exhibit net gain.

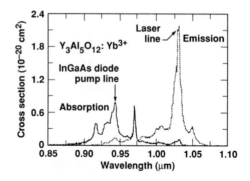

FIGURE 5.17 Absorption and emission of Yb:Y$_3$-Al$_5$O$_{12}$; the potential pump and laser wavelengths are indicated in the figure.

Interestingly, Yb:YAG has recently been "rediscovered" since the advent of InGaAs laser diodes, which can effectively pump the medium directly in the 0.93- to 0.97-μm region.[4,38] Note that flashlamp sources provide extremely poor pumping efficiency for Yb, since the absorption spectra cover such a limited range while the lamps generate light from the uv to the ir. Diode laser pumping, on the other hand, ideally suits this type of laser material. It is worth mentioning that the relatively long emission lifetime of 1.1 ms is a useful storage time for diode pumping, since diode lasers are peak-power-limited devices.

In contrast to Yb^{3+}, dozens of electronic states arise from the $4f^3$ electron configuration of Nd^{3+}, some of which are depicted in Fig. 5.18. The ground state occurs among the ^4I_J manifold, where the ^4I designation describes the $S = 3/2$ spin

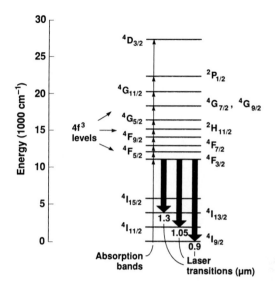

FIGURE 5.18 Energy levels of Nd^{3+} showing the transitions responsible for the absorption bands and laser action. Note that only the major states involved in the transitions have been indicated for clarity of presentation.

and the $L = 6$ orbital angular momentum, and the $J = 9/2$ to $15/2$ states are due to the spin-orbit coupling of these two momenta. The absorption transitions occur from the $^4I_{9/2}$ ground state to the indicated energy levels, and each of the transitions is manifest as a band in the data of Fig. 5.19, which contains the absorption spectrum of LG-750 (Schott Glass Technologies), a Nd-doped phosphate glass. Following the absorption of light energy by one of the pump bands shown in Fig. 5.19, the energy relaxes to the $^4F_{3/2}$ metastable excited state. Importantly, the many absorption bands of Nd^{3+} permit the efficient flashlamp pumping of this ion in most hosts.

The lifetimes of the various electronic states of Fig. 5.18 may be estimated with the aid of the energy-gap law[39]

$$k_{nr} = k_0(T) \cdot \exp(-\alpha E_{gap}) \quad (5.42)$$

where α is a parameter that is specific to each host material and E_{gap} is the energy separation between the state of interest and the next lower electronic state. The temperature dependence of $k_0(T)$ is relatively weak for rare earth ion impurities. Generally an E_{gap} of 3000 cm^{-1} or more is required for the k_{rad} radiative rate to be appreciable com-

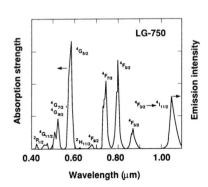

FIGURE 5.19 Absorption and partial emission spectrum of Nd^{3+} in phosphate glass (LG-750, Schott Glass Technologies). The final states of the absorption transitions are noted.

pared with nonradiative decay processes (that is, for $k_{nr} < 10^3$ s^{-1}). Most of the energy gaps for Nd^{3+} are small, resulting in rapid relaxation to the $^4F_{3/2}$ level, which is metastable since it has a larger gap of 5000 cm^{-1}.

Once the $^4F_{3/2}$ state is populated, the Nd laser is capable of lasing at three major wavelengths near 1.3, 1.05, and 0.9 μm, as is depicted in Fig. 5.18. The $^4F_{3/2} \rightarrow$ $^4I_{11/2}$ transition is responsible for the emission band near 1.054 μm in Fig. 5.19 and provides the highest level of gain for Nd-doped glasses and crystals. Nd^{3+} lasers are by far the most technologically important systems, and the three main host materials that are routinely employed are YAG, YLF (LiYF$_4$), and phosphate glass.

Er^{3+} is an example of an ion that possesses numerous metastable levels, thereby allowing 13 different laser transitions to exist for this single center;[16] see Fig. 5.20.

The Er^{3+} ion is illustrative of the disparity that may exist between demonstrated laser transitions and those that are commonly employed, since, of all the laser lines, only the $^4I_{13/2} \rightarrow$ $^4I_{15/2}$ transition at 1.6 μm is utilized in practical systems. The 1.6-μm transition is often realized in flashlamp-pumped glass rod systems, although the metastability of numerous levels does not allow for the quick relaxation to the desired $^4I_{13/2}$ state. Interestingly, the most significant application for Er^{3+} has emerged recently, where the telecommunications industry has found that diode-pumped Er-doped fibers can effectively amplify the signals present in fiber cables. Clearly the market potential for this type of device is enormous.

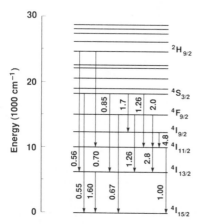

FIGURE 5.20 Energy levels and demonstrated laser transitions of Er^{3+}-doped materials. The wavelengths of laser action are noted in μm.

There is a great gap between the number of RE ion systems that have been lased and the relatively small number of materials that are actually used in practical applications. A summary of all the wavelengths that have been generated by RE ions appears in Fig. 5.21.[16] The wavelengths span the range from 0.18 to 5.2 μm, and both the divalent and trivalent ions are reported. Of all the transitions shown, only three are based on the 5d $\rightarrow$ 4f transition (rather than the 4f $\rightarrow$ 4f), including the 0.18-, 0.3-, and 0.75-μm bands of Nd^{3+}, Ce^{3+}, and Sm^{2+}, respectively. Among the systems that have been demonstrated to have practical utility are the ir transitions of Tm^{3+} and Ho^{3+} near 2 μm; these ions are normally lased in hosts that are sensitized with additional ions (see discussions in Sec. 5.3.6 and 5.3.7). The existence of the many transitions in Fig. 5.21 may be construed as a foreshadowing of the progress that is possible in laser materials and designs in the coming years.

5.3.4 Transition Metal Ion Lasers

The optical properties of transition metal ions are fundamentally different from the rare earth species.[40] This is the case primarily because the 3d $\rightarrow$ 3d electronic

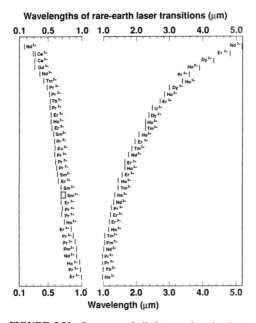

FIGURE 5.21 Summary of all the wavelengths that may be generated by RE ions. Each line represents a particular transition for an RE ion, which may have been lased in numerous host materials.

transitions that are responsible for the absorption and emission features of the transition metal ions interact strongly with the host, in contrast to the relative insensitivity of the rare earth $4f \rightarrow 4f$ transitions. The type of situation that arises is depicted in Fig. 5.22 in terms of the "configuration coordinate" model. The Ti^{3+} ion in Al_2O_3 has been selected for illustrative purposes because its valence shell contains only a single d electron, $3d^1$. The 3d electron is crystal field split into two states, the ^2E and $^2T_2$, by the six nearest-neighbor oxygen anions surrounding the Ti^{3+} ion. The crystal field splitting is roughly an order of magnitude larger than that experienced by the 4f electrons of rare earth ions (as shown in Fig. 5.16 for the Pr^{3+} ion). As depicted in Fig. 5.22, the average TiO distance is slightly larger in the ^2E state compared with the $^2T_2$ state. This enlargement is particularly important, because it produces the wide absorption and emission features that characterize the transition metal ion lasers.

The configuration coordinate diagram of Fig. 5.22 explains how the different Ti-O distances in the ground and excited states give rise to broad spectral features. The Gaussian curve drawn on the ground-state potential energy surface indicates the probabilistic distribution of Ti-O distances that occurs. Since the electronic transition to the excited state occurs rapidly compared with the motion of the Ti-O atoms, this ground-state Ti-O distance distribution is simply "reflected" off the rising side of the upper state energy surface, thereby producing a broad absorption feature. A similar argument applies to the emission process. The actual absorption and emission spectra of Al_2O_3:Ti^{3+} are shown in Fig. 5.23.[41] This material (known as Ti:sapphire) is extremely useful since it may be optically pumped

(a) Splitting due to oxygen neighbors of Al$_2$O$_3$:Ti^{3+}

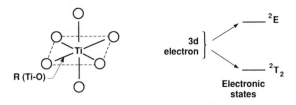

(b) Configuration coordinate model

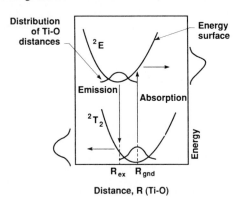

FIGURE 5.22 (*a*) Splitting of the 3d electron of Ti^{3+} into the ^{2}E and ^{2}T$_2$ states due to the interaction with the six nearest-neighbor oxygen anions of Al$_2$O$_3$. (*b*) Configuration coordinate model of the Ti^{3+} impurity depicting how the displacement between the ^{2}T$_2$ and ^{2}E states results in broad absorption and emission features.

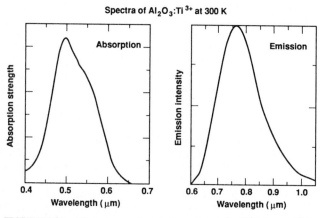

FIGURE 5.23 Absorption and emission spectra of a Ti^{3+}-doped Al$_2$O$_3$ crystal. (*Adapted from Ref 41.*)

with a doubled Nd:YAG laser or an Ar^+ laser, and because the output of the laser is widely tunable from 0.66 to 1.2 μm. Ti:sapphire lasers operate efficiently and are not adversely impacted by the fundamental loss mechanisms shown in Fig. 5.13.

Cr^{3+} lasers[16,23,42,43] are similar to Ti^{3+} lasers, in that these crystals exhibit broad spectral features. Cr^{3+}-doped crystals possess an important advantage, however, in that they have three absorption bands, rather than one, and therefore absorb flashlamp light more efficiently. Another important distinction between Ti^{3+} and Cr^{3+} lasers pertains to the lifetime of the metastable state, which is typically near 1 to 10 μs for Ti^{3+}, and 50 to 300 μs for Cr^{3+}. As a result, Cr lasers can be arranged to store more flashlamp energy than Ti lasers. Finally, because the trivalent oxidation state of Cr is very stable, it may be incorporated into a wide variety of host media. A summary of the tuning ranges achieved by Cr^{3+}-doped materials appears in Fig. 5.24, where it is seen that wavelengths from 0.70 to 1.25 μm can be covered with different host materials. It is crucial to emphasize, however, that many of the reported Cr^{3+}-lasers are not useful because they are flawed in various ways, such as by having low efficiency, or perhaps by permanently coloring under the influence of ultraviolet flashlamp light. The most promising lasers are Cr^{3+}-doped $BeAl_2O_4$ and the $LiCaAlF_6$ family of hosts (known as alexandrite and Cr:LiCAF).

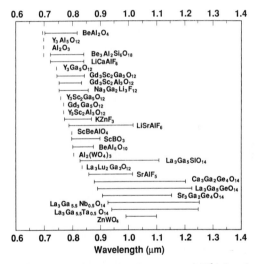

FIGURE 5.24 Reported tuning ranges of Cr^{3+}-doped crystals.

There are, in total, 30 transition metal ions, including the first-, second-, and third-row ions. In spite of this diversity, only four other ions have been lased in addition to Ti^{3+} and Cr^{3+}; they are V^{2+}, Cr^{4+}, Co^{2+}, and Ni^{2+}.[16] These ions exhibit laser output in the 1- to 2.3-μm region, and of these materials, $MgF_2:Co^{2+}$ crystals seem to have the most promise.[44] The Cr^{4+} ion has recently been discovered to lase in forsterite (Mg_2SiO_4), although the practical utility of this system remains uncertain at this time.[45] The limited number of transition metal ion lasers that have been realized is related to several factors. First, the chemical properties of the ions are such that they do not have stable oxidation states and also tend to vaporize rather than dissolve in the host material during crystal growth. Second, many ions turn out to have low emission yields or to have significant excited-state absorption losses, rendering them useless as laser ions.

5.3.5 Color Center Lasers

A number of excellent reviews on color center lasers are currently available. As a consequence, only a brief sketch of salient features of these lasers will be provided here. The reader is directed to several articles by Mollenauer that are particularly informative and thorough.[46-48]

Color center lasers involve active centers that are native defects of the host material. The progenitor to all of these laser species is the F center, although this defect itself is not laser-active. Figure 5.25 contains a reminder of the nature of the F center as well as several of the actual laser-active species (see the text associated with Fig. 5.12 for the initial discussion on this topic). The F_A center of Fig. 5.25 involves the incorporation of a foreign ion, such as Li^+, into the center. The F_2 defect consists of two nearby F centers; F_2^+ contains two vacancies with a single electron, and the $(F_2^+)_A$ incorporates a foreign ion as well. One of the most interesting and useful color center lasers is the Tl^0 atom with a nearby vacancy. This center may also be viewed as involving the sharing of an electron between Tl^+ and a vacancy.

One of the important optical properties of the F center and related species is the broad nature of the absorption and emission features. To understand the origin of these properties, the nature of the electronic structure of the defect centers must be considered. In the simplest sense, the F, F_2, and F_2^+ centers may be modeled as an H atom, H_2 molecule, and H_2^+ molecular ion, respectively, that is embedded in the dielectric continuum of the host medium. As a result the important electronic transitions basically originate from a 1s-2p type of electronic change. The change of the principal quantum number of the orbital occupancy causes the electronic structure of the ground and excited states to be very different. Figure 5.26 depicts the energy levels of the "normal" ground-state configuration (left) and of the relaxed excited state (right). The substantial structural differences between the ground and excited states produce the large shift between the absorption and emission frequencies. The electronic structure of the excited state is also intimately associated with the conduction band states of the host, since the excited states are primarily composed of linear combinations of these orbitals.

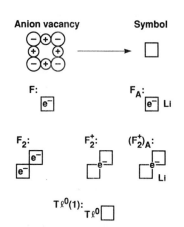

FIGURE 5.25 Schematic illustration of the laser-active color centers are derivable from the F center. *(Reproduced from Ref. 47.)*

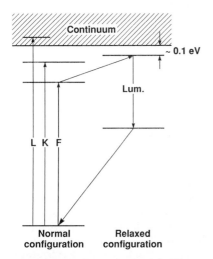

FIGURE 5.26 Energy levels of the F center, showing that the different configurations of the ground and relaxed excited states give rise to the large shift between the absorption and emission features. *(Reproduced from Ref. 48.)*

A tabular listing of the color center lasers that are commonly used in laboratory settings is reproduced in Table 5.3. It is apparent that much of the infrared spectral range is covered by the tunable laser action from these media. The hosts predominantly involve the alkali halides because these crystals afford the most useful properties to the defect centers. Although few other types of lasers effectively and broadly provide tunable radiation in the infrared, the requirement of cryogenic cooling for most of these systems limits their utility to laboratory environments. Finally, it is worth mentioning that the strongly allowed nature of the p-s transitions causes the radiative lifetimes to be rather short, $\tau_{rad} = 100$ to 1000 ns, and the emission cross section to be very high, $\sigma_{em} = 10^{-17} - 3 \times 10^{-16}$ cm^2. In a practical sense, the σ_{em} and τ_{rad} values of color center lasers are closer to the range typical of dye lasers, rather than to the rare earth– and transition metal–based lasers described in this chapter.

TABLE 5.3 Most Used Color Center Lasers

Host	LiF	NaCl:OH	KCl:Tl	KCl:Li	RbCl:Li
Center	F_2^+	$F_2^+:O^{2-}$	$Tl^0(1)$	$F_A(II)$	$F_A(II)$
Pump wavelength, μm	0.647	1.064	1.064	0.61	0.66
Pump power, W	4	6	6	2.0	1.3
Tuning range, μm	0.82–1.05	1.42–1.78	1.4–1.6	2.3–3.1	2.5–3.65
Max. power output, cw	1.8 W	1.2 W	1.1 W	150 mW	90 mW

Reproduced from Ref. 46.

5.3.6 Energy Transfer in Laser Materials

The performance of the laser centers in a material can be enhanced by the presence of additional impurity ions. Figure 5.27 illustrates two possible roles for these extra ions. The sensitizer may increase the general level of white light absorption or provide new absorption features that match the output of a laser pump source. The sensitizer is then anticipated to efficiently transfer the energy to the upper level of the laser ion. Another role that the additional ions may play is that of deactivating the lower laser level, for the case of systems where this level is kinetically bottlenecked. If the lower laser level is not rapidly drained following laser

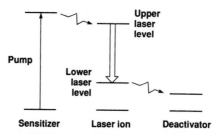

FIGURE 5.27 Illustration showing the role of a sensitizer to excite the upper laser level, or that of a deactivator, to depopulate the lower laser level.

action, the population will build up and introduce an increasing absorption loss exactly at the laser wavelength. The deactivator serves to funnel the energy away from the laser ion. A good example of the need for a deactivator occurs for the $^4I_{11/2} \rightarrow {}^4I_{13/2}$ transition of Er^{3+} at 2.8 µm. Here the $^4I_{13/2}$ level has a lifetime that is about an order of magnitude longer than the $^4I_{11/2}$ state, and the deactivator is required to avert self-quenching of the laser action.[49] Several different rare earth ions have been employed for this purpose.[16]

The physics of energy transfer is complicated and lacks a comprehensive theory by which to describe all phenomena. For the case of dilute dopant concentrations and for dipole-dipole interactions, the energy-transfer rate from a single sensitizer ion to a single laser ion is given by[50]

$$W = \frac{R_0^6}{R^6 \, \tau_{rad}} \tag{5.43}$$

where R is the interion separation, τ_{rad} is the radiative lifetime of the sensitizer S, and R_0 is the critical radius parameter. R_0 is related to the overlap of the sensitizer emission band σ_{em}^S, and the laser ion (L) absorption band σ_{abs}^L, with the relation

$$R_0 = \left(\frac{3c \, \tau_{rad}}{8\pi^4 \, n^2} \int \sigma_{em}^S \sigma_{abs}^L \, d\lambda \right)^{1/6} \tag{5.44}$$

Of course, the statistical ensembles must be suitably averaged in order to interpret the R distances in terms of the actual impurity concentrations.

The nonradiative transfer described in Eqs. (5.43) and (5.44) should not be confused with the radiative transfer process. For the case of radiative transfer, a photon is emitted by the sensitizer, travels through the host medium, and then is absorbed by the laser ion. On the other hand, the sensitizer and laser ions simultaneously undergo transitions during nonradiative transfer. One of the important differences among the effects of the two processes is that only nonradiative transfer leads to a shortening of the sensitizer lifetime by the presence of laser ions. It should be emphasized that nonradiative energy transfer is influenced by many other factors, such as energy migration among the sensitizer ions, nearest neighbor effects, and back transfer. These issues have not been accounted for in the preceding discussion. Several sensitized laser materials have proved to be useful. As an example, consider the material where Cr^{3+} and Nd^{3+} are codoped into $Gd_3Sc_2Ga_3O_{12}$ (GSGG); the Nd^{3+} is the laser ion while Cr^{3+} serves as the sensitizer.[20] The absorption spectrum of GSGG:Cr,Nd is shown in Fig. 5.28. Since the

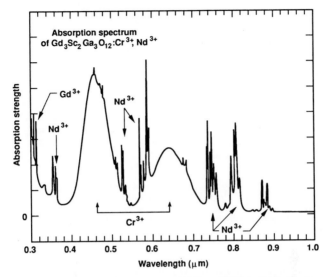

FIGURE 5.28 Absorption spectrum of Cr^{3+} and Nd^{3+} in $Gd_3Sc_2Ga_3O_{12}$. The Cr^{3+} sensitizers provide for more efficient flashlamp absorption and then rapidly transfer the energy to the Nd^{3+} ions.

sharp features are due to the Nd^{3+} ion and the broad bands to Cr^{3+}, it is clear that the Cr^{3+} ion provides greatly enhanced absorption for the flashlamp-pumped system. The $Cr \rightarrow Nd$ energy transfer has been found to be extremely efficient (>90 percent), in part because the Cr emission band centered at 0.8 μm overlaps several Nd absorption bands; see Eq. (5.44). GSGG:Cr,Nd and related garnet host systems have provided the highest flashlamp-pumped efficiencies measured to date.[51,52]

The energy transfer of the Cr^{3+}, Tm^{3+}, Ho^{3+}:YAG crystal is a striking example of an elegant laser system.[53] The Cr^{3+} ions efficiently absorb the flashlamp light. The energy is then transferred to the Tm^{3+} ions, as shown in Fig. 5.29. The Tm^{3+} ions are doped at high enough concentration to allow for efficient cross relaxation in which two Tm^{3+} ions are generated in the $^3F_4$ excited state, for each Tm^{3+} ion initially in the $^3H_4$ state. Lastly, the energy is transferred to the Ho^{3+} ions, which exhibit gain near 2.1 μm. Each of the concentrations of the ions must be carefully chosen to optimize the energy-transfer steps and the laser performance. This system may be operated in both long-pulse and Q-switched modes (since the Ho^{3+} emission cross section is relatively large). The related Cr,Tm:YAG system operates at the long-wavelength tail of the Tm^{3+} emission band, although Cr,Tm:YAG is easiest to operate in a long pulse mode, since the emission cross section is rather low.[54]

5.3.7 Practical Laser Materials

In this section the important optical properties that characterize the potential laser performance of the material are summarized and discussed. All of the materials summarized in Table 5.4 were previously discussed in Sec. 5.3.2 (and also appear in Table 5.2, where the optical, thermal, and physical properties are tabulated). Included in Table 5.4 is the common name of the material, the laser transition,

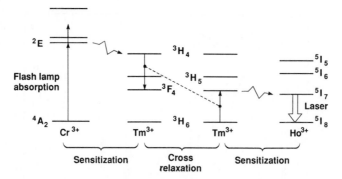

FIGURE 5.29 Energy-transfer dynamics of the Cr^{3+}, Tm^{3+}, Ho_3^+:YAG laser crystal. The Cr^{3+} impurities initially absorb the light and transfer the energy to Tm^{3+}. The Tm^{3+} ions then cross-relax to produce two excited states for each ion initially excited. Lastly the energy is transferred to the Ho^{3+} laser ions.

and the wavelength (λ_l), bandwidth ($\Delta\bar{v}$), cross section (σ_{em}), and saturation fluence (F_{sat}) of the laser transition at hand.[17,20,23,24,27,30,31,32,38,41,43,44,46–48,53–65] The metastable state lifetime is provided (τ_{em}), as is the energy separation of the lower laser level from the ground state (ΔE). ΔE must be $\gg kT = 205$ cm^{-1} in order for the system to operate as a four-level laser at room temperature.

The spectral width of the emission level $\Delta\bar{v}$ is an important quantity for mode-locked pulse generation, since the product of the spectral and temporal full widths must be[66]

$$\Delta\bar{v}\cdot\Delta\tau > 10 \text{ ps·cm}^{-1} \tag{5.45}$$

As a consequence, the width of the laser transition must be greater than the expected width of the laser output pulse.

Among the Nd lasers listed in Table 5.4, Nd:YAG is the most technologically important material. The high cross section at 1064 nm results in straightforward conversion of stored energy density to gain, yet the emission lifetime $\tau_{em} = 0.26$ ms permits adequate energy storage using flashlamps. The saturation fluence $F_{sat} = 0.7$ J/cm^2 allows for efficient extraction of energy from Nd:YAG amplifiers without approaching the damage threshold of the optical elements. The two other transitions at 0.946 and 1.31 μm also lase with reasonable efficiency, although the much higher gain at 1064 nm must clearly be carefully selected against in the cavity; see Sec. 5.2.3. The ground-state splitting ΔE of 857 cm^{-1} is very large in spite of the fact that the 0.946 μm $^4F_{3/2} \rightarrow {}^4I_{9/2}$ transition is resonant (i.e., the $^4I_{9/2}$ is the ground state of Nd^{3+}). The $^4F_{3/2} \rightarrow {}^4I_{9/2}$ transition of Nd:YOS has also been lased at 0.91 μm, although the smaller ΔE value of 356 cm^{-1} requires that a significant level of bleach pumping be exercised in order to achieve net gain. The fluoride medium Nd:YLF is tending to be selected over Nd:YAG more and more often by laser designers, since it has a greatly reduced level of thermal lensing, and because the greater width of the gain feature ($\Delta\bar{v}_l = 15$ vs. 7 cm^{-1}) permits the generation of shorter pulses in a mode-locked Nd:YLF oscillator. The main disadvantage of Nd:YLF lies in the much smaller thermal shock resistance compared with Nd:YAG ($R'_T = 1.1$ vs. 4.6 W/$\sqrt{m}$; see Table 5.2). LG-750 is a Nd-doped phosphate glass that fills an important niche for large fusion laser applications,

TABLE 5.4 Summary of Spectral Properties of Laser Materials

Name	Material	Transition	τ_{em}, ms	λ, μm	$\Delta\bar{\nu}$, cm⁻¹	σ_{em}, 10⁻²⁰ cm²	F_{sat}^{em}, J/cm²	ΔE, cm⁻¹	Pump sources	References
Nd:YAG	Nd:Y₃Al₅O₁₂	$^4F_{3/2} \to {}^4I_{9/2}$	0.26	0.946	14	2.5	9	857	FL, AlGaAs	20, 55–58
		$^4F_{3/2} \to {}^4I_{11/2}$		1.064	7	28	0.7	~2000		
		$^4F_{3/2} \to {}^4I_{13/2}$		1.31	10	9	1.7	~4000		
Nd:YOS	Nd:Y₂SiO₅	$^4F_{3/2} \to {}^4I_{9/2}$	0.21	0.912	50	2.1	10	356	FL, AlGaAs	24, 59
		$^4F_{3/2} \to {}^4I_{11/2}$		1.074	30	10	1.9	~2000		
Nd:YLF	Nd:LiYF₄	$^4F_{3/2} \to {}^4I_{11/2}$	0.48	1.047	15	18	1.1	~2000	FL, AlGaAs	27, 60
LG-750	Nd:phosphate	$^4F_{3/2} \to {}^4I_{11/2}$	0.39	1.054	194	4.0	4.7	~2000	FL, AlGaAs	30, 31
APG-1	Nd:phosphate	$^4F_{3/2} \to {}^4I_{11/2}$	0.38	1.055	206	3.5	5.4	~2000	FL, AlGaAs	31
Cr,Nd:GSGG	Cr,Nd:Gd₃Sc₂Ga₃O₁₂	$^4F_{3/2} \to {}^4I_{11/2}$	0.28	1.061	13	13	1.4	~2000	FL	20, 51, 58
Yb:YAG	Yb:Y₃Al₅O₁₂	$^2F_{5/2} \to {}^2F_{7/2}$	1.1	1.03	50	2.0	10	630	InGaAs	37, 38
Er:glass	Er:phosphate	$^4I_{13/2} \to {}^4I_{15/2}$	8	1.54	200	0.8	16	(200)	FL	61
Er:fiber	Er:SiO₂	$^4I_{13/2} \to {}^4I_{15/2}$	10	1.55	200	0.8	16	(200)	InGaAs(P)	61, 62
Cr,Tm:YAG	Cr,Tm:Y₃Al₅O₁₂	$^3F_4 \to {}^3H_6$	11	2.01	50	0.2	50	590	FL	54
Tm,Ho:YAG	Tm,Ho:Y₃Al₅O₁₂	$^5I_7 \to {}^5I_8$	9	2.09	50	1	10	450	AlGaAs	63, 64
CTH:YAG	Cr,Tm,Ho:Y₃Al₅O₁₂	$^5I_7 \to {}^5I_8$	9	2.09	50	1	10	450	FL	53, 65
Ti:sapphire	Ti:Al₂O₃	$^2E \to {}^2T_2$	0.0032	0.66–1.2	3000	30	0.9	(3000)	D-YAG, ArL	41
Alexandrite	Cr:BeAl₂O₄	$^4T_2 \to {}^4A_2$	0.26	0.70–0.82	2000	0.9	30	(1500)	FL	23
Cr:LiSAF	Cr:LiSrAlF₆	$^4T_2 \to {}^4A_2$	0.067	0.78–1.01	2500	4.8	5	(2000)	FL, AlGaInP	17, 43
Co:MgF₂	Co:MgF₂	$^4T_2 \to {}^4T_1$	0.04	1.5–2.3	1500	0.15	70	(1500)	Nd:YAG	44
LiF, F₂⁺	LiF, 77K	$\sigma_u \to \sigma_g$	10⁻⁴	0.82–1.05	2000	~3 × 10⁴	10⁻³	(2000)	Nd:YAG	46–48
KCl:Tl⁰(I)	Tl:KCl, 77K	$^2P_{3/2} \to {}^2P_{1/2}$	0.0016	1.4–1.6	700	~10³	10⁻²	(700)	Nd:YAG	46–48

FL = flashlamp; AlGaAs, InGaAs, AlGaInP = laser diodes; D-YAG = doubled Nd:YAG laser; ArL = argon ion laser.

where the σ_{em} must be much smaller than Nd:YAG in order to avoid the detrimental effects of ASE, and where large optics must be fabricated at a reasonable cost. Since fusion lasers are fired only every few hours, the low $R'_T = 0.35$ W/$\sqrt{m}$ of LG-750 is not important. APG-1 glass is a similar Nd-doped phosphate glass that has been engineered by Schott Glass Technologies to have a larger R'_T value. The saturation fluences of both LG-750 and APG-1 are near 5 J/cm^2, which still, with some care, permits efficient extraction without exceeding the typical damage thresholds of optical elements. The final Nd laser listed in Table 5.4 is the Cr,Nd:GSGG crystal. Here the Cr^{3+} ions serve as the sensitizers and greatly improve the flashlamp pumping efficiency of the laser.

The Yb:YAG material appears in Table 5.4 because the InGaAs laser diodes that have recently become available are capable of efficiently pumping the single absorption feature of the Yb^{3+} ion, while flashlamps cannot accomplish this task. The τ_{em} of 1.1 ms of Yb:YAG is considerably longer than is typical of Nd-doped materials, thereby allowing better energy storage.

The 1.6-μm laser transition of Er has been operated primarily in glassy hosts. Flashlamp-pumped Er:glass laser rods have been available for some time, although the concept of laser diode-pumped Er-doped fibers is a new idea that may be very significant for the telecommunications industry. Although these systems must be bleach-pumped to a substantial degree since $\Delta E \sim 200$ cm^{-1}, the long lifetime of the metastable excited state of $\sim$10 ms provides a fairly low pump saturation intensity.

Several IR lasers have recently been found to be quite useful, including Cr,Tm:YAG, where Cr serves as the sensitizer and the Tm ions lase. The high saturation fluence renders this system operable only in a long-pulse mode. The Tm,Ho: and Cr,Tm,Ho:YAG crystals lase on the 2.09-μm transition of Ho^{3+} and are pumpable by laser diodes or flashlamps; the Cr is required for better flashlamp absorption.

Several of the transition metal lasers must be considered also. Ti:sapphire has proved to be widely tunable (0.66 to 1.2 μm) and efficient, and may be laser-pumped by either a doubled Nd:YAG or an Ar$^+$ laser. The material is not amenable to either flashlamp or diode laser pumping, however, and the short metastable state lifetime of 3.2 μs renders the material relatively ineffective as a storage device. The Cr lasers, alexandrite and Cr:LiSAF, have wide tuning ranges and can store much more energy for a given pump power than Ti:sapphire because of the longer emission lifetimes (67 and 260 μs). Cr:LiSAF offers a much lower saturation fluence F_{sat} and less thermal lensing than alexandrite but also suffers from a low thermal shock parameter. The Co:MgF$_2$ laser is considered because it is tunable in the IR and can be conveniently pumped by a long-pulse Nd:YAG laser. The favorable thermomechanical attributes of the MgF$_2$ host are also important; see Table 5.2.

The two color center lasers at the end of Table 5.4 are representative examples of these types of systems. Note that the F_{sat} values are very low at 10^{-2} to 10^{-3} J/cm^2, and the excited-state lifetimes are on the order of a microsecond or less. These lasers generate tunable laser output in the important ir region, although most color center lasers must be operated at cryogenic temperatures. It should be noted that the short lifetimes facilitate the generation of ultrashort pulses by the synchronous pumping method (see Sec. 5.2.3).

5.4 FUTURE DIRECTIONS

Although the field of solid-state lasers represents a mature technology in many respects, the field is as much in flux as it has ever been. The eighties, for example, have seen the development of broadly tunable systems, new techniques of mode locking, chirped pulse amplification, miniaturized monolithic architectures, and a resurrection and maturing of slab laser technology and diode pumping. Despite these numerous advances there are three clear areas where solid-state laser architectures will advance significantly in the next few years. The first is to take diode pumped lasers from the minilaser scale to the tens of watts level. This requires development of high average power, two-dimensional diode arrays, which is now well underway. The much smaller heat deposition of the diodes will make high average power systems more feasible, and advanced diode cooling techniques make the diode arrays increasingly more capable. The second is to utilize slab laser technology to operate lasers at the 1 to 10 J per pulse level at repetition rates of about 100 Hz. The issue of *simultaneously* achieving good efficiency and beam quality at high average power remains one of the greatest difficulties of practical system design. Finally, even large fusion lasers are becoming increasingly cheaper on a per joule basis and may ultimately operate at several Hz repetition rate. This last goal requires a marriage of the fields of laser, heat transfer, and flow physics. The resulting laser devices, however, have the potential to operate ultimately at several megawatts of output power per amplifier.

The new laser materials that will be developed in the future will be deployed in systems to generate new wavelengths, operate more efficiently, be produced at lower cost, and have optical properties that are tailored to meet specific technical objectives. Solid-state lasers that operate efficiently in the ultraviolet-blue region are likely to be developed. Many new transition metal lasers that have useful optical properties may be discovered in the next decade. One change that may occur will involve the advent of tailoring some types of laser materials for a specific application. (This has already occurred in the case of Nd-doped glasses for fusion lasers.) This process of designing, rather than discovering, laser materials will be enhanced by a better understanding of the physics and chemistry of solid-state media.

Acknowledgments

Thanks go to our beautiful children, whose laughter can make the worst day worthwhile. This research was performed under the auspices of the U.S. Department of Energy, by Lawrence Livermore National Laboratory under Contract No. W-7405-ENG-48. Portions of this chapter have been reprinted and adapted by permission of the publisher from "Solid-State Lasers," by Stephen A. Payne and Georg F. Albrecht in *Encyclopedia of Lasers and Optical Technology*, edited by R. A. Meyers, copyright 1991 by Academic Press, Inc.

5.5 REFERENCES

1. T. H. Maiman, *Nature*, vol. 187, p. 493, 1960; A. L. Schawlow and C. H. Townes, *Phys. Rev.*, vol. 112, p. 1940, 1958.

2. The following two references are of general interest in the field of lasers, and solid-state lasers in particular. If a section contains no specific references, the reader readily will find further information on the respective subject in one of these two books: W. Koechner, *Solid State Laser Engineering*, 2d ed., Springer Series in Optical Sciences, vol. 1, Springer, New York, 1988. This book, unquestionably the preeminent work on solid-state lasers, is written to be useful for students as well as seasoned practitioners. A. E. Siegman, *Lasers*, University Science Books, Mill Valley, Calif., 1986. Written by one of the ultimate masters on the subject, this voluminous work treats more the theoretical side of lasers in general with added emphasis on beams and resonators.

3. T. Y. Fan and R. L. Byer, "Diode Pumped Solid State Lasers," *IEEE J. Quantum Electron.*, vol. 24, p. 895, 1988.

4. W. F. Krupke, and L. L. Chase, "Ground State Depleted Solid State Lasers: Principles, Characteristics and Scaling," *Opt. Quantum Electron.*, vol. 22, p. 51, 1990.

5. V. M. Ovchinnikov, and V. E. Khartsiev, "Bleach Waves in Two Level Systems," *JETP*, vol. 22, p. 221, 1966.

6. J. E. Swain, et al., "Large-Aperture Glass Disk Laser Systems," *J. Appl. Phys.*, vol. 40, pp. 3973–3977, 1969.

7. T. J. Kane, and R. L. Byer, "Monolithic, Unidirectional Single-Mode Nd:YAG Ring Laser," *Opt. Lett.*, vol. 10, p. 65, 1985.

8. J. J. Degnan, "Theory of the Optimally Coupled Q-Switched Laser," *IEEE J. Quantum Electron.*, vol. 25, p. 214, 1989.

9. D. J. Kuizenga, "Short Pulse Oscillator Development for Nd:Glass Laser Fusion Systems," *IEEE J. Quantum Electron.*, vol. 17, p. 1694, 1981.

10. S. DeSilvestri, V. Magni, O. Svelto, and G. Valentini, et al., "Lasers with Supergaussian Mirrors," *IEEE J. Quantum Electron.*, vol. 26, p. 1500, 1990.

11. G. Albrecht, "Average Power Slab Lasers with Garnet Crystals as the Active Medium," UCRL J-106584, Feb 1991, Available through TID, Lawrence Livermore Ntl. Laboratory, Box 5508, Livermore, CA 94550.

12. D. C. Brown, *High Peak Power Nd:Glass Laser Systems*, Springer Series in Optical Sciences, vol. 25, Springer, New York, 1981, p. 15.

13. G. Mourou, D. Staickland, and S. Williamson, "How Pulse Compression Techniques Can Be Applied to High Energy Laser Amplifiers," *Laser Focus*, June 1986.

14. F. G. Patterson, and M. D. Perry, "Design and Performance of a Multiterawatt, Sub-picosecond Glass Laser," *J. Opt. Soc. Am.* B, accepted for publication.

15. P. F. Moulton, "Paramagnetic Ion Lasers," in M. J. Weber (ed.), *Handbook of Laser Science and Technology*, vol. 1, CRC Press, Boca Raton, Fla., 1982, pp. 21–147.

16. J. A. Caird, and S. A. Payne, "Crystalline Paramagnetic Ion Lasers," in M. J. Weber (ed.), *Handbook of Laser Science and Technology*, suppl. 1, CRC Press, Boca Raton, Fla., 1991, pp. 3–100.

17. See, for example, S. A. Payne, L. L. Chase, H. W. Newkirk, L. K. Smith, and W. F. Krupke, "$LiCaAlF_6$:Cr^{3+}: A Promising New Solid-State Laser Material," *IEEE J. Quantum Electron.*, vol. 24, p. 2243, 1988.

18. H. G. Pfaender, and H. Schroeder, *Schott Guide to Glass*, Van Nostrand, Princeton, N.J., 1983.

19. F. Rosenberger, *Fundamentals of Crystal Growth I*, Springer-Verlag, Berlin, 1979.

20. W. F. Krupke, M. D. Shinn, J. E. Marion, J. A. Caird, and S. E. Stokowski, "Spectroscopic, Optical, and Thermo-mechanical Properties of Neodymium- and Chromium-doped Gadolinium Scandium Gallium Garnet," *J. Opt. Soc. Am.* B, vol. 3, p. 102, 1986.

21. R. Adair, L. L. Chase, and S. A. Payne, "Nonlinear Refractive Index of Optical Crystals," *Phys. Rev.* B, vol. 39, p. 3337, 1989.

22. M. J. Dodge, "Refractive Index," in M. J. Weber (ed.), *Handbook of Laser Science and Technology*, vol. IV, part 2, CRC Press, Boca Raton, Fla., 1986.

23. J. C. Walling, D. F. Heller, H. Samelson, D. J. Harter, J. A. Pete, and R. C. Morris, "Tunable Alexandrite Lasers: Development and Performance," *IEEE J. Quantum Electron.*, vol. 21, p. 1568, 1985.

24. R. Beach, M. D. Shinn, L. Davis, R. W. Solarz, and W. F. Krupke, "Optical Absorption and Stimulated Emission of Neodymium in Yttrium Orthosilicate," *IEEE J. Quantum Electron.*, vol. 26, p. 1405, 1990.

25. H. M. O'Bryan, P. K. Gallagher, and G. W. Berkstresser, "Thermal Expansion of Y_2SiO_5 Single Crystals," *J. Am. Ceram. Soc.*, vol. 71, p. C-42, 1988.

26. B. Woods and S. Velsko, private communication, Lawrence Livermore National Laboratory.

27. T. M. Pollak, W. F. Wing, R. J. Grasso, E. P. Chicklis, and H. P. Jenssen, "CW Laser Operation of Nd:YLF," *IEEE J. Quantum Electron.*, vol. 18, p. 159, 1982; T. M. Pollak, R. C. Folweiler, E. P. Chicklis, J. W. Baer, A. Linz, and D. Gabbe, "Properties and Fabrication of Crystalline Fluoride Materials for High Power Laser Applications," in H. E. Bennett, A. J. Glass, A. H. Guenther, and B. E. Newnam (eds.), *Laser-Induced Damage in Optical Materials*, NBS Special Publication 568, 1980.

28. M. J. Weber, D. Milam, and W. L. Smith, "Nonlinear Refractive Index of Glasses and Crystals," *Opt. Eng.*, vol. 17, p. 463, 1978.

29. B. W. Woods, S. A. Payne, J. E. Marion, R. S. Hughes, and L. E. Davis, "Thermomechanical and Thermooptical Properties of the LiCaAlF6:Cr3+ Laser Material," *Opt. Soc. Am. B*, vol. 8, p. 970, 1991.

30. S. E. Stokowski, "Glass Lasers," in M. J. Weber (ed.), *Handbook of Laser Science and Technology*, vol. I, CRC Press, Boca Raton, Fla., 1982; S. E. Stokowski, R. A. Saroyan, and M. J. Weber, *Laser Glass. Nd-doped Glass Spectroscopic and Physical Properties*, Lawrence Livermore National Laboratory, M-095, Rev. 2.

31. Schott Glass Technologies, published manufacturer's specifications.

32. H. Vanherzeele, "Thermal Lensing Measurement and Compensation in a Continuous-Wave Mode-Locked Nd:YLF Laser," *Opt. Lett.*, vol. 13, p. 369, 1988.

33. W. Koechner, "Thermal Lensing in a Nd:YAG Laser Rod," *Appl. Opt.*, vol. 9, p. 2548, 1970.

34. L. Horowitz, Y. B. Band, O. Kafri, and D. F. Heller, "Thermal Lensing Analysis of Alexandrite Laser Rods by Moiré Deflectometry," *Appl. Opt.*, vol. 23, p. 2229, 1984.

35. Y. R. Shen, *The Principles of Nonlinear Optics*, Wiley, New York, 1984, p. 311.

36. G. M. Dieke, *Spectra and Energy Levels of Rare Earth Ions in Crystals*, Wiley, New York, 1968.

37. L. DeLoach, private communication at Lawrence Livermore National Laboratory.

38. P. Lacovara, H. K. Choi, C. A. Wang, R. L. Aggarwal, and T. Y. Fan, "Room-Temperature Diode-Pumped Yb:YAG Laser," *Opt. Lett.*, vol. 16, p. 1089, 1991.

39. M. J. Weber, "Multiphonon Relaxation of Rare-Earth Ions in Yttrium Orthoaluminate," *Phys. Rev. B*, vol. 8, p. 54, 1973.

40. C. J. Ballhausen, *The Theory of Transition-Metal Ions*, McGraw-Hill, New York, 1962.

41. P. F. Moulton, "Spectroscopic and Laser Characteristics of Ti:Al₂O₃," *J. Opt. Soc. Am. B*, vol. 3, p. 125, 1986.

42. S. A. Payne, L. L. Chase, H. W. Newkirk, L. K. Smith, and W. F. Krupke, "LiCaAlF₆:Cr³⁺: A Promising New Solid-State Laser Material," *IEEE J. Quantum Electron.*, vol. 24, p. 2243, 1988.

43. S. A. Payne, L. L. Chase, L. K. Smith, W. L. Kway, and H. W. Newkirk, "Laser Performance of LiSrAlF₆:Cr³⁺," *J. Appl. Phys.*, vol. 66, p. 1051, 1989.

44. P. F. Moulton, "An Investigation of the Co:MgF₂ Laser System," *IEEE J. Quantum Electron.*, vol. 21, p. 1582, 1985.

45. V. Petricevic, S. K. Gayen, and R. R. Alfano, "Continuous-Wave Laser Operation of Chromium-Doped Forsterite," *Opt. Lett.*, vol. 14, p. 612, 1989.

46. L. F. Mollenauer, "Color Center Lasers," in M. J. Weber (ed.), *Handbook of Laser Science and Technology*, suppl. 1, CRC Press, Boca Raton, Fla., 1991, pp. 101–125.

47. L. F. Mollenauer, "Color Center Lasers," in L. F. Mollenauer and J. C. White, *Tunable Lasers*, Springer-Verlag, Berlin, 1987, pp. 225–277.

48. L. F. Mollenauer, "Color Center Lasers," in M. L. Stitch and M. Bass, *Laser Handbook*, vol. 4, North-Holland, Amsterdam, 1985, pp. 143–228.

49. W. Q. Shi, R. Kurtz, J. Machan, M. Bass, M. Birnbaum, and M. Kokta, "Simultaneous Multiple Wavelength Lasing of (Er,Nd): $Y_3Al_5O_{12}$, *Appl. Phys. Lett.*, vol. 51, p. 1218, 1987.

50. For example, see S. A. Payne and L. L. Chase, "$Sm^{2+} \rightarrow Nd^{3+}$ Energy Transfer in CaF_2," *J. Opt. Soc. Am. B*, vol. 3, p. 1181, 1986.

51. J. A. Caird, M. D. Shinn, T. A. Kirchoff, L. K. Smith, and R. E. Wilder, "Measurements of Losses and Lasing Efficiency in GSGG:Cr,Nd and YAG:Nd Laser Rods," *Appl. Opt.*, vol. 25, p. 4294, 1986.

52. V. A. Smirnov, and I. A. Shcherbakov, "Rare-Earth Scandium Chromium Garnets as Active Media for Solid State Lasers," *IEEE J. Quantum Electron.*, vol. 24, p. 949, 1988.

53. B. M. Antipenko, A. S. Glebov, T. I. Kiseleva, and V. A. Pismennyi, "A New Spectroscopic Scheme of an Active Medium for the 2-μm Band," *Opt. Spectrosc. (USSR)*, vol. 60, p. 95, 1986.

54. G. J. Quarles, A. Rosenbaum, C. L. Marquardt, and L. Esterowitz, "Efficient Room-Temperature Operation of a Flashlamp-Pumped, Cr,Tm:YAG Laser at 2.01 μm," *Opt. Lett.*, vol. 15, p. 42, 1990.

55. T. Y. Fan and R. L. Byer, "Modeling and CW Operation of a Quasi-Three-Level 946 nm Nd:YAG Laser," *IEEE J. Quantum Electron.*, vol. 23, p. 605, 1989.

56. T. Kushida, H. M. Marcos, and G. E. Geusic, "Laser Transition Cross Section and Fluorescence Branching Ratio for Nd^{3+} in Yttrium Aluminum Garnet," *Phys. Rev.*, vol. 167, p. 289, 1968.

57. H. G. Danielmeyer and M. Blätte, "Fluorescence Quenching in Nd:YAG," *Appl. Phys.*, vol. 1, p. 269, 1973.

58. N. P. Barnes, D. J. Gettemy, L. Esterowitz, and R. E. Allen, "Comparison of Nd 1.06 μm and 1.33 μm Operation in Various Hosts," *IEEE J. Quantum Electron.*, vol. 23, p. 1434, 1987.

59. R. Beach, G. Albrecht, R. Solarz, W. Krupke, B. Comaskey, S. Mitchell, C. Brandle, and G. Berkstresser, "Q-switched Laser at 912 nm using Ground-State-Depleted Neodymium in Yttrium Orthosilicate," *Opt. Lett.*, vol. 15, p. 1020, 1990.

60. P. Heinz and A. Laubereau, "Coherent Pulse Propagation in a Nd:YLF Laser Amplifier," *Opt. Commun.*, vol. 82, p. 63, 1991.

61. W. J. Miniscalco and R. S. Quimby, "General Procedure for the Analysis of Er^{3+} Cross Sections," *Opt. Lett.*, vol. 16, p. 258, 1991.

62. E. Desurvire, "Analysis of Erbium-Doped Fiber Amplifiers Pumped in the $^4I_{15/2}$-$^4I_{13/2}$ Band," *IEEE Photonics Technol. Lett.*, vol. 1, p. 293, 1989.

63. T. Y. Fan, G. Huber, R. L. Byer, and P. Mitzscherlich, "Spectroscopy and Diode Laser-Pumped Operation of Tm,Ho:YAG," *IEEE J. Quantum Electron.*, vol. 24, p. 924, 1988.

64. H. Hemmati, "2.07 μm CW Diode-Laser-Pumped Tm,Ho:$YLiF_4$ Room-Temperature Laser," *Opt. Lett.*, vol. 14, p. 435, 1989.

65. G. J. Quarles, A. Rosenbaum, C. L. Marquardt, and L. Esterowitz, "High-Efficiency 2.09 μm Flashlamp-Pumped Laser," *Appl. Phys. Lett.*, vol. 55, p. 1062, 1989.

66. K. L. Sala, G. A. Kenny-Wallace, and G. F. Hall, "CW Autocorrelation Measurements of Picosecond Laser Pulses," *IEEE J. Quantum Electron.*, vol. 16, p. 990, 1980.

Disclaimer

CHAPTER 6
SEMICONDUCTOR LASERS

James J. Coleman

6.1 COMPOUND SEMICONDUCTORS AND ALLOYS

The best-developed semiconductor materials are the elemental semiconductors of column IV of the periodic chart, shown in Fig. 6.1. These include germanium, from which the first transistor was made, and especially silicon, which is the best for most modern commercial microelectronics and electronic integrated circuitry. These materials are perhaps most interesting when formed into *pn* junction diodes by the addition of small, controlled amounts of impurities. The *p*-type regions are formed by adding acceptors, which are column III elements (i.e., B, Al, Ga, In) and lack one electron with respect to the column IV host, and *n*-type regions are formed by adding donors, which are column V elements (i.e., P, As, Sb) and have an extra electron. The lattice structure of the semiconducting forms of the column IV elements is a tetrahedral covalent crystal with a diamond lattice as shown in Fig. 6.2. The lattice constant a_o for these crystalline materials is defined as the size along one edge of the unit cell. All semiconductor materials are characterized by a gap in energy E_g between the conduction and valence bands in the material. These materials are generally transparent to light of lower energy than the energy gap. Shown at the top of Table 6.1 are these parameters for the column IV elemental semiconductors silicon and germanium.

Binary compound semiconductors consist of two elements per unit cell forming a chemical compound. The best developed of these, and the materials of interest for semiconductor lasers, are the III–V compound semiconductors, such as GaAs, AlAs, and InP, each containing one atom from column III and one from column V. As in the case of the elemental semiconductors, the formation of *pn* junction diodes requires impurities including acceptors from the column II elements (i.e., Zn, Cd) and donors from the column VI elements (i.e., S, Se). The column IV elements (i.e., Si, Ge) are amphoteric, meaning they can occupy either a column III site or a column V site and thus may be either donors or acceptors. The lattice structure of the III–V compound semiconductors is a tetrahedral mixed ionic and covalent crystal with a zinc blende lattice, which is the same structure as the diamond lattice but with two alternating atoms. Shown in Table 6.1 are some basic parameters for the most common binary III–V compound semiconductors.

III	IV	V	VI
5 **B** 10.81	6 **C** 12.01	7 **N** 14.01	8 **O** 16.00
13 **Al** 26.98	14 **Si** 28.09	15 **P** 30.97	16 **S** 32.06

	III	IV	V	VI
30 **Zn** 65.38	31 **Ga** 69.72	32 **Ge** 72.59	33 **As** 74.92	34 **Se** 78.96
48 **Cd** 112.40	49 **In** 114.80	50 **Sn** 118.70	51 **Sb** 121.80	52 **Te** 127.60

FIGURE 6.1 Portion of the periodic chart showing the constituents of elemental and compound semiconductors and their dopants.

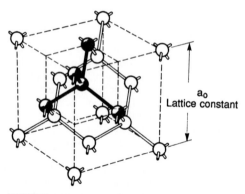

FIGURE 6.2 Schematic of the diamond lattice. *(After James F. Gibbons,* Semiconductor Electronics, *McGraw-Hill, New York, 1966, p. 55.)*

TABLE 6.1 Bandgap Energy and Lattice Constant for Several Elemental and Compound Semiconductor Materials

Material	Energy gap E_g, eV	Lattice constant a_o, Å
Si	1.11	5.431
Ge	0.65	5.658
GaAs	1.43	5.654
AlAs	2.14	5.661
InP	1.35	5.868
GaP	2.26	5.449

6.2 ENERGY BAND STRUCTURE

The principal difference between compound and elemental semiconductors lies in the detailed nature of their energy band structure. Shown in Fig. 6.3 is the momentum (k) space energy band diagram for a "direct" semiconductor material such as GaAs. In a direct semiconductor, the conduction band minimum, labeled Γ in Fig. 6.3, occurs at the same k-space position as the valence band minimum. Note that, for electrons, energy is increasing upward in Fig. 6.3, while the energy for holes increases downward. Since these band minima are at the same k-space position, no momentum transfer process (phonon emission or absorption) is required for efficient optical absorption or recombination. There are subsidiary conduction band minima at slightly higher energies in other directions in k-space, labeled X and L. In indirect semiconductors, one of these subsidiary minima at X or L is lower in energy than the conduction band minimum at Γ, and thus a phonon is required for recombination. This is an *inefficient* second-order optical process. Si and Ge are both indirect semiconductors while many of the III–V compounds are direct semiconductors and suitable for use as *efficient* light emitters and lasers.

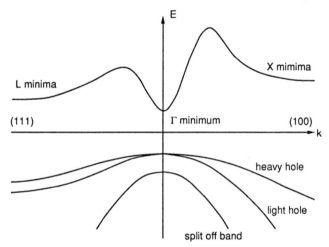

FIGURE 6.3 Momentum space energy dispersion diagram for a direct semiconductor.

In the region of k-space near $k = 0$, the conduction and valence band edges can be approximated as nearly parabolic (as shown in Fig. 6.4a) with the dispersion relation

$$E(k) = \frac{\hbar^2 k^2}{2m^*} \tag{6.1}$$

where $\hbar$ is Planck's constant and m^* is the effective mass of the electron or hole. The density per unit energy interval of electron or hole states $\rho(E)$ at energy E is given by

$$\rho(E) = \frac{1}{2\pi^2} \left(\frac{2m^*}{\hbar^2} \right)^{3/2} E^{1/2} \tag{6.2}$$

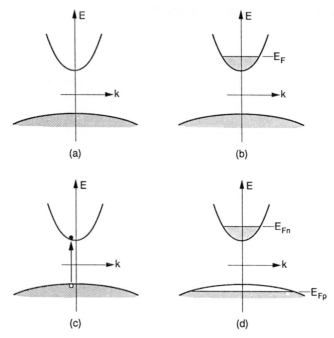

FIGURE 6.4   Conduction and valence band edges at $k = 0$ in the parabolic approximation for (a) an undoped semiconductor, (b) an n-type semiconductor, (c) the absorption process in a direct semiconductor, and (d) population inversion in a semiconductor.

The probability of occupation of any state is given by the Fermi-Dirac function

$$f(E) = \frac{1}{e^{(E-E_F)/kT} + 1} \tag{6.3}$$

where T is temperature, k is Boltzmann's constant, and E_F is the Fermi energy and is defined as the energy where the probability of occupancy is $1/2$. The number of electrons or holes at any energy is the product

$$\rho(E)f(E) \tag{6.4}$$

and the integral over energy of this product is the total density of electrons or holes in the volume. Since the only adjustable parameter is E_F, an n-type semiconductor, for example, with a large density of electrons has a Fermi energy in or near the conduction band, as shown in Fig. 6.4b.

A semiconductor at equilibrium will absorb light having an energy greater than the bandgap energy. Absorption of a photon raises an electron from the valence band to the conduction band in a direct gap semiconductor, as shown in Fig. 6.4c, leaving a hole behind. For fully ionized donors and acceptors, the density of electrons n is given by

$$n = N_d + \delta n \tag{6.5}$$

where N_d is the concentration of ionized donors, and the concentration of holes p is given by

$$p = N_a + \delta p \qquad (6.6)$$

where N_a is the concentration of ionized acceptors. In Eqs. (6.5) and (6.6), δn and δp are the concentrations of excess electrons and holes created by optical absorption or some other process. Of course, δn must equal δp. As larger concentrations (δn, $\delta p \gg N_d, N_a$) of electrons and holes are created, the absorption coefficient decreases until the semiconductor becomes transparent. At this point the population is inverted, as shown in Fig. 6.4d. The carrier concentrations δn and δp define the quasi-Fermi levels E_{Fn} and E_{Fp}. Given sufficient population inversion, stimulated emission and hence gain are present in the semiconductor and laser operation is possible. The transition energy for stimulated emission is given by

$$E = \hbar\omega = E_{Fn} - E_{Fp} \qquad (6.7)$$

where ω is the frequency of the emitted light.

A wider range of bandgap energy and lattice constant can be obtained by using compound semiconductor alloys. Alloys are solid solutions or mixtures of two or more compound semiconductors such as the ternary alloy $Al_xGa_{1-x}As$ and the quaternary alloy $In_{1-y}Ga_yAs_{1-x}P_x$. These are not compounds in the chemical sense, and the composition or mole fraction x, y varies between 0.0 and 1.0. Vegard's law says that the lattice constant of an alloy varies linearly with composition. The energy gap of an alloy varies linearly with composition but has some curvature and can also change from direct to indirect. The bandgap energy vs. composition for $Al_xGa_{1-x}As$ is shown in Fig. 6.5. Above a composition of $x \sim 0.45$ the alloy changes from direct $E_\Gamma(x)$ to indirect $E_X(x)$ with the functional dependence

$$E_\Gamma(x) = 1.424 + 1.247x \qquad (x < 0.45) \qquad (6.8)$$

$$E_X(x) = 1.9 + 0.125x + 0.143x^2 \qquad (x > 0.45) \qquad (6.9)$$

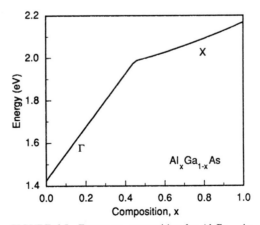

FIGURE 6.5 Energy vs. composition for $Al_xGa_{1-x}As$ showing both direct (Γ) and indirect (X) band edges.

These materials make up part of the optical waveguide in semiconductor lasers, and thus the refractive index variation with composition is also of interest and is given at 1.38 eV and room temperature by

$$n(x) = 3.59 - 0.71x + 0.091x^2 \tag{6.10}$$

6.3 HETEROSTRUCTURES

The physical form of the semiconductor laser diode is similar to most other laser systems and is shown in Fig. 6.6. There is (1) a source of power, which for the semiconductor diode laser is current, (2) an active medium, the semiconductor material, which supports stimulated emission, and (3) a resonant cavity, which is usually a Fabry-Perot cavity formed by cleaving facets along the natural parallel reflecting crystal planes. The mechanism for converting current into an inverted

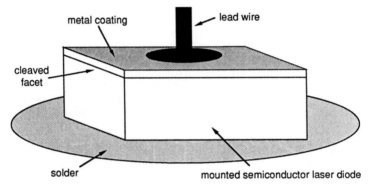

FIGURE 6.6 Schematic diagram for a semiconductor laser diode soldered to a mount and showing the cleaved facets, metal coatings, and lead wire.

carrier population is injection across a forward-biased *pn* electrical junction. The first semiconductor laser diodes were GaAs homostructure lasers formed by diffusing a *p*-type region into an *n*-type GaAs substrate as shown in Fig. 6.7*a*. At zero bias, there is no current flow and electrons and holes are separated by a potential step in the conduction and valence bands. Under forward bias, electrons are injected into the *p*-type GaAs and holes are injected into the *n*-type GaAs forming a region near the *pn* junction, defined by the minority carrier diffusion lengths of the carriers, with a large density of both electrons and holes. If the injected carrier density becomes large enough to reach transparency and overcome the additional losses in the system, which typically is possible only at cryogenic temperatures for homostructure lasers, then laser emission can be obtained.

The development of modern epitaxy, which is the growth of a thin layer of single-crystal material on a similar single-crystal substrate but having different doping or composition, has allowed design of heterostructure lasers. A single heterostructure $Al_xGa_{1-x}As$-GaAs laser diode is shown schematically in Fig. 6.7*b*.

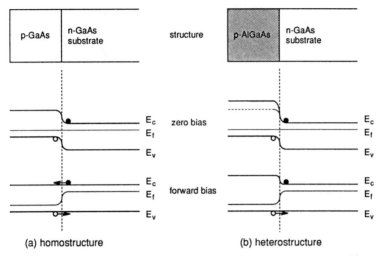

FIGURE 6.7 Structure and band diagrams at zero bias and forward bias for (*a*) the homostructure laser and (*b*) the heterostructure laser.

The $Al_xGa_{1-x}As$ has a larger bandgap energy than GaAs with the heterostructure discontinuity distributed between the conduction and valence bands according to

$$\Delta E_v = 0.35 \, \Delta E_g \tag{6.11}$$

$$\Delta E_c = \Delta E_g - \Delta E_v \tag{6.12}$$

Under forward bias, electrons are confined by the heterostructure discontinuity to the narrower gap *n*-type GaAs while holes are injected into the GaAs to a distance of approximately one diffusion length (1 μm) and recombination takes place on only one side of the junction. This results in a smaller active volume and, correspondingly, a higher injected carrier density for a given current.

There are design constraints associated with heterostructures. The lattice constant of thick epitaxial layers and the substrate must be nearly the same to avoid formation of misfit dislocations ($\Delta a/a \leq 0.2$ percent). An important exception is the strained layer laser structure in which the mismatch strain can be accommodated elastically if the layers are sufficiently thin. Variations in composition result in variations in bandgap energy, index of refraction, and lattice constant. The best-developed semiconductor laser materials are $Al_xGa_{1-x}As$, which is essentially lattice matched to GaAs for any composition, and the quaternary alloy $In_{1-y}Ga_yAs_{1-x}P_x$, which has an additional degree of freedom and can be lattice matched to InP over a wide range of compositions.

6.4 DOUBLE HETEROSTRUCTURE LASER

The double heterostructure laser, shown in Fig. 6.8, consists of a thin GaAs active layer, typically less than 0.1 μm, sandwiched between $Al_xGa_{1-x}As$ layers. The $Al_xGa_{1-x}As$ confining (or cladding) layers have wider bandgap energy and are

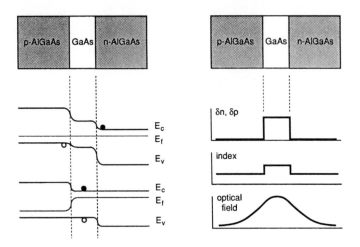

FIGURE 6.8 Structure, band diagrams at zero bias and forward bias, injected carrier profile, index of refraction profile, and optical field profile for the double heterostructure laser.

transparent, and thus exhibit low losses, at the wavelength of laser emission. Under forward bias, injected electrons and holes are confined to a much smaller active volume and their density is much greater for a given current than in a homostructure or single heterostructure laser. In addition, the $Al_xGa_{1-x}As$ has a smaller index of refraction than GaAs, as shown in Fig. 6.8, and the resulting large index step forms a very effective waveguide with a significant portion of optical field extending into confining layers and propagating without significant loss. The double heterostructure laser was the first semiconductor laser to operate at room temperature and laser operation, with threshold current densities of less than 800 A/cm² typical.

The threshold current density J_{th} is defined as the point where, after there is sufficient population inversion for transparency, the gain from stimulated emission exceeds losses from absorption and the ends of the cavity. Equating the gain and losses gives the modal gain $\gamma_t\Gamma$

$$\gamma_t\Gamma = \alpha + \frac{1}{2L} \ln\left(\frac{1}{R_1R_2}\right) \qquad (6.13)$$

where γ_t is the gain (cm⁻¹) at threshold, α is the sum of several different internal loss mechanisms (~5 to 10 cm⁻¹), L is the laser cavity length, and R_1 and R_2 are the reflectivities of the facets (~0.3). Γ is the confinement factor, which is the fraction of the optical mode in the active region d of the waveguide

$$\Gamma = \frac{\displaystyle\int_{-d/2}^{d/2} |E|^2 \, dx}{\displaystyle\int_{-\infty}^{\infty} |E|^2 \, dx} \qquad (6.14)$$

The typical form for a light-current characteristic of a semiconductor laser diode is shown in Fig. 6.9. In the linear region above threshold, the gain γ is given by

$$\gamma = \beta(J - J_0) \tag{6.15}$$

where β is the gain coefficient (cm/A) and J_0 is the current density necessary to reach transparency and is a function of the band structure and effective masses in the active layer. Combining Eqs. (6.13) and (6.15), including the internal quantum efficiency η_i in the semiconductor, and rearranging gives the threshold current density J_{th}

$$J_{th} = \frac{J_0}{\eta_i} + \frac{\alpha}{\eta_i \beta \Gamma} + \frac{(1/2L)\ln(1/R_1 R_2)}{\eta_i \beta \Gamma} \tag{6.16}$$

For AlGaAs-GaAs double heterostructure lasers, the dependence of J_{th} on active region thickness is typically $J_{th}/d \sim 3$ to 5 kA/cm²-μm. The external differential quantum efficiency η (Fig. 6.9) above threshold is given by

$$\eta = \frac{1}{E} \frac{\partial P_0}{\partial I} \tag{6.17}$$

where P is the optical power, I is the current, and E is the emission energy given by

$$E = \hbar\omega = \frac{hc}{\lambda} = \frac{1.239852 \text{ eV-μm}}{\lambda} \tag{6.18}$$

where λ is the emission wavelength. The power generated internally by stimulated emission is

$$P_{total} = \frac{(I - I_{th})\eta_i \hbar w}{q} \tag{6.19}$$

where η_i is the internal quantum efficiency. The power emitted from the facets of the laser P_o is

$$P_o = P_{total} \frac{(1/2L)\ln(1/R_1 R_2)}{\alpha + (1/2L)\ln(1/R_1 R_2)} \tag{6.20}$$

while the external differential quantum efficiency η is given by

$$\eta = \eta_i \frac{(1/2L)\ln(1/R_1 R_2)}{\alpha + (1/2L)\ln(1/R_1 R_2)} \tag{6.21}$$

For many semiconductor lasers η_i can approach unity.

The optical spectrum of a semiconductor laser depends on the gain profile of the medium and the nature of modes in the cavity. The spontaneous emission profile below laser threshold depends on both the absorption spectrum and the occupancy defined by the electron and hole Fermi-Dirac functions, as shown in Fig. 6.10. Above threshold the spontaneous emission spectrum narrows and the laser spectrum is a convolution of the relatively broad emission spectrum and the Fabry-Perot cavity modes as shown in Fig. 6.11. The spacing of the cavity modes is inversely proportional to cavity length.

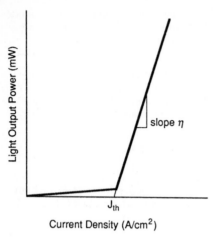

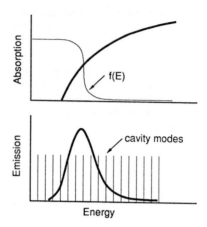

FIGURE 6.9 Light output power vs. current for a semiconductor laser diode.

FIGURE 6.10 Absorption (with the Fermi function) and emission (with cavity modes) spectra for a semiconductor laser.

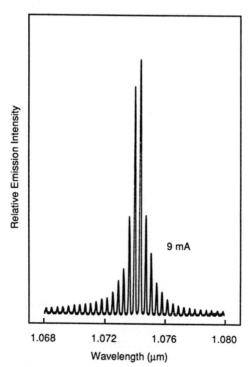

FIGURE 6.11 Typical emission spectrum for a semiconductor laser.

6.5 STRIPE GEOMETRY LASERS

Even the smallest practical broad area double heterostructure lasers require large currents to reach laser threshold. For example, a typical cavity length is ~300 μm and the minimum practical laser width is ~150 μm; thus I_{th} = (800 A/cm^2)(300 × 10^{-4} cm)(150 × 10^{-4} cm) = 360 mA. This is an impractical drive current level. Reasonable drive circuitry and heat sinking require I_{th} < 50 mA, so some method for reducing the area is necessary. The cavity length cannot be greatly reduced without increasing the threshold current density [Eq. (6.13)]; thus reduction of width is the only choice.

The oxide-defined stripe laser structure is formed by first depositing SiO$_2$ over the top surface of the laser wafer and then forming patterned stripes by conventional photolithography and etching. The structure for an oxide-defined stripe laser is shown in Fig. 6.12. The insulating oxide precludes current flow except within the stripe, and typical stripe widths are from 1 to 12 μm or occasionally larger. The

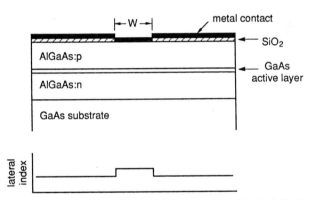

FIGURE 6.12 Schematic cross section and lateral effective index for the oxide-defined stripe laser.

advantages of oxide-defined stripe lasers are that the top surface area is greatly reduced and is precisely defined by the photoprocess. The disadvantages are that SiO$_2$ is a poor thermal conductor, affecting heat dissipation, and that the effective width of the stripe can be much greater than the patterned width, for narrow stripe lasers, because of current spreading in the various layers of the semiconductor heterostructure.

The effective lateral index of refraction of the three-layer slab waveguide is slightly larger under the stripe, as shown in Fig. 6.12, because of the contribution of gain to the complex index in the active layer. Thus there is a net focusing of the laser optical field in the lateral direction and the oxide-defined stripe laser is considered to be gain-guided. Unfortunately, both the magnitude and the width of the lateral effective index step are strong functions of drive current, and a stable emission pattern over a wide range of drive current is unobtainable. The lateral effective index step is typically small and often greater than 10 μm in width, while the transverse index step is usually quite large with an active layer thickness of less than 0.1 μm. This leads to large astigmatism in the near field and its far-field transform. The relationship between the near-field emission pattern of a semiconductor laser, defined as output power vs. position at the laser facet, and the far-field emission pattern, defined as output power vs. angle of radiation at a distance from the facet, is shown schematically in Fig. 6.13. Examples of typical two-di-

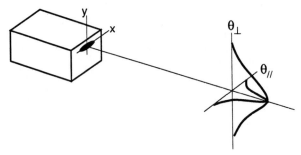

FIGURE 6.13 Schematic diagram showing asymmetry in the near-field and far-field patterns of a semiconductor laser diode.

mensional near-field patterns and corresponding lateral far-field patterns for an oxide-defined stripe laser at two drive levels are shown in Fig. 6.14. The optical field is often in the fundamental mode over a narrow range of currents near laser threshold (upper figures in Fig. 6.14) and then operates in higher-order lateral modes (lower figures in Fig. 6.14) at higher currents.

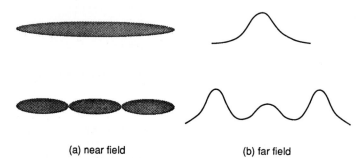

(a) near field   (b) far field

FIGURE 6.14 Optical (*a*) near-field and (*b*) far-field patterns for a laser operating in the fundamental lateral mode (upper) and on a higher-order lateral mode (lower).

6.6 INDEX-GUIDED STRIPE GEOMETRY LASERS

For many applications, the relatively small and variable (with drive current) gain-related step in the lateral index of refraction in gain-guided lasers and the related variable optical properties, such as pulse response, spectra, near-field patterns, and far-field patterns, may be undesirable. Much greater stability in these optical properties can be obtained by designing a structure with a large, built-in, lateral effective index step that is independent of the drive current.

An important example of a common index-guided laser structure is the ridge waveguide laser shown in Fig. 6.15. After epitaxial growth of the structure, narrow mesa stripes are patterned by conventional lithography and etched with wet chemicals or by one of the dry etching methods to near (<0.25 μm), but not through, the active region. The optical field outside the stripe is distorted by the proximity

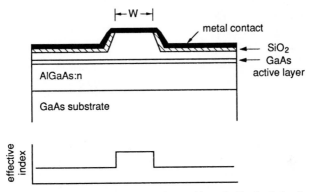

FIGURE 6.15 Schematic cross section and lateral effective index for the ridge waveguide laser.

of the etched surface, oxide, and metallization, resulting in a relatively large step in the lateral effective index. Since the index step is structural, the lateral waveguide that is formed is independent of the drive current and, if the strip width and index step are appropriate with respect to the wavelength, fundamental lateral mode operation can be maintained over a wide range of drive current. In addition, the stripe width necessary for fundamental mode operation is usually only 1 to 3 μm, resulting in threshold currents of less than 20 mA for typical double heterostructure ridge waveguide lasers.

The buried heterostructure (BH) laser shown in Fig. 6.16 is perhaps the most strongly index-guided semiconductor heterostructure laser structure. This laser

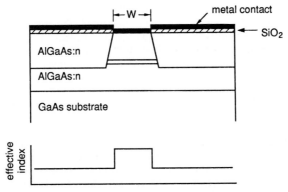

FIGURE 6.16 Schematic cross section and lateral effective index for the buried heterostructure laser.

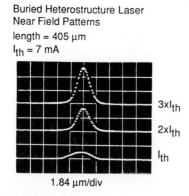

Buried Heterostructure Laser
Near Field Patterns
length = 405 μm
I_{th} = 7 mA

$3 \times I_{th}$

$2 \times I_{th}$

I_{th}

1.84 μm/div

FIGURE 6.17 Fundamental mode near-field patterns at 1, 2, and 3 × I_{th} for a quantum well buried heterostructure laser.

structure is formed by patterning stripes and etching similar to the ridge waveguide laser except that the structure is etched completely through the GaAs active layer. After the processing is completed, an additional wide gap layer is grown outside the stripe region so that the active layer is surrounded by lower index, wider bandgap material. The regrowth layer provides both very strong lateral index guiding and current confinement, since the active layer is now discontinuous. Shown in Fig. 6.17 are the fundamental mode near-field patterns for a buried heterostructure laser at 1, 2, and 3 times laser threshold. A stable fundamental mode near-field pattern is often observed to more than thirty times laser threshold for these structures. Although the buried heterostructure has very low threshold currents and excellent stable optical properties, the complicated processing and epitaxial regrowth requirement are limitations.

6.7 MATERIALS GROWTH

An essential part of any heterostructure laser is the formation of high-optical-quality epitaxial layers having the desired composition and doping. The original and simplest method for obtaining heterostructure laser material is liquid-phase epitaxy (LPE). For this growth method, a graphite slider boat containing a GaAs substrate in the slider and a liquid gallium melt for each layer to be grown is inserted into a furnace at $T > 800°C$ in a hydrogen gas ambient. Each melt is saturated with As and also contains Al or dopants as appropriate, and as the furnace is slowly cooled, the GaAs substrate is slid under each melt in turn for sufficient time to deposit a layer of the desired thickness. The sliding action, tight tolerances, and surface tension of the liquid gallium act to wipe off the liquid between one melt and the next. At first contact, the melt slightly etches the previous grown layer leading to graded heterostructure interfaces. The process is inexpensive and effective on a small scale, but the number of possible layers is limited as a practical matter and scale-up to commercial volume is not feasible.

A more recent epitaxial growth method suitable for the growth of a large number of thin layers is molecular beam epitaxy (MBE). For this method, shuttered effusion cells containing elemental Ga, As, Al, and dopants are connected to an ultrahigh-vacuum chamber and maintained at an elevated temperature suitable for evaporation. A GaAs substrate is mounted and inserted in the chamber, through a vacuum sealed load lock, and raised to the desired growth temperature. A molecular beam of As (or As_2 or As_4) is used to stabilize the surface while the Al, Ga, and dopant source shutters are opened or closed for growth of the appropriate layer. Growth rate, composition, and doping levels are determined by the beam fluxes which are controlled by the effusion cell temperatures. Real-time diagnostics, such as high-energy electron diffraction, are available for in situ monitoring. Abrupt interfaces are possible and very-high-quality GaAs can be grown, but refilling sources can be a problem and scaling to production levels is difficult.

In terms of both control and extension to the production environment, vapor-phase processes are most desirable. A vapor-phase epitaxial growth process that has proved nearly ideal for semiconductor lasers is metalorganic chemical vapor deposition (MOCVD). For this epitaxial growth method, the sources are refrigerated bubblers containing liquid metal alkyls such as trimethylgallium (TMGa) and trimethylaluminum (TMAl) with flowing hydrogen gas for transport and gaseous hydrides, such as arsine, in cylinders. The flows are controlled electronically and valves operated automatically, all under computer control. The gas mixture is transported to the vicinity of a heated susceptor holding a GaAs substrate where a pyrolysis reaction takes place:

$$(1 - x)(CH_3)_3Ga + x(CH_3)_3Al + AsH_3 \longrightarrow Al_xGa_{1-x}As + 3CH_4 \quad (6.22)$$

Growth rate, composition, and doping level are determined by the bubbler temperatures and gas flow rates. Abrupt interfaces are possible and very high optical quality AlGaAs can be grown. Refilling of source materials is simple and scaling to production volume is straightforward. The gaseous hydrides are toxic, however, and require special handling. In addition, real-time diagnostics are unavailable because of the relatively high operating pressure, between 0.1 and 1 atm.

A number of growth and structural parameters are common to semiconductor lasers and independent of growth method. The ideal growth temperature depends on alloy composition with higher Al fractions requiring higher growth temperature. In general, the higher the confining layer composition, the lower the laser threshold current density. The lower the background doping the better, but it is not a strong factor since international doping levels are relatively large, in the high 10^{17} cm^{-3} to low 10^{18} cm^{-3} range. Generally the active layer is not intentionally doped and cap layers for contacts must be heavily doped for low resistance. Abrupt interfaces are most important for quantum well heterostructure lasers. The confining layer thickness, typically 1.0 to 2.0 μm, is determined by the extent to which the optical field extends into confining layers.

Much of the discussion thus far has utilized the $Al_xGa_{1-x}As$-GaAs heterostructure system as an example. In order to access a wavelength range greater than that available from the $Al_xGa_{1-x}As$-GaAs heterostructure system, other III-V compound materials, especially In compounds, must be considered. Indium-compound semiconductor lasers, especially $In_{1-y}Ga_yAs_{1-x}P_x$ lasers, are nearly as well developed and have, in addition, important practical applications. Most characteristics of laser operation are similar to those of the $Al_xGa_{1-x}As$-GaAs heterostructure system. The lattice constant and energy gap of $In_{1-y}Ga_yAs_{1-x}P_x$ are independently variable within limits, and a lattice match is possible to either InP or GaAs substrates. In $In_{1-y}Ga_yAs_{1-x}P_x$-InP long-wavelength lasers are suitable for 1.3 to 1.5-μm low-loss, minimum-dispersion optical fiber systems. $In_{1-y}Ga_yAs_{1-x}P_x$GaAs short-wavelength lasers are suitable for visible emission and plastic fibers. These materials have the advantage that aluminum and its deleterious reactions with oxygen are avoided, but compositional control requirements are more stringent for lattice matching, and the more sophisticated materials growth technologies (MBE, MOCVD) are not as well developed for In compounds.

6.8 QUANTUM WELL HETEROSTRUCTURE LASERS

When the size of the active layer of a semiconductor heterostructure laser is made smaller than an electron wavelength (~200 Å), quantum size effects become im-

portant. The behavior of electrons and holes in the active layer becomes similar
to the particle-in-a-box problem of modern physics. Confined particle states exist
at energies above the bottom of the quantum well, determined by the energy
difference between the well and the confining layers (ΔE_c and ΔE_v), the effective
mass of the particle, and the well width L_z. The bulk parabolic density of states
function described above becomes quantized, as shown in Fig. 6.18a, into a two-
dimensional steplike constant function, independent of energy, and given by

$$\rho(E) = \frac{1}{2\pi^2} \frac{2m^*}{\hbar^2} \frac{\pi}{L_z} \tag{6.23}$$

The single quantum well laser (SQW), which is simply a double heterostructure
laser with a much smaller active layer, has the energy band structure shown in Fig.
6.18b. As in the double heterostructure laser, the discontinuity is split between the
conduction band and the valence band. A quantum state is formed in the conduction
band and separate states are formed in the valence band for the heavy and light

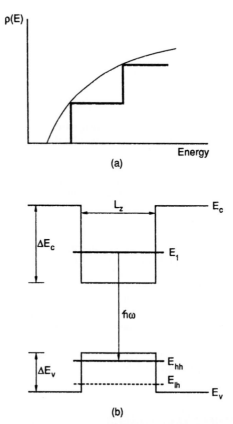

(a)

(b)

FIGURE 6.18 (a) Quantized density of states diagram and (b) energy band diagram for a quantum
well heterostructure laser.

holes. Laser transitions occur between the different quantum states in consideration of the usual quantum-mechanical selection rules.

Quantum well heterostructure lasers were made possible by the modern epitaxial growth methods and require very high quality material and abrupt heterostructure interfaces. The emission wavelength for these lasers is a complex function of size in addition to composition. They have the distinct advantages, however, of very low threshold current density and high efficiency. The transparency current [current necessary to reach population inversion, Eq. (6.16)] scales with total active volume to the quantum limit and can be much smaller than for a conventional double heterostructure laser. The gain coefficient γ can be higher for quantum well heterostructure lasers but the confinement factor Γ is usually much lower, so the modal $\gamma\Gamma$ gain depends on these offsetting effects.

Very thin single well structures are characterized by increasing threshold current density as size is reduced below 100 Å unless the cavity is made unusually long or high-reflectivity coatings are employed. This phenomenon results from the fact that as $L_z \to 0$, the transparency current $J_o \to$ minimum, but $\Gamma \to 0$, causing the threshold current density [Eq. 6.16] to blow up. The solution is to design a structure to minimize these problems. For example, the multiple quantum well laser (MQW) is formed by inserting an increased number of wells, each separated by thin barrier layers, as shown in Fig. 6.19a. Tunneling through the thin barriers allows effective carrier relaxation into the quantum state. Many of the advantages of the single quantum well laser are retained. The active volume is increased, increasing the transparency current density, but the modal gain is also increased since the confinement factor Γ increases. For m wells

$$(\gamma\Gamma)_{MQW} \sim m(\gamma\Gamma)_{SOW} \qquad (6.24)$$

In a multiple quantum well structure, the quantum state energy is somewhat broadened and the index of refraction in the active region is determined by the average composition. In addition, multiple quantum well lasers have a larger differential gain near threshold, which is attractive for modulation.

Another important engineered quantum well heterostructure laser is the graded index separate confinement heterostructure laser (GRIN-SCH), shown in Fig. 6.19b. This structure is formed by growing a compositionally graded region on either side of the well. By shaping the index of refraction in the graded region, a

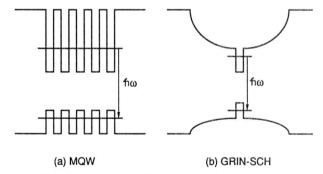

(a) MQW (b) GRIN-SCH

FIGURE 6.19 Energy band diagrams for (a) the multiple quantum well laser and (b) the graded index separate confinement heterostructure quantum well laser.

much higher confinement factor can be obtained while retaining the low-transparency current density of a single thin well structure. The shape of the graded region is not critical, and low threshold lasers have been obtained with simpler, step-graded structures as well. $Al_xGa_{1-x}As$ GRIN-SCH structures with 50-Å GaAs wells have exhibited laser threshold current densities of less than 100 A/cm².

Formation of a combination of low threshold quantum well heterostructure lasers in an index-guided, lateral-waveguide structure, such as the buried heterostructure laser described above, yields very low threshold current, high-efficiency laser diodes with excellent optical field characteristics. Shown in Fig. 6.20 is the light-current characteristic for a quantum well buried heterostructure laser diode. This laser, which is the same device as that in Fig. 6.17, has a very low threshold current (<10 mA), very high external quantum efficiency (>60 percent), and stable fundamental optical modes to high output power (~100 mW per uncoated facet).

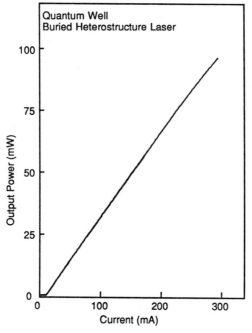

FIGURE 6.20 Light-current characteristic for an uncoated quantum well buried heterostructure laser.

6.9 LASER ARRAYS

For most of the gain-guided and index-guided laser structures described above it is possible to form multiple element arrays with the same processing technology. Simple arrays, consisting of a large number of decoupled elements distributed along a bar as shown in Fig. 6.21, can be utilized to obtain higher total output power than is available from a single element. Electrically, the structure is simply a number

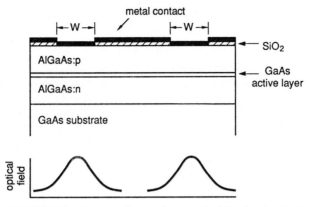

FIGURE 6.21 Schematic diagram for an oxide-defined stripe simple, uncoupled laser array.

of identical diodes in parallel and the optical field pattern is a distributed array of identical single elements. If the laser elements are placed sufficiently close to each other to allow some overlap and sufficient coupling of the optical fields, as shown in Fig. 6.22, a fixed phase relationship is forced between each emitter, and a phase-locked array can result. The extent of phase locking and the phase relationship depend on the structure, spacing, and the resulting lateral effective index of refraction profile.

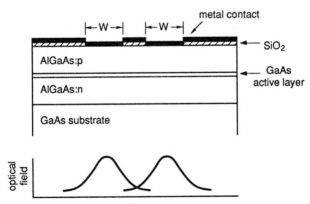

FIGURE 6.22 Schematic diagram for an oxide-defined stripe phase-locked laser array.

In addition to the optical properties of the individual elements, the array itself lends optical properties, known as array supermodes, to the overall structure. These modes arise from the superposition of the fields for N identical weakly coupled emitters. The far-field pattern $F(\phi)$ for the array is the superposition of the far-field pattern for each emitter $|E(\phi)|^2$ and the interference function $G(\phi)$ for the array and is given by

$$F(\phi) = \cos^2(\phi) \, |E(\phi)|^2 \, G(\phi) \tag{6.25}$$

The $\cos^2(\phi)$ term in Eq. (6.25) is near unity for most semiconductor laser arrays. The solution for the interference function allows N array modes from $L = 1$ ($0°$ phase shift) to $L = N$ ($180°$ phase shift). The important parameters in solving for Eq. (6.25) are the wavelength, the center-to-center spacing S, the mode number L, the number of elements N, and the shape of the profile for the individual emitters. The calculated multilobed intensity distributions for a 10-element array for both modes $L = 1$ and $L = 10$ are shown in Fig. 6.23 plotted against the envelope, which is the intensity distribution for a single emitter. The angular separation between peaks in the far-field pattern is given by

$$\Delta\phi = 2 \sin^{-1}\left(\frac{\lambda}{2D}\right) \tag{6.26}$$

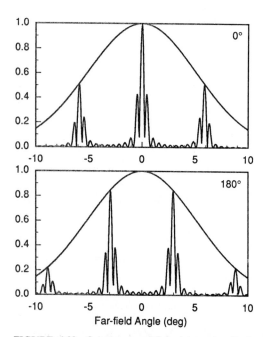

FIGURE 6.23 Calculated multilobed intensity distributions for a 10-element array for both in-phase and out-of-phase modes. The envelope function is the intensity distribution for a single emitter.

Out-of-phase operation ($L = N$) generally dominates since most semiconductor diode laser arrays are unpumped, and thus lossy, in the region between the stripes. Experimental far-field patterns at four drive currents for a 10-element ridge waveguide quantum well heterostructure laser array operating in $180°$ out-of-phase locked mode are shown in Fig. 6.24.

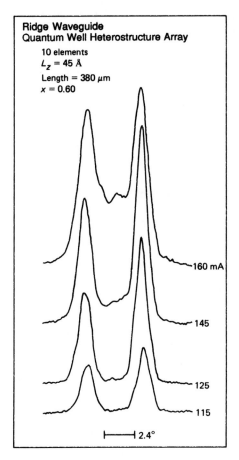

FIGURE 6.24 Far-field emission patterns at four drive current levels for a 10-element ridge wave-guide quantum well heterostructure laser array operating in the 180° out-of-phase mode.

6.10 MODULATION OF LASER DIODES

The simplest and most direct method for modulating a semiconductor laser diode is modulation of the laser drive current, as shown in Fig. 6.25. Direct-current modulation is best described by rate equations for both carriers and photons. If N is the injected electron and hole density and P is the photon density, then the rate equation for carriers is

$$\frac{dN}{dt} = \frac{J}{qd} - \frac{N}{\tau_s} - A(N - N_{\mathrm{tr}})P \qquad (6.27)$$

where d is the active region thickness, τ_s is the spontaneous recombination lifetime, N_{tr} is the transparency carrier density, and A is a constant related to the gain

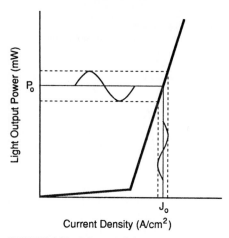

FIGURE 6.25 Schematic light-current characteristic showing direct-current modulation in a semiconductor laser diode.

coefficient. The term J/qd in Eq. (6.27) is the production of carriers, the term N/τ_s is the carriers lost to spontaneous recombination, and the term $A(N - N_{tr})P$ accounts for recombination resulting in stimulated emission. The rate equation for photons is

$$\frac{dP}{dt} = \Gamma A(N - N_{tr})P - \frac{P}{\tau_p} \tag{6.28}$$

where τ_p is the photon lifetime and the term P/τ_p accounts for mirror and internal losses.

For small signal current modulation at frequency ω, the current density J, carrier density N, and photon density P can be approximated by

$$J = J_0 + \Delta J e^{i\omega t}$$

$$N = N_0 + \Delta N e^{i\omega t}$$

$$P = P_0 + \Delta P e^{i\omega t} \tag{6.29}$$

Substitution of Eq. (6.29) into Eqs. (6.27) and (6.28) gives the modulation frequency response

$$\frac{\partial P}{\partial J} = \frac{-(1/qd)\Gamma A P_0}{\omega^2 - i\omega/\tau_s - i\omega A P_0 - A P_0/\tau_p} \tag{6.30}$$

which is shown for two output power levels in Fig. 6.26. The frequency response is flat to frequencies approaching 1 GHz, rises to a peak value at some characteristic frequency, and then quickly rolls off. The peak in the frequency response is given approximately by

$$\omega_{max} \approx \left(\frac{A P_0}{\tau_p}\right)^{1/2} \tag{6.31}$$

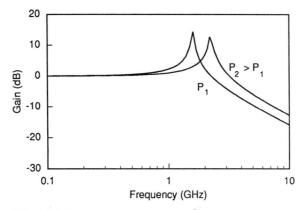

FIGURE 6.26 Small signal frequency response of a semiconductor laser diode at two power levels.

Since A is related to the differential gain of the laser, high-frequency operation requires greater differential gain, greater photon density, and decreased photon lifetime.

Modulation of the laser drive current J modulates the injected carrier density N which, in turn, results in a modulation of the quasi-Fermi levels and the carrier-dependent index of refraction. Since the emission wavelength, and hence frequency, is related to the difference between the quasi-Fermi levels and the index, modulation of the carrier density results in modulation of the emission frequency. Thus amplitude modulation (AM) of the laser power output results in a corresponding frequency modulation (FM). The change (and broadening) of the emission frequency is called "chirp."

6.11 RELIABILITY

The failure mechanisms for semiconductor lasers can be separated into two relatively broad categories. The first category is catastrophic failure associated with exceeding the maximum safe value of some operating parameter. Perhaps the most important example is catastrophic facet damage. When the optical power density at the laser facet reaches a certain value, catastrophic optical damage (COD) occurs. COD for a given material system is a function of the shape of the near-field pattern, the drive current amplitude and pulse length, and the presence or absence of passivation on the facets. COD occurs rapidly and irreversibly at the upper limit for power output of the laser.

If higher power output is required or if the double-ended emission characteristic of a cleaved Fabry-Perot cavity is undesirable, facet coatings allow higher output power from one facet of a given laser structure or array. After the semiconductor laser diode material is completely processed, except for dividing the cleaved bars into individual laser dice, dielectric facet coatings, as shown in Fig. 6.27, can be applied. Typical facet coatings consist of a single-layer dielectric antireflection (AR) coating on one end and a multilayer stacked dielectric, typically two periods of Al_2O_3-Si, for high reflection (HR) on the other end. By adjusting the product of

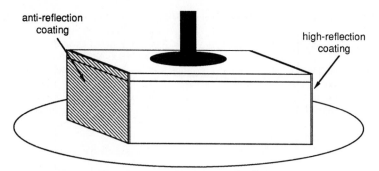

FIGURE 6.27 Schematic diagram showing antireflection (AR) and high-reflection (HR) coatings on the facets of a semiconductor laser diode.

R_1 and R_2 to be the same as for an uncoated laser device, the threshold current density and related properties, such as the emission wavelength, remain the same [see Eq. (6.16)]. The power output, however, is unbalanced toward a single facet. Shown in Fig. 6.28 is the reflectivity on the HR-coated facet as a function of the reflectivity on the AR-coated facet, assuming a constant product $R_1 R_2$. Also shown in Fig. 6.28 is fraction of the total output from the AR-coated facet as a function of the reflectivity on the facet.

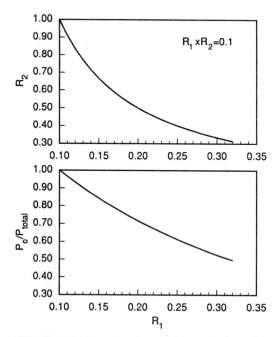

FIGURE 6.28 The reflectivity of the HR-coated facet and the fraction of the total power output from the AR-coated facet as a function of the reflectivity of the AR-coated facet, assuming a constant product $R_1 R_2$.

The second general category of semiconductor laser diode failure is gradual degradation resulting from long-term effects related to materials, such as defects or contaminants, or processing such as handling damage. An estimation of the long-term reliability and mean time between failure (MTBF) of a particular laser design is an important part of commercial laser development, especially for remote or space-based laser systems. The measurement utilized to establish reliability parameters include current constant power measurements, where device failure is defined as the laser output power falling to half its initial value, or constant output power current adjustment, where device failure is defined as the laser drive current rising to twice its initial value.

As semiconductor laser lifetimes have approached those of other solid-state electronic devices, measured in tens of thousands of hours, accelerated life testing methods have become important. By measuring the failure rate or median lifetime as a function of temperature above room temperature (typically 30 to 70°C), as shown in Fig. 6.29, an activation energy E_a can be determined from the slope of the best-fit line. This activation energy allows a preliminary analysis of the dominant failure mechanism, since certain failure mechanisms have characteristic activation energies, and extrapolation of a room-temperature lifetime.

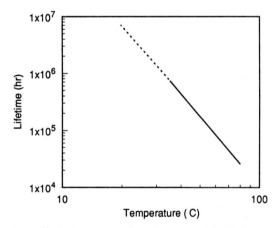

FIGURE 6.29 An example of the laser median lifetime as a function of operating temperature. The dashed line is an extrapolation to room temperature.

6.12 REFERENCES

J. I. Pankove, *Optical Processes in Semiconductors*, Dover, New York, 1971.

C. H. Gooch, *Injection Electroluminescent Devices*, Wiley, New York, 1973.

H. Kressel and J. K. Butler, *Semiconductor Lasers and Heterojunction LEDs*, Academic Press, New York, 1977.

H. C. Casey and M. B. Panish, *Heterostructure Lasers, Part A: Fundamental Principles*, Academic Press, New York, 1978.

H. C. Casey and M. B. Panish, *Heterostructure Lasers, Part B: Materials and Operating Characteristics*, Academic Press, New York, 1978.

J. K. Butler, *Semiconductor Injection Lasers*, IEEE Press, New York, 1979.

G. H. B. Thompson, *Physics of Semiconductor Laser Devices*, Wiley, New York, 1980.

T. P. Pearsall (ed.), *GaInAsP Alloy Semiconductors*, Wiley, New York, 1982.

H. Kressel (ed.), *Topics in Applied Physics*, vol. 39, *Semiconductor Devices for Optical Communication*, Springer-Verlag, Berlin, 1982.

R. G. Hunsberger, *Integrated Optics: Theory and Technology*, 2d ed., Springer-Verlag, Berlin, 1984.

D. K. Ferry (ed.), *Gallium Arsenide Technology*, H. W. Sams, Indianapolis, 1985.

W. T. Tsang (ed.), *Semiconductors and Semimetals*, vol. 22, *Lightwave Communications Technology*, Part A, *Material Growth Technologies*, Academic Press, New York, 1985.

W. T. Tsang (ed.), *Semiconductors and Semimetals*, vol. 22, *Lightwave Communications Technology*, Part B, *Semiconductor Injection Lasers* I, Academic Press, New York, 1985.

W. T. Tsang (ed.), *Semiconductors and Semimetals*, vol. 22, *Lightwave Communications Technology*, Part C, *Semiconductor Injection Lasers* II, Academic Press, New York, 1985.

G. P. Agrawal and N. K. Dutta, *Long Wavelength Semiconductor Lasers*, Van Nostrand Reinhold, New York, 1986.

S. E. Miller and I. P. Kaminow (eds.), *Optical Fiber Telecommunications* II, Academic Press, New York, 1988.

J. T. Verdeyen, *Laser Electronics*, 2d ed., Prentice-Hall, Englewood Cliffs, N.J., 1989.

J. Wilson and J. F. B. Hawkes, *Optoelectronics, An Introduction*, 2d ed., Prentice-Hall, Englewood Cliffs, N.J., 1989.

A. Yariv, *Quantum Electronics*, 3d ed., Wiley, New York, 1989.

P. K. Cheo (ed.), *Handbook of Solid State Lasers*, Marcel Dekker, New York, 1989.

B. G. Streetman, *Solid State Electronic Devices*, 3d ed., Prentice-Hall, Englewood Cliffs, N.J., 1990.

W. Streifer and M. Ettenberg (eds.), *Semiconductor Diode Lasers*, vol. I, IEEE Press, New York, 1990.

CHAPTER 7
INFRARED GAS LASERS

Michael Ivanco and Paul A. Rochefort

7.1 INTRODUCTION

This chapter deals with infrared (ir) gas lasers and their applications. These lasers were the first ones to make significant inroads in industry, and CO_2 lasers, in particular, account for about two-thirds of all industrial laser sales,[1] which comprise the largest share of the laser market. IR gas lasers, which are exclusively molecular lasers,† are unique in that they combine high electric to photonic energy conversion efficiency with high average power. This makes them the easiest lasers to fabricate, and because the active medium is a gas, it is also much easier to scale them to higher power. Indeed, CO_2 lasers in the 2- to 5-kW range are now commonplace in the automobile manufacturing industry.[2]

Recently, CO_2 lasers with average powers of 50 kW have been constructed,[3] and powers of 100 kW appear attainable. No commercial lasers of this size have been constructed to date, however, because a market for them does not yet exist. By contrast, Nd:YAG lasers are not expected to exceed a few kilowatts of average power, although they have advantages in fiber-optic beam delivery. Excimer lasers have much higher efficiencies than Nd:YAG lasers and may some day rival CO_2 lasers in terms of average power; but none have been built that exceed 1 kW.

IR gas lasers are, simply, lasers in which the active medium is a molecular gas, and which lase on vibrational-rotational transitions within the same electronic state. The photon energy of these lasers is therefore limited by the vibrational frequency that a molecule can attain; hence they produce radiation in the infrared. Since the population inversion is typically between pairs of rotational levels, in different vibrational states, the lasers are also line-tunable. CO_2 lasers, for example, can lase on over 200 lines in the 9- to 18-μm range.

In the remainder of this chapter, ir laser resonator theory and design considerations are outlined in Sec. 7.2. The operating principles, characteristics, and applications of specific lasers are discussed in Sec. 7.3, with particular emphasis on CO_2, HF/DF, and CO lasers, which are, by far, the most important ir gas lasers.

---

†Strictly speaking, there are near-ir lines of several neutral and ionic lasers. HeNe lasers, for example, have lines at 1.15, 2.40, and 3.39 μm, but HeNe lasers are most commonly used because of their red and green outputs.

7.2 GAS LASER THEORY

A conceptual model of a laser is that of a positive-feedback saturable light oscillator. The standard laser design is shown in Fig. 7.1. The central section, containing the active medium, which usually consists of an electric gas discharge, is the saturable amplifying section of the laser. The mirrors at either end of the laser tube form the optical resonator and provide the positive feedback of light to the oscillator. One of the mirrors must have some transmission if the light is to be used extracavity.

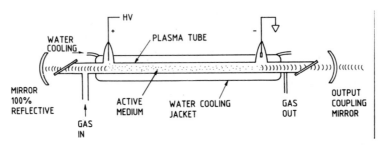

FIGURE 7.1 Standard gas laser design. The discharge is longitudinal and the flow of the gas is axial.

7.2.1 Theory of Light Amplification

Light can be amplified by stimulated emission. That is, a photon will stimulate an excited molecule to emit a photon in the same direction and in phase with the incident one. The energy of the initial photon must equal (or nearly equal) that of the transition (see Fig. 7.2a). However, in the inverse process, a photon will be absorbed when it interacts with a molecule in the lower state (see Fig. 7.2b). For light to be amplified there must be more molecules in the upper state than in the lower state, a situation referred to as population inversion.

As the number of interacting photons increase in a molecular system, the rate of stimulated emission will approach the absorption rate so that the population of the upper and lower states equalizes. As the populations equalize, the gain (or absorption) strength of the system is reduced, and it becomes effectively transparent. This effect is referred to as light-intensity saturation.

In a two-level system, for example, it is impossible to achieve population inversion because of intensity saturation. As the photon density increases, thus increasing the population in the upper state (due to photon absorption), the stimulated-emission rate increases proportionally, thus depopulating the upper level. Hence population equalization can be approached, but inversion can never be created. An inversion can be created between excited states in a multilevel system, however.

For this reason, infrared gas lasers (like most lasers) are four-level systems (see Fig. 7.3). The molecules can be pumped by a variety of means from the ground state to some intermediate upper state (transition $|0> \Rightarrow |1>$). The molecule decays rapidly to the upper lasing level ($|1> \Rightarrow |2>$) and is stimulated to the lower level ($|2> \Rightarrow |3>$) where finally it is deexcited to the ground state ($|3> \Rightarrow |0>$). To

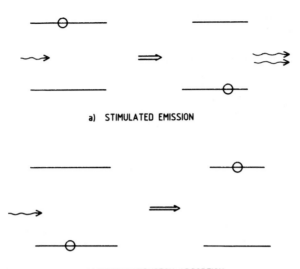

a) STIMULATED EMISSION

b) RESONANT PHOTON ABSORTION

FIGURE 7.2 Photon interaction with a molecule or atom. (*a*) Stimulated emission. In this two-level, one molecule per atom system, there is one photon "in" and two photons "out," on account of the population inversion. (*b*) Resonant photon absorption. In this case, there is one photon "in" and zero photons "out," because there is no inversion.

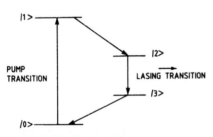

FIGURE 7.3 Four-level laser system.

create the population inversion and to sustain lasing it is necessary that $t_{2\to3} \gg t_{3\to0} \approx t_{1\to2}$,[4] where t is the lifetime of the respective states. In other words, if the lifetime of state $|2\rangle$ is relatively long, then that state represents a bottleneck to energy deactivation; hence population can build up in that state. The $|1\rangle \Rightarrow |2\rangle$ and $|3\rangle \Rightarrow |0\rangle$ transitions are often part of a cascade, as in HF/DF chemical lasers or CO lasers, and may also be lasing transitions.

7.2.2 Gain Broadening and Intensity Saturation

For a given active (laser) medium, there are several mechanisms that frequency-broaden the gain and absorption profiles of the lasing transition. The gain and saturation characteristics of the active medium are dependent on the type of line broadening, as are, ultimately, the mode structure and the output power of the laser.[5]

There are two types of broadening; homogeneous and inhomogeneous. A homogeneously broadened line shape is produced by an ensemble of identical molecules with the same "central" resonant frequency. The energy by levels within each molecule are broadened. An inhomogeneously broadened line shape is produced by an ensemble of molecules, with slightly different resonant frequencies, grouped statistically around the central frequency. Molecules that give rise to an inhomogeneously broadened line are distinguishable, whereas those that give rise to a homogeneously broadened line are not.

In gas lasers there are two mechanisms that give rise to homogeneously broadened line shapes, spontaneous emission and collisions. The broadening from spontaneous emission is uncertainty broadening† and hence is inversely proportional to the lifetime of the transition. Since the lifetimes of infrared transitions are very long, lifetime broadening is not very important in ir gas lasers.

Broadening due to molecular collisions, or pressure broadening, can be visualized by considering the classical model of an exponentially damped oscillator whose phase is periodically interrupted by collisions (see Fig. 7.4). The frequency

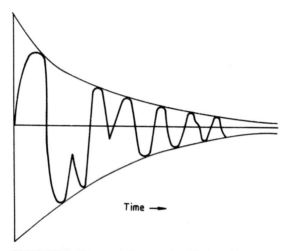

FIGURE 7.4 Exponentially damped oscillation with random phase interruption.

bandwidth of such an oscillator is then given by its Fourier transform, which, because of the phase-interrupting collisions, will have components other than the central resonant frequency ν_0. The frequency bandwidth of a pressure-broadened line is proportional to the mean period between collisions τ_c. The gain profile in the limit of zero electric field is given by the Lorentzian profile

$$g_p(\nu - \nu_0) = C \frac{\Delta\nu_p/2\pi}{(\nu - \nu_0)^2 + (\Delta\nu_p/2)^2} \tag{7.1}$$

where C is a constant proportional to the magnitude of the dipole moment and ν is the lasing frequency. $\Delta\nu_R$ is the full-width half maximum (FWHM) of the transition and is related to the mean collision period by

$$\Delta\nu_p = \frac{1}{\pi\tau_c} \tag{7.2}$$

---

†That is, the linewidths are determined by the Heisenberg uncertainty principle; $\Delta E\,\Delta t \geq h$.

As the radiation intensity within the laser builds up, the gain of the active medium must decrease until it balances optical losses, for steady-state lasing. The intensity-dependent optical gain has the same form as Eq. (7.1) but with intensity broadening and correction for population inversion:

$$G_p(v) = C_p N_0 \left(\frac{\Delta v_p / 2\pi}{(v - v_0)^2 + (\Delta v_p / 2)^2 (1 + I/I_s)} \right) \tag{7.3}$$

where C_p is proportional to the dipole moment of the transition, N_0 is the population inversion, and I is the intensity of the laser electric field. I_s is the saturation intensity and is dependent on the line shape of the transition and the lifetimes of the different levels.[6] When the internal laser intensity equals I_s, the gain in the active medium is reduced by half.

In gas lasers, inhomogeneous broadening is due to the random motion of the molecules. There is a Maxwell-Boltzmann distribution of molecular velocities that is proportional to the square root of the molecule-reduced mass divided by the temperature. The resonant frequency of each molecule is Doppler shifted by v_z/c, such that

$$v = v_0 \left(1 + \frac{v_z}{c} \right) \tag{7.4}$$

where v is the frequency as seen by an observer at rest with respect to the moving molecule, v_z is the velocity component along the line between the molecule and the observer, and c is the speed of light. The line profile of the gas will have a Gaussian shape:

$$g_D(v) = \frac{c}{v_0} \sqrt{\frac{M}{2\pi kT}} e^{-4\ln 2(v - v_0/v_D)^2} \tag{7.5}$$

where M is the mass of the molecule and k is the Boltzmann constant. The width (FWHM) Δv_D of the Doppler broadened line is

$$\Delta v_D = 2v_0 \sqrt{\frac{2kT}{Mc^2} \ln 2} \tag{7.6}$$

The gain of an inhomogeneously broadened line saturates differently than in Eq. (5.3). In this case,

$$G_D(v) = C_D N_0 \frac{1}{\sqrt{1 + I/I_s}} e^{-4\ln 2(v - v_0/\Delta v_D)^2} \tag{7.7}$$

where C_D is proportional to the square of the transition dipole moment and to the inverse of Δv_D. I_s is the same as for pressure broadening.

There are two major consequences of the type of line broadening. In the inhomogeneous case, one resonator mode will interact only with a subset of the molecules.† If the frequency difference between the resonator modes is small enough (see Sec. 7.2.4) then two or more modes can be within the gain bandwidth

---

†At ir frequencies, the bandwidths of the resonator modes are normally much narrower than the linewidth of the transition.

and can easily lase simultaneously. When the active medium is homogeneously broadened, on the other hand, each resonator mode will interact with all the molecules within the gain bandwidth. Therefore, the resonator modes compete for gain within the active medium. It is much easier, therefore, to produce a single-mode laser with homogeneous rather than inhomogeneous line broadening.

The second consequence is that, for typical operating pressures of ir gas lasers, the gain bandwidth of a homogeneously broadened laser is proportional to the total pressure of the active medium. Therefore, to increase the bandwidth only the pressure needs to be increased. In inhomogeneously broadened active media, the gain bandwidth can only be increased by raising the temperature, which normally lowers the gain.

Because the magnitude of the Doppler shift is proportional to the radiation wavelength, inhomogeneous broadening is the dominant mechanism in low-pressure visible and uv lasers. However, for most practical ir gas lasers, inhomogeneous broadening is not a factor because of the relatively long wavelength of ir lasers. For instance, the inhomogeneous width at 10 μm and 300 K with CO_2 lasers is only 31 MHz, whereas for pressure broadening the bandwidth will be about 200 MHz at a typical operating pressure of 40 torr (5.3 kPa).

7.2.3 Gas Laser Inversion Mechanism

Infrared lasing transitions in gases are predominantly neutral, molecular, ro-vibrational transitions in the ground electronic states of molecules. There is a class of noble gas lasers (He, Ne, Ar, Kr, Xe) with infrared electronic transitions that can lase from the near infrared to 125 μm, but these are not widely used. The following section describes the five inversion pumping mechanisms in ir gas lasers; electron impact, resonant energy transfer, chemical reactions, gas-dynamic processes, and optical pumping. The final section discusses depopulation mechanisms of the lower laser level, which are also fundamental for creating and maintaining a population inversion.

Electron Impact. Neutral atoms and molecules can be excited by inelastic scattering of energetic electrons. The probability (cross section) that an energy level will be excited depends, in a first approximation, on the optical transition probability and the energy of the electrons.[8] For a given dipole transition, the probability cross section has an electron energy threshold with a smooth drop-off at higher energies.[9] The electrons, on average, lose a fraction of their energy with each collision with a molecule but regain some energy from the discharge electric field before the next collision. After a few collisions, the distribution of electron energies equilibrates so that they can be characterized by a temperature. Assuming a Gaussian energy distribution, the electron temperature can be shown to be equal to[10]

$$T_e = \frac{e}{\sqrt{2\delta}\,k}\,(El) \qquad (7.8)$$

where e is the electron charge, E is the electric field, l is the mean-free-collision path length, δ is the average fractional energy loss for each collision, and k is the Boltzmann constant.

Since the mean free path is inversely proportional to the gas pressure p, Eq. (7.8) shows that the electron temperature is proportional to the E/p ratio, which is a practical parameter when trying to design or scale up a laser. Buffer gases are often added to the gas mixture, in part to modify the electron temperature, so that there is a better match between the electron energy and that of the molecule to be excited.

For example, in dc discharge, continuous-wave (cw) CO lasers, Xe is added to increase laser output energy. Because Xe has a lower ionization potential than CO, for a given pressure, the discharge can be sustained at a lower voltage, thus lowering the electron temperature. The lower-temperature electrons excite vibrations of the CO molecule more effectively, while reducing electronic excitation and ionization of the molecule.[11]

Two modes of electrical discharge are used to excite gas lasers, pulsed and dc. For dc discharge lasers, the lifetime conditions of the lasing energy levels, described in Sec. 7.2.1, must be met rigorously. As well, the overall input power must be limited so that the lower lasing levels will not be thermally populated, therefore reducing or at worst destroying the population inversion (see Sec. 7.3.1 for more detailed laser gas cooling systems).

The lasing energy lifetimes and the instantaneous (but not average) electrical input power conditions can be relaxed with pulse excitation. Pulse-discharge durations are short, 50 ns to a few microseconds, and create population inversion rapidly. During the lasing period, the gas temperature does not rise appreciably and high gain can be achieved. Pulsed discharges are used mainly for high peak-power generation and for many transitions with low optical gain which will not lase with dc excitation. For high-pressure lasers, i.e., 0.5 to 20 atmospheres, pulsed discharges must be used to create short-duration glow discharges or else filaments will be formed.[12] Filaments are highly ionized low-resistance paths through the gas, similar to lightning, which do not excite the gas volume but can locally damage the electrodes, promoting more filaments.

Gas lasers can also be excited with rf discharges. In most applications, the electron collision frequency is greater than the rf frequency, so that, to the molecule, the rf field is slowly varying and appears similar to a dc discharge. RF excitation is mostly used in industrial lasers because of their compact, efficient power supplies, and their elimination of cataphoresis. In some applications, with reactive or very pure gas systems, rf excitation is advantageous because the electrodes do not have to be in contact with the gas. In waveguide lasers (see Sec. 7.2.4), the electrodes often form part of the optical waveguide and sustain transverse continuous discharges.[13]

Resonant Energy Transfer. The resonant energy process is one in which an excited molecule transfers its energy to another molecule, such that

$$A^* + B \longrightarrow A + B^* \pm \Delta E \tag{1}$$

where * represents an excited state of A or B and ΔE is the energy difference between A^* and B^*. In a simple picture, as the two molecules approach, the time-varying short-range electric field of A^* couples with B, forming a system of two weakly coupled harmonic oscillators. As with two weakly coupled pendulums, energy can be exchanged from one oscillator to the other.[14] Since the interaction between the two particles is resonant, an efficient energy transfer requires that ΔE be not much larger than 0.1 kT, which at 300 K is 20 cm^{-1}. Generally, this means that the process usually excites just one of the vibrational modes of the molecule.

N_2 is the most widely used gas for resonant energy-transfer pumping of lasing media in ir gas lasers. It is used with CO, CO_2, N_2O, and CS_2 lasers.[15]

Excitation by Chemical Reaction.   It is possible to create population inversions between vibrational levels in the course of exothermic chemical reactions.[16] Several exothermic atom transfer reactions are the basis of chemical lasing action. The following reaction, for example,

$$Cl + HI \longrightarrow I + HCl \tag{2}$$

is exothermic by 134 kJ mol^{-1} and there are many ways that this exothermicity can be channeled into the vibrational, rotational, and translational degrees of freedom of the HCl product, 134 kJ mol^{-1} not being sufficient to induce electronic excitation.

In the atom transfer reaction (2), population inversions have been observed between the $v = 3$ and $v = 2$ levels, as well as the $v = 2$ and $v = 1$ levels.[17] Hence, when this reaction is carried out inside a resonant cavity, lasing action can occur. Lasers which use inversions that are created in this way can have very high energy efficiencies and can potentially produce very high powers because of the large energy released in some chemical reactions. Chemical lasers are discussed in more detail in Sec. 7.3.2.

Gas Dynamic Processes.   In gas dynamic processes, population inversion is achieved by the rapid cooling of a hot molecular gas. When a gas is heated, in the stagnation region of an expansion, the energy distribution can be broadened considerably. The inversion is then created downstream from the nozzle orifice, provided that the cooling rate of the gas is faster than the relaxation time of the upper lasing level, but of the same order as that of the lower level.

Typically, vibrations are not cooled as effectively as rotations or translations, in a supersonic expansion, so that the method is ideally suited for producing partial vibrational population inversions. The hot medium can also excite a gas used for resonant energy-transfer processes, thus increasing the efficiency of the process. In practical lasers, the gas is heated by either chemical reaction[18] or an electric discharge.[19] These types of lasers are usually used for high-power industrial applications.

Optical Pumping.   Creating a population inversion with optical pumping is commonly used with a variety of lasing media, mostly solid-state near-ir lasers and visible dye lasers. For infrared gas lasers, the upper lasing level of a molecule is selectively pumped by a high-power cw or pulsed laser (transition 1 in Fig. 7.1) with the lower level left unpopulated. Hence the output wavelength will be longer than the pump wavelength. Although lasing has been demonstrated with wavelengths as short as 2 μm,[20] most optically pumped ir lasers are driven with high-power CO_2 lasers, and lase between 12 and 2000 μm (see Sec. 7.3.4).

Lower-Level Deactivation.   There are two main deactivation channels for the lower laser level, radiative and collisional. Of the two channels, collisional deactivation is the most important one because spontaneous emission lifetimes of infrared transitions are long. This follows from the Einstein coefficient A for spontaneous emission:

$$A = \frac{1}{\tau_{sp}} = \frac{8\pi n |\mu|^2}{3\hbar\epsilon_0 \lambda^3} \tag{7.9}$$

where τ_{sp} is the spontaneous lifetime of the transition, n is the refractive index of the medium, $|\mu|$ is the magnitude of the electric dipole moment, $\hbar$ is Planck's constant, ϵ_0 is the permittivity of free space, and λ is the wavelength of the emitted photon.[21]

From Eq. (7.9) the spontaneous lifetime is inversely proportional to the cubic power of the wavelength. For 10-μm transitions, $\tau_{sp} \approx 100$ μs to 1s. This contrasts with spontaneous lifetimes of 10 to 1000 ns for most visible lasers. This low rate of spontaneous emission for ir transitions is beneficial in some respects. For instance, there is little or no amplified spontaneous emission (ASE), which is a major parasitic loss in gas uv and visible lasers and liquid dye lasers.

Since the τ_{sp}'s are relatively long, collisions, intermolecular or with the laser walls, are the principal molecular energy deactivation mechanism. Buffer gases are used in most gas laser mixes to enhance the deactivation of the lower level as well as to tailor the electron gas temperature. Collisional deactivation rates can vary widely depending on the collision partners. For example, the deactivation rates of the (010) level of CO_2 by Ar, He, and H_2O are 130 $s^{-1}torr^{-1}$, 1.3×10^4, and 1.3×10^5, respectively. Helium is typically used in CO_2 lasers as a buffer gas for this reason. H_2O, which is actually more efficient, would have a very negative effect on the gas discharge. With He as a buffer gas, at pressures of 10 torr (1.3 kPa), collisional deactivation lifetimes will be of the order of 10 μs.[22]

7.2.4 Laser Resonators

The optical resonator in a laser creates a resonating cavity that, in part, determines the lasing frequency† and provides feedback to the amplifying media. The resonator sets up an optical beam that contains light within the laser cavity. There are two types of laser resonators: open resonators, which are used in most lasers, and waveguide resonators.

Open Resonators. Conventional open resonators are formed between two concave mirrors with no walls to contain the radiation perpendicular to the optical axis.‡ The problem of finding a resonator mode between the two mirrors is equivalent to that of finding a beam configuration that is self-consistent through an infinite series of lenses. In other words, the problem is to derive the relative electric-field intensity and phase of the radiation, along all three axes between the two lenses. This field configuration must reproduce itself between each pair of lenses (see Fig. 7.5). The general solution to this problem is a Hermite-Gaussian polynomial.[23] With the optical axis along the z axis, the electric field of the mode can be specified by

$$E_{1,m}(\bar{r}) = E_0 \frac{\omega_0}{\omega(z)} H_1\left(\sqrt{2}\,\frac{x}{\omega(z)}\right) H_m\left(\sqrt{2}\,\frac{y}{\omega(z)}\right)$$
$$\times \exp\left(-\frac{x^2 + y^2}{\omega(z)} - ik\frac{x^2 + y^2}{2R(z)} - ikz + i\eta(l + m + 1)\right) \quad (7.10)$$

H_l and H_m are the different orders of the Hermite polynomial in the x and y directions. The TEM_{00} mode $l = m = 0$ has a Gaussian intensity profile in the

---

†The laser cavity can only support a set of discrete standing waves, for example. In addition, there may be frequency-selective components, gratings, or narrowband reflective mirrors.

‡This contrasts with microwave resonators where the standing electromagnetic wave is confined along all three axes.

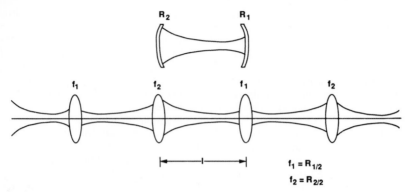

FIGURE 7.5 Self-reproducing beam through an infinite lens series.

radial direction and is usually the most desirable because it has the lowest divergence and produces the highest intensity for a given power. The higher transverse modes $l \neq 0$, $m \neq 0$ are spatially larger and have a frequency offset to the TEM_{00} mode. By adding an aperture to the resonator, transverse-mode lasing can be suppressed.[24] For the TEM_{00} mode and $r^2 = x^2 + y^2$, Eq. (7.10) reduces to[25]

$$E_{00}(r) = E_0 \frac{\omega_0}{\omega(z)} \exp\left[-\frac{r^2}{\omega(z)} - ik\frac{r^2}{2R(z)} - ikz + i\eta \right] \qquad (7.11)$$

where $\omega(z)$ is the $1/e^2$ radial spot size and is equal to

$$\omega(z) = \omega_0 \left[1 + \left(\frac{z}{z_0}\right)^2 \right]^{1/2} \qquad (7.12)$$

ω_0 is the minimum spot size and defines the Rayleigh length z_0 by

$$z_0 = \frac{\pi \omega_0^2 n}{\lambda} \qquad (7.13)$$

k is equal to $2\pi n/\lambda$, n is the index of refraction of the lasing media, and η is the phase factor, which is equal to

$$\eta = \arctan\left(\frac{z}{z_0}\right) \qquad (7.14)$$

From Eq. (7.11), the radius of curvature of the wavefront can be derived and is equal to

$$R(z) = \frac{1}{z}(z^2 + z_0^2) \qquad (7.15)$$

In many lasers one of the resonator mirrors is flat, $R = \infty$; these are hemispherical resonators. If $R = \infty$ for one of the mirrors, then, from Eqs. (7.12) and (7.15), $\omega = \omega_0$ at $z = 0$.

For the TEM_{00} mode the resonator frequency is set by the condition

$$q\pi = \frac{2\pi l}{\lambda_q} - (\eta_1 - \eta_2) \qquad \nu_q = \frac{c}{\lambda_q} \tag{7.16}$$

where q is an integer, η_1 and η_2 are the phase factors at the first and second mirror, and l is the length of the resonator. From Eq. (7.16) the longitudinal mode spacing is equal to

$$\Delta\nu_{\text{mode}} = \frac{c}{2nl} \tag{7.17}$$

To form an optical resonator, the radius of curvature of the mirrors is matched to the curvature of the expanding phase fronts of the laser beam, thus reversing the front onto itself (see Fig. 7.6). There are two approaches to designing a resonator. One is to define the length of the cavity, the minimum spot size, and its position within the cavity, and then, using Eqs. (7.13) and (7.15), set the mirror radius to match the phase fronts. Alternatively, the mirror radius and cavity length can be defined and, again using Eqs. (7.13) and (7.15), the minimum spot size ω_0 can be calculated. With ω_0, the spot size ω can be calculated along the cavity length.

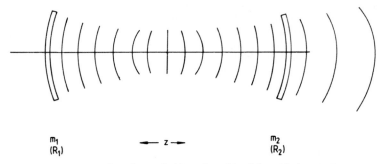

m_1
(R_1)

$\longleftarrow z \longrightarrow$

m_2
(R_2)

FIGURE 7.6 Beam phase-fronts, inside and outside of the optical resonator.

There are limits to the ratio of the cavity length l to mirror curvature (R_1 and R_2 in Fig. 7.6), for the optical cavity to produce a stable, self-consistent mode. The stability criteria are

$$0 \le \left(1 - \frac{l}{R_1}\right)\left(1 - \frac{l}{R_2}\right) \le 1 \tag{7.18}$$

One difficulty with stable resonators is that they are "thin." That is, it is difficult for the resonator mode to fill the active volume, with a relatively short cavity, while still maintaining the TEM_{00} mode. A common parameter that is often used to characterize a mode is the Fresnel number

$$N = \frac{a^2}{l\lambda} \tag{7.19}$$

where $2a^2$ is the effective diameter of the resonator mirrors. Often, for a high-gain laser to achieve single-mode lasing, the Fresnel number must be close to 1, which corresponds to a high-loss resonator.[26]

There is another class of open resonators called unstable resonators that produce near Gaussian modes with relatively large aperture to length ratio. In these lasers, the rear reflector is convex and the front reflector either flat or convex. Unstable resonators have the same optical layout as reflecting telescopes, with the mode separation and losses proportional to the optical magnification. They are usually used with high-gain, large-volume, pulsed lasers, such as TEA CO_2's, to produce single-mode output. This is necessary because, with convex mirrors, there will not be as many round trips inside the laser cavity for building up power. These high-gain lasers are more efficient with low-reflectivity front optics, i.e., large output coupling.[27]

Waveguide Resonators. In open resonators, the optical radiation is kept away from the walls to avoid distortion and to minimize losses. In waveguide lasers, the walls form an integral part of the resonator. The cross section of the cavity can have many shapes but is usually circular or rectangular. Wall materials can be dielectrics, metals, or a combination of both. The waveguide-laser modes are analogous to radio-frequency (rf) waveguide modes except that the channel diameter is many times larger than the radiation wavelength.[28] The laser waveguides are referred to as hollow waveguides because, unlike fiber-optics, the index of refraction of the guiding channel is lower than the media that surrounds it.

Because of the gas laser scaling relationship $P_1 d_1 = P_2 d_2$,[29] where P is gas pressure and d is the diameter of laser bore, most waveguides are operated at much higher pressures than open-resonator lasers. The power per unit length remains the same, but the overall size of the laser is reduced. In addition the lasing bandwidth increases significantly with higher pressure. Single-mode bandwidths as large as 1.2 GHz can be obtained with a CO_2 waveguide laser, compared to the 50-MHz bandwidth of a low-pressure, open-resonator laser operating at 10 torr.[30] These large bandwidths are useful for laser radar, spectroscopy, and laser communications. For a more complete discussion of waveguide lasers see the review article by R. L. Abrams.[31]

Wavelength Tuning. Most infrared gas lasers have many lasing transitions, often closely spaced together. For single-line lasing in the near- to mid-infrared, dielectric-coated mirrors can be made to have high reflectivity over a narrow wavelength range. However, with such a configuration, the mirrors must be replaced in order to change the wavelength, which is usually not a simple task. For the far-infrared there are no good dielectric materials for optical coatings, and other methods must be used.

One of the most convenient and flexible means of selecting a single lasing transition is to use a reflective grating in the Littrow configuration.[32] In the Littrow configuration the first-order diffraction angle is equal to the angle of incidence at a specific wavelength. From the grating equation, the first-order Littrow angle is related to the wavelength such that

$$\lambda = 2d \sin \theta \qquad (7.20)$$

where λ is the wavelength of the incident radiation, d is grating line spacing, and θ is the angle of incidence, measured normal to the grating. Radiation will also diffract in zero order, which is equivalent to specular reflection from the grating.

For laser line selection, the grating replaces one of the reflectors in the optical cavity and acts as a frequency-selective mirror. The various lasing lines can be tuned in by simply rotating the grating with respect to the optical axis. Because of the diffraction properties of gratings, the output beam of the laser will be polarized with the electric vector perpendicular to the grating lines (see Fig. 7.7).[33]

Most gratings used in gas lasers are echelete-type gratings. With these gratings the groove lines are cut in the form of steps and the angle between the long part of the step and the surface of the grating is called the blaze angle. To maximize first-order diffraction and minimize the losses in zero-order diffraction, the blaze angle of the grating should be equal to the Littrow angle.

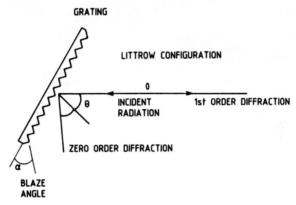

FIGURE 7.7   Grating in the Littrow configuration.

7.3 SPECIFIC GAS LASERS

7.3.1 CO₂ Lasers

The first lasing from CO_2 was reported in 1964 from pure CO_2 in the 10.6-μm band.[34] Only 1 mW of average power was obtained in cw operation with a 5-m tube. At the time, the search for new laser transitions was flourishing and the discovery of this one was "lost in the noise." Within one year, however, a 10^4 increase in average power was obtained[35] (approximately 12 W) using a CO_2-N_2 mixture, and the road to the high-power CO_2 lasers of the present was paved.

Principles of Operation. CO_2 is a linear triatomic molecule (in the symmetry point group $D_{\infty h}$) with four vibrational degrees of freedom. These vibrations are shown schematically in Fig. 7.8. The bending vibration, ν_2 in the nomenclature,† is doubly degenerate. In CO_2, however, the frequency of the ν_2 vibration is almost exactly half that of the ν_1 vibration; hence there is a very strong Fermi resonance between these two levels.‡ They are split by 100 cm⁻¹, and this separation accounts, in part, for the relatively large tuning range (9 to 11 μm) of the CO_2 laser.

---

†The vibrations of a linear triatomic molecule are also routinely labeled $(\nu_1, \nu_2^{\,o}, \nu_3)$, where the number in brackets gives the number of quanta in each vibrational mode. For example, $(2,0,0) = 2\nu_1$.

‡The interaction between the ν_1 and ν_2 modes occurs because the potential experienced by the carbon atom, during the bending motion, depends on the separation between the carbon and oxygen atoms, and hence on the stretching vibration. Further, because $2\nu_2 \approx \nu_1$, then classically, as well as quantum-mechanically, there is enhanced coupling between the two vibrations because of the near resonance.

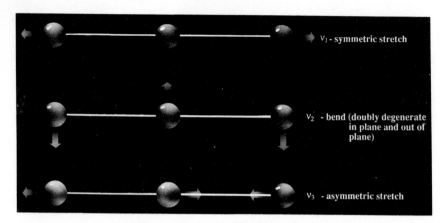

FIGURE 7.8 The normal modes of vibration of a linear triatomic molecule. The other ν_2 vibration has the oxygen atoms (if the molecule is CO_2, for example) moving out of the page and the carbon atom moving into the page. In the absence of any perturbations, such as Fermi resonances, these two vibrations have the same energy. In CO_2 there is a large Fermi resonance, and one of the two bending vibrations is highly perturbed, the $(0,2°,0)$, which has the same symmetry (Σ_g^+) as the $(1,0,0)$ level, with which it is coupled. The $(0,2^2,0)^g$ level (Δ_g symmetry) remains unperturbed. The splitting between the two vibrations is about 50 cm^{-1}.[106]

The CO_2 laser is a four-level system pumped by an electrical discharge (see Sec. 7.2.1). The energy-level diagram for this system is shown in Fig. 7.9. Typically a dc plasma is generated in a mixture of CO_2, N_2, He, and in some cases, a small amount of carbon monoxide.[†] Although inelastic collisions between electrons and CO_2 will lead to some population in the $(0,0,1)$ upper laser level, this is a relatively minor excitation mechanism. Vibrational to vibrational (V-V) energy transfer, between excited N_2 (or N_2^*) and CO_2, is the main process leading to population in the upper laser level. Vibrationally excited N_2 is produced with very high efficiency through the process[36]

$$N_2 + e^- \longrightarrow N_2^- \longrightarrow N_2^* + e^- \tag{3}$$

This energy is readily exchanged in collisions with CO_2 because of the near resonance between the $N_2(v = 1)$ and $CO_2(0,0,1)$ energy levels, thus leading to inversion between the $(0,0,1)$ level and the Fermi-resonance pair, the $(0,2°,0)$ and $(1,0,0)$ levels. The $(0,2°,0)$ and $(1,0,0)$ levels are rapidly relaxed, radiatively and through CO_2-CO_2 collisions, to the $(0,1^1,0)$ level, which helps maintain the inversion. Helium is added to collisionally deactivate the $(0,1^1,0)$ level, which has two beneficial effects. The reduction in the temperature of the bending modes results in a smaller population in the lower laser levels,[‡] which increases the gain. Indeed, if the temperature of the gas rises above 200°C, the lasing will cease.[37]

[†]CO is used primarily because it recombines with O_2, which results from CO_2 dissociation, to reform CO_2, since O_2 tends to promote arcing in gas discharges. Since CO has a large cross section for electron impact excitation, larger even than N_2, it also helps promote the population inversion because its vibrational frequency is also nearly resonant with the $(0,0,1)$ vibration of CO_2.

[‡]Since the $(1,0,0)$ and $(0,2°,0)$ levels are so strongly coupled by the Fermi resonance, any process which leads to the depopulation of one level will also depopulate the other.

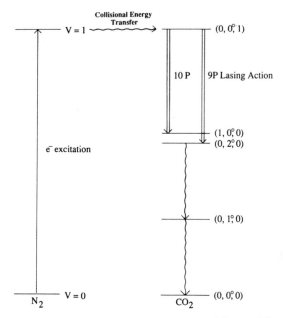

FIGURE 7.9 Pumping scheme and energy-level diagram CO_2 laser. The (0,2^2,0) level lies halfway between the (1,0,0) and (0,$2°$,0) levels but is not shown since it is not radiatively coupled to the (0,$0°$,1) upper laser level.

Second, the collisional deactivation results in a lower temperature of CO_2, which decreases the harmful effects of quantum dilution on gain; i.e., the population is spread over fewer quantum levels.† The theoretical electrical to wall plug efficiency of CO_2 lasers is very high (27 percent,[38] and commercially available cw systems can approach this efficiency.

A list of CO_2 and other common laser wavelength ranges is given in Table 7.1.

Types of CO_2 Lasers. CO_2 lasers can be operated either cw or pulsed. There are a large variety of different cw laser designs which have been developed in response to numerous different applications. The simplest and earliest designs consisted of cylindrical plasma tubes in which a dc discharge was sustained, with electrodes at each end, and axial flow of the gas (see Fig. 7.1, for instance). To achieve good transverse mode quality it is necessary that the resonator have a low Fresnel number [see Eq. (7.19)]. All early designs, therefore, had long, narrow plasma discharge tubes, and the resonators were quite "lossy."

---

†This is of prime importance when the laser is operating on the highest-gain 10.6-μm line [10P(20)], which most industrial lasers do. As the temperature goes up, the population is spread among more rotational levels, thereby decreasing the gain on the 10P(20) line. For some applications, where line tunable operation is desirable, then, on a weaker line, a higher rotational temperature may be an asset. But there is a trade-off between the enhancement of gain due to quantum dilution, and the negative effects of the concomitant increase in vibrational temperature. A higher rotational temperature increases the gain for weak lines [e.g., 10P(40)] at the expense of strong lines; i.e., just because its Boltzmann population is larger. However, the increase in vibrational temperature decreases the gain for all rotational lines.

TABLE 7.1 Wavelength Ranges of Common Infrared
Molecular Gas Lasers*

Molecule	Laser type	Wavelength range, μm
$^{12}CO_2$	Electric discharge	9.13–11.0
CO	Electric discharge	4.75–8.27
HBr	Chemical	4.02–4.65
DBr	Chemical	5.57–6.29
HCl	Chemical	3.57–4.11
DCl	Chemical	5.00–5.61
HF	Chemical	2.41–3.38
DF	Chemical	3.49–4.06

*For a complete list of laser lines, see Ref. 104a, pp. 323–404, 428–483.

Many industrial applications of CO_2 lasers, such as drilling, cutting, and welding, require much higher power, and these early designs were limited. In order to increase the power it is necessary to increase the amount of gas, by raising either the pressure or the volume of the active medium. However, when this is done, the energy that is not converted to laser radiation heats up the gas to the point that population inversion cannot be sustained.

The gas is cooled through collisions with the walls of the plasma tube. Increasing the pressure while keeping the volume constant simply makes cooling less efficient, as well as making a discharge more difficult to sustain. Increasing the volume is not the correct solution, since the surface area of the cylindrical plasma tube scales linearly with a change in radius whereas the volume goes up as r^2. In addition, the Fresnel number becomes larger unless the length of the discharge tube is increased considerably, which can make the laser difficult to house.

The obvious solution to this problem seems to be to increase the axial flow rate of the gas. However, the maximum gain of the laser is near the center of the gas discharge, where water cooling is least efficient. Fast axial flow therefore is necessary but is incompatible with the flow resistance of a long, narrow, cylindrical discharge tube.

Some successful methods of achieving efficient gas cooling while maintaining axial flow have nonetheless been developed to allow scaling of CO_2 lasers to much higher powers. These involve novel electrode geometry, in some cases, such as toroidal gas flow[39] and the use of cylindrical, coaxial, water-cooled electrodes,[40] which result in a doughnut-shaped beam, similar to that of an unstable resonator.

Transverse gas flow can overcome many of these problems, but not without sacrificing some beam quality, since it is extremely difficult to achieve uniform transverse gas flow in a resonator with a low Fresnel number.† However, some very high-power CO_2 lasers have been developed with transverse gas flow, often employing multipass geometry to lower the Fresnel number and obtain reasonable beam quality.[41]

The highest-average-power lasers commercially available employ a hybrid system in order to generate high-power beams with good beam quality. In these

---

†Apertures could be inserted at either end of the resonator so that the laser cavity has a high Q only in the uniform region of the plasma, thus lowering the Fresnel number, but this would be a huge cost in efficiency, which would defeat the purpose of using transverse flow.

systems, a relatively low-power oscillator (≈ 1 kW) is used to pump one or a series of amplifiers. This arrangement is known as a master oscillator, power amplifier (MOPA) chain. It is relatively easy to achieve good beam quality with a 1-kW oscillator, and the problems associated with overheating of laser mirrors and windows can be avoided. The output pulse of the oscillator is then used to pump amplifier sections, usually employing transverse gas flow and gas-dynamic windows.[42] The beam quality is a stronger function of the characteristics of the low-power oscillator than the amplifier. Industrial CO_2 lasers in the 50-kW range have been manufactured using a MOPA chain.[43]

CO_2 lasers can also, as pointed out, be operated as pulsed systems. In some applications high peak powers rather than high average powers are desirable. Lasers have been manufactured which can deliver 100 J of energy in a pulse of 500 ps duration,[44] hence a peak power of 2×10^{11} W, unfocused. However, this laser can only pulse about once a minute;† hence its average power is only 1.67 W.

The most common type of pulsed CO_2 laser is a transversely excited atmospheric (TEA) CO_2 laser. This is similar in design to Fig. 7.1 except that the electrodes are parallel and run the length of the plasma tube. Usually, these tubes are relatively short (about the length of the electrodes), with a high Fresnel number. Because the duty cycle of these lasers is quite low, and large average powers are not usually required for the desired application, there is no need to flow the gas, at least not fast-flow. The electrodes are usually spaced a few centimeters apart and have flat surfaces but are rounded at the edges to avoid instabilities in the electric field. Typically, 40 kV, with a few milliamperes of current, are discharged through a gas mixture at pressures ranging from about 0.01 to 1 atm. A typical CO_2 laser pulse is shown in Fig. 7.10. The pulse has an initial spike of ≈ 200 ns duration, followed by a long tail that is 2 to 3 μs long. The unusual shape of this pulse can be accounted for in the following way.

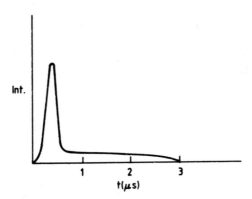

FIGURE 7.10 TEA CO_2 laser pulse with standard (N_2-rich) mixture (mode beating not shown).

In the early portion of the pulse, all of the excitation mechanisms that give rise to population inversion are working in concert; i.e., inelastic electron-N_2 molecule collisions, N_2^*-CO_2 V-V energy transfer, and electron-CO_2 collisions. After the

---

†The main factor limiting the duty cycle of these lasers is the time it takes to recharge the capacitors.

discharge is over, the plasma will persist for several microseconds; but the temperature of the electrons goes down, since there is no electric field remaining to accelerate them; hence their ability to contribute to population inversion decreases.

Lasing quickly starts to equalize the level populations, thus lowering the gain. There is a time lag of the order of a few hundred ns between the initiation of the discharge and the time it takes for the vibrational temperature of the gas to rise, owing to V-V (N_2-CO_2) energy changing collisions. Once the vibrational temperature of the gas rises, the population in the lower laser level, $(0,2°,0)$ or $(1,0,0)$, is not depleted as efficiently, also lowering the gain. A large number of N_2^* molecules are left in excited vibrational levels following electron excitation, however, and these continue to contribute to population in the upper laser level, through V-V energy exchange, until their energy is dissipated. This deactivation process typically takes 2 to 3 μs, which accounts for the long tail on the pulse.

Although the Fresnel number of TEA laser resonators is usually quite large, the laser can be made to lase on the lowest-order transverse mode by putting apertures of the appropriate size in the resonator. This is at the price of a decrease in efficiency; but for most research applications, efficiency is not the prime concern.

For some applications, a temporally smooth, single-longitudinal-mode (SLM) laser pulse is required. Without making any special modifications, the TEA CO_2 laser will lase on several longitudinal modes. This will manifest itself as mode beating, or "spiking," in the temporal profile of the laser which is clearly shown in Fig. 7.11. In addition, in experiments where the CO_2 laser is used to induce photochemical reactions, the long tail of the pulse can seriously complicate the interpretation of the results. The long pulse can be eliminated by lowering the N_2 concentration of the gas. This change in gas composition will give the laser pulse a Gaussian temporal envelope, with mode beating, at the expense of as much as 90 percent decrease in output power.†

The mode-beating frequency [see Eq. (7.17)] is equal to the inverse of the round-trip time in the cavity. The laser can, however, be made to lase SLM by using a low-pressure, quasi-cw gain section, which initiates oscillation on a particular longitudinal mode and enhances the gain of the oscillator on this mode, at the expense of adjacent ones.[45] By using one of these "smoothing tubes," inserted inside the laser oscillator cavity, the laser can be made to operate SLM about 90 percent of the time. The laser pulse shown in Fig. 7.11 (smooth curve) is from the same laser[46] as the mode-beating curve in Fig. 7.11 but with the smoothing tube operational.

Major Industrial Applications. Since CO_2 lasers are high-gain, extremely efficient, and one of the easiest lasers to build, they are the most commonly used industrial lasers for virtually all applications. The interaction of metals in particular with laser light is qualitatively different from their interaction with other forms of energy, mechanical, electrical, or heat. With lasers, it is possible to supply an extremely high energy density to a metal surface. Hence it is possible to perform a number of different functions such as cutting, welding, joining, drilling, and marking of semiconductors and pharamaceuticals, normally accomplished with more traditional tools. CO_2 lasers are becoming commonly used for many of these functions.

Industrial applications of ir gas lasers are dealt with in more detail in Chap. 25. Further information can be found in Ref. 47.

---

†For example, when the gas mix of a Lumonics TEA 840 CO_2 is changed from (16/72/8/4) percent (N_2,He,CO_2,CO), to (4/80/16/0) percent, the power output drops from 3.5 J cm^{-2} to 0.33 J cm^{-2}.

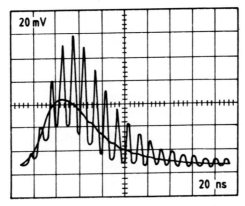

FIGURE 7.11 TEA CO_2 laser pulse with a N_2-lean gas mixture. CO_2 laser pulse with longitudinal mode beating (spiky curve) is shown on the same scale as the single-longitudinal-mode pulse (smooth curve), generated by using a smoothing tube in the same resonator.

Surface Modification.   Since the time required for laser heating of metals is very short, compared to other time scales such as heat conduction, it is possible to perform unique treatments to surfaces that will affect a material's mechanical and corrosion properties. Melting rates with laser processing of metals can be very high and it is possible, by varying power density and interaction time of the laser beam with the surface, to produce extremely large temperature gradients in the surface layer of a metal, with low penetration depths.[48] The resultant cooling rates, into the bulk of the metal, can be extremely high.

A cooling rate of $10^8 \, °C \, s^{-1}$ is not difficult to achieve. Although the temperature at the surface is very high, the actual energy deposited in the bulk metal is not large. Hence it is possible to perform radical modifications to the surface of a metal while leaving the bulk virtually unaffected. Because of these very rapid cooling rates, it is possible to create microstructures at the surface of a metal that would not normally be found in nature. It is, for instance, possible to create a material with a completely amorphous surface layer (laser glazing)[49] but a bulk that is a structured lattice. An amorphous surface layer may make a material more corrosion-resistant.[50]

With somewhat slower cooling rates, lower power densities, and greater penetration depths, it may be possible to "freeze" metastable phases of a material onto a surface.[51] This technique, known as transformation hardening, has been used to harden an alloy of carbon steel and thus make it more resistant to mechanical wear. The technique is used in the auto industry,[52] for example, to make a relatively inexpensive material very hard,† thus avoiding the cost of using special alloys for the construction of some components.

It is possible, with a laser and a metallic powder, to do surface alloying as well.[53] Two-component metals can be produced which utilize the bulk characteristics of

---

†Small regions inside the power steering housings of General Motors cars that are subject to an unusual amount of wear, for instance, are hardened in this way.

one material for its favorable mechanical properties while exploiting the micro-structure of another for its favorable corrosion or wear properties. Cladding of surfaces, one metal to another, can also be done by using a laser.[54] Surface shock hardening[55] using pulsed lasers is another technique that has been used with positive results.

Laser Isotope Separation. The appearance of pulsed, high-peak-power CO_2 lasers unexpectedly gave rise to an explosion of research in the field of laser photochemistry in the mid-1970s[56] which continues today. In 1972, it was discovered[57] that, in many cases, molecules could absorb several, perhaps as many as 40 ir photons successively, and decompose with very high efficiency. The multiphoton excitation process is pictured schematically in Fig. 7.12.

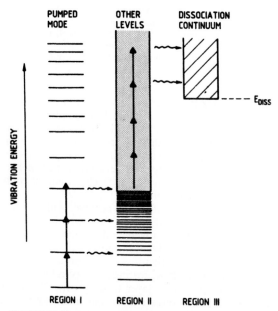

FIGURE 7.12 IR multiphoton absorption in a polyatomic molecule. Region I is sometimes called the "discrete pumping" region. Region II is usually referred to as the quasi-continuum and region III is a "true" continuum.

The IR laser energy initially goes into excitation of a specific bond in the molecule, region I in Fig. 7.12. In region II, the molecule is further excited; but the mode of excitation loses its "uniqueness" as the energy becomes rapidly dissipated among the other vibrational modes. This energy regime is commonly known as the quasi-continuum[58] because the density of states is high enough and the vibrational coupling strong enough that photon absorption is virtually independent of wavelength. Finally, the molecule is excited to region III, above the dissociation limit of the ground state, where the energy levels form a true continuum, and the molecule may dissociate.†

Once the mechanism of photon absorption became slightly better understood, it became clear that ir lasers, CO_2 lasers in particular (since they were the only high-power ones available at the time), could be used to separate isotopes. The principle behind ir laser-induced isotope separation (LIS) relies on the fact that molecules containing different isotopes of the same chemical element will vibrate at different frequencies. The frequency of oscillation is dependent on the number of nuclei that the isotope contains (i.e., its mass) and, depending on the molecule, the way in which the addition of an isotope of different mass affects its symmetry. The ir laser can then be tuned to the same frequency as a (sympathetic) vibrational frequency of the molecule and thus pump energy into that molecular motion.

Examples of both of these effects are shown in the ir spectrum in Fig. 7.13. The molecule in Fig. 7.13 is 1,1,1-trichloroethane (CH_3CCl_3, or TCE, and two of its

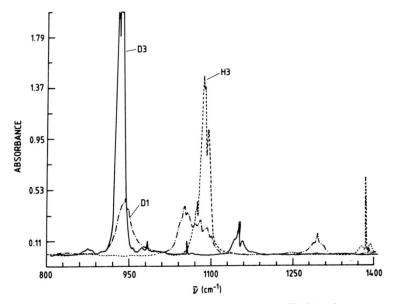

FIGURE 7.13 IR absorption spectrum of 1,1,1-trichloroethane. IR absorption spectrum of CH_3CCl_3 is labeled H_3. The ir absorption spectrum of CD_3CCl_3 is labeled D_3, and the ir absorption spectrum of CH_2DCCl_3 is labeled D_1.

isotopic variants; CD_3CCl_3 and CH_2DCCl_3) and is considered as a potentially suitable molecule for use in separation of deuterium isotopes.[59] The absorption features in the spectrum are due to the symmetrical C-H bending vibration. The CH_3CCl_3

---

†In region III the molecule has enough energy to decompose but will not do so immediately. If there is a significant activation energy, the molecule may have a long unimolecular lifetime. Decomposition of the "superexcited" molecule competes with collisional deactivation by "color," unexcited molecules that can accept some of the excess energy in vibrational-vibrational energy exchanging collisions.

spectrum (see H_3 in Fig. 7.13) shows only one feature below 1250 cm^{-1}; at approximately 1090 cm^{-1}.

When all three H atoms on the molecule are replaced by D atoms, the spectrum shifts considerably to the red (see D_3 in Fig. 7.13), to about 930 cm^{-1}. This illustrates the effect of isotope mass on the spectrum. When only one of the hydrogen atoms is replaced by a deuterium atom, however, there is both a red shift in the spectrum and a change in the number of peaks (see D_1 in Fig. 7.13). The latter occurs because the symmetry of TCE changes from C_{3v} to C_s upon monodeuteration. In C_{3v} symmetry TCE, which has 18 different vibrational modes, has only 6 nondegenerate ones; but in C_s symmetry none of the vibrations are exactly degenerate; hence the spectrum is much more complicated.

Isotope separation by the method of infrared multiphoton absorption and decomposition cannot be carried out on diatomic molecules because they have no "region II"; i.e., there are no other vibrational modes with which to couple. Since all molecules are not perfectly harmonic oscillators, the spacing between vibrational energy levels is not equidistant. In polyatomic molecules, this loss of resonance between the ir photons and the frequency of the "target" vibration is compensated for by the fact that there is a pumping region (II), where excitation is not very sensitive to laser wavelength. Such is not the case in a diatomic molecule.

However, deuterium isotope separation cannot easily be carried out on small polyatomic molecules, such as water. In H_2O, region II occurs at too high an energy, where the effects of loss of the initial resonance between the molecule and the incident photons are already very detrimental. In the case of water, there is the additional complication of free-radical chemistry that occurs after the decomposition pulse,[60] which scrambles the initial isotopic selectivity.

It is clear from Fig. 7.13 that excitation of the fully deuterated form of TCE (D_3), using the 10P(32) line (932 cm^{-1}), in a mixture with fully protonated TCE (H_3), should preferentially dissociate the deuterated form.† In this case, that is exactly what is observed.

TCE decomposes by the following mechanism, following laser irradiation:

$$CD_3CCl_3 + nh\nu \longrightarrow CD_2CCl_2 + DCl \tag{4}$$

(or replace D atoms with H atoms, for the fully protonated form), where n is the number of CO_2 laser photons absorbed. By observing the deuterium concentration in the 1,1-dichloroethene product (CD_2CCl_2 or CH_2CCL_2), it was found that the fully deuterated form of TCE was decomposed with an isotope selectivity of greater than 10,000:1,[59b,59c] where isotope selectivity $\beta_{D/H}$ is defined as

$$\beta_{D/H} = \frac{D_{\text{in products}}}{H_{\text{in products}}} \times \frac{D_{\text{in reactants}}}{H_{\text{in reactants}}} \tag{7.21}$$

This contrasts with typical selectivities of 1 to 4 in conventional physical, or catalytic, isotope separation processes.[56b]

Research toward the application of this technique to the separation of uranium, ^{13}C and D has been extensive. Uranium has been enriched in the ^{235}U isotope using the following scheme:[61]

$$UF_{6 \text{ vapor}} + nh\nu(\approx 16\ \mu m) \longrightarrow UF_{5 \text{ solid}} + F \tag{5}$$

---

†Similarly, excitation of the mono-deuterated form of TCE (Fig. 7–13) using either the 9P(20) line (1044 cm^{-1}) or the 10P(20) line (944 cm^{-1}) of the CO_2 laser should lead to preferential decomposition of CDH_2CCl_3.

If the target molecule is $^{235}UF_6$, which has a small (0.6 cm^{-1}) spectral shift from $^{238}UF_6$, then ^{235}U can be enriched in the product. After decomposition of UF_6, the isotope separation becomes a simple physical one, i.e., the separation of a gas from a solid. The 16-μm light can be generated by optically pumping an ammonia laser,[62] using a high-power, pulsed CO_2 laser. A uranium enrichment pilot plant based on this process has been built. A pilot plant for the enrichment of ^{13}C, using $CHFCl_2$ as a target molecule, has also been constructed.[63]

Medical Applications. Because 10.6-μm light is absorbed very strongly by water, which is present in all living cells, CO_2 lasers are a very useful tool in general surgery, where they are used clinically.[64] Because of the ability of the laser to ablate material, it is ideal for cutting tissue. It also heats the tissue, which means that the laser can cauterize at the same time that it cuts, thus greatly assisting the healing process. Medical applications of ir lasers are dealt with in more detail in Chap. 24.

CO_2 LIDAR. The earliest known application of lasers was in LIDAR,[65] an acronym for "light detection and ranging," which is a generic term encompassing many different forms of laser remote sensing.[66] The basic principles for all forms are the same, however. A pulse of light from a laser is directed toward some target and a bit of it is split off for use as a zero-time reference. A scattered signal is then collected, collimated, and frequency-analyzed. The scattering can be from a solid surface (laser range finding), from a liquid, an aerosol, or a gas. The form of the scattering can be elastic (Rayleigh, resonance, reflection, or Mie) or inelastic (Stokes, or anti-Stokes, Raman, and fluorescence). In another variation, absorption of a laser beam can be measured with remote sensing by looking at the attenuation of a reflected beam.

Most early LIDAR applications employed visible or uv light, primarily because high-speed sensitive detectors were readily available for these wavelengths and because the lasers were more compact than early CO_2 lasers. Near ir, visible and uv lasers all suffer from one major handicap, namely, that they are not eye-safe. Since lasers with high pulse energies are often used in LIDAR, eye safety represents a major concern. The development of compact rf-excited CO_2 lasers, as well as improvements in ir detectors, combined with the sensitivity of heterodyne detection, has made CO_2 lasers much more useful tools for LIDAR applications. Further details concerning LIDAR (and CO_2 LIDAR) can be found elsewhere.[66]

7.3.2 Chemical Lasers

Chemical lasers are "lasers operating on a population inversion produced directly, or indirectly, in the course of an exothermic reaction."[67] The discovery of the chemical laser came about because of fundamental studies in the field of molecular reaction dynamics.[17] They can be highly efficient sources of ir light but have not been exploited as fully as CO_2 lasers. This is primarily because cw-chemical lasers, which would be of primary industrial interest, are much more exotic and are difficult to design. Pulsed chemical lasers are not any more sophisticated than pulsed CO_2 lasers, but the beam quality and frequency characteristics are not usually as good. Hence unless there is a wavelength-specific application, there has not been a large demand for them. There are such applications in the military, where DF chemical lasers, both cw and pulsed, have been in large demand, precisely because of the wavelength at which they lase and because they can be very efficient and scaled to high powers.

Principles of Operation   Early studies of reaction dynamics focused on the energy disposal in exothermic atom-molecule exchange reactions. An example of one of the earliest reactions studied[68] (as mentioned in Section 7.2.3) is that of Cl atoms with HI (hydrogen iodide). This atom transfer reaction is written

$$Cl + HI \longrightarrow I + HCl \tag{6}$$

The relative energetics of the HI reactant and HCl product are shown in Fig. 7.14. Reaction (6) is exothermic by 134 kJ mol^{-1}, but, from Fig. 7.14, it is clear that there are a large number of ways that the reaction exothermicity can be disposed of.

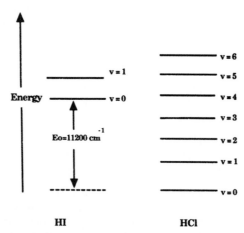

FIGURE 7.14   Relative energies of HI and HCl as well as vibrational energy spacings in the molecules.

The exothermicity is not large enough to result in electronic excitation of HCl; but the energy may be distributed in a large number of ways among the translational, vibrational, and rotational degrees of freedom of the molecule. Eventually, molecular collisions will thermalize the system and lead to equipartitioning of the energy with a Boltzmann statistical distribution. However, if there is some dynamical constraint in the system, which is not uncommon in the course of atom-molecule exchange reactions, then at short times, following reactive collisions between Cl atoms and HI molecules, the product-state distribution may be decidedly nonstatistical. Experimental and theoretical studies have shown that dynamical constraints in three-atom exchange reactions, particularly the existence of a barrier in the entrance channel of the reaction coordinate,[69] can lead to a great deal of vibrational excitation among the reaction products.

The most straightforward way of measuring the vibrational energy distribution following atom-molecule collisions, such as that in reaction (6), is to observe the ir chemiluminescence from HCl. When this is done for reaction (6), the distribution in Fig. 7.15 is observed. It is clear, from this figure, that there is a population inversion between the $v = 3$ and 2 levels of HCl and between the $v = 2$ and 1 levels. If a reaction such as (6) is carried out inside a resonant cavity, then there is a possibility that the chemiluminescence can be amplified and a chemical laser

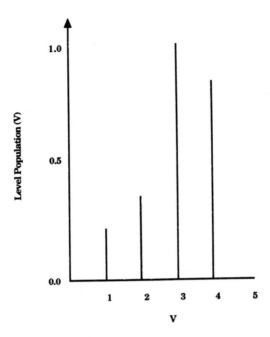

FIGURE 7.15 Nascent vibrational population distributions of HCl following HI + Cl reactive collision.

created. Such has indeed been done for many systems, including the one in reaction (6).[68]

Dynamical effects in atom-molecule collisions can create population inversions between vibrational levels in many processes.[67] In some reactions two vibrationally excited products, with inverted population distributions, can be created. For example:

$$NO_2 \xrightarrow{h\nu_{uv}} NO + O \qquad (7)$$

then

$$CS_2 + O \longrightarrow CS^* + SO^* \qquad (8)$$

Other chemical lasers have been produced following unimolecular reactions. The iodine laser is an example of such a process. A population inversion can be created between the $^2P_{1/2}$ and $^2P_{3/2}$ states in the following reaction:

$$CH_3I \xrightarrow{h\nu_{uv}} CH_3 + I^*(^2P_{1/2}) \qquad (9)$$

The most powerful and most common chemical laser is the HF/DF laser. In this laser collisions between F atoms and H_2, or D_2, produce HF, or DF, with inverted vibrational population distributions. An example of such a reaction is

$$SF_6 \xrightarrow[e^- \text{ discharge}]{h\nu_{uv} \text{ or}} SF_5 + F \qquad (10)$$

then

$$F + H_2 \longrightarrow H + HF^* \tag{11}$$

and

$$HF^* \longrightarrow HF + h\nu_{ir} \tag{12}$$

The energetics of the reaction are shown in Fig. 7.16. A substantial population inversion is created between the $v = 2$ and $v = 1$ states of HF, with a ratio of $(N_{v=2}/N_{v=1}) = 3.2$. The populations were measured by ir chemiluminescence so that no population for the $v = 0$ level could be determined, although experiments on pulsed HF chemical lasers imply that, initially, there is no inversion of population between $v = 1$ and $v = 0$. In the analogous reaction that gives rise to DF, states up to $v = 4$ achieve significant population, and an inversion typically exists between the $v = 3/v = 2$ states and between the $v = 2/v = 1$ states.[70]

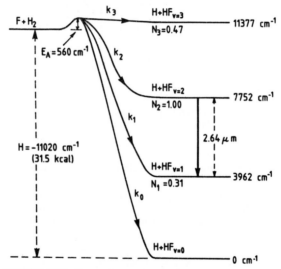

FIGURE 7.16 Energetics of F + H$_2$ reactive collisions. The $v = 2 \rightarrow v = 1$ transition is the first one to achieve lasing in pulsed systems. Lasing also occurs from $v = 3 \rightarrow v = 2$ and $v = 1 \rightarrow v = 0$.

The relative populations given in Fig. 7.16 arise from experiments in which a beam of F atoms is crossed with a beam of H_2 or D_2, which is not the manner in which chemical lasers are normally pumped. However, the experiments are useful in illustrating how a population inversion can be obtained and how large the population inversion can be. In a commercial laser† based on reactions (10) to (12), for instance, in which 80 kV is discharged through a mixture of $SF_6/H_2/He/O_2$ (93/5/1.6/0.4 percent), lasing is observed from $v = 3 \rightarrow v = 2$, $v = 2 \rightarrow v = 1$ and $v = 1 \rightarrow v = 0$.

---

†The laser is a Lumonics series-210 HF/DF laser which operates with a transverse discharge at about 100 torr total pressure.

Lasing in chemical lasers typically takes place in such a cascade process. This cascade is best illustrated in pulsed laser systems, where different lines achieve threshold at different times. As might be expected, from the populations measured for HF* (see Fig. 7.16) the first transitions to achieve threshold are the rotational transitions in the $v = 2 \rightarrow v = 1$ band. Provided that lasing depopulates the $v = 2$ level of HF faster than collisions deactivate the $v = 3$ level, an inversion is created between the $v = 3$ and $v = 2$ levels and lasing between these begins. Finally, both of the above processes eventually build up population in the $v = 1$ level, thus creating an inversion between $v = 1$ and $v = 0$. Lasing then occurs between these levels as well.[70]

Although population inversion is created between pairs of vibrational levels in HF, lasing is not observed in all rotational branches; only in P branches. This is because statistical redistribution of rotational energy is much faster than vibrational redistribution. Hence, with a molecule such as HF, which has a very small moment of inertia, the rotational levels can be far enough apart that energy redistribution among rotational states can reverse a population inversion created in an exothermic reaction, for R-branch transitions,† even though a vibrational inversion may exist.

In addition to reactions (6) to (12), population inversion has also been achieved in abstraction,[71] elimination,[72] and photoelimination[73] reactions. Chemical lasers based on atom-diatomic molecules (also called three-atom exchange) are the most common, and HF/DF lasers are the only ones that are commercially available.‡

Three-atom exchange reactions, of the non-chain-reaction type, have limited efficiencies because HF is collisionally deactivated fairly rapidly. One reaction, which has the potential to have much higher efficiencies, is the H_2/F_2 chain reaction. In addition to reaction (10), one has

$$H + F_2 \longrightarrow HF^* + F \tag{13}$$

which is exothermic by 409 kJ mol^{-1}. Reaction (13) then creates F atoms which can feed back into reaction (11). Such a chain reaction has an advantage over a nonchain reaction because the power output of the laser is not directly related to the power input (electrical, for example), which is required to generate the F atoms that initiate the chain reaction.

In reaction (13), HF is produced with vibrational excitation up to the $v = 10$ level. There is also good evidence that there is a chain branching process in these lasers[74] such that

$$HF_{v=4} + F_2 \longrightarrow HF_{v=0} + 2F \tag{14}$$

This means that far fewer F atoms need to be supplied to initiate the chain reaction, thus making it more efficient. On the negative side, the chemistry of a chain reaction laser is much more difficult to control, because the mixtures of H_2 and F_2 are stable only under certain conditions of temperature and pressure.[75]

Chemical lasers can be operated either pulsed or cw, but operation of cw chemical lasers is a much more difficult engineering problem than is the case with

---

†The rotational constant for HF is 20.94 cm^{-1}. Hence the rotational levels, especially for higher J, can be quite far apart. The $v = 1_{J=6}$ level, for example, lies at an energy of 4841.5 cm^{-1}. The $v = 1_{J=8}$ level, on the other hand, is at 5469.7 cm^{-1}, a difference of 628.2 cm^{-1}, which is very significant for all but extremely high temperatures. At 100°C, for example, the population of the $v = 1_{J=6}$ level is 4.1 times that of the $v = 1_{J=8}$ level. Hence the P(7) and R(7) lines (where these are labeled in emission, not absorption), which both terminate on the $v = 2_{J=7}$ level, will clearly have very different gains. Indeed, R lines are not generally seen with HF lasers.

‡In practice, most laser manufacturers will custom-build virtually any kind of laser system, but no chemical lasers other than HF/DF are advertised for sale.

CO_2 lasers. In CO_2 lasers, once the stimulated emission is over, the molecules eventually (in a few μs) return to the ground vibrational level, where they can be reexcited and contribute to population inversion. Hence, in a CO_2 cw laser, there is a steady-state population of CO_2 molecules in the ground state. With a chemical laser such as HF, however, once the stimulated emission has occurred, one is left with cold HF molecules, which are no longer useful for creating a population inversion and must be eliminated from the laser cavity.

The engineering challenge, then, is to produce an active medium with constant pressure of the products and efficient cooling and product removal. The best method found, capable of achieving this end, was to carry out the mixing of F atoms and H_2 molecules in a supersonic free-jet expansion.[76] Using supersonic flow is a good method of maintaining constant pressure and of removing HF molecules and waste heat from the active medium. In addition, F atoms can be generated in the stagnation region of the nozzle, either by thermal dissociation, which is very simple and efficient, or by arc heating. One of the drawbacks is that such lasers require quite complicated nozzle designs, either banks of nozzles or slit sources, and enormous pumps.

Applications of Chemical Lasers.   Chemical lasers have found applications in the areas of basic research where, for instance, a particular excitation wavelength is required. They have also found some limited applications in medicine,[77] although they suffer from the fact that they are usually quite bulky, and in LIDAR.[78]

The principal applications of chemical lasers have come from the military, specifically for DF chemical lasers. There is strong infrared absorption by water vapor at most principal ir laser wavelengths. This makes transmission of laser beams through the atmosphere quite difficult over long distances. However, there is a transmission window in the ir, between approximately 3.5 and 4.1 μm, which conveniently matches the lasing region of DF lasers, 3.52 to 4.05 μm. Because DF lasers can be scaled up to high power and can be very efficient, there are a number of useful applications, military and nonmilitary.

DF lasers are a natural source of light for use with LIDAR, over large distances in the atmosphere. They cannot be made as compact as CO_2 lasers and are not as widely used; but their superior atmospheric propagation makes them the clear choice for long-distance applications. DF lasers have also been developed, extensively, for use in the Defense Initiative Program (STAR Wars)[79] for the application of disabling ICBMs. The possibility of building a large ground-based laser system for such an application is not feasible with any other high-efficiency laser on account of absorption due to water vapor. Because the energy required to run the lasers can be mostly chemical, rather than electrical, they have also been considered for space-based military applications.[79]

7.3.3 CO Lasers

Since the first demonstration of lasing from carbon monoxide,[80] the wavelength range has been extended from 4 to 8 μm.[81] Population inversion in CO has been achieved using electrical discharges,[82] chemical reactions,[83] or gas-dynamic expansion.[84] The inversion is between rotational states in different vibrational levels of the ground electronic state, and like HF lasers, only P-branch transitions lase. They can be operated both pulsed and cw, although the most common CO laser is an electrically excited cw laser.

CO lasers have very high efficiencies, regardless of the method of excitation. In the chemical reaction:

$$O + CS \longrightarrow CO^* + S + 313 \text{ kJ mol}^{-1} \qquad (15)$$

over 80 percent of the exothermicity is channeled into vibrational excitation of the CO molecule.[85] In electrical-discharge-pumped lasers, an electrical-to-photon conversion efficiency of 63 percent has been obtained.[86] These lasers have also been scaled up to the multikilowatt level.[87]

Principles of Operation. The principles of operation of CO chemical lasers are similar to those of other chemical lasers (see previous section). Similarly the principles of gas-dynamic excitation are not much different from other types of lasers. The inversion processes for electrical pumping, however, are quite different from other ir gas lasers and rely on Treanor pumping.[88]

In an electrical-discharge-excited CO laser, the vibrations in CO are excited by inelastic collisions with relatively low energy (2 eV) electrons. CO has an even larger cross section for excitation by electrons than does N_2, and thus CO lasers are even more efficient than CO_2 lasers. The vibrational energy in the CO laser is quickly redistributed through *V-V* energy exchange; but CO has unusually inefficient *V-T* transfer. Thus, for a short time after excitation, but a much longer time than for other molecules, there is a disequilibrium between the vibrational and translational temperatures in the system. It is during this interval that Treanor pumping can generate at least partial vibrational population inversions in an ensemble of anharmonic oscillators. In fact, the overpopulation of higher energy levels of an anharmonic oscillator is simply a result of "detailed thermodynamic balancing," in the special case that there is a vibrational-translational disequilibrium.[89]

Treanor pumping can be visualized if one considers a pair of two-level molecules in vibrational equilibrium (see Fig. 7.17). The rates of collisional energy transfer in the following process:

$$A^* + B \underset{k_{-1}}{\overset{k_1}{\rightleftharpoons}} A + B^* \qquad (16)$$

are related such that[90]

$$k_1 = k_{-1} e^{(\Delta E/T)} \qquad (7.22)$$

where T is the translational temperature.

MOLECULE A   MOLECULE B

FIGURE 7.17 Two-level system, with two vibrationally excited molecules.

Vibrational energy-exchanging collisions between molecules follow the $\Delta v = \pm 1$ propensity rule. Hence, in collisions between molecules with different vibrational frequencies, energy will flow from the oscillator with the higher frequency to the one with the lower frequency. In an anharmonic oscillator, molecules in the higher v quantum states have much more energy than those in lower v's; but the vibrational frequencies of molecules in high v states are lower on account of the anharmonicity. Because of the $\Delta v = \pm 1$ propensity rule, collisions involving $\Delta v = \pm 2$ or greater are relatively rare. Thus, in V-V exchanging collisions, vibrational energy will tend to flow "uphill," since molecules in low v states have higher vibrational frequencies.

For vibrational energy to flow "downhill," from a high v state in one molecule to a low v state in another, requires that the more highly vibrationally excited collision partner have some threshold amount of extrakinetic energy to make up for its lower vibrational frequency, in order that a vibrational quantum can be exchanged. For energy transfer from low v states to high v states, there is no threshold. This means that, on average, the vibrational energy will tend to pool in higher v states, provided that the V-T cooling mechanism is relatively inefficient.

For instance, if there is vibrational equilibrium, even if there is V-T disequilibrium, it follows that[91]

$$\frac{E_A}{T_A} - \frac{E_B}{T_B} = \frac{\Delta E}{T} \tag{7.23}$$

where T_A and T_B are the vibrational temperatures of the respective molecules. It follows that if the translational temperature $T < T_A$, then $T_B > T_A$. Hence, the energy flows into the oscillator with the smaller energy spacing even though the molecule may be much more highly vibrationally excited.

This is the essence of Treanor pumping, and in theory it should be possible to produce very high vibrational population inversions. In reality, however, V-T processes and spontaneous emission dilute the effect and only partial inversions are created, as in HF lasers.† Unlike HF, the CO molecule has a relatively large moment of inertia.‡ With HF, this made it possible to maintain a partial V-R inversion at temperatures of several hundred degrees Celsius. With CO lasers, by contrast, it is necessary to keep the plasma tube cooled to well below room temperature, i.e., because the rotations are usually well equilibrated with translations and it is beneficial for lasing to keep both temperatures as low as possible. A lower temperature makes the Boltzmann distribution of rotational population narrower; hence the difference in population between rotational levels two quanta apart is much greater, thus permitting partial inversion. The need for CO lasers to be cooled, often to liquid nitrogen temperature, represents a disadvantage in an industrial environment, compared to a CO_2 laser, which only requires room temperature cooling, usually with water.

Applications of CO Lasers.   There has been renewed interest in the industrial uses of CO lasers even though they are more inconvenient to use than CO_2 lasers. This is primarily because 5-μm radiation is coupled more efficiently into the surface of metals; hence there are potential applications in cutting of metals and in laser surface treatment.[87] It has many more potential industrial applications but is un-

---

†Partial inversion means that the population inversion exists only in a small subset of ro-vibrational transitions, P branches in particular.

‡The rotational constant of CO is 1.93 cm^{-1} compared to 19.44 for HF.

likely to supplant CO_2 lasers because the systems suffer from some of the same inconveniences as chemical lasers.

7.3.4 Far-Infrared Lasers

There are a number of polyatomic molecular gases that have lasing transitions in the mid (5 μm) to far-infrared (1000 μm) with some up to 2-mm range. Most of these transitions lase only in pulsed mode, but several will also lase cw. There are two well-characterized excitation methods for far-infrared (fir) lasers, electrical discharge and optical pumping. Both methods are compatible with either pulsed or cw operation. Usually these molecules have very poor energy-conversion efficiencies and are hard to scale to high powers; hence they have few industrial applications. However, fir beams can be used as probes for plasma diagnostics,[92] solid-state studies,[93] and spectroscopy.[94]

Principles of Operation. The first lasers to operate in the far-infrared were pulsed-discharge lasers.[95] The most heavily studied molecules for use as electrical-discharge fir lasers are triatomics. Most of these molecules such as H_2O, D_2O, H_2S, HCN, DCN, and SO_2 will lase from about 40 μm to longer wavelengths. H_2O has the most lasing transitions, with over 60 known lines from 2 to 220 μm.[96] Normally a grating is used for single-line lasing (see Sec. 7.2.4).

Typically, the inversion is created between two normal modes of the molecule, where one of the modes relaxes faster than the other. The probability of radiative transition, from one mode to another, is normally very small. However, when there is an accidental perturbation between the two modes, at some energy level, the states mix and the probability of the ro-vibrational transition increases greatly so that lasing can occur. Pure rotational transitions can also lase when they are connected in cascade with the perturbed levels.[97] Figure 7.18 shows the water vapor lasing transitions connected with the two perturbed levels $(001)6_{33}$ and $(020)6_{61}$.†

Water vapor is probably the most thoroughly studied discharge-fir laser medium. The first lasers were quite large and bulky. However, recent characterization studies[98] together with the use of optical waveguides[99] have reduced their size considerably so that, with a single laser, good output power can be achieved on many lasing transitions.[100]

The number and the range of fir lasing wavelengths was greatly expanded by the advent of optically pumped lasers (see Sec. 7.2.3). Since the molecules have low absorption coefficients for CO_2 laser wavelengths, the fir cavity must be designed for multipassing of the pump radiation.

As an alternative to multipassing, hollow waveguides are often used in fir resonators to minimize optical losses. The waveguides confine the fir mode, thus achieving better pumping efficiencies and more effective cooling of the gas by the walls.[101] One of the impressive features of optically pumped fir lasers is that output power can be scaled with the pumping power. Laser pulse energies as high as 2.6 J at 385 μm, with a D_2O laser, have been reported.[102]

---

†The numbers in brackets label the normal modes of vibration, as in CO_2, except that H_2O is a bent triatomic molecule; hence the bending vibration is not doubly degenerate. The numbers in the subscript are the rotational quantum numbers. For example, $(0,0,1)6_{3,3}$, means that the H_2O molecule has one quantum in the v_3 mode; the total angular momentum quantum number (J), is 6; and the angular momentum quantum numbers K_a and K_c are 3 and 3, respectively. H_2O, from the point of view of rotational spectroscopy, is an asymmetric-top molecule, so that the angular momenta about the three rotation axes are all different; hence they must be specified by separate quantum numbers.

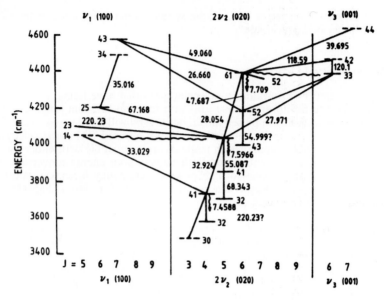

FIGURE 7.18 Water-vapor-laser transitions.

TABLE 7.2 Representative Samples of Far-Infrared Laser Wavelengths

Electrical discharge		Optically pumped	
Molecule	Wavelength, μm	Molecule	Wavelength, μm
H_2O	220.230	D_2O	386
	118.591		356
	89.772		116.6
	78.445		79.9
	67.169	NH_3	281
	55.088		155
	47.224		152
	39.695		148.5
	33.029		124.0
	27.972		105
	26.660		91
	23.365		88.5
	16.932	CH_3F	359.7
D_2O	112.6		320.5
	107.72		288.2
	84.2		250.6
	17.74		230.9
HCN	336.56		221
SO_2	140.8	$^{13}CH_3F$	1202

CH_3F was the first[103] and is still one of the best fir laser molecules to be optically pumped with a CO_2 laser. Other molecules such as NH_3, H_2O, D_2S, CH_3Cl, CH_3Br, and CH_3OH also have optical pumped fir lasing transitions from about 40 to 1200 μm[104] and in one case 2000 μm. NH_3 is a particularly interesting molecule in that it will lase both pulsed and cw very efficiently at 12 μm, when pumped with 9-μm radiation from a CO_2 laser.[105] Table 7.2 gives a list of several common fir lasers and wavelengths of their strongest lines.

7.4 CONCLUSIONS

IR gas lasers are the most widely used lasers in industrial applications, and this situation is unlikely to change. Most of the development in the past has been concentrated on CO_2 lasers because of the unique combination of high efficiency, relative compactness, and ease of construction. Developments of CO_2 lasers, to improve their power per unit area, are ongoing.

Chemical lasers are of more limited use but have found niches in military applications and LIDAR. CO lasers are the most efficient of ir gas lasers, and their shorter output wavelength has advantages over CO_2 lasers in terms of materials processing; but there is a natural resistance to supplanting a proved technology that is cost-effective in many cases. Ironically, the conservatism that kept CO_2 lasers off of the shop floor for many years now helps keep them there. Development of ir gas laser technology is still dynamic, however, and may provide many surprises in the future.

7.5 REFERENCES

1. D. A. Belforte and M. R. Levitt, Annual Technology, Industry and Market Review and Outlook, in D. A. Belforte (ed.), *The Industrial Laser Annual Handbook—1990*, Table 6, PennWell Publishing Co., Tulsa, Okla. 1990.

2. D. M. Roessler, New Laser Processing Developments in the Automotive Industry, in D. A. Belforte (ed.), *The Industrial Laser Annual Handbook—1990*, pp. 109–127, PennWell Publishing Co., Tulsa, Okla. 1990.

3. H. W. Messenger, CO_2 Laser Reaches 50-kW Output, *Laser Focus World*, vol. 27, no. 5, pp. 44–45, May 1991.

4. W. T. Walter, N. Solimene, M. Piltch, and G. Gould, "Efficient Pulsed Gas Discharge Lasers," *IEEE J. Quantum Electron.*, vol. QE-2, pp. 474–479, 1966. (Note: For certain types of pulsed lasers, the lifetime condition between level 2 and 3 can be relaxed. Lasing occurs because of differential pumping between the upper and lower levels.)

5a. W. J. Witteman, "Mode Competition in Lasers with Homogeneous Line Broadening," *IEEE J. Quantum Electron.* vol. QE-5, pp. 92–97, 1969.

5b. M. Sargent III, M. O. Scully, and W. E. Lamb, Jr., *Laser Physics*, chap. IX, pp. 115–136, Addison-Wesley, Reading, Mass., 1974.

6. A. Yariv, *Quantum Electronics*, 3d ed., pp. 176–179, Wiley, New York, 1989.

7. C. C. Davis, "Neutral Gas Lasers," in M. J. Weber (ed.), *Volume* II: *Gas Lasers (CRC Handbook of Laser Science and Technology)*, CRC Press, Inc., Boca Raton, Fla., pp. 55–98, 1982. (Note: This handbook has one of the most comprehensive lists of lasing wavelengths for gas media.)

8. M. J. Seaton, "The Theory of Excitation and Ionization by Electron Impact," in D. R. Bates (ed.), *Atomic and Molecular Processes*, pp. 375–420, Academic Press, New York, 1962.

9. H. S. W. Massey and E. H. S. Burhop, *Electronic and Ionic Impact Phenomena*, Oxford University Press, London, 1969.

10. O. Svelto, *Principles of Lasers* (translated by D. C. Hanna), pp. 83–87, Plenum Press, New York, 1976.

11. M. L. Bhaumik, W. B. Lacina, and M. M. Mann, "High Efficiency Carbon Monoxide Laser at Room Temperature," *IEEE J. Quantum Electron.*, vol. QE-6, p. 575, 1970.

12. M. C. Richardson, K. Leopold, and A. J. Alcock, "Large Aperture CO_2 Laser Discharges," *IEEE J. Quantum Electron.*, vol. QE-9, pp. 934–939, 1973.

13. G. Allcock and D.R. Hall, "An Efficient, RF Excited, Waveguide CO_2 Laser," *Opt. Commun.*, vol. 37, pp. 49–55, 1981.

14. J. B. Marion, "Classical Dynamics of Particles and Systems," pp. 413–415, Academic Press, New York 1970.

15. C. K. N. Patel, Gas Lasers, in A. K. Levine (ed.), *Lasers—Vol. 2*, pp. 101–110, Marcel Dekker, New York, 1968.

16. J. V. V. Kasper and G. C. Pimentel, "HCl Chemical Laser," *Phys. Rev. Lett.*, vol. 14, p. 352, 1965.

17a. J. C. Polanyi, "Proposal for an IR Maser Dependent on Vibrational Excitation," *J. Chem. Phys.*, vol. 34, pp. 347–348, 1961.

17b. J. V. V. Kasper and G. C. Pimentel, "Atomic Iodine Predissociation Laser," *Appl. Phys. Lett.*, vol. 5, pp. 231–233, 1964.

18. E. T. Gerry, "Gasdynamic Laser," *J. Soc. Photo-Opt. Instrum. Eng.*, vol. 9, pp. 61–70, 1971.

19a. H. v. Bülow and E. Zeyfang, "Gas-dynamically Cooled CO-laser with RF-excitation—Design and Performance," Eighth International Symposium on Gas Flow and Chemical Lasers, Madrid, Sept. 10–14, 1990, SPIE vol. 1397, pp. 499–502, 1990.

19b. E. Zeyfang, H. v. Bülow, and M. Stöhr, "Gas-dynamically Cooled CO-laser with RF-excitation—Optical Performance," Eighth International Symposium on Gas Flow and Chemical Lasers, Madrid, Sept. 10–14, 1990, SPIE vol. 1397, pp. 449–452, 1990.

20. N. Skribanowitz, I. P. Herman, and M. S. Feld, "Laser Oscillation and Anisotropic Gain in the $1 \rightarrow 0$ Band of the Optically Pumped HF Gas," *Appl. Phys. Lett.*, vol. 21, pp. 466–470, 1972.

21. O. Svelto, Ref. 10, p. 41.

22. C. K. N. Patel, Ref. 15, pp. 101–110.

23. H. Kogelnik and I. Li, "Laser Beams and Resonators," *Appl Opt.*, vol. 5, pp. 1550–1567, 1966.

24. J. P. Goldsborough, "Design of Gas Lasers," in F. T. Arecchi and E. O. Schulz-Dubois (eds.), *Laser Handbook, Vol. 1*, pp. 605–606, North-Holland, New York, 1972.

25. A. Yariv, Ref. 6, pp. 136–145.

26. G. R. Fowles, *Introduction to Modern Optics*, p. 279, Holt Rinehart and Winston, New York, 1975.

27. A. E. Siegman, *Lasers*, chaps. 22–23, pp. 858–922, University Science Books, Mill Valley, Calif. 1986. (Note: Chapter 22 explains the basic concepts of unstable resonators with hard edge mirrors, and describes a few applications. Chapter 23 introduces more advanced concepts with emphasis on the use of variable-reflectivity mirrors in unstable resonators.)

28. E. A. J. Marcatili and R. A. Schmeltzer, "Hollow Metallic and Dielectric Waveguides for Long Distance Optical Transmission and Lasers," *Bell System Tech. J.*, vol. 43, pp. 1783–1809, 1964.

29. R. L. Abrams and W. B. Bridges, "Characteristics of Sealed-Off Waveguide CO_2, Lasers," *IEEE J. Quantum Electron.*, vol. QE-9, pp. 940–946, 1973.

30. R. L. Taylor and S. Bitterman, "Survey of Vibrational and Relaxation Data for Processes Important in the CO_2-N_2 Laser System," *Rev. Mod. Phys.*, vol. 41, pp. 26–47, 1969.

31. R. L. Abrams, "Waveguide Gas Lasers," in M. L. Stich (ed.), *Laser Handbook—Volume 3*, pp. 41–88, North-Holland, New York, 1979.

32. E. Brannen, "Reflection Gratings as Elements in Far Infrared Masers," *Proc. IEEE*, vol. 53, pp. 2134–2135, 1965.

33. E. Brannen and D. G. Rumbold, "Reflectivity and Polarization Characteristics of Reflection Echelette Gratings," *Appl. Opt.*, vol. 8, pp. 1506–1508, 1969.

34. C. K. N. Patel, W. L. Faust, and R. A. McFarlane, *Bull. Am. Phys. Soc.*, vol. 9, p. 500, 1964.

35. C. K. N. Patel, "Selective Excitation through Vibrational Energy Transfer and Optical Maser Action in N_2-CO_2," *Phys. Rev. Lett.*, vol. 13, pp. 617–619, 1964.

36. F. Kaufman and J. R. Kelso, "Vibrationally Excited Ground State Nitrogen in Active Nitrogen," *J. Chem. Phys.*, vol. 28, pp. 510–511, 1958.

37. C. K. Rhodes and A. Szoke, "Gaseous Lasers—Atomic, Molecular and Ionic," in F. T. Arechi and E. O. Schulz-Dubois (eds.), *Laser Handbook Volume 1*, pp. 265–324, North-Holland, Amsterdam, 1972.

38. J. T. Verdeyen, *Laser Electronics*, p. 281, Prentice-Hall, Englewood Cliffs, N.J., 1981.

39. J. Daniel, "High-Power Lasers," U.S. Patent 4,564,947, Jan. 24, 1986.

40. D. Schuocker, "New Designs for CO_2 Lasers," in D. A. Belforte, *The Industrial Laser Annual Handbook—1989*, pp. 139–148, PennWell Publishing Co., Tulsa, Okla. 1989.

41. J. G. Xin and D. R. Hall, *Opt. Commun.*, vol. 58, p. 420, 1986.

42. *Ibid.*

43. H. W. Messenger, Ref. 3.

44. N. A. Ebrahim, "Lasers, Plasmas and Particle Accelerators—Novel Particle Accelerating Techniques for the 21st Century," *Phys. Can.*, vol. 45, December 1989.

45. A. W. Pasternak, D. J. James, J. A. Nilson, D. K. Evans, R. D. McAlpine, H. M. Adams, and E. B. Selkirk, "Short-Pulse CO_2 Laser for Photochemical Studies," *Appl. Opt.*, vol. 20, pp. 3849–3852, 1981.

46. *Ibid.*

47. D. A. Belforte (ed.), *Industrial Laser Annual Handbook—1990*, PennWell Publishing Co., Tulsa, Okla., 1990.

48. W. W. Duley, *Laser Processing and Analysis of Materials*, Plenum Press, New York, 1983.

49. E. M. Breinan, B. H. Kear, and C. M. Banas, "Processing Materials with Lasers," *Phys. Today*, vol. 29, no. 11, pp. 44–50, 1976.

50. P. Moore, C. Kim, and L. S. Weinman, "Processing and Properties of Laser Surface Melted Titanium Alloys," in Applications of Lasers in Materials Processing, Proceedings of Conference, Washington, D.C., pp. 259–272, Apr. 18–20, 1979.

51. J. S. Eckersley, "Laser Application in Metal Surface Hardening," in A. Niku-Lori (ed.), *Advances in Surface Treatments*, pp. 211–231, Pergamon Press, 1984.

52. *Ibid.*

53. C. W. Draper and J. M. Poate, "Laser Surface Alloying," *Int. Metall. Rev.*, vol. 30, pp. 85–108, 1985.

54a. S. M. Copley, D. Beck, O. Exquivel, and M. Bass, "Laser Solid Interactions," *AIP Conference Proceedings*, vol. 50, American Institute of Physics, New York, 1979.

54b. M. Ikeda, N. Mineta, N. Yusunaga, and S. Fujino, "Ceramic Coating with High-Power CO_2 Lasers," *Mater. Proc.*, vol. 38, pp. 135–140, 1983.

54c. S. J. Mathews, Laser Fusing of Hardfacing Alloy Powders, in E. A. Metzbower (ed.), "Lasers in Materials Processing," *ASM Conference Proceedings*, pp. 166–174, Metals Park, Ohio, 1983.

55. B. P. Fairland, A. H. Clauer, R. G. Tung, and B. A. Wilcox, "Quantitative Assessment of Laser-Induced Stress Waves Generated at Confined Surfaces," *Appl. Phys. Lett.*, vol. 25, pp. 431–433, 1974.

56a. V. S. Letokhov and C. B. Moore, "Laser Isotope Separation," in C. B. Moore (ed.), *Chemical and Biochemical Applications of Lasers*, vol. 3, pp. 1–166, Academic Press, New York, 1977.

56b. R. D. McAlpine and D. K. Evans, "Laser Isotope Separation by the Selective Multiphoton Decomposition Process," *Adv. Chem. Phys.*, vol. 60, pp. 31–98, 1985.

56c. D. M. Golden, M. J. Rossi, A. C. Baldwin, and J. R. Barker, "Infrared Multiphoton Decomposition—Photochemistry and Photophysics," *Acc. Chem. Res.*, vol. 23, pp. 56–62, 1981.

57. N. R. Isenor and M. C. Richardson, "Dissociation and Breakdown of Molecular Gases by Pulsed CO_2 Laser Radiation," *Appl. Phys. Lett.*, vol. 18, pp. 224–226, 1971.

58a. S. Mukamel and J. Jortner, "Multiphoton Dissociation in Intense Infrared Laser Fields," *J. Chem. Phys.*, vol. 65, pp. 5204–5225, 1976.

58b. M. Quack, "Reaction Dynamics and Statistical Mechanics of the Preparation of Highly Excited States by Intense Infrared Radiation," *Adv. Chem. Phys.*, vol. 50, pp. 395–473, 1982.

59a. M. Ivanco, D. K. Evans, R. D. McAlpine, G. A. McRae, and A. B. Yamashita, "Infrared Multiphoton Decomposition and the Possibilities of Laser-Based Heavy Water Processes," *Spectrochim. Acta*, vol. 46A, pp. 635–642, 1990.

59b. G. A. McRae, M. Ivanco, P. E. Lee, and J. W. Goodale, "Laser Isotope Separation of Deuterium-Studies with 1,1,1-Trichloroethane," in Proceedings of the International Symposium on Isotope Separation and Chemical Exchange Uranium Enrichment, Tokyo, Oct. 29–Nov. 1, 1990. Bulletin of the Research Laboratory for Nuclear Reactors-Special Issues 1, 1992, eds. Yasukiho Fujii, Takanobu Ishida and Kazou Takeuchi, pp. 1–2.

59c. G. A. McRae, M. Ivanco and R. A. Back, "Laser Initiated Thermal Reactions and Isotopically Selective Decomposition of 1,1,1-Trichloroethane," *Chem. Phys. Lett.*, vol. 185, pp. 95–100, 1991.

60. R. A. Back, D. K. Evans, R. D. McAlpine, E. M. Verpoorte, M. Ivanco, J. W. Goodale, and H. M. Adams, "Multiphoton Decomposition Studies of Ethanol Vapour," *Can. J. Chem.*, vol. 66, pp. 57–65, 1988.

61. B. B. Snavely, "Separation of Uranium Isotopes by Laser Photochemistry," paper G9, VIII International Quantum Electronics Conference, San Francisco, 1974.

62. T. Yoshida, N. Yamabayashi, N. Miyazaki, and K. Fujisawa, "Infrared and Far-Infrared Laser Emissions from a TE CO_2 Laser Pumped NH_3 Gas," *Opt. Commun.*, vol. 26, pp. 410–414, 1978.

63. J. O'Neill (Ontario Hydro Research), private communication.

64. M. Moretti, "Medical-Laser Technology Responds to User Needs," *Laser Focus World*, pp. 89–102, March 1989 (and references therein).

65. D. A. Buddenhagen, B. A. Lengyel, F. J. McClung, and G. F. Smith, *Proceedings of IRE International Convention*, p. 285, New York, 1961, Part 5 (Institute of Radio Engineers, New York).

66. R. M. Measures, *Laser Remote Sensing—Fundamentals and Applications*, Wiley, New York, 1984.

67. G. C. Pimentel and K. L. Kompa, "What Is a Chemical Laser," in R. W. F. Gross and J. F. Bott (eds.), *Handbook of Chemical Lasers*, pp. 1–31, Wiley, New York, 1976.

68. D. H. Maylotte, J. C. Polanyi, and K. B. Woodall, "Energy Distribution among Reaction Products IV, X + HY (X = Cl, Br; Y = Br, I), Cl + DI," *J. Chem. Phys.*, vol. 57, pp. 1547–1560, 1972 (see also Ref. 17a).

69. R. D. Levine and R. B. Bernstein, "Molecular Reaction Dynamics," p. 93, Oxford University Press, New York, 1974.

70. S. N. Suchard, "Lasing from the Upper Vibrational Levels of a Flash-initiated H_2-F_2 Laser," *Appl. Phys. Lett.*, vol. 23, pp. 68–70, 1973.

71a. M. A. Pollack, "Laser Oscillation in Chemically Formed Carbon Monoxide," *Appl. Phys. Lett.*, vol. 8, pp. 237–239, 1966.

71b. J. H. Parker and G. C. Pimentel, "Hydrogen Fluoride Chemical Laser Emission through H Atom Abstraction from Hydrocarbons," *J. Chem. Phys.*, vol. 48, pp. 5273–5274, 1968.

72a. M. J. Berry and G. C. Pimentel, "Hydrogen Fluoride Elimination Chemical Laser," *J. Chem. Phys.*, vol. 49, pp. 5190–5191, 1968.

72b. M. C. Lin, "Chemical Lasers Produced from $O(^1D)$ Atom Reactions II. A New Hydrogen Fluoride Elimination Laser from the $O(^1D)$ + CH_nF_{4-n} (n = 1, 2 and 3) Reactions," *J. Phys. Chem.*. vol. 75, pp. 3642–3644, 1971.

72c. T. D. Padrick, and G. C. Pimentel, "Addition-Elimination Hydrogen Fluoride Chemical Laser," *Appl. Phys. Lett.*, vol. 20, pp. 167–168, 1972.

73. M. J. Berry and G. C. Pimentel, "Hydrogen Halide Photoelimination Chemical Laser," *J. Chem. Phys.*, vol. 51, pp. 2274–2275, 1969.

74. O. M. Batovskii, G. K. Vasil'ev, E. F. Makarov, and V. L. Tal'rose, "Chemical Laser Operating on Branched Chain Reaction of Fluorine with Hydrogen," *JETP Lett.* (English translation), vol. 9, pp. 200–201, 1969.

75. S. N. Suchard, and J. R. Airey, "Pulsed Hydrogen-Halide Chemical Lasers," in R. W. F. Gross and J. F. Bott (eds.), *Handbook of Chemical Lasers*, Wiley, New York, p. 400, 1976.

76. R. F. W. Gross and D. J. Spencer, "Continuous-Wave Hydrogen-Halide Lasers," in R. W. F. Gross and J. F. Bott (eds.), *Handbook of Chemical Lasers*, p. 206, Wiley, New York, 1976.

77a. J. A. Izatt, N. D. Sankey, F. Partovi, M. Fitzmaurice, R. P. Rava, I. Itzkan, and M. S. Feld, "Ablation of Calcified Biological Tissue Using Pulsed Hydrogen Fluoride Laser Irradiation," *IEEE J. Quantum Electron.*, vol. 26, pp. 2261–2270, 1990.

77b. G. J. Jako and H. K. Herman, "Tissue Cutting and Drilling with the Hydrogen Fluoride Laser," in *Proceedings of the International Society for Optical Engineering*, vol. 12, pp. 193–195, SPIE, Bellingham, Wash., 1987.

78. A. J. Beaulieu, J. A. Nilson, and K. O. Tan, "A Practical DF Laser for Rangefinding Applications," in "Laser Radar Technology and Applications," pp. 8–13, Proceedings of the Meeting, Quebec, Canada, Society of Photo-Optical Instrumentation Engineers, Bellingham, Wash., June 3–5, 1986.

79. J. Grossman, "Military Laser Systems," *Photonics Spectra*, vol. 25, (no. 7), pp. 84–90, July 1991 (and references therein).

80. C. K. N. Patel and R. J. Kerl, "Laser Oscillation on $X^1\Sigma^+$ Vibrational-Rotational Transitions of Carbon Monoxide," *Appl. Phys. Lett.*, vol. 5, pp. 81–83, 1964.

81a. T. X. Lin, W. Rohrbeck, and W. Urban, "Long Wavelength Operation of a Continuous Wave CO-Laser up to 8.18 μm," *Appl. Phys.*, vol. B26, pp. 73–76, 1981.

81b. W. Urban, J. X. Lin, V. V. Subramanian, M. Havenith, and J. W. Rich, "Treanor Pumping of CO, Initiated by CO Laser Excitation," *Chem. Phys.*, vol. 130, pp. 389–399, 1989.

82. N. Legay-sommaire, L. Henry, and F. Legay, *C R Acad. Sci.*, vol. A260, p. 3349, 1965.

83a. M. A. Pollock, Ref. 71a, p. 237.

83b. D. W. Gregg and S. J. Thomas, "Analysis of the CS_2-O_2 Chemical Laser Showing New Lines and Selective Excitation," *J. Appl. Phys.*, vol. 39, pp. 4399–4404, 1968.

83c. G. Hancock and I. W. M. Smith, "IR Chemiluminescence from Vibrationally Excited Carbon Monoxide," *Trans. Faraday Soc.*, vol. 67, pp. 2586–2597, 1971.

84. R. L. McKenzie, "Laser Power at 5 μm from the Supersonic Expansion of Carbon Monoxide," *Appl. Phys. Lett.*, vol. 17, pp. 462–464, 1970.

85. R. E. Center, "High Power: Efficient Electrically-Excited CO Lasers," in M. L. Stitch (ed.). *Laser Handbook*, vol. 3, p. 92, North Holland Amsterdam, New York, 1979.

86. M. M. Mann, D. K. Rice, and R. G. Eguchi, "An Experimental Investigation of High Energy CO_2 Lasers," *IEEE J. Quantum Electron.*, vol. QE-10, p. 682, 1974.

87a. T. Fujioka, "Carbon-monoxide Laser Cuts Thick Steel," *Laser Focus World*, vol. 25, no. 1, pp. 32–33, January 1989.

87b. U. Brinkmann (ed.), "CO Lasers Still Struggle for Acceptance but Are Getting Stronger," *Industrial Laser Rev.*, vol. 5, no. 12, pp. 39–40, May 1991.

88. C. E. Treanor, J. W. Rich, and R. G. Rehm, "Vibrational Relaxation of Anharmonic Oscillators with Exchange Dominated Collisions," *J. Chem. Phys.*, vol. 48, pp. 1798–1807, 1968.

89. *Ibid.*

90. R. E. Center, "High-Power, Efficient Electrically-Excited CO Lasers," in M. L. Stitch (ed.), *Laser Handbook*, vol. 3, p. 95, North Holland Amsterdam, New York, 1979.

91. *Ibid.*

92a. J. B. Gerardo and J. T. Verdeyen, "The Laser Interferometer—Application to Plasma Diagnostics," *Proc. IEEE*, vol. 52, pp. 690–697, 1964.

92b. A. Gondhalekar and F. Keilmann, "Proposal of a New Scheme Using Extreme Forward Light Scattering for Ion Temperature Measurement in Stellarator and Tokamak Plasmas," Max-Planck-Institute für Plasma-physik, Rep. 2/202, 1971, *Opt. Commun.*, vol. 14, pp. 263–266, 1975.

92c. S. Goto, Iwama, N. Satomi, M. Yamanaka, T. Ishimura, and H. Ito, "A 28 μm Water-Vapour Laser Interferometer for Plasma Diagnostics," *Int. J. Infrared Millimetre Waves*, vol. 4, pp. 549–559, 1983.

93a. A. Mayer and F. Keilmann, "Far-Infrared Nonlinear Optics: I. $\chi^{(2)}$ Near Ionic Resonance; II. $\chi^{(2)}$ Contributions from the Dynamics of Free Carriers in Semiconductors," *Phys. Rev.*, vol. B, pp. 6954–6968, 1986.

93b. C. R. Pidgeon, G. D. Holah, F. Al-Berkdar, P. C. Taylor, and U. Strom, "Application of Submillimetre Waveguide Lasers to the Study of Absorption in Elemental Amorphous Solids," *Infrared Phys*, vol. 18, pp. 923–927, 1978.

94. E. A. Rinehart, L. W. Hrubesh, and C. G. Stevens, "Pure Rotational Spectroscopy of SO_2 Using Fix Line Far Infrared Lasers," Eighth International Conference on Infrared and Millimeter Waves, Miami Beach, Fla., 1983, Conference Digest, pp. TH5.6/1–2, 1984.

95. A. Crocker, H. A. Gebbie, M. F. Kimmit, and L. E. S. Mathias, "Stimulated Emission in the Far Infra-red," *Nature*, vol. 202, pp. 169–170, 1964.

96. W. S. Benedict, M. A. Pollack, and W. J. Tomlinson III, "The Water-Vapour Laser," *IEEE J. Quantum Electron.*, vol. QE-5, pp. 108–124, 1968.

97. *Ibid.*

98. P. A. Rochefort, E. Brannen, and Z. Kucerovsky, "Pulsed and CW Operation of Helium-Water Vapour Laser at 28 μm," *Appl. Opt.*, vol. 25, pp. 3838–3842, 1986.

99. P. Belland, "Waveguide CW 118.6 μm H_2O 118.6 μm Laser," *Appl. Phys. B*, vol. 27, pp. 123–128, 1982.

100. P. A. Rochefort, E. Brannen, and Z. Kucerovsky, "Multiple Line and Polarization Control in a Far Infrared Laser with a Compound Grating Resonator," *Appl. Opt.*, vol. 30, pp. 1019–1024, 1991. (Note: Many fir lasers have a large number and range of lasing wavelengths, which allows two-wavelength operation with a modified Littrow configuration.)

101. D. T. Hodges and R. D. Reel, "High-Power Operation and Scaling Behaviour of CW Optically Pumped FIR Waveguide Lasers," *IEEE J. Quantum Electron.*, vol. QE-13, pp. 491–494, 1977.

102. R. Behn, M. A. Dupertuis, I. Kjelberg, P. A. Krug, S. A. Salito, and M. R. Siegrist, "Buffer Gases to Increase the Efficiency of an Optically Pumped Far-infrared D_2O Laser," *IEEE J. Quantum Electron.*, vol. QE-21, pp. 1278–1285, 1985.

103. T. Y. Chang and T. J. Bridges, "Laser Action at 452, 496 and 541 μm in Optically Pumped CH_3F," *Opt. Commun.*, vol. 1, pp. 423–426, 1970.

104a. D. J. E. Knight, "Tables of CW Gas Laser Emissions," in M. J. Weber (ed.), Volume II: *Gas Lasers (CRC Handbook of Laser Science and Technology)*, pp. 421–491, CRC Press, Boca Raton, Fla., 1982 (and references therein).

104b. C. T. Gross, J. Kiess, A. Mayer, and F. Keilmann, "Pulsed High-Power Far-Infrared Gas Lasers—Performance and Spectral Survey," *IEEE J. Quantum Electron.*, vol. QE-23, pp. 377–384, 1987.

104c. R. Wessel, T. Theiler, and F. Keilmann, "Pulsed High-Power Far-Infrared Gas Lasers," *IEEE J. Quantum Electron.*, vol. QE-23, pp. 385–387, 1987.

105a. D. G. Biron, B. G. Danly, R. J. Temkin, and B. Lax, "Far-infrared Raman Laser with High Intensity Laser Pumping," *IEEE J. Quantum Electron.*, vol. QE-17, pp. 2146–2152, 1981.

105b. H. D. Morrison, B. K. Garside, and J. Reid, "Dynamics of the Optically Pumped Mid-infrared NH_3 Laser at High Pump Power—Part I: Inversion Gain," *IEEE J. Quantum Electron.*, vol. QE-20, pp. 1051–1059, 1984.

105c. H. D. Morrison, B. K. Garside, and J. Reid, "Dynamics of the Optically Pumped Mid-infrared NH_3 Laser at High Pump Power—Part II: Raman Gain and ac Stark Shifts," *IEEE J. Quantum Electron.*, vol. QE-20, pp. 1060–1064, 1984.

106. G. Herzberg, *Infrared and Raman Spectra of Polyatomic Molecules*, p. 217, Van Nostrand, New York, 1945.

CHAPTER 8
FREE-ELECTRON LASERS

John A. Pasour

8.1 INTRODUCTION

The free-electron laser (FEL) is different in many ways from other types of lasers. As the name suggests, the radiation from the FEL is produced by a beam of free or unbound electrons. The electrons radiate when they are forced to oscillate in a regular fashion by an appropriate applied field. Thus the FEL has more in common with synchrotron light sources and microwave tubes than with conventional lasers. However, like other lasers, the FEL is capable of generating highly coherent, near-diffraction-limited radiation. Over the past 15 years, great strides have been made in both the theoretical understanding of the FEL interaction and the experimental verification of the many promising predictions of the theory. Experimental research has advanced from the first demonstrations of the physical mechanism to the operation of numerous successful devices at laboratories around the world. These FELs operate at wavelengths from 10^{-5} to 1 cm, and some have power levels in excess of any other source in their spectral regime.

The large amount of research on FELs has been motivated primarily by the unique capabilities of the device and the many applications which require these features. One of the most attractive features of the FEL is its tunability. Unlike typical lasers, the FEL can be described completely by classical mechanics, at least for all of the operating regimes demonstrated to date. The FEL output frequency is determined by continuously variable experimental parameters, such as the kinetic energy of the electrons and the strength of the periodic field which drives the oscillations. Consequently, the FEL may be continuously tuned by varying these parameters. Other important features of the FEL are its high power capabilities and its high efficiency. Because there is no physical lasing medium which must support the radiation field, problems of heating or breakdown which plague conventional solid or gaseous lasers are absent. Enormous powers can be deposited in relativistic electron beams propagating in vacuum. Efficiencies of transferring this electron kinetic energy to FEL radiation have reached 40 percent at millimeter-to-centimeter wavelengths.

These unique features are desirable for a wide range of applications. For example, there is a need for tunable, efficient infrared to ultraviolet sources for biomedical and photochemical applications, laser isotope separation, materials processing, and physics research. High-power sources at these wavelengths have a number of military applications. There are also a number of applications at longer wavelengths. Plasma heating at the electron cyclotron resonance in high-field fusion devices requires efficient millimeter to submillimeter sources at powers >10 MW. High-resolution, long-range radar also needs powerful millimeter to submillimeter sources. FELs are also being examined for various advanced particle accelerator concepts, such as high-frequency, high-accelerating-gradient rf accelerators.

To give a picture of the peak-power capabilities of FEL devices, some representative results are shown in Fig. 8.1. At longer wavelengths, FELs already offer the highest power available from any source. At shorter wavelengths, the peak FEL power is still less than that from conventional lasers, but FELs can operate at frequencies not reachable by other coherent sources. The extremely broad tuning capability of the FEL is evident in some of these devices. The power levels of short-wavelength FELs are also rapidly increasing as more experimental research is completed.

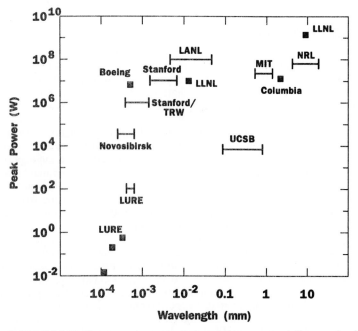

FIGURE 8.1   Peak output power vs. wavelength for a representative sample of FELs.

This chapter begins with a simple theoretical overview of the FEL mechanism, including various operating regimes and efficiency-enhancement schemes. The major components of the FEL are then detailed, and a historical overview of FEL devices is given. A brief discussion of ongoing research and technological challenges remaining to be solved concludes the chapter. For the reader interested in additional details, there are numerous references. In particular, two excellent books[1,2] have recently been written on the subject. A good cross section of early FEL work is contained in Ref. 3. Progress in FEL research has been well documented in a number of special issues of journals[4] and in the proceedings of the annual International Free-Electron Laser Conferences, which have been published recently as special issues of *Nuclear Instruments and Methods in Physics Research.* An extensive list of references is contained in the review by Roberson and Sprangle.[5]

8.2 FEL THEORY

8.2.1 Physical Mechanism

The FEL is conceptually quite simple, consisting only of an electron beam, a periodic pump field, and the radiation field. The most common pump field is a static periodic magnetic field called a wiggler, but any field capable of producing a transverse electron oscillation could in principle be used. A typical configuration of the FEL is illustrated in the simple schematic diagram of Fig. 8.2. The wiggler field is perpendicular to the FEL axis, so electrons injected into the wiggler along the axis begin to oscillate because of the $v \times B$ force. The radiation from the oscillating electrons combines with the wiggler field to produce a beat wave (referred to as the ponderomotive potential), which tends to axially bunch the electrons.

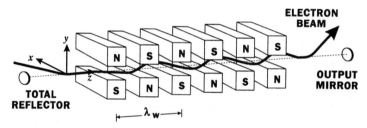

FIGURE 8.2   General configuration of an FEL oscillator.

The bunching caused by the ponderomotive potential provides the coherence that distinguishes the FEL from ordinary synchrotron light sources.

To show explicitly how the ponderomotive force arises and to see how the radiation frequency of the FEL is determined, we assume a particular form for the magnetic wiggler field:

$$B_w = B_w \cos(k_w z)\hat{e}_y \tag{8.1}$$

Here, the wiggler period is $\lambda_w = 2\pi/k_w$, and $\hat{e}_y$ is the unit vector in the y direction. The radiation fields are assumed to have amplitudes E_R and B_R and vary as $\cos(kz - \omega t)$, where the frequency and wavelength are related by the usual vacuum dispersion relation: $\omega = ck = 2\pi c/\lambda$. Electrons injected into the wiggler with axial velocity v_0 acquire a transverse oscillatory velocity (or wiggle velocity) whose amplitude can easily be shown from the Lorentz force equation to be given by

$$v_w = \frac{eB_w}{\gamma_0 m k_w} \sin(k_w z)\hat{e}_x \tag{8.2}$$

where $-e$ is the charge of an electron, $\gamma_0 = (1 - v_0^2/c^2)^{-1/2}$ is the relativistic mass factor, c is the speed of light, and m is the rest mass of the electron. The $v_w \times B_R$ term in the force equation then produces the axial ponderomotive force, which can be shown to vary as $\sin[(k + k_w)z - \omega t]$. The argument of the sine function is just the relative phase between an electron's oscillatory motion and the radiation field, which is usually denoted by ψ. The phase velocity $v_{ph} = \omega/(k + k_w)$ of the ponderomotive potential must be approximately equal to the electron axial velocity in order for the electrons to remain in phase with the potential "wave" long enough to become strongly bunched. Thus we find that to synchronize the ponderomotive wave and the electrons, the frequency must satisfy the resonance condition $\omega = v_z k_w/(1 - \beta_z)$, where $\beta_z = v_z/c$. In terms of the wavelength, this expression is

$$\lambda = \frac{\lambda_w}{\beta_z(1 + \beta_z)\gamma_z^2} \tag{8.3}$$

which for highly relativistic electron beams ($v_0 \approx c$) can be written as $\lambda \approx \lambda_w/2\gamma_z^2$. Using the relation $(\gamma_0/\gamma_z)^2 = 1 + (\gamma_0\beta_\perp)^2$, assuming that $\beta_\perp = \beta_w$, and averaging over a wiggler period, we obtain the well-known FEL resonance condition

$$\lambda = \frac{\lambda_w}{2\gamma_0^2}(1 + a_w^2) \tag{8.4}$$

where $a_w = e\langle B_w\rangle/(mck_w)$ is the normalized wiggler field and $\langle B_w\rangle$ is the rms wiggler field amplitude. It is obvious from Eq. (8.4) that the FEL can be continuously tuned by varying the electron beam energy or the wiggler field amplitude.

Free-electron lasers, like others lasers, can operate either as amplifiers or as oscillators. In the amplifier configuration, an input signal injected into the wiggler along with the electron beam is amplified during a single pass through the wiggler. To be practical, FEL amplifiers require high growth rates, which typically limits their operation to the infrared to millimeter wavelength regime. In an oscillator, a portion of the spontaneously emitted radiation (that having the resonant frequency) is amplified during repeated passes through the interaction region. Although FEL resonators can be similar to those of conventional lasers, there must be a provision for injecting and ejecting the electron beam. Usually this is accomplished by bending the electron beam with an appropriate magnet. Because a high gain per pass is not necessary (assuming the electron beam pulse is of sufficient duration), oscillators are appropriate for short-wavelength operation, where the gain is often relatively low. An important advantage of FEL oscillators is that they can operate at frequencies where other sources do not exist.

8.2.2 FEL Operating Regimes

There are three major operating regimes of FELs, depending on the electron beam and wiggler parameters, in which relatively simple expressions for growth rates and efficiencies can be found. These regimes are referred to as the Compton or single-particle regime, the Raman or collective regime, and the strong-pump or high-gain Compton regime. Free-electron lasers powered by high-energy, low-current, and high-quality beams, such as are produced by rf linacs or storage rings, typically operate in the Compton regime. In the Compton regime, space-charge effects can be neglected because the collective space-charge oscillations are damped by electron thermal motions. The space-charge oscillations occur at the beam plasma frequency $\omega_p = (ne^2/\epsilon_0 m)^{1/2}$, and space-charge effects can be neglected when

$$\frac{\omega_p}{\omega} \gamma^{1/2} \ll \frac{\Delta\gamma}{\gamma} \qquad (8.5)$$

This is the criterion for operation in the Compton regime. The radiation growth rate is usually low owing to the single-particle nature of the interaction, so oscillator operation is preferred. The intrinsic efficiency in the Compton regime is also low, typically 1 percent or less, but efficiency-enhancement techniques (which will be described below) can increase this substantially. If the beam energy and the beam quality are sufficiently high, Compton FELs can operate at short wavelengths, i.e., in the visible or uv.

The other two FEL regimes are characterized by high-current (kA), relatively low-energy (MeV) electron beams. These beams are generally produced by pulse line accelerators, high-current modulators, or induction linacs. The relatively low beam energy limits operation to long wavelengths (millimeter to submillimeter), but impressively large gains on the order of 40 to 50 dB per pass and intrinsic efficiencies on the order of 10 percent have been demonstrated in several experiments. Operation in either the amplifier or the oscillator mode is possible. In the Raman regime, the current is sufficiently large that space-charge waves on the beam are excited, thereby significantly altering the details of the FEL interaction.[1,5,6] In the high-gain Compton or strong-pump regime, the beam current is large, but the wiggler field is so strong that the ponderomotive force dominates the space-charge forces. The distinction between the Raman regime and the high-gain Compton regime can be expressed in terms of a critical plasma density

$$\omega_{\mathrm{crit}} = F \frac{\beta_z^3 \gamma_z^3 a_w^2 c k_w}{2\gamma_0^{3/2}} \qquad (8.6)$$

where F is the filling factor giving the fractional overlap between the electron beam and the radiation field. When $\omega_p \ll \omega_{\mathrm{crit}}$, operation is in the high-gain Compton regime. When $\omega_p \gg \omega_{\mathrm{crit}}$, the Raman regime is indicated.

The growth rates and efficiencies in the various regimes take on relatively simple expressions, tabulated in Table 8.1, when the appropriate approximations are made.

TABLE 8.1 Growth and Efficiency in Various FEL Regimes

Regime	Electric field growth Γ (cm^{-1}) or power gain per pass G	Intrinsic power efficiency
Compton (single-particle, low-gain)	$G = 2\pi F \dfrac{\nu}{\gamma_0} \dfrac{a_w^2}{\gamma_0^2} \dfrac{L_w^3}{r_b^2 \lambda_w} g(\alpha)$	$\dfrac{\lambda_w}{2L_w}$
Raman (collective, high-gain)	$\Gamma = \left(\dfrac{\pi \gamma_z F}{r_b \lambda_w}\right)^{1/2} \left(\dfrac{\nu}{\gamma_0}\right)^{1/4} \dfrac{a_w}{\gamma_0}$	$\dfrac{1}{\pi \gamma_z} \left(\dfrac{\nu}{\gamma_0}\right)^{1/2} \dfrac{\lambda_w}{r_b}$
Strong-pump (single-particle, high-gain)	$\Gamma = 2 \dfrac{F^{1/3}}{r_b} \left(\dfrac{\nu}{\gamma_0}\right)^{1/3} \left(\dfrac{r_b}{\lambda_w}\right)^{1/3} \left(\dfrac{a_w}{\gamma_0}\right)^{2/3}$	$0.18 F^{1/3} \left(\dfrac{\nu}{\gamma_0}\right)^{1/3} \left(\dfrac{\lambda_w a_w}{\gamma_0^2 r_b}\right)^{2/3}$

$\nu = I/17\beta$ kA is the normalized beam current, r_b is the beam radius, F is the filling factor, and $g(\alpha)$ is the gain function discussed below.

The radiation field in the two high-gain regimes grows as $\exp(\Gamma z)$. The Compton power gain is defined as $G = P(L_w)/P(0) - 1$, where L_w is the wiggler length. Note that the scaling of the growth rate and efficiency with various parameters is different in the various regimes. Clearly, not all FELs can fall into one of these simple categories because transition regions must exist. Nevertheless, the simple expressions can be quite useful for back-of-the-envelope calculations to predict FEL performance in particular situations. The remainder of the discussion here emphasizes the Compton regime, which is most relevant to operation in the infrared to ultraviolet regions of primary interest in this handbook. A more detailed discussion of the Raman and high-gain Compton regimes and a complete derivation of the growth rates and efficiencies in all three regimes are given by Sprangle et al.[5-7]

The Compton gain expression contains the gain function, defined as

$$g(\alpha) = -\frac{d}{d\alpha} \frac{\sin^2(\alpha)}{\alpha^2}$$

$$\alpha = k_w L_w \frac{\gamma_0 - \gamma_r}{\gamma_r} \tag{8.7}$$

Here γ_r is the resonant electron energy, and we assume that the laser radiation is at the resonant frequency. The gain function is plotted in Fig. 8.3. Unlike conventional lasers, the gain curve is antisymmetric about the resonance, with zero

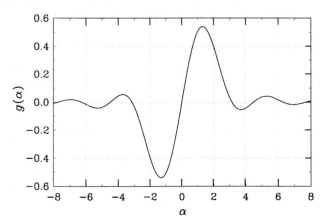

FIGURE 8.3 Gain function for a Compton-regime FEL, as defined by Eq. (8.7).

gain at resonance. From Eq. (8.7) and Fig. 8.3 we see that $g(\alpha) \geqslant 0$ only when $\gamma_0 > \gamma_r$. The width of the positive gain region is $\Delta\alpha = \pi$, corresponding to an energy width $\Delta\gamma/\gamma_r = \lambda_w/2L_w$, which we note from Table 8.1 is simply the intrinsic efficiency of the Compton FEL. From the resonance condition [Eq. (8.4)] the spectral width of the gain curve is then $\Delta\lambda/\lambda = \lambda_w/L_w \equiv 1/N_w$. The maximum value of the gain function is 0.54, which occurs at $\alpha = 1.3$. Thus, for a particular electron energy and frequency, there is an optimum number of wiggler periods $N_{opt} = 0.21\, \gamma_r/(\gamma_0 - \gamma_r)$.

When the gain in an FEL oscillator exceeds the resonator losses, the FEL oscillates and the power increases with time until saturation is reached. To understand gain saturation, it is useful to analyze the interaction in terms of the relative phase between the electrons' oscillatory motion and the radiation field, defined by

$$\psi = (k_w + k)z - \omega t + \phi \tag{8.8}$$

where k, ω, and ϕ are the wave number, frequency, and phase of the optical field. For electrons near resonance in a constant-amplitude radiation field (i.e., at saturation or in the small-gain regime), the evolution of this relative phase is governed by a pendulum-like equation[8]

$$\frac{d^2\psi}{dz^2} = -\Omega^2 \sin(\psi)$$

$$\Omega^2 = \frac{2a_w a_R}{\gamma^2} \tag{8.9}$$

where $a_R = e\langle E_R\rangle/(mc^2)$ is the normalized amplitude of the radiation field. The electrons' motion then can be described in terms of the ponderomotive potential $U = -\Omega^2 \cos \psi$. The FEL interation can therefore be depicted in a phase diagram, as shown in Fig. 8.4, in which the abscissa is the electron position or phase and the ordinate is the momentum. The solid curve represents the potential, whose amplitude is shown increasing unrealistically fast in order to illustrate the concept of saturation and efficiency enhancement. The transfer of energy between the electrons and the optical field is determined from[8]

$$\frac{d\gamma}{dz} = -\frac{\gamma\Omega^2}{2k_w} \sin(\psi) \tag{8.10}$$

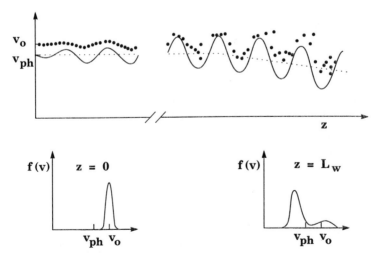

FIGURE 8.4 Schematic diagrams showing electron trapping in the ponderomotive potential wells and efficiency enhancement by decreasing v_{ph} (top) and the initial and final electron distribution functions (bottom).

Thus, depending on their phase with respect to the wave, some electrons gain energy while other lose energy. The electrons must have an initial energy above the resonant value in order to give up net energy to the radiation field. In that case, the radiation field amplitude grows until the ponderomotive potential becomes large enough to trap electrons. Gain saturation occurs as the electrons become trapped and begin to oscillate in the potential wells. In general, only a fraction of the electrons are trapped. Those which enter the FEL with the wrong phase may actually be accelerated and will never become trapped. The motion of a trapped electron is called a synchrotron oscillation, which for an electron with $\psi \ll 1$ has a wavelength

$$\lambda_s = 2\pi \frac{L_w}{\Omega} \tag{8.11}$$

From Eq. (8.9), we see that λ_s varies as the square root of the laser field (fourth root of the laser power). The synchrotron oscillations cause sideband generation, which can become significant when a large portion of the electrons become trapped in the potential wells.[9-11]

8.2.3 Efficiency Enhancement

As discussed above, the intrinsic efficiency of the FEL can be quite small, especially in the Compton regime. Several methods have been developed to enhance the efficiency, however. For example, a tapered wiggler can be used to continue to extract energy from the electrons after they are trapped in the ponderomotive potential wells or "buckets."[7,9] Also, the kinetic energy of the electrons exiting the FEL can be recovered and recycled to the accelerator by using a suitable collector. It turns out that both options are possible and have been demonstrated.

The first method of extending the FEL interaction is to decrease the phase velocity of the ponderomotive wave after the electrons are trapped. If the phase velocity (i.e., the bucket velocity) is decreased gradually as the wave and the trapped electrons travel through the interaction region, the electrons will remain trapped and their velocity will decrease along with the wave velocity. This decrease in electron kinetic energy can be many times larger than can occur during the initial trapping process, as depicted in Fig. 8.4. The phase velocity can be changed in a straight forward way by recalling that $v_{ph} = \omega/(k + k_w) \approx c(1 - \lambda/\lambda_w)$. Hence, by spatially decreasing the wiggler wavelength, the phase velocity of the trapping wave can be decreased to produce the efficiency enhancement.

The second approach to efficiency enhancement is the application of an accelerating force to the trapped electrons. Because the electrons are trapped, they will not be accelerated out of the buckets if the accelerating force is not too large. Instead, the force produces a relative phase shift of the electrons, effectively holding them in the tops of the buckets where they continually lose energy to the wave. The accelerating force can be provided by an axial electric field or by gradually decreasing the amplitude of the wiggler field, thereby decreasing the electron wiggle velocity. By conservation of energy, if the energy corresponding to the wiggle velocity is decreased at the same rate the radiation field energy increases, the axial electron velocity will remain constant and in phase with the ponderomotive potential. In this approach, the energy that would ordinarily appear as an axial acceleration is transferred to the radiation field instead.

Each of the above forms of efficiency enhancement is appropriate only for high-power operation, because a large radiation field amplitude is required to trap the electrons. An alternate approach, called phase-area displacement, uses a reverse-tapered wiggler to accelerate an empty potential well up through the electron beam.[9,12] The electrons are forced to move around the potential wells, resulting in a net negative displacement of all the electrons in phase space. This technique may be quite useful for storage ring FELs, because little energy spread is added to the beam on each pass through the wiggler. However, the gain is reduced because the electrons are not resonant with the ponderomotive potential over much of the wiggler.

Finally, the overall system efficiency can be increased by reusing the electron beam energy which remains after the interaction. This approach is practical in both electrostatic accelerators, in which depressed collectors can be used,[13] and in rf accelerators, in which the spent electrons are passed through a decelerating rf structure.[14] In either case, the recovered energy can be directly applied to the acceleration of additional electrons. A practical problem with this approach is that the energy spread induced by the FEL interaction itself makes beam recovery more difficult, but over 70 percent of the electron energy has been recovered by careful design.

It is clear that in high-power applications, the highest possible overall efficiency will be desired. Consequently, the efficiency-enhancement capability of the FEL is very important. It represents a major advantage over conventional lasers.

8.3 FEL COMPONENTS

Any FEL consists of three major components, each of which is critical to the success of the device: (1) The electron accelerator controls the beam energy, current, pulse duration and repetition rate, and the beam quality. (2) The wiggler, which must be designed to be compatible with the electron beam, is the major contributor to the ponderomotive potential and can provide efficiency enhancement. (3) The radiation optics determine the optical beam quality and must be designed to tolerate the power levels desired. We will briefly address some important issues affecting the choice and/or design of these major components.

8.3.1 Electron Acceleration and Transport

The most important and often most difficult task in any FEL is the generation and transport of a high-quality electron beam. A variation in axial beam energy can greatly impede the electron bunching process, thereby lowering the growth rate and efficiency. A rule of thumb is that the fractional energy spread must be small compared with the intrinsic interaction efficiency, which tends to place a more stringent requirement on beam quality at short wavelengths. Electron energy variations can be produced by the beam emittance (the intrinsic random perpendicular velocity components of the electrons), by gradients in the wiggler field, by the beam's own space charge, and by any voltage variations which occur during acceleration.

The choice of an accelerator is largely determined by the operating energy desired, but the design must be compatible with the other FEL requirements. For operation in the infrared to ultraviolet, electron energies of tens to hundreds of MeV are required. Consequently, appropriate accelerators are rf linacs and storage rings. At longer wavelengths, electrostatic (Van de Graaff) and induction linear accelerators may also be used. Electron beam focusing is usually provided by electromagnets, either solenoids or quadrupoles. Good beam transport is important to maintain high beam quality.

The normalized electron beam emittance $\epsilon_n = \beta\gamma r_b\theta$, where r_b is the beam radius and θ is the angular spread of the electron velocities, is an important quantity in the characterization of beam quality for FELs. Even if all the electrons have identical total energies, the emittance produces a spread in the axial energy component which is given by[15] $\Delta\gamma_z/\gamma_z = \epsilon_n^2/2r_b^2$. Some of the most important contributors to beam emittance are cathode temperature and surface roughness, nonlinear accelerating electric fields or focusing fields, magnetic field aberrations or misalignment, nonlinear space-charge forces, and mismatches between focusing elements. Only by very careful design and fabrication can the beam emittance be kept small enough for efficient FEL operation at short wavelengths.

Another major contributor to axial energy spread in FELs is the wiggler field gradient. The wiggler gradient contribution arises because in any physically realizable wiggler, the magnetic field amplitude must increase away from the axis. In simple terms, the higher off-axis field induces higher wiggle velocities than the lower field on-axis. The higher perpendicular velocity results in a lower axial velocity for the off-axis particles. Quantitatively, $\Delta\gamma_z/\gamma_z = (a_w k_w r_b/2)^2$. Consequently, this effect is more pronounced at higher wiggler fields and larger beam radii. Because the FEL gain and efficiency typically increase with wiggler field and inversely with beam radius, it is usually much better to reduce the beam radius than to reduce the wiggler field amplitude. However, the electron beam radius should not be substantially smaller than that of the radiation beam.

To quantify the reduction in FEL gain due to electron energy variations, it is useful to define a detuning parameter

$$\Theta \equiv \omega \frac{\Delta v_z}{v_z^2} L_w = 4\pi N_w \frac{\Delta\gamma_z}{\gamma_z} \tag{8.12}$$

Jerby and Gover[16] showed that the reduction in gain due to energy spread is very closely approximated by

$$G(\Theta) = \frac{G_{\text{cold}}}{1 + \Theta^2/\pi^2} \tag{8.13}$$

Here, G_{cold} is the cold-beam gain shown in Table 8.1. Typically, the energy spread in short-wavelength FELs is dominated by beam emittance and wiggler gradients. In that case, we can calculate the effective detuning parameter from

$$\Theta^2 = \Theta_\epsilon^2 + \Theta_w^2 = \left(2\pi N_w \frac{\epsilon_n^2}{r_b^2}\right)^2 + (\pi N_w a_w^2 k_w^2 r_b^2)^2 \tag{8.14}$$

It is clear from Eqs. (8.13) and (8.14) that there is an optimum beam radius at which the gain is maximized. This optimum radius is readily found to be[17]

$$r_{\text{opt}} = \frac{1}{a_w k_w} \left[\frac{1}{6N_w^2} (\sqrt{1 + 48(N_w \epsilon_n a_w k_w)^4} - 1) \right]^{1/4} \tag{8.15}$$

However, the optical mode radius, which has a minimum diffraction-limited value of

$$r_{s0} = \left(\frac{\lambda L_w}{2\pi\sqrt{3}} \right)^{1/2} \tag{8.16}$$

imposes a lower limit on the beam radius.[17,18]

The dependence of gain and optimum beam radius on beam current and emittance is shown for a particular case in Fig. 8.5. Here, we have assumed a beam energy of 25 MeV and a 30-period-long wiggler having a period of 2 cm and an rms amplitude of 1 kG. The resulting laser wavelength is about 4 μm. The optimum beam radius decreases with emittance until it becomes equal to the minimum optical mode radius. The emittance range is shown up to only 0.02 cm rad. This value corresponds to the acceptance of the FEL, which Smith and Madey have shown to be[18]

$$A = \frac{1}{4N_w} \sqrt{\lambda\lambda_w} \left(\frac{1 + a_w^2}{a_w^2}\right)^{1/2} \geq \pi \frac{\epsilon_n}{\gamma} \qquad (8.17)$$

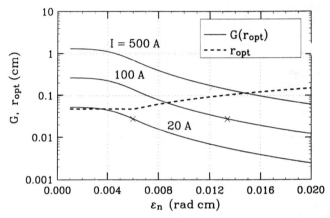

FIGURE 8.5 Optimized FEL gain and optimum beam radius vs. normalized emittance for a particular set of wiggler parameters. Crosses denote Lawson-Penner emittance at the particular peak current levels (assuming a microscopic duty factor of 2 percent).

An empirical relation known as the Lawson-Penner condition relates *average* beam current in rf linacs to the normalized emittance, i.e.,

$$\epsilon_n \approx 0.3\sqrt{\langle I(\text{kA})\rangle} \text{ cm rad} \qquad (8.18)$$

Although no fundamental limitation is suggested by this condition, it is a useful rule of thumb for estimating the emittance levels that can be readily achieved. The Lawson-Penner emittance for the current levels used in Fig. 8.5 is denoted by an x on the two lower-current curves, assuming that the microscopic duty factor is 2 percent, i.e., $I = 50 \langle I \rangle$. The value for the 500-A curve is 0.03 cm rad, which is off the graph. Note that the gain at the Lawson-Penner value of emittance is essentially the same for each of the current values. This suggests the use of the beam brightness, which is defined as $B_n = I/(\pi\epsilon_n)^2$, as a good figure of merit for an electron beam for FEL applications. In fact, for $r_{opt} \geq r_{s0}$, $G(r_{opt})$ varies linearly with B_n, independently of ϵ_n. For the particular parameters used in Fig. 8.5, the constant of proportionality is 5×10^{-7} cm² rad²/A, i.e., B_n must be 10^6 A/(cm rad)² in order to achieve a gain of 0.5. This is 20 times larger than the Lawson-Penner brightness, assuming a 2 percent microscopic duty factor.

It is clear that high-brightness electron beams are required for short-wavelength FEL operation. If emittance is not the dominant contributor to energy spread, however, high brightness is not a sufficient condition. Consequently, Roberson[19] has suggested a generalized version of beam brightness, which he called beam quality and defined as $B_q = I/(\Delta\gamma/\gamma)$, as a more useful figure of merit. Such a definition is particularly appropriate for longer-wavelength FELs, in which there are often a number of important contributors to energy spread.

8.3.2 Wiggler

Nearly all wiggler-based FEL experiments have been performed with either a helical wiggler or a linear wiggler.[1,2] Both permanent magnet and electromagnet wigglers of either type can be fabricated. Helical wigglers produce a perpendicular magnetic field that rotates about the axis. Such a wiggler provides radial focusing for the beam, so that an external focusing field can be avoided.[20] A helical wiggler is also compatible with an external axial magnetic field, which can enhance the FEL interaction via a resonance between the periodic electron oscillations in the wiggler and the gyromotion in the axial field. This effect has been exploited in a number of microwave and millimeter wave FELs.[20] The major disadvantages of the helical wiggler are the difficulty of tapering it for efficiency enhancement and the limited access it provides for diagnosing and aligning the electron beam. It should also be noted that helical wigglers produce circularly polarized radiation. The linear or planar wiggler produces a field in a single transverse direction with a sinusoidally varying amplitude. It therefore produces linearly polarized radiation. However, the linear wiggler provides focusing in only one plane (along the direction of the wiggler field), and it is not compatible with an axial magnetic field, which induces an outward drift of off-axis electrons.[21] Focusing can be provided by quadrupoles, by shaping the wiggler field pole pieces to provide a small degree of field curvature, or by using a "square" wiggler.[21-24] The major advantages of a linear wiggler are the improved access to the beam and the ease of tapering the amplitude or period for efficiency enhancement. The linear wiggler field also has a number of spatial harmonics, which can cause the generation of FEL harmonics.

Schematic diagrams of helical wigglers are shown in Fig. 8.6. The electromagnet shown in Fig. 8.6a consists of two helical windings spaced 1/2 period apart and carrying current in opposite directions. The amplitude of the field on axis is given by[25]

$$B_w = \frac{2\mu_0 I_w}{\lambda_w} \left[\frac{\pi d}{\lambda_w} K_0\left(\frac{\pi d}{\lambda_w}\right) + K_1\left(\frac{\pi d}{\lambda_w}\right) \right] \tag{8.19}$$

where μ_0 is the permeability of free space, d is the diameter of the winding, I_w is the current through the winding, and K_0 and K_1 are Bessel functions. When $d/\lambda_w \geq 1$, this expression may be approximated by

$$B_w[G] \approx 5.6 \frac{I_w[A]}{\lambda_w[cm]} \left(\frac{d}{\lambda_w}\right)^{1/2} \exp\left(-\frac{\pi d}{\lambda_w}\right) \tag{8.20}$$

which is valid to better than 10 percent. One of the problems with helical wigglers is the error field associated with the termination of the windings at each end of the wiggler. A variety of methods have been used to alleviate this problem,[26] including flaring the winding outward to provide an adiabatic entry into the wiggler, using

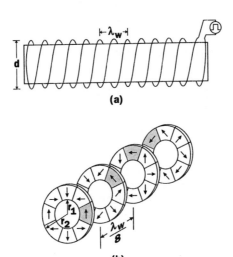

FIGURE 8.6 Helical magnetic wigglers. (*a*) Bifilar helix electromagnet. (*b*) Permanent-magnet helical wiggler.[27]

a graded termination so that the two windings are connected over a number of periods, and providing compensating windings to cancel the error field.

Several versions of permanent-magnet helical wigglers have been built.[27-30] The particular version shown in Fig. 8.6*b* uses eight pie-shaped wedges of permanent-magnet material assembled in an annular disk with the field directions as shown. The assemblies are rotated in eight 45° steps along the wiggler axis to produce one period of the wiggler field. The amplitude of the field on-axis is[27]

$$B_w = 0.8 \, B_r \left[T\!\left(\frac{2\pi r_1}{\lambda_w}\right) - T\!\left(\frac{2\pi r_2}{\lambda_w}\right) \right] \qquad T(x) \equiv K_0(x) + \left(\frac{x}{2}\right)K_1(x) \quad (8.21)$$

Here, r_1 and r_2 are the inner and outer radii of the magnet assemblies, respectively.

Typical linear wiggler configurations are shown in Fig. 8.7. The electromagnet shown in Fig. 8.7*a* uses discrete coils wound around poles in alternating directions. A similar wiggler can be fabricated using a continuous winding which loops back and forth through the poles. The advantage of the discrete coils is that the coil current can be varied to taper the wiggler field amplitude, although the separate power systems required to power the coils in this case add complexity. The field along the axis of the wiggler varies nearly sinusoidally with an amplitude given by[31]

$$B_w = \frac{32\mu_0 N I_w}{\sqrt{2}\pi\lambda_w} \left[\frac{1}{\sinh(\pi h/\lambda_w)} - \frac{1}{3\sinh(3\pi h/\lambda_w)} \right] \qquad (8.22)$$

where h is the distance between the two wiggler faces and NI_w is the number of ampere-turns in the coils.

Several types of permanent-magnet linear wigglers have been built.[31-33] The most common are the all-permanent-magnet design shown in Fig. 8.7*c* and the hybrid wiggler shown in Fig. 8.7*b*. The former of these, often called the Halbach wiggler,[32] uses four magnets per period on each face of the wiggler. The funda-

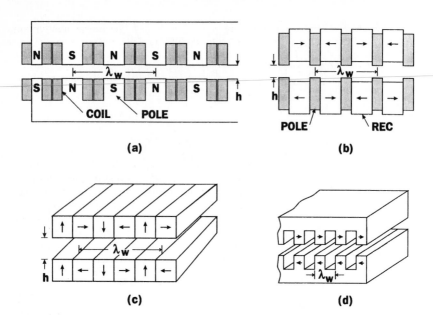

FIGURE 8.7 Planar magnetic wigglers.[31] (*a*) Electromagnet. (*b*) Permanent-magnet hybrid. (*c*) Pure permanent magnet (Halbach configuration). (*d*) Permanent-magnet microwiggler.[33]

mental component of the magnetic field from a general wiggler of this type with M magnets per period, each rotated by an angle $2\pi/M$ from the adjacent one, has an amplitude on axis given by

$$B_w = 2B_r \frac{\sin(\epsilon\pi/M)}{\pi/M} \left[1 - \exp\left(\frac{-2\pi L}{\lambda_w}\right) \right] \exp\left(\frac{-\pi h}{\lambda_w}\right) \qquad (8.23)$$

where B_r is the remanent field of the permanent magnet (usually on the order of 1 T), ϵ is a packing factor denoting the fraction of the axial space occupied by magnets, h is the wiggler gap, and L is the height of the magnets. The maximum field amplitude from a wiggler of this type is obtained when M and L go to infinity and $\epsilon = 1$, in which case

$$B_w(\text{max}) = 2B_r \exp\left(\frac{-\pi h}{\lambda_w}\right) \qquad (8.24)$$

In the typical Halbach configuration, $M = 4$, $L = \lambda_w/4$, and $\epsilon \approx 1$. In this case,

$$B_{w0} = 1.43B_r \exp\left(\frac{-\pi h}{\lambda_w}\right) \qquad (8.25)$$

The hybrid version reduces the field errors due to variations in field strength and direction in the permanent magnets and increases the field on-axis, which is given by[29]

$$B_w = \frac{4B_r}{\sqrt{2}\pi} \left[\frac{1}{\sinh(\pi h/\lambda_w)} - \frac{1}{3\sinh(3\pi h/\lambda_w)} \right] \qquad (8.26)$$

Most permanent magnets are rare-earth-cobalt (REC) or the newer neodymium-iron-boron. These magnets have permeability very near that of vacuum, so they can easily be superimposed.

A number of short-period wigglers have also been built. One example[33] is shown in Fig. 8.7d. This permanent-magnet microwiggler is similar to the Halbach wiggler with $M = 2$, but half the magnets are missing. The maximum value of the field amplitude in this wiggler is a factor of π smaller than that given by Eq. (8.24).

8.3.3 Optical System

The radiation optics required depend on the wavelength of the FEL. Infrared and shorter-wavelength FELs require optical components similar to those found in conventional lasers. A major concern for high-power operation is the damage threshold for these components. At 10 μm, for example, the maximum single-pulse power density that can be tolerated on a copper mirror is about $150/\tau^{1/2}$ MW/cm^2, where τ is the pulse duration in microseconds. Resonator designs capable of handling ultra-high-power levels are being actively studied for the short-wavelength FELs.[34] A useful technique is to use grazing incidence mirrors to spread the radiation over a large surface area. A ring resonator employing such mirrors has the additional advantage that it is relatively insensitive to mirror tilts.[35]

Another important issue for FELs is the design and development of optics suitable for uv and shorter wavelengths. Newnam[36] has reviewed various aspects of the extension of FEL operation into the extreme uv. Although the FEL seems capable of operation at very short wavelengths, the problem of suitable optics is very difficult. To be feasible for FEL operation, it is expected that the mirror reflectivity must be greater than 40 percent. Silicon carbide mirrors might be usable at wavelengths from 60 to 22 nm. Multilayer thin-film mirrors might also be feasible for relatively narrow frequency ranges. Multifaceted metallic mirrors are promising at wavelengths of tens of nm. These mirrors take advantage of the high reflectivity (>95 percent) that can be achieved at large angles of incidence. Such mirrors can also handle relatively high powers. An example of a ring resonator for use in the extreme uv is shown in Fig. 8.8.

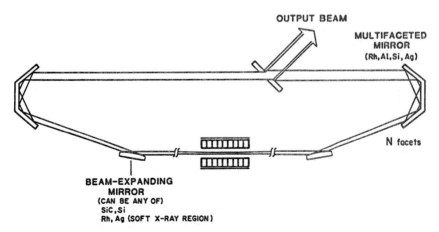

FIGURE 8.8 Ring resonator configuration for an extreme uv FEL. The multifaceted mirrors provide high reflectivity by maintaining a shallow angle of incidence on each facet. *(Reprinted with permission from Brian E. Newnam, in* Free-Electron Lasers, *SPIE, vol. 738, p. 174, 1988.)*

8.4 FEL DEVICES

In many respects, FELs can trace their origins to the work of Motz on undulator radiation in the early 1950s. Motz developed the first magnetic wigglers and used them to generate incoherent radiation at millimeter and optical wavelengths.[37] The first operation of a device employing what we now call the FEL mechanism was reported about 10 years later by Phillips.[23] He called the device a Ubitron and built several microwave-tube-type versions that behaved impressively (>1 MW at >10 percent efficiency) at centimeter wavelengths. It was 15 years later before the experiments leading to the present-day FELs were performed. These experiments have evolved along two separate paths, depending on the type of accelerator used and the operating regime. In the Compton regime, the work was pioneered by Madey and his colleagues at Stanford University using an rf linac operating at 25 to 50 MeV.[38,39] The Raman experiments were pioneered at the Naval Research Laboratory and Columbia University using pulse line accelerators which produced 1-MeV electron beams at currents of tens of kA.[40,41] The first Raman experiments were referred to as stimulated scattering, but after the Stanford group popularized the term FEL, the Raman devices were also referred to by that name.

The early Compton regime experiments were designed primarily to verify some of Madey's important theoretical predictions, such as growth rate and efficiency. Interestingly, Madey's initial analyses[42] used quantum mechanics, making it clear that the device was in fact a laser. The Stanford FEL was operated first as an amplifier at 10.6 μm and later as an oscillator at 3.4 μm. Both experiments used a superconducting rf linac having a peak current of about 2 A together with a 5-m-long helical wiggler. The electron beam consisted of a train of 1-mm-long pulses separated by about 25 m. Even though the fractional energy spread was very small (about 5×10^{-4}), the efficiency was less than 10^{-4} and the gain was so small (a few percent per pass) that it was very difficult to get the device to lase. One of the technical problems characteristic of rf linac FEL oscillators is the extremely tight tolerance on mirror placement. Because the electron bunches are so short and are spaced so far apart, the mirror spacing has to be held to within a micrometer over distances of 10 to 15 m in order to cause the radiation pulses to continue to overlap the electron pulses during repeated passes through the resonator. Because of these difficulties and the marginal gain that could be achieved with the limited-quality electron beams available, it was over 5 years before other Compton FELs were successful.

In the early 1980s, a series of FEL amplifier experiments at 10.6 μm was carried out at Los Alamos, TRW, and Boeing/Spectra Technologies to demonstrate ponderomotive trapping and efficiency enhancement.[3] In these experiments, powerful CO_2 lasers were used as the input source to establish the ponderomotive wave. These experiments also pioneered the use of permanent-magnet wigglers, which were much more amenable to tapering than the Stanford helical electromagnet. Subsequent experiments at Los Alamos, Boeing, and Stanford were aimed at shorter-wavelength operation, higher efficiency, and higher power.[4] A joint TRW-Stanford experiment demonstrated FEL operation in the visible.[43] At Los Alamos, emphasis was placed on efficiency enhancement with a tapered wiggler and rf electron beam recovery.[14] Peak powers of 100 MW have been achieved from Los Alamos' oscillator in the vicinity of 10 μm. At Boeing, a 10-MW visible FEL oscillator has been developed.[44] A key advancement has been the development of much higher peak electron currents in the rf linacs, both by improving the electron guns and by more strongly bunching the electrons before accelerating them. One of the major technical problems thus far has been maintaining the high degree of accelerator stability required for this high-current operation.

The shortest-wavelength Compton FEL experiments have used electron storage rings.[45-47] This work was pioneered at Laboratoire pour l'Utilisation du Rayonment Electromagnetique (LURE) in Orsay, France, and at Novosibirsk in the Soviet Union. These high-energy (hundreds of MeV) beams have achieved uv operation, although the powers and efficiencies are quite low. Even though the electrons make many passes through the wiggler, the thermalization due to the FEL interaction and the synchrotron radiation emitted because of the high electron energy limit the FEL efficiency to about 1 percent.

Other FEL experiments are being performed using cyclic rf accelerators called microtrons.[48] These devices are very compact because the electron beam follows a spiral orbit which brings it through the same rf accelerating cavity many times. Microtrons for FELs have energies of about 20 MeV and peak currents of a few amperes. Their compact size makes them suitable for applications in which enormous power levels are not required and space is at a premium. Unfortunately, the beam quality is not as good as in linear rf accelerators, and thus far the microtron FEL experiments have been disappointing.

Far-infrared FEL experiments are being performed at the University of California at Santa Barbara using an electrostatic accelerator (Van de Graaff generator). These accelerators typically have very low currents, but by recollecting the current (as discussed previously), pulses of several amperes for tens of microseconds can be generated. These accelerators are characterized by extremely good voltage stability and high beam quality. The FEL oscillator has produced 10 kW of radiation at hundreds of microns with very narrow linewidths estimated at 1 part in 10^6.[49]

There have been a larger number of millimeter and centimeter wavelength FELs, but there is insufficient space to describe them here. However, one experiment at Lawrence Livermore National Laboratory (LLNL) is noteworthy because of its successful demonstration of efficiency enhancement. In this experiment,[22] operated at a wavelength of 8 mm, the wiggler amplitude was tapered downward from 3.6 kG, which resulted in the generation of 1 GW of power at an efficiency of about 40 percent. The tapered wiggler increased the output power by over a factor of 5. The experiment used an induction linac operating at ~3 MeV and ~850 A. The output power as a function of wiggler length from this experiment, with and without wiggler tapering, is shown in Fig. 8.9. As can be seen, the experimental results agreed very well with computer simulations.

There are a number of other types of free-electron lasers which we are not able to include in this discussion but which are discussed in the references. One example is the Cerenkov FEL,[50,51] in which an electron beam is propagated near a dielectric surface and interacts with the evanescent optical field which extends into the vacuum. A closely related device is the Smith-Purcell FEL,[52] in which an electron beam is propagated along the surface of a grating. There are also a number of variants on the wiggler-driven FEL. The magnetostatic wiggler can be replaced by an electromagnetic wave,[53] which could be generated by a conventional laser or a high-power microwave source.[17] A special case of an electromagnetic wiggler is the two-stage FEL,[54] in which the same electron beam generates the electromagnetic wave in the first stage and then interacts with the wave in the second stage of the FEL. The advantage of electromagnetic-wiggler-driven FELs is that very short wiggler periods can be achieved, resulting in the use of much lower energy electron beams to generate a particular output wavelength. Unfortunately, it is very difficult to obtain a sufficiently high quality electron beam at the lower energy. Another variant of the wiggler-driven FEL is the optical klystron,[55] in which the wiggler is divided into two sections separated by a drift space. The beam is modulated in the first wiggler section and bunched in the drift space, thereby enhancing the gain in the second wiggler section. The higher gain achievable with optical

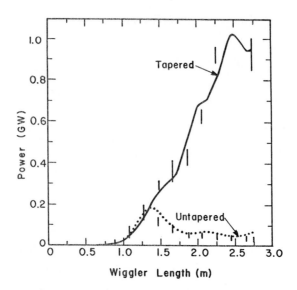

FIGURE 8.9   Output power vs. wiggler length for the LLNL 8-mm FEL amplifier.[20] Vertical lines represent the range of powers measured experimentally, while the solid and dotted lines are the results of computer simulations. *(Courtesy of T. J. Orzechowski.)*

klystrons is typically required with storage ring FELs, because the available length for a wiggler is limited. The optical klystron requires a higher-quality electron beam than a standard FEL.

To conclude this section, it is important to point out that the FEL remains a complicated, expensive device in the optical regime. Of all the Compton FELs that have been built over the past 15 years, there have been about as many failures as successes. Generally, the culprit has been the unacceptable electron beam quality. Although the FEL is conceptually quite simple, the expensive lesson that has been learned time and again is that there is no substitute for a high-quality electron beam.

8.5 FUTURE DIRECTIONS

Ongoing FEL experimental research, especially in the infrared to ultraviolet portions of the spectrum, is directed toward increasing the efficiency and power and/ or decreasing the wavelength of the FEL radiation. Higher-current and better-quality electron beams, high-power optical systems that incorporate sophisticated components such as grazing incidence mirrors and grating rhombs, and electron energy recovery are all being actively pursued in the laboratory. Studies of harmonic generation and sideband instabilities are also important parts of the effort, both theoretically and experimentally. At some laboratories (e.g., Stanford, Duke, and Vanderbilt Universities and the University of California at Santa Barbara), there is a concentration on the development and operation of practical, user-facility FELs for research and medical applications.

In several of these short-wavelength FELs, the interaction length is many times longer than the Rayleigh range of the optical beam ($Z_R = \pi w_0^2/\lambda$, where w_0 is the radiation waist radius). In order for these devices to work, the radiation beam must be guided and focused by the electron beam. Consequently, optical guiding is a subject of theoretical[56-60] and experimental[61-65] interest. There are two separate mechanisms which result in optical guiding: gain guiding and refractive guiding. Gain guiding occurs in any gain medium which is localized in the transverse direction. This mechanism is observed in conventional lasers, and it can be understood simply as light tending to be strongest where it is being amplified. Refractive guiding is analogous to propagation in an optical fiber, in which case a medium having a refractive index larger than unity is used to guide the light. The index of refraction is larger than unity in an FEL because of the electron bunching produced by the ponderomotive force. Refractive guiding can occur even after the FEL interaction has saturated and the radiation has ceased to grow. In the exponential gain regime of the FEL, both mechanisms are present.

The index of refraction in a planar-wiggler FEL with an axially symmetric electron beam having a Gaussian radial density profile can be written as[56,59]

$$n(r, z) = 1 + \frac{\omega_p^2(r, z)}{2\omega^2} \frac{a_w}{a_R} \left\langle \frac{e^{-i\psi}}{\gamma} \right\rangle \tag{8.27}$$

where the angular brackets denote an average over the electron distribution. Clearly, the index can exceed unity so that guiding can occur. Computer simulations show that optical guiding can confine the radiation field to the electron beam over a distance of many Rayleigh lengths.[53,54] Scharlemann[60] has studied the relative importance of gain and refractive guiding by artificially turning off the refractive guiding in numerical simulations, using the parameters shown in Table 8.2. The

TABLE 8.2 Parameters Used in Simulation of Fig. 8.10

Electron energy	38 MeV	Wavelength	21 μm
Beam current	1 kA	Input power	100 kW
Normalized emittance	0.1 rad cm	Initial optical radius	0.25 cm
Beam radius	0.26 cm	Wiggler period	5.5 cm
FEL length	12 m	Initial wiggler field	4.93 kG

results of these simulations are shown in Fig. 8.10. In the exponential gain regime, gain guiding dominates and is strong enough to tightly focus the radiation. However, for $z > 5$ m, the FEL interaction begins to saturate, and gain guiding is no longer strong enough to provide focusing. Refractive guiding, however, continues to focus the radiation. The simulations also show that the output radiation beam quality is much worse without refractive guiding. In fact, the output with refractive guiding is nearly diffraction-limited.

There have also been some experimental observations of optical guiding in FELs, in both the infrared[61-63] and the millimeter wave[64,65] portions of the spectrum.

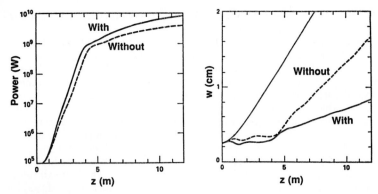

FIGURE 8.10   Computer simulation results of output power (left) and optical beam radius (right) vs. axial distance, with and without refractive guiding. The upper curve on the right is the normal expansion due to free-space diffraction. Simulation parameters are listed in Table 8.2.   *(Reprinted with permission from E. T. Scharlemann, in* Free-Electron Lasers, *SPIE, vol. 738, p. 139, 1988.)*

However, none of the experiments to date have actually involved operation of FELs with interaction lengths of many Rayleigh lengths. Typically, the radiation profile has been compared to that of a vacuum mode and shown to be more tightly focused than could be accounted for by simple diffraction. In the Los Alamos oscillator experiment,[61] optical guiding has been suggested as the mechanism which led to a "walking" of the optical spot over the surface of the resonator mirror. Presumably, a slight misalignment of the electron beam within the resonator steered the optical beam off-axis. Experimental measurements of optical guiding effects will be less ambiguous and much more important as the very long interaction length FELs begin operation.

   Harmonic generation offers a means of achieving shorter-wavelength operation without the need to use higher-energy electron beams and/or shorter-period wigglers. Harmonics are generated naturally with a linear wiggler (odd harmonics only), although the growth rate decreases with increasing harmonic number. Nonetheless, harmonic generation has been experimentally observed.[46] Also, Latham et al.[66] have theoretically compared the operation of a FEL amplifier at the third harmonic with a fundamental-mode FEL at the same frequency. They showed that harmonic operation suppresses the gain of the fundamental. However, the fundamental gain still was larger than the harmonic gain in the cases they considered, so this approach is probably most feasible in an amplifier configuration in which radiation at only the harmonic frequency is injected into the FEL.

8.6 CONCLUSIONS

The FEL has clearly become a major new radiation source. In the last few years, the interest in FELs has grown dramatically as experimental devices have verified the exciting predictions of theory. In the millimeter-wavelength regime, the FEL has the highest power of any available source. When coupled with the inherent tunability and reasonable efficiency of FEL sources, the uniquely high power capability is even more impressive. Not surprisingly, what has until now been an experimental laboratory device is being seriously considered for various specialized applications. Although difficult technical issues remain to be solved before the FEL can be used for some of the more exotic applications, progress is being made rapidly.

8.7 REFERENCES

1. Thomas C. Marshall, *Free-Electron Lasers*, Macmillan, New York, 1985.

2. Charles A. Brau, *Free-Electron Lasers*, Academic Press, New York, 1990.

3. "Free-Electron Generators of Coherent Radiation," S. F. Jacobs et al. (eds.), *Physics of Quantum Electronics*, vols. 7, 8, 9, Addison-Wesley, Reading, MA, 1980, 1982.

4. *IEEE J. Quantum Electron.*, vol. QE-21, no. 7, 1985.

5. C. W. Roberson and P. Sprangle, *Phys. Fluids*, vol. B 1, p. 3, 1989.

6. P. Sprangle, R. A. Smith, and V. L. Granatstein, *Infrared and Millimeter Waves*, vol. 1, K. J. Button (ed.), Academic Press, New York, pp. 279–327, 1979.

7. P. Sprangle, C. M. Tang, and W. Manheimer, *Phys. Rev.* vol. A 21, p. 302, 1980.

8. W. B. Colson, *Phys. Lett.*, vol. 59A, p. 187, 1976; *Phys. Lett.*, vol 64A, p. 190, 1977.

9. N. M. Kroll, P. L. Morton, and M. N. Rosenbluth, vol. 7 of Ref. 3, p. 89; *IEEE J. Quantum Electron.*, vol. QE-17, p. 1436, 1981.

10. M. N. Rosenbluth et al., *Phys. Fluids*, vol. B 2, p. 1635, 1990.

11. R. W. Warren, B. E. Newnam, and J. C. Goldstein, Ref. 4, p. 882.

12. M. N. Rosenbluth et al., in C. A. Brau et al. (eds.), *Free-Electron Generators of Coherent Generation*, SPIE, Bellingham, Wash. vol. 453, pp. 25–40, 1983.

13. L. R. Elias and G. Ramian, vol. 9 of Ref. 3, p. 577, 1982.

14. D. W. Feldman et al., *IEEE J. Quantum Electron.*, vol. QE-23, p. 1476, 1987.

15. V. K. Neil, *JASON Tech. Rept.* JSR-79-10, SRI International, Arlington, Va., December 1979.

16. E. Jerby and A. Gover, *IEEE J. Quantum Electron.*, vol. QE-21, p. 1041, 1985.

17. B. G. Danly et al., *IEEE J. Quantum Electron.*, vol. QE-23, p. 103, 1987.

18. T. I. Smith and J. M. J. Madey, *Appl. Phys.*, vol. B 27, p. 195, 1982.

19. C. W. Roberson, *IEEE J. Quantum Electron.*, vol. QE-21, p. 860, 1985.

20. J. A. Pasour and S. H. Gold, *IEEE J. Quantum Electron.*, vol. QE-21, p. 845, 1985.

21. J. A. Pasour et al., *J. Appl. Phys.*, vol. 53, p. 7174, 1982.

22. T. J. Orzechowski et al., *Phys. Rev. Lett.*, vol 57, p. 2172, 1986.

23. R. M. Phillips, *IRE Trans. Electron Devices*, vol. 7, p. 231, 1960.

24. R. J. Harvey and F. A. Dolezal, *Nucl. Instrum. Methods*, vol. A250, p. 274, 1986.

25. J. P. Blewett and R. Chasman, *J. Appl. Phys.*, vol. 48, p. 2692, 1977.

26. J. Fajans, *J. Appl. Phys.*, vol. 55, p. 43, 1984.

27. M. S. Curtin et al., *Nucl. Instrum. Methods*, vol. A237, p. 395, 1985.

28. K. Halbach, *Nucl. Instrum. Methods*, vol. 187, p. 109, 1981.

29. G. Bekefi and J. Ashkenazy, *Appl. Phys. Lett.*, vol. 51, p. 700, 1987.

30. Paul Diament, *IEEE J. Quantum Electron.*, vol. QE-21, p. 1094, 1985.

31. R. P. Walker, *Nucl. Instrum. Methods*, vol. A237, p. 366, 1985.

32. K. Halbach, *Nucl. Instrum. Methods*, vol. 169, p. 1, 1980; *IEEE Trans. Nucl. Sci.*, vol. NS-28, p. 3136, 1981.

33. G. Ramian et al., *Nucl. Instrum. Methods*, vol. A250, p. 125, 1986.

34. D. M. Shemwell, in Brian Newnam (ed.), *Free-Electron Lasers*, SPIE, Bellingham, Wash., vol. 738, pp. 46–54, 1988.

35. J. M. Eggleston, *Proc. Int. Conf. on Lasers '83*, STS Press, 1983, p. 305.

36. B. E. Newnam, in Brian Newnam (ed.), *Free-Electron Lasers*, SPIE, Bellingham, Wash., vol. 738, pp. 155–175, 1988.

37. H. Motz, *J. Appl. Phys.*, vol. 22, p. 527, 1951; H. Motz et al., *J. Appl. Phys.*, vol. 24, p. 826, 1953.

38. L. R. Elias et al., *Phys. Rev. Lett.*, vol. 36, p. 717, 1976.

39. D. A. G. Deacon et al., *Phys. Rev. Lett.*, vol. 38, p. 892, 1977.

40. V. L. Granatstein et al., *IEEE Trans. Microwave Theory Tech.*, vol. MTT-22, p. 1000, 1974.

41. P. C. Efthimion and S. P. Schlesinger, *Phys. Rev.*, vol. A 16, p. 633, 1977.

42. J. M. J. Madey, *J. Appl. Phys.*, vol. 42, p. 1906, 1971.

43. J. A. Edighoffer et al., *Appl. Phys. Lett.*, vol. 52, p. 1569, 1988.

44. J. L. Adamski et al., *IEEE Trans. Nucl. Sci.*, vol. NS-32, p. 3397, 1985.

45. M. Billardon et al., Ref. 4, p. 805; *Nucl. Instrum. Methods*, vol. A259, p. 72, 1987.

46. R. Prazeres et al., *Nucl. Instrum. Methods*, vol. A272, p. 68, 1988.

47. V. N. Litvinenko, *Synch. Rad. News*, vol. 1, no. 5, p. 18, 1988.

48. E. D. Shaw et al., *Nucl. Instrum. Methods*, vol. A250, p. 44, 1987.

49. L. R. Elias et al., *Phys. Rev. Lett.*, vol. 57, p. 424, 1986.

50. J. E. Walsh, in V. L. Granatstein and I. Alexeff (eds.), *High-Power Microwave Sources*, Artech House, 1987, pp. 421–440.

51. William Case, in V. L. Granatstein and I. Alexeff (eds.), *High-Power Microwave Sources*, Artech House, 1987, pp. 397–420.

52. S. Smith and E. Purcell, *Phys. Rev.*, vol. 92, p. 1069, 1953.

53. R. H. Pantell et al., *IEEE J. Quantum Electron.*, vol. QE-4, p. 905, 1968.

54. L. R. Elias, *Phys. Rev. Lett.*, vol. 42, p. 977, 1979.

55. N. A. Vinokurov and A. N. Skrinsky, Preprint INP 77-59, Novosibirsk, 1977.

56. D. Prosnitz, A. Szoke, and V. K. Neil, *Phys. Rev.*, vol. A 24, p. 1436, 1981.

57. P. Sprangle and C. M. Tang, *Appl. Phys. Lett.*, vol. 39, p. 677, 1981.

58. G. T. Moore, *Opt. Commun.* vol. 52, p. 46, 1984.

59. P. Sprangle, A. Ting, B. Hafizi, and C. M. Tang, *Nucl. Instrum. Methods Phys. Res.*, vol. A272, p. 536, 1988.

60. E. T. Scharlemann, Brian E. Newnam (ed.), *Free-Electron Lasers*, SPIE, Bellingham, Wash., vol. 738, pp. 129–141, 1988.

61. R. W. Warren and B. D. McVey, *Nucl. Instrum. Methods Phys. Res.*, vol. A259, p. 154, 1987.

62. J. E. LaSala, D. A. G. Deacon, and J. M. J. Madey, *Phys. Rev. Lett.*, vol. 59, p. 2047, 1986.

63. T. J. Orzechowski, E. T. Scharlemann, and B. D. Hopkins, *Phys. Rev.*, vol. A35, p. 2184, 1987.

64. J. Fajans, J. S. Wurtele, G. Bekefi, D. S. Knowles, and K. Xu, *Phys. Rev. Lett.*, vol. 57, p. 579, 1986.

65. S. Y. Cai, S. P. Chang, J. W. Dodd, T. C. Marshall, and H. Tang, *Nucl. Instrum. Methods Phys. Res.*, vol. A272, p. 136, 1988.

66. P. E. Latham et al., *Phys. Rev. Lett.*, vol. 66, p. 1442, 1991.

CHAPTER 9
ULTRASHORT LASER PULSES

Li Yan
P. T. Ho and Chi H. Lee

9.1 THEORY OF ULTRASHORT PULSE GENERATION

Since the invention of the laser more than three decades ago, generating short and ultrashort laser pulses has continuously been one of the most important and active areas of laser physics and engineering. Numerous methods have been developed to generate ultrashort laser pulses. Rapid advances have been made, especially in the last decade, during which optical pulses from various lasers have been pushed down to the femtosecond region and the pulse peak power has reached the multi-terawatt level. In general, these methods can be classified into three basic categories: mode locking, gain switching, and shaping and compression of existing pulses. In this section, we briefly describe the general theories of mode locking, gain switching, and pulse compression; detailed discussions are given in Sec. 9.2.

9.1.1 Mode Locking

Mode locking is the first and arguably the most important method of generating ultrashort pulses, and much theoretical and experimental work has been done on this subject since the earliest demonstrations.[1-4]

Mode Locking in Inhomogeneously Broadened Lasers. For an ideal inhomogeneously broadened laser, different species of atoms contribute to different longitudinal modes. (In a laser oscillator, radiation occurs at discrete resonant frequencies which are called the longitudinal modes of the resonator.) Thus, each individual mode saturates independently, and many modes can lase simultaneously. The electric field of the light in the resonator can be written as

$$E(t) = \frac{1}{2} \sum_k A_k e^{i(\omega_k t + \varphi_k)} + \text{c.c.} \tag{9.1}$$

where A_k and φ_k are the amplitude and the phase of the mode designated by k. For any smooth-mode amplitude distribution, the relative phases of modes dom-

9.1

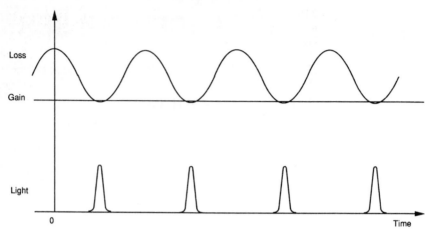

FIGURE 9.1   Active mode locking by amplitude modulation.

inate the temporal structure of light. In the free-running state the relative phases among all the longitudinal modes are random, thus resulting in repetitive noise-like radiation. To generate coherent laser pulses, nonlinear interaction among longitudinal modes has to be introduced to lock the phases of longitudinal modes together.[5] This is mode locking.

The simplest nonlinear interaction is an externally controlled periodic amplitude modulation with the modulation period $1/f_m$ equal to the resonator's round trip transit time T. As shown in Fig. 9.1, in the time domain the modulation corresponds to a periodic shutter which forces light intensity to build up only near the instant of minimum modulation loss. The result is the formation of a light pulse which circulates inside the resonator, producing a train of periodic output pulses. In the frequency domain, sidebands of the longitudinal modes are generated by the modulation. When the modulation frequency is tuned to match the resonator's axial mode spacing, the generated sidebands coincide with other resonator modes. The interactions between the sideband signals and the original signals bring the oscillating modes in phase, and a short pulse is formed. When all longitudinal modes are locked together, the pulse width should approach the inverse of the gain bandwidth.

Mode Locking in Homogeneously Broadened Lasers.   For an ideal homogeneously broadened laser, all lasing atoms share an identical transition line shape. Gain saturation occurs uniformly across the transition line. In the free-running steady state only one single longitudinal mode which experiences the largest net gain can sustain. When a nonlinear modulation is introduced, sidebands of the center mode are generated which expand toward the wings of the line profile to counter the spectral narrowing due to uniform gain saturation. In the steady state, the two processes balance each other. The sidebands thus generated are automatically in phase, resulting in coherent pulses. However, it is difficult to generate enough sidebands to cover the whole gain linewidth.

As shown in Fig. 9.2, a general analysis of the mode locking process considers the effects of gain, loss, spectral filtering, modulation, dispersion, and linear time

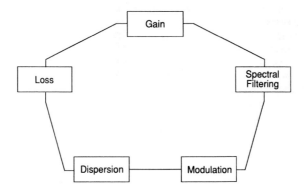

FIGURE 9.2 Self-consistent mode locking process.

delay. A steady-state solution is achieved when the pulse reproduces itself after one round trip. In homogeneously broadened lasers the mode locking process can be analyzed by a master equation in the time domain[6–11] owing to the very nature of automatic phase locking. (Almost all mode-locked lasers which produce ultrashort pulses are homogeneously broadened.) Different mode locking methods differ from one another usually in the way of producing the modulation functions. Once a modulation function is produced, the different mode locking processes are similar. In general, modulation can be classified as amplitude modulation (AM) or phase modulation (FM). When the modulation is driven externally, it is called active mode locking. When the modulation is initiated and influenced by the optical signal itself, it is called passive mode locking. A review of the basic mode locking theories and early experimental works can be found in Ref. 12. In Sec. 9.2 simple descriptions of different mode locking methods will be given.

9.1.2 Gain Switching

Gain switching is another way to generate short pulses.[13,14] The basic mechanism is to turn the gain very rapidly from below threshold to well above threshold, causing very rapid buildup of light intensity. The light intensity then saturates the gain, driving it below threshold, and the light is then quenched. The process is very similar to relaxation oscillation. If pumping continues after the initial pulse, trailing pulses may develop.

Gain switching is different from mode locking. In gain switching, only a single pulse is rapidly developed and generated, and its pulsewidth is longer than, but comparable to, the cavity round trip transit time. In mode locking, a pulse is shuttling inside the cavity, and its width is much shorter than the cavity transit time. Equivalently in the frequency domain, the pulse bandwidth is comparable to, and can be less than, the axial mode spacing in gain switching, but much larger than the axial mode spacing in mode locking. In the extreme transient case, the width of the gain-switched pulse may approach the cavity round trip transit time so that the longitudinal modes of the resonator may not even be defined, distinctly different from mode locking. Similar to Q-switched pulses, the gain-switched pulse may be partially coherent with temporal substructures under the pulse envelope. Coherent gain-switched pulses may be obtained by either limiting the number of longitudinal modes or locking the existing modes. In the latter case, a train of short pulses is generated, but the envelope of the pulse train has about the same width as a single gain-switched pulse.

9.1.3 Pulse Compression

A well-defined phase relationship among different modes may not lead to the shortest pulsewidth allowed by the available bandwidth. A common situation is chirping, where the phase changes in time. The essence of pulse compression is to rearrange the phases of the spectral components of a coherent chirped pulse to minimize the pulse duration.[15,16] Consider a coherent laser pulse whose electric field is

$$\tilde{E}(t) = A(t)e^{i[\omega_0 t + \phi(t)]} \tag{9.2}$$

where the pulse envelop $A(t)$ is a real and smooth function of t. The instantaneous frequency of the pulse is

$$\omega(t) = \omega_0 + \frac{d\phi}{dt} \tag{9.3}$$

Hence for $\phi(t)$ having quadratic and higher-order terms, a frequency chirping is developed across the pulse. For

$$\phi(t) = bt^2 \tag{9.4}$$

the frequency chirping is linear. For $b > 0$, the chirping is positive. The instantaneous frequencies are red-shifted for the pulse leading edge and blue-shifted for the trailing edge. For a strongly chirped pulse, $bt_{p0}^2 \gg 1$, where t_{p0} is the pulsewidth, the pulse bandwidth is determined by

$$\Delta\omega_p \simeq \sqrt{b} \tag{9.5}$$

Pulse compression introduces a group delay in frequency domain to slow down the leading edge and speed up the trailing edge. The outcome is a compressed, bandwidth-limited (chirp-free) pulse.

It is easy to show with a Gaussian pulse that the field spectrum of a linearly chirped pulse has a quadratic phase

$$\phi_p(\omega) = b_p(\omega - \omega_0)^2 \tag{9.6}$$

In order to compress the chirped pulse, this phase must be canceled by an additional phase shift $\Phi_d(\omega)$ introduced by a group delay system. Optimum compression requires

$$\frac{\partial^2\Phi_d}{\partial\omega^2} \equiv \frac{\partial\tau_g}{\partial\omega} = -2b_p \tag{9.7}$$

where

$$\tau_g = \frac{\partial\Phi_d}{\partial\omega} \tag{9.8}$$

is the group delay time. Since b_p follows the sign of b, a negative group delay system is needed to compress a pulse of a positive chirp. For a strongly chirped pulse with a bandwidth $\Delta\omega_p$, the necessary group delay time is

$$|\Delta\tau_g| = \left|\frac{\partial\tau_g}{\partial\omega}\right|\Delta\omega_p \simeq t_{p0} \tag{9.9}$$

9.2 METHODS OF GENERATION

9.2.1 Active Mode Locking

Active mode locking of a laser can be achieved by amplitude modulation or by phase modulation or by gain modulation, which is also called synchronous pumping.

Amplitude Modulation and Phase Modulation. In a classic work[6] Kuizenga and Siegman studied AM and FM mode locking in a homogeneously broadened laser. For amplitude modulation with a round trip amplitude transmission function

$$t_{\text{AM}} = e^{-\Delta_m(1-\cos\omega_m t)} \tag{9.10}$$

where $\omega_m = 2\pi f_m$ is the modulation frequency and Δ_m is the modulation index. The pulsewidth and the bandwidth of the Gaussian pulse are[6,17]

$$t_{p\text{AM}} = \left(\frac{2\sqrt{2}\ln 2}{\pi^2}\right)^{1/2}\left(\frac{g_{\text{sat}}}{\Delta_m}\right)^{1/4}\left(\frac{1}{f_m\Delta f_a}\right)^{1/2} \simeq \frac{0.5}{(f_m\Delta f_a)^{1/2}} \tag{9.11}$$

and

$$\Delta v_{p\text{AM}} = \sqrt{2\ln 2}\left(\frac{\Delta_m}{g_{\text{sat}}}\right)^{1/4}(f_m\Delta f_a)^{1/2} \tag{9.12}$$

where Δf_a is the laser linewidth (FWHM) assuming a Lorentzian shape. The round trip saturated amplitude gain g_{sat} is related to the cavity round trip power loss characterized by an effective mirror reflectance R_{eff} by

$$g_{\text{sat}} = \frac{1}{2}\ln\frac{1}{R_{\text{eff}}} \tag{9.13}$$

Similar expressions can be found for FM mode-locked lasers.[6] The pulse results from the balance between the spectral broadening due to the curvature of modulation $\Delta_m\omega_m^2$ and the spectral narrowing through gain saturation (Δf_a). When an intracavity etalon, with free spectral range Δ_f and finesse $\mathcal{F}$, is present, Δf_a should be modified to

$$\Delta f_a \rightarrow \frac{\Delta f_a}{[1 + (\mathcal{F}^2/2g_{\text{sat}})(\Delta f_a/\Delta_f)^2]^{1/2}} \tag{9.14}$$

The most common device for AM modulation is the acoustooptic modulator in which a standing acoustic wave is excited. By Bragg deflection a portion of light is removed,[18] and consequently the amplitude of the zeroth-order beam is modulated at a frequency equal to twice the frequency of the rf (radio-frequency) driving signal. The electrooptic effect can be used for phase modulation. In a linear electrooptic crystal, the index of refraction is modulated by the voltage of an applied rf electric signal.[18] The light passing through the medium acquires a modulated phase shift at the same frequency of the electrical signal. A phase modulator does not directly remove the energy out of the light beam. By consecutive sideband generation, however, the new frequencies will be shifted away from the gain center and eventually will be removed by gain narrowing. Thus, only the light passing the modulator with minimum phase chirp will survive, resulting in a coherent pulse.

Active AM and FM mode locking techniques are very simple and work in virtually all cw lasers or long-pulse lasers. While in a ring cavity the modulator can be placed in any convenient location, in a linear resonator the modulator is placed close to one end mirror to ensure that the laser pulse passes the modulator only once in every round trip.

Gain Modulation. Modulation of the laser gain is another method of active mode locking. Similar to amplitude modulation, gain modulation favors the growth of light intensity at the instant of maximum gain. When the period of modulation is equal to the cavity round trip time, the process is resonantly enhanced and a pulse is formed inside the laser resonator.

Gain modulation is achieved by modulating the pump source. In semiconductor diode lasers this can be done by directly modulating the driving current.[19,20] In optically pumped lasers, it can be achieved by synchronous pumping using a periodic train of laser pulses from another mode-locked laser. The cavity length of the laser is adjusted equal to the cavity length of the pump laser. As illustrated in Fig. 9.3, upon absorption of a pump pulse the gain increases to a peak value and then is depleted by the laser pulse to below threshold before the arrival of the next pump pulse. Assuming that the relaxation time from the absorption band to the upper laser level is shorter than the pump pulse duration, the modulation function is determined mainly by the upper-level relaxation time of the gain medium. If the relaxation time is fast compared with the pump pulsewidth, the gain modulation follows directly the shape of the pump pulse. In this case, although the curvature of modulation may be sharp, the pump energy cannot be fully integrated for lasing action. Good mode locking occurs when the relaxation time is comparable to the cavity round trip time and the stimulated emission cross section of the gain medium is large. In this case the leading edge of the gain is determined by the pump pulsewidth. The gain, after falling rapidly by saturation, relaxes to a level below the threshold until the arrival of the next pulse. In the other extreme, if the stimulated-emission cross section is small and the relaxation time is much longer than the cavity round trip time, gain modulation is very shallow and mode locking will not be effective and stable.

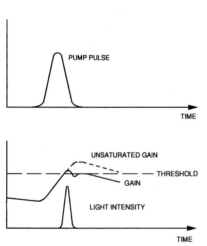

FIGURE 9.3 Dynamics of gain modulation.

9.2.2 Passive Mode Locking

Passive mode locking employs a saturable absorber as the pulse shaping component. A saturable absorber can be modeled on a two-level system and has the property that its absorption decreases with increasing light intensity. Thus the wings of a pulse are attenuated more than the peak, resulting in pulse shortening. Depending on the values of relaxation time, saturable absorbers can be classified into two types, and their corresponding detailed pulse shaping processes are different.

Passive Mode Locking with Fast Saturable Absorbers.   For a fast saturable absorber the relaxation time T_A is much less than the pulsewidth. Thus the population-difference density essentially follows the variation of the pulse intensity and the loss modulation function can be written as[10]

$$L(t) = \frac{L_0}{1 + [|v(t)|^2/P_A]} \tag{9.15}$$

where P_A is the saturation power of the absorber

$$P_A = \frac{\hbar\omega_0 A_A}{\sigma_A T_A} \tag{9.16}$$

and $|v(t)|^2$ equals the sum of powers in the two countertraveling waves inside the resonator. Unlike active mode locking, the modulation function of a fast saturable absorber is governed by the signal itself. The periodicity is preserved when the slightly shaped signal travels back in one round trip. As the signal evolves into the steady state, only the largest transmission peak remains and it is as sharp as the pulse itself. Pulse shortening by the saturable absorber is balanced by broadening due to gain narrowing. Since the effective modulation frequency $(\omega_m)_{\text{eff}}$ is characterized by the curvature of the modulation peak, it is determined by the pulse duration

$$(\omega_m)_{\text{eff}} \propto \frac{1}{t_p} \tag{9.17}$$

By substituting (9.17) into (9.11) we have

$$t_p \propto \frac{1}{\Delta f_a} \tag{9.18}$$

Therefore, in general, passive mode locking is much more efficient in shortening pulsewidth than active mode locking.

Passive Mode Locking with Slow Saturable Absorbers.   Passive mode locking with a slow saturable absorber is somewhat different. For a slow saturable absorber, T_A is much longer than the pulse duration and the absorber is saturated by the integrated pulse intensity [time-dependent energy $E(t)$], and the loss modulation function is given by[9]

$$L(t) = L_0 \exp\left[-\frac{E(t)}{E_A}\right] \tag{9.19}$$

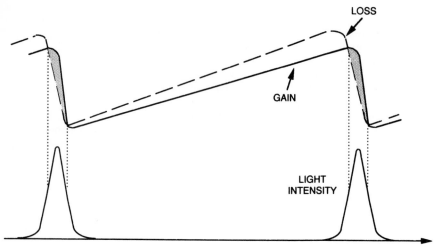

FIGURE 9.4   Passive modeling by a slow saturable absorber and gain saturation.   **TIME**

where

$$E_A = \frac{\hbar\omega_0 A_A}{\sigma_A} \tag{9.20}$$

is the saturation energy of the absorber. Since the absorber is bleached by the pulse energy, by itself it is unstable to the perturbation immediately following the main pulse and satellite pulses may develop. In order to have a clean single pulse, the gain must be saturated after saturation of the absorber, as depicted in Fig. 9.4. Hence the saturable absorber sharpens the leading edge of the pulse and the gain saturation trims the trailing edge. To have effective gain saturation, the lifetime of the gain medium must be shorter than or comparable to the cavity round trip transit time. To ensure that the gain saturation occurs later than the bleaching of the absorber, it is also required that $E_L > E_A$ where E_L is the saturation energy of the gain medium. Together saturations of the absorber and the gain provide the necessary modulation.

Colliding Pulse Mode Locking.   Colliding pulse mode locking (CPM) was invented[21] to enhance the effectiveness of passive mode locking using a saturable absorber. When two coherent pulses traveling in opposite directions overlap at the saturable absorber, the two pulses interfere with each other and form a transient standing wave.[22] At the antinodes of the standing wave the absorber is saturated more completely and results in a maximum transmission or minimum loss. At the nodes of the standing wave the absorber is not bleached. But since there are no energies at the nodes, this again gives minimum loss. Typically the saturable absorber is placed a quarter perimeter away from the gain medium in a ring cavity or at the center of a linear cavity (Fig. 9.5).

Although CPM was first and most successfully employed in ring dye lasers, the method is applicable to other lasers. Siegman[23] proposed using an antiresonant ring (Fig. 9.6) which includes a saturable absorber cell at the end of the optical resonator. When an incoming pulse is split equally and the two parts meet at the saturable absorber cell, CPM is realized. With different saturable absorption dyes, this an-

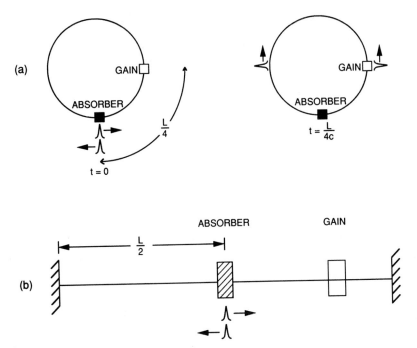

FIGURE 9.5   Colliding pulse mode locking in (*a*) a ring cavity and (*b*) linear cavity.

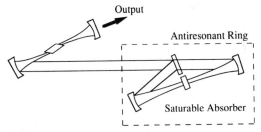

FIGURE 9.6   Colliding pulse mode locking using an antiresonant ring.

tiresonant ring has been applied to synchronously pumped dye lasers, to Nd:YAG lasers, and to Nd:glass lasers.

9.2.3   Additive Pulse Mode Locking

Additive pulse mode locking (APM) has been advancing very rapidly in the last few years. The basic mechanism is the coherent interference of two pulses with a relative phase shift so that the wings destructively interfere.[11,24,25] As shown schematically in Fig. 9.7, a typical laser using additive pulse mode locking is composed of a main cavity and an auxiliary cavity. The main cavity is an ordinary mode-

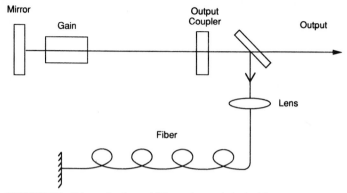

FIGURE 9.7 Schematic of an additive pulse mode-locked laser.

locked oscillator producing repetitive pulses. The auxiliary cavity contains a piece of optical fiber as the nonlinear (Kerr) medium to provide the nonlinear phase shift, and the length of the cavity is nominally equal to the main cavity length or an integer multiple. Part of the output pulse from the main cavity is coupled into the auxiliary cavity and after one round trip returns to the main cavity. In addition to a linear phase retardation, the returned pulse acquires a nonlinear phase retardation due to self-phase modulation. When the length of the auxiliary cavity is adjusted properly such that the peak of the returned pulse is in phase with the main pulse, the central portion of the resultant pulse is constructively enhanced while the wings are reduced by destructive interference. The resulting pulse is thus sharpened.

The output mirror of the main cavity and the auxiliary cavity together can be viewed as a nonlinear mirror with an equivalent reflectivity given by[11]

$$\Gamma = r + L(1 - r^2)\exp[-j(\phi_0 + \Phi_{NL})] \tag{9.21}$$

where ϕ_0 is the linear phase retardation and Φ_{NL} is the nonlinear phase retardation. Although

$$\phi_0 + \Phi_{NL}(0) = 0 \tag{9.22}$$

results in a maximum reflection at the peak of the pulse (antiresonant condition), the largest pulse shortening per pass occurs when

$$\phi_0 = -\pi \tag{9.23}$$

$$\Phi_{NL}(0) = \frac{\pi}{2} \tag{9.24}$$

As the pulse gets narrower, the peak pulse intensity increases; so does the nonlinear phase shift. In the steady state, if the pulse shortening is balanced by the gain narrowing, the shortest pulse is produced when the pulse shortening per pass is largest. For a given pulse energy, the compression ratio equals the ratio of the final peak nonlinear phase shift to the initial peak nonlinear phase shift. For an initial nonlinear phase shift of ~0.01, the compression ratio is about 100. For a larger initial nonlinear phase shift, pulse shortening is easier, but the compression ratio is smaller.

Additive pulse mode locking is passive by its nature. Since pulse shortening is achieved through interference between pulses of different phases and the relaxation time of the nonlinear medium is very fast (<100 fs), additive pulse mode locking can be classified as ultrafast, passive FM mode locking. The nonlinear equivalent reflectivity provides the necessary phase modulation. As with a saturable absorber, mode locking can self-start only under the proper conditions. Ippen et al.[26] have shown that when

$$\frac{\kappa}{g} > \beta \sigma t_p \qquad (9.25)$$

self-starting of additive mode locking is possible. In (9.25) σ is the stimulated-emission cross section of the laser gain medium and κ is proportional to the nonlinear coefficient. Thus for lasers with smaller σ, self-starting of additive mode locking is easier.

9.2.4 Gain Switching

Gain switching is a simple way to generate short pulses. The pulsewidth, borrowed from relaxation oscillation theories,[18] is roughly

$$T_{\text{pulse}} \sim (T_{\text{gain}} T_{\text{photon}})^{1/2} \qquad (9.26)$$

where T_{gain} is the gain turn-on time and T_{photon} is the photon lifetime in the cavity. T_{gain} determines the pulse rise time and T_{photon} the decay time. Thus, to generate very short pulses, the gain medium must have very fast relaxation time and large stimulated-emission cross section, such as dye lasers and semiconductor lasers, for quick switching on and effective quenching. In addition, the cavity length must be short. For semiconductor lasers and dye lasers, $T_{\text{gain}} \simeq 100$ ps, $T_{\text{photon}} \simeq 5$ ps, so $T_{\text{pulse}} \simeq 10$ to 50 ps.

Usually the product of pulsewidth and bandwidth is much larger than unity, with ultrashort substructure under the intensity envelope. To improve the pulse coherence, spectral filtering is needed either by an external cavity grating or by placing a grating inside the cavity in the Littrow configuration. One extreme case of intracavity filtering is the distributed feedback (DFB) laser,[27,28] in which the periodic interference structure is along the longitudinal direction of the cavity. (The longer the periodic structure in the cavity is, the narrower the pulse bandwidth and consequently the higher the degree of pulse coherence will be.) The wavelength of the DFB laser is determined by

$$\lambda_0 = 2n\Lambda \qquad (9.27)$$

where Λ is the groove period and n is the index of refraction of the medium. In a DFB semiconductor laser, periodic grooves are fabricated on the waveguiding sides of the laser cavity to form a phase grating, and, in a DFB dye laser, the gain is modulated spatially along the longitudinal direction of the cavity to form an amplitude grating.[29] Only a very narrow band of light will have net gain from the distributed feedback coupling or spatially modulated gain. As a result the laser output has a very high degree of coherence. In conjunction with gain switching, DFB lasers can produce nearly Fourier-transform-limited pulses from both dye lasers and semiconductor lasers. Since each pulse is generated from below threshold, there is no coherence between pulses.

9.2.5 Pulse Compression

Pulse compression actually consists of two parts: generation of additional frequency components, and rearranging the phases of these components to minimize pulse-width (compression). The generation is by nonlinear phase modulation, and compression with an optical group delay system.

Optical Group Delay Systems.   The most commonly used negative group delay system for optical pulse compression is a pair of gratings parallel to each other[30] as shown in Fig. 9.8a. For first-order diffraction the incident angle θ_{in} and the diffraction angle θ, both relative to the normal of gratings, are related by

$$\frac{\lambda}{d} = \sin \theta_{in} + \sin \theta \tag{9.28}$$

where d is the groove period of the grating. As an optical pulse passes through the grating pair, the longer-wavelength component traverses a longer path than the shorter-wavelength component. Thus a parallel grating pair gives a negative group delay.

A prism sequence, shown in Fig. 9.8b, can also introduce a group delay. As shown by Fork et al.,[31] the angular dispersion gives rise to a negative group delay, while the material portion of the device contributes a positive group delay. Thus, unlike the grating sequence, the prism sequence can be adjusted to produce a net

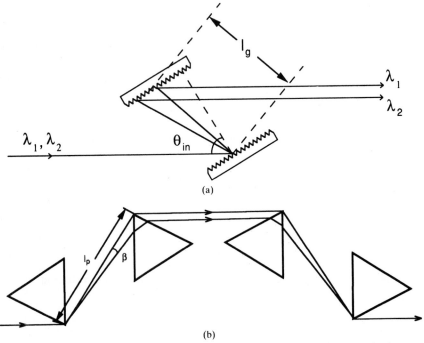

(a)

(b)

FIGURE 9.8   Optical group delay systems by (*a*) a pair of gratings and (*b*) a sequence of prisms.

group delay from positive to negative. For a quartz prism sequence, a separation of about 20 cm can result in a negative group delay.

For pulse compression, an ideal compressor would sum all frequency components in phase. For a linearly chirped Gaussian pulse, a quadratic group delay system is an ideal compressor. However, practical group delay systems are not ideally quadratic. Besides the quadratic phase term, the contribution from the cubic phase term can become significant for large bandwidths. The total group delay $\Delta \tau_g$ is determined by

$$\Delta \tau_g = \Phi''_d \Delta \omega_p + \frac{1}{2} \Phi'''_d \Delta \omega_p^2 \qquad (9.29)$$

Table 9.1 lists the quadratic and cubic contributions to the group delay by a grating pair and by a prism pair, respectively. Because the sign of Φ'''_d of a grating pair is opposite to that of a prism pair, one can use a combination of both group delay systems to cancel their cubic contributions and achieve a nearly ideal compression result.[32,33]

Self-Phase Modulation in Fibers. Self-phase modulation (SPM) in Kerr media[34] is the most widely used means to broaden the pulse bandwidth and introduce frequency chirping. Driven by an intense light field, the index of refraction of a dielectric medium is modified by a nonlinear term. In solids the response times of the nonlinear indexes of refraction, caused by nonlinear electronic polarizabilities, are about 10^{-14} to 10^{-15} s.[35] Thus, except for extremely short optical pulses, the nonlinear effects due to the nonlinear index of refraction γ take place virtually instantaneously, and the index of refraction is given by

$$n = n_0 + \gamma I \qquad (9.30)$$

where γ has the unit of cm^2/W. Optical fibers are nearly ideal Kerr media for pulse spectral broadening. Beam confinement in fibers provides both long interaction

TABLE 9.1 Second and Third Derivatives of Phase with Respect to Frequency for a Double Prism Pair and a Double Grating Pair[32]

Prism	Grating
$\dfrac{d^2\phi_p}{d\omega^2} = \dfrac{\lambda^3}{2\pi c^2}\dfrac{d^2P}{d\lambda^2}$	$\dfrac{d^2\phi_g}{d\omega^2} = \dfrac{-\lambda^3 l_g}{\pi c^2 d^2}\left[1 - \left(\dfrac{\lambda}{d} - \sin\theta_{in}\right)^2\right]^{-1}$
$\dfrac{d^3\phi_p}{d\omega^3} = \dfrac{-\lambda^4}{4\pi^2 c^3}\left(3\dfrac{d^2P}{d\lambda^2} + \lambda\dfrac{d^3P}{d\lambda^3}\right)$	$\dfrac{d^3\phi_g}{d\omega^3} = -\dfrac{d^2\phi_g}{d\omega^2}\dfrac{3\lambda}{2\pi c}\dfrac{\left(1 + \dfrac{\lambda}{d}\sin\theta_{in} - \sin^2\theta_{in}\right)}{\left[1 - \left(\dfrac{\lambda}{d} - \sin\theta_{in}\right)^2\right]}$

Derivatives of the path length P in the prism sequence with respect to wavelength

$$\frac{d^2P}{d\lambda^2} = 4\left[\frac{d^2n}{d\lambda^2} + (2n - n^{-3})\left(\frac{dn}{d\lambda}\right)^2\right]l_p \sin\beta - 8\left(\frac{dn}{d\lambda}\right)^2 l_p \cos\beta$$

$$\frac{d^3P}{d\lambda^3} = 4\frac{d^3n}{d\lambda^3}l_p \sin\beta - 24\frac{dn}{d\lambda}\frac{d^2n}{d\lambda^2}l_p \cos\beta$$

lengths and uniform SPM across the beam profiles. With uniform SPM, self-focusing and spatial chirping will not occur.

Self-Phase Modulation without Group Velocity Dispersion. When the effect of group velocity dispersion (GVD) on the pulse itself is negligible, the pulse temporal envelope does not change as the pulse propagates in the medium. The effect of SPM is to introduce a nonlinear phase shift given by

$$\Phi_{NL} = -\frac{2\pi}{\lambda_0} \frac{\gamma P(t)L}{A} \tag{9.31}$$

The corresponding spectral broadening is

$$\Delta\lambda_{SPM} \simeq \frac{\lambda_0^2}{ct_p} |\Phi_{NL}|_{max} \tag{9.32}$$

For a bell-shaped pulse, since $d^2I(0)/dt^2 < 0$, the coefficient of the quadratic phase term in Φ_{NL} is positive. Thus SPM introduces a positive linear chirping in the central portion of the pulse with the longer-wavelength part in the leading edge of the pulse and the shorter-wavelength part in the trailing edge of the pulse. Since the spectrally broadened pulse is linearly chirped only near the peak of the pulse, small pedestals accompany the compressed central peak.

Self-Phase Modulation with Positive Group Velocity Dispersion. For effective compression, the pulse should be fully linearly chirped. Numerical calculations[36,37] show that, in the presence of positive group dispersion, the pulse eventually will be stretched to a nearly square shape with a linear chirp across the whole pulse. Consequently wings are significantly reduced and are less extensive.[36] (The sign of GVD follows the sign of $d^2n_0/d\lambda_0^2$.[17] Many optical materials exhibit positive GVD in the visible spectrum, turning to negative GVD in the near infrared. For silica optical fibers, the turning point is around 1.3 μm.) Approximate formulas for calculation of parameters for optimal pulse compression using an optical fiber and a pair of gratings were given by Tomlinson et al.[37] For a pulse of peak power P and initial pulsewidth t_{p0}, the optimal fiber length Z_{opt} and the compressed pulsewidth t_p are listed in Table 9.2.

Consideration of Stimulated Raman Scattering. Stimulated Raman scattering (SRS) limits the fiber length for compression of high-power long pulses. For such

TABLE 9.2 Formulas for Optimal Pulse Compression Using Optical Fibers[37]

$$Z_{opt} \simeq \frac{1.6}{A} Z_0 \qquad\qquad t_p \simeq \frac{t_{p0}}{0.63A}$$

$$Z_0 = \frac{t_{p0}^2}{C_1} \qquad\qquad A = \sqrt{\frac{P}{P_1}}$$

$$C_1 = \frac{D(\lambda_0)\lambda_0}{0.322\pi^2 c^2} \qquad\qquad D(\lambda_0) = \frac{\lambda_0\omega^2}{2\pi} k''$$

$$P_1 = \frac{n_0 c\lambda_0 A_{eff}}{16\pi Z_0 n_{2E}} = 792 \left[\frac{\lambda_0(\mu m)A_{eff}(\mu m^2)}{Z_0(cm)}\right] \text{ watt}$$

long pulses, typically tens of picoseconds, the SRS gain G_s and the power of the Stokes pulse P_s at the output of a medium of an interaction length L_{int} are given by[38,39]

$$G_S = \frac{g_r P L_{int}}{A} \tag{9.33}$$

$$P_s(L) = P_{s0} e^{G_s} \tag{9.34}$$

In the above equations, P is the input laser power assuming no significant depletion, g_r is the peak Raman gain coefficient, and A is the transverse beam area. The equivalent input Stokes power P_{s0} comes from the summation of the amplified spontaneous emission along the fiber. Since the Stokes wave grows exponentially, P_s is a rapidly changing function of G_s. When $G_s \simeq 18$ the Stokes power can be comparable to the power of the pump pulse at the fundamental wavelength.

Because of group velocity dispersion the Stokes pulse will eventually walk off from the input pulse and the Raman gain will be reduced significantly. Thus the interaction length is the shorter of the fiber length L and the walk-off distance L_w, given by[39]

$$L_w = \frac{V_s V_l}{V_s - V_l} t_p \tag{9.35}$$

where V_l and V_s are the group velocities of the input laser pulse and the Stokes pulse, respectively, and t_p is the laser pulsewidth. For silica fibers and input pulses at 1.06 μm,

$$L_w \simeq 60 \times t_p(\text{ps}) \qquad \text{cm} \tag{9.36}$$

For the high-power cases such that $L_{int} < L_w$, the fiber length has to be limited by a critical Raman gain, and GVD on the pulse itself is usually negligible. For the low-power cases such that $L_{int} \simeq L_w$, the fiber length can be much longer than L_w, and development of a fully linear chirping is possible.

Soliton Compression in Fibers. Pulse compression relies on creating frequency chirping and balancing the chirping with a proper group delay. For positive chirping such as that produced by self-phase modulation, a negative group delay is needed. In silica fibers, GVD is negative for $\lambda > 1.3$ μm. Thus an optical fiber can serve as both the nonlinear medium and the group delay system. In fact, it is this balance between self-phase modulation and group velocity dispersion that supports the propagation of soliton pulses in a single-mode optical fiber.[40] For a high-order soliton, the pulsewidth varies periodically as the pulse propagates along the fiber.[41] In the region where the pulse is shortening, the mechanism is exactly the same as that of conventional pulse compression except that chirping and compression occur in every infinitesimal step, not separately. Pulse compression can be obtained directly by sending a long pulse through a single-mode fiber of proper length.[42,43]

Self-Phase Modulation in Bulk Media. By simple scaling of area, bulk media can handle much higher powers than single-mode optical fibers. For a practical beam in a bulk medium, however, the spatial intensity distribution creates a non-uniform nonlinear phase retardation across the beam spatial profile. Two effects result. In the near field it causes a spatial chirping effect—a different part of the beam has a different magnitude of frequency chirping. Furthermore, it causes beam wavefront bending, and when the laser power is greater than a critical power, it results in beam self-focusing. The whole beam self-focus length is given by[44,45]

$$z_f = \frac{z_0}{\sqrt{P/P_{cr} - 1}} \tag{9.37}$$

where z_0 is the beam's Rayleigh range and

$$P_{cr} = \frac{c\lambda_0^2}{16\pi^2 n_2} \tag{9.38}$$

is the critical power for beam self-trapping.[46] (n_2 is the nonlinear index of refraction in esu.) Thus the interaction length has to be shorter than z_f. When $P \gg P_{cr}$, $\Phi_{NL} \propto \sqrt{P}$. Usually high peak power is needed, and only the central portion of the beam is selected for pulse compression using bulk media.

To limit the degree of spatial chirping, self-phase modulation in a bulk medium can be obtained through multiple passes in a regenerative amplifier.[47] The stable resonator acts essentially as a large-scale periodic waveguide. With the help of intracavity spatial filtering, the amplified beam can be forced to nearly maintain its beam size and corresponding radius of curvature. Thus the spatial chirping or self-focusing can be confined to a degree corresponding to the pulse peak power at a given pass while the longitudinal phase, which gives self-phase modulation, is accumulated in multiple passes. The reduction of spatial chirping can be about 30 times. This method is especially useful for chirping pulses of medium high powers ($P \sim P_{cr}$), since neither the fiber nor a short bulk medium is suitable.

Chirped Pulse Amplification and Compression. In amplifying short pulses, the laser intensity is limited to about the GW/cm² level in order to avoid self-focusing in the gain medium and optical elements. Chirped pulse amplification and compression[48] is a method to produce ultrashort pulses of ultrahigh powers, especially from high-energy solid-state laser amplifier systems. As illustrated in Fig. 9.9, the pulse can first be temporally stretched to a much longer duration, then amplified to a much higher energy. Assuming that the bandwidth of the amplified

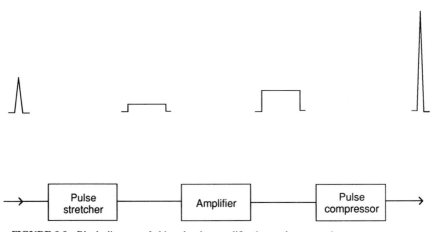

FIGURE 9.9 Block diagram of chirped pulse amplification and compression.

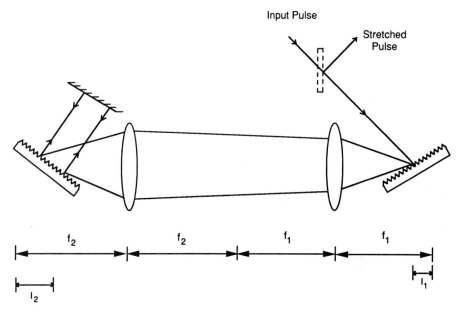

FIGURE 9.10 Pulse stretcher by a pair of antiparallel gratings.

pulse is not substantially narrowed, the pulse can be recompressed to a duration allowed by its spectral width with much higher power.

To stretch pulses of relatively narrow bandwidths such as those from actively mode-locked Nd:YAG and Nd:YLF lasers, a long piece of optical fiber can be used both to broaden the pulse bandwidth by >50 times and to stretch the pulse by about 3 to 5 times.[48] To stretch pulses of large bandwidths, a grating pair in the antiparallel configuration (Fig. 9.10) can be used.[49] A telescope is inserted between the two gratings, which are located inside the two focal planes of the telescope. The grating pair results in a positive group delay whose magnitude can be evaluated by the same formula used for the parallel configuration, but with the distance l_g being replaced by

$$l_g = l_1 + l_2 \tag{9.39}$$

where l_1 and l_2 are, respectively, the distances between the gratings and their nearest focal planes. Such a grating pair can stretch a pulse more than 1000 times,[50] but it requires a relatively large bandwidth to start with. When a pair of gratings of the same grove period and incident angle are used in recompression, the cubic phase distortion will be canceled out automatically.

9.3 ULTRASHORT PULSE LASER SYSTEMS

9.3.1 Neodymium Lasers

Neodymium (Nd) lasers have been mode-locked in a number of ways. Table 9.3 lists some representative results of the mode-locked Nd lasers. The commercial

TABLE 9.3 Short-Pulse Neodymium Lasers

Material	Method	t_p	Power and energy	Rate	Reference
Nd:YAG	Active mode locking	80–100 ps	10–20 W	80–100 MHz	51, 52
	Active mode locking	31 ps	4 W	0.25–1 GHz	54
	Passive mode locking	15–30 ps	~1–5 mJ	5–30 Hz	55, 56
	Active-passive mode locking	10–30 ps	~10 μJ		57, 58
	APM	6 ps	2.4 W	100 MHz	59
	Nonlinear ARR mirror	11 ps	1 W	65 MHz	60
	Diode-pump, FM mode locking	12 ps	65–75 mW	350 MHz	61
	Diode-pump, APM	1.7 ps	25 mW	136 MHz	62
Nd:YLF	Active mode locking	35–50 ps	10–20 W	80–100 MHz	63
	APM	3.7 ps	1–7 W	76 MHz	64
	RPM	4 ps	390 mW	250 MHz	65
	Diode-pump, AM mode locking	7–10 ps	130–160 mW	160 MHz–2 GHz	66, 67
	Diode-pump, APM	1.5–2 ps	10–20 mW	122–136 MHz	68, 69
Nd:glass	Passive mode locking	3–8 ps	~1–5 mJ	1–5 Hz	70, 72
	Active-passive mode locking	3–8 ps	~1 mJ		73, 74
	Active mode locking	5–10 ps	~100 mW	100 MHz	75
	Diode-pump, AM mode locking	7 ps	120 mW	75 MHz	76
	APM	380 fs	60 mW	75 MHz	77
Nd:fiber	NALM* + GVDC†	125 fs	12 mW	~50 MHz	78
Compression:					
Nd:YAG	One-stage compression	~5 ps	~3 W	80–100 MHz	79
Nd:YAG	Two-stage compression	200 fs	~400 mW	82 MHz	80
Nd:YLF	One-stage compression	~3 ps	~3 W	80–100 MHz	
Amplification:					
Nd:YLF	Regenerative amplification	50 ps	~1 mJ	1–10 kHz	63, 85
Nd:YAG	Chirped-pulse amplification	~5 ps	~0.5–1 mJ	5–20 Hz	86
Nd:glass	Regenerative amplification	500 fs	~10 μJ	500 Hz	47
Nd:glass	Chirped-pulse amplification	~1 ps	>1 J	<1 Hz	48, 87, 88

*NALM stands for nonlinear amplifying loop mirror.
†GVDC stands for group velocity dispersion compensation.

flashlamp-pumped, actively mode-locked cw Nd:YAG and Nd:YLF lasers produce ~50-ps and ~100-ps pulses, respectively, with average powers ~20 W. Somewhat shorter pulses have been produced from diode-pumped Nd:YAG[61] and Nd:YLF[66,67] lasers at higher modulation frequencies. In comparison, Nd:glass lasers have much broader linewidths owing to their amorphous structures, and as a result sub-10-ps pulses are generated from an actively mode-locked cw Nd:glass laser.[75]

Since the upper-level lifetimes of Nd gain media are long (~300 μs), pulse shaping by gain saturation is negligible. Thus passively mode-locked Nd lasers use fast saturable absorbers. Because of the relatively small cross-section areas of Nd media and the saturable absorbers, passive mode locking has been used traditionally in high-power Q-switched Nd lasers. In passively mode-locked Nd:YAG lasers, pulses of <30 ps can be generated. Pulses generated from passively mode-locked Nd:glass lasers are typically about 4 to 8 ps.

Additive pulse mode locking produces by far the shortest pulses from neodymium lasers. By resonant-passive mode locking, where a quantum well saturable absorber is placed in the auxiliary cavity, 4-ps pulses have also been produced in a cw Nd:YLF laser.[65]

In a single-stage compression, typically <5-ps and <3-ps pulses with 3 to 4 W average power, respectively, can be obtained from high-power Nd:YAG and Nd:YLF lasers. In two-stage compression, 200-fs pulses have been obtained from a Nd:YAG laser.[80] On the other hand, stimulated Raman scattering can be reduced at a price of small throughput power, on the order of 0.5 W, and subpicosecond pulses can be generated.[81] As short as 30 fs pulses have been generated from a Nd:glass laser system after pulse compression.[82] A soliton compression mechanism has been used to generate ultrashort pulses at 1.32 μm.[83,84]

9.3.2 Titanium:Sapphire Lasers

Because of their very broad linewidth, Ti:Sapphire lasers are very good candidates for tunable, ultrashort pulse sources with high average powers (Table 9.4). Active mode locking using an acoustooptic modulator produces pulses about 1 ps wide with ~1 W, and subsequent fiber-grating compression produced 50-fs pulses with 500 mW power.[89] Passive mode locking using a saturable absorber has produced less than 150 fs pulses directly.[90] Self-starting, additive pulse mode locking using

TABLE 9.4   Short-Pulse Ti:Sapphire Lasers

Method	t_p	Power and energy	Rate	Reference
Active mode locking	1–2 ps	2 W	82 MHz	89
Active mode locking + compression	50 fs	750 mW	82 MHz	89
Passive mode locking + GVDC*	140 fs	300 mW	100 MHz	90
APM + GVDC*	~200 fs	50–100 mW	40–190 MHz	92, 93
RPM	4 ps	100 mW	80–150 MHz	94
Self-mode locking + GVDC*	60–100 fs	100 mW	80 MHz	95, 96
Chirped-pulse amplification	100 fs	~1 mJ		96

*GVDC stands for group velocity dispersion compensation.

an optical fiber produces 200-fs pulses with either external cavity[92] or intracavity[93] GVD compensation. More recently, a Ti:sapphire laser oscillator has been self-mode-locked without the help of the auxiliary fiber cavity, and as short as 60-fs pulses have been generated.[95,96] Mixing of different spatial modes[95] and self-focusing in the Ti:Sapphire crystal itself[97] have been suggested as contributions to the mode locking process. In fact, a self-focusing lens induced in the laser materials produces different beam profiles for different instantaneous powers. In combination with the proper gain or loss apertures inside the resonator, the induced self-focusing lens provides a nonlinear loss modulation, resulting in a preferential oscillation of the peak of the pulse.

9.3.3 Color Center Lasers

Color center lasers form another class of broadband, tunable, ultrashort laser systems (Table 9.5). Although active mode locking by an acoustooptic modulator has produced 6-ps pulses,[98] the main mode locking methods are synchronous pumping, additive mode locking, and passive mode locking. By synchronous pumping with cw mode-locked Nd:YAG lasers, 3- to 10-ps pulses are produced with average powers of hundreds of milliwatts.[99,107,109] Using multiple-quantum-well saturable absorbers, passive mode locking has produced subpicosecond pulses in various color center lasers.[106,110]

Additive pulse mode locking has been successfully employed in a number of lasers systems. In fact, the conceptual root of additive pulse mode locking is from the soliton laser developed by Mollenauer and Stolen.[100] Pulses as short as 60 fs have been produced from a soliton laser.[101] However, soliton formation is only one of many ways to achieve additive mode locking which operates in both the positive and negative group velocity dispersion region in the optical fibers.[102] Using APM, pulses as short as 75 fs have been generated from KCl:Tl[105] and NaCl[108] lasers. Color center lasers often need to be synchronously pumped by other mode-locked lasers, and APM acts to shorten the pulse durations.

9.3.4 Semiconductor Lasers

The advantages of semiconductor lasers are compactness, high repetition rates, and high efficiency (Table 9.6). For semiconductor lasers with external cavities, the repetition rates are about 1 to 15 GHz, and for monolithic semiconductor lasers the repetition rates range from 30 to 300 GHz. The semiconductor laser has been mode-locked with active mode locking (gain modulation) and passive mode locking and has been gain-switched. Changing the gain of the semiconductor laser is particularly easy, since the carrier lifetime in the active region is $\sim$ ns, and relatively small electrical power is used. Active mode locking by gain modulation is usually done in an external cavity and in the microwave region; lower modulation frequencies allow gain recovery between pulses. Short electrical pulses also have been applied to modulate the gain. Typical pulsewidths from active mode locking are in the 10-ps range, although increasing the modulation frequency can decrease the pulsewidth to the subpicosecond range.[124] Passive mode locking has been achieved with semiconductor saturable absorbers, either in a separate chip from the laser chip[119] or integrated on the same chip to form a colliding pulse mode-locked system.[131] Gain switching can be done with either a sinusoidal or pulse waveform, the latter generated with a comb generator. Typically 10- to 50-ps pulses are obtained.

TABLE 9.5 Short-Pulse Color Center Lasers

Material	λ_0	Method	t_p	Power and energy	Rate	Reference
KCl:Tl	~1.5 μm	Active mode locking	6 ps	150 mW	41 MHz	98
		Sync. pump	3–10 ps	100 mW–1 W	82 MHz	99
		Sync. pump + soliton	60–130 fs			100, 101
		AM mode locking + soliton	380 fs			98
		Sync. pump + APM	75–260 fs	20–50 mW	82–100 MHz	103–105
		Passive mode locking	22 ps		83 MHz, chopped	106
NaCl	~1.6 μm	Sync. pump	5–10 ps	300–700 mW	82–164 MHz	107
		Sync. pump + APM	75–150 fs	300 mW	164 MHz	108
		Passive mode locking	275 fs	3.7-kW peak	83 MHz, chopped	106
KCl:Li	~2.7 μm	Sync. pump	11 ps	10–20 mW	82 MHz	109
		Passive mode locking	120 fs	470-W peak	100 MHz, chopped	110
RbCl:Li	~2.9 μm	Sync. pump	8 ps	10–20 mW	82 MHz	109
		Passive mode locking	190 fs	415-W peak	137 MHz, chopped	110
LiF:F$_2^+$	~870 nm	Sync. pump	0.7 ps	15–20 mW	82 MHz, chopped	111
		CPM	390 fs	10 mW	77 MHz, chopped	112
		CPM + GVDC*	180 fs	16 mW	77 MHz, chopped	113
Amplification:						
NaCl:F$_2^+$	1.55 μm	Q-switch YAG pump, multipass	200 fs	1–2 μJ	1 kHz	114

*GVDC stands for group velocity dispersion compensation.

TABLE 9.6 Short-Pulse Semiconductor Lasers

Material	Method	t_p	Power and energy	Rate	Reference
~0.8 μm	Active mode locking	5–20 ps	1–5 mW	~1 GHz	19, 115, 116
	Active + passive mode locking	5 ps	0.8 mW	0.3 GHz	117
	Passive mode locking	0.6–2 ps			118, 119
	G. S. and Q. S.	5 ps	10-W peak	18.5 GHz	120
	Passive mode locking, monolithic	2.4 ps	30 mW	108 GHz	121
	G. S., monolithic surf. emit.	4 ps			122
	Compression	0.46 ps	70-W peak	0.3 GHz	117
~1.3 μm	Active mode locking	<1 ps, 5–15 ps	~0.5 mW	1–16 GHz	123, 124
	G. S.	1–30 ps			125, 126
	Compression	4–7 ps		1–8 GHz	127, 128
~1.5 μm	Active mode locking, monolithic	4 ps		40 GHz	129
	G. S.	10–30 ps			126, 130
	CPM, monolithic	0.64–1.1 ps	10-mW peak	30–300 GHz	131
	Amplification + compression	23 ps	1 pJ	1 GHz	132

Although subpicosecond pulses can be obtained from semiconductor lasers, coherence between pulses is achieved only with low average power output, no more than mW. There are efforts to obtain coherent pulses via amplification with semiconductor diode amplifiers.[116,132]

9.3.5 Dye Lasers

Thus far, mode-locked dye lasers consistently produce the shortest pulses (Table 9.7). Passive mode locking and synchronous pumping are the two most-used methods. Passive mode locking using slow saturable absorbers has been particularly successful in dye lasers, being the first to generate subpicosecond pulses,[133] sub-100-fs pulses,[21] and the shortest optical pulses directly from a laser oscillator.[134,135] The most-often-used gain medium is rhodamine 6G with 3,3'-diethyloxadicarbocyanine iodide (DODCI) as the saturable absorber.[21] With compensation of GVD by an intracavity prism sequence, a ring cavity CPM laser has generated 27-fs pulses, the shortest pulses directly from a laser oscillator.[134] Consistently, CPM lasers produce ≤50-fs pulses. CPM can also be realized in a linear cavity terminated with an antiresonant ring.[136] Typically, a CPM laser provides about ≤50 mW average power.

Because of the relatively large stimulated cross section and short gain relaxation time, dye lasers are also successfully mode-locked by synchronous pumping. Both cw mode-locked argon ion lasers and the frequency-doubled, cw mode-locked Nd:YAG and Nd:YLF lasers have been used as pump lasers and produced about 2- to 5-ps pulses. By compressing the Nd:YAG and Nd:YLF pump laser pulses, synchronously pumped dye lasers can generate pulses of ~200 fs. Hybrid mode locking, which combines synchronous pumping and passive mode locking, is also used to produce sub-100-fs pulses. Typical average powers from synchronously pumped dye lasers are about 100 to 300 mW. Cavity dumping can increase the single pulse energy by 30 to 50 times at a reduced pulse repetition rate of several MHz.[133]

Although cw mode-locked dye lasers generate femtosecond pulses, the peak powers (a few kilowatts) are low for many spectroscopic experiments. There are several methods of amplification of femtosecond pulses. Different methods use different pump lasers. In general the trade-off between pulse energy and pulse repetition rate is such that the average power remains around 10 mW. Reference 145 provides a review. For high-repetition-rate (1 to 5 MHz) applications, the cavity-dumped argon ion laser is used as the pump laser for amplification of cavity-dumped CPM laser pulses.[146] In the intermediate frequency range (5 to 10 kHz), femtosecond pulses can be amplified by multiple passes through a single dye amplifying jet pumped by a copper vapor laser and followed by a grating pair or prism pair to compensate the additional group velocity dispersion encountered in the dye amplifier components.[147,148] An alternate method uses a frequency-doubled Nd:YAG regenerative amplifier to pump a multistage dye amplifier chain.[149] The output of a master cw mode-locked Nd:YAG laser is split to synchronously pump a dye oscillator and to inject the Nd:YAG regenerative amplifier. One advantage of this method is that the pump pulses of ~90 ps greatly reduce amplified spontaneous emission (ASE) and increase the extraction efficiency. To further increase pulse peak power to the gigawatt level, a frequency-doubled, Q-switched Nd:YAG laser can be used to pump a multistage dye amplifier with saturable absorbers placed between each stage to suppress ASE, and a grating pair is used to compensate the extra dispersion of the amplified pulses.[150] Another multistage gigawatt dye

TABLE 9.7 Short-Pulse Dye Lasers

Material	Pump	Method	t_p	Power and energy	Rate	Reference
Rhodamine 6-G $\lambda_0 \sim 630$ nm	cw argon	CPM + GVDC*	<40 fs	50–100 mW	100 MHz	134, 135
	cw argon	ARR	130 fs	100 mW	100 MHz	136
	ml argon	Sync. pump	3–5 ps			137
	ml argon	Sync. pump + ARR	130 fs	60 mW	80 MHz	138
	ml YAG	Sync. pump	2 ps, 70 fs	30 mW	100 MHz	139
	ml YAG	Sync. pump + ARR	85 fs	mW	82 MHz	140
	Comp. ml YAG	Sync. pump	~200 fs	150–300 mW	82 MHz	141, 142
	ml YAG	Sync. pump + PM + GVDC*	~60 fs	100 mW	125 MHz	143, 144
Amplification	Cavity-dump. argon		<100 fs	10–50 nJ	1–5 MHz	146
	Copper vapor laser		~50 fs	1–5 µJ	5–10 kHz	147, 148
	Nd:YAG regen. ampl.		85 fs	5 µJ	1–5 kHz	149
	Q-switch Nd:YAG		70 fs	0.5–1 mJ	10 Hz	150
	XeCl laser		70 fs	7 GW	100 Hz	151
Compression:						
$\lambda_0 \sim 630$ nm		Fiber + grating	<20 fs			153–155
$\lambda_0 \sim 630$ nm		Fiber + grating + prism	6 fs			32
$\lambda_0 \sim 800$ nm		Fiber + grating + prism	9 fs			198
$\lambda_0 \sim 500$ nm		Fiber + grating + prism	10 fs			199
$\lambda_0 \sim 620$ nm		Bulk medium + grating	<25 fs	100 µJ	6 Hz	156

*GVDC stands for group velocity dispersion compensation.

amplifier uses a XeCl excimer laser as the pump source, and pulses of 70 fs are amplified to 7 GW at 100 Hz.[151]

Pulse compression has been used for dye lasers to obtain the shortest optical pulses.[152–155] Because of the initial ultrashort pulse durations, short fibers are used and the group velocity dispersion helps to linearize the chirping. Formulas in Table 9.2 can be used to calculate the fiber lengths and the compressed pulsewidths. Using the combination of a grating sequence and a prism sequence to compensate the cubic phase delays, the record for shortest pulses, 6 fs, has been produced.[32] Using bulk quartz, amplified 100-fs pulses at 630 nm have been compressed to less than 25 fs with an energy output of ~100 μJ.[156]

Although most of the advances in generation and amplification of ultrashort pulses were first achieved in dye lasers based on rhodamine 6G centered at about 620 nm, the same techniques have been applied to a wide range of organic dyes. Ultrashort pulses of comparable durations have been generated in the spectral regions from blue to near infrared.[157–171]

Gain switching of short cavity dye lasers is a simple and effective way to generate picosecond and femtosecond pulses.[172] A short cavity length results in a short photon lifetime, while the large stimulated-emission cross section of a typical dye provides rapid gain switching. Typical cavity lengths are from 100 μm to a few millimeters, and the main pump sources are nitrogen or excimer lasers with pulse durations of several nanoseconds. The generated short pulses, sometimes after amplification, are often used to pump a second short cavity laser at a slightly longer wavelength. By cascade pumping, pulses of a few picoseconds can be easily generated in wide wavelength ranges,[173–175] particularly in the short-wavelength region where efficient cw pump sources are not available. This technique is simple, but the pulses are broadband. Gain-switched distributed feedback dye lasers (DFDLs) can produce coherent ultrashort pulses. Several methods have been used to create a high-visibility grating in the active medium. One can use two pump beams to interfere in the gain medium,[176] but a high degree of coherence is required of the pump beams. In order to utilize pump pulses which have poor spectral or temporal coherence, several techniques were developed.[177,178] Reflection[179] or transmission[180] holographic gratings were used in place of beam splitters. Direct projection and reduction of the image of a coarse grating using a high-quality microscope objective was reported to produce excellent results.[181] Typically 500-fs pulses have been generated from 400 to 760 nm.[181] After chirping and compression, 60-fs pulses were generated at 248 nm.[182]

9.3.6   Other Short Pulse Laser Systems

In some cases, noticeably in the uv and mid-ir, ultrashort pulses cannot be generated directly for lack of broadband laser media. To produce ultrashort laser pulses at these spectral regions, nonlinear conversions and other pulse shaping techniques are used.

In the uv region, it is difficult to generate ultrashort pulses directly from excimer lasers since the gain duration is too short for a short pulse to develop in a mode locking scheme. (The best results are 120 ps.[183]) An electro-optic modulator, overdriven with a high-voltage pulse, has been used to slice long pulses from a KrF laser into a train of short pulses. The secondary pulses are then quenched with gain saturation and fast optical breakdown in xenon, leading to a single pulse of ~30 ps.[184]

High-repetition-rate uv femtosecond pulses have been obtained by intracavity second harmonic generation (SHG) from CPM dye lasers at about 620 nm.[185,186] Pulses as short as 43 fs with 3 mW average power have been obtained using a beta barium borate (β-BaB_2O_4) crystal.[187] Femtosecond pulses at 308, 248, and 193 nm for XeCl, KrF, and ArF amplifiers, respectively, have been generated by SHG and sum frequency generation.[188–190] Picosecond pulses in 0.95 to 1.2 μm[191] and femtosecond pulses in 0.7 to 1.8 μm[192] were generated by parametric oscillation using SHG from Q-switched short pulse Nd lasers as the pump lasers. Tang and his coworkers have demonstrated a cw mode-locked optical parametric oscillator (OPO) which consists of a CPM laser at 620 nm and an OPO intersecting at the KTP crystal.[193] The tuning range is 0.72 to 4.5 μm; 105-fs pulses with 2 mW average power per direction have been obtained at 840 nm, and similar pulse durations are expected in 0.72 to 1.6 μm. Pulsewidths in 1.6 to 4.5 μm are expected to be 100 to 300 fs.

Generation of ultrashort pulse continuum provides tunable ultrashort pulses in a broad spectral region. By amplifying sub-100-fs pulses from a mode-locked dye laser at 620 nm to approximately several microjoules and focusing them on a thin liquid jet, supercontinuum spanning 0.4 to 1.0 μm can be generated[171,194–197] by self-phase modulation and self-steepening effects. Since the liquid jet is thin enough, group velocity dispersion is negligible and the continuum retains the same pulse duration. By compression of selectively amplified continuum, 9-fs pulses at ~800 nm[198] and 10-fs pulses at ~500 nm,[199] respectively, have been obtained.

Ultrashort pulses can be generated by soliton-like compression in an optical fiber with negative group velocity dispersion. One special case is soliton Raman generation in which the pump has positive GVD, while the Stokes wave has negative GVD. The strong pump pulse induces a nonlinear index change which chirps and broadens the Stokes pulse through cross-phase modulation (XPM).[200,201] When XPM is balanced by the negative velocity dispersion, a shorter Stokes pulse is produced, similar to the soliton compression which utilizes self-phase modulation instead. Using 1.32-μm pump pulses from mode-locked Nd:YAG lasers, Stokes pulses near 1.4 μm have been compressed in a single pass to as short as 100 fs.[202] With injection of cw diode-laser signals, tunable, femtosecond soliton pulses of 200 fs have been produced from a fiber Raman amplifier.[203] Synchronously pumped fiber Raman lasers[204] have also produced femtosecond pulses.[205–207] In erbium-doped fiber lasers, where the wavelength ~1.5 μm is in the negative GVD region, soliton pulse shaping plays a key role in the formation of ultrashort pulses. Pulses of 4 ps have been generated.[208]

Because of the narrow gain linewidths at low pressures, ultrashort pulses cannot be generated directly by mode locking CO_2 lasers. Instead, pulse shaping techniques are used. Using optical free induction decay induced by laser breakdown, pulses of 100 ps were produced.[209] Subpicosecond pulses, as short as 130 fs, at 10 μm have been produced[210,211] by ultrafast slicing of a long pulse CO_2 beam by semiconductor switches[212,213] controlled by high-power femtosecond pulses from a CPM dye laser-amplifier system.

9.4 METHODS OF PULSEWIDTH MEASUREMENTS

To measure ultrashort optical pulses, special techniques and devices with picosecond to femtosecond time resolution are needed. They can be classified into electronic and optical techniques. References 214 and 215 provide detailed discussions of picosecond pulse measurements.

9.4.1 Electronic Techniques

Sampling Oscilloscopes. For a pulse train of high repetition rate, the intensity envelope of optical pulses can be converted to electrical pulses by a fast photodetector and displayed on a sampling scope. The width of the optical pulses t_p is given to a good approximation by

$$t_p \simeq \sqrt{\tau_{\text{dis}}^2 - \tau_{\text{res}}^2} \tag{9.40}$$

where τ_{dis} is the displayed pulsewidth and τ_{res} is the combined electronic response time of the measurement device. For $t_p > \tau_{\text{res}}$, electronic techniques give direct and linear display of the envelope of optical pulses. Typical response times of ultrafast pin-diode detectors are about 20 to 30 ps. At room temperature, response times of sampling scopes are <20 ps. The response time of a superconductor sampling scope cooled to low temperatures can be as fast as ~5 ps. Optical pulses with durations less than 20 ps are difficult to measure with good resolution by a sampling scope.

Streak Cameras. A streak camera is an electron device with fast time resolution. A description of its structure and principles is given in Ref. 216. It can be operated on a single-shot mode provided that the peak power of the optical pulse is large enough. The time resolution is limited by the variation of photoelectron emission time, the electron space charge effect, and electron beam aberrations. Commercially available streak cameras can have a resolution as short as <1 ps. For pulse trains of high repetition rates, but low peak powers, synchronous-scan streak cameras with a time resolution of about 5 to 10 ps can be used. The streak camera is limited by its photocathode response, which ranges from the uv to near-ir regions. Infrared pulses must be converted to the shorter-wavelength region by nonlinear conversions such as second harmonic generation.

9.4.2 Optical Correlation Methods

Most ultrashort optical pulsewidth measurements are performed by using optical correlation methods. Among them the second-order autocorrelation is most often used.[214,215] Higher-order nonlinear methods are discussed in Refs. 214 and 217.

Autocorrelation. Autocorrelation is done by interfering two replicas of an optical pulse with a relative time delay. Mathematically the background-free intensity autocorrelation is given by[214,218]

$$G_0^2(\tau) = \frac{\int_{-\infty}^{\infty} I(t)I(t + \tau)dt}{\int_{-\infty}^{\infty} I^2(t)dt} \tag{9.41}$$

and the with-background intensity autocorrelation is given by[214,218]

$$G_B^2(\tau) = 1 + 2G_0^2(\tau) \tag{9.42}$$

where $I(t) = \frac{1}{2} A^2(t)$ is the optical time-averaged pulse intensity.

For coherent pulses, the autocorrelation trace exhibits a single peak. The pulsewidth t_p (FWHM) can be deconvolved from the width, τ_{auto} (FWHM), of the autocorrelation peak by assuming a certain pulse shape. Table 9.8 lists the deconvolution factors for some common pulse shapes. For partially coherent pulses,

TABLE 9.8 Intensity Autocorrelation Widths and Bandwidths of Four Different Transform-Limited Pulse Shapes. All Are FWHM Values

$I(t)$	τ_{auto}/t_p	$t_p \Delta \nu_p$
$1 \ (0 \le t \le t_p)$	1	0.886
$\exp\left[-\dfrac{(4 \ln 2)t^2}{t_p^2}\right]$	$\sqrt{2}$	0.441
$\mathrm{sech}^2\left(\dfrac{1.76t}{t_p}\right)$	1.55	0.315
$\exp\left[-\dfrac{(\ln 2)t}{t_p}\right] \ (t \ge 0)$	2	0.11

which are essentially noise bursts, the autocorrelation trace exhibits a coherent spike sitting on top of a broad pedestal which is a measure of the overall pulse duration (Fig. 9.11). The width of the spike is determined by the bandwidth of the pulses. In the case of with-background-type autocorrelation, the ratio of peak to background is 3 to 1 for coherent pulses, and the ratio of peak to pedestal to background is 3 to 2 to 1 for partially coherent pulses. For continuous noises, the pedestal becomes infinitely broad, and the peak-to-background ratio is 3 to 2. For

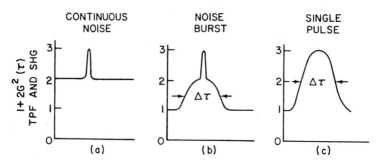

FIGURE 9.11 Theoretical autocorrelation traces of pulses with different degrees of coherence. *(From Ippen and Shank.[214])*

the background-free-type autocorrelation, the traces have similar signatures of pulse coherence, but without the background.

Slow Scan Autocorrelation. If the variation of the time delay is slow enough, interference fringes can be resolved. The autocorrelation function with background is given by[218]

$$g_B^2(\tau) = \frac{\int_{-\infty}^{\infty} [E(t) + E(t + \tau)]^4 dt}{2 \int_{-\infty}^{\infty} E^4(t) dt} \tag{9.43}$$

where $E(t) = A(t) \cos[\omega t + \phi(t)]$ is the electric field. For a coherent pulse, the peak-to-background ratio is 8 to 1. Such slow autocorrelation is especially useful to reveal the phase information of the pulse.[219]

Autocorrelators for Repetitive Pulses *Background-Free Autocorrelator.* Figure 9.12*a* shows the schematic of a background-free autocorrelator using second harmonic generation. The incoming pulse is split into two beams. Of the two beams, one experiences a fixed time delay and the other a variable time delay by moving the mirror in that arm. The two beams then recombine in a second harmonic crystal at an angle determined by the phase matching condition. When the two pulses overlap at the crystal, the second harmonic signal is generated in the bisecting direction and is detected by a photomultiplier. An aperture is used to block the fundamental beams, and if the second harmonic signal is weak, a lock-in amplifier can be used with one beam chopped periodically. The second harmonic signal as a function of the relative time delay gives the autocorrelation of the pulse measured.

With-Background Autocorrelator. Figure 9.12*b* shows the schematic of a with-background-type autocorrelator using second harmonic generation. The two beams

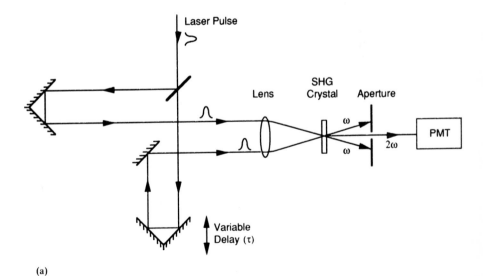

(a)

FIGURE 9.12 Setup of (*a*) background-free autocorrelator and (*b*) with background autocorrelator.

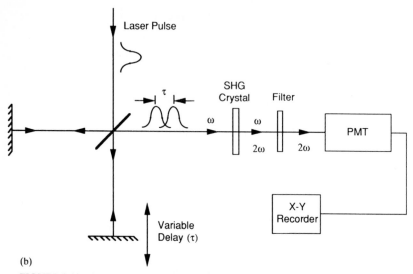

(b)

FIGURE 9.12 (*Continued*) Setup of (*a*) background-free autocorrelator and (*b*) with background autocorrelator.

recombine and propagate collinearly. Using a type I second harmonic crystal, each beam by itself can generate a second harmonic signal which also propagates collinearly with, but polarizes orthogonally to, the fundamental pulse. The total second harmonic signal results from the sum of electric fields of the two pulses. Thus, at large relative time delay, a background second harmonic signal is contributed from the sum of intensities of second harmonic signals generated by the two separated pulses. A filter is used to absorb the fundamental signal.

Single Shot Autocorrelators.   For laser pulses with low repetition rates, autocorrelation via stepping delay becomes impractical. Single-shot intensity autocorrelation is needed to measure pulsewidths. Originally two-photon fluorescence (TPF) was used.[214] The two split beams were made to collide in a liquid dye cell. While transparent at the fundamental wavelength, the dye molecules are lifted to an upper level by two-photon absorption, and then emit fluorescence in the visible spectrum. The spatial distribution of the fluorescence intensity gives the with-background intensity autocorrelation of the pulse. By photographing the fluorescence in the direction perpendicular to the colliding axis, pulsewidths can be deduced.

A more efficient way to produce a well collimated beam is by noncollinear second harmonic generation.[220,221] As shown in Fig. 9.13, two synchronized beams overlap in the second harmonic crystal at an angle determined by phase matching. When the beam size is much larger than the spatial extent of the optical pulse, the spatial distribution of the time-integrated SHG is proportional to the background-free intensity autocorrelation of the pulse. The pulsewidth can be deconvolved from the spatial width of the second harmonic signal by

$$t_p = \frac{2n W_{\text{FWHM}} \sin(\phi_m/2)}{\eta c} \tag{9.44}$$

In (9.44) W_{FWHM} is the spatial width of the second harmonic signal, ϕ_m is the angle between the two beams, n is the index of refraction of the crystal, c is the speed

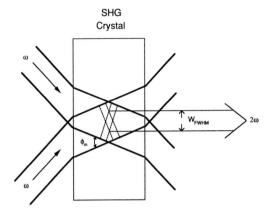

FIGURE 9.13   Setup of a single-shot autocorrelator using second harmonic generation.

of light in vacuum, and η is the same deconvolution factor, τ_{auto}/t_p, as in autocorrelation. With this single-shot autocorrelator, pulsewidths of about 50 fs have been resolved.[222]

9.5 APPLICATION OF ULTRAFAST OPTICS TO HIGH-FREQUENCY ELECTRONICS

9.5.1 Introduction

There are many scientific applications of picosecond and femtosecond lasers in physics, chemistry, and electronics. In this section we discuss only the application of ultrafast optics to electronics, in particular microwave electronics.

Rapid advances in both microwave and optical technologies make it possible to merge them to form a new class of devices that can perform high-speed, high-frequency electronic functions. With the advent of femtosecond lasers we envision such devices to work in the picosecond time domain. We refer to the new technology as picosecond optical electronics, which combines the fields of optics and electronics. The link between these two fields is picosecond photoconductivity.[223]

Optical techniques for controlling the microwave devices offer unique advantages in (1) near-perfect isolation between controlling and controlled devices, (2) elimination of rf feed in a large array system, (3) immunity from electromagnetic interference, (4) light weight and compact size, (5) extremely fast response, (6) high power-handling capability, and (7) possibility for monolithic integration.

The conversion of optical ultrashort pulses to electrical pulses can be understood as follows. An electrical pulse train is instantaneously generated via the picosecond photoconductivity effect when a photoconductor is illuminated with an optical pulse train. Through this effect one strips off the optical carrier frequency while retaining the ultrawide bandwidth of the optical pulses. Since the electrical pulses thus generated have picosecond time duration, it is a natural match to millimeter waves whose periods are in the tens of picosecond range.

Picosecond photoconductors can now be engineered to generate electric pulses as short as 1 ps[224] with jitter-free switching. All subsequent electronic events can be slaved to this initial pulse. Picosecond optical techniques can be utilized again to generate other pulses that can be precisely time-controlled. Using picosecond optical electronic techniques, it is possible now to generate microwave and millimeter-wave signals that are in complete time synchronization with the exciting optical pulses. Radiofrequency waveform generation, ranging from one monocycle[225] to continuous wave, with peak power up to a few megawatts in the pulsed case and frequency up to 20 GHz in the cw case,[226] has been achieved. Phase control of the microwave signals can be accomplished by simply controlling the arrival time of the exciting optical pulses, giving potential for real-time-delay phased array radiation.

The progress in the application of ultrafast optics to microwave and millimeter-wave technology through picosecond optoelectronics will be reviewed in this section. The term "picosecond optoelectronics" is defined as electronic function performed by picosecond optical pulses via photoconductivity. In addition to the millimeter-wave generation and control, the picosecond optical pulse can also be used to characterize monolithic millimeter-wave integrated circuits and to phase lock free-running microwave oscillators to optical pulses. For a detailed recent review in ultrafast optoelectronics the reader is referred to the articles in *Picosecond Optoelectronic Devices*,[227] the proceedings of the 1985, 1987, 1989, and 1991 Picosecond Electronics and Optoelectronics Conferences,[228–231] and Ref. 232.

9.5.2 Picosecond Photoconductors

A photoconductor exhibiting picosecond response time is referred to as a picosecond photoconductor. This phenomenon is referred to as the picosecond photoconductivity effect, which was first reported in 1972.[223] Picosecond photoconductors used as ultrafast switching and gating devices have been demonstrated. In this application, the photoconductor is dc-biased at the input terminal and a fast rise electrical output signal is obtained instantaneously when it is irradiated by a picosecond optical pulse. Thus in a sense it is a three-terminal device where the control terminal is an optical port.

A large number of devices are based on the picosecond photoconductivity effect. They include switches, gates, samplers, electronic impulse function correlators. A/D converters, optical detectors, dc to rf converters, coherent microwave generators, and terahertz radiation source generators.[233,234] In this chapter we concentrate on optical techniques for the characterization of monolithic microwave integrated circuits (MMIC) and on phase locking of MMICs to the optical pulses. A detailed review in the general area on the application of ultrafast optics to microwaves can be found in Refs. 235 and 236.

9.5.3 On-Wafer Optoelectronic Characterization of Monolithic Millimeter-Wave Integrated Circuits

Microwave measurements are traditionally performed in the frequency domain, where desired results are typically expressed in the form of scattering parameters. Circuit components are usually mounted in a 50-Ω microstrip network and the measurements performed with a network analyzer, such as the HP 8510. For the GaAs monolithic microwave and millimeter-wave integrated circuit (MMIC) manufacturer, it is desirable to be able to characterize the devices before the wafer is diced, i.e., to perform on-wafer measurements.

In an effort to reduce the cost of large-scale on-wafer measurements and to achieve a more diverse technique, the use of optoelectronic techniques has been proposed. Several approaches for making these measurements have been studied. Frequency-domain measurements have been performed by electro-optic probing of the electrical signal on a microstrip line.[237] An alternative is to generate a very short electrical signal on-wafer by illuminating a biased photoconductor with a picosecond optical pulse.[238,239] The electrical signal on a line can then be sampled by electro-optic (EO) or photoconductive (PC) sampling. Some results for a GaAs FET mounted in a silicon-on-sapphire test circuit have been presented using this approach with PC sampling.[240]

In the optoelectric characterization of MMICs, it is clearly advantageous to be able to generate the characterization signal on the wafer with only dc bias required, thereby avoiding the difficult problem of launching the microwave or millimeter-wave signal. The technique described here differs from that reported in Ref. 237. Upon a comparison of optoelectronic time-domain measurements with frequency-domain measurements, a number of trade-offs are evident. When measurements are performed in the time domain, broadband information can be obtained with a single measurement. The time-domain approach can be used to characterize nonlinear effects and the current optical technique can be simply extended for multiport measurements.

The measurement system for obtaining two-port scattering parameters using PC sampling is shown in Fig. 9.14. The pulse generation occurs at ports 1 or 4, with the pulse traveling away from the device under test (DUT) either being terminated in the matched load or being windowed out by the sampling time duration. If the

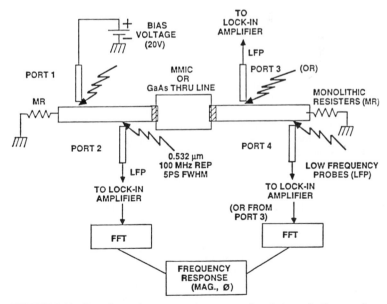

FIGURE 9.14 Optoelectronic measurement system using photoconductive sampling.

DUT is assumed linear, the broadband scattering parameters can be reconstructed from the measured time-domain response by appropriately windowing the measured data at the sampling ports and using the fast Fourier transform (FFT).

The implementation of both the PC and EO sampling of a 28-GHz MMIC and the comparison of the optical sampling results with that obtained by a conventional network analyzer measurement have been reported in Ref. 236. The reader is referred to that paper for detail. Figure 9.15 shows the magnitude of S_{21} parameter

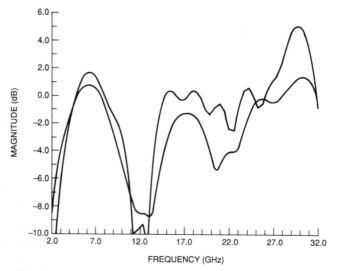

FIGURE 9.15 S_{21} magnitude determined by EO sampling (light) and PC sampling (bold).

measured by the EO and PC sampling techniques, and the corresponding data as determined by the network analyzer measurements (HP 8510) are shown in Fig. 9.16. In addition to the complete phase data of S_{21}, the measurement of both magnitude and phase of S_{11} has also been reported in Ref. 241 and 242. The magnitude of S_{21} measured by the PC sampling technique agrees well with that obtained by the conventional network analyzer. The magnitude measured by the EO technique is low and may be due to calibration problems. Calibration is more difficult with EO sampling because the signal is very sensitive to the local substrate thickness and thus position of the sampling beam. This is because of the interference effect. In order to avoid the interference effect, one has to use laser pulses with durations shorter than the round trip time for the optical pulse to travel through the substrate.

Another important performance criterion of the S-parameter measurement technique is the dynamic range of the measured data. This issue has recently been investigated. In particular it has been found that in photoconductive sampling, the dynamic range is intrinsically limited by the time span of the sampling. This effectively imposes a limited time window on the measurement. This effect has been

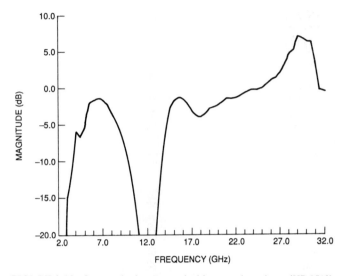

FIGURE 9.16 S_{21} magnitude measured with network analyzer (HP 8510).

demonstrated in the measurement of the S_{21} parameter of a three-stage 12-GHz MMIC amplifier with the photoconductive sampling technique. The measured S_{21} parameter with different measurement times is shown in Fig. 9.17. A good agreement with network analyzer measurement is obtained up to 40-dB dynamic range when the measurement time is 2 ns. The dynamic range reduced to 30 dB when the measurement time only spans over 1.1 ns. From Fig. 9.17, two distinct features are evident. First, the separation of the peaks of the ripples is approximately equal to the inverse of the length of the time span of measurement. Second, the dynamic range increases as the sampling range increases. These observations can be explained by the following simple model. Let's consider a sine wave with period t_0 measured by PC sampling with a time window of T. The measured waveform and

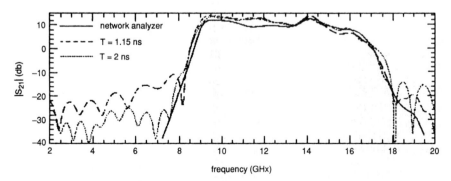

FIGURE 9.17 Effect of time windowing.

the magnitude of its Fourier transform can be expressed as follows:

$$A(t) = \begin{cases} \sin \dfrac{2\pi t}{t_0} & ; \ -\dfrac{T}{2} < t < \dfrac{T}{2} \\[2mm] 0 & ; \ \text{elsewhere} \end{cases} \tag{9.45}$$

$$|A(f)| = \left| \frac{\sin(\pi f T)}{\pi} \ \frac{t_0}{1 - f^2 t_0^2} \right| \tag{9.46}$$

First, it is clear from Fig. 9.18 that time windowing generates ripples, which limits the dynamic range. Second, the distance between two minima is $1/T$ and independent of t_0. A computer simulation, using a 7-GHz bandwidth waveform with a center frequency of 12 GHz and 2-ns time window, shows that the dynamic range at 6 GHz is 38 dB, which is in excellent agreement with the experimental data. This model and the experimental data show that the limitation is primarily on the length of sampling time, not the noise created by laser fluctuations.

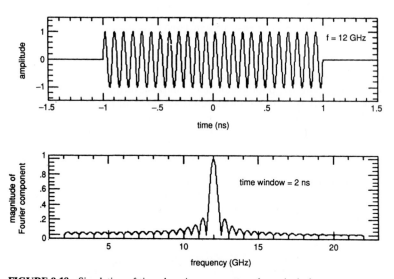

FIGURE 9.18 Simulation of time-domain measurement for a single frequency.

9.5.4 Microwave Phase Locking Using Electro-optic Harmonic Mixing

As mentioned in the previous section, the electro-optic (E-O) sampling technique has potential applications in microwave measurement including MMIC characterizations. However, in the particular method to measure the waveform at microwave frequency by the E-O sampling technique, there is a requirement for stable synchronization between the microwave source and the laser pulses. Even in the case of amplitude measurement alone (excluding phase), the signal-to-noise ratio is proportional to the frequency stability of the microwave source. In order to achieve high-sensitivity measurement, synchronization is essential.

A new phase locking technique, which can potentially lock any MMIC oscillator for frequency up to 100 GHz at the wafer level before the wafer is diced and the device is rf packaged, will be discussed. This technique can be accomplished without requiring mechanical contacts to supply microwave and millimeter wave signal to the wafer.

The schematic of the phase-locked loop (PLL) is shown in Fig. 9.19. The harmonic mixer consists of a GaAs microstrip line, a polarizer (optional), and a

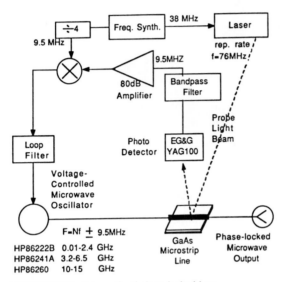

FIGURE 9.19 Schematic of phase-locked loop.

photodetector. The E-O effect can be detected by a conventional scheme which measures the rotation of laser beam polarization with a polarizer[237] or by the multiple-beam interference effect from the substrate without using a polarizer. The electro-optic effect results in the mixing of the microwave signal on the stripline with one of the harmonic frequencies of the laser repetition rate, thus producing an intermediate frequency f_{IF}. The upper frequency limit of this mixer is determined by the thickness of the GaAs substrate, and the laser pulsewidth. The output signal from the photodetector is amplified by a 10-MHz amplifier with a gain of 80 dB, and then mixed with a 10-MHz reference signal derived from the driver of the A-O modulator. The resultant error signal is applied to the voltage-controlled oscillator (VCO), such as HP86222B, which can be controlled by an external dc voltage through an active loop filter. The presence of the loop filter is essential in the PLL. The locked frequency of the VCO has to be $f = 100N \pm 10$ (MHz), where N is an integer and the laser pulse repetition rate is 100 MHz.

The phase-locked condition can be determined by monitoring the output waveform of the intermediate-frequency amplifier and the output of the 10-MHz reference oscillator by using an oscilloscope. The time base of the scope is triggered by one of the signals. If phase locking is achieved, very clean waveforms from both signals can be observed. If the output signal is not phase-locked, only one trace stays clear. Without phase locking, the output of the microwave source has a wide

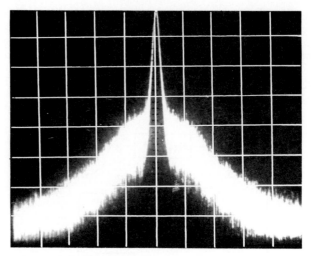

FIGURE 9.20   Spectrum of a phase-locked 10-GHz signal using PC
sampler.

spectrum and is very unstable. After the signal has been phase locked, the measured
spectrum becomes narrow and stable. Figure 9.20 shows a phase-locked 10-GHz
signal. Phase-locked signals up to 15 GHz have been obtained.

Progress in the application of ultrafast optical techniques using photoconductive
switching to the microwave and millimeter-wave technology has been reviewed. A
time-domain network analyzer using a photoconductive sampling method for on-
wafer MMIC characterization has also been described. A new microwave phase
lock scheme which synchronizes the microwave output with picosecond laser pulses
has been discussed. With this synchronized microwave source, it is possible to
measure the microwave waveform. Since this technique allows the conversion of
rf phase signal to dc error signal without requiring the rf probe physically contacting
(or connecting) with the microwave circuit, this makes on-wafer phase locking and
characterization of a MMIC circuit possible, prior to the wafer's being diced and
the individual MMIC packaged.

9.6   CONCLUSIONS

We have surveyed the most important types of ultrashort pulse lasers and methods
of measurement with ultrashort pulses. The progress in time resolution in the past
two decades has been most remarkable. Today, it is reaching the limit of optics.
We have not had the space to mention all the scientific advances made possible by
these ultrashort pulse techniques. The application to engineering problems, on the
other hand, has not kept pace to exploit fully the available bandwidth. Even so,
optical techniques have enabled electronics to increase its speed to the subpicosecond regime, as demonstrated by the examples in Sec. 9.5. We anticipate that
the inherent wider bandwidth of optics will play an increasingly important role in
various engineering applications and continuously advance scientific research.

Acknowledgment

We are pleased to acknowledge partial support by the U.S. Air Force Office of Scientific Research, the National Science Foundation, and DARPA. We also thank J. Goldhar for helpful comments.

9.7 REFERENCES

1. L. E. Hargrove, R. L. Fork, and M. A. Pollack, "Locking of He-Ne Laser Modes Induced by Synchronous Intracavity Modulation," *Appl. Phys. Lett.*, vol. 5, pp. 4–5, 1964.

2. S. E. Harris and R. Targ, "FM Oscillation of the He-Ne Laser," *Appl. Phys. Lett.*, vol. 5, pp. 202–204, 1964.

3. H. W. Mocker and R. J. Collins, "Mode Competition and Self-Locking Effects in a Q-Switched Ruby Laser," *Appl. Phys. Lett.*, vol. 7, pp. 270–273, 1965.

4. A. J. DeMaria, C. M. Ferrar, and G. E. Danielson, Jr., "Mode Locking of a Nd^{3+}-Doped Glass Laser," *Appl. Phys. Lett.*, vol. 8, pp. 22–24, 1966.

5. M. J. DiDomenico, Jr., "Small-Signal Analysis of Internal (Coupling-Type) Modulation of Lasers," *J. Appl. Phys.*, vol. 35, pp. 2870–2876, 1964.

6. D. J. Kuizenga and A. E. Siegman, "FM and AM Mode-Locking of the Homogeneous Laser—Parts I and II," *IEEE J. Quantum Electron.*, vol. QE-6, pp. 694–715, 1970.

7. H. A. Haus, "A Theory of Forced Mode Locking," *IEEE J. Quantum Electron.*, vol. QE-11, pp. 323–330, 1975.

8. E. M. Garmire and A. Yariv, "Laser Mode Locking with Saturable Absorbers," *IEEE J. Quantum Electron.*, vol. QE-3, pp. 222–226, 1967.

9. H. A. Haus, "Theory of Mode Locking with a Slow Saturable Absorber," *IEEE J. Quantum Electron.*, vol. QE-11, pp. 736–746, 1975.

10. H. A. Haus, "Theory of Mode Locking with a Fast Saturable Absorber," *J. Appl. Phys.*, vol. 46, pp. 3049–3058, 1975.

11. E. P. Ippen, H. A. Haus, and L. Y. Liu, "Additive Pulse Mode Locking," *J. Opt. Soc. Am.*, vol. B 6, pp. 1736–1745, 1989.

12. G. H. C. New, "The Generation of Ultrashort Laser Pulses," *Rep. Prog. Phys.*, vol. 46, pp. 877–971, 1983.

13. P. P. Sorokin, J. R. Lankard, E. C. Hammond, and V. L. Moruzzi, "Laser-Pumped Stimulated Emission from Organic Dyes: Experimental Studies and Analytical Comparisons," *IBM J. Res. Develop.*, vol. 11, pp. 130–147, 1967.

14. C. Lin, "Studies of Relaxation Oscillations in Organic Dye Lasers," *IEEE J. Quantum Electron.*, vol. QE-11, pp. 602–609, 1975.

15. F. Gires and P. Tournois, "Inferféromètre utilisable pour la compression d'impulsions lumineuses modulées en fréquence," *C. R. Acad. Sci. (Paris)*, vol. 258, pp. 6112–6115, June 1964.

16. J. A. Giordmaine, M. A. Duguay, and J. W. Hansen, "Compression of Optical Pulses," *IEEE J. Quantum Electron.*, vol. QE-4, pp. 252–255, 1968.

17. A. E. Siegman, *Lasers*, University Science, Mill Valley, Calif., 1986.

18. See, for example, A. Yariv, *Quantum Electronics*, 3d ed., Wiley, New York, 1989.

19. P.-T. Ho, L. A. Glasser, E. P. Ippen, and H. A. Haus, "Picosecond Pulse Generation with a cw GaAlAs Laser Diode," *Appl. Phys. Lett.*, vol. 33, pp. 241–242, 1978.

20. K. Y. Lau, N. Bar-Chaim, I. Ury, Ch. Harder, and A. Yariv, "Direct Amplitude

Modulation of Short-Cavity GaAs Lasers up to X-Band Frequencies," *Appl. Phys. Lett.*, vol. 43, pp. 1–3, 1983.

21. R. L. Fork, B. I. Greene, and C. V. Shank, "Generation of Optical Pulses Shorter than 0.1 psec by Colliding Pulse Mode Locking," *Appl. Phys. Lett.*, vol. 38, pp. 671–672, 1981.

22. R. L. Fork, C. V. Shank, R. Yen, and C. A. Hirlimann, "Femtosecond Optical Pulses," *IEEE J. Quantum Electron.*, vol. QE-19, pp. 500–506, 1983.

23. A. E. Siegman, "Passive Mode Locking Using an Antiresonant-Ring Laser Cavity," *Opt. Lett.*, vol. 6, pp. 334–335, 1981.

24. K. J. Blow and D. Wood, "Mode-Locked Lasers with Nonlinear External Cavities," *J. Opt. Soc. Am.*, vol. B 5, pp. 629–632, 1988.

25. M. Morin and M. Piché, "Interferential Mode Locking: Gaussian Pulse Analysis," *Opt. Lett.*, vol. 14, pp. 1119–1121, 1989.

26. E. P. Ippen, L. Y. Liu, and H. A. Haus, "Self-Starting Condition for Additive-Pulse Mode-Locked Lasers," *Opt. Lett.*, vol. 15, pp. 183–185, 1990.

27. H. Kogelnik and C. V. Shank, "Stimulated Emission in a Periodic Structure," *Appl. Phys. Lett.*, vol. 18, pp. 152–154, 1971.

28. H. Kogelnik and C. V. Shank, "Coupled-Wave Theory of Distributed Feedback Lasers," *J. Appl. Phys.*, vol. 43, pp. 2327–2335, 1972.

29. See, for example, H. A. Haus, *Waves and Fields in Optoelectronics*, Prentice-Hall, Englewood Cliffs, N.J., 1984.

30. E. B. Treacy, "Optical Pulse Compression with Diffraction Gratings," *IEEE J. Quantum Electron.*, vol. QE-5, pp. 454–458, 1969.

31. R. L. Fork, O. E. Martinez, and J. P. Gordon, "Negative Dispersion Using Pairs of Prisms," *Opt. Lett.*, vol. 9, pp. 150–152, 1984.

32. R. L. Fork, C. H. Brito Cruz, P. C. Becker, and C. V. Shank, "Compression of Optical Pulses to Six Femtoseconds by Using Cubic Phase Compensation," *Opt. Lett.*, vol. 12, pp. 483–485, 1987.

33. C. H. Brito Cruz, P. C. Becker, R. L. Fork, and C. V. Shank, "Phase Correction of Femtosecond Optical Pulses Using a Combination of Prisms and Gratings," *Opt. Lett.*, vol. 13, pp. 123–125, 1988.

34. R. A. Fisher, P. L. Kelley, and T. K. Gustafson, "Subpicosecond Pulse Generation Using Optical Kerr Effect," *Appl. Phys. Lett.*, vol. 14, pp. 140–143, 1969.

35. T. Y. Chang, "Fast Self-Induced Refractive Index Changes in Optical Media," *Opt. Eng.*, vol. 20, pp. 220–232, 1981.

36. D. Grischkowsky and A. C. Balant, "Optical Pulse Compression Based on Enhanced Frequency Chirping, *Appl. Phys. Lett.*, vol. 41, pp. 1–3, 1982.

37. W. J. Tomlinson, R. H. Stolen, and C. V. Shank, "Compression of Optical Pulses Chirped by Self-Phase Modulation in Fibers," *J. Opt. Soc. Am.*, vol. B 2, pp. 139–149, 1984.

38. R. G. Smith, "Optical Power Handling Capacity of Low Loss Optical Fiber as Determined by Stimulated Raman and Brillouin Scattering," *Appl. Opt.*, vol. 11, pp. 2489–2494, 1972.

39. R. H. Stolen and A. M. Johnson, "The Effect of Pulse Walkoff on Stimulated Raman Scattering in Fibers," *IEEE J. Quantum Electron.*, vol. QE-22, pp. 2154–2160, 1986.

40. A. Hasegawa and F. Tappert, "Transmission of Stationary Nonlinear Optical Pulses in Dispersive Dielectric Fibers I Anomalous Dispersion," *Appl. Phys. Lett.*, vol. 23, pp. 142–144, 1973.

41. See, for example, G. P. Agrawal, *Nonlinear Fiber Optics*, Academic Press, San Diego, 1989.

42. L. F. Mollenauer, R. H. Stolen, and J. P. Gordon, "Experimental Observation of Picosecond Pulse Narrowing and Solitons in Optical Fibers," *Phys. Rev. Lett.*, vol. 45, pp. 1095–1098, 1980.

43. L. F. Mollenauer, R. H. Stolen, J. P. Gordon, and W. J. Tomlinson, "Extreme Picosecond Pulse Narrowing by Means of Soliton Effect in Single-Mode Optical Fibers," *Opt. Lett.*, vol. 8, pp. 289–291, 1983.

44. O. Svelto, "Self-Focusing, Self-Trapping, and Self-Phase Modulation of Laser Beams," in E. Wolf (ed.), *Progress in Optics XII*, North-Holland, 1974.

45. S. A. Akhmanov, R. V. Khokhlov, and A. P. Sukhorukov, "Self-Focusing, Self-Defocusing, and Self-Modulation of Laser Beams," in F. T. Arecchi and E. O. Schulz-Dubois (eds.), *Laser Handbook*, North-Holland, 1972.

46. R. Y. Chiao, E. Garmire, and C. H. Townes, "Self-Trapping of Optical Beams," *Phys. Rev. Lett.*, vol. 13, pp. 479–482, 1964.

47. L. Yan, P.-T. Ho, C. H. Lee, and G. L. Burdge, "Generation of High-Power, High Repetition Rate, Subpicosecond Pulses by Intracavity Chirped Pulse Regenerative Amplification," *Appl. Phys. Lett.*, vol. 54, pp. 690–692, 1989.

48. P. Maine, D. Strickland, P. Bado, M. Pessot, and G. Mourou, "Generation of Ultrahigh Peak Power Pulses by Chirped Pulse Amplification," *IEEE J. Quantum Electron.*, vol. QE-24, pp. 398–403, 1988.

49. O. E. Martinez, "Design of High-Power Ultrashort Pulse Amplifiers by Expansion and Recompression," *IEEE J. Quantum Electron.*, vol. QE-23, pp. 1385–1387, 1987.

50. M. Peesot, P. Maine, and G. Mourou, "1000 Times Expansion/Compression of Optical Pulses for Chirped Pulse Amplification," *Opt. Commun.*, vol. 62, pp. 419–421, 1987.

51. See, for example, W. Koechner, *Solid-State Laser Engineering*, Springer-Verlag, New York, 1984.

52. S. De Silesrtri, P. Laporta, and V. Magni, "14-W Continuous-Wave Mode-Locked Nd:YAG Laser," *Opt. Lett.*, vol. 11, pp. 785–787, 1986.

53. A. M. Johnson and W. M. Simpson, "Continuous-Wave Mode-Locked Nd:YAG-Pumped Subpicosecond Dye Lasers," *Opt. Lett.*, vol. 8, pp. 554–556, 1983.

54. T. Sizer, II, "Mode Locking of High-Power Neodymium:Yttrium Aluminum Garnet Lasers at Ultrahigh Repetition Rates," *Appl. Phys. Lett.*, vol. 55, pp. 2694–2695, 1989.

55. H. Graener and A. Laubereau, "Shorter and Bandwidth-Limited Nd:YAG Laser Pulses," *Opt. Commun.*, vol. 37, pp. 138–142, 1981.

56. H. Vanherzeele, J. L. Van Eck, and A. E. Siegman, "Colliding Pulse Mode Locking of a Nd:YAG Laser with an Antiresonant Ring Structure," *Appl. Opt.*, vol. 20, pp. 3484–3486, 1981.

57. H. P. Kortz, "Characterization of Pulsed Nd:YAG Active/Passive Mode-Locked Laser," *IEEE J. Quantum Electron.*, vol. QE-19, pp. 578–584, 1983.

58. A. Del Corno, G. Gabetta, G. C. Reali, V. Kubecek, and J. Marek, "Active-Passive Mode-Locked Nd:YAG Laser with Passive Negative Feedback," *Opt. Lett.*, vol. 15, pp. 734–736, 1990.

59. L. Y. Liu, J. M. Huxley, E. P. Ippen, and H. A. Haus, "Self-Starting Additive-Pulse Mode Locking of a Nd:YAG Laser," *Opt. Lett.*, vol. 15, pp. 553–555, 1990.

60. T. F. Carruthers and I. N. Dulling III, "Passive Laser Mode Locking with an Antiresonant Nonlinear Mirror," *Opt. Lett.*, vol. 15, pp. 804–806, 1990.

61. G. T. Maker and A. I. Ferguson, "Frequency-Modulation Mode Locking of a Diode-Pumped Nd:YAG Laser," *Opt. Lett.*, vol. 14, pp. 788–790, 1989.

62. J. Goodberlet, J. Jacobson, J. G. Fujimoto, P. A. Schulz, and T. Y. Fan, "Self-Starting Additive-Pulse Mode-Locked Diode-Pumped Nd:YAG Laser," *Opt. Lett.*, vol. 15, pp. 504–506, 1990.

63. P. Bado, M. Bouvier, and J. S. Coe, "Nd:YLF Mode-Locked Oscillator and Regenerative Amplifier," *Opt. Lett.*, vol. 12, pp. 319–321, 1987.

64. J. K. Chee, E. C. Cheung, M. N. Kong, and J. M. Liu, "Passive Mode Locking of a cw Nd:YLF Laser with a Nonlinear External Coupled Cavity," *Opt. Lett.*, vol. 15, pp. 685–687, 1990.

65. U. Keller, T. K. Woodward, D. L. Sivco, and A. Y. Cho, "Coupled-Cavity Resonant Passive Mode-Locked Nd:Yttrium Lithium Fluoride Laser," *Opt. Lett.*, vol. 16, pp. 390–392, 1991.

66. U. Keller, K. D. Li, B. T. Khuri-Yakub, D. M. Bloom, K. J. Weingarten, and D. C. Gerstenberger, "High-Frequency Acousto-optic Mode Locker for Picosecond Pulse Generation," *Opt. Lett.*, vol. 15, pp. 45–47, 1990.

67. K. J. Weingarten, D. C. Shannon, R. W. Wallace, and U. Keller, "Two-Gigahertz Repetition-Rate, Diode-Pumped, Mode-Locked Nd:YLF Laser," *Opt. Lett.*, vol. 15, pp. 962–964, 1990.

68. J. Goodberlet, J. Jacobson, J. Wang, J. G. Fujimoto, T. Y. Fan, and P. A. Schulz, "Ultrashort Pulse Generation with Additive Pulse Modelocking in Solid State Lasers: Ti:Al$_2$O$_3$, Diode Pumped Nd:YAG and Nd:YLF," in C. B. Harris, E. P. Ippen, G. A. Mourou, and A. H. Zewail (eds.), *Ultrafast Phenomena VII*, Springer-Verlag, New York, 1990, pp. 11–13.

69. G. P. A. Malcolm, P. F. Curley, and A. I. Ferguson, "Additive-Pulse Mode Locking of a Diode-Pumped Nd:YLF Laser," *Opt. Lett.*, vol. 15, pp. 1303–1305, 1990.

70. D. J. Bradley and W. Sibbett, "Streak-Camera Studies of Picosecond Pulses from a Mode-Locked Nd:Glass Laser," *Opt. Commun.*, vol. 9, pp. 17–20, 1973.

71. W. Zinth, A. Laubereau, and W. Kaiser, "Generation of Chirp-Free Picosecond Pulses," *Opt. Commun.*, vol. 22, pp. 161–164, 1977.

72. L. S. Goldberg, P. E. Schoen, and M. J. Marrone, "Repetitively Pulsed Mode-Locked Nd:Phosphate Glass Laser Oscillator-Amplifier System," *Appl. Opt.*, vol. 21, pp. 1474–1477, 1982.

73. W. Seka and J. Bunkenburg, "Active-Passive Mode-Locked Oscillator at 1.054 μm," *J. Appl. Phys.*, vol. 49, pp. 2277–2280, 1978.

74. L. S. Goldberg and P. E. Schoen, "Active-Passive Mode Locking of an Nd:Phosphate Glass Laser Using #5 Saturable Dye," *IEEE J. Quantum Electron.*, vol. QE-20, pp. 628–630, 1984.

75. L. Yan, J. D. Ling, P.-T. Ho, and Chi H. Lee, "Picosecond-Pulse Generation from a Continuous-Wave Neodymium:Phosphate Glass Laser," *Opt. Lett.*, vol. 11, pp. 502–503, 1986.

76. F. Krausz, T. Brabec, E. Wintner, and A. J. Schmidt, "Mode Locking of a Continuous Wave Nd:Glass Laser Pumped by a Multistripe Diode Laser," *Appl. Phys. Lett.*, vol. 55, pp. 2386–2388, 1989.

77. C. Speilmann, F. Krausz, E. Wintner, and A. J. Schmidt, "Subpicosecond Pulse Generation from a Nd:Glass Laser Using a Nonlinear External Cavity," *Opt. Lett.*, vol. 15, pp. 737–739, 1990.

78. M. E. Fermann, M. Hofer, F. Haberl, A. J. Schmidt, and L. Turi, "Additive-Pulse-Compression Mode Locking of a Neodymium Fiber Laser," *Opt. Lett.*, vol. 16, pp. 244–246, 1991.

79. J. D. Kafka, B. H. Kolner, T. Baer, and D. M. Bloom, "Compression of Pulses from a Continuous-Wave Mode-Locked Nd:YAG Laser," *Opt. Lett.*, vol. 9, pp. 505–506, 1984.

80. B. Zysset, W. Hodel, P. Beaud, and H. P. Weber, "200-femtosecond Pulses at 1.06 μm Generated with a Double-Stage Pulse Compressor," *Opt. Lett.*, vol. 11, pp. 156–158, 1986.

81. A. M. Johnson, R. H. Stolen, and W. H. Simpson, "80 × Single-Stage Compression of Frequency Doubled Nd: Yttrium Aluminum Garnet Laser Pulses," *Appl. Phys. Lett.*, vol. 44, pp. 729–731, 1984.

82. L. Yan, P.-T. Ho, Chi H. Lee, and G. L. Burdge, "Generation of Ultrashort Pulses from a Neodymium Glass Laser System," *IEEE J. Quantum Electron.*, vol. QE-26, pp. 2431–2440, 1989.

83. A. S. Gouveis-Neto, A. S. L. Gomes, and J. R. Taylor, "Generation of 33-fsec Pulses at 1.32 μm through a High-Order Soliton Effect in a Single-Mode Optical Fiber," *Opt. Lett.*, vol. 12, pp. 395–397, 1987.

84. S. J. Keen and A. I. Ferguson, "Subpicosecond Pulse Generation from an All Solid-State Laser," *Appl. Phys. Lett.*, vol. 55, pp. 2164–2166, 1989.

85. X. D. Wang, P. Basséras, R. J. D. Miller, J. Sweetser, and I. A. Walmsley, "Regenerative Pulse Amplification in the 10-kHz Range," *Opt. Lett.*, vol. 15, pp. 839–841, 1990.

86. D. F. Voss and L. S. Goldberg, "Simultaneous Amplification and Compression of Continuous-Wave Mode-Locked Nd: YAG Laser Pulses," *Opt. Lett.*, vol. 11, pp. 210–212, 1986.

87. C. Sauteret, D. Husson, G. Thiell, S. Seznec, S. Gary, A. Migus, and G. Mourou, "Generation of 20-TW Pulses of Picosecond Duration Using Chirped-Pulse Amplification in a Nd: Glass Power Chain," *Opt. Lett.*, vol. 16, pp. 238–240, 1991.

88. M. D. Perry, F. G. Patterson, and J. Weston, "Spectral Shaping in Chirped-Pulse Amplification," *Opt. Lett.*, vol. 15, pp. 381–383, 1990.

89. J. D. Kafka, M. L. Watts, D. J. Roach, M. S. Keirstead, H. W. Schaaf, and T. Baer, in C. B. Harris, E. P. Ippen, G. A. Mourou, and A. H. Zewail (eds.), *Ultrafast Phenomena VII*, Springer-Verlag, New York, 1990, pp. 66–68.

90. N. Sarukura, Y. Ishida, and H. Nakano, "Generation of 50-fsec Pulses from a Pulse-Compressed, cw, Passively Mode-Locked Ti: Sapphire Laser," *Opt. Lett.*, vol. 16, pp. 153–155, 1991.

91. P. M. W. French, J. A. R. Williams, and J. R. Taylor, "Femtosecond Pulse Generation from a Titanium-Doped Sapphire Laser Using Nonlinear External Cavity Feedback," *Opt. Lett.*, vol. 14, pp. 686–688, 1989.

92. J. Goodberlet, J. Wang, J. G. Fujimoto, and P. A. Shulz, "Femtosecond Passively Mode-Locked Ti: Al$_2$O$_3$ Laser with a Nonlinear External Cavity," *Opt. Lett.*, vol. 14, pp. 1125–1127, 1989.

93. W. Sibbett, "Hybrid and Passive Mode Locking in Coupled-Cavity Lasers," in C. B. Harris, E. P. Ippen, G. A. Mourou, and A. H. Zewail (eds.), *Ultrafast Phenomena VII*, Springer-Verlag, New York, 1990, pp. 2–7.

94. U. Keller, W. H. Knox, and H. Roskos, "Coupled-Cavity Resonant Passive Mode-Locked Ti: Sapphire Laser," *Opt. Lett.*, vol. 15, pp. 1377–1379, 1990.

95. D. E. Spence, P. N. Kean, and W. Sibbett, "60-fsec Pulse Generation from a Self-Mode-Locked Ti: Sapphire Laser," *Opt. Lett.*, vol. 16, pp. 42–44, 1991.

96. J. Squier, F. Salin, and G. Mourou, "100-fs Pulse Generation and Amplification in Ti: Al$_2$O$_3$," *Opt. Lett.*, vol. 16, pp. 324–326, 1991.

97. M. Piché, N. McCarthy, and F. Salin, in *Digest of Annual Meeting of the Optical Society of America*, Optical Society of America, Washington, D.C., 1990, paper MB8.

98. J. F. Pinto, C. P. Yakymyshyn, and C. R. Pollock, "Acousto-optic Mode-Locked Soliton Laser," *Opt. Lett.*, vol. 13, pp. 383–385, 1988.

99. L. F. Mollenauer, N. D. Vieira, and L. Szeto, "Mode Locking by Synchronous Pumping Using a Gain Medium with Microsecond Decay Times," *Opt. Lett.*, vol. 7, pp. 414–416, 1982.

100. L. F. Mollenauer and R. H. Stolen, "The Soliton Laser," *Opt. Lett.*, vol. 9, pp. 13–15, 1984.

101. F. M. Mitschke and L. F. Mollenauer, "Ultrashort Pulses from the Soliton Laser," *Opt. Lett.*, vol. 12, pp. 407–409, 1987.

102. K. J. Blow and B. P. Nelson, "Improved Mode Locking of an F-Center Laser with a Nonlinear Nonsoliton External Cavity," *Opt. Lett.*, vol. 13, pp. 1026–1028, 1988.

103. P. N. Kean, X. Zhu, D. W. Crust, R. S. Grant, N. Langford, and W. Sibbett, "Enhanced Mode Locking of Color-Center Lasers," *Opt. Lett.*, vol. 14, pp. 39–41, 1989.

104. J. Mark, L. Y. Liu, K. L. Hall, H. A. Haus, and E. R. Ippen, "Femtosecond Pulse Generation in a Laser with a Nonlinear External Resonator," *Opt. Lett.*, vol. 14, pp. 48–50, 1989.

105. X. Zhu, P. N. Kean, and W. Sibbett, "Coupled-Cavity Mode Locking of a KCl:Tl Laser Using an Erbium-Doped Optical Fiber," *Opt. Lett.*, vol. 14, pp. 1192–1194, 1989.

106. M. N. Islam, E. R. Sunderman, C. E. Soccolich, I. Bar-Joseph, N. Sauer, T. Y. Chang, and B. I. Miller, "Color Center Lasers Passively Mode Locked by Quantum Wells," *IEEE J. Quantum Electron.*, vol. QE-25, pp. 2454–2463, 1989.

107. J. F. Pinto, E. Georgiou, and C. R. Pollock, "Stable Color-Center Laser in OH-Doped NaCl Operating in the 1.41- to 1.81-μm Region," *Opt. Lett.*, vol. 11, pp. 519–521, 1986.

108. C. P. Yakymyshyn, J. F. Pinto, and C. R. Pollock, "Additive-Pulse Mode-Locked NaCl:OH$^-$ Laser," *Opt. Lett.*, vol. 14, pp. 621–623, 1989.

109. M. N. Islam, L. F. Mollenauer, and K. R. German, "Tunable Picosecond Pulses Near 3 μm from Mode-Locked RbCl:Li and KCl:Li F_A(II) Color Center Lasers," in *Digest of the Conference on Lasers and Electro-Optics*, Optical Society of America, Washington, D.C., 1989, paper MD2.

110. C. L. Cesar, M. N. Islam, C. E. Soccolich, R. D. Feldman, and R. F. Austin, "Femtosecond KCl:Li and RbCl:Li Color-Center Lasers Near 2.8 μm with a HgCdTe Multiple-Quantum-Well Saturable Absorber," *Opt. Lett.*, vol. 15, pp. 1147–1149, 1990.

111. N. Langford, K. Smith, and W. Sibbett, "Subpicosecond-Pulse Generation in a Synchronously Mode-Locked Ring Color-Center Laser," *Opt. Lett.*, vol. 12, pp. 817–819, 1987.

112. N. Langford, K. Smith, and W. Sibbett, "Passively Mode-Locked Color-Center Laser," *Opt. Lett.*, vol. 12, pp. 903–905, 1987.

113. N. Langford, R. S. Grant, C I. Johnston, K. Smith, and W. Sibbett, "Group-Velocity-Dispersion Compensation of a Passively Mode-Locked Ring LiF:F$_2^+$ Color-Center Laser," *Opt. Lett.*, vol. 14, pp. 45–47, 1989.

114. G. Sucha, M. Wegener, S. Weiss, and D. S. Chemla, "Kilohertz Amplification of Femtosecond Pulses Near 1.55 μm to Microjoule Energies," in C. B. Harris, E. P. Ippen, G. A. Mourou, and A. H. Zewail (eds.), *Ultrafast Phenomena VII*, Springer-Verlag, New York, 1990, pp. 32–34.

115. J. Kuhl, M. Serenyi, and E. O. Gobel, "Bandwidth-Limited Picosecond Pulse Generation in an Actively Mode-Locked GaAs Laser with Intracavity Chirp Compensation," *Opt. Lett.*, vol. 12, pp. 334–336, 1987.

116. P. J. Delfyett, C.-H. Lee, G. A. Alphonse, and J. C. Connolly, "High Peak Power Picosecond Pulse Generation from AlGaAs External Cavity Mode-Locked Semiconductor Laser and Traveling-Wave Amplifier," *Appl. Phys. Lett.*, vol. 57, pp. 971–973, 1990.

117. P. J. Delfyett, C.-H. Lee, L. T. Florez, N. G. Stoffel, T. J. Gmitter, N. C. Andreadakis, G. A. Alphonse, and J. C. Connolly, "Generation of Subpicosecond High-Power Optical Pulses from a Hybrid Mode-Locked Semiconductor Laser," *Opt. Lett.*, vol. 15, pp. 1371–1373, 1990.

118. H. Yokoyama, H. Ito, and H. Inaba, "Generation of Subpicosecond Coherent Optical Pulses by Passive Mode Locking of an AlGaAs Diode Laser," *Appl. Phys. Lett.*, vol. 40, pp. 105–107, 1982.

119. Y. Silberberg and P. W. Smith, "Subpicosecond Pulses from a Mode-Locked Semiconductor Laser," *IEEE J. Quantum Electron.*, vol. QE-22, pp. 759–761, 1986.

120. P. P. Vasil'ev, "Picosecond Injection Laser: A New Technique for Ultrafast Q-Switching," *IEEE J. Quantum Electron.*, vol. QE-24, pp. 2386–2391, 1988.

121. S. Sanders, L. Eng, J. Paslaski, and A. Yariv, "108 GHz Passive Mode Locking of a Multiple Quantum Well Semiconductor Laser with an Intracavity Absorber," *Appl. Phys. Lett.*, vol. 56, pp. 310–311, 1990.

122. J. R. Karin, L. G. Melcer, R. Nagarajan, J. E. Bowers, S. W. Corzine, P. A. Morton, R. S. Geels, and L. A. Coldren, "Generation of Picosecond Pulses with a Gain-Switched GaAs Surface-Emitting Laser," *Appl. Phys. Lett.*, vol. 57, pp. 963–965, 1990.

123. J. T. K. Chang and J. I. Vukusic, "Active Mode Locking of InGaAsP Brewster Angled Semiconductor Lasers," *IEEE J. Quantum Electron.*, vol. QE-23, pp. 1329–1331, 1987.

124. S. W. Corzine, J. E. Bowers, G. Przybylek, U. Koren, B. I. Miller, and C. E. Soccolich, "Actively Mode-Locked GaInAsP Laser with Subpicosecond Output," *Appl. Phys. Lett.*, vol. 52, pp. 348–350, 1988.

125. H. F. Liu, M. Fukazawa, Y. Kawai, and T. Kamiya, "Gain-Switched Picosecond Pulse (<10 ps) Generation from 1.3 μm InGaAsP Laser Diodes," *IEEE J. Quantum Electron.*, vol. QE-25, pp. 1417–1425, 1989.

126. C. Lin and J. E. Bowers, "Measurement of 1.3 and 1.55 μm Gain-Switched Semiconductor Laser Pulses with a Picosecond IR Streak Camera and High-Speed InGaAs PIN Photodiode," *Electron. Lett.*, vol. 21, pp. 1200–1202, 1985.

127. A. Takada, T. Sugie, and M. Saruwatari, "High-Speed Picosecond Optical Pulse Compression from Gain-Switched 1.3-μm Distributed Feedback-Laser Diode (DFB-LD) through Highly Dispersive Single-Mode Fiber," *J. Light Wave Tech.*, vol. 5, pp. 1525–1533, 1987.

128. A. S. Hou, R. S. Tucker, and G. Eisenstein, "Pulse Compression of an Actively Modelocked Diode Laser Using Linear Dispersion in Fiber," *IEEE Photon. Tech. Lett.*, vol. 2, pp. 322–324, 1990.

129. R. S. Tucker, U. Koren, G. Raybon, C. A. Burrus, B. I. Miller, T. L. Koch, and G. Eisenstein, "40 GHz Active Mode-Locking in a 1.5 μm Monolithic Extended-Cavity Laser," *Electron. Lett.*, vol. 25, pp. 621–622, 1989.

130. I. H. White, D. F. G. Gallagher, M. Osinski, and D. Bowley, "Direct Streak-Camera Observation of Picosecond Gain-Switched Optical Pulses from a 1.5 μm Semiconductor Laser," *Electron. Lett.*, vol. 21, pp. 197–199, 1985.

131. Y. K. Chen, M. C. Wu, T. Tanbun-Ek, R. A. Logan, and M. A. Chin, "Subpicosecond Monolithic Colliding-Pulse Mode-Locked Multiple Quantum Well Lasers," *Appl. Phys. Lett.*, vol. 58, pp. 1253–1255, 1991.

132. G. P. Agrawal and N. A. Olsson, "Amplification and Compression of Weak Picosecond Optical Pulses by Using Semiconductor-Laser Amplifiers," *Opt. Lett.*, vol. 14, pp. 500–502, 1989.

133. C. V. Shank and E. P. Ippen, "Subpicosecond Kilowatt Pulses from a Mode-Locked cw Dye Laser," *Appl. Phys. Lett.*, vol. 24, pp. 373–375, 1974.

134. J. A. Valdmanis, R. L. Fork, and J. P. Gordon, "Generation of Optical Pulses as Short as 27 femtoseconds Directly from a Laser Balancing Self-Phase Modulation, Group-Velocity Dispersion, Saturable Absorption, and Saturable Gain," *Opt. Lett.*, vol. 10, pp. 131–133, 1985.

135. A. Finch, G. Chen, W. Sleat, and W. Sibbett, "Pulse Asymmetry in the Colliding-Pulse Mode Locked Dye Laser," *J. Mod. Opt.*, vol. 35, pp. 345–354, 1988.

136. H. Vanherzeele, J.-C. Diels, and R. Torti, "Tunable Passive Colliding Pulse Mode-Locking in a Linear Laser," *Opt. Lett.*, vol. 9, pp. 549–551, 1984.

137. C. K. Chan and S. O. Sari, "Tunable Dye Laser Pulse Converter for Production of Picosecond Pulses," *Appl. Phys. Lett.*, vol. 25, 403–406, 1974.

138. H. Vanherzeele, R. Torti, and J.-C. Diels, "Synchronously Pumped Dye Laser Passively Mode-Locked with an Antiresonant Ring," *Appl. Opt.*, vol. 23, pp. 4182–4184, 1984.

139. T. Sizer II and G. Mourou, "Picosecond Dye Laser Pulses Using a cw Frequency Doubled Nd:YAG as the Pump Source," *Opt. Commun.*, vol. 37, pp. 207–210, 1981.

140. P. Bado, I. N. Duling III, T. Sizer II, T. B. Norris, and G. A. Mourou, "Generation of 85-fsec Pulses by Synchronous Pumping of a Colliding-Pulse Mode-Locked Dye Laser," *J. Opt. Soc. Am.*, vol. B 2, pp. 613–615, 1985.

141. J. D. Kafka and T. Baer, "A Synchronously Pumped Dye Laser Using Ultrashort Pump Pulses," in *SPIE* 533 *Ultrashort Pulse Spectroscopy and Applications*, pp. 38–45, 1985.

142. A. M. Johnson and W. M. Simpson, "Tunable Femtosecond Dye Laser Synchronously Pumped by the Compressed Second Harmonic of Nd:YAG," *J. Opt. Soc. Am.*, vol. B 2, pp. 619–625, 1985.

143. M. Nakazawa, T. Nakashima, H. Kubota, and S. Seikai, "65-femtosecond Pulse Generation from a Synchronously Pumped Dye Laser without a Colliding-Pulse Mode-Locking Techniques," *Opt. Lett.*, vol. 12, pp. 681–683, 1987.

144. W. T. Lotshaw, D. McMorrow, T. Dickson, and G. A. Kenney-Wallace, "Synchronously Pumped, Femtosecond Dye Laser Insensitive to Cavity-Length Variations of up to 15 μm," *Opt. Lett.*, vol. 14, pp. 1195–1197, 1989.

145. W. H. Knox, "Femtosecond Optical Pulse Amplification," *IEEE J. Quantum Electron.*, vol. QE-24, pp. 388–397, 1988.

146. M. C. Downer, R. L. Fork, and M. Islam, "3 MHz Amplifier for Femtosecond Optical Pulses," in D. H. Austin and K. B. Eisenthal (eds.), *Ultrafast Phenomena IV*, Springer-Verlag, New York, 1984, pp. 27–29.

147. W. H. Knox, M. C. Downer, R. L. Fork, and C. V. Shank, "Amplified Femtosecond Optical Pulses and Continuum Generation at 5-kHz Repetition Rate," *Opt. Lett.*, vol. 9, pp. 552–554, 1984.

148. W. H. Knox, M. C. Downer, R. L. Fork, C. V. Shank, and J. A. Valdmanis, "35-fs 5 kHz Pulse Amplifier," in *Digest of the Conference on Lasers and Electro-Optics*, Optical Society of America, Washington, D.C., 1985, paper TuE3.

149. I. N. Duling III, T. Norris, T. Sizer II, P. Bado, and G. A. Mourou, "Kilohertz Synchronous Amplification of 85-femtosecond Optical Pulses," *J. Opt. Soc. Am.*, vol. B 2, pp. 616–618, 1985.

150. R. L. Fork, C. V. Shank, and R. T. Yen, "Amplification of 70-fs Optical Pulses to Gigawatt Powers," *Appl. Phys. Lett.*, vol. 41, pp. 223–225, 1982.

151. C. Rolland and P. B. Corkum, "Amplification of 70 fs Pulses in a High Repetition Rate XeCl Pumped Dye Laser Amplifier," *Opt. Commun.*, vol. 59, pp. 64–68, 1986.

152. C. V. Shank, R. L. Fork, R. Yen, and R. H. Stolen, "Compression of Femtosecond Optical Pulses," *Appl. Phys. Lett.* vol. 40, pp. 761–763, 1982.

153. J. G. Fujimoto, A. M. Weiner, and E. P. Ippen, "Generation and Measurement of Optical Pulses as Short as 16 fs," *Appl. Phys. Lett.*, vol. 44, pp. 832–834, 1984.

154. J. M. Halbout and D. Grischkowsky, "12-fs Ultrashort Optical Pulse Compression at a High Repetition Rate," *Appl. Phys. Lett.*, vol. 45, pp. 1281–1283, 1984.

155. W. H. Knox, R. L. Fork, M. C. Downer, R. H. Stolen, C. V. Shank, and J. A. Valdmanis, "Optical Pulse Compression to 8 fs at a 5-kHz Repetition Rate," *Appl. Phys. Lett.*, vol. 46, pp. 1120–1121, 1985.

156. C. Rolland and P. B. Corkum, "Compression of High-Power Optical Pulses," *J. Opt. Soc. Am.*, vol. B 5, pp. 641–647, 1988.

157. P. M. W. French and J. R. Taylor, "Generation of Sub-100-fsec Pulses Tunable Near 497 nm from a Colliding-Pulse Mode-Locked Ring Dye Laser," *Opt. Lett.*, vol. 13, pp. 470–472, 1988.

158. P. M. W. French, M. M. Opalinska, and J. R. Taylor, "Passively Mode-Locked cw Coumarin 6 Ring Dye Laser," *Opt. Lett.*, vol. 14, pp. 217–218, 1989.

159. P. M. W. French and J. R. Taylor, "Passive Mode-Locked Continuous-Wave Rhodamine 110 Dye Laser," *Opt. Lett.*, vol. 11, pp. 297–299, 1986.

160. H. Kubota, K. Kurokawa, and M. Nakazawa, "29-fsec Pulse Generation from a Linear-Cavity Synchronously Pumped Dye Laser," *Opt. Lett.*, vol. 13, pp. 749–751, 1988.

161. M. Mihailidi, Y. Budansky, X. M. Zhao, Y. Takiguchi, and R. R. Alfano, "Quasi-linear Ring Colliding-Pulse Mode-Locked Femtosecond Laser Using Binary Energy-Transfer Gain Dye Mixture," *Opt. Lett.*, vol. 13, pp. 987–989, 1988.

162. P. M. W. French and J. R. Taylor, "The Passive Modelocking of the Continuous Wave Rhodamine B Dye Laser," *Opt. Commun.*, vol. 58, pp. 53–55, 1986.

163. P. M. W. French and J. R. Taylor, "Passive Mode Locking of the Continuous Wave DCM Dye Laser," *Appl. Phys.*, vol. B 41, pp. 53–55, 1986.

164. M. D. Dawson, T. F. Boggess, and A. L. Smirl, "Femtosecond Synchronously Pumped Pyridine Dye Lasers," *Opt. Lett.*, vol. 12, pp. 254–256, 1987.

165. P. M. W. French, J. A. R. Williams, and J. R. Taylor, "Passive Mode Locking of a Continuous-Wave Energy-Transfer Dye Laser Operating in the Near Infrared around 750 nm," *Opt. Lett.*, vol. 12, pp. 684–686, 1987.

166. P. Georges, F. Salin, and A. Brun, "Generation of 36-fsec Pulses Near 775 nm from a Colliding-Pulse Passively Mode-Locked Dye Laser," *Opt. Lett.*, vol. 14, pp. 940–942, 1989.

167. P. Georges, F. Salin, G. Le Saux, G. Roger, and A. Brun, "Femtosecond Pulses at 800 nm by Passive Mode Locking of Rhodamine 700," *Opt. Lett.*, vol. 15, pp. 446–448, 1990.

168. J. A. R. Williams, P. M. W. French, and J. R. Taylor, "Passive Mode Locking of a cw Energy-Transfer Dye Laser Operating in the Infrared Near 800 nm," *Opt. Lett.*, vol. 13, pp. 811–813, 1988.

169. M. D. Dawson, T. F. Boggess, and A. L. Smirl, "Picosecond and Femtosecond Pulse Generation Near 1000 nm from a Frequency-Doubled Nd:YAG-Pumped cw Dye Laser," *Opt. Lett.*, vol. 12, pp. 590–592, 1987.

170. F. S. Choa and P. L. Liu, "Broadband Ultrafast Pulse Generation from Synchronously Pumped Dye Lasers," *Opt. Lett.*, vol. 13, pp. 743–745, 1988.

171. W. H. Knox, "Generation and Kilohertz-Rate Amplification of Femtosecond Optical Pulses around 800 nm," *J. Opt. Soc. Am.*, vol. B 4, pp. 1771–1776, 1987.

172. F. P. Schäfer, "Principles of Dye Laser Operation," in F. P. Schäfer (ed.), *Dye Lasers*, Springer-Verlag, New York, 1990, pp. 1–89.

173. P. H. Chiu, S. C. Hsu, S. J. C. Box, and H.-S. Kwok, "A Cascade Pumped Picosecond Dye Laser System," *IEEE J. Quantum Electron.*, vol. QE-20, pp. 652–658, 1984.

174. K. Bohnert, T. F. Boggess, K. Mansour, D. Maxson, and A. L. Smirl, "Tunable Near-Infrared Picosecond Pulses from a Short-Cavity Dye Laser," *IEEE J. Quantum Electron.*, vol. QE-22, pp. 2195–2199, 1986.

175. S. Szatmári, "Pulse Shortening of 5×10^3 by the Combined Pulse Forming of Dye Oscillators, Saturated Amplifiers and Gated Saturable Absorbers," *Opt. Quantum Electron.*, vol. 21, pp. 55–61, 1989.

176. C. V. Shank, J. E. Bjorkholm, and H. Kogelnik, "Tunable Distributed-Feedback Dye Laser," *Appl. Phys. Lett.*, vol. 18, pp. 395–396, 1971.

177. Z. Bor and A. Müller, "Picosecond Distributed Feedback Dye Lasers," *IEEE J. Quantum Electron.*, vol. QE-22, pp. 1524–1533, 1986.

178. J. Hebling and Z. Bor, "Distributed Feedback Dye Laser Pumped by a Laser Having a Low Degree of Coherence," *J. Phys. E*, vol. 17, pp. 1077–1080, 1984.

179. Z. Bor, "A Novel Pumping Arrangement for Tunable Single Picosecond Pulse Gen-

erated with N$_2$ Laser Pumped Distributed Feedback Laser," *Opt. Commun.*, vol. 29, pp. 103–108, 1979.

180. A. N. Rubinov and T. Sh. Éfendiev, "Holographic DFB Dye Lasers," *Optica Acta*, vol. 32, pp. 1291–1301, 1985.

181. S. Szatmári and F. P. Schäfer, "Subpicosecond, Widely Tunable Distributed Feedback Dye Laser," *Appl. Phys.*, vol. B 46, pp. 305–311, 1988.

182. S. Szatmári and F. P. Schäfer, "Simplified Laser System for the Generation of 60 fs Pulses at 248 nm," *Opt. Commun.*, vol. 68, pp. 196–202, 1988.

183. T. M. Shay, R. C. Sze, M. Maloney, and J. F. Figueira, "120-ps Duration Pulses by Active Mode Locking of an XeCl Laser," *J. Appl. Phys.*, vol. 64, pp. 3758–3760, 1988.

184. J. J. Curry, S. T. Feng, and J. Goldhar, "Generation of KrF Laser Pulses on a Picosecond Time Scale Using Electro-optic Modulation," *Opt. Lett.*, vol. 14, pp. 782–784, 1989.

185. F. Laermer, J. Dobler, and T. Elsaesser, "Generation of Femtosecond UV Pulses by Intracavity Frequency Doubling in a Modelocked Dye Laser," *Opt. Commun.*, vol. 67, pp. 58–62, 1988.

186. G. Focht and M. C. Downer, "Generation of Synchronized Ultraviolet and Red Femtosecond Pulses by Intracavity Frequency Doubling," *IEEE J. Quantum Electron.*, vol. QE-24, pp. 431–434, 1988.

187. D. C. Edelstein, E. S. Wachman, L. K. Cheng, W. R. Bosenberg, and C. L. Tang, "Femtosecond Ultraviolet Pulse Generation in β-B$_a$B$_2$O$_4$," *Appl. Phys. Lett.*, vol. 52, pp. 2211–2213, 1988.

188. J. H. Glownia, J. Misewick, and P. P. Sorokin, "160-fsec XeCl Excimer Amplifier System," *J. Opt. Soc. Am.*, vol. B 4, pp. 1061–1065, 1987.

189. S. Szatmári, F. P. Schäfer, E. Muller-Horsche, and W. Muchenheim, "Hybrid Dye-Excimer Laser System for the Generation of 80 fs, 900 GW Pulses at 248 nm," *Opt. Commun.*, vol. 63, pp. 305–309, 1987.

190. S. Szatmári and F. P. Schäfer, "Generation of Input Signals for ArF Amplifiers," *J. Opt. Soc. Am.*, vol. B 6, pp. 1877–1883, 1989.

191. A. S. Piskarskas, V. J. Smilgevičius, A. P. Umbrasas, J. P. Juodišius, A. S. L. Gomes, and J. R. Taylor, "Picosecond Optical Parametric Oscillator Pumped by Temporally Compressed Pulses from a Q-Switched, Mode-Locked, cw-Pumped Nd:YAG Laser," *Opt. Lett.*, vol. 14, pp. 557–559, 1989.

192. R. Laenen, H. Graener, and A. Laubereau, "Broadly Tunable Femtosecond Pulses Generated by Optical Parametric Oscillation," *Opt. Lett.*, vol. 15, pp. 971–973, 1990.

193. E. S. Wachman, D. C. Edelstein, and C. L. Tang, "Continuous-Wave Mode-Locked and Dispersion-Compensated Femtosecond Optical Parametric Oscillator," *Opt. Lett.*, vol. 15, pp. 136–138, 1990.

194. R. L. Fork, C. V. Shank, C. Hirlimann, and R. Yen, "Femtosecond White-Light Continuum Pulses," *Opt. Lett.*, vol. 8, pp. 1–3, 1983.

195. P. Bado, I. N. Duling III, T. Sizer II, T. B. Norris, and G. A. Mourou, "Generation of White Light at 1 kHz," in *SPIE 533 Ultrashort Pulse Spectroscopy and Applications*, pp. 59–62, 1985.

196. G. R. Olbright and G. R. Hadley, "Generation of Tunable Near-Infrared Amplified Femtosecond Laser Pulses and Time-Correlated White-Light Continuum," *J. Opt. Soc. Am.*, vol. B 6, pp. 1363–1369, 1989.

197. J. H. Glownia, J. Misewich, and P. P. Sorokin, "Subpicosecond Time-Resolved Infrared Spectral Photography," *Opt. Lett.*, vol. 12, pp. 19–21, 1987.

198. P. C. Becker, H. L. Franitor, R. L. Fork, F. A. Beisser, and C. V. Shank, "Generation of Tunable 9 Femtosecond Optical Pulses in the Near Infrared," *Appl. Phys. Lett.*, vol. 54, pp. 411–412, 1989.

199. R. W. Schoenlein, J.-Y. Bigot, M. T. Portella, and C. V. Shank, "Generation of Blue-

Green 10 fs Pulses Using an Excimer-Pumped Dye Amplifier," *Appl. Phys. Lett.*, vol. 58, pp. 801–803, 1991.

200. M. N. Islam, L. F. Mollenauer, R. H. Stolen, J. R. Simpson, and H. T. Shang, "Cross-Phase Modulation in Optical Fibers," *Opt. Lett.*, vol. 12, pp. 625–627, 1987.

201. P. L. Baldeck, P. P. Ho, and R. R. Alfano, "Cross-Phase Modulation: A New Technique for Controlling the Spectral, Temporal, and Spatial Properties of Ultrashort Pulses," in R. R. Alfano (ed.), *The Supercontinuum Laser Source*, Springer-Verlag, New York, 1989, pp. 117–183.

202. A. S. Gouveia-Neto, A. S. L. Gomes, and J. R. Taylor, "Femtosecond Soliton Raman Generation," *IEEE J. Quantum Electron.*, vol. QE-24, pp. 332–340, 1988.

203. E. J. Greer, D. M. Patrick, P. G. J. Wigley, J. I. Vukusic, and J. R. Taylor, "Tunable, Femtosecond Soliton Generation from Amplified Continuous-Wave Diode-Laser Signals," *Opt. Lett.*, vol. 15, pp. 133–135, 1990.

204. R. H. Stolen and C. Lin, "Fiber Raman Lasers," in M. J. Weber (ed.), *Handbook of Laser Science and Technology*, vol. I, CRC Press, Boca Raton, Fla., 1982, pp. 265–273.

205. J. D. Kafka and T. Baer, "Fiber Raman Soliton Laser Pumped by a Nd:YAG Laser," *Opt. Lett.*, vol. 12, pp. 181–183, 1987.

206. M. N. Islam, L. F. Mollenauer, R. H. Stolen, J. R. Simpson, and H. T. Shang, "Amplifier/Compressor Fiber Raman Lasers," *Opt. Lett.*, vol. 12, pp. 814–816, 1987.

207. E. A. Golovvchenko, E. M. Dianov, P. V. Mamyshev, A. M. Prokhorov, and D. G. Fursa, "Theoretical and Experimental Study of Synchronously Pumped Dispersion-Compensated Femtosecond Lasers," *J. Opt. Soc. Am.*, vol. B 7, pp. 172–181, 1990.

208. J. D. Kafka, T. Baer, and D. W. Hall, "Mode-Locked Erbium-Doped Fiber Laser with Soliton Pulse Shaping," *Opt. Lett.*, vol. 14, pp. 1269–1271, 1989.

209. E. Yablonovitch, J. Goldhar, "Short CO_2 Laser Pulse Generation by Optical Free Induction Decay," *Appl. Phys. Lett.*, vol. 25, pp. 580–582, 1974.

210. P. B. Corkum, "High Power, Subpicosecond 10-μm Pulse Generation," *Opt. Lett.*, vol. 8, pp. 514–516, 1983.

211. C. Rolland and P. B. Corkum, "Generation of 130-fsec Midinfrared Pulses," *J. Opt. Soc. Am.*, vol. B 3, pp. 1625–1629, 1986.

212. A. J. Alcock, P. B. Corkum, and D. J. James, "A Fast Scalable Switching Technique for High-Power CO_2 Laser Radiation," *Appl. Phys. Lett.*, vol. 27, pp. 580–582, 1975.

213. P. B. Corkum and D. Keith, "Controlled Switching of 10-micrometer Radiation Using Semiconductor Étalons," *J. Opt. Soc. Am.*, vol. B 2, pp. 1873–1879, 1985.

214. E. P. Ippen and C. V. Shank, "Techniques for Measurements," in S. L. Shapiro (ed.), *Ultrashort Light Pulses: Picosecond Techniques and Applications*, Springer-Verlag, Berlin, 1977, pp. 83–122.

215. D. J. Bradley and G. H. C. New, "Ultrashort Pulse Measurements," *Proc. IEEE*, vol. 62, pp. 313–345, 1974.

216. D. J. Bradley, "Methods of Generation," in S. L. Shapiro (ed.), *Ultrashort Light Pulses: Picosecond Techniques and Applications*, Springer-Verlag, Berlin, 1977, pp. 17–81.

217. H. Schulz, H. Schüler, T. Engers, and D. von der Linde, "Measurement of Intense Ultraviolet Subpicosecond Pulses Using Degenerate Four-Wave Mixing," *IEEE J. Quantum Electron.*, vol. QE-25, pp. 2580–2585, 1989.

218. K. L. Sala, G. A. Kenney-Wallace, and G. Hall, "CW Autocorrelation Measurements of Picosecond Laser Pulses," *IEEE J. Quantum Electron.*, vol. QE-16, pp. 990–996, 1980.

219. J.-C. Diels, J. J. Fontaine, I. C. McMichael, and F. Simoni, "Control and Measurement of Ultrashort Pulse Shapes (in Amplitude and Phase) with Femtosecond Accuracy," *Appl. Opt.*, vol. 24, pp. 1270–1282, 1985.

220. J. Janszky, G. Corradi, and R. N. Gyuzalian, "On a Possibility of Analyzing the Temporal Characteristics of Short Pulses," *Opt. Commun.*, vol. 23, pp. 293–298, 1977.

221. C. Kolmeder, W. Zinth, and W. Kaiser, "Second Harmonic Beam Analysis, A Sensitive Technique to Determine the Duration of Single Ultrashort Laser Pulses," *Opt. Commun.*, vol. 30, pp. 453–457, 1979.

222. F. Salin, P. Georges, G. Roger, and A. Brun, "Single-Shot Measurement of a 52 fs Pulse," *Appl. Opt.*, vol. 26, pp. 4528–4531, 1987.

223. S. Jayaraman and C. H. Lee, "Observation of Two-Photon Conductivity in GaAs with Nanosecond and Picosecond Light Pulse," *Appl. Phys. Lett.*, vol. 20, pp. 392–395, 1972.

224. M. B. Ketchen, D. Grischkowsky, T. C. Chen, C-C. Chi, I. N. Duling, III, N. H. Halas, J.-M. Halbout, J. A. Kash, and G. P. Li, "Generation of Subpicosecond Electrical Pulses on Coplanar Transmission Lines," *Appl. Phys. Lett.*, vol. 48, pp. 751–753, 1986.

225. H. A. Sayadian, M.-G. Li, and Chi. H. Lee, "Generation of High-Power Broad-Band Microwave Pulses by Picosecond Optoelectronic Technique," *IEEE Trans. Microwave Theory Tech.*, vol. MTT-37, pp. 43–50, 1989.

226. D. Butler, E. A. Chauchard, K. J. Webb, K. A. Zaki, Chi H. Lee, P. Polak-Dingels, H.-L. A. Huang and H. C. Huang, "A cw 20 GHz Optoelectronic Source with Phased-Array Applications," *Microwave Opt. Tech. Lett.*, vol. 1, pp. 119–123, 1988.

227. Chi H. Lee (ed.), *Picosecond Optoelectronics Devices*, Academic Press, New York, 1984.

228. G. A. Mourou, D. M. Bloom, Chi H. Lee (eds.), *Picosecond Electronics and Optoelectronics*, Springer Ser. Electrophys., vol. 21, Springer-Verlag, Berlin, 1985.

229. F. J. Leonberger, C. H. Lee, F. Capasso, and H. Morkoc (eds.), *Picosecond Electronics and Optoelectronics*, Springer Ser. Electron & Photonics, vol. 24, Springer-Verlag, Berlin, 1987.

230. G. Sollner and D. M. Bloom (eds.), *Picosecond Electronics and Optoelectronics*, OSA Proceedings, 1989.

231. Technical Digest of OSA Topical Meeting on Picosecond Electronics and Optoelectronics, Salt Lake City, 1991.

232. D. H. Auston, "Ultrafast Optoelectronics," in W. Kaiser (ed.), *Ultrafast Laser Pulses*, Springer Ser. Appl. Physics, vol. 60, Springer-Verlag, Berlin, 1988.

233. M. Van Exter and D. R. Grischkowsky, "Characterization of an Optoelectronic Terahertz Beam System," *IEEE Trans. Microwave Theory Tech.*, vol. MTT-38, pp. 1684–1691, 1990.

234. J. T. Darrow, B. B. Hu, X-C Zhang, and D. H. Auston, "Subpicosecond Electromagnetic Pulses from Large Aperture Photoconducting Antennas," *Opt. Lett.*, vol. 15, pp. 323–325, 1990.

235. Chi H. Lee, "Picosecond Optics and Microwave Technology," *IEEE Trans. Microwave Theory Tech.*, vol. MTT-38, pp. 596–607, 1989.

236. Chi H. Lee, "Optical Control of Semiconductor Closing and Opening Switches," *IEEE Trans. Electron Devices*, vol. 37, pp. 2426–2438, 1990.

237. K. J. Weingarten, M. J. Rodwell, and D. M. Bloom, "Picosecond Optical Sampling of GaAs Integrated Circuits," *IEEE J. Quantum Electron.*, vol. QE-24, pp. 198–220, 1988.

238. D. H. Auston, "Picosecond Photoconductors: Physical Properties and Applications," in Chi H. Lee (ed.), *Picosecond Optoelectronic Devices*, Academic Press, Orlando, Fla., 1984.

239. D. H. Auston, "Impulse Response of Photoconductor in Transmission Lines," *IEEE J. Quantum Electron.*, vol. QE-19, pp. 639–648, 1983.

240. D. E. Cooper and S. C. Moss, "Picosecond Optoelectronic Measurement of the High

Frequency Scattering Parameters of a GaAs FET," *IEEE J. Quantum Electron.*, vol. QE-22, pp. 94–100, 1986.

241. H. L. A. Hung, P. Polak-Dingels, K. J. Webb, T. Smith, H. C. Huang, and Chi H. Lee, "Millimeter-Wave Monolithic Integrated Circuit Characterization by a Picosecond Optoelectronic Technique," *IEEE Trans. Microwave Theory Tech.*, vol. MTT-37, pp. 1223–1231, 1989.

242. K. J. Webb, E. A. Chauchard, P. Polak-Dingels, C. H. Lee, H. L. Hung, and T. Smith, "A Time Domain Network Analyzer Which Uses Optoelectronic Technique," in *Digest of 1989 IEEE MTT-S International Microwave Symp.*, vol. 1, p. 220, 1989.

CHAPTER 10
OPTICAL MATERIALS— UV, VUV

Jack C. Rife

This chapter discusses properties and selection of ultraviolet (uv) window, mirror, and coating materials. The uv range of the electromagnetic spectrum extends from energies (wavelengths) of about 3 eV (400 nm) just outside the visible to a vague boundary near 6000 eV (0.2 nm), the start of the x-ray range. This chapter primarily discusses spectral ranges in terms of energy. An energy scale better serves spectroscopy, since prominent spectral features across the whole range have the imprint of a relatively narrow range of unoccupied final electronic states. A wavelength scale will be used at times, however, to emphasize the longer wavelengths and provide a useful gauge for thin film dimensions.

The uv and x-ray ranges of the electromagnetic spectrum divide into subregions that are overlapping and whose boundaries are not commonly agreed on. Figure 10.1 shows the approximate ranges of the various named regimes and physical phenomena that determine experimental ranges. The near uv extends from just outside the visible at 3 eV to the beginning of the vacuum uv at 6.7 eV, where air is no longer transparent and vacuum is required. This is the region where radiation begins to be energetic enough to be ionizing. The vacuum uv extends to the beginning of the x-ray region proper at 2000 to 6000 eV where He and other gases become sufficiently transparent. This is also where Be windows are transparent and thick enough to support one atmosphere of differential pressure. Other regions include the extreme ultraviolet or xuv that extends from the approximately 11.9-eV cutoff of the largest bandgap window, LiF, to the x-ray region. The grazing incidence region begins at 30 eV where instruments must be designed with grazing incidence optics (excluding diffraction spectrometers using multilayers and crystals). Here the normal incidence reflectance of homogeneous mirror materials falls off approximately as $1/E^4$. The soft x-ray region normally starts at the grazing incidence boundary at 30 eV and extends to 6000 eV. An additional subregion receiving particular attention now for in vivo soft x-ray microscopy is the water window from the carbon K edge at 285 eV to the oxygen K edge at 540 eV, where water is transparent relative to the absorption in organic materials. This chapter emphasizes the spectral region from 3 to 40 eV, which is more likely to be of interest for electro-optics. XUV and soft x-ray materials are covered as well, however, in recognition of the potential of free electron lasers, laser-driven xuv lasers, and multilayer optics in that region.

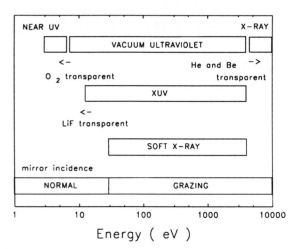

FIGURE 10.1 Ultraviolet regimes.

Currently, a variety of scientific fields are demanding and achieving new levels of uv optical system performance. The next generation of synchrotron radiation sources will require the initial beam-line mirror to maintain arcsecond figure accuracy under continuous, nonuniform heat loads of up to 100 W over square millimeters of mirror surface. Applications of excimer lasers, free electron lasers, and harmonics of various solid-state lasers must deal with radiation damage problems of high instantaneous uv power as well. Astronomy, microscopy, and microelectronic projection lithography are all demanding higher wavefront accuracy and lower scattering to improve imaging. Improvements in fabrication and metrology are helping to meet these needs, as are new knowledge of standard materials and the development of new materials. In addition, concepts such as multilayer reflective coatings, which permit normal incidence designs with greater light-gathering power and less wavefront distortion in what was the grazing incidence regime, offer new possibilities. Developments in uv optics appear in a variety of conference proceedings; but the national and international conferences on Synchrotron Radiation Instrumentation, the NIST Boulder Laser Induced Damage in Optical Materials symposia, and many SPIE (International Society for Optical Engineering) meetings, in particular, are sources of the latest work.

Samson[1] has written the most frequently used handbook on uv techniques. It provides useful graphs of standard uv window and xuv thin film filter transmission, discussion of polarization effects, and graphs of standard uv reflective coatings. General information on optics is also contained in the book by Born and Wolf,[2] the AIP[3] and the OSA[4] Handbooks, or in the two journals, *Applied Optics* and *Optical Engineering* on instrument design;[5] but these works are focused on longer wavelengths. Laser-related optical material issues are covered extensively in a CRC Handbook.[6]

This chapter is intended as a guide to fundamental properties, application issues, and sources of information on uv optical materials. It is not a collection of graphs and tables available elsewhere. This chapter does not cover organic materials such as polymers, liquid crystals, or Bragg optics (crystals or Langmuir-Blodgett films) except for multilayers.

10.1 FUNDAMENTAL PHYSICAL PROPERTIES

10.1.1 Optical Constants

Forms. In the uv, the dominant photon-material interactions are bound-to-bound and ionizing transitions of atomic electrons. These interactions lead to incoherent photoabsorption or photoemission and associated coherent, elastic Rayleigh scattering and reflection. Phonon processes or lattice vibration effects modify band-to-band transitions in this spectral region slightly and are usually neglected. For solids and molecular systems, the atoms are considered frozen on the time scale of absorption or scattering. In absorption measurements, the intensity transmitted I through a sample of thickness x of an incident beam of intensity I_0 is given by

$$I = I_0\, e^{-\mu x} \tag{10.1}$$

where μ is the absorption coefficient most often known by α at longer wavelengths. μ is given by $\rho\sigma$, the product of the number density of atoms and molecules and the atomic cross section. Typical absorption cross sections per atom are 10^{-15} to 10^{-18} cm^2. Inelastic processes such as the Raman and Compton effects are less likely. Raman cross sections are about 10^{-29} cm^2, and the Compton effect begins to be significant only above 2000 eV.[7]

Bulk optical properties are determined by Maxwell's equations, where the photon-solid interaction is incorporated in the displacement field $D = \varepsilon E$, the product of the complex dielectric coefficient and the applied electric field. The complex dielectric coefficient is directly connected to atomic polarizability and oscillator strengths of atomic transitions. The dielectric coefficient ε is given by the density of atoms ρ and the atomic polarizability α as

$$\varepsilon = 1 + 4\pi\rho\alpha \tag{10.2}$$

The atomic polarizability can be approximated on the basis of a collection of Lorentz oscillators

$$\alpha = \frac{e^2}{m} \sum_j \frac{f_j}{(\omega_j^2 - \omega^2) - i\omega\Gamma_j} \tag{10.3}$$

where f_j, ω_j, and Γ_j are the oscillator strength, frequency, and half width of the jth transition. For solids, a correction needs to be made to account for the polarization of the surrounding medium. In addition, the band structure will spread out the distribution of atomic oscillator strength for bound transitions.

Optical constants of the various elements and solid materials completely determine ideal, linear transmittance and reflectance. Formulas and characteristic behavior in transmission and reflection are discussed more extensively in Secs. 10.2 and 10.3, respectively. The complex constants represent the real and imaginary parts of the linear response of the medium or polarization and scattering and absorption. Various interrelated forms in the uv are used depending on the spectral region and physical processes of interest.

In the visible and near uv, the complex dielectric coefficient $\varepsilon = \varepsilon_1 + i\varepsilon_2$ is directly related to internal fields and band structure calculations, as indicated above. The complex index of refraction $N = n + ik$, where n is the index of refraction and k the extinction coefficient, makes a clearer connection with physical measurements. The phase velocity in the medium is c/n, the wavelength in the medium is λ_0/n (λ_0 in vacuum), and the absorption coefficient is given by

$$\mu = \frac{4\pi k}{\lambda_0} \tag{10.4}$$

The index of refraction takes the form $N = 1 - \delta - i\beta$ in the soft x-ray region, where n is very close to one and k very close to zero.

The complex atomic scattering factor or amplitude $f = f_1 + f_2$ is more often used in the x-ray region, where f_1 is proportional to Thomson scattering off of the Z atomic electrons, as if they were free, plus a term due to scattering associated with ionizing transitions or "anomalous" dispersion.[8,9]

The forms of optical constants are related by the formulas

$$\varepsilon = N^2 \tag{10.5}$$

$$\varepsilon_1 = n^2 - k^2 \tag{10.6}$$

$$\varepsilon_2 = 2nk \tag{10.7}$$

$$\delta = \frac{r_e \lambda^2}{2\pi} \sum_q \rho_q f_{1q} \tag{10.8}$$

$$\beta = \frac{r_e \lambda^2}{2\pi} \sum_q \rho_q f_{2q} \tag{10.9}$$

where r_e is the classical radius of the electron e^2/mc^2 and ρ_q is the number of atoms per unit volume of type q. The real and imaginary parts of the optical constants are rigorously related via a Kramers-Kronig intergral.[10] For the index of refraction

$$n(\omega) - 1 = \frac{2}{\pi} P \int_0^\infty \frac{\omega' k(\omega')}{\omega'^2 - \omega^2} \, d\omega' \tag{10.10}$$

$$k(\omega) = -\frac{2\omega}{\pi} P \int_0^\infty \frac{n(\omega') - 1}{\omega'^2 - \omega^2} \, d\omega' \tag{10.11}$$

where P stands for the principal value integral.

Connection with Electronic Structure. In the regions of transitions from the valence band or core levels to bound final states or the conduction band, the optical constants are sensitive to bonding in the material. Band structure calculations predict the magnitude and shape of spectral features relatively well but do not predict energy locations accurately. Lynch has recently summarized interband phenomena such as critical points, discontinuities in the joint density of states, and excitons at the fundamental edge.[11] For theoretical verification, band-to-band transitions can be located very accurately at critical points by modulation spectroscopy. Just above core level edges, structures in the optical constants primarily map the densities of conduction band states of appropriate atomic symmetry, since the core levels are well defined in energy. Near-edge structure may also be analyzed with the aid of molecular orbital and cluster calculations.[12] Above threshold regions, the spectral dependences of the optical constants of materials are fairly well understood in terms of atomic processes. The strongest electronic features in n and k occur in the valence to conduction band transitions below 30 eV. The bunching of oscillator strength in this region leads to values of n less than one above about 30 eV. At higher energies, n rises toward one as $1/(\hbar\omega)^2$ with the entries of further absorption edges appearing as perturbations. Absorption rises abruptly at core level edges and then falls to zero toward higher energy approximately as $1/(\hbar\omega)^{7/2}$, considerably above the edges.[13] The distribution of oscillator strength above each absorption edge is also strongly affected by overlap of initial and final state wavefunctions leading to delayed onset of higher angular momenta transitions and the Cooper minimum.[14,15,16] Additional modifications of the expected distribution of oscillator strength include shape resonances and autoionization.[17] More than 30 eV above core level edges, backscattering of the outgoing photoelectron from nearby atoms and interference at the absorption site or extended absorption fine structure (EXAFS) modulates the absorption by several percent over a few hundred eV. Inversion of the EXAFS modulation yields nearest-neighbor distances to an accuracy as good as 0.001 nm.[18]

The integrity of a set of optical constants covering the entire range of electronic transitions is often verified by applying a form of the Thomas-Reiche-Kuhn sum rule for atomic oscillator strength. The most commonly used sum rule is

$$n_{\text{eff}}(\omega) = \frac{m}{2\pi^2 \rho e^2} \int_0^\omega \omega' \varepsilon_2(\omega') d\omega' \tag{10.12}$$

where n_{eff} is the number of effective electrons contributing at each frequency per atom or molecule and ρ is the atomic or molecular number density. Here n_{eff} must equal the total number of electrons per atom or molecule at energies above the $1s$ edge.[10]

Sources.   Palik[19] gives optical constants for a variety of often used metals, semiconductors, and insulators for infrared to x-ray energies. In particular, the source provides fine energy scale constants in the valence band to conduction band region, and some critical analysis of the measurement sources. Chapters are included on the origins and measurement of optical constants.

The compilation of Henke et al.[20,21] of the atomic scattering factors for the elements above 30 eV provides data to determine reasonably well the optical properties of any material, given the density, at energies away from the absorption edges. The data do not give a good picture at the edges of bound unoccupied states that are sensitive to the chemical environment. The constants have been derived from absorption measurements, theoretical extrapolations, and the Kramers-Kronig relations. An IBM-compatible computer program is freely available to graphically display the optical constants and the transmission and reflectance from any molecular solid.[22]

The state of xuv optical constant theory, measurement, and data bases has been covered recently by a variety of SPIE papers.[23] Soft x-ray electron binding edges, characteristic x-rays, and standard filter transmissions and mirror reflectivities along with useful constants and formulas have been collected in a handy pocket guide issued by the Center for X-ray Optics.[24]

Measurement Techniques.   UV optical constants are experimentally determined by measurements of absorption, reflectance, ellipsometry, or direct refraction and Kramers-Kronig inversion of the data where needed.

Absorption is measured in transmission, primarily. Corrections for reflectance, thickness, and oxide or contaminating layers can present problems for accurate measurements. Use of the Kramers-Kronig relations to obtain the index of refraction depends on careful extrapolation of the absorption to low and high energies. Errors will not appreciably affect the shape of optical constant structure but will alter the magnitude. Sum rules such as Eq. (10.12) are helpful in gauging the suitability of extrapolations. Weak absorption in transparent materials is measured by laser calorimetry[25] or photoacoustic methods. Multiphoton processes put a limit on sensitivity, however.

Optical constants can also be determined from reflectance measurements. Below the start of the grazing incidence regime at 40 eV, reflectance measurements can be made with tolerable flux in normal incidence. Beyond 40 eV, grazing incidence must be used for reasonable signal levels, and measurements must be made in orthogonal directions to the incident photon beam to determine polarization effects. Measurements made at a variety of grazing angles at a single energy can be used to derive the optical constants by intersection of isoreflector curves in n and k space or by simultaneous solution.[26] Alternatively, reflectance data over a wide energy range at normal incidence or a given grazing angle with a highly polarized source can be Kramers-Kronig inverted to obtain both optical constants.[27] The surface sensitivity of reflectivity measurements limits the accuracy of optical constants. In the region below 20 eV the reflectivity sampling depth is typically 10 to 50 nm, so that a monolayer can have 1 percent effects. Aspnes has observed a surface sensitivity of 1 percent in the polarized reflectance of (110) Si in the near uv.[28] Optical constants based on thin film measurements should often be called pseudo-optical constants because they can vary significantly with deposition conditions owing to roughness, oxide or other overlayers, and void fraction or packing density. Aspnes has discussed effective medium approximations for dealing with the analysis of measurements of imperfect films.[29] Scattering from rough surfaces can remove light from the specularly reflected beam and alter measured values. Scattering has a greater effect at normal angles and shorter wavelengths, and is discussed more extensively in Secs. 10.3 and 10.5. For critical telescope applications, measurement of the pseudo-optical constants of a witness plate near the grazing angle of use is desirable for instrument performance predictions.

Ellipsometry provides a way to simultaneously determine both optical constants in an intensity-independent way.[29] Quick determinations can be made of ideal specimens, but correcting for overlayers requires careful work. Recently, accurate uv measurements have been made using polarized synchrotron radiation by Johnson et al.[30] using MgF_2 prisms or three-mirror polarizers as analyzers. The measurements lose accuracy at energies larger than 20 to 30 eV, however, owing to the constraints of polarizers and the fact that the optimum specimen grazing angle for maximum phase shift moves to 45° at higher energies where p-polarized reflectance is very small.

Direct refraction measurements on transparent specimens can be made in the visible and near uv by measuring deviation angles or phase shifts interferometrically.[31] In the xuv, the index of refraction can be determined by refractive corrections to Bragg angles of multilayer coatings[32] and by direct measurement of the phase shift with x-ray interferometers.[33,34] Optical constants have also been determined, in the xuv, from the diffraction efficiency of transmission gratings.[35]

10.1.2 Structure and Thermophysical Properties

Bulk physical properties of materials are often limiting factors in selection. Standard texts on solid-state physics by Kittel[36] and materials science by Van Vleck[37] are useful for general data and background. Lattice structures of inorganic crystals are available in the series *Crystal Structures*.[38] Thermal radiation, conductivity, diffusivity, and expansion, together with specific heat and viscosity of materials, are available from the national source TPRC/CINDAS and its compilation of material properties.[39] Reference 3 is a useful condensed guide to a wide range of thermal and mechanical properties. Reference 6 lists thermal properties and elastics constants for electro-optic materials. Extensive data on metallic alloys are available in the *Metals Handbook*.[40] A chart of the vapor pressure of the elements is available in the *RCA Review*.[41]

10.2 TRANSMISSIVE UV OTPICS

This section discusses linear and nonlinear effects and formulas followed by consideration of some standard bulk, coating, and filter materials. It primarily covers transparent optics from 3 to 11.9 eV, the room temperature cutoff of the largest bandgap material, LiF, but will also discuss xuv thin film filters.

10.2.1 Transmission Formulas and General Linear Behavior

Absorption. Linear transmission in normal incidence through a window, assuming that no interference occurs between reflections from the front and back surfaces and that the same transparent medium exists before and after the window, is given by

$$T = \frac{(1 - R)^2 \tau}{1 - R^2 \tau^2} \tag{10.13}$$

where τ is the internal transmission determined by Eq. (10.1) and R is the normal incidence reflectance. Here absorption losses are assumed to include any scattering as well as electronic absorption. The formula includes multiple reflections; but if the reflectivity is weak, it becomes

$$T \approx (1 - R)^2 \tau \tag{10.14}$$

which represents single reflections off the front and back surfaces.

UV absorption in transmissive optics can be divided into an impurity or disorder plateau that limits transparency and the rising fundamental absorption edge. In the high-transparency region, the density of defects can be roughly determined by Smakula's equation (ignoring effective field effects)

$$\rho f = \frac{m}{\pi^2 e^2} \int \omega n \, \Delta k \, d\omega \tag{10.15}$$

where ρ is the density of defects; f is the oscillator strength of the defect transition, usually between 0 and 1; n is the index of refraction of the host; and Δk is the absorption induced by the defect.[10,42,43] The basic shape of the fundamental absorption edge for direct-gap materials is given by single-electron band-structure calculations to be a M_0 critical point, rising like $(E - E_g)^{1/2}$ where E_g is the bandgap.[11] The absorption edge for indirect gap materials is correspondingly given by $\mu \propto (E - E_g)^n$, where the exponent is 2 for allowed transitions and 3 for forbidden. Beyond the single-electron picture, however, the absorption is modified to a considerable degree by the screened electron-hole interaction. This many-body effect leads to the formation of excitons or bound states of the electron-hole pair. The interaction is stronger in low dielectric constant, large bandgap insulators leading to a more localized excitation and strong spectral features. Electron-hole exchange interaction can also modify the characteristic critical point shape at the fundamental edge and throughout the interband region.

In addition to the above characteristics, the absorption coefficient of many semiconductors and insulators rises initially from the impurity plateau at the fundamental absorption edge with the characteristic exponential form of the Urbach tail. The functional form is given by

$$\mu = Ae^{-B(\omega_0 - \omega)} \tag{10.16}$$

where A is a constant.[11] For some materials B is also constant while for others B is given by

$$B = B_0 \frac{2kT}{E_p} \tanh\left(\frac{E_p}{2kT}\right) \tag{10.17}$$

where E_p is an effective phonon energy. The exponential dependence can be caused by the effects of electric fields induced by defects and phonons or by exciton-phonon coupling.

The location of the cutoff energy of the fundamental absorption edge is temperature-dependent, as is the general broadening. The shift with temperature is in the range from 1 to 4 meV/K near room temperature but drops to zero near the liquid nitrogen boiling point for a variety of insulators.[44] It has been explained in terms of phonon-assisted interband transitions in SiO_2 and Al_2O_3[45] and in terms of excitons for fluorides.

Refraction. Refraction at the interface between two isotropic media of different indexes of refraction n_1 and n_2 is given by Snell's law

$$n_1 \sin i_1 = n_2 \sin i_2 \tag{10.18}$$

where i_1 and i_2 are the angles of incidence and exit relative to the interface normal. Both the incident and refracted rays lie in the plane of incidence. Snell's law fails for the extraordinary ray in anisotropic media.

The spectral dependence or dispersion of the index of refraction in transparent materials is usually cast as a polynomial in powers of the wavelength by Sellmeir's equation

$$n^2 = 1 + \sum_i \frac{A_i \lambda^2}{\lambda^2 - \lambda_i^2} \tag{10.19}$$

where three terms are sufficient to fit most materials.[46] This is entirely equivalent to a set of Lorentz oscillators as in Eq. (10.3). Abbe's constant v_d is used for convenience to quantify the dispersion of some materials, particularly silica. It is given by

$$v_d = \frac{n_d - 1}{n_f - n_c} \tag{10.20}$$

where the characters d, f, and c indicate the wavelengths 589, 486, and 656 nm, respectively. Tables of the coefficients of the Sellmeir equation and other polynomials and their temperature dependences are available in Ref. 4 and given by Dodge.[46] Detailed tables of the index for several insulators are given in Palik.[19] Values of the index for nonlinear materials are also given by Singh.[47] The variation of index with temperature leads to thermal lensing in high-power lasers, a limitation in laser rod and cavity design.

Polarization. Polarization properties of optical materials and polarizer design have been covered extensively by Bennett and Bennett.[48] The polarization of light can be altered at the interface of two media by reflection or throughout a medium by birefringence. Reflection will be covered in Sec. 10.3. Birefringence, or polarization dependence of the index of refraction in anisotropic media, has been tabulated for a variety of materials by Dodge,[46] by Bennett and Bennett,[48] and by Singh for nonlinear materials.[47] Birefringence can be induced in materials by stress and strain and by electric or magnetic fields. Simple uv quarter-wave plates can be made by applying uniaxial stress to crystals of quartz, MgF_2, and LiF. Piezo-optic and elasto-optic constants are used to predict the effects of stress and strain, respectively, on the index of refraction. The constants for a variety of materials are given in several sources,[49,50] although little data is available for the ultraviolet. Birefringence induced by electric and magnetic fields is discussed in the next section.

10.2.2 Applied Field and Nonlinear Behavior

Strong electric fields or intense radiation fields can modify the optical properties of a medium. The polarization P of the medium is given by

$$P_j = \chi_{jk}^{(1)} E_k + \chi_{jkl}^{(2)} E_k E_l + \chi_{jklm}^{(3)} E_k E_l E_m \qquad (10.21)$$

where E is the electric field and $\chi^{(1)}$, $\chi^{(2)}$, and $\chi^{(3)}$ are the first-, second-, and third-order susceptibilities. The χ's can be complex, and summation over the duplicated indexes is assumed in Eq. (10.21). Effects of the first-order susceptibility were discussed in the previous section. The second-order susceptibility applies only to materials without inversion symmetry. It leads to phenomena such as second harmonic and difference frequency generation, the linear electro-optic (Pockel's) effect, and the photorefractive effect. Essentially all materials show a third-order effect. Third-order susceptibility leads to two-photon absorption and the associated change in the index of refraction n_2 and to the quadratic Kerr effect. The alteration of the index can occur with a large dc electric field or in an intense laser beam itself, which leads to self-focusing and damage. Third-order susceptibility also leads to third harmonic generation and stimulated Raman and Brillouin scattering.

Two-photon absorption can be described as a modification of the normal absorption [Eq. (10.1)]. Thus

$$\frac{dI}{dx} = -[\mu + \beta I(x)]I(x) \qquad (10.22)$$

The second-order cross section is given by

$$\sigma_2 = \frac{\hbar\omega}{\rho}\beta \approx 10^{-50}\,\frac{cm^4 s}{photon\text{-}molecule} \qquad (10.23)$$

for most materials. Correspondingly, the nonlinear index of refraction n_2 for materials with inversion symmetry is given by

$$n = n_0 + n_2 E^2 \qquad (10.24)$$

Recently, it has been shown that n_2 is related by Kramers-Kronig inversion to the two-photon absorption coefficient just as the linear index of refraction is related to linear absorption.[51]

Tables of electro-optic coefficients with wavelength and temperature dependence are available in *Landolt-Börnstein*.[50] Tables of the linear electro-optic coefficients in a more abbreviated form are in Kaminow.[52] Tables of $\chi^{(2)}$ for second harmonic generation and $\chi^{(3)}$ for a variety of nonlinear processes for a variety of materials for 1 μm and longer are given by Singh.[47] Tables of β and n_2 are given by Smith.[53] Taylor et al. give β for fused silica, LiF, BaF_2, SrF_2, CaF_2, and MgF_2 at 248 nm.[54] Reintjes has discussed uv and xuv third, fifth, and higher harmonic generation in rare gases and metal vapors.[55]

Magneto-optical effects such as Faraday rotation are due to the splitting of atomic energy levels in magnetic fields. Tables of magneto-optic coefficients such as the Verdet constant are available in Chen.[56]

10.2.3 Transparent UV Materials

This section discusses the specific absorption and materials properties of a number of standard uv optical materials used as windows, uv low-pass cutoff filters, polarizing materials, or in harmonic generation. Emphasis is on pure crystalline materials. Pure silica glass has been discussed more extensively as the most widely used window material. UV absorption in silicate, borate, and phosphate glasses has been treated in detail by Sigel.[57] Fluoride glasses are covered in a recent book.[58] The materials discussed below have been ordered primarily by high-energy cutoff, with the highest first. The high-energy cutoff is taken as that energy where the transmission falls below 10 percent. All cutoffs given are for room temperature, except where noted. Figure 10.2 shows the uv transmission cutoff of some standard uv windows, and Table 10.1 provides a few useful thermophysical constants and

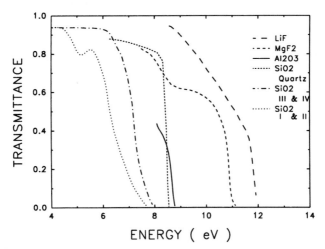

FIGURE 10.2 Transmittance vs. energy at 20 °C of some typical uv windows of LiF, MgF_2, and Al_2O_3,[1] three types of SiO_2-quartz,[76] types III or IV fused silica, and types I and II fused silica.[71] Thicknesses for the measurements were 1.95, 1.0, 0.32, 10, 3.0, and 6.0 mm, respectively. Transmission at energies below the cutoff is determined by reflectance and impurity absorption.

TABLE 10.1 UV Transmissive Materials

		UV cutoff, eV	K, W/cm-K	α, 10^{-6}/K	C, J/g-K	E, GPa	ρ_m, g/cm³	T_m, K	K/α thermal stability	E/ρ_m specific stiffness	$K/\rho_m C$ thermal diffusion
LiF		11.9	0.113	34	1.62	112	2.64	1140	0.0033	42.42	0.026
MgF$_2$	p	11.1	0.21	14	0.97	140	3.18	1528	0.0150	44.03	0.068
	s			8.9					0.0236	44.03	0.068
CaF$_2$	p	10.2	0.10	18.7	0.912	165	3.18	1630	0.0053	51.89	0.035
Al$_2$O$_3$	p	8.8	0.251	5.6	0.774	350	3.98	2300	0.0448	87.94	0.082
	s		0.230	5.0					0.0460	87.94	0.075
SiO$_2$ quartz	p	8.6	0.133	7.4	0.736	87	2.65	1740	0.0180	32.83	0.068
	s		0.074	14.0					0.0053	32.83	0.038
SiO$_2$ fused	s	5.4–7.3	0.014	0.42	0.765	70	2.21	1450	0.0333	31.67	0.008
β-BBO	p	6.5	0.008	36		124	3.84	1198	0.0002	32.29	
	s		0.0008	4					0.0002	32.29	
KDP	p	6.2	0.020	42	0.879	71	2.34	526	0.0005	30.34	0.010
	s		0.020	27					0.0007	30.34	0.010
CaCO$_3$	p	5.8	0.055	25	0.828	144	2.71	1610	0.0022	53.14	0.025
	s		0.046	−5.8					−0.0079	53.14	0.020
C diam		5.5	6.6	1.0	0.510	1035	3.52	3770	6.6000	294.03	3.677

K = thermal conductivity, α = thermal expansion, C = specific heat, E = Young's modulus (C_{11} for anisotropic crystals), ρ_m − mass density, all at 273 or 293 K. T_m = transformation and melting temperature, p parallel, s perpendicular to c axis. Values may vary according to material purity and preparation. Consult original references mentioned in the text.

figures of merit, including thermal stability, specific stiffness, and thermal diffusivity. The figures of merit are discussed in Secs. 10.3 and 10.4. Similar information is available for some standard reflecting materials in Sec. 10.3 and Table 10.2.

LiF. Lithium fluoride has the largest high-energy cutoff available, 11.9 eV. Optical constants are given in Palik.[19] LiF is somewhat hygroscopic (solubility 0.225 g/cm^2-24 h[59] or 0.27 g/100 g[60]). This can cause aging and a loss in transmission, which is most significant in thin films.[61] LiF is soft (Knoop hardness 110 kg/mm^2 [60]) and difficult to polish. It cleaves much more easily at liquid nitrogen temperatures. Radiation damage leads to the formation of brownish-color centers that can often be annealed out at 200 to 400°C.[62] See Sec. 10.4.

MgF$_2$. Magnesium fluoride has a cutoff of 10.9 eV. MgF_2 is weakly birefringent in the uv (1.2 to 1.5 percent[48]). Lynch has given the uv optical constants up to 50 eV.[63] The low-energy refractive index and temperature dependence is given by Bennett and Glassman.[64] It is used for wide-gap uv windows and for single antireflection or enhanced reflection coatings. It is less subject to radiation damage than LiF (see Sec. 10.4.3). MgF_2 is used for antireflection coatings because its index of refraction is intermediate between vacuum or air and many transparent substrates. It is often used as protective overlayer for normal incidence mirrors coated with aluminum.[65,66] The aluminum is protected from oxidation, which lowers the reflectance, and the overlayer thickness is adjusted for constructive interference and maximum reflectance near the cutoff. Magnesium fluoride is difficult to polish, but not as difficult as LiF (Knoop hardness 415 kg/mm^2 [60]).

CaF$_2$. Calcium fluoride has a high-energy cutoff of 10.2 eV. It can be used as a temperature tuned window for hydrogen Lyman-α radiation.[44] The refractive index and temperature dependence are given by Bennett and Glassman.[64] It can be cleaved. Knoop hardness is 120 kg/mm^2.[60] King and Nestor have recently reported fabrication of increased size and radiation damage resistant bulk crystals suitable for high-power laser windows.[67]

BeF$_2$. Beryllium fluoride exists in glassy and crystalline forms isometric with SiO_2. Optical constants from 10 to 50 eV are given by Bedford et al.[68] Their observation suggests that glassy BeF_2 has a uv cutoff as large as 9.5 eV, and crystalline BeF_2 may have a cutoff as large as LiF, 11.9 eV. Be toxicity is a problem in fabrication. BeF_2 is only slightly hygroscopic (0.066 g/cm^2-24 h[59]). Fluoroberyllate glasses are potential high-power laser materials because of their low n_2.[69]

Al$_2$O$_3$. Sapphire is the crystalline form of Al_2O_3 and is used more often for windows than the glassy form, alumina. Lynch has given the uv optical constants up to 50 eV.[63] Sapphire is hard (Knoop hardness 1370 kg/mm^2 [60]) and has a high thermal conductivity. It has a cutoff of 8.7 eV and is often used for windows because its large Young's modulus permits a thinner window when standing off a vacuum. MgF_2 is an almost perfect antireflection coating on sapphire. Sapphire is only weakly birefringent in the uv (0.9 to 1.3 percent[48]). Absorption from 1 to 8.5 eV in the Urbach tail and impurity plateau has been discussed by Innocenzi et al.[70] Major absorption features at 4.8 and 7.0 eV in the highest-purity samples seem due to trace impurities, although the 7.0-eV feature may be a result of oxygen vacancies.

SiO$_2$. Silica, SiO_2, or in its crystalline form, quartz, is the most prevalent window material. Quartz will be treated separately as a polarizing material below. Silica is broadly discussed by Brückner[71] and in *Glass Science and Technology*,[72] and the optical constants are given by Philipp.[73] SiO_2 has a Knoop hardness of 741 kg/mm[2].[60]

Fused silica or amorphous SiO_2 is classified into four types. Metallic impurities at the ppb level primarily determine the high-energy cutoff between 5.4 and 7.6 eV, although the water or OH content affects the impurity plateau at 7.5 eV via the concentration of oxygen vacancies.[74] OH content strongly affects the infrared transmission at 2.78 μm. Types I (Schott Infrasil) and II (Schott Herasil, Homosil, and Optosil) are forms of natural quartz fused in an electric arc and in flame, respectively. Types I and II have 5 to 20 ppm and 150 to 400 ppm by weight of OH, respectively. Both have Na and Al impurities. Impurities limit the uv cutoff to 5.4 to 7.2 eV, while pure quartz has a high-energy cutoff of about 8.5 eV. An oxygen treatment of type II fused silica (Schott Ultrasil) eliminates the transmission dip near 5.2 eV shown in Fig. 10.2. Type III (Corning 7940, Schott Supersil 2) is a synthetic fused silica formed by flame hydrolysis and has 200 to 1200 ppm OH. Type IV (Corning 7943, Schott Suprasil W) is a synthetic fused silica formed by plasma deposition and has less than 3 ppm OH. Type III and IV uv cutoffs are as high as 7.6 eV, with variations due to ppb metallic impurities. The highest-energy transmitting optics, or "uv grade," are made of selected type III or IV fused silica. The standard optical glass is borosilicate crown (e.g., Schott BK-7). It has less OH than type III fused silica but contains various network modifying cations that shift the uv cutoff to about 3.9 eV.

Luminescence bands at 1.9, 2.8, 4.3, and 6.7 eV in fused silica and quartz appear to be related to intrinsic defects, although the last may be due to band-to-band recombination.[75] Luminescence efficiencies vary in silica from type to type, and radiation damage can enhance the luminescence. This can limit the usefulness of silica optics with excimers or other uv lasers. See. Sec. 10.4.3.

Low-expansion silica ceramics such as Zerodur (Schott) have been developed and are discussed in Sec. 10.3.2.

Quartz has a high-energy cutoff of 8.5 eV, as mentioned above. The optical constants are given by Philipp.[76] It is weakly birefringent (1 to 2 percent[48]) and often used for quarter wave plates. It is also optically active. Along the optical axis, the slightly different phase velocities of right- and left-hand circularly polarized light lead to a rotation of linear polarization by 21.7°/mm.[48]

KDP and BBO. KDP (KH_2PO_4) and β-barium borate or BBO ($\beta\text{-}BaB_2O_4$) are two examples of nonlinear materials useful out into the uv, with cutoffs of 6.2 and 6.5 eV, respectively. Chen et al. have compared their usefulness, together with urea and KTP, in second through fifth harmonic generation with a Nd-YAG laser.[77] BBO has only recently become available and may be the material of choice for high-power applications below 4.9 eV. Eimerl et al. have determined the optical, mechanical, and thermal properties.[78] BBO is only slightly hygroscopic. KDP is moderately birefringent in the ultraviolet (4 to 6 percent). The indexes of refraction in the transparent region are given by Singh.[47] KDP is hygroscopic (solubility 33 g/100 g[60]).

Diamond. Crystalline diamond and synthetic diamond films are attractive because of diamond's extreme properties. It is the hardest material (Knoop hardness 5700 to 10,400 kg/mm[2] [60]), has the highest thermal conductivity, and has a very wide spectral range into the infrared.[79] The high-energy cutoff of pure diamond (type IIA) is about 5.5 eV. The optical constants are given by Edwards and Philipp.[80] New synthetic diamond deposition capabilities are offering tough, thin film windows (discussed with xuv filters below), hard machine tool and optical coatings, and the possibility of diamond electronics. But microstructure, adhesion, and doping remain problems in some cases.[81,82]

CaCO₃. Calcite is most often used as a polarizing material. It has a high-energy cutoff of 5.8 eV. Calcite is strongly birefringent in the uv (10 to 20 percent) but is not optically active. Higher-energy uv performance can be attained with quartz or magnesium fluoride, but they are much less birefringent so the design is made more difficult. Most uv polarizers are so-called pile-of-plates polarizers that take advantage of Brewster's angle. Calcite is relatively soft (Vickers hardness 75 to 135 kg/mm^2 [60]).

10.2.4 Coatings and Filters

Design of uv coatings and thick filters becomes progressively more difficult at higher energies because fewer transparent materials are available. Filters above 12 eV must be free-standing thin films because of the high absorption. Optical constants of uv coatings are similar to those discussed for bulk materials in the previous section, but the transmission is much less, owing to inhomogeneities such as columnar growth that occur during deposition. This will be discussed in more detail in Sec. 10.5.2. High-reflectivity coatings and polarizing beamsplitters are covered in Sec. 10.3, and the design of coatings to prevent laser damage, such as rugate filters, is reserved for Sec. 10.4.

Many aspects of coatings and filters are dealt with in the series *The Physics of Thin Films*[83] and by MacLeod,[84] Dobrowolski,[85] and Rancourt.[86] Cook and Stokowski[87] and Costich[88] have also discussed thick and thin film filters, respectively.

Antireflection Coatings. At angles of incidence from normal to grazing, reflection losses in the ultraviolet can be minimized with suitable antireflection (AR) coatings. Full transmission and zero reflectance of linearly polarized light can be achieved with uncoated windows in *p*-polarized light only at Brewster's angle.

The simplest AR coating is a single $\lambda/4$ coating with an index equal to the geometric mean of the indexes of the incident and substrate media. Reflections from the front and back surfaces interfere destructively. Such coatings are broadband and work well over a range of incident angles. MgF$_2$ is the most often used material because its index is close to the geometric mean for a variety of substrates in vacuum or air.

Multilayer AR coatings can be designed to reduce the reflection of a single wavelength at a single angle, to cover a wide spectral range, or to cover an intermediate range with a flatter response. The most common multilayer ultraviolet AR coatings are two-layer $\lambda/4$ designs such as Al$_2$O$_3$/MgF$_2$, but non-$\lambda/4$ high-index designs ZrO$_2$/SiO$_2$ are useful with high-index substrates and high fluences. ZrO$_2$ multilayers are limited to energies below 4.4 eV and usually have significant inhomogeneities but a high laser damage resistance. Other useful uv coating materials include HfO$_2$, Sc$_2$O$_3$, Y$_2$O$_3$, ThF$_4$, Na$_3$AlF$_6$, and AlF$_3$.[89,90,91]

The ideal AR coating is a graded index material that makes a smooth transition from the index of the incident medium to the index of the substrate over a thickness larger than $\lambda/4$. Unfortunately, no solid materials with an index of nearly unity are available for windows in air or vacuum. Graded index coatings have been discussed by Lowdermilk.[92] Recently, considerable development work has been done on porous or graded solid sol-gel coatings, in particular. Graded index coatings have the advantage of lower peak field at the final surface and a higher laser damage threshold, as is discussed in Sec. 10.4.

Transmission Band Filters.   Band filters composed of a multilayer dielectric stack are widely available, but primarily for the visible. Narrowband filters can be obtained with a Fabry-Perot stack or with multiple Fabry-Perot stacks. Malherbe discusses filters using MgF_2 for energies of 10.2 and 6.2 eV.[93] Polarization interference filters of several types are also available.[48]

Chromaphore filters, made by dispersing transition metal or rare earth ions in liquids, have a behavior similar to doped glasses.[94] Water has the highest-energy cutoff of room temperature liquids that can serve as host, 6.7 eV.

Molecular oxygen has a high-energy cutoff of about 6.4 eV that limits uv spectroscopy in air. But it also has a useful narrow window with a low cross section of 1×10^{-20} cm^2 at the Lyman-α line of hydrogen at 10.20 eV.[1] Purging spectrometers with N_2 in place of air increases the uv cutoff to 7.3 eV.

Christiansen filters depend on the scattering of inclusions in a solid or liquid matrix. At a specific energy the scattering goes to zero as the index of the inclusion matches that of the surrounding medium.[94] Wojak et al. discuss Christiansen filters for energies from 3.3 to 5.4 eV.[95]

Neutral Density Filters.   Neutral density sets provide a number of filters with fixed attenuation over a broad wavelength range. They are usually made of Ni in various complexes in glass or in thin alloy films deposited on fused silica or quartz. The host or substrate determines the uv transmission range. Neutral density filters are discussed by Dobrowolski.[85]

Thin Film XUV Filters.   Above the LiF cutoff energy and below the hard x-ray region, transmission filters must be thin films owing to large absorption coefficients. Above 500 eV, broadband high-pass filters with thicknesses of several micrometers can be made of the lowest atomic number elements, such as Be, that have little remaining oscillator strength and absorption. Above 400 eV, synthetic diamond films are tough enough to withstand an atmosphere of differential pressure over a considerable area and still be thin enough to transmit adequately. In the xuv below 400 to 500 eV, thin film filters must be used. Their windows of transmission occur at energies below core absorption edges where the absorption is at a minimum. For transmissions greater than 1 percent, thicknesses range between 50 and 200 nm. These fragile filters are often mounted on grids to support large areas and are subject to pinholes, particularly materials such as In. The transmission of xuv filters can be determined roughly from the optical constants of the elements given by Henke.[20,21] Transmission measurements of a number of xuv filters are illustrated in Samson.[1] Recently, Powell et al. have collected absorption measurements and discussed oxidation and other aging effects for B, C, Al, Ti, Sb, Sn, and In thin films in the ranges from 7 to 500 ev.[96] Figure 10.3 illustrates the transmission of a few standard xuv thin film filters; In, Sn, Ti, Aged Al, C, and Be.[96,20,21] Several groups have also fabricated thin multilayer films for special uses such as soft x-ray polarizing beamsplitters.[97,98]

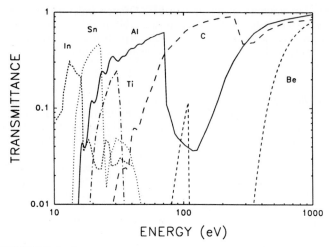

FIGURE 10.3 Transmittance vs. energy of a few standard xuv thin film filters. In, Sn, Ti, and C of 100 nm thickness. Aged Al of 90 nm pure Al and 1.5 nm oxide thickness. Be of 2 μm thickness. Transmittance of In, Ti, and Sn above 50 eV omitted for clarity. Optical constant data: In, Sn, and Ti;[96] Al, C, and Be.[20,21]

10.3 REFLECTIVE UV OPTICS

This section examines reflectance features and thermomechanical behavior of typical ultraviolet mirror materials. The term "reflectance" should be applied to real measured values including scattering and other losses, and the term "reflectivity" should be reserved for the ideal case. Damage from external effects during use is reviewed in Sec. 10.4, and fabrication issues are discussed in Sec. 10.5. Section 10.3.1 covers the connection of reflectivity to optical constants and electronic structure and the nature of scattering losses due to rough surfaces. Section 10.3.2 will address the limitations of substrate properties on performance. Section 10.3.3 discusses the reflectance of standard elemental materials, and the final section, 10.3.4, considers the behavior of multilayer reflective coatings.

10.3.1 Reflectivity Formulas and General Behavior

Formulas.   Specular reflectivity vs. angle at an interface with vacuum as a function of the optical constants and polarization is given by Fresnel's laws as

$$R_s = \frac{(a - \cos i)^2 + b^2}{(a + \cos i)^2 + b^2} \tag{10.25}$$

and

$$R_p = R_s \frac{(a - \sin i \tan i)^2 + b^2}{(a + \sin i \tan i)^2 + b^2} \tag{10.26}$$

with

$$2a^2 = \sqrt{c^2 + 4n^2k^2} + c \qquad (10.27)$$

$$2b^2 = \sqrt{c^2 + 4n^2k^2} - c \qquad (10.28)$$

and

$$c = n^2 - k^2 - \sin^2 i \qquad (10.29)$$

Here i is the angle of incidence relative to the surface normal and R_s and R_p are the reflectivities in s-polarized and p-polarized light with the electric vector of the light perpendicular and parallel to the plane of incidence, respectively. The term s-polarized comes from the German word for perpendicular, *senkrecht*. At higher energies and sometimes in the ultraviolet, the terms σ and π are often used instead. Figure 10.4 plots the s, p, and average reflectivity vs. angle of incidence, for a set of optical constants n and k, to illustrate the general behavior. R_s and R_p are equal at normal and extreme grazing angles; but in between, R_p passes through a mini-

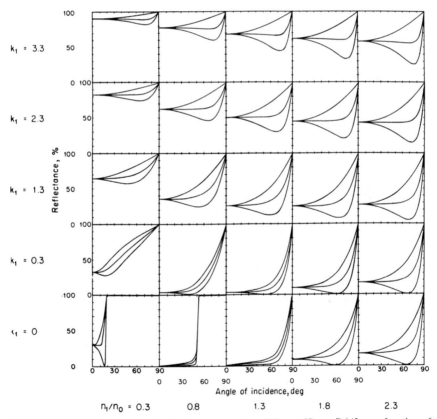

FIGURE 10.4 R_s (upper curves), R_p (lower curves), and $R_{av} = (R_s + R_p)/2$ as a function of angle of incidence for various values of the refractive index ratio n_1/n_0 and k_1. The incident medium, having refractive index n_0, is assumed to be nonabsorbing. *(Courtesy of J. M. Bennett.[48])*

mum, a Brewster's angle where the oscillator's dipole is oriented toward the reflected beam. This angle approaches 45° at higher energies as n approaches 1 and k zero. In normal incidence, the equations for either polarization reduce to

$$R = \frac{(n - 1)^2 + k^2}{(n + 1)^2 + k^2} \tag{10.30}$$

Reflectivity is more completely expressed as a complex number including the magnitude as discussed above and a phase shift. In ellipsometry, the ratio of the complex reflection coefficients r_s and r_p is expressed as

$$\frac{r_p}{r_s} = \tan \psi \, e^{i\Delta} \tag{10.31}$$

with an amplitude $\tan \psi$ and a phase shift Δ. The phase shift in normal incidence reflectance from a nonabsorbing medium is π on reflection from a higher-index medium and zero on reflection from a lower-index medium. In general, the difference in phase shifts of s and p polarizations upon reflection is given by

$$\tan \Delta = \frac{-2b \sin i \tan i}{a^2 + b^2 - \sin^2 i \tan^2 i} \tag{10.32}$$

The phase shift difference is $\pi/2$ at the "principal angle of incidence," near but not equal to the minimum in R_p. Because of the minimum in R_p and not R_s, a quarter-wave retarding reflector to convert linear to purely circular polarization is not efficient. Performance of a three-mirror linear vuv polarizer and a single-mirror circular vuv polarizer has been measured by Westerveld et al.[99] General polarization properties in reflection have been discussed by Bennett and Bennett.[48]

Normal and Grazing Incidence Behavior.   Normal incidence reflectance of essentially all materials is large in the near ultraviolet region with the bunching of oscillator strength due to transitions from valence to conduction band transitions. The normal reflectance of metals tails off relatively smoothly from high values in the infrared and visible. The normal reflectance of semiconductors shows peaks due to the band structure character. As an example, Fig. 10.5 shows the normal incidence reflectance of a chalcopyrite semiconductor, $CuGaS_2$.[100] The normal reflectance of insulators is even more peaked and atomic-like with less dispersive bands.

Toward higher energies, normal incidence reflectance falls off smoothly as $1/E^4$, except for jumps and structure due to core level transitions. The reflectance falls below 10 percent above 30 to 40 eV for all materials. But, because $n < 1$ in the soft x-ray region, high reflectivity can be restored by using grazing angles of incidence, since there is total external reflectance below some critical angle. At higher energies, n rises back to one as $1/E^2$, the critical angle decreases, and more grazing angles are required to maintain reasonable reflectivity. The critical angle away from absorption edges is roughly given by

$$\theta_c \approx 3.7 \times 10^{-11} \frac{\sqrt{\rho \, Z}}{E} \tag{10.33}$$

where θ_c is the critical grazing angle in radians, ρ is the number of atoms per cm^3, Z is the atomic number, and E is the energy in eV.[101] This formula is more accurate above 100 eV, where absorption is small and so the critical angle is more sharply

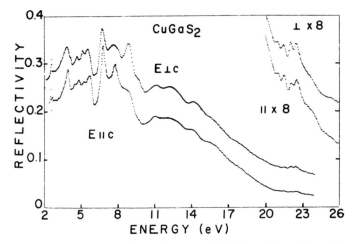

FIGURE 10.5   Normal incidence reflectance of $CuGaS_2$, a chalcopyrite semiconductor, at 80 K for light polarized perpendicular and parallel to the c axis. Outside
of the fundamental valence to conduction band transitions from 3 to 20 eV, excitons
can be seen at about 2.5 eV and transitions from the Ga 3d core levels near 22 eV
(expanded ordinate).[100]

defined. Clearly, higher-Z elements are preferred mirror coatings at higher
energies.

Retroreflecting systems for use above 30 eV can be constructed from a set of
grazing angle reflections in a multiple mirror[102,103] or "whispering gallery"[104] approach. Such systems have been investigated for end mirrors in a soft x-ray free-
electron laser cavity. A nine-mirror alumium retroreflector has been fabricated in
ultra-high vacuum with a total reflectance of 89 percent at 21.2 eV.[105] An alternative
approach, multilayer coatings, can offer reflectances as high as 60 percent in normal
incidence in the soft x-ray range (Mo/Si multilayer at 95 eV), although they are
narrowband and absorption limits reflectance at lower energies. Multilayer coatings
are discussed in Sec. 10.3.4.

Scattering and Surface Sensitivity.   Reflectance is sensitive to surface preparation
and condition. In the normal incidence regime and below the critical angle, the
sampling depth can be as small as 10 nm, so contaminants and oxide layers can
have a significant effect. This has already been discussed in Sec. 10.1.1 with the
sources and measurement of optical constants.

Particularly troublesome for many applications is scattering from surface roughness. Scattering theory and measurements have been reviewed in a recent monograph by Bennett and Mattsson.[106] Mie scattering or absorption losses by external
contaminants, such as dust, are mentioned in Sec. 10.4; and techniques of measuring
surface roughness are reviewed in Sec. 10.5. The simplest theory of diffractive
scattering from correlated surface features is the "scalar" theory, which applies to
unpolarized light and does not address the angular dependence of the scattered
light. It predicts that the intensity of specularly reflected light I_s is given by

$$I_s = I_0\, e^{-(4\pi\sigma\sin\theta/\lambda)^2} \tag{10.34}$$

where I_0 is the intensity reflected by an ideal smooth surface, σ is the rms surface roughness (dependent on the bandwidth of the measurement), and θ is the grazing angle. This formula properly predicts greater scattering at shorter wavelengths and more normal angles and often gives a reasonable estimate of soft x-ray scattering. The more complete angular dependent "vector" theories depend on the polarization, the scattering angle, and the roughness and transverse spatial periodicity of the roughness. Such scattering can be considered to arise from an ensemble of gratings of different spatial periods, each period being smaller than the transverse coherence length. Measurements of angle-dependent scattering agree well with theory except at angles near backscattering in s polarization.[107,108,109]

Wavefront distortions due to roughness with spatial periods longer than the transverse coherence length simply lead to geometrical scattering from slope errors. This is because the diffracted light falls within the image width. The coherence length naturally divides spatial periods into the regions of mirror finish and figure. For focusing mirrors used with synchrotron radiation, for example, the coherence length for a quasi-monochromatic, extended incoherent source with image of radius σ_i is given by

$$x_c \simeq \frac{\lambda f}{2\pi \, \sigma_i \, \theta} \tag{10.35}$$

where f is the mirror to image distance.[110]

10.3.2 Substrates

Substrates for uv mirrors are often a limiting consideration. Mirror thermal conductivity, thermal expansion, size, weight, stiffness, and ability to figure and polish can be critical factors. Thermal and elastic deformations are discussed in this section. They are often analyzed independently or jointly with commercial finite-element computer programs. Structural stability of bulk materials and thin film coatings is considered in Sec. 10.4.1. Figuring and polishing are reserved for discussion in Sec. 10.5.1.

Thermal Distortion. *High Heat Load.* Dealing with thermal distortion is particularly difficult for high-heat-load laser and synchrotron radiation applications. Free-electron lasers and new wiggler and undulator synchrotron sources are predicted to deliver continuous heat loads of 100 W/mm^2 and higher, and arcsecond figure accuracies are desired. Damage by ionizing radiation often precludes the use of materials such as glasses that can shrink, or wide-bandgap crystalline materials that can expand. Radiation damage phenomena are reviewed in Sec. 10.4. For continuous power loads, the standard figure of merit for low thermal distortion substrate selection is thermal stability or the ratio of thermal conductivity to the coefficient of thermal expansion K/α. For pulsed laser power loads, where the pulse duration is greater than 1 ns, a better figure of merit is the thermal diffusivity or thermal conductivity, divided by the density, times the specific heat per unit mass $K/\rho_m C$. Table 10.2 lists the parameters at 273 to 293 K for some typical mirror materials. The parameters vary considerably with temperature, which is generally unfavorable but can be used to advantage in certain situations. Previously mentioned compilations of thermophysical properties should be consulted.

Approaches to dealing with the continuous thermal distortion problem at the new generation of synchrotron radiation sources have recently been summarized.[111]

With the thick substrates required to hold figure accuracy and nonuniform source illumination, distortions include a front-to-back bending of the whole mirror and a bump along the surface due to thermal gradients. Both can lead to 10-arcsecond slope errors even with water cooling from the back or sides. One approach to reducing distortion is to use more efficient liquid gallium cooling through channels near the reflecting surface. Another approach is to use silicon substrates cooled to near 125°K, where the silicon expansion coefficient crosses zero and the thermal conductivity is four times better than at room temperature.

Low Heat Load.   For telescopes and other applications where low heat loads and radiation damage are not a problem, low-expansion glass ceramics are used for mirror substrates to minimize figure distortions. The thermal expansion coefficient of fused silica is 0.42×10^{-6}, but several low-expansion SiO_2 glasses have been developed with a thermal expansion at least a factor of 5 smaller near room temperature. These include Schott Zerodur, Corning 9600, and ULE (Corning 7971). The thermal expansion advantage may be lost at other temperatures. Variation of the thermal expansion coefficients and simple dimensional stability with temperature of Zerodur, ULE, fused silica, BK-7, Invar and Super-Invar, silicon, and several metals have been measured by Jacobs.[112] The dimensional stability of high-precision optics made of Zerodur, in particular, must be questioned if the optics have been exposed to temperatures exceeding 150°C.[113] Recently, low-expansion glass ceramics (Schott Zerodur M and Corning 9600) that are more stable under thermal cycling have been developed.[114,115]

Elastic Deformation and Adaptive Optics.   Elastic deformation is a consideration for large, heavy, and active or adaptive optics. The standard figure of merit for mirrors subject to weight distortions is the specific stiffness or Young's modulus divided by the density E/ρ_m. This is given for various materials in Table 10.2. A crude rule of thumb for minimal solid substrate bending under gravitational forces is a 6 to 1 maximum ratio of width to thickness. Honeycomb or other integral substrate ribbing can decrease weight while retaining stiffness and minimizing figure distortion for a given mirror size. SiC, Be, and low-expansion glass telescope mirrors are often manufactured in such forms. Stiffness is also a concern for active or adaptive mirror optics. These mirrors are designed to either be bent infrequently or to be dynamically adjusted on the time scale of seconds or shorter to adjust focus. Examples of the former are relatively inexpensive, plane or singly-curved grazing angle x-ray mirrors that are bent to large radii of curvature. These have been used extensively at synchrotron radiation sources.[116] Examples of the latter are adaptive mirrors being used in ground-based telescopes to minimize the effects of atmospheric turbulence and improve angular resolution.[117]

10.3.3 Reflective UV Materials

Standard uv mirror materials, primarily thin film coatings but also bulk mirrors, fall into several regions of use. In the regime of normal incidence, 3 to 40 eV, low-Z materials have the best reflectivity from 3 to 20 eV owing to bunching of oscillator strength at low energies; but high-Z materials generally have higher reflectivity above 20 eV. In the grazing incidence regime, above 30 eV, low-Z materials offer high reflectivity below their critical energies and a sharp cutoff; but high-Z materials generally have higher reflectivity out to higher energies, for a given grazing angle, owing to their greater density of effective electrons. These spectral behaviors are, of course, modified in each element by the occurrence of absorption edges. As illustrations, Fig. 10.6 plots the normal incidence reflectance of low- and high-Z materials: aged Al, Au, Ag, Os, diamond, and SiC; Fig. 10.7 shows the grazing reflectivity above 30 eV of a low-Z material, SiO_2; and Fig. 10.8 plots the grazing reflectivity above 30 eV of a high-Z material, Au.

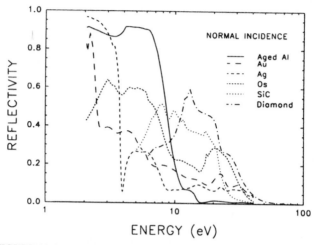

FIGURE 10.6 Normal incidence reflectivity of some of the largest magnitude normal incidence reflecting materials in the ultraviolet.[19,20,21,120]

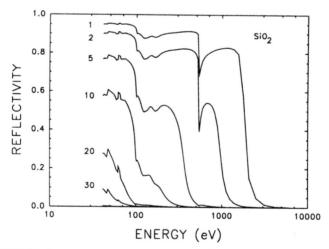

FIGURE 10.7 Grazing reflectivity vs. energy of a low-Z material, SiO_2, at grazing angles in degrees as indicated. Si $L_{2,3}$ and O K absorption edges appear at 100 and 525 eV, respectively. Critical energy cutoff is sharper at higher energies where absorption is lower.[20,21]

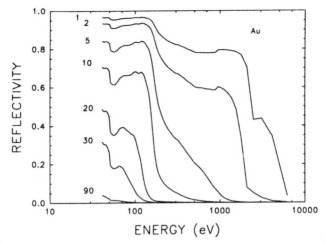

FIGURE 10.8   Grazing reflectivity vs. energy of a high-Z material, Au, at grazing angles in degrees as indicated. Au $O_{2,3}$, $N_{2,3}$, and $M_{4,5}$ absorption edge effects appear near 50, 550, and 2200 eV, respectively. Critical energy cutoffs are more gradual than lower-Z materials because of higher absorption.[20,21]

The materials discussed will be ordered roughly by energy range of use. Table 10.2 gives some thermophysical data for the materials covered. Palik[19] has given optical constants for many of these materials. Those data and Henke's optical constants can be used to generate reflectivities with Eqs. (10.25) through (10.29). Besides these sources, Madden has reviewed early evaporated thin film reflectance measurements in the vuv below 30 eV.[65] Hunter has reviewed vuv optics materials with comments on coating quality.[61] Weaver et al. have published optical constants for 43 metals from 0.1 to 500 eV.[118] And Windt et al. have recently reported reflectance measurements and optical constants for 16 metals[119] as well as C, diamond, Al, Si, and CVD SiC[120] from 10.2 to 517 eV. Polishing of bulk mirrors and thin film morphology are discussed in Sec. 10.5. Multilayer mirror coatings are reserved for Sec. 10.3.4, and damage phenomena are discussed in Sec. 10.4.

Ag Coating.   Freshly evaporated silver has the highest normal incidence reflectance in the range from 0.4 to 3.3 eV. Because of its tendency to degrade with the formation of AgS on the surface, it must have a protective transparent coating. The relationship between visible scattering and microstructure of thin Ag films has been studied.[107]

Al Coating.   Freshly evaporated aluminum has the highest normal incidence reflectance from 3.3 to 10 eV. An oxide layer of 0.5 nm forms in 1 h in air and a terminal thickness of about 2 nm forms in about 10 days.[65] Vacuum evaporated reflectances of 63 percent at 10.2 eV correspondingly fall to values of 41 and 27 percent. Overcoating with MgF_2 to a thickness of 25 nm before air exposure yields reflectances of about 80 percent at 10.2 eV. Values above 90 percent could, in principle, be achieved if the coating had single-crystal optical constants.[66] Aluminum filters are reported to form a terminal oxide thickness of 7.5 nm on each

TABLE 10.2 UV Reflective Materials

	K, W/cm-K	α, 10^{-6}/K	C, J/g-K	E, GPa	ρ_m, g/cm^3	T_m, K	K/α thermal stability	E/ρ_m specific stiffness	$K/\rho_m C$ diffusivity
Be	1.8	12.3	1.88	300	1.86	1560	0.146	161.29	0.515
C diam	6.6	1.0	0.510	1035	3.52	3770	6.600	294.03	3.677
Al	2.3	23.0	0.945	69	2.70	933	0.100	25.56	0.901
SiC-CVD	1.6	2.0	0.70	460	3.21	2830	0.800	143.30	0.712
SiC bonded	1.6	2.6	0.75	311	2.92	2380	0.615	106.51	0.731
Si	1.7	2.5	0.736	120	2.33	1685	0.680	51.50	0.991
SiO$_2$ fused	0.014	0.42	0.765	70	2.21	1450	0.033	31.67	0.008
SiO$_2$ zerodur	0.016	0.05	0.82	91	2.53	420	0.320	35.97	0.008
Ni	0.9	12.8	0.439	210	8.90	1726	0.070	23.60	0.230
Cu	4.0	16.7	0.385	120	8.96	1358	0.240	13.39	1.160
Mo	1.1	5.0	0.272	340	10.2	2890	0.220	33.33	0.397
Ag	4.3	19.0	0.233	75	10.5	1234	0.226	7.14	1.758
Os	0.88	4.7	0.130	560	22.4	3300	0.187	25.00	0.302
Pt	0.73	8.9	0.136	150	21.4	2045	0.082	7.01	0.251
Au	3.1	14.2	0.130	75	19.3	1338	0.218	3.89	1.236

K = thermal conductivity, α = thermal expansion, C = specific heat, E = Young's modulus, ρ_m = mass density, all at 273 or 293 K. T_m is transformation and melting temperature. Values may vary according to material purity and preparation. Consult original references mentioned in the text.

surface.[96] Visible half-wave Al_2O_3 and SiO_x overcoats are often added to improve surface protection for aluminum mirrors to be used below 6.2 eV.[121] Aged aluminum mirrors with a native oxide coating or aluminum overcoated with MgF_2 have reasonably high reflectances from the overlayer at normal or grazing angles above 12 eV as well. Normal reflectance of aged aluminum is, in fact, better than Pt up to 17 eV. The reflectance of aluminum at higher energies and grazing angles of less than 40° is better than platinum because the low-Z material has lower values of n and k.

SiC Bulk or Coating.   From 8-22 eV, polished, chemically vapor deposited (CVD) SiC has a considerably higher reflectance than platinum. SiC also has high figures of merit for high-power applications. Normally β-SiC (cubic), which can be highly polished, is deposited on SiC or graphite substrates by CVD at high temperatures for mirror blanks, but monolithic CVD blanks can be made.[122] Powders of α-SiC (hexagonal) are more commonly high-temperature-processed into bulk substrates. The substrates are classed as recrystallized or hot-pressed, sintered, and reaction-bonded or siliconized according to fabrication and vary in porosity and thermo-mechanical properties.[123] Large blanks up to 60 cm in diameter are now available. Highly reflective thin films have also been deposited on other substrates at low temperatures, using ion-beam and other deposition methods.[124] Grazing reflectance and optical constants of CVD-SiC have been determined by several groups in the ranges 40 to 180 eV,[101] 80 to 1000 eV,[125] and 10 to 500 eV.[120]

Au, Pt, Ir, Rh, Re, Os, W, Ru Coatings.   These metals are the best reflectors in the full 10 to 30 eV range at normal incidence;[61,119] and because of their high Z, they have the highest critical energies for a given grazing angle. Their reflectances in the 10 to 30 eV range are similar but vary in magnitude. Except for gold, they all require high-temperature e-beam evaporation or other deposition method. Au, Re, Ru, and Os exhibit their highest reflectance when evaporated onto room-temperature substrates, but the others require heated substrates. Energetic deposition techniques, described in Sec. 10.5, would perhaps not need heated substrates for smooth coatings.

Gold is the most common coating for normal or grazing incidence mirrors. Gold is relatively inert, although there is some evidence of a thin oxide overlayer.[126] Evaporated gold tends to roughen as the layer thickens. Layers of 15 nm are the smoothest and have the best overall vuv reflectance.[127] Smooth layers can also be formed electrolytically or by sputter deposition. Aspnes has obtained the near-uv pseudo-optical constants of gold up to 5.8 eV with ellipsometry and has correlated the variations with void fraction, crystallinity, and roughness arising from differences in deposition.[128]

Osmium has the highest normal incidence reflectance of evaporated metal coatings from 10 to 30 eV. Osmium has been observed, however, to oxidize and evaporate in a low earth orbit.[129] See Sec. 10.4.2. Ir falls between Os and Pt in reflectance, but Pt is most often evaporated for reasons of expense. Pt has also been deposited electrolytically in dense and smooth films up to 100 nm thick. Pt is used at grazing angles to attain a slightly higher critical energy than gold. W and Re form oxide overlayers on air exposure.

SiO₂ Bulk.   Fused silica or low-expansion glass ceramics (see Sec. 10.3.2) can be highly polished and offer high reflectivity uncoated in the xuv at grazing angles. Applications include grazing incidence Wolter telescopes. The reflectance and optical constants of two low-expansion glasses from 11.8 to 155 eV have been reported by Rife and Osantowski.[130] Shrinkage up to 3 percent and distortion can occur in silicas in ionizing radiation. See Sec. 10.4.

Be Bulk or Coatings.   Beryllium substrates offer light weight and thermal stability. Low Z yields low absorption and sharp critical energy or angle cutoffs. Only a very small oxide layer, if any, is present. Near net shape parts are available with hot and cold isostatic pressed powders.[131] Toxicity is a factor for thin film deposition and often requires a dedicated deposition chamber.

Ni Bulk or Coatings.   Completely metal mirrors can be fabricated on aluminum or copper substrates with a thick overcoating of electroless Ni. Electroless Ni coatings are chemically deposited out of solution, and they are discussed in Sec. 10.5.2. Electroless nickel can be diamond-turned to nearly the correct figure and cosmetically polished to a supersmooth finish. As with Be, the low Z offers high reflectance and sharp cutoffs in grazing incidence. Rife and Osantowski have measured the grazing reflectance of electroless nickel from 40 to 150 eV.[101]

10.3.4 High-Reflectance Multilayer Coatings

Sub-XUV Dielectric.   High-reflectance (HR) uv dielectric coatings typically consist of a stack of alternating low and high index, $\lambda/4$ thick transparent layers. Stack thickness can be adjusted to select transmission for laser cavity resonator mirrors.

Much of the discussion and references in Secs. 10.2.3 and 10.2.4 on transparent uv materials and coatings can be carried over and will not be duplicated. UV HR coating materials have been discussed by Rainer et al.,[89] Lowdermilk,[90] and DeBell et al.[91] Malherbe has also discussed a few MgF_2-based HR coatings for the range from 5.6 to 7.3 eV.[93] Damage by ionizing radiation and laser radiation is discussed in Secs. 10.4.3 and 10.4.4, respectively.

XUV.   Interest in xuv multilayer coatings has grown rapidly since the first materials combinations were developed that formed smooth, durable interfaces. Spiller[132] and Barbee[133] have reviewed the techniques of multilayer fabrication and calculations of multilayer performance. Multilayer development has been spurred by the benefits of extending normal incidence optics designs to the spectral domain of grazing incidence optics and of engineering synthetic crystals to serve in the x-ray region. Recent advances include multilayer coated gratings that offer the potential for higher resolving power soft x-ray spectrometers, efficient imaging spectrographs for soft x-ray telescopes, and robust replacements for diffracting crystal optics.[134,135,136]

Multilayers are synthetic crystals for which Bragg's law, $m\lambda = 2D \sin \theta$, is obeyed, where D is the multilayer period, m the order, and θ the grazing angle. Scattering layers of selected thickness alternate with relatively transparent spacer layers to form the stack. Selection of materials for the scattering and spacer layers depends on the optical constants and material compatibility. Absorption edges, particularly in the spacer layer, dramatically reduce the refractive index contrast necessary for high reflectance. For wavelengths longer than 10 nm, large absorption for all materials limits useful stack thickness. Minimum practical 2D spacings are about 3 nm, so higher multilayer orders must be used to attain shorter wavelengths in normal incidence. Layer materials must not interdiffuse for maximum reflectance and must be smooth to minimize scattering losses. Several standard coatings such as ReW/C, W/C, W/Si, and Mo/Si have been developed, but many possibilities have not yet been explored.

Peak reflectances for multilayers typically are 10 to 20 percent in normal incidence, although about 60 percent has been achieved for Mo/Si at 95 eV where Si is very transparent. Multilayer reflectance rises at more grazing angles. The bandwidth and angular acceptance are determined by the effective number of periods N contributing, which depends on the penetration depth of the radiation. Bandwidths are roughly given by $\Delta\lambda/\lambda \approx 1/mN$. Bandwidths can be as small as 0.3 to 1.0 percent at the shortest wavelengths.

10.4 DAMAGE AND DURABILITY

This section covers durability and damage of ultraviolet materials in two parts. First, questions of mechanical stability and interdiffusion at interfaces as well as practical considerations of surface contamination are discussed. Second, the more specific problems of radiation damage are addressed. They are divided, somewhat arbitrarily, into ionizing radiation damage and laser-induced damage. Defects due to fabrication are discussed in Sec. 10.5.

10.4.1 Structural Stability

Stress and Adhesion.   Stress relief can be a long-term problem for bulk materials, particularly those that are metastable or not properly heat-treated. Length variations and distortions over time proscribe the use of materials such as cast iron for precision mounts and favor materials such as granite. But ppm length variations can occur over years or days at 60 °C even for apparently stable, low-expansion materials like ULE, Invar, and Super-Invar.[137] Heat treatment of precision metal mirror substrates must be carefully considered if figure is not to be destroyed at elevated temperatures.

Stress in optical thin films is not necessarily a problem but can cause distortions with thin substrates. Thin film breakdowns such as delamination, "orange peel," and crazing result from differences in thermal expansion coefficients, stresses, and adhesion. Doerner and Nix have reviewed stresses and deformation in thin films.[138] Pulker has examined the stress, adherence, hardness, and density of metal and dielectric optical thin films.[139,140] Thin film morphology, columnar and nodule growth, strongly affects these factors. Morphology varies most rapidly in the smaller thicknesses. Tensile stress builds up in depositions on unheated substrates owing to forces acting to collapse residual voids of columnar microstructure. Packing density can be increased, stresses lowered, and adhesion improved with energetic deposition processes such as ion-assisted sputtering, which is discussed in Sec. 10.5. Packing density can also be increased by heating the substrate during deposition, but this may lead to unacceptable stresses at ambient conditions. Matching expansion coefficients is desirable but not always possible. The thin film designer may choose to alter the thicknesses of individual layers away from the ideal optical performance values in order to match stresses.

Adhesion is determined by bonding, lattice mismatch, and interdiffusion. Tabor has reviewed adhesion.[141] Metal-metal adhesion and interface segregation energies have recently been obtained using photoemission.[142] Adhesion can sometimes be improved by undercoating. For example, gold coatings on glass are often under-coated with Cr to improve adhesion. The Cr diffuses into the gold along grain boundaries and forms a chromium oxide interface with the SiO_2.[143]

Interdiffusion and Crystallization. Coating and substrate interdiffusion can lead to formation of alloys that alter roughness or destroy structural stability. Interdiffusion in thin films has been reviewed by Weaver,[144] Nakahara,[143] and Greer and Spaepen.[145] A catastrophic interdiffusion example is the attempt to use a combination of Ga/In eutectic with some mirror substrates for nonstress cooling in ultrahigh vacuum.[146] In the case of an aluminum substrate, the mirror crumbles to dust. Interdiffusion and crystallization in x-ray multilayer metal reflectors are a potential concern for aging and operation at elevated temperatures. Piecuch has discussed effects,[147] and Knight et al. have reviewed material selection for high x-ray flux applications.[148] Insoluble metal combinations can be selected from binary phase diagram data to minimize interdiffusion at interfaces.[149,150] Crystallization of amorphous layers leading to increased roughness, destruction of the layers, and much reduced reflectivity has been observed in W/C, Co/C, and Cr/C multilayers at temperatures between 650 and 750 °C.[151]

10.4.2 Surface Degradation

Physisorption. Various volatile contaminants can adsorb on and absorb into surfaces with relatively weak bonding. H_2O is commonly physisorbed at room temperature and in ordinary atmospheric conditions. Water has a partial pressure that limits the vacuum in ultrahigh vacuum systems and is normally baked away at temperatures above 150 °C. A thin water layer is not detrimental, if the material is not hygroscopic, but large amounts can be absorbed into the cavities in low packing density thin films, leading to lower laser damage thresholds. Laser damage thresholds are discussed in Sec. 10.4.4.

Chemisorption/Oxidation. Many materials have a native oxide coating that grows to a thickness limited by the diffusion of oxygen through the oxide to the interface. Several specific examples have already been discussed in Sec. 10.3. Native oxide coatings on aluminum and silicon are roughly 2.0 and 0.5 to 1.0 nm thick. Thicknesses can be morphology-dependent. Terminal oxide thicknesses are reported to be 2.0 nm for pure, quickly evaporated aluminum mirror coatings[65] and 7.5 nm for each surface of aluminum filters.[96] Scott et al. have measured the growth rate and reflectance at 21.2 eV of oxide layers on silicon and aluminum in a clean vacuum system.[152] As mentioned in Secs. 10.2 and 10.3, MgF_2 overcoatings are applied to aluminum to prevent oxidation and obtain higher reflectivity in the vuv. Very uniform anodic coatings of selected thickness can also be formed on aluminum with applied potentials in solution.[153] Some oxides have low vapor pressure. Osmium has been observed to oxidize and evaporate in a low earth orbit.[129]

Overlayers, such as cracked carbon, can be formed from volatile compounds by the action of ionizing radiation. This is discussed in Sec. 10.4.3.

Dust. Dust can be a problem for transparent and reflective optics as sources of scattering and absorption. In grazing incidence, in particular, the footprint of the surface shadowed increases as $(\sin \theta)^{-1}$. Estimates of Mie scattering and absorption for soft x-ray grazing telescopes put the maximum fractional area of dust coverage from 5×10^{-5} to 10^{-3}.[154,155] Dust can also act as sites for high-power laser damage.

Protective Strip Coatings. Total integrated scattering and x-ray photoelectron measurements indicate that strip coatings used to protect optical surfaces can often leave a residue of carbon contamination and particulate matter.[156]

10.4.3 Ionizing Radiation Damage

This section discusses creation and accumulation of isolated defects due to ionizing radiation in the bulk and on the surface of uv materials. The photon or particle must be energetic enough to cause bound electrons to be excited to the conduction band or free electron state. Equivalent isolated ionizing damage processes can occur through multiphoton transitions in laser fields. Laser damage effects resulting from increased absorption of transiently populated excited states or lattice heating and melting will be reserved for the next section. This section focuses on ionizing radiation damage effects in alkali halides and silica, where the effects have received the greatest attention. Radiation damage in a variety of materials has been reviewed by Williams and Friebele.[62]

The absorption of uv light or the inelastic scattering of energetic particles lead to the creation of electron-hole pairs that recombine in times from femtoseconds to microseconds depending on band structure and the density and cross section of recombination sites. High-energy excitation will lead to a cascade of electron-hole pairs as excited photo- and Auger electrons and holes produce secondaries. The process terminates when the hot electrons and holes no longer have the energy necessary to excite an electron-hole pair. The number of electron-hole pairs created by a photon of energy E is given by $E/2.8E_g$, where E_g is the bandgap. This relationship holds for many materials. Lattice displacements can occur from excited, localized states or from an incident particle imparting enough kinetic energy.[157]

Displacement damage further depends on bonding, direction, and lattice distortion and can lead to cascades if the displaced particle has sufficient energy.

Bulk. Bulk radiation damage can lead to increased absorption, luminescence, and dimensional changes. The damage phenomena are complicated with many different defect types whose concentrations vary with material, temperature, dose rate, and radiation quality.

Alkali halides are most susceptible to point defect damage. Williams and Friebele,[62] Williams,[158] and Tanimura and Itoh[159] have reviewed sources and forms of damage for alkali and alkaline earth halides. The primary, stable room temperature damage is F centers, which are electrons trapped at halide vacancies. The absorption band associated with F centers occurs in the normally transparent region from visible to the uv. In LiF, for example, the band is centered around 5.0 eV. The damage can often be annealed away at temperatures of 200 to 400 °C.

Radiation damage in silicate glasses has been reviewed by Williams and Friebele[62] and by Friebele.[75] The primary radiation induced defects leading to absorption in pure amorphous silica are the E′ center, a hole trapped at an intrinsic oxygen vacancy, and the E center, correlated with peroxyl radicals, Si-O-O. Absorption bands of these defects occur at 5.85 and 7.6 eV, respectively. Aluminum and alkali dopants or impurities in silicate glasses or quartz can increase their susceptibility to radiation, which produces color centers and a brownish tinge.[62] Fluorescence associated with the defects in pure silica occurs at energies of 1.9, 2.8, 4.3, and 6.7 eV. Such fluorescence can cause spurious signals and backgrounds for equipment using short-wavelength lasers and silica optics.

Dimensional changes in windows or lenses due to radiation damage can lead to wavefront distortion. For example, crystalline materials such as the alkali halides and quartz expand under prolonged neutron irradiation. At the limit, quartz can expand 14 percent in volume, while amorphous silica can contract 3 percent to the same equilibrium volume.[160]

There are only a few measurements of bulk uv radiation damage of thin film coatings. Most are associated with HR coatings for visible free-electron lasers. Elleaume et al.[161] and Velghe et al.[162] have measured damage in SiO_2/TiO_2 and SiO_2/Ta_2O_5 thin film coatings at the Orsay FEL; Bakshi et al.[163] have observed damage in SiO_2/TiO_2 coatings at the NSLS TOK undulator; and Early et al.[164] have measured He-Ne reflectivity losses in SiO_2/ZrO_2, Al_2O_3, Ta_2O_5 and HfO_2 thin film coatings on exposure to 248- and 351-nm laser radiation.

Surface. Ionizing radiation can cause ion desorption from various materials, notably alkali halides.[165,166] This can lead to an increased surface roughness and enhanced scattering and even a change of surface properties from insulating to metallic.

In ionizing radiation environments, volatile physisorbed compounds can decompose to molecules that are nonvolatile or chemisorbed. Carbon contamination, for example, can limit the performance of synchrotron radiation and free-electron laser optics.[161,163] It results from uv cracked hydrocarbons or carbon monoxide, always present in stainless-steel ultrahigh vacuum systems. Several studies indicate that carbon cracking is driven by primary and secondary photoelectrons and that the growth rate is reduced at elevated temperatures.[167,168] In situ methods of removing carbon contamination by oxygen glow discharges have been developed.[169,170]

Selective removal of fluorine from MgF_2 thin films under 250-keV α particle radiation drastically reduces the reflectance of MgF_2 overcoated Be in the visible.[171] This may present serious problems for fluoride-coated optics to be used in space.[172]

10.4.4 Laser Radiation Damage

Catastrophic radiation damage in intense laser fields is generally caused by linear absorption at surfaces, impurities, or defects. This absorption produces runaway heating, leading to lattice melting, vaporization, or thermally induced fracture. Intrinsic linear or nonlinear absorption in the bulk or at the interfaces can provide destructive lattice heating. The temperature rise is governed by host thermal conductivity, fluence, and laser pulse duration. At the laser damage threshold (LDT), melting and pitting with explosive plasma blowoff and crazed cracking are observed on surfaces. The complicated phenomena are material- and morphology-dependent. The discussion below is divided into separate considerations of mirrors, windows, and coatings.

The best source of information on laser damage and approaches to raising the LDT in optical materials are the proceedings of the NIST Boulder Laser Damage Symposia. Processes in absorptive and transparent materials have also been discussed in a number of Material Research Society symposia.

Mirrors. LDT values for absorptive materials range from 0.05 to 2.0 J/cm^2 in fluence and from 10^8 to 10^4 W/cm^2 in intensity. Bloembergen has summarized the processes for surfaces in absorptive materials.[173]

For pulse durations longer than about 1 ns, the heat diffusion length during the pulse is typically larger than the absorption length of the radiation, which is about 30 to 100 nm. The system is in quasi-steady state, and for longer pulses the heat diffuses away. In this case, the temperature rise is governed by the heat diffusion length l_d, which is given by

$$l_d = \sqrt{t_p\,\kappa} \tag{10.36}$$

where t_p is the pulse duration and κ is the thermal diffusivity. Thermal diffusivities for a number of window and mirror materials are given in Tables 10.1 and 10.2. The temperature rise ΔT is given approximately by

$$\Delta T \approx \frac{(1 - R)F}{l_d\,\rho C} \tag{10.37}$$

where F is the fluence. For more accurate estimation of ΔT, one would take into account variations of the absorptance and thermal constants with temperature.[174] As an example, a material which requires a fluence of 1 J/cm^2 to melt with a 100-ns pulse (10^7 W/cm^2) may require a fluence of 10 J/cm^2 with a 10,000-ns pulse (10^4 W/cm^2). At elevated temperatures, vaporized material takes away some of the heat, and at about 10^8 W/cm^2 ionized plasmas form and begin to absorb the radiation. For comparison, soft x-ray laser plasma sources are normally generated with laser intensities of 10^{12} to 10^{15} W/cm^2, and laser electric fields are comparable to atomic binding fields at about 10^{14} W/cm^2.

For laser pulses shorter than about 1 ns, the diffusion length becomes shorter than the absorption length of the radiation, and the pulse duration is no longer a factor. For these shorter pulses, the temperature rise is approximately given by

$$\Delta T \approx \frac{(1 - R)F\,\mu}{C} \tag{10.38}$$

Measurements of uv laser mirror damage are sparse, but the theory and measurements at longer wavelengths are in good agreement. One example is the measurements of Kurosawa et al., who measured the damage thresholds for laser cavity mirrors of Al, Si, Mo, SiO_2, and SiC in a Ar_2^* excimer laser operating at a wavelength of 126 nm and a pulse duration of 10 ns.[175]

Windows. Bulk transparent materials damage catastrophically for LDTs from 1 to 30 J/cm^2 and intensities above about 10^9 W/cm^2. The single-pulse, bulk damage mechanism in pure, bulk, crystalline, wide bandgap materials has been controversial, but it now seems clear that absorption proceeds through multiphoton absorption across the gap or at defect sites. Initial creation of electron-hole pairs is followed by single photon excitation and heating of the free carriers and energy transfer to the lattice by phonon creation.[176] Avalanche breakdown arising from acceleration of the free carriers by the electric field of the light and subsequent impact ionization, as in dc breakdown, appears to play a minor role at most.

Laser absorption in transparent materials is at a maximum where the electric field peaks, at interfaces and in defects such as cracks and voids. Subsurface defects caused by polishing can act as damage sites. Polishing techniques are discussed in Sec. 10.5. For pure, transparent windows, damage often occurs first on the exit surface where the field has an antinode. Surface damage phenomena in wide-gap materials has been reviewed by Reif.[177] A few sources of 2-photon absorption coefficients were given in Sec. 10.2.2.

The LDT can decrease with multiple subthreshold laser pulses. Clearly, this is due to permanent or relatively permanent defects created by multiphoton-generated electron-hole pairs, as mentioned in the previous section. At some level of defect accumulation, catastrophic laser absorption and thermal damage occur.

Coatings. Thin film LDTs are lower than bulk and can vary up to 20 percent even for coatings applied using the same materials and coating design. This is a result of impurity content and film morphology, which are sensitive to deposition conditions. Fabrication and morphology are examined in Sec. 10.5. The lower LDTs of thin films are partly due to lower thermal conductivities, which can be one to two orders of magnitude smaller than bulk values as a result of dendritic and columnar growth.[178,179] Energetic deposition techniques that increase packing density, such as ion-assisted deposition, can improve thermal conductivities. Water absorbed into the voids in thin films can also lower the LDT. Subthreshold conditioning of such films can desorb the water and raise operating levels. Damage can also be caused by diffusion along grain boundaries at temperatures roughly one-half of melting.[143] A variety of thermomechanical failure modes at elevated temperatures, outside of bulk modes already mentioned, have been discussed in Sec. 10.4.1. Rainer et al.[89] and Lowdermilk[90] have examined the LDT at 248 nm of a number of transparent single-layer AR and HR thin film uv coatings.

Increased LDTs in multilayer coatings can be achieved with reduced electric fields at graded index interfaces such as in sol-gel coatings[92] or rugate filters[180] or by moving the antinodes in non-$\lambda/4$ designs away from the interfaces, where various defects are likely to occur[181] and into the less absorbing, low-index layers. Single $\lambda/2$ coatings under multilayer coatings also seem to raise LDTs. In addition, durable single $\lambda/2$ coatings are often deposited over multilayer stacks for surface protection. Synthetic diamond films seem a viable candidate for lowered LDTs due to high thermal conductivity and melting point, but low thermal expansion may be a problem.[182]

10.5 FABRICATION

This section examines some techniques for fabricating uv window and mirror materials. Limited consideration is given to purifying methods to generate starting stock, although impurities clearly can be a performance-limiting factor. Many fabrication techniques apply equally well to visible and infrared optical materials, but wavelength-dependent problems of scattering, absorption, and LDT are often greater at shorter wavelengths and force greater care in obtaining pure, homogeneous low-scattering materials.

10.5.1 Bulk Windows and Mirrors

Material Quality.   Bulk quality is usually specified in terms of uniformity, impurity content, crystallinity, and defect concentration—quantities not all independent. Glasses typically have the indexes of refraction specified to within ±0.1 percent and dispersion to ±0.08 percent. Striae or index variations, bubble content, and annealing are stipulated as well. Impurity content can be a significant problem for laser optics, limiting the LDT. Platinum inclusions in phosphate laser glasses acquired from the glass melting crucibles made of platinum limited the LDT in the NOVA laser.[183] Metal microstructure determines toughness and strongly affects subsequent machining and polishing. Specifications on processing and heat treatment of standard metals, as well as machinability and impurity content, are available in Ref. 40.

Figuring and Polishing.   Optical flats and spheres have been made as accurately as 0.1 arcsecond or $\lambda/100$ to $\lambda/400$ figure accuracy, and 0.02-nm rms roughness, where λ is the 632.8-nm He-Ne laser wavelength. Aspheres have been made as accurately as about 1 arcsecond or $\lambda/4$ to 2λ figure accuracy and 0.5- to 1.0-nm rms surface roughness.

The primary techniques for form control and final finish are loose abrasive polishing, diamond turning, and plasma–ion beam erosion, and each will be discussed in turn. All the techniques depend on feedback from surface metrology to achieve final figure and finish.

Loose Abrasive Grinding and Polishing.   Abrasive polishing is the oldest and still most often used method of obtaining simply figured and finely finished optical surfaces. It lends itself to simple surfaces such as planes and spheres because it is an area-averaging technique. Subaperture tools or laps are needed for aspherics, and roll-off figure errors occur at part edges. Brown has reviewed the complex abrasive polishing processes.[184] For metals, standard polishing is a gouging. Final roughness height is roughly proportional to particle size and lap pressure and inversely proportional to the Young's modulus of the part and a complex pitch-penetration function. Microstructure of metals such as molybdenum can limit finish quality.[185] For glasses, chemistry is an integral part of the abrasive polishing process. For both metals and glasses, close fit of the lap and part is crucial to achieving superpolishes with roughness better than 0.5 nm rms. Relative lap and part rotation, slurries, and temperatures must be tightly controlled.

Generation of aspheres with abrasive polishing requires subaperture tools and computer control. Computer position control and predictable tool wear profile minimize figure errors in the difficult midspatial period range of 0.5 to 5 mm.[186]

In the last 15 years, several low lap pressure, submerged slurry methods have been developed that significantly reduce subsurface damage due to lap or tool pressure. However, surfaces that have been superpolished with standard abrasive techniques may also have satisfactory levels of subsurface damage. Subsurface damage can lead to lower LDTs or increased light scattering. Subsurface damage can often be detected by direct light-scattering techniques but may be missed entirely by contact profilometry. Low subsurface damage polishing techniques include the float polishing method of Namba and Tsuwa[187] and the elastic emission machining of Mori et al.[188,189] Recent improvements in the ductile grinding of glass, through precise control of grinding wheel penetration and other parameters, may lead to rapid fabrication of smooth, low subsurface damage optics, bypassing expensive polishing.[190,191]

Diamond Turning. Ultraprecision diamond machining of optics has been developed in the last 15 years.[192] Diamond turning generates surfaces directly as smooth as 25 nm rms and contours as accurate as $\lambda/10$ to 2λ in the visible, depending on machine, part size, and desired contour. However, it requires sharp diamond tool tips, air-bearing lathe spindles, interferometric position control, high thermal stability, and careful adjustment of computer controlled figure generation. Aspheres with high curvature and an axis of rotation that are difficult to do by other methods can be more easily achieved with diamond turning. Production rates are also faster than with labor-intensive grinding and polishing. Compatible materials are somewhat limited but include aluminum, copper, electroless nickel, zinc sulfide, cadmium telluride, alkali halides, and many plastics. Accuracies are sufficient for final form and finish in the infrared, but uv and soft x-ray optics require a critical, cosmetic final polish to remove the diamond turning marks.

Plasma-Ion Erosion. Plasma-assisted chemical[193] and ion[194] etching techniques remove materials without mechanical contact over a subaperture area that can be computer controlled to sweep selectively over a surface to generate the final contour. The techniques are only recently being used. They offer no roll-off part error, and minimal part heating. The ion erosion technique operates with higher-energy ions, so it may lead to subsurface damage. Both techniques promise high figure accuracy and finish and rapid fabrication, but midspatial period errors in the 0.5- to 5-mm range can be a problem.

Measurement of Figure and Finish. Process control and final quality of figure and finish are intimately connected with the sensitivity and consistency of surface curvature and roughness metrology. Bennett and Mattsson[106] and Franks et al.[195] have discussed the sensitivity of various surface metrology instrumentations. Takacs and Church have analyzed the figure, finish, and performance of a variety of synchrotron radiation mirrors.[196] They have observed that the height errors of polished mirrors measured as a power spectral density function increase approximately as the 4/3 power of the spatial period. Scattering theory has been reviewed in Sec. 10.3.1.

Instruments to determine figure and finish can be best characterized by height sensitivity and spatial period bandwidth. Figure 10.9 shows the ranges of peak-to-peak height and spatial period for several types of instruments. The diagonal dashed line drawn indicates a 1-arcsec slope error. Sinusoidal surface errors below the line have maximum slopes less than 1 arcsec, approximately the state of the art for aspheres. A vertical line could also be drawn at spatial periods of about 0.1 mm that separates the regions of geometrical scattering and diffractive scattering or figure and finish, as discussed in Sec. 10.3.1.

Depth sensitivity of the various roughness measuring techniques, relative to the sampling depth of the application wavelength, also may have to be considered owing to subsurface scattering. Scattering measurements at the wavelength of use are always preferable.

Figure. Standard optical figure tests during fabrication, including the simple knife-edge, Ronchi, and Hartmann tests and interferometry, have been reviewed in *Optical Shop Testing.*[197] Phase measuring interferometers[198] are commonly used

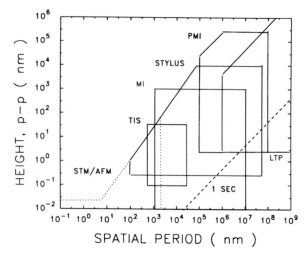

FIGURE 10.9 Approximate height vs. spatial period range of surface profilometers. For surface figure: LTP, long trace profilometer; PMI, 2D phase measuring interferometers. For surface finish: standard STYLUS; STM/AFM, scanning tunneling or atomic force microscopes; MI, microfocus interferometers; and TIS, total integrated scattering for He-Ne laser radiation (specular losses due to roughness over the indicated range). Stylus instruments are limited in slope by stylus tip radius. Dashed line indicates typical mirror figure specification or amplitude of sinusoidal waves for a given spatial period that yield a maximum slope error of 1 arcsecond, about the state of the art in nonspherical optics. For spatial periods less than the coherence length, scattering dominates geometrical errors and the mirror finish specification should be determined correspondingly. See text.

and have been developed to high degree with rapid acquisition of digitized two-dimensional contour maps, up to 10 by 10 cm in size with height accuracies of 10 to 20 nm, and spatial periods of 0.1 to 100 mm. Reference surfaces are needed, and complicated figures and grazing optics are not easily or accurately tested. Grazing incidence optics that are diffraction-limited in the visible can be coarsely tested by analyzing the deviations of the point spread function from the expected diffraction image of a point source. Becker and Heynacher have reported a mechanically contacting, three-axis coordinate measuring machine capable of measuring slope errors as small as 0.1 arcsec (10 nm over any 20 mm) for optics as large as $400 \times 600 \times 200$ mm³.[199] Recently, Takacs has developed a noncontacting, long-trace profilometer for testing grazing angle cylinders and other figures. This profilometer has a height sensitivity of 1.8 to 1 mm and spatial period bandwidth of 1 to 1000 mm. It is easy to use and at least as sensitive as phase measuring interferometers.[196]

Finish. Finish is characterized by the rms surface roughness or surface height vs. spatial period variations for spatial periods less than about 0.1 mm.

Instruments with the finest spatial wavelength range are based on stylus profilometers. Commercial roughness-measuring instruments can measure rms roughness as small as 0.1 nm and spatial wavelengths as small as the smallest tip radius, or about 100 nm. They have a typical travel range of about 100 mm. Scanning tunneling microscopes or atomic force microscopes can measure spatial wavelengths down to atomic levels on flat surfaces, but the piezoelectric crystals used for scanning limit spatial accuracy. Actual scattering at the wavelength of use may not agree with predictions based on theory and the very surface-sensitive, near-atomic-size stylus measurements because of additional subsurface scattering.

Commercially available noncontacting, microfocus interferometers (Wyco or Zygo) with reference surface or using the sample as a reference surface can measure surface roughnesses of 0.1 nm to 1 μm or 0.01 to 100 nm, respectively, and spatial wavelengths of 1 μm to 10 mm and 2 to 200 μm, respectively.

Total integrated scattering measurements collect scattered light and compare it to the incident flux. Bandwidth-limited rms surface roughness is then determined from Eq. (10.34). For a typical He-Ne laser instrument, the rms surface roughness sensitivity is 0.1 to 35 nm over a spatial period of 0.5 to 30 μm. Angle-resolved scattering instruments can selectively sample surface and subsurface damage, forming a two-dimensional image of the surface. In the infrared, the angle-dependent scattering is often referred to using the popular bidirectional reflectance distribution function (BRDF) nomenclature.

10.5.2 Coating Deposition

This section is concerned with physical or chemical thin film deposition, although the distinction blurs in several cases. Transmissive and reflective thin film filter design has been discussed in Secs. 10.2.4 and 10.3.4, respectively. Ionizing and high-power laser damage have been discussed in Secs. 10.4.3 and 10.4.4. Thin film deposition has been generally discussed in the handbook edited by Maissel and Glang[200] and the book edited by Vossen and Kern.[201]

Thin Film Morphology. The microstructure of thin vacuum deposited films is affected by film growth rate, substrate temperature, and the evaporant or sputtered atom energy, as well as materials and substrate preparation. Residual gases adsorbed on the substrate can also nucleate and alter growth. Since the development of high-vacuum techniques in the 1960s, quality and impurity levels in thin films have considerably improved.

The most common thin film nucleation and growth mode is by islands or the Volmer-Weber mode.[202,203] It leads to poly- or noncrystalline thin films. The Movchan-Demchishin structure zone model of thin film growth relates morphology most closely to substrate temperature. Because the energy of thermally evaporated atoms is less than 0.4 eV, such adatoms striking a substrate whose temperature is less than 0.45 of the deposited coating melting temperature will stick and not diffuse.[204,205] These adatoms will shadow subsequent atoms that arrive, leading to the growth of dendritic chains several atoms wide. As these low-temperature films grow thicker, the dendritic chains will coalesce; and the voids will move to the surface of the larger-diameter columns, with the packing density remaining at 80 to 90 percent. A hierarchy of columnar microstructures and nodules develops and eventually becomes large enough to be visible in optical microscopes. Lower substrate temperatures and faster growth lead to more amorphous films. At substrate temperatures higher than about 0.45 of the melting temperature of the thin film, adatoms will diffuse and the film will become crystalline.

The packing density and the LDT of coatings thermally deposited on low-temperature substrates are improved by energetic deposition techniques. These techniques enhance diffusion or atomic motion by increasing the energy of adatoms or with an additional beam of ions striking the substrate.[206] Depending on ion energy, there can be some ion penetration and damage to the surface. Bennett et al. have compared the optical properties and structure of TiO_2 thin films prepared by a variety of techniques including e-beam, ion-assisted, and rf and ion sputtered deposition as well as reactive evaporation, ion plating, and dip coating.[207] Techniques to achieve denser crystalline growth will be discussed with the deposition methods below.

Thermal Evaporation—Resistive and e-Beam. Thermal evaporation from resistively heated boats or filaments of refratory materials is the oldest method of vacuum coating. E-beam heating with a high-current arc of electrons in a magnetic field focused into the center of the source material is a more recent modification. E-beam evaporation eliminates contamination from the boat and can evaporate refractory materials. Both methods yield a stream of vapor with thermal energies less than 0.4 eV and deposition rates of 0.2 to 10 nm/s. Since the vapor pressure is a strong function of temperature, direct control of the deposition rate and thickness is normally achieved only within ± 5 percent. The large thermal mass and power heating circuits also lead to a slow rate-control response time of about 1 s. Multiple sources and shutters are often used. Deposition thicknesses are measured separately by quartz crystal thickness monitors or ionization gauge monitors. Because of the directionality of the evaporant sources, source-to-substrate distance is often somewhat large, and substrate motion is used to yield more uniform coatings.

Recently, higher packing density films on low-temperature substrates have been obtained with ion-assisted deposition. In this process, material for the coating is supplied by evaporation and the extra energy to compact the material, reduce voids, and improve adherence to the substrate is supplied by a simultaneous ion beam with energy from 50 to 1000 eV.[206,208,209] Ion plating, which consists of thermal evaporation with ions generated by an associated plasma discharge, can also produce higher-density films.

Thermal Evaporation—MBE. Molecular beam epitaxy (MBE) is basically slow thermal evaporation onto heated substrates. It is primarily used for semiconductor fabrication but deserves separate treatment because it can achieve control of crystal growth. MBE has been discussed by various authors.[210,211] MBE growth takes place in ultrahigh vacuum to minimize contamination and to permit electron diffraction and other diagnostics to analyze the growing crystalline surface. Growth rates are about one monolayer per second. Evaporation sources are stable, thermocouple-controlled Knudsen effusion cells. Shutters with 0.1 s actuation time can select among multiple sources. Uniformities as good as 0.5 percent across 100 mm have been achieved.

Sputtering. Sputtering deposition techniques have been discussed in Vossen and Kern.[201] A plasma discharge of inert gas leads to sputtering of target material. Adatom energies are about 10 eV tailing to 50 to 80 eV. This leads to high-density films and facilitates the use of low-temperature substrates. Modern magnetron and triode sources increase the deposition rate up to 1 nm/s. They also confine the plasma near the sputtering target and away from the substrate, thus minimizing mixing due to ion bombardment, which can affect material within 1 to 2 nm of the interface.[133] Deposition rate is determined by the discharge power, and thickness control can be as good as 0.1 nm. Alloys can be directly deposited, and multisource configurations are often used with the substrate moving past them. Thermally isolated substrate temperatures rise as high as 180 °C.

Pulsed Laser Deposition.   Pulsed laser deposition is a new technique still being developed. It has been reviewed by Cheung and Sankur.[212] It has advantages of energetic deposition with adatom energies as high as 100 eV, a point source, and good thickness control proportional to the total energy of a number of pulses. Drawbacks can include directionality of the vaporized stream and splattered material due to nonuniform vaporization.

Chemical—Solution.   Deposition of inorganic films from solutions has been discussed by Lowenheim.[201] This can be further divided into chemical reactions and electrochemical reactions. An example of the former is electroless nickel used for surfaces to be diamond-turned. Electroless nickel is really an alloy with P, B, or Cu depending on the autocatalytic chemical reduction. Electrodeposited or electroplated materials include a broad range of metals. Electrodeposited Au or Pt can form smooth mirror coatings of 100 nm thickness over polished conductive substrates.

Chemical—CVD.   Chemical vapor deposition (CVD) has been discussed by Kern and Ban.[201] Low-pressure CVD growth is normally regulated by reaction rates on the heated substrate surfaces. Condensation and evaporation from the supersaturated vapor proceed in equilibrium until a nucleus exceeds a given size. Lower temperatures and higher gas phase concentrations favor polycrystalline growth. Plasma-assisted CVD dissociates the vapor molecules away from the substrate providing high-energy molecular fragments and more latitude in substrate film formation.[213] High-quality Si and Si_3N_4 polycrystalline films are often grown by CVD at growth rates up to 0.3 nm/s. Recently, great strides are being made in the development of CVD growth of diamond thin films.[81,82]

10.6 REFERENCES

1. J. A. R. Samson, *Techniques of Vacuum Ultraviolet Spectroscopy*, Wiley, New York, 1967.

2. M. Born and E. Wolf, *Principles of Optics*, 3d ed., Pergamon Press, Oxford, 1965.

3. *American Institute of Physics Handbook*, McGraw-Hill, New York, 1982.

4. W. G. Driscoll (ed.), *Handbook of Optics*, McGraw-Hill, New York, 1978.

5. *Applied Optics and Optical Engineering*, vols. I–X, Academic Press, New York, 1965–1983.

6. *CRC Handbook of Laser Science and Technology*, vols. III, IV, V, CRC Press, Boca Raton, Fla., 1986.

7. L. Kissel and R. H. Pratt, in B. Crasemann (ed.), *Atomic Inner-Shell Processes*, Plenum Press, New York, 1985.

8. A. H. Compton and S. K. Allison, *X-rays in Theory and Experiment*, 2d ed., Van Nostrand, New York, 1935.

9. R. W. James, *The Optical Principles of Diffraction of X-Rays*, Ox Bow Press, Woodbridge, Conn., 1982.

10. D. Y. Smith, in E. D. Palik (ed.), *Handbook of Optical Constants of Solids*, Academic Press, New York, 1985.

11. D. W. Lynch, in E. D. Palik (ed.), *Handbook of Optical Constants of Solids*, Academic Press, New York, 1985.

12. Proceedings 5th International Conference on X-ray Absorption Fine Structure, *Physica*, B158, 1989.

13. V. B. Berestetskii, E. M. Lifshitz, and L. P. Pitaevskii, *Quantum Electrodynamics*, vol. 4 of *Course of Theoretical Physics*, 2d ed., Pergamon Press, Oxford, 1982, p. 211.

14. C. Kunz, in B. O. Seraphin (ed.), *Optical Properties of Solids—New Developments*, North-Holland, Amsterdam, 1976.

15. J. W. Cooper, *Phys. Rev.* vol. 128, p. 681, 1962.

16. U. Fano and J. W. Cooper, *Rev. Mod. Phys.*, vol. 40, p. 441, 1968.

17. J. L. Dehmer, A. C. Parr, and S. H. Southworth, in G. V. Marr (ed.), *Handbook of Synchrotron Radiation*, vol. 2, North-Holland, Amsterdam, 1987.

18. E. A. Stern and S. M. Heald, in E. E. Koch (ed.), *Handbook on Synchrotron Radiation*, vol. 1b, North-Holland, Amsterdam, 1983.

19. E. D. Palik (ed.), *Handbook of Optical Constants of Solids*, Academic Press, New York, 1985.

20. B. L. Henke, P. Lee, T. J. Tanaka, R. L. Shimabukuro, and B. J. Fujikawa, *Atom. Data Nucl. Data Tables*, vol. 27, no. 1, 1982.

21. B. L. Henke, J. C. Davis, E. M. Gullikson and R. C. C. Perera, *Lawrence Berkeley Laboratory Report* LBL-26259, 1988.

22. M. M. Thomas, J. C. Davis, C. J. Jacobsen, and R. C. C. Perera, *Nucl. Instrum. Methods*, vol. A291, p. 107, 1990.

23. N. K. Del Grande, P. Lee, J. A. R. Samson, and D. Y. Smith (eds.), "X-Ray and Vacuum Ultraviolet Interaction Data Bases, Calculations, and Measurements," *Proc. SPIE*, vol. 911, 1988.

24. D. Vaughan (ed.), *X-Ray Data Booklet*, Lawrence Berkeley Laboratory, 1986.

25. P. A. Temple, in E. D. Palik (ed.), *Handbook of Optical Constants of Solids*, Academic Press, New York, 1985.

26. W. R. Hunter, in E. D. Palik (ed.), *Handbook of Optical Constants of Solids*, Academic Press, New York, 1985.

27. D. M. Roessler, *Br. J. Appl. Phys.*, vol. 16, p. 1359, 1965.

28. D. E. Aspnes and A. A. Studna, *Phys. Rev. Lett.*, vol. 54, p. 1956, 1985.

29. D. E. Aspnes, in E. D. Palik (ed.), *Handbook of Optical Constants of Solids*, Academic Press, New York, 1985.

30. R. L. Johnson, J. Barth, M. Cardona, D. Fuchs, and A. M. Bradshaw, *Rev. Sci. Instrum.*, vol. 60, p. 2209, 1989.

31. J. Shamir, in E. D. Palik (ed.), *Handbook of Optical Constants of Solids*, Academic Press, New York, 1985.

32. T. W. Barbee, Jr., *Proc. SPIE*, vol. 911, p. 169, 1988.

33. U. Bonse, I. Hartmann-Lotsch, and H. Lotsch, in A. Bianconi, L. Incoccia, and S. Stipcich (eds.), *EXAFS and Near Edge Structure*, Springer-Verlag, Berlin, 1983, p. 376.

34. U. Bonse, H. Lotsch, and A. Henning, *J. X-Ray Sci. Tech.*, vol. 1, p. 107, 1989.

35. R. Tatchyn, I. Lindau, E. Källne, and E. Spiller, *Phys. Rev. Lett.*, vol. 53, p. 1264, 1984.

36. C. Kittel, *Introduction to Solid State Physics*, Wiley, New York, 1967.

37. L. H. Van Vleck, *Elements of Materials Science and Engineering*, Addison-Wesley, Reading, Mass., 1975.

38. R. W. G. Wyckoff, *Crystal Structures*, vols. 1–4, 2d ed., Interscience, New York, 1965.

39. Y. S. Touloukian and C. Y. Ho (eds.), *Thermophysical Properties of Matter*, vols. 1–13, IFI/Plenum, New York, 1979.

40. *Metals Handbook*, 9th ed., American Society for Metals, Metals Park, Ohio, 1978–1989.

41. R. E. Honig and D. A. Kramer, *RCA Rev.*, vol. 21, p. 360, 1960; 30, p. 285, 1969.

42. A. Smakula, *Z. Phys.*, vol. 59, p. 603, 1930.

43. W. B. Fowler, in W. B. Fowler (ed.), *Physics of Color Centers*, Academic Press, New York, 1968.

44. W. R. Hunter and S. A. Malo, *J. Phys. Chem. Sol.*, vol. 30, p. 2739, 1969.

45. M. Reilly, *J. Chem. Phys. Sol.*, vol. 31, p. 1041, 1970.

46. M. J. Dodge, in *CRC Handbook of Laser Science and Technology*, vol. IV, part 2, CRC Press, Boca Raton, Fla., 1986.

47. S. Singh, in *CRC Handbook of Laser Science and Technology*, vol. III, CRC Press, Boca Raton, Fla., 1986, p. 3.

48. J. M. Bennett and H. E. Bennett, in W. G. Driscoll (ed.), *Handbook of Optics*, McGraw-Hill, New York, 1978.

49. A. Feldman, in *CRC Handbook of Laser Science and Technology*, vol. IV, part 2, CRC Press, Boca Raton, Fla., 1986.

50. W. R. Cook, R. F. S. Hearmon, H. Jaffe, and D. F. Nelson, Piezooptic and Electrooptic Constants, *Landolt-Börnstein*, Springer-Verlag, New York, p. 495.

51. M. Sheik-Bahae, D. J. Hagan, and E. W. Van Stryland, *Phys. Rev. Lett.*, vol. 65, p. 96, 1990.

52. I. P. Kaminow, in *CRC Handbook of Laser Science and Technology*, vol. IV, CRC Press, Boca Raton, Fla., 1986, p. 253.

53. W. L. Smith, in *CRC Handbook of Laser Science and Technology*, vol. III, CRC Press, Boca Raton, Fla., 1986, pp. 229, 259.

54. A. J. Taylor, R. B. Gibson, and J. P. Roberts, *Opt. Lett.*, vol. 13, p. 814, 1988.

55. F. J. Reintjes, in M. Bass and M. L. Stitch (eds.), vol. 5, *Laser Handbook*, North-Holland, Amsterdam, 1985.

56. D. Chen, in *CRC Handbook of Laser Science and Technology*, vol. IV, CRC Press, Boca Raton, Fla., 1986, p. 287.

57. G. H. Sigel, in *Treatise on Materials Science and Technology*, vol. 12, Academic Press, New York, 1977, p. 5.

58. *Fluoride Glass Optics*, Academic Press, New York 1990.

59. C. F. Cline, D. D. Kingman, and M. J. Weber, *J. Non-Cryst. Sol.*, vol. 33, p. 417, 1979.

60. M. J. Weber, in *CRC Handbook of Laser Science and Technology*, vol. IV, CRC Press, Boca Raton, Fla., 1986, p. 5.

61. W. R. Hunter, *SPIE*, vol. 140, p. 122, 1978.

62. R. T. Williams and E. J. Friebele, in *CRC Handbook of Laser Science and Technology*, vol. III, CRC Press, Boca Raton, Fla., 1986, p. 229.

63. D. W. Lynch, in M. J. Weber (ed.), *CRC Handbook of Laser Science and Technology*, vol. IV, CRC Press, Boca Raton, Fla., 1986.

64. J. M. Bennett and A. T. Glassman, in *CRC Handbook of Laser Science and Technology*, vol. IV, CRC Press, Boca Raton, Fla., 1986.

65. R. P. Madden, *Physics of Thin Films*, vol. 1, 1963, p. 123.

66. O. R. Wood, II, P. J. Maloney, H. G. Craighead, and J. E. Sweeney, in *Basic Properties of Optical Materials*, NBS Special Publication 697, p. 181, 1985.

67. C. W. King and O. H. Nestor, *SPIE*, vol. 1047, p. 80, 1989.

68. K. L. Bedford, R. T. Williams, W. R. Hunter, J. C. Rife, M. J. Weber, D. D. Kingman, and C. F. Cline, *Phys. Rev.*, vol. B27, p. 2446, 1983.

69. M. J. Weber, *NBS Spec. Publ.*, 574 p. 3, 1980.

70. M. E. Innocenzi, R. T. Swimm, M. Bass, R. H. French, A. B. Villaverde, and M. R. Kokta, *J. Appl. Phys.*, vol. 67, p. 7542, 1990.

71. R. Brückner, *J. Non-Crystal. Sol.*, vol. 5, pp. 123, 177, 1970.

72. *Glass Science and Technology*, vols. I–IV, Academic Press, Boston, vol. V, American Ceramics Society, 1991.

73. H. R. Philipp, in *Handbook of Optical Constants of Solids*, Academic Press, New York, 1985, p. 749.

74. A. Appleton, T. Chiranjivi, and M. Jafaripour-Ghazvini, in S. T. Pantelides (ed.), *The Physics of SiO$_2$ and Its Interfaces*, Pergamon Press, New York, 1978, p. 94.

75. E. J. Friebele, in D. R. Uhlman and N. J. Kreidl (eds.), *Glass: Science and Technology*, vol. V, American Ceramics Society, 1991.

76. H. R. Philipp, in E. D. Palik (ed.), *Handbook of Optical Constants of Solids*, Academic Press, New York, 1985.

77. C. Chen, Y. X. Fan, R. C. Eckardt, and R. L. Byer, *SPIE*, vol. 681, p. 12, 1986.

78. D. Eimerl, L. Davis, S. Velsko, E. K. Graham, and A. Zalkin, *J. Appl. Phys.*, vol. 62, p. 1968, 1987.

79. M. Seal and W. J. P. van Enckevort, *SPIE*, vol. 969, p. 144, 1988.

80. D. F. Edwards and H. R. Philipp, in *Handbook of Optical Constants of Solids*, Academic Press, New York, 1985, p. 665.

81. J. C. Angus and C. C. Hayman, *Science*, vol. 241, p. 913, 1988.

82. W. A. Yarbrough and R. Messier, *Science*, vol. 247, p. 688, 1990.

83. *Physics of Thin Films*, vols. I–XIV, Academic Press, New York, 1963–1989.

84. H. A. MacLeod, *Thin Film Optical Filters*, American Elsevier, New York, 1986.

85. J. A. Dobrowolski, *Handbook of Optics*, McGraw-Hill, New York, 1978.

86. J. D. Rancourt, *Optical Thin Films*, Macmillan, New York, 1987.

87. L. M. Cook and S. E. Stokowski, in W. J. Weber (ed.), *CRC Handbook of Laser Science and Technology*, vol. IV, CRC Press, Boca Raton, Fla., 1986.

88. V. R. Costich, in W. J. Weber (ed.), *CRC Handbook of Laser Science and Technology*, vol. V, CRC Press, Boca Raton, Fla., 1987.

89. F. Rainer, W. H. Lowdermilk, D. Milam, C. K. Carniglia, T. T. Hart, and T. L. Lichtenstein, *Appl. Opt.*, vol. 24, p. 496, 1985.

90. W. H. Lowdermilk, *SPIE*, vol. 541, p. 124, 1985.

91. G. DeBell, L. Mott, M. von Genten, *SPIE*, vol. 895, p. 254, 1988.

92. W. H. Lowdermilk, in *CRC Handbook of Laser Science and Technology*, vol. V, CRC Press, Boca Raton, Fla., 1987, p. 431.

93. A. Malherbe, *Appl. Opt.*, vol. 13, pp. 1275, 1276, 1974.

94. L. M. Cook and S. E. Stokowski, in *CRC Handbook of Laser Science and Technology*, vol. IV, CRC Press, Boca Raton, Fla., 1986, p. 93.

95. U. Wojak, U. Czarmetzki, and H. F. Dobele, *Appl. Opt.*, vol. 26, p. 4788, 1987.

96. F. R. Powell, P. W. Vedder, J. F. Lindblom, and S. F. Powell, *Opt. Eng.*, vol. 29, p. 614, 1990.

97. J. B. Kortright and J. H. Underwood, *Nucl. Instrum. Methods*, vol. A291, p. 272, 1990.

98. C. K. Malek, J. Susini, A. Madouri, M. Ouahabi, R. Rivoira, F. R. Ladan, Y. Lepetre, and R. Barchewitz, *Opt. Eng.*, vol. 29, p. 597, 1990.

99. W. B. Westerveld, K. Becker, P. Zetner, J. J. Corr, and J. W. McConkey, *Appl. Opt.*, vol. 24, p. 2256, 1985.

100. J. C. Rife, R. N. Dexter, P. M. Bridenbaugh, and B. W. Veal, *Phys. Rev.*, vol. B16, p. 4491, 1977.

101. J. C. Rife and J. F. Osantowski, *SPIE*, vol. 315 p. 103, 1981.

102. B. E. Newnam, *NBS Spec. Pub.*, 746, p. 261, 1988.

103. D. R. Gabari and D. L. Shealy, *Opt. Eng.*, vol. 29, p. 641, 1990.

104. O. I. Tolstikhin and A. V. Vinogradov, *Appl. Phys.*, vol. B50, p. 213, 1990.

105. M. L. Scott, in R. W. Falcone and J. Kirz (eds.), *OSA Proc. on Short Wavelength Coherent Radiation Generation and Applications*, vol. 2, 1988, p. 322.

106. J. M. Bennett and L. Mattsson, *Introduction to Surface Roughness and Scattering*, Optical Society of America, Washington, D.C., 1989.

107. J. M. Bennett, H. H. Hurt, J. P. Rahn, J. M. Elson, K. H. Guenther, M. Rasigni, and F. Varnier, *Appl. Opt.*, vol. 24, p. 2701, 1985; H. H. Hurt and J. M. Bennett, *Appl. Opt.*, vol. 24, p. 2712, 1985.

108. H. Hogrefe R-P. Haelbich, and C. Kunz, *Nucl. Instrum. Methods*, vol. A246, p. 198, 1986.

109. E. Marx and V. Vorburger, *Appl. Opt.*, vol. 29, p. 3613, 1990.

110. M. Born and E. Wolf, *Principles of Optics*, 3d ed., Pergamon Press, Oxford, 1965, p. 520.

111. R. K. Smither, *Nucl. Instrum. Methods*, vol. A291, p. 286, 1990.

112. S. F. Jacobs, in *Applied Optics and Optical Engineering*, vol. X, Academic Press, New York, 1987, p. 71.

113. O. Lindig and W. Pannhorst, *Appl. Opt.*, vol. 24, p. 3330, 1985.

114. S. F. Jacobs and D. Bass, *Appl. Opt.*, vol. 28, p. 4045, 1989.

115. R. Haug, A. Klass, W. Pannhorst, and E. Rodek, *Appl. Opt.*, vol. 28, p. 4052, 1989.

116. S. Heald, *SPIE*, vol. 315, p. 79, 1981.

117. H. W. Babcock, *Science*, vol. 249, p. 253, 1990.

118. J. H. Weaver, C. Krafka, D. W. Lynch, and E. E. Koch, *Physik Daten, Physics Data, Optical Properties of Metals*, vol. 18–1, Fach-information zentum, Karlsruhe, 1981.

119. D. L. Windt, W. C. Cash, Jr., M. Scott, P. Arendt, B. Newnam, R. F. Fisher, and A. B. Swartzlander, *Appl. Opt.*, vol. 27, p. 246, 1988.

120. D. L. Windt, W. C. Cash, Jr., M. Scott, P. Arendt, B. Newnam, R. F. Fisher, A. B. Swartzlander, P. Z. Takacs, and J. M. Pinneo, *Appl. Opt.*, vol. 27, p. 279, 1988.

121. G. Hass, J. B. Heaney, and W. R. Hunter, *Physics of Thin Films*, vol. 12, p. 1, 1982.

122. M. A. Pickering, R. L. Taylor, J. T. Keeley, and G. A. Graves, *Nucl. Instrum. Methods*, vol. A291, p. 95, 1990.

123. S. Sato, A. Iijima, S. Takeda, M. Yanaghihara, T. Miyahara, A. Yagashita, T. Koide, and H. Maezawa, *Rev. Sci. Instr.*, vol. 60, p. 1479, 1989.

124. R. A. M. Keski-Kuha, J. F. Osantowski, H. Herzig, J. S. Gum, and A. R. Toft, *Appl. Opt.*, vol. 27, p. 2815, 1988.

125. M. Yanagihara, M. Niwano, T. Koide, S. Sato, T. Miyahara, Y. Iguchi, S. Yamaguchi, and T. Sasaki, *Appl. Opt.*, vol. 25, p. 4586, 1986.

126. K. Platzoder and W. Steinmann, *J. Opt. Soc. Am.*, vol. 58, p. 588, 1968.

127. L. R. Canfield, G. Hass, and W. R. Hunter, *J. Phys*, vol. 25, p. 124, 1964.

128. D. E. Aspnes, E. Kinsbron, and D. D. Bacon, *Phys. Rev.*, vol. B21, p. 3290, 1980.

129. T. R. Gull, H. Herzig, J. F. Osantowski, and A. R. Toft, *Appl. Opt.*, vol. 24, p. 2660, 1985.

130. J. Rife and J. Osantowski, *J. Opt. Soc. Am.*, vol. 70, p. 1513, 1980.

131. T. Parsonage, *Laser Focus*, June 1990, p. 77.

132. E. Spiller, *Mat. Res. Soc. Symp.*, vol. 56, p. 419, 1986.

133. T. W. Barbee, Jr., *Opt. Eng.*, vol. 25, p. 898, 1986.

134. J. C. Rife, W. R. Hunter, T. W. Barbee, Jr., and R. G. Cruddace, *Appl. Opt.*, vol. 28, p. 2984, 1989.

135. J. C. Rife, T. W. Barbee, Jr., W. R. Hunter, and R. G. Cruddace, *Phys. Scr.*, vol. 41, p. 418, 1990.

136. R. G. Cruddace, T. W. Barbee, Jr., J. C. Rife, and W. R. Hunter, *Phys. Scr.*, vol. 41, p. 396, 1990.

137. S. F. Jacobs, *Applied Optics and Optical Engineering*, vol. X, Academic Press, New York, 1987, p. 71.

138. M. F. Doerner and W. D. Nix, *CRC Crit. Rev. Solid State Mater. Sci.*, vol. 14, p. 225, 1988.

139. H. K. Pulker, *SPIE*, vol. 325, p. 84, 1982.

140. H. K. Pulker, *Coatings on Glass*, Elsevier, Amsterdam, 1984.

141. D. Tabor, in J. M. Blakely (ed.), *Surface Physics of Materials*, vol. 2, Academic Press, New York, 1975, p. 475.

142. M. Martensson, A. Stenborg, O. Bhorneholm, A. Nilsson, and J. N. Andersen, *Phys. Rev. Lett.*, vol. 60, p. 1731, 1988.

143. S. Nakahara, *SPIE*, vol. 346, p. 39, 1982.

144. C. Weaver, *Physics of Thin Films*, vol. 6, p. 301, 1971.

145. A. L. Greer and F. Spaepen, in L. L. Chang and B. C. Giessen (eds.), *Synthetic Modulated Structures*, Academic Press, New York, 1985, p. 419.

146. W. R. Hunter and R. T. Williams, *Nucl. Instrum. Methods*, vol. 222, p. 359, 1984.

147. M. Piecuch, *Rev. Phys. Appl.*, vol. 23, p. 1727, 1988.

148. L. V. Knight, J. M. Thorne, A. Toor, and T. W. Barbee, Jr., *Rev. Phys. Appl*, vol. 10, p. 1631, 1988.

149. *Binary Alloy Phase Diagrams*, vols. I, II, ASM, Metals Park, Ohio, 1986.

150. W. G. Moffatt, *The Handbook of Binary Phase Diagrams*, Genium Publishing Corp., 1986.

151. E. Ziegler, Y. Lepetre, I. K. Schuller, and E. Spiller, *Appl. Phys. Lett.*, vol. 48, p. 1354, 1986.

152. M. L. Scott, P. N. Arendt, B. J. Cameron, J. M. Saber, and B. E. Newnam, *Appl. Opt.*, vol. 27, p. 1503, 1988.

153. L. R. Canfield, R. G. Johnston, and R. P. Madden, *Appl. Opt.*, vol. 12, p. 1611, 1973.

154. L. Van Speybroeck, *Dust*, Smithsonian Astrophysical Observatory Internal Memorandum, 1987.

155. P. Slane, E. R. McLaughlin, D. A. Schwartx, L. P. Van Speybroeck, J. W. Bilbro, and B. H. Nerren, *SPIE*, vol. 1113, p. 12, 1989.

156. J. M. Bennett, L. Mattsson, M. P. Keane, and L. Karlsson, *Appl. Opt.*, vol. 28, p. 1018, 1989.

157. E. Sonder and W. A. Sibley, *Point Defects in Solids*, chap. 4, Plenum Press, New York, 1972.

158. R. T. Williams, *Opt. Eng.*, vol. 28, p. 1024, 1989.

159. K. Tanimura and N. Itoh, *Nucl. Instrum. Methods*, vol. B46, p. 207, 1990.

160. A. Franks and M. Stedman, *Nucl. Instrum. Methods*, vol. 172, p. 249, 1980.

161. P. Elleaume, M. Velghe, M. Billardon, and J. M. Ortega, *Appl. Opt.*, vol. 24, p. 2762, 1985.

162. M. F. Velghe, M. E. Couprie, and M. Billardon, *Nucl. Instrum. Methods*, vol. A296, p. 666, 1990.

163. M. H. Bakshi, M. A. Cecere, D. A. G. Deacon, and A. M. Fauchet, *Nucl. Instrum. Methods*, vol. A296, p. 677, 1990.

164. J. Early, V. Sanders, and W. Lemon, *NIST Spec. Pub.*, 775, p. 233, 1989.

165. N. Itoh, *Nucl. Instrum. Methods Phys. Res.*, vol. B27, p. 155, 1987.

166. R. F. Haglund, Jr., *SPIE*, vol. 895, p. 182, 1988.

167. K. Boller, R.-P. Haelbich, H. Hogrefe, W. Jark, and C. Kunz, *Nucl. Instrum. Methods*, vol. 208, p. 273, 1983.

168. R. A. Rosenberg and D. C. Mancini, *Nucl. Instrum. Methods*, vol. A291, p. 101, 1990.

169. E. D. Johnson and R. F. Garrett, *Nucl. Instrum. Methods*, vol. A266, p. 381, 1988.

170. T. Koide, T. Shidara, K. Tanaka, A. Yagishita, and S. Sato, *Rev. Sci. Instr.*, vol. 60, p. 2034, 1989.

171. M. H. Mendenhall and R. A. Weller, *Appl. Phys. Lett.*, vol. 57, p. 1712, 1990.

172. R. Cowen, *Sci. News.*, vol. 138, p. 276, 1990.

173. N. Bloembergen, *Mat. Res. Soc. Symp.*, vol. 51, p. 3, 1985.

174. M. Sparks and E. Loh, Jr., *J. Opt. Soc. Am.*, vol. 69, pp. 847, 859, 1979.

175. K. Kurosawa, W. Sasaki, M. Okuda, Y. Takigawa, K. Yoshida, E. Fujiwara, and Y. Kato, *Rev. Sci. Instr.*, vol. 61, p. 728, 1990.

176. S. C. Jones, P. Braulich, R. T. Casper, X. Shen, and P. Kelly, *Opt. Eng.*, vol. 28, p. 1039, 1989.

177. J. Reif, *Opt. Eng.*, vol. 28, p. 1122, 1989.

178. A. H. Guenther and J. K. McIver, *SPIE*, vol. 895, p. 246, 1988.

179. J. C. Lambropoulos, M. R. Jolly, C. A. Amsden, S. E. Gilman, M. J. Sinicropi, and D. Diakomihalis, *J. Appl. Phys.*, vol. 66, p. 4230, 1989.

180. M. Zukic and K. H. Guenther, *SPIE*, vol. 895, p. 271, 1988.

181. D. H. Gill, B. E. Newman, and J. McLeod, *NBS Spec. Publ.*, 509, p. 206, 1977.

182. S. Albin, A. Cropper, L. Walkins, C. E. Byvik, A. M. Buoncristiani, K. V. Ravi, and S. Yokota, *SPIE*, vol. 969, p. 186, 1988.

183. J. S. Hayden, D. L. Sapak, and A. J. Marker III, *SPIE*, vol. 895, p. 176, 1988.

184. N. J. Brown, *Ann. Rev. Mat. Sci.*, vol. 16, p. 371, 1988.

185. J. M. Bennett, S. M. Wong, and G. Krauss, *Appl. Opt.*, vol. 19, p. 3562, 1989.

186. G. Hull-Allen and G. Gunnarsson, *Photonics Spectra*, March 1990, p. 147.

187. Y. Namba and H. Tsuwa, *Ann. CIRP*, vol. 27, p. 511, 1978.

188. Y. Mori, T. Okuda, K. Sugiyama, and K. Yamaguchi, *J. Jap. Soc. Precis. Eng.*, vol. 51, p. 1033, 1985.

189. Y. Higashi, T. Meike, J. Suzuki, Y. Mori, K. Yamauchi, K. Endo, and H. Namba, *Rev. Sci. Instr.*, vol. 60, p. 2120, 1989.

190. T. G. Bifano, T. A. Dow, and R. O. Scattergood, *SPIE*, vol. 966, p. 100, 1988.

191. K. L. Blaedel, P. J. Davis, and D. J. Nikkel, *Opt. Eng. OE Repts.*, December 1990, p. 7.

192. G. M. Sanger and S. D. Fantone, in M. J. Weber, (ed.), *CRC Handbook of Laser Science and Technology*, vol. V, CRC Press, Boca Raton, Fla., 1987.

193. L. D. Bollinger and C. B. Zarowin, *SPIE*, vol. 966, p. 82, 1988.

194. S. R. Wilson, D. W. Reicher, and J. R. McNeil, *SPIE*, vol. 966, p. 74, 1988.

195. A. Franks, B. Gale, and M. Stedman, *SPIE*, vol. 830, p. 2, 1989.

196. P. Z. Takacs and E. L. Church, *Nucl. Instrum. Methods*, vol. A291, p. 253, 1990.

197. D. Malacara (ed.), *Optical Shop Testing*, Wiley, New York, 1978.

198. M. Schaham, *SPIE*, vol. 306, p. 183, 1981.

199. K. Becker and E. Heynacher, *SPIE*, vol. 733, p. 149, 1986.

200. L. I. Maissel and R. Glang (eds.), *Handbook of Thin Film Technology*, McGraw-Hill, New York, 1970.

201. J. L. Vossen and W. Kern (eds.), *Thin Film Processes*, Academic Press, New York, 1978.

202. R. W. Vook, *SPIE*, vol. 346, p. 2, 1982.

203. J. A. Venebles, G. D. T. Spiller, and M. Hanbucken, *Rep. Prog. Phys.*, vol. 47, p. 399, 1984.

204. K. H. Guenther, D. J. Smith, and L. Bangjun, *SPIE*, vol. 678, p. 2, 1986.

205. R. P. Netterfield, *SPIE*, vol. 678, p. 14, 1986.

206. P. J. Martin and R. P. Netterfield, in E. Wolf (ed.), *Progress in Optics*, vol. XXIII, 1986, p. 113.

207. J. M. Bennett, E. Pelletier, G. Albrand, J. P. Borgogno, B. Lazarides, C. K. Carnigilia, R. A. Schmell, T. H. Allen, T. Tuttle-Hart, K. H. Guenther, and A. Saxer, *Appl. Opt.*, vol. 28, p. 3303, 1989.

208. D. VanVechten, G. K. Hubler, E. P. Donovan, and F. D. Correll, *J. Vac. Sci. Technol.*, vol. A8, p. 821, 1990.

209. G. K. Hubler, D. VanVechten, E. P. Donovan, and C. A. Carosella, *J. Vac. Sci. Technol.*, vol. A8, p. 831, 1990.

210. E. H. C. Parker, (ed.), *The Technology and Physics of Molecular Beam Epitaxy*, Plenum Press, New York, 1985.

211. C. T. Foxon, *CRC Crit. Rev. Solid State Mater. Sci.*, June 1981, p. 235.

212. J. T. Cheung and H. Sankur, *CRC Rev. Solid State Mater. Sci.*, vol. 15, p. 63, 1988.

213. S. M. Ojha, *Physics of Thin Films*, vol. 12, Academic Press, New York, 1982, p. 237.

CHAPTER 11
OPTICAL MATERIALS: VISIBLE AND INFRARED

W. J. Tropf, T. J. Harris, and M. E. Thomas

11.1 INTRODUCTION

This chapter presents intrinsic properties of optical materials useful in the visible and infrared spectral regions. Materials discussed include insulators, semiconductors, and metals. Crystalline, polycrystalline, and amorphous (glassy) materials are considered. Only materials with a reasonably complete and consistent set of property data are included. Applications for these materials include bulk optics (windows, lenses, prisms, beamsplitters, etc.), optical fibers, mirrors, and thin film coatings (spectral and neutral density filters, protective and antireflective coatings).

Both commonly used and relatively new optical materials are included. Emphasis is on data, formulas, and references. Physical, thermal, mechanical, and optical properties are given. Optical properties are summarized as parameters for index of refraction and absorption coefficient formulas that allow calculation of these quantities over broad spectral regions. Room-temperature properties are emphasized. The data are complemented with a short section on the origin of properties (especially optical) that includes a descriptive explanation of the underlying physics.

Previous compilations of optical property data are given in Refs. 1 to 12. Extensive references are also provided.

11.2 TYPES OF MATERIALS

Materials used for transparent optical elements (e.g., windows, lenses, beamsplitters, fibers, nonlinear crystals) are essentially fully dense and pure. Fully dense means the material is essentially at theoretical (or x-ray) density. Significant deviations from full density (e.g., <99.9 percent of theoretical density) mean that voids in the material are sufficiently large and numerous to make the material highly scattering, hence translucent. High purity is necessary to ensure low absorption throughout the intrinsic transmission region. (Intentional impurities are used to make devices such as lasers, or to add color to make broadband filters.)

Transparent materials may be classified in several ways. The materials may be crystalline or noncrystalline. Crystalline materials include both single crystals and polycrystalline forms. Single-crystal materials, particularly those with high isotopic purity, are used in applications requiring very low scatter, very low absorption (high purity), substrates for epitaxy, or in devices making use of birefringence (wave-mixing crystals, acoustooptical and polarization devices). Polycrystalline optical materials are used in applications requiring strength or the need to manufacture into near-final-form shape (e.g., by hot pressing or deposition).

Noncrystalline optical materials include glasses and plastics. Glass is the most commonly used optical material, accounting for over 90 percent of the optical elements produced. The term glass is applied to a material that retains an amorphous state upon solidification. More accurately, glass is an undercooled, inorganic liquid with a very high room-temperature viscosity. The gradual change of viscosity with temperature is characterized by several temperatures below the melting point, especially the glass transition temperature and the softening point temperature. The transition temperature defines a second-order phase transition; the material is a stable glass below this temperature.

Under proper conditions, glass can be formed from many different inorganic mixtures. Several hundred different optical glasses are commercially available. Primary glass-forming compounds include oxides, halides, and chalcogenides, with the most common being the oxides of silicon, boron, and phosphorus. Varying the composition varies the glass properties, especially refractive index and dispersion. The continuous variation of these properties makes glass suitable for making achromatic optical systems and gradient index devices.

Plastics have several important properties that make them attractive for visible applications. These advantages include low cost (through ease of manufacture and assembly), low weight (half that of glass), and high impact strength. Disadvantages are low tolerance to heat and abrasion, high temperature coefficients of the refractive index (80 to 150 $\times$ 10^{-6}/K), greater index of refraction inhomogeneity (an order of magnitude greater than glass), and the inability to adjust index of refraction easily by varying composition. The most important optical plastic is PMMA or poly{methyl methacrylate} (also called plexiglas or lucite) because of its superior weathering and scratch resistance. Polycarbonate and polystyrene are also important optical plastics. Polystyrene also can be copolymerized to make optical plastics such as SAN poly{styrene-*co*-acrylonitrile}) or NAS (poly{styrene-*co*-methacrylate}).

Optical materials can also be classified as insulators, semiconductors, and conductors (metals). Insulators have few free carriers and high electrical resistivity (10^{14} to 10^{22} Ω-cm) while metals have very low resistivity ($\sim 10^{-6}$ Ω-cm) and large free carrier concentration. Semiconductors have an intermediate resistivity (10^{-2} to 10^9 Ω-cm) and can be characterized as having a modest bandgap (<3 eV). (On the other hand, some materials are called large-bandgap semiconductors (diamond, AlN, SiC) because of their structural similarity to more conventional semiconductors.) Typical insulators are the oxides and halides (including glasses) while elemental or binary semiconductors are typically formed of elements from the IV, III to V, and II to VI columns of the periodic table. Semiconductors used for optical elements typically have high purity (are intrinsic or undoped) and high resistivity. A bandgap of 0.5 eV or greater is desired for bulk transmissive elements at room temperature because thermally-populated conduction electrons in low-bandgap materials produce excessive absorption.

11.3 APPLICATIONS

11.3.1 Windows and Optical Elements

Materials for windows and optical elements must be made with sufficient purity and uniformity to have low absorption and scatter and good index of refraction homogeneity. For most low-power window applications, the absorption and scatter attenuation coefficients typically need to be less than 0.1 cm^{-1} (and known down to 10^{-3} cm^{-1}). Transparency for long-length optical fibers (1-km length) and high-power laser windows requires low-level absorption coefficients, i.e., well below 10^{-3} cm^{-1}. For low-power-loss applications, higher-order intrinsic processes and extrinsic impurities and defects become important.

Uniformity of the refractive index throughout an optical element is a prime consideration in selecting materials for high-performance lenses, elements for coherent optics, laser harmonic generation, and acoustooptical devices. In general, highly-pure single crystals achieve the best uniformity, followed by glasses (especially those selected by the manufacturer for homogeneity), and lastly polycrystalline materials.

Practical manufacturing techniques limit the size of optics of a given material (glasses are typically limited by the moduli, i.e., deformation caused by the element's weight). Some manufacturing methods, such as hot pressing, also produce significantly lower-quality material (especially when the thinnest dimension is significantly increased). Cost of finished optical elements is a function of size, raw material cost, and the difficulty of machining, polishing, and coating the material. Any one of these factors can dominate cost.

References 8 to 10 contain general information on the manufacturing methods for glasses and crystalline materials. Information on cutting and polishing of optical elements can be found in Refs. 2, 13, and 14.

11.3.2 Optical Fibers

Several material types are available as optical waveguides: glasses, plastics, crystalline materials, and reflective hollow tubes. Glass fibers are inexpensive and operate in the visible and near infrared (0.4 to 2.3 μm). Very inexpensive plastic fibers (e.g., PMMA or polystyrene) are available for visible wavelengths. Relatively new fluoride-glass fibers are now available for infrared applications up to 6 or 7 μm. Chalcogenide glass fibers have the longest infrared cutoff wavelength (up to 20 μm or more). For short-wavelength operation, silica fibers are available. (BeF$_2$ glass offers very short wavelength capability but is not commercially available because it is very hygroscopic and highly toxic.)

Fibers used for long-distance data transmission require low loss (attenuation coefficients less than 10^{-5} cm^{-1}) and low dispersion. Both the lowest loss and minimum dispersion wavelengths lie in the middle of the transparent region between the electronic transitions (and scatter contributions to loss) and the lattice vibrations, and fortunately are not very different. For pure silica fibers, the minimum loss is at 1.6 μm (but in practice is closer to 1.5 μm owing to impurity absorption). The minimum dispersion wavelength (defined as $d^2n/d\lambda^2 = 0$) of silica is at 1.272 μm. The location of these ideal wavelengths has spurred the development of laser diodes operating at 1.3 to 1.5 μm.

Longer-wavelength fibers offer the potential of lower attenuation. As the infrared absorption edge moves to longer wavelengths, the minimum loss wavelength (intersection of Rayleigh scatter loss and multiphonon absorption) has lower attenuation. The minimum loss in the best silica-glass fibers is 0.3 dB/km at 1.5 μm. Heavy-metal-fluoride glass fibers have the potential of less than 10^{-3} dB/km loss at 3.5 μm.

Single crystals make excellent fibers because of low scatter, typically one-tenth that of amorphous materials. High-purity single crystals (e.g., the alkali halides) also have very low absorption and low dispersion, especially in the infrared, making such materials attractive for low-loss fibers. Fibers of Al_2O_3, alkali and silver halides, and KRS-5 are commercially available for infrared applications.

Short-distance fiber applications do not have the loss and dispersion constraints of data transmission. Typical noncommunication applications include directing laser or lamp energy, illuminating or viewing hard-to-reach locations, and energy coupling (i.e., from image plane to detectors). "Ordered" (or "coherent") fiber bundles are used to couple image planes and have been made to magnify (or reduce) image linear dimensions tenfold. Transmission of fibers is improved by cladding with material of different index or by gradual changes in index ("gradient index") with radius. Fibers with specially shaped, birefringent cores to maintain polarization are also available.

Further information on optical fibers and fiber materials can be found in Refs. 15 and 16.

11.3.3 Coatings, Filters, and Mirrors

Spectral filters can be made in several ways. The long- and short-wavelength transmission limits of bulk optical components can be used to limit spectral response. Typical glasses and crystalline materials can be modified by adding various "dopants" or impurities that modify lattice vibrations or form color centers or second phases that modify the spectral response of the base material. Organic dyes, in a gelatin, plastic, vinyl, or other host material, are other examples.

Thin-film interference filters, made of layers of different index materials, are particularly important for precision applications. Layer thickness is accurately controlled to create the desired spectral response. Such a filter can be designed for very narrow bandpasses (i.e., laser line filters with half widths < 0.01 percent of center wavelength) and for broadband applications (e.g., 20 percent bandpass) that may require sharp spectral cutoffs. In interference filters, the absorption characteristics of the substrate are important for blocking higher harmonics of the filter bandpass. Interference coatings are also used to reduce reflection, either to enhance optical transmission (e.g., of a lens or window) or to increase absorption (e.g., as a light trap or to increase efficiency of a semiconductor detector).

A number of references detail the theory of interference filters,[1,17] and several computer tools are available for their design and analysis. For many applications, manufacturers have stock filters that are suitable. Custom filters can also be designed and made by a filter manufacturer to customer specification to satisfy unique requirements.

Gratings or prisms can be used as tunable spectral filters (i.e., monochromators) as can interferometers (including etalons) and acoustooptical devices. Resonant filters have also been developed for use in the infrared.

Neutral density filters are made in several ways: coated as broadband multilayer transmission filters, as attenuating filters (e.g., partially developed silver halide emulsion or other absorbing material in a thin film or bulk matrix), or as a partially coated surface (e.g., using evaporated metal). (Variable neutral density filters also can be achieved using crossed polarizers.)

Coatings also can be used to protect the optical components. Coatings may be applied to increase scratch resistance (hardness), strength, and resistance to moisture or other environmental effects (e.g., to prevent oxidation). In the past decade, carbon coatings have been developed to produce good-quality, ultrahard diamond films.

Metals are used as coatings to produce highly reflective mirrors for use as optical elements. The large free-carrier concentration of metals results in a high plasma frequency and therefore high reflectivity for the visible throughout the far-infrared and microwave regions. Aluminum is the most commonly used metal for mirror coatings. It is usually overcoated with MgF_2 or a silicon oxide to protect the aluminum from oxidation or scratching. Gold offers very high reflectance in the infrared. Solid copper mirrors are sometimes used when heat dissipation is required.

Magnesium and beryllium are used as lightweight mirror blanks (beryllium also has higher elastic modulus and lower thermal expansion compared to magnesium, making it attractive for some applications despite high cost and manufacturing difficulties). Invar or other low-expansion alloys are also used as low-expansion mirror bases or structural elements where thermal expansion must be minimized (i.e., for laser cavities or etalons).

Most mirror blanks are glass or ceramic. Some are made of a low-expansion material such as fused silica or a specially formulated low-expansion ceramic (e.g., Zerodur, Cer-Vit, or ULE fused silica).

11.3.4 High-Power Devices

Heating of optical elements by high-intensity light creates special problems. Heating of elements by absorption, and the subsequent thermal expansion and change in refractive index can significantly affect optical performance. A typical high-power beam has a large radial gradient in light intensity, resulting (through absorption) in a radial temperature profile that creates an effective gradient index in lenses or distorts the surface of mirrors. This effect is called "thermal lensing," and its effects differ from those of uniformly raising the temperature of an optical system. The fundamental approach to alleviate thermal lensing effects is selection of materials with very low absorption coefficients. If significant absorption heating remains, high-thermal-conductivity materials can reduce thermal gradients, and materials with low expansion and low (or negative) index of refraction temperature coefficients are possible approaches to minimizing thermal lensing effects.

Materials used for high-power and nonlinear optics are often subject to very intense light. These materials are frequently characterized by a "damage threshold," the power density that damages the crystal. Damage usually begins with the ionization of a single molecule, typically a contaminant, defect, or inclusion. The damage threshold is usually expressed in power density, and typical values range from 0.01 to 10 GW/cm^2.

11.4 MATERIAL PROPERTIES

11.4.1 Origin of Properties

Material properties arise from composition, structural order, and bonds. Because all properties arise from the same basic factors, they are highly interrelated. It is unrealistic to develop a set of desired properties and then attempt to discover a material that meets them. Rather, one needs to understand the origin of the properties, and design around the available range of the interrelated properties of real materials.

Composition includes consideration of the size and mass of the atoms, the electronic structure of the atoms (including chemical valence), and stoichiometry (ratio of the constituent atoms). Structural order factors include the bonding arrangements (short-range order) and the long-range order (i.e., whether a material is crystalline or amorphous). The number of direction-dependent components in a property of a crystalline material depends on the particular crystalline structure.

The chemical bonds have great influence over properties. Three basic bond types are recognized: covalent, ionic, and metallic. Covalent bonding is moderately strong and directional. Diamond, silicon, and germanium are prototypical covalent compounds. The characteristics of these prototypical covalent materials belie the normal covalent bond; the nonfilled outer shells of these materials allow electron overlap without infrared vibrational excitations. This charge overlap creates an attractive interaction, that, combined with high coordination number and symmetry of these materials, creates a very strong binding force. In other covalent materials, the electronic overlap and coordination are reduced; hence the bonds become weaker (and also have some ionic character).

Alkali halides are representative ionic bonded compounds. Ionic bond strength arises from the electrostatic attraction between positive and negative ions. The strength of the bond is highly dependent on the interatomic distance (i.e., on the atomic number of the constituents, particularly the negative ion). Strong nonmetallic bonds manifest themselves in large bandgaps, high melting point, high strength and hardness, and high-frequency lattice vibrations. Metallic bonds are characterized by variable strength and a large number of free (conduction) electrons (one or more per atom). Alkali metals have relatively weak bonds because of large interatomic distance.

Symmetry of crystalline materials gives rise to anisotropic (tensor) properties. For some crystals, there is significant difference between the property components, and knowledge of this difference may be required in analyzing the performance of the material. Index of refraction is a second-rank tensor. Usually the index is given along the principal lattice directions. With this convention, cubic (and isotropic) materials are characterized by one index of refraction, tetragonal and hexagonal (uniaxial) materials by two independent indexes of refraction, and all other (biaxial) materials by three indexes. For uniaxial materials, the following nomenclatures for refractive index are equivalent:

Optical nomenclature	E-field nomenclature
Ordinary ray (n_0, n_ω, or ω)	E-field perpendicular to c axis ($n\perp$)
Extraordinary ray (n_e, n_ϵ, or ϵ)	E-field parallel to c axis ($n\|$)

Properties of polycrystalline and glassy materials are usually treated as isotropic, greatly simplifying characterization, particularly the stress-strain relationships and optical characteristics dependent on stress (or strain).

Properties can be characterized as intrinsic or extrinsic. Intrinsic properties are those characteristic of the pure, stoichiometric, defect-free, ideal material. Extrinsic properties are those that are attributable to impurities (dopants), composition changes, and defects. Many optical materials are intentionally modified to produce extrinsic properties. Examples of deliberate material modification include the addition of dopants (impurities) to create laser transitions in host materials, to color materials, to modify bandgaps in conductors in semiconductors, to strengthen materials, or to change crystalline phase (e.g., cubic zirconia). Stoichiometry can be modified in the manufacture of materials, adding defects (e.g., color centers) that modify the electronic structure and add additional lattice vibrations, hence increase optical absorption. Even the most carefully prepared material will exhibit some extrinsic effects. Properties first to be affected include optical absorption, strength, and thermal conductivity.

11.4.2 Optical Properties

Intrinsic optical properties of a material are determined by three basic physical processes: free carriers, lattice vibrations, and electronic transitions.[5,6,18,19] However, the dominant physical process depends on the material and spectral region of interest. All materials have contributions to the complex index of refraction from electronic transitions. Insulators and semiconductors also require the characterization of the lattice vibrations (or phonons) to fully understand the optical properties. Semiconductors are additionally influenced by free-carrier effects. The strength of free-carrier effects depends on the carrier concentration; thus it is the dominant process determining the optical properties of metals in the visible and infrared.

In the range of transparency of a bulk material, more subtle processes such as multiphonon processes, impurity and defect absorption, and scattering become important. Intrinsic atomic (Rayleigh) scattering is a very weak effect but is important in long-path optical fibers and uv-transparent materials. Extrinsic scattering, caused by density (local composition) variations, defects, or grains in polycrystalline solids, is typically much larger than Rayleigh scattering in the visible and infrared spectral regions. Impurity and defect (electronic or vibrational) absorption features can be of great concern depending on the spectral region, incident radiation intensity, and material temperature.

Both intrinsic and extrinsic optical properties are represented by a complex index of refraction $\hat{n}^{\dagger}$ (which is the square root of the complex relative permittivity or dielectric constant $\hat{\epsilon}_r$):

$$\hat{n} = n + ik = \sqrt{\hat{\epsilon}_r} \qquad i = \sqrt{-1} \qquad (11.1)$$

where n (the real index of refraction) is physically related to the (phase) velocity of light in a material by $n = c/v$ (where c is the speed of light in a vacuum and v the speed of light in the material). The (real) index of refraction is the most fundamental optical characteristic. In transparent materials, the index determines the refractivity, surface reflectivity, and optical path length. The imaginary part of the complex index primarily determines the absorption coefficient. The absorption coefficient β is related to k (called the index of extinction) by

$$\beta = 4\pi\nu k \qquad (11.2)$$

---

†This definition of $\hat{n}$ assumes a harmonic field of the form $e^{-i\omega t}$. Other definitions use a field of the form $e^{i\omega t}$. These conventions lead a different form for the complex index, i.e., $\hat{n} = n - i \cdot k$. Care must be taken to assure sign consistency in all resulting equations, especially those from different sources (which may use a different convention). See Ref. 20 for more information on sign conventions.

where ν is frequency (in wavenumbers). In regions of high absorption (e.g., $k >$ 0.1), the imaginary part of the index of refraction begins to contribute to reflectivity and refraction.

The temperature T and frequency dependence of the real part of the index of refraction $n(\nu,T)$ is determined by the dominant physical processes previously mentioned. Figure 11.1a illustrates the frequency dependence for an insulating polar crystal. Minor contributions near the infrared edge of transparency come from multiphonon transitions. The value of $n(\nu,T)$ is essentially the sum of the strengths of all electronic and lattice vibration resonances, and is dominated by those with fundamental oscillation frequencies above ν. Figure 11.1a also indicates regions of validity for the popular Sellmeier model [see Sec. 11.5.4 and Eq. (11.11)].

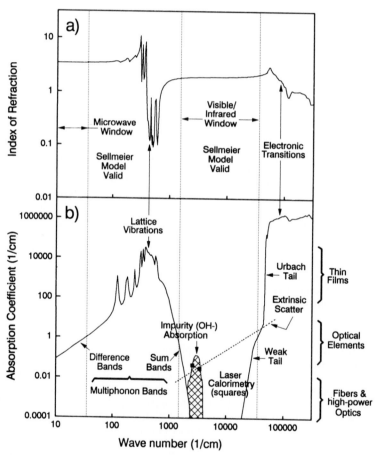

FIGURE 11.1   The index of refraction (a) and the absorption coefficient (b) for a typical insulator (yttrium oxide). The microwave and visible and infrared transparent regions are shown by vertical dotted lines (1 cm^{-1} absorption coefficient criterion). These transparent regions are separated by the lattice vibrations, and high-frequency transparency is terminated by the electronic transitions. Attenuation in the visible and infrared is dominated by scatter and impurity absorption. Tolerable attenuation for a variety of applications is shown on the right side of the figure.

Temperature dependence of the real index of refraction is attributable to two effects: thermal expansion and the temperature dependence of the polarizability.[21] Positive thermal expansion (usual case) decreases the refractive index while a positive change in polarizability (usual case) increases refractive index. Materials with high thermal expansion (and weak bonding, low melting point, e.g., alkali halides) have negative dn/dT while materials with low expansion (strong bonds, high melting point, e.g., oxides) have positive dn/dT. Since the magnitude of the polarizability increases near the electronic and infrared resonances, dn/dT increases (becomes less negative or more positive) at the edges of transparent region.

Temperature and frequency dependence of the imaginary part of the index of refraction $k(v,T)$ is more involved and requires consideration of not only the dominant physical processes but also higher-order processes, impurities, and defects as illustrated in Fig. 11.1b. The spectral regions of the fundamental resonances are opaque. Transparent regions for insulators are divided into two regions: microwave, and visible-infrared. For a nonconducting thin film (thickness on the order of 1 μm), microwave (or far-infrared) transparency lies below the minimum active transverse optical vibrational mode frequency, and the visible and infrared transparent region extends from the maximum longitudinal optical vibrational mode frequency up to the material electronic bandgap. Transparency of a bulk insulator (typical thickness 0.1 to 1 cm) is determined by the absorption edges associated with these three processes. Microwave transparency ends in the region where absorption arises from two-phonon difference bands and the one-phonon red wing (the low-frequency side of the absorption band). Infrared transparency of a bulk solid typically begins in the region of three-phonon sum bands (essentially the third harmonic of the fundamental infrared-active, or one-phonon, vibrations). Visible-ultraviolet transparency ends at the "Urbach tail," which can be phenomenologically interpreted as thermal fluctuations in the bandgap energy.[18]

Transmission, reflection, and emission characteristics of a material are directly determined from the complex index of refraction. For a medium with no surface roughness and no bulk scatter, the total power law, a statement of conservation of energy, is given by

$$\tau_s + \rho_s + \alpha_{abs} = 1 \qquad (11.3)$$

where τ_s is the specular transmittance, ρ_s is the specular reflectance, and α_{abs} is the absorbance. These specular terms can be expressed as a function of the complex index of refraction. For an infinite slab of thickness L, these terms are, ignoring interference effects,

$$\tau_s = \frac{[1 - R(\theta,v)]^2 \exp(-\beta(v)L/\cos\theta_a)}{1 - R^2(\theta,v)\exp(-2\beta(v)L/\cos\theta_a)} \qquad (11.4)$$

$$\rho_s = R(\theta,v) + \frac{R(\theta,v)[1 - R(\theta,v)]^2 \exp(-2\beta(v)L/\cos\theta_a)}{1 - R^2(\theta,v)\exp(-2\beta(v)L/\cos\theta_a)} \qquad (11.5)$$

and

$$\alpha_{abs} = \frac{[1 - R(\theta,v)][1 - \exp(-\beta(v)L/\cos\theta_a)]}{1 - R(\theta,v)\exp(-\beta(v)L/\cos\theta_a)} \qquad (11.6)$$

where θ_a is the angle from the normal to the ray propagating inside the material and $\beta(v)$ is the absorption coefficient. The magnitude of the single-surface Fresnel power reflection coefficient for unpolarized light $R(\theta,\nu)$ is a function of the angle of incidence, wavenumber, and index of refraction, and is given by

$$R(\theta\nu) = \tfrac{1}{2}(R_s + R_p) \tag{11.7}$$

$$R_s = \left| \frac{\hat{n}_0\cos(\theta) - \hat{n}_a\cos(\theta_a)}{\hat{n}_0\cos(\theta) + \hat{n}_a\cos(\theta_a)} \right|^2 \tag{11.8}$$

where

$$R_p = \left| \frac{\hat{n}_a\cos(\theta) - \hat{n}_0\cos(\theta_a)}{\hat{n}_a\cos(\theta) + \hat{n}_0\cos(\theta_a)} \right|^2 \tag{11.9}$$

Here θ is the angle of incidence, θ_a is the refracted angle inside the material (found from the complex form of Snell's law), $\hat{n}_0$ is the complex index of refraction in the region outside the material (usually $\hat{n}_0 \approx 1$), and $\hat{n}_a$ is the complex index of refraction of the material. R_s and R_p are the linear polarization components. If the light is polarized, the net reflection coefficient must be appropriately modified.

Kirchhoff's law states that the absorptance of a material is equal to the emittance. For a smooth, flat surface and no bulk scattering the emittance (or emissivity) ϵ is

$$\epsilon(\Omega_i,\nu,T) = \alpha_{\text{abs}}(\Omega_i,\nu,T) \tag{11.10}$$

where Ω_i is the angular position of an observer.

11.5 PROPERTY DATA TABLES

Property data for many optical materials are given in Tables 11.1 through 11.7. The tables contain physical, mechanical, thermal, and optical properties of selected optical materials. Materials selected for the data tables include both well-known, widely used, and relatively new substances. We have chosen to group similar materials together for ease of comparing potential candidates and conveniently accessing all properties for a given application. Much care has been exercised to ensure that the property data are for optical-grade material (fully dense, highly pure, and as close to intrinsic as possible) and that the property data are consistent and reasonable. Looking at the tables, one can see that there is wide variation in the quality and completeness of the property data. We emphasize optical properties and give appropriate formula constants and extensive references. Previous compilations of optical property data are given in Refs. 1 to 4, 11, and 12.

We have, in the main, selected materials that are well characterized. Of the multitude of optical materials used for all purposes, few are satisfactorily characterized to meet the needs of all applications.

11.5.1 Physical Properties

All materials are characterized by name for identification, a chemical formula (crystalline materials) or approximate composition (glasses, amorphous substances), and a density (ρ, in g/cm^2). Crystalline materials are further identified by crystal class, space group, unit cell lattice parameters (a, b, and c), molecular weight (of a formula unit in amu), and number of formula units per unit cell (Z).

(See Refs. 22 and 23 for compilations of crystallographic data.) These terms allow calculation of theoretical density. Cold water solubility (in g per 100 g water) is also given for most crystalline materials.[24] Characteristic temperatures are also given: primarily melting and phase change temperatures for crystals[24] and glass transition and softening temperatures for amorphous (glassy) materials.[9,25,26] A high melting temperature indicates strong bonding.

Crystals are completely identified by both the chemical formulation and the space group. Chemical formulation alone is insufficient for identification because many formulations have several different structures (called polymorphs) that have different properties. Properties in the data table pertain only to the specific structure listed. Materials in the data tables having several stable polymorphs at room temperature include GeO_2, SiO_2 (eight polymorphs), C (diamond, graphite, and amorphous forms), and ZnS (cubic and hexagonal). Often literature property values do not adequately identify the material. Space group, lattice constant, and other crystallographic data are given in Refs. 22 and 23.

The space group also identifies the appropriate number of independent terms that describe a physical quantity. Noncubic materials require two or more values to fully describe thermal expansion, thermal conductivity, refractive index, and other properties. Often, scalar quantities are given in the literature when a tensor characterization is needed. Such a characterization may be adequate for polycrystalline materials but is unsatisfactory for single crystals that require directional properties.

Glasses are identified by traditional names characteristic of their composition and their dispersion relative to their index of refraction. Loosely, crown glasses have low dispersion [Abbe number (see below), $v_d > 50$] and flint glasses have high dispersion ($v_d < 50$). A more specific identifier is a six-digit number representing the first three digits of ($n_d - 1$) and the first three digits of v_d (defined in military standard MIL-G-174). Each manufacturer also has its own designator that uniquely identifies each glass. For example, the glass with code 517624 has the following manufacturers' designations:

Manufacturer	Designation for glass 517624
Schott	BK-7
Corning	B-16-64
Pilkington	BSC-517642
Hoya	BSC-7
O'Hara	BSL-7 (glass 516624)

Properties of glass are primarily determined from the composition but also depend on the manufacturing process, specifically the thermal history. In general, both composition and processes are proprietary, so compositions given are only illustrative. For detailed work, the manufacturer can supply a "melt data sheet" providing accurate optical properties for a specific glass lot. To reduce the effect of thermal history, optical glasses are annealed. Annealing removes stress and reduces index of refraction variations.

Since glasses can be viewed as supercooled liquids, they are characterized by a gradual softening with temperature, and a hysteresis between glass and crystalline properties. The glass transition temperature is the approximate point of transition from liquid to a glassy state. Above this glass transition temperature, the glass may phase separate or crystallize. The glass transition temperature is typically two-thirds of the melting (or liquidus) temperature.[27]

Viscosity at the glass transition temperature is approximately 10^{13} poise. Glasses should be annealed just under this temperature and kept well below (e.g., 100 to 150 K below) the transition temperature to avoid distortion. A second important temperature is the softening temperature, usually defined as having a viscosity of $10^{7.6}$ poise. This is the minimum temperature for forming optical elements; at this temperature glass quickly deforms under its own weight.

11.5.2 Thermal Properties

The principal thermal properties given in the tables are room-temperature heat capacity, Debye temperature, thermal conductivity, and room-temperature thermal expansion. The high-temperature oxide materials are also characterized by the temperature derivative of thermal expansion and total hemispherical emissivity. The latter, while an optical property, is most often used for radiative transfer calculations and is therefore included in the thermal property tables.

Heat Capacity. The constant-pressure heat capacity [c_p, J/(g·K), a scalar property] is the energy needed to raise the temperature of a unit mass of a substance by one Kelvin. Above the Debye temperature (see below), the heat capacity of crystalline materials usually rises slowly with temperature. The data tables give room-temperature (300 K) heat capacity.[4,28,29]

Debye Temperature. The Debye temperature (Θ_D, Kelvins) is a measure of the maximum lattice vibrational frequency and can be used to model the low-temperature heat capacity. At temperatures above the Debye temperature, heat capacity rises slowly, primarily thermal expansion. A high Debye temperature is indicative of strong bonds, hence high melting temperature, low thermal expansion, high strength, and elastic moduli.[30] The average infrared vibrational frequency of solids is proportional to the Debye temperature.[31] Debye temperature is not applicable to amorphous materials.

Thermal Conductivity. Thermal conduction (second rank tensor) in a material [κ, W/(m·K)] is the rate of heat flow between opposite sides of an infinite slab with a 1 Kelvin temperature difference. This property is especially important in relieving thermal stress and optical distortions caused by rapid heating or cooling. For most nonmetallic crystalline substances, thermal conductivity increases proportional to heat capacity near 0 K, reaches its highest value at a low temperature (usually <100 K), and falls gradually thereafter (nominally as T^{-1}). Therefore, thermal conductivity for crystalline materials is given for several temperatures. Generally, amorphous materials have lower thermal conductivity and much less temperature dependence.

Thermal conductivity is highly dependent on purity and order. Mixed crystals, second-phase inclusion, nonstoichiometry, voids, and defects can all lower the thermal conductivity of a material. Values given in the data tables are for the highest-quality material. A compilation of thermal conductivity data is given in Ref. 28 and a review of theory and data in Ref. 32.

TABLE 11.1a Physical Properties of Selected Oxides

Material	Crystal structure and space group	Unit cell lattice parameters, Å	Formulas per unit cell, Z	Molecular weight, amu	Density, g/cm³	Melting point,* K	Cold water solubility, g/100 g
ALON (5AlN·9Al$_2$O$_3$) (aluminum oxynitride spinel)	Cubic Fd3m (227)	7.948	1	1122.59	3.71	$T_m = 2323$ (2438 ± 15)	insol.
Beryllium oxide (BeO) (Bromellite)	Hexagonal P6mc (186)	$a = 2.693$ $c = 4.395$	2	25.01	3.01	$T_p = 2373$ $T_m = 2843$	0.00002
Calcium carbonate (CaCO$_3$) (Calcite)	Hexagonal R3c (167)	$a = 4.9898$ $c = 17.060$	6	100.09	2.7102	$T_p = 323$ $T_m = 1612$	0.0014
Germanium dioxide (GeO$_2$) (low germania)	Tetragonal P4$_2$/mnm (136)	$a = 4.396$ $c = 2.863$	2	104.59	6.30	$T_m = 1359$	insol.†
Sapphire (Al$_2$O$_3$) (aluminum oxide, α-alumina, Corundum)	Hexagonal R3c (167)	$a = 4.759$ $c = 12.989$	6	101.96	3.987	$T_m = 2319$	insol. (0.000098)
Spinel (MgAl$_2$O$_4$) (magnesium aluminum spinel)	Cubic Fd3m (227)	8.084	8	142.27	3.577	$T_m = 2408$	insol.

TABLE 11.1a (Continued)

Material	Crystal structure and space group	Unit cell lattice parameters, Å	Formulas per unit cell, Z	Molecular weight, amu	Density, g/cm^3	Melting point,* K	Cold water solubility, $g/100\ g$
Magnesium oxide (MgO) (Periclase)	Cubic Fm3m (225)	4.2117	4	40.30	3.584	$T_m = 3073$	0.00062
Fused silica (SiO_2) (vitreous silica, fused quartz)	Amorphous	—	—	60.08	2.202	$T_m = 1983$	insol.
α-Quartz (SiO_2) (natural quartz, rock crystal)	Hexagonal $P3_121$ (154)‡	$a = 4.9136$ $c = 5.4051$	3	60.08	2.648	$T_p = 860$ $T_m = 1986$	insol.
Titanium dioxide (TiO_2) (rutile)	Tetragonal P4/mmm (136)	$a = 4.5937$ $c = 2.9618$	2	79.88	4.245	$T_m = 2128$	insol.
YAG ($Y_3Al_2[AlO_4]_3$) (yttrium aluminum aluminate)	Cubic Ia3d (230)	12.008	8	593.7	4.55	$T_p = 2193$ $T_m = 2223$	insol.
Yttrium oxide (Y_2O_3) (yttria)	Cubic Ia3 (206)	10.603	16	225.81	5.033	$T_p = 2640$ $T_m = 2710$	0.00018
Cubic zirconia (ZrO_2:Y_2O_3) (typically 12% yttria)	Cubic Fm3m (225)	5.148	4	(121.0)	5.891	$T_m = 3000$	insol.

*Notation: T_m is the melting temperature; T_p is the phase change temperature.
†A water-soluble (hexagonal) form of GeO_2 exists.
‡Right-handed quartz; left-handed quartz is P3₁21 (152).

11.14

TABLE 11.1b Mechanical Properties of Selected Oxides

Material	Moduli, GPa			Poisson's ratio	Strength, MPa		Knoop hardness, kg/mm²	Fracture toughness, MPa·m$^{1/2}$	Elastic constants Refs.
	Elastic	Shear	Bulk		Tension	Flexure			
ALON	317	128	203	0.24	220	310	1800–1900	1.4	*
Beryllium oxide	395	162	240	0.23	140	275	1250		3, 35, 36
Calcite	83	32	73.2	0.31			75–135		3, 35, 36
Germanium oxide	378	150	259	0.26					35
Sapphire									
∥ to c axis	400	162	250	0.23	900	1200	2000–2250	3.0–5.0	3, 4, 35, 36
60° to c axis	450				310	500			
⊥ to c axis	360								
Spinel	275	109	197	0.28_5	105	170	1500–1800	1.2–1.9	3, 35, 36
Magnesium oxide	195	75	160	0.29_5		130	640–690		3, 4, 35, 36
Fused silica	72.6	31.2	36	0.16_4		110	635	0.8	†
α-Quartz	95	44	38	0.08			955		3, 4, 35, 36
∥ to c axis							900		
⊥ to c axis							1000		
Titanium dioxide	293	115	215	0.27			880		3, 35, 36
YAG	283	113	186	0.25			1350		3, 35
Yttrium oxide	173	67	145	0.30	100	140	650–750	0.7	‡
Cubic zirconia	232	88.6	205	0.29	200		1150	2.0	35, 49

*ALON stiffness and compliance estimated from the engineering moduli: $c_{11} = 393$ GPa, $c_{12} = 108$ GPa, $c_{44} = 119$ GPa, $s_{11} = 2.89$ TPa^{-1}, $s_{12} = -0.62$ TPa^{-1}, $s_{44} = 8.40$ TPa^{-1}.

†Fused silica elastic constants given in Ref. 48 as $c_{11} = 77.6$ GPa, $c_{44} = 31.2$ GPa, and therefore $c_{12} = 2c_{11} - 2c_{44} = 15.2$ GPa, $s_{11} = 13.77$ TPa^{-1}, $s_{12} = -2.26$ TPa^{-1}, $s_{44} = 32.05$ TPa^{-1}.

‡Yttria stiffness and compliance estimated from the engineering moduli: $c_{11} = 233$ GPa, $c_{12} = 101$ GPa, $c_{44} = 67$ GPa, $s_{11} = 5.82$ TPa^{-1}, $s_{12} = -1.76$ TPa^{-1}, $s_{44} = 14.93$ TPa^{-1}.

TABLE 11.1c Thermal Properties of Selected Oxides

Material	Heat capacity, J/g·K	Debye temperature, K	Thermal conductivity, W/m·K				Thermal expansion (1/K) at 300 K		Total hemispherical emissivity*		
			250 K	300 K	500 K	1000 K	$\alpha \cdot 10^6$	$d\alpha/dT \cdot 10^9$	300 K	500 K	1000 K
ALON	0.830			12.6	7.0	5.8	5.66	3.86	0.66	0.67	0.40
Beryllium oxide	1.0285	1280	420	350	200	80	6.32	6.05			
|| to c axis							7.47	2.51			
⊥ to c axis							5.64	8.08			
Calcite	0.8220						(4.0)				
|| to c axis			6.2	5.4	(4.2)		8.8				
⊥ to c axis			5.1	4.5	(3.4)		−2.8				
Germanium oxide	0.5341						4.51	4.65			
|| to c axis							1.61	8.01			
⊥ to c axis							7.01	5.57			
Sapphire	0.777	1030	58	46	24.2	10.5	6.77	3.97			
|| to c axis				(35)			7.15	3.88	0.53(c||s)		
⊥ to c axis				(33)			6.65	3.79	0.54(c⊥s)		

Spinel											
Single crystal	0.8191	850	(30)	(25)	9–10	5–6	6.97	3.36	0.61	0.62	0.35
Polycrystalline				16							
Magnesium oxide	0.9235	950					10.6	4.6?			
Single crystal			73	59	32	13					
Polycrystalline			58	48	27	9.7					
Fused silica	0.7458	—	1.28	1.38	1.62	2.87	0.51	0.30	0.76		
α-Quartz	0.7458	470					10.49	24.12			
∥ to c axis		(572)	12.7	10.4	6.0	—	6.88	18.89			
⊥ to c axis			7.5	6.2	3.9	—	12.38	26.50			
Titanium dioxide	0.6910	760	9.3	8.4	5.9	—	7.56	4.75			
∥ to c axis			11.8	10.4	(8.0)	—	8.97	6.07			
⊥ to c axis			8.3	7.4	(5.5)	—	6.86	4.06			
YAG	0.625	754	16.7	13.4	9.2		7.7				
Yttrium oxide	0.4567	465	(30)	13.5	6.4	4.5	6.56	2.66	0.59	0.50	
Cubic zirconia	0.46	563		1.8	1.9	2.1	10.23 (20–1500°C)				0.23

*0.5-cm nominal thickness.

11.17

TABLE 11.1d Optical Properties of Selected Oxides

Material	Bandgap, eV	Transparent regions* Optical, μm	Transparent regions* Microwave, cm⁻¹	Maximum ν_{LO}	Refractive index Optical n_x	Refractive index DC n_o	Refractive index DC $dn_o/dT \cdot 10^4$	References
ALON	6.2	0.23–4.8	<19	969	1.771	3.063	2.2	19, 50, 51
Beryllium oxide	10.2	0.21–[3.5]	<30					53
o-ray				1095	1.709	2.71		52–54
e-ray				1085	1.724	2.77		52–54
Calcite	(7)							
o-ray		0.24–2.2	<11.5	1549	2.6	2.942		55–57
e-ray		0.21–3.3	<10	890	2.0	2.850		55–57
Germanium oxide	5.6	0.3–5.0						
o-ray				852	2.15	4.6		58
e-ray				816	2.02	4.1		58

Material								
Sapphire								
o-ray	8.8	0.19–5.0	<30	914	1.7555	3.19		59
e-ray	8.7	0.19–0.52	<28	871	1.7478	3.067		19, 57, 59–62
						3.408		57, 59–63
Spinel	≈8	0.2–5.3	<21	869	1.701	2.88	2.5	19, 50, 64
Magnesium oxide	7.8	0.35–6.8	<40	725	1.7196	3.135	1.70	21, 59, 65, 66
Fused silica	8.4(7.7)	0.16–3.8		1263	1.4506	1.957		67
α-Quartz								
o-ray	8.4	0.155–4.0	<98	1215	1.5352	2.105		57, 59, 61
e-ray			<120	1222	1.5440	2.152		57, 59, 61
Titanium dioxide								
o-ray	3.03	0.42–4.0		806	2.432	9.15		54, 68
e-ray	3.04			811	2.683	12.85		54, 68
YAG	8.0	0.21–5.2	<38	921	1.815	3.226		69, 273
Yttrium oxide	5.6	0.29–7.1	<24	620	1.892	3.43	1.05	19, 50, 70
Cubic zirconia	3.6	0.38–6.0	<4.3	668	2.0892	5.1	~3?	57, 71, 72

*Region with absorption coefficient <1 cm^{-1}.

TABLE 11.1e Sellmeier Dispersion Formula Parameters for Selected Oxides

Material	Sellmeier equation parameters						Range, μm	Temp., °C	Ref.
	λ_1	λ_2	λ_3	A_1	A_2	A_3			
ALON	0.10256	18.868		2.1375	4.582		0.4–2.3	Room	51
Beryllium oxide									
o-ray	0.07908	9.7131		1.92274	1.24209		0.44–7.0	22.4	73*
e-ray	0.08590	10.4797		1.96939	1.67389		0.44–7.0	22.4	73*
Calcite									
o-ray	0.05580 $\lambda_4 = 7.005$	0.14100	0.197	0.8559 $A_4 = 0.6845$	0.8391	0.0009	0.35–2.2	Room	74
e-ray	0.07897	0.14200	11.468	1.0856	0.0988	0.3170	0.2–2.1	Room	74
Germanium dioxide (glass)	0.068972606	0.15396605	11.841931	0.80686642	0.71815848	0.85416831	0.36–4.28	22–26	75
Sapphire									
o-ray	0.0726631	0.1193242	18.028251	1.4313493	0.65054713	5.3414021	0.2–5.5	20	76
e-ray	0.0740288	0.1216529	20.072248	1.5039759	0.55069141	6.5927379	0.2–5.5	20	76

								Room	64‡
Spinel	0.09942	15.826	1.8938	3.0755			0.35–5.5	Room	64‡
Magnesium oxide	0.0712465	0.1375204	26.89302	1.111033	0.8460085	7.808527	0.36–5.4	20	77
Fused silica	0.0684043	0.1162414	9.896161	0.6961663	0.4079426	0.8974794	0.21–3.71	20	78
α-Quartz									
o-ray	0.0600 $\lambda_4 = 8.844$	0.1060 $\lambda_5 = 20.742$	0.1190	0.663044 $A_4 = 0.565380$	0.517852 $A_5 = 1.675299$	0.175912	0.185–0.71	18	79
e-ray	0.0600 $\lambda_4 = 8.792$	0.1060 $\lambda_5 = 19.70$	0.1190	0.665721 $A_4 = 0.539173$	0.503511 $A_5 = 1.807613$	0.214792	0.185–0.71	18	79
Titanium dioxide									
o-ray	0.0	0.2834		1.873	3.040		0.43–1.5	RT	80
e-ray	0.0	0.2903		2.256	3.941		0.43–1.5	RT	80
YAG	0.1095	17.825		2.293	3.705		0.4–4.0	RT	81†
Yttrium oxide	0.1387	22.936		2.578	3.935		0.2–12	20	82
Cubic zirconia	0.062543	0.166739	24.320570	1.347091	2.117788	9.452943	0.36–5.1	25	71

*Our fit to the referenced data.
†Our fit to the referenced data. Also see Ref. 269, p. 314, for another dispersion equation.
‡New fit: infrared data from Alpha Optical Systems, Inc., added to reference data.

11.21

TABLE 11.1f Temperature Change of Refractive Index (dn/dT in 10^{-6}/K) of Selected Oxides

Material	Wavelength, μm													Ref.
	0.254	0.365	0.405	0.458	0.546	0.589	0.633	0.768	1.06	1.15	1.32	3.39	10.6	
ALON							11.6							83
Beryllium oxide														
o-ray				8.2	8.2	8.2	8.2							53
e-ray				13.4	13.4	13.4	13.4							53
Calcite														
o-ray		3.6	[3.4]	3.2		2.4	2.1							4
e-ray		14.4	[13.8]	13.1		12.1	11.9							4
Germanium oxide														
Sapphire														$a = 84$
o-ray				11.7[a]		13.6[b]	12.6[c]	12.3[c]			7.5[?d]			$b = 85$
e-ray				12.8[a]		14.7[b]								$c = 86$
														$d = 87$

											e = 83
---	---	---	---	---	---	---	---	---	---	---	f = 85
Spinel						9.0[f]	13.0[e]				
Magnesium oxide	19.5	18.9	17.5	15.9	15.3	14.8	13.6				77
		18.6	17.6	16.5							79
											85
Fused silica		9.5	9.6	9.8	16.0	9.9	10.1		10.3		88
											87
α-Quartz											
o-ray	−2.9	−5.4	−5.7	−6.0	−6.2						79
e-ray	−4.0	−6.2	−6.4	−6.7	−7.0						79
Titanium dioxide											
o-ray		−4									80
e-ray		−9									80
YAG			11.9	10.7	10.4	9.4	9.3	9.1	9.9		89
											89
Yttrium oxide			10.0	9.0	8.2	8.3		8.6			83
											90
Cubic zirconia	16.0	13.7	10.0	9.0	8.2	7.9	7.3	6.5	6.3		71

TABLE 11.2a Physical Properties of Selected Halides

Material	Crystal class and space group	Unit cell lattice parameters, Å	Formulas per unit cell, Z	Molecular weight, amu	Density, g/cm³	Melting point,* K	Cold water solubility, g/100 g†
Lithium fluoride (LiF)	Cubic Fm3m (225)	4.0173	4	25.94	2.66	$T_m = 1115$	0.27
Sodium fluoride (NaF) (Valliaumite)	Cubic Fm3m (225)	4.6342	4	41.99	2.806	$T_m = 1266$	4.22
Sodium chloride (NaCl) (Halite, common or rock salt)	Cubic Fm3m (225)	5.63978	4	58.44	2.164	$T_m = 1074$	35.7
Sodium bromide (NaBr)	Cubic Fm3m (225)	5.9732	4	102.89	3.207	$T_m = 1028$	116
Sodium iodide (NaI)	Cubic Fm3m (225)	6.475	4	149.89	3.668	$T_m = 934$	184
Potassium fluoride (KF)	Cubic Fm3m (225)	5.347	4	58.10	2.52	$T_m = 1131$	92.3
Potassium chloride (KCl) (Sylvite)	Cubic Fm3m (225)	6.293	4	74.55	1.987	$T_m = 1043$	34.4
Potassium bromide (KBr)	Cubic Fm3m (225)	6.600	4	119.00	2.750	$T_m = 1007$	53.5
Potassium iodide (KI)	Cubic Fm3m (225)	7.065	4	166.00	3.127	$T_m = 954$	127.5
Cesium bromide (CsBr)	Cubic Pm3m (221)	4.286	1	212.81	4.49	$T_m = 908$	124.3

Material	Structure (space group)	Lattice constant (Å)	Z	Molecular weight	Density	Temperature (K)	Solubility
Cesium iodide (CsI)	Cubic Pm3m (221)	4.566	1	259.81	4.53	$T_m = 898$	44.0
Magnesium fluoride (MgF$_2$) (Sellaite)	Tetragonal P4$_2$/mnm (136)	a = 4.623 c = 3.053	2	62.30	3.152	$T_m = 1536$	0.0076 (<0.0002)
Calcium fluoride (CaF$_2$) (Fluorite)	Cubic Fm3m (225)	5.46295	4	78.08	3.181	$T_m = 1633$ $T_p = 1424$	0.0016 (0.0001)
Strontium fluoride (SrF$_2$)	Cubic Fm3m (225)	5.7996	4	125.62	4.277	$T_m = 1710$	0.011 (0.0001)
Barium fluoride (BaF$_2$)	Cubic Fm3m (225)	6.2001	4	175.33	4.887	$T_m = 1553$	0.12 (0.004)
β-Lead fluoride (PbF$_2$)	Cubic Fm3m (225)	5.951	4	245.20	7.750	$T_m = 1094$	0.064
YLF (LiYF$_4$) (lithium yttrium fluoride)	Tetragonal I4$_1$/a (88)	a = 5.175 c = 10.74	4	171.84	3.968	$T_m = 1092$	insol.
Silver bromide (AgBr) (Bromyrite)	Cubic Fm3m (225)	5.7745	4	187.78	6.478	$T_m = 705$	8.4×10^{-6}
Silver chloride (AgCl) (Cerargyrite)	Cubic Fm3m (225)	5.547	4	143.32	5.578	$T_m = 728$	8.9×10^{-6}
Thallium chloride (TlCl)	Cubic Pm3m (221)	3.8425	1	239.82	7.0183	$T_m = 703$	0.29
Thallium bromide (TlBr)	Cubic Pm3m (221)	3.9846	1	284.29	7.462	$T_m = 740$	0.05
KRS-5 (Tl[Br,I]) (Thallium bromoiodide)	Cubic Pm3m (221)	4.108	1	[307.79]	7.371	$T_m = 687$	<0.05

*Notation: T_m is the melting temperature, T_p is a phase change temperature.
†Solubilities in parenthesis refer to single crystals.

TABLE 11.2b Mechanical Properties of Selected Halides

Material	Moduli, GPa			Poisson's ratio	Flexure strength, MPa*	Knoop hardness, kg/mm²	Fracture toughness, MPa·m$^{1/2}$	Elastic constants References
	Elastic	Shear	Bulk					
Lithium fluoride	110	45	65.0	0.22_5	10.9 (27)†	102–133		3, 4, 35, 36
Sodium fluoride	76	30.7	48.5	0.24		60		3, 4, 35, 36
Sodium chloride	37	14.5	25.3	0.26	2.4 (9.6)†	15–18		3, 4, 35, 36
Sodium bromide	29	11.6	19.9	0.26				3, 4, 35, 36
Sodium iodide	22	8.4	16.1	0.28				3, 4, 35, 36
Potassium fluoride	41	16	31.8	0.28				3, 4, 35, 36
Potassium chloride	22	8.5	18.4	0.29	2.3 (9.6)†	7–9		3, 4, 35, 36
Potassium bromide	18	7.2	15.2	0.30	1.1 (11.0)†	6–7		3, 4, 35, 36
Potassium iodide	14	5.5	11.9	0.30		5.0		3, 4, 35, 36

								References
Cesium bromide	22	8.6	16.7	0.28	8.4	16–20		3, 35, 36
Cesium iodide	18	7.3	12.6	0.26	5.6			3, 35, 36
Magnesium fluoride	137	53.9	99.1	0.26_9	60 (100)†	415–580	1.0	3, 35, 36
Calcium fluoride	110	42.5	95.7	0.29	40 (90)†	160–180	0.5	3, 4, 35, 36
Strontium fluoride	89	34.5	69.9	0.29		140–154		3, 35, 36
Barium fluoride	65.6	25.2	58.4	0.31	27	65–90		3, 35, 36
β-Lead fluoride	60.6	22.7	60.7	0.33		200		3, 35, 36
YLF	85	32	81	0.32	35	260–325		91
Silver bromide	24.7	8.8	40.6	0.39_5				3, 4, 35, 36
Silver chloride	22.9	8.1	44.0	0.41	26	9.5		3, 4, 35, 36
Thallium chloride	24	9.0	23.6	0.33		13		3, 4, 35, 36
Thallium bromide	23	8.7	22.4	0.33		12		3, 4, 35, 36
KRS-5	19.6	7.3	20.4	0.33	26.2	33–40		3, 4, 35, 36

*Elastic limit values in some cases.
†Strength values for polycrystalline materials; other values are for single crystals.

TABLE 11.2c Thermal Properties of Selected Halides

Material	Heat capacity, J/g·K	Debye temperature, K	Thermal conductivity, W/m·K					Thermal expansion (1/K) at 300 K α·10⁶
			250 K	300 K	500 K	1000 K		α·10⁶
Lithium fluoride	1.6200	735	19	14	7.5	3.5		34.4
Sodium fluoride	1.1239	492	22	16	8	5		33.5
Sodium chloride	0.8699	321	8	6.5	4	1.5		41.1
Sodium bromide	0.5046	225		5.6		0.54		41.8
Sodium iodide	0.3502	164		4.7				44.7
Potassium fluoride	0.8659	336		8.3				31.4
Potassium chloride	0.6936	235	8.5	6.7	3.8	1.5		36.5
Potassium bromide	0.4400	174	5.5	4.8	2.4	0.8		38.5
Potassium iodide	0.3192	132		2.1		0.4(m)		40.3
Cesium bromide	0.2432	145		0.85				47.2

Cesium iodide	0.1983	124		1.05		48.6
Magnesium fluoride	1.0236	535				10.7
∥ to c axis				(21)		14.2
⊥ to c axis				(30)		8.9
Calcium fluoride	0.9113	510	13	9.7	5.5	18.9
Strontium fluoride	0.6200	378	11	8.3		19.0
Barium fluoride	0.4474	283		12		18.4
β-Lead fluoride	0.3029	218		(28)	(2.2)	29
YLF	0.79			6.3		
∥ to c axis						8.3
⊥ to c axis						13.3
Silver bromide	0.2790	145	1.11	0.93	0.57	33.8
Silver chloride	0.3544	180	1.25	1.19	4(m)	32.4
Thallium chloride	0.2198	126		0.75		52.7
Thallium bromide	0.1778	116		0.59		52.5
KRS-5	(0.16)	(110)		0.54		58

TABLE 11.2d Optical Properties of Selected Halides

| Material | Bandgap, eV | Transparent regions* | | | Refractive index | | | |
| | | Optical, μm | Microwave, cm^{-1} | Maximum ν_{LO} | Optical | DC | | References |
					n_∞	n_o	$dn_o/dT \cdot 10^4$	
Lithium fluoride	13.7	0.120–6.60	<16.5	662	1.388	3.018	5.66	21, 54, 57, 92, 93
Sodium fluoride	11.5	0.135–11.2		414	1.320	2.25		54, 92, 93
Sodium chloride	8.75	0.174–18.2	<8.5	264	1.555	2.43	3.84	21, 54, 57, 92, 93
Sodium bromide	7.1	0.20–(24)		209	1.615	2.45		4, 54, 92, 93
Sodium iodide	5.9	0.26–(24)		181	1.73	2.57		4, 54, 92, 93
Potassium fluoride	10.8	0.14–15.8		330	1.357	2.46		4, 54, 92, 93
Potassium chloride	8.7	0.18–23.3	<9.4	214	1.475	2.18	3.20	21, 54, 57, 92, 93
Potassium bromide	7.4	0.200–30.3	<9.5	165	1.537	2.21	3.42	21, 54, 57, 92, 93
Potassium iodide	6.34	0.250–38.5	<10	139	1.629	2.252		54, 92, 93
Cesium bromide	7.3	0.230–43.5		112	1.669	2.53		54, 92, 93

Material								References
Cesium iodide	6.1	0.245–62	<7	85	1.743	2.38	1.3	54, 57, 92, 93
Magnesium fluoride	10.8	0.13–7.7	<26.0			2.35	1.0	59
o-ray		0.13–7.7	<27.8	621	1.3734	2.20		57, 59, 93, 94
e-ray		0.13–7.7	<25.7	625	1.3851	2.30		57, 59, 93, 94
Calcium fluoride	10	0.135–9.4	<12	463	1.4278	2.607	3.21	66, 93, 95
Strontium fluoride	9	(0.13)–11.0	<11	374	1.4316	2.543	2.65	66, 93, 95
Barium fluoride	9.1	0.14–12.2	<10	326	1.4663	2.713	2.90	66, 93, 95
β-Lead fluoride	~5	(0.29)–12.5		337	1.731	5.41		275
YLF	~9			566		2.18		69, 96
o-ray		0.18–6.7			1.447	2.74		274
e-ray		0.19–6.7			1.469			
Silver bromide	2.69	0.49–35		138	2.166	3.53		93, 97, 98
Silver chloride	2.98	0.42–23		196	2.002	3.34		93, 98
Thallium chloride	3.6	0.38–30		158	2.136	5.62		54, 93, 99
Thallium bromide	(3.1)	0.44–38		101	2.271	5.46	−8.5	21, 54, 93, 99
KRS-5	2.37	0.58–42		~90	2.38	(5.5)		100

*Region with absorption coefficient <1 cm^{-1}.

TABLE 11.2e Sellmeier Dispersion Formula Parameters for Selected Halides

Material		Sellmeier equation parameters					Range, μm	Temp., °C	Ref.
	λ_1	λ_2	λ_3	A_1	A_2	A_3			
Lithium fluoride	0.07291	28.24699		0.9255630	5.128197		0.4–6.0	23.6	101
	0.07376	32.790		0.92549	6.96747		0.1–11.0	20	92
Sodium fluoride	0.0	0.117	40.57	0.41572	0.32785	3.18248	0.15–17.0	20	92
Sodium chloride	0.0	0.050	0.100	0.00055	0.19800	0.48398	0.20–30.0	20	92
	$\lambda_4 = 0.128$ $\lambda_7 = 60.98$	$\lambda_5 = 0.158$ $\lambda_8 = 120.34$	$\lambda_6 = 40.50$	$A_4 = 0.38696$ $A_7 = 3.17064$	$A_5 = 0.25998$ $A_8 = 0.30038$	$A_6 = 0.08796$			
Sodium bromide	0.0	0.125	0.145	0.06728	1.10463	0.18816	0.21–34.0	20	92
	$\lambda_4 = 0.176$	$\lambda_5 = 0.1884$	$\lambda_6 = 74.63$	$A_4 = 0.00243$	$A_5 = 0.24454$	$A_6 = 3.7960$			
Sodium iodide	0.0	0.170	86.21	0.478	1.532	4.27	0.25–40.0	20	92
Potassium fluoride	0.0	0.126	51.55	0.55083	0.29162	3.60001	0.15–22.0	20	92
Potassium chloride	0.0	0.100	0.131	0.26486	0.30523	0.41620	0.18–35.0	20	92
	$\lambda_4 = 0.162$	$\lambda_5 = 70.42$		$A_4 = 0.18870$	$A_5 = 2.620$				
Potassium bromide	0	0.146	0.173	0.39408	0.79221	0.01981	0.20–40.0	20	92
	$\lambda_4 = 0.187$	$\lambda_5 = 60.61$	$\lambda_6 = 87.72$	$A_4 = 0.15587$	$A_5 = 0.17673$	$A_6 = 2.06217$			
Potassium iodide	0.0	0.129	0.175	0.47285	0.16512	0.41222	0.25–50.0	20	92
	$\lambda_4 = 0.187$ $\lambda_7 = 98.04$	$\lambda_5 = 0.219$	$\lambda_6 = 69.44$	$A_4 = 0.44163$ $A_7 = 1.92474$	$A_5 = 0.16076$	$A_6 = 0.33571$			
Cesium bromide	0.0905643	0.1671517	119.0155	0.9533786	0.8303809	2.847172	0.36–39.0	27	102
	0.0	0.119	0.137	0.33013	0.98369	0.00009	0.18–40.0	20	92
	$\lambda_4 = 0.145$	$\lambda_5 = 0.162$	$\lambda_6 = 100.50$	$A_4 = 0.00018$	$A_5 = 0.30914$	$A_6 = 4.320$			

							Range (μm)		Ref.
Cesium iodide	0.0229567	0.1466	0.1810	0.34617251	1.0080886	0.28851800	0.29–50.0	24	103
	$\lambda_4 = 0.2120$	$\lambda_5 = 161.0$		$A_4 = 0.39743178$	$A_5 = 3.3605359$				
Magnesium fluoride									
o-ray	0.04338408	0.09461442	23.793604	0.48755108	0.39875031	2.3120353	0.2–7.0	19	94
e-ray	0.03684262	0.09076162	23.771995	0.41344023	0.50497499	2.4904862	0.2–7.0	19	94
Calcium fluoride	0.050263605	0.1003909	34.649040	0.5675888	0.4710914	3.8484723	0.23–9.7	24	104
	0.0127821	0.0936663	34.8259	0.3439319	0.694269	3.8902192	0.21–8.7	21.7	84
Strontium fluoride	0.05628989	0.10801027	39.906666	0.67805894	0.37140533	3.3485284	0.21–11.5	20	84
Barium fluoride	0.057789	0.10968	46.3864	0.643356	0.506762	3.8261	0.26–10.3	25	105
β-Lead fluoride	0.00034911	0.17144455	0.28125513	0.6695342	1.3086319	0.01670641	0.3–11.9	19	106
	$\lambda_4 = 796.67469$			$A_4 = 2007.8865$					
YLF									
o-ray	0.0	0.09649	7.14125	0.38757	0.70757	0.18849	0.225–2.6		96
e-ray	0.0	0.09359	11.61708	0.31021	0.84903	0.53607	0.225–2.6		96
Silver bromide	0.0	0.29185		2.45000	1.24168		0.49–0.67		98*
Silver chloride	0.1039054	0.2438691	70.85723	2.062508	0.9461465	4.300785	0.54–21.0	23.9	107
Thallium chloride	0.0	0.28129		1.91159	1.65125		0.43–0.74		99*
	0.0	0.30097		2.43600	1.24408		0.43–0.66		98†
Thallium bromide	0.0	0.30632		2.31028	1.84788		0.43–0.77		99*
	0.0	0.32556		2.53300	1.66885		0.54–0.65		98†
KRS-5	0.150	0.250	0.350	1.8293958	1.6675593	1.1210424	0.58–39.4		100
	$\lambda_4 = 0.450$	$\lambda_5 = 164.59$		$A_4 = 0.04513366$	$A_5 = 12.380234$				

*Our fit to the referenced data.
†Our fit to the index calculated by the dispersion formula given in Ref. 98.

11.33

TABLE 11.2f Temperature Change of Refractive Index (dn/dT in 10^{-6}/K) of Selected Halides

Material	Wavelength (μm)										Ref.
	0.254	0.365	0.458	0.546	0.633	0.768	1.15	3.39	10.6	30	
Lithium fluoride	−13.3		−16.0		−16.7		−16.9	−14.5			84
Sodium fluoride		−16.4	−17.3	−17.8	−18.0	−18.3	−18.4	−16.4	[25]		92
Sodium chloride	−12.4		−11.9		−12.8		−13.2	−12.5			84
		−15.4	−16.2	−16.6	−16.8	−17.0	−17.1	−16.6	−7.8		92
	−14.3		−34.2		−35.4		−36.4	−36.3			84
Sodium bromide	16.5	−28.4	−30.8	−31.7	−32.3	−32.6	−33.1	−33.1	−29.3	[10.8]	92
Sodium iodide	48.8	−30.4	−35.7	−37.7	−38.6	−39.4	−40.2	−40.4	−37.9	18.8	92
Potassium fluoride	−19.9	−21.0	−38.4	−43.6	−46.0	−47.9	−49.6	−50.5	−48.6	−15.7	92
Potassium chloride		−22.3	−22.8	−23.1	−23.2	−23.3	−23.4	−23.1	−17.0		92
			−34.9		−35.8		−36.2	−36.2	−34.8		84
Potassium bromide	−22.1	−30.0	−31.1	−31.5	−31.7	−31.9	−31.1	−32.0	−30.1	[15.8]	92
			−39.3		−41.2		−41.9	−42.1	−41.1		84
Potassium iodide	−10.2	−33.1	−35.5	−36.2	−36.6	−36.9	−37.2	−37.2	−36.1	−14.2	92
	88.9	−36.0	−41.5	−43.0	−43.8	−44.2	−44.7	−44.8	−44.0	−30.8	92
Cesium bromide	−82.0	−86.1	−85.5	−85.1	−84.7	−84.5	−84.1	−83.8	−83.1	−75.8	92
Cesium iodide	−57.4	−87.5	−98.6	−99.0	−99.3	−99.2	−98.5	−95.1	−91.6	−88.0	103
		−97.8	−97.4	−96.7	−96.3	−95.8	−95.2	−94.8	−94.4	−90.6	92
Magnesium fluoride											
o-ray					1.12		0.88	1.1			84
e-ray					0.58		0.32	0.6			84

Material										
Calcium fluoride	−7.5	−9.6	−9.8	−10.4	−11.3	−10.6	−11.5	−11.2	[−5.2]	84
					−10.4		−10.3	−8.1		104
					−11.8		−12.0	−11.5		89
					−13.1		−13.4	−12.8		108
Strontium fluoride					−12.4		−12.6	−12.4	−9.8	84
					−12.0		−12.7	−13.0		89
Barium fluoride		−15.0	−15.2		−16.0	−15.5	−16.2	−15.9	−14.5	84
					−15.2		−17.1	−16.8		105
					−16.7		−16.8	−16.3		89
					−16.4					108
β-Lead fluoride YLF										
o-ray				−6.7						96
e-ray				−23.0						96
Silver bromide								−61	−50	109
Silver chloride					−61.0			−58	−35	107
Thallium chloride										109
Thallium bromide									−233	
KRS-5					−250	−245	−240	−237	−195	100

TABLE 11.3a Physical Properties of Selected Semiconductors

Material	Crystal class and space group	Unit cell lattice parameters, Å	Formulas per unit cell, Z	Molecular weight, amu	Density, g/cm³	Melting point,* K	Cold water solubility,† g/100 g
Diamond (C) (type IIa)	Cubic Fd3m (227)	3.56696	8	12.011	3.515	$T_p = 1770$ $T_m > 3823$	insol.
Silicon (Si)	Cubic Fd3m (227)	5.43085	8	28.0855	2.329	$T_m = 1680$	insol.
Germanium (Ge)	Cubic Fd3m (227)	5.65741	8	72.59	5.326	$T_m = 1211$	insol.
β-Silicon carbide (SiC) (carborundum)	Cubic F$\bar{4}$3m (216)	4.3590	4	40.10	3.216	$T_d \approx 2050$	insol.
Aluminum arsenide (AlAs)	Cubic F$\bar{4}$3m (216)	5.6611	4	101.89	3.730	$T_m = 2113$	decomp.
Aluminum nitride (AlN)	Hexagonal P6₃mc (186)	$a = 3.1127$ $c = 4.9816$	2	40.988	3.257	$T_v = 2500$	decomp.
Boron nitride (BN)	Cubic F$\bar{4}$3m (216)	3.615	4	24.828	3.491	$T_p = 1100$ $T_m > 3000$	insol.
Boron phosphite (BP)	Cubic F$\bar{4}$3m (216)	4.538	4	41.795	2.971	$T_p > 1400$ $T_m > 2300$	insol.
Gallium arsenide (GaAs)	Cubic F$\bar{4}$3m (216)	5.6533	4	144.64	5.3169	$T_m = 1511$	insol.
Gallium nitride (GaN)	Hexagonal P6₃mc (186)	$a = 3.186$ $c = 5.178$	2	83.73	6.108	$T_d = 1160?$	insol.

Compound	Crystal structure	Lattice constant (Å)	Z	Molecular weight	Density	Melting temp.	Solubility
Gallium phosphide (GaP)	Cubic F4̄3m (216)	5.4495	4	100.695	4.133	$T_m = 1740$	0.000086
Indium arsenide (InAs)	Cubic F4̄3m (216)	6.0584	4	189.74	5.667	$T_m = 1216$	
Indium phosphide (InP)	Cubic F4̄3m (216)	5.8688	4	145.795	4.757	$T_m = 1345$	
Lead sulfide (PbS) (Galena)	Cubic Fm3m (225)	5.935	4	239.26	7.602	$T_m = 1390$	insol.
Lead selenide (PbSe) (Clausthalite)	Cubic Fm3m (225)	6.122	4	286.16	8.284	$T_m = 1338$	insol.
Lead telluride (PbTe) (Altaite)	Cubic Fm3m (225)	6.443	4	334.79	8.314	$T_m = 1190$	
β-Zinc sulfide (ZnS) (zincblende, sphalerite)	Cubic F4̄3m (216)	5.4094	4	97.44	4.088	$T_p = 1293$ $T_m = 2100$	0.000065
Zinc selenide (ZnSe)	Cubic F4̄3m (216)	5.6685	4	144.34	5.263	$T_m = 1790$	insol.
Zinc telluride (ZnTe)	Cubic F4̄3m (216)	6.1034	4	192.98	5.636	$T_m = 1510$	decomp.
Cadmium sulfide (CdS) (Greenockite)	Hexagonal P6₃mc (186)	$a = 4.1367$ $c = 6.7161$	2	144.47	4.819	$T_m = 1560$	0.00013
Cadmium telluride (CdTe)	Cubic F4̄3m (216)	6.4830	4	240.01	5.849	$T_m = 1320$	insol.
Calcium lanthanum sulfide ($CaLa_2S_4$)	Cubic I4̄3d (220)	8.685	4	446.16	4.524	$T_m = 2083$	(slight)

*Notation: T_m is the melting temperature, T_p is the phase change temperature, T_d is the decomposition temperature.
†Solubilities in parenthesis refer to single crystals.

11.37

TABLE 11.3b Mechanical Properties of Selected Semiconductors

Material	Moduli, GPa			Poisson's ratio	Flexure strength, MPa	Knoop hardness, kg/mm²	Fracture toughness, MPa·m^(1/2)	Elastic constants references
	Elastic	Shear	Bulk					
Diamond	1100	500	460	0.10	2940	7500–9200	2.0	4, 35, 36
Silicon	165	66.2	97.7	0.22_6	120–140	1150	0.95	4, 35, 36
Germanium	132	54.8	75.0	0.20_9	100	850	0.66	4, 35, 36
Silicon carbide	436	186	220	0.18	250	2880	4.0	35
Aluminum arsenide	116	44	75.3	0.26		480–500		110
Aluminum nitride	323	130	205	0.23_8	280–370	1230	2.7–3.5	*
Boron nitride	833	375	358	0.11		>4600		†
Boron phosphide	324	136	172	0.19		4700		113
Gallium arsenide	116	46.6	75.0	0.25	55	700–720	0.8	35, 36
Gallium nitride	91	36	75	0.29	70	750		4, 35
Gallium phosphide	140	56.5	89.3	0.24_4	100	845–910	0.9	3, 35, 36
Indium arsenide	74	28	61	0.30		375–410		3, 36

Material							Ref.
Indium phosphide	86	33	72.5	0.30		510–535		3, 35, 36
Lead sulfide	71.5	27.5	59.3	0.30				3, 4, 35, 36
Lead selenide	66.9	25.9	54.0	0.25				4, 35
Lead telluride	49.4	19.0	41.1	0.30				3, 35, 36
β-Zinc sulfide	82.5	31.2	76.6	0.32	60	160–190	0.5	3, 4, 35, 36
					100§	250§	0.88	
Zinc selenide	75.4	29.1	61.8	0.31	55	100–130	0.33	3, 35, 36
Zinc telluride	61.1	23.5	51.0	0.31	24	82		3, 35, 36
Cadmium sulfide	42	15	59	0.38?	28	122		3, 4, 35, 36
Cadmium telluride	38.4	14.2	42.4	0.37	22–31	45–56		3, 35, 36
Calcium lanthanum sulfide	96	[38.4]	[64]	0.25	81	570	0.68	‡

*Aluminum nitride stiffness and compliance estimated from the engineering moduli assuming isotropy:[111] $c_{11} = 379$ GPa, $c_{12} = 119$ GPa, $c_{44} = 130$ GPa, $s_{11} = 3.10$ TPa^{-1}, $s_{12} = -0.74$ TPa^{-1}, $s_{44} = 7.67$ TPa^{-1}.

†Cubic boron nitride stiffness and compliance from:[112] $c_{11} = 783$ GPa, $c_{12} = 146$ GPa, $c_{44} = 418$ GPa, $s_{11} = 1.36$ TPa^{-1}, $s_{12} = -0.214$ TPa^{-1}, $s_{44} = 2.39$ TPa^{-1}.

‡Calcium lanthanum sulfide stiffness and compliance estimated from the engineering moduli, assuming isotropy:[114] $c_{11} = 98.4$ GPa, $c_{12} = 46.8$ GPa, $c_{44} = 50.0$ GPa, $s_{11} = 14.7$ Pa^{-1}, $s_{12} = -4.7$ Pa^{-1}, $s_{44} = 20.0$ Pa^{-1}.

§CVD FLIR grade material (β-zinc sulfide).

TABLE 11.3c Thermal Properties of Selected Semiconductors

Material	Heat capacity, J/g·K	Debye temperature, K	Thermal conductivity, W/m·K					Thermal expansion (1/K) at 300 K $\alpha \cdot 10^6$
			250 K	300 K	500 K	1000 K		
Diamond	0.5169	2240	≤2800	≤2000				1.25
Silicon	0.7139	645	191	140	73.6	30.6		2.618
Germanium	0.3230	380	74.9	59.9	33.8	17.4		5.7
Silicon carbide	0.59	(1000)		490		(100)		2.8
Aluminum arsenide	0.452	416		(80)				3.52
								3.06
								3.75
Aluminum nitride ‖ to c axis ⊥ to c axis	0.796	950	500	320	150	50		2.7
Boron nitride	0.513	1900		1300				3.5
Boron phosphide	0.71	985	460	360				2.9
Gallium arsenide	0.345	344	(65)	54	27	(16)		5.73

Material							
Gallium nitride							
∥ to c axis				130			5.59
⊥ to c axis							3.17
Gallium phosphide	0.435	(600)	120	100			5.3
Indium arsenide	0.2518	460	(50)	27.3	15	10 (m)	5.2
Indium phosphide	0.3117	251		68	33	17 (800)	4.5
Lead sulfide	0.209	302		2.5			19.0
Lead selenide	0.175	227	2	1.7	1		19.4
Lead telluride	0.151	138	2.5	2.3	1.8		(20)
β-Zinc sulfide	0.4732	125		16.7	10	6.5	6.8
Zinc selenide	0.339	340		13	8	5	7.1
Zinc telluride	0.218	270		11			8.8
Cadmium sulfide	0.3814	225		27	13		4.2
∥ to c axis							3.5
⊥ to c axis							5.6
Cadmium telluride	0.210	(300)	8.2	6.3			5.0
Calcium lanthanum sulfide	(0.36)	160		1.7	1.5		14.6

TABLE 11.3d Optical Properties for Selected Semiconductors

Material	Bandgap, eV		Transparent regions*			Refractive index			References
						Optical	DC		
	300 K	0 K	Optical, μm	Microwave, cm^{-1}	Maximum ν_{LO}	n_x	n_o	$dn_o/dT \cdot 10^4$	
Diamond									
Type IIa	5.47 I	5.48 I			1332				93
Type I or CVD	7.4 D		0.24–2.7	<1600		2.38	2.38	0.096	72, 115, 116
			0.34–2.7	<1000			2.42		54, 117
Silicon	1.107 I	1.205 I	1.1–6.5	<125	520	3.415	3.455	1.35	61, 93, 118
	3.07 D								
Germanium	0.665 I	0.746 I	1.8–(15)	<95	301	4.001	4.001	3.40	61, 93, 118
	0.805 D	0.898 D							
Silicon carbide	2.2 I	2.390 I	(0.5)–(4)		970	2.563	3.0		46, 119, 120
Aluminum arsenide	2.13 I	2.25 I			402	2.857	3.188		35, 121, 122
	2.95 D								
Aluminum nitride			0.36–4.4	<100			3.02		3, 121
o-ray	5.88 D				895	2.20			123–126
e-ray	5.74 D				888	2.17			
						2.22			
Boron nitride	6.4 I		[0.2]–		1340	2.12	2.66		93, 121
Boron phosphide	2 I	2 I	[0.5]–		834	2.78			93, 113, 121
		6 D							

Gallium arsenide	1.428 D	1.522 D	0.90–17.3	<46	292	3.32	3.606	54, 93, 121
Gallium nitride								127, 128
o-ray	3.7				746	2.35	3.1	
e-ray	3.5				744	2.31	3.2	
Gallium phosphide	2.261 I / 2.78 D	2.338 I / 2.88 D	0.54–10.5	<20	403	3.01	3.332	35, 93, 129–131
Indium arsenide	0.359 D	0.4105 D	3.9–(20)		241	3.44	3.814	93, 132
Indium phosphide	1.351 D / 2.25 I	1.4205 D	0.93–20	(<8)	345	3.09	3.52	35, 93, 121
Lead sulfide	0.41 D	0.28 D	2.9–	<20	212	4.1	12.25	54, 119, 133
Lead selenide	0.27 D	0.15 D	4.6–	—	116	4.7	14.3	4, 54, 133
Lead telluride	0.31 D	0.20 D	4.0–20†	—	112	5.67	36	119, 133
β-Zinc sulfide	3.58 D	3.84 D	0.40–12.5	<20	352	2.258	2.85	35, 50, 54, 84, 93, 118
Zinc selenide	2.58 D	2.80 D	0.51–19.0	<14	250	2.435	3.02	35, 50, 54, 84, 93, 118, 134
Zinc telluride	2.26 D	2.39 D	(0.6–25)	<12	205	2.70	3.18	35, 54, 93, 118, 134
Cadmium sulfide	2.42 D	2.582 D						
o-ray			0.52–(14.8)		304	1.7085	2.95	35, 43, 54, 93
e-ray			0.52–(14.8)		306	1.7234	3.04	35, 43, 54, 93
Cadmium telluride	1.45 D	1.61 D	0.85–29.9	<17	169	2.68	3.206	35, 93, 134
Calcium lanthanum sulfide	2.7		0.65–14.3		314	~2.6	~4.4	114, 135

*Region with absorption coefficient <1 cm^{-1}.
†Fuzzy edge.

TABLE 11.3e Sellmeier Dispersion Formula Parameters of Selected Semiconductors

Material	Sellmeier equation parameters						Range, μm	Temp., °C	Ref.
	λ_1	λ_2	λ_3	A_1	A_2	A_3			
Diamond	0.1060	0.1750		4.3356	0.3306		0.225–∞		115
Silicon	0.301516485	1.13475115	1104.0	10.6684293	0.0030434347484	1.54133408	1.36–11	27	43
Germanium	0.0	0.66411	62.21	9.28156	6.72880	0.21307	2–12	20	136
β-Silicon carbide	0.1635			5.5705			0.47–0.69		120*
Aluminum arsenide	0.0	0.2822	27.62	1.0792	6.0840	1.900	0.56–2.2	27	122*
Aluminum nitride									
o-ray	0.0	0.1715	[15.03]	2.1399	1.3786	[3.861]	0.22–0.59		137†
e-ray	0.0	0.1746	[15.03]	2.0729	1.6173	[4.139]	0.22–0.57		137†
Boron nitride									
Boron phosphide	0.267			6.841			0.45–0.63		113*
Gallium arsenide	0.0	0.408	37.17	2.5	7.497	1.935	1.4–11		138
Gallium nitride									
o-ray	0.0	0.256	17.86	2.60	1.75	4.10	<10		127
e-ray	0.0	—	18.76	(4.35)	—	5.08			127
Gallium phosphide	0.172 $\lambda_4 = 27.52$	0.234	0.345	1.390 $A_4 = 2.056$	4.131	2.570	0.8–10	27	129
	0.25650	0.40644		7.55036	0.54107		0.55–0.70	24.5	130*
	0.234	0.334		2.635	5.404		0.44–4.0	17	131

Material							Range (μm)	Temp (°C)	Ref.
Indium arsenide	0.0	2.551	45.66	10.1	0.71	2.75	3.7–31.3	RT	132
Indium phosphide	0.0	0.6263	[32.935]	7.255	2.316	[2.765]	0.95–2.1	25	139‡
Lead sulfide	0.77	[140.85]		15.9	[133.2]		3.5–10	27	133
Lead selenide	1.37			21.1			5.0–10	27	133
Lead telluride	1.26			31.2			4.5–10	27	133
	1.563			30.046			4.0–12.5	Room	140
β-Zinc sulfide	0.3142303	0.1759417	33.88656	0.3390403	3.760687	2.731235	0.55–10.5	21.6	84
	0.0	0.23979	36.525	1.57299	2.52873	3.23924	0.50–14	20	141
Zinc selenide	0.190630	0.3787826	46.99456	4.298015	0.6277656	2.895563	0.55–18.0	20.3	84
	0.0	0.336155		3.00	1.90		0.45–2.5	Room	134
Zinc telluride	0.0	0.29934	48.38	2.19424	2.73228	3.08889	0.55–18	20	141
	0.0	0.376829		3.27	3.01		0.57–2.5	Room	134
	0.0	0.37766	56.5	3.30230	2.98190	2.63580	0.55–30	20	141
Cadmium sulfide									
o-ray	0.23622804	0.48285199		3.96582820	0.18113874		0.51–1.4		43
e-ray	0.22426984	0.46693785	0.50915139	3.97478769	0.26680809	0.00074077	0.51–1.4		43
Cadmium telluride	0.0	0.605		4.68	1.53		0.86–2.5	Room	134
	0.317069	72.0663		6.1977889	3.2243821		6–22	27	142
	0.0	71.43		6.05	3.55		7.5–31	Room	132
Calcium lanthanum sulfide									

*Our fit to the referenced data.
†Our fit to the referenced data, a better fit than given in Ref. 124. The infrared terms are from Ref. 276 and extend the index to about 5 μm.
‡The infrared term is from Ref. 277 and extends the index to about 10 μm.

TABLE 11.3f Temperature Change of Refractive Index (dn/dT in 10^{-6}/K) of Selected Semiconductors

Material	Wavelength, μm										Ref.
	0.405	0.546	0.633	1.15	2.5	3.39	5.0	10.6	20	30	
Diamond	[12]	10.1	[9.7]							9.6	143 116
Silicon				[200]	166	162	159	157			118
Germanium					462	434	416	404	[401]		118
β-Silicon carbide											
Aluminum arsenide											
Aluminum nitride											
Boron nitride											
Boron phosphide											
Gallium arsenide			250	200	200	200 206	200 216 188	200 202 187	182		278 144 109
Gallium nitride				61							128
Gallium phosphide		200	160	140	100	100	100				131 278

11.46

Material									
Indium arsenide						450	300	250	278
Indium phosphide						83	82	77	109
Lead sulfide					−2100	−1900	−1700		133
Lead selenide					−2300	−1400	−860		133
						−2900	−1500		140
Lead telluride					−2100	−1500	−1200		133
						−2750	−1600		140
β-Zinc sulfide	62	55	46	43	43	43	41		141
			46		42		41		92
		63.5	49.8		45.9		46.3		108
Zinc selenide	143	106	69	64	63	63	61	[59]	141
		106	70		62		61		92
		91.1	59.7		53.4		52		108
Zinc telluride				100					
Cadmium sulfide									
o-ray							58.6		145
e-ray							62.4		145
Cadmium telluride			147	100	98.2		98.0		108
					100		100		278
Calcium lanthanum sulfide						100			

TABLE 11.4a Physical Properties of Selected Nonlinear Crystals

Material	Crystal structure and space group	Unit cell lattice parameters, Å	Formulas per unit cell, Z	Molecular weight, amu	Density, g/cm³	Melting point,* K	Cold water solubility, g/100 g
ADP ($NH_4H_2PO_4$) (ammonium dihydrogen phosphate)	Tetragonal I$\bar{4}$2d (122)	$a = 7.4997$ $c = 7.5494$	4	115.03	1.800	$T_p = 148$ $T_m = 463$	22.7
KDP (KH_2PO_4) (potassium dihydrogen phosphate)	Tetragonal I$\bar{4}$2d (122)	$a = 7.452$ $c = 6.959$	4	136.09	2.339	$T_p = 123$ $T_p = 450$ $T_m = 526$	20
KTP ($KTiOPO_4$) (potassium titanyl phosphate)	Orthorhombic Pna2_1 (33)	$a = 12.840$ $b = 6.396$ $c = 10.584$	8	197.97	2.945	$T_c = 1209$ $T_d = 1423$	
Lithium niobate ($LiNbO_3$)	Hexagonal R3c (161)	$a = 5.1483$ $c = 13.8631$	6	147.842	4.628	$T_c = 1480$ $T_m = 1523$	insol.
Potassium niobate ($KNbO_3$)	Orthorhombic Bmm2 (38)	$a = 5.6946$ $b = 3.9714$ $c = 5.7203$	2	180.01	4.621	$T_p = 498$ $T_c = 703$ $T_m = 1460$	

Material	Crystal system (space group)	Lattice parameters (Å)	Z	M	ρ	Temperatures (°C)	
Lithium iodate (LiIO$_3$)	Hexagonal P6$_3$ (173)	$a = 5.4815$ $c = 5.1709$	2	181.84	4.487	$T_p = 520$ $T_m = 693$	80.3
LBO (LiB$_3$O$_5$) (lithium triborate)	Orthorhombic Pna2$_1$ (33)	$a = 8.4473$ $b = 5.1395$ $c = 7.3788$	2	119.37	2.993	$T_m = 1107$	
BBO (Ba$_3$[B$_3$O$_6$]$_2$) (β-barium metaborate)	Hexagonal R3 (146)	$a = 12.547$ $c = 12.736$	6	668.88	3.84	$T_p = 1170$	
Tellurium (Te)	Hexagonal P3$_1$21 (152)	$a = 4.44693$ $c = 5.91492$	3	127.60	6.27	$T_p = 621$ $T_m = 723$	insol.
Cadmium selenide (CdSe)	Hexagonal P6$_3$mc (186)	$a = 4.30$ $c = 7.02$	2	191.36	5.653	$T_m > 1625$	insol.
Urea ([NH$_2$]$_2$CO) (carbamide, carbonyl diamide)	Tetragonal P42$_1$m (113)	$a = 5.661$ $c = 4.712$	2	60.06	1.321	$T_m = 408$	Very high

*Notation: T_m is the melting temperature; T_p is the phase change temperature; T_c is the Curie temperature; T_d is the decomposition temperature (no melt).

11.49

TABLE 11.4b Thermal and Mechanical Properties of Selected Nonlinear Crystals

Material	Heat capacity, J/g·K	Thermal conductivity, W/m·K	Thermal expansion coeff. (10^{-6}/K)	Moduli, GPa			Poisson's ratio	Strength, MPa	Knoop hardness, kg/mm²	Elastic constants references
				Elastic	Shear	Bulk				
ADP	1.26			29	11	27.9	0.32_5			3, 4, 35, 36
‖ to c axis		0.71	10.7							
⊥ to c axis		1.26	27.2							
KDP	0.88	2.1		(38)	(15)	(28)	(0.26)			3, 4, 35, 36
‖ to c axis			39.2							
⊥ to c axis			22.0							
KTP	0.728									
		2 (a)	11							
		3 (b)	9							
		3.3 (c)	0.6							
Lithium niobate	0.63	5.6		170	68	112	0.25		~5	3, 35, 36, 146
‖ to c axis			(9.0)*							
⊥ to c axis			4.1							
			14.8							
Potassium niobate			(37)							35, 147

Lithium iodate				55	22.4	33.5	0.23			3, 35
∥ to c axis			48							
⊥ to c axis			28							
LBO									~500	
BBO				30	11	60.6	0.41			148
∥ to c axis			36							
⊥ to c axis			4							
Tellurium	0.202			39	16	25	0.24	11	18	3, 35
∥ to c axis		3.5	−1.6							
⊥ to c axis		4.0	27.5							
Cadmium selenide	0.252	(9)		42	15.3	53	0.37		44–90	35
∥ to c axis										
⊥ to c axis										
Urea	1.551			~9	~3	17	0.41			35
∥ to c axis										
⊥ to c axis										

(a) (b) (c) Crystal axes.
*(9,0) Polycrystalline average.

11.51

TABLE 11.4c Optical Properties of Selected Nonlinear Crystals

Material	Optical transparent region,* μm	Refractive index				Nonlinear optical coefficients		Damage threshold, GW/cm²
		Optical			DC			
		n_x	$dn_x/dT \cdot 10^6$	Ref.	n_o	d_{ij}, pm/V	Ref.	
ADP	0.185–1.45					At 1.06 μm:	35, 44	0.5
o-ray		1.516	−47.1	149, 272	3.74	$d_{14} = 0.72$		
e-ray		1.470	−4.3	149, 272	7.56	$d_{36} = 0.76$		
KDP	0.176–1.42					At 1.06 μm:	35, 44	0.5
o-ray		1.502	−39.6	149, 272	6.8	$d_{14} = 0.58$		
e-ray		1.460	−28.3	149, 272		$d_{36} = 0.63$		
KTP x	[0.35–4.5]	1.734	11	150	3.41	At 1.06 μm:	151	1
y		1.741	13	150	3.32	$d_{15} = 6.1$ $d_{31} = 6.5$		
z		1.820	16	150	3.92	$d_{24} = 7.6$ $d_{32} = 5.0$		
						$d_{33} = 13.7$		
Lithium niobate	[0.5–5.0]					At 1.06 μm:	35, 44	0.1
o-ray		2.214	44	152, 279	9.2	$d_{15} = 6.9$ $d_{31} = -5.0$		
e-ray		2.140	37.9	152, 279	5.3	$d_{22} = 2.3$ $d_{33} = -29.7$		

Material	Region	n	angle	Ref.	value	d coefficients		Ref.	value
Potassium niobate x	[0.4–5.0]	2.19	23	153, 280	6.1	At 1.06 μm:	$d_{31} = -12.9$	35, 182	0.35
y		2.22	−34	153, 280	27.9	$d_{15} = -12.4$	$d_{32} = 11.3$		
z		2.09	63	153, 280	4.9	$d_{24} = 11.9$	$d_{33} = -19.6$		
Lithium iodate						At 1.06 μm:			
o-ray	0.38–[5.5]	1.883	−8.9	44	2.6	$d_{14} = 0.3$	$d_{31} = -5.0$	35, 44	0.1
e-ray	0.38–[5.5]	1.737	−7.8	44	2.5	$d_{15} = 5.0$	$d_{33} = -5.0$		
LBO x	0.17–[2.5]	1.566		154		At 1.06 μm:	$d_{31} = -1.5$	154	7.5
y		1.590		154		$d_{15} = -1.4$	$d_{32} = 1.7$	270	
z		1.607		154		$d_{24} = 1.6$	$d_{33} = 0.1$		
BBO	0.205–3.0					At 1.06 μm:			
o-ray		1.540	−16.6	148	2.59	$d_{15} \approx d_{31}$	$d_{31} < 0.1$	148	5.0
e-ray		1.655	−9.3	148	2.85	$d_{22} = 1.6$	$d_{33} < 0.1$		
Tellurium	3.5–[32]					At 10.6 μm:			
o-ray		4.778		155	5.74	$d_{11} = 920$		35	0.05
e-ray		6.222		155	7.28	$d_{14} =$			
Cadmium selenide	[0.53–20]					At 10.6 μm:			
o-ray		2.448		155	3.09	$d_{15} = 31.0$	$d_{31} = -29$	35	0.06
e-ray		2.467		155	3.19		$d_{33} = 55$		
Urea	[0.21–1.4]					At 1.06 μm:			
o-ray		1.464		156		$d_{14} = 1.4$		44	1.5
e-ray		1.581		156		$d_{36} = 1.35$			

*Region with absorption coefficient <1 cm^{-1}.

TABLE 11.4d Sellmeier Dispersion Formula Parameters for Selected Nonlinear Crystals

Material	λ_1	λ_2	λ_3	A_1	A_2	A_3	Range, μm	Temp., °C	Ref.
				Sellmeier equation parameters					
ADP									
o-ray	0.0944632	33.47508		1.298990	43.17064		0.4–1.06	33	149
e-ray	0.0926997	28.84136		1.162166	12.01997		0.4–1.06	33	149
KDP									
o-ray	0.0919120	33.37520		1.256618	33.89909		0.4–1.06	33	149
e-ray	0.0902551	28.49129		1.131091	5.75675		0.4–1.06	33	149
KTP									
n_x	0.0	0.21473	15.281	2.16747	0.83733	4.0	0.4–1.06		157
n_y	0.0	0.22293	15.709	2.19229	0.83547	4.0	0.4–1.06		157
n_z	0.0	0.23422	13.672	2.25411	1.06543	4.0	0.4–1.06		157
Lithium niobate									
o-ray	0.0	0.217	16.502	1.39198	2.51118	7.1333	0.4–3.1	RT	152
e-ray	0.0	0.210	25.915	1.32468	2.25650	14.503	0.4–3.1	RT	152
Potassium niobate									
n_x	0.2118			3.9328			0.49–1.06	RT	153
n_y	0.1969			3.7936			0.49–1.06		153
n_z	0.1857			3.3836			0.49–1.06		153

Lithium iodate									
o-ray	0.0	0.0	0.18731	13.0	1.03132	1.37623	1.06745	0.5–5	RT	158
e-ray	0.0	0.0	0.17715	12.6	0.83086	1.08807	0.55458	0.5–5	RT	158
LBO									
n_x	0.0	0.0	0.1065	10.0	0.4630	0.9912	1.388	0.25–1.06	RT	159
n_y	0.0	0.0	0.1090	10.0	0.4650	0.0740	1.848	0.25–1.06	RT	159
n_z	0.0	0.0	0.1106	10.0	0.5154	1.0711	1.861	0.25–1.06	RT	159
BBO									
o-ray	0.0	0.0	0.1338	10.0	0.7126	1.0279	1.535	0.2–1.06	RT	148
e-ray	0.0	0.0	0.1249	10.0	0.5525	0.8205	0.423	0.2–1.06	RT	148
Tellurium									
o-ray	0.0	0.0	1.9952	108.69	17.5346	4.3289	3.7800	4.0–14.0		155
e-ray	0.0	0.0	1.6052	116.28	28.5222	9.3068	9.2350	4.0–14.0		155
o-ray	0.0	0.0	1.0757	100.0	3.0164	18.8133	7.3729	8.5–30.3	RT	155
e-ray	0.0	0.0	1.0394	100.0	0.9041	36.8133	6.2456	8.5–30.3	RT	155
Cadmium selenide									
o-ray	0.0	0.0	0.4764	58.14	3.2243	1.7680	3.1200	1.0–22.0	RT	155
e-ray	0.0	0.0	0.4659	60.24	3.2009	1.8875	3.6461	1.0–22.0	RT	155
Urea									
o-ray	0.0	0.0	0.1732	10.0	0.7256	0.4167	1.451	0.3–1.06	RT	156
e-ray	0.0	0.0	0.1732		0.7007	0.8000		0.3–1.06	RT	156

TABLE 11.5a Physical Properties of Selected Metals

Material	Crystal structure and space group	Unit cell lattice parameters, Å	Formulas per unit cell, Z	Molecular weight, amu	Density, g/cm³	Melting point,* K	Oxidation potential, V
Aluminum (Al)	Cubic Fm3m (225)	4.04958	4	26.98154	2.699	$T_m = 933.4$	+1.66
Beryllium (Be)	Hexagonal P6$_3$/mmc (194)	$a = 2.2680$ $c = 3.5942$	2	9.01218	1.857	$T_m = 1550$	+1.85
Copper (Cu)	Cubic Fm3m (225)	3.61496	4	63.546	8.934	$T_m = 1356.2$	−0.337
Gold (Au)	Cubic Fm3m (225)	4.07825	4	196.9665	19.288	$T_m = 1336.2$	−1.50 Does not oxidize
Iron (Fe)	Cubic Im3m (229)	2.8664	2	55.847	7.875	$T_c = 1043$ $T_p = 1185$ $T_p = 1667$ $T_m = 1810$	+0.44
Magnesium (Mg)	Hexagonal P6$_3$/mmc (194)	$a = 3.20927$ $c = 5.21033$	2	24.305	1.737	$T_m = 923$	+2.37
Nickel (Ni)	Cubic Fm3m (225)	3.524	4	58.69	8.903	$T_c = 631$ $T_m = 1728$	+0.250
Platinum (Pt)	Cubic Fm3m (225)	3.9237	4	195.09	21.452	$T_m = 2042$	−1.2 Does not oxidize
Silver (Ag)	Cubic Fm3m (225)	4.0862	4	107.8682	10.502	$T_m = 1234.0$	−0.7991 Oxide marginally stable at 300 K, unstable above

*Notation: T_m is the melting temperature; T_p is the phase change temperature; T_c is the Curie temperature.

TABLE 11.5b Mechanical Properties of Selected Metals

Material	Moduli, GPa			Poisson's ratio	Yield strength, MPa		Knoop hardness, kg/mm²	Fracture toughness, MPa·m$^{1/2}$	Elastic constants references
	Elastic	Shear	Bulk		Ultrapure*	Alloy†			
Aluminum	70.3	26.2	75.2	0.345	10	47.4	32		13, 27, 28
Beryllium	310	150	115	0.046	120–250	250–260		9–13	13, 27, 28
Copper	129.8	48.3	137.8	0.343	34	220	48		13, 27, 28
Gold	78	27	217	0.44	125		60		13, 27, 28
Iron	205	81.0	165	0.282					13, 27, 28
Magnesium (polycrystalline)	44.7	17.3	35.6	0.291		180			13, 27, 28
Nickel	197	78.5	190.3	0.296		130			13, 27, 28
Platinum	168	61	228	0.377		140			27, 28
Silver	82.7	30.3	103.6	0.367		130			13, 27, 28

*Polycrystalline; yield strength of single crystals can be extremely low.
†High purity (greater than 99 percent) annealed alloy.

TABLE 11.5c Thermal Properties of Selected Metals

Material	Heat capacity, J/g·K	Debye temperature, K	Electronic heat capacity, μJ/g·K²	Thermal conductivity (W/m·K)			Thermal expansion coefficient, K⁻¹, α·10⁶	Total hemispherical emissivity		
				200 K	500 K	Melt		Polished, 300 K	Oxidized, 500 K	Oxidized, 1000 K
Aluminum	0.9030	428	50.03	237	237	211	23.1	0.02	0.12 (0.03)*	0.2 (0.05)*
Beryllium	1.9736	1440	18.86	(350)	140		11.4			
Copper	0.3858	343	10.94	413	388	330	16.5	0.03	0.45 (0.02)*	0.8 (0.03)*
Gold	0.1288	165	3.70	327	309	247	14.2	0.02	(0.035)*	(0.056)*
Iron	0.4462	470	89.17	94	61.5	32.6†	11.8	0.06	0.50	0.62
Magnesium ∥ c axis ⊥ c axis	1.0244	410	53.5	159	151	145	24.8 25.1 24.4	0.07	0.12	
Nickel	0.4411	450	119.6	106	72.1	63.8†	13.4	0.054	0.3–0.5	0.55–0.6
Platinum	0.1326	240	34.9	70	70	80+	8.8	0.04	(0.05)*	(0.104)*
Silver	0.2367	225	5.99	430	413	355	18.9	0.01	(0.02)*	(<0.03)*

*Polished, not oxidized.
†Phase transition temperature.

TABLE 11.5d Optical Properties of Selected Metals

Material	Complex index of refraction								Free carrier parameters		
	0.40 μm	0.60 μm	0.80 μm	1.0 μm	2.5 μm	5.0 μm	10.0 μm	Ref.	$\nu_p \cdot 10^{-4}$	$\nu_c \cdot 10^{-2}$	Ref.
Aluminum	0.40 + 3.92i	0.97 + 6.00i	1.99 + 7.05i		3.3 + 26i	8.6 + 48i	30 + 96i	160 161	11.9	6.60	171
Beryllium	2.98 + 2.20i	2.64 + 2.27i	2.7 + 2.8i	2.6 + 3.1i	2.75 + 7.25i	6.85 + 14.3i	11.9 + 20i	4	5.96	0.732	171
Copper	0.85 + [2i] 1.48 + 2.0i	0.17 + 3.07i 0.34 + 2.47i	0.12 + 5.07i	0.35 + 6.8i	1.71 + 17.6i	4.0 + 30i 2.21 + 31.7i	11.0 + 55i 11.6 + 60.3i	160 162 163 164 165			
Gold	1.45 + [1.9i]	0.23 + 2.97i	0.16 + 4.84i	0.22 + 6.71i	0.82 + 17.3	3.27 + 35.2	11.5 + 67.5i	160 166	7.28	2.15	171
Iron		[2.84 + 3.3i] 19.8 + 3.9i	3.0 + 3.7i	3.24 + 4.26i	4.14 + 8.02i	4.97 + 15i	6.3 + 30i	167	3.30	1.47	171
Nickel		2.48 + 4.38i	2.85 + 5.10i	4.03 + 9.64i	4.25 + 17.7i	7.0 + 37i		168 169	3.94	3.52	171
Platinum		[2.3 + 4i]	2.8 + 5i	3.5 + 5.8i	3.9 + 7.7i	5.8 + 23i	10 + 38i	170	4.15	5.58	171
Silver	0.075 + 1.93i	0.060 + 3.75i	0.90 + 5.45i	0.28 + 7.8i 0.22 + 6.71i	0.67 + 18.3i 0.82 + 17.3i	3.27 + 35.2i	11.5 + 67.5i	160 165 166	7.27	1.45	171

TABLE 11.6a Physical Properties of Selected Glasses

Glass type	Selected glass code	Density, g/cm^3	Example composition (for the general type)	Temp., K	
				Glass	Soften
Deep crown	479587 TiK1	2.39	Alkali alumo-borosilicate glass	613	
Fluor crown	487704 FK5	2.45	(boro)Phosphide glass w. high fluoride content	737	945
Titanium flint	511510 TiF1	2.47	Titanium alkali alumoborosilicate glass	716	[892]
Borosilicate	517642 BK7	2.51	70%SiO$_2$ 10%B$_2$O$_3$ 8%Na$_2$O 8%K$_2$O 3%BaO 1%CaO	836	989
Phosphate crown	518651 PK2	2.51	70%P$_2$O$_5$ 12%K$_2$O 10%Al$_2$O$_3$ 5%CaO 3%B$_2$O$_3$	841	994
Crown	522595 K5	2.59	74%SiO$_2$ 11%K$_2$O 9%Na$_2$O 6%CaO	816	993
Crown flint	523515 KF9	2.71	67%SiO$_2$ 16%Na$_2$O 12%PbO 3%ZnO 2%Al$_2$O$_3$	718	934
Light barium crown	526600 BaLK1	2.70	Borosilicate glass	782	954
Antimony flint	527511 KzF6	2.54	Antimony borosilicate glass	717	
Zinc crown	533580 ZK1	2.71	71%SiO$_2$ 17%Na$_2$O 12%ZnO	835	1005
Extra light flint	548458 LLF1	2.94	63%SiO$_2$ 24%PbO 8%K$_2$O 5%Na$_2$O	721	901
Dense phosphate crown	552635 PSK3	2.91	60%P$_2$O$_5$ 28%BaO 5%Al$_2$O$_3$ 3%B$_2$O$_3$	875	1009
Barium crown	573575 BaK1	3.19	60%SiO$_2$ 19%BaO 10%K$_2$O 5%ZnO 3%Na$_2$O 3%B$_2$O$_3$	875	1019

Name	Glass	Density	Composition		
Light barium flint	580537 BaLF4	3.17	51%SiO$_2$ 20%BaO 14%ZnO 6%Na$_2$O 5%K$_2$O 4%PbO	842	1004
Light flint	581409 LF5	3.22	53%SiO$_2$ 34%PbO 8%K$_2$O 5%Na$_2$O	692	858
Special long crown	586610 LgSK2	4.15	Alkali earth aluminum fluoroborate glass	788	
Fluor flint*	593355 FF5	2.64		788	843
Dense barium crown	613586 SK4	3.57	39%SiO$_2$ 41%BaO 15%B$_2$O$_3$ 5%Al$_2$O$_3$	916	1040
Special short flint	613443 KzFSN4	3.20	Aluminum lead borate glass	765	867
Extra dense barium crown	618551 SSK4	3.63	35%SiO$_2$ 42%BaO 10%B$_2$O$_3$ 8%ZnO 5%Al$_2$O$_3$	912	1064
Flint	620364 F2	3.61	47%SiO$_2$ 44%PbO 7%K$_2$O 2%Na$_2$O	705	866
Dense barium flint	650392 BaSF10	3.91	43%SiO$_2$ 33%PbO 11%BaO 7%K$_2$O 5%ZnO 1%Na$_2$O	757	908
Barium flint*	670472 BaF10	3.61	46%SiO$_2$ 22%PbO 16%BaO 8%ZnO 8%K$_2$O	853	908
Lanthanum crown	720504 LaK10	3.81	Silicoborate glass w. rare earth oxides	893	976
Tantalum crown*	741526 TaC2	4.19	B$_2$O$_3$/La$_2$O$_3$/ThO$_2$ rare earth oxides	928	958
Niobium flint*	743492 NbF1	4.17		863	898
Lanthanum flint	744447 LaF2	4.34	Borosilicate glass w. rare earth oxides	917	1013
Dense flint	805254 SF6	5.18	33%SiO$_2$ 62%PbO 5%K$_2$O	696	811
Dense tantalum flint*	835430 TaFD5	4.92	B$_2$O$_3$/La$_2$O$_3$/ThO$_2$/T$_2$O$_3$	943	973

*Hoya glasses; others are Schott glasses.

TABLE 11.6b Thermal and Mechanical Properties of Selected Glasses

Glass type	Selected glass code	Heat capacity, J/g·K	Thermal conductivity, W/m·K	Thermal expansion, 10^{-6}/K	Elastic modulus, GPa	Poisson's ratio	Knoop hardness, kg/mm^2
Deep crown	479587 TiK1	0.842	0.773	10.3	40	0.254	330
Fluor crown	487704 FK5	0.808	0.925	9.2	62	0.205	450
Titanium flint	511510 TiF1	[0.81]	[0.953]	9.1	58	0.239	440
Borosilicate	517642 BK7	0.858	1.114	7.1	81	0.208	520
Phosphate crown	518651 PK2	[0.80]	1.149	6.9	84	0.209	520
Crown	522595 K5	0.783	0.950	8.2	71	0.227	450
Crown flint	523515 KF9	[0.75]	[1.01]	6.8	67	0.202	440
Light barium crown	526600 BaLK1	0.766	1.043	9.1	68	0.234	430
Antimony flint	527511 KzF6	[0.82]	[0.946]	5.5	52	0.212	380
Zinc crown	533580 ZK1	[0.77]	[0.894]	7.5	68	0.240	430
Extra light flint	548458 LLF1	[0.71]	[0.960]	8.1	60	0.210	390
Dense phosphate crown	552635 PSK3	[0.72]	[1.004]	8.6	84	0.226	510
Barium crown	573575 BaK1	0.687	0.795	7.6	74	0.253	460

Light barium flint	580537 BaLF4	0.670	0.827	6.4	76	0.244	460
Light flint	581409 LF5	0.657	0.866	9.1	59	0.226	410
Special long crown	586610 LgSK2	[0.51]	0.866	12.1	76	0.290	340
Fluor flint*	593355 FF5	[0.80]	[0.937]	8.6	[65]	[0.238]	500
Dense barium crown	613586 SK4	0.582	0.875	6.4	82	0.268	500
Special short flint	613443 KzFSN4	[0.64]	[0.769]	5.0	60	0.276	380
Extra dense barium crown	618551 SSK4	[0.57]	[0.806]	6.1	79	0.265	460
Flint	620364 F2	0.557	0.780	8.2	58	0.225	370
Dense barium flint	650392 BaSF10	[0.54]	[0.714]	8.6	67	0.256	400
Barium flint*	670472 BaF10	0.569	0.967	7.2	95	0.277	610
Lanthanum crown	720504 LaK10	[0.53]	[0.814]	5.7	111	0.288	580
Tantalum crown*	741526 TaC2	[0.48]	[0.861]	5.2	117	0.299	715
Niobium flint*	743492 NbF1	[0.48]	[0.845]	5.3	108	0.308	675
Lanthanum flint	744447 LaF2	[0.47]	[0.695]	8.1	93	0.289	480
Dense flint	805254 SF6	0.389	0.673	8.1	56	0.248	310
Dense tantalum flint*	835430 TaFD5	[0.41]		6.4	126	0.299	790

Values in brackets are estimated using methods given in Ref. 9.
*Hoya glasses; others are Schott glasses.

TABLE 11.6c Optical Properties of Selected Glasses

Glass type	Selected glass code	Transparent region,* μm	Refractive index n_d	v_d	$dn_o/dT \cdot 10^6$	Refs.
Deep crown	479587 TiK1	0.35–	1.47869	58.70	−3.5	25
Fluor crown	487704 FK5	0.29–(2.5)	1.48749	70.41	−2.9	25
Titanium flint	511510 TiF1	0.36–	1.51118	51.01	−0.8	25
Borosilicate	517642 BK7	0.31–2.6	1.51680	64.17	1.3	25
Phosphate crown	518651 PK2	0.31–(2.5)	1.51821	65.05	1.4	25
Crown	522595 K5	0.315–2.76	1.52249	59.48	0.2	25
Crown flint	523515 KF9	0.33–(2.5)	1.52341	51.49	2.7	25
Light barium crown	526600 BaLK1	0.31–(2.6)	1.52642	60.03	−1.1	25
Antimony flint	527511 KzF6	0.325–	1.52682	51.13	5.0	25
Zinc crown	533580 ZK1	0.31–	1.53315	57.98	3.7	25
Extra light flint	548458 LLF1	0.32–	1.54814	45.75	1.8	25
Dense phosphate crown	552635 PSK3	0.315–(2.5)	1.55232	63.46	−1.6	25
Barium crown	573575 BaK1	0.31–2.7	1.57250	57.55	1.2	25

Light barium flint	580537 BaLF4	0.335–(2.6)	1.57957	53.71	3.7	25
Light flint	581409 LF5	0.32–(2.5)	1.58144	40.85	1.3	25
Special long crown	586610 LgSK2	0.35–	1.58599	61.04	-4.9	25
Fluor flint†	593355 FF5	0.36–	1.59270	35.45	-0.9‡	26
Dense barium crown	613586 SK4	0.325–2.68	1.61272	58.63	1.2	25
Special short flint	613443 KzFSN4	0.325–2.38	1.61340	44.30	1.6	25
Extra dense barium crown	618551 SSK4	0.33–2.9	1.61765	55.14	1.5	25
Flint	620364 F2	0.33–2.73	1.62004	36.37	2.7	25
Dense barium flint	650392 BaSF10	0.34–(2.6)	1.65016	39.15	2.3	25
Barium flint†	670472 BaF10	0.35–2.7	1.67003	47.20	3.0‡	26
Lanthanum crown	720504 LaK10	0.345–(2.5)	1.72000	50.41	3.2	25
Tantalum crown†	741526 TaC2	0.33–	1.74100	52.59	5.1‡	26
Niobium flint†	743492 NbF1	0.325–	1.74330	59.23	6.2‡	26
Lanthanum flint	744447 LaF2	0.36–(2.5)	1.74400	44.72	-0.8	25
Dense flint	805254 SF6	0.37–2.74	1.80518	25.43	9.3	25
Dense tantalum flint†	835430 TaFD5	0.34–	1.83500	42.98	2.8‡	26

*Region with absorption coefficient <1 cm^{-1}.
†Hoya glasses; others are Schott glasses.
‡Measured at 632.8 μm.

TABLE 11.6d Schott Dispersion Formula Parameters for Selected Glasses

Selected glass code	A_0	A_1	A_2	A_3	A_4	A_5
				Schott equation coefficients		
479587 TiK1	2.1573978	$-8.4004189 \cdot 10^{-3}$	$1.0457582 \cdot 10^{-2}$	$2.1822593 \cdot 10^{-4}$	$-5.5063640 \cdot 10^{-6}$	$5.4469060 \cdot 10^{-7}$
487704 FK5	2.1887621	$-9.5572007 \cdot 10^{-3}$	$8.9915232 \cdot 10^{-3}$	$1.4560516 \cdot 10^{-4}$	$-5.2843067 \cdot 10^{-6}$	$3.4588010 \cdot 10^{-7}$
511510 TiF1	2.2473124	$-8.9044058 \cdot 10^{-3}$	$1.2493525 \cdot 10^{-2}$	$4.2650638 \cdot 10^{-4}$	$-2.1564809 \cdot 10^{-6}$	$2.6364065 \cdot 10^{-6}$
517642 BK7	2.2718929	$-1.0108077 \cdot 10^{-2}$	$1.0592509 \cdot 10^{-2}$	$2.0816965 \cdot 10^{-4}$	$-7.6472538 \cdot 10^{-6}$	$4.9240991 \cdot 10^{-7}$
518651 PK2	2.2770533	$-1.0532010 \cdot 10^{-2}$	$1.0188354 \cdot 10^{-2}$	$2.9001564 \cdot 10^{-4}$	$-1.9602856 \cdot 10^{-6}$	$1.0967718 \cdot 10^{-6}$
522595 K5	2.2850299	$-8.6010725 \cdot 10^{-3}$	$1.1806783 \cdot 10^{-2}$	$2.0765657 \cdot 10^{-4}$	$-2.1314913 \cdot 10^{-6}$	$3.2131234 \cdot 10^{-7}$
523515 KF9	2.2824396	$-8.5960144 \cdot 10^{-3}$	$1.3442645 \cdot 10^{-2}$	$2.7803535 \cdot 10^{-4}$	$-4.9998960 \cdot 10^{-7}$	$7.7105911 \cdot 10^{-7}$
526600 BaLK1	2.2966923	$-8.2975549 \cdot 10^{-3}$	$1.1907234 \cdot 10^{-2}$	$1.9908305 \cdot 10^{-4}$	$-2.0306838 \cdot 10^{-6}$	$3.1429703 \cdot 10^{-7}$
527511 KzF6	2.2934044	$-1.0346122 \cdot 10^{-2}$	$1.3319863 \cdot 10^{-2}$	$3.4833226 \cdot 10^{-4}$	$-9.9354090 \cdot 10^{-6}$	$1.1227905 \cdot 10^{-6}$
533580 ZK1	2.3157951	$-8.7493905 \cdot 10^{-3}$	$1.2329645 \cdot 10^{-2}$	$2.6311112 \cdot 10^{-4}$	$-8.2854201 \cdot 10^{-6}$	$7.3735801 \cdot 10^{-7}$
548458 LLF1	2.3505162	$-8.5306451 \cdot 10^{-3}$	$1.5750853 \cdot 10^{-2}$	$4.2811388 \cdot 10^{-4}$	$-6.9875718 \cdot 10^{-6}$	$1.7175517 \cdot 10^{-6}$
552635 PSK3	2.3768193	$-1.0146514 \cdot 10^{-2}$	$1.2167148 \cdot 10^{-2}$	$1.1916606 \cdot 10^{-4}$	$6.4250627 \cdot 10^{-6}$	$-1.7478706 \cdot 10^{-7}$
573575 BaK1	2.4333007	$-8.4931353 \cdot 10^{-3}$	$1.3893512 \cdot 10^{-2}$	$2.6798268 \cdot 10^{-4}$	$-6.1946101 \cdot 10^{-6}$	$6.2209005 \cdot 10^{-7}$

580537 BaLF4	2.4528366	$-9.2047678 \cdot 10^{-3}$	$1.4552794 \cdot 10^{-2}$	$4.3046688 \cdot 10^{-4}$	$-2.0489836 \cdot 10^{-5}$	$1.5924415 \cdot 10^{-6}$
581409 LF5	2.4441760	$-8.3059695 \cdot 10^{-3}$	$1.9000697 \cdot 10^{-2}$	$5.4129697 \cdot 10^{-4}$	$-4.1973155 \cdot 10^{-6}$	$2.3742897 \cdot 10^{-6}$
586610 LgSK2	2.4750760	$-5.4304528 \cdot 10^{-3}$	$1.3893210 \cdot 10^{-2}$	$2.2990560 \cdot 10^{-4}$	$-1.6868474 \cdot 10^{-6}$	$4.3959703 \cdot 10^{-7}$
593355 FF5*	2.4743324	$-1.0955338 \cdot 10^{-2}$	$1.9293801 \cdot 10^{-2}$	$1.4497732 \cdot 10^{-3}$	$-1.1038744 \cdot 10^{-4}$	$1.1136008 \cdot 10^{-5}$
613586 SK4	2.5585228	$-9.8824951 \cdot 10^{-3}$	$1.5151820 \cdot 10^{-2}$	$2.1134478 \cdot 10^{-4}$	$3.4130130 \cdot 10^{-6}$	$1.2673355 \cdot 10^{-7}$
613443 KzFSN4	2.5293446	$-1.3234586 \cdot 10^{-2}$	$1.8586165 \cdot 10^{-2}$	$5.4759655 \cdot 10^{-4}$	$-1.1717987 \cdot 10^{-5}$	$2.0042905 \cdot 10^{-6}$
618551 SSK4	2.5707849	$-9.2577764 \cdot 10^{-3}$	$1.6170751 \cdot 10^{-2}$	$2.7742702 \cdot 10^{-4}$	$1.2686469 \cdot 10^{-7}$	$4.504790 \cdot 10^{-7}$
620364 F2	2.5554063	$-8.8746150 \cdot 10^{-3}$	$2.2494787 \cdot 10^{-2}$	$8.6924972 \cdot 10^{-4}$	$-2.4011704 \cdot 10^{-5}$	$4.5365169 \cdot 10^{-6}$
650392 BaSF10	2.6531250	$-8.1388553 \cdot 10^{-3}$	$2.2995643 \cdot 10^{-2}$	$7.3535957 \cdot 10^{-4}$	$-1.3407390 \cdot 10^{-5}$	$3.6962325 \cdot 10^{-6}$
670472 BaF10*	2.7324621	$-1.2490460 \cdot 10^{-2}$	$1.8562334 \cdot 10^{-2}$	$9.9990536 \cdot 10^{-4}$	$-6.8388552 \cdot 10^{-5}$	$4.9257931 \cdot 10^{-6}$
720504 LaK10	2.8984614	$-1.4857039 \cdot 10^{-2}$	$2.0985037 \cdot 10^{-2}$	$5.4506921 \cdot 10^{-4}$	$-1.7297314 \cdot 10^{-5}$	$1.7993601 \cdot 10^{-6}$
741526 TaC2*	2.9717137	$-1.4952593 \cdot 10^{-2}$	$2.0162868 \cdot 10^{-2}$	$9.4072283 \cdot 10^{-4}$	$-8.8614104 \cdot 10^{-5}$	$5.3191242 \cdot 10^{-6}$
743492 NbF1*	2.9753491	$-1.4613470 \cdot 10^{-2}$	$2.1096383 \cdot 10^{-2}$	$1.1980380 \cdot 10^{-3}$	$-1.1887388 \cdot 10^{-4}$	$7.3444350 \cdot 10^{-6}$
744447 LaF2	2.9673787	$-1.0978767 \cdot 10^{-2}$	$2.5088607 \cdot 10^{-2}$	$6.3171596 \cdot 10^{-4}$	$-7.5645417 \cdot 10^{-6}$	$2.3202213 \cdot 10^{-6}$
805254 SF6	3.1195007	$-1.0902580 \cdot 10^{-2}$	$4.1330651 \cdot 10^{-2}$	$3.1800214 \cdot 10^{-3}$	$-2.1953184 \cdot 10^{-4}$	$2.6671014 \cdot 10^{-5}$
835430 TaFD5*	3.2729098	$-1.2888257 \cdot 10^{-2}$	$3.3451363 \cdot 10^{-2}$	$-6.8221381 \cdot 10^{-5}$	$1.1215427 \cdot 10^{-4}$	$-4.0485659 \cdot 10^{-6}$

*Hoya glasses; others are Schott glasses.

TABLE 11.7a Physical Properties of Selected Specialty Glasses and Plastics

Glass type	Density, g/cm^3	Typical composition	Temp., K Glass	Temp., K Soften	Temp., K Melt
Fused silica (SiO$_2$) (e.g., Corning 7940)	2.648	100% SiO$_2$	1273		1983
Calcium aluminate glass, BS-39B	3.1	50% CaO, 34% Al$_2$O$_3$, 9% MgO	1023	(970)	
CORTRAN 9754	3.581	33% GeO$_2$, 20% CaO, 37% Al$_2$O$_3$, 5% BaO, 5% ZnO	786	873	
ULTRAN 30 (548743)	4.02				
ZBL	4.78	62% ZrF$_4$, 33% BaF$_2$, 5% LaF$_3$	580		820
ZBLA	4.61	58% ZrF$_4$, 33% BaF$_2$, 5% LaF$_3$, 4% AlF$_3$	588		820
ZBLAN	4.52	56% ZrF$_4$, 14% BaF$_2$, 6% LaF$_3$, 4% AlF$_3$, 20% NaF	543		745
ZBT	4.8	60% ZrF$_4$, 33% BaF$_2$, 7% ThF$_4$	568	723	
HBL	5.78	62% HfF$_4$, 33% BaF$_2$, 5% LaF$_3$	605		832
HBLA	5.88	58% HfF$_4$, 33% BaF$_2$, 5% LaF$_3$, 4% AlF$_3$	580		835
HBT	6.2	60% HfF$_4$, 33% BaF$_2$, 7% ThF$_4$	593		853
Arsenic trisulfide (As$_2$S$_3$)	3.43	100% As$_2$S$_3$	436	573	
Arsenic triselenide (As$_2$Se$_3$)	4.69	100% As$_2$Se$_3$		345	
AMTIR-1/TI-20	4.41	55% Se, 33% Ge, 12% As	635	678	
AMTIR-3/TI-1173	4.70	60% Se, 28% Ge, 12% Sb	550	570	
Plexiglas/Lucite (acrylic)	1.19	Poly(methyl methacrylate) (PMMA)	375		
Lexan 101	1.25	Polyester carbonate copolymer	420	410	608
Polystyrene	1.05	Poly(styrene)	360		516

TABLE 11.7b Thermal and Mechanical Properties of Selected Specialty Glasses and Plastics

Glass type	Heat capacity, J/g·K	Thermal conductivity, W/m·K	Thermal expansion, 10^{-6}/K	Elastic modulus, GPa	Poisson's ratio	Strength, MPa	Knoop hardness, kg/mm^2
Fused silica (SiO$_2$)	0.746	1.38	0.51	72.6	0.16$_4$	110	635
Calcium aluminate glass, BS-39B	0.865	1.23	8.0	104	0.29	90	760
CORTRAN 9754	0.54	~0.8	6.2	84.1	0.29	50	560
Ultran 30	0.58	0.667	11.9	76	0.297		380
ZBL	0.538		18.8		0.31		228
ZBLA	0.534		18.7	60.2	0.25	11	230–240
ZBLAN	0.633	0.4	17.5	60	0.31		225
ZBT	0.511		4.3	60	0.279	62	250
HBL	0.413		18.3				228
HBLA	0.414		17.3	56	0.3		240
HBT	0.428		6.0			62	250
Arsenic trisulfide (As$_2$S$_3$)	0.473	0.17	26.1	15.8	0.296	16.5	180
Arsenic triselenide (As$_2$Se$_3$)	0.349	0.205	24.6	18.3	0.288		120
AMTIR-1/TI-20	0.293	0.25	12.0	21.9	0.266	18.6	170
AMTIR-3/TI-1173	0.276	0.22	14.0	21.7	0.265	17.2	150
Plexiglas/Lucite	1.40	0.193	67.9	3.30	0.36		
Lexan 101	1.25	0.19	67.5	2.40	0.41	62	
Polystyrene	1.20	0.154	70.0	3.20	0.325	35	

TABLE 11.7c Optical Properties of Selected Specialty Glasses and Plastics

Glass type	Transparent region,* μm	Refractive index			Refs.
		Wavelength, μm	n	$dn/dT \cdot 10^6$	
Fused silica (SiO$_2$)	0.16–3.8	0.5893	1.438	10	88
Calcium aluminate glass, BS-39B	0.38–4.9	0.5893	1.676	7.4	172
		3.0	1.672		
CORTRAN 9754	0.36–4.8	0.5893	1.664		9, 173
		1.3	1.66		
Ultran 30	0.22–3.95	0.5893	1.54830	–5.8	25
ZBL	0.25–7.0	0.5893	1.523		174, 175
ZBLA	0.29–7.0	0.5893	1.521		174–176
ZBLAN	0.24–6.9	0.5893	1.480	14.7	176
ZBT	0.32–6.8				177, 178
HBL	0.25–7.3	0.5893	1.498		175
HBLA	0.29–7.3	0.5893	1.504		175, 179
HBT	0.22–7.7				178
Arsenic trisulfide (As$_2$S$_3$)	0.62–11.0	0.60	2.63646	85	180, 181
		1.00	2.47773	17	
		10.6	2.37698	~0	
		∞	2.73		

Material	Region*	Wavelength (µm)	n	dn/dT (×10⁻⁶/°C)	Ref.
Arsenic triselenide (As₂Se₃)	0.87–17.2	1.0	2.93	40	182, 183
		8–10	2.7789		
		12	2.7728		
		∞	3.29		
AMTIR-1/TI-20	0.75–(14.5)	1.0	2.6055	101	144, 182
		3.0	2.5187	77	
		10.0	2.4976	72	
AMTIR-3/TI-1173	0.93–16.5	3.0	2.6366	98	144, 182
		5.0	2.6173	92	
		8.0	2.6088	87	
		12.0	2.5983	93	
Plexiglas/Lucite	0.36–1.63	0.4861	1.5014	-105	184, 185
		0.5893	1.4950		
		0.6328	1.4934		
		0.6563	1.4928		
Lexan 101	0.35–1.63	0.4861	1.5995	-107	184, 185
		0.5893	1.5854		
		0.6328	1.5816		
		0.6563	1.5800		
Polystyrene	0.39–1.63	0.5893	1.595	-142	184
		0.4861	1.604		186
		0.5893	1.590	-120	
		0.6563	1.585		

*Region with absorption coefficient < 1 cm⁻¹.

TABLE 11.7d Sellmeier Dispersion Formula Parameters for Various Glasses

Material	Sellmeier equation parameters						Range, μm	Temp., °C	Ref.
	λ_1	λ_2	λ_3	A_1	A_2	A_3			
Fused silica	0.0684043	0.1162414	9.896161	0.6961663	0.4079426	0.8974794	0.21–3.71	20	78
BS-39B	0.1155	14.981		1.7441	1.6465		0.43–4.5	RT	172*
ULTRAN-30	0.089286	7.9103		1.36689	0.34711		0.36–2.3	RT	25*
ZBL	0 $\lambda_4 = 37.9$	0.182	20.7	1.03 $A_4 = 1.97$	0.265	1.22	1–5		187
ZBLA	0.0969	26.0		1.291	2.76		0.64–4.8		176†
ZBLAN	0.0954	25		1.168	2.77		0.50–4.8		176†
HBL	0 $\lambda_4 = 41.5$	0.172	20.8	0.96 $A_4 = 2.22$	0.299	0.86	1–5		187
Arsenic trisulfide	0.150 $\lambda_4 = 0.450$	0.250 $\lambda_5 = 27.3861$	0.350	1.8983678 $A_4 = 0.1188704$	1.9222979 $A_5 = 0.9569903$	0.8765134	0.56–12.0	25	180
AMTIR-1/TI-20	0.29007	32.022		5.298	0.6039		1.0–14.0	25	144*
AMTIR-3/TI-1173	0.29192	42.714		5.8505	1.4536		3.0–14.0	25	144*
	0.29952	38.353		5.8357	1.064		0.9–14.0	Room	188†
FK5 glass	0.0776227030	0.138959626	9.93162512	1.03630719	0.15210770	0.91316627	0.37–2.33		43
BaLF4 glass	0.0856548405	0.173243878	10.8069635	1.25385390	0.19811351	1.01615191	0.37–2.33		43
KzFSN4	0.0948292206	0.201806158	8.28807544	1.38374965	0.16462681	0.85913757	0.37–2.33		43
SF13 glass	0.117985319	0.249318140	2.67826498	1.68311631	0.22881310	0.05174839	0.40–1.06		43

*Our fit to referenced data.
†Sellmeier parameters from Wemple index formula.

Thermal Expansion. Thermal expansion (second rank tensor) is expressed in terms of the fractional change in length for a temperature rise of 1 Kelvin. This measure is called the thermal expansion coefficient (α, $1/K$), and relates strain (second rank tensor) to temperature. Since the (linear dimension) thermal expansion coefficient is small, the volume expansion coefficient is 3α [and density varies as $\rho/(1 + 3\alpha\Delta T)$]. Thermal expansion arises from the anharmonic nature of the interatomic potential and generally increases with temperature. The magnitude of the thermal expansion coefficient is generally inversely proportional to the melting temperature. Higher-order thermal expansion terms are necessary to accurately model expansion over a wide temperature range. For the high-temperature oxide materials a second expansion term, $d\alpha/dT$, is given. Compilations of thermal expansion data are given in Refs. 28 and 33.

Total Hemispherical Emissivity. Total hemispherical emissivity (scalar) is estimated for the high-temperature oxides and metals. For oxides, the total emissivity can become a contributor to radiation relief, For metals, oxidation significantly changes the optical properties. Compilations of thermal radiative properties are given in Refs. 28 and 34.

11.5.3 Mechanical Properties

The primary mechanical properties are elastic properties, strength, and hardness. There is a hierarchy of elastic properties: interatomic force constants, the elastic constants (stiffness and compliance), and the technical (macroscopic) elastic moduli. The elastic constants (fourth rank tensor) relate stress (second rank tensor) to strain (second rank tensor). For nonisotropic materials (particularly single crystals), one must use the elastic constants (c_{xx} and s_{xx}), and the data tables give appropriate references (or estimates of the elastic constants) for some materials. For most other applications, the engineering moduli (elastic, shear, and bulk moduli plus Poisson's ratio) are sufficient. Isotropic materials have only two independent engineering moduli; the elastic (Young's) modulus and Poisson's ratio are reported. Engineering moduli for crystalline materials are either taken from measurements or estimated using the elastic constants and the Voigt and Reuss methods (noncubic materials) or the Haskin and Shtrickman method (cubic materials).[3,4,35,36]

Engineering Moduli. Elastic (or Young's) modulus (E, GPa), shear modulus (K, GPa), and bulk modulus (B, GPa) are given for all materials. Volume compressibility is the reciprocal of the bulk modulus. For a material in tension, Poisson's ratio (ν, dimensionless) is the negative ratio of transverse strain over longitudinal strain.

Strength. Strength values quoted are measured in flexure (units are MPa unless otherwise noted). Tensile strength values are typically 50 to 90 percent of flexure strength values. Strength is determined, in large part, by material imperfections and therefore is dependent on purity, preparation, surface quality, and the size of the tested sample (larger samples have a greater likelihood of a flaw, hence lower strength). Strength values, therefore, should be used only as a guide in material selection; design must be based on the characteristics of the specific supplier's material. (Strength is a tensor quantity, but seldom reported as such.)

Hardness and Fracture Toughness. These measures are empirical and relative. Hardness and toughness are related to both bond strength and material defects. They give an indication of a material's resistance to surface damage (Knoop indent test hardness, kg/mm^2) and crack propagation with applied stress (fracture toughness, K_{IC}, MPa·m$^{1/2}$). High values indicate greater resistance. Materials with Knoop hardness greater than 750 kg/mm^2 are quite hard; those with hardness of 100 kg/mm^2 are relatively soft, difficult to polish, and very susceptible to handling damage. Many alkali halides have a hardness less than 10 kg/mm^2 and must be handled with great care.

11.5.4 Optical Properties

Optical properties are presented in several tables. The first gives summary optical parameters, the second gives Sellmeier or Schott dispersion parameters for modeling the room-temperature index of refraction in the visible and infrared. For selected materials, additional tables give dn/dT, Sellmeier parameters for microwave index of refraction, and the infrared absorption edge.

Bandgap. The bandgap (a second rank tensor) is usually defined as the minimum energy for an electronic transition to occur, hence the onset of electronic absorption.[37] In practice, the bandgap is not well defined and may be found by extrapolating absorption or using an arbitrary criterion. We have tried to provide a consistent set of bandgap values (E_g, eV) corresponding to the approximate onset of "significant" absorption. Even thin films will not be transparent at energies (frequencies) above the bandgap. A large bandgap is frequently indicative of low refractive index. A compilation of bandgap energies is given in Ref. 37.

Range of Transparency. To provide a useful measure of transparency, we have selected an absorption coefficient less than 1 cm^{-1} as the criterion that defines the useful transparent range. This criterion allows meaningful comparison of materials and also defines the practical range of transparency for optical elements. The visible and infrared range of transparency is given in micrometers. Many materials also are transparent in the microwave (radar) region and may serve as a window or optical element for either optical or microwave or combined applications. The microwave window is determined using the same criterion of 1 cm^{-1} to define the practical range of transparency. This range is defined in terms of the maximum wavenumber (reciprocal of wavelength ν, cm^{-1}). Compilations of transparency data are given in Refs. 1 to 3, 11, 38, and 39.

Maximum Longitudinal Frequency. The maximum longitudinal frequency (ν_{LO}, wavenumbers) defines the onset of "significant" absorption beyond the fundamental lattice vibrations. Even thin films will not be transparent at frequencies (wavenumbers) below the maximum longitudinal frequency. This frequency is also significant in modeling the infrared (multiphonon) absorption edge.[19,40]

Index of Refraction. Refractive index is summarized in terms of $n\infty$, the net electronic contribution to the index of refraction and therefore a typical value in the middle of the transparent region, or for glasses by n_d (index at 587.56 nm) or other index value. For materials useful in the microwave region, n_0 (the square root of the static dielectric constant) and its temperature derivative are included. Complex index compilations are found in Refs. 5 to 7; real index data are also found in Refs. 1 to 4 and 11.

Dispersion is represented by Sellmeier (most crystalline materials) or Schott (most glasses) formulas. The Sellmeier or Drude (or Maxwell-Helmholtz-Drude[41]) dispersion model arises from treating the absorption like simple mechanical or electrical resonances. Sellmeier proposed the following dispersion formula in 1871 (although Maxwell had also considered the same derivation in 1869[42]):

$$n^2 - 1 = \sum_{i=1} \frac{A_i \lambda^2}{\lambda^2 - \lambda_i^2} \qquad (11.11)$$

An often used slight modification of this formula puts the first wavelength at zero ($\lambda_1 = 0$); i.e., the first term is a constant. Several other dispersion equations are modified forms of the Sellmeier equation or approximations to it. Power series approximations to the Sellmeier equation are expressed in many forms. One common form is the Schott glass formula:

$$n^2 = A_0 + A_1 \lambda^2 + A_2 \lambda^{-2} + A_3 \lambda^{-4} + A_4 \lambda^{-6} + A_5 \lambda^{-8} \qquad (11.12)$$

This equation is typically accurate to $\pm 3 \times 10^{-6}$ in the visible (400 to 765 nm) and within $\pm 5 \times 10^{-6}$ from 365 to 1014 nm (i.e., within the lot-to-lot variability of glasses). It is commonly used by glass manufacturers to characterize their products and has gained acceptance in optical design codes. A comparison of the Schott power series formula with a three-term Sellmeier formula showed equivalent accuracy of the range of the Schott fit, but that the Sellmeier model was accurate over a much wider wavelength range.[43]

For many glasses, the Abbe number v_d is also given. The Abbe number is a measure of dispersion in the visible and is defined as $v_d = (n_d - 1)/(n_F - n_C)$ where n_F and n_C are the refractive index at 486 and 656 nm. The temperature coefficient is given as a function of wavelength; for most glasses only dn_e/dT (temperature change of the index at 546 nm) is given.

For materials transparent at microwave wavelengths, Table 11.8 gives the low-frequency index of refraction in the form of a Sellmeier model. The data tables for insulators and high-bandgap semiconductors (Tables 11.1 to 11.3) list the region of microwave transparency.

Table 11.10 gives references for additional indexes of refraction, dispersion formulas, and temperature dependence of refractive index data, including many materials not listed in the data tables.

Absorption Coefficient.   Absorption at the infrared edge of insulators (especially highly ionic insulators) can be characterized by an exponential absorption coefficient[40] α of the form

$$\alpha = A \exp\left(-\gamma \frac{v}{v_o}\right) \qquad (11.13)$$

where A is a constant (dimensions same as the absorption coefficient), γ is a dimensionless constant typically around 4, v_0 is frequency or wavenumber of the maximum transverse optical frequency (units are cm^{-1} for wavenumbers; values are given in the property data tables) and v is the frequency or wavenumber of interest (units are cm^{-1} for wavenumbers). This formula works reasonably well for ionic materials in the range of absorption coefficients from 0.001 to 10 cm^{-1}. Table 11.9 lists these parameters for selected materials at room temperature. Calculated absorption values are good to 20 percent for ionic materials and within a factor of 2 for covalent materials. More accurate models for the infrared absorption edge, including temperature dependence, are available (see Ref. 19).

TABLE 11.8 Sellmeier Parameters for Microwave Index of Refraction

Material	$\epsilon(0)$	$\epsilon(\infty)$	ν_{min}, cm^{-1}	Refs.
ALON	9.278	4.867	403	51
Beryllium oxide				
o-ray	6.94	2.95	724	52
e-ray	7.65	2.99	680	52
Sapphire				
o-ray	9.4073	4.9240	428.4	62
e-ray	11.6155	4.5460	387.6	62
Spinel	8.1141	3.7992	440	64
Magnesium oxide	9.807	3.007	396	189
Quartz				
o-ray	4.3261	3.1878	403.6	62
e-ray	4.5648	2.1286	522	62
YAG	10.403	9.42	123	69
Yttrium oxide	11.422	8.067	194	70
Lithium fluoride	8.84	2.04	306	190
Sodium chloride	5.95	2.71	164	190
Potassium bromide	4.60	2.54	114	190
Potassium chloride	4.80	2.18	142	190
Potassium iodide	5.02	2.69	101	191
Cesium iodide	6.56	3.22	62	192
Magnesium fluoride				
o-ray	5.86	3.26	247	190
e-ray	4.41	1.92	420	190
Calcium fluoride	6.63	2.04	257	193
Strontium fluoride	6.20	2.07	217	193
Barium fluoride	6.96	2.15	184	193
Aluminum arsenide	10.1	8.2	361.8	194
Gallium arsenide	13.001	11.034	268.7	195
	12.8	10.9	268.8	196
Gallium phosphide	11.100	9.091	363.4	129
Zinc sulfide	8.41	5.10	271	197
Zinc selenide	9.12	5.93	203	197
Zinc telluride	10.10	7.28	178	197
Cadium sulfide				
o-ray	9.6	5.3	241	198
e-ray	10.7	5.4	234	198
Cadmium telluride	10.294	7.194	141	199
	9.4	6.7	140	196
Indium phosphide	12.40	9.61	303.7	200
Tellurium				
o-ray	30.0	23.0	92.4	201
e-ray	43.2	36.0	88.1	201
Arsenic trisulfide	7.46	6.49	180	202

Sellmeier formula for the microwave index of refraction:

$$n^2(\nu) = \epsilon(\infty) + \frac{[\epsilon(0) - \epsilon(\infty)] \cdot \nu_{min}^2}{\nu_{min}^2 - \nu^2}$$

11.76

TABLE 11.9 Parameters for Modeling Infrared Absorption Edge

Material	A, cm^{-1}	γ	ν_0, cm^{-1}	Refs.
ALON	41,255	4.96	969	19
Beryllium oxide	36,550	5.43	1090	*
Calcite				
o-ray	1,633	2.58	1549	1†
e-ray	456	1.88	890	1†
Sapphire				
o-ray	55,222	5.03	900	40
	94,778	5.28	914	19
e-ray	33,523	4.73	871	63
Spinel	147,850	5.51	869	64†
Magnesium oxide	41,420	5.29	725	189†
Quartz				
o-ray	107,000	4.81	1215	1†
e-ray	196,000	5.16	1222	1†
Yttrium oxide	184,456	5.36	620	19
Cubic zirconia	226,390	4.73	668	203
Lithium fluoride	21,317	4.39	673	40, 204
Sodium fluoride	41,000	5.0	425	191
Sodium chloride	24,273	4.79	268	40, 204
Potassium chloride	8,696	4.19	213	40, 204
Potassium bromide	6,077	4.25	166	40, 204
Potassium iodide	8,180	3.86	139	191†
Cesium iodide	12,800	3.88	85	205†
Magnesium fluoride	11,213	4.29	617	40
Calcium fluoride	105,680	5.1	482	40
Barium fluoride	49,641	4.5	344	40
Strontium fluoride	22,548	4.4	395	40
Thallium chloride	6	1.58	158	206
KRS-5	5,400	4.0	100	206
Gallium arsenide	2,985	3.69	292	195†
ß-Zinc sulfide	227,100	5.31	352	207†
Zinc selenide	179,100	5.33	250	208†
Cadmium telluride	5,460	4.25	168	209†
Calcium lanthanum sulfide	3,910	3.58	314	*
AMTIR-3	10,320	3.13	235	210†
Fused silica	54,540	5.10	1263	*
ZBT	$2.97 \cdot 10^5$	4.4	500	187
Arsenic trisulfide	108,650	4.10	350	202†
Arsenic triselenide	7,730	3.52	250	183†

*Estimated from our measurements.
†Our fit to the referenced data.

TABLE 11.10 Refractive Index Data for other Materials

Material	Wavelength range, μm	Dispersion formula	Thermooptic data	Reference	Notes
ADA and AD*A	0.4–1.06	Sellmeier	No	149	Also Zernike fit
ADP ($NH_4H_2PO_4$)	0.21–1.53	Zernike	No	211	$T = 24.8°C$
AD*P ($NH_4D_2PO_4$)	0.4–1.06	Sellmeier	No	149	Also Zernike fit
Aluminum oxide (Al_2O_3)	0.25–0.69	Cornu	Yes	65	Sapphire, hex. cryst., $T = 24°C$
	0.27–5.6	Sellmeier	No	212	o-ray
	1.0–5.6	Herzberger	No	213	o-ray; see Ref. 212
Aluminum phosphide ($AlPO_4$)	0.4–2.6	No	No	81	Hexagonal crystal
Ammonium oxalate ($C_2O_4(NH_4)_2 \cdot H_2O$)	0.45–1.4	No	No	268	Orthorhombic crystal
Arsenic selenide glass (As_2Se_3)	0.82–1.15	Sellmeier	Yes	214	
Arsenic sulfide glass (As_2S_3)	0.6–12.0	Herzberger	No	213	Also see Ref. 78
Barium fluoride (BaF_2)	0.5–11.0	Herzberger	No	213	
	0.55–1.8	No	Yes	215	$T = -200–23°C$
Barium formate ($Ba(COOH)_2$)	Visible	Yes	No	216	Orthorhombic crystal
Barium sulfate($BaSO_4$)	Visible	Sellmeier	Yes	79	Orthorhombic crystal
Barium titanate ($BaTiO_3$)	0.4–1.0	Sellmeier	No	217	
Beryllium oxide (BeO)	0.43–0.69	Sellmeier	Yes	53	$T = 22.4°C$
Bismuth germanate ($Bi_4Ge_3O_{12}$)	0.4–0.64	Sellmeier	No	218	Cubic crystal
Bismuth silicate ($Bi_4Si_3O_{12}$)					
Bismuth germanate ($Bi_{12}GeO_{20}$)	1.2–5.0	Herzberger	No	219	Cubic crystal
	0.4–0.7	Sellmeier	(Yes)	220	Several dispersion fits
Bismuth silicate ($Bi_{12}SiO_{20}$)	0.4–0.7	No		221	Same index as $Bi_{12}GeO_{20}$
BGZA glasses	0.4–5.3	Polynomial	Yes	222	$T = 25°C$ and others

Material	Range (μm)	Formula		Ref.	Comments
Cadmium selenide (CdSe)	0.8–4.0	No	No	81	Hexagonal crystal
	1.0–22	Sellmeier	No	172	Hexagonal crystal
	10–25	Sellmeier	No	223	Hexagonal crystal
Cadmium sulfide (CdS)	0.51–14.0	No	No	224	Hexagonal crystal; $T = 25°C$
Cadmium telluride (CdTe)	2–30	No	Yes	225	Polycrystalline; $T = 24$ and 298 K
	2–30	Sellmeier	Yes	226	Fit to data of Ref. 225
	7.5–31.3	Sellmeier	No	132	
Calcium aluminate glass	0.6–4.3	Herzberger	No	213	Bausch & Lomb sample
Calcium fluoride (CaF₂)	0.6–8.3	Herzberger	No	213	Data from Ref. 253
	0.55–1.8	No	Yes	215	$T = -180$ and 25°C
Calcium molybdate (CaMoO₄)	0.45–3.8	No	No	81	Tetragonal crystal
Calcium tungstate (CaWO₄)	0.45–2.4	No	No	81	Tetragonal crystal
	0.45–1.5	No	Yes	215	$T = -180$–20°C
CDA and CD*A	0.4–1.06	Sellmeier	No	149	Also Zernike fit
Cerium fluoride (CeF₃)	0.35–0.64	Sellmeier	No	227	Biaxial crystal
Cesium bromide (CsBr)	2–30	Sellmeier	Yes	228	Also Herzberger fit; 80–300 K
Cesium chloride (CsCl)	0.18–40.0	Sellmeier	Yes	92	Critique
Cesium fluoride (CsF)	0.15–30.0	Sellmeier	Yes	92	Critique
Cesium iodide (CsI)	2–30	Sellmeier	Yes	228	Also Herzberger fit; 80–300 K
Chalcopyrite compounds*	Transparent region	Polynomial	No	229	11 tetragonal compounds
		Sellmeier	No	155	5 tetragonal compounds
		Sellmeier	Yes	230	3 tetragonal compounds
Copper(I) bromide (CuBr)	0.44–10	Sellmeier	Yes	231	Cubic crystal
Diamond (C)	0.23–∞	Herzberger	No	232	Analysis of data
Gadolinium gallium garnet (Gd₃Ga₅O₁₂ or GGG)	0.40–1.06	Zernike	No	90	$T = 26°C$
Gadolinium scandium gallium garnet (Gd₃Sc₂Ga₃O₁₂ or GSGG)	0.4–1.06	Zernike	Yes	90	$T = 26°C$

TABLE 11.10 (Continued)

Material	Wavelength range, μm	Dispersion formula	Thermooptic data	Reference	Notes
Gallium arsenide (GaAs)	0.9–1.7	No	Yes	244	$T = 103, 187, 300$ K
Gallium phosphide (GaP)	0.5–4.0	No	No	81	Cubic crystal
Garnets	0.50–1.0	Sellmeier	No	233	10 garnet compounds
Germania/silica glasses (GeO_2/SiO_2)	0.4–4.0	Sellmeier	No	75	Various compositions; $T = 22$–$26°C$
Germanium (Ge)	1.9–18.0	Cauchy	Yes	118	Critique
	2.0–13.5	Herzberger	No	213	
Glass. Corning 1723	0.36–2.7	No	Yes	234	Aluminosilicate glass
Corning 7913	0.26–2.6	No	Yes	234	Vycor, optical grade
Corning 7940	0.23–3.4	No	Yes	234	Fused silica
Glass. Schott	0.365–1.01	Herzberger	No	235	24 types
Glass. doped silica	0.47–2.4	Sellmeier	No	236	9 compositions
	0.44–1.1	Sellmeier	No	237	8 compositions
	0.44–1.5	Sellmeier	No	238	6 compositions
Hafnia. cubic (HfO_2:Y_2O_3)	0.36–5.0	Sellmeier	Yes	239	$T = 20°C$; 9.6% yttria
IR-20 glass	0.5–5.0	Herzberger	No	213	Bausch & Lomb sample
KDA	0.4–1.06	Sellmeier	No	149	Zernike fit also
KDP (KH_2PO_4)	0.21–1.53	Zernike	No	211	$T = 24.8°C$
KD*P	0.4–1.06	Sellmeier	No	149	Zernike fit also
Lanthanum fluoride (LaF_3)	0.25–0.55	Cornu	No	240	Biaxial crystal
	0.35–0.68	Sellmeier	No	227	Biaxial crystal

Material	Wavelength range (μm)	Dispersion formula		Ref.	Comments
Lead molybdate (PbMoO$_4$)	0.45–0.67	No	No	241	Tetragonal crystal
Di-lead molybdate (Pb$_2$MoO$_5$)	0.4–0.7	Sellmeier	No	242	Monoclinic crystal
Lead telluride (PbTe)	4–12.5	No	Yes	140	T = 80, 130, 300 K
Lead titanate (PbTiO$_3$)	0.45–1.15	Sellmeier	No	243	Trigonal crystal
Lithium bromide (LiBr)	0.21–20.0	Sellmeier	Yes	92	Critique
Lithium chloride (LiCl)	0.17–16.0	Sellmeier	Yes	92	Critique
Lithium fluoride (LiF)	0.105–0.200	Damped	No	245	VUV data
	0.5–6.0	Herzberger	No	213	See Ref.101
Lithium indium sulfide (LiInS$_2$)	0.42–11.0	No	No	246	Orthorhombic crystal
Lithium iodate (LiIO$_3$)	0.46–1.06	Sellmeier	Yes	44	Hexagonal crystal
Lithium iodide (LiI)	0.25–25.0	Sellmeier	Yes	92	Critique
Lithium sulfate (Li$_2$SO$_4$ · H$_2$O)	0.36–1.7	No	No	268	Monoclinic crystal
Lithium tantalate (LiTaO$_3$)	0.45–4.0	No	No	81	Hexagonal crystal
Magnesium barium fluoride (MgBaF$_4$)	Visible	Yes	No	216	Orthorhombic crystal
Magnesium fluoride (MgF$_2$)	0.15–10.0	Sellmeier	Yes	247	Critique
	0.115–0.50	No	No	248	(o-ray index?)
	1.0–6.7	Herzberger	No	213	Kodak IRTRAN 1
Magnesium oxide (MgO)	0.5–5.5	Herzberger	No	213	See Ref. 77
	0.25–0.70	No	Yes	249	T = 23°C
α-Mercury sulfide (HgS)	0.62–11.0	Sellmeier	No	155	Uniaxial crystal
Neodymium fluoride (NdF$_3$)	0.35–0.64	Sellmeier	No	227	Biaxial crystal
Potassium chloride (KCl)	0.25–15.0	Sellmeier	Yes	84	T near 20°C; several grades
Potassium malate (K$_2$C$_4$H$_4$O$_5$· 1½H$_2$O)	0.25–0.70	Sellmeier	No	250	Monoclinic crystal
Potassium tartrate (K$_2$C$_4$H$_4$O$_6$· ½H$_2$O)	0.36–1.4	No	No	268	Monoclinic crystal
Praseodymium fluoride (PrF$_3$)	0.35–0.64	Sellmeier	No	227	Biaxial crystal

TABLE 11.10 (Continued)

Material	Wavelength range, μm	Dispersion formula	Thermooptic data	Reference	Notes
RDA and RD*A	0.4–1.06	Sellmeier	No	149	Also Zernike fit
RDP and RD*P	0.4–1.06	Sellmeier	No	149	Also Zernike fit
Rubidium bromide (RbBr)	0.21–50.0	Sellmeier	Yes	92	Critique
Rubidium chloride (RbCl)	0.18–40.0	Sellmeier	Yes	92	Critique
Rubidium fluoride (RbF)	0.15–25.0	Sellmeier	Yes	92	Critique
Rubidium iodide (RbI)	0.24–64.0	Sellmeier	Yes	92	Critique
Selenium (Se)	1.06–10.6	No	No	251	Trigonal crystal, $T = 23°C$
Silicon (Si)	1.2–14.0	Cauchy	Yes	118	Critique
	1.3–11.0	Herzberger	No	213	
Silicon carbide (SiC)	0.45–0.69	Cauchy	No	120	Uniaxial forms
	0.47–0.69	No	No	252	Cubic (ß) form
Silicon oxide (SiO₂)	0.5–4.3	Herzberger	No	213	Fused quartz: GE sample
	0.55–3.1	No	No	253	o-ray only
	0.14–0.23	No	No	254	Quartz
Silver antimony sulfide (Ag₃SbS₃)	1.5–10.6	Sellmeier	No	255	Trigonal crystal Pyrargyrite
Silver arsenic sulfide (Ag₃AsS₃)	0.59–4.62	Yes	No	256	Trigonal crystal Proustite
Sodium bromate (NaBrO₃)	0.30–0.77	Sellmeier	No	257	Cubic crystal
Sodium chlorate (NaClO₃)	0.23–0.72	Sellmeier	No	257	Cubic crystal
Strontium fluoride (SrF₂)	0.15–14	Sellmeier	Yes	247	Critique
Strontium molybdate (SrMoO₄)	0.45–2.4	No	No	81	Tetragonal crystal
Strontium titanate (SrTiO₃)	1.0–5.3	Herzberger	No	213	Unknown orientation

Material	Range (μm)	Formula		Ref.	Comments
Tellurium dioxide (TeO$_2$)	0.4–1.0	Sellmeier	No	258	Tetragonal tellurite crystal
	0.47–1.15	Sellmeier	No	259	
Thallium bromide (TlBr)	10–24	No	No	260	$T = 45°C$
Thallium chloride (TlCl)	10–18	No	No	260	$T = 45°C$
Titanium dioxide (TiO$_2$)	0.45–2.4	No	No	81	Tetragonal crystal
	0.44–1.53	Cauchy	No	80	Tetragonal rutile crystal
Yttrium aluminum garnet (Y$_3$Al$_5$O$_{12}$)	0.4–4.0	No	No	81	Cubic crystal
Yttrium iron garnet (Y$_3$Fe$_5$O$_{12}$)	1.4–5.5	Sellmeier	No	261	Y-Gd iron garnet similar
Yttrium lithium fluoride (YLiF$_4$)	0.225–2.6	No	No	262	Tetragonal crystal
Yttrium orthoaluminate (YAlO$_3$)	0.40–1.10	Sellmeier	No	159	Biaxial crystal
Zinc germanium phosphide (ZnGeP$_2$)	0.64–12.0	Sellmeier	Yes	263	See Refs. 155, 229, and 230
Zinc oxide (ZnO)	0.8–4.0	No	No	81	$T = 80, 300$ K
Zinc selenide (ZnSe)	2–8	No	Yes	140	Hexagonal crystal; $T = 25°C$
α-Zinc sulfide (ZnS)	0.36–1.4	No	No	224	Cubic crystal
β-Zinc sulfide (ZnS)	0.45–2.4	No	No	81	Natural sphalerite crystal
	0.37–1.53	Cauchy	No	80	Kodak IRTRAN 2 sample
	1.0–13.5	Herzberger	No	213	CVD Cleartran sample
	0.4–13.0	Power	No	264	Based on data from Ref. 141
	0.5–14.0	Yes	No	265	IRTRAN-2
	1.0–10.5	No	Yes	266	
Zinc tungstate (ZnWO$_4$)	0.42–3.6	No	No	81	Monoclinic crystal
Zirconia, cubic (ZrO$_2$:Y$_2$O$_3$)	0.36–5.5	Sellmeier	No	267	11–33% yttria

Reference 229 gives index data for the following compounds of space group Ī42d: AgGaS$_2$,, AgGaSe$_2$,*, CdGeAs$_2$,*, AgInSe$_2$, CdGeP$_2$,*, CuAlS$_2$, CuGaS$_2$, CuGaSe$_2$, CuInS$_2$, ZnGeP$_2$*, and ZnSiAs$_2$. Compounds marked with asterisk also found in Ref. 155. Reference 230 gives the temperature dependence of the index of refraction for CdGeP$_2$, CuGaS$_2$, and ZnGeP$_2$.

Free-carrier contributions to the optical constants can be accurately modeled by a plasma frequency and a relaxation frequency. A complex dielectric constant that includes conduction (free charge) effects $\hat{\epsilon}_c$ can be defined as

$$\hat{\epsilon}_c = \hat{\epsilon}_r(\text{bound charge}) - \frac{v_p^2}{v^2 + i \cdot v \cdot \gamma_c} \tag{11.14}$$

where v_p is the plasma frequency (strength) term (usually in wavenumbers, cm^{-1}) and γ_c is the relaxation frequency (or width) term (also usually in wavenumbers). For metals, free-carrier effects dominate the infrared (typically below 1000 cm^{-1}) optical constants. Table 11.5d gives these parameters for selected metals. Semiconductors and insulators also exhibit free-carrier absorption, albeit at much longer wavelengths. Unlike metals, the free-carrier concentration is highly dependent on purity and temperature. Absorption is proportional to the free-carrier concentration.

Nonlinear Optical Coefficients. There are many higher-order coefficients that describe the optical behavior of materials. While a complete catalog of these properties is well beyond the scope of this chapter, it is necessary to at least allude to an important parameter, the second-order nonlinear susceptibility. With the high electric fields generated by lasers, the nonlinear susceptibility gives rise to important processes such as second-harmonic generation, optical rectification, parametric mixing, and the linear electrooptic (Pockels) effect. The nonlinear optical coefficient d, the usual term used to describe these nonlinear properties, is a third-rank tensor equal to one-half of the second-order nonlinear susceptibility $\chi^{(2)}$. Nonlinear optical coefficients are universally written in reduced index notation d_{ij}, where the index $i = 1$, 2, or 3 and the index j runs from 1 to 6.[44] (The usual electrooptic coefficient r_{ij} is proportional to d_{ij}.) Table 11.4c lists values of the nonlinear optical coefficient in mks units (m/V) for common nonlinear optical materials.

Scatter. Scatter is both an intrinsic and extrinsic property. Rayleigh, Brillouin, Raman, and stoichiometric (index variation) contributions to scatter have been derived in simple form and used to estimate scatter loss in several fiber-optical materials.[45] These contributions to scatter typically have a λ^{-4} dependence. Polycrystalline materials have additional grain-boundary scatter that has a λ^{-m} dependence, where the parameter m typically lies between 1 and 2.[46,47]

11.6 REFERENCES

1. W. G. Driscoll (ed.), *Handbook of Optics*, McGraw-Hill, New York, 1978.

2. W. L. Wolfe and G. J. Zissis (eds), *The Infrared Handbook*, Environmental Research Institute of Michigan, 1985.

3. M. J. Weber (ed.), *Handbook of Laser Science and Technology*, vol. IV: *Optical Materials*, Part II, CRC Press, Boca Raton, Fla., 1986.

4. D. E. Gray (ed.), *American Institute of Physics Handbook*, 3d ed. McGraw-Hill, New York, 1972.

5. E. D. Palik (ed.), *Handbook of Optical Constants of Solids*, Academic Press, Orlando, 1985.

6. E. D. Palik (ed.), *Handbook of Optical Constants of Solids II*, Academic Press, Orlando, 1991.

7. B. O. Seraphin and H. E. Bennett, "Optical Constants," in R. K. Willardson and A. C. Beer (eds.), *Semiconductors and Semimetals*, vol. 3: *Optical Properties of III-V Compounds*," Academic Press, New York, 1967.

8. S. Musikant (ed.), *Optical Materials: A Series of Advances*, Marcel Dekker, New York, 1990.

9. F. V. Tooley (ed.), *The Handbook of Glass Manufacture*, Ashlee Publishing Co., New York, 1985.

10. S. Musikant, *Optical Materials: An Introduction to Selection and Application*, Marcel Dekker, New York, 1985.

11. L. N. Durvasula and N. P. Murarka, *Handbook of the Properties of Optical Materials*, GACIAC HB 84-01, IIT Research Institute, Chicago, Ill. (January 1984).

12. I. W. Donald and P. W. McMillan, "Review of Infrared Transmitting Materials," *J. Mater. Sci.*, vol. 13, pp. 1151–1176, 1978.

13. G. W. Fynn and W. J. A. Powell, *Cutting and Polishing Optical and Electronic Materials*, 2d ed. Adam Hilger, Bristol, 1988.

14. W. Zschommler, *Precision Optical Glassworking*, SPIE, Bellingham, Wash., 1986.

15. T. Izawa and S. Sudo, *Optical Fibers: Materials and Fabrication*, KTK Scientific Publishers, Tokyo, 1980.

16. P. A. Tick and P. L. Bocko, "Optical Fiber Materials," in S. Musikant (ed.), *Optical Materials: A Series of Advances*, Marcel Dekker, New York, pp. 147–322, 1990.

17. P. Yeh, *Optical Waves in Layered Media*, Wiley, New York, 1988.

18. T. Skettrup, "Urback's Rule Derived from Thermal Fluctuations in the Band Gap Energy," *Phys. Rev. B*, vol. 18, pp. 2622–2631, 1978.

19. M. E. Thomas, R. I. Joseph, and W. J. Tropf, "Infrared Transmission Properties of Sapphire, Spinel, Yttria, and ALON as a Function of Temperature and Frequency," *Appl. Opt.*, vol. 27, pp. 239–245, 1988.

20. R. T. Holm, "Convention Confusions," in E. D. Palik (ed.), *Handbook of Optical Constants of Solids II*, Academic Press, Orlando, pp. 21–55, 1991.

21. A. J. Bosman and E. E. Havinga, "Temperature Dependence of Dielectric Constants of Cubic Ionic Compounds," *Phys. Rev.*, vol. 129, pp. 1593–1600, 1963.

22. J. D. H. Donnay and H. M. Ondik (eds.), *Crystal Data Determination Tables*, 3d ed. U.S. Department of Commerce, 1973.

23. R. W. G. Wyckoff, *Crystal Structures*, Wiley, New York, 1963.

24. R. C. Weast (ed.), *CRC Handbook of Chemistry and Physics*, 71st ed., CRC Press, Boca Raton, Fla., 1990.

25. Schott Glass Technologies, Duryea, Pa.

26. Hoya Optics, Inc., Fremont, Calif.

27. S. Sakka and J. D. MacKenzie, "Relation between Apparent Glass Transition Temperature and Liquidus Temperature for Inorganic Glasses," *J. Non-Cryst. Sol.*, vol. 6, pp. 145–162, 1971.

28. Y. S. Touloukian (ed.), *Thermophysical Properties of Matter*, IFI/Plenum, New York, 1970.

29. I. Barin and O. Knacke, *Thermophysical Properties of Inorganic Substances*, Springer-Verlag, Berlin, 1973.

30. H. M. Ledbetter, "Estimation of Debye Temperature by Averaging Elastic Coefficients," *J. Appl. Phys.*, vol. 44, pp. 1451–1454, 1973.

31. J. N. Plendl, "Some New Interrelations in the Properties of Solids Based on Anharmonic Cohesive Forces," *Phys. Rev.*, vol. 123, pp. 1172–1180, 1961.

32. G. A. Slack, "The Thermal Conductivity of Nonmetallic Crystals," in H. Ehrenreich, F. Seitz, and D. Turnbull (eds.), *Solid State Physics*, vol. 34, Academic Press, New York, 1979.

33. R. S. Krishnan, R. Srinivasan, and S. Devanarayanan, *Thermal Expansion of Crystals*, Pergamon Press, Oxford, 1979.

34. Aleksader Sala, *Radiant Properties of Materials*, Elsevier, Amsterdam, 1986.

35. K.-H. Hellwege and A. M. Hellwege (eds.), *Landolt-Börnstein Numerical Data and Functional Relationships in Science and Technology, New Series, Group III: Crystal and Solid State Physics*, vol. 11: *Elastic, Piezoelectric, Pyroelectric, Piezooptic, Electrooptic Constants and Nonlinear Susceptibilities of Crystals*, Springer-Verlag, Berlin, 1979.

36. G. Simmons and H. Wang, *Single Crystal Elastic Constants and Calculated Aggregate Properties: A Handbook*, MIT Press, Cambridge, Mass., 1971.

37. W. H. Stehlow and E. L. Cook, "Compilation of Energy Band Gaps in Elemental and Binary Compound Semiconductors and Insulators," *J. Phys. Chem. Ref. Data*, vol. 2, pp. 163–199, 1973.

38. D. E. McCarthy, "Reflection and Transmission of Infrared Materials: 1. Spectra from 2–50 microns," *Appl. Opt.*, vol. 2, pp. 591, 1963; "2. Bibliography," *Appl. Opt.*, vol. 2, pp. 596, 1963; "3. Spectra from 2 μ to 50 μ," *Appl. Opt.*, vol. 4, p. 317, 1965; "4. Bibliography," *Appl. Opt.*, vol. 4, p. 507, 1965.

39. A. Smakula, "Synthetic Crystals and Polarizing Materials," *Opt. Acta*, vol. 9, pp. 205–222, 1962.

40. T. F. Deutsch, "Absorption Coefficient of Infrared Laser Window Materials," *J. Phys. Chem. Solids*, vol. 34, pp. 2091–2104, 1973.

41. P. G. Nutting, "Dispersion Formulas Applicable to Glass," *J. Opt. Soc. Am.*, vol. 2–3, pp. 61–65, 1919.

42. J. W. S. Rayleigh, "The Theory of Anomalous Dispersion," *Phil. Mag.*, vol. 48, pp. 151–152, 1889.

43. B. Tatian, "Fitting Refractive-Index Data with the Sellmeier Dispersion Formula," *Appl. Opt.*, vol. 23, pp. 4477–4485, 1984.

44. S. Singh, "Nonlinear Optical Properties," in M. J. Weber (ed.), *Handbook of Laser Science and Technology*, vol. III: *Optical Materials*, Part I, CRC Press, Boca Raton, Fla. 1986.

45. M. E. Lines, "Scattering Loss in Optic Fiber Materials. I. A New Parametrization," *J. Appl. Phys.*, vol. 55, pp. 4052–4057, 1984; "II. Numerical Estimates," *J. Appl. Phys.*, vol. 55, pp. 4058–4063, 1984.

46. J. A. Harrington and M. Sparks, "Inverse-Square Wavelength Dependence of Attenuation in Infrared Polycrystalline Fibers," *Opt. Lett.*, vol. 8, pp. 223–225, 1983.

47. D. D. Duncan and C. H. Lange, "Imaging Performance of Crystalline and Polycrystalline Oxides," *Proc. SPIE*, vol. 1326, pp. 59–70, 1990.

48. I. J. Fritz and R. A. Graham, "Second Order Elastic Constants of High-Purity Vitreous Silica," *J. Appl. Phys.*, vol. 45, pp. 4124–4125, 1974.

49. H. M. Kandil, J. D. Greiner, and J. F. Smith, "Single Crystal Elastic Constants of Yttria-Stabilized Zirconia in the Range 20° to 700°C," *J. Am. Ceram. Soc.*, vol. 67, pp. 341–346, 1984.

50. M. Stead and G. Simonis, "Near Millimeter Wave Characterization of Dual Mode Materials," *Appl. Opt.*, vol. 28, pp. 1874–1876, 1989.

51. W. J. Tropf and M. E. Thomas, "Aluminum Oxynitride (ALON) Spinel," in E. D. Palik (ed.), *Handbook of Optical Constants of Solids II*, Academic Press, Orlando, pp. 775–785, 1991.

52. E. Loh, "Optical Phonons in BeO Crystals," *Phys. Rev.*, vol. 166, pp. 673–678, 1968.

53. H. W. Newkirk, D. K. Smith, and J. S. Kahn, "Synthetic Bromellite. III. Some Optical Properties," *Am. Mineralogist*, vol. 51, pp. 141–151, 1966.

54. K. F. Young and H. P. R. Frederikse, "Compilation of the Static Dielectric Constant of Inorganic Solids," *J. Phys. Chem. Ref. Data*, vol. 2, pp. 313–410, 1973.

55. K. H. Hellwege, W. Lesch, M. Plihal, and G. Schaack, "Two Phonon Absorption Spectra and Dispersion of Phonon Branches in Crystals of Calcite Structure," *Z. Physik*, vol. 232, pp. 61–86, 1970.

56. R. K. Vincent, "Emission Polarization Study on Quartz and Calcite," *Appl. Opt.*, vol. 11, pp. 1942–1945, 1972.

57. A. A. Volkov, G. V. Kozlov, and A. M. Prokhorov, "Progress in Submillimeter Spectroscopy of Solid State," *Infrared Phys.*, vol. 29, pp. 747–752, 1989.

58. A. Kahan, J. W. Goodrum, R. S. Singh, and S. S. Mitra, "Polarized Reflectivity Spectrum of Tetragonal GeO_2," *J. Appl. Phys.*, vol. 42, pp. 4444–4446, 1971.

59. J. Fontanella, "Low Frequency Dielectric Constants of α-quartz, Sapphire, MgF_2, and MgO," *J. Appl. Phys.*, vol. 45, pp. 2852–2854, 1974.

60. A. S. Barker, "Infrared Lattice Vibrations and Dielectric Dispersion in Corundum," *Phys. Rev.*, vol. 132, pp. 1474–1481, 1963.

61. E. V. Loewenstein, D. R. Smith, and R. L. Morgan, "Optical Constants of Far Infrared Materials," *Appl. Opt.*, vol. 12, pp. 398–406, 1973.

62. E. E. Russell and E. E. Bell, "Optical Constants in the Far Infrared," *J. Opt. Soc. Am.*, vol. 57, pp. 543–544, 1967.

63. M. E. Thomas, "Infrared Properties of the Extraordinary Ray Multiphonon Processes in Sapphire," *Appl. Opt.*, vol. 28, pp. 3277–3278, 1989.

64. W. J. Tropf and M. E. Thomas, "Magnesium Aluminum Spinel ($MgAlO_4$)," in E. D. Palik (ed.), *Handbook of Optical Constants of Solids II*, Academic Press, Orlando, pp. 881–895, 1991.

65. M. A. Jeppesen, "Some Optical, Thermo-optical, and Piezo-optical Properties of Synthetic Sapphire," *J. Opt. Soc. Am.*, vol. 48, pp. 629–632, 1958.

66. M. Wintersgill, J. Fontanella, C. Andeen, and D. Schuele, "The Temperature Variation of the Dielectric Constant of 'Pure' CaF_2, SrF_2, BaF_2, and MgO," *J. Appl. Phys.*, vol. 50, pp. 8259–8261, 1979.

67. W. W. Ho, "High-Temperature Dielectric Properties of Polycrystalline Ceramics," in *Material Research Symposium Proceedings*, vol. 124, pp. 137–148, 1988.

68. G. Gervais and B. Piriou, "Temperature Dependence of Transverse- and Longitudinal-optic Modes in TiO_2 (Rutile)," *Phys. Rev. B*, vol. 10, pp. 1642–1654, 1974.

69. Y. Zhen and P. D. Coleman, "Far-IR Properties of $Y_3Al_5O_{12}$, $LiYF_4$, $Cs_2NaDyCl_6$, and Rb_2NaYF_6," *Appl. Opt.*, vol. 23, pp. 548–551, 1984.

70. W. J. Tropf and M. E. Thomas, "Yttrium Oxide (Y_2O_3)," in E. D. Palik (ed.), *Handbook of Optical Constants of Solids II*, Academic Press, Orlando, pp. 1081–1098, 1991.

71. D. L. Wood and K. Nassau, "Refractive Index of Cubic Zirconia Stabilized with Yttria," *Appl. Opt.*, vol. 21, pp. 2978–2981, 1982.

72. M. E. Thomas and R. I. Joseph, "Optical Phonon Characterization of Diamond, Beryllia, and Cubic Zirconia," *Proc. SPIE*, vol. 1326, pp. 120–126, 1990.

73. D. F. Edwards and R. H. White, "Beryllium Oxide," in E. D. Palik (ed.), *Handbook of Optical Constants of Solids II*, Academic Press, Orlando, pp. 805–814, 1991.

74. Calcite index data compiled from many sources as originally given in Ref. 4 and repeated in Refs. 1 and 2. These data are compiled from several sources and are not easily fit to a simple Sellmeier model. Short wavelength data have wide variations. This fit is good in the visible and near infrared.

75. J. W. Fleming, "Dispersion in GeO_2-SiO_2 Glasses," *Appl. Opt.*, vol. 23, pp. 4486–4493, 1984.

76. I. H. Malitson and M. J. Dodge, "Refractive Index and Birefringence of Synthetic Sapphire," *J. Opt. Soc. Am.*, vol. 62, pp. 1405A, 1972. Also see M. J. Dodge, "Refractive Index," in *Handbook of Laser Science and Technology*, vol. IV, *Optical Materials: Part 2*, p. 30, CRC Press, Boca Raton, 1986.

77. R. E. Stephens and I. H. Malitson, "Index of Refraction of Magnesium Oxide," *J. Natl. Bur. Stand.*, vol. 49, pp. 249–252, 1952.

78. I. H. Malitson, "Interspecimen Comparison of the Refractive Index of Fused Silica," *J. Opt. Soc. Am.*, vol. 55, pp. 1205–1209, 1965.

79. T. Radhakrishnan, "Further Studies on the Temperature Variation of the Refractive Index of Crystals," *Proc. Indian Acad. Sci.*, vol. A33, pp. 22–34, 1951.

80. J. R. DeVore, "Refractive Index of Rutile and Sphalerite," *J. Opt. Soc. Am.*, vol. 41, pp. 416–419, 1951.

81. W. L. Bond, "Measurement of the Refractive Index of Several Crystals," *J. Appl. Phys.*, vol. 36, pp. 1674–1677, 1965.

82. Y. Nigara, "Measurement of the Optical Constants of Yttrium Oxide," *Jap. J. Appl. Phys.*, vol. 7, pp. 404–408, 1968.

83. C. H. Lange and D. D. Duncan, "Temperature Coefficient of Refractive Index for Candidate Optical Windows," *SPIE Proc.*, vol. 1326, pp. 71–78, 1990.

84. A. Feldman, D. Horowitz, R. M. Walker, and M. J. Dodge, "Optical Materials Characterization Final Technical Report, Feb. 1, 1978–Sept. 30, 1978," *NBS Technical Note 993*, February 1979.

85. K. Vedam, J. L. Kirk, and B. N. N. Achar, "Piezo- and Thermo-Optic Behavior of Spinel ($MgAl_2O_4$)," *J. Sol. State Chem.*, vol. 12, pp. 213–218, 1975.

86. J. Tapping and M. L. Reilly, "Index of Refraction of Sapphire Between 24 and 1060°C for Wavelengths of 633 and 799 nm," *J. Opt. Soc. Am.*, A., vol. 3, pp. 610–616, 1986.

87. N. C. Fernelius, R. J. Harris, D. B. O'Quinn, M. E. Gangl, D. V. Dempsey, and W. L. Knecht, "Some Optical Properties of Materials Measured at 1.3 μm," *Opt. Eng.*, vol. 22, pp. 411–418, 1983.

88. W. S. Rodney and R. J. Spindler, "Index of Refraction of Fused-quartz Glass for Ultraviolet, Visible, and Infrared Wavelengths," *J. Res. Natl. Bur. Stand.*, vol. 53, pp. 185–189, 1954.

89. H. G. Lipson, Y. F. Tsay, B. Bendow, and P. A. Ligor, "Temperature Dependence of the Refractive Index of Alkaline Earth Fluorides," *Appl. Opt.*, vol. 15, pp. 2352–2354, 1976.

90. C. S. Hoefer, K. W. Kirby, and L. G. Deshazer, "Thermo-optic Properties of Gadolinium Garnet Laser Crystals," *J. Opt. Soc. Am. B.*, vol. 5, pp. 2327–2332, 1988.

91. P. Blanchfield and G. A. Saunders, "The Elastic Constants and Acoustic Symmetry of $LiYF_4$," *J. Phys. C*, vol. 12, pp. 4673–4689, 1979.

92. H. H. Li, "Refractive Index of Alkali Halides and Its Wavelength and Temperature Derivatives," *J. Phys. Chem. Ref. Data*, vol. 5, pp. 329–528, 1976.

93. S. S. Mitra, "Infrared and Raman Spectra Due to Lattice Vibrations," in S. Nudelman and S. S. Mitra (eds.), *Optical Properties of Solids*, pp. 333–451, Plenum, New York, 1969.

94. M. J. Dodge, "Refractive Properties of Magnesium Fluoride," *Appl. Opt.*, vol. 23, pp. 1980–1985, 1984.

95. R. P. Lowndes, "Dielectric Response of Alkaline Earth Fluorides," *J. Phys. C*, vol. 2, pp. 1595–1605, 1969.

96. N. P. Barnes and D. J. Gettemy, "Temperature Variation of the Refractive Indices of Yttrium Lithium Fluoride," *J. Opt. Soc. Am.*, vol. 70, pp. 1244–1247, 1980.

97. D. E. McCarthy, "Refractive Index Measurements of Silver Bromide in the Infrared," *Appl. Opt.*, vol. 12, p. 409, 1973.

98. H. Schröter, "Über die Brechungsindizes einiger Schwermetallhalogenide im Sichtbaren und die Berechnung von Interpolationsformeln für den Dispensionverlauf," *Z. Phys.*, vol. 67, pp. 24–36, 1931 (in German).

99. T. Barth, "Some New Immersion Melts of High Refraction," *Am. Mineral.*, vol. 14, pp. 358–361, 1929.

100. W. S. Rodney and I. H. Malitson, "Refraction and Dispersion of Thallium Bromide Iodide," *J. Opt. Soc. Am.*, vol. 46, pp. 956–961, 1956.

101. L. W. Tilton and E. K. Plyler, "Refractivity of Lithium Fluoride with Application to the Calibration of Infrared Spectrometers," *J. Res. Natl. Bur. Stand.*, vol. 47, pp. 25–30, 1951.

102. W. S. Rodney and R. J. Spindler, "Refractive Index of Cesium Bromide for Ultraviolet, Visible, and Infrared Wavelengths," *J. Res. Natl. Bur. Stand.*, vol. 51, pp. 123–126, 1953.

103. W. S. Rodney, "Optical Properties of Cesium Iodide," *J. Opt. Soc. Am.*, vol. 45, pp. 987–992, 1955.

104. I. H. Malitson, "A Redetermination of Some Optical Properties of Calcium Fluoride," *Appl. Opt.*, vol. 2, pp. 1103–1107, 1963.

105. I. H. Malitson, "Refractive Properties of Barium Fluoride," *J. Opt. Soc. Am.*, vol. 54, pp. 628–632, 1964.

106. I. H. Malitson and M. J. Dodge, "Refraction and Dispersion of Lead Fluoride," *J. Opt. Soc. Am.*, vol. 59, pp. 500A, 1969. Also see M. J. Dodge, "Refractive Index," in *Handbook of Laser Science and Technology*, vol. IV, *Optical Materials*: Part 2, p. 31, CRC Press, Boca Raton, Fla., 1986.

107. L. W. Tilton, E. K. Plyler, and R. E. Stephens, "Refractive Index of Silver Chloride for Visible and Infra-red Radiant Energy," *J. Opt. Soc. Am.*, vol. 40, pp. 540–543, 1950.

108. R. J. Harris, G. T. Johnson, G. A. Kepple, P. C. Krok, and H. Mukai, "Infrared Thermooptic Coefficient Measurement of Polycrystalline ZnSe, ZnS, CdTe, CaF_2 and BaF_2, Single Crystal KCl, and TI-20 Glass," *Appl. Opt.*, vol. 16, pp. 436–438, 1977.

109. Y. Tsay, B. Bendow, and S. S. Mitra, "Theory of the Temperature Derivative of the Refractive Index in Transparent Crystals," *Phys. Rev. B*, vol. 5, pp. 2688–2696, 1972.

110. N. Chetty, A. Muñoz, and R. M. Martin, "First-Principles Calculation of the Elastic Constants of AlAs," *Phys. Rev. B*, vol. 40, pp. 11934–11936, 1989.

111. F. P. Skeele, M. J. Slavin, and R. N. Katz, "Time-Temperature Dependence of Strength in Aluminum Nitride," in *Ceramic Materials and Components for Engines*; *Proceedings of the Third International Symposium*, American Ceramic Society, pp. 710–718, 1989.

112. V. A. Pesin, "Elastic Constants of Dense Modification of Boron Nitride," *Sverkhtverd. Mater.*, vol. 6, pp. 5–7, 1980 (in Russian).

113. W. Wettling and J. Windscheif, "Elastic Constants and Refractive Index of Boron Phosphide," *Solid State Comm.*, vol. 50, pp. 33–34, 1984.

114. M. E. Hills, "Preparation, Properties, and Development of Calcium Lanthanum Sulfide as an 8- to 12-micrometer Transmitting Ceramic," *Naval Weapons Center Report* TP 7073, September 1989.

115. F. Peter, "Über Brechungsindizes und Absorptionkonstanten des Diamanten zwishen 644 and 226 mμ," *Z. Phys.*, vol. 15, pp. 358–368, 1923 (in German).

116. J. Fontanella, R. L. Johnston, J. H. Colwell, and C. Andeen, "Temperature and Pressure Variation of the Refractive Index of Diamond," *Appl. Opt.*, vol. 16, pp. 2949–2951, 1977.

117. T. M. Hartnett and R. P. Miller, "Potential Limitations for Using CVD Diamond as LWIR Windows," *Proc. SPIE*, vol. 1307, pp. 474–484, 1990.

118. H. H. Li, "Refractive Index of Silicon and Germanium and Its Wavelength and Temperature Derivatives," *J. Phys. Chem. Ref. Data*, vol. 9, pp. 561–658, 1980.

119. A. S. Barker and A. J. Sievers, "Optical Studies of the Vibrational Properties of Disordered Solids," *Rev. Mod. Phys.*, vol. 47, suppl. 2, 1975.

120. P. T. B. Schaffer, "Refractive Index, Dispersion, and Birefringence of Silicon Carbide Polytypes," *Appl. Opt.*, vol. 10, pp. 1034–1036, 1971.

121. M. Neuberger, *Handbook of Electronic Materials*, vols. II, V, IFI/Plenum, New York, 1971.

122. R. E. Fern and A. Onton, "Refractive Index of AlAs," *J. Appl. Phys.*, vol. 42, pp. 3499–3500, 1971.

123. S. Bloom, "Band Structures of GaN and AlN," *J. Phys. Chem. Solids*, vol. 32, pp. 2027–2032, 1971.

124. L. Roskovcová, J. Pastrňák, and R. Babušková, "The Dispersion of the Refractive Index and the Birefringence of AlN," *Phys. Stat. Sol.*, vol. 20, pp. K29–K32, 1967.

125. J. Pastrňák and L. Roskovcová, "Optical Absorption Edge of AlN Single Crystals," *Phys. Stat. Sol.*, vol. 26, pp. 591–597, 1968.

126. A. T. Collins, E. C. Lightowlers, and P. J. Dean, "Lattice Vibration Spectra of Aluminum Nitride," *Phys. Rev.*, vol. 158, pp. 833–838, 1967.

127. A. S. Barker and M. Ilegems, "Infrared Lattice Vibrations and Free-Electron Dispersion in GaN," *Phys. Rev. B*, vol. 7, pp. 743–750, 1973.

128. E. Ejder, "Refractive Index of GaN," *Phys. Stat. Sol (a)*, vol. 6, pp. 445–447, 1971.

129. D. F. Parsons and P. D. Coleman, "Far Infrared Optical Constants of Gallium Phosphide," *Appl. Opt.*, vol. 10, pp. 1683–1685, 1971.

130. D. F. Nelson and E. H. Turner, "Electro-optic and Piezoelectric Coefficients and Refractive Index of Gallium Phosphide," *J. Appl. Phys.*, vol. 39, pp. 3337–3343, 1968.

131. D. A. Yas'kov and A. N. Pikhtin, "Optical Properties of Gallium Phosphide Grown by Float Zone I. Refractive Index and Reflection Coefficient," *Mat. Res. Bull.*, vol. 4, pp. 781–788, 1969. Also see D. A. Yas'kov and A. N. Pikhtin, "Dispersion of the Index of Refraction of Gallium Phosphide," *Sov. Phys. Sol. State*, vol. 9, pp. 107–110, 1967.

132. O. G. Lorimor and W. G. Spitzer, "Infrared Refractive Index and Absorption of InAs and CdTe," *J. Appl. Phys.*, vol. 36, pp. 1841–1844, 1965.

133. J. N. Zemel, J. D. Jensen, and R. B. Schoolar, "Electrical and Optical Properties of Epitaxial Films of PbS, PbSe, PbTe, and SnTe," *Phys. Rev.*, vol. 140A, pp. 330–342, 1965.

134. D. T. F. Marple, "Refractive Index of ZnSe, ZnTe, and CdTe," *J. Appl. Phys.*, vol. 35, pp. 539–542, 1964.

135. P. L. Provenzano, S. I. Boldish, and W. B. White, "Vibrational Spectra of Ternary Sulfides with the Th_3P_4 Structure," *Mat. Res. Bull.*, vol. 12, pp. 939–946, 1977.

136. N. P. Barnes and M. S. Piltch, "Temperature-Dependent Sellmeier Coefficients and Nonlinear Optics Average Power Limit for Germanium," *J. Opt. Soc. Am.*, vol. 69, pp. 178–180, 1979. Also see H. W. Icenogle, B. C. Platt, and W. L. Wolfe, "Refractive Indexes and Temperature Coefficients of Germanium and Silicon," *Appl. Opt.*, vol. 15, pp. 2348–2351, 1976.

137. J. Pastrňák and L. Roskovcová, "Refractive Index Measurements on A1N Single Crystals," *Phys. Stat. Sol.*, vol. 14, pp. K5–K8, 1966.

138. A. H. Kachare, W. G. Spitzer, and J. E. Fredrickson, "Refractive Index of Ion-Implanted GaAs," *J. Appl. Phys.*, vol. 47, pp. 4209–4212, 1976.

139. G. D. Pettit and W. J. Turner, "Refractive Index of InP," *J. Appl. Phys.*, vol. 36, p. 2081, 1965.

140. F. Weiting and Y. Yixun, "Temperature Effects on the Refractive Index of Lead Telluride and Zinc Sulfide," *Infrared Phys.*, vol. 30, pp. 371–373, 1990.

141. H. H. Li, "Refractive Index of ZnS, ZnSe, and ZnTe and Its Wavelength and Temperature Derivatives," *J. Phys. Chem. Ref. Data*, vol. 13, pp. 103–150, 1984.

142. A. G. DeBell, E. L. Dereniak, J. Harvey, J. Nissley, J. Palmer, A. Selvarajan, and W. L. Wolfe, "Cryogenic Refractive Indices and Temperature Coefficients of Cadmium Telluride from 6 μm to 22 μm," *Appl. Opt.*, vol. 18, pp. 3114–3115, 1979.

143. G. N. Ramachandran, "Thermo-optic Behavior of Solids, I. Diamond," *Proc. Ind. Acad. Sci.*, vol. A25, pp. 266–279, 1947.

144. Amorphous Materials, Inc., Garland, Tex.

145. R. Weil and D. Neshmit, "Temperature Coefficient of the Indices of Refraction and the Birefringence in Cadmium Sulfide," *J. Opt. Soc. Am.*, vol. 67, pp. 190–195, 1977.

146. R. W. Weiss and T. K. Gaylord, "Lithium Niobate: Summary of Physical Properties and Crystal Structure," *Appl. Phys.*, vol. A37, pp. 191–203, 1985.

147. S. D. Phatak, R. C. Srivastava, and E. C. Subbarao, "Elastic Constants of Orthorhombic $KNbO_3$ by X-ray Diffuse Scattering," *Acta Cryst.*, vol. A28, pp. 227–231, 1972.

148. D. Eimerl, L. Davis, and S. Velsko, "Optical, Mechanical, and Thermal Properties of Barium Borate," *J. Appl. Phys.*, vol. 62, pp. 1968–1983, 1987.

149. K. W. Kirby and L. G. DeShazer, "Refractive Indices of 14 Nonlinear Crystals Isomorphic to KH_2PO_4," *J. Opt. Soc. Am. B*, vol. 4, pp. 1072–1078, 1987.

150. J. D. Bierlein and H. Vanherzeele, "Potassium Titanyl Phosphate: Properties and New Applications," *J. Opt Soc. Am.*, B., vol. 6, pp. 622–633, 1989.

151. F. C. Zumsteg, J. D. Bierlein, and T. E. Gier, "$K_xRb_{1-x}TiOPO_4$: A New Nonlinear Optical Material," *J. Appl. Phys.*, vol. 47, pp. 4980–4985, 1976.

152. D. F. Nelson and R. M. Mikulyak, "Refractive Indices of Congruently Melting Lithium Niobate," *J. Appl. Phys.*, vol. 45, pp. 3688–3689, 1974.

153. Y. Uematsu, "Nonlinear Optical Properties of $KNbO_3$ Single Crystal in the Orthorhombic Phase," *Jap. J. Appl. Phys.*, vol. 13, pp. 1362–1368, 1974.

154. C. Chen, Y. Wu, A. Jiang, B. Wu, G. You, R. Li, and S. Lin, "New Nonlinear-Optical Crystal: LiB_3O_5," *J. Opt. Soc. Am.*, B., vol. 6, pp. 616–621, 1989.

155. G. C. Bhar, "Refractive Index Interpolation in Phase-matching," *Appl. Opt.*, vol. 15, pp. 305–307, 1976.

156. M. J. Rosker, K. Cheng, and C. L. Tang, "Practical Urea Optical Parametric Oscillator for Tunable Generation Throughout the Visible and Near-Infrared," *IEEE J. Quant. Elect.*, vol. QE-21, 1600–1606, 1985.

157. T. Y. Fan, C. E. Huang, B. W. Hu, R. C. Eckardt, Y. X. Fan, R. L. Byer, and R. S. Feigelson, "Second Harmonic Generation and Accurate Index of Refraction Measurements in Flux-grown $KTiOPO_4$," *Appl. Opt.*, vol. 26, pp. 2390–2394, 1987.

158. M. M. Choy and R. L. Byer, "Accurate Second-Order Susceptibility Measurements of Visible and Infrared Nonlinear Crystals," *Phys. Rev. B*, vol. 14, pp. 1693–1706, 1976.

159. K. Kato, "Tunable UV Generation to 0.2325 μm in LiB_3O_5," *IEEE J. Quantum Electron.*, vol. QE-26, pp. 1173–1175, 1990.

160. L. G. Schulz, "The Optical Constants of Silver, Gold, Copper, and Aluminum. I. The Absorption Coefficient k," *J. Opt. Soc. Am.*, vol. 44, pp. 357–361; "II. The Index of Refraction n," *J. Opt. Soc. Am.*, vol. 44, pp. 362–368, 1954.

161. E. Shiles, T. Sasaki, M. Inokuti, and D. Y. Smith, "Self-Consistency and Sum-Rule Tests in the Kramers-Kronig Analysis of Optical Data: Applications to Aluminum," *Phys. Rev. B*, vol. 22, pp. 1612–1628, 1980.

162. S. Mattei, P. Masclet, and P. Herve, "Study of Complex Refractive Indices of Gold and Copper Using Emissivity Measurements," *Infrared Phys.*, vol. 29, pp. 991–994, 1989.

163. P. F. Robusto and R. Braunstein, "Optical Measurements of the Surface Plasmon of Copper," *Phys. Stat. Sol. (b)*, vol. 107, pp. 443–449, 1981.

164. A. P. Lenham and D. M. Treherne, "Application of the Anomalous Skin-Effect Theory to the Optical Constants of Cu, Ag and Au in the Infrared," *J. Opt. Soc. Am.*, vol. 56, pp. 683–685, 1966.

165. H.-J. Hagemann, W. Gudat, and C. Kunz, "Optical Constants from the Far Infrared to the X-ray Region: Mg, Al, Cu, Ag, Au, Bi. C, and AL_2O_3," *J. Opt. Soc. Am.*, vol. 65, pp. 742–744, 1975.

166. G. P. Motulevich and A. A. Shubin, "Influence of Fermi Surface Shape in Gold on the Optical Constants and Hall Effect," *Soviet Physics JETP*, vol. 20, pp. 560–564, 1965.

167. J. H. Weaver, E. Colavita, D. W. Lynch, and R. Rosei, "Low-Energy Interband Absorption in bcc Fe and hcp Co," *Phys. Rev. B*, vol. 19, pp. 3850–3856, 1979.

168. D. W. Lynch, R. Rosei, and J. H. Weaver, "Infrared and Visible Optical Properties of Single Crystal Ni at 4K," *Solid State Comm.*, vol. 9, pp. 2195–2199, 1971.

169. P. B. Johnson and R. W. Christy, "Optical Constants of Transition Metals: Ti, V, Cr, Mn, Fe, Co, Ni, and Pd," *Phys. Rev. B*, vol. 9, pp. 5056–5070, 1974.

170. J. H. Weaver, "Optical Properties of Rh, Pd, Ir, and Pt," *Phys. Rev. B*, vol. 11, pp. 1416–1425, 1975.

171. M. A. Ordal, R. J. Bell, R. W. Alexander, L. L. Long, and M. R. Querry, "Optical Properties of Fourteen Metals in the Infrared and Far Infrared: Al, Co, Cu, Au, Fe, Pb, Mo, Ni, Pd, Pt, Ag, Ti, V, and W," *Appl. Opt.*, vol. 24, pp. 4493–4499, 1985.

172. Barr & Stroud, Ltd., Glasgow, Scotland (UK).

173. J. W. Fleming, "Optical Glasses," in M. J. Weber (ed.), *Handbook of Laser Science and Technology*, vol. IV: *Optical Materials*, Part II, CRC Press, Boca Raton, Fla, pp. 69–83, 1986.

174. C. T. Moynihan, M. G. Drexhage, B. Bendow, M. S. Boulos, K. P. Quinlan, K. H. Chung, and E. Gboji, "Composition Dependence of Infrared Edge Absorption in ZrF_4 and HfF_4 Based Glasses," *Mat. Res. Bull.*, vol. 16, pp. 25–30, 1981.

175. R. N. Brown, B. Bendow, M. G. Drexhage, and C. T. Moynihan, "Ultraviolet Absorption Edge Studies of Fluorozirconate and Fluorohafnate Glass, *Appl. Opt.*, vol. 21, pp. 361–363, 1982.

176. R. N. Brown and J. J. Hutta, "Material Dispersion in High Optical Quality Heavy Metal Fluoride Glasses," *Appl. Opt.*, vol. 24, pp. 4500–4503, 1985.

177. B. Bendow, M. G. Drexhage, and H. G. Lipson, "Infrared Absorption in Highly Transparent Fluorozirconate Glass," *J. Appl. Phys.*, vol. 52, pp. 1460–1461, 1981.

178. M. G. Drexhage, C. T. Moynihan, and M. Saleh, "Infrared Transmitting Glasses Based on Hafnium Fluoride," *Mat. Res. Bull.*, vol. 15, pp. 213–219, 1980.

179. M. G. Drexhage, O. H. El-Bayoumi, C. T. Moynihan, A. J. Bruce, K.-H. Chung, D. L. Gavin, and T. J. Loretz, "Preparation and Properties of Heavy-Metal Fluoride Glasses Containing Ytterbium or Lutetium," *J. Am. Ceram. Soc.*, vol. 65, pp. C168–C171, 1982.

180. W. S. Rodney, I. H. Malitson, and T. A. King, "Refractive Index of Arsenic Trisulfide," *J. Opt. Soc. Am.*, vol. 48, pp. 633–636, 1958.

181. A. R. Hilton and C. E. Jones, "The Thermal Change in the Nondispersive Infrared Refractive Index of Optical Materials," *Appl. Opt.*, vol. 6, pp. 1513–1517, 1967.

182. J. A. Savage, "Optical Properties of Chalcogenide Glasses," *J. Non-Cryst. Sol.*, vol. 47, pp. 101–116, 1982.

183. D.J. Treacy, "Arsenic Selenide (As_2Se_3)," in E. D. Palik (ed.), *Handbook of Optical Constants of Solids*, Academic Press, Orlando, pp. 623–639, 1985.

184. J. Brandrup and E. H. Immergut (eds.), *Polymer Handbook*, 3d ed., Wiley, New York, 1989.

185. R. M. Waxler, D. Horowitz, and A. Feldman, "Optical and Physical Parameters of Plexiglas 55 and Lexan," *Appl. Opt.*, vol. 18, pp. 101–104, 1979.

186. J. I. Kroschwitz (ed.), *Concise Encyclopedia of Polymer Science and Engineering*, Wiley, New York, 1990.

187. B. Bendow, R. N. Brown, M. G. Drexhage, T. J. Loretz, and R. L. Kirk, "Material Dispersion of Fluorozirconate-type Glasses," *Appl. Opt.*, vol. 20, pp. 3688–3690, 1981.

188. P. Klocek, and L. Colombo, "Index of Refraction, Dispersion, Bandgap and Light Scattering in GeSe and GeSbSe Glasses," *J. Non-Cryst. Sol.*, vol. 93, pp. 1–16, 1987.

189. D. M. Roessler and D. R. Huffman, "Magnesium Oxide (MgO)," in E. D. Palik (ed.), *Handbook of Optical Constants of Solids II*, Academic Press, Orlando, pp. 919–955, 1991.

190. M. E. Thomas and R. I. Joseph, "A Comprehensive Model for the Intrinsic Transmission Properties of Optical Windows," *Proc. SPIE*, vol. 929, pp. 87–93, 1988.

191. J. I. Berg and E. E. Bell, "Far-Infrared Optical Constants of KI," *Phys. Rev.*, vol. B4, pp. 3572–3580, 1971.

192. P. Vergnat, J. Claudel, A. Hadni, P. Strimer, and F. Vermillard, "Far Infrared Optical Constants of Cesium Halides at Low Temperatures," *J. Phys.*, vol. 30, pp. 723–735, 1969 (in French).

193. D. R. Bosomworth, "Far-Infrared Optical Properties of CaF_2, SrF_2, BaF_2, and CdF_2," *Phys. Rev.*, vol. 157, pp. 709–715, 1967.

194. E. D. Palik, O. J. Glembocki, and K. Takarabe, "Aluminum Arsenide (AlAs)," in E. D. Palik (ed.), *Handbook of Optical Constants of Solids II*, Academic Press, Orlando, pp. 489–499, 1991.

195. E. D. Palik, "Gallium Arsenide (GaAs)," in E. D. Palik (ed.), *Handbook of Optical Constants of Solids*, Academic Press, Orlando, pp. 429–443, 1985.

196. C. J. Johnson, G. H. Sherman, and R. Weil, "Far Infrared Measurement of the Dielectric Properties of GaAs and CdTe at 300 K and 8 K," *Appl. Opt.*, vol. 8, pp. 1667–1671, 1969.

197. T. Hattori, Y. Homma, A. Mitsuishi, and M. Tacke, "Indices of Refraction of ZnS, ZnSe, ZnTe, CdS, and CdTe in the Far Infrared," *Optics Comm.*, vol. 7, pp. 229–232, 1973.

198. L. Ward, "Cadmium Sulphide (CdS)," in E. D. Palik (ed.), *Handbook of Optical Constants of Solids II*, Academic Press, Orlando, pp. 579–595, 1991.

199. E. J. Danielewicz and P. D. Coleman, "Far Infrared Optical Properties of Selenium and Cadmium Telluride," *Appl. Opt.*, vol. 13, pp. 1164–1170, 1974.

200. O. J. Glembocki and H. Pillar, "Indium Phosphide (InP)," in E. D. Palik (ed.), *Handbook of Optical Constants of Solids*, Academic Press, Orlando, pp. 503–516, 1985.

201. E. D. Palik, "Tellurium (Te)," in E. D. Palik (ed.), *Handbook of Optical Constants of Solids II*, Academic Press, Orlando, pp. 709–723, 1991.

202. D. J. Treacy, "Arsenic Sulfide (As_2S_3)," in E. D. Palik (ed.), *Handbook of Optical Constants of Solids*, Academic Press, Orlando, pp. 641–663, 1985.

203. J. A. Cox, D. Greenlaw, G. Terry, K. McHenry, and L. Fielder, "Comparative Study of Advanced IR Transmissive Materials," *Proc. SPIE*, vol. 683, pp. 49–62, 1988.

204. L. L. Boyer, J. A. Harrington, M. Hass, and H. B. Rosenstock, "Multiphonon Absorption in Ionic Crystals," *Phys. Rev. B*, vol. 11, pp. 1665–1680, 1975.

205. J. E. Eldridge, "Cesium Iodide (CsI)," in E. D. Palik (ed.), *Handbook of Optical Constants of Solids II*, Academic Press, Orlando, pp. 853–874, 1991.

206. T. Hidaka, T. Morikawa, and J. Shimada, "Spectroscopic Small Loss Measurements on Infrared Transparent Materials," *Appl. Opt.*, vol. 19, pp. 3763–3766, 1980.

207. E. D. Palik and A. Addamiano, "Zinc Sulfide (ZnS)," in E. D. Palik (ed.), *Handbook of Optical Constants of Solids*, Academic Press, Orlando, pp. 597–619, 1985.

208. P. A. Miles, "Temperature Dependence of Multiphonon Absorption in Zinc Selenide," *Appl. Opt.*, vol. 16, pp. 2891–2896, 1977.

209. E. D. Palik, "Cadmium Telluride (CdTe)," in E. D. Palik (ed.), *Handbook of Optical Constants of Solids*, Academic Press, Orlando, pp. 409–427, 1985.

210. A. R. Hilton, D. J. Hayes, and M. D. Rechtin, "Infrared Absorption of Some High-Purity Chalcogenide Glasses," *J. Non-Cryst. Sol.*, vol. 17, pp. 319–338, 1975.

211. F. Zernike, "Refractive Indices of Ammonium Dihydrogen Phosphate and Potassium Dihydrogen Phosphate between 2000 Å and 1.5 μ," *J. Opt. Soc. Am.*, vol. 54, pp. 1215–1220, 1964. Errata: *J. Opt. Soc. Am.*, vol. 55, pp. 210E, 1965.

212. I. H. Malitson, F. V. Murphy, and W. S. Rodney, "Refractive Index of Synthetic Sapphire," *J. Opt. Soc. Am.*, vol. 48, pp. 72–73, 1958. Also see I. H. Malitson, "Refraction and Dispersion of Synthetic Sapphire," *J. Opt. Soc. Am.*, vol. 52, pp. 1377–1379, 1962.

213. M. Herzberger and C. D. Salzberg, "Refractive Indices of Infrared Optical Materials and Color Correction of Infrared Lenses," *J. Opt. Soc. Am.*, vol. 52, pp. 420–427, 1962.

214. Y. Ohmachi, "Refractive Index of Vitreous As_2Se_3," *J. Opt. Soc. Am.*, vol. 63, pp. 630–631, 1973.

215. T. W. Houston, L. F. Johnson, P. Kisiuk, and D. J. Walsh, "Temperature Dependence of the Refractive Index of Optical Maser Crystals," *J. Opt. Soc. Am.*, vol. 53, pp. 1286–1291, 1963.

216. P. S. Bechthold and S. Haussühl, "Nonlinear Optical Properties of Orthorhombic Barium Formate and Magnesium Barium Fluoride," *Appl. Phys.*, vol. 14, pp. 403–410, 1977.

217. A. R. Johnston, "Dispersion of Electro-optic Effect in $BaTiO_3$," *J. Appl. Phys.*, vol. 42, pp. 3501–3507, 1971.

218. D. P. Bortfeld and H. Meier, "Refractive Indices and Electro-optic Coefficients of the Eulitities $Bi_4Ge_3O_{12}$ and $Bi_4Si_3O_{12}$," *J. Appl. Phys.*, vol. 43, pp. 5110–5111, 1972.

219. E. Burattini, G. Cappuccio, M. Grandolfo, P. Vecchia, and Sh. M. Efendiev, "Near-Infrared Refractive Index of Bismuth Germanium Oxide ($Bi_{12}GeO_{20}$)," *J. Opt. Soc. Am.*, vol. 73, pp. 495–497, 1983.

220. K. Vadam and P. Hennessey, "Piezo- and Thermo-optical Properties of $Bi_{12}GeO_{20}$, II. Refractive Index," *J. Opt. Soc. Am.*, vol. 65, pp. 442–445, 1975.

221. R. E. Aldrich, S. L. Hou, and M. L. Harvill, "Electrical and Optical Properties of $Bi_{12}SiO_{20}$," *J. Appl. Phys.*, vol. 42, pp. 493–494, 1971.

222. S. Mitachi and T. Miyashita, "Refractive-Index Dispersion for BaF_2-GdF_3-ZrF_4-AlF_3 Glasses," *Appl. Opt.*, vol. 22, pp. 2419–2425, 1983.

223. G. C. Bahr, D. C. Hanna, B. Luther-Davies, and R. C. Smith, "Tunable Down-Conversion from an Optical Parametric Oscillator," *Opt. Comm.*, vol. 6, pp. 323–326, 1972.

224. T. M. Bieniewski and S. J. Czyzak, "Refractive Indexes of Single Hexagonal ZnS and CdS Crystals," *J. Opt. Soc. Am.*, vol. 53, pp. 496–497, 1963. See also S. J. Czyzak, W. M. Baker, R. C. Crane, and J. B. Howe, "Refractive Indexes of Single Synthetic Zinc Sulfide and Cadmium Sulfide Crystals," *J. Opt. Soc. Am.*, vol. 47, pp. 240–243, 1957.

225. J. E. Harvey and W. L. Wolfe, "Refractive Index of Irtran 6 (Hot-pressed Cadmium Telluride) as a Function of Wavelength and Temperature," *J. Opt. Soc. Am.*, vol. 65, pp. 1267–1268, 1975.

226. N. P. Barnes and M. S. Piltch, "Temperature-Dependent Sellmeier Coefficients and Coherence Length for Cadmium Telluride," *J. Opt. Soc. Am.*, vol. 67, pp. 628–629, 1977. Also see Ref. 255.

227. R. Laiho and M. Lakkisto, "Investigation of the Refractive Indices of LaF_3, CeF_3, PrF_3, and NdF_3," *Phil. Mag. B*, vol. 48, pp. 203–207, 1983.

228. A. Selvarajan, J. L. Swedberg, A. G. DeBell, and W. L. Wolfe, "Cryogenic Temperature IR Refractive Indices of Cesium Bromide and Cesium Iodide," *Appl. Opt.*, vol. 18, pp. 3116–3118, 1979.

229. S. H. Wemple, J. D. Gabbe, and G. D. Boyd, "Refractive-Index Behavior of Ternary Chalcopyrite Semiconductors," *J. Appl. Phys.*, vol. 46, pp. 3597–3605, 1975.

230. G. C. Bhar and G. Ghosh, "Temperature-Dependent Sellmeier Coefficients and Coherence Lengths for Some Chalcopyrite Crystals," *J. Opt. Soc. Am.*, vol. 69, pp. 730–733, 1979.

231. E. H. Turner, I. P. Kaminow, and C. Schwab, "Temperature Dependence of Raman Scattering, Electro-optic, and Dielectric Properties of CuBr," *Phys. Rev. B*, vol. 9, pp. 2524–2529, 1974.

232. D. F. Edwards and E. Ochoa, "Infrared Refractive Index of Diamond," *J. Opt. Soc. Am.*, vol. 71, pp. 607–608, 1981.

233. S. H. Wemple and W. J. Tabor, "Refractive Index Behavior of Garnets," *J. Appl. Phys.*, vol. 44, pp. 1395–1396, 1973.

234. J. H. Wray and J. T. Neu, "Refractive Index of Several Glasses as a Function of Wavelength and Temperature," *J. Opt. Soc. Am.*, vol. 59, pp. 774–776, 1969.

235. M. Herzberger, "Colour Correction in Optical Systems and a New Dispersion Formula," *Opt. Acta*, vol. 6, pp. 197–215, 1959.

236. S. Kobayashi, N. Shibata, S. Shibata, and T. Izawa, "Characteristics of Optical Fibers in Infrared Wavelength Region," *Rev. Elect. Commun. Lab.*, vol. 26, pp. 453–467, 1978.

237. J. W. Fleming, "Material and Mode Dispersion in $GeO_2 \cdot B_2O_3 \cdot SiO_2$ Glasses," *J. Am. Ceram. Soc.*, vol. 59, pp. 503–507, 1976.

238. J. W. Fleming, "Material Dispersion in Lightwave Glasses," *Elect. Lett.*, vol. 14, pp. 326–328, 1978.

239. D. L. Wood, K. Nassau, T. Y. Kometani, and D. L. Nash, "Optical Properties of Cubic Hafnia Stabilized with Yttria," *Appl. Opt.*, vol. 29, pp. 604–607, 1990.

240. M. P. Wirick, "The Ultraviolet Optical Constants of Lanthanum Fluoride," *Appl. Opt.*, vol. 12, pp. 1966–1967, 1966.

241. D. A. Pinnow, L. G. Van Uitert, A. W. Wagner, and W. A. Bonner, "Lead Molybdate: A Melt-Grown Crystal with a High Figure of Merit for Acousto-optic Device Applications," *Appl. Phys. Lett.* vol. 15, pp. 83–86, 1969.

242. N. Uchide, S. Miyazawa, and K. Ninomiya, "Refractive Indices of Pb_2MoO_5 Single Crystal," *J. Opt. Soc. Am.*, vol. 60, pp. 1375–1377, 1970.

243. S. Singh, J. P. Remeika, and J. R. Potopowicz, "Nonlinear Optical Properties of Ferroelectric Lead Titanate," *Appl. Phys. Lett.*, vol. 20, pp. 135–137, 1972.

244. D. T. F. Marple, "Refractive Index of GaAs," *J. Appl. Phys.*, vol. 35, pp. 1241–1242, 1964.

245. P. Laporte and J. L. Subtil, "Refractive Index of LiF from 105 to 200 nm," *J. Opt. Soc. Am.*, vol. 72, pp. 1558–1559, 1982.

246. G. D. Boyd, H. M. Kasper, and J. H. McFee, "Linear and Nonlinear Optical Properties of $LiInS_2$," *J. Appl. Phys.*, vol. 44, pp. 2809–2812, 1973.

247. H. H. Li, "Refractive Index of Alkaline Earth Halides and Its Wavelength and Temperature Derivatives," *J. Phys. Chem. Ref. Data*, vol. 9, pp. 161–289, 1980.

248. M. W. Williams and E. T. Arakawa, "Optical Properties of Crystalline MgF_2 from 115 nm to 400 nm," *Appl. Opt.*, vol. 18, pp. 1477–1478, 1979.

249. J. Strong and R. T. Brice, "Optical Properties of Magnesium Oxide," *J. Opt. Soc. Am.*, vol. 25, pp. 207–210, 1935.

250. L. Schüler, H. Betzler, H. Hesse, and S. Kapphan, "Phase-Matched Second Harmonic Generation in Potassium Malate," *Opt. Comm.*, vol. 43, pp. 157–159, 1982.

251. L. Gample and F. M. Johnson, "Index of Refraction of Single-Crystal Selenium," *J. Opt. Soc. Am.*, vol. 59, pp. 72–73, 1969.

252. P. T. B. Schaffer and R. G. Naum, "Refractive Index and Dispersion of Beta Silicon Carbide," *J. Opt. Soc. Am.*, vol. 59, p. 1498, 1969.

253. W. W. Coblentz, "Transmission and Refraction Data on Standard Lens and Prism Material with Special Reference to Infra-red Spectroradiometry," *J. Opt. Soc. Am.*, vol. 4, pp. 432–447, 1920.

254. V. Chandrasekharan and H. Damany, "Dispersion of Quartz in the Vacuum Ultraviolet from Interference in a Thin Parallel Plate," *Appl. Opt.*, vol. 7, pp. 687–688, 1968.

255. J. D. Feichtner, R. Johannes, and G. W. Roland, "Growth and Optical Properties of Single Crystal Pyrargyrite," *Appl. Opt.*, vol. 9, pp. 1716–1717, 1970.

256. K. F. Hulme, O. Jones, P. H. Davies, and M. V. Hobden, "Synthetic Proustite (Ag_3AsS_3): A New Crystal for Optical Mixing," *Appl. Phys. Lett.*, vol. 10, pp. 133–135, 1967.

257. S. Chandrasekhar and M. S. Madhava, "Optical Rotary Dispersion of Crystals of Sodium Chlorate and Sodium Bromate," *Acta Cryst.*, vol. 23, pp. 911–913, 1967.

258. N. Uchida, "Optical Properties of Single Crystal Paratellurite (TeO_2)," *Phys. Rev. B*, vol. 4, pp. 3736–3645, 1971.

259. S. Singh, W. A. Bonner, and L. G. van Uitert, "Violation of Kleinmann's Symmetry Condition in Paratellurite," *Phys. Lett.*, vol. A38, pp. 407–408, 1972.

260. D. E. McCarthy, "Refractive Index Measurements of Thallium Bromide and Thallium Chloride in the Infrared," *Appl. Opt.*, vol. 4, pp. 878–879, 1965.

261. B. Johnson and A. K. Walton, "The Infra-Red Refractive Index of Garnet Ferrites," *Br. J. Appl. Phys.*, vol. 16, pp. 475–477, 1965.

262. D. E. Castleberry and A. Linz, "Measurement of the Refractive Indices of $LiYF_4$," *Appl. Opt.*, vol. 14, p. 2056, 1975.

263. G. D. Boyd, E. Buehler, and F. G. Storz, "Linear and Nonlinear Optical Properties of $ZnGeP_2$ and CdSe," *Appl. Phys. Lett.*, vol. 18, pp. 301–304, 1971.

264. M. Debenham, "Refractive Indices of Zinc Sulfide in the 0.405–13-μm Wavelength Range," *Appl. Opt.*, vol. 23, pp. 2238–2239, 1984.

265. C. A. Klein, "Room-Temperature Dispersion Equations for Cubic Zinc Sulfide," *Appl. Opt.*, vol. 25, pp. 1873–1875, 1986.

266. W. L. Wolfe and R. Korniski, "Refractive Index of Irtran-2 as a Function of Wavelength and Temperature," *Appl. Opt.*, vol. 17, pp. 1547–1549, 1978.

267. D. L. Wood, K. Nassau, and T. Y. Kometani, "Refractive Index of Y_2O_3 Stabilized Zirconia: Variation with Composition and Wavelength," *Appl. Opt.*, vol. 29, pp. 2485–2488, 1990.

268. M. V. Hobden, "Phase-Matched Second-Harmonic Generation in Biaxial Crystals," *J. Appl. Phys.*, vol. 38, pp. 4365–4372, 1967.

269. L. G. DeShazer, S. C. Rand, and B. A. Wechsler, "Laser Crystals," in *Handbook of Laser Science and Technology*, vol. V, *Optical Materials*: Part III, pp. 281–338, CRC Press, Boca Raton, Fla., 1987.

270. S. Lin, Z. Sun, B. Wu, and C. Chen, "The Nonlinear Optical Characterization of a LiB_3O_5 Crystal," *J. Appl. Phys.*, vol. 67, pp. 634–638, 1990.

271. K. W. Martin and L. G. DeShazer, "Indices of Refraction of the Biaxial Crystal $YAlO_3$," *Appl. Opt.*, vol. 12, pp. 941–943, 1973.

272. C. S. Hoefer, "Thermal Variations of the Refractive Index in Optical Materials," *Proc. SPIE*, vol. 681, pp. 135–142, 1986.

273. J. P. Hurrell, S. P. S. Porto, I. F. Chang, S. S. Mitra, and R. P. Bauman, "Optical Phonons of Yttrium Aluminum Garnet," *Phys. Rev.*, vol. 173, pp. 851–856, 1968.

274. S. A. Miller, H. E. Rast and H. H. Caspers, "Lattice Vibrations of $LiYF_4$," *J. Chem. Phys.*, vol. 52, pp. 4172–4175, 1970.

275. J. D. Axe, J. W. Gaglianello, and J. E. Scardefield, "Infrared Dielectric Properties of Cadmium Fluoride and Lead Fluoride," *Phys. Rev.*, vol. 139, pp. A1211–A1215, 1965.

276. D. A. Yas'kov and A. N. Pikhtin, "Refractive Index and Birefringence of Semiconductors with the Wurtzite Structure," *Sov. Phys. Semicond.*, vol. 15, pp. 8–12, 1981.

277. A. N. Pikhtin and A. D. Yas'kov, "Dispersion of the Refractive Index of Semiconductors with Diamond and Zinc-blende Structures," *Sov. Phys. Semicond.*, vol. 12, pp. 622–626, 1978.

278. M. Bertolotti, V. Bogdanov, A. Ferrari, A. Jascow, N. Nazorova, A. Pikhtin, and L. Schirone, "Temperature Dependence of the Refractive Index in Semiconductors," *J. Opt. Soc. Am., B.*, vol. 7, pp. 918–922, 1990.

279. D. S. Smith, H. D. Riccius, and R. P. Edwin, "Refractive Indices of Lithium Niobate," *Optics Commun.*, vol. 17, pp. 332–335, 1976.

280. B. Zysset, I. Biaggio, and P. Günter, "Refractive Indices of Orthorhombic $KNbO_3$. I. Dispersion and Temperature Dependence," *J. Opt. Soc. Am. B.*, vol. 9, pp. 380–386, 1992.

CHAPTER 12
OPTICAL FIBERS

Carlton M. Truesdale

INTRODUCTION

The progress of optical fiber technology is embedded in dreams of revolutionizing the way we communicate with one another. Systematically, the building blocks have been created and mass-produced so that those dreams are becoming real. The basic receiver was in existence in the 1950s. The laser appeared in 1960 and the demonstration of the first room-temperature continuous-wave semiconductor laser by Hayashi and Panish[1] at Bell Laboratories in 1970 made possible the transmission and reception of information using signals of light. This technology is being used to rewire America and the world.

Kapany and Simms announced the fabrication of the first infrared optical fiber in 1965.[2] This chalcogenide glass fiber had transmission loss of more than 10,000 dB/km. By 1970 Corning researchers Kapron, Keck, and Maurer had fabricated silica-based optical fibers that reached the 20 dB/km milestone.[3] By reducing the impurities in the source materials and improving the homogeneity of the glass processing, an attenuation of 0.2 dB/km was obtained by Miya, Hosaka, and Miyashita.[4] In 1978, Pinnow,[5] Van Uitert,[6] and Goodman et al.[7] proposed that an ultra-low-loss fiber with a loss less than 0.01 dB/km for non-silica-based fibers was theoretically possible. This announcement motivated many researchers to study and discover other infrared materials.

Today standard silica-based optical fiber for optical communications typically has less than 0.2 dB/km attenuation as measured at 1550 nm. Optical fibers have come a long way and are becoming a more refined technology.

12.1 THEORY OF FIBER TRANSMISSION

The following section describes the theory of light propagating in an optical fiber. The concepts of reflection and refraction are reviewed to help confer the notion of total internal reflection in a fiber, and also the concepts of numerical aperture, modal properties of single-mode fibers, the solution of the wave equation for single-mode fibers, and description of the modal propagation for multimode fibers.

12.1.1 Basic Concepts

An optical fiber consists of at least two distinct regions known as the "core" and "cladding." The refractive index of a material n is shown by Eq. (12.1).

$$n = \frac{v(\text{medium})}{c} \tag{12.1}$$

where $v(\text{medium})$ is the velocity of light in the medium and c is the speed of light in a vacuum ($c = 3 \times 10^8$ m/s). The refractive index is a measure of the relative speed of light in the material as compared with light traveling in a vacuum. The larger the refractive index of a material, the more slowly light travels within it.

The refractive index of the core n_1, when made higher than the refractive index of the cladding n_2, can confine a light beam to the core region of a fiber. The light beam is confined in the core by total internal reflection.

There are two classes of optical fiber. These classes are separated both by construction and by their propagation properties. They are called single-mode and multimode fibers. Figure 12.1 shows that the usual dimensions of a single-mode fiber core are on the order of 3 to 10 μm while the cladding diameter is 125 μm.

INDEX PROFILE

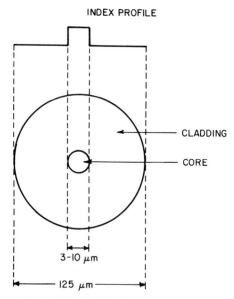

FIGURE 12.1 Typical dimensions and refractive index profile of a single-mode fiber.

To protect this fragile "wire," optical fibers are coated with a soft primary and a harder secondary polymer coating. Single-mode fibers allow only one pathway for a bundle of rays to be transmitted. Multimode fiber, on the other hand, usually has a 40- to 300-μm core. Multimode fiber cladding diameter ranges from 125 μm to larger than 300 μm, and it is also protected with coating materials. Multimode fibers can contain up to several hundred paths for ray bundles.

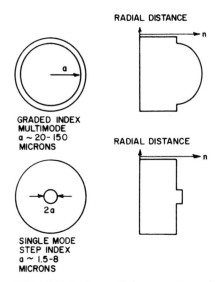

GRADED INDEX
MULTIMODE
$a \sim 20\text{-}150$
MICRONS

SINGLE MODE
STEP INDEX
$a \sim 1.5\text{-}8$
MICRONS

FIGURE 12.2 Geometrical cross sections and graded and step refractive index profiles of multimode and single mode fiber, respectively.[44]

The profile of the refractive index as a function of radial position $n(r)$ is most commonly classified as step or graded. Figure 12.2 illustrates these two types of refractive index profiles.

12.1.2 Definition of Reflection and Refraction

Light rays are electromagnetic waves and must obey Maxwell's equations. Reflection and refraction occur at the interface of two regions of refractive index. As shown by Fig. 12.3, total internal reflection is the process by which light rays traveling through material with an index n_1, which is greater than the index n_2 beyond the interface, are directed back into the original material because the propagation angle α_1 made with the normal of the interface surface is greater than the critical complementary angle θ_c given by Eq. (12.2).[8,9]

$$\theta_c = \cos^{-1}\left(\frac{n_2}{n_1}\right) \tag{12.2}$$

Figure 12.4 shows that if $n_1 < n_2$, the ray can be refracted and reflected. The angle at which the ray is refracted α_1 and reflected α_2 is given in Eqs. (12.3) and (12.4),

$$\theta_1 = \text{complementary launch angle} \tag{12.3}$$

$$\theta_2 = \sin^{-1}\left(\frac{n_1 \sin \theta_1}{n_2}\right) \tag{12.4}$$

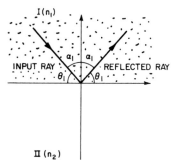

FIGURE 12.3 Reflection of a light ray for $n_1 > n_2$, which leads to total internal reflection.[10]

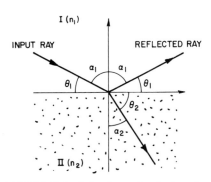

FIGURE 12.4 Refraction and reflection of a light ray for $n_1 < n_2$.[10]

12.1.3 Guiding Property of Optical Fiber

Using the above concepts of reflection and refraction, Fig. 12.5 shows graphically how light is confined to the axial portion of the fiber by total internal reflection.

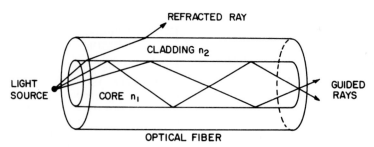

FIGURE 12.5   Light propagation in fiber due to frustrated internal reflection.

In Fig. 12.6, the cone of the critical angle encloses the launch angle θ' from air to the optical fiber core. The critical angle is found by substituting $n_2 = 1$ in Eq. (12.2). The light is guided in the core at an angle θ. If a ray is launched with an angle greater than θ_c, the ray enters the fiber core and is refracted into the fiber cladding. (The ray picture is really only appropriate for the discussion of multimode fibers.)

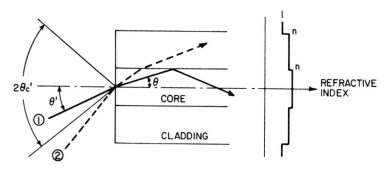

FIGURE 12.6   Light transmission in an optical fiber.[10]

The numerical aperture, or NA, of a fiber is defined as

$$NA = (n_1^2 - n_2^2)^{1/2} \tag{12.5}$$

The maximum acceptance angle of a fiber is given by the critical angle described by

$$\theta_c' = 2 \sin^{-1}(n_1 \sin \theta_c) \tag{12.6}$$

12.1.4 Modal Classification of Light Propagation in Optical Fibers

Groups of bundles of rays with the same path are classified as modes. This allows a more thorough treatment of the wave nature of light rays. The wave vector defined by k characterizes the direction and magnitude of a plane wave. The magnitude of $|k|$ for a plane wave in a vacuum is given in Eq. (12.7).

$$k = \frac{2\pi}{\lambda} \tag{12.7}$$

where λ is the wavelength of the light. Where the refractive index is not unity, this phase constant becomes kn_1 and the wavelength becomes λ/n_1. See Fig. 12.7 for a schematic.

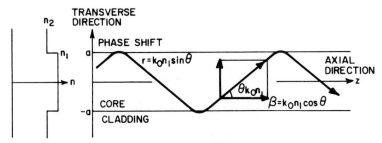

FIGURE 12.7   Formation of a mode in a waveguide in relation to the propagation direction.[10]

The plane wave has two components that propagate in the axial and transverse directions.[10] The axial propagation constants β and the transverse propagation constant γ are given below in Eqs. (12.8) and (12.9), respectively.

$$\beta = kn_1 \cos \theta \tag{12.8}$$

$$\gamma = \pm kn_1 \sin \theta \tag{12.9}$$

The standing wave of the electric field is created by the reflection of the plane wave in the transverse direction at the two core-cladding interfaces. After two reflections the total phase change is a multiple of 2π. The axial propagation constant does not change as the plane wave travels axially.

To determine the total number of modes present in an optical fiber, the normalized frequency or modal volume V provides convenient information. V is defined as

$$V = \frac{2\pi n_1 \rho (2\Delta)^{1/2}}{\lambda} = \rho k(n_1^2 - n_2^2)^{1/2} \tag{12.10}$$

where ρ is the core radius and Δ is called the profile height parameter, which is given by

$$\Delta = \frac{n_1^2 - n_2^2}{2n_1^2} = \frac{\sin^2 \theta_c}{2} \tag{12.11}$$

For single-mode fibers $V < 2.4$, which means there is one fundamental mode that is degenerate. The V value of multimode fibers is defined to be $V > 2.4$, which means several modes are possible. The fundamental mode for a single-mode fiber is shown in Fig. 12.8. The cutoff wavelength λ_c for single-mode operation is given by

$$\lambda_c = \frac{2\pi n_1 \rho (2\Delta)^{1/2}}{2.4} \tag{12.12}$$

The λ_c corresponds to the wavelength at which only one fundamental mode will propagate. Below this cutoff value the optical fiber will allow other higher-order modes to be guided.

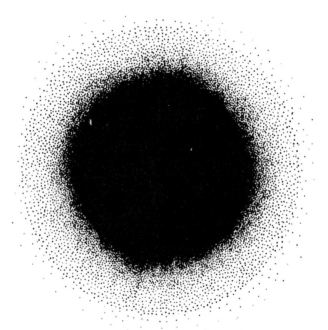

FIGURE 12.8 Qualitative representation of the light intensity for the core region of a step profile single-mode fiber.

12.1.5 Scalar Wave Equation

If one wishes to construct the electric and magnetic fields for the bound modes on weakly guiding optical fiber, the vectorial relationship of the equations from Maxwell's equations is reduced to the solution of the transverse electric field Ψ by the scalar wave equation given below.[11]

$$\left[\frac{\partial^2}{\partial r^2} + \frac{1}{r}\frac{\partial}{\partial r} + \frac{1}{r^2}\frac{\partial^2}{\partial \phi^2} + k^2 n(r)^2 - \beta^2 \right] \Psi = 0 \tag{12.13}$$

where r is the radial position, ϕ is the azimuthal position, β is the transverse propagation constant, and Ψ is more commonly called the wavefunction.

For a step index waveguide the solution to Eq. (12.13) is found using Bessel functions. Equation (12.13) can be transformed into Bessel's equation by making the substitution $\Psi = A\Psi(r)\exp(iv\phi)$ and obtaining the result

$$\left[\frac{\partial^2}{\partial r^2} + \frac{1}{r}\frac{\partial}{\partial r} + k^2 n(r)^2 - \beta^2 - v^2\right]\Psi(r) = 0 \qquad (12.14)$$

where v represents the azimuthal mode. The light travels axially along the z direction. The fundamental mode has two components: one polarized along the x direction and one along the y direction. Without making polarization corrections, the fundamental mode has degenerate x and y polarized modes.

The electric field solutions for the core and cladding are Bessel functions with arguments related to the phase constants or wavevector magnitudes of the optical fiber given by

$$E_z = \begin{cases} AJ_v(k_1 r)e^{iv\phi} \; r > \rho \; \text{(core)} & (12.15a) \\ BK_v(k_2 r)e^{iv\phi} \; r > \rho \; \text{(cladding)} & (12.15b) \end{cases}$$

where J_v and K_v are the ordinary and hyperbolic Bessel functions of the vth kind, and the magnitude of the core and cladding wavevectors k_1 and k_2 is given by

$$k_1 = (n_1^2 k^2 - \beta^2)^{1/2} \qquad (12.16)$$

$$k_2 = (\beta^2 - n_2^2 k^2)^{1/2} \qquad (12.17)$$

The wavevectors are solved by using the fact that at the core-clad interface ($r = \rho$) $\Psi(r)$ and its first derivative $\overline{d\Psi} \; dr = \Psi'(r)$ are continuous. This means that $\Psi(r = \rho; \text{core}) = \Psi(r = \rho; \text{cladding})$ and $\Psi'(r = \rho; \text{core}) = \Psi'(r = \rho; \text{cladding})$. This leads to the solution as a result of solving the eigenvalue equation given below.

$$k_1 J_{v+1}(k_1\rho)K_v(k_2\rho) = k_2 J_v(k_1\rho)K_{v+1}(k_2\rho) \qquad (12.18)$$

There is an interesting relationship between the modal volume V and the core and cladding parameters U and W, which are represented by $k_1\rho$ and $k_2\rho$, respectively. The relationship is given by the following equation:

$$U^2 + W^2 = V^2 \qquad (12.19)$$

Table 12.1 displays $V, U,$ and W for the fundamental mode ($v = 0$). A plot of U and W vs. V is shown in Fig. 12.9.

The propagation constant β is related to $U, V, \Delta,$ and ρ as

$$\beta = \frac{1}{\rho}\left(\frac{V^2}{2\Delta} - U^2\right)^{1/2} \qquad (12.20)$$

For single-mode operation ($0 < V < 2.4$) the fractional power in the core η is given by[11]

$$\eta = \frac{\displaystyle\int_0^\rho \int_0^{2\pi} \Psi^2(r)r \; dr \; d\phi}{\displaystyle\int_\infty^\rho \int_0^{2\pi} \Psi^2(r)r \; dr \; d\phi} = \frac{U^2}{V^2}\left[\frac{W^2}{U^2} + \frac{K_0^2(W)}{K_1^2(W)}\right] \qquad (12.21)$$

A plot of the fractional power η as a function of V is shown in Fig. 12.10.

TABLE 12.1 Modal Parameters U, V, and W for Single-Mode Operation

V, modal volume	U, core parameter	W, cladding parameter
6.007	2.045	5.648
5.801	2.036	5.433
5.610	2.027	5.231
5.430	2.016	5.043
5.221	2.014	4.818
4.991	1.995	4.575
4.814	1.989	4.384
4.617	1.968	4.177
4.407	1.953	3.952
4.216	1.936	3.745
4.016	1.908	3.535
3.813	1.883	3.316
3.610	1.861	3.094
3.411	1.830	2.879
3.201	1.803	2.646
3.003	1.767	2.429
2.805	1.735	2.204
2.600	1.691	1.976
2.407	1.644	1.758
2.203	1.592	1.525
2.002	1.530	1.292
1.800	1.455	1.061
1.600	1.368	0.832
1.402	1.262	0.611
1.300	1.201	0.498
1.200	1.134	0.393
1.100	1.060	0.294

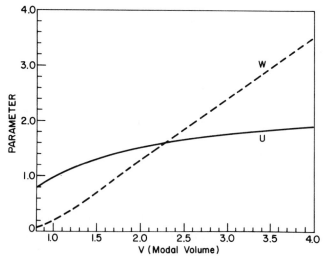

FIGURE 12.9 W and U modal parameters vs. the modal volume parameter V.

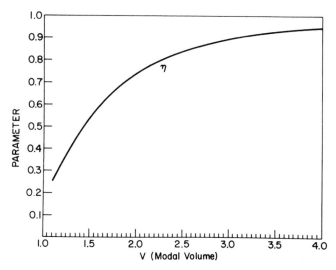

FIGURE 12.10 Fractional core power vs. the modal volume parameter V.

If the scalar wave equation is used to solve for the transverse components of the electric field, it cannot be used to account for polarization effects. To account for polarization effects, the full vector wave equation must be used. Mathematically, terms of the form $\vec{\nabla} \ln n^2(r)$ must be computed. The definition of $\vec{\nabla}$ is given in the following equation:

$$\vec{\nabla} = \hat{x}\,\frac{\partial}{\partial x} + \hat{y}\,\frac{\partial}{\partial y} = \hat{r}\,\frac{\partial}{\partial r} + \hat{\phi}\,\frac{\partial}{r\partial\phi} \tag{12.22}$$

These terms account for hybrid modes that are formed because the action of the $\vec{\nabla} \ln n^2(r)$ operator creates a rotation of the electric vector. The corrections to the propagation constant $\delta\beta$ and the core parameter δU are given by

$$\delta\beta = \frac{-(2\Delta)^{3/2}}{2\rho}\,\frac{U^2 W}{V^3}\,\frac{K_0(W)}{K_1(W)} \tag{12.23}$$

$$\delta U = \frac{\Delta U W}{V^2}\,\frac{K_0(W)}{K_1(W)} \tag{12.24}$$

For graded index profiles, the solution of the radial differential equation is solved approximately using Hermite-Gaussian functions.[4] The modes are more complex, but their general description is essentially described by the step index formalism given above.

12.1.6 Multimode Fiber Approximations

Because multimode fibers will transmit a large number of modes, it is not possible to describe modal propagation in as much mathematical detail as shown above. It is more appropriate to use the concepts of rays to understand this type of optical fiber. As shown in Secs. 12.1.2 and 12.1.3, the maximum angle accepted by a fiber launched from air is the angle

$$\theta_c \simeq \sin \theta = (n_1^2 - n_2^2)^{1/2} \tag{12.25}$$

In the small angle approximation, the free-space modes entering the front of the core are given by πa^2.[12] The pair of modes that are polarized orthogonal to each other occupy a cone of solid angle $\pi\delta^2$, where [12]

$$\delta = \frac{\lambda}{\pi a} \tag{12.26}$$

The total number of modes N accepted by the multimode fiber is[13]

$$N \simeq \frac{2\theta_c^2}{\delta^2} \simeq \frac{V^2}{2} \tag{12.27}$$

The first mode is composed of two modes and all others contain four. When a fiber has a large number of modes, we can relate the mode v to the cutoff value of the core parameter U by the equation[14]

$$U_c \simeq (2v)^{1/2} \tag{12.28}$$

Knowing that U above cutoff is nearly U_c, the ratio of the power in the core to the total power for the vth mode η_v is approximated by[14]

$$\eta_v = \frac{U^2}{V^2}(V^2 - U_c^2)^{1/2} = 1 - \frac{v}{N(2N - 2v)^{1/2}} \tag{12.29}$$

The average η over all modes is given by the integration of Eq. (12.25) over all modes which results in the following equation.[14]

$$\eta = 1 - \frac{4}{3}\left(\frac{2}{N}\right)^{1/2} \tag{12.30}$$

12.2 MATERIALS FOR THE FABRICATION OF OPTICAL FIBER

A large variety of materials are suitable for making infrared fibers. Glass fiber has been the most promising fiber material, and it can be constructed from materials like metal oxides, metal halides, and chalcogenides. A review of these materials will be discussed in connection with the fabrication of the core and cladding regions and the outer coating process. A representation of the attenuation vs. time for silicate, heavy metal oxides, fluoride glasses, chalcogenide glasses, and halide crystals is shown in Fig. 12.11.

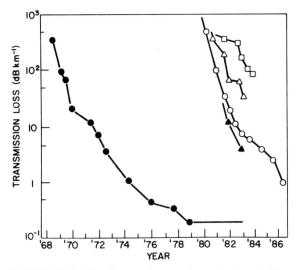

FIGURE 12.11 The history of attenuation reduction of silica, heavy-metal oxides, fluoride glasses, chalcogenide glasses, and halide crystals represented by the data shown using the filled circles, encircled points, open circles, open triangles, and open squares.[45]

12.2.1 Metal Oxide Glass Fibers

Silicon dioxide is by far the most used material in the construction of optical fiber. SiO_2 has superior transmission and mechanical properties compared with other materials. GeO_2 is good for ultra-low-loss transmission at longer wavelengths in the infrared compared with silica. This is because the infrared absorption due to Ge-O lattice vibrations is shifted toward the red owing to the heavier weight of Ge. Examples of the effect of other metal oxide glass dopants on the refractive index are shown in Fig. 12.12.

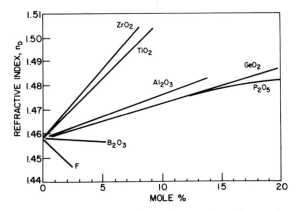

FIGURE 12.12 Index modifying effect of GeO_2, P_2O_5, F, B_2O_3, and other common dopants for high-silica waveguides.[46]

12.2.2 Metal Halides

Baldwin has provided a review of metal halides.[15] Materials known as "heavy metal" fluoride glasses are a subset of this larger family. These materials incorporate anions of the halogens (Cl_2, F_2) in place of oxygen. The application of multicomponent fluoroberylatte glasses to low-loss infrared fibers has been explored at Corning. Michel Poulain discovered fluorozirconate glasses in 1974–1975.[16] With intense research activity there are only a few metal fluoride glass compositions that have been found to be practical for optical fiber application. Foremost among them is the ZrF_4-BaF_2-LaF_3-AlF_3-NaF (ZBLAN) system first reported by Ohsawa and colleagues of the Furikawa Electric Company in 1981.[17]

12.2.3 Chalcogenides

Glasses formed from metals such as arsenic, germanium, and antimony combined with the heavier elements in the oxygen family (the "chalcogens") like S, Se, and Te are infrared-transparent and are called chalcogenide glasses. Usually, the electronic absorption edges of these glasses are in the middle ir, visible, or near-ir region. These materials have high refractive indexes ($\sim$2.5) and low glass transition temperatures ($\sim$200 to 300°C). They are prepared by mixing and melting purified elemental materials in vacuum. Chalcogenide glasses are easily damaged by moisture. The most pervasive extrinsic loss is due to hydrogen and oxygen impurities. The most serious problem that chalcogenide fibers have when used for low-loss telecommunication applications is the existence of a weak intrinsic absorption tail at around 5 μm which limits the minimum attainable attenuation to 1 to 10 dB/km.[18] Short-length fiber applications in the ir are more attractive.

12.2.4 Rare-Earth Dopants

There is a keen interest in doping silicates or fluoride glasses with rare-earth ions like erbium, neodymium, and praesodymium. Because the valence shell of the lanthanides involves 4f orbitals, the matrix or host glass does not modify their chemical nature too drastically. The f-orbitals are shielded by the 5d electrons.[19] By appropriately exciting certain electronic levels, fluorescence occurs in the 1.3- to 1.5-μm region. Erbium-doped glasses have been shown to lase at around 1.53 μm. Fiber cores doped with erbium have been shown to amplify 1.55-μm optical signals.[20] This novel fiber amplifier device may revolutionize the transmission of optical signals by directly amplifying optical signals without converting to an electrical signal and back.

12.3 FABRICATION METHODS

Several processing techniques have been used to produce glass fibers for transmission applications. These techniques have been successful in making low-loss optical fiber for optical transmission. The outside vapor deposition process (OVD), modified chemical vapor deposition process (MCVD or IVD), vapor-phase axial deposition process (VAD), plasma chemical vapor deposition process (PCVD and PMCVD), and double-crucible melting process will be summarized.

12.3.1 Outside Vapor Deposition (OVD)

The outside vapor deposition (OVD) process steps are illustrated in Fig. 12.13. The process was first developed by Corning and is sometimes referred to as the "soot process."[21-23] The first process step involves the delivery of vapors of $SiCl_4$, $GeCl_4$, O_2, CH_4, etc., to a burner where soot particles (glass spheres) are produced. The soot is collected on a revolving bait or target rod that traverses the burner. At this point the preform is porous.

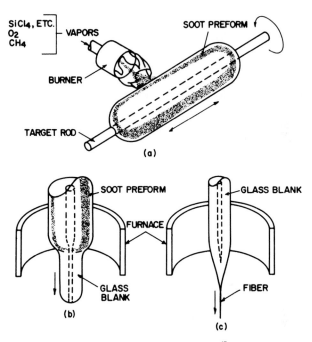

FIGURE 12.13 Schematic of the OVD process.[47]

Compositional control is achieved by varying the ratio of dopants like $GeCl_4$ or BCl_3 to $SiCl_4$. In the second process step the preform is consolidated in a furnace at a temperature near 1500°C under an atmosphere of Cl_2 with other drying agents to remove traces of water. In the third and final step, the preform is drawn into a fiber by feeding the glass preform into a furnace heated to 2000°C. The miniaturized glass fiber preserves its original geometrical and material structure.

12.3.2 Modified Chemical Vapor Deposition (MCVD)

The modified chemical vapor deposition (MCVD) process, practiced by Bell Laboratories, injects vapors of materials like $SiCl_4$, $GeCl_4$, and BCl_3 inside a silica tube that is heated externally by an oxyhydrogen burner to produce glass.[24,25] This process is also called the inside vapor deposition process (IVD). A schematic of the process is shown in Fig. 12.14. The rotating fused silica tube is heated to about 1400°C by the burner that travels the length of the tube. At each pass of the burner

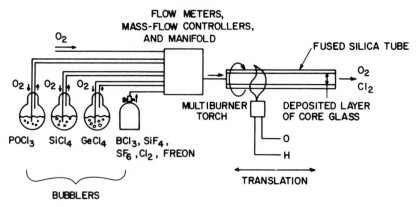

FIGURE 12.14 Schematic of the MCVD process.[48]

a new layer of glass is formed. About 50 layers are usually deposited by multiple passes of the burner. After the glass is deposited on the silica tube, the burner temperature is raised to collapse the tube into a solid preform. Fiber is drawn from the preform in the same way as the OVD process.

12.3.3 Vapor-Phase Axial Deposition (VAD)

The vapor-phase axial deposition (VAD) is a variation of the OVD process.[26–28] It also produces soot particles in a flame. The main difference from OVD is that the soot particles are collected from a vertically hung bait rod. The consolidation can be done simultaneously but is usually done in a different step. The core composition is deposited by a small torch at the tip of the preform, and the cladding is formed from multiple burners along the axis of the bait. Figure 12.15 illustrates the type of processing.

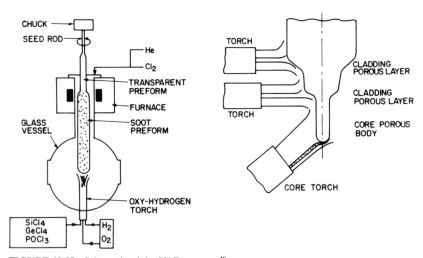

FIGURE 12.15 Schematic of the VAD process.[49]

12.3.4 Plasma Chemical Vapor Deposition

The variations of the MCVD process use a plasma instead of a flame to initiate the soot-forming reaction. These are called plasma-enhanced modified chemical vapor deposition (PMCVD)[29,30] (see Fig. 12.16) and plasma chemical vapor dep-

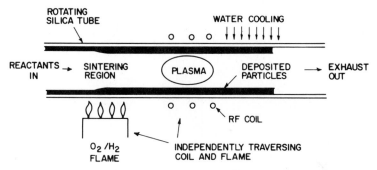

FIGURE 12.16 Schematic of the PMCVD process.[50]

osition (PCVD—see Fig. 12.17).[31,32] These differ in the type of plasma used. The temperature of the plasma is approximately 10,000°C. Because of the high temperatures present in the silica tube, a complete conversion of the metal halides into soot occurs. Up to 2000 thin layers of glass can be deposited on the silica tube.

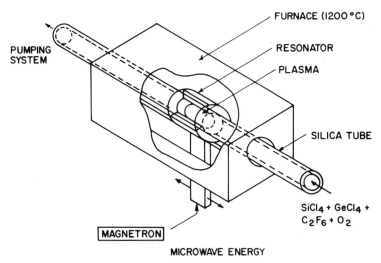

FIGURE 12.17 Schematic of the PCVD process.[51]

12.3.5 Direct Melt Processes

For fiber applications requiring compositions for which no volatile source compounds are available, the double-crucible technique is useful. The double-crucible technique consists of two concentric crucibles which contain separate molten core and clad glasses.[33] Depending upon the speed of the draw process, either step index fiber (fast) or graded index fiber (slow) can be produced. Usually, alkali ions like Na^+ and K^+ are exchanged between the core and the cladding to manufacture graded index fiber. Figure 12.18 illustrates this manufacturing process.

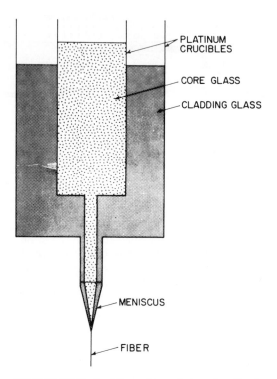

FIGURE 12.18 Double-crucible technique for directly melting glasses.[33]

12.4 FIBER LOSSES

In this section the losses that occur as a result of fiber material scattering, absorption, and bending will be discussed. This introduces the concept of Rayleigh scattering, impurity absorption, and losses attributable to a perturbation in the optical path by bending the fiber. Other mechanisms for scattering light are a result of the existence of inhomogeneities in the materials based on compositional fluctuations or the presence of bubbles and strains introduced in the process of jacketing or cabling the fiber. All absorption or scattering processes induce a loss of the form

$$a = \text{loss(dB)} = 10 \log_{10}\left[\frac{p(z)}{p(0)}\right] \tag{12.31}$$

where $p(z)$ is the power transmitted over a distance z in kilometers that is given by

$$p(z) = p(0)e^{-\alpha z} \tag{12.32a}$$

where α is the power loss coefficient in units of inverse kilometers and $p(0)$ is the initial power of the light.

To convert the power loss coefficient to units of dB/km, one uses Eq. (12.31) to calculate a and the $\log_{10}(e) = 0.434$.

$$\frac{\text{Loss(dB)}}{z} = \frac{a}{z} = 4.34(2\alpha) \tag{12.32b}$$

12.4.1 Rayleigh Scattering

Rayleigh scattering is caused by the interaction of light and the granular appearance of atoms and molecules on a microscopic scale. The scattering sites are much smaller than the wavelength of the light. Rayleigh scattering is proportional to λ^{-4}.

In the case of a one-component glass, the scattering loss coefficient α_s is given by[34]

$$\alpha_s = \frac{8\pi^3}{3\lambda^4}(n^8 p^2)kT\beta_T \tag{12.33}$$

where n is the refractive index of the glass, p is the photoelastic coefficient, k is the Boltzmann constant, T is the glass transition temperature in Kelvin, and β_T is the isothermal compressibility.

For multicomponent glasses the Rayleigh scattering loss coefficient α_{sc} is written as[34]

$$\alpha_{sc} = \frac{16\pi^3 n}{3\lambda^4}\left(\frac{2N^2}{2C}\right)\overline{\Delta C^2}\,\delta V \tag{12.34}$$

where N is ratio of indexes of refraction of the inclusion or region of compositional change to the medium, $\overline{C}$ is the density, $\overline{\Delta C^2}$ is the mean square of the density fluctuation, and δV is the volume of the compositional varying region.

12.4.2 Absorption Losses

Absorption losses arise from the presence of impurities in the glass fiber. These impurities may add strong absorption features in the infrared region between 1.2 and 3.5 μm. The presence of OH allows fundamental vibration of the Si-OH that occurs at 2.73 μm. The reduction of OH impurity is very critical in reducing absorptions between 1.2 and 2 μm. The occurrence of transition metals like Fe, Co, Ni, Cu, and V also adds additional absorption losses in the infrared. Table 12.2 is a compilation of the absorption for the impurities listed above that were taken from the data of Ohishi,[35] Osanai,[36] and Schultz[37] for the impurity level being 1 part in 10^{-6}. The observation that can be made from Table 12.5 is that to be able to fabricate low-loss fiber, the OH and transition metal impurities must be kept $\leq$ 10 ppb (10 parts in 10^{-9}). Figure 12.19 graphs the effects of metal impurities on the absorption of fibers.

TABLE 12.2 Absorption Losses in Units of dB/km for Various Transition Metals and OH

Material	1.3 μm	1.5 μm	2.0 μm	3.0 μm
Fe, 1 ppm	100	190	92	2
Co, 1 ppm	100	200	130	4
Ni, 1 ppm	100	190	90	0.5
Cu, 1 ppm	250	200	3	0.01
V, 1 ppm	600	200		
OH, 1 ppm	90 (1.4 μm)	10		

Data taken from Refs. 34, 35, and 36.

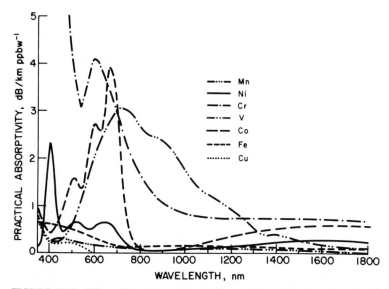

FIGURE 12.19 Effect of metal impurities in silica on absorptivity in optical fibers.[37]

12.4.3 Bending Losses

Two classes of fiber losses arise from either large-radii bends or small fiber curvatures with small periods. These bending effects are called macrobending and microbending losses, respectively. It is crucial to be able to characterize these losses because they are important if one wishes to be able to wrap fiber around a mandrel or storage spool.

If a fiber is bent from the straight position, the light may be radiated away from the guide, causing optical leakage. As the radius of curvature of the fiber bend decreases, the bending loss will increase exponentially. The pure macrobending loss α_c of the fundamental mode of a single-mode fiber to the radius of curvature of the bend R is given by[38]

$$\alpha_c = A_c R^{-1/2} \exp(-UR) \qquad (12.35)$$

where

$$A_c = \frac{1}{2} \left(\frac{\pi}{\rho V^3} \right)^{1/2} \left[\frac{U}{V K_1(V)} \right]^2 \qquad (12.36)$$

and

$$U = \frac{4 \Delta n V^3}{3 \rho V^2 n_2} \qquad (12.37)$$

Δn is the difference in refractive index between the fiber core and cladding, ρ is the core radius, V is the modal volume, n_2 is the cladding index, and K_1 is the hyperbolic Bessel function of first order.

A fiber with small randomly oscillating deviations around the straight position with periods that may be small will lead to microbending losses. Loss occurs because of coupling between the fundamental mode and higher-order modes that radiate power. Figure 12.20 presents theoretical results for fiber with a Δn in the range of 0.825 percent, 0.55 percent, and 0.275 percent.[39] The observation that the magnitude of the loss decreases as Δn increases holds for macrobending and microbending effects.

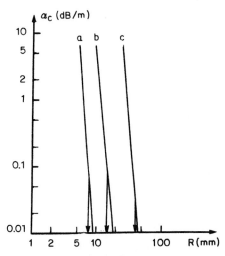

FIGURE 12.20 Bending loss of a function of radius of curvature for fibers operating at 1300 nm and having a 1180 cutoff wavelength. The index differences Δn are (a) 0.825 percent, (b) 0.55 percent, and (c) 0.275 percent. The arrows indicate the corresponding critical radius of curvature where the loss rises exponentially with wavelength.[39]

12.5 PULSE BROADENING

During fiber transmission pulse broadening can occur owing to intermodal dispersion. Intermodal dispersion is a combination of material dispersion and waveguide dispersion. It depends in part on the refractive index profile of the fiber. The intermodal dispersion occurs because the optical pulse transit time through a fiber depends on the optical path or the particular mode that is launched. The refractive index profile controls the amount of intermodal dispersion. Material dispersion occurs because the refractive index of the materials used to make the fiber vary as a function of wavelength. Waveguide dispersion occurs because the propagation constant β varies with wavelength. The profile of the refractive index affects both intermodal and material dispersion and causes pulse spreading or broadening that limits the bandwidth of the transmission system.

The theory used to calculate the pulse broadening for single-mode fibers is taken from a discussion of pulse broadening for planar waveguides. There are two types of rays that propagate in a fiber. The meridional rays cross the fiber axis between reflections, but skew rays are rays that never cross the fiber axis. This theory is appropriate for fibers because meridional rays of a fiber correspond to the rays of a planar waveguide. Skew rays of a fiber are independent of the azimuthal or transverse invariant $\overrightarrow{\gamma}$, which is given by γ/k, where γ is defined in Eq. (12.9). As a result, skew rays will behave the same as meridional rays in a fiber.

The bandwidth determines the maximum frequency of optical pulses that can be transmitted before severe degradation of the pulse train results. A calculation of the optimum profile for limiting the pulse broadening for single-mode fibers is also included. The dispersion in multimode fibers with a power law refractive index will be described. The optimum profile exponent that limits the pulse broadening for this class of fibers will be presented.

12.5.1 Intermodal Dispersion

How much dispersion occurs is related to the path a light ray travels and the refractive index profile of the fiber. The step index, parabolic, and power clad law refractive index profiles will be discussed. For this discussion only the fundamental mode is considered.

The distance traveled for half a wave period results from solving the ray path equations given by the eikonal equation for $\overrightarrow{\nabla} n(r) = 0$, which is for ray trajectories that are straight lines.[8,9]

$$\frac{d}{ds}\left[n(r)\,\frac{dr}{ds} \right] = \frac{dn(r)}{dr} \qquad (12.38a)$$

$$\frac{d}{ds}\left[n(r)\,\frac{dz}{ds} \right] = 0 \qquad (12.38b)$$

where r is the displacement from the center of the core of the fiber, $n(r)$ is the refractive index profile, s is the distance along the ray path, and z is the direction of the propagating ray.

Examples of refractive index profiles $n(r)$ are shown in Fig. 12.21. The ray invariant $\overrightarrow{\beta}$ equals β/k, where β is defined in Eq. (12.8). The ray invariant $\overrightarrow{\beta}$ determines if a ray is bound or is refracted. For bound rays, $n_2 < \overrightarrow{\beta} < n_1$, and for refracting rays $0 \le \overrightarrow{\beta} < n_2$. The relationship between the ray invariant $\overrightarrow{\beta}$ and the propagation direction z is given by[40]

$$\overrightarrow{\beta} = n(r)\cos\theta(r) = n(r)\,\frac{dz}{dr} \qquad (12.39)$$

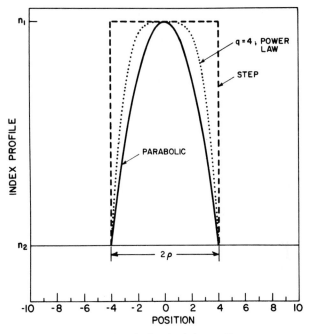

FIGURE 12.21 Examples of refractive index profiles.

Substituting Eq. (12.39) into Eq. (12.38a) leads to a rearrangement of the ray equation given by the following equation:

$$\vec{\beta}^2 \frac{d^2r}{dz^2} = \frac{1}{2} \frac{dn^2(r)}{dr} \tag{12.40}$$

A single integration of this equation leads to

$$\vec{\beta} \frac{dr}{dz} = (n^2(r) - \vec{\beta}^2)^{1/2} \tag{12.41}$$

Integration of this equation determines the ray propagation as a function of radial position for bound rays when $0 \le r \le r_{tp}$ and for refracting rays when $0 \le r \le \rho$.

$$z(r) = \vec{\beta}^2 \int_0^r \frac{dr}{\sqrt{(n_1^2 - \vec{\beta}^2)}} \tag{12.42}$$

The position where the ray turns direction is called the turning point r_{tp}. It is defined by

$$n(r_{tp}) = \vec{\beta} \tag{12.43}$$

TABLE 12.3 Parameters for Calculating Pulse Dispersion

Profile	$n(r)$	r_{tp}	z_p	L_0	t_p
step index	$\begin{array}{l} n_1;\ |r| \leq \rho \\ n_2;\ |r| > \rho \end{array}$	ρ	$\dfrac{2\rho\vec{\beta}}{(n_1^2 - \vec{\beta}^2)^{1/2}}$	$\dfrac{2\rho n_1^2}{(n_1^2 - \vec{\beta}^2)^{1/2}}$	$\dfrac{z n_1^2}{c\vec{\beta}}$
Parabolic	$n_1\sqrt{1 - 2\Delta\left(\dfrac{r}{\rho}\right)^2}$	$\dfrac{\rho}{n_1}\sqrt{\dfrac{n_1^2 - \vec{\beta}^2}{2\Delta}}$	$\dfrac{\pi\rho\vec{\beta}}{n_1\sqrt{2\Delta}}$	$\dfrac{z_p}{2}\left(\dfrac{n_1^2}{\vec{\beta}} + \vec{\beta}\right)$	$\dfrac{z n_1}{2c}\left(\dfrac{\vec{\beta}}{n_1} + \dfrac{n_1}{\vec{\beta}}\right)$
Clad power law	$n_1\sqrt{1 - 2\Delta\left(\dfrac{r}{\rho}\right)^q}$	$\dfrac{\rho}{n_1}\left(\dfrac{n_1^2 - \vec{\beta}^2}{2\Delta}\right)^{1/q}$	$\sqrt{\dfrac{2\pi}{\Delta}\left(\dfrac{\rho_q}{r}\right)^q}\ \dfrac{\vec{\beta} r_{tp}}{q n_1}\ x\ \dfrac{\Gamma\left(\dfrac{1}{q}\right)}{\Gamma\left(\dfrac{1}{2} + \dfrac{1}{q}\right)}$	$\dfrac{z_p}{q+2}\left(\dfrac{q n_1^2}{\vec{\beta}} + \dfrac{2\vec{\beta}}{n_1}\right)$	$\dfrac{z n_1}{c(q+2)}\left(\dfrac{2\vec{\beta}}{n_1} + \dfrac{q n_1}{\vec{\beta}}\right)$

For a step index profile $r_{tp} = \rho$. The optical path length L_0 is defined as the integral given by the following equation:

$$L_0 = \int_{-r_{tp}}^{r_{tp}} n(r)ds \tag{12.44}$$

where ds is the differential arc length.

Substituting $ds = \dfrac{ds}{dz}\dfrac{dz}{dr}\,dr = \vec{\beta}\,n(r)dr/[n^2(r) - \vec{\beta^2}]^{1/2}$ into Eq. (12.41) leads to the optical path length given by

$$L_0 = \int_{-r_{tp}}^{r_{tp}} \frac{n^2(r)dr}{[n^2(r) - \vec{\beta^2}]^{1/2}} \tag{12.45}$$

The ray half period follows, using Eq. (12.42) to be

$$z_p = \vec{\beta} \int_{-r_{tp}}^{r_{tp}} \frac{dr}{[n^2(r) - \vec{\beta^2}]^{1/2}} \tag{12.46}$$

To determine the time t_p to travel a distance z with an optical path L_0 and ray half period z_p, it is calculated as

$$t_p = \frac{zL_0}{cz_p} \tag{12.47}$$

Finally, to determine the dispersion or pulse spread t_p, because the front of the wave travels with $\vec{\beta} = n_1$ and the back of the wave travels with $\vec{\beta} = n_2$, one uses

$$t_p = t(\vec{\beta} = n_2) - t(\vec{\beta} = n_1) \tag{12.48}$$

Table 12.3 displays the r_{tp}, L_0, and t_p for various refractive index profiles. The dispersion times for the step index, parabolic, and power clad law refractive index profiles are given in Table 12.4.

12.5.2 Optimum Intermodal Dispersion Profile for Clad Power Law

As seen in Table 12.3, the transit time t_p for the clad power law refractive index profile is given by[40]

$$t_p(\vec{\beta}) = \frac{zn_1}{c(q+2)}\left(\frac{qn_1}{\vec{\beta}} + \frac{2\vec{\beta}}{n_1}\right) \tag{12.49}$$

TABLE 12.4 Dispersion Times as a Function of Refractive Index

Refractive index	t_d	$t_d\ (\Delta \ll 0)$
Step index	$\dfrac{zn_1}{c}[(1 - 2\Delta)^{-1/2} - 1]$	$\dfrac{zn_1\Delta}{c}$
Parabolic	$\dfrac{zn_1}{2c}[(1 - 2\Delta)^{-1/2} + (1 - 2\Delta)^{1/2} - 2]$	$\dfrac{zn_1\Delta^2}{2c}$
Clad power law	$\dfrac{zn_1}{c(q+2)}[q(1 - 2\Delta)^{-1/2} + 2(1 - 2\Delta)^{1/2} - 2]$	$\dfrac{zn_1\Delta^2}{c(q+2)}(q\Delta^2 + q\Delta - 2\Delta)$

A minimum transit time occurs by performing the derivative of Eq. (12.46) with respect to $\vec{\beta}$ and setting the result equal to zero. The optimum $\vec{\beta}_m$ is defined so that transit time $t_p(\vec{\beta}_m)$ is a minimum. The values of $\vec{\beta}_m$ and $t_p(\vec{\beta}_m)$ are given by

$$\vec{\beta}_m = \left(\frac{q}{2}\right)^{1/2} n_1 \qquad t_p(\vec{\beta}_m) = \frac{z(8q)^{1/2}n_1}{c(q+2)} \tag{12.50}$$

The optimum power law exponent q_m is determined by setting the dispersion or pulsewidth t_d in Table 12.4 equal to zero. This operation is written as follows:

$$q_m \frac{n_1}{n_2} + 2\frac{n_2}{n_1} = q_m + 2 \tag{12.51}$$

so that

$$q_m = 2(1 - 2\Delta)^{1/2} \tag{12.52}$$

where Δ is defined by Eq. (12.11) and the minimum pulsewidth is given by[41]

$$t_d = \frac{z(\sqrt{n_1} - \sqrt{n_2})^2 n_1}{c(n_1 - n_2)} \simeq \frac{z\Delta^2 n_1}{8c} \tag{12.53}$$

Figure 12.13 shows the typical bandwidth of a power law refractive index. The maximum bandwidth occurs near $q = 2$.

12.5.3 Material Dispersion

Material dispersion is due to the fact that rays with different wavelengths, even though they take the same path, travel at different speeds. Although material dispersion is usually a small effect, when it combines with intermodal dispersion, it can be quite significant.

For example, the step index fiber has a transit time t_p for $\vec{\beta} = n_1$ given by $t_p = zn_1/c$. If n_1 varies with wavelength, the incremental time differential Δt is written as $\Delta t = z\Delta n_1/c$, where Δn_1 is given by

$$\Delta n_1 = n_1(\lambda) - \lambda \frac{dn_1}{d\lambda} \tag{12.54}$$

Δn_1 is called the on axis group index n_a. The pulsewidth t_d assuming that $\delta\lambda \ll \lambda$ leads to

$$t_d = \left|\frac{\partial t}{\partial \lambda}\right| \delta\lambda = \frac{z\lambda}{c}\left|\frac{d^2 n_1}{d\lambda^2}(\lambda)\right| \delta\lambda \tag{12.55}$$

The group index $n_g(r,\lambda)$ for a general refractive index $n(r,\lambda)$ is written as

$$n_g(r,\lambda) = n(r,\lambda) - \lambda \frac{dn}{d\lambda}(r,\lambda) \tag{12.56}$$

In the description of dispersive fiber materials, we limit our discussion to linear dispersion. In other words, $n_g = a(\lambda)n + b(\lambda)$, where a and b are functions of λ only.

12.5.4 Profile Dispersion

Profile dispersion is the combination of intermodal and material dispersion. The equations for the optical path length L_m that accounts for material dispersion with group index $n_g(r,\lambda)$ are modified.[11]

$$L_m = \int_{-r_{tp}}^{r_{tp}} \frac{n(r,\lambda)n_g(r,\lambda)dr}{[n2(r,\lambda) - \overrightarrow{\beta^2}]^{1/2}} \qquad (12.57)$$

For a step index profile $n_g(r,\lambda)$ is given by the expression given in Eq. (12.54). The pulse spread t_d is then written as

$$t_d = \frac{zn_1}{c}(\lambda)\left[\frac{n_1}{n_2} - 1\right]\left[1 - \frac{\lambda dn_1}{n_1(\lambda)d\lambda}(\lambda)\right] \qquad (12.58)$$

The first and second terms in square brackets are related to the intermodal and material dispersion, respectively.

Using the approach given above, the group index $n_g(r,\lambda)$ for a power law profile is written as[11]

$$n_g(r,\lambda) = \left[n_1(\lambda) - \lambda\frac{dn_1^{(\lambda)}}{d\lambda}\right]\left[1 - 2\Delta\left(\frac{r}{\rho}\right)^q\right]2n_1(\lambda)\left(\frac{r}{\rho}\right)^q\frac{dn_1^{(\lambda)}}{d\lambda} \qquad (12.59)$$

The ray transit time of the power law profile is defined by

$$t_p = \frac{zn_a}{c(q + 2)}\left[(p + q)\frac{n_1}{\overrightarrow{\beta}} + (2 - p)\frac{\overrightarrow{\beta}}{n_1}\right] \qquad (12.60)$$

where n_a is the on axis group index defined by Eq. (12.54) and p is given by

$$p = \frac{n_1(\lambda)}{n_a}\frac{\lambda}{\Delta(\lambda)}\frac{d\Delta(\lambda)}{d\lambda} \qquad (12.61)$$

12.5.5 Optimum Profile Dispersion for Power Law Refractive Index

The minimum transit time occurs by taking the derivative of Eq. (12.57) with respect to $\overrightarrow{\beta}$ and setting the result to zero. The values of $\overrightarrow{\beta}_m$ and $t_p(\overrightarrow{\beta}_m)$ are given by

$$\overrightarrow{\beta}_m = \left(\frac{p + q}{2 - p}\right)^{1/2} n_1 \qquad t_p(\overrightarrow{\beta}_m) = \frac{2zn_a(p + q)^{1/2}}{c(q + 2)}(2 - p)^{1/2} \qquad (12.62)$$

The optimum power law exponent q_m is determined by setting the dispersion or pulsewidth $t_d = t(n_2) - t(n_1)$ equal to zero. This operation is written as follows:

$$(p + q)\frac{n_1}{n_2} + (2 - p)\frac{n_2}{n_1} = p + q_m + 2 - p \qquad (12.63)$$

So that[40]

$$q_m = 2(1 - 2\Delta)^{1/2} - p[1 + (1 - 2\Delta)^{1/2}] \approx 2(1 - p) - \Delta(2 - p) \qquad (12.64)$$

The minimum pulse spread is given by

$$t_d = \frac{z(\sqrt{n_1} - \sqrt{n_2})^2}{c(n_1 - n_2)} \, n_a \simeq n_a z \Delta^2 \tag{12.65}$$

Equation (12.61) shows that for most cases where $p \gg \Delta$ the optimum profile parameter q_m is controlled more by material dispersion than by intermodal dispersion.

12.5.6 Waveguide Dispersion

The propagation constant β, which is determined by the eigenvalue equation given by Eq. (12.18), is a function of k because of material dispersion and because it is a function of the v value. It is very difficult to separate the material and waveguide dispersion.

The delay time is related to the group velocity of the optical pulse and the speed of light as given by

$$t = \frac{z}{v_g} = z \frac{d\beta}{d\omega} \tag{12.66}$$

where ω is the frequency of the light and v_g is the group velocity.[42] The group velocity for a given refractive index can be written as[11]

$$v_g = \frac{(c\beta/k) \int_0^{2\pi} \int_0^\infty \Psi^2(r) r \, dr \, d\phi}{\int_0^{2\pi} \int_0^\infty \Psi^2(r) r \, dr \, d\phi} \tag{12.67}$$

The pulse spread is given by the derivative of the delay time t with respect to the frequency and is written by

$$t_d = z \frac{d^2\beta \delta\omega}{d\omega^2} \tag{12.68}$$

For a single-mode fiber Eq. (12.67) can be put in a form that is dependent upon modal parameters U and V and distortion parameter D defined as[11]

$$D = \frac{1}{(2\Delta)^{1/2}} \frac{d^2(\rho\beta)}{dV^2} \tag{12.69}$$

Table 12.5 displays the group velocity and distortion parameter for a step and power law refractive index profiles. The pulse spread can be written as

$$t_d = \frac{2zn_1 DV\Delta\delta\omega}{c\omega} = \frac{2zn_1 DV\Delta\delta\lambda}{c\lambda} \tag{12.70}$$

TABLE 12.5 Group Velocity and Distortion Parameters

Profile	v_g	D
Step index	$\dfrac{c}{n_1}\left[1 + \dfrac{\Delta U^2}{V^2}(1 - 2\sigma^2)\right]$	$\dfrac{2U^2 W\sigma}{V^3}$
Power law	$\dfrac{c}{n_1}\left[1 + \Delta\dfrac{(2-q)}{(2+q)}\left(\dfrac{G}{V}\right)^{2q/q+2}\right]$	$\left[\dfrac{U^2\sigma^4}{W^2} + \dfrac{(U^2 - W^2)\sigma^3}{W^3} + \dfrac{(W^2 - U^2)\sigma^2}{W^2} + \dfrac{3\sigma}{2W} - 1\right]$ $\dfrac{q(2-q)}{(2+q)^2}\left(\dfrac{G^{2q}}{V^{3q+2}}\right)^{1/(q+2)}$

$$\sigma = \frac{K_0(W)}{K_1(W)}$$

$$G = \frac{\Gamma(1/q + 1/2)}{2\Gamma(1/q)}(q + 2)\pi^{1/2}$$

12.5.7 Bandwidth

The previous subsection addressed the pulse spreading produced by intermodal and chromatic dispersion as a function of material and waveguide dispersion. The total pulse spread (rms) σ_t^2 is given by[43]

$$\sigma_t^2 = \sigma_i^2 + \sigma_c^2 \tag{12.71}$$

where subscripts i and c refer to the intermodal and chromatic dispersion.

The dispersion coefficient D_λ in units of picoseconds per nanometer-kilometer accounts for the source spread in bandwidth $\Delta\lambda$ and is given as[43]

$$D_\lambda = \frac{\sigma_t}{\Delta\lambda L} \tag{12.72}$$

where L is the length of the fiber. The bandwidth is defined as the frequency at which the Fourier amplitude of the transformed pulse response falls to half of its peak value. For Gaussian pulse shapes, the optical bandwidth is[43]

$$\Delta f_{-3\mathrm{dB(optical)}} \approx \frac{1.87}{\sigma_t} \tag{12.73}$$

One might think that at zero dispersion wavelength σ_t becomes zero and the bandwidth can approach infinity, but at this wavelength the second-order dispersion effect that has been neglected becomes important. The optical bit rate B can be used in place of the frequency bandwidth, since they are identical.

12.6 REFERENCES

1. I. Hayashi, M. B. Panish, P. W. Foy, and S. Sumski, *Appl. Phys. Lett.*, vol. 17, p. 109, 1970.
2. N. S. Kapany and R. J. Simms, *Infrared Phys.*, vol. 5, p. 69, 1965.

3. F. P. Kapron, D. B. Keck, and R. D. Maurer, *Appl. Phys. Lett.*, vol. 17, p. 423, 1970.

4. T. Miya, Y. Terunuma, T. Hosaka, and T. Miyashita, *Electron. Lett.*, vol. 15, p. 107, 1979.

5. D. A. Pinnow, A. L. Gentile, A. G. Stanlee, A. J. Timper, and L. M. Hobrock, *Appl. Phys. Lett.*, vol. 33, p. 28, 1978.

6. L. G. Van Uitert and S. H. Wemple, *Appl. Phys. Lett.*, vol. 33, p. 57, 1978.

7. C. H. L. Goodman, *Solid-State Electron. Dev.*, vol. 2, p. 129, 1978.

8. M. Born and E. Wolf, *Principles of Optics*, Pergamon Press, Oxford, 1970.

9. D. Marcuse, *Light Transmission Optics*, Van Nostrand, New York, 1972.

10. T. Katsuyama and H. Matsumura, *Infrared Optical Fibers*, Adam Higler, Philadelphia, 1989.

11. A. W. Snyder and J. D. Love, *Optical Waveguide Theory*, Chapman and Hall, New York 1983.

12. G. Toraldo di Francia, *J. Opt. Soc. Am.*, vol. 59, p. 799, 1969.

13. S. E. Miller, E. A. J. Marcatili, and T. Li, *Proc. IEEE*, vol. 61, p. 1703, 1973.

14. D. Gloge, *Appl. Opt.*, vol. 10, p. 2252, 1971.

15. C. M. Baldwin, R. M. Almeida, and J. D. Mackenzie, *J. Non-Cryst. Solid*, vol. 43, p. 309, 1981.

16. M. Poulain, M. Poulain, and J. Lucas, *Mater. Res. Bull.*, vol. 10, p. 243, 1975.

17. K. Ohsawa, T. Shibata, N. Nakamura, and S. Yoshida, in Technical Digest, 7th European Conference on Optical Communication, Bella Center, Copenhagen, Denmark, 1981.

18. T. Kanamori, Y. Terunuma, S. Takahashi, and T. Miyashita, *J. Lightwave Technol.*, vol. 3, p. 607, 1984.

19. B. E. Douglas and D. H. McDaniel, *Concepts and Models of Inorganic Chemistry*, Blaisdell, London, 1965.

20. M. Shimizu, M. Yamada, T. Takeshia, and M. Horiguchi, in Technical Digest, Optical Amplifiers and Their Applications (Optical Society of America), Washington, D.C., vol. 13, p. 12, 1991.

21. D. B. Keck, P. C. Shultz, and F. Zimar, U.S. Patent 3,737,292.

22. P. C. Schultz, *Proc. IEEE*, vol. 68, p. 1187, 1980.

23. M. G. Blankenship and C. W. Deneka, *IEEE J. Quantum Electron.*, vol. QE-18, p. 1918, 1982.

24. J. B. MacChesney, *Proc. IEEE*, vol. 68, p. 1181, 1980.

25. S. R. Nagel, J. B. MacChesney, and K. L. Walker, *Optical Fiber Communications*, Academic Press, New York, 1985.

26. T. Izawa, S. Kobayashi, S. Sudo, and E. Hanawa, in Technical Digest, 2nd International Conference on Integrated Optics and Optical Fiber Communications, Tokyo, p. 375, 1971.

27. T. Izawa and W. Inagaki, *Proc. IEEE*, vol. 68, p. 1184, 1980.

28. S. Tomaro, S. Kawachi, M. Yasu, and M. Edahira, *Trans. IEEE* Japan, vol. E65, p. 717, 1982.

29. J. W. Fleming and V. R. Raju, *Electron. Lett.*, vol. 17, p. 867, 1981.

30. J. W. Flemming, J. B. MacChesney, and P. B. O'Connor, U.S. Patent 4,331,462, 1982.

31. P. Geittner, D. Kuppers, and H. Lydtin, *Appl. Phys. Lett.*, vol. 28, p. 645, 1971.

32. P. Bachman, P. Geittner, and H. Lydtin, in Technical Digest, Conference on Optical Communications, Washington, D.C., WA1, 1986.

33. G. J. Koel, in Technical Digest, Eighth European Conference on Optical Communication, p. 21, 1982.

34. D. A. Pinnow, T. C. Rich, F. W. Ostermayer, and M. DiDomenico, *Appl. Phys. Lett.*, vol. 22, p. 527, 1973.

35. Y. Ohishi, S. Mitachi, T. Kanamori, and T. Manabe, *Phys. Chem. Glasses*, vol. 24, p. 135, 1983.

36. H. Osanai, *J. Inst. Electronics Commun. Eng. Japan*, vol. 63, p. 385, 1980.

37. P. C. Schultz, *J. Am. Ceram. Soc.*, vol. 57, p. 309, 1974.

38. W. A. Gambling, H. Matsumura, and C. M. Ragdale, *Opt. Quantum Electron.*, vol. 11, p. 43, 1979.

39. Luc. B. Jeunhomme, *Single-Mode Fiber Optics*, Marcel Dekker, New York, 1990.

40. A. Ankiewicz and C. Park, *Opt. Quantum Electron.*, vol. 9, p. 87, 1977.

41. R. Olshansky and D. B. Keck, *Appl. Opt.*, vol. 15, p. 483, 1976.

42. J. D. Jackson, *Classical Electrodynamics*, Wiley, New York, 1967.

43. C. Yeh, *Handbook of Fiber Optics*, Academic Press, New York, 1990.

44. T. G. Giallorenzi, *Proc. IEEE*, vol. 66, p. 744, 1978.

45. S. R. Nagel, In Technical Digest, Conference on Optical Fiber Communication, Optical Society of America, Washington, D.C., 1985, paper TuR1.

46. S. Yoshida, Kougaku Gijutsu Contact, vol. 24, pp. 681–691, 1986.

47. P. A. Tick and P. L. Bocko, "Optical Fiber Materials," in *Optical Materials*, Marcel Dekker, New York, 1990.

48. J. B. MacChesney and P. B. O'Connor, U.S. Patent 4,217,027.

49. N. Niizeki, N. Inagaki, and T. Edahiro, in T. Li (ed.), *Optical Fiber Communications*, vol. 1, Academic Press, New York, p. 97, 1985.

50. S. R. Nagel, J. B. MacChesney, and K. L. Walker, in vol. 1, T. Li (ed.), *Optical Fiber Communications*, Academic Press, New York, p. 97, 1985.

51. A. Kats, *Philips Tech. Rev.* 10/11/12, p. 36, 1986.

CHAPTER 13
NONLINEAR OPTICS

Gary L. Wood and Edward J. Sharp

13.1 INTRODUCTION

The introduction of the visible laser, first demonstrated[1] experimentally on May 15, 1960, marked the beginning of the rapid expansion of nonlinear optics.[2] Today, the field of nonlinear optics remains very active in both a basic and an applied sense. New nonlinear effects have been discovered and are being explored both experimentally and theoretically. The technology has continued to mature, and refinements in the development of lasers and high-quality optical materials have resulted in a variety of commercially available nonlinear optical devices. A number of good references are devoted to introducing the reader to the broad field of nonlinear optics.[3–9]

The interaction of light with matter includes both a linear as well as a nonlinear component. Even though the nonlinear component is usually weak and often not noticeable, it can become dominant at optical resonances or for large incident intensities, such as those produced with laser beams. For instance, it has been observed that the optical behavior of many materials deviates from the linear behavior found in Snell's and Beer's law at high intensities. This is due to the fact that these laws are approximations and are only valid at low intensities. In general, the index of refraction n and the absorption coefficient α are intensity-dependent,

$$n = n_0 + \Delta n(I) \qquad (13.1a)$$

$$\alpha = \alpha_0 + \Delta \alpha(I) \qquad (13.1b)$$

where I is the incident intensity.

A full quantum-mechanical treatment is necessary in order to describe the detailed behavior of nonlinear optics. A number of nonlinear effects, such as the description of the initial buildup of stimulated scattering, have no classical analog. A number of nonlinear effects that occur on time scales over which the material state is interacting coherently with the photon field also require quantum-mechanical analysis. Because of the somewhat limited space and in the interest of clarity, this chapter is concerned mostly with nonlinear optical effects that can be understood without a full quantum-mechanical treatment. Fortunately, many nonlinear optical effects can be described quite adequately without a detailed understanding of quantum physics, and we begin by considering the classical electric dipole.

13.2 LINEAR OPTICS: THE HARMONIC POTENTIAL WELL

The classical picture begins by considering an electron attached by a spring to an infinitely massive positive charge (a nucleus) under the influence of a sinusoidally oscillating electric field $E(t)$. Considering only one dimension, Newton's equation of motion for the system depicted in Fig. 13.1 is

$$m\frac{d^2x}{dt^2} + \eta\frac{dx}{dt} + m\omega_0^2 x = eE(t) \tag{13.2}$$

where x is the distance the electron has moved from its equilibrium position, m is the mass of the electron, η is some resistance term or damping term, and ω_0 is the

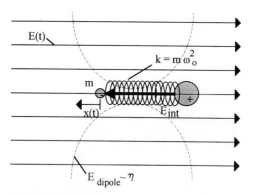

FIGURE 13.1   This figure shows an electron (negative charge) of mass m bound to a positive charge of infinite mass through the internal field E_{int}. For linear systems this internal field is represented by a spring with spring constant $k = m\omega_0^2$. The displacement of the electron from equilibrium is given by $x(t)$. This system is located within an oscillating optical electric field $E(t)$ and dampening η is represented as a reradiated dipole field given by the dotted lines as E_{dipole}.

resonant frequency of the electron. The oscillating macroscopic polarizability is then given by the solution $x(t)$ of Eq. (13.2) above as

$$P(t) = Nd(t) = Nex(t) \tag{13.3}$$

where N is the number of oscillators per unit volume and d is the dipole moment. For $E = 0.5\, E_0\, (e^{i\omega t} + e^{-i\omega t})$, where ω is the optical frequency, the solution to x in the linear equation above can be found by substituting $x = 0.5\, x_0\, (e^{i\omega t} + e^{-i\omega t})$ into Eq. (13.2) to get

$$x_0 = \frac{eE_0}{m}\,(\omega_0^2 - \omega^2 + i\Gamma\omega)^{-1} \tag{13.4}$$

where $\Gamma = \eta/m$. This gives a complex polarizability of

$$P(t) = \frac{Ne^2 E(t)}{m} (\omega_0^2 - \omega^2 + i\Gamma\omega)^{-1} \tag{13.5}$$

The index of refraction and the absorption coefficient can be approximated from the above equation if it is assumed that only one resonance is near the optical frequency and that the local field the electron sees is close to the incident field. The polarizability is linear in the electric field and can be written as $P = \chi E$ where χ is the complex linear susceptibility tensor. This linear relationship between P and E results in an index of refraction and absorption coefficient that is independent of the electric field. That is, for weak absorption, small local field correction, and only one dominant nearby resonance[10] ω_0, the index of refraction and the absorption coefficient are independent of the optical electric field, as is observed at low optical intensities, and given as

$$n^2 \sim 1 + \frac{Re(\chi)}{\epsilon_0} \tag{13.6}$$

$$\alpha \sim \frac{\omega Im(\chi)}{nc\epsilon_0} \tag{13.7}$$

Although this simple oscillating spring model seems to predict field-independent indexes and absorption coefficients, and many other observed behaviors of the index and absorption (such as the correct dispersion and Lorentzian linewidth), it is not immediately clear why it should be successful. To provide a clearer understanding it is necessary to investigate the spring model in greater detail.

The spring model uses a linear (Hooke's law) restoring force $F = kx$, where k is the spring constant or $k = m\omega_0^2$. In general, the charges responding to the oscillating electric field have been moved from some equilibrium point and try to return to that point when the field is removed. This is called stable equilibrium and can be thought of as a potential well that has a minimum point. The potential V can be expanded about the minimum point located at $r = 0$.

$$V(r) = V|_{r=0} + \frac{dV}{dr}\bigg|_{r=0} r + \frac{d^2V}{dr^2}\bigg|_{r=0} \frac{r^2}{2} + \cdots \tag{13.8}$$

Since the first two terms can be set to zero at $r = 0$, the potential becomes

$$V(r) = \frac{d^2V}{dr^2}\bigg|_{r=0} \frac{r^2}{2} + \frac{d^3V}{dr^3}\bigg|_{r=0} \frac{r^3}{3!} + \cdots \tag{13.9}$$

which yields a force of $dV = \mathbf{F} \cdot d\mathbf{r}$, or

$$F = \frac{d^2V}{dr^2}\bigg|_{r=0} r + \frac{d^3V}{dr^3}\bigg|_{r=0} \frac{r^2}{2} + \cdots \tag{13.10}$$

The first term in Eq. (13.10) is Hooke's law and is the dominant term for small deviations of the electron from the equilibrium position. With larger perturbations, associated with larger electric fields, the restoring force must include higher-order terms and results in a more complicated nonlinear behavior. A comparison of anharmonic and harmonic potential wells is illustrated in Fig. 13.2.

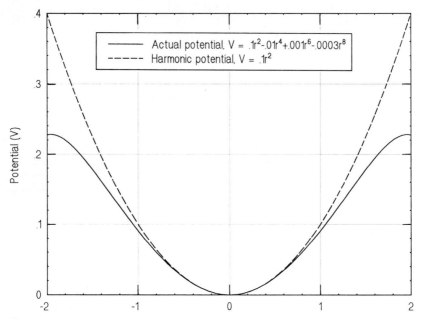

FIGURE 13.2 A symmetric potential well is shown for a molecular system by the solid line. The equation for this potential is given as $V = 0.1r^2 - 0.01r^4 + 0.001r^6 - 0.0003r^8$. A harmonic potential of $V = 0.1r^2$ is also shown by the dotted line. The harmonic potential closely follows the actual potential until $r > 1$. Above $r = 1$ higher-order terms must be kept in the expansion of the potential.

13.3 NONLINEAR OPTICS: THE ANHARMONIC POTENTIAL WELL

For a general restoring force derived from a nonparabolic stable equilibrium potential well, the force equation about the minimum potential is

$$m\frac{d^2x}{dt^2} + \eta\frac{dx}{dt} + \omega m_0^2 x + {}_{i=1}\Sigma^N D_i x^{i+1} = eE(t) \tag{13.11}$$

and the solution for small anharmonic contributions is of the form

$$x(t) = \sum_{j=1}^{N} \frac{x_j \exp(ij\omega t) + \text{c.c.}}{2} \tag{13.12}$$

By equating the coefficients of $e^{i\omega t}$, the second-order term x_2 can be found oscillating at 2ω and is given as

$$x_2 = \frac{-De^2E_0^2}{2m^2[(\omega_0^2 - \omega^2) + i\omega\eta]^2(\omega_0^2 - 4\omega^2 + 2i\omega\eta)} \tag{13.13}$$

In addition to the $\omega = \omega_0$ resonance there is also a resonance at $\omega = \omega_0/2$. The anharmonic terms result in a nonlinear polarizability which, in general, for three dimensions is

$$P_i(x_p) = \chi_{ij}^{(1)}E_j + \chi_{ijk}^{(2)}E_jE_k + \chi_{ijkl}^{(3)}E_jE_kE_l + \cdots \tag{13.14}$$

where E is the total electric field present. The first term is the result of a linear restoring force (harmonic potential well) and describes linear, low-intensity, optical interactions with materials. The second term, which includes the second harmonic term of Eq. (13.13), has a second-order susceptibility $\chi_{ijk}^{(2)}$ and two electric fields E_j and E_k. This term results from an anharmonic potential well and is only found in crystals without inversion symmetry. In fact, all the even-numbered, higher-order susceptibilities are eliminated in materials that have a center of inversion. The third-order term, and all higher odd-ordered terms, are found in all materials, assuming that the material will not suffer laser-induced damage first.

In the spring model discussed above, the electron is held in place by the attraction of the positive nucleus. Large deviations from the electron equilibrium position with the application of an applied electric field result in the charges experiencing the anharmonic region of the potential well and determine the relative importance of the nonlinear terms. The linear susceptibility is proportional to the square of the expectation value of the dipole, or $\chi^{(1)} \propto \langle 2|d|1\rangle^2 = d_{21}^2$, where $|2\rangle$ is the final state wavefunction and $|1\rangle$ is the initial state. In fact, the nth-order susceptibility is proportional to the dipole multiplied by $n + 1$, or $\chi^{(n)} \propto d_{21}^{n+1}$. Since the dipole moment multiplied by an electric field is the energy $\hbar\omega$, the dipole moment can be written as a resonance frequency divided by some characteristic field strength which binds the charges $d_{21} \propto \omega_0/E_{int}$. Comparing terms in the polarization vector yields

$$\frac{P^{n+2}}{P^n} = \frac{\chi^{(n+2)}E^{n+2}}{\chi^{(n)}E^n} = d_{21}^2E^2 \propto \left(\frac{E}{E_{int}}\right)^2 \tag{13.15}$$

It can be seen that when $(E/E_{int})^2 \ll 1$ the polarization vector will rapidly converge. For materials transparent to optical radiation, $E_{int} \approx e/4\pi\epsilon_0(1 \text{ Å})^2 \approx 10^9$ V/cm and, since most laser pulses cannot achieve this electric field, the expansion is in general valid. Therefore, away from resonance, the third-order susceptibility is much smaller in magnitude than the second-order susceptibility, which in turn is much smaller than the linear susceptibility.

In addition to index of refraction and absorption coefficients that depend on the optical electric field, nonlinear effects often introduce new radiation frequencies. The polarization vector frequency ω_p describes the material frequency response to the incident electric fields. When the susceptibility tensor is independent of the optical electric field, that is, in the linear optical regime, the polarization vector frequency ω_p is equal to the incident frequency ω_1. This is the material response at normally encountered optical intensities. As an example, in the solution to the classical dipole problem above, with the introduction of an anharmonic potential well, the response included a term oscillating at twice the incident frequency. Many combinations of frequencies are possible with the nonlinear terms of the susceptibility tensors, and this is a convenient way to characterize the different types of optical nonlinearities.

Equation (13.14) is valid when the response of the media is instantaneous. In general, the polarizability P is the result of the electric fields located throughout space and time. The polarizability is given as the product of the susceptibility and the fields integrated over space and time. However, for slowly varying field amplitudes only the susceptibility is integrated and the susceptibility is given in terms of the frequency response, as we have already shown above.

The quantum-mechanical picture of linear optics can be described by one photon absorption followed by reemission to some final state. Nonlinear optics arises in the classical picture from high-intensity radiation (large optical fields), which causes the charges to oscillate anharmonically in the potential well. The quantum-mechanical picture of nonlinear optics involves higher-order interactions of the photon field and the matter fields. Feynmann diagrams are often employed to catalog the various absorptions and emissions, with the final atomic susceptibility being the sum of all combinations yielding the same number of absorptions and emissions. This involves various intermediate states that may be real or virtual depending on the photon energy and the excited material state (see Fig. 13.3). There are many good sources that derive and describe the nonlinear susceptibilities using quantum field theory.[11,12] In general it can be stated that the susceptibilities are all made up of two parts. The susceptibilities all have dipole terms d and resonant frequency terms. For instance, the *atomic* third-order nonlinear susceptibility that yields a polarizability oscillating at the incident frequency can be written as[13]

$$\chi_a^{(3)}(\omega;\omega,-\omega,\omega) = \Sigma_{mpq} d_{nm} d_{mp} d_{pq} d_{qn}$$

$$\times \{[(\omega_{mn} - \omega)(\omega_{pn} - 2\omega)(\omega_{qn} - \omega)]^{-1}$$

$$+ [(\omega_{mn} - \omega)\omega_{pn}(\omega_{qn} - \omega)]^{-1}$$

$$+ [(\omega_{mn} - \omega)\omega_{pn}(\omega_{qn} + \omega)]^{-1}$$

$$+ [(\omega_{mn} + \omega)(\omega_{pn} + 2\omega)(\omega_{qn} + \omega)]^{-1}$$

$$+ [(\omega_{mn} + \omega)\omega_{pn}(\omega_{qn} + \omega)]^{-1}$$

$$+ [(\omega_{mn} + \omega)\omega_{pn}(\omega_{qn} - \omega)]^{-1}\} \qquad (13.16)$$

where $\chi(\omega)^{(3)} = N_n\chi_a^{(3)}$, N_n is the number density of oscillators in the nth state. The initial and final state is designated as n, while m, p, q are intermediate states. In general, the states consist of both electronic (fast) and nuclear or ionic (slow) components, which by the adiabatic approximation can be separated and treated independently. Associated with each resonant term is a linewidth (related to the resistance due to radiative damping) appearing as $i\Gamma$ [not included in Eq. (13.16) when far from resonance]. If the linewidth of the real material states lies outside the excited photon states, then all the intermediate states are virtual and the interaction is instantaneous with no population redistribution. Otherwise, at or near resonance, there is a population redistribution in time and the nonlinear susceptibility is enhanced because a term or terms in the denominator become small as $\Delta\omega \rightarrow 0$. Since resonant enhancement can dramatically enhance the susceptibility, even insignificant nonlinearities (i.e., non-phase-matched) can become important and dominant. For significant resonant enhancement, absorption can cause the transmission to be small. However, a coherent mixture of two states can result in a large resonant nonlinearity and yet have weak linear absorption.[14] Often resonant enhancement allows higher-order processes to be described as a product of two or more lower-order processes called a cascaded event.

Third Harmonic Generation (3HG)

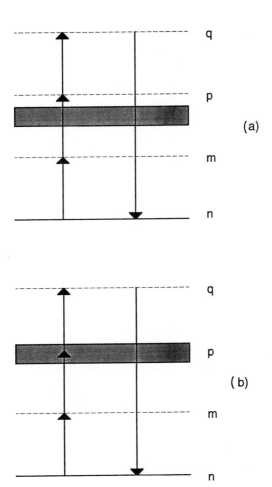

FIGURE 13.3 One set of photon absorptions and emissions that contribute to third-harmonic generation (3HG). Part (*a*) shows a linewidth broadened real material state below virtual state *p* and the total transition can be considered virtual. Part (*b*) shows a two-photon absorption resonantly enhanced 3HG with the real state at *p*.

13.4 SECOND-ORDER NONLINEARITIES: $\chi^{(2)}$

The second term in Eq. (13.14) describes second-order nonlinearities and can be written for two complex electric fields $E(\omega_1)$ and $E(\omega_2)$ as

$$P_i^{(2)}(\omega_p) = \chi_{ijk}^{(2)}E_jE_k$$

$$= \frac{\chi_{ijk}^{(2)}}{4}[E(\omega_1) + E(\omega_1)^* + E(\omega_2) + E(\omega_2)^*]_j$$

$$\times [E(\omega_1) + E(\omega_1)^* + E(\omega_2) + E(\omega_2)^*]_k \qquad (13.17)$$

which results in

$$P^{(2)} = \chi_{ijk}^{(2)}(0;\omega_1,-\omega_1)A_{j1}A_{k1}^*/2 + \chi_{ijk}^{(2)}(0;\omega_2,-\omega_2)A_{j2}A_{k2}^*/2$$

$$+ \chi_{ijk}^{(2)}(2\omega_1;\omega_1,\omega_1)(A_{j1}A_{k1}e^{i2\omega_1 t} + \text{c.c.})/4$$

$$+ \chi_{ijk}^{(2)}(2\omega_2;\omega_2,\omega_2)(A_{j2}A_{k2}e^{i2\omega_2 t} + \text{c.c.})/4$$

$$+ \{[\chi_{ijk}^{(2)}(\omega_1 + \omega_2;\omega_1,\omega_2)A_{j1}A_{k2}e^{i(\omega_1+\omega_2)t} + \text{c.c.}]$$

$$+ [\chi_{ikj}^{(2)}(\omega_2 + \omega_1;\omega_2,\omega_1)A_{k2}A_{j1}e^{i(\omega_2+\omega_1)t} + \text{c.c.}]\}/4$$

$$+ \{[\chi_{ijk}^{(2)}(\omega_1 - \omega_2;\omega_1,-\omega_2)A_{j1}A_{k2}^*e^{i(\omega_1-\omega_2)t} + \text{c.c.}]$$

$$+ [\chi_{ikj}^{(2)}(\omega_2 - \omega_1;\omega_2,-\omega_1)A_{k2}A_{j1}^*e^{i(\omega_2-\omega_1)t} + \text{c.c.}]\}/4 \qquad (13.18)$$

where $E = Ae^{i\omega t}$ has been used. Each $\chi^{(2)}$ has a unique frequency dependence and the corresponding polarization frequency is given in $\chi^{(2)}(\omega_p;\ldots)$ as the first frequency. The material response to the incident optic fields can be very fast if only virtual transitions occur such as the interaction of off-resonance electrons. The response for these cases is essentially instantaneous, with the response field following the incident field.

The first two terms in Eq. (13.18) yield a zero polarization frequency $\omega_p = 0$ and give rise to the nonlinear effect of optical rectification.[15] Essentially, the incident field generates a polarization in the medium that appears as an applied dc electric field.

If one frequency is zero or close to it, $\omega_2 \sim 0$, then the polarization appears as

$$P^{(2)}(\omega_1) = \chi^{(2)}(\omega_1;\omega_1,0)A_1A_2e^{i\omega_1 t} \qquad (13.19)$$

which describes the Pockel's effect. This susceptibility can include slow ionic, as well as fast electronic mechanisms, and will in general be different from a susceptibility that has only optical frequency contributions. The electric field A_2 is a dc field applied across the material while A_1 is the optical beam which passes through the material and experiences an electric-field-dependent change in phase or index of refraction, but no change in frequency. The index change can be seen by writing

$$P_i = \chi_{ij}E_j = (\chi^{(1)}_{ij} + \chi^{(2)}_{ijk}E_k)E_j$$

$$= (\chi_{ij} + \Delta\chi_{ij})E_j \qquad (13.20)$$

so that

$$n^2 = \frac{\epsilon}{\epsilon_0} = \frac{\chi_{ij} + \Delta\chi_{ij}}{\epsilon_0} \sim (n_0 + \Delta n)^2 \qquad (13.21)$$

or $\Delta n \approx \Delta\chi_{ij}/2n_0\epsilon_0$ (for $\Delta n^2 \to 0$) contains the applied low-frequency field dependence. This induced birefringence was one of the first experiments in nonlinear optics performed in the late 1800s by Pockel. In the literature this second-order susceptibility is replaced with the electro-optic coefficients

$$\chi^{(2)}{}_{ijk} = -\frac{\epsilon_{ii}\epsilon_{jj}}{4\epsilon_0} r_{ijk} \qquad (13.22)$$

where the electro-optic tensor is usually written in contracted notation as $r_{ij,k} = r_{ij}$, $i = 1–6$, $j = 1–3$.

In addition to the electric-field-induced index change, there is a corresponding absorption change which can be expressed through the Kramers-Kronig relationship.[16] Only the real part of the complex polarization vector is related to the index change. The imaginary part of this susceptibility is related to electroabsorption. The applied electric field can be large enough to move an absorption edge in some materials, resulting in a change in the absorption spectrum.

The third and fourth terms of Eq. (13.18) have a component of the second-order susceptibility that exhibits a polarization frequency twice the incident frequency which gives rise to second-harmonic generation (SHG). This is an important means for converting a long-wavelength laser to a shorter wavelength. For instance, the output of a Nd:YAG laser at 1064 nm is outside the visible spectrum, but with second-harmonic generation the beam can be converted to a green laser output at 532 nm. The efficient conversion of a significant fraction of the incident beam energy to a new wavelength requires that the incident frequency beam and the generated frequency beam travel in phase (at the same velocity) throughout the material. This process, known as phase matching, can be accomplished in some materials at particular angles and temperatures. For example, Fig. 13.4 shows how a negative birefringent crystal can be used for SHG. In this example an ordinary

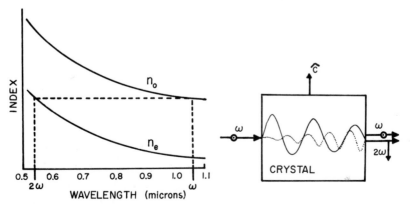

FIGURE 13.4 A negative birefringent material showing type I phase matching for an input beam at $\lambda = 1.06$ μm and a second harmonic at $\lambda = 0.532$ μm. The crystal is shown with optimum coupling, that is, noncritical phase matching.

polarized pump beam is incident perpendicular to the crystal c axis. Phase matching is achieved at the second harmonic of the pump for the extraordinary polarized wave $(n_\omega^0 = n_{2\omega}^e)$. For positive birefringent crystals the roles of the ordinary and extraordinary waves are reversed. Notice that the fundamental and second harmonic are copropagating throughout the crystal. This is the preferred arrangement for maximum coupling and is known as noncritical phase matching. In general the input beam is not incident at 90° to the c axis $[n_\omega^0 = n_{2\omega}^e(\theta)]$ and the ordinary and extraordinary polarized beams diverge. This may limit the interaction to less than the crystal length and hence reduce the SHG efficiency. This constitutes critical phase matching. There are two types of phase matching using birefringent crystals, and Fig. 13.4 is an example of type I where two incident photons of one polarization are converted to a single photon of the other polarization at the second harmonic. Type II phase matching occurs when the input beam consists of both polarization states and the second harmonic is polarized either ordinary or extraordinary depending on positive or negative birefringence, respectively. For example, type II phase matching for a positive birefringent crystal obeys the condition $n_{2\omega}^0 = [n_\omega^e(\theta) + n_\omega^0]/2$. A popular technique to investigate new materials for their SHG capabilities is the well-known Kurtz powder method.[17]

The last two terms of Eq. (13.18) involve, respectively, sum and difference frequency mixing. This is a convenient method to mix different laser frequencies to generate a coherent source that is not restricted to the second harmonic. This phenomenon forms the basis for optical parametric oscillators which can be used to fabricate tunable laser sources.[18]

The various components of the susceptibility tensors are determined by the material symmetry. The tensor elements corresponding to the various crystal symmetries can be found in Chap. 11 and in many books[19] and handbooks[20] on nonlinear optics. In the above analysis the order of the electric fields is taken into account and the susceptibility is dependent on the order of the frequencies. However, far from resonance, i.e., a lossless medium with instantaneous response, the order is not important and the susceptibility does not depend on the order of the frequencies,

$$\chi_{ijk}^{(2)} E_{1j} E_{2k} = \chi_{ikj}^{(2)} E_{2k} E_{1j} \tag{13.23}$$

so that, far from resonance, the polarizability can be written as

$$P_i(\omega_p) = \chi_{ij}^{(1)} E_j + 2\chi_{ijk}^{(2)} E_j E_k + 4\chi_{ijkl}^{(3)} E_j E_k E_l + \cdots \tag{13.24}$$

where the factors 2 and 4 in the second and third terms are due to the degeneracy.

13.5   THE THIRD-ORDER SUSCEPTIBILITIES: $\chi^{(3)}$

The third-order term $P_i^{(3)}(\omega_p)$ has 22 distinct frequencies. The third-order susceptibility is a fourth-rank tensor and contains 81 terms. However, because of symmetry, most optical materials have considerably fewer elements and isotropic materials have only two independent elements. For isotropic materials, with i not equal to j and $i, j = 1$–3; $\chi_{iiii} = \chi_{jjjj} = \chi_{kkkk}$, $\chi_{iijj} = \chi_{jjii}$, $\chi_{iijj} = \chi_{jiji}$, $\chi_{ijji} = \chi_{jiij}$, and $\chi_{iiii} = \chi_{iijj} + \chi_{ijij} + \chi_{ijji}$.

Table 13.1 lists some of the many possible mechanisms that give rise to a real third-order susceptibility and the time scale over which it is considered significant.

TABLE 13.1 Microscopic Mechanisms That Can Generate $Re\ \chi^{(3)}$

Mechanisms	Comments
Electronic Kerr effect (EKE)[a]	Electron cloud distortion, nonharmonic potential well ($\sim$fs), observed in all materials
Orientational Kerr effect (OKE)[b]	Reorientation of asymmetric molecules (ns–ps), observed mainly in liquids
Electrostriction[c]	Electric field gradient pressure (ns), observed in all materials
Thermal[d]	Density changes ($>$ns) or bandgap changes (fs), observed in absorbing materials
Saturation of a resonance[e]	Reduction in the number of oscillators (ns), observed in all materials near resonance
Creation of free changes[f]	Change in the number and type of oscillators (ps–fs), usually observed in semiconductors
Nonparabolic bandgap[g]	Deviation from harmonic potential (fs), observed in narrow-bandgap semiconductors

[a]R. W. Hellwarth, "Third-Order Optical Susceptibilities of Liquids and Solids," in J. H. Sanders and S. Stenholm (eds.), *Progress in Quantum Electronics*, Pergamon Press, New York, 1977.
[b]O. Svelto, "Self-Focusing, Self-Trapping, and Self-Phase Modulation of Laser Beams," in E. Wolf (ed.), *Progress in Optics XII*, North-Holland, 1974.
[c]E. L. Kerr, *IEEE J.Quantum Electron.*, vol. QE-6, p. 616, 1970.
[d]S. A. Akhmanov, D. P. Krindach, A. V. Migulin, A. P. Sukhorukov, and R. V. Khokhlov, *IEEE J. Quantum Electron.*, vol. QE-4, p. 568, 1968.
[e]A. Javan and P. L. Kelley, *IEEE J. Quantum Electron.*, vol. QE-2, p. 470, 1966.
[f]A. Miller, D. A. B. Miller, and S. D. Smith, *Adv. Phys.*, vol. 30, p. 697, 1981.
[g]P. A. Wolff and G. A. Pearson, *Phys. Rev. Lett.*, vol. 17, p. 1015, 1966.

The total observed third-order susceptibility is the sum of all these many possible contributions. The third-order susceptibility can be written as

$$P_i(\omega_p) = \chi_{ijkl}^{(3)} E_j E_k E_l \tag{13.25}$$

$$P_{0i}(\omega_p) = (1/4)\delta\chi_{ijkl}^{(3)}(\omega_p;\omega_1,\omega_2,\omega_3)\ A\dagger_j(\omega_1)A\dagger_k(\omega_2)A\dagger_l(\omega_3)e^{\pm i\omega_1 t}e^{\pm i\omega_2 t}e^{\pm i\omega_3 t} \tag{13.26}$$

where $\dagger = *$ when $\omega < 0$; otherwise $\dagger = +1$, $\omega_p = \omega_1 + \omega_2 + \omega_3$, and P_0 is the complex polarizability. All combinations of ω and jkl are possible. Since the ordering of the electric fields is arbitrary, the degeneracy factor δ is included.[21] For three different frequencies, where a minus sign is considered a different frequency, the degeneracy factor is $\delta = 6$. If two frequencies are alike, then $\delta = 3$. If all the frequencies are alike, then $\delta = 1$.

The most general case is known as four-wave mixing and has three different frequencies mixing to generate a fourth frequency. Three incident beams interact on the material and generate a fourth beam. All four beams are then present to interact with each other. The intensity of the generated beam is related to the intensities of the mixing beams and the magnitude squared of the third-order susceptibility, $I_4 \propto |\chi^{(3)}|^2 I_1 I_2 I_3$. Notice the time-averaged intensity, which is the observed intensity for optical frequencies, is defined to be $\langle I \rangle = c\epsilon_0 n A A^*/2$. Four-wave mixing is a valuable tool for investigating resonances within a material and, by delaying the probe beam, investigating time responses as well. A particularly simple four-wave mixing geometry, degenerate four-wave mixing, uses three mutually coherent beams of equal frequency and polarization. As shown in Fig. 13.5 two beams, a forward pump beam ($\mathbf{k}_f$) and a backward pump beam ($\mathbf{k}_b$) are arranged to be counterpropagating through the material. A third beam, known as the probe beam ($\mathbf{k}_p$), is incident at a small angle to one of the pump beams. In this arrangement it is possible to observe a generated beam ($\mathbf{k}_c$), also at the frequency of the incident beams, but counterpropagating to the probe beam ($\mathbf{k}_c = -\mathbf{k}_p$). In addition, this generated beam is the phase conjugate of the incident probe beam. More about this subject can be found in the following chapter.

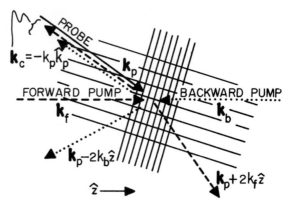

FIGURE 13.5 Four-wave mixing spatial gratings for an isotropic medium showing the scattering (dotted and dashed lines) of the forward and backward pump beams $\mathbf{k}_f$ and $\mathbf{k}_b$, respectively where $\bar{\mathbf{k}}_f = k_f\hat{z}$, $\bar{\mathbf{k}}_b = -k_b\hat{z}$ and $|\bar{\mathbf{k}}_b| = |\bar{\mathbf{k}}_f|$. The dotted lines correspond to scatter from a transmission grating and the dashed lines scatter from a reflection grating. Scattered beams traveling opposite the probe beam $\mathbf{k}_p$ are the observed conjugate beam while the other scattered beams are not phase-matched and can only be observed in thin media.

Third-harmonic generation[22] (THG) results when all three incident frequencies add to the polarization vector frequency.

$$P_{0i}(\omega_p = 3\omega) \sim \chi^{(3)}{}_{ijkl}(3\omega;\omega,\omega,\omega)A_j e^{i\omega t}A_k e^{i\omega t}A_l e^{i\omega t} \tag{13.27}$$

Only electronic mechanisms can participate, since a fast response is required. It is very difficult to phase-match the third harmonic and, as a consequence, it is not nearly as important for frequency up-conversion as is second-harmonic generation. If one of the electric fields is a low frequency or zero frequency, the resulting change in index is known as the dc Kerr effect, or

$$P_i(\omega) \longrightarrow \chi^{(3)}{}_{ijkl}(\omega;\omega,0,0)A_j e^{i\omega t}A_k^* A_l \tag{13.28}$$

where A_k and A_l have frequencies near or equal to zero. John Kerr first noticed a birefringence in materials with an applied electric field in the late 1800's that was proportional to the square of the applied field. The effect which bears his name has been found in all materials if the applied field does not damage the material first.

If only a single frequency beam is incident on the material, a number of third-order nonlinear effects can occur, such as nonlinear absorption, refraction, and scattering. The frequency of the generated beam can be at the frequency of the incident beam with no applied dc fields with the following combinations of fields:

$$P_{0i}(\omega) \propto \chi^{(3)}_{ijkl}(\omega;\omega,\omega,-\omega) A_j e^{i\omega t} A_k e^{i\omega t} A_l^* e^{-i\omega t} \tag{13.29}$$

This appears as a change in the linear susceptibility away from resonance. From Eq. (13.20)

$$\Delta\chi_{ij} = 3\chi^{(3)}_{ijkl}(\omega;\omega,\omega,-\omega) A_k e^{i\omega t} A_l^* e^{-i\omega t} \tag{13.30}$$

This change in susceptibility is related to the third-order nonlinear index n_2 if the third-order susceptibility is real, and it is related to the nonlinear absorption β if the third-order susceptibility is imaginary, where $\Delta n = n_2|A|^2/2$ and $A\alpha = \beta I$. In fact, the two are related through the Kramers-Kronig relationships for nonlinear optics.[23] Since the same mechanism that gives rise to nonlinear absorption also produces nonlinear refraction (index change), a value of the nonlinear index can be extracted from a measurement of nonlinear absorption vs. wavelength through the use of the Kramers-Kronig relationships, and vice versa.

Stimulated scattering is observed when the incident beam photon is absorbed to some virtual level where a previously scattered photon stimulates a return to some excited state (Stokes emission) or to a state below the initial energy level (anti-Stokes emission). By this process a new frequency is generated, where for Stokes emission,

$$P_{0i}(\omega_S) \propto \chi^{(3)}_{ijkl}(\omega_S;\omega_L,-\omega_L,\omega_S) A_{Lj} e^{i\omega_L t} A^{*Lk} e^{-i\omega_L t} A_{Sl} e^{i\omega_S t} \tag{13.31}$$

where ω_L denotes the laser angular frequency and ω_S denotes the Stokes angular frequency. The imaginary polarizability is negative for Stokes emission ($\omega_S < \omega_L$), and positive for anti-Stokes emission ($\omega_S > \omega_L$). The presence of a negative imaginary polarization leads to an exponentional gain of the Stokes generated frequency and the anti-Stokes emission decreases exponentially. Two of the most common types of stimulated scattering mechanisms that can arise in solids, liquids, and gases are vibrational and acoustic scattering. If the stimulated scattering involves excited vibrational states, then $\omega_L - \omega_S = \omega_V$, and the scattering is known as stimulated Raman scattering. If the scattering excites acoustic vibrations it is known as stimulated Brillouin scattering.

Higher-order susceptibilities such as $\chi^{(4)}$ and $\chi^{(5)}$ are usually very small and not important unless extremely high intensities are being used or the laser frequency is near a resonance. Fifth-order nonlinearities are sometimes encountered with large incident intensities when the third-order susceptibility begins to saturate. Many semiconductors with two-photon absorption (2PA) are observed to produce a $\chi^{(5)}$ index change which can be viewed as (2PA) followed by a change in index. At large intensities, before these higher-order terms become important, plasma breakdown often occurs, usually leading to damage in solid materials. Photoionization, which can lead to plasma formation, is a form of a higher-order nonlinearity. Often multiphoton absorption is required to excite an electron to ionization energies. The number of photons, and hence the order of the susceptibility, is given as $E/\hbar\omega + 1$, where E is the ionization energy. Also, very high harmonics are possible by paying careful attention to phase matching by mixing gases at the proper pressures near resonances where dispersion is dramatic.

13.6 PROPAGATION THROUGH NONLINEAR MATERIALS

It is clear from the above discussion that high intensities from laser sources are capable of affecting a nonlinear (in the optical electric field, E) material response. This material response will, in turn, affect the propagation of the laser beam through the material. To understand the way in which this occurs, Maxwell's wave equation must be modified to include the material response. Assuming a nonmagnetic material, Ohm's law $\mathbf{J} = \sigma\mathbf{E}$, and writing the electric displacement as $\mathbf{D} = \epsilon_0\mathbf{E} + \mathbf{P}$, the wave equation is

$$\nabla^2\mathbf{E} = \mu_0\sigma\frac{\partial\mathbf{E}}{\partial t} + \mu_0\epsilon_0\frac{\partial^2\mathbf{E}}{\partial t^2} + \mu_0\frac{\partial^2\mathbf{P}}{\partial t^2} \qquad (13.32)$$

where $\mathbf{E}$ is the component of the optical field transverse to the direction of propagation given by $\mathbf{k}$. The polarizability can be described as $\mathbf{P} = \mathbf{P}' + \Delta\mathbf{P}$. The wave equation can then be written as

$$\nabla^2\mathbf{E} = \mu_0\sigma\frac{\partial\mathbf{E}}{\partial t} + \mu_0\epsilon\frac{\partial^2\mathbf{E}}{\partial t^2} + \mu_0\frac{\partial^2\,\Delta\mathbf{P}}{\partial t^2} \qquad (13.33)$$

where $\mathbf{D} = \epsilon\mathbf{E} = \epsilon_0\mathbf{E} + \mathbf{P}'$. Assuming a wave propagating in the $+z$ direction of the form $E(x,y,z,t) = 0.5[A(x,y,z,t)e^{i(kz-\omega t)} + \text{c.c.}]$, a nonlinear response of the form $\Delta P = 0.5[\Delta P_0(x,y,z,t)e^{i(k_pz-\omega_p t)} + \text{c.c.}]$, employing the slowly varying envelope approximation (SVEA), and time averaging over an optical cycle yields

$$\nabla_t^2 A + 2ik\left(\frac{\partial A}{\partial z} + v^{-1}\frac{\partial A}{\partial t}\right) = -\mu_0\omega_p^2\,\Delta P_0\exp(i\Delta kz) \qquad (13.34)$$

where $v = [\mu_0\epsilon(1 + i\sigma/\omega\epsilon)]^{-1/2}$, $\Delta k = k_p - k$, and $\omega_p = \omega$. A change of coordinates to the retarded time of $z' = z$, $T = t - z/v$ eliminates the derivative with respect to time. The field A is actually the total electric field amplitude present including the field generated by the change in polarizability. The change in polarizability is whatever combination yields the field frequency. The single propagation equation can be broken up into a set of coupled propagation equations describing the incident and generated fields. For instance, with three different incident beams with distinct frequencies and directions, the complete description of their interaction in a thick $\chi^{(3)}$ material will involve four coupled wave equations. One equation describes the generation and propagation of the new beam, which acts back on the incident beams, whose interactions lead to three additional equations. The frequency of the generated beam could be one of the following: $\omega_4 = \omega_1 + \omega_2 + \omega_3$, $\omega_4 = \omega_1 + \omega_2 - \omega_3$, or $\omega_4 = \omega_1 - \omega_2 - \omega_3$. These processes are known as four-wave difference-frequency mixing.

If the wave is traveling as a pulse in time, the phase velocity can be reinterpreted as the group velocity $v_g = d\omega/dk$, where k is the centerline wavenumber. The change in coordinates to retarded time then represents a coordinate system that rides along with the pulse. As mentioned earlier, the material response is not always instantaneous, and for these cases an equation that describes the time response of the material to the incident beams is coupled to the propagation equations. For a third-order susceptibility the time-dependent polarizability is written as

$$\Delta P_i(t) = 0.5[\Delta P_{0i}\exp(-i\omega_p t) + \text{c.c.}] \qquad (13.35)$$

where

$$\Delta P_{0i}(t) = \int_{-\infty}^{\infty} \exp(i\omega_p t'_1)\,dt_1 \int_{-\infty}^{\infty} \exp(i\omega_p t'_2)\,dt_2 \int_{-\infty}^{\infty} \exp(i\omega_p t'_3)\,dt_3$$
$$\times\; \chi^{(3)}(t'_1,t'_2,t'_3)E(t_1)E(t_2)E(t_3) \quad (13.36)$$

where $t'_s = t - t_s$, $s = 1\text{–}3$. For instance, a nonlinear change in index is represented by the general expression as

$$\Delta n(t) = \int_{-\infty}^{\infty} n_2\, R(t'_1)\, A(t_1)\, A(t_1)^*\, dt_1 \quad (13.37)$$

where R is the response function. For changes in index that arise from virtual transitions, which are practically instantaneous, the response function is a delta function, $R(t'_1) = \delta(t'_1)$ and $\Delta n(t) = n_2 A(t) A(t)^*$. Notice the factor of one-half is missing because the change in index has not been time-averaged. For many situations the response of the material Δn is given by an equation that satisfies

$$\tau\, \frac{d\Delta n}{dT} + \Delta n = n_2|A(T)|^2 \quad (13.38)$$

where τ is the nonlinear response time. Since the pulse is time-dependent, this equation has a solution

$$\Delta n(T) = \frac{n_2}{\tau} \exp\left(\frac{-T}{\tau}\right) \int_{0}^{\infty} \exp\left(\frac{T'}{\tau}\right) |A(T')|^2\, dT' \quad (13.39)$$

which appears similar to the form of the polarization response.

In order to see the coupled equations more clearly, suppose there is only one incident beam E_i and we are interested in investigating third-harmonic generation (3HG), $\omega_3 = 3\omega_i$. The total electric field becomes $E_T = E_i + E_3$, and the coupled equations far from resonance with isotropic symmetry for linear polarization become

$$\nabla_t^2 A_3 + 2ik_3\frac{dA_3}{dz'} = -\frac{\delta_3(3\omega_i)^2}{\epsilon_0 c^2}\,\omega\chi^{(3)}_{1111}(3\omega_i;\omega_i,\omega_i,\omega_i)A_i^3\exp(-i\Delta kz') \quad (13.40)$$

$$\nabla_t^2 A_i + 2ik_i\frac{dA_i}{dz'} = -\frac{\delta_i\omega_i^2}{\epsilon_0 c^2}\,\omega\chi^{(3)}_{1111}(\omega_i;3\omega_i,-\omega_i,-\omega_i)A_3 A_i^{*2}\exp(i\Delta kz') \quad (13.41)$$

where $\delta_i = 3$, $\delta_3 = 1$, $\Delta k = k_3 - 3k_i$. Equation (13.40) describes the propagation of the generated third-harmonic wave and Eq. (13.41) describes the effect of the third-harmonic wave back on the incident beam. If the beam is a plane wave, or if diffraction can be ignored because the beam spot size does not change significantly within the material, then the ∇_t^2 term can be set to zero, $\nabla_t^2 \to 0$. For Gaussian beam propagation this approximation is valid in the focal region when the material thickness L is much less than the Rayleigh range, $z_0 = \pi r_0^2 n/\lambda$, or $L \ll z_0$, where r_0 is the radius of the minimum spot measured at e^{-2} of the maximum intensity. Figure 13.6 shows the focal region of a Gaussian beam using diffraction-limited optics where the confocal beam parameter b is defined as twice the Rayleigh range. Materials where $L \ll z_0$ which are positioned in the region of the minimum waist experience nearly collimated light (plane phase fronts). In this chapter the condition $L \ll z_0$ refers to an optically thin material and an optically thick material is the

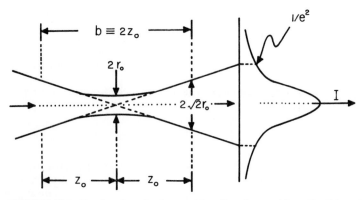

FIGURE 13.6   Focal region of a beam with a Gaussian spatial profile $I(r) = I_0 \exp(-2r^2/r_0^2)$.

opposite case or $L \gg z_0$. If only a small fraction of the incident beam is converted into the third harmonic, the effect on the incident beam is minimal and the change in the incident beam can be set to zero or $dA_i/dz' = 0$. This approximation is known as the undepleted pump approximation because the driving, or pumping, beam does not lose much energy to the generated beam and the equations are decoupled. For this very special set of conditions the weak third-harmonic beam amplitude can be calculated from

$$2ik_3 \frac{dA_3}{dz'} = -\frac{\delta_3(3\omega_i)^2}{\epsilon_0 c^2} \, \omega\chi^{(3)}{}_{1111}(3\omega_i;\omega_i,\omega_i,\omega_i) A_i^3 \, \exp(-i\Delta k z') \quad (13.42)$$

which for no linear absorption has the solution

$$A_3(L) - A_3(0) = i \frac{3\pi}{\epsilon_0 k_i n_i n_3} L \, \chi^{(3)}{}_{1111}(3\omega_i;\omega_i,\omega_i,\omega_i) A_i^3 \, \exp\left(\frac{-i\Delta k L}{2}\right) \frac{\sin(\Delta k L/2)}{\Delta k L/2}$$

$$(13.43)$$

For focused Gaussian beams $\Delta k \to \Delta k + 2/z_0$ owing to the phase of a Gaussian beam. Since the intensity is defined as $I_3 = cn_3\epsilon_0|A_3|^2/2$ and the field of the third-harmonic wave is zero at the front of the material, the intensity for $I_3 \ll I_i$ is

$$I_3(z) = \frac{36\pi^2}{\epsilon_0^4 k_i^2 n_i^5 n_3 c^2} \times |\chi^{(3)}{}_{1111}|^2 \, L^2 \, I_i^3 \, \mathrm{sinc}^2\left(\frac{\Delta k L}{2}\right) \quad (13.44)$$

Notice that the maximum third-harmonic intensity occurs when $\Delta k L/2 = 0$ or $n_3/\lambda_3 = 3n_i/\lambda_i$. This is known as Bragg matching or the phase matching condition for 3HG, which was discussed earlier. For materials much thicker than the wavelength, the generated intensity is negligible when the beams are not Bragg matched. Also, notice that the intensity of the third harmonic grows as the incident intensity cubed. This is because we have considered the initial buildup of the third harmonic only when it is much weaker than the incident beam. If pump depletion is considered (the solution to the coupled set of equations), then the third harmonic is seen to saturate at large incident intensities.

The general case of four different frequencies that mix in all possible combinations involves three separate generated frequencies $\omega_1 + \omega_2 + \omega_3$, $\omega_1 + \omega_2 - \omega_3$, and $\omega_1 - \omega_2 - \omega_3$. Each will have a unique phase-matching condition and each has an identical generated intensity which, in the undepleted pump approximation for plane waves, is proportional to the length squared and the multiplication of the three incident intensities.

If we are interested in $\chi^{(3)}$ processes with generated frequencies that correspond to the incident frequency ω and assume two incident fields are present with frequencies ω_a and ω_b, the phase-matched equations become

$$\nabla_t^2 A_a + 2ik_a \frac{dA_a}{dz'} = -\frac{(\omega_a)^2}{\epsilon_0 c^2} \chi^{(3)}_{1111}(3|A_a|^2 + 6|A_b|^2)A_a \qquad (13.45)$$

$$\nabla_t^2 A_b + 2ik_b \frac{dA_b}{dz'} = -\frac{(\omega_b)^2}{\epsilon_0 c^2} \chi^{(3)}_{1111}(3|A_b|^2 + 6|A_a|^2)A_b \qquad (13.46)$$

This type of two-beam coupling through the third-order susceptibility is referred to as cross-phase modulation in the literature. If $\omega_a = \omega_b$, the situation can be used to describe a strong pump beam propagating along with a weak probe beam. It can also be used to describe two beams of degenerate frequency but different polarization, i.e., $a = x$, $b = y$. The first term on the right-hand side of the equations describes a self-induced change in the susceptibility while the second term describes a change in the susceptibility due to the presence of the other beam. It is the presence of the second term on the right-hand side of the equations that couples the equations. Assuming plane waves and letting A_a be the undepleted pump beam, the equations for the undepleted pump beam propagation are given as

$$\frac{\partial I_b}{\partial z} = -2\eta\chi''^{(3)}\left(I_b^2 + 2\frac{n_b}{n_a}I_aI_b\right) - \alpha_0 I_b \qquad (13.47)$$

$$\frac{\partial \phi_b}{\partial z} = \eta\chi'^{(3)}\left(I_b + 2\frac{n_b}{n_a}I_a\right) \qquad (13.48)$$

where $A(z) = a(z)e^{i\phi(z)}$, $\chi^{(3)} = \chi'^{(3)} + i\chi''^{(3)}$, $\eta = 3k_b'/cn_b^2\epsilon_0^2$, k_b' is real, and $I_m(k_b) = \alpha_0/2I_m(n)$. Equations (13.47) and (13.48), respectively, describe the changes in amplitude and phase of the plane wave with position as it propagates through the material. The solution to the first equation can be determined by letting $4\eta\chi''^{(3)}(n_b/n_a)I_a = \alpha_{eff}$, which acts as a constant linear absorption term. The equation appears as

$$\frac{\partial I_b}{\partial z} = -\alpha I_b - \beta_1 I_b^2 \qquad (13.49)$$

where $\alpha = \alpha_0 + \alpha_{eff}$ and the nonlinear absorption term is $\Delta\alpha = \beta_1 I_b$ where $\beta_1 = 2\eta\chi''^{(3)}$. It is therefore the imaginary contribution to the third-order susceptibility which produces the intensity-dependent absorption. A quantum-mechanical derivation of 2PA (the simultaneous absorption of two photons) yields a nonlinear absorption of this form. For this mechanism β is called the 2PA coefficient. The intensity at any position z and transverse distance r is given in terms of the initial intensity $I_{ob}(z = z'',r,t')$ as

$$I_b(z,r,t') = I_{0b}(z = z'',r,t')e^{-\alpha L}[1 + \beta_1 L_{eff}I_{0b}(z = z'',r,t')]^{-1} \qquad (13.50)$$

where $L_{eff} = (1 - e^{-\alpha L})/\alpha$, z'' is the location of the front face of the material, and $z > z'' + L$. The energy transmitted through the nonlinear material is

$$T = \frac{\iiint I_b(z'' + L, r, t')\, r\, dr\, d\theta\, dt'}{\iiint I_{ob}(z'', r, t')\, r\, dr\, d\theta\, dt'} = e^{-\alpha L}\,[\beta_1 L_{eff} I_{ob}]^{-1}\, ln(1 + \beta_1 L_{eff} I_{ob})$$

(13.51)

where the intensity is assumed to have a Gaussian spatial profile of e^{-2} radius r_0 and a square temporal pulsewidth of τ. If the pump beam I_a is set to zero, the phase equation yields

$$\phi_b(z = L, r, t') = \frac{\eta \chi'^{(3)}}{\beta_1\, ln[1 + \beta_1 L_{eff} I_{ob}(z'', r, t')]}$$

(13.52)

If there is no two-photon absorption, $\beta_1 \to 0$, then the change in phase at the end of the material $z = L$ is given as

$$\phi(z = L, r, t') = \eta \chi'^{(3)} I_{ob}(z = 0, r, t')\, L_{eff}$$

(13.53)

where for small linear absorption $L_{eff} \approx L$ or

$$\phi(z = L, r, t') = \left[\frac{3k'_{ob}\chi'^{(3)} L |A_b(z'', r, t')|^2}{2n_b \epsilon_0} \right]$$

(13.54)

where k_{ob} is the vacuum wavevector for beam b. This phase term can be written in terms of a change in index of refraction, $\exp(i\phi) = \exp(ik_{ob}\Delta nz)$ where the change in index is given as $\Delta n = (3\chi'^{(3)} z/n_b\epsilon_0)(|A_b|^2/2)$ or $n_2 = 3\chi'^{(3)}/n_b\epsilon_0$. The term n_2 is known as the nonlinear index of refraction. In the literature the nonlinear index is often given in esu units, in which case $n_2 = 12\pi\chi'^{(3)}_{1111}/n$ for an isotropic material. The conversion from mks to esu is given as $(n_2)|_{mks} = 1/9 \times 10^{-8}\,(n_2)|_{esu}$. Since the nonlinear index is a function of intensity, the largest index change will occur at the highest intensity. For a Gaussian beam this means the index change is largest at the center of the beam and falls off to no index change in the wings of the Gaussian. Phase fronts propagating through the nonlinear medium will be bent, which will affect the direction of beam propagation. If the nonlinear index is positive, $n_2 > 0$, plane wave phase fronts (such as at the focus of a Gaussian beam) will become concave about the direction of propagation and the beam will focus. Likewise, for a negative nonlinear index the phase fronts will be convex about the direction of propagation and the beam will defocus. Although the phase fronts have a Gaussian shape, for paraxial rays (rays close to the axis of propagation), a Gaussian phase front is similar to the spherical phase front introduced by a spherical lens. Therefore, the thin nonlinear material acts similar to thin concave and convex lenses for paraxial rays. An equation relating the focal length of the lens to the change in index is

$$f^{-1} = L \frac{\partial^2 \Delta n}{\partial r^2}\bigg|_{r=0}$$

(13.55)

For a Gaussian spatial profile $I = I_0 \exp(-2r^2/r_0^2)$ the focal length is given by $f = r_0^2 c \epsilon_0 n_0 / n_2 I_0 L$.

Another effect of this phase change is to shift the frequency of the incident beam. This is known as self-phase modulation and is the result of the time dependence of the electric field amplitude $A(t') = A(t - zn/c)$. The change in frequency is determined by $\exp(i\omega t + i\delta\omega t)$ or $\delta\omega = \partial\phi/\partial t$ or $\delta\omega = ik_0n_2z\,\partial|A(t - zn/c)|^2/\partial t$. For a Gaussian temporal intensity, the maximum frequency shift occurs at the two inflection points and no frequency shift occurs at the maximum intensity. The pulse is down-shifted (red-shifted) at the front end of the pulse and up-shifted (blue-shifted) at the back of the pulse. The result of these shifts is a spread of frequency components in time (see Fig. 13.7). Self-phase modulation can be useful in pulse compression if the material in which it propagates has a negative group velocity dispersion so that the red-shifted frequency components travel more slowly than the blue-shifted components. Since this is not the dispersion condition usually encountered in most materials, combinations of prisms are used which yield an effective anomalous dispersion.

In semiconductors the propagation equations (13.47) and (13.48) are often modified to include terms that describe the change in amplitude and phase due to the change in the number density N of free carriers. Free carriers can be generated in transparent materials by two-photon absorption across the bandgap or by linear absorption from defects and dopants in the bandgap. In fact, the charges can be generated indirectly through local heating by the laser. If this process is significant, thermal runaway followed by damage can occur. Once in the bands the charges diffuse and recombine, reducing the local charge density, but this reduction is only significant for pulses longer than the diffusion and recombination times.

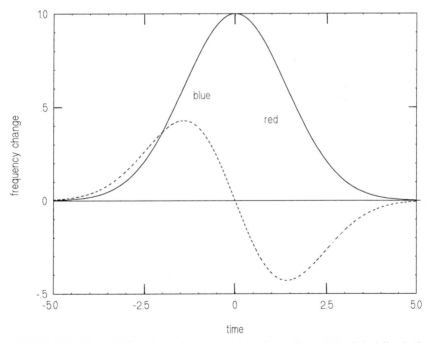

FIGURE 13.7 The solid line shows the temporal intensity profile and the dotted line is the corresponding frequency shift. The leading edge of the pulse is red-shifted while the trailing edge is blue-shifted.

From the above analysis, beam propagation through a thin nonlinear material will experience a phase change if $|\chi'^{(3)}| > 0$ and/or nonlinear absorption if $\chi''^{(3)} > 0$. A focused Gaussian beam has an electric field after propagating through the nonlinear material of length L at $z = z''$,

$$E(xz_0 + L,r,t') = Y\, E(xz_0,r,t')$$

$$= \sqrt{I}\,(1 + x^2)^{-1/2}\, e^{i(kL - \omega t')} e^{-\alpha L/2}$$

$$\times \exp\{-r^2/[r_0^2(1 + x^2)]\}\, \exp[ikr^2/2R(x)]\, \exp[i\phi] \quad (13.56)$$

where Y is a transmission function, $x = z''/z_0$ and $x = 0$ occur at the minimum waist (at focus), $R(x)$ is the radius of curvature of the phase $R(x) = xz_0(1 + x^{-2})$, $I(z = L, z'' = 0, r = 0, t')$ is given by Eq. (13.50), and $\phi(z = L, z'' = 0, r, t')$ is given by Eq. (13.52). The general transmission function is

$$Y = (1 + q)^{i\Xi - 1/2} \quad (13.57)$$

where $q = \beta I_0 L_{\text{eff}} \exp[-2r^2/r_0^2(1 + x^2)]/(1 + x^2)$, I_0 is the peak on-axis incident intensity, and $\Xi = \eta \chi'^{(3)}/\beta_1$. Propagation of the beam from a nonlinear material to the far field can be significantly different from a beam that encounters a thin material without any nonlinear response. Using a beam with a Gaussian spatial profile and a thin nonlinear material placed at $z = z''$, the observed fluence (energy density) can be determined at any location z by propagating the beam using the Huygens-Fresnel formalism. The observed fluence at z is

$$F(z,r) = \frac{c\epsilon_0}{2} \int_{-\infty}^{\infty} |E(z,r,t)|^2\, dt \quad (13.58)$$

where

$$E(z,r,t) = \frac{2\pi \exp(i\pi r^2/\lambda z_d)}{i\lambda z_d} \int_0^{\infty} E(L,r',t') \exp(i\pi r'^2/\lambda z_d) J_0\left(\frac{2\pi r r'}{\lambda z_d}\right) r'\, dr' \quad (13.59)$$

where J_0 is the zeroth-order Bessel function, r' is the transverse beam profile within the nonlinear material and $z_d = z - (z'' + L)$. The solution to this equation is found by numerical integration. Another method which provides an analytic solution is to expand the exponential containing the phase term into a series summation. For the case of a single beam, $I_a = 0$, the on-axis electric field with no two-photon absorption propagated to the far field is

$$E(z,r,t) = \sqrt{I_0(t')}\, \frac{z_0}{z}\, e^{i(kL - \omega t')}\, e^{-\alpha L/2}$$

$$\times \sum_{n=0}^{\infty} i^{n+1} (-\phi'(t'))^n [(2n + 1) - ix]$$

$$\times \{n!\,(1 + x^2)^{n-1/2} [(2n + 1)^2 + x^2]\}^{-1} \quad (13.60)$$

where $\phi' = \eta \chi'^{(3)} I_0(t') L_{\text{eff}}$. The transmission in the far field is given in general as

$$T(t) = \int_0^{\infty} E(z,r,t) E^*(z,r,t) r\, dr \left[\int_0^{\infty} E_L(z,r,t) E_L^*(z,r,t) r\, dr\right]^{-1} \quad (13.61)$$

where E_L is the linear field or the field when $\phi' = 0$.

The transmission converges rapidly if the phase is small or if $\phi' \ll \pi$ but requires more terms be kept for larger phase changes. For instance, in the case where the thin nonlinear medium is placed at the focus, $z'' = 0$, when $\phi' \approx 5\pi$ for $n = 55$ the accuracy is to the seventh decimal position; 50 terms, 4 places; 45 terms, 3 places; and 40 terms, 1 place. For an instantaneous phase change of $\phi' = 4\pi$ the on-axis intensity is reduced 20 times in the far field compared with the linear case.

The z scan[24] is a popular technique that uses a thin material to determine the sign and magnitude of the index change and the magnitude of the nonlinear absorption. The experimental setup to determine the index change is relatively simple. As shown in Fig. 13.8, a small aperture is placed on-axis in front of a detector in the far field. For an incident laser beam of fixed energy, the normalized transmittance (ratio D_2/D_1 in Fig. 13.8) is measured at different positions along the z

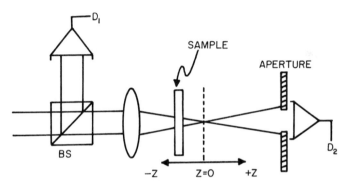

FIGURE 13.8 Experimental setup for the z-scan measurement. The transmission through the thin sample and aperture is determined by the ratio of the energy measured by detector D_2 to that measured by detector D_1, $T = D_2/D_1$.

direction while the sample is translated symmetrically through the focused beam waist. If the nonlinear index of the thin sample is positive, and the sample is positioned before the focus, as shown in Fig. 13.9a, the nonlinear medium will focus the beam earlier and to a smaller waist. This situation is represented in Fig. 13.9a by the beam profile shown with dotted lines. Since the waist is made smaller (greatly exaggerated for clarity) the beam expands more rapidly owing to diffraction; remains collimated over a shorter distance in the near field; and diverges at a larger beam angle in the far field, which reduces the irradiance at detector D_2. When the same sample passes into the postfocal position, as shown in Fig. 13.9b, the positive self-lensing of the nonlinear material tends to reduce the beam divergence (dotted lines), which results in increased irradiance at detector D_2. If the nonlinear index of the sample is negative and the sample is placed in the prefocal region, the beam waist at the focus will be increased and the focus will be closer to the aperture, as shown by the dashed lines in Fig. 13.9a. As a result, more radiation will pass through the aperture, producing an increased signal on the detector. When the material passes into the postfocal region the negative lensing effect of the material will spread the already diverging rays even more (dashed lines in Fig. 13.9b), so that the irradiance will be significantly decreased at the detector.

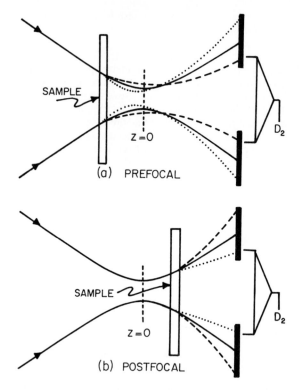

FIGURE 13.9 Nonlinear refraction: (*a*) prefocal ($z < 0$) diffraction, (*b*) postfocal ($z > 0$) diffraction. The solid line in both figures is linear (low-intensity) diffraction. The dotted line shows $n_2 > 0$ and the dashed line corresponds to $n_2 < 0$.

The difference in behavior shown by negative and positive nonlinear refraction therefore provides a unique signature of the sign of the nonlinearity. As Fig. 13.10 shows, a prefocal transmittance minimum followed by a postfocal maximum is the signature of a positive nonlinearity while a prefocal maximum followed by a minimum is the signature of a negative nonlinearity.

The magnitude of the nonlinearity can be determined in a straightforward manner if the phase change is small, $|\phi| \leq 1$, and the aperture is small resulting in a small transmission, $T < 0.05$, so that only on-axis beams are passed. Under these conditions the peak-to-valley transmission change ΔT_{p-v} is proportional to the nonlinear index. The relationship is given as

$$\Delta T_{p-v} = 0.406|\langle \Delta\phi \rangle| \tag{13.62}$$

$$\langle \Delta\phi \rangle = \frac{2\pi}{\gamma\lambda_0} n'_2(1 - R)I_0(z = 0, r = 0)L_{\text{eff}} \tag{13.63}$$

where $\gamma = \sqrt{2}$ if the n'_2 is instantaneous and $\gamma = 2$ if the n'_2 is accumulative in time as in Eq. (13.38). Here n'_2 has the units (cm²/kW), $L_{\text{eff}} = (1 - e^{-\alpha l})/\alpha$, l is

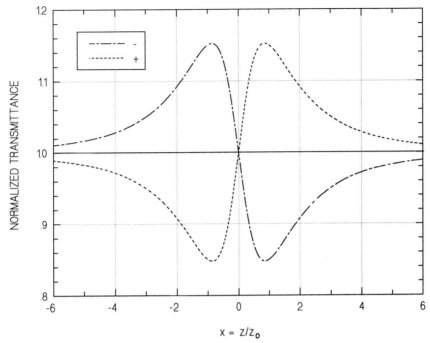

FIGURE 13.10 Theoretical plots of the z-scan transmission normalized so that the transmission is 100 percent with a semiclosed aperture and low intensity. The dashed line is the z scan for $n_2 > 0$ and the dot-dashed line for $n_2 < 0$. These plots were generated with an aperture such that the transmission is 5 percent and the phase change $|\langle\Delta\phi\rangle| = 0.75$.

the material length, R is the front surface reflection coefficient, and $\langle\ \rangle$ is a time-average quantity. The separation of the peak-to-valley is given by $\Delta z_{p-v} = 1.7z_0$. The presence of nonlinear absorption can be determined by opening the shutter and capturing all the scattered radiation on the detector during the z scan. If only an index change occurs, Im $\chi^{(3)} = 0$, then there is no corresponding change in the transmission with position. However, a drop in transmission with a minimum at the focus is the signature of nonlinear absorption. In practice both may be present, so that, by subtracting an apertured z scan from an open-aperture z scan, the nonlinear absorption can be separated from the nonlinear refraction.

For weak nonlinear refraction, as required by the z scan, the time-averaged apertured on-axis transmission can be written as

$$T = Y^*Y = 1 + 4\langle\Delta\phi\rangle x\,(1 + x^2)^{-1}\,(9 + x^2)^{-1}. \qquad (13.64)$$

The open-apertured transmission (sensitive to nonlinear absorption only) for weak nonlinear absorption, $\beta I_0 L \ll 1$ and a Gaussian temporal profile is given as

$$T = 1 - \frac{\beta(1 - R)I_0(z = 0, r = 0)\,L_{\mathrm{eff}}}{2\sqrt{2}(1 + x^2)} \qquad (13.65)$$

By fitting these equations to the z-scan data a better determination of the nonlinear parameters can be found than by using Eqs. (13.62) and (13.63). However, the signal-to-noise for small apertures [where Eqs. (13.64) and (13.65) are applicable] may increase. A larger aperture can be used to decrease the signal-to-noise; however, Eqs. (13.64) and (13.65) are no longer applicable, but Eq. (13.62) can be used if the right-hand side is multiplied by $(1 - T_L)^{0.25}$ and $T_L < 0.7$.[25]

Up to this point the solutions have not involved diffraction effects so the beams have been plane waves or Gaussian beams propagating in thin materials placed at the focus. Analytic solutions that include diffractive effects can be found if some approximations are used.[26,27] For instance, the general equations for a single incident Gaussian beam in cylindrical coordinates $\nabla_t^2 = \partial^2/\partial r^2 + r^{-1}\,\partial/\partial r$ is given by equating the real and imaginary parts and denoting the electric field amplitude as $A = a(z,r)e^{ik\xi(z,r)}$

$$I\frac{\partial^2\xi}{\partial r^2} + r^{-1}\,I\frac{\partial\xi}{\partial r} + \frac{\partial I}{\partial r}\frac{\partial\xi}{\partial r} + \frac{\partial I}{\partial z} = -\alpha I - \beta I^2 \tag{13.66}$$

$$(8kI^2)^{-1}\left(\frac{\partial I}{\partial r}\right)^2 - (4kI)^{-1}\frac{\partial^2 I}{\partial r^2} - (4krI)^{-1}\frac{\partial I}{\partial r} + \frac{k}{2}\left(\frac{\partial\xi}{\partial r}\right)^2 + k\frac{\partial\xi}{\partial z} = k\,n_2'I \tag{13.67}$$

where β is the nonlinear absorption coefficient $\beta = 3k\chi''^{(3)}/cn^3\epsilon_0^2$ and $n_2' = 3\chi'^{(3)}/cn^3\epsilon_0^2 = n_2/ce\epsilon_0 n^2$. Notice that when the derivatives with respect to radial position r are ignored, the equations become the familiar equations (13.47) and (13.48). If diffraction is assumed to be small, $k \to \infty$, and the nonlinearities are also assumed to be negligible, then $\xi = r^2/2(z + R)$ is a solution to Eq. (13.67) and represents a spherical wave of radius R when $z = 0$. By assuming weak diffraction effects and small nonlinearities, a solution for the phase is of the form $\xi = r^2\zeta(z)/2$. This phase is inserted into Eqs. (13.66) and (13.67) to generate a new set of wave equations which have a Gaussian intensity of the form

$$I = I_0 \exp(-\alpha z)f^{-2}(z)\exp[-r^2/r_0^2f^2(z)] \tag{13.68}$$

as a solution when $\beta = 0$. Therefore, the beam is assumed to propagate through the nonlinear medium as a Gaussian which is correct only for no nonlinear absorption and small nonlinearities. Equation (13.66) then yields the solution $\zeta = f^{-1}\,df/dz$. Notice that the intensity at $f = 1$ describes the intensity of the beam at the material front surface $z = 0$. Also notice that $f^{-1}df/dz = R^{-1}$ or the inverse radius of curvature of the phase front. Using the modified version of Eq. (13.67) with the Gaussian expanded in a Taylor's series and retaining only the lowest-order terms in r^2 (the paraxial ray approximation) yields the equation with no absorption

$$f^{-1}\frac{d^2f}{dz^2} = (k^2r_0^4f^4)^{-1} - n_2'I_0(f^4r_0^2)^{-1} \tag{13.69}$$

The first term on the right side describes diffraction which causes the beam to diverge and the second term describes the effect of the nonlinear index on the beam. The nonlinear index term has the opposite sign from the diverging diffraction term and causes the beam to converge for positive n_2'. The solution to this equation describes the propagation of the beam waist in the z direction and is given as

$$r^2 = r_0^2f^2 = r_0^2\left\{1 + \frac{2z}{R} + \left[R^{-2} + (k^2r_0^4)^{-1} - \frac{n_2'I_0}{r_0^2}\right]z^2\right\} \tag{13.70}$$

If the beam at the front face of the material has a plane phase front (as for a beam at focus) then $R^{-1} \to 0$. Since the total power in the beam is $P = I_0 \pi r_0^2$, the equation for a plane initial phase front is

$$r^2 = r_0^2 f^2 = r_0^2 \left[1 + (k^2 r_0^4)^{-1} z^2 - \frac{n_2' P z^2}{\pi r_0^4} \right] \tag{13.71}$$

When the nonlinear index term is as large as the diffraction term, the beam waist will not diverge or converge but remains constant. This is known as self-trapping, a form of spatial soliton. However, this situation is unstable, and for powers at this threshold or above, the beam will collapse. A somewhat stable self-trapping solution is found if the nonlinear index saturates above some intensity before damage occurs. The critical power for beam collapse, or self-focusing, occurs when

$$P_{cr} = \frac{c \epsilon_0 \lambda_0^2}{4 \pi n_2} \text{ (mks)} = \frac{c \lambda_0^2}{32 \pi^2 n_2} \text{ (esu)} \tag{13.72}$$

The distance over which this collapse occurs is a function of the power above critical power. By setting the beam radius to zero, this distance is found to be

$$z_{SF} = \frac{k r_0^2}{(P/P_{cr} - 1)^{1/2}} \tag{13.73}$$

For a beam propagating in a thick material even a small fraction of power above the critical power is enough to collapse the beam to zero radius within the material. Of course, the beam does not really achieve zero radius, but the radius does decrease in many materials until the energy of the beam is dissipated through either stimulated scattering, plasma generation, heat, etc. In solids the result of catastrophic self-focusing is often a damaged material. The collapse of the beam to a point or singularity does not occur in more exact solutions, i.e., where the SVEA is not used. A major criticism of this method of analysis is that it does not include the whole beam and therefore cannot be entirely accurate. In fact, the critical power for catastrophic self-focusing derived above is indeed off by a numerical factor of 3.77 or $P_{cr} = 3.77 P_{paraxial}$.[28] However, this method does yield analytic solutions that do predict the general behavior and close agreement with numerical solutions over the region where these equations are approximately valid, namely, for small nonlinearities where the beam still propagates as a Gaussian.

Another popular method to propagate a beam in a thick nonlinear material involves a modified ABCD rule valid only for small nonlinearities and Gaussian beams as well.[29] Since these methods involve paraxial rays and predict the whole beam sharply focuses to a point they are called aberration-less self-focusing theories. In reality, rings are observed on beams propagating in nonlinear materials above the critical power. This is the result of the interference from different portions of the beam beyond the paraxial region. Different radial positions along the beam have different phases and beam divergences. The center of the beam undergoes the largest phase change, but the wings of the beam are changed very little. At π phase intervals away from the center, a ring can appear. An estimate for the change in nonlinear index in the beam center from the number of rings N is given by[30]

$$\Delta n L \sim N \lambda_0 \tag{13.74}$$

To this point, an analytic expression for the transmission through a thick cell has not been given when nonlinear absorption is present. When the dominant nonlinearity is nonlinear absorption, the change in transmission, to the first order, is given as

$$\Delta T = \frac{\beta P_0 n}{4\lambda} \tan^{-1}\left(\frac{L}{z_0}\right) \tag{13.75}$$

For the thin cell, $L \ll z_0$, then $\Delta T = -\beta L I_0/2$ as can be seen from Eq. (13.51). For the thick cell, $L \gg z_0$, $\Delta T = -(\beta P_0 n/4\lambda)(\pi/2)$. Note that the change in transmission is power-dependent.

Earlier the point was made that the phase velocity could be reinterpreted as the group velocity and the wave equations would describe a pulse in time. This is valid only as long as the group velocity dispersion (GVD) can be ignored. This is usually a good approximation except for ultrashort pulses and/or for pulses that travel a long distance in a material such as pulses traveling along a dielectric fiber cable.[31] In fact, GVD can be ignored only if the propagation distance L is such that $L_D \gg L$, where $L_D = T_p^2/|\beta_2|$, T_p is the pulsewidth, $\beta_2 = d(v_g^{-1})/d\omega = d(v_g^{-1})/d\omega$, and v_g = group velocity. For silica glass, the material commonly used today in optical fibers, β_2 can range from $+60$ ps^2/km in the visible to -20 ps^2/km at $\lambda = 1.55$ μm. For a $L = 1$-cm-thick material with the GVD coefficient of silica fiber at visible wavelengths, the GVD must be included for pulses less than about 80 fs. For a 50-km-long fiber of silica glass, GVD must be included for visible pulses less than about 170 ps. To describe the propagation of pulses that require the inclusion of GVD, the wave equation is derived in the Fourier transformed frequency space with the propagation wavevector $k(\omega)$ expanded in terms of the frequency. The equation is then transformed back to the time domain. The equation describing the propagation of a beam that encounters a nonlinear index change due to the third-order susceptibility is

$$-\frac{i}{2\lambda}\nabla_t^2 A + \frac{\partial A}{\partial z'} + \frac{i}{2}\beta_2\frac{\partial^2 A}{\partial T^2} + \frac{\alpha A}{2} = i\psi|A|^2 A \tag{13.76}$$

to the second order in $k(\omega)$, which is the order that includes the GVD. For an optical fiber $\psi = n_2 k_0/2A_{\text{eff}}$, and for a nonlinear material $\psi = n_2 k_0/2$. For the optical fiber the electric field above is only a function of the propagation distance z' and the retarded time T, i.e., $A(z',T)$. The beam is assumed to propagate along the fiber without any change in the launched mode; therefore, there is no diffraction and the first term in the equation is zero, $\nabla_t^2 A \to 0$. (For propagation in nonfiber nonlinear materials, dropping the diffraction term will require plane incident waves or a thin material placed within the focal volume.) In order to simplify the description of fiber propagation, assume the beam propagates in the fundamental mode. The effective core area is given as $A_{\text{eff}} \sim \pi r_0^2$ for the fundamental mode approximated as a Gaussian of width r_0 ($HW e^{-2}M$). This width is approximately the core radius (the high index central part of the fiber) as long as $V = k_0 a_0(n_1^2 - n_2^2)^{1/2} > 2.4$, where a_0 is the core radius, n_1 the core index, and n_2 the cladding index. However, for values of V greater than $V > 2.405$ more than one mode can propagate in the fiber.

The effect of GVD without any nonlinearity is to spread the pulse in time with propagation distance z much the same way diffraction spreads the transverse spatial profile. It is actually possible to narrow the temporal pulse by appropriate frequency chirping. However, if the pulse is propagated far enough it will broaden again. This is analogous to a lens that will focus a spatial beam, but after the focal point, the beam again spreads in space.

The group velocity β_1^{-1} and the GVD term β_2 are usually the only terms in the expansion of $k(\omega) = k_0 + {}_{j=1}\Sigma [d^j/d\omega^j (v_g^{-1})]|_{\omega=\omega_0} (\omega - \omega_0)^j/_j!$, where $[d^j/d\omega^j (v_g^{-1})]|_{\omega=\omega_0} = \beta_j$ required to describe pulse propagation along fibers or for short pulses in materials. When $\beta_2 \sim 0$ (for silica glass $\beta_2 \approx 0$ at $\lambda \sim 1.3$ µm) or the pulses are tens of femtoseconds, higher-order terms such as β_3 may be required.

If linear absorption is ignored and the $\nabla_t^2 A$ term is zero, the equation of propagation for a temporal pulse becomes

$$i \frac{\partial A}{\partial z'} = \frac{\beta_2}{2} \frac{\partial^2 A}{\partial T^2} - \psi |A|^2 A \qquad (13.77)$$

which is known in the literature as the nonlinear Schrödinger equation. This equation can describe the propagation of radiation in a nonlinear fiber or a nonlinear material with no diffraction. If GVD is not important but the material can support the propagation of a beam in 2-spatial dimensions, such as a slab waveguide, the above equation can also describe the beam propagation provided $T \rightarrow x$ or y and $\beta_2 \rightarrow -(k)^{-1}$. Analytic solutions to the nonlinear Schrödinger equation do exist and are derived from a method called the inverse scattering technique, which is described elsewhere.[32] In the temporal regime, a beam propagating in the negative $\beta_2 < 0$ region, known as anomalous dispersion, and with a positive n_2, will excite a propagation eigenmode called a bright soliton.[33] (A bright soliton is also excited if $n_2 < 0$ and the dispersion is normal or $\beta_2 > 0$.) In the spatial regime, a beam propagating in a medium with a positive n_2 can excite a bright spatial soliton.[34] In the temporal regime, propagation in the normal dispersion region can result, with the proper initial conditions, in the propagation eigenmode called a dark soliton.[35] Likewise, it is possible to excite a spatial dark soliton with $n_2 < 0$.[36] Table 13.2 summarizes the conditions required to observe bright or dark solitons.

TABLE 13.2 Solitons

	Temporal	Spatial
Bright	$\beta_2 < 0, n_2 > 0$ or $\beta_2 > 0, n_2 < 0$	$n_2 > 0$
Dark	$\beta_2 > 0, n_2 > 0$ or $\beta_2 < 0, n_2 < 0$	$n_2 < 0$

The normal solution for the lowest-order bright temporal soliton is

$$A(z',T) = P_0^{1/2} \, \text{sech}\left(\frac{T}{T_0}\right) \exp(iz'|\beta_2|/2T_0^2) \qquad (13.78)$$

where P_0 is the incident peak power and T_0 is a measure of the width of the soliton. The normalized lowest-order bright soliton intensity is shown as the solid line in Fig. (13.11a). This width can be related to half the temporal width at half of the maximum intensity, HWHM, by $T_{HWHM} = 0.88 \, T_0$. The soliton order is given by

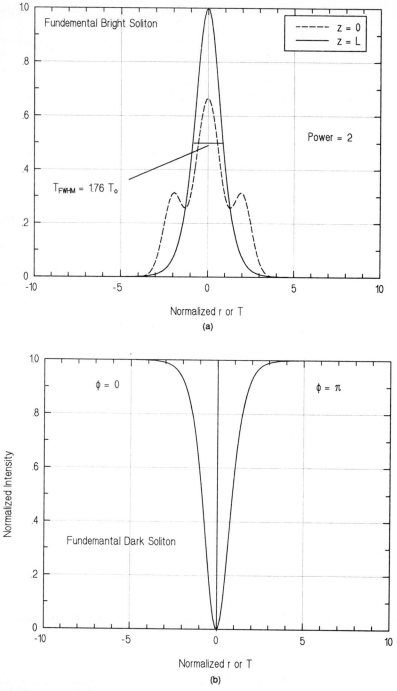

FIGURE 13.11 (*a*) The normalized fundamental bright spatial or temporal soliton mode (solid line), $I = \text{sech}^2 av$ where $v = r$ or T. The dashed line shows the input beam. (*b*) A normalized dark spatial or temporal soliton mode, $I = 1 - \text{sech}^2 av$.

$N^2 = \psi P_0 T_0^2 / |\beta_2|$ which becomes an integer during propagation. The lowest-order soliton $N = 1$ propagates without changing shape or width with propagation distance. For $N > 1$ this is no longer true, but the pulsewidth varies periodically with distance. An incident power of $P_0 = 1$ W at $\lambda = 1.55$ μm in a silica fiber will yield a lowest-order soliton for $T_0 \sim 3$ ps. (For the spatial soliton the power required to launch the lowest-order soliton is given as $P_1 = 2n_0 w / n_2 a_0 k^2$, where a_0 replaces T_0 as the spatial beamwidth, w is the slab thickness, and it is assumed that $w \ll a_0$.) Given the incident power and pulsewidth, the soliton order can be determined. As the power is increased the soliton order can be increased and vice versa. In fact, with absorption losses the power will decrease so the pulsewidth will increase to maintain a constant N. However, the increase in the pulsewidth is usually less than in the linear case. By periodically resupplying the soliton with the energy loss due to absorption and scattering, the original width can be recovered. In this manner the pulse can be refreshed along a transmission line. For input pulse field amplitudes of the form $\text{sech}(T/T_0)$ a propagation eigenmode exists. If the input pulse has a different form than $A = A_0 \text{sech}(T/T_0)$, it will evolve into the appropriate eigenmode soliton for the appropriate initial N of the pulse (see Fig. 13.11(a)). The higher-order solitons, $N > 1$, are not as simple as the $N = 1$ given in Eq. (13.78). These higher-order modes share one thing in common, a periodic behavior. The pulsewidth changes with propagation distance but returns to the original width after some distance. The period is found to be $z_p = \pi L_D/2 = 0.322\pi T^2_{\text{FWHM}}/2|\beta_2|$.

For pulses launched that do not correspond to some integer multiple of the order, they will after some propagation distance. Typically the distance required is about three soliton periods. If a beam is launched such that $N \le 0.5$, no soliton is formed. In fact, for $0.5 < N \le 3/2$ the soliton will become $N = 1$ after some distance.

A physical picture of the formation of the bright temporal solitons is the interplay between the GVD and self-phase modulation. Self-phase modulation (for $n_2 > 0$) separates the red frequencies to the leading edge of the pulse and the blue frequencies to the trailing edge. Through the anomalous GVD, the red light is slowed down more than the blue light so that the pulse contracts. For the fundamental soliton mode the two effects exactly balance and the pulse travels undistorted with distance. The higher-order solitons oscillate because the two effects are out of phase with each other, with the self-phase modulation leading the GVD.

An important consideration for optical communication regarding soliton transmission is the interaction of solitons along the fiber length. It has been found that two solitons separated by some distance initially can attract or repel one another depending on the relative phase difference of the soliton electric field envelopes. In fact, if there is a $2n\pi$ phase difference between the two solitons, where n is some integer, they appear to attract each other. However, a $(2n + 1)\pi$ phase difference repels the two solitons. The amount of attraction or repulsion depends on the initial separation, the relative soliton order, and the phase difference. Two solitons separated in phase by $2n\pi$ will actually move together and apart periodically along the length of the material. The distance at which the two solitons collapse to one is given by $z_c \sim z_p \exp(\Delta T/T_0)$, which is approximately valid for $\Delta T/T_0 > 3$.

For a spatial soliton the physical explanation of this behavior is that the $2n\pi$ phase difference allows constructive interference between the two solitons and increases the refractive index between them. An increased index bends both beams toward each other and they collapse. Likewise, a $(2n+1)\pi$ phase difference causes interference between the two solitons and the index is lower between them than on the opposite side of each soliton, so they move apart.

For normal GVD ($\beta_2 > 0$), such as found in silica glass in the visible, dark temporal solitons can be observed. (A spatial dark soliton corresonds to $n_2 < 0$.) A dark soliton can be launched by allowing the pulse to have a constant amplitude as time extends to positive and negative infinity, but a dip at t, $(x) = 0$. The dark soliton has the solution

$$|A(z',T)|^2 = \frac{P_0}{N^2} \left\{ 1 - v^2 \operatorname{sech}^2 \left[v \left(\frac{T}{T_0} - \frac{\lambda z'}{L_D} \right) \right] \right\} \qquad (13.79)$$

where λ determines the visibility or contrast, $v^2 = 1 - \lambda^2 = (A_0^2 - |A|^2_{min})/A_0^2$, and N is not necessarily an integer. For $\lambda = 0$, $|A|^2_{min} = 0$ and the dip is completely dark at $T = 0$. Figure 13.11b shows the normalized dark soliton intensity in space or time for $v = 1$, $|A|^2_{min} = 0$, $\lambda = 0$. The solution for this case is $|A| = (\sqrt{P_0}/N) \tanh (T/T_0)$ and does not change width or shape as the pulse is propagated. The so-called gray soliton occurs when the dip in cw intensity does not go to zero, $|A|^2_{min} \neq 0$. The velocity of the dark soliton relative to the retarded time coordinates is given as 2λ. That is, moving along with a quasi-cw pulse (long pulse compared with the intensity dip) the dark soliton advances or retards, depending on the velocity. For a completely dark soliton the velocity is zero. For spatial dark solitons the velocity is interpreted as the angle at which the soliton propagates relative to the optic axis z. This angle is given as $\tan \theta = \lambda(n_2 E_0^2/2n_0)^{1/2}$. The soliton has a width given as $v^{-1}(2n_0/n_2 E_0^2 k^2)^{1/2}$. The use of a phase mask, and the resulting interference, is one popular method to create dark solitons that transmit the full incident intensity. The parameter λ, which determines the contrast and the velocity, or angle, of the dark soliton, is a function of the phase difference across a step phase mask.[37] For instance, $\lambda = -\cos \Delta\varphi$, where $\Delta\varphi = (\varphi_{left} - \varphi_{right})/2$. If the phase difference $\Delta\varphi = \pi$, then $\lambda = 0$ and the velocity is zero (the spatial soliton does not travel at an angle to the propagation axis) and the soliton is completely dark at the center of time or space (see Fig. 13.11b).

The full solution for 3-spatial and 1-temporal dimension does not yield analytic solutions. However, there are a number of numerical techniques that can be effective in describing beam propagation. One of the more popular is the slit-step Fourier method.[38] The material length is divided into a number of sections. The larger the sections, the larger the error, but the shorter the computing time. This is a trade-off that must be determined in order to determine the section size. The electric field of the beam is propagated over a section with only diffraction, or GVD, using a fast Fourier transform (FFT) algorithm. At the center of the section the electric field is multiplied by the nonlinearity that would be present over the whole section. The beam is then propagated to the end of the section, again with only diffraction or GVD. The beam at the end of the run can then be compared with a second run with different step sizes to determine the acceptibility of the error.

13.7 ACKNOWLEDGMENTS

We gratefully acknowledge helpful discussions with W. W. Clark, III, G. J. Salamo, T. F. Boggess, G. A. Swartzlander, Jr., and E. W. Van Stryland.

13.8 REFERENCES

1. T. H. Maiman, "Stimulated Optical Radiation in Ruby," *Nature*, vol. 187, p. 493, 1960.
2. It is generally understood that the term "optics" refers to the portion of the electromagnetic spectrum that includes the visible as well as the infrared wavelengths.
3. N. Bloembergen, *Nonlinear Optics*, W. A. Benjamin, Inc., London, 1965.
4. G. C. Baldwin, *An Introduction to Nonlinear Optics*, Plenum Press, New York, 1969.
5. A. Yariv, *Quantum Electronics*, Wiley, New York, 1975.
6. H. Rabin and C. L. Tang (eds.), *Quantum Electronics: A Treatise*, Academic Press, New York, 1975.
7. Y. R. Shen, *The Principles of Nonlinear Optics*, Wiley, New York, 1984.
8. F. A. Hopf and G. I. Stegeman, *Applied Classical Electrodynamics*, vol. II, Wiley, New York, 1986.
9. R. W. Boyd, *Nonlinear Optics*, Academic Press, San Diego, Calif., 1991.
10. J. D. Jackson, *Classical Electrodynamics*, Wiley, New York, 1975.
11. T. K. Yee and T. K. Gustafson, *Phys. Rev.*, vol. A18, p. 1597, 1978.
12. R. Loudon, *The Quantum Theory of Light*, Clarendon Press, Oxford, 1985.
13. N. B. Delone and V. P. Krainov, *Fundamentals of Nonlinear Optics of Atomic Gases*, Wiley, New York, 1988.
14. S. E. Harris, J. E. Field, and A. Imamoglu, "Nonlinear Optical Properties Using Electromagnetically Induced Transparency," in *Nonlinear Optics: Materials, Phenomena and Devices*, IEEE, New York, p. 317, 1990.
15. M. Bass, P. A. Franken, J. F. Ward, and G. Weinreich, *Phys. Rev. Lett.*, vol. 9, p. 446, 1962.
16. D. C. Hutchings, M. Sheik-Bahae, D. J. Hagan, and E. W. Van Stryland, *Opt. Quantum Electron.*, vol. 24, p. 1, 1992.
17. S. K. Kurtz, *IEEE J. Quantum Electron.*, vol. QE-4, p. 578, 1968.
18. R. L. Byer, "Parametric Oscillators and Nonlinear Materials," in P. G. Harper and B. S. Wherrett (eds.), *Nonlinear Optics*, Academic Press, New York, 1977.
19. A. Yariv and P. Yeh, *Optical Waves in Crystals*, Wiley, New York, 1984.
20. S. Singh, "1.1 Nonlinear Optical Materials," in M. J. Weber (ed.), *Handbook of Laser Science and Technology*, vol. III, *Optical Materials*: Part I, CRC Press, Boca Raton, Fla., 1986.
21. P. D. Maker and R. W. Terhune, *Phys. Rev.*, vol. 137, p. A801, 1965.
22. J. F. Reintjes, *Nonlinear Optical Parametric Processes in Liquids and Gases*, Academic Press, Orlando, Fla., 1984.
23. M. Sheik-Bahae, D. J. Hagan, and E. W. Van Stryland, *Phys. Rev. Lett.*, vol. 65, p. 96, 1990.
24. M. Sheik-Bahae, A. A. Said, and E. W. Van Stryland, *Opt. Lett.*, vol. 14, p. 955, 1989.
25. M. Sheik-Bahae, A. A. Said, T. H. Wei, D. J. Hagan, and E. W. Van Stryland, *IEEE J. Quantum Electron.*, vol. 26, p. 760, 1990.
26. A. K. Ghatak and K. Thyagarajan, *Contemporary Optics*, Plenum Press, New York, 1978.
27. S. A. Akhmanov, R. V. Khokhlov, and A. P. Surkhorukov, "Self-Focusing, Self-Defocusing and Self-Modulation of Laser Beams," in F. T. Arecchi and E. O. Schultz-DuBois (eds.), *Laser Handbook*, North Holland, New York, 1972.
28. J. H. Marburger, "Self-Focusing: Theory," in *Progr. Quantum Electron.*, vol. 4, p. 35, 1975.

29. M. Sheik-Bahae, A. A. Said, D. J. Hagan, M. J. Soileau, and E. W. Van Stryland, *Opt. Eng.*, vol. 30, p. 1228, 1991.

30. S. D. Durbin, S. M. Arkalian, and Y. R. Shen, *Opt. Lett.*, vol. 6, p. 411, 1981.

31. G. P. Agrawal, *Nonlinear Fiber Optics*, Academic Press, New York, 1989.

32. R. K. Dodd, J. C. Eilbeck, J. D. Gibbon, and H. C. Morris, *Solitons and Nonlinear Wave Equations*, Academic Press, New York, 1984.

33. A. Hasegawa and F. Tappert, *Appl. Phys. Lett.*, vol. 23, p. 142, 1973.

34. J. S. Aitchison, Y. Silverberg, A. M. Weiner, D. E. Leaird, M. K. Oliver, J. L. Jackel, E. M. Vogel, and P. W. E. Smith, *J. Opt. Soc. Am.*, vol. B8, p. 1290, 1991.

35. A. Hasegawa and F. Tappert, *Appl. Phys. Lett.*, vol. 23, p. 171, 1973.

36. G. A. Swartzlander, Jr., *Opt. Lett.*, vol. 17, p. 493, 1992.

37. Y. S. Kivshar and S. A. Gredeskul, *Opt. Commun.*, vol. 79, p. 285, 1990.

38. C. R. Menyuk, *Opt. Lett.*, vol. 12, p. 614, 1987.

CHAPTER 14
PHASE CONJUGATION

Gary L. Wood

Real time phase conjugation from nonlinear media was first observed and explained as a conjugate beam by Zel'dovich and his coworkers in the Soviet Union in 1972.[1] It was noticed that a high-intensity ruby laser beam (λ = 694 nm) focused into methane gas caused a retrobeam to emerge with unusual properties. The retrobeam seemed to propagate from the medium back to the source without diverging but actually converging (see Fig. 14.1). In other words, this retrobeam retraced the incident beam. In addition, it was observed that the retrobeam would restore any phase aberrations (see Fig. 14.2). The retrobeam became known as a phase conjugate beam. Phase conjugation has since been observed in many media and from several different physical mechanisms; for example, see Ref. 2 for a list of materials in which phase conjugation has been observed. Although phase conjugation is mostly observed in the visible to infrared part of the spectrum (λ = 400 nm to 10.6 μm) there is no reason it should be limited to this region. Phase conjugation should be observable from microwave[3] to x-ray[4] wavelengths in the near future with suitable sources and nonlinear materials. The many implications of phase conjugation are still not known and could affect basic physical processes in unexpected ways.[5,6] There also seem to be many potential applications of phase conjugation.

This chapter is divided into three sections. The first section describes what phase conjugation means. The second section describes some of the more usual methods employed to generate phase conjugation. Finally, the third section outlines some of the applications of phase conjugation.

14.1 PHASE CONJUGATION: WHAT IT IS

Perhaps the best way to understand phase conjugation is to begin with the wave equation in vacuum or

$$\nabla^2 E = \mu_0 \epsilon_0 \frac{\partial^2 E}{\partial t^2} \tag{14.1}$$

In spherical coordinates with a point source emitter the wave equation becomes

$$\frac{1}{r^2} \frac{\partial(r^2\, \partial \mathbf{E}/\partial r)}{\partial r} = \frac{1}{c^2} \frac{\partial^2 \mathbf{E}}{\partial t^2} \tag{14.2}$$

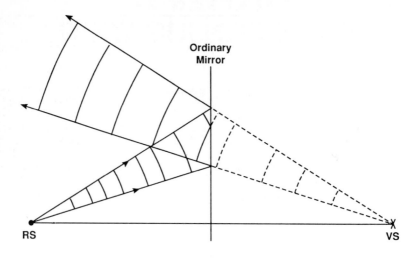

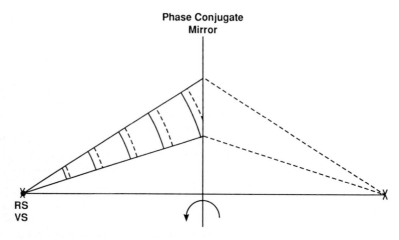

FIGURE 14.1 Comparison of an ordinary mirror (*a*) and a phase conjugate mirror (*b*). The ordinary mirror obeys the law of reflection and the reflected beam continues to diverge as if from a virtual source located the same distance behind the mirror as the actual source is in front of the mirror. The phase conjugate mirror returns the reflected beam to the source and exactly reverses the divergence to become a converging beam. It is as if the plane from the mirror to the virtual source is rotated so the virtual source and real source lie on top of one another.

The solution to this equation is $\mathbf{E} = \mathrm{Re}\{(\mathbf{E}_0/r)\exp[i(\mathbf{k}\cdot\mathbf{r} - \omega t)]\}$, where $c = \omega/k$. This solution suggests that the electric field at the source E_0 falls off in magnitude as the inverse of the radius and the electric field propagates as a wave with velocity $c = \omega/k$ away from the source. If the wave is confined to the z direction $[(x^2 + y^2) \ll z^2$, where x and y are transverse coordinates to the z direction] then the solution can be approximated and rearranged to yield $E = (E_0/z)\exp\{ik[(x^2 + y^2)/2z]\}\exp[i(kz - \omega t)]$. The first term in the above solution describes the magnitude of the electric field as a function position in the z direction. The second

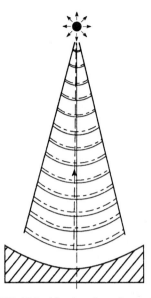

FIGURE 14.1 (*Continued*) *c* shows that a phase conjugate mirror behaves as if the reflector takes on the exact curvature of the incident beam over its entire surface. A conjugate beam can be generated this way with a flexible mirror.

term describes the phase of the electric field in the transverse plane as a function of the position along the z axis. This is the term that describes the diffraction of the beam. Rays drawn perpendicular to this phase front indicate the direction of the beam propagation.[7] Notice that the radius of curvature of the phase in the transverse direction is $+z$, indicating that the beam is spreading out away from the origin. The final term is the wave propagation term that describes the movement of the wave in the positive z direction. If the solution is now broken up into a spatial part and a temporal part and the spatial part is conjugated, the result is $E = E^*(r)f(t) = (E_0/z) \exp\{-ik[(x^2 + y^2)/2z]\} \exp[-i(kz + \omega t)]$. This wave is traveling in the negative z direction (or back toward the source) as seen by the third component of the electric field. Also the radius of curvature is negative but negative relative to a source opposite the original source. The curvature appears the same, but the direction is changed. Instead of diverging, the rays are pointing inward and converging to the orig-

inal source (see Fig. 14.3). This is exactly the behavior of a phase conjugate beam. Therefore, the phase conjugation process involves some way of conjugating only the spatial component of an electric field.[8,9] A conventional mirror (metallic or dielectric) changes the propagation direction but does not affect the phase of the transverse components. This results in the familiar beam that continues to spread in the direction of propagation upon reflection, as was shown in Fig. 14.1.

Another way to view this phase conjugate process is to think of a reversal of the time axis upon reflection.[10] Since the actual electric field is the real component of the complex field or Re $E = (E + E^*)/2$, it is easy to see that taking the complex conjugate of the spatial component is equivalent to taking the complex conjugate of the time part only. This is the same as allowing the beam to diverge to the phase conjugate mirror, at which point time is run backward ($t \rightarrow -t$, for phase only) and the beam is seen to converge back to the source. In the literature phase conjugation is often referred to as time reversal. However, this picture is not accurate for any amplitude variations in time.

In the laboratory one very reliable test for phase conjugation is to distort the phase of the incident beam to be conjugated. A phase distorter can be made with a frosted glass plate or with an HCl acid etched glass slide. (If plastic is used the polarization of the transmitted beam may be affected, which could affect the conjugation process.) The incident beam propagates through the distorter and into the phase conjugate mirror. The return beam is then allowed to pass back through the distorter and the distortions are removed. A return beam from a beam splitter can verify the undistorted phase conjugate beam. There are some practical considerations when performing this experimental test. First, the distortion should only

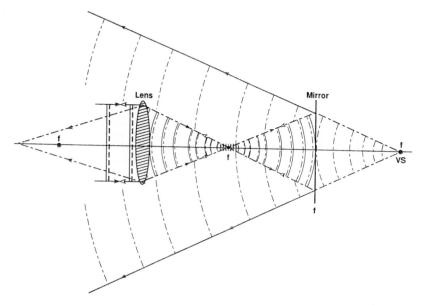

FIGURE 14.2 A plane wave entering to the left of the lens propagates to the right in the figure with phase fronts depicted as solid lines. The beam is focused at f and strikes a mirror at $2f$ from the lens. A phase conjugate mirror will return the beam exactly as it entered as shown by the dashed phase fronts. The ordinary mirror returns a diverging beam (dashed phase fronts) as if from a virtual source located f inside the mirror. Notice some of this beam is refocused at $1.5f$ to the left of the lens.

be of a phase nature because the phase conjugation will not undistort amplitude variations. Also, all of the scattered light that propagates through the distorter should be introduced into the phase conjugate mirror or the phase correction will not be completely faithful.

14.2 PHASE CONJUGATION: HOW TO GENERATE IT

As discussed in the previous section, in order to generate a phase conjugate beam, some means of conjugating the spatial components of the beam is necessary. There are many ways this can be accomplished and new methods are continuing to be developed. This section outlines some of the more popular methods available. Section 14.2.1 discusses four-wave mixing and $\chi^{(3)}$ nonlinear optical susceptibilities. Section 14.2.2 outlines the photorefractive effect and the many techniques for phase conjugation using this phenomenon. Section 14.2.3 is a discussion of stimulated scattering processes. This is the original physical mechanism for phase conjugate generation and continues to be very popular. Section 14.2.4 discusses some unconventional methods that are possible because of state-of-the-art technology. It should be mentioned that this list is by no means exhaustive. For instance, phase conjugation has been demonstrated from photon echoes,[11] three-wave mixing,[12] forward four-wave mixing,[13] and superluminescence,[14] but these methods are not as commonly employed as the four mentioned for discussion.

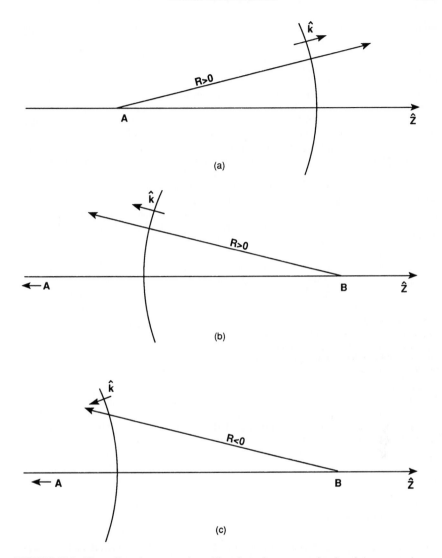

FIGURE 14.3 The radius of curvature is positive about the source point that the curvature is concave. For a source to the left radiating a spherical phase wave the radius is positive (*a*). When a beam reflects from a conventional mirror the source is now to the right and the radius of curvature remains positive (*b*) and continues to diverge. The phase conjugate mirror (*c*) has a negative radius of curvature so the phase is convex about the source. The phase conjugate therefore has the phase of the incident beam.

14.2.1 Four-Wave Mixing

Four-wave mixing is an attractive and frequently used method of generating a phase conjugate beam.[15] This is due partly to the ability to describe the process exactly and the ability to generate the phase conjugate beam at low incident beam powers. Four-wave mixing is a nonlinear optical effect that describes the generation of a phase conjugate beam when three incident beams interact on a material through the third-order nonlinear susceptibility.

Chapter 13 describes the third-order susceptibility; however, the following is a brief review. In order to describe the effect of an electric field on a material classically (many photons) the dielectric vector $\mathbf{D}$ is separated into two parts, the vacuum contribution and the material contribution. This is written as $\mathbf{D} = \mathbf{E} + 4\pi\mathbf{P}$ in Gaussian units. The material contribution comes from the polarization vector $\mathbf{P}$. This vector can be written as $P_i = \chi_{ij}E_j$ where a summation is implied over repeated indexes. The effect of the electric field is to change the polarization of the material and this in turn will affect the propagation of optical beams through the material. The effect is to reduce the phase velocity. This type of material response occurs when the force required to move a charge increases linearly with an applied electric field or for a charge in a parabolic potential well. If the potential well is not parabolic, this will be evident in a nonlinear force on the charge. This nonlinear force can be expressed in the material response by expanding the real polarization vector in terms of the total real electric fields as $P_i(\omega_1) + P^{(2)}_i(\omega_3) + P^{(3)}_i(\omega_4) = \chi^{(1)}_{ij}(\omega_1)E_j(\omega_1) + 2\chi^{(2)}_{ijk}(\omega_3;\omega_1,\omega_2)E_j(\omega_1)E_k(\omega_2) + 4\chi^{(3)}_{ijkl}(\omega_4;\omega_1,\omega_2,\omega_3)E_j(\omega_1)E_k(\omega_2)E_1(\omega_3) + \cdots$, where $\chi^{(n)}$ is the susceptibility tensor of rank $n + 1$ and n is the order, and the 2 and 4 are due to the order of $ijkl$, having no physical significance away from resonance.[16,17] The polarization vector responding at the initial frequency is the linear term here. The second-order polarization vector responds at the frequency $\omega_3 = \omega_1 + \omega_2$ and the third order at frequency $\omega_4 = \omega_1 + \omega_2 + \omega_3$. It is these higher-order susceptibilities that change the usual Maxwell's equations and lead to nonlinear optics.[18] The even-order susceptibilities are present only for asymmetric materials, so the first nonlinear term to appear for symmetric materials is of the third order.

The third-order susceptibility can be found quantum-mechanically from electrodynamic quantum field theory.[19] Stimulated scattering and two-photon absorption are both attributed to the third-order susceptibility and involve the absorption and emission or absorption at two separate space-time points, respectively. Nonlinear refraction, multiwave mixing, and third-harmonic generation are also described by third-order susceptibilities. The latter processes involve fourth-order perturbation theory with the virtual absorption and the emission of four photons under all possible combinations. The frequencies of the photons and the amount of absorption to emission determine which type of third-order process will occur. All of these are electronic processes and are fast (~femtoseconds) and typically small in magnitude. However, for large conjugated organic molecules, or near resonances, the dipole moment can be large. The range of magnitudes for the nonlinear refraction susceptibility off resonance is probably 10^{-15} to 10^{-9} esu.[20] In addition to the electronic third-order susceptibility, which all materials possess, most materials can exhibit a slower effective third-order susceptibility which is often larger than the electronic contribution.[21] A list of possible physical mechanisms and the materials in which these mechanisms are most likely to be observed is given in Table 13.1 along with the appropriate references.

Usually four-wave mixing is performed with degenerate (or nearly degenerate) frequencies and the generated fourth beam is also the same frequency as the incident beams. The degenerate case is chosen in order to ensure phase matching (conservation of momentum), as will be discussed later. By writing the real fields in terms of complex fields, $e = (E + E^*)/2$ the third-order polarizability $P^{(3)}_i = \chi^{(3)}_{ijkl}e_je_ke_l$ can have all possible combinations of E and E^*. However, to ensure $P^{(3)}$ has the same frequency as the incident electric fields, only combinations of two complex fields and a complex conjugate field can exist. The amplitude of the generated fourth field is described by the mixing of the three incident fields or $P_4^{(3)} \propto \chi^{(3)} E_1 E_2 E_3^*$. The total nonlinear polarization vector that mixes all combinations of the fields present is given by

$$P_i^{(3)} (\omega) = D\chi^{(3)}_{ijkl}(\omega;\omega,\omega,-\omega)(E_1 + E_2 + E_3 + E_4)_j$$
$$\times (E_1 + E_2 + E_3 + E_4)_k(E_1 + E_2 + E_3 + E_4)^*_l \quad (14.3)$$

where $D = 6$ for a completely nondegenerate case, $D = 3$ for the partially degenerate case (when $\omega_k = \omega_l$), and $D = 1$ for the completely degenerate case.[22] This is the same $\chi^{(3)}$ responsible for nonlinear index of refraction n_2 and nonlinear absorption β. The third-order susceptibility has been characterized in many materials, for instance, see Refs. 18 and 23. The magnitude of the third-order susceptibility is often compared to CS_2 because CS_2 has a fairly large nonlinearity and the dominant nonlinear mechanism (orientational Kerr effect) is well understood in this material. The magnitude of $\chi_{1111}^{(3)}$ for CS_2 for linear polarization in the visible is $\chi_{1111}^{(3)} = 5.5 \times 10^{-13}$ esu.[24] Nonlinear absorption, the term characterized by the imaginary part of $\chi^{(3)}$,[25] can alone be responsible for four-wave mixing.[26] However, the phase conjugate gain is at the same time reduced by the absorption so this may not be the most efficient way to generate a phase conjugate beam. The following analysis assumes that $\chi^{(3)}$ is real unless otherwise stated.

The steady-state Maxwell's equation within the context of the slowly varying envelope approximation (SVEA) for plane waves is (Gaussian units)

$$\frac{dA}{dz} = -\frac{\alpha}{2} A + i\frac{2\pi\omega}{nc} \Delta P \, e^{i(\mathbf{k}_p - \mathbf{k})\cdot \mathbf{r}} \quad (14.4)$$

where A is the complex amplitude of the electric field wave ($\mathbf{E} = \mathbf{A}e^{i(\mathbf{k}\cdot\mathbf{r} - \omega t)}$), ΔP, in this case, is the change in polarization due to third-order susceptibility, and $\mathbf{k}_p$ is the wave vector of the polarized media. In order to Bragg match (conservation of momentum) the difference between the wavevectors must be zero, $\mathbf{k}_p - \mathbf{k} = 0$.[27] Bragg matching usually gives the largest response, especially for thick media (media has many grating periods). For the four-wave mixing situations considered here the medium is considered thick. For thin media Bragg matching is not required as in Ref. 28. One typical geometry used in degenerate four-wave mixing to ensure Bragg matching is to allow two of the incident beams to be counterpropagating. The third beam is then Bragg matched at any incident angle with the generated beam propagating opposite to it. In this geometry the first two counterpropagating beams are referred to as the pumping beams, the third incident beam is the probe beam, and the generated beam will turn out to be the phase conjugate of the probe beam (see Fig. 14.4). The four complex electric fields propagating close to the z axis can be written as

$$\mathbf{E}_1 = A_1 \, e^{i(k_1 z - \omega t)}\mathbf{e}_1$$

$$\mathbf{E}_2 = A_2 \, e^{-i(k_2 z + \omega t)}\mathbf{e}_2$$

$$\mathbf{E}_3 = A_3 \, e^{i(k_3 z - \omega t)}\mathbf{e}_3$$

$$\mathbf{E}_4 = A_4 \, e^{-i(k_4 z - \omega t)}\mathbf{e}_4 \quad (14.5)$$

where $\mathbf{e}$ is the polarization vector. Beams 1 and 2 are the pump beams and 4 is the probe beam E_p. Beam 3 will be the generated beam. Notice that beam 1 and beam 3 travel along the $+z$ axis while beam 2 and beam 4 travel along the $-z$

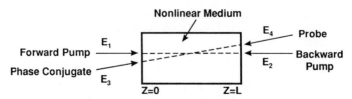

FIGURE 14.4 The typical geometry for four-wave mixing that ensures phase matching. Two pump beams counterpropagating along the z axis interact along a nonlinear material of length L. Pump beam 1 propagates along the positive z axis and pump beam 2 propagates along the negative z axis where $|\mathbf{k}_1| = |\mathbf{k}_2|$. A probe beam traveling at a shallow angle to the z axis propagates in the negative z direction. The phase conjugate beam is Bragg matched by traveling in the positive z direction and counterpropagating to the probe beam.

axis. Since the propagation is assumed to take place along the z axis, the angle between the probe beam 4 and the pump beams is assumed to be small. This also ensures a large interaction length.

If the medium is isotropic there is a wonderful physical picture that can be used to describe four-wave mixing. With this high degree of symmetry it is possible to write the third-order polarization vector responsible for the amplitude of the conjugate beam as a vector equation[29]

$$P^{(3)}(\mathbf{k}_3) = A'(\mathbf{E}_2 \cdot \mathbf{E}^*_p)\mathbf{E}_1 + B'(\mathbf{E}_1 \cdot \mathbf{E}^*_p)\mathbf{E}_2 + C(\mathbf{E}_1 \cdot \mathbf{E}_2)\mathbf{E}^*_p$$
$$+ A''(\mathbf{E}_2^* \cdot \mathbf{E}_p)\mathbf{E}_1 + B''(\mathbf{E}_1^* \cdot \mathbf{E}_p)\mathbf{E}_2 \quad (14.6)$$

where the A,B,C terms are combinations of the third-order susceptibilities that depend on the angles and polarizations of the interfering fields. If the medium is transparent and all three incident beams are polarized perpendicular to the plane of incidence then $A' = B' = 3(\chi_{1122} + \chi_{1212})$, $C = 6\chi_{1221}$, and $\chi_{\text{eff}} = \chi_{1111} = A' + B' + C/2$. As a vector equation it can be seen that there are six interference (dot product) terms. The first term above describes the interference of pump beam 2 and the probe beam. Since the frequency term cancels out, this is a static grating. Pump beam 1 sees (or reads) this grating and scatters into direction $+k_p$ (opposite the original k_p direction). The grating written by the A term is $k_{Ag} = |\mathbf{k}_p - \mathbf{k}_2|$. The magnitude of the A grating is given as $|\mathbf{k}_g| = 2|\mathbf{k}|\sin(\theta/2)$, where θ is the total angle between the wavevector difference, $\mathbf{k}_p - \mathbf{k}_2$, and $\mathbf{k}_g = 2\pi/\lambda_g$, where λ_g is the grating period. Only the A' contribution will be phase matched (Bragg matched, $\mathbf{k}_{A'3} = -k_2\mathbf{z} + \mathbf{k}_p + k_1\mathbf{z} = \mathbf{k}_p$, while $\mathbf{k}_{A''3} = k_2\mathbf{z} - \mathbf{k}_p + k_1\mathbf{z} = 2|k_1|\mathbf{z} - \mathbf{k}_p$, where $\mathbf{z}$ is the unit vector) and able to contribute to the scattering of pump beam 1 into the direction opposite the probe beam in thick media. The phase conjugate in this picture is due to the reflection of pump beam 1 off the grating written by pump beam 2 and the probe beam. Because pump beam 1 is scattered in a forward direction this grating is known as a transmission grating. The second grating term involves the grating written by the B terms. This grating arises from the interference of pump beam 1 and the probe beam. This grating will have a larger angle θ and will have a correspondingly smaller grating period. The B grating involves the pump beam 2 reading this grating and scattering opposite to the probe beam $+k_p$. Because pump beam 2 is backscattered, this grating is known as a reflection grating. Again B' is Bragg matched while B'' is not. These gratings are summarized in Fig. 14.5.

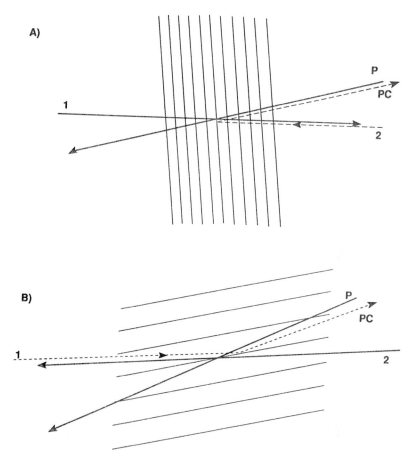

FIGURE 14.5 Four-wave mixing can produce two spatial gratings shown as (*a*) and (*b*). The dashed line is the read beam and the solid lines are the write beams. The small-period grating (*a*) is written by pump beam 1 and the probe beam. Pump beam 2 backscatters off this grating to become the phase conjugate beam. This grating is known as the reflection grating. The large-period grating (*b*) is written by pump beam 2 and the probe beam. Pump beam 1 scatters off this grating to become the phase conjugate beam. This grating is known as the transmission grating.

If one of the pump beams has a different polarization or is not coherent with the other pump and the probe beam, but is still at the same frequency, then one of these gratings can be selected over the other. In this way the material response to the grating period can be determined. The smaller grating period B is likely to diffuse away faster than the long-period grating A, and this could be measured with grating discrimination. However, the magnitude of the phase conjugate beam will not be as large as when both gratings are present with equal modulation.

The writing of the first two gratings, A and B in Eq. (14.6), is analogous to real-time holography.[30] The small grating period, the B grating, is a reflection hologram, and the large grating period, the A grating, is a transmission hologram. The transmission hologram is written by the interference of a beam reflected off an object, the probe beam, and a plane wave reference beam, the pump beam, with both incident on the same side of a holographic material. This hologram is then read by a plane wave beam, second pump, incident on the opposite side of the hologram opposite the direction of the first reference beam. The holographic image is then formed where the object was, i.e., the phase conjugate beam. The non-Bragg matched scattering will form a virtual image if the nonlinear material is thin. With real-time holography, via nonlinear materials, there is no waiting for the holographic material to develop. This allows the object to move (as long as the response time of the nonlinear medium is fast enough) and the phase conjugate image to follow.

So far the C grating in Eq. (14.6), the grating written by the two pump beams, has been ignored. This grating would be important in the case when the probe beam has a different polarization or coherence than the two pump beams. This grating is not a spatial grating but a temporal grating that oscillates at twice the incident frequency 2ω. This is known as a "breathing" grating from which the probe beam alone could scatter. (It should be noted, however, that the pump beams could scatter radiation and this scattered radiation might act like a probe beam from which four-wave mixing could occur.) This grating is usually not significant but can be if the material has a resonant two-photon absorption.[31] Two photons from the pump beams drive the system to a real excited energy state and a generated phase conjugate beam plus the probe beam deexcite the material back to its ground state. In this way phase matching occurs automatically. The phase conjugate reflectivity is then independent of the incident angle to the pump beams. It is interesting that the buildup of a phase conjugate beam usually begins from noise or stray photons but with this situation the buildup can begin from the vacuum. Pumping a sample with the two-photon resonance builds the breathing grating at 2ω. Fields in the vacuum can then be encouraged to amplify at the frequency ω. This vacuum cavity can be a set of mirrors placed at any angle to the pump beams. In this way an empty resonant cavity will build up a field. This process leads to squeezed light and the noise of these fields is reduced from the coherent state over half a cycle.

Bragg matching only ensures that the propagation of the generated beam reflects back opposite to the probe beam; it is still necessary to show that the phase of the generated beam is the conjugate of the probe beam. This will involve deriving the form of the field amplitude $\mathbf{A}(r)$ for the phase conjugate beam from the coupled SVEA equations. Keeping only those terms in ΔP that are Bragg matched yields the following set of coupled equations;[32]

$$
\begin{aligned}
\frac{dA_1}{dz} = {} & -\frac{\alpha}{2} A_1 + iC_1(A_1A_3{}^* + A_2{}^*A_4)A_3 + iC_2(A_1A_4{}^* + A_2{}^*A_3)A_4 \\
& + iC_3(A_1A_2{}^*)A_2 + iC_4(|A_1|^2 + |A_2|^2 + |A_3|^2 + |A_4|^2)A_1 + iC_5(A_3A_4 \\
& + A_1A_2)A_2{}^* + iC_6(A_1A_4)A_4{}^* + iC_7(A_1A_3)A_3{}^* + iC_8(A_1A_1)A_1{}^*
\end{aligned}
$$

$$
\begin{aligned}
\frac{dA_2}{dz} = {} & +\frac{\alpha}{2} A_2 - iC_1(A_1{}^*A_3 + A_2A_4{}^*)A_4 - iC_2(A_1{}^*A_4 - A_2A_3{}^*)A_3 \\
& - iC_3(A_1{}^*A_2)A_1 + iC_4(|A_1|^2 + |A_2|^2 + |A_3|^2 + |A_4|^2)A_2 \\
& - iC_5(A_3A_4 - A_1A_2)A_1{}^* - iC_6(A_2A_3)A_3{}^* - iC_7(A_2A_4)A_4{}^* \\
& - iC_8(A_2A_2)A_2{}^*
\end{aligned}
$$

$$\frac{dA_3}{dz} = -\frac{\alpha}{2} A_3 + iC_1(A_1^*A_3 + A_2A_4^*)A_1 + iC_2(A_1A_4^* + A_2^*A_3)A_2$$

$$+ iC_3(A_4^*A_3)A_4 + iC_4(|A_1|^2 + |A_2|^2 + |A_3|^2 + |A_4|^2)A_3$$

$$+ iC_5(A_3A_4 + A_1A_2)A_4^* + iC_6(A_2A_3)A_2^* + iC_7(A_1A_3)A_1^*$$

$$+ iC_8(A_3A_3)A_3^*$$

$$\frac{dA_4}{dz} = +\frac{\alpha}{2} A_4 - iC_1(A_1A_3^* - A_2^*A_4)A_2 - iC_2(A_1^*A_4 - A_2A_3^*)A_1$$

$$- iC_3(A_4A_3^*)A_3 - iC_4(|A_1|^2 + |A_2|^2 + |A_3|^2 + |A_4|^2)A_4$$

$$- iC_5(A_3A_4 - A_1A_2)A_3^* - iC_6(A_1A_4)A_1^* - iC_7(A_2A_4)A_2^*$$

$$- iC_8(A_4A_4)A_4^*$$

$$(14.7)$$

where the $C's$ contain the appropriate $\chi^{(3)}$ terms. Notice that C_3, C_4, C_6, C_7, C_8 affect the phase of the transmitted beam but not the amplitude. The case of arbitrary polarization in a cubic material can be found in Ref. 33. If the nonlinearity does not depend on the vector properties of the fields (non-Kerr medium), such as a thermal nonlinearity, then the above equations reduce to[34]

$$\frac{dA_1}{dz} = -\frac{\alpha}{2} A_1 + i\gamma A_1(|A_1|^2 + 2|A_2|^2 + 2|A_3|^2 + 2|A_4|^2) + 2i\gamma A_3A_4A_2^*$$

$$\frac{dA_2}{dz} = +\frac{\alpha}{2} A_2 - i\gamma A_2(2|A_1|^2 + |A_2|^2 + 2|A_3|^2 + 2|A_4|^2) - 2i\gamma A_3A_4A_1^*$$

$$\frac{dA_3}{dz} = -\frac{\alpha}{2} A_3 + i\gamma A_3(2|A_1|^2 + 2|A_2|^2 + |A_3|^2 + 2|A_4|^2) + 2i\gamma A_1A_2A_4^*$$

$$\frac{dA_4}{dz} = +\frac{\alpha}{2} A_4 - i\gamma A_4(2|A_1|^2 + 2|A_2|^2 + 2|A_3|^2 + |A_4|^2) - 2i\gamma A_1A_2A_3^*$$

$$(14.8)$$

The coupling constant γ is given by

$$\gamma = \frac{3\pi\omega\chi^{(3)}}{cn} \qquad (14.9)$$

Notice that the coupling coefficient is the same for all combinations. Equation (14.8) is also valid for a Kerr medium if all four beams are linear polarized parallel to each other. The solution of these coupled equations will be dependent on the boundary conditions. The nonlinear material is assumed to extend from $z = 0$ to $z = L$.

To show the generated beam is the phase conjugate of the probe beam assume the probe and conjugate beams are small in magnitude compared with the pump beams. Also assume the pumps are not depleted; that is, they do not change with their amplitude noticeably with propagation through the medium. The second term on the right side of Eq. (14.8) will only change the phase of those beams and can be ignored for now. The solution of beam 3 with no absorption is found to be[35]

$$A_3(z) = A_3(0) \frac{\cos[g(z - L)]}{\cos(gL)} - iA_4(L)^* \frac{\sin(gz)}{\cos(gL)} \tag{14.10}$$

where the gain $g = 2\gamma(8\pi/cn)(I_{10}I_{2L})^{1/2}$. The boundary condition requires that the generated beam be produced within the material from zero initial amplitude; therefore, $A_3(0) = 0$. With this condition the generated beam at $z = L$ is $A_3 = i \tan(gL)$ $A_4(L)^*$. The generated beam field is the phase conjugate of the incident beam 4 field with a magnitude of $\tan(gL)$. The phase conjugate reflectivity R is defined as the ratio of the intensity of the phase conjugate return beam to the intensity of the incident probe beam $R = I_3/I_4$. Under the above assumptions, the reflectivity will be $R = \tan^2(gL)$. [If the third-order susceptibility is complex, $R = \tan^2(|g|L)$, where $|g| = (g^*g)^{1/2}$. Notice that a measurement of the phase conjugate reflectivity depends on the magnitude of the contributions from both the real and imaginary third-order susceptibility. Therefore, it is not possible to separate the contribution of the nonlinear index of refraction and nonlinear absorption by measuring the phase conjugate reflectivity alone.] This shows that not only is the generated beam the phase conjugate of the incident beam but, for four-wave mixing, the reflectivity can be greater than 1 when $gL > \pi/4$.

Notice that the reflectivity becomes infinite when $gL = n\pi/2$, where n is any integer. This situation is known as the phase conjugate oscillation condition. The medium is so unstable that stray photons directed into the pumped medium will have gain and a phase conjugate output can occur. To obtain a gain-length product of $\pi/2$ requires

$$\frac{\pi}{2} = 2\gamma \frac{8\pi}{cn} L (I_{10}I_{2L})^{1/2} \tag{14.11}$$

If the pump intensities are assumed to be equal, the length is assumed to be 1 cm, and the third-order susceptibility is assumed to be that of CS_2, then the pump intensities need to be $I = 356$ MW/cm^2 to achieve a gain-length product of $\pi/2$. In reality the phase conjugate signal does not go to infinity, of course, but it can get very large. Often with large intensities there is some beam distortion and beam steering that limit the amount of beam overlap. These effects did not come out of the analysis performed here because the beams were assumed to be plane waves. Also with large gains it is easy to get nonoptimized oscillations that build up from scattering centers. Finally, as the phase conjugate beam builds up it is no longer accurate to consider the pump beams as undepleted so the approximation used to derive the reflectivity is not correct. Most often, for the reasons cited above, the phase conjugate reflectivity falls below $R = 100$ unless special care is taken.

Keeping all the phase terms, the magnitude of the phase conjugate beam with no absorption and no pump depletion ($dA_{1,2}/dz = 0$) can also be found from the above equations. (The changes in beams 1 and 2 are still zero but it is no longer assumed that the pump beam intensities are much greater than the probe and conjugate beams.) With the pump ratio defined as $r = I_{2L}/I_{10}$ and the probe ratio as $q = I_{4L}/(I_{10} + I_{2L})$, the phase conjugate reflectivity is

$$R = \frac{I_{pc}}{I_4} = \frac{\sin^2(gLh)}{h^2 + \sin^2(gLh)} \tag{14.12}$$

where $h^2 = 1 + (1 - r)^2/16r$.[36] Notice that r can be replaced with $1/r$ with no noticeable difference, indicating that the reflectivity is independent of the probe direction with regard to the pump beams. If $r = 1$, the pump beams are of equal incident intensity; then the phase conjugate reflectivity R is given as $R = \tan^2(gL)$, as was derived when the phase terms from beam 3 and beam 4 were ignored. If the pump ratio does not equal 1, it is not possible to reach the phase conjugate oscillation condition.

With the assumption of undepleted pump beams and keeping only the pump beams in the phase terms, the solution with absorption can be found. With absorption the phase conjugate reflectivity becomes[37]

$$R = \left| \frac{2g \sin(HL_{\text{eff}}/2)}{[H \cos(HL_{\text{eff}}/2) + \alpha \sin(HL_{\text{eff}}/2)]} \right|^2 \tag{14.13}$$

where the effective interaction length is given by $L_{\text{eff}} = \{[1 - \exp(-\alpha L)]/\alpha\}$, L is the overlap length of the probe beam with the pump beams, and $H = (4|g|^2 e^{-\alpha L} - \alpha^2)^{1/2}$. Notice that the reflectivity goes to infinity when $\tan(HL_{\text{eff}}/2) = -H/\alpha$, which reduces to $gL_{\text{eff}} = \pi/2$ when $\alpha = 0$. If the combined intensity of the forward and backward pump beams is a significant fraction of the saturation intensity, then nonlinear absorption and refraction will become manifest. This situation can be described by Eq. (14.13) with the appropriate modifications to g and α.[38,39] For instance, at line center with a two-level atomic system,

$$g = i \, \alpha_0 \left(\frac{I_1 I_2}{I_s^2}\right)^{1/2} \left[\left(1 + \frac{I_1 + I_2}{I_s}\right)^2 - \frac{4 I_1 I_2}{I_s^2}\right]^{-3/2} \tag{14.14}$$

and

$$\alpha = \alpha_0 \left(1 + \frac{I_1 + I_2}{I_s}\right) \left[\left(1 + \frac{I_1 + I_2}{I_s}\right)^2 - \frac{4 I_1 I_2}{I_s^2}\right]^{-3/2}$$

where α_0 is the low-intensity absorption. The saturation intensity I_s is given by $I_s = h\nu/2\sigma T_1$, where σ is the absorption cross section and T_1 is the energy relaxation time assuming homogeneous line broadening.

For high reflectivities pump depletion must be considered.[40] Unfortunately there is no closed-form solution for this situation, but rather the solution is of the form of an elliptic integral of the first kind which is easily solved numerically. The reflectivity still oscillates when plotted versus the gain-length product, but the maximum reflectivity no longer occurs at integer intervals of $\pi/2$. One distinguishing feature of the pump depletion case is that the reflectivity is multivalued with respect to the gain length. In other words, there can appear more than one value of reflection at certain gain-length values (see Fig. 14.6). When this occurs usually one branch of the reflection value will be stable.

The gain length at which the reflectivity becomes multivalued for a pump ratio of 1 and small q (not much depletion) occurs when the gain length is slightly larger than 2. It increases steadily to reach about 3 at $q = 0.3$ and then continually decreases for larger q values. With no absorption, the transmission of the pump beam going in the $+z$ direction (the forward pump) is given as $T_f = 1 - q(1 + r)R$. The transmission of the backward-going pump beam is $T_b = 1 - q(1 + r)R/r$, and the probe transmission is given as $T_p = 1 + R$. The maximum reflectivity is approximately $R_{\text{max}} \sim r/q(1 + r)$ for $r < 1$, and $R_{\text{max}} \sim 1/q(1 + r)$ for $r > 1$.

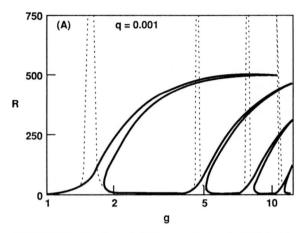

FIGURE 14.6 The theoretical phase conjugate reflectivity is plotted versus the gain-length product for the small probe to pump beam ratio $q = 0.001$. The broken line curves are the solutions for the undepleted pump approximation with no absorption. The solid curves are the solutions for the depleted pump situation with no absorption. The most striking feature of the depleted pump solutions is their bivalued nature. Notice that although q is small and the undepleted pump approximation should be fairly accurate, the two solutions depart at gain-length products long before the reflectivity saturates. For larger values of q, the reflectivity will be smaller, the curvature will become less pronounced, and the maximum will move to smaller gain-length products. *(After reference S. Guha and P. Conner, Opt. Commun., vol. 89, p. 107, 1992.)*

Also, with pump depletion the symmetry of the reflectivity with r and $1/r$ is lost. The larger the pump depletion, the larger the asymmetry.

With absorption and pump depletion the solution becomes very difficult. Numerical solutions have been found out to the point when the reflectivity becomes multivalued.[41] For small depletion there can be gain lengths where the reflectivity is higher with absorption than without it. The threshold for multivalued reflectivities is smaller with absorption and the maximum reflectivity is smaller. The maximum reflectivity is approximately given by $R_{max} \sim re^{-2\alpha L}/q(1 + r)$.

So far, four-wave mixing has been with beams of equal frequency and this is known as degenerate four-wave mixing (DFWM). Four-wave mixing can occur for different frequency beams with some restrictions. If the two pump beams are of the same frequency ω, but different from the probe beam $\omega + \delta$, the conjugate beam will be $\omega - \delta$. The change in the polarizability that leads to the conjugate beam is proportional to $P_{pc}(\omega_{pc} = \omega_p^*) \propto E_1 E_2 E_p^*$ so the frequency response is $\omega + \omega - (\omega + \delta) = \omega - \delta$. Since the total difference in frequency between the probe beam and its conjugate is 2δ, there will not be exact cancellation of the conjugate wave with the probe. A frequency mismatch implies a wavevector mismatch of magnitude $\Delta k = 2n\delta/c$. The phase conjugate reflectivity for the nondegenerate case with small reflectivities because of undepleted pumps is[42]

$$R = \frac{(gL)^2 \tan^2(\beta L)}{(gL)^2 + \left(\dfrac{\Delta kL}{2}\right)^2 \sec^2(\beta L)} \tag{14.15}$$

where $\beta = [g_{pc}g_{pc}^* + (\Delta k/2)^2]^{1/2}$. In general the reflectivity falls off rather sharply from the Bragg matched condition $\Delta k = 0$ and has a series of oscillations that decay in maximum amplitude as Δk gets larger. (In fact, the minima appear at $\Delta k L/2\pi = n$, where n is an integer.) If the probe beam is slightly offset in frequency from the pumps, it is possible to ensure phase matching by arranging the normally counterpropagating pump beams to be at a slight angle to each other.[43]

Equation (14.15) assumes that the medium response is instantaneous. This is a good assumption for third-order susceptibilities that exhibit only the EKE (see Table 13.1) but may need modification otherwise. The interference of two different frequencies results in a moving intensity grating. If this grating moves along faster than the material can respond, then a material grating will not form. For an exponential decay in the change of susceptibility, the value of the third-order susceptibility is $\chi^{(3)} = \chi^{(3)}/(1 + i\Delta\omega\tau)$, where τ is the decay time of the nonlinear susceptibility.[44] Notice that as the frequency difference and/or the time response becomes large the magnitude of $\chi^{(3)}$ decreases. By comparing this theoretical expression with the experimental data it is possible to determine the response time of the nonlinearity for a frequency mismatch.

The above analysis for the mixing of different frequencies is generally valid for media that respond locally. For media that launch material waves, such as electrostriction (acoustic waves) or thermally (thermal waves), a different type of phase conjugation can occur. The phase mismatch still limits the reflectivity, but the frequency mismatch can significantly increase the nonlinear index or gain. To understand this, suppose the two pump beams have the same frequency, the probe is assumed to be shifted up in frequency (anti-Stokes shifted), and the phase conjugate will then be down-shifted. (The probe could have been shifted down with a reverse in the following pump intensities but up shifting works better.) Also, suppose the pump beams have unequal intensities with the larger-intensity pump and the probe propagating in the $+z$ direction. The weak pump and conjugate will travel in the $-z$ direction opposite the probe and strong pump. The frequency mismatch creates an intensity beat frequency written by the probe and the weak pump that can move at the same velocity as an acoustic wave in the $+z$ direction of the reflection pump. The scatter of the strong pump off the grating moving away from it down-shifts the conjugate to a Stokes beam (so $\Delta\omega_{\text{total}} = 0$ or energy is conserved overall, but Δk will not equal zero). Conjugate beam interference further increases the modulation of the moving grating which increases the intensity of the conjugate. This process is unstable when the net gain can overcome losses due to acoustic dampening and scattering of the weak pump into an anti-Stokes phase conjugate. Once above the threshold gain (gain > losses) there appears a near exponential buildup of the conjugate beam in time until significant pump depletion occurs. The name commonly given to this type of phase conjugation for acoustic waves is Brillouin-enhanced four-wave mixing (BEFWM).[45] Because of the frequency shift, buildup from noise is reduced (especially for the anti-Stokes probe) and very small probe intensities can be significantly amplified. For example, in CS_2 it has been observed at 1064 nm and for a 6.4-ns pulse that a 0.5 nJ/0.03 cm^2 probe was amplified to a 1.4-mJ conjugate with a strong pump beam at 100 mJ/0.47 cm^2.[46] This corresponds to a reflectivity of 2.8×10^6 and the efficiency of conversion from weak pump to conjugate of 1 percent. For larger probe intensities the reflectivity fell off to $R = 1000$ at $E_p = 10^{-5}$ J, but the conversion efficiency increased to 15 percent. These values were achieved by slightly adjusting the pump beams so they were not exactly counterpropagating and the frequency shift was not exactly the acoustic shift. In order to observe BEFWM it is necessary for the pump beams to be above some threshold value. The minimum threshold is predicted for a phase mismatch of $\Delta k L = 5$ and the pump ratio of 0.3.

14.2.2 Photorefraction (MKS units)

The photorefractive effect describes a light-induced index of refraction change and was first observed in $LiNbO_3$, $LiTaO_3$, $BaTiO_3$,[47] and KTN.[48] It was noticed that these materials seemed to undergo a change in index of refraction that persisted for some time after illumination with milliwatt cw laser beams. The index change occurred within the same volume as the incident laser beam. This effect is detrimental for many of the applications of these crystals (such as second harmonic generation and electro-optic modulators) and these early investigators viewed photorefraction as a new type of "optical damage." Unlike catastrophic damage usually associated with laser damage, this index change, and hence the "damage," can be eliminated by heat treatment or uniform illumination of the crystals.

The basic physical mechanism underlying photorefraction has since been determined and is well understood today.[49-51] (However, the finer details of the mechanism and the identification of the dopant-defect donors continue to undergo revisions.)[52,53] There are only a small number of crystals in which the traditional photorefractive effect, as will be described below, has been observed. Most of these materials can be grouped into semiconductors, perovskites, sillenites, and tungsten-bronzes. It has been observed that the photorefractive response is usually larger when the crystals are doped. The photorefractive process starts when light incident on a transparent crystal of asymmetric symmetry excites trapped charges. These charges then either diffuse or drift (if an applied electric field is present) to new locations where they are retrapped into empty sites. This process continues until the freed charges are trapped in dark (no illumination) areas of the crystal (for pulses longer than the diffusion time); see Fig. 14.7. As the freed charges diffuse a space charge field develops. Eventually the diffusion force is exactly compensated by this space charge field and no more charges diffuse to dark areas. This static space charge field can distort the material lattice, causing an index change in the asymmetric crystals. The process of a static electric field changing the index is the well-known Pockels effect used in electro-optic modulators.[54] The Pockels effect involves the second-order susceptibility. From the total polarization vector the change in susceptibility can be found from the static field $E_k(0)$ to be

$$P_i(\omega) = \chi^{(1)}{}_{ij} E_j(\omega) + \chi^{(2)}{}_{ijk} E_j(\omega) E_k(0)$$

$$= (\chi^{(1)} + \Delta\chi)_{ij} E_j(\omega) \tag{14.16}$$

where $\Delta\chi_{ij} = \chi^{(2)}{}_{ijk} E_k(0) = 2n\epsilon_0\Delta n$. Usually the Pockels effect is characterized by the electro-optic coefficients r_{ij}, which are related to the change in index of refraction in mks units as

$$\Delta n_{ij} = -\epsilon_{ii}\epsilon_{jj}r_{ijk}\frac{E_k}{2n} \tag{14.17}$$

where ϵ is the high-frequency dielectric constant in the principal axis system. The electro-optic coefficient is usually written in contracted notation as $r_{ij,k} = r_{mk}$.

If the incident light consists of two coherent beams which combine at the crystal to produce a spatial sinusoidal interference pattern, the freed charges will diffuse to form a similar periodic charge distribution pattern, at least for a small interference modulation.[55] The interference pattern of the electric fields is given by

$$I = I_0[1 + \text{Re}(m\, e^{i\mathbf{k_g}\cdot\mathbf{r}})] \tag{14.18}$$

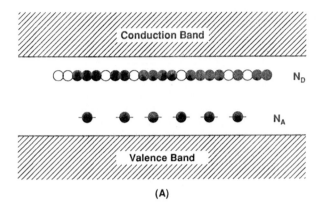

(A)

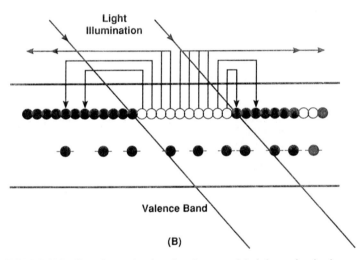

(B)

FIGURE 14.7 Here the two-level semiconductor model of photorefraction is presented with the vertical axis representing energy and the horizontal axis space. The initial charge distributions are shown in (a). In (a) the donor level N_D is partially filled with electrons and the acceptor level N_A is completely filled (fully compensated). It is assumed that the conduction band is initially empty. (b) The presence of optical radiation on part of the crystal causes the excitation of charges from the donor level to the conduction band. In the conduction band the electrons are free to diffuse (or drift if an external electric field is applied), and eventually the electrons recombine in empty donor sites in dark regions of the crystal.

where $I_0 = I_1 + I_2$, $\mathbf{k}_g = \mathbf{k}_2 - \mathbf{k}_1$, I_1 is the intensity of beam 1 while I_2 is the intensity of beam 2, $\mathbf{k}_1$ the incident wave vector of beam 1, $\mathbf{k}_2$ the incident wave vector of beam 2, and $\mathbf{k}_g$ the grating wave vector with a magnitude given by

$$|\mathbf{k}_g| = 2|\mathbf{k}| \sin(\theta) \qquad (14.19)$$

where the two incident waves are assumed to have the same wavelength and θ is the half angle between $\mathbf{k}_2$ and $\mathbf{k}_1$. The grating strength or modulation m is given in mks units by

$$m = \frac{(c\epsilon_0 n/2)(2\mathbf{E}_1 \cdot \mathbf{E}_2^*)}{I_0} \tag{14.20}$$

where $\mathbf{E}_1$ and $\mathbf{E}_2^*$ are the complex incident electric field strengths.

The material response to the incident light can be found by solving four coupled equations. These four basic equations govern the transport of charges as a function of time and applied electric field E_0; they are the equation of continuity of charge, the rate equation, the charge current equation, and Poisson's equation. If it is assumed that the incident crossing beams are copropagating at a small equal but opposite angle (θ) to the $\mathbf{z}$ axis and the beams propagate in the $\mathbf{x}$-$\mathbf{z}$ plane, the equations in mks units are

$$\frac{\partial n}{\partial t} - \frac{\partial N_D^+}{\partial dt} = \frac{1}{q} \nabla j$$

$$\frac{\partial N_D^+}{\partial t} = (sI + \beta)(N_D - N_D^+) - \gamma_R n N_D^+ \tag{14.21}$$

$$j = q\mu n E - k_B T \mu \nabla n$$

$$\nabla E = \frac{q}{\epsilon}(n + N_A - N_D^+)$$

where n is the electron number density, N_D is the dopant number density, N_D^+ is the ionized dopant density, N_A is the number density of negative charges which electrically compensate for N_D^+ in the dark (number density of acceptor sites), j is the current density, s is the ionization cross section, I is the total intensity, μ is the mobility, E is the electric field in the crystal, $k_B T$ is the Boltzmann temperature, ϵ is the static dielectric constant, β is the dark contribution due to a Maxwell-Boltzmann distribution in the bandgap, q is the elementary charge, and γ_R is the recombination coefficient. The intensity here is a reduced intensity, i.e., $I = I/h\nu$ (the number of photons per unit area per second) and the dimensions of the recombination coefficient γ_R are m^3/s. These equations are valid only under the following approximations: (1) in the bandgap only two levels are involved and one is inactive; (2) there is only one dominant charge carrier (electrons); and (3) there is no two-photon absorption or excited state absorption.

These equations are highly nonlinear and cannot be solved in closed form as they stand. However, they can be solved in the linear approximation of small grating modulation index $m \ll 1$, for times longer than the recombination time, and for low intensities (no two-photon absorption) to find the resulting space-charge field. The steady-state space-charge field with no applied electric field is given as[56]

$$\mathbf{E}_{sc} = \hat{\mathbf{e}}_x \, \mathrm{Re}\left\{ -im'k_g \frac{k_B T}{q} \left[1 + \left(\frac{k_g}{k_0}\right)^2 \right]^{-1} (1 - e^{-t/\tau})e^{ik_g x} \right\}$$

$$= -\frac{im'\mathbf{E}_{0sc}e^{ik_g x}}{2} + \text{c.c.} \tag{14.22}$$

where k_0 is the inverse Debye screening length given by $k_0^2 = q^2 N_{\mathrm{eff}}/\epsilon k_B T$ and the effective number of charge carriers is given as $N_{\mathrm{eff}} = (N_A/N_D)(N_D - N_A)$, $\mathbf{e}_x$ is unit vector in the x direction, and $m' = m/(1 + I_d/I_0)$ where $I_d = \beta/s$ is due to the dark current and I_0 is the total incident intensity. Notice the magnitude of the space-charge field depends on the crossing angle of the two beams. The inverse time response τ^{-1} is given by

$$\tau^{-1} = \frac{qI_0}{\hbar\omega\epsilon}\left[1 + \left(\frac{k_g}{k_0}\right)^2\right]\left[\frac{\alpha_{ph}\mu}{\gamma N_A}\left\{1 + \left(\frac{k_g}{K}\right)^2\right\}^{-1}\right] + \frac{\sigma_d}{\epsilon} \qquad (14.23)$$

where α_{ph} is the photorefractive absorption $\alpha_{ph} = s(N_D - N_A)$, $K^{-2}/k_B = T\mu/q\gamma N_A$, and σ_d is the dark conductivity. The time dependence with no dark conductivity has a I_0^{-1} dependence and for a Strontium Barium Niobate (SBN) crystal at $\lambda = 442$ nm, $\tau(s) \sim 0.1/I_0(\text{W/cm}^2)$.[57] The buildup of phase conjugation in time cannot be shorter than the space-charge buildup time shown here. The space-charge field grating is shifted from the intensity grating by 90°. The 90° phase shift comes about by considering diffusion only. With an applied electric field (and drift), or a photovoltaic field, the static space-charge field will be different from above. For instance, the phase shift will not be 90°, the amplitude of the space-charge field can be increased, and the time response will change.

It is not the space-charge field but the index grating that is important for describing the interaction of the material with light. With the aid of Eqs. (14.17), (14.20), and (14.22) the index change from two interfering beams is

$$\Delta n_{ij}(x) = n_{1ij} (\text{Re})e^{-i\varphi}\frac{\mathbf{A}_1 \cdot \mathbf{A}_2^*}{I_0} e^{ik_g x}\Big|\hat{\mathbf{e}}_x \qquad (14.24)$$

where φ is the phase of the grating shift (for a diffusion current only, $\varphi = \pi/2$), and $n_{1ij} = (cn\epsilon_0/2)\, \epsilon_i\epsilon_j r_{mk}E_{0k}{}^{sc}/n$. The magnitude of the index change can be as large as $\delta n/n \sim 10^{-4}$ and independent of intensity (for $I \gg I_d$). It is now possible, with four fields present in the counterpropagating geometry, to have these fields couple via four-wave mixing. The SVEA equations for plane waves can be written as

$$\frac{dA_i}{dz} = -\frac{\alpha}{2}A_i + i\left[\frac{2\pi\Delta n}{\lambda_0}\cos(\theta')\right]A_j \qquad (14.25)$$

where θ' is the half crossing angle inside the crystal. For crossing angles close to the z axis, $\cos(\theta') \sim 1$. Only standing-wave terms can become index gratings so the phase matched, coupled equations describing four-wave mixing for the same linear polarization in photorefractive media are[58]

$$\frac{dA_1}{dz} = -\frac{\alpha}{2}A_1 + i\frac{\pi}{\lambda_0}\left[\frac{n_1}{I_0}e^{-i\varphi_1}(A_1A_4^* + A_2^*A_3)A_4\right.$$

$$\left. + \frac{n_2}{I_0}e^{i\varphi_2}(A_1A_3^* + A_2^*A_4)A_3 + \frac{n_3}{I_0}e^{i\varphi_3}(A_1A_2^*)A_2\right]$$

$$\frac{dA_2}{dz} = +\frac{\alpha}{2}A_2 - i\frac{\pi}{\lambda_0}\left[\frac{n_1}{I_0}e^{i\varphi_1}(A_1^*A_4 + A_2A_3^*)A_3\right.$$

$$\left. - \frac{n_2}{I_0}e^{-i\varphi_2}(A_1^*A_3 + A_2A_4^*)A_4 - \frac{n_3}{I_0}e^{-i\varphi_3}(A_1^*A_2)A_1\right]$$

$$(14.26)$$

$$\frac{dA_3}{dz} = +\frac{\alpha}{2} A_3 - i \frac{\pi}{\lambda_0} \left[\frac{n_1}{I_0} e^{-i\varphi_1} (A_1 A_4^* + A_2^* A_3) A_2 \right.$$

$$\left. - \frac{n_2}{I_0} e^{-i\varphi_2} (A_1^* A_3 + A_2 A_4^*) A_1 - \frac{n_4}{I_0} e^{-i\varphi_4} (A_4^* A_3) A_4 \right]$$

$$\frac{dA_4}{dz} = -\frac{\alpha}{2} A_4 + i \frac{\pi}{\lambda_0} \left[\frac{n_1}{I_0} e^{i\varphi_1} (A_1^* A_4 + A_2 A_3^*) A_1 \right.$$

$$\left. + \frac{n_2}{I_0} e^{i\varphi_2} (A_1 A_3^* + A_2^* A_4) A_2 + \frac{n_4}{I_0} e^{i\varphi_4} (A_4 A_3^*) A_3 \right]$$

Beam 1 is a pump beam traveling in the $+z$ direction; beam 2 is a pump beam traveling in the $-z$ direction. Beam 3 is the phase conjugate beam traveling in the $-z$ direction, and beam 4 is the probe beam traveling in the $+z$ direction. In order to simplify these equations one grating can be selected out, say n_1. This is possible experimentally by adjusting the coherence or polarization so only beams 1 and 4, and/or beams 2 and 3 can write a grating. This grating is known as the transmission grating. The simplified, coupled equations can be solved assuming undepleted pumps and no absorption[59] to yield a phase conjugate reflectivity

$$R = \left| \frac{A_3(0)}{A_4^*(0)} \right|^2 = \left| \frac{\sinh^2(gL/2)}{\cosh^2[(gL/2) + \frac{1}{2} \ln(r)]} \right| \tag{14.27}$$

where r is the pump ratio $r = I_2(L)/I_1(0)$, and the gain is $g = i\pi n_1 e^{-i\varphi_1}/\lambda_0$. If the pump ratio is 1 and the photorefractive grating is local, $\varphi_1 = 0$, then the reflectivity is $R = \tan^2(\pi n_1 L/2\lambda_0)$, as was derived in the four-wave mixing from $\chi^{(3)}$ suscep-tibilities. The difference is that $3(\mu_0/\epsilon_0)^{1/2} \chi_{\text{mks}}^{(3)} (I_{10} I_{2L})^{1/2}/20\pi n^2 \epsilon_0$, has been replaced with n_1. This indicates that the photorefractive gain coefficient is independent of the intensity, at least for $I_0 \gg I_d$. This introduces important differences in the behavior of these materials. For instance, the photorefractive gain saturates in most of the ferroelectric crystals with milliwatt beam powers. The saturated change in index at these beam powers can yield large gain coefficients so that large-reflectivity phase conjugation can occur with milliwatt beams. This is useful in that less expensive, cw lasers are often able to provide all the necessary beam power. In order to maximize the reflectivity with a phase shifted grating the pump beams should not be equal. Also, the maximum reflectivity for a phase shift of $\varphi = \pi/2$, which is the shift with no applied electric field, is finite so self-oscillation does not occur for this case.

If the writing beams are only beams 1 and 3, and/or beams 2 and 4, then the grating is referred to as a reflection grating. The only difference between these two types of gratings, in terms of the phase conjugate reflectivity, is that the gain term for the reflection grating has $-n_2 e^{i\varphi_2}$ instead of $+n_1 e^{-i\varphi_1}$.

The depleted pump solutions have been derived exactly and different reflectivities exist for the transmission and reflection gratings. The solutions involve solving algebraic equations with more than one root indicating multivalued reflectivities (as was found for the un-phase-shifted $\chi^{(3)}$ type four-wave mixing solutions).

Because of the nonlocal index grating some very interesting four-wave mixing geometries are possible with a single incident beam.[60] Notice that for these cases the reflectivity cannot exceed unity by conservation of energy and the solutions require the depleted pump equations. Figure 14.8a shows the two pump beams replaced with conventional mirrors aimed at one another. This geometry is called the passive phase conjugate mirror (PPCM). Initially, the pump intensity of beam

Passive Phase Conjugate Mirrors

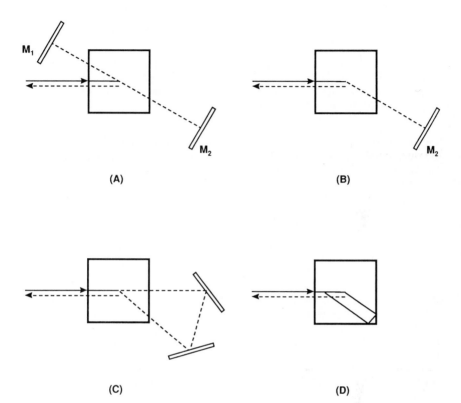

FIGURE 14.8 Four passive phase conjugate mirror geometries are shown. A linear passive phase conjugate mirror (LPPCM) is shown in (*a*). Two conventional mirrors aimed at each other are placed off axis on opposite ends of a photorefractive crystal. (*b*) shows the semilinear PPCM where only mirror M_2 remains. (*c*) shows the ring PPCM where two conventional mirrors redirect the incident beam back into and across the incident beam. (*d*) shows the two-interaction region, total internal reflection, self-pumped PCM (sometimes just called the self-pumped phase conjugate mirror). The incident beam has two regions of four-wave mixing with a corner cube reflector. Each of these configurations requires a single incident beam (probe) from which the pump #1 off M_1 and pump #2 off M_2 are derived.

1 off mirror 1 at $z = 0$ is zero, $I_1(0) = 0$, and the pump beam 2 off mirror 2 is zero at $z = L$, $I_2(L) = 0$ (no beams other than the probe beam are entering the crystal). In order for a phase conjugate reflectivity to exist with these initial conditions, the gain-length product must satisfy the following threshold condition, $M_1 M_2 = \exp[(g + g^*)L]$, where the mirror reflectivities are given as $M_1 = I_1(0)/I_2(0)$ and $M_2 = I_2(L)/I_1(L)$. The threshold condition is another way of stating that the gain in one round trip must equal or exceed the losses due to the mirror reflectivities. If both mirrors are assumed to be of unity reflectivity, then the threshold gain-length product goes to zero. The exponential is twice the real part of the gain coefficient. Therefore, it is required that the real part of the gain be

of sufficient magnitude in order to begin an oscillation and establish a conjugate beam. Only the nonlocal grating contributes to the real part of the gain so an unshifted grating will not build up a phase conjugate signal. Physically the situation can be described by the incident probe beam scattering radiation off imperfections into all directions. Some of this scattered radiation (which is usually much less intense than the incident beam) is the initial beam 1. Beam 1 will see gain as a result of the shifted grating, if the crystal is orientated properly, and bounces off mirror 2 to become beam 2. This process of beam amplification, known as two-beam coupling, occurs only with shifted gratings. Beam 2 strikes mirror 1 and the process is repeated until steady state is achieved.

An even more interesting geometry is to take away mirror 1. This resonator is known as the semilinear mirror and is depicted in Fig. 14.8b. It can be shown that only the transmission grating can build the conjugate signal and only if a seed beam between the crystal and mirror 2 is initially provided. Therefore, this resonator is not self-starting. In addition, the threshold gain-length product for this resonator, the semilinear mirror, is given by

$$gL = -(1 + M_2)^{1/2} \ln\left[\frac{(1 + M_2)^{1/2} - 1}{(1 + M_2)^{1/2} + 1}\right] \tag{14.28}$$

If mirror 2 is assumed to be unity, the threshold gain length becomes $gL = 2.49$.

Another popular resonator cavity is the ring cavity shown in Fig. 14.8c. An incident beam passes through the crystal, reflects off two different mirrors external to the crystal, and on the second reflection is reintroduced into the back of the crystal crossing the incident beam. It can be shown that this device is self-starting and also has a threshold given by

$$gL = -\left(\frac{M + 1}{M - 1}\right) \ln\left(\frac{M + 1}{2M}\right) \tag{14.29}$$

where M is the reflectivity of the external mirrors assumed to be equal. If the mirrors in the ring resonator are assumed to be unity, the threshold gain length is $gL = 1$.

A completely self-contained geometry was found experimentally and involves two interaction regions within the crystal.[61,62] Figure 14.8d shows the typical pattern of light intensity observed when looking down on the top of the crystal. The beam enters the crystal and breaks into two beams that are totally internally reflected from the corners back into the incident beam. There appear to be two regions within the crystal where four-wave mixing occurs, and these regions also appear to be coupled. One way to think about this resonator is that the first interaction region derives from a ring resonator and the second interaction region is a double-phase conjugate mirror (which is discussed below). The two-interaction-region model is not self-starting; it requires a seed beam as the semilinear mirror. The threshold gain length for this resonator has been found to be $gL = 4.68$. It has been observed that uniform illumination of a photorefractive crystal with an established two-interaction-region conjugate can have a larger reflectivity than without illumination.[63] This suggests competition between gratings within the crystal that reduces the optimum reflectivity. It should be mentioned that there are other models for the phase conjugation from a single incident beam that is corner cube reflected.[64,65]

Photorefractive crystals offer a unique ability to provide strong seed beams from the amplified scattering of many two-beam couplings that occur over the width of the incident beam. Typically the radiation couples energy toward ($BaTiO_3$) or away

(SBN) from the c axis of a uniaxial single crystal. (The direction of the c axis is defined as the arrow pointing from the positive poling voltage crystal face to the negative face.) This amplified scattered radiation is known as fanned light and can be significant.[66] For instance, the transmitted light can be depleted by OD = 3 because of this scattered radiation. Therefore, the geometries that require seed beams can be made to oscillate as long as the gain length is above threshold by careful attention to the direction of this fanned light.

Resonators that use total internal reflection are known as self-pumped phase conjugate mirrors. These oscillators still have a close analogy to a holographic process. An incident plane wave can be passed through an image bearing transparency. This image is then the incident beam in the crystal. The crystal will fan this image, effectively scrambling this light and causing the image to go back into plane waves. The fanned light strikes the corner of the crystal and reflects back onto itself. Therefore, the pump beams are the fanned light and the probe is the image beam. The conjugate becomes the Bragg scattered pump or fanned light.

There are many other geometries that can be observed with photorefractive crystals mainly owing to the large gain lengths available with milliwatt laser beams. For instance, the gain-length product in transparent Ce-doped SBN with extraordinary light is typically 10 without optimization.[67] Some of the latest research in this area is the incoherent double-phase conjugate mirrors. These resonators involve two or more sources that are incoherent to each other but are themselves coherent sources and of the same wavelength, for instance, two He-Cd lasers. These lasers can be aimed into a single crystal from opposite sides, producing two-phase conjugate beams. If one beam is turned off, the other beam also decays away but there is no steady-state cross talk between the two beams in thick crystals. To date, there are several different geometries,[68] but they all seem to operate with the same underlying principle. For instance, the bridge conjugator shown in Fig. 14.9 fans the two incident beams into each other within the crystal. This particular geometry works well for crystals with large r_{33} electro-optic coefficients, such as SBN. As a result, many gratings are written, but owing to the incoherence of the different sources, most of these gratings are erased within a short time. However, there is one direction of scattering that allows a grating to build up because both sources will cooperate to build the same grating. This grating is further enhanced by further scattering and is reinforced. Four-wave mixing then occurs over at least two regions, which depending on the fanning, may extend over a large portion of the crystal. These two regions contain the incident beam, the scattered beam, and the scattered beam from the second source from which the phase conjugate is derived. These conjugators appear as ring resonators with twice the interaction length. Therefore, the threshold for these resonators should be $gL_{\text{eff}} = g2L = 2$ (where L is the length of one interaction region), with perfect feedback.[69]

14.2.3 Stimulated Scattering[70,71] (Gaussian Units)

Phase conjugation from stimulated scattering was the first nonlinear process to generate a phase conjugate beam. Stimulated scattering is a particularly simple way to generate a conjugate beam but usually requires a large power laser. A single beam from the laser is directed into the medium (usually focused) which may be almost any type of material. The backscattered beams usually become a strong phase conjugate above some threshold intensity. Figure 14.10 shows a typical stimulated scattering phase conjugate geometry. Because there is only one incident beam, the reflectivity cannot exceed unity for these processes. The physical process

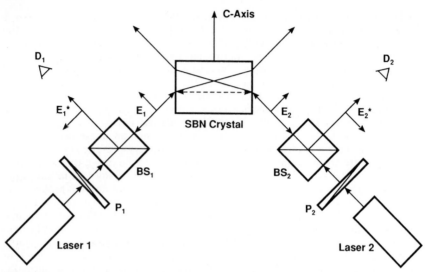

FIGURE 14.9 This shows the experimental configuration for the bridge double-phase conjugate mirror in an SBN crystal which fans opposite to the c axis. Two separate laser sources, laser 1 and laser 2, of the same wavelength pass through a polarizer P and a beamsplitter BS and then enter the crystal and fan into each other (dotted line). A cooperative grating is built up and incident beam 1 produces a phase conjugate shown reflecting off the beam splitter BS_1 and into the detector D_1. The same also occurs for beam 2, and a phase conjugate is recorded at detector D_2. If beam 1(2) is interrupted between the laser and BS, beam 2(1) will erase the grating that has been established and the phase conjugate of beam 1(2) will decay with a typical photorefractive time response while pc2 decays instantaneously.

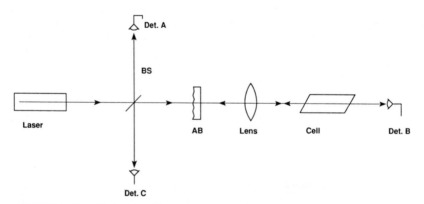

FIGURE 14.10 This is a typical experimental configuration for stimulated scattering. The incident beam passes through a beamsplitter where the reflected part of the incident beam is sampled by detector A. The beam then passes through a phase aberration plate AB and onto a lens. The lens focuses the incident beam into a cell. The cell is filled with a nonlinear medium from which a phase conjugate stimulated scattered beam builds up. Detector B monitors the forward-scattered light and/or the incident beam. The phase conjugate beam passes back through the phase distortion plate, and the phase is restored to the original incident beam at that point. The phase conjugate beam is sampled from reflection off the beamsplitter by detection at detector C. The phase conjugate reflectivity is simply the ratio of detector C to detector A.

of stimulated scattering is discussed next before discussion of how this process leads to a phase conjugate beam.

Stimulated scattering is always present in scattered radiation but is usually very weak. Spontaneous scattering usually dominates at low intensities, but at high intensities stimulated scattering can be significant. Stimulated scattering is the process that describes an excited material state becoming deexcited because of the presence of a real photon with the subsequent increase by one in that photon field (same k and polarization). Elastic scattering leads to stimulated Rayleigh scattering and inelastic scattering can lead to stimulated Brillouin or Raman scattering.

The scattering process in the photon picture is proportional to $(n + 1)$ where n is the photon number. The photon number generates the stimulated part of the scattering and the $+1$ is related to the spontaneous scattering. The photon number is related to the product of two electric fields which is proportional to the intensity. Therefore, the scattered beam has an intensity-dependent term associated with the stimulated process. Because many photons are involved in these processes, a classical approach can be taken. This can be written as a third-order process from the polarization vector for the generated field as $\mathbf{P}(\omega_{new}) = \chi^{(3)}: \mathbf{E}(\omega_i)\,\mathbf{E}(-\omega_i)^*\,\mathbf{E}(\omega_{new})$, where ω_i is the incident-beam angular frequency. The stimulated scattering process will then involve a gain in the scattered (generated) frequency that is intensity-dependent. The following analysis will only consider inelastic processes; however, even so-called elastic processes have a bandwidth from which different frequencies could mix.

The scattering process begins when two separate beams (the incident laser beam and the generated stimulated scattered beam) interfere within some material that responds by changing the index of refraction. If the index modulation is 90° out of phase with the intensity modulation, the amplitudes of the two beams are coupled. This process is known as two-beam coupling and involves the simultaneous solution to three equations, two optical frequency electric field propagation equations and a material-light interaction equation. This is the same two-beam coupling, requiring a nonlocal index grating, that was important in the photorefractive crystals. The nonlocal index grating for most materials, other than photorefractive crystals, originates from a noninstantaneous material response to a moving grating produced by the interference of two different frequencies. Assuming SVEA and no transverse components (plane waves) these coupled equations can be written as[72]

$$\left(\frac{d}{dz} + \frac{\alpha}{2}\right) A_1 + \frac{n}{c}\frac{dA_1}{dt} = i\gamma_1 \Delta n A_2\, e^{i\Delta k z}\, e^{i\Delta \omega t}$$

$$\left(\frac{d}{dz} + \frac{\alpha}{2}\right) A_2 + \frac{n}{c}\frac{dA_2}{dt} = i\gamma_2 \Delta n A_1\, e^{-i\Delta k z}\, e^{-i\Delta \omega t}$$

$$\frac{d^2 \Delta\Lambda}{dt^2} + R\frac{d\Delta\Lambda}{dt} + q^2 \Delta\Lambda = \Phi A_1 A_2^*\, e^{-i\Delta k z}\, e^{-i\Delta \omega t} \qquad (14.30)$$

$\Delta k = k_2 - k_1$, $\Delta\omega = \omega_1 - \omega_2$, α is the absorption coefficient, E_T (the total electric field) $= 1/2\, A_1 e^{i(k_1 z - \omega_1 t)} + 1/2\, A_2 e^{i(k_2 z - \omega_2 t)} +$ c.c. for an incident beam E_1 and a scattered beam E_2, γ is a constant given by

$$\gamma_{1,2} = 2\pi(\lambda_{01,2}\cos\theta_{1,2})^{-1} \qquad (14.31)$$

where θ is the angle the field makes with the propagation axis (z axis). $\Delta\Lambda$ is some parameter in the material modulated by the interfering fields (which is not stationary unless $\omega_2 = \omega_1$) and can be related to the change in the complex index of refraction by $\Delta n = (\partial n/\partial\Lambda)\Delta\Lambda = \text{Re}[12\pi\chi^{(3)}\mathbf{E}_1\cdot\mathbf{E}_2^*]$ (for linear polarization),[73] q is the angular frequency of the material wave, R is some resistance term, and Φ is some material interference constant.

As shown above, the change in index can be determined from a knowledge of the material modulation factor $\Delta\Lambda$. The material-induced wave amplitude can be found from the solution to the material equation. The solution is

$$\Delta\Lambda = \Phi A_1 A_2^* \, e^{-i\Delta kz} \, e^{-i\Delta\omega t} \, [q^2 - (\omega_2 - \omega_1)^2 + i(\omega_2 - \omega_1)R]^{-1} \quad (14.32)$$

The resistance term R is the Lorentzian linewidth full-width half-maximum (FWHM) of the generated, scattered wave. When solving the coupled wave equations, it is necessary to Bragg match, or phase match, which requires the complex conjugate of $\Delta\Lambda$ for the SVEA equation for field 2. The steady-state, plane-wave, SVEA-coupled amplitude equations without absorption can be rewritten as

$$\frac{dI_1}{dz} = 8\pi \frac{dn}{d\Lambda} \frac{\gamma_1\Phi\Delta\omega R \, I_1 I_2}{cn_2[(q^2 - \Delta\omega^2)^2 + \Delta\omega^2 R^2]}$$

$$\frac{dI_2}{dz} = -8\pi \frac{dn}{d\Lambda} \frac{\gamma_2\Phi\Delta\omega R \, I_1 I_2}{cn_1[(q^2 - \Delta\omega^2)^2 + \Delta\omega^2 R^2]} \quad (14.33)$$

where only the imaginary part of Λ contributes to the change in amplitude and $\Delta\omega = \omega_2 - \omega_1$. The exact solutions to these equations can be found in Ref. 74. If the incident beam is not depleted much, I_1 remains relatively constant and I_2 with absorption becomes

$$I_2(z) = I_2(0)e^{-\alpha z} \exp(-8\pi(dn/d\Lambda)\gamma_2\Phi\Delta\omega RzI_1\{cn_1[(q^2 - \Delta\omega^2)^2 + \Delta\omega^2 R^2]\}^{-1})$$

$$= I_2(0) \exp[(g - \alpha)z] \quad (14.34)$$

and at line center or $q = \Delta\omega$,

$$I_2(z) = I_2(0)e^{-\alpha z} \exp[-8\pi(dn/d\Lambda)\gamma_2\Phi I_1 z/cn_1\Delta\omega R] \quad (14.35)$$

If $\omega_2 < \omega_1$, field 2 has exponential gain with the incident laser intensity I_1, at least until pump depletion occurs. Notice that at line center the gain is proportional to R^{-1}, which is also related to the time response of the medium τ. The generation of frequencies less than the incident frequency is known as Stokes emission. This equation shows there is an initial exponential gain in intensity in the positive z direction of the Stokes frequency and a corresponding exponential decrease in the anti-Stokes frequency ($\omega_2 > \omega_1$). However, generation of stimulated Stokes radiation is often accompanied by anti-Stokes radiation.[75] The reason for the emergence of anti-Stokes radiation is the presence of four-wave mixing and stimulated scattering coupling to the Stokes emission.

If the incident intensity is gradually increased the stimulated scattering is seen to rise sharply above some threshold incident intensity and then level off at some saturation intensity where the undepleted incident beam approximation is certainly not valid. This transition from undepleted to depleted solutions occurs at the threshold laser intensity. This threshold can be understood from the following analysis. The Stokes intensity can be written as $I_S(z) = I_S(0)\exp[(g - \alpha)L]$, where L is the length of the material (or the focal volume length), g is the gain, α is the sum of the losses, and $I_S(0)$ is the intensity of scattered Stokes radiation initially present from spontaneous emission. For Rayleigh scatter of Stokes emission into the focal volume, the initial Stokes beam is $I_S(0) = I_L(dS/d\Omega)\Delta\Omega\, L$, where I_L is the laser intensity present, $dS/d\Omega$ is the probability for Rayleigh spontaneous Stokes scattering per unit length per steradian ($\sim 10^{-7}\ cm^{-1}\ r^{-1}$), and L is the confocal length. For significant buildup, $I_S \sim I_L$, or $1 \sim (dS/d\Omega)\Delta\Omega\, L\, \exp[(g - \alpha)L]$, where $\Delta\Omega$ is the angle over which the effective length is the largest ($\sim 10^{-4}$ to 10^{-6} sr). Typically, the gain-length product needs to be on the order of approximately 30 (for the stimulated Stokes intensity) to achieve the threshold condition. The threshold intensity can be found more accurately in cgs units from

$$I_{Lt} \sim \left\{ -\frac{1}{L} ln\left[\frac{dS}{d\Omega} \Delta\Omega L \right] + \alpha \right\} \left\{ \frac{8\pi(dn/d\Lambda)\Phi\Delta\omega R\gamma_2}{cn_1[(q^2 - \Delta\omega^2)^2 + \Delta\omega^2 R^2]} \right\}^{-1} \quad (14.36)$$

Notice that the threshold intensity is inversely proportional to the linewidth at line center. Deviations from the threshold intensity derived above can occur if the initial Stokes radiation is increased within the focal volume owing to enhanced feedback in the backward direction from Rayleigh scattering (for thick materials) or reflection from the walls (for a thin material).[76]

The buildup of stimulated scattering in time can be determined by solving the coupled equations with the time derivative included. According to Zel'dovich et al., the undepleted pump solution for the Stokes intensity is proportional to the exponent of $-R(t_2 - t_1) + [4Rg'L \int_{t_1}^{t_2} I_L(t')dt]^{1/2}$, where g' is the modified gain (gain without the incident laser intensity I_L) and t_1 and t_2 are the instances of time to be investigated. In order to have a stimulated scatter buildup comparable to the incident laser, the Stokes intensity must be at or above threshold. This requires the exponential to reach approximately 30 before the pulse, $\Delta t = t_p$, is over. So the above expression must be equal to 30 or more during the time interval of the pulsewidth. For long pulses such that $t_{pc} = t_p R = t_p/\tau \geq 30$, the gain length can be 30, the steady-state threshold condition. However, for shorter pulses, $t_p/\tau < 30$, the gain length needs to be larger than 30 to achieve the threshold during the pulse. When t_{pc} is less than 30, the condition on $g'IL$ is $g'IL > (15 + Rt_p/2)^2/Rt_p$ for a square pulse. For instance, if $Rt_p = 10$, $g'IL \sim 4$ or 1.33 times the steady-state condition.

With the coherent Stokes and coherent anti-Stokes radiation occurring simultaneously, it is often possible to observe higher-order terms. The higher-order Stokes terms emerge when material in the ground-state absorbs the Stokes field. The second-order Stokes field is then $\omega_s - \omega_v = \omega_i - 2\omega_v$. The same thing happens for the anti-Stokes field.

To summarize, stimulated scattering has been shown to arise in materials that have an intensity-dependent index of refraction. The scattered frequency will be different from the incident frequency, and this new frequency will have gain throughout the material. Many of the properties of stimulated scattering depend on the resistance of the material (or inertia) to the incident beam. The Lorentzian shaped linewidth (FWHM) of the generated beam is given by R. The gain at line center was also found to depend inversely on R. Finally, the buildup time of the stimulated beam was found to be inversely proportional to the resistance R. Therefore, a large resistance term can generate a large linewidth with a small gain and a short response time.

Stimulated scattering has been shown to be a generated beam from a single incident beam, but it has not yet been shown that it is directed back at the source or that the spatial components have been phase conjugated. In order to do that, it is necessary to rewrite the SVEA equations without the assumption of plane waves or

$$\frac{\nabla_{\perp}^2 A_s}{i2k_s} + \frac{dA_s}{dz} = \frac{g'}{2} |A_L(r,z)|^2 A_s \tag{14.37}$$

where g' is a modified gain, $\nabla_{\perp}^2$ is the Laplacian with respect to the transverse components, and r is the radius in cylindrical coordinates. With the Laplacian it is now possible to describe the propagation of the beam waist and divergence. This equation assumes no absorption, and the laser field is assumed to be undepleted. With the definition of power given as $P_s = \int |A_s(r,z)|^2 d^2r$, the above equation can be rewritten as[77]

$$\frac{dP_s}{dz} = g(z)P_s(z) \tag{14.38}$$

where the new gain coefficient is given as

$$g(z) = G[\int |A_L(r,z)|^2 |A_s(r,z)|^2 d^2r][\int |A_s(r,z)|^2 d^2r]^{-1} \tag{14.39}$$

It is assumed the transverse field is zero at r equal to infinity. Now it is seen that the gain is sensitive to the overlap, or correlation, of the laser intensity and the Stokes intensity in space. There are many possible amplitudes and directions of the generated Stokes beams, and spontaneous scattering will generate most of them. The formula above suggests that only those beams generated with the largest transverse correlation over the interaction length will have the largest gain and produce the highest Stokes beam intensity. There is a full range of possible incident intensity spatial patterns. For uncomplicated patterns (a few transverse modes or image elements) that remain uncomplicated in the high-intensity region of a material the highest correlation will occur for beams propagating along the path of the incident beam. Stimulated scattering will then occur in the forward and backward directions. These directions are also favored because of the long interaction length. If the incident intensity pattern is complicated (multimode) and/or changes its spatial profile significantly in a deterministic way throughout the high-intensity region, then the highest correlation will occur for a backward-propagating phase conjugate beam. This is the only beam that can exactly retrace the incident beam so as to maintain a high degree of spatial correlation over the longest path length.

The length of a material can play a significant role in determining if stimulated scattering will be a phase conjugate beam. A material that is of similar length or shorter than the confocal beam length will not have good discrimination of the phase conjugate beam over noise beams as would a material which is much longer than the confocal beam length. An optical waveguide can scramble the incident beam modes. A waveguide is ideal for the observation of stimulated scattering phase conjugation not only because of the mode scrambling but also because the beams are maintained at a high intensity over a long interaction length. In order to phase conjugate an image with a waveguide the guide must at least have as many modes as the resolution elements of the image. The number of modes that must be present for a phase conjugate beam to appear has been determined.[78] There must be no fewer than 10 modes present and there cannot be more than 10^4 to 10^7 modes in steady state. The limitation of 10 modes is due to the number of modes needed for adequate discrimination of the phase conjugate beam over background noise beams. The upper limit on the number of modes is due to the ability of the noise beams at low gain (which are always present and taking energy from the incident beam) to significantly deplete the incident intensity.

The detailed theory of the derivation of phase conjugation from stimulated scattering is difficult (because of the need for a large number of modes) and has several versions.[79-81] Analytic solutions can be obtained under some simplifying assumptions; however, for most practical cases these assumptions are not accurate. Numerical solutions exist and begin with a complex incident transverse beam profile. The number of points in the transverse profile that adequately represent the incident beam is given as N. This will then require the simultaneous solution of $2N$ coupled nonlinear differential equations with $2N$ boundary conditions. These can be solved on a computer, but it takes long run times and requires a large computer memory. There is a three-dimensional solution that involves a perturbative technique that works well for many practical situations with less time and memory.[82]

Zel'dovich's original argument can be used here to support the premise of a phase conjugate return beam from stimulated scattering. Equation (14.39) can be rewritten as

$$g = \frac{G\langle E_L{}^* E_L E_s{}^* E_s\rangle}{\langle E_s{}^* E_s\rangle} \qquad (14.40)$$

where $\langle \ \rangle$ is the correlation integral and E is the electric field. If the incident laser beam is aberrated so that it appears as a random speckle pattern, then statistical analysis can be used to describe the speckle. The use of Gaussian statistics, and in particular joint Gaussian statistics, allows the pairwise correlation of four laser fields to be written as

$$\langle E_L{}^* E_L E_s{}^* E_s\rangle = \langle E_L{}^* E_L\rangle\langle E_s{}^* E_s\rangle + \langle E_L{}^* E_s{}^*\rangle\langle E_L E_s\rangle + \langle E_L{}^* E_s\rangle\langle E_L E_s{}^*\rangle \qquad (14.41)$$

and usually $\langle E_L{}^* E_s{}^*\rangle = \langle E_L E_s\rangle = 0$ because they are uncorrelated. The case of the Stokes beam being the phase conjugate of the incident laser beam would then yield $g_c = 2G\langle I_L\rangle = g_c{}'\langle I_L\rangle$, where $g_c{}'$ is the modified conjugate gain. A Stokes beam that remains uncorrelated to the laser beam has $\langle E_{L,s}{}^* E_{s,L}\rangle = 0$ so that $g_{uc} = G\langle I_L\rangle = g_{uc}{}'\langle I_L\rangle$, where $g_{uc}{}'$ is the uncorrelated modified gain. The correlated Stokes beam has a gain twice the uncorrelated beam, $g_c = 2g_{uc}$, so the phase conjugate beam is favored. The undepleted solutions for the two competing fields have the form $I_{uc} = I_{0uc}\exp(g_{uc}L)$, and $I_c = I_{0c}\exp(2g_{uc}L)$, where I_0 is the initial intensity. In order for the uncorrelated beam to have a similar Stokes intensity with the correlated beam, it is necessary for the initial scattering of the uncorrelated beam to be $I_{0uc} = I_{0c}\exp(g_{uc}L)$. For the threshold case where $2g_{uc}L = 30$, the initial uncorrelated noise intensity would need to be $I_{0uc} = I_{0c}e^{15}$ to compete with the conjugate signal.

In order to suppress the noise beams it is desirable to have a small angle over which the uncorrelated beams can have gain. The initial scattering can be estimated by assuming spontaneous scattering into 4π sr, but the uncorrelated beam has gain only over some smaller angle $\Delta\theta$ and the correlated beam only at the diffraction solid angle $\sim\lambda^2/S$, where S is the area of the acceptance aperture. The condition for competition of uncorrelated beams to correlated beams then becomes $\Delta\theta \sim \lambda^2 e^{15}/S$. The number of different transverse modes can be given by $\Delta\theta S/\lambda^2$ which must be much less than $e^{15} \sim 3 \times 10^6$ for conjugate discrimination. The angle for gain in the media can be found from the deviation of the Stokes beam from the laser beam for focused beams and from diaphragms that define the beam propagation path. In a waveguide the defining aperture S is related to the deviation from linear polarization by reflection off the guide walls or the greater losses at larger angles. The aperture S could be the waveguide opening or the lens area and this could be adjusted to change the uncorrelated, noise beam contribution. A larger aperture introduces more uncorrelated intensity and reduces the phase conjugate intensity but also reduces the probability of damage due to high intensities. A large aperture can be used if a high-intensity seed beam is introduced so that the buildup of the conjugate is not from noise, but from this seed beam.[83]

The polarization properties of phase conjugate beams from stimulated scattering are like ordinary mirrors. For instance, a right-handed polarized beam will become left-handed just as an ordinary mirror will. This is in contrast to the phase conjugate generated from four-wave mixing. For linear polarization, the phase conjugate beam will have the same polarization as the linear polarization of the incident beam. In order to generate a phase conjugate beam from stimulated scattering, the incident beam must be polarized.[84]

Stimulated Scattering Mechanisms. There are two very common physical mechanisms that generate stimulated scattering, Brillouin and Raman scattering. Brillouin scattering can be thought of as the scattering of an incident photon $\mathbf{k}_i$ by an acoustic phonon $\mathbf{k}_a$ and redirected with a wavelength $\mathbf{k}_s$ where $\mathbf{k}_i - \mathbf{k}_s = \mathbf{k}_a$. Brillouin scattering differs from Raman scattering in that the induced vibrational mode within the material is an acoustic phonon rather than an optical phonon.[85] An acoustic vibrational mode has a much longer wavelength, so the energy removed by the material in the Brillouin scattering process is substantially less than for the Raman case. The frequency shift of the generated radiation is typically much smaller than a Raman-shifted frequency and on the order of 1 to 10 GHz. As a result $|\mathbf{k}_i| \sim |\mathbf{k}_s|$ so that the angle of scattering can be given as $\Delta k = 2k_i \sin(\Theta/2)$, or $\Theta = 2 \sin^{-1}(k_a/2k_1)$, where Θ is the angle between the incident and scattered beams. In order to describe stimulated Brillouin scattering (SBS), Eq. (14.30) can be modified so that Λ is the density ρ, $\Delta\Lambda \rightarrow$ density amplitude of acoustic wave or $\rho - \rho_0$ where ρ_0 is the mean density, $R \rightarrow$ Brillouin linewidth and related to the viscosity, $q \rightarrow$ the frequency of acoustic wave, and $\xi \rightarrow \Delta k^2\rho_0(\partial n^2/\partial\rho)/8\pi$. The resonant ($q = \Delta\omega$) gain becomes $g = 2(\partial n/\partial\rho)^2 2\pi\Delta k I_L n_1/c\lambda_{02}Rv_a$, where v_a is the acoustic velocity.

Stimulated Brillouin scattering occurs when the incident photon excites the material and a Stokes shifted photon deexcites the medium, generating two identical Stokes photons. In the classical picture of SBS, the square of the electric field changes the density of the material, which leads to an intensity-dependent index. In SBS the incident field creates acoustic waves of many different wavelengths and in many different directions from which the incident field scatters. The Doppler-shifted scattered wave interferes with the incident wave, and some acoustic waves at the velocity of sound in the material are reinforced by the beat frequency of this interference. Two-beam coupling can transfer energy from the incident beam into these scattered beams, resulting in the depletion of the incident beam. The gain length is largest along the longest length (directly forward or backward), but the sound relaxation is largest in the backward direction so the gratings developed from backscattering respond the most quickly (stimulated scattering buildup time is inversely proportional to the linewidth) and get the largest gain. Also, the gain is largest in the backward direction since $\Delta k|_{mx} = 2k$ when $\Theta = \pi$. The stimulated scattering equations can describe backscatter if the Stokes wave is $E_2 = A_2 \exp[-i(k_2z + \omega_2t)]$. A typical threshold value for SBS is 10 to 100 MW/cm^2.[86] The gain is typically 0.01 to 0.02 cm/MW, making the gain an order of magnitude larger than Raman gains.

Raman scattering is the relaxation radiation that is different from the initial frequency radiation by one or more phonons.[87] The shift in wavelength can occur over a range of values depending on the material, but these shifts tend to be large, typically 60 nm. Stimulated Raman scattering differs from spontaneous Raman scattering in that it is a coherent effect. A phase relationship is maintained between the absorbed incident photon and the scattered photon.

Stimulated Raman scattering (SRS) can be observed in many materials at 100 MW/cm^2.[88] The gain for typical materials may be 2×10^{-3} cm/MW. CS_2 has a fairly large gain at about 24×10^{-3} cm/MW. In order to observe a significant fraction of stimulated scattering the gain threshold requires $gL \sim 30$, or for a 1 cm interaction region, $I = 1.25$ GW/cm^2 in CS_2. If the interaction length was 10 cm instead, then the intensity would have been in the 100 MW/cm^2 range. For this reason a waveguide is often employed to maintain a high intensity over a long path length. SRS can be described by Eq. (14.30) if $\Lambda \rightarrow NX$, where N is the number of vibrational oscillators per volume and X is the normal vibrational coordinate $\xi \rightarrow N(\partial\alpha/\partial X)_0/4m$ and $R \rightarrow \Delta\nu/2\pi$, where $\Delta\nu$ is the spontaneous Raman linewidth.

Stimulated Raman scattering has been observed in both forward and backward directions. Phase conjugation from backscattered stimulated Raman scattering has been demonstrated in a waveguide configuration and with multimode incident beams.[89] Forward-stimulated Raman scattering from a multimode beam has also been shown to produce a poor conjugate beam (based on image restoration).[90] If the incident beam is not multimode or sent into a waveguide, stimulated Raman scattering is most often observed in the forward direction; however, this beam is not a conjugate beam.

SBS and stimulated Raman scattering can also occur as a result of self-focusing.[91] The self-focusing threshold is power-dependent. Therefore, self-focusing can occur before the scattering threshold is reached, given enough power. As the beam collapses the intensity becomes large and the threshold is reached. Actually it is much more complicated than this, but the main point is that experimental thresholds will not match theoretical predications if self-focusing is involved.[92] Achieving threshold in this way has some disadvantages for generating phase conjugate beams. For instance, the polarization and phase of the scattered beam can be altered so that the phase conjugate beam is not of good quality.[93]

In addition to acoustic gratings it is possible to stimulate many other types of gratings as well. SBS has a large gain coefficient but is still most often employed in a gas medium to avoid damage due to the high intensities needed. These other stimulated sources will need even higher intensities to go above threshold. Waveguides are helpful in achieving the gain length needed to approach threshold and should be used if possible. Stimulated thermal scattering is often seen with long pulses.[94] The gain for thermal scattering is usually much less than for Brillouin scattering. Stimulated Rayleigh scattering which has a small gain[95] and the larger gain Rayleigh wing scattering, due to the beating of closely separated frequencies, has been observed in liquids.[96,97] Stimulated scattering in photorefraction has been observed with and without seed noise.[98,99] All these scattering mechanisms, and the phase conjugates that can result, are not as common as stimulated Brillouin scattering because most have a smaller gain coefficient.

14.2.4 Adaptive Optics[100]

Adaptive optics is the correction of phase distortions by sampling the propagation path and properly adjusting the phase for the transmission or reflection optics of a signal beam. In this way a phase-corrected signal beam is the phase conjugate of the sampling probe beam. There are many possible ways this could be done. The adjustable optics could consist of a grid of movable mirrors,[101] membranes, or electro-optic elements. Spatial light modulators offer a variety of possible mechanisms and are being made fairly compact.[102] Unlike the earlier sections, this approach does not need to involve the incident beam directly in the generation of a phase conjugate signal. This allows the possibility of using different wavelengths for the sampling and propagation beams. It also allows much smaller intensities to be used, but a limit is imposed by the transmission of the phase-measuring devices. Figure 14.11 shows a generic scheme for sampling the medium with one source and then, with the aid of an adaptive optical device, sending a phase-corrected beam back along the identical path. The phase-corrected beam will not suffer spot size distortions due to propagating in an inhomogeneous index medium. Rather, the beam will propagate with a diffraction-limited spot size. There are some disadvantages to using this approach over the nonlinear optics approach. For example, adaptive optics works well only if the adjustable optics are much smaller than the spatial extent over which there is a significant phase change. In addition, if phase correction is desired through the atmosphere or through water, adjustments need to be made on the order of the phase coherence time. For visible light propagating down through the entire atmosphere, the adjustments need to be made on the order of milliseconds (1 to 0.1 kHz). The control of many individual elements makes these systems more complex, costly, and larger than the nonlinear approach. Adaptive optics has been used successfully for static as well as dynamic (propagation through the atmosphere) applications. For instance, astronomers can use stars of visual magnitude 8 or brighter to sample the atmosphere and provide correction to visible telescope lenses or mirrors as long as the sampling star image is within 2 seconds of the image of the interested star.[103] Another method that is free from the constraint of having a $< +8$ magnitude star in the field of view is to excite a thin layer of sodium atoms about 100 km high in the atmosphere with a strong

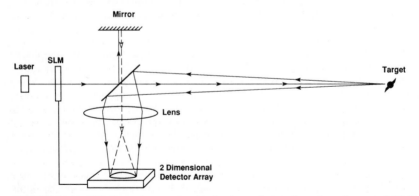

FIGURE 14.11 This figure shows a generic adaptive optical configuration using a spatial light modulator (SLM). A probe beam samples the medium between the source laser and the target. The return probe is split off by a beamsplitter at the receiver and focused on a detector array. Part of the incident probe beam is also reflected off the beamsplitter, then off a mirror and focused at the detector array. The interference of the two beams reveals the phase delay across the transverse beam profile, and this information is fed to the SLM. The SLM then delays the phase at the appropriate spatial position to phase compensate for the atmospheric distortion. Therefore, the beam after the SLM is a phase conjugate version of the returning probe beam.

laser beam. The atmospheric distortion due to turbulence is then measured by measuring the relative phase of the sodium fluorescence that has traveled back to the ground near the source laser. Phase correction of an incident plane wave from a star eliminates atmospheric turbulence effects and allows for diffraction-limited spot sizes. Phase correction through the atmosphere for laser sources could also be important for some military applications and is being pursued for the strategic defense initiative.[104]

14.3 APPLICATIONS[105]

The most obvious application of phase conjugation is to provide distortion correction. The propagation of phase conjugate beams through nonlinear and/or absorbing media has raised some questions regarding the ability to produce perfect phase conjugate beams. Many nonlinear mechanisms will yield a global phase change between the incident and conjugate beam; however, this phase difference is not usually important for beam propagation. Yariv[106] has recently shown theoretically that a linearly absorbing media can propagate a perfect conjugate if the losses depend only on z (the propagation direction). Perfect conjugation with propagation in media with a nonlinear index of refraction and/or absorption depends on the presence of gratings written by the incident and conjugate beams.

Distortion-free images possible with phase conjugation are usually desirable and critical for photolithography.[107] Typical optical imaging devices can never produce a "perfect" image because of diffraction. Without using very expensive optics to produce diffraction-limited resolution, optical aberrations, imperfect polishing, and misalignment will cause some degree of image distortion. In addition, nonlinear optical phase distortions and laser speckle, when they occur, will further reduce the image quality. For instance, it is often desired to amplify an image or a single transverse laser oscillator mode. The amplifier will introduce distortions on the incident beam for all the reasons mentioned above. If a phase conjugate mirror is placed after the amplifier and the image is reflected from it, as it passes back through the amplifier, these distortions can be removed. Amplifiers could consist of anything from the traditional pumped gain medium to photorefractive crystals that couple energy by way of two-beam coupling.[108,109] Second-harmonic generation also induces distortions which can be corrected with a double-pass geometry.[110]

An optical beam that must be transmitted large distances through the atmosphere and reflected back to the source will flicker (scintillation). Atmospheric turbulence will cause the reflected beam to wander about the desired path. A phase conjugate beam returning through the same turbulence will undo the phase distortions and be a perfect mirror.[111] This will occur even if there is thermal blooming caused by an intense laser source.[112] Images carried by optical fibers become severely distorted due to dispersion. A phase conjugate mirror at the end of a fiber can unscramble this distortion and present a distortion-free image back at the origin. However, a more useful application is to carry the image away from the origin. This can be done if there is an exactly matching fiber that the phase conjugate is sent through.[113]

The fidelity of a strong phase conjugate beam is excellent from most methods of generation. (A weak conjugate beam that is competing with noise will obviously have noise distortion.) Stimulated scattering produces frequency-shifted, nearly conjugate beams. Although not perfect conjugates, the backscattered beams are usually close enough to also produce excellent fidelity. High fidelity from wave-guided stimulated scattering requires a polarization preserving waveguide at low powers. Most distortions in the phase conjugated beam can be attributed to radiation from the image not being collected and sent to the phase conjugate mirror. Special care must be taken to collect as much of the image as possible. This is especially true when taking the Fourier transform of an image. The higher-order Fourier transform spatial modes which carry high spatial frequency information, so necessary for good edge quality, can easily miss the phase conjugate mirror. Some nonlinear phase shifts, such as introduced by self-focusing, can cause imperfect conjugate return signals and lead to image degradation. Saturation of the nonlinear material, slow response time of the nonlinearity, and large diffusion of the gratings can also lead to poor image quality.

High-power solid-state lasers suffer from thermal effects which degrade performance. Thermal lensing reduces the laser efficiency and can be somewhat compensated for with a lens of opposite effect placed within the laser cavity. However, this lens is not dynamic or able to exactly compensate the thermal lens. As a result the thermal problem is somewhat reduced for lower-power lasers but still present at high powers. By replacing the back mirror of a laser resonator with a phase conjugate mirror, this problem can be corrected along with the addition of some other desirable laser resonator features.[114,115] The phase conjugate mirror will correct the dynamic thermal distortions in the cavity as well as any other distortions present. Some of the other features of this laser cavity include the fact that it will not possess longitudinal modes that depend on the laser cavity length. A phase conjugate resonator can support any frequency that falls within the bandwidth and the resonator length itself. This means if the cavity length changes, the output frequency will remain the same. Also, unlike a conventional resonator, the phase conjugate resonator has a stable configuration regardless of the mirror curvature or the cavity length.

The fact that the phase conjugate beam is retroreflected makes it attractive as a beacon.[116] Since most phase conjugate mirrors have some acceptance angle, the laser source and beacon could be in relative motion. Therefore, the conjugate mirror can be accurately tracked, since any portion of the incident beam is reflected and concentrated back onto the source. Another tracking scheme involves a laser that illuminates some distant object as in Fig. 14.12. Scattered light from the object is amplified and sent to a conjugate mirror, reamplified, and this amplified light will then converge onto the scattering object. When the phase conjugate reflectivity becomes larger than 1, the phase conjugate mirror can couple to some reflecting surface to form a phase conjugate resonator. This means the finder laser can be removed after an oscillation builds up and the reflecting surface will automatically be tracked by the phase conjugate mirror. In order to track this object, the phase conjugate reflectivity R_{pc} times the reflecting surface reflectivity R must be greater than 1, $R_{pc} \times R > 1$. Also, the distance moved between scatter and conjugate return must be less than the original spot size. Another type of tracking, the novelty tracking filter, tracks objects only if there is some difference between a reference pattern and an image. Phase conjugation can be used to look for phase difference information.[117]

Phase conjugation can be used in laser communications. A beam incident on a phase conjugate mirror can be encoded with temporal information.[118] By modulating one of the pump beams in degenerate four-wave mixing or applying a uniformly modulated electric field across a photorefractive crystal, the modulation can be transferred to the conjugate beam. The double-phase conjugate mirrors of photorefractive crystals can also transmit modulated beams to one another. One beam can be modulated and the other will receive the signal. These communication schemes can have a fairly large field of view and strong, high-quality return signals.

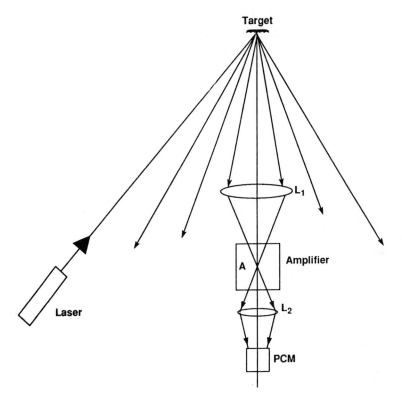

FIGURE 14.12   A phase conjugate tracking scheme that tracks the object target. A laser illuminates the object, which then scatters the laser radiation. Some fraction of this laser radiation is collected by lens L_1 and amplified in A. The amplified light is then focused by lens L_2 into a phase conjugate mirror. The phase conjugate beam will then be reamplified and converge back onto the source, thereby tracking the object target. If the phase conjugate reflectivity times the reflectivity of the object is greater than 1, the laser can be eliminated because the object and the phase conjugate mirror form a resonator.

The ability of more than one incident beam to be connected to the conjugate mirror allows for an optical relay station. This may be most appropriate for the double-phase conjugate mirrors if a good way of connecting one at a time (beam steering) is found.

Since phase conjugation is closely related to holography (if the generation mechanism is degenerate four-wave mixing), it should be possible to generate a moving three-dimensional hologram in real time. The method of generation would essentially be the same as the tracking scheme in Fig. 14.12. This would have a wide range of possible applications. It should also be possible to produce these holograms in color by using a multifrequency laser source (such as an argon-ion and a krypton laser), separating the wavelengths and writing the images at separate locations within the material. Multiwavelength (color) phase conjugation has been demonstrated in self-pumped phase conjugate mirrors.[119]

Manipulation of images such as image subtraction,[120,121] exclusive or,[122] correlation, and convolution are all possible with phase conjugation. Subtraction can be used to produce the negative of an image.[123] Correlation and convolution can be achieved from DFVM by imaging the Fourier transforms of three images with three separate input beams in a nonlinear material. The conjugate beam when propagated the distance of the objects, 2 times the focal length, yields the field E_{pc} $\sim \langle (E_1 * E_2)E_p \rangle$ where * denotes the convolution and $\langle \ \rangle$ the correlation.[124] Using a delta function for one of the pump beams and a series of point images for the other pump beam the conjugate is the probe image placed at each point image. If the point images become a solid image and the probe is a solid object, the conjugate consists of point images of any solid object that is the same in the pump and probe beams.[125] This last operation could be used as an image recognition scheme. In actual practice many image recognition schemes employ an optical correlator that involves a matched filter placed at the Fourier transform plane (FTP) of an imaging system.[126] The matched filter is a spatial light modulator capable of matching the phase, amplitude, or both to some object held in a memory. A second lens is placed at a focal distance f away from the FTP which focuses to a detector array that measures the intensity (correlation) and position at f behind the second lens. Good correlation peaks (recognition) are observed when the image and filter match phase and amplitude. To understand the phase aspect of the correlation peak consider a phase front that passes through the FTP with some curvature. The second lens will not be able to focus this image well and the correlation peak will not be large. However, a phase front that is exactly compensated (phase conjugated) by the match filter will have a plane wave curvature. This phase front will focus on the detector plane, as a result of the second lens, leading to a high degree of correlation.

A more versatile image recognition scheme than the matched filter device is the use of neural networks.[127,128] Neural networks try to emulate the function of biological neurons, which are the basic building blocks for the mental processing of information. The neuron itself has an input, an output, feedback, gain, and thresholding, and is highly connected. It is believed that the neural networks allow a greater amount of parallel processing which is important for handling large amounts of information quickly. Optics offers a unique capability in this area because information can be processed in parallel quite naturally. Optical neural networks with feedback and thresholding have been designed and built to perform the task of image recognition and associative memory.[129,130]

One version of an optical neural network consists of an input image that is directed onto a hologram that has previously recorded overlapping images written at different angles as shown in Fig. 14.13. The incident image may be similar to, but not an exact copy of, one of the prerecorded images. For instance, the incident image may be an incomplete version of one of the prerecorded images or it may be distorted. In any case this incident image scatters off the hologram and into the directions of the recorded images. The amount of light scattered into these directions is proportional to the likeness of the incident image to one of the recorded images. Next, the scattered light is directed onto a thresholding device such as a self-pumped phase conjugate mirror. The most intense scattered beams build up a conjugate signal and get redirected back to the hologram which bends the light back to the source. A mirror is used to feed this light back into the system. This feedback increases the intensity at the phase conjugate mirror and reinforces the conjugation of the strongest-intensity scattered beams. Eventually most of the optical energy is directed into the scattered beams that had the initial high scattering because that phase conjugate signal is most strongly established. The output image will then be the prerecorded hologram that most closely resembles the input image. Hence, the image-recognition optical neural network depends on phase conjugation to provide a flexible, automatic threshold.[131–133]

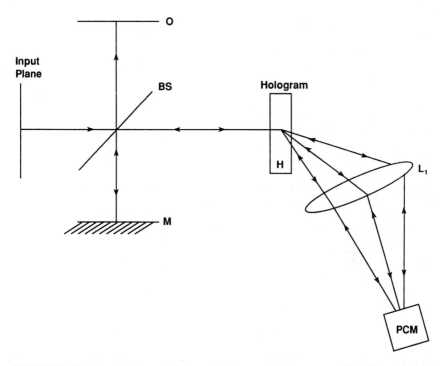

FIGURE 14.13 A typical schematic diagram depicting associative memory. The hologram labeled *H* contains a number of overlapping images, each written at different angles before appearing in this optical arrangement. The hologram is presented with an input image. This input image may or may not be a previously stored image. The input image may also be only part of a stored image. The hologram will scatter the input image into the various angles with which it was written. These scattered beams are collected by a lens L_1 and focused onto a phase conjugate mirror. The scattered components that have the largest intensity will phase conjugate and return to the hologram and back to a beamsplitter *BS*. The beamsplitter sends the reflected radiation to a mirror, forming a cavity. The larger the correlation of the input image to the recorded image, the larger the incident intensity scattered at that previously recorded image writing angle. Because of the thresholding of the conjugator only the most correlated images will resonate and that image can be viewed at plane *O*. The image at *O* will be the recorded image that most resembles the input image.

There are many other applications for phase conjugation including laser gyros,[134] optical limiting,[135] temporal pulse compression,[136] temporal pulse reversal,[137] spatial light modulators,[138] and bistable switches.[139] The applications and potential applications of phase conjugation continue to grow, and it has already spread into almost all areas of modern optics. It remains to be seen how commercially successful phase conjugation will become. It was discovered as a scientific curiosity, but its potential applications in so many different optical areas seem to assure its future success.

14.4 REFERENCES

1. B. Ya. Zel'dovich, V. I. Popovichev, V. V. Ragulsky, and F. S. Faizullov, "On Relationship between Wavefronts of Reflected and Exciting Radiation in Stimulated Brillouin Scattering," *JETP Lett.*, vol. 15, p. 109, 1972.

2. D. M. Pepper, "Nonlinear Optical Phase Conjugation," *Opt. Eng.*, vol. 21, p. 155, 1982.

3. R. McGraw, D. N. Rogovin, W. W. Ho, B. Bobbs, R. Shih, and H. R. Fetterman, "Nonlinear Response of a Suspension Medium to Millimeter-Wavelength Radiation," *Phys. Rev. Lett.*, vol. 61, p. 943, 1988.

4. P. L. Shkolnikov and A. E. Kaplan, "Feasibility of X-Ray Resonant Nonlinear Effects in Plasmas," *Opt. Lett.*, vol. 16, p. 1153, 1991.

5. P. W. Milonni, E. J. Bochove, and R. J. Cook, "Quantum Theory of Spontaneous Emission and Excitation Near a Phase-Conjugating Mirror," *J. Opt. Soc. Am. B.*, vol. 6, p. 1932, 1989.

6. N. F. Andreev, V. I. Bespalov, M. A. Dvoretsky, and G. A. Pasmanik, "Phase Conjugation of Single Photons," *IEEE J. Quantum Electron.*, vol. QE-25, p. 346, 1989.

7. A. K. Ghatak and K. Thyagarajan, *Contemporary Optics*, Plenum Press, New York, 1978, chap. 1.

8. D. M. Pepper, D. A. Rockwell, and H. W. Bruesselbach, "Phase Conjugation: Reversing Laser Aberrations," *Photonics Spectra*, August 1986.

9. C. R. Giuliano, "Applications of Optical Phase Conjugation," *Phys. Today*, vol. 27, April 1981.

10. D. M. Pepper, "Applications of Optical Phase Conjugation," *Sci. Am.*, vol. 74, January 1986.

11. N. C. Griffen and C. V. Heer, "Focusing and Phase Conjugation of Photon Echoes in Na Vapor," *Appl. Phys. Lett.*, vol. 33, p. 865, 1978.

12. A. Yariv, "Compensation for Atmospheric Degradation of Optical Beam Transmission by Nonlinear Optical Mixing," *Opt. Commun.*, vol. 21, p. 49, 1977.

13. Yu, I. Kucherov, S. A. Lesnik, M. S. Soskin, and A. I. Khizhnyak, "Copropagating Four-Beam Interaction in Slowly-Responding Media," in V. I. Bespalov (ed.), *Phase Conjugation in Nonlinear Media*, Gorky, USSR, p. 111, 1982.

14. V. G. Koptev, A. M. Lazaruk, I. P. Petrovich, and A. S. Rubanov, "Optical Phase Conjugation at Superluminescence," *JETP Lett.*, vol. 28, p. 434, 1979.

15. R. W. Hellwarth, "Generation of Time-Reversed Wavefront by Nonlinear Reflection," *J. Opt. Soc. Am.*, vol. 67, p. 1, 1977.

16. N. Bloembergen, *Nonlinear Optics*, W. A. Benjamin, Inc., Reading, Mass., 1964, chap. 1.

17. Y. R. Shen, *The Principles of Nonlinear Optics*, Wiley, New York, 1984, chaps. 1, 2.

18. J. F. Reintjes, *Nonlinear Optical Parametric Processes in Liquids and Gases*, Academic Press, New York, 1984, chaps. 1, 2.

19. R. Loudon, *The Quantum Theory of Light*, Clarendon Press, Oxford, 1985, chap. 9.

20. M. Sheik-Bahae, D. J. Hagan, and E. W. Van Stryland, "Dispersion and Band-Gap Scaling of the Electronic Kerr Effects in Solids Associated with Two-Photon Absorption," *Phys. Rev. Lett.*, vol. 65, p. 96, 1990.

21. W. L. Smith, "Nonlinear Optical Properties," in M. J. Weber (ed.), *Handbook of Laser Science and Technology*, CRC Press, Boca Raton, Fla., 1986.

22. P. D. Maker and R. W. Terhune, "Study of Optical Effects Due to an Induced Polarization Third Order in the Electric Field Strength," *Phys. Rev.*, vol. 137, p. A801, 1965.

23. T. Y. Chang, "Fast Self-Induced Refractive Index Changes in Optical Media: A Survey," *Opt. Eng.*, vol. 20, p. 220, 1981.

24. W. E. Williams, M. J. Soileau, and E. W. Van Stryland, "Optical Switching and n_2 Measurements in CS_2," *Opt. Commun.*, vol. 50, p. 256, 1984.

25. J. H. Bechtel and W. L. Smith, "Two-Photon Absorption in Semiconductors with Picosecond Laser Pulses," *Phys. Rev.*, vol. B13, p. 3515, 1976.

26. G. Rivoire, J. L. Ferrier, J. Gazengel, J. P. Lecoq, and N. Phu Xian, "Influence of the Non Linear Refraction Index on the Phase Conjugation Efficiency in Stimulated Scattering Effects," *Opt. Commun.*, vol. 48, p. 143, 1983.

27. A. Yariv and P. Yeh, *Optical Waves in Crystals*, Wiley, New York, 1984, chap. 6.

28. I. Khoo and Y. Zhao, "Probe Beam Amplification and Phase Conjugation Self-Oscillation in a Thin Media," *IEEE J. Quantum Electron.*, vol. QE-25, p. 368, 1989.

29. Y. R. Shen, *The Principles of Nonlinear Optics*, Wiley, New York, 1984, chap. 14.

30. A. Yariv, "Phase Conjugate Optics and Real-Time Holography," *IEEE J. Quantum Electron.*, vol. QE-14, p. 650, 1978.

31. M. L. Claude, L. L. Chase, D. Hulin, and A. Mysyrowicz, "Optical Phase Conjugation in Semiconductors," in H. Haug (ed.), *Optical Nonlinearities and Instabilities in Semiconductors*, Academic Press, Boston, 1988.

32. J. F. Reintjes, *Nonlinear Optical Parametric Processes in Liquids and Gases*, Academic Press, New York, 1984, chap. 5.

33. R. K. Jain and M. B. Klein, "Degenerate Four-Wave Mixing in Semiconductors," in R. A. Fisher (ed.), *Optical Phase Conjugation*, Academic Press, New York, 1983.

34. A. Maruani, "Propagation Analysis of Forward Degenerate Four-Wave Mixing," *IEEE J. Quantum Electron.*, vol. QE-16, p. 558, 1980.

35. D. M. Pepper and A. Yariv, "Optical Phase Conjugation Using Three-Wave and Four-Wave Mixing via Elastic Photon Scattering in Transparent Media," in R. A. Fisher (ed.), *Optical Phase Conjugation*, Academic Press, New York, 1983.

36. J. H. Marburger and J. F. Lam, "Effect of Nonlinear Index Changes on Degenerate Four-Wave Mixing," *Appl. Phys. Lett.*, vol. 35, p. 249, 1979.

37. R. G. Caro and M. C. Gower, "Phase Conjugation by Degenerate Four-Wave Mixing in Absorbing Media," *IEEE J. Quantum Electron.*, vol. QE-18, p. 1376, 1982.

38. R. L. Abrams and R. C. Lind, "Degenerate Four-Wave Mixing in Absorbing Media," *Opt. Lett.*, vol. 2, p. 94, 1978.

39. R. L. Abrams, J. F. Lam, R. C. Lind, D. G. Steel, and P. F. Liao, "Phase Conjugation and High-Resolution Spectroscopy by Resonant Degenerate Four-Wave Mixing," in R. A. Fisher (ed.), *Optical Phase Conjugation*, Academic Press, New York, 1983.

40. R. Lytel, "Pump-Depletion Effects in Noncollinear Degenerate Four-Wave Mixing in Kerr Media," *J. Opt. Soc. Am.*, vol. B 3, p. 1580, 1986.

41. S. Guha and P. Conner, "Degenerate Four-Wave Mixing in Kerr Media in the Presence of Nonlinear Refraction, Pump Depletion, and Linear Absorption," *Opt. Commun.*, vol. 89, p. 107, 1992.

42. A. Yariv and D. M. Pepper, "Amplified Reflection, Phase Conjugation, and Oscillation in Degenerate Four-Wave Mixing," *Opt. Lett.*, vol. 1, p. 16, 1977.

43. N. F. Pilipetsky and V. V. Shkunov, "Narrowband Four-Wave Reflecting Filter with Frequency and Angular Tuning," *Opt. Commun.*, vol. 37, p. 217, 1981.

44. J. P. Huignard and A. Marrackchi, "Coherent Signal Beam Amplification in Two-Wave Mixing Experiments with Photorefractive BSO Crystals," *Opt. Commun.*, vol. 33, p. 249, 1981.

45. A. M. Scott and K. D. Ridley, "A Review of Brillouin-Enhanced Four-Wave Mixing," *IEEE J. Quantum Electron.*, vol. QE-25, p. 438, 1989.

46. J. R. Ackerman and P. S. Lebow, "Improved Performance from Noncollinear Pumping in a High-Reflectivity Brillouin-Enhanced Four-Wave Mixing Phase Conjugator," *IEEE J. Quantum Electron.*, vol. QE-25, p. 479, 1989.

47. A. Ashkin, G. D. Boyde, J. M. Dziedzic, R. G. Smith, A. A. Ballman, H. J. Levenstein, and K. Nassau, "Optically-Induced Refractive Index Inhomogeneities in $LiNbO_3$ and $LiTaO_3$," *Appl. Phys. Lett.*, vol. 9, p. 72, 1966.

48. F. S. Chen, "A Laser-Induced Inhomogeneity of Refractive Indices in KTN," *J. Appl Phys.*, vol. 38, p. 3418, 1967.

49. N. V. Kukhtarev, "Kinetics of Hologram Recording and Erasure in Electrooptic Crystals," *Sov. Tech. Phys. Lett.*, vol. 2, p. 438, 1976.

50. J. Feinberg, D. Heiman, A. R. Tanguay, Jr., and R. W. Hellwarth, "Photorefractive Effects and Light Induced Charge Migration in Barium Titanate," *J. Appl. Phys.*, vol. 51, p. 1297, 1981.

51. P. Gunter, "Holography, Coherent Light Amplification and Optical Phase Conjugation with Photorefractive Materials," *Phys. Rep.*, vol. 93, p. 199, 1982.

52. F. P. Strohkendl, J. M. C. Jonathan, and R. W. Hellwarth, "Hole-Electron Competition in Photorefractive Gratings," *Opt. Lett.*, vol. 11, p. 312, 1986.

53. D. Mahgerefteh and J. Feinberg, "Explanation of the Apparent Sublinear Photoconductivity of Photorefractive Barium Titanate," *Phys. Rev. Lett.*, vol. 64, p. 2195, 1990.

54. A. Yariv, *Quantum Electronics*, Wiley, New York, 1975, chap. 14.

55. G. C. Valley and M. B. Klein, "Optimal Properties of Photorefractive Materials for Optical Data Processing," *Opt. Eng.*, vol. 22, p. 704, 1983.

56. S. Ducharme and J. Feinberg, "Altering the Photorefractive Properties of $BaTiO_3$ by Reduction and Oxidation at 650°C," *J. Opt. Soc. Am.*, vol. B 3, p. 283, 1986.

57. M. D. Ewbank, R. R. Neurgaonkar, W. K. Cory, and J. Feinberg, "Photorefractive Properties of Strontium Barium Niobate," *J. Appl. Phys.*, vol. 62, p. 374, 1987.

58. M. Cronin-Golomb, B. Fischer, J. O. White, and A. Yariv, "Theory and Applications of Four-Wave Mixing in Photorefractive Media," *IEEE J. Quantum Electron.*, vol. QE-20, p. 12, 1984.

59. For exact solution with absorption see M. R. Belic and M. Lax, "Exact Solution to the Stationary Holographic Four-Wave Mixing in Photorefractive Crystals," *Opt. Commun.*, vol. 56, p. 197, 1985.

60. M. Cronin-Golomb, B. Fischer, J. O. White, and A. Yariv, "Passive Phase Conjugate Mirror Based on Self-Induced Oscillation in an Optical Ring Cavity," *Appl. Phys. Lett.*, vol. 42, p. 919, 1983.

61. J. Feinberg, "Self-Pumped, Continuous-Wave Phase Conjugator Using Internal Reflection," *Opt. Lett.*, vol. 7, p. 486, 1982.

62. K. R. MacDonald and J. Feinberg, "Theory of a Self-Pumped Phase Conjugator with Two Coupled Interaction Regions," *J. Opt. Soc. Am.*, vol. 73, p. 548, 1983.

63. G. J. Dunning, D. M. Pepper, M. B. Klein, "Control of Self-Pumped Phase-Conjugate Reflectivity Using Incoherent Erasure," *Opt. Lett.*, vol. 15, p. 99, 1990.

64. P. Yeh, "Theory of Unidirectional Photorefractive Ring Oscillators," *J. Opt. Soc. Am.*, vol. B 2, p. 1924, 1985.

65. J. F. Lam, "Origin of Phase Conjugate Waves in Self-Pumped Photorefractive Mirrors," *Appl. Phys. Lett.*, vol. 46, p. 909, 1985.

66. G. Salamo, M. J. Miller, W. W. Clark, III, G. L. Wood, and E. J. Sharp, "Strontium Barium Niobate as a Self-Pumped Phase Conjugator," *Opt. Commun.*, vol. 59, p. 417, 1986.

67. G. L. Wood, W. W. Clark, III, M. J. Miller, E. J. Sharp, G. J. Salamo, and R. R. Neurgaonkar, "Broadband Photorefractive Properties and Self-Pumped Phase Conjugation in Ce-SBN:60," *IEEE J. Quantum Electron*, vol. QE-23, p. 2126, 1987.

68. E. J. Sharp, W. W. Clark, III, M. J. Miller, G. L. Wood, G. J. Salamo, and R. R. Neurgaonkar, "Double Phase Conjugation in Tungsten Bronze Crystals," *Appl. Opt.*, vol. 29, p. 743, 1990, and references therein, also A. A. Zozulya and A. V. Mamaev, "Mutual Phase Conjugation of Incoherent Light Beams in a Photorefractive Crystal," *JETP Lett.*, vol. 49, p. 553, 1989. P. Ye, D. Wang, Z. Zhang, and X. Wu, "Mutually Coherent Beam Induced Self-Pumped Phase Conjugate Reflection in $BaTiO_3$," *Appl. Phys. Lett.*, vol. 55, p. 830, 1989.

69. M. Cronin-Golomb, "Almost All Transmission Grating Self-Pumped Phase Conjugate Mirrors Are Equivalent," *Opt. Lett.*, vol. 15, p. 897, 1990.

70. Y. R. Shen, *The Principles of Nonlinear Optics*, Wiley, New York, 1984, chaps. 10, 11.

71. A. Yariv, *Quantum Electronics*, Wiley, New York, 1975, chap. 18.

72. B. Ya. Zel'dovich, N. F. Pilipetsky, and V. V. Shkunov, *Principles of Phase Conjugation*, Springer-Verlag, Berlin, 1985, chap. 2.

73. S. Y. Auyang and P. A. Wolff, "Free-Carrier-Induced Third-Order Optical Nonlinearities in Semiconductors," *J. Opt. Soc. Am.*, vol. B 6, p. 595, 1989.

74. P. Yeh, "Exact Solution of a Nonlinear Model of Two-Wave Mixing in Kerr Media," *J. Opt. Soc. Am.*, vol. B 3, p. 747, 1986.

75. B. Bobbs and C. Warner, "Raman-Resonant Four-Wave Mixing and Energy Transfer," *J. Opt. Soc. Am.*, vol. B 7, p. 234, 1990.

76. M. Sparks, "Stimulated Raman and Brillouin Scattering: Parametric Instability Explanation of Anomalies," *Phys. Rev. Lett.*, vol. 32, p. 450, 1974.

77. B. Ya. Zel'dovich, N. F. Pilipetsky, and V. V. Shkunov, *Optical Phase Conjugation*, Springer-Verlag, Berlin, 1985, chap. 4.

78. R. W. Hellwarth, "Phase Conjugation by Stimulated Scattering," in R. A. Fisher (ed.), *Optical Phase Conjugation*, Academic Press, New York, 1983.

79. R. W. Hellwarth, "Theory of Phase Conjugation by Stimulated Scattering in a Waveguide," *J. Opt. Soc. Am.*, vol. 68, p. 1050, 1978.

80. B. Ya. Zel'dovich and V. V. Shkunov, "Limits of Existence of Wavefront Reversal in Stimulated Light Scattering," *Sov. J. Quantum Electron.*, vol. 8, p. 15, 1978.

81. P. Suni and J. Falk, "Theory of Phase Conjugation by Stimulated Brillouin Scattering," *J. Opt. Soc. Am.*, vol. B 3, p. 1681, 1986.

82. P. H. Hu, J. A. Goldstone, and S. S. Ma, "Theoretical Study of Phase Conjugation in Stimulated Brillouin Scattering," *J. Opt. Soc. Am.*, vol. B 6, p. 1813, 1989.

83. N. G. Basov, V. F. Efimkov, I. G. Zubarev, A. V. Kotov, and S. I. Mikhailov, "Control of Characteristics of Phase-Conjugate Mirrors in the Amplification Regime," *Sov. J. Quantum Electron.*, vol. 11, p. 1335, 1981.

84. V. N. Blaschuk, V. N. Krasheninnikov, N. A. Mel'nikov, N. F. Pilipetsky, V. V. Ragulsky, V. V. Shkunov, and B. Ya. Zel'dovich," SBS Wavefront Reversal for the Depolarized Light (Theory and Experiment)," *Opt. Commun.*, vol. 28, p. 137, 1978.

85. C. Kittel, *Introduction to Solid State Physics*, Wiley, New York, 1976, chaps. 10, 11.

86. Y. R. Shen, *The Principles of Nonlinear Optics*, Wiley, New York, 1975, chap. 10.

87. N. Bloembergen, "The Stimulated Raman Effect," *Am. J. Phys.*, vol. 35, p. 989, 1967.

88. Y. R. Shen, *The Principles of Nonlinear Optics*, Wiley, New York, 1975, chap. 10.

89. B. Ya. Zel'dovich and V. V. Shkunov, "Wavefront Reproduction in Stimulated Raman Scattering," *Sov. J. Quantum Electron.*, vol. 7, p. 610, 1977.

90. A. I. Sokolovskaya, G. L. Brekhovskikh, and A. D. Kudryavtseva, "Light Beam Wavefront Reconstruction and Real Volume Image Reconstruction of the Object at the Stimulated Raman Scattering," *Opt. Commun.*, vol. 24, p. 74, 1978.

91. J. H. Marburger, "Self Focusing: Theory," *Progr. Quantum Electron.*, vol. 4, p. 35, 1975.

92. M. M. Loy and Y. R. Shen, "Study of Self-Focusing and Small-Scale Filaments of Light in Nonlinear Media," *IEEE J. Quantum Electron.*, vol. QE-9, p. 409, 1973.

93. F. A. Hopf, A. Tomita, and T. Liepmann, "Quality of Phase Conjugation in Silicon," *Opt. Commun.*, vol. 37, p. 72, 1981.

94. R. M. Herman and M. A. Gray, "Theoretical Prediction of the Stimulated Rayleigh Scattering in Liquids," *Phys. Rev. Lett.*, vol. 19, p. 824, 1967.

95. I. L. Fabellinskii, D. I. Mash, V. V. Morozov, and V. S. Starunov, "Stimulated Scattering of Light in Hydrogen Gas at Low Pressures," *Phys. Lett.*, vol. 27A, p. 253, 1968.

96. A. D. Kudriavtseva, A. I. Sokolovskaia, J. Gazengel, N. Phu Xuan, and G. Rivoire, "Reconstruction of the Laser Wave-Front by Stimulated Scattering in the Pico-second Regime," *Opt. Commun.*, vol. 28, p. 446, 1978.

97. M. Denariez and G. Bret, *Phys. Rev.*, vol. 171, p. 160, 1968.

98. T. Y. Chang and R. W. Hellwarth, "Optical Phase Conjugation by Backscattering in Barium Titanate," *Opt. Lett.*, vol. 10, p. 408, 1985.

99. R. A. Mullen, D. J. Vickers, and D. M. Pepper, "Stimulated Photorefractive Scattering Phase-Conjugators Back-Seeded with Retro-Reflector Arrays," Conference on Lasers and Electro-Optics, Anaheim, Calif., May 21–25, 1990.

100. Special issue on adaptive optics, *J. Opt. Soc. Am.*, vol. 67, 1977.

101. W. R. Wu, R. O. Gale, L. J. Hornbeck, and J. B. Sampsell, "Electro Optical Performance of an Improved Deformable Mirror Device," *SPIE Spatial Light Modulators and Applications*, vol. 825, p. 24, 1987.

102. A. D. Fisher and J. N. Lee, "The Current Status of Two-Dimensional Spatial Light Modulator Technology," *SPIE Optical and Hybrid Computing*, vol. 634, p. 352, 1986.

103. C. A. Beichman and S. Ridgway, "Adaptive Optics and Interferometry," *Phys. Today*, vol. 44, p. 48, April 1991.

104. C. K. N. Patel and N. Bloembergen, "Strategic Defense and Directed-Energy Weapons," *Sci. Am.*, vol. 257, p. 39, 1987.

105. T. R. O'Meara, D. M. Pepper, and J. O. White, "Applications of Nonlinear Optical Phase Conjugation," in R. A. Fisher (ed.), *Optical Phase Conjugation*, Academic Press, New York, 1983.

106. A. Yariv, "Fundamental Media Considerations for the Propagation of Phase-Conjugate Waves," *Opt. Lett.*, vol. 16, p. 1376, 1991.

107. M. D. Levenson, K. M. Johnson, V. C. Hanchett, and K. Chiang, "Projection Photolithography by Wave-Front Conjugation," *J. Opt. Soc. Am.*, vol. 71, p. 737, 1981.

108. D. Yu. Nosach, V. I. Popovichev, V. V. Ragul'skii, and F. S. Faizullov, "Cancellation of Phase Distortions in an Amplifying Medium with a Brillouin Mirror," *Sov. Phys. JETP*, vol. 16, p. 435, 1972.

109. A. E. Chiou and P. Yeh, "Laser-Beam Cleanup Using Photorefractive Two-Wave Mixing and Optical Phase Conjugation," *Opt. Lett.*, vol. 11, p. 461, 1986.

110. L. M. Frantz, "Theory of Phase Conjugation in Frequency Doubling," *J. Opt. Soc. Am.*, vol. B 7, p. 335, 1990.

111. A. Yariv, "Compensation for Atmospheric Degradation of Optical Beam Transmission," *Opt. Commun.*, vol. 21, p. 49, 1977.

112. C. J. Wetterer, L. P. Schelonka, and M. A. Kramer, "Correction of Thermal Blooming by Optical Phase Conjugation," *Opt. Lett.*, vol. 14, p. 874, 1989.

113. G. J. Dunning and R. C. Lind, "Demonstration of Image Transmission through Fibers by Optical Phase Conjugation," *Opt. Lett.*, vol. 7, p. 558, 1982.

114. D. M. Pepper, D. A. Rockwell, and H. W. Bruesselbach, "Phase Conjugation: Reversing Laser Aberrations," *Photonics*, August 1986.

115. A. E. Siegman, P. A. Belanger, and A. Hardy, "Optical Resonators Using Phase-Conjugate Mirrors," in R. A. Fisher (ed.), *Optical Phase Conjugation*, Academic Press, New York, 1983.

116. J. AuYeung and A. Yariv, "Phase-Conjugate Optics," *Opt. News*, vol. 13, spring 1979.

117. D. Z. Anderson, D. M. Lininger, and J. Feinberg, "Optical Tracking Novelty Filter," *Opt. Lett.*, vol. 12, p. 123, 1987.

118. D. M. Pepper, "Phase Conjugation and Beam Combining and Diagnostics," *SPIE Remote Sensors Using Hybrid Phase-Conjugator/Modulators*, vol. 739, p. 71, 1987.

119. G. J. Salamo, M. J. Miller, W. W. Clark, III, G. L. Wood, E. J. Sharp, and R. R. Neurgaonkar, "Photorefractive Rainbows," *Appl. Opt.*, vol. 27, p. 4356, 1988.

120. Y. Tomita, R. Yahalom, and A. Yariv, "Real-Time Image Subtraction with the Use of Wave Polarization and Phase Conjugation," *Appl. Phys. Lett.*, vol. 52, p. 425, 1988.

121. P. Yeh, T. Y. Chang, and P. H. Beckwith, "Real-Time Optical Image Subtraction Using Dynamic Holographic Interference in Photorefractive Media," *Opt. Lett.*, vol. 13, p. 586, 1988.

122. S. Kwong, G. A. Rakuljic, and A. Yariv, "Real Time Image Subtraction and Exclusive Or Operation Using a Self-Pumped Phase Conjugate Mirror," *Appl. Phys. Lett.*, vol. 48, p. 201, 1986.

123. E. J. Sharp, W. W. Clark, III, M. J. Miller, G. L. Wood, B. Monson, G. J. Salamo, and R. R. Neurgaonkar, "Double Phase Conjugation in Tungsten Bronze Crystals," *Appl. Opt.*, vol. 29, p. 743, 1990.

124. D. M. Pepper, J. AuYeung, D. Fekete, and A. Yariv, "Spatial Convolution and Correlation of Optical Fields via Degenerate Four-Wave Mixing," *Opt. Lett.*, vol. 3, p. 7, 1978.

125. J. O. White and A. Yariv, "Real-Time Image Processing via Four-Wave Mixing in a Photorefractive Medium," *Appl. Phys. Lett.*, vol. 37, p. 5, 1980.

126. M. A. Flavin and J. L. Horner, "Average Amplitude Matched Filter," *Opt. Eng.*, vol. 29, p. 31, 1990.

127. K. Fukushima, S. Miyake, and T. Ito, "Neocognitron: Neural Network Model for a Mechanism of Visual Pattern Recognition," *IEEE Trans. Syst. Man Cybern.*, vol. SMC-13, p. 826, 1983.

128. J. Y. Jau, Y. Fainman, and S. H. Lee, "Comparison of Artificial Neural Networks with Pattern Recognition and Image Processing," *Appl. Opt.*, vol. 28, p. 302, 1989.

129. T. Kohonen, *Self-Organization and Associative Memory*, Springer-Verlag, New York, 1984.

130. For a review see *Proceedings*, *Neural Networks for Computing Conference*, Snowbird, Utah, Apr. 13–16, 1986; or K. Kyuma, Optical Neural Networks—A Review," *NLO*, vol. 1, p. 39, 1991.

131. Y. Owechko, G. J. Dunning, E. Marom, and B. H. Soffer, "Holographic Associative Memory with Nonlinearities in the Correlation Domain," *Appl. Opt.*, vol. 26, p. 1900, 1987.

132. Y. Fainman and S. H. Lee, "Applications of Photorefractive Crystals to Optical Signal Processing," *SPIE Optical and Hybrid Computing*, vol. 634, p. 380, 1986.

133. A. Yariv, S. Kwong, and K. Kyuma, "Optical Associative Memories Based on Photorefractive Oscillators," *SPIE Nonlinear Optics and Applications*, vol. 613, p. 2, 1986.

134. I. McMichael and P. Yeh, "Self-Pumped Phase-Conjugate Fiber-Optic Gyro," *Opt. Lett.*, vol. 11, p. 686, 1986.

135. M. Cronin-Golomb and A. Yariv, "Optical Limiters Using Photorefractive Nonlinearities," *J. Appl. Phys.*, vol. 57, p. 4906, 1985.

136. D. T. Hon, "Pulse Compression by Stimulated Brillouin Scattering," *Opt. Lett.*, vol. 5, p. 516, 1980.

137. D. A. B. Miller, "Time Reversal of Optical Pulses by Four-Wave Mixing," *Opt. Lett.*, vol. 5, p. 300, 1980.

138. A. Marrakchi, A. R. Tanguay, Jr., J. Yu, and D. Psaltis, *Opt. Eng.*, vol. 24, p. 124, 1985.

139. G. P. Agrawal and C. Flytzanis, "Bistability and Hysteresis in Phase Conjugated Reflectivity," *IEEE J. Quantum Electron.*, vol. QE-17, p. 374, 1981.

CHAPTER 15
ULTRAVIOLET AND X-RAY DETECTORS

George R. Carruthers

15.1 OVERVIEW OF ULTRAVIOLET AND X-RAY DETECTION PRINCIPLES

The basic principles of ultraviolet and x-ray detectors are largely similar to those of detectors used for visible light; however, the similarities are greater at the longer wavelengths (closest to the visible). Especially in the x-ray portion of the spectrum, a number of unique detection techniques (not used in the visible) are applicable.

One detection technique, the use of photographic film, is applicable at all wavelengths shortward of the near-infrared, including x-rays. It is also the oldest and simplest technique, and still has advantages in some applications. However, it has been replaced by other techniques for most applications where higher sensitivity, more quantitatively accurate photometric information, or more immediate data availability is required.

In common with visible-light detectors, most ultraviolet detectors and many x-ray detectors are based on the principles of photoelectric emission or of photoconductivity. Detectors based on a third process, gas photoionization, are also used in the far-ultraviolet and x-ray wavelength ranges. Scintillation detectors are used for high-energy x-ray and gamma-ray detection.

The basic detection principles are the same for nonimaging detectors (in which the objective is simply to measure the intensity and/or spectral distribution of an incoming beam of radiation) and for imaging detectors (in which one also seeks to preserve the intensity versus position information in a two-dimensional field of view). However, imaging detectors often involve the use of additional components or techniques in order to preserve and record the spatial intensity distribution information.

15.2 PHOTOGRAPHIC FILM

Photographic emulsions depend on the energy of incoming photons to convert crystals of silver halides to a developable form, i.e., so that they are decomposed and leave pure silver behind in the development process. Since this is a threshold process, any photon of energy greater than the threshold can cause halide grains to become developable. Therefore, any films which are sensitive to visible light can, in principle, also be used to detect ultraviolet (uv) or x-ray radiation. In fact, photographic film was used by W. Roentgen in his initial discovery of x-rays and was used in the first detection of solar x-rays from rockets in the late 1940s.

In much of the ultraviolet and low-energy x-ray range, conventional photographic films have low sensitivity because the gelatin binder and overcoat used in the emulsion is opaque to the radiation. As a result, very few of the silver halide grains are exposed to the radiation and made developable. One method of circumventing this difficulty is to coat the front surface of the film with a phosphor which gives off visible light when exposed to uv or x-rays. The film then detects the emitted visible light. Another, more efficient (but more complex) technique is to use special emulsions (known as *Schumann emulsions*) in which the amount of gelatin used is kept to an absolute minimum. This allows the silver halide grains to be directly exposed to incoming uv and x-ray radiation. However, such emulsions are difficult to manufacture and handle; they are extremely sensitive to abrasion, mechanical pressure, and chemical contamination.

Photographic films have the major advantages of simplicity and low cost, in comparison to other detection techniques (especially imaging detectors). They can be used for imaging very large fields of view with very high resolution, a capability not yet matched by electronic detectors. However, the quantitative accuracy of photography is not as good as is achieved with electronic sensors. This is because the blackening of the film is nonlinear with integrated radiation flux and is subject to threshold effects and reciprocity failure, and the response is not necessarily the same from one film sample to the next, even in the same batch of emulsion. The absolute sensitivity of films (measured as a quantum efficiency, or number of detectable blackened grains per 100 incident photons), even when optimized for the uv and x-ray wavelength ranges, is rather low (typically a few percent). Since films are also sensitive to visible light, filters and/or spectrographs must be used which efficiently reject the unwanted longer-wavelength radiations. Nevertheless, in applications where quantitative accuracy and very high sensitivity are not important, photographic film is still very useful and cost-effective.

15.3 NONIMAGING PHOTOIONIZATION DETECTORS

Gas photoionization is a useful technique for detecting far-ultraviolet and x-ray radiations whose photon energies exceed the photoionization thresholds of the gases used. Suitable gases are available for detection of wavelengths below about 170 nm.[1,2] For example, nitric oxide, NO (ionization potential = 9.15 eV), can be used to detect the uv below 135 nm, and has a photoionization quantum efficiency of 81 percent at 121.6 nm.[3] The noble gases He, Ne, Ar, Kr, and Xe have 100 percent photoionization quantum efficiencies below their threshold wavelengths of 50.4, 57.5, 78.7, 88.5, and 102.2 nm, respectively.

15.3.1 Unity-Gain Photoionization Chambers

The simplest photoionization detector is the unity-gain photoionization chamber (ion chamber). It consists of a metal container for the gas, an input window for the radiation to be detected, and a collecting electrode passing into (but electrically insulated from) the chamber. Figure 15.1 shows diagrams of two types of ion chambers used in space science investigations.

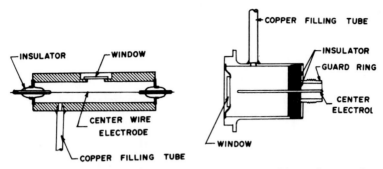

FIGURE 15.1 Diagrams of two far-UV ionization chambers (also used as gas gain or as Geiger counters) used by the Naval Research Laboratory in space-based astronomy and upper-atmosphere studies.

When ionizing radiation enters the chamber, gas atoms or molecules are broken up into positive ions and electrons. If a potential difference is applied between the shell and the internal electrode, positive ions flow to the negative electrode and electrons flow to the positive electrode, in equal numbers. Hence, an electric current will flow in direct proportion to the rate of production of electron-ion pairs, which in turn is proportional to the intensity of the incoming radiation. The current is independent of the applied voltage, as long as the voltage is high enough to result in efficient collection of the electron-ion pairs, but not so high as to cause collision-induced secondary ionizations.

In the design of an ion chamber, one must consider other factors in addition to the photoionization threshold and photoionization quantum efficiency of the gas used. Another important aspect is the absorption coefficient of the gas: the path length of radiation in the gas should be adequate for the absorption of nearly all of the incoming photons. The absorption efficiency of the gas is proportional to the absorption coefficient (per atom or molecule) and the gas pressure. A highly absorbing gas requires a smaller pressure $\times$ path length product than a weakly absorbing one. The absorption coefficient of a gas varies with wavelength, and in a manner not necessarily related to the variation of photoionization yield with wavelength.

The spectral response of an ion chamber is determined by the photoionization yield versus wavelength of the gas filling and the transmission versus wavelength of the input window. Therefore, different combinations of gas fillings and windows can be used to tailor the response of the detector over a wide range. Figures 15.2 and 15.3 show some typical detection efficiency versus wavelength curves for uv and x-ray ion chambers. If the energy of a photon exceeds twice the ionization potential of the gas, it is possible to produce two or more electron-ion pairs per photon absorbed. This is a particularly important effect in x-ray detection, and is the basis of the proportional counter to be discussed below.

15.3.2 Gas-Gain Photoionization Chambers and Proportional Counters

If the voltage between the shell and the central electrode of an ionization chamber is increased, so as to increase the electric field in the space between, electrons and ions produced by photoionization will gain increasing amounts of kinetic energy between collisions with neutral gas molecules. Eventually, this energy becomes

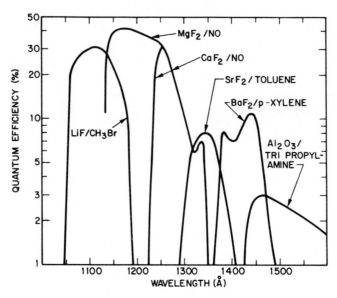

FIGURE 15.2 Quantum efficiency vs. wavelength for various combinations of ionization chamber windows and gas fillings useful in the far ultraviolet.

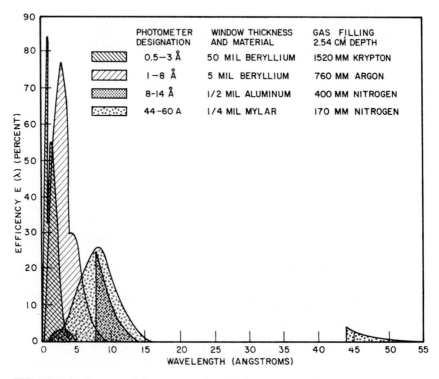

FIGURE 15.3 Quantum efficiency vs. wavelength in the x-ray spectral range for ionization chambers with various combinations of window materials and gas fillings. (Courtesy of R. Taylor, NRL)

sufficient to cause collisional ionization (secondary ionization) of the neutral gas, so that more than one electron-ion pair is produced per primary photoionization. The resulting *gas gain* increases rapidly above the secondary-ionization threshold voltage, in a manner dependent on the nature of the gas filling (see Fig. 15.4). Of practical importance is that the gain versus voltage characteristic should not be too steep, or the detector will tend to be unstable and small voltage changes can cause large output fluctuations.

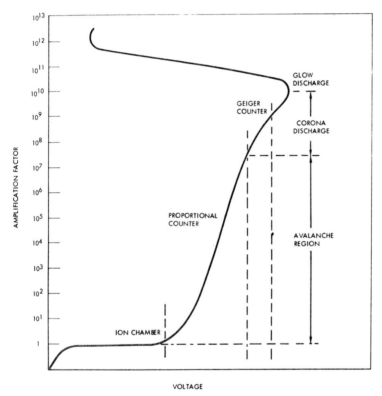

FIGURE 15.4   Gas photoionization chamber amplification vs. voltage, showing the three main useful ranges (unity gain, proportional counting, and Geiger counting). (Courtesy of H. Friedman, NRL)

In proportional counters and Geiger counters, it is usually necessary to include other gases in the detector filling besides that needed for initial photon detection, in order to produce appropriate gain versus voltage characteristics. For example, uv ion chambers using gases such as nitric oxide or organic gases also contain inert gases such as neon or argon. X-ray ion chambers which use inert gases for primary photon detection must also include small amounts of organic or electronegative "quench" gases, to prevent instabilities and high-voltage discharges in the detector and to reduce the slope of the gain versus voltage curve.

Gain factors as high as 10^6 to 10^7 can be utilized in proportional counters with suitable gas fillings. With sensitive pulse-counting electronics, the individual photoionization events can then be detected as discrete counts, thus providing much better sensitivity than a similar unity-gain ion chamber with the same gas filling, in which only photocurrent can be measured. Typically, current-measuring techniques are limited to currents above about 10^{-14} A (about 10^5 electrons/s), whereas photon-counting techniques allow measurement of 1 photoevent/s or less.

As mentioned above, extreme-ultraviolet and x-ray photons can produce more than one primary electron-ion pair in the gas. In the x-ray range, typically the number of electron-ion pairs produced by a photon of energy E (eV) is $E/30$. In the gas-gain mode of operation, the final output pulse is proportional to the number of primary photoionization events, as well as the gas gain factor; hence gas-gain ion chambers are also known as *proportional counters*. This is an important feature, because measurements of the output pulse amplitudes can allow a means for determining the energies of the detected photons, as well as their fluxes. That is, it provides a spectroscopic capability. The *spectral resolution*—i.e., the degree to which photons of varying energy can be distinguished—is generally rather low for extreme-uv and soft x-ray photons, but improves toward higher energies. The resolution of a proportional counter is given, approximately, by

$$\frac{\delta E}{E} \simeq \frac{0.2}{E^{1/2}} \tag{15.1}$$

where δE is the full width at half maximum (FWHM) of the line profile and E is the energy in keV.

Because of their spectroscopic capabilities, high quantum efficiencies, and wider dynamic range than Geiger counters (discussed below), proportional counters have been the most widely used x-ray detectors in x-ray astronomy. Very large area, multiple-wire counters have been constructed to provide a large collecting area without the need for x-ray-focusing optics. Figure 15.5 is a diagram of a typical detector used by the Naval Research Laboratory (NRL). Note that the detector consists of two layers; the bottom layer does not see the x-rays of interest and serves as an anticoincidence shield against energetic-charged-particle events. Any event detected simultaneously in both the upper and lower layers is rejected. Very large (2000 cm² open area) proportional counters were developed by NRL for its instrument package on NASA's first High Energy Astronomy Observatory (HEAO-1), launched in 1977.

15.3.3 Geiger Counters

If the applied voltage on an ion chamber is continually increased in the gas-gain range, a point is reached in which pulse-amplitude saturation occurs—that is, all output pulses are about the same size, regardless of the initial number of electron-ion pairs produced by an input photon. This is known as the *Geiger counting region* of the gain versus voltage curve.

The advantage of the Geiger counting mode is that the count rate is only weakly dependent on voltage, whereas, in the gas-gain mode, the count rate depends strongly on voltage. Hence, Geiger counters are easier to use in practical applications. Also, with a given pulse-counting circuit, the Geiger counter mode will give better sensitivity than the proportional counter mode because, in the latter, some counts will fall below the detection threshold. However, it is usually true (especially in far-uv Geiger counters) that the counting efficiency (counts per incident photon) is considerably lower than the detection efficiency of a unity-gain ion chamber with the same gas filling. An advantage in comparison to the use of

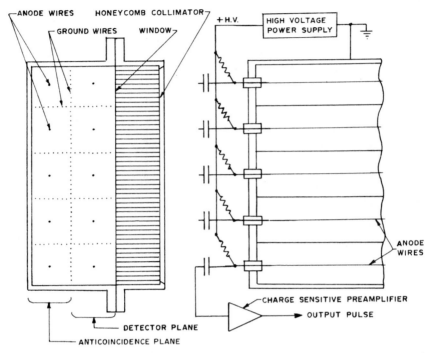

FIGURE 15.5 Diagram of a large area multi-wire proportional counter used by the Naval Research Laboratory for x-ray astronomy. The anode wires can be connected to individual output amplifiers, or two or more can be connected to a single amplifier.

photomultipliers in the uv and x-ray ranges is that narrower bands with sharper long-wavelength cutoffs can often be obtained.

A disadvantage of Geiger counters is that they have relatively long dead times following pulses in which new photoevents are not detected; hence they provide poor photometric accuracy except at very low count rates (less than a few thousand counts/s). Another disadvantage, in comparison to proportional counters, is that Geiger counters cannot discriminate among detected photons having different energies, since all photoevents produce pulses of the same size. In comparison to photomultipliers, especially in the far-uv, the detection efficiencies of Geiger counters are usually lower and less stable.

15.4 IMAGING PROPORTIONAL COUNTERS

A large-area multiwire proportional counter can be made to provide imaging in one dimension, if each of the wires is connected to a separate amplifier and counting circuit. The position of a photoionization event is indicated by which wire gives the largest signal; however, by comparing the signals on wires adjacent to that giving the largest signal, the centroid of the event can be determined to an accuracy greater than the spacing between wires in the counter.

This approach can be extended to two dimensions if two crossed arrays of collector wires are used, one placed slightly above the other so that the x and y

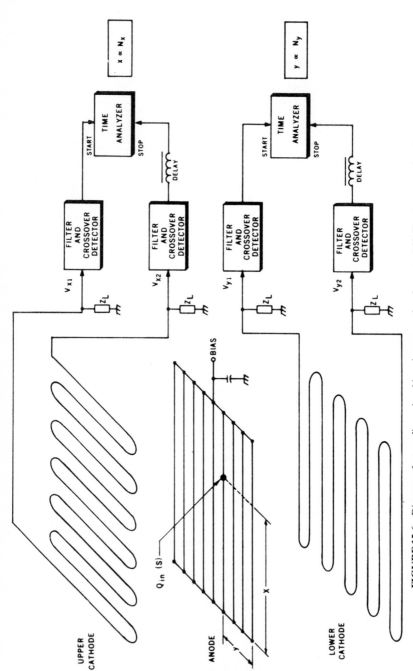

FIGURE 15.6 Diagram of a two-dimensional imaging proportional counter shown in Figure 15.5, here the signal is derived from the ion current to cathodes placed in close proximity to the anode, rather than from the electron current to the anode. The location of the photoevent relative to each of the cathode wires is determined from the time difference in the arrival of a charge pulse at the two ends of the wire.

arrays share nearly equally the signals produced by photoevents in the counter gas. Of course, this approach becomes complex because of the large number of amplifiers and counting circuits needed for high resolution in two dimensions.

If a single high-resistivity wire is used, the position of an event along the wire can also be determined, by comparing the amplitudes of signals, or the times at which signals are received, at two amplifiers—one attached to each end of the wire. Either the inverse amplitude or the delay time of the pulse is proportional to the length of wire (impedance) in series between the event and the amplifier; hence the ratio of the output signals due to an event at position x along a wire of length L is proportional to $x/(L - x)$. This ratio is independent of the pulse amplitude, and depends only on the location of the event.

Figure 15.6 shows a method for obtaining two-dimensional imaging using two resistive collector wires and the time-delay method, with four output amplifiers.[4,5] It can provide about 1 part in 100 resolution in each dimension. A device of this type was used in the second High Energy Astronomical Observatory (HEAO-2, named *Einstein*), which was launched in 1978.[6,7]

Another approach to x-ray imaging is provided by the gas-scintillation proportional counter. Here, the detection of an event is by way of the ultraviolet radiation emitted as a result of a photoionization event in a gas scintillation chamber, such as shown in Fig. 15.7.[8] The chamber is filled with a noble gas (such as argon or xenon) and contains parallel grids of electrodes, which provide for acceleration and multiplication of the photoelectrons. In the low-voltage-gradient drift region between the entrance window and the first grid, the x-rays are absorbed and photoelectrons produced are transported (without multiplication) through the grid into the high-voltage-gradient scintillation region between the two grids. In this region, photoelectrons are accelerated to sufficient energy that they produce uv emission when they collide with the noble gas atoms. The pulses of uv light, or scintillations, produced have amplitudes proportional to the number of photoelectrons in a photoionization event and increase with the accelerating voltage.

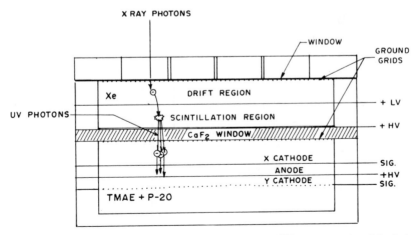

FIGURE 15.7 Diagram of an imaging gas scintillation counter.[8] The upper portion of the device is a gas scintillator, in which x-ray produced photoelectrons excite uv emission in a noble gas. The lower portion is an imaging proportional counter, as per Fig. 15.6, whose gas filling contains a component which is photoionized by the uv light from the gas scintillator.

The detection of the event, however, requires a second stage of the device to convert the flash of uv light into an electrical signal. This is typically done by making the rear window of the scintillation chamber of a uv-transmissive material, such as calcium fluoride, and interfacing it with a uv-sensitive imaging detector. This latter can be an imaging proportional counter, whose fill gas contains a component which is photoionized by the uv radiation, such as triethylamine (TEA) or tetrakis(dimethylamino)ethylene (TMAE) (see Fig. 15.7). Alternatively, a large-area microchannel plate with a CsI photocathode and two-dimensional imaging anode can be used[9] (as will be discussed in later sections).

The advantages of imaging gas-scintillation proportional counters over conventional imaging proportional counters include better energy resolution and better rejection of charged-particle background events.

15.5 PHOTOEMISSIVE DETECTORS

Photoelectric emission, or photoemission, is the basis for a large fraction of the detector technologies (imaging and nonimaging) useful in the wavelength range extending from the very near infrared (ir) through the ultraviolet. It is also used in many types of x-ray detectors, particularly imaging types.

Photoemission is the process in which a photon incident on a solid surface causes an electron (photoelectron) to be ejected from that surface. The photoelectron can then be collected, or amplified by other processes such as secondary emission, for detection. The photoemissive efficiency (photoelectrons per incident photon) and its variation with wavelength are properties of the photoemissive surface material (photocathode).[10] The energy of the photon must exceed a threshold energy, or work function, $E_{th} = E_A + E_G$ (see Fig. 15.8) of the photocathode material, in order to result in emission of a photoelectron. Here, E_A is the electron affinity and E_G is the bandgap energy (difference between valence band and conduction band energies) of the photocathode material. In general, the smaller E_{th} is, the longer the wavelength to which the photocathode is sensitive. Also, the quantum efficiency of the photocathode for photons of energy exceeding E_{th} is higher if the electron affinity E_A is minimized (i.e., the ratio E_A/E_G should be minimized).

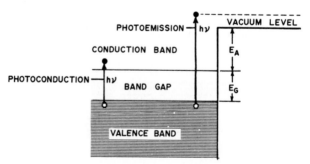

FIGURE 15.8 Energy-level diagram of a semiconductor showing processes of photoemission and of photoconduction.

A wide variety of photocathode materials have been developed which are useful in both nonimaging and imaging detectors. The long-wavelength limit of sensitivity ranges from above 1000 nm in the near ir to near 100 nm in the far uv. For most uv and x-ray applications, photocathodes which are insensitive to visible and infrared radiation are preferred, because (1) interference due to stray long-wavelength radiation is eliminated; (2) thermal emission (dark current) is usually lower for photocathodes having higher threshold energies; and (3) photocathodes sensitive in the far uv and x-ray regions are less chemically reactive, and hence less susceptible to degradation by exposure to air or poor vacuum, than are photocathodes sensitive in the near uv and visible.

Photocathodes used for the near- and middle uv are typically compounds of alkali metals, usually cesium or rubidium, with tellurium (for middle uv) or antimony (for near uv and visible).[10,11] These compounds must be prepared in ultrahigh-vacuum conditions and cannot be subsequently exposed to air or poor vacuum. Photocathodes for the far uv are typically alkali halides such as CsI or KBr, and can be prepared by evaporation of the salt in vacuum onto the desired substrates. These photocathodes can be exposed to dry air or nitrogen, but are degraded (to varying degrees) by exposure to water vapor. For the extreme uv through x-ray regions of the spectrum, almost any material (including common metals) will provide some sensitivity. However, the alkali halides, alkaline-earth halides (such as MgF_2 and BaF_2), and metal oxides (such as BeO and Al_2O_3) are typically the most efficient.[12-14]

Two major types of photocathodes are widely used, distinguished by whether the photoelectrons are ejected from the same surface onto which the radiation is incident (opaque or reflective photocathodes), or from the opposite surface of a thin layer on which the radiation is incident (semitransparent photocathodes), as shown in Fig. 15.9. Semitransparent photocathodes are normally deposited on the rear surface of a transparent window, such as the faceplate of a photomultiplier or image intensifier tube. Opaque photocathodes typically have higher quantum efficiencies than do semitransparent photocathodes of the same material, especially in the ultraviolet (see Fig. 15.10). This is because an opaque photocathode can be made thick enough to absorb most of the incoming radiation, while still not inhib-

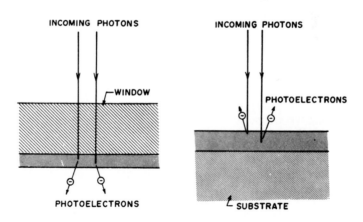

SEMITRANSPARENT   OPAQUE   PHOTOCATHODE
PHOTOCATHODE

FIGURE 15.9 Two main types of photoemissive cathodes. The photocathode surface (from which the photoelectrons are emitted) is normally in a vacuum environment.

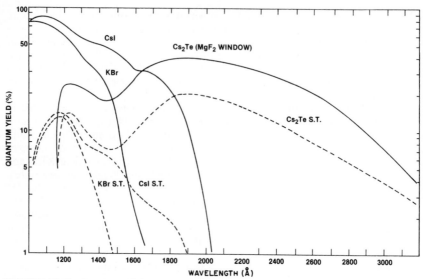

FIGURE 15.10 Quantum efficiency vs. wavelength is the middle- and far-ultra-violet wavelength range, of commonly used semitransparent and opaque photocathodes.

iting the escape (from the front surface) of those photoelectrons produced close to the light-incident surface. However, particularly for imaging applications, semi-transparent photocathodes are more often used in the wavelength range longward of 105 nm because they are more compatible with most electron-optical geometries. On the other hand, shortward of 105 nm, the only suitable window materials are very thin metal and/or plastic films, which usually cannot withstand 1-atm pressure differentials. Therefore, in the extreme uv and x-ray wavelength ranges, opaque photocathodes in windowless configurations are most often used.

In the extreme uv and x-ray ranges, the quantum efficiencies of opaque photocathodes remain high, on the average, but show marked variations with wavelength (see Fig. 15.11 and Refs. 12 and 13) and are strongly dependent on the angle of incidence of the radiation onto the surface. This is because the depth of penetration of the radiation into the material can be large compared with the maximum escape depth of photoelectrons. Hence, large angles of incidence increase quantum yield by providing a larger path length for incoming photons without further inhibiting the escape of photoelectrons. However, for each wavelength, there is a limit to the angle of incidence which can be usefully employed, because the reflectance of the photocathode material increases with angle of incidence, tending to offset the increase in efficiency expressed as photoelectrons per photon absorbed. In general, the maximum efficiency occurs at larger (more nearly grazing) angles of incidence as the wavelength decreases in the extreme-uv and soft x-ray range. For x-rays at normal incidence, semitransparent photocathodes made of low-density, "fluffy" CsI deposited on an x-ray-transparent thin-film substrate can have higher quantum efficiencies than full-density opaque CsI photocathodes deposited on microchannel plates.[15]

15.5.1 Nonimaging Photoemissive Detectors

Photoemissive devices without intrinsic imaging capability are normally used to make photometric measurements of radiation in a defined field of view and in a

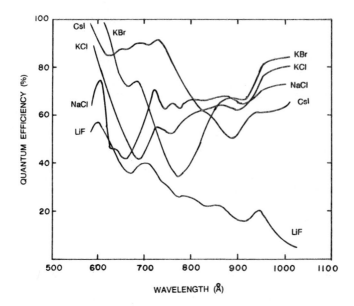

FIGURE 15.11 Quantum efficiency vs. wavelength of opaque alkali halide photocathodes in the extreme-ultraviolet wavelength range.[12] The apparent quantum efficiency can exceed 100% of photon energies sufficient to release two photoelectrons (i.e., twice the threshold energy).

wavelength range determined by the photoemissive device in combination with auxiliary filters and other optical components. However, they can be used to generate images by scanning a small field of view in one or two dimensions, although the intensity measurements at each point in the scanned field are not simultaneous as in a true imaging detector or photographic exposure.

Photodiodes. Photodiodes are the simplest form of photoemissive detector, consisting of a photocathode surface and a collecting electrode (which is at a positive potential relative to the photocathode). Versions of photodiodes using opaque and semitransparent photocathodes are shown in Fig. 15.12. Normally, the space between the photocathode and collector is maintained at high vacuum, so that pho-

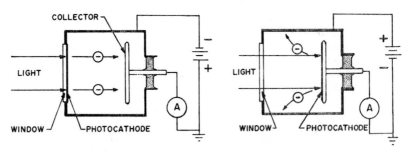

FIGURE 15.12 Diagrams of two simple photodiodes, using semitransparent (left) and opaque (right) photocathodes. The window on the opaque photocathode device is optional, provided the detector is operated in vacuum.

toelectrons do not collide with gas molecules in transiting the space to the collector (which could result in degraded time response) and to protect the photocathode from degradation by active gases. In the far-uv and x-ray ranges, a vacuum is also required to avoid absorption of the radiation of interest. Because of their simplicity, and the fact that the collected photocurrent is independent of applied voltage above a minimum threshold, photodiodes are preferred for accurate photometric measurements if light levels are adequately high (i.e., if the photocurrent produced is above about 10^{-13} A). Photodiodes are often used as photometric standards in the calibration of other types of uv and x-ray sensors.

Photomultipliers. The principle of secondary emission is used in photomultipliers to increase the primary photocurrent from a photocathode and thereby increase the ease and accuracy of measurement of low-intensity radiation. If an electron is accelerated and allowed to impact the surface of a suitable material, two or more *secondary* electrons may be emitted from the surface (i.e., geometrically similar to the photoemission from an opaque photocathode). Materials are chosen which have a high coefficient of secondary emission (average number of secondary electrons produced per incident primary electron). This coefficient is also dependent on the energy of the primary electron and its angle of incidence on the surface. Materials which are useful secondary emitters include alkali and alkaline-earth halides, metal oxides (BeO, MgO, etc.), and special semiconducting glasses.

Since the coefficient of secondary emission is usually a small number (less than 10), typically several stages of secondary emission are used in photomultiplier tubes. Figure 15.13 shows diagrams of two types of photomultipliers, including their secondary-emitting surfaces (dynodes). The total gain of a photomultiplier with n dynodes with gain G per dynode is G^n. If a photomultiplier has 14 dynodes with a gain of 3 per stage, a gain of 5×10^6 is produced. Gains of 10^6 to 10^7 are high enough that individual photoelectron events can be detected and counted (as in Geiger counters). Thus, the sensitivity of a photomultiplier is much greater than that of a photodiode using the same photocathode.

The *photon-counting* mode of operation gives the highest possible sensitivity in any type of radiation detector. In fact, the accuracy of measurement is limited mainly by the statistical fluctuation in the number of photons detected; if, in a given time interval, N photoevents occur, the statistical accuracy is $\pm \sqrt{N}$ and the signal-to-noise ratio is $N/!N = !N$. If there are *dark counts*, due to thermionic emission from the photocathode, cosmic rays, etc., constituting a background count rate N_b, in addition to a signal count N_s, the signal-to-noise ratio is $N_s/\sqrt{(N_s + N_b)}$. Since N_s is increased (for a given light level) by a higher photocathode quantum efficiency, it is seen that, at the very lowest light levels, quantum efficiency is a major determinant of detectivity and signal-to-noise ratio.

Photomultipliers are most often used with semitransparent, "end-on" photocathodes, as shown in Fig. 15.13; however, opaque photocathode configurations are also available. For devices sensitive in the wavelength range longward of 200 nm, a window must still be provided to prevent degradation of the photocathode by exposure to air; but windowless photomultipliers are practical at shorter wavelengths, especially below 120 nm.

Channel Multipliers and Microchannel Plates. As shown in Fig. 15.13, conventional photomultiplier tubes make use of several discrete dynodes, each held at a fixed potential relative to the photocathode, with increasing potential along the dynode string to the final collecting electrode. In the channel multiplier[16] (see Fig. 15.14), the dynode string is replaced by a single, continuous tube of semiconducting glass whose inside surface is specially processed to have a high secondary emission

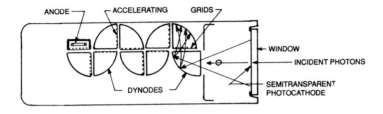

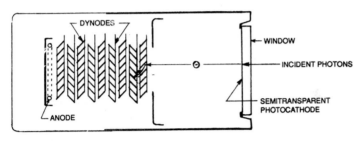

FIGURE 15.13 Diagrams of two photomultipliers, using different types of secondary-emitting dynodes for photoelectron multiplication.

coefficient. If a voltage is applied along the length of this tube, a primary electron entering one end collides with the wall and is multiplied, as shown in Fig. 15.14. The advantage of this approach is that the device can be made much simpler and more compact than a discrete-dynode multiplier chain. A potential disadvantage, relative to photomultipliers, is that channel multipliers are usually limited to operation at lower light levels than can be handled by photomultipliers. This is because the output current must be small compared to the current conducted in the semiconducting walls of the channel multiplier.

If a straight-channel electron multiplier as shown in Fig. 15.14 is operated at very high gains (greater than about 10^4), a problem known as *ion feedback* is encountered. Positive ions are produced by electron bombardment of the channel walls near the high-potential end of the channel; these in turn are attracted to the low-potential end of the tube (i.e., they travel in the opposite direction from that of the electrons). If the positive ions strike the wall near the front end, they can

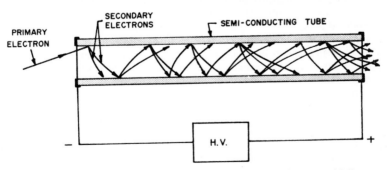

FIGURE 15.14 Principle of operation of a continuous channel electron multiplier.

also produce secondary electrons which result in *false counts*. To avoid this problem, channel multipliers are usually curved, into a C or helical shape, as shown in Fig. 15.15. This impedes the travel of the heavy ions back up the channel, without significantly impeding the flow of electrons down the channel. Curved channel multipliers can be used at gains as high as 10^7 without significant ion-feedback problems. To increase their utility as light detectors, channel multipliers can be equipped with conical inputs, as shown in Fig. 15.15, which serve as photocathode surfaces (below about 120 nm), or can be coated with materials such as CsI to enhance their far-uv and/or x-ray sensitivities.

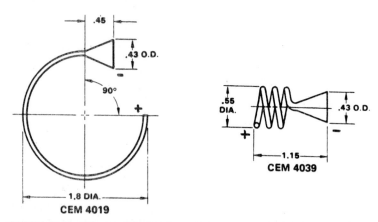

FIGURE 15.15 Two channel multipliers which are curved to inhibit "ion feedback" when operated at high gain, and which have funnel-shaped entrance apertures to facilitate their use as photon or electron detectors. (Courtesy of Galileo Electro-Optics, Inc.)

Since the gain versus voltage of a channel multiplier is relatively insensitive to scale, these devices can be made very small and still have high gains. Therefore, an extension of the channel multiplier approach is to make a large two-dimensional array of channel multipliers,[17] in the form of holes in a plate of semiconducting glass (see Fig. 15.16). The individual channels can be made as small as 10 μm in diameter, with 12- to 15-μm center-to-center spacing. Hence, a 40-mm-diameter microchannel plate (MCP) can have of the order of 7×10^6 separate channels. This allows the use of channel multipliers for imaging as well as nonimaging applications, since the outputs of each of the channels can, in principle, be collected and analyzed independently. However, even if all the outputs are collected by a single anode, the use of MCPs allows for a more compact photomultiplier design in cases where a large-diameter photocathode is required, since the dimension along the optical axis can be made much shorter than in a conventional photomultiplier. Also, the time response can be much shorter because of the short distance from the photocathode to final collecting electrode.

As mentioned above, ion feedback is a problem if straight channels are used with gains above about 10^4. It is difficult to make microchannel plates with curved channels, although this has been accomplished with satisfactory results and curved MCPs have been operated with gains[18,19] as high as 10^6. A simpler approach, where high gain is needed, is to stack two or more straight-channel MCPs in series, each operated with less than 10^4 gain, but with their channel axes tilted relative to each

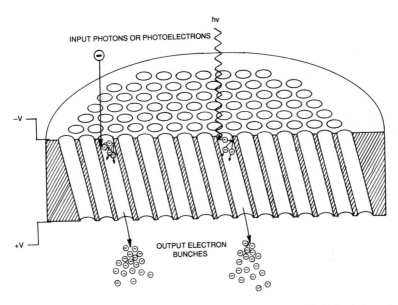

FIGURE 15.16 Diagram of a microchannel plate, a large array of individual channel multipliers fused together. The amplified output retains the spatial intensity distribution of the input.

other to inhibit ion feedback at the interfaces (see Fig. 15.17). This is called a *chevron* or "*Z*" configuration, dependent on whether two or three plates are used.

Microchannel plate sensors can be used with semitransparent or opaque photocathodes, as in conventional photomultipliers. However, a unique feature is that opaque photocathodes sensitive in the far-uv and x-ray ranges can be deposited directly on the front surface of the MCP, or (if the highest possible sensitivity is not essential) the front surface of the MCP can itself serve as the photocathode (shortward of about 120 nm). It should be noted, however, that the effective quantum efficiencies of opaque photocathodes of materials such as CsI, when deposited on the front surfaces of MCPs, are typically considerably lower than when deposited on flat substrates. Also, there is marked dependence on the angle of incidence of the radiation relative to the microchannel axes as well as to the front surface of the MCP, especially at extreme uv and x-ray wavelengths.[20-28]

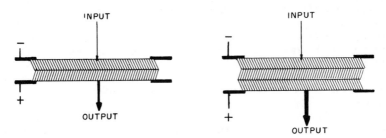

FIGURE 15.17 Use of two microchannel plates, stacked in series ("chevron" configuration) or three stacked together ("Z" configuration) to obtain higher electron multiplication gain than possible with a single, straight-channel plate.

15.5.2 Imaging Photoemissive Detectors

Image Intensifiers. Image intensifiers, or image converters, are among the simplest and first-developed electronic imaging devices.[29,30] They consist of a photocathode surface, onto which an optical image is projected, and a phosphor screen which emits light when the photoelectrons, accelerated to several kilovolts energy by a potential difference between the photocathode and phosphor, impact the phosphor. Since a phosphor screen can emit several hundred photons when impacted by an electron having 10 to 20 keV energy (see Fig. 15.18), a net gain in the number of photons (image intensification) can result. Also, if the spectral distribution of the light incident on the photocathode is different from that which is emitted by the phosphor, this allows viewing of radiation (such as ultraviolet or x-ray radiation) to which the eye is not directly sensitive (image conversion).

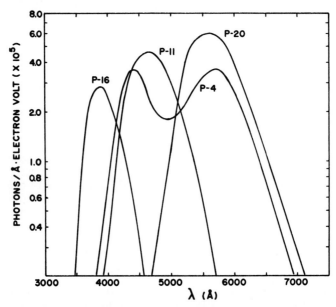

FIGURE 15.18 Light intensity output (photons/A) of commonly used phosphor screens versus wavelength, per electron volt of energy deposited by bombarding electrons. (Courtesy of ITT Electron Tube Division).

Image intensifiers use three basic techniques for focusing the photoelectrons emitted by the photocathode onto the phosphor screen, so that the visible light image output by the phosphor is as close as possible in resolution and geometric fidelity to the optical image projected on the photocathode. These are proximity focusing, electrostatic focusing, and magnetic focusing (see Fig. 15.19).

Proximity focusing, the simplest of the three, is actually a misnomer, in that no active focusing of photoelectrons occurs—the objective is to minimize the spreading of photoelectrons which naturally occurs when a plane-parallel photocathode and phosphor screen are placed facing each other and a potential difference is applied (see Fig. 15.19a). Although the photoelectrons are accelerated by the potential difference in the direction perpendicular to the plane-parallel electrodes, spreading of photoelectrons emitted from a point on the photocathode occurs because they

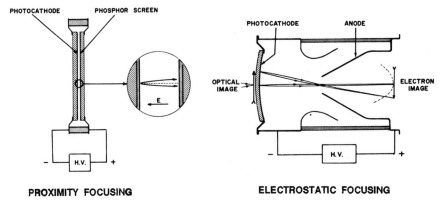

PROXIMITY FOCUSING   ELECTROSTATIC FOCUSING

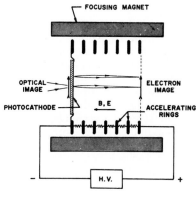

MAGNETIC FOCUSING

FIGURE 15.19 Three electron-optical techniques used in photoemissive electronic imaging devices. Proximity "focusing" does not produce an electron image, as do electrostatic and magnetic focusing.

are released from the photocathode with nonzero velocities (due to the difference between the energy threshold of photoelectric emission and the actual energy of the absorbed photon) and in random directions. Hence, the initial emission velocity usually includes a transverse, as well as an axial, component of velocity. The transverse component is not affected by the electric field between the electrodes, and hence results in divergence of the photoelectron trajectories and image blurring.

The spread of a point image is proportional to the maximum initial transverse or radial velocity v_r and the time of flight t from cathode to anode. Only the latter can be readily altered, and can be minimized by decreasing the cathode-anode spacing L and by increasing the accelerating potential V_L (which increases the axial velocity of the photoelectrons). The lateral displacement δ of a photoelectron is given by

$$\delta = v_r t = \left(\frac{2eV_r}{m}\right)^{1/2} 2L\left(\frac{2eV_L}{m}\right)^{-1/2} = 2L\left(\frac{V_r}{V_L}\right)^{1/2} \quad (15.2)$$

and the maximum diameter of a point image is therefore $D_{max} = 2\delta$. In practice, the minimum spacing L and the maximum potential V_L are limited by the risk of high-voltage breakdown between the electrodes due to field emission. Typically, the electric field $E = V_L/L$ should not exceed about 5 kV/mm. For a given electric field E, it is seen that resolution increases as $1/V_L$, but that the accelerating voltage (and hence phosphor gain) are reduced with closer spacings.

Most practical proximity-focused image intensifiers have relatively low resolutions [on the order of 15 to 20 line pairs per millimeter (lp/mm)]. However, practical advantages of such devices are their simplicity and compactness (in the axial direction). They can easily be made in very large diameter formats, which can in some cases compensate for low spatial resolution. They also have the advantage, in some applications, of being relatively unaffected by external magnetic fields.

Electrostatic focusing uses electric fields in a manner to actively focus photoelectrons in transit from the photocathode to the phosphor screen (see Fig. 15.19b). Electrostatic focusing is based on the fact that a particle in a potential field follows laws similar to those followed by light rays in a medium of varying refractive index. The "index of refraction" is here proportional to the velocity of the electron at a given point in space, and equipotential surfaces are analogous to lens surfaces in ordinary optics.

In a typical electrostatically focused image tube, as shown in Fig. 15.19b, the photocathode and phosphor screen are curved toward each other, and a conical accelerating electrode with an aperture is used to accelerate and focus the photoelectrons. The resolution achieved at the center of the field can be much better than that typically obtained with proximity-focused devices (typically 40 to 50 lp/mm); however, the resolution decreases toward the edges of the field.

The photocathode curvature in electrostatically focused image tubes is a problem for uv imaging, because fiber-optic faceplates cannot be used to match the flat focal surfaces of most optical systems as in the visible. Correction of a flat focal surface to the curved photocathode, in general, requires a double-concave faceplate or a separate correcting lens. The outer radius is determined by the index of refraction at the wavelength of interest and the inner radius. It is noteworthy, however, that electrostatic tube photocathodes better match the convex focal surfaces of Schmidt cameras than do flat photocathodes.

A unique feature of electrostatic focusing is that (de)magnification ratios (image size on the photocathode/image size on the phosphor screen) other than unity can be readily achieved. Demagnification of the photocathode image of 2:1 to 10:1 is often used, especially in applications involving large-area, low-resolution images (such as in many medical or industrial x-ray applications). The demagnification results in a brighter image on the phosphor (in proportion to the square of the demagnification); however, a given image resolution on the phosphor corresponds to lower resolution as measured at the photocathode surface (in inverse proportion to demagnification).

Magnetic focusing uses a combination of electric and magnetic fields to accelerate and focus photoelectrons. Plane-parallel photocathode and anode are used, with the fields coaxial (see Fig. 15.19c). The magnetic field corrects for the spreading of the photoelectrons because of their randomly oriented initial emission velocities, by confining the radial components of velocities to circles centered on magnetic field lines. The condition for focusing is that the electrons complete one (or an integral number) of loops (in the radial direction) in the same time that it takes them to travel from the photocathode to the phosphor screen. This condition can be shown to be satisfied when the distance L, magnetic field B, and accelerating potential V are related by

$$L = \frac{\pi}{B}\left[\frac{2mV}{e}\right]^{1/2} \tag{15.3}$$

for single-loop focusing. To first order, this expression is independent of the initial emission velocity of the photoelectrons.

Magnetic focusing can produce very high resolution (100 lp/mm or better) with uniform resolution over a large flat format. Its main disadvantage is the requirement for a large focusing magnet or solenoid coil.

In many practical applications of x-ray imaging, such as in medicine, it may be difficult to use x-ray-sensitive photocathodes in image tubes because of the unavailability of windows which can stand atmospheric pressure but still transmit the x-rays. In these cases, it is common practice to use an x-ray phosphor to convert the x rays to visible light. The phosphor is deposited on the front surface of a fiber-optic faceplate, which is the input window and photocathode substrate of a visible-light-sensitive image intensifier. A thin aluminum film is used on the outer surface of the phosphor to block visible light and to reflect forward-emitted phosphor light back toward the faceplate. Although less efficient than use of an x-ray-sensitive photocathode, this approach is relatively simple and flexible.

Although originally developed for direct viewing applications, image intensifiers can also be used with photographic recording of their light output. This provides an increase in the effective recording sensitivity and/or a broader spectral range of the photographic process. The overall gain is greatly increased if the phosphor screen is deposited on a fiber-optic output window, with the recording film pressed into contact with the outside surface of the fiber-optic plate, instead of using lens coupling of the phosphor light to the film. In scientific applications such as astronomy, in practice, the benefits of image intensifiers are primarily due to the higher quantum efficiencies of photocathodes as compared to photographic emulsions, rather than the light amplification factor per se.

Image intensifiers can also be coupled to television camera tubes and solid-state video sensors, such as charge-coupled devices (CCDs), to improve their low-light-level sensitivity and/or broaden their spectral range. As in photographic recording, the highest gain (and the most compact overall device) is realized by the use of fiber-optic coupling of the phosphor to the final sensing device.

Electrographic Detectors. Electrographic detectors, like image intensifiers, are very simple in principle. They use film- or plate-based emulsions, very similar to ordinary photographic emulsions, to detect accelerated photoelectrons directly. The electron-sensitive emulsion takes the place of the phosphor screen in an image intensifier, as in Fig. 15.19.

In electrography, each photoelectron (when accelerated to the order of 20-keV energy) can produce one or more blackened grains in the processed emulsion. Hence, in comparison to direct photography, the gain factor is the ratio of the photocathode quantum efficiency to that of a photographic emulsion exposed to the same photon image. However, other advantages accrue from electrography in comparison to photography (or an image intensifier with photographic recording).

Typical electron-sensitive emulsions have much finer grains and much lower fog density than high-speed photographic emulsions. Also, the relationship between processed film density and exposure is linear over a very wide range, and there are no threshold effects or reciprocity failure.

A practical difficulty with electrography which has hindered its widespread use in the visible and near- and middle-uv ranges is that the recording emulsion outgasses in vacuum. The gases released (primarily water vapor) tend to react with the photocathode and degrade its quantum efficiency, unless elaborate techniques for protecting the photocathode are employed. However, this is much less of a problem in the far-uv and x-ray wavelength ranges, because photocathodes useful in this spectral region are much less susceptible to degradation by emulsion out-

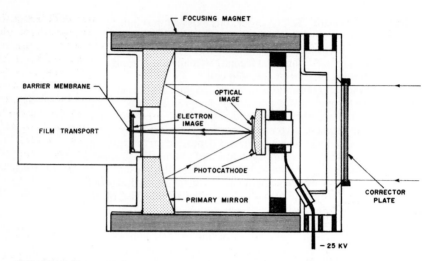

FIGURE 15.20 Diagram of an electrographic Schmidt camera used for far-ultraviolet studies at the Naval Research Laboratory. Opaque alkali-halide photocathodes are used to provide high overall detection efficiency. The barrier membrane shown is a thin, aluminized plastic film which is transparent to energetic electrons but prevents exposure of the film by stray visible light.

gassing. Electrographic detectors have been widely used in far-uv cameras and spectrographs by NRL, for space-based astronomical and upper-atmosphere observations.[14] Figure 15.20 shows an NRL-developed electrographic Schmidt camera, which makes use of the high efficiencies of opaque alkali-halide photocathodes in the far uv (as shown in Fig. 15.10).

Imaging Microchannel Plate Detectors. As discussed in Sec. 15.4.1, microchannel plates provide a convenient means for amplifying photoelectrons while still retaining the spatial distribution information. Here, we will discuss their applications to image intensifiers and electrographic detectors and introduce several new types of imaging detectors with direct, electrical-signal outputs (i.e., which do not utilize phosphor screens or film).

When a microchannel plate is introduced into any of the three basic types of image intensifiers (as shown in Fig. 15.21), it greatly increases the overall photon gain, because of the multiplication of photoelectrons within the MCP. Normally, a single-stage image intensifier does not provide enough gain for the ultimate low-light-level capability (with direct viewing of the phosphor, photographic recording, or video recording) and therefore requires amplification to the level where single photoelectron events can be individually detected. A single-stage MCP can provide an electron gain of several hundred, and a two-stage MCP can provide gains of 10^5 or more, before the MCP output electrons impact the phosphor screen. When an MCP image intensifier is coupled to a CCD or other video sensor, this allows true photon-counting imaging systems whose signal-to-noise ratio is limited only by the statistical fluctuations in the number of photoevents at low light levels.

An MCP can also be used as an intermediate stage in an electrographic detector, such as that shown in Fig. 15.20. This is especially useful for observations of very faint, diffuse sources, because individual photoelectron events are amplified to the extent that they can be individually detected and measured with a microdensitometer, the device normally used for quantitative analysis of electrographic (and

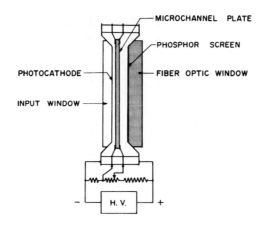

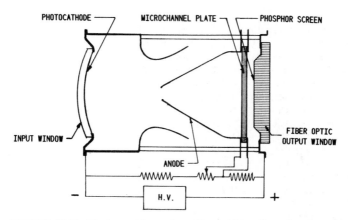

FIGURE 15.21 Versions of image intensifiers which utilize microchannel plates to increase the overall gain of the devices.

photographic) exposures.[31] Hence, the signal-to-noise ratio is determined by photoevent statistics, rather than the readout noise of the microdensitometer, at low light levels.

A number of electronic imaging devices are based on the direct detection and processing of the amplified photoelectron events which exit the back surface of an MCP or stack of MCPs. These include various schemes for determining the spatial location (x, y coordinates) of the individual events. Examples include crossed-grid arrays of collecting anodes, continuous resistive anodes, and various "coded anode" devices such as the wedge-and-strip multianode microchannel array (MAMA), Codacon, and delay-line anode devices.

For use in x-ray or far-uv detection, a photocathode material (such as CsI) can be deposited on the front surface of the upper MCP. Alternatively, the MCP can be fed with photoelectrons from a separate opaque or semitransparent photocathode, using proximity, electrostatic, or magnetic focusing.

The simplest electronic readout technique for MCPs which retains one- or two-dimensional spatial resolution involves the use of a one- or two-dimensional array of discrete collecting electrodes, each of which is connected to a separate amplifier and counting circuit.[32] This approach allows very high count rates and high reliability (the failure of one readout circuit affects only one pixel in the array), but has the disadvantage that the number of pixels is limited by the number of amplifiers and counting circuits which can reasonably be accommodated. Therefore, in practice, this approach is limited to relatively small arrays (a few hundred pixels total).

The crossed-grid readout technique (see Fig. 15.22) uses two orthogonal arrays of parallel wire collectors, connected to each other by divider resistors.[33] A photoelectron event injects an electrical signal into one or more of the x- and y-grid

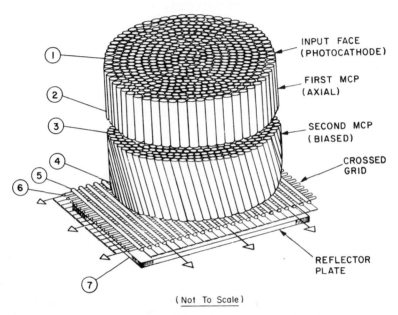

INPUT FACE
(PHOTOCATHODE)

FIRST MCP
(AXIAL)

SECOND MCP
(BIASED)

CROSSED
GRID

REFLECTOR
PLATE

(Not To Scale)

FIGURE 15.22 A crossed-grid readout scheme for microchannel plate detectors.[33] The anode wires in each axis are resistively coupled, so the number of amplifiers needed is considerably less than the number of wires.

electrodes. The position of the event in each axis is determined from the ratio of the amplitudes of the signals recorded by two amplifiers on opposite sides of the event. (The amplifier closest to the event receives the largest signal.) Array sizes of 1000×1000 pixels or more are practical, using $100 + 100$ amplifiers. This type of detector was used in the high-resolution imager for soft x-rays (0.25 to 1.5 keV) on the second High Energy Astronomical Observatory (HEAO-2), or Einstein, satellite, launched in 1978.

In the resistive-anode readout technique,[34-37] the crossed-grid array of collecting anodes is replaced by a continuous strip of semiconducting material (see Fig. 15.23). Amplifiers are attached to two ends of the strip (for a two-dimensional readout, such as may be used in spectroscopy) or to four sides (for two-dimensional imaging). For the one-dimensional case, the position x along the strip (length L) is determined from the ratio of the charge pulse amplitudes input to the two amplifiers (proportional to $x/(L - x)$. Advantages of the resistive strip readout technique are simplicity and need for fewer amplifiers. However, the spatial resolution and maximum count rate are less than for some other approaches. The resolution is limited by thermal noise in the resistive strip, although it can be improved by using higher MCP gain. Resolutions of 0.2 to 1 percent of the active length are typically achieved.

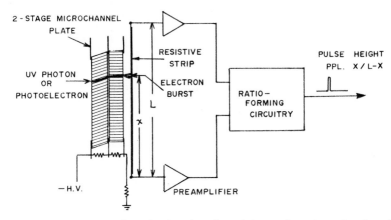

FIGURE 15.23   A one-dimensional version of a resistive-anode readout microchannel plate detector. The location of a photoevent is determined from the ratio of signal amplitudes output by each end of the anode.

Another class of coded anodes consists of two or more sets of electrodes, connected to different amplifiers but electrically isolated from each other, whose geometries are such that the position of a charge pulse can be determined from the ratios of charge picked up by the anodes.[38,39] A one-dimensional example consists of two sets of triangular electrodes, interleaved with each other (see Fig. 15.24a). Provided the triangle pattern cycle width is smaller than the diameter of an MCP output charge pulse, the ratio of charges in the upper set of triangles to that in the lower set is simply $y/(L - y)$ (as in the one-dimensional resistive anode). A two-dimensional version, a wedge-and-strip anode scheme, is illustrated in Fig. 15.24b. Here, only three amplifiers are needed, and the spatial coordinates of the photoevent are determined from the ratios of signals in the three amplifiers (A, B, and C).

Although the electrode patterns in typical anode arrays repeat on a scale of order 1 mm, the centroid of the charge pulse can be determined to much better accuracy. In fact, a rather large gap is normally used between the back of the MCP stack and the anode to assure that the charge cloud spreads to a diameter at least equal to the pattern width, to allow accurate ratioing. Since the electrodes are metallic, the thermal noise inherent in resistive anodes is eliminated, and resolutions of 0.2 percent of the image dimensions or better have been demonstrated.

The multianode microchannel array coded-anode technique[40] is illustrated in Fig. 15.25. Although a crossed grid of discrete electrodes is used, the technique allows an array of $a \times b$ pixels in one dimension to be read out using $a + b$ readout amplifiers, where a = coarse-encoding electrodes and b = fine-encoding electrodes, as shown in Fig. 15.25b for a one-dimensional array. A photoevent is sensed simultaneously on at least one fine and one coarse electrode; the coarse electrodes remove the ambiguity which results from using a fine electrode to sense positions

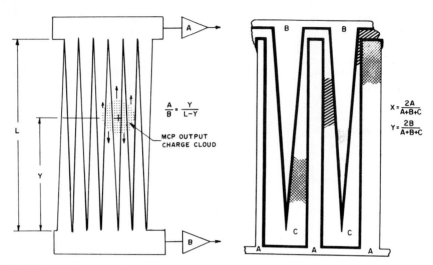

FIGURE 15.24   A one-dimensional version (left) of a "partitioned anode" readout for a micro-channel plate detector. As for the resistive-anode detector, the ratio of outputs to amplifiers connected to the upper and lower sets of electrodes determines the location of the photoevent. One version (right) of a "wedge and strip" anode[38] which requires only three amplifiers to determine the location of a photoevent in two dimensions.

at different segments of the array. A two-dimensional image of $(a \times b)^2$ pixels can be read out using $2(a + b)$ electrodes. MAMAs with up to 1024×1024 pixels are currently in use.

The MAMA typically uses a curved-channel MCP with a gain of 10^5 to 10^6, with the anode layer in close proximity (about 50 mm) to the back surface of the MCP to minimize spreading of the charge cloud before it is detected. Most of the other coded-anode devices use two or more stacked MCPs, with total gains of 10^7 or more, and with a larger gap between the rear of the MCP and the sensing anodes. Unlike most of the previously described coded-anode devices, the MAMA does not depend on centroiding; hence its electrode patterns must have cycle lengths smaller than the desired resolution limit (e.g., 25-μm spacing for 50-μm resolution). The MAMA has the advantage of a higher maximum count rate, and hence wider dynamic range, compared to the centroiding, high-gain-MCP coded-anode detectors.

The Codacon is somewhat similar to the MAMA, except for the details of the collecting electrode structure and readout electronics.[41] Figure 15.26 illustrates a typical anode structure for a one-dimensional array. Here, vertical charge spreader electrodes collect the output charge from the MCP. These electrodes are insulated from, but capacitively coupled to, a series of horizontal electrodes with alternating wide and narrow segments, which result in either a large or small pulse in the connected output amplifier. The combination of large and small pulses in the series of six electrodes (connected to three amplifiers) produces a 3-bit Gray code address of the photoevent, uniquely locating at which of the eight horizontal locations it occurred. This technique can be extended to larger numbers of addresses by using more than 3 bits worth of horizontal coding electrodes; also it can be extended to two-dimensional arrays.

The most recent development in MCP coded-anode readouts is the delay-line approach.[42,43] This is similar to the resistive anode readout in that the signals at

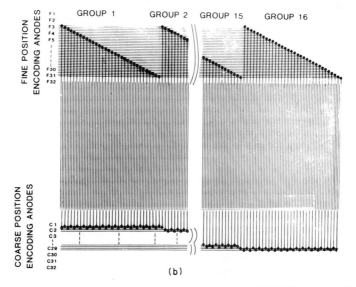

FIGURE 15.25 (a) Diagram of a Multi-Anode Microchannel Array (MAMA) detector. (b) Illustration of a one-dimensional array of coarse and fine encoding anodes, which can localize photoevents in a × b positions with only a + b amplifiers. (Courtesy of J. G. Timothy, Stanford University).

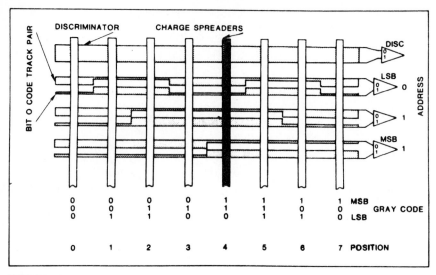

FIGURE 15.26 Diagram of the Gray-code encoding scheme and anode configuration used in the Codacon microchannel plate detector.[41]

two ends of a collecting anode are compared. However, in this case, it is the pulse arrival times, not their amplitudes, which are compared. This is the same approach used in the imaging proportional counter (IPC) discussed previously (see Fig. 15.6), but the electrode patterns are in the form of flat, metallic collectors (as in the wedge and strip detector). As shown in Fig. 15.6 for the IPC, a timer measures the time difference between the start and stop pulses, with an additional fixed delay introduced at the bottom end of the delay line to ensure that, regardless of the position of the charge pulse, the signal is always received first from the top end. The time interval recorded is $\Delta_t = t_0 + 2x/v$, where t_0 is the fixed time delay, x is the location of the charge pulse centroid, and v is the delay line propagation speed. Time (and hence space) resolution are improved by slow propagation velocity; however, this acts to increase the detector dead time and hence reduce dynamic range.

Since high resistivity is not required (impedance can be primarily capacitive), the resolution is not limited by thermal noise as in resistive anode detectors. A unique feature of the delay line technique is that the spatial resolution is, in principle, nearly independent of the total length of the collector—hence, very large format, high-resolution MCP sensors are possible.

15.6 SOLID-STATE DETECTORS

15.6.1 Nonimaging Solid-State Detectors

Photoconductive sensors, based on silicon or other semiconducting materials, can be used to detect ultraviolet and x-ray radiation, as is done in the visible and infrared. In the uv and x-ray range, the advantages are the compactness and simplicity in comparison to photomultipliers and proportional counters; however, in some applications, a disadvantage is the fact that most photoconductive materials (such as silicon) have equal or greater sensitivity in the visible and near ir, which can cause problems in making accurate uv or x-ray measurements if the source in question is also a strong emitter in the visible and ir. Except for energetic x-rays, the signal produced by the absorption of a single photon is not adequate for photon-counting sensitivity. Also, the sensors must be cooled to cryogenic temperatures to minimize thermal dark current.

In the x-ray wavelength range, an advantage of solid-state sensors is that they can provide spectral information, in a manner analogous to proportional counters. This is because each x-ray photon produces a discrete photoevent (consisting of a large number of electron-hole pairs) in the material, the size of which is proportional to the absorbed photon energy. For silicon, each electron-hole pair requires 3.7 eV to be created; hence theoretically the absorption of a photon with energy E (eV) produces $E/3.7$ electron-hole pairs. The detection efficiency, wavelength range of sensitivity, and spectral resolution achievable depend on the type of semiconducting material, its thickness, details of its processing, and the temperature at which it is operated, as well as the sensitivity of the associated electronics. Cooling the device allows better sensitivity and spectral resolution, by eliminating the noise associated with thermal dark current.

Figure 15.27 illustrates a typical solid-state sensor used for astronomical x-ray nondispersive spectroscopy.[44] It is a reverse-biased silicon p-n junction, which when cooled is essentially nonconducting, unless x-rays or other photons create electron-hole pairs in the carrier-depleted junction zone (photoconductivity). It was used in the HEAO-2 satellite (*Einstein*) and was sensitive in the 0.4- to 4-keV energy range. It had an energy resolution of about 120 eV (versus 400 to 500 eV for a proportional counter in the 1 keV range) and high quantum efficiency (greater than 80 percent in the energy range above 1 keV).

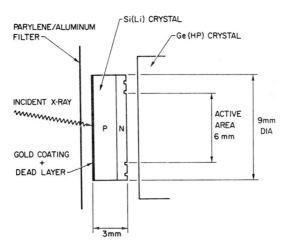

FIGURE 15.27 Diagram of a solid-state x-ray detector, used in the *Einstein* High Energy Astronomy Observatory for non-dispersive spectroscopy.[44] The filter in front blocks visible and uv light, whereas the Ge(HP) crystal is used as an anticoincidence detector to reject charged-particle events.

15.6.2 Imaging Solid-State Detectors

Two-dimensional solid-state sensor arrays, such as charge-coupled devices, can be used for ultraviolet and x-ray detection, although the devices must be specially prepared for use in the wavelength range of interest. The comments regarding solid-state nonimaging devices in Sec. 15.6.1 also apply to imaging devices. In addition to direct detection of uv and x-ray radiation, solid-state array sensors can be used to detect the visible-light outputs of image intensifiers, whose photocathodes are in turn sensitive to uv and/or x-rays. CCD arrays are now available in formats as large as 4096 × 4096 pixels, and with pixel sizes ranging from about 30 μm to as small as 7 μm. Therefore, they offer significantly better resolution capability than most other electronic imaging devices we have discussed. Reference 45 gives a good introduction to the operating principles of CCD arrays.

CCD arrays can be operated in an analog readout mode (as in TV cameras), giving a signal proportional to the integrated photon flux (which can be digitized for further processing). However, if operated with an image intensifier, in electron-bombarded mode, or with x-ray photon inputs, the charge pulses due to single photoevents can be individually detected, above the dark and readout noise of the CCD. In this case, photon-counting modes similar to those used with coded-anode MCP detectors or proportional counters may be used.

Figure 15.28 shows that there is an extremely wide range of photon penetration depths in silicon, ranging from less than 10 nm in the near and middle uv to 100 μm for 10-keV x-rays.[46] Most CCDs tend to have low sensitivity in the middle and far uv (300 to about 70 nm) because of (1) the high absorptivity of silicon in this

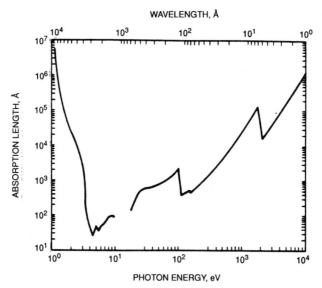

FIGURE 15.28   Characteristic absorption depth vs. wavelength in silicon.[46]

wavelength range, which causes photons to be absorbed in the gate structure or very close to the surface of the device, and (2) the high absorptivity of silicon oxide (in the far uv below 150 nm), which is normally present on the surfaces of CCDs and other silicon-based sensors, and tends to prevent the far-uv photons from reaching the silicon device at all.[47–51]

An obvious advantage of direct detection of ultraviolet and x-ray radiation with a CCD, versus using it to observe the output of a uv- or x-ray-sensitive image tube, is that the total system is smaller, lighter, less complex, and requires no high voltages. However, especially in the middle uv, disadvantages are that, since the CCD has comparable or greater sensitivity to visible and near-ir radiation, it is difficult to reject unwanted response to these longer wavelengths. Also, since the CCD lacks the gain factor of an image intensifier, its sensitivity at low light levels is limited by the device readout noise (typically 5 to 50 electrons per pixel) rather than by photon statistics.

A simple method for providing uv sensitivity, applicable to both frontside- and backside-illuminated CCDs, is to coat the device with a phosphor, such as coronene or liumogen. The phosphor converts uv radiation into visible light, to which the device is more sensitive. This approach was used in the Wide Field/Planetary camera, which is one of the primary scientific instruments on board the Hubble Space Telescope. However, the detection efficiencies of phosphor-coated CCDs, especially at wavelengths shortward of about 150 nm, are lower than what is possible with direct detection of the photons within the silicon, using specially prepared CCDs.

CCDs can be made directly sensitive to uv and soft x-rays by special preparation of the devices.[46-52] To avoid absorption in the frontside polysilicon gate electrodes and silicon oxide insulating layer, the CCD must be operated with backside illumination. The device also must usually be thinned (typically, to about 10 μm or less) so that electrons produced near the back surface can be efficiently collected by the frontside electrodes with minimal lateral spreading. Figure 15.29 shows theoretical and measured internal quantum efficiencies of properly prepared silicon

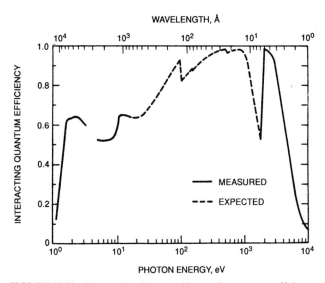

FIGURE 15.29   Measured and expected interacting quantum efficiency (excluding surface reflection losses) vs. wavelength in a thinned, backside-illuminated CCD optimized for use in the uv and soft x-ray wavelength ranges.[46-49]

devices, exclusive of surface reflection losses.[46-49] It is seen that very high efficiencies (more than 50 percent at most wavelengths) can be achieved.

Absorption of uv photons close to the surface of the silicon, even in thinned, backside-illuminated devices, can still be a problem because of trapping of photogenerated electrons by positive charges in surface defects. These produce a potential gradient in the surface layer which is opposite to that in the bulk of the silicon; i.e., electrons are attracted toward the back surface (where they are lost to recombination), rather than to the collecting electrodes (see Fig. 15.30). This can be corrected in a number of ways.

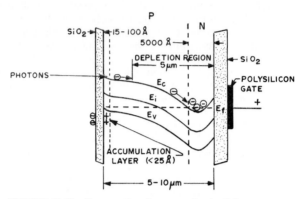

FIGURE 15.30 Cross section diagram and potential energy curves in a thinned, backside-illuminated CCD, with the backside accumulated to avoid trapping of photoelectrons there.[48]

One method is to implant, by means of an ion beam incident on the backside of the thinned CCD, p-type impurities to create a potential maximum, about 100 nm from the back surface.[51] This is followed by a second thinning, to just beyond the potential maximum, so that electrons photogenerated near the backside see an electric field directing them away from the backside and toward the frontside potential well. Another method is to charge the back surface oxide layer with photoelectrons, produced by strong illumination of the device with ultraviolet radiation.[46–49] This negative surface charge compensates for the positive surface defect charges, and repels electrons photogenerated within the silicon away from the backside. Yet another method is to coat the back surface with a thin film of platinum (flash gate).[52] The platinum has a higher work function than the silicon, so that its Fermi level is lower, relative to the vacuum energy level, than in silicon. When a layer of platinum is brought nearly into contact with the silicon (separated by only a thin oxide layer), electrons pass through the oxide to the platinum layer, to equalize the Fermi levels on each side (see Fig. 15.31). Since this means lowering

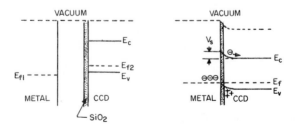

FIGURE 15.31 Illustration of the use of a "flash gate" (a thin film of a high work function metal, such as platinum) to provide accumulation of the back surface of a thinned CCD.[52]

the energy levels on the silicon side, a potential maximum exists at the surface which repels photogenerated electrons away from the interface.

For energetic x-ray-sensing applications, where a relatively thick CCD is required to efficiently absorb the photons, use of *deep-depletion* devices based on high-resistivity *p*-type silicon is beneficial.[51,53] The high resistivity allows a steep potential gradient, extending throughout the bulk of the device (see Fig. 15.32), which improves charge collection efficiency and minimizes lateral diffusion of photoelectrons. In turn, this increases spatial resolution and detection efficiency.

An important benefit of direct detection of extreme-uv and x-ray photons is

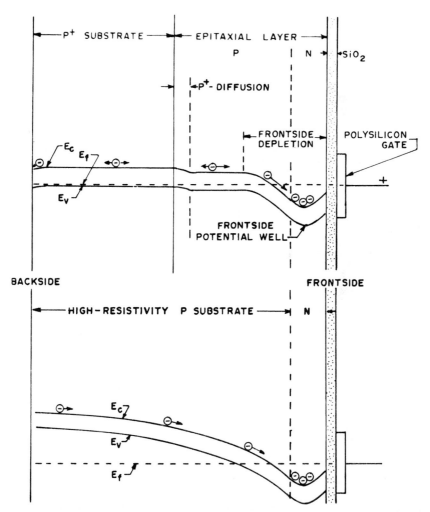

FIGURE 15.32 (Top) Diagram of a typical, unthinned CCD and potential curves, showing poor collection efficiency for electrons photogenerated outside of the frontside depletion region.[46] (Bottom) Illustration of how the depletion region can be enlarged, by the use of high-resistivity silicon (and a higher gate voltage), to increase the efficiency of photoelectron collection in a thick CCD (as may be required for efficient detection of energetic x-rays).

that each photon can produce more than one electron-hole pair. Theoretically, one electron-hole pair results from each 3.7 eV of absorbed photon energy, yielding an effective gain relative to detection of visible-light photons with the same quantum efficiency. For example, an x-ray photon of 1.2-nm wavelength (1000-eV energy) can produce 270 electron-hole pairs, whereas the readout noise of a good scientific-grade CCD can be 10 electrons per pixel or less. Hence, in detection of high-energy photons, signal-to-noise ratios approaching the statistical limit for true photon-counting systems can be achieved.

Of particular interest is that a CCD used to directly detect an x-ray image can also provide spectral information on each photon event, as discussed above for nonimaging solid-state detectors. That is, the CCD can provide three dimensions of spatial and spectral information. The amplitude of each photoevent is proportional to the energy of the absorbed photon (provided there are no losses either in a frontside dead layer, important at low energies, or less than total absorption within the active thickness of the device, important at high energies). In principle, the limit on energy resolution can approach the 3.7 eV required for each electron-hole pair, but in practice is limited by the CCD readout noise and Fano noise. As an example of current state of the art, a resolution (line FWHM) of about 120 eV has been demonstrated at the Fe^{55} 5.9-keV x-ray line.[50,52,54]

In general, the spectral resolving power $\delta E/E$ is usually better for shorter wavelengths (more energetic photons) because of the larger charge pulse each photon produces (which allows its absolute size to be measured more accurately). However, thicker devices are required for efficient detection of the more energetic x-rays, which tends to degrade the spectral (as well as spatial) resolution because the charge is split among several pixels, rather than concentrated in a single pixel, degrading the accuracy of measuring the total. This effect can be minimized by using a deep-depletion CCD, in which a high and uniform potential gradient is maintained through the thickness of the device to minimize charge spreading.

Several spaceflight experiments involving extreme-uv and x-ray imaging and spectroscopy are currently under development. They use CCD arrays for direct detection of the photons and take advantage of their capability for simultaneous nondispersive spectrometry and two-dimensional imaging.

15.6.3 Imaging Electron-Bombarded Silicon Detectors

When a high-energy electron is incident on the sensitive area of a solid-state silicon sensor, such as a CCD, many electron-hole pairs can be produced in the collisions of the electron with silicon atoms in the deceleration process. As is the case for absorption of high-energy photons in silicon, one electron-hole pair can be produced for each 3.7 eV lost by the incident electron. Therefore, neglecting any energy loss in a surface dead layer, it can be seen that a 10-keV electron can produce a charge pulse of about 2700 electron-hole pairs. Since this is far in excess of the typical readout noise of a CCD, single energetic electrons are easily detected and counted.

An electron-bombarded CCD (EBCCD) sensor consists of an image tube structure (see Fig. 15.20) with a semitransparent or opaque photocathode, sensitive in the wavelength range of interest, with a CCD array at the electron focus. There are two major advantages of this approach for uv and x-ray sensing. First, as mentioned above, is the high gain factor resulting from the electron-bombarded mode. Secondly, and in some cases more importantly, the spectral response of the device is determined by the image tube photocathode, not by the CCD itself; hence it is easy to make a device which is totally insensitive to visible and near-uv light while retaining high sensitivity in the far-uv and/or x-ray ranges.

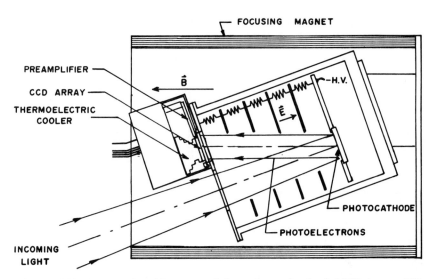

FIGURE 15.33   Diagram of an oblique magnetic focus, electron-bombarded CCD detector. This device makes use of the high quantum efficiency and short-wavelength utility of windowless, opaque alkali-halide photocathodes.

As was shown in Figs. 15.10 and 15.11, opaque photocathodes typically have much higher quantum efficiencies than semitransparent photocathodes in the middle-uv through x-ray wavelength ranges; also, they can be used without windows, which can limit the ranges of spectral sensitivity. Figure 15.33 shows one type of magnetically focused EBCCD sensor, developed at Princeton University and the Naval Research Laboratory,[55–57] which takes advantage of these benefits of opaque photocathodes in addition to those of EBCCD sensing. The oblique-focus sensor has been used by Princeton in sounding rocket astronomy experiments. NRL has also developed an EBCCD version of the opaque-photocathode Schmidt camera shown in Fig. 15.20, and both types of EBCCD sensors are planned for use by NRL in a future satellite mission.

15.7 SCINTILLATION DETECTORS

Scintillation detectors make use of the visible or uv radiation emitted by a transparent solid when a high-energy photon interacts with it. The light flash is typically detected by a photomultiplier tube (see Fig. 15.34), although an imaging detector could also be used with an appropriate optical system. In a sense, this can be considered an extension of the use of conversion phosphors (discussed earlier for use in uv and x-ray imaging) to higher energies, by use of a thicker fluorescent medium.

As is true of proportional counters and solid-state sensors, the amplitude of the signal pulse of a scintillation counter is proportional to the energy of the incident photon. Therefore, spectroscopic as well as intensity information about the incoming photons can be obtained. The spectral resolution, $\delta E / E$, improves at higher energies (roughly as $1/\sqrt{E}$). A typical NaI scintillator can yield about 8 percent resolution (FWHM) at 660 keV.

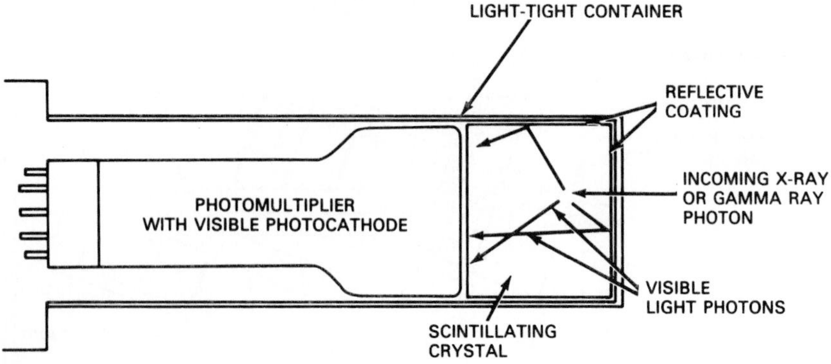

DIAGRAM OF SCINTILLATION
DETECTOR

FIGURE 15.34   Diagram illustrating the principle of a scintillation detector, as used for detection and non-dispersive spectrometry of high-energy x-rays and gamma rays.

Scintillation detectors are normally used only for high-energy x-rays and gamma rays (above about 20 keV), where proportional counters become inefficient (because the photons are not efficiently absorbed in the active gas volume). Scintillation detectors make use of high-density solid materials, such as NaI or CsI, which provide better stopping power for the energetic photons. These materials, however, must usually be doped with a small amount of an impurity, such as thallium, to provide efficient conversion of the photon energy into visible light. The detection efficiency is nearly 100 percent, until energies are reached at which the photon being detected can pass completely through the scintillation crystal (more than several hundred keV). For example, a typical NaI crystal 7.5 cm in diameter and thickness is only 40 to 50 percent efficient near 500 keV, but the sensitivity is improved by using a larger crystal.

Large scintillation crystals can be obtained which, when viewed by an array of photomultiplier tubes, can provide high sensitivity to low-intensity sources (such as in high-energy astronomy). They can also be obtained in various shapes, and combined to provide anticoincidence shielding for rejection of background events (such as those caused by cosmic rays). Scintillation detectors have been used in the High Energy Astronomical Observatories and Gamma Ray Observatory (GRO) space missions, among others, as well as in ground-based laboratory work. Very large scintillation detectors were used by NRL in its Oriented Scintillation Spectrometer Experiment (OSSE) on GRO, launched in April 1991.

15.8   REFERENCES

1. Matsunaga, F. M., "Photoionization Yield of Several Molecules in the Schumann Region," Contribution No. 27, Hawaii Institute of Geophysics, University of Hawaii, November 1961.

2. Watanabe, K., T. Nakayama, and J. Mottl, "Ionization Potentials of Some Molecules," *J. Quant. Spectrosc. Radiat. Transfer*, vol. 2, p. 369, 1962.

3. Watanabe, K., F. M. Matsunaga, and H. Sakai. "Absorption Coefficient and Photoionization Yield of NO in the Region 580–1350 A," *Applied Optics*, vol. 6, p. 391, 1967.

4. Borkowski, C. J., and M. K. Kopp, "Some Applications and Properties of One- and Two-Dimensional Position-Sensitive Proportional Counters," *IEEE Trans. Nucl. Sci.*, vol. NS-17, no. 3, p. 340, 1970.

5. Borkowski, C. J., and M. K. Kopp, "Proportional Counter Photon Camera," *IEEE Trans. Nucl. Sci.*, vol. NS-19, no. 3, p. 161, 1972.

6. Harvey, P., et al., "The HEAO-B Imaging Proportional Counter Design," *IEEE Trans. Nucl. Sci.*, vol. NS-23, no. 1, p. 487, 1976.

7. Humphrey, A., et al., "Imaging Proportional Counter for HEAO," *IEEE Trans. Nucl. Sci.*, vol. NS-25, no. 1, p. 445, 1978.

8. Hailey, C. J., W. H.-M. Ku, and M. H. Vartanian, "An Imaging Gas Scintillation Proportional Counter for Use in X-Ray Astronomy," *Nucl. Instrum. Methods*, vol. 213, p. 397, 1983.

9. Simons, D. G., et al., "Performance of an Imaging Gas Scintillation Proportional Counter with Microchannel Plate Readout," *IEEE Trans. Nucl. Sci.*, vol. NS-32, no. 1, p. 345, 1985.

10. Sommer, A. H., *Photoemissive Materials: Preparation, Properties, and Uses*, Wiley, New York, 1968.

11. Fisher, G. B., et al., "A Standard for Ultraviolet Radiation," *Applied Optics*, vol. 12, p. 799, 1973.

12. Metzger, P. H., "On the Quantum Efficiencies of Twenty Alkali Halides in the 12–21 eV Region," *J. Phys. Chem. Solids*, vol. 26, p. 1879, 1965.

13. Lukirskii, A. P., E. P. Savinov, I. A. Brytov, and Yu. F. Shepelev, "Efficiency of Secondary-Electron Multipliers with Au, LiF, MgF_2, SrF_2, BeO, KCl, and CsI Photocathodes in the 23.6 to 113 A Wavelength Region," *U.S.S.R. Acad. Sci. Bulletin Physics*, ser. 28, p. 774, 1964.

14. Carruthers, G. R., "Magnetically Focused Electronographic Image Converters for Space Astronomy Applications," *Applied Optics*, vol. 8, p. 633, 1969.

15. Kowalski, M. P., et al., "Quantum Efficiency of Cesium Iodide Photocathodes at Soft X-Ray and Extreme Ultraviolet Wavelengths," *Applied Optics*, vol. 25, p. 2440, 1986.

16. Kurz, E. A., "Channel Electron Multipliers," *American Laboratory*, March 1979.

17. Wiza, J. L., "Microchannel Plate Detectors," *Nucl. Instrum. Methods*, vol. 162, p. 587, 1979.

18. Timothy, J. G., and R. L. Bybee, "Preliminary Results with Microchannel Array Plates Employing Curved Microchannels to Inhibit Ion Feedback," *Rev. Sci. Instrum.*, vol. 48, p. 292, 1977.

19. Timothy, J. G., "Curved-Channel Microchannel Array Plates," *Rev. Sci. Instrum.*, vol. 52, no. 8, p. 1131, 1981.

20. Martin, C., and S. Bowyer, "Quantum Efficiency of Opaque CsI Photocathodes with Channel Electron Multiplier Arrays in the Extreme and Far Ultraviolet," *Applied Optics*, vol. 21, p. 4206, 1982.

21. Fraser, G. W., "The Soft X-Ray Quantum Detection Efficiency of Microchannel Plates," *Nucl. Instrum. Methods*, vol. 195, p. 523, 1982.

22. Carruthers, G. R., and C. B. Opal, "Detection Efficiencies of Far-Ultraviolet Photon-Counting Detectors," in *Advances in Electronics and Electron Physics*, vol. 64B, Academic Press, London, p. 299, 1985.

23. Siegmund, O. H. W., et al., "High Quantum Efficiency Opaque CsI Photocathodes for the Extreme and Far Ultraviolet," in "Ultraviolet Technology," *Proc. SPIE*, vol. 687, p. 117, 1986.

24. Simons, D. G., et al., "UV and XUV Quantum Detection Efficiencies of CsI-Coated Microchannel Plates," *Nucl. Instrum. Methods in Physics Research*, vol. A261, p. 579, 1987.

25. Carruthers, G. R., "Quantum Efficiencies of Imaging Detectors with Alkali Halide Photocathodes—1. Microchannel Plates with Separate and Integral CsI Photocathodes," *Applied Optics*, vol. 26, p. 2925, 1987.

26. Siegmund, O. H. W., et al., "Ultraviolet Quantum Detection Efficiency of Potassium Bromide as an Opaque Photocathode Applied to Microchannel Plates," *Applied Optics*, vol. 26, p. 3607, 1987.

27. Carruthers, G. R., "Further Investigation of CsI-Coated Microchannel Plate Quantum Efficiencies," *Applied Optics*, vol. 27, p. 5157, 1988.

28. Siegmund, O. H. W., et al., "Soft X-Ray and Extreme Ultraviolet Quantum Detection Efficiency of Potassium Chloride Photocathode Layers on Microchannel Plates," *Applied Optics*, vol. 27, p. 4323, 1988.

29. Carruthers, G. R., "Electronic Imaging Devices in Astronomy," *Astrophysics and Space Science*, vol. 14, p. 332, 1971.

30. Coleman, C. I., and A. Boksenberg, "Image Intensifiers," *Contemp. Phys.*, vol. 17, p. 209, 1976.

31. Heckathorn, H. M., and G. R. Carruthers, "Microchannel Intensified Electrography," in "Instrumentation in Astronomy IV," *Proc. SPIE*, vol. 331, p. 415, 1982.

32. Timothy, J. G., and R. L. Bybee, "One-Dimensional Photon-Counting Detector Array for Use at EUV and Soft X-Ray Wavelengths," *Applied Optics*, vol. 14, p. 1632, 1975.

33. Kellogg, E., P. Henry, S. Murray, and L. Van Speybroeck, "High-Resolution Imaging X-Ray Detector," *Rev. Sci. Instrum.*, vol. 47, p. 282, 1976.

34. Lampton, M., and F. Paresce, "The Ranicon: A Resistive Anode Image Converter," *Rev. Sci. Instrum.*, vol. 45, p. 1098, 1974.

35. Weiser, H., R. C. Vitz, H. W. Moos, and A. Weinstein, "Sensitive Far UV Spectrograph with a Multispectral Element Microchannel Plate Detector for Rocket-Borne Astronomy," *Applied Optics*, vol. 15, p. 3123, 1976.

36. Opal, C. B., P. D. Feldman, H. A. Weaver, and J. A. McClintock, "Two-Dimensional Ultraviolet Imagery with a Microchannel-Plate/Resistive-Anode Detector," in "Instrumentation in Astronomy III," *Proc. SPIE*, vol. 172, p. 317, 1979.

37. Firmani, C., et al., "High-Resolution Imaging with a Two-Dimensional Resistive Anode Photon Counter," *Rev. Sci. Instrum.*, vol. 53, no. 5, p. 570, 1982.

38. Martin, C., et al., "Wedge-and-Strip Anodes for Centroid-Finding Position-Sensitive Photon and Particle Detectors," *Rev. Sci. Instrum.*, vol. 52, no. 7, p. 1067, 1981.

39. Siegmund, O. H. W., et al., "Application of Wedge and Strip Image Readout Systems to Detectors for Astronomy," in "Instrumentation in Astronomy VI," *Proc. SPIE*, vol. 627, p. 660, 1986.

40. Timothy, J. G., and R. L. Bybee, "Photon-Counting Array Detectors for Space and Ground-Based Studies at Ultraviolet and Vacuum Ultraviolet (VUV) Wavelengths," in "Ultraviolet and Vacuum Ultraviolet Systems," *Proc. SPIE*, vol. 279, p. 129, 1981.

41. McClintock, J. E., et al., "Rocket-Borne Instrument with a High-Resolution Microchannel Plate Detector for Planetary UV Spectroscopy," *Applied Optics*, vol. 17, p. 3071, 1982.

42. Lampton, M., O. Siegmund, and R. Raffanti, "Delay Line Anodes for Microchannel-Plate Spectrometers," *Rev. Sci. Instrum.*, vol. 58, no. 12, p. 2298, 1987.

43. Siegmund, O. H. W., M. L. Lampton, and R. Raffanti, "A High Resolution Delay Line Readout for Microchannel Plates," in "EUV, X-Ray, and Gamma-Ray Instrumentation for Astronomy and Atomic Physics," *Proc. SPIE*, vol. 1159, p. 476, 1989.

44. Holt, S. S., "Si(Li) X-Ray Astronomical Spectroscopy," *Space Sci. Instrum.*, vol. 2, p. 205, 1976.

45. Janesick, J., and M. Blouke, "Sky on a Chip: The Fabulous CCD," *Sky and Telescope*, vol. 74, no. 3, p. 238, 1987.

46. Janesick, J., et al., "Backside Charging of the CCD," in "Solid State Imaging Arrays," *Proc. SPIE*, vol. 570, p. 46, 1985.

47. Janesick, J., et al., "CCD Advances for X-Ray Scientific Measurements in 1985," in "X-Ray Instrumentation in Astronomy," *Proc. SPIE*, vol. 597, p. 364, 1985.

48. Janesick, J. R., et al., "The Potential of CCDs for UV and X-Ray Plasma Diagnostics," *Rev. Sci. Instrum.*, vol. 56, no. 5, p. 796, 1985.

49. Janesick, J. R., et al., "Present and Future CCDs for UV and X-Ray Scientific Measurements," *IEEE Trans. Nucl. Sci.*, vol. NS-32, no. 1, p. 409, 1985.

50. Stern, R. A., R. C. Catura, M. M. Blouke, and M. Winzenread, "EUV Astronomical Spectroscopy with CCD Detectors," in "Instrumentation in Astronomy VI," *Proc. SPIE*, vol. 627, p. 583, 1986.

51. Bosiers, J. T., et al., "CCDs for High Resolution Imaging in the Near and Far UV," in "Ultraviolet Technology," *Proc. SPIE*, vol. 687, p. 126, 1986.

52. Janesick, J., T. Elliott, T. Daud, and D. Campbell, "The CCD Flash Gate," in "Instrumentation in Astronomy VI," *Proc. SPIE*, vol. 627, p. 543, 1986.

53. Walton, D., R. A. Stern, R. C. Catura, and J. L. Culhane, "Deep-Depletion CCDs for X-Ray Astronomy," in "State-of-the-Art Imaging Arrays and Their Applications," *Proc. SPIE*, vol. 501, p. 306, 1984.

54. Chowanietz, E. G., D. H. Lumb, and A. A. Wells, "Charge-Coupled Devices (CCDs) for X-Ray Spectroscopy Applications," in "X-Ray Instrumentation in Astronomy," *Proc. SPIE*, vol. 597, p. 381, 1985.

55. Lowrance, J. L., P. Zucchino, G. Renda, and D. C. Long, "ICCD Development at Princeton," in *Advances in Electronics and Electron Physics*, vol. 52, Academic Press, London, p. 441, 1979.

56. Carruthers, G. R., et al. "Development of EBCCD Cameras for the Far Ultraviolet," in *Advances in Electronics and Electron Physics*, vol. 74, Academic Press, London, p. 181, 1988.

57. Jenkins, E. B., et al., "IMAPS: a High-Resolution, Echelle Spectrograph to Record Far-Ultraviolet Spectra of Stars from Sounding Rockets," in "Ultraviolet Technology II," *Proc. SPIE*, vol. 932, p. 213, 1988.

CHAPTER 16
VISIBLE DETECTORS

Suzanne C. Stotlar

16.1 INTRODUCTION

The physical universe is made up of objects with temperatures greater than absolute zero, which means that the atoms and molecules of an object are in motion. These motions result in interactions with other atoms and molecules (via bonds and collisions). Therefore, these elementary particles are subject to accelerations which result in electromagnetic radiation.

Radiometry is the science of measurement of the electromagnetic spectrum. Information is obtained and evaluated by detection[1] of the electromagnetic radiation on which it is impressed. This chapter discusses those detectors which provide a response to that portion of the electromagnetic spectrum which is visible to the human eye. Applications for detectors in this region are often related to human response, i.e., turning on lights when the sky is dark (to us), verifying the color purity of a paint or cloth sample, reading a printed page, or aiding in exploratory surgery. The visible spectral region lies between violet light of wavelength 400 nm to red at 700 nm. The detectors encompassed here include those most important by application and availability.

16.1.1 Performance Parameters

Responsivity is the response per unit of incident power. Generally, the (current) responsivity is given in amperes per watt, since more detectors are current sources. Voltage responsivity requires a knowledge of load (and frequency) conditions. *Spectral responsivity* is the responsivity at a specific wavelength. Most detectors, especially semiconductor detectors, exhibit some variation with wavelength.

Quantum efficiency is the fraction of photoelectrons created per incident photon, or

$$QE = \frac{1.24\, R_l}{\lambda} \times 100\% \tag{16.1}$$

where λ is the wavelength (μm) and R_l is the current responsivity (A/W).

This results from the fact that an electron has an energy of 1 eV = 1.6×10^{-19} J, and a photon has an energy

$$h\nu = \frac{hc}{\lambda} \tag{16.2}$$

where h is Planck's constant (6.626×10^{-34} W·s²), c is the speed of light (2.998×10^{10} cm/s), and λ is the wavelength (μm).

Since surface defects or reflectance can prevent a photon from entering the detector, only those photons absorbed by the material should be counted. This may be called the *internal quantum efficiency*.

Dynamic range is the variation in incident power the detector can experience while maintaining acceptable linearity.

Linearity is the change in signal relative to the original value as the incident power is varied or scanned across the surface of the absorbing (active) area.

Speed, or *10–90 percent rise time*, is the detector's ability to follow the rise (fall) of an incident energy pulse. Depending on the application and detector type, 10–90 percent values can range from seconds to femtoseconds [1 femtosecond (fs) = 10^{-15} s].

Spectral matching factor (SMF) is a measure of the overlap between the spectral distribution of the incident power on a detector and the spectral sensitivity (responsivity).

Active area is the primary light-collecting area of the detector surface. In some detectors the nonactive area can absorb light and also generate output signal.

Response uniformity describes the variation in output as a small spot or beam of incident energy is moved across the detector.

Angular response is the responsivity as the angle of the incident power is varied from the direction normal to the plane surface of the detector receiving area. When the angle of incidence varies from the normal, the output is generally reduced by the cosine of the incident angle.

Field of view (FOV) is the solid angle cone from which energy can reach the detector. Most detectors have a window or a lens between the active area of the detector and the source of energy, and the FOV is determined by the distance of the detector surface from the front surface of the window or by the curvature of the lens.

Optical absorption coefficient is the fraction of light absorbed per unit thickness of an optical material. The *absorption depth* is the thickness required to absorb 63 percent ($1 - e^{-1}$) of the incident energy. For many detector materials this is wavelength-dependent.

Antireflection coatings are applied to detector materials and, where applicable, windows or lenses to reduce signal loss due to reflection.

Noise limits the range and sensitivity to low levels of light of all detectors. Not all detectors, however, experience the same sources or same degree of noise. Major noise sources in the detector include:

Shot noise, which is generated in thermal and semiconductor detectors by the random emission of electrons, can be calculated from

$$i_s = (2qi_D \, \Delta f)^{1/2} \qquad \text{A (rms)} \tag{16.3}$$

where q is the electron charge (1.6×10^{-19} C), i_D is the dark current (A), and Δf is the operating bandwidth (Hz).

Flicker or f^{-1} noise sources are not well-characterized, but decrease rapidly with frequency.

Temperature noise is also present in thermal detectors due to fluctuations in the detector which are not due to a change in signal. Proper detector selection can ensure that this noise contribution is negligible.

Thermal Johnson noise is the noise in the equivalent resistor, or

$$i_t = (4kT\,\Delta f/R_{\mathrm{SH}})^{1/2} \quad \mathrm{A} \tag{16.4}$$

where k is Boltzmann's constant (1.38×10^{-23} J/K), T is the temperature (K), R_{SH} is the shunt or equivalent resistance (Ω), and Δf is the operating bandwidth (Hz).

Radiation noise results from statistical fluctuation in the number of incident photons, generally small for most applications.

Amplifier noise, or noise in the associated electronics, may be the dominant noise source for the application, as each type of noise described above may be present, depending on the design.

Microphonic noise results from mechanical displacement of wiring or components when the system experiences vibration or shock. Some detection systems are particularly sensitive to microphonics.

Total noise current is the quadrature sum of all (significant) noise currents.

Noise equivalent power (NEP) is the incident light power for which the signal-to-noise ratio is equal to one, and thus is a measure of the minimum detectable power:

$$\mathrm{NEP} = \frac{I_N}{R_I} \quad \mathrm{W} \tag{16.5}$$

where I_N is the total noise current and R_I is the responsivity. (The normalized NEP for unit bandwidth is often reported as $\mathrm{NEP}^* = \mathrm{NEP}/\Delta f^{1/2}$ with units $\mathrm{W \cdot Hz^{-1/2}}$.)

Detectivity, the inverse of the NEP, is used as a figure of merit for comparing detectors.

Area normalized detectivity (D^*) permits comparison of detectors of different area.

$$D^* = \frac{(A\,\Delta f)^{1/2}}{\mathrm{NEP}} \quad \mathrm{cm \cdot Hz^{1/2} \cdot W^{-1}} \tag{16.6}$$

where A is the active area of the detector ($\mathrm{cm^2}$) and Δf is the bandwidth of the associated electronics (Hz). This figure of merit is most appropriate for comparing detectors of similar spectral response and physical type.

Multielement arrays provide multiple detectors on a single piece of the detective material. *Hybrid arrays* package more than one detector and/or detector electronics together.

Optical crosstalk refers to the signal measured in a nonilluminated array element when its nearest neighbor is illuminated (usually 80 percent of the active area of the element).

Electrical crosstalk is the current measured in a detector array element when the current in its nearest neighbor is varied. *Element-to-element isolation* and/or *interelement leakage* are similar characteristics.

16.2 THE HUMAN EYE AS A DETECTOR

The visible region of the energy spectrum is defined by the human eye, and, therefore, it is appropriate to discuss the eye first. Many applications for this region are designed to replace, evaluate, or satisfy visual response. Figure 16.1 shows the typical spectral composition of skylight and sunlight combined to form daylight. (Atmospheric conditions and solar altitude result in modest variation of the spectral distribution.) The human retina is well-adjusted to the peak spectral distribution of daylight. The retina is actually a compound detector since it has three sensitive layers [nerve cells, light-sensitive cells (rods and cones), and pigment-containing cells]. Because it converts light to an electrical signal (to the optic nerve), the retina is the detector. Although it is packaged with other optical and functional components, including the lens, iris, pupil, and cornea, into a single assembly, the eye is frequently called a detector. Like the eye, other detectors are commonly packaged with optical components which modify or enhance their inherent characteristics.

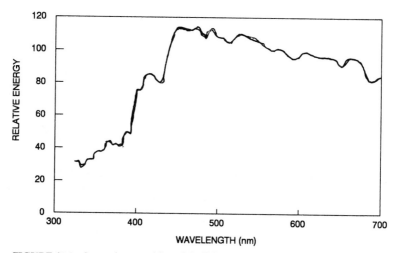

FIGURE 16.1   Spectral composition of daylight.

The retina is the inner layer of the eyeball. It occupies about four-fifths of the eye toward the rear of the eyeball. As shown in Figure 16.2, the retina touches the choroid. The nerve cells form a layer toward the center of the central cavity. The light-sensitive long, thin rods and wider cones compose the middle layer. The rods lie toward the edges of the retina, and the cones are near the center. Nerve fibers attached in front of the rods and cones come together from all parts of the retina to form the optic nerve, which passes through the retina and other layers of the back of the eyeball and goes to the brain. Pigment-containing cells form a third layer toward the outside near the choroid. Light must pass through the layer of nerve cells to reach the light-sensitive cells.

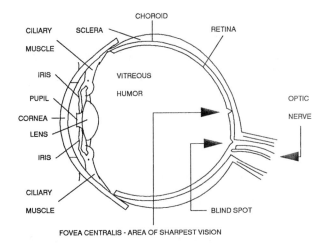

FIGURE 16.2 Cross-sectional diagram of the human eye.

Vision is the ability of the eye to form an image of an object and send the image to the visual centers of the brain. Light rays from the object pass through the cornea, the aqueous humor, the pupil, the lens, and the vitreous humor to reach the retina. When light rays strike the light-sensitive cells of the retina, they stimulate the nerve cells. The optic nerve carries the message to the visual cortex of the brain. The cones lie mainly in the macula, a small area near the center of the retina. They are less sensitive than rods and respond only to fairly strong light. The cones in the macula do most of our seeing in daylight or fairly bright artificial light and are responsible for color vision. There may be three kinds of cones which respond separately to red, green, and blue light. In the center of the macula is a small depression called the fovea centralis. The layer of nerve cells is very thin here, permitting the sharpest vision.

The area of the retina outside the macula contains the rods. These cells are extremely sensitive, even to very dim light. They produce a somewhat blurred image and do not provide color vision. They saturate and lose sensitivity in bright light.

The retina contains two light-sensitive pigments. Rods contain visual purple, and cones contain iodopsin. When light hits the retina, chemical changes in the pigments cause electrical impulses to pass along nerves attached to the rods and cones. The pigmented cells of the retina behind the rods and cones absorb stray light.

Like the image formed on the film of a camera, the image formed on the retina is inverted. The brain processes the image so that it is seen in the correct orientation. There are no rods or cones where the optic nerve leaves the eye. The brain fills in the image around this blind spot.

A detection system contains not only the detector (and generally additional components to collect and transport signal) but a means of processing the information. It may produce a report or provide a means of controlling other components. The eye and brain form a highly sophisticated detection system. The human visual detection system contains two eyes. The optic nerves from the two eyes fuse together at the optic chiasma, near the brain. The brain forms a single image from the images of the two eyes. The slight differences in the images give humans stereoscopic vision. This system is highly sophisticated and capable of broad dynamic range in input power (to 10^6), broad color discrimination (greater than 10^3), self-correction to some degree, and long operating life (10^2 years).

By using a series of filters, the response of other detectors (usually silicon photodiodes) can be modified to have a spectral response similar to the eye, as shown in Fig. 16.3. Such a detector is called a *CIE standard photometric observer*.[2]

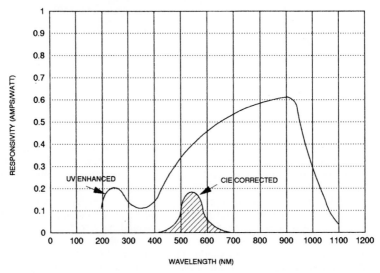

FIGURE 16.3 Photodiode spectral response showing CIE-corrected (filtered) response.

16.3 PHOTOGRAPHIC FILM

Photographic film is the standard of comparison for many other detector types in terms of sensitivity, dynamic range, and (for arrays) resolution. The dynamic range of (high-quality) film is six orders of magnitude (10^6), the smallest resolvable spot or pixel is about 10 μm in diameter, and the minimum level of detection is 1 to 10 photons.

Photographic film consists of one or more (usually three for color) light-sensitive emulsions deposited onto a carrier, usually paper or plastic. While post-light-exposure processing is necessary to provide a permanent record, the (chemical) change is irreversible.

Although the most common application of visible film is to provide a photographic image, film has other uses in detection.

Photographic film generally responds to exposure in a nonlinear manner. By measuring the transmittance of the processed transparency, the quantitative result of exposure may be determined using the following expression for density, as shown in Fig. 16.4,

$$D = \log_{10}\left(\frac{1}{\tau}\right) \tag{16.7}$$

where τ is the transmittance.

Black-and-white film is easier and more accurate to use for this type of application than color film. The densitometer and the photographic camera (with appropriate filters) are useful for examining the spatial distribution of laser beams and focused spots. For high-power lasers, the burn pattern may be examined.

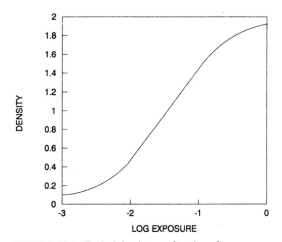

FIGURE 16.4   Typical density as a function of exposure.

16.4 PHOTOELECTRIC DETECTORS

In general, the most sensitive types of detectors are fabricated from semiconductor materials. Photoemissive, photoconductive, and photodiode (photovoltaic) detectors all rely on the photoelectric effect. Figure 16.5 gives a diagram of the energy level structure in a semiconductor. In (pure) semiconductor materials, electrons can only occupy energy levels separated by forbidden bands. In order to contribute to the electronic signal, an electron must be excited to an energy level in the conduction band. If the electron is originally in the valence band, the required photon energy is greater than the bandgap e_G. If the electron is in a donor level, corresponding to the energy of an electron in a donor impurity atom intentionally introduced into the semiconductor, the required photon energy must be only e_d or greater. Acceptor impurities, which can accept an electron excited from the valence band with a minimum required energy e_a, can also produce energy levels within the forbidden band. A hole, or a missing electron, is produced in the valence band. This hole can move through the semiconductor material and contribute to the signal current as a positively charged carrier. N-type materials are those in which the

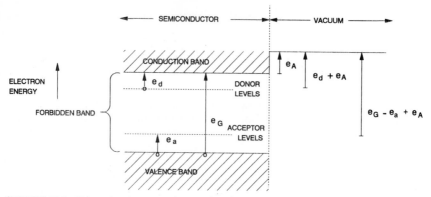

FIGURE 16.5   Energy level structure in a semiconductor.

majority carriers are electrons, while holes are the majority carrier in *p*-type materials. Pure semiconductors, in which the dominant mechanism of detection is the excitation of electrons from the valence band to the conduction band, are called *intrinsic* semiconductors. *Extrinsic* semiconductors are those for which the dominant mechanism is excitation from either donor or acceptor levels.

In the vacuum region outside a semiconductor surface, an electron has an electrostatic attraction to the surface. The energy required to overcome this attraction is the electron affinity e_A, or work function, of the material. If an absorbed photon from an incident beam of energy contains energy greater than $e_G + e_A$ (to overcome both the bandgap and the electron affinity), then an electron can escape from the surface of an intrinsic semiconductor material and can be collected as a signal current. This is the external photoelectric effect, usually called *photoemission*. In an extrinsic material the energy required for photoemission can be as small as $e_d + e_A$.

If an electron is excited to the conduction band but remains within the material to contribute to the signal current, the phenomenon is called the *internal photoelectric effect*. For an intrinsic material, the required photon energy for the internal photoelectric effect is $e > e_G$. For an extrinsic material, the energy is $e > e_d$. Consequently, the internal effect can be used to detect lower-energy (further-infrared) photons than the external effect. Detectors employing both the internal and external photoelectric effects are available for the visible spectral region.

16.4.1   Vacuum Phototubes

A vacuum phototube consists of an evacuated envelope (usually glass or quartz), a photocathode, and an anode. Figure 16.6 shows a schematic diagram of a simple operating circuit of a vacuum phototube and a typical configuration of a phototube with a cylindrical photocathode. The negative terminal of a power supply is connected to the photocathode. The positive terminal is connected to the anode through an anode resistor as shown. Radiation which satisfies the energy conditions described in Sec. 16.4 causes electrons to be emitted by the photocathode. The electrons are attracted by the positive potential of the anode, and an electrical current passes through the resistor. The voltage drop $V_o = i_p R_a$ across the anode resistor is a measure of the photocurrent.

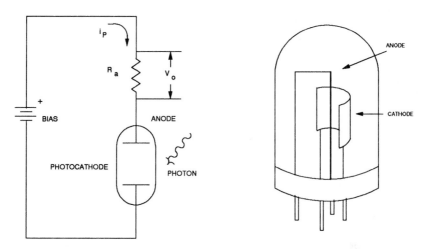

FIGURE 16.6 Simple operating circuit of a vacuum phototube with a typical configuration of a phototube with cylindrical photocathode.

The application and expected performance characteristics determine the overall shape and configuration of the phototube components. Vacuum phototubes are available in both head-on and side-on package styles. For the visible spectral region there are many material choices, including gallium arsenide (cesium-doped) and sodium-potassium-antimony-cesium. The application, including spectral region, temperature, signal level, and other parameters, must be considered in selecting the type and material, which should be done in consultation with the supplier. Low bias voltage (50- to 250-V) phototubes are typically 1 cm² but may be as large as 2 cm². They have low capacitance and can be used for high-speed applications with a rise time as low as 1 ns. High-voltage biplanar phototubes are designed for subnanosecond rise times and peak anode currents for short pulses of several amperes, making them suitable for use with pulsed lasers. Typical performance characteristics are shown in Table 16.3 in Sec. 16.7.1. The responsivity varies with wavelength, so the spectral composition of the radiant source must be known if measurement of the incident radiation is required.

Phototubes experience a number of effects which can degrade performance: fatigue of the photocurrent with time or overheating, sensitivity to microphonics, and electromagnetic impulse and electrostatic discharge (EMI/ESD). Packaging is generally bulky and difficult to integrate with other optical or electrical components. Most vacuum phototubes exhibit poor linearity of response as the incident radiation is varied in power or position across the collecting surface. It is advisable to illuminate all the photocathode in order to reduce effects of spatial nonuniformity. The current responsivity of a photocathode is

$$R_i = \frac{q\eta\lambda}{hc} \qquad (16.8)$$

where η is the fraction of photons which produce electrons. A vacuum photocathode is capable of producing signal currents too low to be measured by available electronics. The photomultiplier described in the next section provides electron multiplication. This gain feature makes the photomultiplier more useful in low-light-level applications.

16.4.2 Photomultipliers

Photomultipliers provide high internal current gain by combining the external photoelectric effect with low-noise secondary electron multiplication, and are approximately linear over a wide dynamic range.

A material which has a low work function will emit a stream of electrons when bombarded with photons of frequency ν such that

$$h\nu > \phi_o \qquad (16.9)$$

where ϕ_o is the work function of the photocathode. Excess photon energy is converted into kinetic energy. A photomultiplier tube (PMT) is illustrated schematically in Fig. 16.7.

For an uncooled PMT, thermionic emission is often the dominant source of dark current. The thermionic current of a photocathode is greater than that of the dynodes and is given by

$$i_T = A_d S T^2 e^{-\phi_o/kT} \qquad A \qquad (16.10)$$

where A_d is the photocathode area, S is a constant ($S = 4\pi mqk^2/h^3$, where m is the mass of the electron), T is the photocathode temperature (K), and ϕ_o is the work function of the photocathode. For high values of gain and R_{eff} (effective resistance of the resistor chain), the PMT will be shot-noise limited. Figure 16.7 is a simplified diagram. Most PMTs have a wrap-around feature and much more complex dynode structures. The PMT consists of a photoemissive detector and a low-noise amplifier contained in the same vacuum jacket. When an incident photon is absorbed by the photocathode to produce a photoelectron, the electron is accelerated toward the first dynode by the voltage drop across R_1. If this potential is 100 V, then the electron will have a kinetic energy of 100 eV plus the excess of the photon energy over the work function of the photocathode. This kinetic energy will cause a number of electrons to be emitted from the dynode. Each of these electrons will be accelerated toward the second dynode by the voltage drop across R_2. In this way, a large number of electrons are collected at the anode for each electron emitted by the photocathode. If three electrons are emitted by each dynode

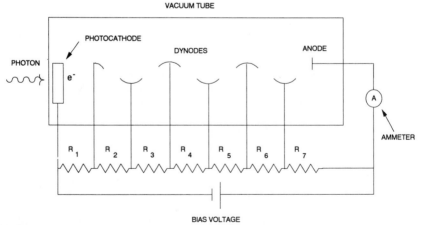

FIGURE 16.7   Photomultiplier tube (PMT) and connection diagram.

for one incident electron, the gain of the six dynodes is $3^6 = 729$. It is more likely that 11 dynodes would be used in practice, for a total gain of 3^{11} or about 2×10^5. This gain is sufficient to raise a photocathode current of 8×10^{-19} A to 1.6×10^{-13} A. This signal level is compatible with electronic ammeters, as discussed in Ref. 2.

Photomultipliers are generally used in low-light-level and deep-ultraviolet (uv) applications which cannot be addressed by silicon photodiodes. They are expensive, delicate, and bulky compared to other detectors like photodiodes.

PMTs require magnetic shielding and are difficult to integrate with other optical and electrical components. They saturate easily and require large power supplies. Photoemissive materials are also used in a variety of imaging devices including those which utilize microchannel plates (an array of small glass tubes 5 to 20 μm in diameter which can operate as electron multipliers). Each channel can be used to produce one intensified picture element, or *pixel*. The silicon, lead, and alkali oxide compounds of which the tubes are made provide a gain of 2 electrons per single wall collision.

16.4.3 Photoconductors

A photoconductor is a semiconductor device which exhibits a change in conductance (resistance) when radiant energy is incident on it. Table 16.1 lists some examples of intrinsic semiconductor materials used in photoconductors.

TABLE 16.1 Typical Intrinsic Detector Materials

Detector	e_G, eV	T, K (typical operation)	λ max., μm
Ge	0.67	193	1.9
Si	1.12	300	1.1
GaAs	1.5	300	0.9
CdSe	1.8	300	0.69
CdS	2.4	300	0.52

Figure 16.8 shows a typical connection diagram for the photoconductive detector. Radiant energy incident on the detector produces an electron-hole pair which lowers the detector resistance by producing more carriers. The change in the photoconductor resistance R_p produces a change in the voltage drop across R_a.

As with photoemissive detectors, the photon energy must be greater than the bandgap e_G. At wavelengths less than $\lambda_{\max} = hc/e_G$, the current responsivity is

$$R_i = \frac{G(1 - \rho)\eta q \lambda}{hc} \quad \text{A/W} \tag{16.11}$$

where G is the photoconductive gain, ρ is the reflectance of the front surface, η is the quantum efficiency of the electron charge, and $hc = 1.9856 \times 10^{-25}$ W·s·m. The photoconductive gain is given by

$$G = \frac{\tau}{T_\tau} \tag{16.12}$$

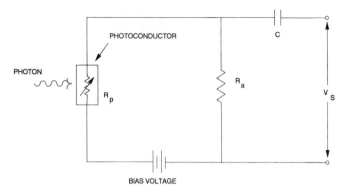

FIGURE 16.8 Connection diagram for a photoconductive detector.

where τ is the lifetime of charge carriers and T_τ is the travel time between the electrodes. The travel time is inversely proportional to the bias voltage. The gain can have values greater than 10^3. This gain results from the fact that the lifetime of the charge carriers is longer than the transit time between electrodes, permitting new carriers to be generated before the previously generated ones have been collected at the electrodes.

The most common photoconductors are made of cadmium selenide (CdSe) or cadmium sulfide (CdS).These photoconductors are fabricated in a continuous pattern. They are found along sidewalks, home entrances, and other areas where they turn on lights as the sun goes down. Both CdS and CdSe have a spectral sensitivity similar to that of the human eye. Photoconductors can serve as optical switches. Since such detectors have high values of resistance when not illuminated and low resistance when (sufficiently) illuminated, they can provide control signals which can be rapid and decoupled from noise sources that might otherwise be present. Silicon, germanium, and compound semiconductors such as gallium arsenide are used for switching when fabricated with narrowly separated electrodes deposited on the same surface (usually the side toward the source).

Optical switches usually require illumination of the entire sensitive region between the electrodes. Linear position sensing, however, does not require the entire photoconductor to be illuminated. Ohmic (low-resistance) contacts are formed on the same surface of the semiconductor (usually silicon) material. By comparing the signals between the electrodes and a third contact, the centroid of an incident beam can be determined.

Photoconductor detectors can be linear if care is taken in irradiating and collecting the signal. The requirements to support linearity need to be reviewed with the supplier.

16.4.4 Photodiodes

Silicon photodiodes perform countless operations daily. Applications range from counting syringes in a hospital supply room to measuring atomic transitions in uv-visible spectroscopy for blood analysis to directing satellites or operating a mouse for a personal computer. The silicon photodiode will be the detector of choice in any application which its characteristics permit it to perform. It benefits from over

40 years of technological development. For applications such as radiometry, pyrometry, colorimetry, and optical spectroscopy, this device offers the broadest options in terms of cost, reliability, packaging, linearity, ruggedness, and ability to be integrated with other optical and electronic components. Silicon is not the only type of photodiode, but it is the most important for the visible spectral region.

Photodiodes employ the photovoltaic effect. The photovoltaic effect requires a potential barrier with an electric field, in this case a *p-n* junction. Figure 16.9 illustrates the energy-band diagram of a *p-n* junction. An incident photon of energy greater than or equal to the bandgap e_G creates a hole-electron pair as shown.

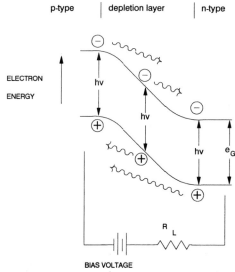

FIGURE 16.9   Energy-band diagram of a *p-n* junction.

The electric field of the junction will not allow the hole-electron pair to recombine. The photoelectrons are thus available to produce a current through an external circuit.

Electron-hole pairs can be produced by photon absorption in three regions. In the *p* region, the positively charged hole is immobile, but the electron can diffuse toward the junction region, called the *depletion layer*. Once in the depletion layer, it drifts rapidly across the junction under the influence of the junction field and contributes to the signal current. In the *n* region, the electron is immobile, but the hole can move toward the junction by diffusion (and then drift rapidly across it). In the depletion layer itself, both the electron and the hole are mobile and can move in opposite directions, both contributing to the signal current. Since drift in the junction field is faster than the diffusion process, the best time response (and highest current) is produced when the photons are absorbed in the depletion layer.

Electrons and holes swept across the junction make up the current which generates a voltage across the load resistor. This signal voltage is the output of the semiconductor photodiode. The electrical characteristics of the photodiode are illustrated in Fig. 16.10. In this figure the uppermost curve shows the relationship

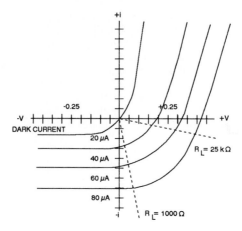

FIGURE 16.10 Typical current voltage characteristics of junction photodiodes.

between current and voltage when no radiant energy is incident on the photodiode
detector.

The bend in the curve occurs at $V = 0$ in this case. If a positive bias voltage is
applied across the junction, the detector is said to be *forward-biased*, and if the
bias voltage is negative, then the detector is said to be *reverse-biased*. A photodiode
with no external bias is in photovoltaic mode. The depletion layer is very thin, but
the detector noise is very low. When operated with a reverse bias, the photodiode
is in photoconductive mode. Although the noise level is higher due to the shot-
noise contribution (not present in photovoltaic mode), the speed of response is
improved. When silicon p-n junctions are used in solar energy conversion applications, maximum power transfer occurs when the load resistance $R_L = R_{SH}$, the
shunt resistance. The other curves in Fig. 16.10 represent the i-v relationship when
increasing levels of radiant power are incident on the detector.

Photovoltaic devices are commonly operated as current generators with a current
responsivity expressed as the current in amperes divided by the incident power in
watts required to generate that current. A photodiode in photovoltaic mode is
usually amplified using a transimpedance-mode amplifier as shown in Fig. 16.11.
This figure also shows a photoconductive-mode circuit in which the detector is
reverse biased.

A p-n junction is formed by diffusing a shallow layer of an acceptor, or p-type,
impurity into the surface of a donor, or n-type, wafer so that the concentration of
acceptors in the layer exceeds that of the donors. The p-type region is transparent,
therefore light can reach the junction. Ideally, all light is absorbed in the depletion
region, but in reality it penetrates further. Reverse biasing increases the depletion
region, reducing the number of photoelectrons created outside the depletion region.
Additional donor impurities are diffused into the rear surface in order to provide
an ohmic contact with the metal electrode. As the bias is increased, few photoelectrons are created outside the depletion region. Increasing the depletion region
decreases the capacitance (at least until the device is fully depleted) and makes
the detector faster.

Photodiodes are generally large (in active area) compared to integrated circuits,
transistors, and low-power diodes. The junction is usually deeper and more likely

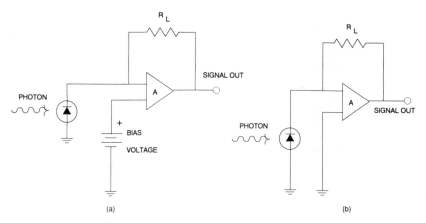

FIGURE 16.11 Connection diagrams for (*a*) photoconductive and (*b*) photovoltaic bias mode photodiodes with transimpedance amplifier.

to be provided with passivation. Consequently, they can exhibit stable performance characteristics without hermetic sealing. This results in an enormous variety of packaging options. Photodiode manufacturers can provide single chips very cheaply for use in complex multielement arrays with internal signal modification (amplifiers, clock circuitry, etc.), coatings to enhance or modify signal response, and optics to increase or control the effective aperture. These sophisticated subsystems may cost thousands of times more.

Because of the nature of the photoelectric effect, nearly all device parameters are temperature-dependent. The spectral response is both wavelength- and temperature-dependent.

In most applications, the primary noise sources are shot noise (for photoconductive mode) and Johnson noise. Semiconductors, however, are sensitive to EMI field and physical strain. Silicon, in fact, is widely used in strain transducers. For many devices, the dark current is dominated not by shot noise but by mechanical stresses applied to the semiconductor material or chip in packaging it. If the chip is attached to a substrate material with a different coefficient of thermal expansion, the degree of stress may vary with temperature. Most semiconductor detectors are sensitive to vibration and microphonics, although appropriate packaging can reduce the effect of these noise sources.

There are many variations on the *p-n* photodiode described above. By varying the purity (resistivity) of the bulk material, the manufacturer can change or improve the characteristics of the device. Silicon photodiodes fabricated of 400 Ω·cm or higher *n*-type bulk material are often called *p-i-n*, as the higher resistivity material is nearly intrinsic (free of impurities and defects). *P*-type bulk material is used for *n-p* junctions. More important in the infrared spectral region, *p*-type devices exhibit higher noise than *n*-type. Guard rings are diffused around the active areas to control leakage in most *p*-type detectors and in a few *n*-type.

16.4.5 Avalanche Photodiodes

When high speed is the primary design consideration, an avalanche photodiode may be used. Avalanche photodiodes (APDs) are reverse-biased junction photodiodes in which the reverse bias produces a field such that charge carriers accelerated by the bias create more carriers by impact ionization. The APD is the semiconductor analog to the gas-filled phototube. Above a certain voltage bias, an uncontrolled current flows in the semiconductor detector. This bias is called the breakdown voltage. APDs are designed to have high breakdown voltages (several hundred volts or more in some cases). The operational reverse bias is close to the breakdown voltage as shown in Fig. 16.12. The gain, or multiplication factor, M is highly dependent on the reverse voltage and temperature. Minor defects in the crystal or variations in the junction and uneven strain during mounting limit the size, uniformity, and linearity of these devices. In the visible region, silicon is the most common APD material. Noise in APDs includes shot noise due to either leakage current or the incident radiation, multiplication noise from the avalanche process, and thermal noise (Johnson) from the photodiode.

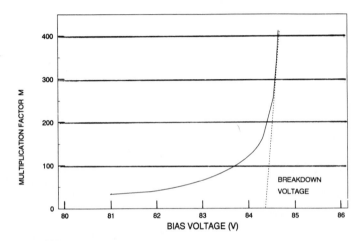

FIGURE 16.12 Multiplication factor versus bias for a typical avalanche photodiode at 25°C.

16.5 THERMAL DETECTORS

A detector which produces an electrical signal (or a change in output) in response to a change in its bulk temperature is classified as a *thermal detector*. Thermocouples, thermistors, bolometers, and thermopiles are thermal detectors. Pyroelectric detectors are also grouped with thermal detectors. Thermal detectors generally lack the sensitivity of semiconductor detectors, but they can provide flatter wavelength response and a variety of options in packaging, size, and cost. Most of them are passive devices, requiring no bias. Because of their stable properties, thermopiles are frequently used as calibration standards. Thermal detectors absorb radiation, which produces a temperature change that in turn changes a physical or electrical property of the detector. Since a change in temperature takes place, thermal detectors are generally slow in response and have relatively low sensitivity compared to other detectors. The thermal detector works by heating the detector chip. Therefore, a heat equation needs to be examined to model the thermal detector mathematically.

Primary noise sources for a thermal detector are white noise (the noise associated with a blackbody) and noise due to random thermal fluctuations (Johnson noise).

16.5.1 Thermocouples and Thermopiles

A thermocouple consists of two dissimilar metals connected in series. As the temperature of this junction varies, the electromotive force (emf) developed at the output terminal varies. If two dissimilar metal wires are connected at both ends, current flows in the electrical loop. The voltage needed to stop current flow at the measuring terminals in Fig. 16.13 is the emf developed by the thermocouple.

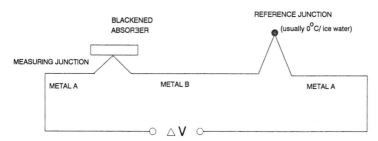

FIGURE 16.13 Thermocouple connection diagram.

Thermocouples are generally connected with a twin so that a reference junction can be established. As shown, 0°C (the temperature of ice) is the most commonly used reference temperature. Standard tables allow a user to determine the temperature from the (millivoltage) readings. As detectors, thermocouples are not very sensitive.

A quantity of N thermocouples in series increases the responsivity by a factor of N. Such a device is called a thermopile. The most commonly used thermopile materials are bismuth and antimony as a series of wires or evaporated thin films. Energy-absorbing black paints are used to coat thermopiles to make them absorb uniformly over a broad wavelength range.

16.5.2 Bolometers and Thermistors

The bolometer has been in use since it was invented by Langley in 1881 to detect both visible and infrared radiation. A temperature change produced by the absorption of radiation causes a change in electrical resistance of the material used to fabricate the bolometer. This change in resistance can be used to sense radiation just as in the photoconductor. However, the basic detection mechanisms are different. In the bolometer, radiant power produces heat within the material, which in turn produces the resistance change. There is not a direct photon-lattice interaction. Figure 16.14 shows a typical connection diagram for the bolometer. Although bolometers are generally fabricated of materials which will absorb visible light naturally, they are usually made with a black coating to reduce reflection. In practice they are often as thin as a few micrometers and are cooled to reduce noise. Their design can make them very sensitive to microphonics.

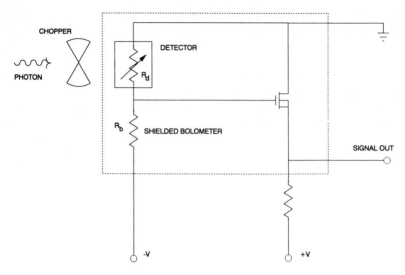

FIGURE 16.14   Bolometer connection diagram.

The most commonly used bolometers are metal, thermistor, and semiconductor bolometers.

The thermistor is a second-generation bolometer and perhaps the most popular. It has found wide application, ranging from burglar alarms to fire-detection systems to industrial temperature measurement to spaceborne horizon sensors and radiometers. When hermetically sealed, this type of detector can resist such environmental extremes as vibration, shock, temperature variations, and high humidity. Table 16.2 gives some typical values of parameters for a thermistor operating at room temperature.

TABLE 16.2   Typical Thermistor Characteristics

Size	1–3 mm^2
Resistance	250 kΩ–2.5 MΩ
Time constant	1–100 ms
NEP	10^{-10}–10^{-8} W
Spectral response	0.4–10 μm

Commercially available thermistors are fabricated with a bead which increases the effective aperture of the detector while providing protection for the sensor.

As shown in Fig. 16.15, the ray directed toward the center of the thermistor is not refracted (bent) at the lens interface; however, the rays directed to the edge of the detector are refracted by the lens; giving the detector an appearance of being larger by a factor of n, where n is the index of refraction of the lens. Since the detector is a two-dimensional device, the virtual area increase is n^2. As a result,

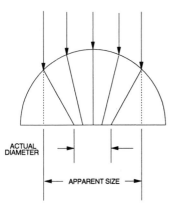

ACTUAL
DIAMETER

←— APPARENT SIZE —→

FIGURE 16.15 Immersion lens on thermistor.

the signal energy collected on the thermistor is increased by n^2, as is the signal-to-noise ratio.

Germanium, silicon, and arsenic triselenide are among the most useful thermistor materials.

16.5.3 Pyroelectric Detectors

In 2300 BC, the Greeks considered tourmaline a magical material because, when heated, it produced a charge which shocked the unwary. This effect was pyroelectricity, and the observation is one of the earliest on record. Pyroelectric detectors are frequently classified with thermal detectors, which can be confusing, because the pyroelectric speed of response is limited only by the phonon vibration rate and not by bulk thermal characteristics.

The effect results from disruption (absorption) of a photon by the outer electron cloud of the molecule. Pyroelectric detectors sense changes and *only* changes in incident flux resulting in a displacement current. They are inherently broadband in response and can be high-speed. They can be used to detect particles or far-infrared photons to the microwave region. 50-GHz detectors (10^{-12} rise time) have been built.

They are passive devices, which means they do not require biasing and can utilize smaller packaging as well as being rugged and linear in output (greater than 6 decades, or orders of magnitude, change in incident response). However, nature has a way of giving and taking. In spite of all these positive features, they have low sensitivity compared to most semiconductor types of detectors. Their ruggedness and low sensitivity suggest operation in high flux or noxious environments. One of the reasons pyroelectric detectors are frequently classified with thermal detectors is that their high dielectric constants can make them slow, especially where maximum sensitivity is required.

Figure 16.16 shows the basic construction of a pyroelectric detector. Pyroelectric detectors generally have opposing electrodes. If radiation strikes the crystal between the electrodes, the device is edge-type. If it is absorbed by one electrode or transmitted through an electrode, it is face-type. The coplanar configuration places both contacts on the same surface of the sensing material.

Pyroelectric materials are those for which the molecule exhibits a net dipole moment. Energy absorbed by the outer electron cloud results in a shift, or change, in this dipole moment. In order to return to its original state, the material emits phonons. If the molecules in a pyroelectric material are aligned, the phonons are absorbed by the nearby molecules. An increase in temperature of the bulk material interferes with the effect because it changes the spacing between the molecules. There are thousands of pyroelectric materials, including human bone and skin, but only a few which exhibit a response large enough to be useful. Triglycerine sulfide (TGS), triglycerine selenate (TGSe), lithium tantalate ($LiTaO_3$), lead zirconate (PZT), strontium barium niobate (SBN), and polyvinyl fluoride (PVF) are the most common materials. Except for $LiTaO_3$, they are available with doping to change their properties or in various stoichiometric ratios. The properties of each version

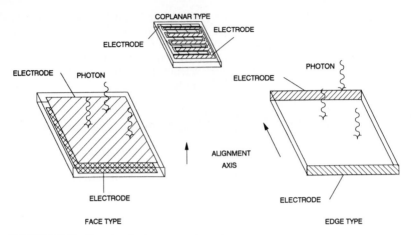

FIGURE 16.16 Pyroelectric detector configurations.

vary and need to be discussed with the supplier. The dipole moments of these materials are not naturally aligned. So that the effect can be transmitted and collected, a high electric field (poling) is applied between the electrodes. Those pyroelectric materials which are also ferroelectric experience a phase transition near the Curie temperature which permits the dipoles to move more freely, requiring a lower field. When practical, ambient temperature is raised to the Curie temperature for poling.

$LiTaO_3$ is the most common pyroelectric material because it has application in other areas, can be grown in crystalline form readily, and has stable characteristics. SBN is more difficult to grow (as a crystal) than $LiTaO_3$. It has limited commercial availability in high-speed detection. PVF (and similar materials) is a plastic film which exhibits pyroelectricity when stretched and poled. It is relatively inexpensive, thin, and available in large area. When used in the visible spectral region, pyroelectric detectors are coated with an absorbing medium. TGS and TGSe are hygroscopic and fragile. The most sensitive of the pyroelectric materials and the only ones with material absorption in the visible spectral region, TGS and TGSe families are sensitive to microphonics and may depole easily.

Pyroelectric detectors can also be fabricated in a coplanar structure which may be easier and cheaper to use. The coplanar structure behaves as a thin, fast sensor because it has low capacitance. Narrow separation of the electrodes gives it high sensitivity. Coplanar-electrode pyroelectric detectors are generally used for research purposes and are not available commercially. This configuration is most successful with materials which can be poled after fabrication of the detector.

In general, pyroelectric detectors do not compare to semiconductors (photovoltaics) in sensitivity. The current responsivity of a silicon photodiode can be typically 0.5 A/W as opposed to 0.5 μA/W for a high-speed pyroelectric detector or 50 μA/W for a sensitive one. Face-type configurations are used for low-light-level applications, and edge-type configurations are used for high speed. Because it is so low, the quantum efficiency of pyroelectric detectors is not discussed. On the other hand, semiconductors generally exhibit wavelength-dependent response and are sensitive over only limited spectral regions, whereas pyroelectric detectors are broadband in response, although many materials are transparent in visible and near-ir regions. The maximum D^* for photovoltaic detectors is approximately 10^{15} cm·$Hz^{1/2}$·W^{-1}, as opposed to 10^{10} cm·$Hz^{1/2}$·W^{-1} for sensitive and 10^7 cm·$Hz^{1/2}$·W^{-1} for high-speed pyroelectric detectors. Because of their higher sensitivity (and subsequent D^*), photovoltaic detectors are used in most applications where suitable wavelength response is obtained. Pyroelectric detectors are used when broadband, room-temperature, low-cost, high-energy, unbiased (passive), or very high frequency operation is desirable. Chapter 17 discusses the operation of pyroelectric detectors in greater detail.

16.6 OTHER DETECTORS

Although many detector types are covered here, a few more should be mentioned:

Phototransistors are *p-n-p* devices in which the collector current is proportional to incident radiation. The internal gain is typically 50. Generally small and inexpensive, they are nonlinear and have poor uniformity.

Photodarlingtons are phototransistors followed by amplifier stages in one package.

Photo-FETs are field-effect transistors in which the output current is proportional to the incident radiation.

A transistor exposed to light is by no means similar in performance to a photodiode in combination with an amplifier, since the design requirements for each are different, but these devices can satisfy switching or modulation applications where low cost, gain, and speed are more important than linearity or uniformity.

16.7 DETECTION SYSTEMS AND SELECTION GUIDE

The detector is only one element of a detection system. A means of collecting light, which may include lenses, filters, apertures, gratings, polarizers, integrating spheres, fiber optics, diffusers, attenuators, and other devices may determine the spectral character, distribution, intensity, and duration of the light reaching the detector. The detector itself may be coated to enhance absorption (with an antireflection coating, for example) or to select certain wavelengths. The detector signal itself must be processed. Whether the output is displayed on a meter or an oscilloscope or used to control other systems, the signal processing system places both requirements and limitations on the performance of the detector. The collection optical elements, the detector, and the signal processing components comprise a detection system. Frequently, some portion of the optics, the detector, and the first stages of the electronics will be housed together as an optohybrid. Typical optohybrid devices have several elements. Each element has its own amplifier. Depending on the application, the gain can be varied and individual filters can be applied to the silicon active areas.

TABLE 16.3 Typical Performance Characteristics of Visible Detectors

Detector type	Area, cm²	Responsivity at λ peak, A/W	Operating bias, V	Maximum signal, μA	Current density, μA/cm²	Dark current, pA	Capacitance, pF	Rise time, ns	Detectivity *D, cm·Hz$^{1/2}$·W^{-1}
Low-voltage vacuum phototube	1–20	0.030	50–250	<1	<0.05	1–100	2	1	10^8
High-voltage biplanar phototube	1–200	0.040	1000–2500	1–1000	5	5000	2	0.1–1	
Photomultiplier tube (5–16 stages, 10^5–10^7 gain)	0.1–1	0.030	1000–5000	1	5	50–5000 (anode)	1–10	0.1–1	10^{12}–10^{16}
Photoconductor (low–high illumination)	0.1–1	10^3–10^6 V/W	10–350	10^5				10^6–10^9	10^8–10^{11}
Photodiode	0.01–1	0.6	0–200	100	100	1–1000	1–10^4	1–10^6	10^{12}–10^{15}
Avalanche photodiode	10^{-4}–0.2	50	350	20	100	1000	5	0.5–2	10^{12}
Thermopile	0.008–0.7	40 V/W	0				5	10^6–10^8	10^8
Bolometer	0.01–0.3	5000 V/W	10–1500					10–10^8	10^7–10^{10}
Pyroelectric	0.01–1	10^{-6}–10^{-5}	0	>10^{-6}	>10^6	1–10^3 (Johnson noise)	1–10^3	10^{-2}–10^8	10^7–10^{10}

16.7.1 Detector Selection

In the visible spectral region, the silicon junction photodiode is the detector of choice whenever the application is compatible with its characteristics. Silicon photodiodes are available in a broad range of sizes, packages, and performance characteristics.

- If the spectral character of the source is not known, either an array with spectrally selective filters on the active elements or a broadband response detector such as a pyroelectric detector may be necessary.
- If the application is at low temperature (− 100°C, for example), germanium might be the material of choice.
- If the application requires fast response, the detector *and* the signal processing system must be faster unless the response of the system is well-characterized.
- If the incident power density is high, attenuation of the input power may be required to keep the detector in the linear portion of its response curve, usually $250 \ mW/cm^2$ for silicon photodiodes, for example. Large amounts of attenuation may be difficult to calibrate so another physical type may be desirable.
- If a detector is to be operated near the limit of its performance range, the potential effect on the signal should be discussed with the manufacturer.

Table 16.3 compares the typical performance characteristics of some of the detectors discussed in this chapter.

16.8 REFERENCES AND FURTHER READING

Detection and Radiometry

1. Grum, F., and R. J. Becherer, *Radiometry, Optical Radiation Measurements*, vol. 1, Academic Press, New York, 1979.
2. Dereniak, E. L., and D. G. Crowe, *Optical Radiation Detectors*, Wiley, New York, 1984.
3. Wolfe, W. L., and G. J. Zissis, eds., *The Infrared Handbook*, Office of Naval Research, Department of the Navy, Washington, D.C.

Semiconductor Detectors

4. Shur, M., *Physics of Semiconductor Devices*, Prentice Hall, Englewood Cliffs, N.J., 1990.
5. Joshi, N. V., *Photoconductivity: Art, Science, and Technology*, Optical Engineering Series, vol. 25, Marcel Dekker, New York, 1990.

Detectors

6. Budde, W., *Physical Detectors of Optical Radiation*, Optical Radiation Measurements, vol. 4, Academic Press, Orlando, Fla. 1983.
7. *Laser Focus World Buyer's Guide*, Pennwell, Tulsa, Okla.
8. *Photonics Buyer's Guide*, Laurin, Pittsfield, Mass.

CHAPTER 17
INFRARED DETECTORS

Suzanne C. Stotlar

17.1 INTRODUCTION

The physical universe is made up of objects with temperatures greater than absolute zero, which means that the atoms and molecules of an object are in motion. These motions result in interactions with other atoms and molecules (via bonds and collisions). Therefore, these elementary particles are subject to accelerations which result in electromagnetic radiation.

Radiometry is the science of measurement of the electromagnetic spectrum. This chapter discusses those detectors which provide a response to that portion of the electromagnetic spectrum which is not visible to the human eye because the energy of the photon is too low to elicit a response. The lower energy results in a longer wavelength. This region of the spectrum is called *infrared* (ir). Light of 700 to 3000 nm is generally called *near infrared* and longer than 20 μm (20,000 nm) to 1000 μm is called *far infrared*. There are many detector types for the infrared region. No one type has a response broad enough to cover the entire region without modification, although pyroelectric detectors and thermopiles come close. This chapter is a companion to Chap. 16. This review of ir detectors cannot be comprehensive, but addresses those currently most important by application and availability.

17.1.1 Performance Parameters

Terminology and performance parameters common to ir detectors are the same as those for visible detectors. Section 16.1.1 reviews those most commonly used.

17.2 PHOTOGRAPHIC FILM

Infrared film is similar to visible film and plates. Infrared emulsions respond to visible and ultraviolet light as well as infrared. Color infrared film is available with sensitivity to wavelengths approaching 1 μm. Filters are frequently used to enhance the longer wavelengths. In visible color film, the top layer is blue-sensitive with a yellow filter (to remove the blue) below it. This prevents the red and green layers from responding to the shorter wavelength light. In color infrared film, the layers are green-sensitive, red-sensitive, and near-infrared-sensitive. If a yellow (minus blue) filter is not built into the film, the user may need to add one. Chemical processing of visible color film results in blue dye colorants where only blue light exposed the film. In color infrared film, a color translation scheme is used so that where near-ir light exposed the film, the image appears red; where the red band only was excited, the image is green; and where radiation excites the green band only, the image appears blue. Extending the response of film to longer wavelengths involves limiting fogging caused by background radiation as well as using longer-wavelength-sensitive materials. Cooled apertures and enclosures lower background radiation.

A densitometer and photographic camera (with appropriate filter) using black-and-white or single-emulsion film is useful for examining the spatial distribution of laser beams and focused spots. For high-power lasers the "burn" pattern may be examined.

17.3 PHOTOELECTRIC DETECTORS

See Sec. 16.4 for a description of photoelectric detector physics. The internal photoelectric effect is more important in the infrared spectral region than the external photoelectric effect, or photoemission. Although cathodes fabricated from Ag-O-Cs (S-1), NaKCsSb (multialkali), and gallium arsenide, plus a few other materials, have a spectral response which reaches slightly into the ir, vacuum phototubes and photomultipliers are not considered infrared devices. Silicon is generally the material of choice in the near ir to 1.1 μm, with many detector types competing at longer wavelengths.

17.3.1 Photoconductors

See Sec. 16.4.3 for a description of photoconductor physics. Table 17.1 lists some examples of intrinsic materials used in photoconductors.

TABLE 17.1 Typical Intrinsic Photoconductor Detector Material

Detector	e_G, eV	Typical operating temperature, K	λ max., μm	Usable wavelength range, μm
Ge	0.67	193	1.9	0.9–1.9
Si	1.12	300	1.1	0.2–1.1
GaAs	1.5	300	0.9	0.4–0.9
PbS	0.42	300	2.4	1.1–3.5
PbSe	0.23	300	3.9	1.0–5.0
InSb	0.18	77	4.9	2.0–5.5
$Hg_{0.8}Cd_{0.2}Te$		77*	11.0	8.0–14.0
$Hg_{0.8}Cd_{0.2}Te$		195*	4.5	3.0–5.0

*Also at 300 K with reduced performance.

Because infrared light is more deeply penetrating than visible light, the transit time T_r of the charge carriers is

$$T_r = \frac{d^2}{\mu V_b} \qquad (17.1)$$

where d is the distance between the electrodes, μ is the charge mobility, and V_b is the bias voltage. The photocurrent is

$$i_{\text{ph}} = (1 - \rho) \frac{q\Phi\lambda}{\eta hc} \frac{\tau\mu V_b}{d^2} \qquad (17.2)$$

where ρ = reflectance of the front surface
$\quad \eta$ = quantum efficiency of the electron charge
$\quad hc = 1.9856 \times 10^{-23}$ W·s·cm
$\quad q$ = electron charge
$\quad \tau$ = lifetime of the charge carrier
$\quad \Phi$ = incident flux
$\quad \lambda$ = wavelength

Size, operating temperature, and physical configuration as well as illumination and other environmental conditions, such as vibration or background electromagnetic radiation, influence the actual performance of infrared photoconductors. Cooling in particular is useful in shifting the response curve further into the ir. Thermal (Johnson) noise limits the infrared wavelength response. Cooling minimizes thermal noise by reducing the number of free charge carriers and maximizes responsivity, which is most effectively done by cooling the detector to a temperature such that $kT \ll e_G$. Cooling does have a limitation in that for each material there is a range above absolute zero in which the semiconductor properties remain the same with further cooling.

As the wavelength increases, the energy of the photon decreases. Eventually, the photon lacks sufficient energy to create free carriers by causing electrons to cross the energy gap. The minimum optical frequency which will produce a free electron from the covalent bond is $\nu = e_G/h$, or, in terms of wavelength,

$$\lambda_{\text{max}} = \frac{1.24}{e_G} \qquad (17.3)$$

where λ_{max} is the maximum wavelength of radiation to produce an electronic transition, and e_G is the bandgap in electron-volts.

Another technique used to improve the responsivity is external biasing. This reduces the energy necessary to create an electron transition, extending the wavelength response near the bandgap. However, since the number of thermally excited electrons (noise) also increases, biasing may be performed in conjunction with cooling.

Intrinsic semiconductor materials have no impurities which can reduce the bandgap or alter physical properties. However, doping an intrinsic semiconductor can be used to control the wavelength response, producing an extrinsic semiconductor. As shown in Fig. 17.1, the material becomes p type if acceptor impurities are used and n type if donor impurities are used. P-type materials have positive majority carriers (holes), and n-type have negative majority carriers (electrons). In intrinsic semiconductors, both electrons and holes produce a change in conduction while only majority carriers contribute to the current in an extrinsic semiconductor. The

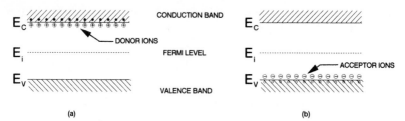

FIGURE 17.1 Energy-band diagram of extrinsic semiconductors with (*a*) donor and (*b*) acceptor ions.

result of doping to produce a *p*-type material is that less energy is required for an electron to jump from the valence band to the acceptor level. The electrons are trapped in the acceptor level while the holes (majority carriers) are mobile and can cause conduction changes. The energy necessary to produce this change is less than the original bandgap, so the long wavelength cutoff has been increased. An *n*-type semiconductor is produced by introducing impurity atoms resulting in weakly bound electrons which reside in a donor level within the bandgap and can be excited into the conduction band by the absorption of energy. While a photon causes an electron to move from the donor level to the conduction band, resulting in conduction by electrons (majority carriers), a corresponding hole is trapped in the donor level. Table 17.2 lists some extrinsic semiconductors used as ir detectors. Extrinsic semiconductors are generally operated at low temperatures in order to reduce the effects of thermal carrier generation.

Control of thermally ionized carrier conduction is provided for temperatures lower than

$$T_C = \frac{e_G}{k \ln (\nu \sigma_c N_v / \sigma_{ph} E_P^B)} \tag{17.4}$$

where ν is the thermal drift velocity
    σ_c = free-electron capture cross section
    σ_{ph} = photoionization cross section
    N_v = acceptor/donor concentration
    E_p^B = photon irradiance

TABLE 17.2 Extrinsic Photoconductor Materials

Material	e_G, eV	Conductor type	Typical operating temperature, K	Wavelength range, μm
Ge:Cu	0.041	*p*	4.2	1.5–27
Ge:Cd	0.06	*p*	4.2	1.5–100
Si:As	0.0537	*n*	<30	0.9–28
Si:Bi	0.0706	*p*	<30	0.9–22
Si:In	0.165	*p*	≤50	0.9–7
Si:Mg	0.087	*p*	<50	0.9–12

For large irradiance values

$$T_c \le \frac{e_G}{k} \tag{17.5}$$

where k is Boltzmann's constant 1.380662×10^{-23} W·s·K. Since there is also a relationship between the optical cutoff frequency and e_G,

$$\lambda_c = \frac{1.24}{e_G} \tag{17.6}$$

$\lambda_c T_c$ is less than or equal to a constant. As the wavelength response becomes longer, the cooling temperature becomes lower. A 10-μm wavelength cutoff, for example, requires liquid nitrogen (77 K) cooling.

Photoconductors exhibit $1/f$ noise, generation-recombination (G-R) noise, and Johnson (thermal) noise. Flicker, or $1/f$, noise dominates at low frequencies and results from material and manufacturing defects. G-R noise dominates at midrange frequencies in photoconductors and is a type of shot noise caused by fluctuation in generation, recombination, or trapping rates in a semiconductive photoconductor. It is influenced by both thermal and photon excitation. Cooling the detector reduces the expression for G-R noise to

$$i^2_{G-R} = 4q^2 G^2 \eta E_p A_d \tag{17.7}$$

where q = the electron charge
G = photoconductive gain
η = quantum efficiency
E_p = photon irradiance
A = sensing area of the detector

At high frequencies Johnson noise is dominant.

If the performance of an infrared detector is limited by the fluctuation in arrival rate of incident photons (photon noise) from the background, the detector is described as BLIP (background-limited infrared performance). BLIP is sometimes used to describe any photon-limited ir detector even if it is not background- but signal-limited. BLIP is applied properly only to those detectors with a field of view which contains more background than signal photons. The noise equivalent power (NEP) of a BLIP detector is

$$\text{NEP}(\lambda, f) = \frac{i_{G-R}}{R_i} = 2\frac{hc}{\lambda} \sqrt{\frac{E_p^B A_d \, \Delta f}{\eta}} \tag{17.8}$$

where i_{G-R} = photon current noise
R_i = current responsivity
h = Planck's constant
c = speed of light
λ = wavelength
E_p^B = background photon irradiance
A_d = detector sensitive area
Δf = electrical bandwidth
η = quantum efficiency

The BLIP area normalized detectivity is

$$D^*(\lambda, f) = \frac{\lambda}{2hc} \left(\frac{\eta}{E_p^B} \right)^{1/2} \tag{17.9}$$

BLIP photoconductor detectors generally have the load resistor (see Fig. 16.8, where R_a is the load resistor) mounted on the detector heat sink in order to minimize Johnson noise. Extrinsic BLIP photoconductors have load resistors with values much less than that of the detector, while for intrinsic BLIP photoconductors, the effective resistance is that of the detector.

17.3.2 Photodiodes

Photodiodes (p-n) are formed by diffusing a thin layer of acceptor (p-type) impurity into the surface of an n-type-doped semiconductor so that the number of acceptors exceeds that of the donors. The p-n junction is where the p-type layer joins the n-type bulk. The layer is thin so that photons can reach the depletion region.

Conversely, a n-p junction is formed by diffusing a donor impurity layer into bulk p-type material. Near the junction a depleted region exists which separates photogenerated electron-hole pairs.

The photovoltaic effect is described in Sec. 16.4.4. Most photodiodes exhibit a fairly abrupt ir cutoff (λ_{max}) in response due to a rapid increase in the photon penetration depth near the bandgap. Silicon (and to some extent germanium) declines more gradually in response. Because they are the least expensive, are the most readily available, and generally offer the best performance options, including 300 K operating temperature, silicon photodiodes are commonly used for applications between 0.7 and 1.1 μm. Silicon's responsivity increases between 0.7 μm and approximately 0.9 μm. Beyond 1 μm, its sensitivity decreases as the absorption coefficient falls. Light penetrates farther into the material, and charge collection becomes less efficient. The infrared response can be improved somewhat by increasing the bias (to widen the depletion region) and using high purity (resistivity) p-type silicon as the bulk material. Increasing the bias also decreases the capacitance until a minimum value is reached. At this point the detector is described as *fully depleted*. Pulsed yttrium aluminum garnet (YAG) laser (1.06 μm) applications typically use fully depleted p-type silicon photodiodes [full width at half maximum (FWHM) response time less than 16 ns]. The increased bias also results in increased shot noise, which is higher in p-type detectors than in n-type anyway, so a guard ring is used to reduce the surface leakage current. Figure 17.2 is a schematic diagram of a planar diffused p-n junction photodiode and a planar diffused n-p junction photodiode with a guard ring. Planar diffusion is a manufacturing technique which results in termination of the junction on the active surface of the device. A heavy oxide is used to protect the junction. Heavy doping (n^+ or p^+) is used to improve the ohmic (nonrectifying) characteristics of the contacts. A high resistivity (greater than 400 Ω·cm for n-type) bulk material will result in a p-i-n diode. This region is generally termed *intrinsic*, although *lightly doped with fewer impurities* might be more accurate. Some manufacturers do not describe their product as p-i-n because intrinsic material has neither doping nor defects. For silicon, the resistivity of intrinsic material would be greater than 50,000 Ω·cm, so 400 Ω·cm is only relatively intrinsic compared to 1 Ω·cm material. The capacitance per unit area is a better comparison than the p-n or p-i-n designation.

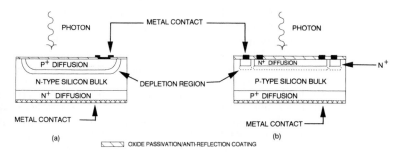

FIGURE 17.2 Schematic diagram of planar diffused photodiodes: (*a*) *p-n* junction; (*b*) *n-p* junction with guard ring.

Photodiodes are operated in photovoltaic or photoconductive mode as shown in Fig. 16.11. Photovoltaic, or unbiased, operation has slower response but no shot noise. Photoconductive, or reverse-biased, operation has shot noise but faster response and slightly higher sensitivity. Silicon (and germanium) used for photodiodes has a long lifetime. Light absorbed outside the diffused area may contribute to the optical current. This is particularly true in the near-ir spectral region as photons become more deeply penetrating as the wavelength is increased. This contribution can increase fall time or produce other results which are difficult to interpret, particularly in multielement or array patterns. A typical silicon multielement pattern for ir beam (usually a 1.06-μm YAG laser) guidance is shown in Fig. 17.3. A quadrant pattern is formed by planar diffusion on high-resistivity *p*-type silicon. The high resistivity provides more response at 1.06 μm and lower capacitance for improved rise time. A reverse bias (usually about 200 V) is used to further enhance the responsivity and speed. The guard ring, which must also be

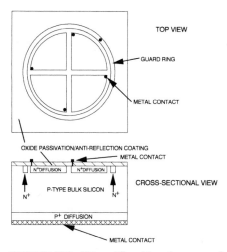

FIGURE 17.3 Silicon diffused junction *n-p* quadrant detector with guard ring.

reverse-biased to the same level as the active areas in order to be fully effective, reduces dark current contribution from the margin (surface and bulk) of the device. Hole-electron pairs generated in the margin areas are also collected by the guard ring. Although quadrant detectors are used in several ways, a common technique employs a laser spot approximately the size of the quadrant. By comparing the signal of each quadrant, the centroid of the beam can be located. Silicon detectors typically offer six or more decades of linear response to incident light, so very fine resolution can be achieved. Silicon has a high temperature coefficient at this wavelength, so a heater (or other temperature control mechanism) is frequently used to improve operational linearity. Applications for this type of device include laser-guided weapons and robotics. More sophisticated versions (including other materials) are used in machine vision applications. Chapter 19 discusses imaging detectors and systems.

At wavelengths beyond the response of silicon, other materials are used. Germanium photodiodes are available for some applications. The low bandgap (0.67 eV) produces high (thermal) noise at room temperature so that cooling is desirable, particularly for low-light-level applications. Since the development of fiber-optic technology, a high-speed near-ir detector which can operate near 300 K has become important to communication applications. (Infrared fibers can deliver information faster with lower losses than silica fibers.) The heterojunction photodiode is an ir detector in which the junction is formed between two different bandgap semiconductor materials. The larger bandgap material is more transparent to light and can act as a window to transmit optical radiation to the junction.

Proper material selection for heterojunction combinations can be used to optimize specific performance parameters. The lattice constants of the two materials must be closely matched to obtain low leakage current. The ternary III-V semiconductor $Al_xGa_{1-x}As$ exhibits direct bandgap transitions for $x < 0.4$. When epitaxially grown on a gallium arsenide substrate, heterojunctions with perfectly matched lattices are formed. These devices are used in the 0.65- to 0.85-μm wavelength range. Multiple heterojunctions can be formed on the same substrate. Ternary compounds such as $Ga_{0.47}In_{0.53}As$, which has a bandgap of 0.73 eV, and quaternary compounds such as $Ga_xIn_{1-x}As_yP_{1-y}$ can be used. Both are a perfect lattice match in InP. Figure 17.4 shows a typical InGaAs detector and its responsivity curve. The wide bandgap of the InP substrate makes it transparent to near-ir light. This rear entrance configuration permits the most efficient absorption of the desired

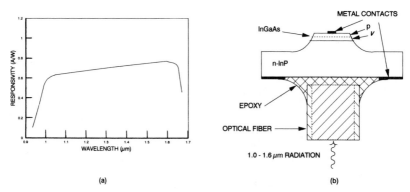

(a)

(b)

FIGURE 17.4 Typical InGaAs detector and its responsivity curve.

wavelength light. The optical fiber is coupled to the rear surface. Diffusion technologies are not well-developed for compound semiconductor materials. Since the layers are deposited epitaxially anyway, mesa fabrication techniques are common with heterojunction photodiodes. With the mesa structure, the device pattern is defined by etching. The perimeter of the junction must be passivated, usually with a deposited coating.

The manufacturing technology, device geometry, and illumination and environmental conditions determine actual performance characteristics. Availability of some ir semiconductor detectors is a continuing problem because of issues related to the materials. The material may be difficult to grow, ohmic contact formation may be limited, or material properties may be highly variable. Maximum performance may also be difficult to achieve because of packaging characteristics. HgCdTe and PbSnTe at 77 K, for example, are commonly mounted on cold "fingers" in order to minimize the mass which must be cooled. The resulting assembly is not only bulky (because it uses liquid nitrogen) but also microphonic. Consequently, thermoelectrically cooled detectors may be more practical for some applications. Common infrared photovoltaic detector materials and their operating parameters are listed in Table 17.3.

TABLE 17.3 Semiconductor Materials for IR Photovoltaic Devices

Material	Typical operating temperature, K	Peak wavelength, μm	Usable range, μm
Si	300	0.9	0.2–1.1
GaInAs	300	1.6	0.9–1.7
InGaAsP	300	1.3	1–1.6
Ge	300	1.5	0.9–1.9
InAs	77	3.1	1.8–3.8
InSb	77	4.9	2–5.5
$Hg_{0.7}Cd_{0.3}Te$	195	4.5	1–5.5
$Hg_{0.8}Cd_{0.2}Te$	77	10.5	8–11.5
PbSnTe	77	11	8–11.5

17.3.3 Avalanche Photodiodes (APDs)

Avalanche photodiodes provide internal gain in the number of electron-hole pairs created per absorbed photon through an effect called *avalanche multiplication*. A high reverse bias results in a high electric field within the semiconductor which accelerates the electrons to kinetic energy levels large enough to produce new electron-hole pairs by collision. These new carriers are also accelerated and produce further collisions. Since charge collection occurs almost instantaneously, APDs are useful in high-speed applications. Their gain-bandwidth product can exceed several hundred gigahertz, suitable for microwave frequencies. The most common application for APDs is in optical-fiber communication. As in fabrication of junction photodiodes, the adage "no gain without pain" is certainly applicable to APDs. The high reverse bias necessary to produce avalanche multiplication results in an operational bias near the reverse bias breakdown voltage V_{BR}. This parameter varies to some degree with each device, as it depends not only on the design, but on (1) defects in the material, (2) contamination introduced during junction formation and anomalies or defects in passivation layers, (3) strain resulting from packaging, (4) nonuniform strain resulting from packaging or mounting of the package into the test system, and (5) internal stress resulting from field effects within the device. This bend, or knee, in the I-V curve also tends to be quite temperature-sensitive. Figure 16.12 shows the multiplication as a function of electric field for a typical APD. The APD photocurrent is similar to that of the junction photodiode except for the multiplication factor M, which represents the avalanche gain:

$$i_{prms} = q\eta m\Phi_p \frac{M}{\sqrt{2hc}} \tag{17.10}$$

where q = electron charge
η = quantum efficiency
m = modulation index
Φ_p = average optical power
$M/\sqrt{2}$ = rms gain

The rms value is used because APDs are commonly found in frequency-modulated applications. The shot-noise current produced by the optical signal, the background, and the detector dark current constitute the rms shot noise after multiplication:

$$\langle i_s^2 \rangle = 2q(i_p + i_B + i_D)\langle M^2 \rangle B \tag{17.11}$$

where B = bandwidth
i_p = photon noise current
i_B = background noise current
i_D = dark current
and $\langle M^2 \rangle$ is the mean-square internal gain, given by

$$\langle M^2 \rangle = M^2 F(M) \tag{17.12}$$

where $F(M)$ is the noise factor, a measure of the increase in the shot noise compared to an ideal noiseless multiplier. Johnson (or thermal) noise is the other APD noise source.

The optimum performance of an APD is obtained when the shot noise is approximately equal to the Johnson noise for a given level of incident power. For high-frequency and large-bandwidth operation, the minimum detectable power is limited by the Johnson noise of the load resistor and the noise of the amplifier. The noise equivalent power is given by

$$\text{NEP} = \sqrt{2}\,\frac{hc}{\lambda\eta}\left(\frac{i_{eq}}{qF(M)^2}\right)^{1/2} \tag{17.13}$$

where $F(M)$ usually follows a curve similar to Fig. 17.5 and

$$i_{eq} = (i_B + i_D)F(M) + 2\,\frac{kT}{qR_{eq}M^2} \tag{17.14}$$

The gain and noise factor behavior of the avalanche photodiode are complex and explained fairly well by Sze.[10]

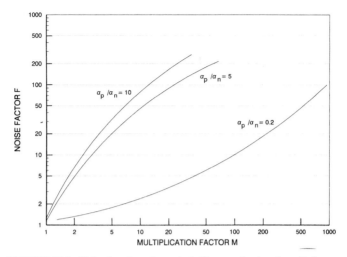

FIGURE 17.5   Noise function of a typical silicon avalanche photodiode.

The spatial field distribution, both in terms of uniformity and shape, is critical to the performance of an APD. Figure 17.6 shows some typical device configurations used to control leakage current along the junction edge due to junction curvature or high field concentration. The beveled structure is used to control the shape of the field, provided that damage-free surfaces can be maintained (usually a chemical-mechanical polish followed by a deposited passivation). The guard ring (usually diffused or ion-implanted) is made deeper than the primary diffusion in order to provide a barrier. The mesa and beveled configurations are easier to package in high-frequency applications, as their geometry can facilitate impedance matching. APDs require defect-free junction formation (at each level), as microplasmas (small areas in which the breakdown voltage is lower) limit performance as well as device operational life. The illumination and responsivity also need to be uniform in order to maintain uniform avalanche gain. If the radiation is not monochromatic, variation in the penetration depth must also be considered. Consequently, APDs are rarely larger than a few millimeters in diameter and nearly always round.

Because they are difficult to manufacture reproducibly, APDs are not offered by many suppliers in silicon and germanium. APDs can be fabricated from any semiconductor material for which reasonable material uniformity can be achieved. Heterojunction avalanche photodiodes fabricated from III-V alloys are particularly interesting, since the wavelength response can be changed or tuned by varying the alloy composition. The device structure of Fig. 17.6a is common. The substrate material (shown here as p^+) is actually part of the contact, and the device consists of epitaxially grown layers. These include AlGaAs/GaAs, AlGaSb/GaSb, GaInAsP/InP, and GaAsIn/InP, the most common of which is probably the last.

Metal-semiconductor (Schottky-barrier) devices, which will be discussed in Sec. 17.3.4, can also achieve avalanche gain. Platinum silicide (PtSi) in particular has received interest because it responds in both the uv and, when cooled, ir. The configuration shown in Fig. 17.6b has a diffused guard ring. In the ir, it is not only cooled, but also rear- (back-) illuminated.

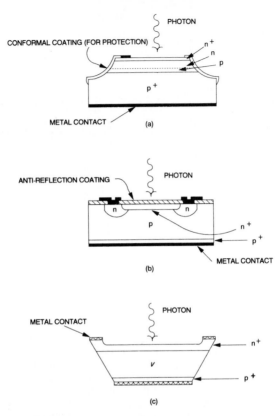

FIGURE 17.6 Typical avalanche photodiode device structures: (*a*) mesa structure; (*b*) guard ring structure; (*c*) beveled *p-i-n* structure.

17.3.4 Internal Photoemissive Detectors (Schottky)

A Schottky diode is formed by depositing a metal of an appropriate work function onto the clean surface of a semiconductor. The work function is the energy necessary for an electron to escape into vacuum from an initial energy at the Fermi level. The Fermi level is occupied by the highest-energy electron at absolute zero. The work function is defined as $\phi = e_i - e_F$, where e_i is the ionization energy and e_F is the energy of the Fermi level. Work functions for metals vary from 2 to 6 eV. When an electron is at a distance x from the metal, a positive charge is induced on the surface of the metal. The charge (equal to that on a positive charge at $-x$) is called the *image charge*. The image force is given by

$$F = \frac{-q^2}{16\pi\epsilon_0 x^2} \tag{17.15}$$

where ϵ_0 is the permittivity of free space. Integrating this function between infinity and x provides the potential energy of an electron at a distance x from the metal surface. When an external field ξ is applied, the total potential energy as a function of distance (downward from the x axis) is given by

$$\text{PE}(x) = \frac{q^2}{16\pi\epsilon_0 x} + q\xi x \quad \text{eV} \tag{17.16}$$

The image-force-induced reduction of the potential energy for charge carrier emission when an electric field is applied is called the *Schottky effect*. Figure 17.7 compares this effect for a metal-vacuum system with a metal-semiconductor. The

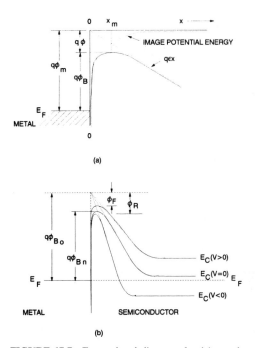

FIGURE 17.7 Energy-band diagrams for (*a*) metal-vacuum and (*b*) metal-semiconductor Schottky interfaces.

Schottky barrier reduction, or image force lowering, $\Delta\phi$ and the location of x_m as shown are given by

$$x_m = \left(\frac{q}{16\pi\epsilon_0\xi}\right)^{1/2} \quad \text{cm} \tag{17.17}$$

and

$$\Delta\phi = \left(\frac{qE}{4\pi\epsilon_0}\right)^{1/2} = 2\xi x_m \quad \text{V} \tag{17.18}$$

For a metal-semiconductor system, the field is the maximum field at the interface, and ϵ_0 is replaced by the permittivity of the semiconductor material (the product of the dielectric constant k and ϵ_0), or

$$\Delta\phi = \left(\frac{q\xi}{4\pi k\epsilon_0}\right)^{1/2} \quad V \tag{17.19}$$

Figure 17.7a shows the energy-band diagram for a metal-vacuum interface when an electric field is applied. The barrier reduction results from the combined effects of the image force and the electric field. Figure 17.7b shows the energy-band diagram for a metal-semiconductor system incorporating the Schottky effect for an n-type semiconductor. The intrinsic barrier height is $q\phi_{Bo}$, and the height at thermal equilibrium is $q\phi_{Bn}$. The change in the barrier height for forward bias ($V > 0$) is less than that for $V = 0$. The greatest change is for $V < 0$ under reverse bias. Although the dielectric constant results in a small value of $\Delta\phi$, there is a significant effect on current transport processes. Again, Sze[10] describes the Schottky diode.

Schottky photodiodes must have thin, semitransparent metal layers if the photon is to be absorbed in the semiconductor ($h\nu > e_G$). For $V < V_B$, the diode behaves like a p-i-n photodiode. For $V \approx V_B$ (high reverse bias voltage), the diode behaves like an avalanche photodiode.

For $e_G > h\nu > q\phi_{Bn}$ and $V < V_B$, photoexcited electrons (hot electrons) in the metal can cross the barrier into the semiconductor. This internal photoemission effect is important in the ir region where the detector can be back-illuminated.

Schottky monolithic focal-plane arrays are formed by evaporating platinum onto a p-type silicon substrate followed by heat treatment which is then mounted onto a charge-coupled device (CCD) readout in an interline transfer configuration. Imaging is discussed in Chap. 18. Dereniak and Crowe[3] also provide a good explanation of imaging arrays. This array is useful between 1.1 and 5 μm and is operated at 77 K, as are single-element PtSi Schottky photodiodes. As might be expected, the Schottky effect is temperature-dependent.

In some cases temperature cycling, exposure to elevated cycling, or exposure to elevated temperatures (greater than 373 K, for example) can cause the device to fail prematurely. PtSi, the most likely Schottky detector choice for the ir, is able to tolerate some variation in the storage temperature.

Gold-silicon Schottky photodiodes with antireflection-coated semitransparent gold metalization or back illumination are also used in the near ir from 0.2 to 1.1 μm. However, a Si junction photodiode is preferred for most applications.

17.4 THERMAL DETECTORS

A detector which produces an electrical signal (or a change in output) in response to a change in its bulk temperature is classified as a *thermal detector*. Thermocouples, thermistors, bolometers, and thermopiles are thermal detectors. Pyroelectric detectors are also grouped with thermal detectors. Thermal detectors generally lack the sensitivity of semiconductor detectors, but they can provide flatter wavelength response and a variety of options in packaging, size, and cost. Most of them are passive devices, requiring no bias. Because of their stable properties, thermopiles are frequently used as calibration standards. Thermal detectors absorb radiation, which produces a temperature change that in turn changes a physical or electrical property of the detector. Since a change in temperature takes place, thermal detectors are generally slow in response and have relatively low sensitivity compared to other detectors. The thermal detector works via the heating of the detector chip. Therefore, a heat equation needs to be examined to model the thermal detector mathematically.

In the ir spectral region, a semiconductor detector with an appropriate wavelength response may not be available or, if available, may be difficult to operate, especially at low temperature. The lack of semiconductors with broadband response and suitability for heat-sinking, accounts for the greater importance of thermal detectors in the infrared than in the visible.

Primary noise sources for a thermal detector are *white noise* (the noise associated with a blackbody) and *Johnson noise* (noise due to random thermal fluctuations).

17.4.1 Thermocouples and Thermopiles

Thermocouples and thermopiles are discussed in Sec. 16.5.1. Optical blocks are used to absorb radiation in the near ir and middle IR. In the far ir (greater than 20 μm), flat wavelength response may not be feasible. In any case, the reflectance and absorption of the coating as the wavelength varies must be known in order to calibrate the sensitivity of the thermocouple or thermopile.

17.4.2 Bolometers and Thermistors

Bolometers and thermistors are described in Sec. 16.5.2. A bolometer can be a thin blackened strip of metal, semiconductor, or superconductor material. The resistance varies with temperature as given by

$$R(G) = R_0(1 + \alpha \Delta T) \tag{17.20}$$

where R_0 is the strip resistance when $\Delta T = 0$ and α is the temperature coefficient of resistance. In metals α is positive, while in semiconductors it is negative. The bolometer is very sensitive to change in temperature and generally is used with chopped or modulated sources of incident radiation.

Semiconductor bolometers have much higher voltage responsivity than metal ones because of higher absolute values of α. Thus, it is the semiconductor bolometer which will be considered here. Germanium and thin-film semiconductors are highly useful in the ir. They are usually operated at temperatures below 4.2 K. At such cryogenic temperatures, near-theoretical D^* (detectivity) can be achieved.

The germanium bolometer is useful over the 5- to 100-μm wavelength region. Bolometers are particularly useful beyond 15 μm, where there are few alternatives except pyroelectric detectors.

In a bolometer, a temperature change produced by the absorption of radiation causes a change in the electrical resistance of the sensing element. A typical low-temperature bolometer is packaged, or housed, in a dual cryogenic dewar which can be maintained at liquid helium (4.2 K) or pumped helium (less than 4.2 K) temperatures. The bolometer element (or elements, if a dual circuit is used for reference) may be a thin slice of germanium mounted in a strainfree manner to the two electrical leads which provide thermal contact with both. (The housing is evacuated.) The thermal conductance of germanium ensures that the temperature is uniform throughout the element. The load resistor is placed inside the housing to reduce Johnson noise. The temperature of the element is determined by the heat equilibrium equation, given by

$$H \frac{dT}{dt} + K_0(T_d - T_0) = I^2 R_0 \tag{17.21}$$

when the thermal conduction of the leads just equals the current heating effects,

where H = thermal capacity of the element (mass m times the specific heat C_p)
  K_0 = average thermal conductance from the element to the heat sink at initial temperature T_0
  T_d = detector temperature
  I = bias current

The heat conductance K can be determined by

$$K = 4\sigma\eta A T^3 \tag{17.22}$$

where σ = Stefan-Boltzmann constant
  η = emissivity
  A = surface area
  T = temperature

Incident radiation causes additional heating according to

$$H \frac{dT}{dt} + K(T - T_0) = I^2 R_0 + \Phi_e^b + \eta\Delta\phi_e \tag{17.23}$$

where Φ_e^b is the background radiant energy, $\Delta\phi_e$ is the signal radiant energy, and η is the emissivity. Bear in mind that T_0 is the initial temperature of the detector *and* the temperature of the heat sink. Equation (17.23) is valid only if $\Delta\phi_e$ is much less than the electrical power dissipation ($\Delta\phi_e < I^2 R_0$) and the background radiation power is less than or equal to the electrical power dissipation ($\Phi_e^b \leq I^2 R_0$). Bolometers are intended for low-light-level applications.

The voltage responsivity is

$$\mathfrak{R}_V = \eta I \alpha R_{\text{eff}} (K^2 + \omega^2 H)^{-1/2} \tag{17.24}$$

where ω is the angular frequency and R_{eff} is the effective resistance. For a shielded bolometer such as that shown in Fig. 16.14, $R_{\text{eff}}^{-1} = R_D^{-1} + R_L^{-1}$. The relationship for the resistance versus cryogenic temperature for a semiconductor bolometer ($\alpha \Delta T \gg 1$) is

$$R(T) = R_0 \left(\frac{T_0}{T}\right)^A \tag{17.25}$$

where R_0 is the resistance of the detector at temperature T_0 and is an empirically derived constant.[3] This constant is approximately 4, but must be determined for each bolometer. The temperature coefficient of resistance α is given by

$$\alpha(T) = \frac{1}{R}\left(\frac{dR}{dT}\right) = \frac{-A}{T} \tag{17.26}$$

The temperature coefficient increases as the temperature is reduced. Lowering the temperature (below 4 K) also results in a lower specific heat and a higher resistance

R. Of course, R may become too large for the amplifier so there may be an optimum operating temperature. The voltage responsivity is

$$\mathfrak{R}(B) = -\left[\frac{A(B-1)}{(A+1)B - A^2 B^A}\right]^{1/2}\left(\frac{R_0}{T_0 K}\right)^{1/2} \tag{17.27}$$

where $B \approx T/T_0$ and K is the heat conductance. For a given T_0, R_v has a maximum value determined by

$$\mathfrak{R}_{max} = -0.7\left(\frac{\mathfrak{R}_0}{T_0 K}\right)^{1/2} \tag{17.28}$$

The optimum value of the (dark) electrical dissipation is $0.1 T_0 K$. Using Eq. (17.28) as an expression for $\mathfrak{R}_{max}$ gives the Johnson NEP as

$$NEP_J = \left(\frac{4k T_0^2 K \Delta f}{0.7}\right)^{1/2} \tag{17.29}$$

where Δf is the equivalent noise bandwidth and the photon noise power is

$$NEP_P = (16k T_0^2 K \Delta f)^{1/2} \tag{17.30}$$

These can be added to estimate the total NEP:

$$NEP \approx 4.92 T_0 (k K \Delta f)^{1/2} \tag{17.31}$$

The thermal time constant is given by

$$\tau = \frac{H}{K} \tag{17.32}$$

Bolometers cannot be operated at high frequencies without sacrificing responsivity. Reducing the heat capacity H by reducing the mass can increase frequency response. Thin-film bolometers and composite bolometers such as blackened bismuth or nichrome deposited on sapphire (heat capacity 1/60th that of germanium) and then attached to a small germanium chip have been fabricated with reported rise times of 1 ns (10^{-9} s). However, such devices are susceptible to microphonics and other effects which may interfere with fast response. Furthermore, the housing and detector configuration are difficult to impedance-match to associated electronics. Bolometers are not recommended for high-frequency applications. Although bolometers can also be fabricated from superconductor materials, they are still developmental.

Thermistors are more frequently the bolometer of choice. The sensing element is typically a sintered wafer of manganese, nickel, and cobalt oxides mounted on an electrically insulating but thermally conductive substrate such as sapphire. An absorbing coating is often applied to enhance or ensure broadband wavelength response. Usually fabricated as a pair, one thermistor provides a shielded reference while the other responds with a temperature coefficient as high as 5%/°C. This coefficient also varies as $1/T^2$. Although the thermistor lacks the usually high responsivity of the bolometer, it is cheaper, easier to use (generally 300 K), and physically smaller as no housing is required. Johnson noise is usually the dominant noise source.

17.4.3 Pyroelectric Detectors

Pyroelectric detectors are used in many infrared applications, often replacing more sensitive, but more difficult- (and expensive-) to-use cooled detectors. Because of their relative lack of detectivity when compared to semiconductor devices, they require considerable energy for use. However, the speed of response of pyroelectric detectors is potentially subpicosecond and is currently limited by available electronics.

The characteristics of pyroelectric devices which make them so desirable in the infrared, i.e., stability, ruggedness, speed, economy, broadband response, room-temperature operation, and large area, are also useful in other regions of the energy spectrum. Pyroelectric detectors are frequently classified with thermal detectors, which can be confusing since the speed of response is limited only by the phonon vibration rate and not by bulk thermal characteristics.

Pyroelectric materials are those for which the molecule exhibits a net dipole moment. Energy absorbed by the outer electron cloud results in a shift, or change, in this dipole moment. In order to return to its original state, the material emits phonons. If the molecules in a pyroelectric material are aligned, the phonons are absorbed by the nearby molecules. An increase in temperature of the bulk material interferes with the effect because it changes the spacing between the molecules.

Pyroelectric materials exhibit a net dipole moment (polarization is the dipole moment per unit volume) which is temperature-sensitive. Although encompassing a great many materials, including human bone and skin, only a handful exhibit sufficient pyroelectric current to be useful. Some of these are triglycerine sulfate (TGS) and triglycerine selenate (TGSe), lithium tantalate ($LiTaO_3$), and lithium niobate ($LiNbO_3$), strontium barium niobate ($Sr_xBa_{1-x}Nb_2O_3$ or SBN), lanthanum-doped lead zirconate (PLZT), lead zirconate (PZT), and polyvinylfluoride (PVF).

Except for $LiTaO_3$, they are available with doping to change their properties or in various stoichiometric ratios. The properties of each version vary and need to be discussed with the supplier. The dipole moments of these materials are not naturally aligned. So that the effect can be transmitted and collected, a high electric field is applied (poling) between the electrodes. Those pyroelectric materials which are also ferroelectric experience a phase transition near the Curie temperature which permits the dipoles to move more freely, requiring a lower field. When practical, the ambient temperature is raised to the Curie temperature.

$LiTaO_3$ is the most common pyroelectric material because it has application in other areas, can be grown in crystalline form readily, and has stable characteristics. SBN is more difficult to grow (as a crystal) than $LiTaO_3$. It has limited commercial availability in high-speed detection. Plastic films such as PVF and similar materials exhibit pyroelectricity when stretched and poled. They are relatively inexpensive, thin, and available in large area. When used in the visible spectral region, pyroelectric detectors are coated with an absorbing medium. TGS and TGSe detector materials are hygroscopic and fragile. Although the most sensitive of the pyroelectric materials and the only ones with material absorption in the visible spectral region, TGS and TGSe (families) are sensitive to microphonics and may depole easily.

Characteristics of some pyroelectric detector materials are given in Table 17.4.

Figure 16.16 shows the basic construction of a pyroelectric detector. Pyroelectric detectors generally have opposing electrodes. If radiation strikes the crystal between the electrodes, the device is edge-type. If it is absorbed by one electrode or transmitted through an electrode, it is face-type. Pyroelectric detectors can also be fabricated in a coplanar structure[9] which may be easier and cheaper to use. The coplanar structure behaves as a thin, fast sensor because it has low capacitance.

TABLE 17.4 Pyroelectric Materials and Their Properties

Material	$p(T)$ at 300 K, $\mu C/cm^2 \cdot K$	ϵ at 300 K, 1 Mhz	Curie temperature, K	Resistivity, $\Omega \cdot cm$	Characteristics
LiTaO$_3$	0.0176	43	600	10^{13}	Must be poled in the boule, polarity difficult to maintain through fabrication, poor absorption of 10.6 μm.
TGS	0.035	50	49	10^{12}	Low damage threshold (250 mW/cm^2 continuous wave), microphonic, fractures easily, hygroscopic.
TGSe			23		Used at low temperatures only, similar to TGS.
PVF, PVF$_2$	0.002	10	*	4×10^{13}	Low damage threshold, plastic film.
PLZT	0.030	1000	200	10^{13}	Various levels of lanthanum doping, as with SBN, many possible compositions.
LiNbO$_3$	0.0083	28	>800 (at 1200 K)	10^{13}	Boule-poled, 800°C is used for poling.
50/50 SBN	0.065	380	117	10^{11}	Relatively easy to grow.
BaTiO$_3$	0.02	160	126		Fractures easily.

*Elevated temperature (to 50°C) eases poling.

Narrow separation of the electrodes gives them high sensitivity. Coplanar electrode pyroelectric detectors are generally used for research purposes and are not available commercially. This configuration is most successful with materials which can be poled after fabrication of the detector.

The pyroelectric detector element is often coated to enhance or control optical response. The least expensive and most common type of coating is an optical black, usually a metallic oxide or a paint. Such coatings are thick, therefore slow, and frequently fragile. La Delfe and Stotlar[6] have developed thin, rugged coatings for high-speed applications or where a narrowband response is desired without the use of additional optical elements.

When a photon is absorbed by the crystal lattice, the lattice spacings change, changing the value of the electric polarization. A charge is developed at the surface of the crystal normal to the axis of polarization. If these surfaces are electroded and connected through an external circuit, free charge will be brought to the electrodes to balance the surface charge, generating a current in the circuit. The induced current is proportional to the rate of the change of the crystal temperature, or

$$I = \lambda_p(T)A\,\frac{dT}{dt} \tag{17.33}$$

where $\lambda_p(T)$ is the pyroelectric coefficient at temperature T, A is the electrode area, and dT/dt is the rate of change of the temperature.

The responsivity of a pyroelectric detector is inversely proportional to the electrode separation. The rise time, on the other hand, is proportional to the electrode area divided by the electrode separation. High responsivity has meant slow detectors. The edge-type (radiation impinges between the electrodes) configuration is useful for high-speed devices because radiation-electrode interaction can be ignored and because the large electrode separation provides fast rise times. Edge-type detectors with high responsivity are very small and difficult to illuminate without striking the electrodes. Those with large size have very low responsivity. The current responsivity is

$$R_I = \frac{\eta\lambda_p(T)}{sC_pl} \tag{17.34}$$

where η = emissivity
$\quad s$ = material density
$\quad C_p$ = specific heat
$\quad l$ = electrode separation

The voltage responsivity is

$$R_V = \frac{\eta\lambda_p(T)}{sC_pl} \frac{R_L}{[1 + (2\pi f R_E C_E)^2]^{1/2}} \tag{17.35}$$

where f = frequency
$\quad R_E$ = effective resistance ($R_E^{-1} = R_L^{-1} + R_p^{-1}$) for R_p the detector
$\quad\quad$ resistance and $R_L = \rho l/A$, where ρ is the resistivity
$\quad C_E$ = effective capacitance ($C_E = C_p + C_L$)

For high-speed detectors, C_E is minimized by reducing stray or input capacitance and using large electrode separation in the detector. The detector capacitance is

$$C_P = \frac{\epsilon(T, \omega)\epsilon_0 A}{l} \tag{17.36}$$

where $\epsilon(T, \omega)$ is the dielectric constant and ϵ_0 is the permittivity of free space. Since $R_L \ll R_p$, $R_E = R_L$ and the voltage responsivity for high-speed detectors is

$$R_V = \frac{\eta\lambda_p(T)R_L}{sC_pl} \tag{17.37}$$

for $(2\pi f R_E C_E)^2 \ll 1$. R_L is chosen to provide impedance matching with the external circuit. The primary detector noise source in high-speed pyroelectric detectors is Johnson noise,

$$V_J = (4kTR_L\Delta f)^{1/2} \tag{17.38}$$

where k is the Boltzmann constant, T is the temperature (K), and Δf the frequency bandwidth. When used with high-speed oscilloscopes, the minimum detectable power is generally determined by the minimum detectable voltage of the oscilloscope. In order to operate in a linear region, the devices are generally operated at a power density one order of magnitude lower.

Very sensitive pyroelectric detectors can have a minimum detectable power (or noise equivalent power) of 10^{-10} W.

The area-normalized detectivity is

$$D^*(\lambda, f) = \frac{(A_d \Delta f)^{1/2}}{\text{NEP}_\lambda} \tag{17.39}$$

where A_d is the sensitive area, Δf is the bandwidth, and NEP_λ is the noise equivalent power at wavelength λ (NEP_λ = noise/responsivity).

Again, the actual performance of pyroelectric detectors is often limited by associated electronics. Figure 17.8 shows the voltage (*a*) and the current (*b*) responsivity of a typical pyroelectric detector. The voltage responsivity peaks near a fraction of a hertz to a few (up to 10) Hertz. It decreases as $1/f$ for low to moderate frequencies. Relaxation in the dielectric constant results in equivalent capacitance limited by stray capacitance at high (greater than MHz) frequencies. The voltage responsivity versus frequency is flattened by selection of a load resistor. The current responsivity varies little with frequency.

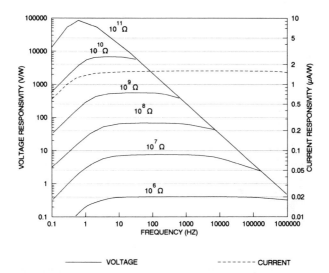

FIGURE 17.8   Typical pyroelectric detector responsivity (voltage responsivity for various external load resistors).

The figure of merit used to compare pyroelectric materials at low frequencies is

$$M_L = \frac{\lambda_p(T)}{\epsilon s C_P} \tag{17.40}$$

At high frequencies it is

$$M_F = \frac{\lambda_p(T)}{s C_P}$$

Because semiconductor detectors are discussed widely in current literature, a comparison of pyroelectric and semiconductor photovoltaic detectors is given in Table 17.5.

TABLE 17.5 Pyroelectric Versus Photovoltaic Detectors

Parameter sensitivity	Sensitive pyroelectric detectors	High-speed pyroelectric detectors	Typical photovoltaic
Current responsivity	10 μA/W	0.1 μA/W	0.5 μA/W
Quantum responsivity, $QE = \dfrac{1.24R}{\lambda} \times 100\%$	$10^{-4}\%$ at 10 μm	$10^{-6}\%$ at 10 μm	$\geq 70\%$ at 1 μm
Spectral response	Generally broadband (especially if coated) unless material transmits visible and near ir	Generally broadband except where material is transparent (thin coatings only)	Wavelength-dependent
Noise	Johnson noise/amplifier noise	Johnson noise/amplifier noise	Shot, thermal, photon, fabrication recombination, etc.
Detectivity D^*	$\leq 10^{10}$ cm·Hz$^{1/2}$·W^{-1}	10^{7} cm·Hz$^{1/2}$·W^{-1}	10^{14} cm·Hz$^{1/2}$·W^{-1}

In general, pyroelectric detectors do not compare to semiconductors (photovoltaics) in sensitivity. The current responsivity of a silicon photodiode can be typically 0.5 μA/W for a high-speed pyroelectric detector or 50 μA/W for a sensitive one.

The quantum efficiency of a typical silicon photodiode would be 70 percent at 0.9 μm as opposed to 5×10^{-6} percent for a pyroelectric detector. Consequently, the quantum efficiency of pyroelectric detectors is not discussed. On the other hand, semiconductors generally exhibit wavelength-dependent response and are sensitive over only limited spectral regions whereas pyroelectric detectors are broadband in response, although many materials are transparent in visible and near-ir regions.

Photovoltaic detectors experience thermal, shot, photon, Johnson, recombination, modulation $(1/f)$, and fabrication-induced noise generation while pyroelectric detectors exhibit thermal, amplifier-related noise and Johnson noise, which usually dominates. The minimum detectable power of photovoltaic detectors is generally detector-limited, while pyroelectric detectors are usually electronics-limited.

The maximum D^* for photovoltaic detectors is approximately 10^{15} cm·Hz$^{1/2}$·W^{-1} as opposed to 10^{10} cm·Hz$^{1/2}$·W^{-1} for sensitive and 10^{7} cm·Hz$^{1/2}$·W^{-1} for high-speed detectors.

Because of their higher sensitivity (and subsequent D^*), photovoltaic detectors are used in most applications where suitable wavelength response is obtained. Pyroelectric detectors are used when broadband, room-temperature, low-cost, high-energy, unbiased (passive), or very high frequency operation is desirable.

17.5 OTHER DETECTORS

Section 16.6 describes some of the other detectors of interest to the ir spectral region. The detectors described in this section are also commercially available.

Photon drag detectors are doped germanium crystals in which a change in voltage drop (across the crystal) occurs when a laser beam induces an electric field. The photon drag effect has a rapid response time but fairly slow fall time.

Golay cells are thermal detectors consisting of small gas-filled cells with an absorptive face. The other side of the cell is a flexible mirror which moves a reflected beam on a photocell. The motion is proportional to the incident energy.

17.6 DETECTION SYSTEMS AND SELECTION GUIDE

The detector is only one element of a detection system. The means of collecting light, which may include lenses, filters, apertures, gratings, polarizers, integrating spheres, fiber optics, diffusers, attenuators, and other devices may determine the spectral character, distribution, intensity, and duration of the light reaching the detector. The detector itself may be coated to enhance absorption (with an anti-reflection coating, for example) or to select certain wavelengths. The detector signal itself must be processed. Whether the output is displayed on a meter or an oscilloscope or used to control other systems, the signal processing system places both requirements and limitations on the performance of the detector. The collection optical elements, the detector, and the signal processing components comprise a detection system.

17.6.1 Detector Selection

In the near-infrared spectral region, the silicon junction photodiode is the detector of choice whenever the application is compatible with its characteristics. Silicon photodiodes are available in a broad range of sizes, packages, and performance characteristics.

InGaAs detectors bridge the gap between 1.1 and 2.6 μm. PtSi and InSb are useful in the 3- to 5-μm region. InSb and HgCdTe are commonly used in the 5- to 12-μm spectral region for low light applications. Pyroelectric detectors are common in broadband applications and those where low cost of the detector or room-temperature operation are critical requirements. Far-ir applications are addressed by bolometers and pyroelectric detectors.

The infrared spectral region encompasses too many material choices to be included here. Generally, a semiconductor device is used unless the application precludes it. However, most of the non-single-element semiconductor materials currently available exhibit material-related limitations such as poor uniformity which should be discussed with the manufacturer.

- If the spectral character of the source is not known, either an array with spectrally selective filters on the active elements or a broadband response detector such as a pyroelectric detector may be necessary.

- If the application requires fast response, the detector *and* the signal processing system must be faster unless the response of the system is well-characterized.

- If the incident power density is high, attenuation of the input power may be required to keep the detector in the linear portion of its response curve, usually 250 mW/cm² for silicon photodiodes, for example. Large amounts of attenuation may be difficult to calibrate, so another physical type may be desirable.

- If a detector is to be operated near the limit of its performance range, the potential effect on the signal should be discussed with the manufacturer.

TABLE 17.6 Typical Performance Characteristics of Infrared Detectors

Detector type	Area (cm²)	Responsivity at λ peak (A/W)	Operating bias (V)	Maximum signal (μA)	Current density (μA/cm²)	Dark current (pA)	Capacitance (pF)	Risetime (ns)	Detectivity D (cm Hz$^{1/2}$ w^{-1})
Photoconductor (low–high illumination)	.1–1	10^3–10^6 V/W	10–350	10^5				10^6–10^9	10^8–10^{11}
Photodiode	.01–1	.6	0–200	100	100	1–1000	1–10^4	1–10^6	10^{12}–10^{15}
Avalanche Photodiode	10^{-4}–.2	50	350	20	100	1000	5	.5–2	10^{12}
Thermopile	.008–.7	40 V/W	0				5	10^6–10^8	10^8
Bolometer	.01–.3	5000 V/W	10–1500					10–10^8	10^7–10^{10}
Pyroelectric	.01–1	10^{-6}–10^{-5}	0	$>10^{-6}$	$>10^6$	1–10^3 (Johnson Noise)	1–10^3	10^{-2}–10^8	10^7–10^{10}

17.24

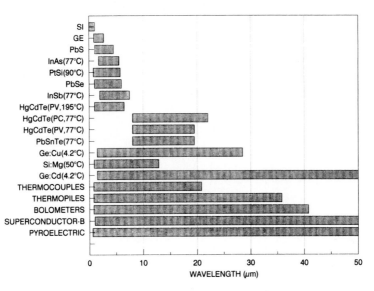

FIGURE 17.9   Typical wavelength ranges for common ir detectors.

Table 17.6 compares the typical performance characteristics of some of the detectors discussed in this chapter while Figure 17.9 provides their typical performance regions.

17.7 REFERENCES AND FURTHER READING

1. Astheimer, R. W., and S. Weiner, *Application Notes for Pyroelectric Detectors*, Barnes Engineering, 1971.

2. Buddle, W., *Physical Detectors of Optical Radiation*, Optical Radiation Measurements, vol. 4, Academic Press, Orlando, 1983.

3. Dereniak, E. L., and D. G. Crowe, *Optical Radiation Detectors*, Wiley, New York, 1984.

4. Grum, F., and R. J. Becherer, Optical Radiation Measurements, vol. 1, *Radiometry*, Academic Press, Arco, 1979.

5. Joshi, N. V., *Photoconductivity: Art, Science, and Technology*, Marcell Optical Engineering Series, vol. 25, Dekker, New York, 1990.

6. La Delfe, P. C., and S. C. Stotlar, U.S. Patent No. 4,595,832, "Thermal Sensor with an Improved Coating," U.S. Department of Energy, 1986.

7. *Laser Focus World Buyer's Guide*, Laurin, Pittsfield, Mass., 1992.

8. Middlehock, S., and Ardet, S. A., *Silicon Sensors*, Academic Press, San Diego, 1989.

9. Stotlar, S. C., "Pyroelectric Devices with Coplanar Electrodes," U.S. Patent No. 3,932,753, Harshaw Chemical Co., 1976.

10. Sze, S. M., *Physics of Semiconductor Devices*, Wiley, New York, 1981.

11. Sze, S. M., *Semiconductor Devices, Physics and Technology*, Wiley, New York, 1985.

12. Wolfe, W. L., and G. J. Zissis, eds., *The Infrared Handbook*, IRIA Center, Ann Arbon, Mich., 1978.

13. *1992 Photonics Buyer's Guide*, Laurin, Pittsfield, Mass., 1992.

CHAPTER 18
IMAGING DETECTORS

Frederick A. Rosell

18.1 INTRODUCTION

Although television imagery was demonstrated in the late 1920s, it has become a household word since the end of World War II. Basically, television uses imaging sensors to convert photon images, which have been focused by lenses onto a photosensitive surface, to electron images. These electron images are reconverted to visible light images for direct or remote viewing by an observer. While the displayed image must be in the visible, the input image may be formed in any spectral band from the ultraviolet to the far infrared. In the broad sense, these sensors should be called *image converters*. In this chapter, the sensors considered will be restricted to those operating in the ultraviolet to near-infrared portion of the spectrum (0.2 to 1.1 μm).

Direct-view imagers convert a photon image incident on a photoemitter to an electron image that is then accelerated to a phosphor which creates a visible image directly. The World War II sniperscope used a scene illuminator filtered to exclude the visible spectrum and a near-infrared direct-view image converter for nighttime infantry actions. After World War II, direct-view light amplifiers (multiple-stage intensifiers) were developed which can image at light levels down to natural starlight without the aid of auxiliary scene illuminators.

The TV broadcast industry developed camera tubes for entertainment use. The standard camera tube used was the image orthicon, or IO, which was of moderately high sensitivity because of a prestorage electron image gain mechanism and a very low noise preamplifier. To image at low light levels, the standard glass face plates of the IO were replaced by fiber-optic face plates which could be coupled directly to image intensifiers with fiber-optic end plates.[1] Improved IOs, with higher prestorage gains and the image isocon, which eliminates most of the noise in the scene lowlights, were developed to further extend sensitivity. Even further sensitivity improvements were obtained by camera tubes with higher prestorage gains. The most notable of these was the secondary electron conduction (SEC) camera tube and a tube with a silicon diode matrix gain-storage target. The SEC camera tube achieves higher gain by secondary-electron-induced conduction while the silicon diode matrix target tube, which goes by several trade names such as SIT or EBS, obtains gain by electron-bombardment-induced conductivity.[2,3] These new devices also resulted in systems of smaller size and weight and of greater versatility. Active, range-gated cameras using pulsed lasers provided outstanding high-contrast imagery under darkest night conditions. However, with the advent of passive far-infrared imagers, interest in low-light-level TV (LLLTV) waned, and government support for further development all but ceased except for advanced image intensifiers.

In 1950, RCA announced the development of the vidicon, whose photosensitive surface is photoconductive.[4] Lacking prestorage gain, the vidicon was of low sensitivity and its image storage target exhibited lag effects. However, with the advent of transistorized circuitry, small portable cameras could be built at low cost. Later, new photosurfaces became available and one, lead oxide, improved performance to the point where tubes using it displaced the IO as the broadcast industry standard.

Many variants of the above camera tubes were devised and still have considerable interest in scientific applications, but the current trend appears to be toward solid-state imagers which, only now, are approaching and, in some cases, even exceeding certain performance parameters of their vacuum-tube counterparts. The current approach to low-light-level imaging appears to be the coupling of very high gain second- and third-generation intensifiers to the solid-state imagers.

18.2 PHOTOSURFACES

The three general types of photosensitive surfaces used in imaging devices in the uv to near-ir spectrum are classified as *photoemissive*, *photovoltaic*, and *photoconductive*. The primary parameters are usually quantum efficiency and the limiting noise mechanism. However, in many cases, the photosurface also serves as the image storage mechanism and other parameters such as integration time, storage capacity, and frame-to-frame image retention make compromises in sensitivity necessary. The principles of the above photoconversion processes are well-known and will not be discussed except as they affect imaging performance.

18.2.1 Photoemitters

When light impinges on a photoemitter, electrons are emitted spontaneously. For camera tube use, the photoemitters are generally semitransparent, which limits their quantum efficiency to about 20 percent. As shown in Fig. 18.1, photoemitters are available for nearly the whole 0.1- to 1.1-μm spectral band. Only the S-1, used in the World War II sniperscopes, has appreciable response beyond 0.9 μm, and it is characterized by high thermionic emission* (10^{-11} to 10^{-12} lm/cm^2) at room temperature as opposed to 10^{-13} to 10^{-14} lm/cm^2 for the S-10 and 10^{-15} to 10^{-16} lm/cm^2 for the S-20. The thermionic emission of the S-1 can be reduced by thermoelectric cooling to acceptable levels. An integrated value for sensitivity is often quoted, assuming that the illumination source is a blackbody at 2854 K. Some representative values are given in Table 18.1. The most commonly used laboratory source for these measurements is an ordinary tungsten bulb.

---

*The units for thermionic emission are in terms of an equivalent background illumination as measured with the lens cap on.

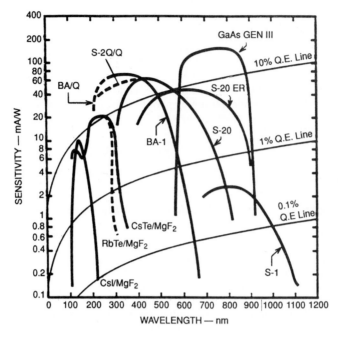

FIGURE 18.1 Spectral sensitivity versus wavelength for various photo-emitters (1 μm = 1000 nm).

Table 18.1 assumes that the sensor's spectral bandpass is unrestricted by spectral filters. Analytically, luminous sensitivity is found from

$$S_L = \frac{\int_{\lambda_1}^{\lambda_2} S(\lambda)E_s(\lambda)\,d\lambda}{680 \int_{0.40}^{0.76} \overline{y}(\lambda)E_s(\lambda)\,d\lambda} \quad \frac{A}{\text{lm}} \tag{18.1}$$

where $\overline{y}(\lambda)$ is the relative photospectral response of the human eye, $S(\lambda)$ is the spectral responsivity of the sensor (A/W), and $E_s(\lambda)$ is the spectral radiance due

TABLE 18.1 Integrated Luminous Sensitivity of Selected Photoemitters for a 2854 K Source

Photoemitter type	Luminous sensitivity, μA/lm
S-1	20
S-10	40
S-20	150
S-20 ER	200–450
GaAs	1000–2000

to the source (W/m²). Note also that if this spectral response is given in $S(\lambda)$ A/W, the quantum efficiency $Q(\lambda)$ can be found from

$$Q(\lambda) = \frac{1.24\ S(\lambda)}{\lambda} \quad \text{electrons/photon} \quad (18.2)$$

where λ is the wavelength (μm). The ultimate limit to any imaging system's ability is photon-to-electron conversion noise (photon noise is nonexistent). This noise, for a uniformly lighted area of a photoemitter, in terms of a root mean square (rms) photocurrent i_n is

$$i_n = (2ei\,\Delta f)^{1/2} \quad \text{A} \quad (18.3)$$

where e is the charge of an electron and Δf is the measuring bandwidth (Hz).

18.2.2 Photoconductive and Photovoltaic Photosurfaces

The development of the vidicon, which was based on an antimony trisulfide photocathode with an S-18 spectral response, led to the development of a large number of photoconductors which are too numerous to list in detail. The S-18 response is similar to that of the eye. The vidicon was simple, very small, and inexpensive, which accounts for its popularity over a long time period. Its primary faults were its sublinear response to input light and frame-to-frame lag in scene lowlights. Vidicons have been constructed with usable response out to beyond 2 μm. Usually, photoconductors used in camera tubes result in the formation of electron-hole pairs and thus are subject not only to photoelectron noise but also to generation-recombination noises such that the rms noise current, in analogy to that for photoemissive surfaces, becomes

$$i_n = (4ei\,\Delta f)^{1/2} \quad (18.4)$$

The noise for photovoltaic surfaces is like that for photoemitters. A lead oxide surface, which fits this category, was developed. It is used in a vidicon structure,

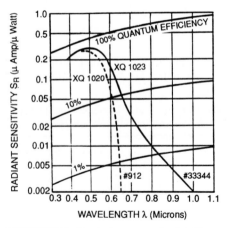

FIGURE 18.2 Spectral responses for two Plumbicon© (lead oxide) photosurfaces versus wavelength.

and has displaced the IO as the standard for commercial broadcast use.[5] The spectral responses for two such surfaces, one of which is optimized for improved red response, is shown in Fig. 18.2. Note that, in general, vidicons are preamp- rather than photoelectron-noise-limited.

The most popular photosurface for military use is based on silicon. This surface is constructed of a matrix of diodes with up to 2000 diodes/in. One of its virtues is an extended near infrared as shown by its spectral response in Fig. 18.3. Note that the spectral response of any photosurface may be modified by coatings and other manufacturing processes.

Photoconductive and photovoltaic photosurfaces usually serve not only to convert scene images to electron images but also to store the image for subsequent sequential electronic readout.

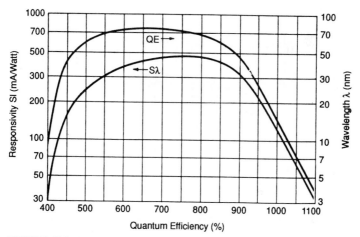

FIGURE 18.3   Spectral response of a typical silicon diode matrix photosurface.

18.3 IMAGING TUBES

While solid-state imaging devices will probably supplant vacuum-tube imagers for most industrial and military applications in the next decade, TV camera tubes will still offer a degree of versatility that the solid-state devices cannot match. For example, camera tubes are analog devices which are sampled in only one direction by the raster rather than in both, the raster line number is variable, tubes can be more easily gated for use in active systems or for exposure control, and view fields can be electronically zoomed.

18.3.1 Vidicons

Although the image orthicon preceded the vidicon's invention by a decade or more, the vidicon is simple, and a vidiconlike readout is often used in camera tubes with prestorage image gain.

Most vidicons employ a photoconductive or photovoltaic photosurface consisting of a material such as antimony trisulfide, lead oxide, or silicon. The quantum efficiencies of such materials can be very high but not greater than unity when used in an imaging application. Nevertheless, vidicons are of low sensitivity because of noise generated in the readout process and lag. Therefore, their application has been restricted to daylight levels.

The schematic of a typical vidicon is shown in Fig. 18.4. A signal electrode is deposited on the inner surface of the faceplate. The faceplate can be either glass or a fiber-optic plate. The photosurface, or target, is deposited on the signal electrode. Ordinarily, the signal electrode is biased 15 to 40 V positive with respect to

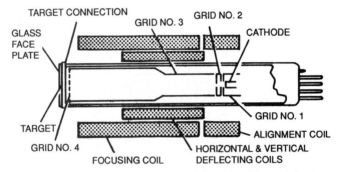

FIGURE 18.4   Schematic of the vidicon and associated focus, deflection, and alignment coils.

the electron gun. The side of the photocathode facing the electron gun is periodically charged to the electron gun potential by the action of the raster-scanning electron beam. If light is incident on a resolution element of the photosurface during the interval between successive scans, electrons are generated within the photosurface and move in an appropriate direction so as to discharge the charge stored on it. However, current cannot flow in the external target lead resistor until the beam once again passes the illuminated point. The function of the beam is to recharge the photocathode point by point back to gun-cathode potential, and the resulting charging current flowing through the target lead constitutes the video signal. Signal storage results from the fact that scene light impinging on a point on the photosurface continually discharges it between successive passes of the beam. The charge replaced by the beam is the total integrated amount of charge discharged during the period between scans. This period is called the *frame time*. In commercial practice, the frame time is $\frac{1}{30}$ s.

There are usually three important time constants in the operation of the tube. First, there is a signal-storage time constant, or dielectric relaxation time, consisting of the RC time constant of the layer. This time constant should be long compared to the frame time. A second time constant is associated with the photoprocess itself and has to do with the mobility of charge carriers. The third time constant, the readout time constant, is a product of the beam resistance times the photoconductive capacitance. Both the second and third time constants, which collectively limit the speed and efficiency of signal readout, should be short.

18.3.2 Camera Tubes with Prestorage Gain

The principal noise limiting the sensitivity of a vidicon is not photo-to-electron conversion noise but noise due to the preamplifier. The solution is to employ a photoemissive photocathode and to amplify the electron image prior to storing the image for readout by the electron beam. Usually the target of the camera tube, in conjunction with image electron acceleration, serves as both the gain mechanism and the image storage medium.

18.3.3 Image Orthicons and Isocons

One of the earliest camera tubes with a gain storage target was the image orthicon, shown schematically in Fig. 18.5. This magnetically focused and deflected camera tube employed a photoemissive photosurface that creates an electron image which is accelerated to the target. The electrons that strike the target (which originally was of glass) generate secondary electrons which are collected by the target mesh. The charge distribution created on the target is stored for a frame time and read out by an electron beam. The electron-beam current returns to an electron multiplier. The signal current consists of the original beam current less the amount deposited on the target to recharge it. While the target gain was only about 4, this gain, called *prestorage gain*, together with the low noise associated with the electron multiplier, resulted in an image tube that could be used under studio broadcast lighting conditions and was a favorite with broadcasters for many years. Later on, a new target material consisting of a thin film of magnesium oxide with a gain of about 20 was developed which improved sensitivity further. Another tube based on the image orthicon was the image isocon. In an ordinary image orthicon, the greatest noise is generated in the scene blacks because the full unmodulated beam

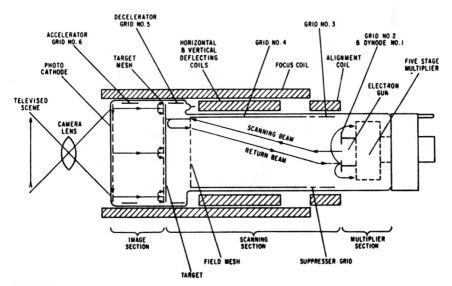

FIGURE 18.5 Schematic of image orthicon showing functions of the elements.

returns to the electron multiplier. In the isocon, unmodulated electrons in the electron readout beam are blocked.

The IO and the isocon are magnetically focused, and this results in images of high geometric fidelity and uniformity.

18.3.4 SEC and EBS Camera Tubes

Camera tubes were developed in the 1960s that had an electrostatically focused image section as shown in Fig. 18.6. Typically, the electron image formed by the image section is accelerated by a high voltage (about 10 kV) to the gain-storage target. This image is inverted, top for bottom, and reversed, left for right. The advantage of the electrostatic focus is that it is light in weight and permits electronic image zoom, electronic light level control, and gating (shuttering).

Two types of targets have been used in the above configuration. The first was composed of a highly porous layer of potassium chloride deposited on a thin film of aluminum oxide. The photoelectron image impinging on the target gives rise to secondary electrons within the KCl layer, which constitute gain. The gain of this type of tube, which is called the *secondary electron conduction* (SEC) camera tube, is typically 120 to 150. Another target which is employed in the same basic tube structure is composed of a silicon oxide diode array with a density of about 2000 diodes per inch. The gain obtainable with this target is typically 1500 to 2000. This tube has been generically known as the SEBIR (secondary electron bombardment induced response) camera tube, but is better known as the SIT (silicon intensified tube) or the EBS (electron-bombarded silicon) camera tube. The EBS and SIT tubes are the most commonly used for military applications.

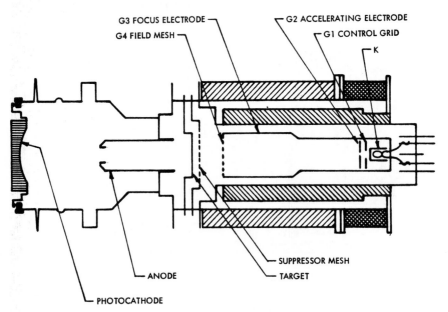

FIGURE 18.6 Schematic of a typical camera tube with an electrostatically focused image section and a high-gain-storage target.

All of the above tubes are in the medium-low-light-sensitive category, although the SIT tube approaches true low-light-level capability. Often, camera tube imagers, which have readout time constants, require prestorage gains beyond that required to become photoelectron-noise-limited to prevent lag effects. All of the above camera tubes can be constructed with fiber-optic faceplates for the purpose of coupling them to other light-amplifying devices such as image intensifiers.

18.3.5 Image Intensifiers

The function of image intensifiers, or image converters, is to amplify the image of the scene. This amplified image can be viewed directly or coupled to other intensifiers or imaging tubes by fiber optics or lenses. The simplest image intensifiers, which include those used in the World War II sniperscope, are simple diodes (or triodes, if a gating electrode is included) and are called *first-generation intensifiers*. More recently, intensifiers have been developed which include a microchannel plate (MCP) amplifier to provide even higher gains. The second-generation tube includes the MCP amplifier and can be used with any photoemissive photocathode, while the third-generation tube is basically a second-generation tube with a gallium arsenide input photoemitter.

First-Generation Intensifiers. A schematic of an electrostatically focused first-generation intensifier is shown in Fig. 18.7. The fiber-optic faceplate serves two functions; one is to permit coupling to yet another intensifier or image device and

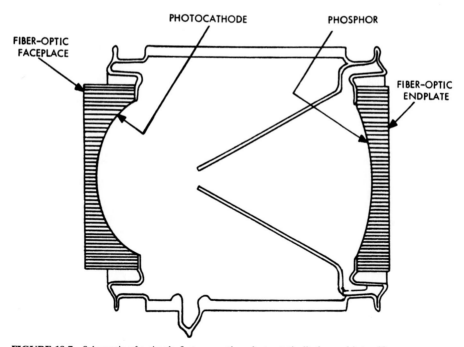

FIGURE 18.7 Schematic of a simple first-generation electrostatically focused intensifier.

the second is to simplify the electron lens design by correcting for image curvature. The device shown in Fig. 18.7 is a diode. If desired, a grid can be incorporated for the purpose of electronic shuttering at voltage levels below that needed if the entire voltage across the device (about 15 kV) is to be cut off in very short times. The output phosphor is usually a conventional P-20 surface and has a fairly narrow spectrum centered at 0.53 to 0.55 μm which is near the spectral wavelength at which the eye peaks. The input photocathode is usually an S-20 or S-20VR surface. The S-20 photoemitter is superior for coupling to another intensifier, while the S-20VR has enhanced sensitivity in the near infrared, which is desirable for many military applications.

The gain due to high-energy electrons striking the phosphor is approximately 1000 photons/electron but, because of the somewhat low quantum efficiency of the photosurface and transmission losses, the luminous gain is in the neighborhood of 100. Part of this gain results from the fact that the phosphor is green. For hand-held direct-viewing devices for night vision, three first-generation intensifiers are coupled in cascade by butting the output of one intensifier to the faceplate of the next, etc.

Second- and Third-Generation Intensifiers.   A second-generation image intensifier, as shown in Fig. 18.8, employs a microchannel plate amplifier interposed between the phosphor and the photocathode but near the phosphor.[6] The MCP is a two-dimensional array of hollow glass fibers similar to a fiber-optic faceplate with the cores of each fiber etched out. The inside of each fiber is coated with a thin film which is a secondary-electron emitter. The electron image generated by the input photocathode is amplified by the MCP and further accelerated after amplification to the phosphor.

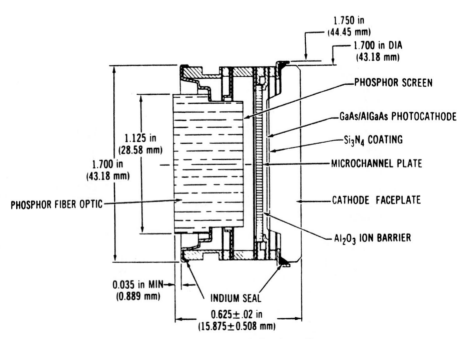

FIGURE 18.8   Schematic of a wafer-type microchannel plate intensifier.

18.4 SOLID-STATE IMAGING DEVICES

Solid-state imaging devices have been under development for over two decades and can now come close to camera tubes in overall performance.[7] Within the next decade, solid-state imagers will probably dominate. There are currently three basic types of solid-state staring arrays:

1. Interline-transfer charge-coupled device (CCD) arrays
2. Frame-transfer CCD arrays
3. Charge injection devices (CIDs)

The three types are shown schematically in Figs. 18.9 to 18.11.

18.4.1 Interline-Transfer CCD Imager

A typical *interline-transfer CCD* area array is shown schematically in Fig. 18.9. Every other vertical column consists of a two-phase line array of image cells, and the in-between columns are analog shift or charge-transport registers, which are light-shielded. Since the light-shielded area is approximately equal to the light-sensitive area, the effective photosurface area is about half the total picture area. Assuming a $\frac{1}{30}$ s frame time, it is found that the integration time for both fields 1 and 2 is $\frac{1}{30}$ s. In the first field, photoelectrons are collected for $\frac{1}{30}$ s and are then rapidly shifted into the shielded transport register. This image is then transported, a line at a time, into the horizontal charge-transport register. The portion of the image in the vertical transport register is completely read into the horizontal register in a field time of $\frac{1}{60}$ s. After the first field is read, the second field, which has been integrating image for $\frac{1}{30}$ s, is shifted into the vertical transport register and read in $\frac{1}{60}$ s. As a result, the integration time for both fields is equal to the frame time of $\frac{1}{30}$ s.

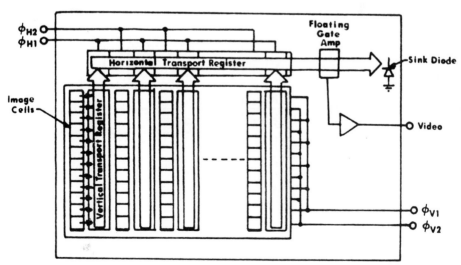

FIGURE 18.9 Image cell and readout register organization for a typical interline-transfer CCD area array.

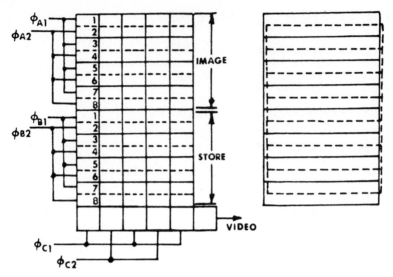

FIGURE 18.10 Typical two-phase frame-transfer CCD area array with interlace. Line overlap is one-half line in the vertical direction.

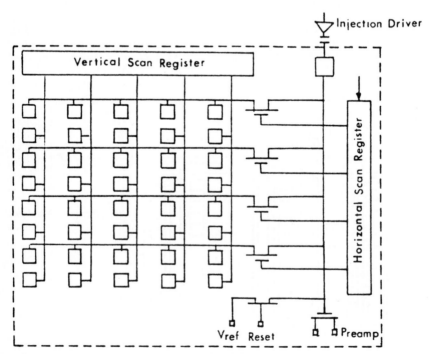

FIGURE 18.11 Typical charge-injected imaging device.

Since only half of the scene image is converted to electron charge, it is common in many applications to use two arrays with the scene image split by a 45° half-silvered mirror so that the second array fills in the gaps of the first array.

18.4.2 Frame-Transfer CCD

The *frame-transfer CCD array* is depicted in Fig. 18.10. The top half of the array is used to image for a field or a frame time. At the end of the integration period, the entire photoconverted image is very rapidly shifted into the storage area, which is identical to the image area but is shielded from light. During the next integration period, the storage image is read by the horizontal register sequentially from the video signal. An interlace feature can be incorporated by shifting the light collection center of an image cell. In the array of Fig. 18.10, the light collection center in phase 1 would be between cells 1 and 2 for phase A-1 and between cells 2 and 3 for phase A-2. Thus the horizontal-scan line area shifts up or down one-half scan line after each field. In this way the number of scan lines is effectively doubled in the vertical direction and the Nyquist frequency, or sampled data frequency, limit is equal to twice the number of cells in units of TV lines per picture height (TVL/PH).

18.4.3 Charge-Injection Devices

The *CID solid-state image sensors* use an x-y-addressed array of charge-storage capacitors which store photon-generated charge in metal-oxide-semiconductor (MOS) inversion regions. Readout of the first self-scanned arrays was effected by sequentially injecting the stored charge into the substrate and detecting the resultant displacement current to create a video signal. The charge-storage sites can be read out in any arbitrary order. Arrays can be designed with integral digital MOS decoders for x and y line selection to allow random access. The integration time as well as the scan sequence could then be externally programmed for special applications.

An array designed for raster scan which includes integral shift registers is diagrammed in Fig. 18.11. Each sensing site consists of two MOS capacitors with their surface inversion regions coupled such that charge can readily transfer between the two storage regions. A larger voltage is applied to the row-connected electrodes so that photon-generated charge collected at each site is stored under the row electrode, thereby minimizing the capacitance of column lines. A line is selected for readout by setting its voltage to zero by means of the vertical scan register. Signal charge at all sites of that line is transferred to the column capacitors, corresponding to a row enable condition. The charge is then injected by driving each column voltage to zero, in sequence, by means of the horizontal scan register and the signal line. The net injected charge is measured by integrating the displacement current in the signal line, over the injection interval. Charge in the unselected lines remains under the row-connected electrodes during the injection pulse time (column voltage pulse). This corresponds to a half select condition. The array is constructed on an epitaxial layer so that the reverse-biased epitaxial junction can act as a collector for the injected charge. This effectively prevents the charge injected at any site from being collected by neighboring sites.

Solid-state imagers are characterized by high quantum efficiencies and low read-out noise but do not fall into the low-light-level category. Most are considered to be daylight imagers.

18.5 IMAGING SYSTEM PERFORMANCE MODEL

Imaging system performance is generally specified in terms of the ability of a sensor-augmented observer to resolve scene detail on the display as a function of the scene light level and contrast. Performance is usually measured in the laboratory by using bar patterns as test objects. It has been found that the laboratory measurements can be analytically predicted with high accuracy for low-light-level sensors and adequate accuracy for daylight sensors.[8]

The basic premise in the analytical model is that the photodetection process of any phototransducer, including the human eye, is subject to statistical fluctuation. This notion has been experimentally verified. In the sensor-augmented observer case, it is assumed that a signal-to-noise ratio can be associated with an image on the display and that this image signal-to-noise ratio will be identical to that on the observer's retina after taking into account the eye-brain's ability to integrate in space and time. This signal-to-noise ratio is designated SNR_D for display signal-to-noise ratio. The noise is assumed to be entirely generated by the sensor and not by the eye.

18.5.1 Ideal Sensor System

An ideal sensor is defined here as one whose spatial frequency response is unity at all spatial frequencies. Spatial frequency is defined as the number of half cycles which may be fitted into the picture height, and the units are television lines per picture height.

Consider the schematic of a very simple imaging system, shown in Fig. 18.12. The phototransducer converts the photon image incident on it to a photoelectron image. The incident photon image can be considered to be noisefree since the existence of a noisy coherent electron emission from a photosurface has yet to be demonstrated. Continuing the description, the phototransduced image is passed to the signal processor, whose main purpose is to amplify and magnify the image signals and noises alike. If the signals and noises are not further degraded or filtered by the spatial frequency response of the processor or the display and if no further noise is added, then the signal-to-noise ratio of the image appearing on the display will be identical to that at the output of the input photosurface. Furthermore, if the gains and magnifications of the signal processor and display combination are

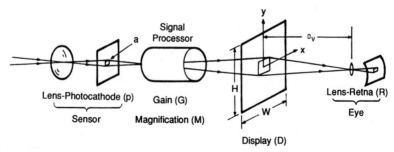

FIGURE 18.12 Schematic of the electro-optical imaging process.

large enough that the observer's eye is neither light-level- nor image-size-limited, then the image's signal-to-noise ratio at the output of the eye's retina will be identical to that on the display if due account is taken of the eye-brain combination's ability to integrate in space and time. These conditions can be achieved in practice for a range of image sizes, signal amplifications, display luminances, image magnifications, and observer-to-display viewing distances. In other cases, noise added within the signal processor or even in the observer's retinal photoprocessor may be a factor.

Suppose the input image is a rectangular image of area a amid uniform background. The phototransduced image area a is converted to n_0 electrons and an equivalent area of background is converted to n_b electrons. The incremental signal $\Delta n = n_0 - n_b$. Also, if $n_0 = \dot{n}_0' at$ and $n_b = \dot{n}_b' at$, where $\dot{n}'$ is the generation rate of electrons in electrons/m^2·s, the incremental signal is

$$\Delta \dot{n}' at = (\dot{n}_0' - \dot{n}_b')at \qquad (18.5)$$

The rms fluctuation noise associated with converting photons to electrons is taken to be the average of that due to object and background and is given by $[(\dot{n}_0' + \dot{n}_b')at/2]^{1/2}$. Then, the peak-to-peak SNR_D can be written as

$$SNR_D = \frac{\Delta \dot{n}'(at)^{1/2}}{[(\dot{n}_0' + \dot{n}_b')/2]^{1/2}} \qquad (18.6)$$

In a television camera tube, the two-dimensional picture is sequentially read out by a scanning electron beam to provide a one-dimensional, time-varying electrical signal. The video signal current can be obtained from the relation

$$i = \dot{n}'eA \qquad (18.7)$$

Where e is the charge of an electron (C) and A is the total effective scanned area of the focal plane. With Eq. (18.7), Eq. (18.6) becomes

$$SNR_D = \sqrt{\frac{a}{A}} \, t \, \frac{\Delta i}{(ei_{av})^{1/2}} \qquad (18.8)$$

Note that video bandwidth does not appear in the above equation and is not fundamental to SNR_D. The real bandwidth is determined by the eye's ability to integrate in space and time. However, multiplying and dividing the above equation by $(2 \, \Delta f_V)^{1/2}$, where Δf_V is the video bandwidth, gives

$$SNR_D = \sqrt{2 \, \Delta f_v \frac{a}{A}} \, \frac{\Delta i}{(2ei_{av} \, \Delta f_V)^{1/2}} \qquad (18.9)$$

The quantity $(2ei_{av} \, \Delta f_V)^{1/2}$ can be recognized as the rms shot-noise current normally associated with photoelectron noise, so

$$SNR_D = \sqrt{2 \, \Delta f_v \frac{a}{A}} \, SNR_{VO} \qquad (18.10)$$

where SNR_{VO} is the video signal-to-noise ratio if the noise is white. Once more, video bandwidth Δf_V is not fundamental to SNR_D. The observer is unaffected by its magnitude.

The most popular test pattern for measuring resolution is a bar pattern with alternating, equally spaced light and dark bars. The basic modeling assumption is that, to detect a bar pattern, one must detect one bar in the target. Thus, the area a in Eqs. (18.5) to (18.10) becomes

$$a = xy = \epsilon y^2 \tag{18.11}$$

where y is the width of the bar and $x = \epsilon y$ is the length, ϵ being the length-to-width ratio. The total focal plane area is αY^2 where α is the picture aspect ratio (length-to-width) and Y is the picture width. The ratio a/A can then be written as

$$\frac{a}{A} = \frac{\epsilon y^2}{\alpha Y^2} \tag{18.12}$$

The spatial frequency N in TVL/PH for a repetitive bar pattern is

$$N = \frac{Y}{y} \tag{18.13}$$

and

$$\frac{a}{A} = \frac{\epsilon}{\alpha N^2} \tag{18.14}$$

and Eq. (18.8) becomes

$$\mathrm{SNR}_D = \sqrt{\frac{\epsilon t}{\alpha}} \frac{1}{N} \frac{\Delta i}{(el_{av})^{1/2}} \tag{18.15}$$

This is the basic equation for the modeling of ideal photoelectron-noise-limited sensors. If there are additional independent white noise sources, they can be added in quadrature; i.e.,

$$\mathrm{SNR}_D = \sqrt{\frac{2\epsilon \, \Delta f_v t}{\alpha}} \frac{1}{N} \frac{\Delta i}{[ei_{av} + i_1^2 + \cdots + i_n^2]^{1/2}} \tag{18.16}$$

18.5.2 Effect of Modulation Transfer Functions

Scene details are blurred by sensor optical elements, intensifiers, camera tube targets and scanning beams, electrical filters, displays, etc. The effects of these components can usually be described in terms of spatial frequency response. If the components are linear, the frequency responses are called *modulation transfer functions* (MTFs) $R_0(N)$. The effect of MTFs is to smear signals from light and dark bars together, causing incremental signal levels to decrease and the MTFs of components following points of noise insertion to filter the noise so that it is no longer white. Rewriting Eq. (18.16) to include these effects gives

$$\mathrm{SNR}_D = \sqrt{\frac{\epsilon t}{\alpha}} \frac{8}{\pi^2} \frac{R_0(n)}{N} \frac{\Delta i}{[J_1^2 \beta_1(N) + \cdots + J_n^2 \beta_n(N)]^{1/2}} \tag{18.17}$$

where the $8/\pi^2$ comes from considering only the fundamental of the square-wave input pattern, J_n^2 is the mean square noise current density (A^2/half-cycle) computed as if the noise were white in the video and divided by $2\ \Delta f_V$ to obtain a noise current density, and $\beta(N)$ represents a noise filtering function defined by

$$\beta(N) = \frac{1}{N} \int_0^N R_f^2(N)\ dn \tag{18.18}$$

Here, $R_f(n)$, which represents the product of all the MTFs of components following a point of noise insertion, serves the purpose of changing a uniform noise density spectrum to a nonwhite spectrum.

18.5.3 Threshold or Limiting-Resolution Characteristic

To determine the limiting resolution versus incremental signal current characteristic, a graphical plot of SNR_D is constructed as shown in Fig. 18.13. The observer's display threshold signal-to-noise ratio SNR_{DT} is usually set equal to a constant equal to about 2.5, and the intersection of the SNR_D curve at any given input current level with SNR_{DT} gives the threshold resolution. The curves shown in Fig. 18.13, were drawn for a 100 percent contrast pattern using the definition

$$C = \frac{i_h - i_l}{i_h} \tag{18.19}$$

where i_h is the current due to scene highlights and i_l is that due to lowlights,

$$\Delta i = C i_h \tag{18.20}$$

so that curves for contrasts less than 1.0 can be generated.

Threshold Resolution versus Photocathode and Scene Illuminance. Using the luminous sensitivity expression of Eq. (18.1) and making note of the prestorage signal gain G, gives the incremental signal current

$$\Delta i = \frac{G S_L A\ \Delta E_{\mathrm{pc}}}{e_v e_h} \quad \mathrm{A} \tag{18.21}$$

where $e_v e_h$ is the product of the vertical and horizontal scan efficiencies.

This can be referred to the scene luminous excitance ΔM_s by the well-known lens equation at infinity focus as

$$\Delta E_{\mathrm{pc}} = \frac{T_0\ \Delta M_S}{4f^2} \tag{18.22}$$

where T_0 is the transmittance of the lens and

$$f = \frac{(F_L^2 + D_0^2/4)^{1/2}}{D_0} \tag{18.23}$$

for a lens of focal length F_L and effective diameter D_0.

Noise Current Density. Most imaging sensors have several sources of noise although, in some cases, one may be dominant. For low-light-level TV camera tubes, the dominant noise is photoelectron when operated at high prestorage gain levels

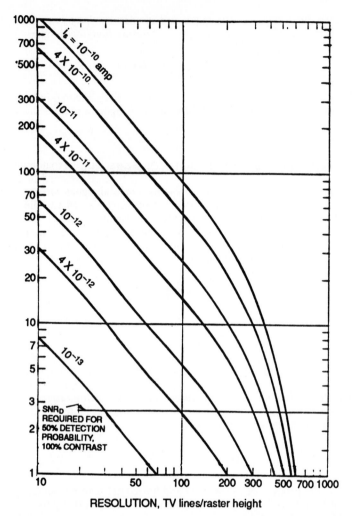

FIGURE 18.13 SNR_D versus bar pattern spatial frequency for a typical low-light-level camera as a function of highlight signal current level.

and a mixture of photoelectron and preamp noise at low gains. Daylight cameras using vidicons or solid-state sensors are usually preamp-limited. However, other noises such as dark current and fixed-pattern may be a factor. The mean square photoelectron noise density for photoemissive photocathodes or photovoltaic photosurfaces is

$$J_{pe}^2 = \frac{(2 - C)G^2 ei_h}{2e_v e_h} \quad (18.24)$$

For photoconductors,

$$J_{pe}^2 = \frac{(2 - C)ei_h}{2e_v e_h} \quad (18.25)$$

when the intensifier is of the second- or third-generation variety, the microchannel plate is noisy, and, in effect, doubles the J_{pe}^2 of Eq. (18.24).

Fixed-pattern noise due to either variations in gain in MCP devices or cell-to-cell sensitivity variations in solid-state devices has been almost totally ignored, and, while the experimental data needed for analysis is lacking, this noise could be ultimately limiting as it is for many mid-infrared and far-infrared imagers. The suggested form for mean square pattern noise is

$$J_f^2 = \frac{t}{e_v e_h} \cdot \frac{[(2 - c)M^i b/2]}{N_x N_y} \tag{18.26}$$

where M is the modulation of signal and background current due to nonuniformities in either gain or sensitivity, N_y is the number of pixels in the vertical in the case of solid-state sensors or the effective number of raster lines if the sensor is a camera tube, and N_x is the limiting resolution in the horizontal. This limit may be due to the number of pixels or the video bandwidth employed.

The preamp noise for camera tubes is nonlinear, and, overall, the noise current is given by

$$I_{pa}^2 = 4KT \left[\frac{\Delta f_v}{R_L} + \frac{R_{EQ}}{3 f_0^2 R_L^2} \Delta f^3 \right] \tag{18.27}$$

where R_L is the load resistance, R_{EQ} is the equivalent anode resistance, K is Boltzmann's constant, and $f_0 = 1/2\pi C R_L$ where C is the camera tubes interelectrode capacitance. For first-order calculations, this noise can be considered to be white-averaged over the video bandwidth Δf_V.

In the case of solid-state imagers, the noises associated with various dark currents, readout noises, etc., are not usually specified in data sheets, but the signal-to-noise ratio at a given operating current level is given and the rms noise I_{nr} can be determined. The noise current density in this case is

$$J_{nr}^2 = \frac{I_{nr}^2}{2 \Delta f_v} \tag{18.28}$$

where the bandwidth is given by

$$\Delta f_v = \frac{N_x N_y F_r}{2 e_v e_h} \tag{18.29}$$

where $N_x N_y$ is the product of the number of effective horizontal and vertical pixels and F_r is the frame rate in s^{-1}.

Operating Signal Current Level. Due to signal storage capacity limits, the light level incident on the input photosurface and/or the prestorage gain, if any, must be adjusted so that the imaging device does not saturate because of scene highlights. As a practical matter, the photocurrent due to the scene is averaged over the scene or some fraction of it and this average is used to control either the incident light level through iris or neutral density filters or the prestorage signal gain, or both. As a rule of thumb, the average current level is set equal to half the maximum the sensor can deliver, and this current is called the *operating current* I_{op}. Suppose that the control parameter is prestorage gain G and that the current at the input photosurface is i_{av}, where

$$i_{av} = \frac{T_0 S_p A_e E_p}{4 f^2 e_v e_h} \tag{18.30}$$

Then the output current $I_{av} = Gi_{av}/e_v e_h$. If

$$I_{av} \leq I_{op}$$

then $G = G_{max}$ and

$$I_{av} > I_{op} \qquad G = \frac{I_{op}e_v e_h}{i_{av}} \qquad (18.31)$$

and, finally, if the calculated value of $G \leq G_{min}$, then assume $G = G_{min}$, since the maximum sensor performance has been realized.

18.6 MODULATION TRANSFER FUNCTIONS

It has become common to specify imager resolution in the spatial frequency rather than the space domain. Furthermore, these characterizations of resolution are in terms of one-dimensional spatial frequency responses, while in reality images are two-dimensional. The use of one-dimensional frequency responses implies that spatial responses are independent functions of x and y. Also, the use of Fourier frequency transform theory implies linearity. However, even when independence and linearity conditions are not met, the concepts are still used as if they were.

18.6.1 Overall Imaging Sensor Modulation Transfer Functions

A typical low-light-level imaging sensor block diagram is shown in Fig. 18.14. Most images in the ultraviolet to near-infrared portion of the spectrum can be used with a variety of lenses. Typically manufacturer's specifications address only the imager, and do not consider performance of the device when coupled to a lens or image line-of-sight instabilities. The input photosurface generates noise which is filtered by the intensifier phosphors and all of the modulation transfer functions which

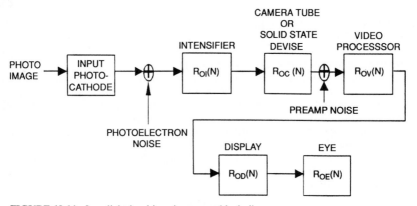

FIGURE 18.14 Low-light-level imaging sensor block diagram.

follow. The preamp noise is filtered only by the video processor, the display, and the eye. The overall MTF, $R_{OS}(N)$, is given by

$$R_{OS}(N) = R_{OI}(N) \cdot R_{OC}(N) \cdot R_{OV}(N) \cdot R_{OD}(N) \cdot R_{OE}(N) \tag{18.32}$$

The noise-filtering function for photoelectron noise is

$$\beta_{pe}(N) = \frac{1}{N} \int_0^N R_{ope}^2(N) \, dN \tag{18.33}$$

where

$$R_{ope}(N) = R_{OS}(N) \tag{18.34}$$

For preamp or readout noise

$$\beta_{pa}(N) = \frac{1}{N} \int_0^N R_{opa}^2(N) \, dN \tag{18.35}$$

where

$$R_{opa}(N) = R_{ov}(N) \cdot R_{OD}(N) \cdot R_{OE}(N) \tag{18.36}$$

18.6.2 Intensifier MTFs

The MTF $R_{OI}(N)$ of a first-generation intensifier results primarily from the two fiber-optic endplates, if used, and the phosphor. The fiber diameters in the faceplate are usually about 6 μm in diameter and usually have only a small effect on MTF. The phosphor most commonly used in intensifiers is the P-20, which has a peak spectral output in the green, but its particles are the prime cause of resolution loss. Averaged over several hundreds of first-generation intensifiers, the MTF of a single stage was found to be well-approximated by

$$R_{OI}(N) = \exp -\left(\frac{N}{48.8 \, Y}\right)^{1.42} \tag{18.37}$$

where Y is the effective phosphor height, mm.

Second- and third-generation intensifiers using microchannel plate amplifiers are still in active development. The primary limit to the MTF of these devices is the MCP, whose fiber diameters can range from 10 to 25 μm. Generally, the MTF of most imaging sensors can be curve-fitted by an equation of the form

$$R_0(N) = \exp -\left(\frac{N}{N_{OI}}\right)^k \tag{18.38}$$

where

$$k = k_1 \qquad N \le N_{OI}$$

$$k = k_2 \qquad N \ge N_{OI}$$

For one intensifier with an MCP fiber pitch of 14 μm, an effective focal plane height (and phosphor height) of 9.525 mm, $k_1 = 1.578$, $k_2 = 1.770$, and $N_{OI} = 374.3$ TVL/PH.

18.6.3 Camera Tube MTFs

At one time an almost bewildering array of camera tubes of different types and sizes was available, but the number has been decreasing yearly. The MTFs of almost any camera tube can be fitted to the function given by Eq. (18.25). Typically N_O is between 300 and 700 TVL/PH and k_1, k_2 fall between 1.4 and 1.7. Observe that the data supplied by most manufacturers is in the form of a square-wave amplitude response which must be converted to a sine-wave response as shown in Ref. 8.

Invariably, a camera tube is read out by a raster process in the vertical direction. This is a sampling process. When sampled, the sensor's resolution, as measured, will be a function of the phase angle θ between the bars in the test pattern and the raster lines. Thus, the MTF will be of the form

$$G_{OC}(N) = R_{OC}(N) \cos \theta \qquad (18.39)$$

In the worst case, where the number of lines in the pattern equals the number of raster lines and each raster line is displaced by one-half line from the pattern line, $\theta = 90°$ and $G_{OC}(N) = 0$. In the best case, where raster and pattern lines are equal and coincide, $\theta = 0°$ and $G_{OC}(N) = R_{OC}(N)$. For practical purposes, in calculating overall performances, a good approximation is to let $G_{OC}(N) = R_{OC}^2(N)$.

18.6.4 Solid-State Imagers

TV camera tubes usually have photosurfaces and targets which are nearly continuous in nature but are sampled in the vertical by the scanning electron beam. Solid-state imagers have well-defined picture elements, or pixels, which are quantized in both directions and, therefore, are sampled in both directions. A single pixel in the focal plane is of size Δx by Δy, and the MTF is given by

$$G_0(N) = \frac{\sin \pi N/2N_p}{\pi N/2N_p} \qquad (18.40)$$

where $N_p = n_p$ with n_p being the number of pixels which can be fitted into a picture height in the appropriate direction (x or y). $G_0(N)$ is used rather than $R_0(N)$ to indicate that the function is nonlinear because of sampling. It is recommended, because of the sampling, that $G_0(N)$ be replaced by $G_0^2(N)$ when a more realistic performance estimate is desired. The highest frequency which can be reproduced at the output of a sampler is half the number of samples per cycle, or 1 per half-cycle. This frequency is sometimes called the *Nyquist frequency*, and, for a solid-state sensor, it will be N_p half-cycles or lines per picture height.

18.6.5 Video Processors

Usually, the MTFs of video processors act as either filters to limit video bandwidth or aperture correctors. Video bandwidth filters are used only in the horizontal direction. Aperture correction is possible in both x and y, but is seldom used in the vertical.

Bandwidth Filters. The MTF of most passive filters can be written as

$$R_{OV}(N) = \left[1 - \left(\frac{N}{N_v} \right)^{2n} \right]^{-1} \tag{18.41}$$

where N_v is the line number at which the response is down 3 dB and n is the number of poles. When $n = 1$, the filter is a simple RC circuit.

Aperture Correctors. Aperture correction in the horizontal (cross-scan direction for camera tubes) can take a variety of forms. One of the most common is the delay-line corrector whose frequency response $G_A(N)$ is given by

$$G_A(N) = \frac{1}{2} \left[(1 + G_p) + (1 - G_p) \cos \left(\frac{\pi N}{N_A} \right) \right] \tag{18.42}$$

where G_p is the peak gain at frequency N_A TVL/PH.

18.6.6 Display MTF

The function of the display is both to filter the image and reconstruct the image for viewing. The filtering function is very important. In the sampled data directions, input image frequencies which are higher than the Nyquist frequency result in spurious responses which are reflected back into the image signal spectrum. A further condition is that the raster line structure should not be distracting. When a uniformly lighted scene is viewed, it is desirable that no raster lines be visible.[10] This is called the flat field condition. However, as can be shown, this can result in substantial image detail loss, therefore a compromise is indicated. A commercial broadcast standard allows 2.5 percent response at the raster line frequency, which is twice the number of raster lines SNR. Usually, the display MTF is Gaussian and given by

$$R_{OD}(N) = \exp -\left(\frac{N}{N_D} \right)^2 \tag{18.43}$$

and

$$N_D = \frac{2N_R}{\sqrt{-\ln 0.025}} = 1.04 N_R \tag{18.44}$$

18.6.7 Eye MTF

The observer's eye, like other elements in imaging systems, has limits to its resolving capability. However, in making threshold resolution measurements, the observer is usually allowed to adjust viewing distance to optimize the ability to resolve the test pattern in use. Thus the eye's frequency response is not a factor. In some applications, the viewing distance is fixed and the eye's ability to resolve must be taken into account. This subject has not been sufficiently researched at this time.

18.6.8 Other MTFs

There are numerous other MTFs which are involved, such as those due to charge transfer in CCDs, wavelength-dependent diffusion effects, and temporal lag, which are too specialized to be treated here.

18.7 APPLICATIONS

The basics of television evolved during the late 1920s and '30s, but development of commercial applications were placed on hold during World War II. At the end of the war,the demand for television by the public exploded and led to very rapid advances during the 1950s and '60s. The television broadcast and home television receivers of the day were entirely based on vacuum-tube technology. Broadcast cameras, with the advent of color, often weighed in at a ton or more. However, with the advent of solid-state circuitry, the weight was reduced to the point where even broadcast-quality imagery could be generated by portable cameras. The home TV receiver gradually incorporated solid-state circuitry as well, with consequent improvement in quality and reliability. About the only vacuum-tube components now in use are the camera tube and the cathode-ray-tube (CRT) display, and the end may be in sight for these. Digital video processors can now provide a bewildering array of special effects.

Video cassette recording is commonplace among consumers, and video cassette cameras, or camcorders, may soon replace the photographic camera as the medium of choice for recording family events. While the Department of Defense sponsored many television developments, including solid-state circuitry, it was the great popularity of TV that led to the reduction in cost of sensors, receivers, and recorders, and it is this reduction in cost that will spur a high increase in applications during the next decade. These applications will not be limited to the public consumer arena.

High-definition TV will become a reality, producing imagery of photographic quality. Low costs have brought TV imaging well within the budgets of security and law-enforcement agencies. Courtroom trials will be recorded for use in appeals which may occur years later when the original witnesses may be unavailable. Very powerful video processing techniques will expand the use of imagers in fully automated quality control and process testing.

Classroom lectures by noted teachers will improve education. Business efficiency will increase through face-to-face meetings via satellite-communicated televised meetings. The list is endless. While television imaging has already added a new dimension to everyday life and has a potential to do great good, it also has the power to do great harm by such effects as stimulating publicity-seeking terrorists and invasion of privacy.

18.8 REFERENCES

1. Rosell, F. A., "Limiting Resolution of Low-Light-Level Imaging Sensors," *J. Opt. Soc. Am.*, vol. 59, no. 5, May 1969.

2. Rosell, F.A., "Television Camera Tube Performance Data and Calculations," Chap. 5 in *Photoelectronic Imaging Devices*, vol. 1, Bibermand and Nudelman, eds., Plenum Press, New York, 1971.

3. Rosell, F. A., "The Limiting Resolution of Low-Light-Level Imaging Sensors" and "Television Camera Tube Performance Data and Calculations," Chaps. 14 and 22 in *Photoelectronic Imaging Devices*, vol. 2, Biberman and Nudelman, eds., Plenum Press, New York, 1971.

4. Weimer, P. K., et al., "The Vidicon Photoconductive Camera Tube," *Electronics*, May 1950.

5. Levitt, R. S., "The Performance and Capabilities of Recently Developed Plumbicon® TV Camera Pickup Tube," *Proc. 105th Technical Conference*, April 20–25, 1969.

6. Csorba, I. P., *Image Tubes*, Sams, Indianapolis, 1985.

7. Barbe, D. F., "Imaging Devices using the Charge Coupled Concept," *Proc. IEEE*, vol. 63, no. 1, January 1975.

8. Rosell, F., and G. Harvey, "Fundamentals of Thermal Imaging Systems," NRL report 8311, EOTPO report no 46, May 1979.

9. Coltman, J. W., "The Specification of Imaging Properties by Response to a Sine Wave Input," *J. Opt. Soc. Am.*, vol. 44, no. 6, June 1984.

10. Schade, Otto, Sr., "Image Reproduction by a Line Raster Process," *Perception of Displayed Information*, L. M. Biberman, ed., Plenum Press, New York, 1973.

CHAPTER 19
HOLOGRAPHY

Tung H. Jeong

19.1 INTRODUCTION

In an attempt to correct the spherical aberration of the electron microscope, Dennis Gabor[1,2] gave birth to the basic concepts of holography in 1948, for which he was awarded the Nobel Prize in physics in 1971. However, it was not until after the invention of the laser and further contributions by E. N. Leith and J. Upatnieks[3,4,5] and independently by Yu. N. Denisyuk[6,7,8] that holography was developed to the present stage.

In the discussion that follows, we will concentrate on optical holography, although the concepts involved are applicable to any type of coherent waves. In particular, we will analyze the recording and reconstruction of the wavefronts of a three-dimensional object. These analyses are applicable to hybrid developments summarized in Sec. 19.5.

Generally, the process of *holography* can be defined as the recording of the interference pattern between two mutually coherent radiation fields on a two- or three-dimensional medium. The result is said to be a *hologram*. It is a complex diffraction grating. When one of the fields is directed at the hologram, the diffraction reconstructs the wavefronts of the other field.

Holography can be studied and appreciated at many levels. It is sufficiently simple for those without a technical background to produce beautiful works of art. Indeed, it can now be pursued at the same level as photography.

19.2 THEORY OF HOLOGRAPHIC IMAGING

The most elegant exposition on holography is through the mathematics of communication theory[3] and Fourier transformations.[9] However, for the sake of practicality, we will present two complementary treatments, one through rigorous mathematics which leads to numerical results, and the other through a graphic model which offers a conceptual understanding of all major features of holography.

19.2.1 Basic Mathematics of Hologram Recording and Reconstruction

Figure 19.1 depicts a generalized system for recording and playing back a hologram. Let the origin of the x, y, z coordinate system be located in the plane of a hologram.

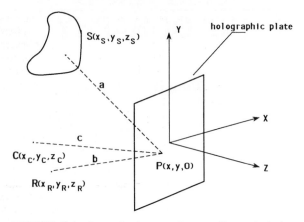

FIGURE 19.1 Generalized system for recording and playing back a hologram.

A point $S(x_S, y_S, z_S)$ on the object, which is illuminated by an ideal coherent source, scatters light toward the hologram being recorded. It interferes with a coherent wave from a point source located at $\mathbf{R}(x_R, y_R, z_R)$. The subsequent interference pattern is recorded all over the surface of the hologram, which is in the x-y plane.

During the playback, a point source $\mathbf{C}(x_C, y_C, z_C)$ located in the vicinity of $\mathbf{R}$ with a wavelength λ', which can be different from the recording wavelength λ, is directed at the hologram. We wish to study the nature of the resultant diffraction.

In particular, let us concentrate on point $P(x, y, 0)$. Let a be the distance between $\mathbf{S}$ and P, and let b be the distance between $\mathbf{R}$ and P, where

$$a = \sqrt{(x - x_S)^2 + (y - y_S)^2 + z_S^2} \tag{19.1}$$

and

$$b = \sqrt{(x - x_R)^2 + (y - y_R)^2 + z_R^2} \tag{19.2}$$

Furthermore, let the instantaneous amplitudes of the object and reference waves at P be $\mathbf{S}(x, y, t)$ and $\mathbf{R}(x, y, t)$, respectively, having duly noted that they both have spherical wavefronts. We have thus

$$S(x, y, t) = \frac{S_0}{a} \exp\{j\phi_S + j\omega t\} \tag{19.3}$$

and

$$R(x, y, t) = \frac{R_0}{b} \exp\{j\phi_R + j\omega t\} \tag{19.4}$$

where

$$\phi_S = -\left(\frac{2\pi}{\lambda}\right)a + \psi_S \tag{19.5}$$

and

$$\phi_R = -\left(\frac{2\pi}{\lambda}\right)b + \psi_R \tag{19.6}$$

Here ω is the angular frequency and S_0, R_0, ψ_S, ψ_R are constants.

From Eqs. (3) and (4) the total amplitude at P is

$$H(x, y, t) = S + R \tag{19.7}$$

$$= \left\{ \frac{S_0}{a} \exp(j\phi_S) + \frac{R_0}{b} \exp(j\phi_R) \right\} \exp(j\omega t)$$

and the total intensity is

$$I(x, y, t) = |H(x, y, t)|^2 \tag{19.8}$$

At this point, the nature of the recording medium as well as the processing procedure play an important role to the amplitude transmittance $T(x, y)$ at P. For simplicity, let

$$T(x, y) = KI(x, y) \tag{19.9}$$

where K is a positive constant. From Eqs. (7), (8), and (9), we obtain

$$T(x, y) = K\left\{ \frac{S_0^2}{a^2} + \frac{R_0^2}{b^2} + \left(\frac{S_0 R_0}{ab} \right)[\exp\{j(\phi_S - \phi_R)\} + \exp\{-j(\phi_S - \phi_R)\}] \right\} \tag{19.10}$$

This is the hologram at P.

Next we consider illuminating P with a spherical wave front originating from $C(x_C, y_C, z_C)$ with a wavelength $\mu\lambda$, where μ is the ratio of the illuminating and the recording wavelengths.

At the left side of P, we can denote the amplitude of the illuminating light as

$$C(x, y, z = -0, t) = \frac{C_0}{c} \exp\left\{ (j\phi_C) + j\left(\frac{\omega}{\mu} t \right) \right\} \tag{19.11}$$

where

$$\phi_C = -\left(\frac{2\pi}{\mu\lambda} \right) c + \psi_C \tag{19.12}$$

and

$$c = \sqrt{(x_C - x)^2 + (y_C - y)^2 + z_C^2} \tag{19.13}$$

Thus, at the right side of P, the amplitude is

$$C(x, y, z = +0, t) = T(x, y)C(x, y, z = -0, t)$$

$$= K\left(\frac{S_0^2}{a^2} + \frac{R_0^2}{b^2} \right) \frac{C_0}{c} \exp\left[j\phi_C + j\left(\frac{\omega}{\mu} \right) t \right]$$

$$+ \frac{KS_0 R_0 C_0}{abc} \exp\left[j(\phi_S - \phi_R + \phi_C) + j\left(\frac{\omega}{\mu} \right) t \right] \tag{19.14}$$

$$+ \frac{KS_0 R_0 C_0}{abc} \exp\left[j(-\phi_S + \phi_R + \phi_C) + j\left(\frac{\omega}{\mu} \right) t \right]$$

$$= I + II + III$$

Notice that for a usual case where $\mu = 1$, $\mathbf{C} = \mathbf{R}$, and $b = c$, we have the essence of holography. The II and III terms become

$$II \propto \frac{S_0}{a} \exp(j\phi_S + j\omega t) \tag{19.15}$$

and

$$III \propto \frac{S_0}{a} \exp(-j\phi_S + j\omega t) \tag{19.16}$$

Except for a constant, Eq. (19.15) is identical to the object. This is called the *direct image*. Equation (19.16) represents a conjugate image.

Note the direct or the conjugate images can be real, virtual, or a combination of the two. This can occur when the "object" is actually a real image projected onto the hologram plane. As depicted in Fig. 19.1, however, the direct image is virtual and the conjugate is real. The first term in Eq. (19.14) represents the directly transmitted wave from $\mathbf{C}$, and is sometimes called the *dc component*.

19.2.2 Properties of Holographic Images[10]

For designing holographic optical elements, as well as for displaying holograms which will be illuminated by nonlaser sources, it is useful to understand some optical characteristics peculiar to holography.

Although a more general treatment is possible,[11] a useful understanding can be derived with a paraxial approximation, as is the case with simple lenses. Assuming $\mathbf{S}$, $\mathbf{R}$, and $\mathbf{C}$ are located relatively closely to the z axis, we can perform an expansion on a and b as follows:

$$
\begin{aligned}
a &= [z_S^2 + (x - x_S)^2 + (y - y_S)^2]^{1/2} \\
&= z_S \left[1 + \frac{(x - xS)^2 + (y - y_S)^2}{z_S^2} \right]^{1/2} \\
&\approx z_S \left[1 + \frac{(x - x_S)^2 + (y - y_S)^2}{2z_S^2} \right] \\
&\approx \frac{x^2 + y^2 - 2x_S x - 2y_S y}{2z_S} + z_S + \frac{x_S^2 + y_S^2}{2z_S}
\end{aligned}
\tag{19.17}
$$

and similarly

$$b \approx \frac{x^2 + y^2 - 2x_R x - 2y_R y}{2z_R} + z_R + \frac{x_R^2 + y_R^2}{2z_R} \tag{19.18}$$

Using these approximations on Eqs. (19.5) and (19.6) yields

$$\phi_S = -\frac{2\pi}{\lambda} \left\{ \frac{1}{2z_S} (x^2 + y^2) - \frac{x_S}{z_S} x - \frac{y_S}{z_S} y \right\} + const. \tag{19.19a}$$

and
$$\phi_R = -\frac{2\pi}{\lambda}\left\{\frac{1}{2z_R}(x^2 + y^2) - \frac{x_R}{z_R}x - \frac{y_R}{z_R}y\right\} + constant \qquad (19.19b)$$

therefore,
$$\phi_S - \phi_R = -\frac{2\pi}{\lambda}\left\{\frac{1}{2}\left(\frac{1}{z_S} - \frac{1}{z_R}\right)(x^2 + y^2) \right.$$
$$\left. - \left(\frac{x_S}{z_S} - \frac{x_R}{z_R}\right)x - \left(\frac{y_S}{z_S} - \frac{y_R}{z_R}\right)y\right\} + const. \quad (19.20)$$

We can now represent the phase factors from the second and third terms of Eq. (19.14) as

$$\phi_1 = \phi_S(x, y) - \phi_R(x, y) + \phi_C(x, y)$$

$$\approx -\frac{2\pi}{\mu\lambda}\left\{\frac{1}{2}\left(\frac{\mu}{z_S} - \frac{\mu}{z_R} + \frac{1}{z_C}\right)(x^2 + y^2) - \left(\frac{\mu x_S}{z_S} - \frac{\mu x_R}{z_R} + \frac{x_C}{z_C}\right)x \qquad (19.21)\right.$$

$$\left. - \left(\frac{\mu y_S}{z_S} - \frac{\mu y_R}{z_R} + \frac{y_C}{z_C}\right)y\right\} + const.$$

and similarly

$$\phi_2 \approx -\frac{2\pi}{\mu\lambda}\left\{\frac{1}{2}\left(-\frac{\mu}{z_S} + \frac{\mu}{z_R} + \frac{1}{z_C}\right)(x^2 + y^2) - \left(-\frac{\mu x_S}{z_S} + \frac{\mu x_R}{z_R} + \frac{x_C}{z_C}\right)\right.$$
$$\left. \times x - \left(-\frac{\mu y_S}{z_S} + \frac{\mu y_R}{z_R} + \frac{y_C}{z_C}\right)y\right\} + const. \qquad (19.22)$$

Just as in Eq. (19.19), we readily recognize the form $A(x^2 + y^2) + Bx + Cy$, which represents spherical wavefronts. By finding the centers of these spheres, we can find the locations of the real and the virtual images from the hologram.

Let the centers of the direct and conjugate images be located at (x_1, y_1, z_1) and (x_2, y_2, z_2), respectively. It should be noticed that Eq. (19.19) is a general expression of the phase distribution of a point source in the x-y plane (in the paraxial approximation). Therefore, the phase distribution of a spherical wavefront originating from (x_1, y_1, z_1) on the x-y plane can be written in the similar form

$$\phi_1 = -\frac{2\pi}{\mu\lambda}\left\{\frac{1}{2z_1}(x^2 + y^2) - \frac{x_1}{z_1}x - \frac{y_1}{z_1}y\right\} + const. \qquad (19.23)$$

By comparing Eq. (19.23) and Eq. (19.21), it is easy to solve for the coordinates and arrange them in a form similar to lens equations:

$$\frac{1}{\mu z_1} - \frac{1}{z_S} = \frac{1}{\mu z_C} - \frac{1}{z_R} \qquad (19.24a)$$

$$\frac{x_1}{z_1} = \frac{\mu x_S}{z_S} + \left(\frac{x_C}{z_C} - \frac{\mu x_R}{z_R}\right) \qquad (19.24b)$$

$$\frac{y_1}{z_1} = \frac{\mu y_S}{z_S} + \left(\frac{y_C}{z_C} - \frac{\mu y_R}{z_R}\right) \qquad (19.24c)$$

The location of the conjugate image can be solved in the same fashion

$$\frac{1}{\mu z_2} + \frac{1}{z_S} = \frac{1}{\mu z_C} + \frac{1}{z_R} \tag{19.25a}$$

$$\frac{x_2}{z_2} = -\frac{\mu x_S}{z_S} + \left(\frac{x_C}{z_C} + \frac{\mu x_R}{z_R}\right) \tag{19.25b}$$

$$\frac{y_2}{z_2} = -\frac{\mu y_S}{z_S} + \left(\frac{y_C}{z_C} + \frac{\mu y_R}{z_R}\right) \tag{19.25c}$$

For a special case where $\mu = 1$ and $C = R$, Eqs. (19.24a) and (19.25a) show that the hologram is a weird lens with two focal lengths.

Also from Eqs. (19.24) and (19.25), we can derive the following useful lateral, longitudinal, and angular magnifications for both the direct and conjugate images. For direct image:

$$M_{lat} = \frac{dx_1}{dx_S} = \frac{dy_1}{dy_S} = \mu\frac{z_1}{z_S} = \left\{1 + \frac{1}{\mu}\frac{z_S}{z_C} - \frac{z_S}{z_R}\right\}^{-1} \tag{19.26a}$$

$$M_{long} = \frac{dz_1}{dz_S} = \frac{1}{\mu} M_{lat}^2 \tag{19.26b}$$

$$M_{ang} = \frac{d(x_1/z_1)}{d(x_S/z_S)} = \mu \tag{19.26c}$$

For conjugate image:

$$M_{lat} = -\mu\frac{z_2}{z_S} = \left\{1 - \frac{1}{\mu}\frac{z_S}{z_C} - \frac{z_S}{z_R}\right\}^{-1} \tag{19.27a}$$

$$M_{long} = \frac{1}{\mu} M_{lat}^2 \tag{19.27b}$$

$$M_{ang} = -\mu \tag{19.27c}$$

19.3 VOLUME HOLOGRAMS—A GRAPHIC MODEL

The theoretical analysis presented above assumes a two-dimensional recording medium. Like "thin" lens treatments in geometric optics, the results are generally useful. In reality, all media have a thickness of at least 1 μm. When holograms are recorded on photorefractive crystals, the medium is truly three-dimensional.

Mathematical treatments for volume holograms are a great deal more complex[12] and beyond the scope of this chapter. However, we present here a graphic model[13] which offers intuitive understanding of all the major characteristics.

Figure 19.2 depicts in two dimensions the constructive interference pattern formed by A and B, two point sources of coherent radiation. It consists of a family of hyperbolic lines. A figure of revolution around axis AB is a corresponding family

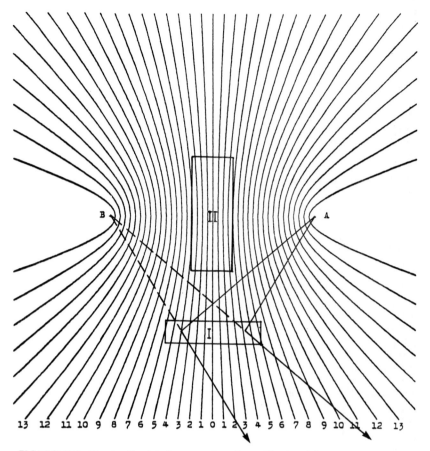

FIGURE 19.2   Constructive interference pattern formed by *A* and *B*.

of hyperboloidal surfaces. The difference in optical path between rays from *A* and *B* on any point of a given surface is constant, in integral numbers of wavelengths.

The zeroth order, for example, is a flat plane that perpendicularly bisects *AB*. Any point on it is equidistant from *A* and *B*. Likewise, any point on the *n*th-order surface represents an optical path difference of *n* wavelengths.

19.3.1   Transmission Holograms

Region I of Fig. 19.2 shows the edge of a thick recording medium. In the case of silver halide, such as the red-sensitive 8E75 emulsion from Agfa, the thickness is approximately 6 μm.

Light from *A* and *B* arrives at the plane of the emulsion from the same side, and forms a family of narrow hyperboloidal surfaces throughout the volume on chemical processing. In between these surfaces of interference maxima are the minima.

The essential feature of the graphic model is to regard these hyperboloidal surfaces as if they were partially reflecting mirrors. Rays shown by dotted lines in Fig. 19.2 show a unique property of hyperbolic mirrors: Any light from A incident on any part of any mirror is reflected to form a virtual image of B, and vice versa. Of course, retracing the reflected rays results in a real image of A.

Suppose a photoplate is exposed in region I and properly processed. Source A, now called the *reference beam*, is directed alone at the finished hologram placed at its original position. Some light will be directly transmitted through the hologram, but the remainder will be reflected to create a virtual image of B. An observer viewing the reflected light will see a bright spot located at B.

If a converging beam is directed through the hologram in a reversed direction and is focused at A (now called the *conjugate beam*) along the same path, a real image of B is created at the precise original location. Thus we have created a transmission hologram of an object consisting of a single point.

If we substitute a three-dimensional object illuminated by coherent light in the vicinity of B, each point on the object interferes with A to form a unique set of hyperbolic mirrors. Any elementary area on the photoplate receives light from different points on the object and records a superposition of all sets of mirrors. Each elementary area thus records a distinctly different set of points holographically. All together, the photoplate records all the views that can be seen through the plate at the object.

When the finished hologram is illuminated by A, light is "reflected" by the mirrors and recreates the virtual image of the object. By moving one's eye to different locations, different views are recreated to form a completely three-dimensional image.

Because the "reflected" light is actually transmitted from source A, through the hologram, to the viewer on the other side, this is a *transmission hologram*. Such a hologram can be broken into small pieces, and each piece can recreate a complete view of the object.

If a narrow laser beam is directed in a conjugate direction toward A, a real image is projected onto a screen located at B. By moving the beam around (but always directing it toward A), a different view of the object is projected.

Figure 19.3 shows a realistic microscopic cross section of an emulsion after exposure to two beams, A and B, each at 45° from opposite sides of the normal. Because the emulsion is about 10 wavelengths thick, the "mirrors" are very close together. When A alone is directed through the hologram, the rays penetrate through not one but several partial mirrors.

This volume grating exhibits diffraction that is wavelength-selective, a phenomenon explored by Lippmann[14] and Bragg,[15] for which each was awarded a Nobel Prize. If red light from a He-Ne laser were used for the recording and a point source of white (incandescent) light is located at A, a predominantly red image is recreated. But because there are only a few planes in this quasi-thick hologram, other colors still come through, showing a spectral smear. By using photopolymer 50 μm thick, we were able to make A transmission holograms viewable by incandescent light.

Multichannel Hologram. Using this volume effect, multiple exposures can be made of different objects with corresponding reference beams from different angles. MCC (Byte) is currently using this property for computer storage.

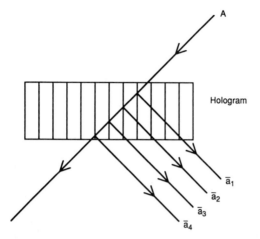

FIGURE 19.3   Cross section of an emulsion after exposure.

19.3.2 Reflection Hologram

As shown in Fig. 19.2, region II contains standing waves formed by light from A and B in opposite directions. Nodal planes are one-half wavelength apart.

By placing a photoplate in region II, it is exposed to hyperboloidal planes that are practically flat and parallel to the plate. An emulsion layer 10 wavelengths thick would create 20 partially reflecting surfaces. Reconstruction can be accomplished readily by using an incandescent point source.

When a hologram is made in this manner, the reconstructed image is viewed from the side of the reference beam, and is called a *reflection hologram*. A particularly simple configuration for making such a hologram was discovered by Denisyuk.[6,7,8] The object is placed directly on one side of the photoplate and the exposure is made by a single source from the opposite side. The direct incidence is the reference beam, and the transmitted and subsequently scattered light is the object beam.

19.4 MATERIAL REQUIREMENTS

Holography can be practiced at the amateur level by using inexpensive equipment and materials, with a total expenditure similar to that for popular photography.[16] What follows is a more complete listing suitable for a research laboratory.

19.4.1 Lasers

The helium-neon laser was first used by Leith and Upatnieks[3,4,5] and remains the most popular source today. However, there are numerous alternatives, both continuous wave (cw) and pulsed. Table 19.1 summarizes lasers now available, with their wavelengths and typical outputs.

TABLE 19.1 Available Lasers

Type	Wavelength, nm	Typical power outputs, mW
He-Ne	633	1–80
Argon	458	400
	477	800
	488	2000
	514	3000
Krypton	476	200
	521	300
	647	1000
He-Cd	442	50
Semiconductor	780	25

For direct portraiture of human subjects[17] and other mechanically unstable objects, the most popular choice has been the Q-switched ruby laser with a 694-nm output in the vicinity of 1 J per pulse and a pulse duration of about 20 ns. Lately, the frequency-doubled Nd:YAG laser[18] with a 530-nm output is frequently used.

In choosing the laser, the most important factors are

1. *Coherence*[19]: Determines the ranges of optical path differences, i.e., the depth of the scene, for the recording.
2. *Power*: Determines the exposure duration.
3. *Wavelength*: Determines the recording material and the reconstruction color.

19.4.2 Optics

In general, with few exceptions, holography requires high-quality optics. The profile of the Gaussian beam must be preserved.

Front-surfaced mirrors: Since the laser beam diameters are small, flat mirrors need not be large. Flatness should be one-fourth wavelength or better. For efficiency and for use with high-power lasers, special coatings for the wavelengths chosen are sometimes necessary.

Beam splitters: The simplest beam splitter is a piece of glass. However, there are two parallel surfaces, resulting in multiple reflections. Disks with a graduated coating on one side and an antireflection coating on the other side are available. Cubes made by combining two prisms with a coated diagonal surface are superior for fixed beam ratios. Since most laser outputs are plane-polarized, cubes made by a judicious combination of two calcite prisms afford the possibility of variable beam ratios.

Polarization rotators: Half-wave plates are most common. Two are necessary for use with polarization beamsplitters—one to orient the plane of the input, and the other to rotate one of the outputs to render its plane parallel to the other.

Beam expanders: For high-power lasers, one must avoid focusing the beam to a small point where ionization of the air may take place. Here, very clean plano-concave lenses, double concave lenses, or convex front surface mirrors must be used. For cw lasers, a microscope objective focusing the beam through a pinhole of appropriate size is ideal; this combination is called a *spatial filter*.

Large collimators: In making high-quality holographic optical elements (HOEs) as well as master holograms, one often needs to create a conjugate beam through a large recording plate. The most obvious element to be used is a large positive lens. A better alternative is a concave mirror. It serves as a single surfaced reflector, which folds the beam to minimize the table size, and collimates or focuses as well. For display holography, a telescope mirror does well. For HOEs, off-axis elliptical or parabolic mirrors are sometimes necessary.

Diffusers: Sometimes "soft" light for the object is necessary. An expanded beam over 2 cm in diameter directed on opal or ground glass usually serves the purpose.

HOEs: With a basic holographic laboratory, one can create HOEs for more efficient and exotic work.[20] For example, the conjugate image of a transmission hologram of a diffused beam (on opal glass) recreates a similar beam, but preserves the plane of polarization and localizes the light. A hologram of a plane beam is in fact a beam splitter. When the diffraction efficiency is varied across the hologram, it serves as a variable beam splitter.

Optical fibers[21]: It is possible to launch the output of a laser directly into fiber-optic elements containing fiber couplers as beam splitters and to perform all the functions provided by the above components with the exceptions of collimation and conjugation.[22] Single-mode fibers are necessary for the reference beam. The only disadvantages at this time are insertion losses and higher requirements for mechanical and temperature stability.

19.4.3 Hardware

Except for pulsed work, the most dominant piece of hardware in a holography laboratory is the vibration isolation table. Other equipment includes holders for each item of optical equipment as well as various sizes of holders for photoplates and vacuum platens for film.

For long exposures, fringe lockers[23] should be considered a necessity. This device detects the movement of interference fringes and provides feedback to control a mirror (typically the reference beam) with speaker or piezoelectric movements. Sometime more than one system is required.

Electronic shutters are not necessary for low-power lasers, when the exposure exceeds 1 s. With argon and krypton ion lasers, however, a commercial shutter is advisable.

Power meters, especially those calibrated for each laser wavelength, are necessary for professional work. On the other hand, a simple photographic meter, such as the Luna-pro or the Minolta, will do well.

19.4.4 Recording Material and Processing Chemicals

Silver halide emulsion coated on glass or flexible backing (triacetate or polyester) has been, and will remain for a long time, the workhorse of holographic recording.[24]

The most commonly used are the Agfa 8E75 and 8E56 emulsions, which are sensitized for He-Ne and argon lasers, respectively. Both are capable of resolving 5000 lines per millimeter and thus are used for both transmission and reflection holograms.[25]

The processing of silver halide is a well-explored technology.[26] The reader is well-advised to begin by using the simplest technique[16,27] and progress from there.

Dichromated gelatin (DCG)[28] is a popular medium if one has high output lasers in the blue region. It offers high diffraction efficiency, low scatter, and environmental stability when correctly sealed in glass. Ready-coated plates are presently not available in the market. Currently, DCG is used in most HOE applications such as head-up displays (HUDs) and common jewelry items in gift shops.

Photoresist-coated plates[29] are used for making surface relief master holograms. After proper processing, the plates can be electroformed to create metallic embossing masters.[30]

Thermoplastic[31] material is amenable to electronic processing and can be erased and recorded up to 1000 times. It is used in holographic nondestructive testing systems.

Photopolymers[32] are the material of the future. It offers high diffraction efficiency and low scatter as does DCG, but is capable of being sensitized to all regions of the visible spectrum. It has promising portents for HOE and full-color holographic applications.[33]

19.5 GENERAL PROCEDURES

Like photography, holography can be practiced at all levels of sophistication and cost. In general, one should begin with the simplest procedures[16] by making single beam reflection and transmission holograms, using the JD-2 processing kit.[25] From there, one can progress to a more professional level by consulting detailed texts.[34-36]

Figure 19.4 represents a typical configuration for creating a master hologram H_1 suitable for transferring to H_2. The latter is made by placing a photoplate in the vicinity of the conjugate image, with a reference beam from the side of H_1 or the opposite side. The result is a focused image transmission or reflection hologram. This example shows the object being illuminated by two object beams from two sides. One is a point source, giving sharp shadows and the other is diffused for soft lighting.

The general procedure for making H_1 is

1. On a vibration-free table, lay out the optical configuration approximately as shown in Fig. 19.4. Pinholes for spatial filters need not be inserted at this point.

2. This step is optional if the laser used has an internal etalon for selecting a single frequency, thereby emitting a beam of long coherent length. In general, beam path lengths, beginning from the first beam splitter to the photoplate, must be equalized well within the coherent length available. Tie one end of a long string to the base of the first beam splitter, and measure the three paths. Equalize them by judicious relocation of beam-folding mirrors.

3. Measure the relative intensity of light arriving at the photoplate from the reference beam and from the illuminated object. The ratio should be at least 4:1. Adjust the beam ratio by rotating the quarter-wave plates.

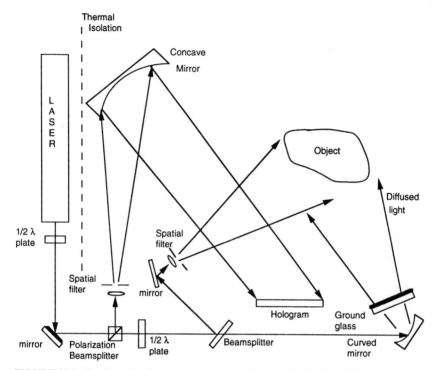

FIGURE 19.4 Configuration for creating a master hologram suitable for transfer.

4. Measure the total intensity at the photoplate and determine the exposure time. In general, a strip test is advisable. This is done by covering the plate with an opaque mask and exposing one vertical strip at a time. With each adjacent strip having twice the exposure, the best choice can then be made with the processed hologram.

Often, air movement alone can be a critical deterrent. The system should be covered without introducing further disturbance.

19.6 CURRENT APPLICATIONS

With few exceptions, the current applications of holography can be generally classified into the following catagories. Because of space limitations, they are simply listed with pertinent references.

19.6.1 Display

This is the most obvious use of the three-dimensional medium. There is now a small but active group of professional artists around the world creating highly imaginative images.[37] Most large cities now have galleries exhibiting and selling their works.

Custom portraitures[38] are recorded routinely by using pulsed lasers. Recently, group portraits of 200 people in motion (doing "the wave") were made during the Fourth International Symposium on Display Holography in Lake Forest, Ill.[39]

For commercial applications, embossed holograms are most popular, because of their mass-produceability. *National Geographic Magazine* has used it in no less than three covers.

Because it is difficult to duplicate a hologram without the original master, anticounterfeiting application has become a large segment of the business. Holograms are now used on credit cards, postage stamps, paper currency, baseball cards, clothing labels, superbowl tickets, among many other products. Currently, real-time holographic TV is being developed.[40]

19.6.2   Holographic Interferometry[41-43]

Holograms can reveal microscopic changes on a specimen, yielding quantitative data. The three general methods are (1) double exposure, with the change occurring between exposures; (2) time average, a single exposure made while the subject undergoes many normal mode vibrations; and (3) real time, viewed through a hologram at the specimen.

19.6.3   Holographic Optical Elements (HOEs)

As shown in a previous section, a hologram behaves like a lens and a grating simultaneously. Volume holograms are effective notched filters. Among other current applications are scanning,[44] head-up display for aircrafts (HUD),[45] and general optical elements.[46] Many current HOEs can be generated by computers.[47]

19.6.4   Precision Measurements

With the development of powerful and coherent lasers, deep three-dimensional scenes can be recorded in a short time, even below one picosecond. Particle sizing[48] was an early application. A more recent example is the recording of bubble-chamber tracks[49] at Fermilab. By using picosecond lasers, "light-in-flight" measurements are possible.[50,51]

19.6.5   Optical Information Processing and Computing[52]

Besides three-dimensional displays, the first dramatic application of holography was in complex spatial filtering.[53] That was the first use of holography in the emerging field of optical computing.

The fact that holograms are formed by the interference of two wavefronts, and that one would recall the other, is directly related to the phenomenon of associated memory in the field of neural networks.[54]

When a hologram is addressed, a parallel rather than serial output is obtained. Thus information can be stored and/or transferred at a rate not possible for electronic computers. Furthermore, holograms can be duplicated more easily and economically.

A major limitation of electronic computers is in interconnections. Thus, connecting two $n \times n$ arrays of processors would require n^4 wires.[55] This is not a problem with holograms.

Using new exotic recording materials such as photorefractive crystals, read and write computer memory systems capable of handling terabytes of information are at the horizon.[56]

With the development of exotic photorefractive materials, the subject of phase conjugation[57] promises parallel-processing optical amplifiers.

Judged by the current activities and funding in research, optical computing may be the most important application of holography, both in extent and economics.

19.6.6 Education

Holography is one subject that can be learned at all levels, by students of all ages and of divergent interests.[58] Basic holograms can be made with a budget below that for photography and without previous training. Yet the basic ideas involved relate directly to many Nobel Prizes in physics. Those won by Michelson, Lippmann, Bragg, Zernicke, and Gabor are a few examples. Increasingly, this worthy subject is being integrated into elementary science educational programs with great success.[59] It is already an integral part of basic university physics texts.[60]

19.7 REFERENCES

1. Gabor, D., "A New Microscopic Principle," *Nature*, no. 161, pp. 777–779, 1948.
2. Gabor, D., "Microscopy by Reconstructed Wavefronts, *Proc. Phys. Soc.*, vol. A194, pp. 454–487, 1949.
3. Leith, E. N., and J. Upatnieks, "Reconstructed Wavefronts and Communication Theory," *J. Opt. Soc. Am.*, vol. 53, no. 12, pp. 1377–1381, 1963.
4. Leith, E. N., and J. Upatnieks, "Reconstruction with Continuous Tone Objects," *J. Opt. Soc. Am.*, vol. 53, no. 12, pp. 1377–1381, 1963.
5. Leith, E. N., and J. Upatnieks, "Wavefront Reconstruction with Diffused Illumination and Three-Dimensional Objects," *J. Opt. Soc. Am.*, vol. 54, no. 11, 1295–1301, 1964.
6. Denisyuk, Yu. N., "Photographic Reconstruction of the Optical Properties of an Object in Its Own Scattered Radiation Field," *Sov. Phys. Dokl.*, vol. 7, pp. 543–545, 1962.
7. Denisyuk, Yu. N., "On the Reproduction of the Optical Properties of an Object by the Wave Field of Its Own Scattered Radiation," *Opt. Spectrosc.*, vol. 15, pp. 279–284, 1963.
8. Denisyuk, Yu. N., "On the Reproduction of the Optical Properties of an Object by the Wave Field of Its Own Scattered Radiation II, *Opt. Spectrosc.*, vol. 18, pp. 152–157, 1965.
9. Goodman, J., *Introduction to Fourier Optics*, McGraw-Hill, New York, 1968.
10. Meier, R. W., "Magnification and Third-Order Aberration in Holography," *J. Opt. Soc. Am.*, vol. 55, pp. 987–992, 1965.
11. Champagne, E. B., "Nonparaxial Imaging, Magnification and Aberration Properties in Holography, *J. Opt. Soc. Am.*, vol. 57, pp. 51–55, 1967.
12. Syms, R. R. A., *Practical Volume Holography*, Oxford University Press, 1991.
13. Jeong, T. H., "A Geometric Model for Holography," *Am. J. Phy.*, pp. 714–717, August 1975.

14. Lippmann, G., "Sur la Theorie de la Photographie des Couleurs Simples et Composées par la Methode Interferentielle," *J. Phys.*, Paris, vol. 3, 1894.

15. Bragg, W. L., "A New Type of X-ray Microscope," *Nature*, vol. 143, pp. 678, 1939.

16. Jeong, T. H., *Laser Holography—Experiments You Can Do*, Integraf, Lake Forest, Ill., 1988.

17. Bjelkhagen, H., "Holographic Portraits: Transmission Master Making and Reflection Copying Techniques, *Proc. ISDH*, T. H. Jeong, ed., Lake Forest College, Ill., vol. I, pp. 45–54, 1982.

18. Bates, H. E., "Burst-mode Frequency Doubled YAG:Nd^{3+} Laser for Time-Sequenced High Speed Photography and Holography," *Appl. Opt.*, vol. 12, pp. 1172–1178, 1973.

19. Jeong, T. H., Z. Qu, E. Wesly, and Q. Feng, "Coherent Length and Holography," *Proc. of the Lake Forest Fourth International Symposium on Display Holography*, T. H. Jeong and H. Bjelkhagen, eds., SPIE, vol. 1600, pp. 387–401, 1992.

20. Jeong, T. H., "HOE for Holography," *Holography '89*, Varna, Bulgaria, Yu. N. Denisyuk and T. H. Jeong, eds., *Proc. SPIE*, vol. 1183, 1989.

21. Gilbert, J. A., and T. D. Dudderar, "Uses of Fiber Optics to Enhance and Extend the Capabilities of Holographic Interferometry," *SPIE Inst. for Advanced Technology*, P. Greguss and T. H. Jeong, eds., vol. IS 8, pp. 146–159, 1991.

22. Jeong, T. H., "Demonstration in Holometry Using Fiber Optics," *Proc. SPIE*, R. J. Pryputniewicz, ed., vol. 746, pp. 16–19, 1987.

23. Phillips, N., "Fringe Locking Devices for the Stabilization of Holographic Interference," *Lake Forest International Symposium on Display Holography* (ISDH), T. H. Jeong, ed., vol. II, pp. 111–130, 1985.

24. Phillips, N., "Bridging the Gap Between Soviet and Western Holography," *SPIE Inst. for Advanced Technology*, P. Greguss and T. H. Jeong, eds., vol. IS 8, pp. 206–214, 1991.

25. For updates in availability and current prices of emulsions and chemicals, send stamped and self-addressed envelope to HoloInfo Center, P.O. Box 586, Lake Forest, IL 60045.

26. Phillips, N., *The Silver Halides—The Workhorse of the Holography Business*, ISDH, T. H. Jeong, ed., vol. III, pp. 35–74, 1988.

27. Saxby, G., *Manual of Practical Holography*, Focal Press, London, 1991.

28. Coblijn, A. B., "Theoretical Background and Practical Processing for Art and Technical Work in Dichromated Gelatin Holography," *SPIE Inst. for Advanced Technology*, P. Greguss and T. H. Jeong, eds., vol. IS 8, pp. 305–325, 1991.

29. Cvetkovich, T. J., *Technique in Using Photoresist*, ISDH, vol. III, pp. 501–510, 1988.

30. Burns, J. R., "Large Format Embossed Holograms," *Proc. SPIE*, L. Huff, ed., vol. 523, pp. 7–14, 1984.

31. Lin, L. H., and H. L. Beauchamp, "Write and Erase in situ Optical Memory Using Thermoplastic Holograms," *Applied Optics*, vol. 9, pp. 2088–2092, 1970*b*.

32. Smothers, W. K., T. J. Trout, A. M. Weber, and D. J. Mickish, "Hologram Recording in DuPont's New Photopolymer Materials," *IEEE 2d Intl. Conf. on Holographic Systems, Components, and Applications*, University of Bath, U.K., 1989.

33. Jeong, T. H., and E. Wesly, "Progress in True Color Holography," *Three-Dimensional Holography: Science, Culture, Education*, Kiev, USSR, V. Markov and T. H. Jeong, eds., *SPIE*, vol. 1238, pp. 298–305, 1989.

34. Collier, R. J., C. B. Burckhardt, and L. H. Lin, *Optical Holography*, Academic Press, New York, 1971.

35. Hariharan, P., *Optical Holography, Principles, Techniques and Applications*, Cambridge University Press, Cambridge, U.K., 1984.

36. Saxby, G., *Practical Holography*, Focal Press, London, 1988.

37. Section 5 of ISDH, vol. III, "Artistic Techniques and Concepts" (10 articles by artists).

38. Unterseher, F., "Integrating Pulse Holography with Varied Holographic Techniques," ISDH, T. H. Jeong, ed., vol. III, pp. 403–420, 1988.

39. Smith, S., and T. H. Jeong, "Method and Apparatus for Producing Full Color Stereographic Holograms, U. S. patent 5022727, June 11, 1991.

40. Benton, S., *Experiments in Holographic Video Imaging, SPIE Inst. for Advanced Optical Technology*, P. Greguss and T. H. Jeong (eds.), vol. IS 8, pp. 247–267, 1991.

41. Vest, C., *Holographic Interferometry*, Wiley Interscience, New York, 1979.

42. Abramson, N., "The Making and Evaluation of Holograms," Academic Press, New York 1981.

43. Pryputniewicz, R. J., Automatic System for Quantitative Analysis of Holograms, *SPIE Institute for Advanced Technology*, vol. IS 8, pp. 215–246, 1991.

44. Beiser, L., *Holographic Scanning*, Wiley, New York, 1988.

45. McCauley, D. G., and C. E. Simpson, "Holographic Optical Element for Visual Display Applications, *Appl. Opt.*, vol. 12, pp. 232–242, 1973.

46. Chang, B. J., and C. D. Leonard, "Dichromated Gelatin for the Fabrication of Holographic Optical Elements," *Appl. Opt.*, vol. 15, pp. 2407, 1979.

47. Cindrich, I. N., and S. H. Lee, eds., "Holographic Optics: Optically and Computer Generated," *Proc. SPIE*, vol. 1052, 1989.

48. Thompson, B. J., "Holographic Particle Sizing Techniques," *J. Phys. E: Scientific Instruments*, vol. 7, pp. 781–788, 1974.

49. Akbari, H., and H. Bjelkhagen, "Big Bubble Chamber Holography," *ISDH II*, T. H. Jeong, ed., pp. 97–110, 1988.

50. Abramson, N., "Light-in-Flight Recording: High-Speed Holographic Motion Pictures of Ultrafast Phenomena," *Proc. SPIE*, vol. 22, pp. 215–222, 1983.

51. Pettersson, S. G., H. Berstrom, and N. Abramson, "Light-in-Flight Recording with View-Time Expansion Using a Skew Reference Wave or Multiple Reference Pulses," *ISDH III*, T. H. Jeong, ed., pp. 315–325, 1988.

52. Caulfield, H. J., "Holograms in Optical Computing," *SPIE Institute for Advanced Optical Technology*, vol. IS 8, pp. 54–61, 1991.

53. VanderLugt, A., "Signal Detection by Complex Spatial Filtering," *IEEE Trans. Information Theory*, vol. IT-10, pp. 139–145, 1964.

54. Caulfield, H. J., J. Kinser, and S. K. Rogers, "Optical Neural Netowrks," *Proc. IEEE*, vol. 77, pp. 1573–1590, 1989.

55. Caulfield, H. J., "Parallel N^4 Weighted Optical Interconnections, *Appl. Opt.*, vol. 26, pp. 4039–4040, 1987.

56. Parish, T., "Crystal Clear Storage," *Byte*, pp. 283–288, November 1990.

57. Fisher, R., *Optical Phase Conjugation*, Academic Press, New York, 1983.

58. Jeong, T. H., "Holography and Education," *SPIE Institute for Advanced Optical Technology*, vol. IS 8, P. Greguss and T. H. Jeong, eds., pp. 360–369, 1991.

59. Tomaszkiewicz, F., "A Continuing Laser-Imaging Program of Instruction in a Public Middle School," T. H. Jeong and H. Bjelkhagen, eds., ISDH IV, SPIE, vol. 1600, 263–267, 1992.

60. Halliday, D., and R. Resnick, *Fundamental of Physics*, 3d ed., essay 18, Wiley, New York, 1988.

CHAPTER 20
LASER SPECTROSCOPY AND PHOTOCHEMISTRY

G. Rodriguez, S. B. Kim, and J. G. Eden

20.1 INTRODUCTION

In its most general sense, spectroscopy encompasses the study of the structure of matter by the observation of its interaction with electromagnetic radiation. Not only are those frequencies at which a species absorbs radiation of interest, but, once one or more photons are consumed, detailed information regarding the structure of an atom or molecule is revealed by the manner in which it disposes of the additional energy. At first glance, then, the various branches of spectroscopy at optical wavelengths can be categorized on the basis of (1) the number of photons absorbed by the atom or molecule under study and (2) the experimental technique chosen to detect a specific product generated by the optical process.

Spectroscopy and photochemistry are both undergirded by the characteristic of matter to absorb radiation at specific wavelengths in the electromagnetic spectrum. As shown qualitatively in Fig. 20.1 for crown glass (after Ref. 1), absorption by a molecule (or atom) at different points in the spectrum corresponds to energizing or exciting various degrees of freedom of the molecule. The tantalizing opportunity that awaits the spectroscopist or photochemist, therefore, is the ability to excite specific states or modes of a molecule and to observe its subsequent behavior. One may wish to isolate the molecule and determine the way in which it responds to the excitation (i.e., radiating another photon, dissociating, ejecting an electron, etc.) or it may be useful to allow the excited molecule to transfer some fraction of its energy to a different species through a collision. A critical factor in the almost overwhelming success of spectroscopy is that the absorption spectra of molecules (and atoms) are unique to a species—a signature or fingerprint of its structure. Note, too, from Fig. 20.1, that, not surprisingly, decreasing the wavelength of the radiation allows one to access increasingly energetic modes of the species under study.

The demonstration of the laser in 1960 changed forever the face of spectroscopy and photochemistry. Lasers quickly displaced lamps and arcs as optical sources as it was recognized that the intensities and narrow spectral linewidths available with laser radiation enabled one to examine atoms and molecules and their participation in chemical reactions to a level of detail that was previously unimaginable. Three decades later, lasers now available range in wavelength from the submillimeter

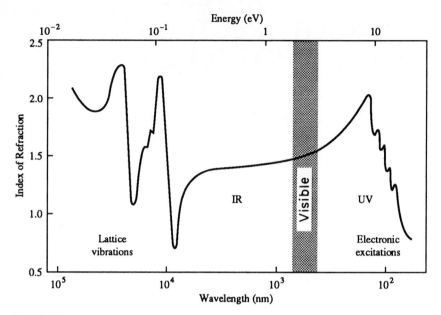

FIGURE 20.1 Variation with wavelength of the index of refraction for a crown glass containing approximately 74 percent SiO_2, 21 percent Na_2O, and 5 percent CaO by weight. *(Reprinted from Ref. 1, by permission.)*

region ($\lambda \sim 500$ µm lasers such as CO_2-pumped CH_3F) to the x-ray systems recently developed at the Naval Research Laboratory, Lawrence Livermore National Laboratory, and Princeton University that emit in the realm below 5 nm. Viewed from a different perspective, the frequency spans five orders of magnitude ($\sim 10^{12}$ to 10^{17} Hz). During this intervening period, the development of new lasers or methods for stabilizing laser frequency or narrowing linewidth, for example, was invariably and closely followed by the introduction of spectroscopic techniques which offered (1) an increasingly detailed view of atomic and molecular structure and (2) the ability to perform precise measurements of fundamental constants. The reverse viewpoint is that applications envisioned for the laser set demanding specifications that, in turn, drove its further development. In any case, the diversity of spectroscopic techniques available today makes a detailed review of them impractical, but it is the aim of this chapter to give the reader a broad overview of the more versatile techniques and their extraordinary capabilities. Toward that end, several of the methods most often employed by the laser spectroscopic community, such as laser-induced fluorescence and multiphoton spectroscopy, are reviewed here, and the interested reader will be referred to the literature for more details concerning each technique.

Emphasis will be placed on the spectral region extending from the infrared to the vacuum ultraviolet ($\lambda < 200$ nm). Several examples drawn from the literature and the work of the authors' laboratory will illustrate the application of prominent spectroscopic techniques to unraveling the structure of atoms and small molecules. Although most of the specific instances cited here involve the gas phase, virtually all of the spectroscopic tools and methods described later have been extensively applied to solids, liquids, and plasmas as well.

Although many of the lasers most often used for laser spectroscopy have been discussed in Chaps. 3 to 8, a few words concerning the spectroscopist's primary tool are in order. Most of the laser systems employed in laboratory spectroscopy are gas, dye, and (insulating) solid-state oscillators. Only recently have semiconductor lasers made significant inroads into spectroscopy in the visible and near infrared (nir), but that trend is expected to accelerate in the next 5 years.

Despite the wide array of commercial laser systems that is presently available, a relative handful have dominated the major spectroscopic techniques. For continuous-wave (cw) experiments, the argon ion laser-pumped dye laser has been the workhorse of high-resolution ($\Delta \nu < 100$ MHz) spectroscopy for almost two decades but the recently developed titanium-doped sapphire ($Ti:Al_2O_3$) system is already supplanting dyes in several spectral regions, particularly in the nir itself (~ 700 to 900 nm), and in the wavelength interval accessible by second-harmonic generation. The wide tunability of $Ti:Al_2O_3$ lasers when pumped in the green (500 to 540 nm), the relative ease of handling the crystal, and its rugged mechanical and optical characteristics make it quite attractive to the spectroscopist for whom lasers are of secondary interest and reliability is imperative.

When peak intensity is an important parameter, dye lasers pumped by pulsed neodymium-doped yttrium aluminum garnet ($Nd:YAG$) or rare gas-halide excimer lasers are the systems of choice. Used in conjunction with nonlinear crystals such as potassium titanyl phosphate (KTP) or β-BaB_2O_4, the spectral region extending from about 200 to 900 nm and beyond can be covered continuously with these lasers and peak pulse intensities exceeding 1 MW/cm^2 are readily attainable.

20.2 LASER-INDUCED FLUORESCENCE AND ABSORPTION SPECTROSCOPY

Laser-induced fluorescence (LIF) and laser absorption spectroscopy are, perhaps, the simplest and most prevalent of the laser spectroscopic methods. Both are, nevertheless, powerful techniques drawing on the narrow linewidth of laser radiation to access specific excited quantum states of an atom or molecule.

20.2.1 LIF

As illustrated by the diagram of Fig. 20.2, LIF involves tuning the wavelength (photon energy) of a laser until it coincides with that for a transition of the atom or molecule of interest.[2] Coupling of the two states by a resonant optical field results in the upper state of the transition being populated, thereby depleting the lower state number density. Once the excited species is produced, it can radiate to lower levels, producing a characteristic and readily identifiable emission spectrum. Often, though, the pumped excited state can relax through collisions with background atoms or molecules in the gas phase or by phonon-assisted processes in crystals. In this situation, nearby electronic states also fluoresce. Such secondary emissions can be suppressed to an extent by lowering the gas pressure or crystal temperature but, if their presence is unavoidable, they can generally be isolated by sequentially pumping several adjacent excited states.

The overriding characteristic of laser-induced fluorescence is its selectivity—namely, the ability it offers to examine specific states of a particular species. Further, since the absorption spectra for all atoms and molecules are unique, LIF is a

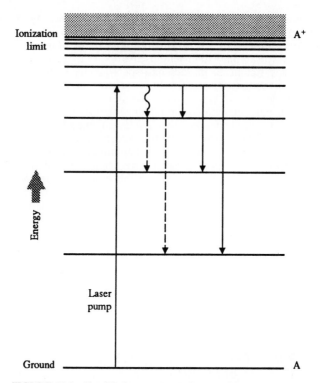

FIGURE 20.2 Simplified energy-level diagram of an atom or molecule (denoted A), schematically illustrating the probing of excited states by laser-induced fluorescence. The lower level of the transition pumped by the laser need not be ground, but is pictured as shown for convenience. The dashed vertical lines denote fluorescence on transitions not directly pumped by the laser but whose upper states are populated by collisions.

convenient means for identifying one constituent in a mixture. Although emission spectroscopy thrived long before the advent of the laser, other approaches to exciting fluorescence in an atom or molecule are considerably less discriminating than laser techniques. The spectra emitted by a gas discharge or arc, for example, are often terribly congested and difficult or impossible to analyze. The reason for this is rooted in the nature of electron impact excitation, which produces a wide array of excited states, primarily because the electron energy distribution function for a discharge is far from monoenergetic. As an example, see Fig. 4 of Ref. 2, which compares the emission of the sulfur diatomic molecule produced by a discharge with an LIF spectrum in the same wavelength region. In contrast, the energy of photons produced by a laser is well-defined, and the worst of lasers generate beams for which the ratio of spectral linewidth ($\Delta\lambda$) to center wavelength λ_0 is 10^{-3} to 10^{-4}. With modest effort, this ratio can be reduced by another 3 to 4 orders of magnitude. In any case, LIF has repeatedly simplified the spectra of atoms and molecules and quite literally revolutionized our notions of chemical structure and microscopic dynamical processes.

LIF experiments generally involve an arrangement similar to that shown in Fig. 20.3. A laser beam is directed through the sample, and the resulting fluorescence is collected by lenses or optical fibers, dispersed by a spectrograph or monochromator, and detected by a photomultiplier or diode array. Usually, the pump laser is tunable in order to access different quantum states, but occasionally a coincidence exists between the fixed wavelength of an available laser and the transition of interest.

With such an experiment, one has the option of (1) fixing the laser wavelength and scanning the detection system to observe all fluorescence produced by that particular pump wavelength or (2) holding the detector wavelength constant and scanning the pump laser with the intent of determining the relative efficiency of producing a *particular* "product" state as a function of wavelength. These complementary methods of acquiring spectra are known as *fluorescence* and *laser-excitation* spectroscopy, respectively. While there exist a number of variations on these themes, most LIF experimental techniques can be broadly classified into one of the two camps.

A simple, but interesting, example of the principles represented in Fig. 20.2 is provided by the tellurium dimer Te$_2$. If tellurium vapor is irradiated with a blue dye laser beam, the diatomic molecule will produce superradiant laser emission at

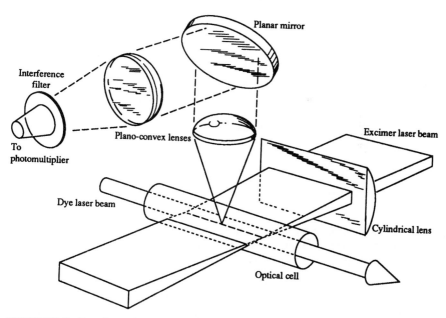

FIGURE 20.3 Experimental arrangement representative of those used in LIF studies. The dye laser is tuned to a resonance of the atom or molecule under study, and the resulting fluorescence is collected by lenses and detected by a photomultiplier. It is generally advantageous to be able to scan this laser in wavelength. A second laser (an ultraviolet excimer laser, in this case) is often provided as well to produce a desired transient species for subsequent probing by the dye laser. The diatomic radical xenon monofluoride (XeF), for example, lives only briefly in the gas phase and can be produced by photodissociating XeF$_2$ with a uv laser. The electronic states of the radical can subsequently be probed by the time-delayed dye laser. *(After Ref. 3.)*

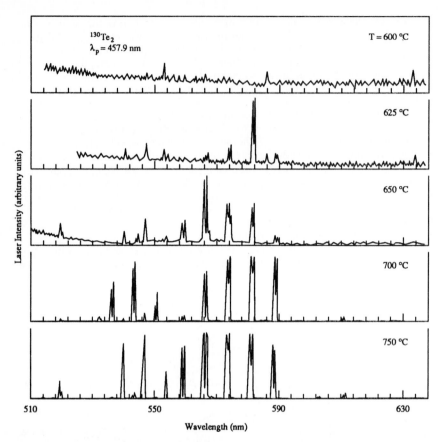

FIGURE 20.4 Laser emission from the $^{130}Te_2$ dimer produced upon irradiating tellurium vapor with a pulsed (~10-ns) dye laser beam at 457.9 nm (λ_p). The number of oscillating lines and the distribution of intensity among them changes dramatically with temperature. Note also that several of the most intense laser lines (particularly those at the highest temperatures) are broadened artificially because of detector saturation.

a wide range of visible wavelengths. Figure 20.4 shows the laser lines produced in the green to red region of the spectrum (510 to 640 nm) when a dye laser tuned to 457.9 nm illuminates about 2 to 6-torr of Te vapor. The dye laser serves to excite ground state (XO^+) Te_2 molecules to various vibrational levels of the excited electronic B state. Molecules in these levels can subsequently fluoresce via transitions back to ground and will readily produce coherent (stimulated) emission when the vapor is pumped by pulsed or cw lasers. As the temperature of the optical cell containing the ^{130}Te metal source is raised, the number of lasing lines and the distribution of intensity among them changes rapidly. These data reflect the fact that increasing temperature populates ever more energetic vibrational levels in the ground molecular state. In conjunction with the blue laser pump, this results in additional B state vibrational levels being populated and, hence, new laser lines.

Further detail concerning this molecule can be gleaned from Fig. 20.5, which illustrates the laser excitation spectra obtained by scanning the dye laser wavelength

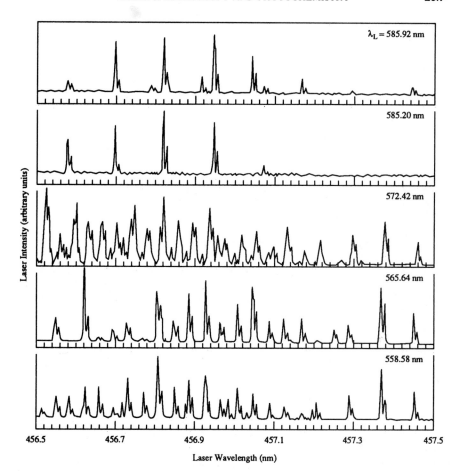

FIGURE 20.5 Excitation spectra obtained by scanning the dye laser (pump) wavelength while recording the intensity of specific $^{130}Te_2$ laser lines (denoted λ_L) in the yellow or orange.

while measuring the intensity of a *particular* Te_2 laser line. Discussing the structure of this molecule in detail is beyond the scope of this chapter but, clearly, the varying complexity of the spectra of Fig. 20.5 demonstrates the degree to which the upper state vibrational and rotational levels are coupled to one another.

As a second, more detailed, example, consider the partial energy level diagram for the indium monoiodide (InI) diatomic molecule shown in Fig. 20.6. The abscissa represents the separation in angstroms (1 Å = 10^{-8} cm) between the indium and iodine atomic nucleii and the ordinate is given in wave numbers (cm^{-1}), a unit of energy for which 1 eV ≈ 8065 cm^{-1}. If an InI molecule absorbs a violet photon (~25,000 cm^{-1}), an excited state is produced which can subsequently re-radiate, as indicated by the vertical arrow labeled *molecular fluorescence*.

However, the excited molecule is also able to absorb a second photon *from the same laser pulse*. The combined energies of the two absorbed photons are sufficient to leave the molecule in an unstable state—one which results in the dissociation

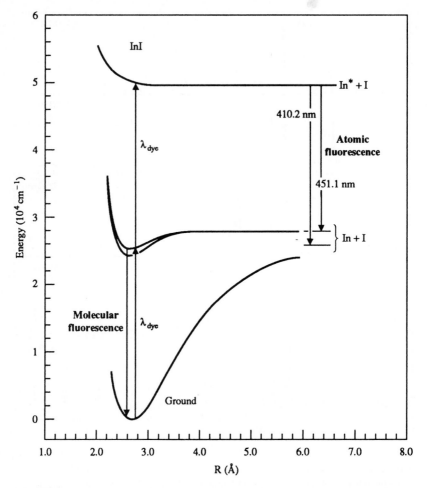

FIGURE 20.6   Partial energy-level diagram for the indium monoiodide molecule (InI), qualitatively illustrating both laser-induced molecular and atomic fluorescence. An asterisk denotes an electronic excited state of an atom or molecule.

of the molecule. The products (fragments) are a ground state iodine atom and an electronically excited In atom that subsequently fluoresces in the violet (451.1 or 410.2 nm).

A block diagram of the experimental arrangement for these studies is given in Fig. 20.7. For a fixed dye laser wavelength (λ_{dye}, Fig. 20.6) tuned to a transition of the molecule, much can be learned concerning the structure of all three InI states involved in the two-step process by simply scanning the monochromator over the wavelengths emitted by the diatomic molecule and the atom. Conversely, excitation spectra are obtained by holding the detection system wavelength constant and scanning the dye laser. The temporal histories of the atomic indium and molecular emissions produced in such experiments are shown by the fluorescence

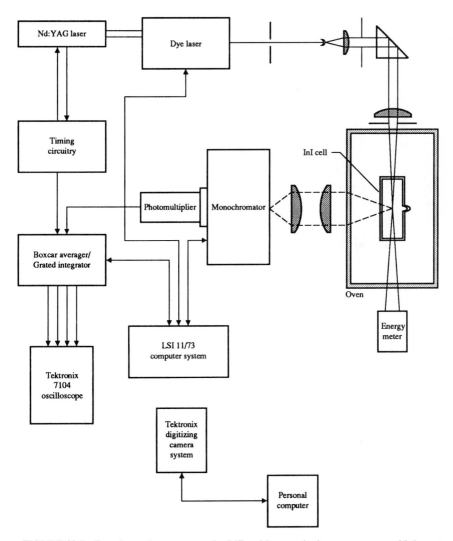

FIGURE 20.7   Experimental arrangement for LIF and laser excitation spectroscopy of InI.

waveforms of Fig. 20.8, and Fig. 20.9 presents representative excitation spectra. The molecular spectrum is considerably simpler than its atomic counterpart and serves to identify some of the structure in the atomic excitation spectrum. Consequently, excitation or conventional LIF spectra are capable of probing a molecule at critical points in its electronic structure diagram, allowing for the flow of energy within the molecule (or atom) to be determined to an unprecedented degree of accuracy.

Going one step further with Fig. 20.6, it is straightforward to add a *second* dye laser to the experimental system for the purpose of learning more about the struc-

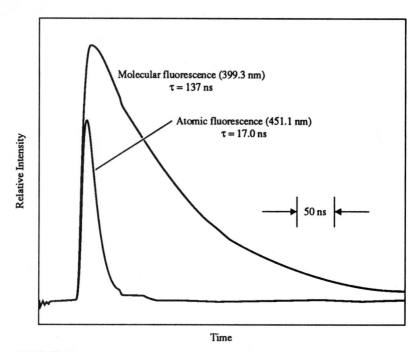

FIGURE 20.8   Atomic In and molecular InI fluorescence waveforms recorded in the experiments depicted in Fig. 20.6. Note the difference in atomic and molecular excited-state lifetimes
(denoted τ) that are determined from the fall time of the emission waveforms.

ture of atomic indium. In this case, the first dye laser pulse is fixed in wavelength
and the atomic In emission is detected as before. The second dye laser pulse arrives
an adjustable time delay following pulse 1 and is scanned in wavelength. Imagine
the situation that occurs when the wavelength of laser pulse 2 is coincident with a
transition of In for which the state denoted In* in Fig. 20.6 is its *lower* level; that
is, dye laser 2 excites the In* atoms to still-higher-lying states. At such a wavelength,
the In 451-nm violet fluorescence will be *suppressed*, since the In* population has
been depleted by pulse 2. Therefore, if the In 451-nm emission is recorded as the
wavelength of laser 2 is varied, a *fluorescence suppression spectrum*, such as that
given in Fig. 20.10, is obtained. The clear suppressions in the spectrum correspond
to transitions of atomic In from the 6s $^2S_{1/2}$ (In*) state to various members of three
Rydberg series, as indicated on the figure. These brief examples serve to illustrate
the versatility of laser-induced fluorescence and closely related techniques in gaining
insight into the structure and optical and chemical behavior of atoms and molecules.

20.2.2   Bound $\rightleftarrows$ Free Processes

It is fair to say that molecular spectroscopic studies have predominantly dealt with
electronic states that are bound (i.e., stable). The prominent aspects of the examples
of Te$_2$ and InI discussed in the last section certainly fall into that category. Often,
however, one of the electronic states involved in an optical transition is, as illus-

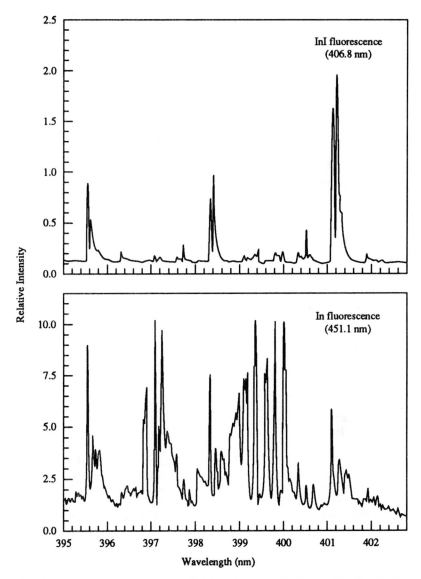

FIGURE 20.9 Segment of the laser excitation spectra obtained by scanning the dye laser wavelength in the 395- to ~403-nm region. The top spectrum was recorded by detecting molecular emission at 406.8 nm and the lower spectrum was obtained by monitoring atomic fluorescence at 451 nm. A number of features appear in both spectra, and the molecular spectrum serves to identify several peaks in the atomic excitation spectrum.

trated in Fig. 20.11, repulsive (dissociative). The interaction of the discrete states of the bound level with the ground state continuum has the effect of producing gently undulating, rather than highly structured, emission and absorption spectra. Emission from a limited number of vibrational levels of the $^1\Sigma_u^+$ excited state of

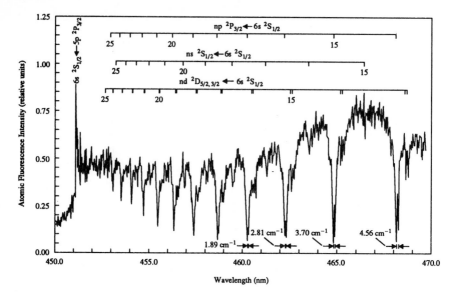

FIGURE 20.10 Fluorescence suppression spectrum for atomic indium showing the depletion of the In* ($6^2S_{1/2}$) state's population as a dye laser is tuned into resonance with Rydberg transitions of the atom. The three easily identifiable Rydberg series that are observed are indicated.

Zn_2, for example, is shown by the top spectrum in Fig. 20.12. Instead of the sharp features characteristic of Figs. 20.4, 20.5, and 20.9, weak oscillations in the spectral envelope attributable to the upper state wavefunction are observed. Similar results are obtained if the reverse process—bound ← free absorption or photoassociation—is observed. As displayed in Fig. 20.12b for the KrF molecule, the absorption of a photon by a colliding pair of ground-state atoms having a repulsive interaction potential yields a spectrum once again characterized by deeply modulated (Franck-Condon) oscillations. The comparative lack of structure that is characteristic of bound ⇄ free spectra precludes the precise determination of molecular constants that are readily deduced from bound ⇄ bound spectra, but numerical quantum simulations of bound → free spectra have, nevertheless, provided structural information concerning transient molecules that is otherwise difficult to obtain.

20.2.3 Absorption Spectroscopy

Laser absorption spectroscopy is similar to LIF in that a laser is once again scanned through transitions of an atom or molecule. In this case, however, the primary parameter of interest is the depletion of the pump beam by the absorbing medium. Recall that a laser tuned to a molecular or atomic transition is a double-edged sword in the sense that it populates the upper state but also depletes (at least momentarily) the lower level. Therefore, one determines those optical frequencies at which the laser beam is attenuated by the medium by simply scanning the laser wavelength and recording the intensity of the transmitted beam with a photodetector. It must be emphasized that the resulting spectrum provides little information regarding the products generated when the medium absorbs a photon of a given

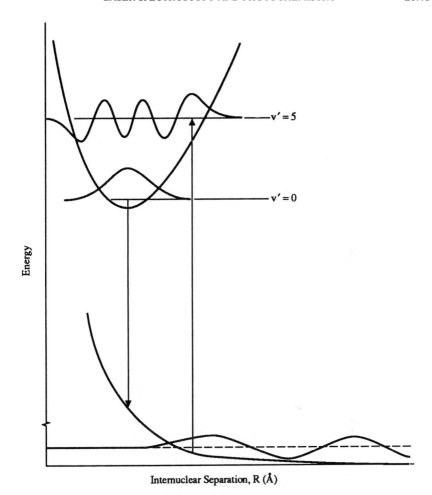

FIGURE 20.11 Partial energy-level diagram of a diatomic molecule having an unstable (dissociative) ground state. Bound → free emission from the excited state $v' = 0$ level and bound ← free absorption terminating at $v' = 5$ are both indicated.

energy. Nevertheless, when combined with laser excitation spectra in the same wavelength region, absorption spectra can yield considerable detail regarding branching ratios for populating specific energy levels and the structure of the electronic states involved.

Part *a* of Fig. 20.13 is a schematic diagram of an experiment for measuring absorption spectra. Either a tunable laser or a broadband lamp can provide the source (or reference) spectrum that is propagated through the sample. For illustrative purposes, the figure shows a sample of low volatility and, therefore, requires heating to obtain an adequate vapor pressure for measurements to be made. Both the spectrum transmitted by the sample and the source spectrum can be displayed simultaneously by a dual diode array detector. This is a desirable feature because,

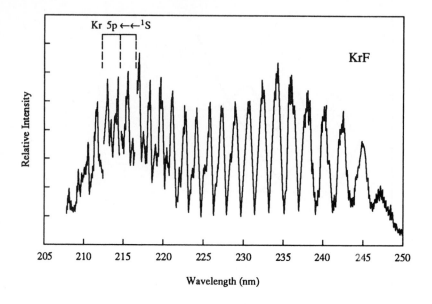

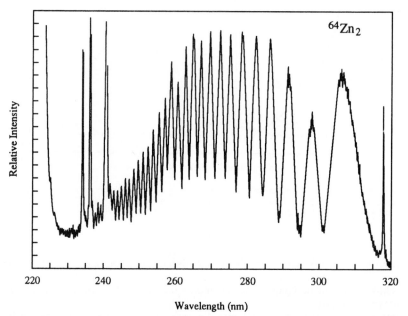

FIGURE 20.12 (*a*) Bound → free emission spectrum of the $v' = 85$ level of the Zn_2 ($^1\Sigma_u^+$) state. (*b*) Bound ← free excitation spectrum for the KrF molecule recorded by scanning a dye laser over the 205- to 250-nm region and monitoring KrF B → X and C → A fluorescence.

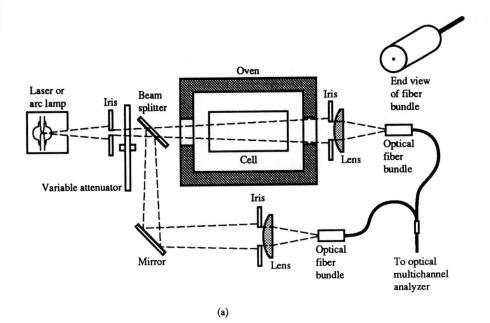

(a)

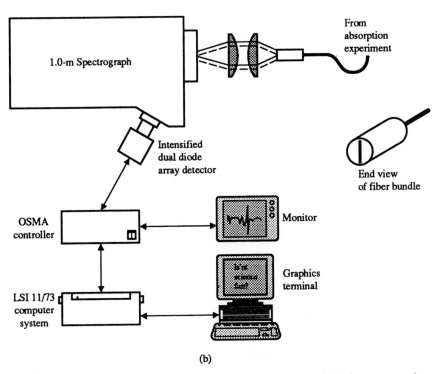

(b)

FIGURE 20.13 (*a*) Diagram of an experimental setup for measuring the absorption spectrum of an atom or molecule. While this layout is applicable to the study of a wide range of gaseous, liquid, and solid samples, the measurement of the spectrum for a metal (or metal salt) having a low volatility, and thus requiring an oven to obtain reasonable sample vapor pressure, is illustrated here. (*b*) Optical detection and electronic analysis equipment for absorption experiments. The source and transmitted spectra are displayed simultaneously by a dual diode array detector.

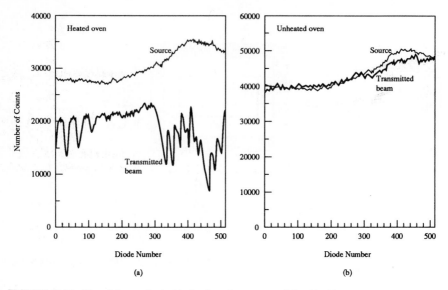

FIGURE 20.14 Raw data acquired with the detection system of Fig. 20.13*b* showing (*a*) the source (reference) beam from a Xe arc lamp and the beam transmitted by the metal-halide vapor in the heated optical cell and (*b*) the analogous spectra recorded with the optical cell at room temperature.

as indicated in Fig. 20.14, the reference signal is rarely constant as a function of wavelength. The primary advantage of acquiring absorption spectra with a laser, as opposed to classical spectroscopy using a lamp source, is that the spectral resolution in the former situation is dictated by the laser's narrow linewidth, leading to resolving powers several orders of magnitude higher than that available with a spectrograph.

Figure 20.15 shows the absorption spectrum of the iodine molecule I_2 over a small portion of the visible that was recorded with a tunable (dye) laser having a linewidth of 0.04 cm^{-1} (~1.3 GHz). This well-known spectrum frequently serves as a wavelength calibration in high-resolution spectroscopic experiments. A second example is illustrated in Fig. 20.16, which displays the absorption spectrum of a thin (5-mm) section of a fluoride crystal ($LiYF_4$, known by the acronym YLF) that has been doped with 1 atomic percent of erbium (Er) atoms. Because of the weak interaction of this rare-earth atom with the crystalline lattice, the spectrum shown is that for the trivalent ion, Er^{3+}, and ion-lattice interactions account for the relative breadth of the absorption lines. It is clear that, as the temperature of the crystal is increased from 77 K (liquid nitrogen temperature) to 300 K, the absorption lines broaden and new features appear. As shown by the inset to Fig. 20.16, the peaks in the spectrum are associated with transitions of the ion from its ground state, $^4I_{15/2}$, to sublevels of the $^4F_{7/2}$manifold.

20.3 PHOTOIONIZATION AND PHOTOELECTRON SPECTROSCOPY

The spectroscopic techniques described in the last section are effective in those situations in which optical radiation is produced by a chemical reaction or when

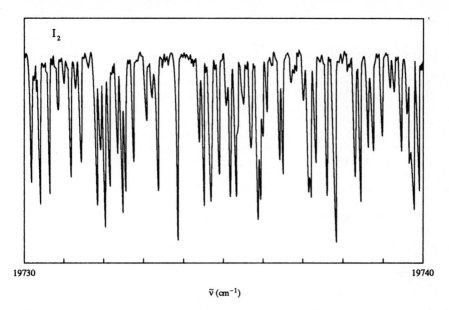

FIGURE 20.15 Absorption spectrum for I_2 vapor in the green (19,730 to 19,740 cm^{-1}) with a dye laser having a linewidth of about 1.3 GHz (0.04 cm^{-1}).

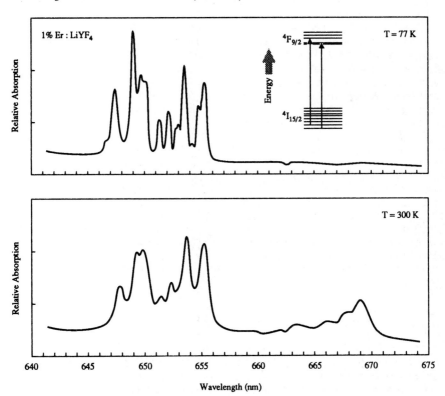

FIGURE 20.16 Absorption spectrum at 77 K and 300 K for a 5-mm-long YLiF$_4$ crystal doped with 1 atomic percent Er. The features shown arise from transitions of Er^{3+} from ground ($^4I_{15/2}$) to various Stark sublevels of the $^4F_{7/2}$ manifold.

an atom or molecule, upon absorbing a photon at one wavelength, radiates into another portion of the spectrum. Often, however, a photon is not released as a product and an entirely different perspective on atomic and molecular structure is afforded by photoelectron spectroscopy (PES). Instead of detecting the fluorescence produced when a species absorbs one or more photons, the focus of PES is the electron freed by ionization. Of particular interest are the energies, angular distribution, and relative number of electrons produced by a sample illuminated with optical radiation of a known frequency. Knowledge of photoelectron energies is especially useful in determining ionization potentials and, frequently, the identity and nature of electronic states of the *neutral* species. Figure 20.17 portrays in a generalized manner several processes by which the photoionization of a molecule *MX* can occur. If an optical source is available that generates photons having energies exceeding the ionization potential of *MX*, then the molecule is photoionized in a single step as illustrated in Fig. 20.17a. Since the ionization potentials of most atoms and molecules exceed 5 eV, single-photon ionization spectroscopy requires a laser in the vacuum ultraviolet (vuv; $100 \leq \lambda \leq 200$ nm) or extreme ultraviolet (xuv; $20 \leq \lambda < 100$ nm)—spectral regions in which few practical lasers exist at present.

A well-developed *incoherent* source that has proven useful in photoionization spectroscopy is the helium lamp, which produces strong resonance radiation at 58.4 nm ($h\nu = 21.2$ eV). Combining the lamp with an electron energy analyzer has been applied successfully to more than 200 molecules,[4] and the method and apparatus are known collectively as HeI PES.

The availability of commercial lasers in the visible and ultraviolet ($1.8 \leq h\nu < 6$ eV) that are capable of generating peak pulse intensities exceeding 50 MW has driven efforts over the last decade, in particular, to develop photoionization techniques requiring the absorption of several photons. Only a few of the possible schemes for the multiphoton ionization (MPI) of the molecule *MX* are illustrated in diagrams *b–d* of Fig. 20.17. Although Sec. 20.4 will discuss these processes further, it is clear from Fig. 20.17 that the electron energies produced by the various

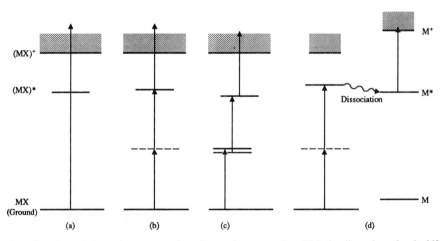

FIGURE 20.17 Schematic representation of several processes by which the diatomic molecule *MX* can be photoionized. In (*a*), *MX* is photoionized in a single step whereas in (*b*) to (*d*), the energy of available photons requires a multistep process involving intermediate states.

processes are dependent upon the photon energy and the locations (and symmetry) of the intermediate excited states of the electrically neutral molecule (denoted MX^*) or atomic fragment (M^*). A more detailed picture of the multistep ionization of MX is provided by Fig. 20.18, which shows qualitatively the distribution of energies expected in the electron spectrum. The discrete energies at which the peaks in the distribution appear reflect the vibrational state structure of the molecular ion ground state (MX^+). Figure 20.18 also suggests the flexibility the experimentalist has in selecting a specific intermediate state for excitation with one laser and, subsequently, photoionizing the molecule with a laser pulse of a different wavelength. Such "two-color" experiments enable one to probe both the excited states of MX, the structure of its ion state, and processes occurring in the continuum

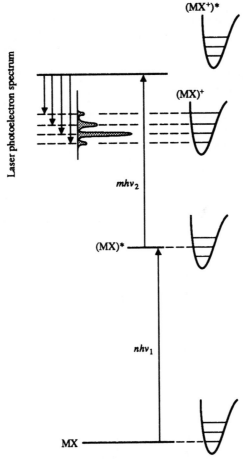

FIGURE 20.18 Generalized energy-level diagram for the heteronuclear molecule MX showing the energies of the electrons released by multiphoton ionization of the molecule. Note that m and n are integers. *(After Ref. 4.)*

in a detailed manner that is generally not accessible with fluorescence methods. Note, too, from Fig. 20.18, that, by decreasing the wavelength of the second laser (i.e., increasing $h\nu_2$), more energetic states of the molecular ion [such as $(MX^+)^*$ in Fig. 20.18] can be accessed.

A straightforward (and inexpensive!) approach to photoionization spectroscopy is to measure the time-integrated photoelectron (and ion) current with the simple diode configurations[5] depicted in Fig. 20.19. A tunable, and generally pulsed, laser beam is directed along the axis of an optical cell containing the gas or vapor of interest and two electrodes, one of which may be cylindrical. As the laser wavelength is tuned through a multiphoton resonance, free electrons are produced and collected by imposing a weak electric field on the region between the electrodes.

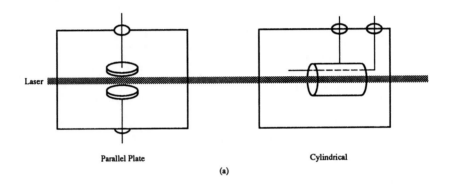

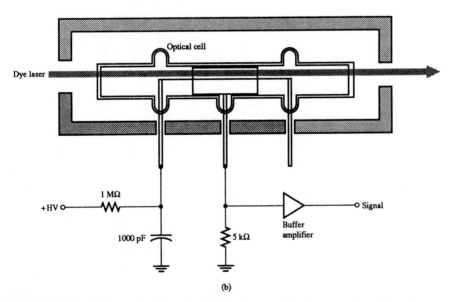

FIGURE 20.19 (*a*) Representative cell and electrode geometries for measuring relative electron yields in photoionization experiments. *(After Ref. 5, by permission.)* (*b*) Ionization cell employing the cylindrical configuration, showing a simple biasing and signal extraction circuit.

The resulting current in the external biasing circuit (refer to Fig. 20.19*b*) is a measure of the relative photoelectron yield.

More detailed information regarding the photoionization process can be obtained with an electron energy analyzer, of which there are two basic configurations: the spherical electrostatic and time-of-flight systems. The latter has the advantage of being able to record the entire energy spectrum with each laser pulse. A schematic diagram of the time-of-flight analyzer developed by Kruit and Read[6] is shown in Fig. 20.20. Photoelectrons produced by a focused laser beam in the region between the pole pieces of a 1-Tesla (T) magnet enter a 50-cm-long drift region through a small hole in one of the pole pieces. The 1-T magnetic field allows the spectrometer to collect electrons emitted into a solid angle of 2π steradians without disturbing the energy distribution. From the time required for the electrons to drift the length of the flight tube (measured by a multichannel detector), the *inital* energy distribution of the electrons—that is, the energies of the electrons "at birth"—can be determined. Part *b* of Figure 20.21 gives the electron energy spectrum observed when xenon is photoionized by the five-photon process shown in Fig. 20.21*a*. Since the Xe ion ground state has two spin-orbit split components ($^2P_{3/2}$ and $^2P_{1/2}$), two photoelectron peaks appear in the energy spectrum,[7] and the widths of the peaks are determined by the resolution of the instrument ($\sim$15 meV).

Photoelectron spectra of molecules are considerably more complex than their atomic counterparts, and most experiments have concentrated on the smallest diatomics. An example of the detail that can be gleaned from a molecule, given the ability to resolve photoelectron energies, is illustrated in Fig. 20.22. In two-color experiments reported by O'Halloran and coworkers,[8] a highly-excited state of H_2 was pumped by a two-photon process ($2\,h\nu_1$) and ionized with a third photon ($h\nu_2$). By recording the relative intensities of different peaks (i.e., at specific energies) in the photoelectron energy spectrum as the wavelength of the second laser was scanned, the elegant details of the photoionization process were examined closely.

As suggested by Fig. 20.18, photoionization is a complicated process in which several "exit channels" interact and each contributes to the total photoelectron

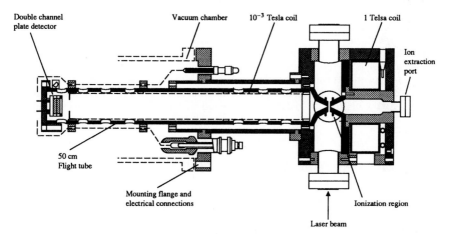

FIGURE 20.20 Schematic diagram of the "magnetic bottle" time-of-flight electron energy analyzer described in Ref. 6. *(Reprinted by permission.)*

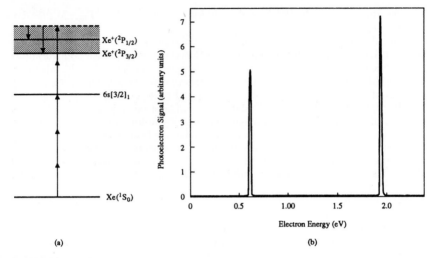

(a)   (b)

FIGURE 20.21 (a) Partial energy level diagram of xenon showing the five-photon ionization of the atom that was first reported in Ref. 7. Because the Xe ion ground state has two-spin-orbit split components, two groups of photoelectrons are expected in the electron energy distribution, as illustrated. (b) Photoelectron energy spectrum for the five-photon scheme of part a that was measured with the analyzer of Fig. 20.20.

yield. Consider, for example, the two spectra shown in Fig. 20.22 (reproduced from Ref. 8). The top spectrum was obtained by recording the intensity of a peak that correlates with the ultimate production of the H_2^+ ion in its third vibrational level ($v^+ = 2$). Similarly, the lower trace reflects the production of H_2^+ ($v^+ = 1$) ions as a function of the wavelength ($h\nu_2$) of the second laser. Peaks and valleys (suppressions) in the two spectra reveal the presence of Rydberg states that are accessed by the third (and final) photon, and result in the prompt ionization of the molecule because these states lie above the H_2^+ ($v^+ = 1$) limit. Note, however, the interesting coincidence of the dips in the $v^+ = 2$ spectrum with peaks in the $v^+ = 1$ spectrum. This phenomenon is a result of the competition between the two ionization processes (known as rotational and vibrational autoionization) that yield molecular ions in the $v^+ = 2$ and 1 states. Further details regarding these and similar experiments can be found in Refs. 4, 8, and 9, but the examples cited above demonstrate the new horizons in atomic and molecular physics that can be explored when narrow bandwidth lasers are combined with electron energy spectrometers.

20.4 MULTIPHOTON SPECTROSCOPY

Multiphoton processes are those involving the simultaneous absorption of two or more photons by an atom or molecule. Consider, for example, Fig. 20.23a, which shows two states $|i>$ and $|f>$ coupled by three photons having energies of $E/3$. Since no real energy levels of the species exist between initial and final states $|i>$ and $|f>$, respectively, the transition is said to be a nonresonant, three-photon

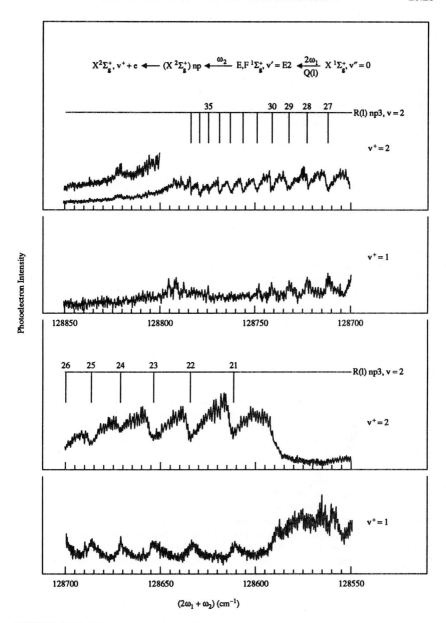

FIGURE 20.22 Photoelectron spectra of H_2 obtained by photoionizing the molecule in a two color experiment. The two spectra were recorded by observing separate exit channels in the photoelectron spectrum. *(Reprinted by permission from Ref. 8.)*

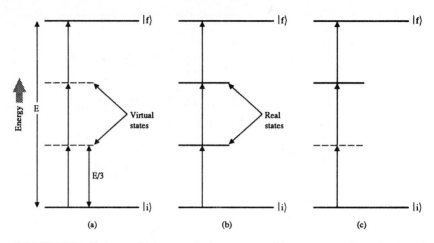

FIGURE 20.23 Various multiphoton excitation processes: (*a*) nonresonant, three-photon excitation via virtual states; (*b*) excitation by the sequential absorption of three photons; (*c*) two-photon resonant, three-photon excitation. For each of the three cases, the states $|i\rangle$ and $|f\rangle$ are of opposite parity.

excitation process. In general, the transition rate for an N-photon process is given by time-dependent perturbation theory as

$$R = \sigma^{(N)}\left(\frac{I}{h\nu}\right)^N \qquad (20.1)$$

where I is the optical intensity (expressed in W/cm^2) and $\sigma^{(N)}$ is the generalized multiphoton absorption or excitation cross section expressed in units of cm^{2N} s^{N-1} (i.e., cm^2 for $N = 1$, cm^4·s for $N = 2, \ldots$). Therefore, although much of spectroscopy entails driving optically allowed transitions of an atom by photons having energies equal to the separation between the two states in question, photons of considerably less energy are also capable of inducing such transitions.

While it is not necessary for the energies of the absorbed photons to be equal (as in Fig. 20.23*a*), the *sum* of the energies of the absorbed photons must equal the energy difference between the initial and final states. The implications of this are profound, since it, in effect, relaxes the stringent requirements that single-photon spectroscopy places on the photon energy. Multiphoton spectroscopy has proven to be a powerful tool in the hands of the physicist or chemist since: (1) high-lying electronic states can now be accessed with intense laser sources operating at longer wavelengths and (2) it provides information that complements conventional one-photon spectroscopy, particularly if N is even.[10] If N is even, states that cannot be populated by single-photon transitions because of selection rules are now accessible.

Because of the nonlinear dependence of the multiphoton transition rate on laser intensity, it is frequently advantageous to drive the process at the highest available intensities. For this reason, prior to the advent of high-intensity laser sources, multiphoton transitions were rarely observed. The last two decades, however, have witnessed the extensive development of a wide range of systems capable of driving nonlinear transitions in atomic and molecular systems. The conventional Nd:YAG-pumped dye laser, for example, routinely produces ~1-mJ, 10-ns pulses that can be focused to intensities exceeding 10 MW/cm^2—more than ample to observe a wide variety of multiphoton processes. Recent advances in the development of lasers emitting pulses as short as 10 fs have led to the generation of focal intensities beyond 10^{14} W/cm^2, thus opening entirely new realms of highly nonlinear multiphoton phenomena in liquid, gaseous, and solid media for investigation.

Multiphoton absorption which terminates at energies above the ionization limit is referred to as *multiphoton ionization* (MPI), and when an MPI sequence involves an intermediate resonance, the entire process is known as *resonantly enhanced, multiphoton ionization* (REMPI). Figure 20.21, discussed in Sec. 20.3 in connection with photoelectron spectroscopy, shows an example of REMPI of the Xe atom.

A more recent demonstration of the attractiveness of MPI for elucidating the structure of small molecules is the five-photon ionization of the iodine dimer (I$_2$) in the spectral region lying between about 567 and 587 nm. A portion of the MPI spectrum obtained by detecting the photoelectrons produced when the molecule is ionized is shown in Fig. 20.24. Despite the fact that molecular states lying more than 8 eV above ground are being probed, the signal-to-noise ratio of the data, a

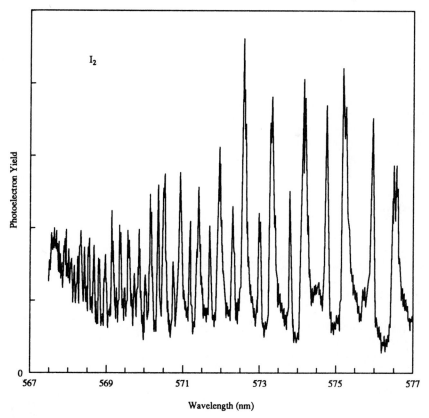

FIGURE 20.24 MPI spectrum of I$_2$ in the yellow (567.5 $\leq \lambda \leq$ 577 nm) obtained with the photoelectron spectrometer of Fig. 20.20.

direct result of the sensitivity of the photoelectron spectrometer, is greater than 10:1 over most of the spectrum. While it is convenient to detect MPI processes by the electron(s) ejected when the atom or molecule under study is ionized, other detection methods, such as monitoring the final state by the fluorescence it may produce, are usually also available. Time-of-flight *mass* spectrometers are also often used in MPI experiments to detect the product ions and often are an invaluable tool in identifying the species generated by the ionization process.

20.5 NONLINEAR LASER SPECTROSCOPY

20.5.1 Introduction

Nonlinear laser spectroscopy draws on the nonlinear response that many media exhibit toward intense optical fields. In particular, resonances that exist in the nonlinear components of the medium's susceptibility provide one with several powerful and versatile options for examining atomic or molecular structure with unprecedented resolution.

The response of a specific medium to incident laser radiation can be expressed in terms of its dielectric polarization, P. Expanded in a Taylor series involving the optical beam's electric field E, the induced polarization can be written

$$P_i = \sum_j \chi_{ij}^{(1)} E_j + \sum_{jk} \chi_{ijk}^{(2)} E_j E_k + \sum_{jkl} \chi_{ijkl}^{(3)} E_j E_k E_l + \cdots \qquad (20.2)$$

where the subscripts, $i, j, k, l, \ldots$, refer to a Cartesian coordinate system, and the coefficients χ are the susceptibility tensor elements of the medium. Note, too, that the summations are taken over all the spatial components of the electric field. The susceptibility tensor elements of nonlinear media are frequency-dependent, and it is usually more convenient to expand the polarization into a Fourier series:

$$P_i(t) = \sum_\infty \frac{1}{2} P_i(\omega) e^{-i\omega t} + cc \qquad (20.3)$$

which yields

$$
\begin{aligned}
P_i(\omega_3) = &\sum_j \chi_{ij}^{(1)} (-\omega_3, \omega_3) E_j(\omega_3) \\
&+ \sum_{jk} \chi_{ijk}^{(2)} (-\omega_3, \omega_2, \omega_1) E_j(\omega_2) E_k(\omega_1) \\
&+ \sum_{jkl} \chi_{ijkl}^{(3)} (-\omega_3, \omega_2, \omega_1, \omega_0) E_j(\omega_2) E_k(\omega_1) E_l(\omega_0) + \cdots \qquad (20.4)
\end{aligned}
$$

for the first three polarization terms in an isotropic medium.

The first term in Eq. (20.4) describes the linear or dipole response of the medium, and is dependent only on the frequency of the incident field, as is the case for conventional absorption spectroscopy (discussed in Sec. 20.2) for which the frequency of the probe laser $= \omega_3$. The second term represents processes such as second harmonic generation (SHG: $\omega_3 = \omega_1 + \omega_2$, $\omega_1 = \omega_2$), sum frequency generation (SFG: $\omega_3 = \omega_1 + \omega_2$) and difference frequency generation (DFG: $\omega_3 = \omega_1 - \omega_2$) that require the presence of *two* optical fields. In isotropic media with inversion symmetry, $\chi^{(2)}$ becomes zero, and all other even-order processes have zero susceptibility as well. This leaves the third-order susceptibility $\chi^{(3)}$ as the lowest nonzero, nonlinear component in isotropic media. The third-order susceptibility is responsible for processes such as two-photon absorption, four-wave mixing, third-harmonic generation, Raman effects, and self-focusing, several of which are depicted schematically in Fig. 20.25. Notice that many of the processes do not require resonance with the energy states of the medium.

In summary, nonlinear laser spectroscopic studies involve the use of a nonlinear interaction as a sensitive measurement of the properties of the medium being studied. Although a more complete treatment of the subject of nonlinear laser

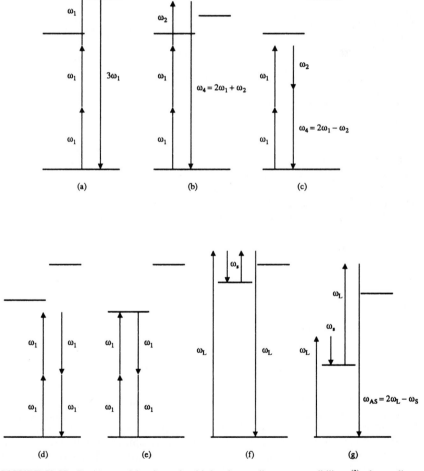

FIGURE 20.25 Processes arising from the third-order nonlinear susceptibility $\chi^{(3)}$ of a medium: (a) third-harmonic generation; (b) four-wave, sum-frequency mixing; (c) four-wave, difference-frequency mixing; (d) nonlinear index of refraction; (e) two-photon absorption; (f) stimulated Stokes Raman scattering; and (g) anti-Stokes Raman scattering. *(Adapted from Ref. 11.)*

spectroscopy can be found elsewhere,[11,12] the following sections describe a few of the most widely used spectroscopic techniques.

20.5.2 Coherent Anti-Stokes Raman Scattering (CARS)

The first experimental demonstration of CARS spectroscopy was reported by Maker and Terhune[13] in 1965. In the intervening years, CARS has been extensively applied to combustion diagnostics, nonlinear dispersion measurements in gases, liquids, and solids, the determination of Raman cross sections, and high-resolution spectroscopy of molecules. The technique is powerful because CARS signals are (as its name implies) coherent and thus easily distinguishable from incoherent radiation. CARS spectroscopy is a form of four-wave mixing and is also dependent on the third-order term of the susceptibility, $\chi^{(3)}$. Consequently, the phase-matching condition requires combining three frequencies: ω_1, ω_1, and ω_2, to yield a fourth wave at a frequency of $\omega_3 = 2\omega_1 - \omega_2$. Since ω_3 is equivalent to the anti-Stokes Raman frequency of the medium, CARS provides a convenient method for obtaining the Raman spectrum of the medium. The Raman active modes can be molecular electronic, vibrational, or rotational levels in gases or liquids, or phonon modes in solids. Figure 20.26a is a generalized energy-level diagram illustrating a typical CARS experiment, and part b of the figure shows four different CARS geometries which satisfy the phase-matching condition.[14] The simplest geometry, collinear, generally yields the most intense CARS signals since the interaction length between the beams is maximized. The collinear arrangement, however, is an inconvenient one for separating the CARS signal from the pump beams, and it is often necessary to include spectral filters (such as prisms, gratings, or dichroic mirrors) to discriminate the CARS signal from background radiation. Also, the collinear geometry significantly reduces the spatial resolution that can be a priority in several applications. For this reason, one of the noncollinear geometries is frequently chosen since the interaction length is now reduced to the region of overlap between the laser beams. For a noncollinear CARS experiment, the phase matching angles are calculated from the wave vector relation: $\Delta k = k_{3AS} - (2k_1 - k_2)$.

The CARS signal intensity is given by the expression:[15]

$$I_3 = C_1 I_1^2 I_2 \mid \chi^{(3)} \left(-\omega_{3AS}, \omega_1, \omega_1, -\omega_2 \right) \mid^2 N^2 L_z^2 \qquad (20.5)$$

where N = number density of the Raman medium
I_i = laser beam intensity at frequency ω_i
L_z = interaction length

and C_1 is a constant defined as

$$C_1 = \frac{24\pi^3}{\lambda_{AS} \, c \, n_{AS} n_1^2 n_2} \qquad (20.6)$$

where n_i is the refractive index at frequency ω_i. From Eq. (20.5), I_3 is proportional to the square of the nonlinear susceptibility, $\chi^{(3)}$, which is frequency-dependent and consists of both real and imaginary parts associated with the electronic structure of the molecules (or atoms) in the medium. The susceptibility itself can be written as

$$\chi^{(3)} = \frac{N \, C_2 \, \Delta j}{\omega_s^4} \left[\frac{\partial \sigma}{\partial \Omega} \right]_k \left[\frac{1}{2\delta\omega_k - i\Gamma_k} \right] + \chi_{nr} \qquad (20.7)$$

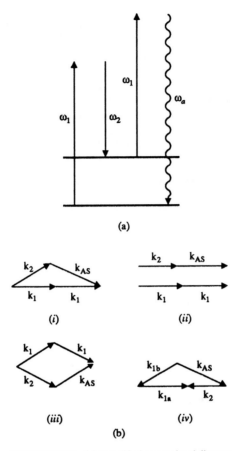

FIGURE 20.26 (*a*) Simplified energy-level diagram depicting the CARS process. *(After Ref. 14.)* (*b*) *k*-vector phase-matching conditions for several CARS schemes: (*i*) noncollinear, (*ii*) collinear, (*iii*) BOXCARS, and (*iv*) ARCS. *(After Ref. 11, by permission.)*

The first term on the right-hand side of Eq. (20.7) is the portion of the susceptibility that dominates near a Raman-active resonance. The second term, χ_{nr}, is the nonresonant contribution that manifests itself as a background superimposed onto the resonant signal. Occasionally, weak Raman resonances are obscured by χ_{nr} and its effect must be suppressed by polarization filtering techniques. For the generally strong resonant term, C_2 is a constant, ω_s is the Stokes frequency, N is again the medium number density, Δj is the temperature-dependent population difference factor, $\left(\dfrac{\partial\sigma}{\partial\Omega}\right)_k$ is the differential Raman scattering cross section, $\delta\omega_k$ is the detuning from the Raman transition resonance frequency [i.e., $\delta\omega_k = (\omega_1 - \omega_2) - \omega_{AS}$], Γ_k is the linewidth of the Raman transition, and k is the index identifying a specific transition. Determining the CARS spectrum temperature is often imperative and is extracted from the Boltzmann factor in Δj. Thus, critical parameters such as linewidth, temperature, and various molecular constants can be readily determined from analysis of CARS spectra.

A typical CARS experimental setup[16] for the study of gases in a free jet expansion is given in Fig. 20.27. A frequency-doubled, pulsed Nd:YAG laser, operating at 532 nm and a pulse repetition frequency of 10 Hz, serves as the source of ω_1. The third harmonic of the Nd:YAG laser at 355 nm pumps a dye laser which provides the second wave at ω_2. After combining ω_1 with ω_2 in the BOXCARS geometry (see Fig. 20.26b) with a focusing lens, the beams are focused onto the free jet gas expansion, and the emerging CARS wave, ω_3, is spatially filtered from ω_1 and ω_2 with a pinhole. Any scattered light at 532 nm can be removed spectrally from the CARS signal with an I_2 cell filter, and the CARS signal is detected by a photomultiplier (PMT) mounted onto a monochromator. The spectrum of the CARS signal can be acquired by scanning the monochromator or by tuning the dye laser. The primary limitation on the spectral resolution of a CARS spectrum is imposed by the linewidths of the lasers.

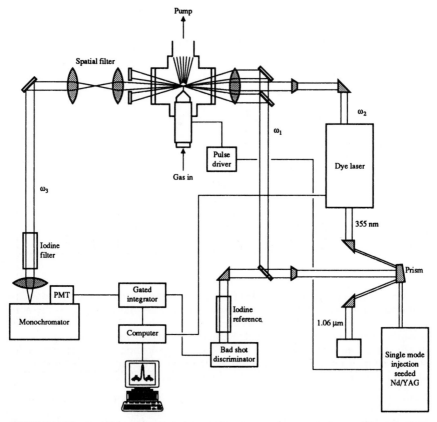

FIGURE 20.27 Experimental setup for the CARS spectroscopy of gases in a free jet expansion. The BOXCARS geometry is well suited for detecting a small Raman shift because it simplifies discriminating against undesirable background radiation. *(After Ref. 16.)*

A typical rotational spectrum for N_2 cooled in a free jet expansion is shown in Fig. 20.28. By replacing the jet expansion chamber with a simple Bunsen burner, the CARS spectrum of N_2 in a flame can be obtained, and the results are illustrated in Fig. 20.29 for various temperatures. This straightforward example demonstrates the value of CARS for *in-situ* diagnostics.

The application of CARS spectroscopy to *solids* has yielded the values of the third-order nonlinear susceptibilities in centrosymmetric crystals.[17] As illustrated in Fig. 20.30, a representative experiment for studying CARS in crystals involves intersecting two dye laser beams in the sample and, again, detecting the CARS signal with a photomultiplier. The dye lasers are scanned in such a manner as to vary $\omega_1 - \omega_2$ linearly while keeping $\omega_3 = 2\omega_1 - \omega_2$ constant. A sample CARS spectrum for BaF_2 is shown in Fig. 20.31. The difference in frequency between the observed maximum and minimum in the spectrum can be related to the effective third-order nonlinear susceptibility of the crystal and the full width at half maximum (FWHM) of the peak in the signal at $\omega_1 - \omega_2 \simeq 240$ cm^{-1} is proportional to twice the linewidth of the Raman transition.[17]

Several variations on the basic CARS theme, such as CW CARS, CARS with ultrafast lasers, and Raman-induced Kerr-effect spectroscopy (RIKES CARS) are

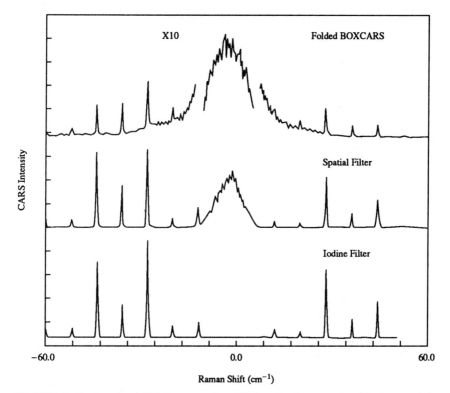

FIGURE 20.28 Rotational CARS spectra of neat N_2 in a free jet expansion. The large peak in the center is scatter that is gradually filtered by a combination of spatial and spectral approaches. *(After Ref. 16, reprinted by permission.)*

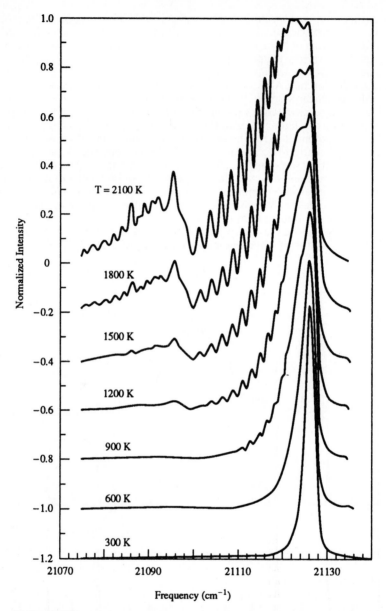

FIGURE 20.29 Rovibrational CARS spectra of N_2 in a flame observed as a function of temperature. At lower temperatures, only the $\Delta J = 0$, $(v', v'') = (1, 0)$ transition is observed but, at elevated temperatures, the (2, 1) hot band appears. *(After Ref. 15.)*

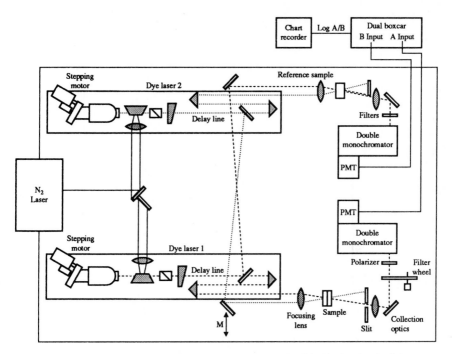

FIGURE 20.30 Experimental setup of CARS spectroscopy of solids. Notice that both lasers are tunable dye systems. *(Reprinted from Ref. 17 by permission.)*

all well-developed. For a more extensive treatment of CARS spectroscopy, the reader is referred to the literature cited.

20.5.3 Saturation Spectroscopy

Saturation spectroscopy capitalizes on the narrow linewidth of a laser in order to selectively study atoms or molecules having inhomogeneously broadened linewidths larger than that of the laser. Since only those atoms or molecules that are in resonance with the laser frequency are excited, a second laser is then able to probe the excited species in a manner that is free of inhomogeneous broadening.

The discovery of the *Lamb dip* by McFarlane, Bennett, and Lamb[18] in 1963 demonstrated that the hole-burning effect in the gain spectrum of a 1.15-μm He-Ne laser is a measure of the Doppler-free linewidth of the Ne transition. Similar Lamb-dip saturation experiments were performed by Freed and Javan[19] on several P- and R- branch transitions of a CO_2 laser at 10.6 μm. Figure 20.32 shows the Lamb dip in the P(20) line of the (001)-(000) transition of CO_2 that was obtained by adjusting the cavity length to tune the laser frequency and by monitoring the spontaneous emission signal at 4.3 μm.

To describe saturation spectroscopy, consider first the dynamics of a simple, two-level system pumped at resonance. For a gaseous medium, Doppler broadening is the dominant inhomogeneous broadening mechanism, and a laser tuned to a resonance pumps only those atoms or molecules whose velocity component v_z lies

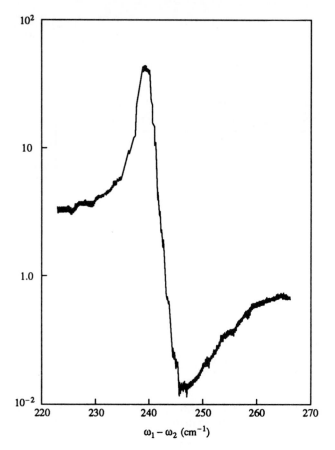

FIGURE 20.31 CARS spectrum of BaF_2 at 77 K. *(After Ref. 17.)*

along the direction of propagation of the laser beam. That is, only that fraction of the atoms within the Doppler-broadened linewidth whose resonant frequency is given by $\omega = \omega_{21} - kv_z$ (where ω_{21} is the transition frequency) are excited to the upper state. The result is a hole-burning effect in the velocity distribution of the gas at $v_z = (\omega - \omega_{21})/k$, where the atoms have been selectively removed from the velocity distribution. If a second beam of frequency ω' is counterpropagated with respect to the first laser (ω) for the purpose of probing those atoms that participate in the hole-burning process, the result is the formation of a second saturation hole at a frequency $\omega' = 2\omega_{21} - \omega$, provided the probe beam is weak compared to the pump. This second resonant dip (at ω') occurs because the hole created at $\omega' = \omega_{21} + kv_z$ by the probe beam correlates with the dip at $\omega = \omega_{21} - kv_z$ produced by the counterpropagating pump. In the weak saturation limit ($I/I_s \ll 1$), the absorption coefficient at ω' can be expressed as

$$\alpha(\omega') \approx \alpha_o(\omega') \left\{ 1 - \frac{2\Gamma^2 \, I/I_s}{[(\omega' - \omega) + 2(\omega - \omega_{21})]^2 + 4\,\Gamma^2} \right\} \qquad (20.8)$$

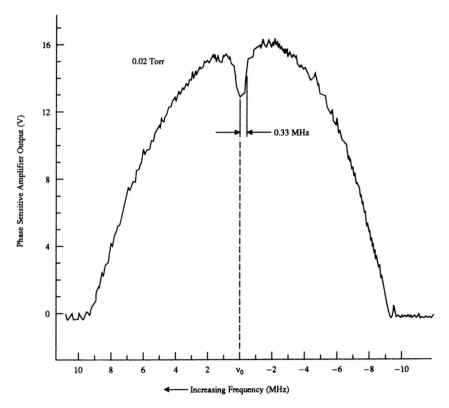

FIGURE 20.32 Lamb dip in the P(20) (001)-(000) transition in CO_2 at 20 mTorr. The transition occurs at 10.6 μm, and the signal is detected by monitoring spontaneous emission at 4.3 μm. Also, the CO_2 laser is tuned by dithering the mirrors with a piezoelectric crystal to adjust the cavity length. The spectral width (FWHM) of the dip is a result of pressure broadening but indicates a Doppler-free full-width of 660 kHz. *(After Ref. 19.)*

where $\alpha_o(\omega')$ is the absorption coefficient for the probe beam alone at $\omega' = \omega_{21} + kv_z$, $2\Gamma = $ FWHM of the homogeneous lineshape, and the probe beam intensity I is given by $I = c|E(\omega')|^2(2\pi n)^{-1}$. The saturation intensity I_s is expressed as

$$I_s = \frac{c\Gamma \ |E|^2}{8\pi n \ \Omega^2 \ T_1} \tag{20.9}$$

where T_1 is the relaxation time for the excited state population, n is the refractive index of the medium, and $\Omega = (2\pi/h) \ |\langle 1|er\cdot E(\omega')|2\rangle|$ is the Rabi frequency for the transition (for the pump tuned onto resonance). Notice that Eq. (20.8) reflects the fact that the change in the hole absorption coefficient (α) is proportional to the intensity I, or $|E|^2$, which is consistent with a third-order nonlinear optical effect. When $\omega' = 2\omega_{21} - \omega$, the halfwidth of the dip is Γ, the homogeneous halfwidth of the transition. Hence, saturation spectroscopy allows one to extract the homogeneous linewidth for the transition even though the system is inhomogeneously broadened. Other resonances (ω_{ij}) can also be probed with counterpropagating laser beams, provided that $\omega' = 2\omega_{ij} - \omega$. It must be emphasized that the above result holds only for cases where $I/I_s \ll 1$. Situations where the probe beam intensity is comparable to that for the pump beam require a treatment of coherence and atom-field effects, and the reader is referred to Ref. 20 for a complete mathematical treatment of the subject.

A typical experimental setup for saturation spectroscopy with counterpropagating beams[21,22] is shown schematically in Fig. 20.33. A laser beam is split into two components—which serve as the pump beam and a weaker probe—and are counterpropagated through the absorbing sample. Mechanically chopping the pump beam simplifies adjusting the experiment such that a signal arising from the probe beam will be observed at the detector only when the pump beam is present at the sample. Thus, the probe beam excites those molecules having the proper velocity component along the beam, $+v_z$, only when the pump beam excites those molecules having a velocity component of $-v_z$. With this technique and a narrow-linewidth dye laser, Hänsch, Shahin, and Schawlow[22] studied the fine structure of the hydrogen Balmer series in experiments in which a hydrogen discharge provided the absorbing medium. The application of saturation spectroscopy to resolving the fine structure of the H_α transition is illustrated in Fig. 20.34. The Doppler-broadened profile that would be measured otherwise is also shown.

The discussion above can be extended to a three-level system provided that the two transitions involved for the pump and probe beams share a common level. The probe beam must again be weak with respect to the pump intensity for the results discussed above to be valid. If the pump and probe beams are of the same frequency ($\omega' = \omega$), and yet the resonant frequencies differ ($\omega_{01} \neq \omega_{02}$), the Lamb

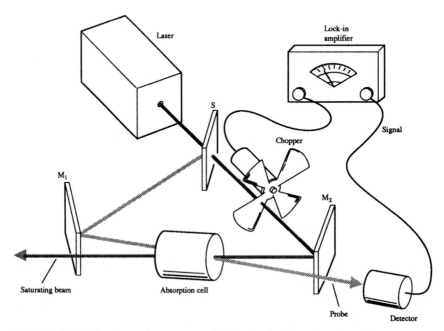

FIGURE 20.33 Experimental arrangement for Doppler-free saturation absorption spectroscopy with counterpropagating beams. *(After Ref. 21, by permission.)*

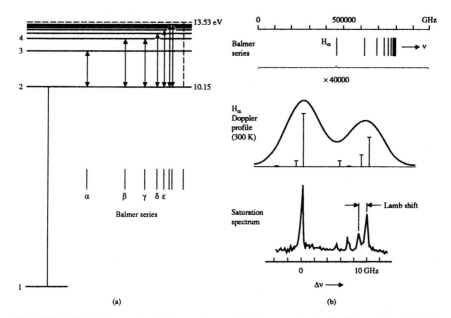

FIGURE 20.34   (*a*) Energy-level diagram for atomic hydrogen showing the Balmer lines; (*b*) Saturation spectrum of the H_α line in the red showing fine structure components that are normally not resolvable with other methods due to Doppler broadening.   *(After Ref. 22, reprinted by permission.)*

dip will occur at $\omega = \frac{1}{2}(\omega_{01} + \omega_{02})$. An example of multilevel saturation spectroscopy is given in Fig. 20.35, which shows the hyperfine structure of the ($6s6p$ $^2D^0_{3/2}$ ← $6s^2$ $^2D_{3/2}$) resonance transition of lutetium (Lu) observed by resonance ionization mass spectrometry (RIMS) saturation spectroscopy.[23] The two obvious Lamb dips arise from conventional two-level, Doppler-free saturation but the local maximum near 22,124.45 cm^{-1} is a result of three-level saturation. Known as a *crossover resonance*, this peak occurs exactly halfway between the normal Lamb dips [i.e., $\omega = \frac{1}{2}(\omega_{01} + \omega_{02})$]. An energy-level diagram of the hyperfine components of lutetium is given in Fig. 20.36. The Lamb dips of Fig. 20.35 are correlated with the $(F', F'') = (5, 4)$ and $(5, 5)$ hyperfine transitions, and the crossover resonance associated with both the $(5, 4)$ and $(5, 5)$ transitions lies at $\omega = \frac{1}{2}(\omega_{45} + \omega_{55})$.

20.5.4   Coherent Transient Spectroscopy

Optical Nutation.   When a two-level system is resonantly excited with a laser pulse, the population difference between the two levels becomes coupled to the frequency of the electric field produced by the laser. This characteristic frequency, known as the *Rabi frequency*, is given by the expression

$$\Omega^* = \left[(\omega - \omega_0)^2 + \left(\frac{\gamma}{\hbar} 2\, E \right)^2 \right]^{1/2} \tag{20.10}$$

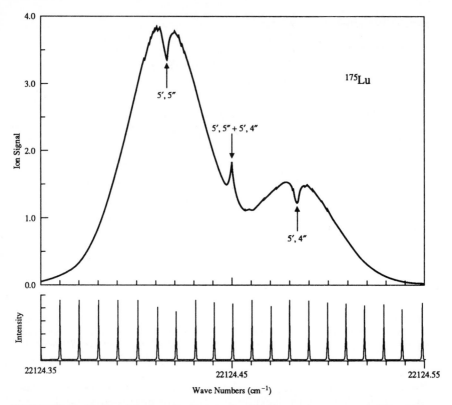

FIGURE 20.35   Portion of the hyperfine Doppler-free RIMS saturation spectrum in ^{175}Lu. The transition of interest is the $6s6p$ $^2D^0_{3/2} \leftarrow 6s^2$ $^2D_{3/2}$ resonance line. The arrows indicate the Doppler-free hyperfine transitions which have linewidths of approximately 100 MHz. The Lamb dips are two-level saturation transitions, whereas the narrow peak corresponds to a three-level crossover feature. The Doppler width is 1.5 GHz. *(After Ref. 23, reprinted by permission.)*

where $(\omega - \omega_0)$ is the detuning from resonance, $\gamma = \langle + |e \cdot r| - \rangle = \langle - |e \cdot r| + \rangle$ is the dipole matrix element, and E is the amplitude of the electric field from a circularly polarized laser source. The optical nutation signal is usually detected by observing the intensity modulation of the transmitted laser beam. The temporal behavior of the oscillations in the optical nutation signal can be understood by considering an ensemble of coherently excited dipoles interacting with the laser at the Rabi frequency. Relaxation mechanisms in the medium dampen the oscillations and the decay of the optical nutation signal is described by a characteristic time T_2, which is known as the *dephasing* or *transverse relaxation* time. To observe optical nutation, the Rabi frequency must be greater than the characteristic frequency of the decay, T_2^{-1}, and in an homogeneously broadened system, T_2 is inversely proportional to the FWHM of the transition. Therefore, from the observed oscillation frequency, one can determine the Rabi frequency, and thus the value of the matrix element γ.

A representative optical nutation signal[24,25] for methyl fluoride (CH_3F) is shown at the top of Fig. 20.37. To obtain this waveform, methyl fluoride molecules are

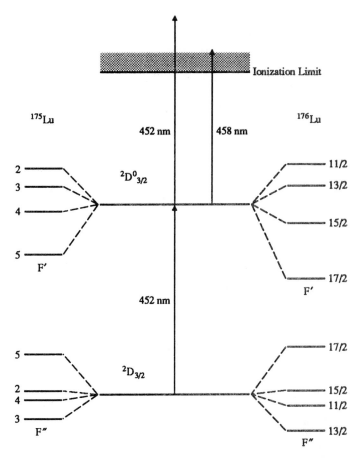

FIGURE 20.36 Partial energy-level diagram for ^{175}Lu and ^{176}Lu, also showing the atomic hyperfine states. The laser wavelengths employed in obtaining the RIMS saturation spectrum of Fig. 20.35 are also indicated. *(After Ref. 23.)*

switched into resonance with the frequency of a cw dye laser by applying a dc Stark field that is controlled externally by a pulse generator. The field is applied to decouple those molecules from an *inhomogeneously* broadened transition that were in resonance with the laser *before* the field was applied. Once the field is present, a new set of molecules comes into resonance with the laser, and the optical nutation signal appears. Interference between the radiation from the ensemble of the dipoles in the medium and the transmitted laser beam produces the oscillatory signal having a period of Ω^{*-1} and a decay constant determined by T_2.

Free Induction Decay. Let us now consider a system that has come into equilibrium after a resonant laser field has been applied. If the laser's optical field is suddenly removed, the system does not immediately return to thermal equilibrium. Rather, the medium continues to radiate a coherent optical field that weakens with time as the system dephases. The observation of the temporal decay of the co-

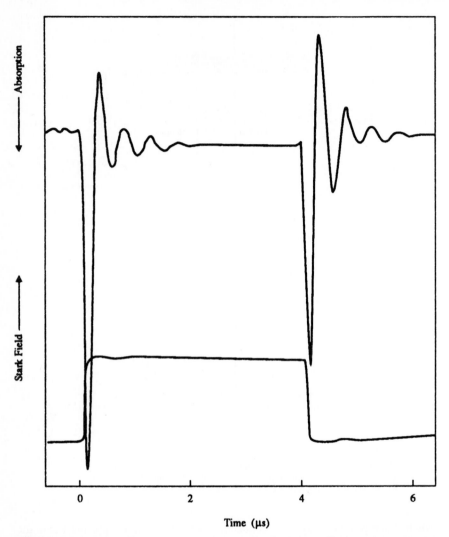

FIGURE 20.37 Optical nutation signal (upper trace) in methyl fluoride ($^{13}CH_3F$) pumped by a CO_2 laser at 9.7 μm. The oscillation frequency is a measure of the Rabi frequency, and the time constant for the decay of the oscillations is T_2^{-1}. The temporal behavior of the applied dc Stark field is shown by the lower trace. *(After Refs. 24 and 25—see text.)*

herence signal is known as free-induction decay (FID). If the exciting laser light has a linewidth narrower than T_2^{-1}, then the FID intensity decays with a characteristic time constant given by

$$\tau_{FID}^{-1} = T_2^{-1} + \Gamma_p \qquad (20.11)$$

where T_2 is the dephasing time and Γ_p is the contribution to the overall linewidth due to power broadening, where

$$\Gamma_p = T_2 (1 + (\gamma/h)^2 \, 4 \, E^2 \, T_1 \, T_2)^{1/2}. \qquad (20.12)$$

As before, γ is the dipole matrix element, E is the electric field amplitude of the laser, and T_1 is the population difference or longitudinal relaxation time.

An FID signal generated in molecular I_2 is depicted in Fig. 20.38. Experimentally,[26] a cw dye laser is tuned to the $(v, J) = (15, 60) \leftarrow (2,59)$ transition of the $B^3\Pi_{0u}{}^+ \leftarrow X^1\Sigma_g^+$ band of I_2 in the visible. A pulsed electro-optical modulator, incorporating an ammonium dideuterium phosphate (AD*P) crystal, is inserted in the beam path to switch the laser frequency in and out of resonance with the transition. When the laser is switched off resonance, the resulting FID signal is detected by measuring the absorption of the laser beam through the I_2 cell. The oscillation frequency of the FID signal of Fig. 20.38 is 32.83 MHz and the decay

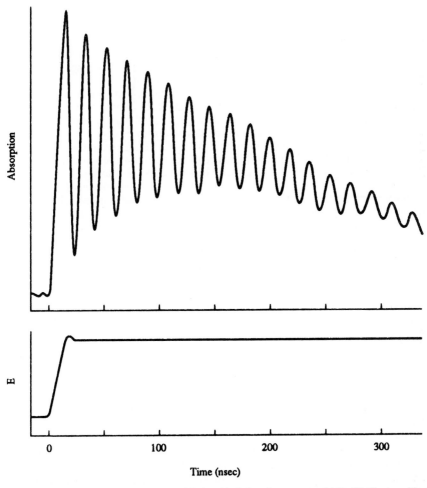

Time (nsec)

FIGURE 20.38 Free induction decay (FID) signal in I_2 for a laser power of 1.9 mW. The transition of the molecule being excited is $(v,J) = (15,60) \leftarrow (2,59)$ of the $B^3\Pi_{0u}{}^+ \leftarrow X^1\Sigma_g^+$ band which lies at 16,956.43 cm^{-1}. The FID decay time (extrapolated as a function of laser power) yields $\tau_{\text{FID}} = T_2/2 = 243$ nsec, where T_2 is the dephasing time. *(After Ref. 26, reprinted with permission.)*

time of the waveform depends on laser power in the manner specified by Eqs. (20.11) and (20.12). In the low power limit, the decay time of the FID signal is related to the dephasing time T_2 by $\tau_{FID} = T_2/2$. Therefore, in the (extrapolated) zero power limit, $\tau_{FID} = 243$ ns and $T_2 = 486$ ns.

20.6 PHOTOCHEMISTRY

It has long been recognized that optical radiation is capable of initiating chemical reactions, particularly in the gas and liquid phases. Since the strengths of chemical bonds in many molecules are less than 5 to 6 eV, the absorption of a single photon of the proper visible or uv wavelength has sufficient energy to rupture one or more bonds within the "precursor" molecule, thus freeing the desired atom or molecular radical. Through the careful choice of laser wavelength and precursor, specific products can often be produced. In addition to such *photodissociation* processes—that is, photon-induced fragmentation of a molecule—other mechanisms, such as photoionization, can be exploited to address a particular species against a background of other atoms or molecules that may be present.

In short, laser photochemistry is built on the premise that specific reactions can be initiated at the expense of other reaction channels or desired chemical products produced by matching the laser wavelength(s) to a combination of reactants. The introduction of photons provides an additional, external degree of freedom in controlling reactions and enables one to drive a reaction far from thermal equilibrium. Being able to utilize photochemistry to the full, however, presumes an understanding of the processes that ensue when a photon is absorbed by one of the constituents in a mixture of reactants. Furthermore, since the distribution and identity of products of photochemical reactions change with the wavelength, it is necessary to explore the chemistry in detail if one wishes to optimize the formation of a particular product. Seldom is this information available and much of the versatility of photochemistry remains to be explored. Nevertheless, the large photon fluences (photons/cm²·s) and narrow bandwidths available with lasers has accelerated efforts to develop efficient, laser-driven chemical processes suitable for commercial production. This section briefly describes two examples of laser-initiated reactions that illustrate the potential specificity of photochemical reactions—namely, photochemical vapor deposition and the synthesis of vitamin D. Also, the application of photochemistry in the gas and liquid phases to purification and elemental separation are briefly discussed.

20.6.1 Photochemical Vapor Deposition

In 1932, Romeyn and Noyes[27] reported " . . . the photochemical decomposition of germane by radiation transmitted by thin layers of quartz" and observed deposition of germanium when germane vapor was illuminated with uv photons from a mercury arc or hydrogen lamp. Since the late 1970s, more than 20 elemental and compound films have been deposited (or etched) by the photodissociation of polyatomic molecules. While lamps provide sufficient optical intensities to allow several elements to be deposited, most films have required either a visible or uv laser. One well-studied example is aluminum. The alkyl trimethylaluminum $(Al(CH_3)_3)$ is absorbing for wavelengths below about 255 nm and, for $230 \leq \lambda \leq 255$ nm, a single photon has sufficient energy to free an aluminum atom:

$$Al(CH_3)_3 + h\nu \ (\geq 4.9 \ eV) \longrightarrow Al(CH_3)_2 + CH_3 \longrightarrow Al + C_2H_6 + CH_3.$$

$$(20.13)$$

Below 230 nm, $AlCH_3$ radicals, rather than aluminum atoms, are the predominant product which ultimately results in the incorporation of carbon into the metal film. This serves as a simple example of the influence of the optical source wavelength on the reaction products and film quality. Photodissociating trimethylaluminum with a KrF excimer laser ($\lambda = 248$ nm) has been found to be capable of depositing Al films more than 0.5 μm in thickness at deposition rates up to 1000 Å/min. The resistivities of films deposited from both $Al(CH_3)_3$ and trimethylaluminum hydride (TMAH) compare favorably with bulk values and show the films to be suitable as ohmic contacts and interconnects.

Similarly, silicon thin films have been deposited on a variety of substrates by photodissociating the silanes SiH_4, Si_2H_6, or Si_3H_8 with an excimer laser (typically ArF, 193 nm). Growth rates generally range from 100 to 300 Å/min and the film properties are comparable to those for films deposited by plasma processes. With care given to surface preparation, gas mixture flow, and substrate temperature, epitaxial (crystalline) films have also been grown at temperatures 200 to 400°C *lower* than those normally required by existing (thermal) deposition processes. Deposition of compound thin films, such as the dielectrics SiO_2 or Si_3N_4, is accomplished by irradiating mixtures of disilane (Si_2H_6) and N_2O (or O_2) or NH_3, respectively, with vuv photons from a lamp or laser. Silicon dioxide growth rates exceeding 1000 Å/min have been demonstrated at temperatures below 300°C, and the quality of photochemically deposited films with respect to pinhole density and dielectric strength is comparable or superior to films deposited by other processes.

A similar approach has succeeded in etching semiconductor materials such as Si and GaAs, except, in this case, photodissociation is employed to free a chemically *reactive* species which results in the removal of surface atoms. Photodissociating CH_3Br in the uv, for example, produces free bromine atoms which will effectively etch both metals and semiconductors. Although these processes are still in the early developmental stages, several specific reactions have already been implemented in integrated circuit (IC) mask repair.

20.6.2 Vitamin D Synthesis

Vitamin D is produced naturally in humans by the interaction of sunlight with cholesterol in the skin.[28] It has long been known that a synthetic analog to this process is the single-step photochemical conversion of 7-dehydrocholesterol (7-DHC) to previtamin D (P). As illustrated in Fig. 20.39, the isomers 7-DHC, lumisterol (L) and tachysterol (T), as well as vitamin D itself (D), all absorb throughout the ultraviolet. Direct conversion of 7-DHC to previtamin D occurs when 7-DHC absorbs a single, near-uv ($\lambda \sim 300$ nm) photon (see Fig. 20.40a). This process has been developed commercially with ~330-nm radiation from mercury arc lamps, but the yield of previtamin D is marginal—in the 20 to 50 percent range. Hackett and coworkers[28,29] have demonstrated that irradiating 7-DHC with two separate wavelengths improves the previtamin D yield considerably. When 7-DHC absorbs a 5-eV photon, for example, the predominant product is tachysterol, rather than previtamin D, and Fig. 20.39 shows that the tachysterol absorption

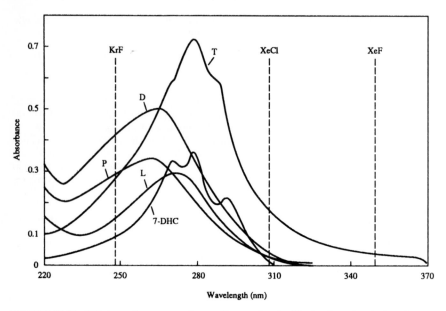

FIGURE 20.39 UV absorption spectra of several isomers involved in the photochemical production of vitamin D. The positions of several excimer laser wavelengths are also indicated. *(Reprinted from Ref. 28, by permission.)*

spectrum peaks at longer wavelengths (i.e., to the red) than do those for all the other molecules in the synthesis chain. Tachysterol can subsequently be transformed into previtamin D by illuminating the molecule with $\lambda \sim 350$ nm photons. In the experiments of Refs. 28 and 29, the combination of KrF (248 nm) and N_2 (337 nm) lasers produced previtamin D yields of about 80 percent, or roughly double those representative of the more conventional approach. Therefore, by the use of two or more separate optical wavelengths, the chemistry of a synthetic chain can be tailored to favor the formation of a particular product.

20.6.3 Purification and Elemental Separation

Other processes have received considerable attention for possible commercial applications, one of which is purification. An example is the purification of 2,4,5-trichlorophenoxyacetic acid (2,4,5-T) containing trace amounts of 2,3,7,8-tetrachlorodibenzo-p-dioxin (TCDD).[30] Figure 20.41 shows the ultraviolet absorption spectra for the two molecules and suggests that, for wavelengths beyond roughly 300 nm, photolysis of the dioxin could be accomplished while affecting 2,4,5-T in only a minimal way. Table 20.1 (after Ref. 30) substantiates this premise and shows that the spectral region around 320 nm is the most effective in terms of selectively decomposing TCDD. At longer wavelengths, TCDD is photodissociated (although its absorption cross section is falling rapidly) but an undesirable side effect is that a growing fraction of the 2,4,5-T is also decomposed.

Silane (SiH_4) has also been purified by Clark and Anderson[31] who showed that, since impurities such as PH_3 and B_2H_6 absorb more strongly at $\lambda = 193$ nm than

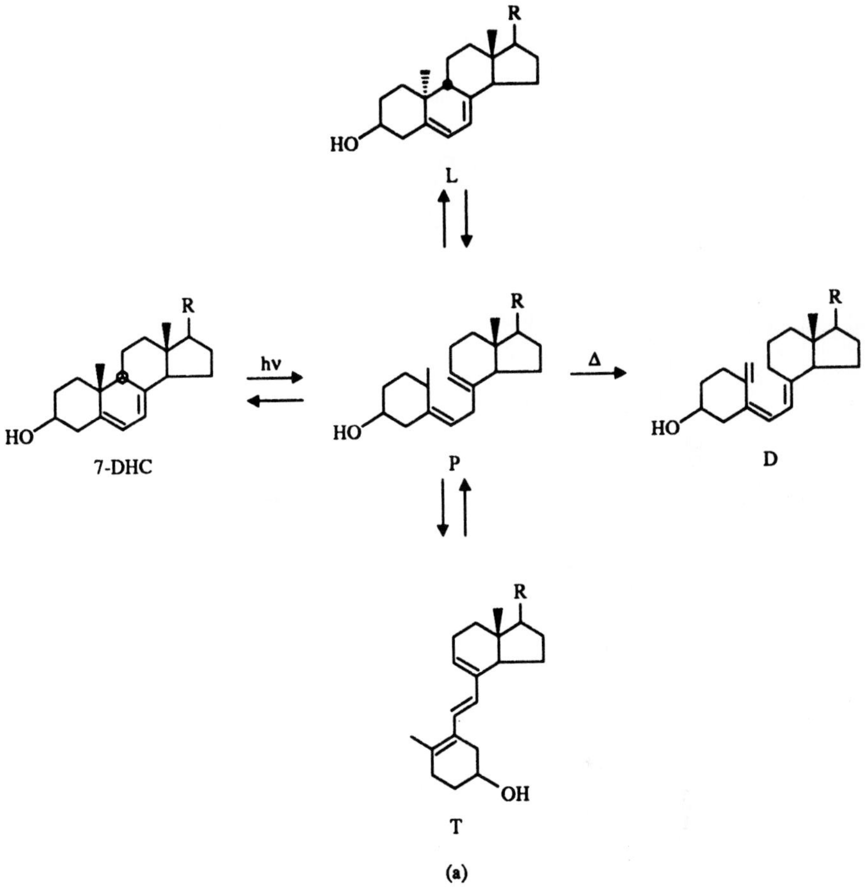

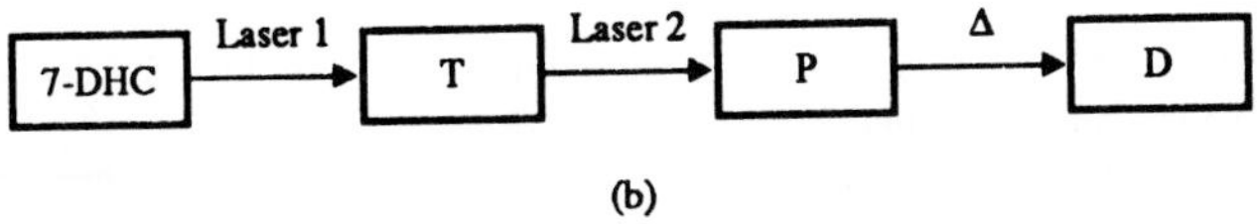

FIGURE 20.40 (*a*) Reactions involved in the conventional synthesis of vitamin D (denoted D) with a lamp as the optical source. (*b*) Simplified block diagram of the two-laser synthesis scheme for producing vitamin D with higher yields; the first laser is typically KrF (248 nm: $h\nu = 5.0$ eV) and the second is N_2 (337 nm: $h\nu \approx 3.7$ eV). *(After Ref. 28.)*

does SiH_4, more than 40 percent of contaminant PH_3 and B_2H_6 and 99 percent of AsH_3 molecules were photolyzed in a 100:1 (SiH_4:contaminant) mixture. In contrast, less than 6 percent of the SiH_4 was affected by the presence of the vuv laser beam. Also, the number of SiH_4 molecules photolyzed per contaminant molecule removed from the gas mixture ranged between 1.1 and 1.3.

Donohue[32] has noted that "the vast majority of conventional industrial processes involve the liquid phase," and, while photochemistry in liquids has been less ex-

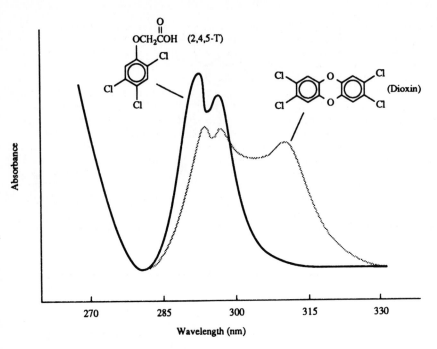

FIGURE 20.41 The ultraviolet absorption spectra of TCDD (dashed curve) and 2,4,5-T. *(After Ref. 30.)*

tensively studied than gas-phase processes, one prominent success in the former category is the photostimulated separation of the rare-earth elements in solution. The lanthanides are critical to a variety of scientific, technological, and commercial applications (including samarium in magnets and europium in color television picture tubes), and separating the elements from one another in naturally occurring

TABLE 20.1 Data Summarizing the Photolysis of TCDD in 2,4,5-T at Various UV Wavelengths

Optical source (λ, nm)	Power, W	Time, h	2,4,5-T, %	TCDD, ppm
None			100	0.058
XeCl (308)	0.1	2	101	0.052
XeCl (308)	0.05	4	102	0.048
N$_2$ (337)	0.05	4	96	0.057
Hg arc (360)	400	6	92	0.007†
Hg arc (360)	400	20.00*	76	†
Hg arc (360)	400	7.25*	39	0.033‡

*CH$_3$OH solvent
†Init. = 0.033
‡Init. = 0.390
Source: After Ref. 30.

mixtures is generally a challenging and expensive proposition. Donohue, however, has successfully demonstrated photochemical schemes for removing Eu and Ce from solutions containing mixtures of lanthanide elements. Europium, for example, has been photoreduced by irradiating artificially prepared aqueous or alcohol solutions with ultraviolet radiation from an excimer laser (193 or 248 nm) or a low-pressure mercury lamp.[32] The result is that Eu precipitates out of the solution as $EuSO_4$ (aqueous) or $EuCl_2$ (alcohol solutions). The efficiency of the separation process is described by Φ, the quantum yield for the photochemical process in the overall separation scheme, and β, the separation factor. The latter is a measure of the degree to which the desired product is enriched relative to its concentration in the reactant mixture. For the Eu separation experiments described in Ref. 32 (methanol solutions), the photochemical yield ranged from 50 to 60 percent and values of β greater than 1000 were obtained.

Similarly, cerium has been separated from binary Ce/lanthanide solutions by photo-oxidation of the Ce ion, again followed by precipitation and solvent extraction. Figure 20.42 shows the values for the separation factor that were obtained

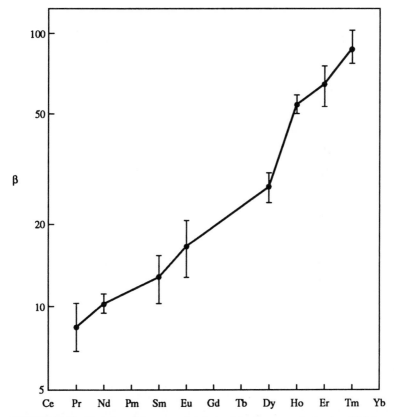

FIGURE 20.42   Variation of the separation factor β for the removal of cerium (Ce) from binary solutions containing Ce and one of the above rare earths. The optical source was an XeCl (308-nm) laser, the solvent was dilute HCl, and Ce precipitated out as ceric iodate.   *(After Ref. 32, reprinted by permission.)*

for eight different Ce/rare earth solutions when an XeCl excimer laser (308 nm) was the optical source. Cerium precipitated as ceric iodate. Note that β increases rapidly with the mass of the *other* rare earth.

20.6.4 Summary

The foregoing applications of photochemistry are diverse examples of the ability of optical radiation to drive a desired reaction. As laser-based studies of fundamental reaction kinetics and product yields for a wide range of chemical processes continue, more reactions having commercial potential will undoubtedly be developed.

20.7 CONCLUDING COMMENTS

Space limitations have precluded a comprehensive discussion of many powerful, but less widespread, laser spectroscopic techniques that have been developed in recent years, and the reader is encouraged to consult the references for more details concerning a particular method of interest. Photoacoustic, laser-induced breakdown, stimulated emission pumping, and optical-optical double resonance (OODR) spectroscopy are examples of the myriad of techniques that have been effectively applied to elucidating atomic and molecular structure.[5,9,33] Their exclusion here is in no way intended to reflect their importance to spectroscopy. The spectroscopic techniques described in this chapter are, however, representative of the scope of laser diagnostic methods presently available and are clearly complementary in the sense that no one approach provides a complete view of the structure of the atom or molecule under study and a careful choice of a combination of techniques is generally required to realistically assess its optical and chemical properties.

Probing the structure of atoms and molecules and initiating chemical reactions with lasers have become ever more sophisticated in the past 10 to 15 years as the tools available to the spectroscopist and photochemist have improved. Narrower laser linewidths, increased pulse intensities, and newly developed uv and vuv lasers continue to deepen our insight into the fundamental nature of matter and to suggest avenues for applying that heightened understanding to new or more efficient chemical syntheses. Furthermore, there is no reason to expect that this rapidly evolving and exciting period is in its twilight. On the contrary, it appears that several recent developments, including the advent of lasers in the x-ray region, will give birth to entirely new fields of photophysics and spectroscopy. Also, the introduction of laser-driven chemistry to commercial applications will likely accelerate as the engineering of ultraviolet lasers approaches the level enjoyed by older systems such as the infrared CO_2 system.

20.8 ACKNOWLEDGMENTS

The authors thank K. V. Reddy and T. Donohue of Amoco Technology and P. M. Dehmer of the Argonne National Laboratory for valuable discussions. We also wish to express our gratitude to K. K. King of Ohio State University for supplying the indium monoiodide spectra, to H. Tran for the $Er:YLiF_4$ absorption data, and to Y. Gu for the I_2 photoelectron spectrum. Support for this work was provided by the Air Force Office of Scientific Research and the National Science Foundation.

20.9 REFERENCES

1. Nassau, K., *The Physics and Chemistry of Color*, p. 211, Wiley, New York, 1983.

2. For a well-written overview of this technique, see D. R. Crosley, *J. Chem. Education*, vol. 59, pp. 446–455, 1982.

3. Ediger, M. N., and J. G. Eden, *J. Chem. Phys.*, vol. 85, pp. 1757–1769, 1986.

4. Kimura, K., in *Photodissociation and Photoionization*, pp. 161–199, K. P. Lawley, ed., Wiley, New York, 1985.

5. Parker, D. H., in *Ultrasensitive Laser Spectroscopy*, pp. 233–309, D. S. Kliger, ed., Academic, New York, 1983.

6. Kruit, P., and F. H. Read, *J. Phys. E: Scientific Instruments*, vol. 16, pp. 313–324, 1983.

7. Compton, R. N., J. C. Miller, A. E. Carter, and P. Kruit, *Chem. Phys. Lett.*, vol. 71. pp. 87–90, 1980.

8. O'Halloran, M. A., S. T. Pratt, F. S. Tomkins, J. L. Dehmer, and P. M. Dehmer, *Chem. Phys. Lett.*, vol. 146, pp. 291–296, 1988.

9. Svanberg, S., *Atomic and Molecular Spectroscopy*, Springer, New York, 1991.

10. For more details, the reader is referred to Ref. 9 and R. Bruzzese, A. Sasso, and S. Solimeno, "Multiphoton Excitation and Ionization of Atoms and Molecules," *La Rivista del Nuovo Cimento*, vol. 12, pp. 1–105, 1989.

11. Reintjes, J. F., *Nonlinear Optical Parametric Processes in Liquids and Gases*, Academic, Orlando, 1984.

12. Levenson, M. D., *Introduction to Nonlinear Laser Spectroscopy*, Academic, New York, 1982.

13. Maker, P. D., and R. W. Terhune, *Phys. Rev.*, vol. 137, pp. A801–818, 1965.

14. Shen, Y. R., *The Principles of Nonlinear Optics*, Wiley, New York, 1984.

15. Vewrdieck, J. F., R. J. Hall, J. A. Shirley, and A. C. Eckbreth, *J. Chem. Education*, vol. 59, p. 495–503, 1982.

16. Nibler, J. W., in *Applied Laser Spectroscopy*, pp. 313–328, W. Demtröder and M. Inguscio, eds., Plenum, New York, 1990.

17. Levenson, M. D., and N. Bloembergen, *Phys. Rev. B.*, vol. B10, pp. 4447–4463, 1974.

18. McFarlane, R. A., W. R. Bennett, and W. E. Lamb, *Appl. Phys. Lett.*, vol. 2, pp. 189–190, 1963.

19. Freed, C., and A. Javan, *Appl. Phys. Lett.*, vol. 17, pp. 53–56, 1970.

20. Letokhov, V. S., and V. P. Chebotayev, *Nonlinear Laser Spectroscopy*, Springer-Verlag, Berlin, 1977.

21. Schawlow, A. L., *Rev. Mod. Phys.*, vol. 54, pp. 697–707, 1982.

22. Hänsch, T. W., I. S. Shahin, and A. L. Schawlow, *Nature*, vol. 235, pp. 63–65, 1972.

23. Feary, B. L., D. C. Parent, R. A. Keller, and C. M. Miller, *J. Opt. Soc. Am.*, vol. B7, pp. 3–8, 1990.

24. Brewer, R. G., R. L. Shoemaker, and S. Stenholm, *Phys. Rev. Lett.*, vol. 33, pp. 63–65, 1974.

25. Brewer, R. G., *Phys. Today*, vol. 30, pp. 50–59, 1977.

26. Genack, A. Z., and R. G. Brewer, *Phys. Rev. A*, vol. 17, pp. 1463–1473, 1978.

27. Romeyn, H., Jr., and W. A. Noyes, Jr., *J. Am. Chem. Soc.*, vol. 54, pp. 4143–4154, 1932.

28. Hackett, P. A., C. Willis, M. Gauthier, and A. J. Alcock, *Proc. SPIE.*, vol. 458, pp. 65–73, 1984.

29. Malatesta, V., C. Willis, and P. A. Hackett, *J. Am. Chem. Soc.*, vol. 103, pp. 6781–6783, 1981.

30. Perettie, D. J., S. M. Khan, J. B. Clark, and J. M. Grzybowski in *Laser Applications in Chemistry*, pp. 245–249, K. L. Kompa and J. Wanner, eds., Plenum, New York, 1984.

31. Clark, J. H., and R. G. Anderson, *Appl. Phys. Lett.*, vol. 32, pp. 46–49, 1978.

32. Donohue, T., "Applied Laser Photochemistry in the Liquid Phase" in *Laser Applications in Physical Chemistry*, pp. 89–172, D. K. Evans, ed., Marcel Dekker, New York, 1989.

33. Steinfeld, J. I., *Molecules and Radiation*, 2d ed., MIT Press, Cambridge, Mass., 1985.

CHAPTER 21
FIBER-OPTIC SENSORS

Charles M. Davis and Clarence J. Zarobila

21.1 INTRODUCTION

Various transduction mechanisms using optical fibers to sense physical parameters have been demonstrated. The mechanisms discussed in this chapter exploit the modulation of optical intensity, phase, and state of polarization. The light may be modulated either inside or outside the fiber. With the exception of microbending, amplitude modulation is performed by some means outside the fiber. Alternatively, phase and polarization modulation generally are performed within the fiber.

All of these transduction approaches may employ multimode or single-mode components. However, intensity modulation usually employs multimode components while phase modulation typically uses single-mode components. As such, intensity-modulated sensors tend to be less expensive and less sensitive.

In Sec. 21.2, various transduction methods are discussed. A description of the optical components is given in Sec. 21.3. Techniques for fabricating temperature sensors are given in Sec. 21.4. Static and dynamic pressure sensors are considered in Sec. 21.5. Accelerometers and rate-of-rotation sensors are discussed in Secs. 21.6 and 21.7, respectively. Finally, electric/magnetic field sensors are described in Sec. 21.8. References are included in Sec. 21.9. The approach taken below is to describe the basic principles in a simple form and to give appropriate references for those who desire to pursue a specific topic in more detail.

21.2 FIBER-OPTIC SENSOR TRANSDUCTION

Various modulation techniques used for transduction are described in this section.

21.2.1 Intensity Modulation

Light is guided by the core and cladding of an optical fiber. A detailed discussion of the mechanism is given in Chap. 12. In multimode fibers, the "ray picture" is commonly used.[1] In this case, light is considered to propagate down the fiber at angles (measured with respect to the axis) whose maximum value is limited by the refractive indices of the core and cladding. This fact is exploited in the case of amplitude sensors. Changes in the parameter being measured result in variations in optical intensity. While the intensity modulation takes place in a transducer, additional intensity variations may occur in the optical path external to the transducer. This is a serious disadvantage that directly affects the accuracy of the measurement.

Reflection/Transmission.   When the light exits the fiber, it emerges in a cone of radiation whose cross section is a function of the distance from the end of the fiber. Thus if a second fiber, aligned coaxially with the first, is allowed to intercept the light,[2] the fraction of light trapped by the second fiber is a function of the separation (see Fig. 21.1a). An alternative approach,[3,4,5] shown in Fig. 21.1b, replaces the second fiber by a reflecting surface perpendicular to the axis of the optical fiber. In this case, a portion of the light reflected by the surface is recaptured by the fiber and propagates in the opposite direction. Finally, a second fiber, parallel to the first (see Fig. 21.1c) may be employed[4,5,6] to capture a portion of the back-reflected light. In all cases, a fiber bundle or bundles may be used. Typical plots of the

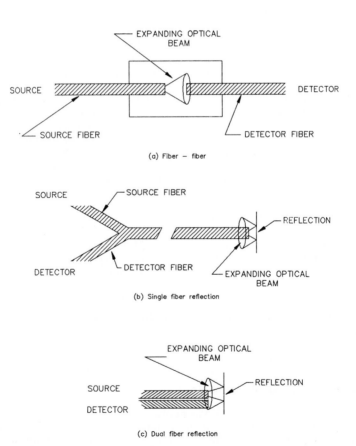

FIGURE 21.1   Various fiber geometries for amplitude sensing. (*a*) Fiber-fiber; (*b*) single fiber; (*c*) dual fiber reflection.

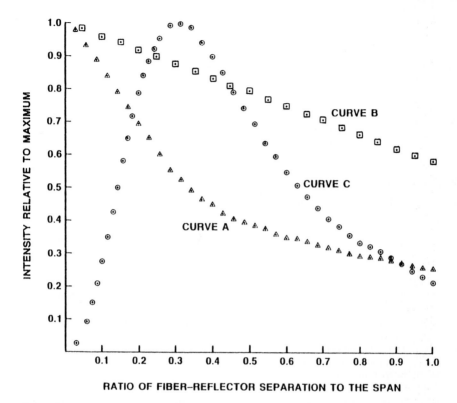

FIGURE 21.2   Reflection sensor response.

detected light versus separation are given in Fig. 21.2. Calibration is required to relate the variation in intensity to the measured parameter. An example of a pressure sensor which uses this approach is described in Sec. 21.5.1. Sources, detectors, and other optical components required are discussed in Sec. 21.3.

Microbending.   A second transduction mechanism leading to intensity modulation is considered here. In this approach, microscopic deformations of the fiber are employed to couple a portion of the light from the core to the cladding, where it is lost. This phenomenon is known as *microbend* loss and should not be confused with macrobend loss.[4] In the microbend sensing approach, microscopic deformations, whose periodicity is optimized to approximate the mode-group spacing in the fiber, are used. In the communication industry, extensive effort has been directed at developing optical fibers relatively insensitive to microbending.[7,8,9] By means of reverse engineering, optical fibers which exhibit high microbend sensitivity have been fabricated.[10] With these fibers, a variety of sensors have been demonstrated.[11,12,13]

Microbend sensors generally employ multimode components. In Fig. 21.3, a schematic of a generic microbend sensor is shown. Light from the optical source [e.g., a light-emitting diode (LED)] is introduced into the optical fiber, where it enters both the core and cladding. For long transmission lines, the latter component

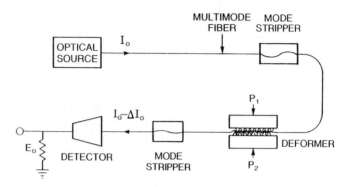

FIGURE 21.3 Bright-field microbend sensor.

is lost and can be neglected. In the case of sensors, the leads are generally much shorter (on the order of 100 m) and some residual cladding light may reach the sensor. This light must be mode-stripped. Mode stripping is often accomplished by the fiber jacket; otherwise, an appropriate index-matching material applied to the cladding should be used. A mechanical deformer (usually a pair of toothed or serrated plates) whose deformation spacing has been optimized for the fiber is used to couple light from the core to the cladding. For small deformations (e.g., less than 1 μm), the degree of coupling is linearly related to the modulation depth. The sensor mechanical design is such that the physical parameter to be measured causes a force to be exerted on the deformer. After deformation, it is advisable to mode-strip in order to ensure that only core light reaches the photodetector. A discussion of microbending theory and the mechanical design of such sensors is given below. Specific examples of sensors are given in Secs. 21.4, 21.5, and 21.6.

Microbend Theory. A detailed study of the solutions to Maxwell's equations for a cylindrical waveguide shows that only certain modes of propagation, β, are allowed. These modes fall in the range

$$k_2 \leq \beta \leq k_1 \qquad (21.1)$$

where k_2 and k_1 are the wave numbers $2\pi n/\lambda$ in the cladding and core, respectively. Exact solutions for β are obtained by satisfying the boundary conditions for the electric and magnetic fields at the core-cladding interface. It is found that these solutions can be spaced into *mode groups*, which, in the case of a step-index waveguide, are separated by

$$\delta\beta = \left(\frac{2\sqrt{\Delta}}{a}\right)\left(\frac{p}{P}\right) \qquad (21.2)$$

In Eq. (21.2), Δ is the relative difference $(n_c - n_{cl})/n_c$ in refractive index between the core and cladding, a is the core radius, p is the order of the mode group, and P is the total number of mode groups. An equivalent expression has been derived for graded-index fiber.[13] However, the present discussion will be restricted to step-index fiber, since such fiber is generally employed in microbend sensors. In this case, the primary mode coupling occurs between the highest-order core modes and the cladding modes. Thus p/P is taken to be unity. If the microbends contain spatial frequencies[14] corresponding to $\delta\beta$ in Eq. (21.2), then maximum mode coupling results. The optimum separation L_c between microbends is found from

$$\frac{2\pi}{L_c} = \delta\beta = \sqrt{2}\,\frac{NA}{(n_c a)} \qquad (21.3)$$

where NA is the numerical aperture discussed in Sec. 21.3.1. Equation (21.3) is the design equation for microbend mode coupling. A typical value of L_c is 2 mm.

Mechanical Design. It has been shown[13] that the change in transmitted light ΔT for a change in microbend amplitude ΔX can be related to the applied force ΔF in the form

$$\frac{\Delta T}{\Delta X} = \left(\frac{\Delta T}{\Delta F}\right)\left(\frac{\Delta F}{\Delta X}\right) \qquad (21.4)$$

where $\Delta F/\Delta X$ is the stiffness of the sensor. The stiffness is a function of the number of deformations and the diameter and elastic moduli of the fiber. An example of a microbend accelerometer is given in Sec. 21.5.1.

Grating. A conceptually simple intensity-type sensor is the moving-grating transducer shown in Fig. 21.4. Two axially aligned fibers are separated by a small gap. A pair of optical gratings is placed in the gap. These gratings consist of a cyclic grid of transmissive and opaque parallel line elements of equal width. When the gratings are moved relative to one another, there is a change in the transmitted intensity. The diverging light beam from the input fiber on the left in Fig. 21.4 is collimated by a short graded-index self-focusing (SELFOC) lens and then partially transmitted through the gratings in a collimated beam that is focused into the output fiber by a second SELFOC lens. Assuming the two gratings each consist of 5-μm-wide grating elements that are spaced 5 μm apart, the transmitted light intensity will vary cyclically with a 10-μm repetition rate. The sensitivity is greatest when

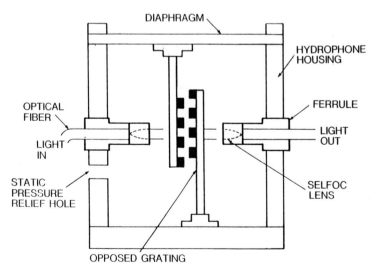

FIGURE 21.4 Moving-grating optical hydrophone.

the quiescent, or bias, point is set at a relative displacement of 2.5 μm, 7.5 μm, 12.5 μm, etc. In addition, decreasing the width of the grating elements will increase the sensitivity at the expense of a decrease in the dynamic range.

If one grating is attached to a plate or diaphragm onto which the force resulting from the parameter being measured is applied and the other plate remains stationary, the variation in optical intensity is proportional to the applied force and therefore to the parameter of interest. If an air-backed diaphragm is employed to deform the optical fiber, a pressure sensor results. The most serious disadvantage of the grating approach is the difficulty in maintaining the optimum bias point. This type of transducer has been employed in a hydrophone[15] where one of the gratings was mounted on the rigid base plate of the liquid-filled housing while the other was attached to a flexible diaphragm exposed to the sound pressure in the water. The fill liquid is employed to compensate for changes in external static pressure. In this application, the considerably lower compressibility of the fill liquid significantly reduces the acoustic sensitivity.

Phase Modulation. The most sensitive fiber-optic sensors are based on optical phase modulation. They generally require single-mode fiber and other single-mode optical components such as coherent sources (lasers). The source optical frequency is approximately 10^{14} Hz. The frequency response of photodetectors is significantly less, and therefore they cannot be used to directly measure phase modulation. However, the modulation frequency associated with the signal of interest is many orders of magnitude lower than the optical frequency. Thus by using optical interferometers which convert phase modulation to intensity modulation, the detection of the signals of interest is accomplished. A variety of interferometers is shown in Fig. 21.5 and discussed in Ref. 3.

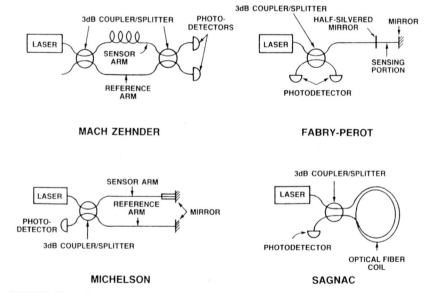

FIGURE 21.5   Fiber-optic interferometers. (*a*) Mach-Zender; (*b*) Michelson; (*c*) Fabry-Perot; (*d*) Sagnac.

As an example, consider the case of the Mach-Zehnder interferometer shown in Fig. 21.5. The output of a single-mode laser is coupled into the optical-fiber lead. The light coupled into the lead fiber is split by a 3-dB coupler/splitter (C/S) into two arms of the interferometer. The C/S serves the same role in fiber-optic interferometers as the half-silvered beam splitter in conventional optical interferometers and is commercially available from a variety of manufacturers. The light in one arm of the interferometer serves as a reference, while the light in the other arm is modulated by the parameter of interest. The light through the arms is recombined in the second C/S, where interference occurs. At this point, any modulation of the phase difference between the light propagating in the two interferometer arms is converted to an amplitude modulation at the output ports of the second C/S. The photodetectors convert the optical modulation to an electric current modulation which is subsequently detected. In the case of the Michelson and Sagnac interferometers shown in Fig. 21.5, a single C/S serves identically as the input and output C/S. Finally, in the case of the Fabry-Perot interferometer, a partially reflecting mirror is used to split and recombine the light.

Interference. If the coherence length of the optical source exceeds the Optical Path length Difference (OPD) between the sensing and reference arms, interference occurs at the output C/S. The light reaching a photodetector varies, resulting in an equivalent variation in output current I. A plot of output current versus phase is shown in Fig. 21.6. The relationship between the current and the phase is given by

$$\Delta I = I_o \pm A \cos (\Delta \phi) \qquad (21.5)$$

where I_o is the average current, A is $\leq I_o$, and A/I_o is defined as the visibility or depth of modulation (DM). The light reaching the upper photodiode of the Mach-Zehnder shown in Fig. 21.5 is 180° out of phase with that reaching the lower photodiode, thereby maintaining conservation of energy.

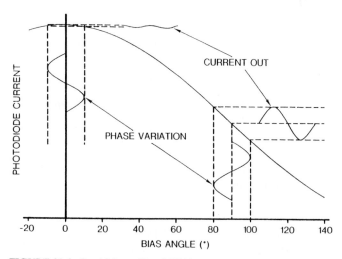

FIGURE 21.6 Sensitivity at 0° and 90° bias.

In Fig. 21.6, bias points corresponding to $\Delta\phi = 0$ and $90°$ are shown in order to demonstrate sensitivity. A sinusoidal signal of amplitude $\pm10°$ is superimposed about the bias points. The amplitude of the resulting output current is obtained by projecting the phase oscillation (input signal) upward onto the solid curve and plotting the resulting optical current about a horizontal line, as is normally done for any transfer function. For the case of the $90°$ bias, the resulting current is large and of the same frequency as the input signal. For the $0°$ bias, however, the amplitude is small and exhibits a frequency equal to twice the excitation frequency. Consider such a signal, initially at the $90°$ bias point, which drifts toward the $0°$ bias point. The amplitude of the current will decrease and for a bias at less than $10°$, a second harmonic will appear. The current amplitude becomes a minimum at the $0°$ bias point, where the fundamental component is 0, leaving only the second harmonic. This process is referred to as *fading*. The $90°$ bias condition is known as *quadrature*. Techniques for maintaining quadrature are considered in Sec. 21.3.3.

Optomechanical Effects.   The transduction mechanism by which a fiber-optic axial strain produces a phase change in the optical path length is discussed in this section. The phase ϕ of the light can be expressed by the equation

$$\phi = knL \qquad \text{rad} \qquad (21.6)$$

where ϕ is in radians, k is the wavenumber $2\pi/\lambda$, λ is the free-space wavelength of the light, and L is the length of fiber in the sensor. Changes in k, n, or L result in changes in ϕ. A mechanical force applied to the fiber results in changes in n and L and, therefore, in ϕ. The corresponding expression relating these changes is

$$\Delta\phi = k\Delta(nL) = knL\left(\frac{\Delta n}{n} + \frac{\Delta L}{L}\right) \qquad (21.7)$$

where $\Delta L/L$ is the axial strain S_{11}, and, for mechanical strain, $\Delta n/n$ is given by

$$\frac{\Delta n}{n} = \frac{n^2[(P_{11} + P_{12})S_{12} + P_{12}S_{11}]}{2} \qquad (21.8)$$

In Eq. (21.8), P_{11} and P_{12} are Pockels coefficients and S_{12} is the radial strain due to an axial stress. The strains S_{11} and S_{12} are related by

$$S_{12} = -\mu S_{11} \qquad (21.9)$$

where μ is Poisson's ratio. Substituting Eqs. (21.8) and (21.9) into Eq. (21.7) yields

$$\Delta\phi = knL\left\{\frac{1 + n^2[(1 - \mu)P_{12} - \mu P_{11}]}{2}\right\}S_{11} \qquad (21.10)$$

In fused silica $P_{11} = 0.12$, $P_{12} = 0.27$, $n = 1.46$, and $\mu = 0.17$. Substituting these values into Eq. (21.10) results in

$$\Delta\phi = 0.796knLS_{11} = 0.796\phi S_{11} \qquad (21.11)$$

Thus, a strain in the sensing fiber caused by the physical parameter of interest induces a proportional phase change in the optical fiber. Equation (21.11) is the fundamental design equation for many interferometric fiber-optic sensors.

Thermal Effects. Thermal effects introduce an added term to Eq. (21.10). An expression which considers the effect of temperature[16] is

$$\frac{\Delta\phi}{\phi\Delta T} = \left(\frac{1}{n}\right)\left(\frac{\partial n}{\partial T}\right)_\rho + \frac{S_{11} - (n^2/2)[(P_{11} + P_{12})S_{12} + P_{12}S_{11}]}{\Delta T} \quad (21.12)$$

where subscript ρ indicates the change at constant density.

21.2.3 Polarization Effects

Strain-Induced Birefringence. In an ideal circular, strain free, single-mode fiber, the orthogonal states of polarization are degenerate (i.e., each direction of polarization propagates with the same phase velocity and thus is characterized by the same refractive index). If mechanical perturbations occur, the degeneracy is removed and the two states no longer exhibit the same propagation constant $\beta = nk$. The states propagate at different velocities and thus interfere within the fiber. This leads to a beat length corresponding to the length of fiber required for that state which propagates at the greatest velocity to travel one wavelength farther than the distance traveled by the slower state. Such a fiber is said to have a *modal birefringence B*, defined by

$$B = \frac{(\beta_x - \beta_y)}{2\pi} \quad (21.13)$$

where the x direction is taken to be the direction of the electric field vector corresponding to the fastest state. The Poynting vector is along the z axis. The beat length L can be shown[3] to be equal to λ/B. In the case of fiber exhibiting large B, the value of L is small. Thus a mechanical disturbance will generally occur over a length of fiber corresponding to many beat lengths. This tends to average out the effect of such disturbances. Alternatively, for large L (small B), the effect of such mechanical disturbances does not average out and noise may result if optical components which discriminate between the states are employed. Thus, by the use of a polarizer/analyzer combination, it is possible to fabricate a sensor. Alternatively, in the case of interferometric sensors, optical interference between the polarization states may result in system noise.

Faraday Effect. The Faraday effect[17] refers to the magneto-optic phenomenon wherein the plane of polarization of linearly polarized light propagating through a material is rotated by an angle θ in the presence of a magnetic field. This effect is present to some degree in all transparent materials, but when the Faraday effect is employed in a sensor, glass (either bulk or fiber) is generally used. θ may be expressed as

$$\theta = \mu VHL \quad (21.14)$$

where μ = magnetic permeability
  V = Verdet constant
  H = amplitude of the magnetic field
  L = length over which H acts

The use of optical fiber as the medium permits the value of L to be varied. The Faraday effect is used in sensing electric current and magnetic fields. In the former case, the current I gives rise to a magnetic field H where $H = I/2\pi R$ and R is the distance from the axis of the current-carrying wire.

Pockels Effect. The Pockels effect refers to the phenomenon wherein the difference in refractive index Δn between the two polarization states in a crystalline material is varied by the application of an electric field. The corresponding expression is

$$\Delta n = n^3 T_{ij} E \qquad (21.15)$$

where n is the index of refraction in the absence of an applied field, T_{ij} is a tensor element relating the directions of light propagation and the electric field to the crystallographic axes, and E is the applied electric field. While this effect does not occur in amorphous materials (e.g., optical fibers), it nevertheless is employed in "fiber-optic" voltage sensors. In that case, the crystalline material and the polarizer/ analyzer combination are situated in the electric field being measured. Two lengths of optical fiber are used to transmit the light to and from the measurement region.

21.3 FIBER-OPTIC SENSOR COMPONENTS

The various components used in fiber-optic sensors are discussed in this section. They include optical fiber, couplers, connectors, sources, detectors, and various modulation and demodulation electronics.

21.3.1 Optical Fiber, Couplers, and Connectors

In Chap. 12 of this book, optical fibers and their properties are dealt with in detail. The requirements for optical fibers used in sensors differ somewhat from those used in optical communication. In particular, the buffering material, normally used only for protection, may serve an added role of enhancing the strain produced by the measured physical parameter. Thus, a compliant buffer[13] renders the fiber more sensitive to static and dynamic pressure. On the other hand, metal-buffered fiber has been demonstrated to be less sensitive to pressure than bare fiber.[18] In many applications, the optical fibers are wound around a mandrel such as a compliant plastic or metallic cylinder.[19] This is an alternative approach to the compliant buffer. Such a winding procedure introduces a tensile stress, and therefore a small-diameter fiber that has been overstressed by spooling under a high tension is often used. In the case of microbend sensors, a noncompliant buffer is required in order to prevent creep and to allow the glass fiber to experience a maximum mechanical deflection. Metal-jacketed fibers are therefore generally used for microbend applications.[11,12] Finally, the NA is generally greater in sensing applications in order to prevent macrobend losses and to achieve increased sensitivity.[3,4]

In many interferometric applications, strain birefringence in the optical leads results in deleterious effects at the output of the interferometer, notably, a change in visibility (see Sec. 21.2.2, under "Optomechanical Effects") and excess phase noise. To minimize these effects, polarization-maintaining (PM) fiber[20] is often used. This is especially true in the case of rate-of-rotation sensors.[21]

Interferometers require the light to be divided into a sensor and a reference path (occasionally both paths are used in a differential sensing configuration). Optical couplers accomplish this function. These devices replace the bulk beamsplitters used in nonfiber interferometers. Optical couplers bring two or more fibers together in such a manner that the light in the core of one overlaps the core of the others.[22,23] Splitting occurs by means of evanescent mode coupling. This is accomplished with a minimum of optical loss (typically less than 0.1 dB). These same couplers are also required for optical multiplexing and in multimode applications. Most manufacturers provide custom splitting ratios for a small additional cost.

As a rule, standard connectors are used in fiber-optic sensors. However, on demating/mating, the optical losses often vary. This requires amplitude-modulated sensors to be rezeroed and possibly recalibrated. Connectors may also be a source of drift. Finally, care must be taken to reduce back-reflections, especially if single-mode lasers are employed as the optical source, for reasons discussed in the next section.

21.3.2 Sources and Detectors

Fiber-optic sensors generally take advantage of the small size and low power consumption offered by solid-state optical sources, e.g., light-emitting diodes and injection laser diodes (ILD). The former are most often used with amplitude-modulated sensors, while the latter may be used with both amplitude-modulated and phase-modulated sensors. Such devices are discussed in Chap. 6. For the purpose of optical detection (see Chaps. 16 and 17) *p-i-n* diodes or avalanche photodiodes (APDs) are employed.

Of particular importance for phase-modulated sensors are the noise characteristics of ILDs and photodetectors. Detection resolution depends on the noise present at the frequencies of interest. The noise is equivalent to a minimum detectable phase shift (resolution) $\Delta\phi_{min}$ and, therefore, to the strain through Eq. (21.11). The functional dependence of these optical noise sources are listed in Table 21.1.

In general, if sufficient power is available, photodetector noise, which is proportional to $P_o^{-1/2}$ or P_o^{-1}, is not a limitation. However, in the case of multiplexing, the photodetector noise is a factor in determining the total number of sensors which may be powered by a single source.

Laser noise may result in either amplitude noise $\Delta\phi_{amp}$ or phase noise $\Delta\phi_{phase}$. As seen in Table 21.1, amplitude noise is inversely proportional to the square root of the detection frequency. For measurements made above a few kilohertz, amplitude noise is generally not a problem. For lower-frequency measurements, various heterodyning techniques have been used to move the measurements to a higher frequency.[24] Alternatively, common-mode rejection techniques[3] have been employed. Optical phase noise can be eliminated by matching the optical path length[25]

TABLE 21.1 Noise Characteristics

ILDs	Photodetectors
$\Delta\phi_{Amp.} \approx 1/f^{1/2}$	$\Delta\phi_{Shot} \approx (\Delta B/P_o)^{1/2}/DM$
$\Delta\phi_{Phase} \approx OPD(\Delta\nu\Delta B)^{1/2}$	$\Delta\phi_{Dark} \approx NEP/P_o$

f = Measurement Frequency
OPD = Optical Path-Length Difference
$\Delta\nu$ = Laser Bandwidth
ΔB = Measurement Bandwidth
P_o = Optical Power at Photodetector
DM = Depth of Modulation

of the interferometer arms, thereby causing OPD to approach 0. As can be seen in Table 21.1, this causes $\Delta\phi_{phase}$ to $\to$ 0. In those cases where an optical path length difference is required for the purpose of demodulation,[26] a laser having a small $\Delta\nu$ minimizes $\Delta\phi_{phase}$. Finally, it has been demonstrated that optical noise can be reduced by up to 30 dB if the output of a noise-compensating interferometer having a path length mismatch equal to or greater than that in the sensing interferometer is fed back to the laser.[27]

To maintain a narrow line width,[28] it is necessary to reduce or eliminate optical feedback[29] to the ILD. To accomplish this, optical isolators are routinely used. At the common communication wavelengths of 1300 and 1550 nm, these isolators are integral to the laser package. Isolators have the added advantage of eliminating the tendency of the laser to mode-hop (in the case of single-longitudinal-mode devices). Mode hopping is the phenomenon where the wavelength of the laser rapidly undergoes a finite change which appears as noise in the system output. Distributed feedback lasers (DFBs) have been developed[30] to eliminate mode-hopping. However, the linewidth of DFB lasers is very sensitive to optical feedback and integral isolators are employed to maintain a small $\Delta\nu$. At other wavelengths (e.g., 830 nm), bulk isolators may be employed. Finally, a new class of solid-state lasers having extremely narrow linewidths ($\approx$5 kHz) consists of diode-laser-pumped neodymium-doped yttrium aluminum garnet (Nd:YAG) lasers.[31] Pigtailed versions of these devices are not readily available; therefore bulk isolators and optical alignment fixtures are required. For specific interferometers having very small path length differences (e.g., Sagnac interferometers), superluminescent diodes[32] (SLD) can be used. These devices exhibit coherence lengths of $\approx$50 μm or less, compared to 200 μm for multi-longitudinal-mode ILDs, several meters for single-longitudinal-mode ILDs, and 100 m or more for gas and Nd:YAG lasers.

ILDs are commonly provided with integral thermoelectric coolers. These maintain constant-temperature operation for the purpose of wavelength and output stabilization. The forward current of a typical ILD is on the order of 50 mA. The thermocooler requires a current less than 1 A. ILDs are easily damaged by transient current. Thus it is essential that a laser driver incorporating current regulation and transient suppression is used. Alternatively, battery power may be employed. Furthermore, ILDs are sensitive to static electricity and appropriate handling techniques must be employed.

In most applications, *P-I-N* diodes are used for photodetection. These have the advantage of not requiring back-biasing. For wavelengths between the visible and $\approx$1.1 μm, silicon detectors are generally used. Above 1.1 μm, germanium, GaAs, and InGaAsP are used. In general, the noise equivalent power (NEP) of the photodetector is an important parameter that establishes system resolution (see Table 21.1). Typical values of NEP are in the range 10^{-14} to 10^{-12} W/$\sqrt{\text{Hz}}$.

21.3.3 Demodulation

Amplitude-modulated sensors make use of relatively simple demodulation techniques. Transimpedance amplifiers, bandpass filters, and appropriate amplifiers are used. It is important to stabilize the output amplitude of the optical source, since any variation in output appears as a signal. Appropriate signal conditioning is required to interface the demodulated output to the data collection equipment. Also, digital techniques may be used.

Phase-modulated devices require significantly more complex demodulation techniques, including passive homodyning,[33] phase-locked-loop (PLL),[34] phase-generated carrier (PGC),[35] and synthetic heterodyning.[36] The output of a fiber-optic interferometer is highly nonlinear away from the quadrature point as well as for large phase shifts (e.g., greater than 0.05 rad). In the case of passive homodyning, the measurement span is determined by the difference between the maximum nonlinearity that can be tolerated and the minimum detectable phase shift. Typically, this span is less than 80 dB. Temperature differences between the interferometer arms are primarily responsible for drift away from quadrature. If free-running, a typical interferometer employing long fiber arms may drift through hundreds of radians in a laboratory environment. This causes signal fading and distortion which render measurements almost impossible. For this reason, passive homodyne detection is seldom used. The other three demodulation techniques eliminate signal fading.

In the case of PLL demodulation, active feedback to either a phase-shifting element located in the interferometer or to the single-mode laser source is used to maintain quadrature[34] (see Fig. 21.7). In the former case, a fiber-wound piezoelectric element is commonly used. Integrated-optic phase shifters[37] may also be used as they become available. The PLL demodulator integrates the output of the transimpedance amplifier (or difference amplifier, if common-mode rejection is used) and feeds it back to either the phase shifter or the laser. Thus a signal that drives the output of the interferometer away from quadrature results in a voltage at the output of the integrator proportional to the signal. This voltage is fed to the PZT or laser (see discussion below), resulting in an optical phase shift equal to that produced by the signal, thereby rebalancing the interferometer. In this case, the measurement span is set by the difference between the maximum feedback voltage and the voltage corresponding to the minimum detectable phase shift. A span as large as 120 dB may be achieved. A disadvantage of feedback to the laser is that a laser is required for each interferometer/sensor.

Feedback to the laser is employed in those cases where electrical leads to the interferometer are not allowed (e.g., in an explosive environment). This can be accomplished because single-longitudinal-mode ILDs may be frequency- (v) or wavelength-modulated. If the laser forward current is modulated, the output fre-

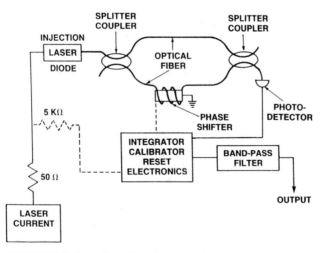

FIGURE 21.7 Phase-locked loop demodulation.

quency varies linearly for a small modulation. Typical values of $\Delta\nu/\Delta i$ are 0.5 to 3.0 GHz/mA for low modulation frequencies (i.e., less than 50 kHz). Operation at higher frequencies results in a reduced response. In either a Mach-Zehnder or Michelson interferometer, the change in phase with modulation current can be shown to be directly proportional to the product of the OPD and $\Delta\nu/\Delta i$. As a consequence, laser modulation results in phase modulation in the interferometer. Therefore, feedback to the laser has the same effect as feedback to a phase-shifting element in the interferometer. As can be seen in Table 21.1, an OPD can result in phase noise, in which case a narrow-linewidth laser is advantageous.

In the PGC approach, a sinusoidal carrier frequency, usually a factor of 3 or more greater than the highest signal frequency of interest, is used to modulate the laser or phase shifter. When the laser is modulated, the need for electrical leads to the interferometer is eliminated, but an OPD is required as discussed above. A typical carrier amplitude[35] is 2.65 rad peak. This drives the interferometer well into the nonlinear region.

Because of the resulting nonlinearity, the signal of interest is mixed with the carrier. Sidebands occur about both the carrier frequency and twice the carrier frequency (as well as about higher harmonics). These are electronically processed to extract the signal. In the PGC approach, a measurement span near 100 dB can be achieved. The primary advantage of the PGC approach is the ability to utilize a single laser for powering several interferometers. Up to 16 such interferometers have been demonstrated.[38]

The synthetic heterodyne approach introduces a separate modulation frequency into each arm of the interferometer. As above, the nonlinearity of the interferometer results in mixing between the signal and the difference between the modulation frequencies. This approach is complicated and therefore not generally used.

21.4 TEMPERATURE SENSORS

A variety of techniques for the measurement of temperature using optical fibers have been demonstrated.[39-48] Several of these are discussed below.

21.4.1 Interferometric Temperature Sensors

Phase-modulated fiber-optic sensors have been shown to be a highly sensitive means for measuring temperature. In most cases, it is desirable to employ a single insensitive fiber-optic lead from the sensing region to the demodulator. In addition, for many applications, "point" sensors are required. These requirements are best met by the use of Fabry-Perot–type interferometers (refer to Fig. 21.5). Fabry-Perot interferometric temperature sensors generally employ a sensing region a few millimeters long or less, located between a pair of reflecting surfaces. Researchers at Texas A&M University[39] have developed fusion-splicing techniques for incorporating dielectric mirrors directly into a continuous length of single-mode fiber. Using such mirrors, they have fabricated Fabry-Perot interferometers approximately 1.5 mm long. Temperatures ranging between -200 and $+1050°C$ have been measured by fringe-counting techniques. At temperatures above 100°C, a sensitivity $\Delta\phi/\Delta T$ on the order of 0.1 rad/°C was shown. In another demonstration[40] of a Fabry-Perot–type sensor embedded in a graphite-epoxy composite, the normalized sensitivity $\Delta\phi/\Delta T$ was measured to be $8.0 \times 10^{-6}/°C$. These sensors employed single-longitudinal-mode ILDs and optical isolators. The latter is used to eliminate problems associated with back-reflection into the laser, as discussed above.

An alternative approach employing multimode ILDs has been demonstrated by researchers at Optical Technologies Inc.[41] In this approach, the Fabry-Perot cell was attached to the end of a lead fiber. The cell was fabricated using a short length (2 to 4 mm) of sensing fiber. Dielectric mirrors were deposited on each end of the cell. Since the coherence length (≈ 0.2 mm) of the multimode laser is significantly shorter than the optical path length (twice the fiber length) of the Fabry-Perot cell, interference does not occur at the Fabry-Perot cell. However, the relative phase shift between the light reflected from the near and far mirrors is maintained. A second Fabry-Perot cell of approximately the same length is contained in the demodulator (see Fig. 21.8). This second cell is used to adjust the path lengths of the two beams to within the coherence length of the multimode ILD. Interference occurs at this compensation Fabry-Perot cell. Thus a "divided" Fabry-Perot geometry, shown in Fig. 21.8, is formed. Since multimode ILDs are not adversely affected by back-reflected light, the need for isolators is eliminated, and cost is thereby reduced. Furthermore, multimode ILDs are less expensive than their single-mode counterparts. Finally, this approach lends itself to the use of PLL demodulation by varying the length of the compensating cell. With this technique, a resolution and accuracy of 0.01°C and 0.1°C, respectively, were achieved.[42] Using a similar divided Fabry-Perot geometry and employing fringe counting techniques, researchers[43] at Texas A&M demonstrated a resolution of ≈ 4°C.

Such Fabry-Perot cells, each having a unique length, may be serially located along a single fiber.[44] A number of compensating cells equal to the number of sensing cells and having approximately the same length as the corresponding sensing

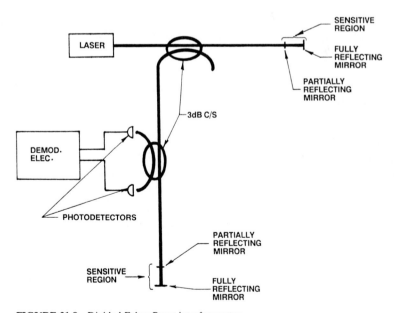

FIGURE 21.8 Divided Fabry-Perot interferometer.

element are required. In addition, alternative configurations employing an input 3 × 3 fiber-optic coupler have been demonstrated.[45]

Temperature sensors employing Fabry-Perot interferometers whose dimensions are sufficiently small to allow the use of very short coherence length sources (e.g., LEDs) are available from Metricor.[46] These sensors employ a Fabry-Perot cavity containing a material whose optical refractive index is large and temperature-dependent. Changes in temperature vary the refractive index and, by Eq. (21.6), the optical phase. Since a thin film of material is employed, the thermal time constant is small, allowing rapidly changing temperatures to be tracked (e.g., a 50°C change in 3 ms). Alternatively, the use of an evacuated cavity between the end of the optical fiber and a diaphragm permits the measurement of pressure.

21.4.2 Intensity-Based Temperature Sensors

A technique for the measurement of temperature employs the phosphorescent decay[47] of a material after illumination by an ultraviolet light pulse. A fiber-optic lead carries the ultraviolet pulse to the phosphorescent material located at the fiber end. The ultraviolet energy absorbed by the material is radiated as visible light. The optical fiber collects a portion of the visible light and returns it to the detector. For properly chosen materials, the decay time is a function of temperature. The round-trip distance between the source and the phosphorescent material is limited by the high attenuation of optical fiber to ultraviolet light and the poor conversion efficiency of the phosphorescent material. This technique has been employed from below −180°C to above 450°C with resolutions of less than 1°C. Such sensors are available from Luxtron, Mountain View, Calif.

Finally, a blackbody radiation technique[48] has been used to measure temperatures up to several thousand degrees Celsius. These sensors employ a sapphire fiber in the high-temperature region and conventional fiber in the low-temperature region. As is well-known, the wavelength at which the blackbody radiation curve exhibits a peak is inversely proportional to temperature. By measuring the intensity of the blackbody radiation from the sensor at several wavelengths, the blackbody radiation curve is determined and the temperature is calculated.

21.5 STATIC AND DYNAMIC PRESSURE SENSORS

Various approaches for measuring pressure with optical fibers have been demonstrated.[4] These may be broadly grouped as static and dynamic sensors. Static sensors are generally intensity-modulated devices and, as such, suffer from lead sensitivity. Interferometric techniques are generally restricted to dynamic measurements (acoustic). This is due primarily to low-frequency noise arising from temperature and $1/f$-type noise. For high-frequency and acoustic measurements, mechanical resonances limit the response.

21.5.1 Static Pressure Sensors

Some of the techniques for the measurement of static pressure are transmissive, moving grating, near total internal reflection, reflection, and microbending.[4] The last two techniques are described below.

Static pressure sensors utilizing the reflection technique are conceptually simple and can be grouped into two general classes: one employing a single lead fiber and the other, a pair of fibers (or bundle) as shown in Figs. 21.1*b* and 21.1*c*, respectively. The corresponding intensity versus displacement curves are given in Fig. 21.2. Calibration is required to relate the displacement of the reflector to the pressure. A typical sensor head design is shown in Fig. 21.9. The pressure port is at the left and the optical port is at the right. A diaphragm or some other pressure-sensitive transducer (e.g., bourdon tube or bellows) separates the pressure and optical regions. In order to isolate the optical side of the diaphragm from the environment, a permanent plug is used in the optical connector. A section of optical fiber may serve as the plug.

Since intensity fluctuations are interpreted as pressure fluctuations, it is important to use a well-regulated optical source. Without such regulation, fluctuations of 10 percent or more may occur, thereby limiting the accuracy of the sensor. Lead perturbations due to microbending or temperature fluctuations can cause errors as large as several percent. Referencing techniques[5] have been shown to reduce such errors by an order of magnitude. In Ref. 5, a dual-wavelength approach was employed. In this case, the permanent plug (connector) shown in Fig. 21.9 incorporates a dichroic mirror that passes one wavelength and reflects the other. Figure 21.10 illustrates the dichroic cutoff and the relative spectral distributions of two commercially available LEDs. The reference spectrum is almost completely reflected, while the measured spectrum is transmitted. An example of the corrected outputs with and without lead bending is shown in Fig. 21.11. Such bending was in excess of that expected in practice. See Ref. 5 for details.

Researchers[49] at Babcock-Wilcox have demonstrated a microbend pressure sensor. In this case, a fused-silica diaphragm deforms under the influence of pressure, causing microbending. Referencing is accomplished by means of a second optical fiber configured along the same path, but not passing through the microbender. The reference fiber is used to correct for environmental influences.

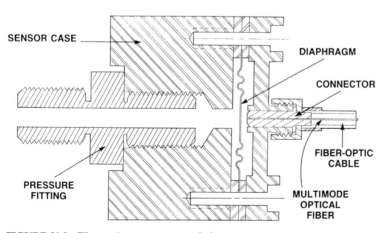

FIGURE 21.9 Fiber-optic pressure sensor design.

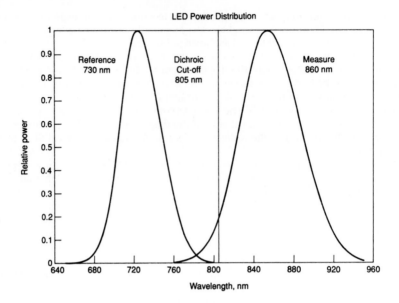

FIGURE 21.10 Dichroic cutoff and spectral distributions of LEDs.

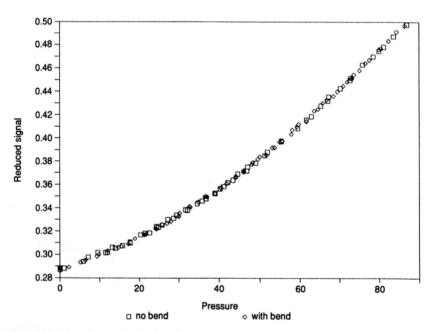

FIGURE 21.11 Processed reflection data.

21.5.2 Dynamic Pressure Sensors

Dynamic pressure sensors have been developed for acoustic applications. Most employ interferometric techniques because of their high sensitivity and large dynamic range. The optical fiber in the sensing arm of the interferometer is caused to undergo a mechanical strain arising from the acoustic pressure. This strain can be significantly increased over that of a bare fiber by the use of either a compliant jacket on the fiber or a compliant mandrel around which the fiber is tightly wound. The resulting mechanical strain causes a phase shift, given by Eq. (21.11). The response of bare, plastic-jacketed, and metal-jacketed fibers as well as fiber-wound compliant mandrels has been reported.[13] Plastic jacketing increases the acoustic response compared to the bare fiber, while metal-jacketing decreases the response. The design approach most often used involves a fiber-wound compliant mandrel. A simplified calculation for an isotropic plastic is given below.

S_{11} in Eq. (21.11) can be expressed in the form

$$S_{11} = \left(\frac{\partial L}{\partial P}\right)\frac{\Delta P}{L} \tag{21.16}$$

where ΔP is the acoustic pressure and L is the sensing fiber length. For an isotropic material, S_{11} is

$$S_{11} = \left(\frac{\partial V}{\partial P}\right)\frac{\Delta P}{3V} \tag{21.17}$$

where V is the volume of the material. Since $V(\partial P/\partial V)$ is the bulk modulus K, the right-hand side of Eq. (21.17) becomes $\Delta P/3K$ and therefore Eq. (21.11) becomes

$$\Delta\phi = \frac{0.26\phi\Delta P}{K} \tag{21.18}$$

For a nylon mandrel, $K \approx 6 \times 10^9$ Pa. Substituting this value of K and rearranging Eq. (21.18) gives

$$\frac{\Delta\phi}{\phi\Delta P} = 4.3 \times 10^{-11} \text{ Pa}^{-1} \tag{21.19}$$

$\Delta\phi/\phi\Delta P$ is the normalized pressure sensitivity, which serves as a design parameter. If a given minimum detectable acoustic pressure ΔP_{min} is to be measured, and the minimum detectable phase shift $\Delta\phi_{min}$ is known, then the length of fiber can be determined from Eq. (21.19) ($\phi = knL$). $\Delta\phi_{min}$ is generally established by the laser (see Table 21.1). A typical value is 20 μrad/$\sqrt{\text{Hz}}$ at 1 kHz, assuming an ILD operating at 0.83 μm and an OPD of 4 cm. $\Delta\phi_{min}$ at 1 kHz has been measured[27] to be 1 μrad/$\sqrt{\text{Hz}}$, using a diode-pumped Nd:YAG laser and an OPD of 1 m. Increased sensitivity can be achieved by substituting a thin-walled, air-backed metallic cylinder for the nylon.[50] The normalized pressure sensitivity can be doubled by the use of a push-pull arrangement[51] wherein the acoustic pressure causes the fiber in one arm of the interferometer to expand while that in the other arm contracts. Instead of using a mandrel, a very thick jacketed fiber wound in a planar spiral and embedded in a thin polyurethane layer[52] resulted in a normalized pressure sensitivity of 1.25×10^{-10} Pa^{-1}.

An early approach[53] to the measurement of acoustic signals employed a Mach-Zehnder interferometer incorporating a pair of moving mirrors in the sensing arm. One mirror remained fixed and the other was attached to a diaphragm subjected to the acoustic pressure. The sensor light was introduced between the mirrors at an angle, thereby producing multiple reflections. With air as a medium between the mirrors, a minimum detectable pressure of 48 dB relative to 1 μPa ($\approx$240 μPa) was achieved for frequencies near 1 kHz. If the air is replaced by a compliant material (e.g., silicone rubber) for operation in the ocean at elevated pressures, a reduction in sensitivity of up to 76 dB may occur.

Single-mode fibers allow two polarization states to propagate. This fact has been used for the detection of acoustic waves. In one approach,[54] a highly birefringent fiber is wound on a compliant cylinder. Linearly polarized coherent light is injected into the fiber. In order to excite both eigenmodes equally, the light is launched at 45° relative to the predominant eigenmode direction. When subjected to an acoustic wave, the compliant cylinder deforms, modulating the state of polarization. The light exiting the fiber is analyzed to determine the relative percentage of light in each of the two eigenmodes. The degree of cross-coupling is linearly related to the amplitude of the acoustic wave.

An approach to the measurement of high-frequency ultrasonic waves using polarized light in bare fiber has been demonstrated.[55] When the acoustic wavelength is less than the fiber diameter, strain birefringence is induced in the fiber, modulating the state of polarization. It was shown that, in the frequency range 10 MHz to 30 MHz (corresponding to acoustic wavelengths approximately one-half to one-sixth the fiber diameter), the transduction mechanism is linear with frequency.

A variety of optical intensity modulation techniques for acoustic measurements have been demonstrated. These are reviewed in Ref. 13, where detailed discussions are given. Two of these have been extensively investigated. These are microbend-modulated and moving-grating sensors. In the case of microbend modulation, the acoustic pressure is caused to modulate the mechanical deformer as described in Sec. 21.2.1 under "Microbending." A minimum detectable pressure of 60 dB relative to 1 μPa (1 mPa) for frequencies above 500 Hz was achieved.[13] In the case of the moving grating approach (see Sec. 21.2.1 under "Grating"), the acoustic wave modulates the relative position of the grating. A minimum detectable pressure in the range of 50 dB relative to 1 μPa at frequencies between 100 Hz and 2 kHz was reported.[56]

21.6 ACCELEROMETERS

Fiber-optic accelerometers detect the inertial force exerted on an accelerating mass. The force may be caused to directly strain a fiber as, for example, in microbending. Alternatively, the force may be converted into pressure-induced strains and detected. Finally, a variety of other intensity-modulation techniques such as reflection or moving-grating may be used.

21.6.1 Microbend-Modulated Accelerometer

The inertial force can be caused to modulate the microbender described in Sec. 21.2.1 under "Microbending." A highly sensitive horizontal microbend fiber-optic accelerometer has been demonstrated.[12,57] In a horizontal configuration, the deformer is not subjected to the weight of the mass, and therefore the tendency to creep, with a resultant baseline shift, is eliminated. The device uses a cantilever beam to which is attached the inertial mass. One set of deforming teeth is attached to the case, and the other is attached to the mass (see Fig. 21.12). This constitutes a mass-spring system with the stiffness of the fiber and the cantilever beam determining the spring constant. In order to detect acceleration frequencies up to 100 Hz, a resonant frequency of 300 Hz was selected. Accelerations as small as 5 μg at 1 Hz were detected in the direction perpendicular to the cantilever beam. Minimum detectable accelerations in the cross-axis directions were 27 dB higher. The measurement span was observed to be in excess of 90 dB. A prototype model has been compared to conventional sensors, all located at the University of Southern California Los Angeles Basin seismic network.[58] The optical fiber accelerometer exhibited better low-frequency (below 10 Hz) response than a velocity-coupled geophone (Mark Products L4C seismometer). This result is promising for low-frequency borehole seismic sensing.

21.6.2 Phase-Modulated Seismometers

Interferometric techniques have been used to measure acceleration.[59] A simple approach makes use of plastic-jacketed single-mode optical fiber. The fiber is configured in a loosely wound coil, located at the bottom of a container filled with a liquid (e.g., water or mercury, depending on the desired sensitivity). A liquid is used to reduce cross-axis effects and to eliminate spurious resonances. The coil

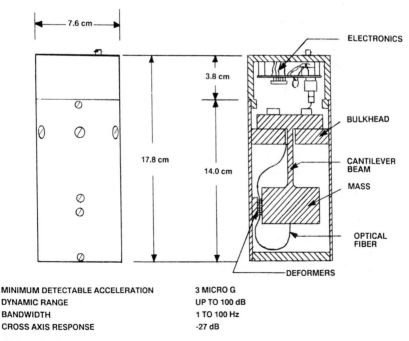

MINIMUM DETECTABLE ACCELERATION	3 MICRO G
DYNAMIC RANGE	UP TO 100 dB
BANDWIDTH	1 TO 100 Hz
CROSS AXIS RESPONSE	-27 dB

FIGURE 21.12 Horizontal microbend seismometer.

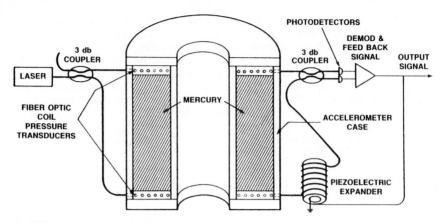

FIGURE 21.13 Fiber-optic seismometer.

constitutes one arm of the interferometer. Another coil external to the liquid serves as a reference arm. When the device (shown in Fig. 21.13) is accelerated upward, the pressure at the bottom of the container where the sensor coil is located increases, thereby decreasing the optical path length of the fiber. Similarly, if the container is accelerated downward, the pressure on the coil decreases, thereby increasing the optical path length.

An example calculation of the acceleration resolution achievable by this method is shown below. The change in pressure by the sensing coil is

$$\Delta P = \rho h \Delta a \qquad (21.20)$$

where ρ is the liquid density, h is the height of the liquid column, and Δa is the acceleration. The normalized pressure sensitivity $\Delta\phi/\phi\Delta P$ ($= C$) of a plastic-jacketed fiber [≈ 400 μm outside diameter (OD)] has been measured[60] to be approximately 2.5×10^{-11} Pa^{-1}. Utilizing Eq. (21.20) and this result for $\Delta\phi/\phi\Delta P$, the acceleration is found to be

$$\Delta a = \frac{\Delta\phi}{C\rho h\phi} = \frac{\Delta\phi}{C\rho hknL} \qquad (21.21)$$

Assuming the use of a diode-pumped Nd:YAG laser emitting at a wavelength of 1319 nm, $\Delta\phi_{min}$ at 1 Hz is on the order of 10^{-6} rad. If $h = 10$ cm, $\rho = 13.6$ g/cm^3 (mercury), $L = 100$ m, $n = 1.46$, $k = 2\pi/\lambda$, and $C = 2.5 \times 10^{-11}$ Pa^{-1}, it is found that $\Delta a_{min} = 4.2 \times 10^{-8}$ m/s^2, or $\approx 4 \times 10^{-9}$ g. Results from ongoing research by the present authors indicate that values of Δa_{min} an order of magnitude less than that calculated herein can be expected. In addition, the fundamental resonance frequency is calculated to be greater than 5 kHz. This yields a useful measurement bandwidth in excess of 1 kHz.

Another phase-modulated approach makes use of two fiber-wound, silicone rubber mandrels separated by a 520 g inertial mass.[61] In this case, the device acts as a velocimeter above the fundamental resonant frequency (≈ 300 Hz). The optical path length in each arm was approximately 13 m. Below resonance, the reported sensitivity is 10,500 rad/g. However, the use of silicone rubber mandrels is expected to cause a variation in sensitivity with temperature. Furthermore, the design is expected to result in significant cross-axis effects.

A miniature phase-modulated accelerometer employing a multimode diode laser as the optical source has been demonstrated.[62] A divided Fabry-Perot geometry was employed (see Sec. 21.4.1). The sensor was 2.5 cm long and 0.25 mm in diameter. The frequency range investigated was 100 Hz to 25 kHz. Accelerations in the range $10^{-3}g$ to more than $3g$ were measured. The device was designed for measurements to above $10g$. An undesirable cross-axis resonance occurred at 2.8 kHz.

21.7 *RATE-OF-ROTATION SENSORS*

Fiber-optic rate-of-rotation sensors (gyroscopes) employing the Sagnac effect exhibit performances equivalent to those achieved by ring-laser gyroscopes. Because of the geometric flexibility and somewhat lower cost of the fiber-optic gyroscopes, these devices appear ideal for use in a wide range of navigation applications, both commercial and defense-related. Compared to ring-laser gyroscopes, the warm-up time of fiber-optic gyroscopes is short. Comprehensive discussions of these devices and the optical components required are given in Refs. 3 and 21. A highly simplified discussion of their operation is given below.

The Sagnac effect is used to generate an optical path length difference ΔL, which is proportional to the rate of rotation Ω about an axis normal to the plane of the interferometer coil (see Fig. 21.5). Consider a circular optical path of radius R where light is injected simultaneously in opposite directions around the perimeter. For a stationary loop, the clockwise and counterclockwise beams will arrive at the injection point in phase. However, if the loop is rotating in a clockwise direction, the beam traveling in the counterclockwise direction arrives earlier than the beam traveling in the clockwise direction. This leads to a difference in phase between the two counterpropagating beams. In view of the fact that the beams travel identical paths, their respective phases are not influenced by reciprocal effects (i.e., those which affect both beams identically), such as temperature, vibration, and electromagnetic fields. Rotation, on the other hand, is a nonreciprocal effect. The phase difference arising from the rotation is directly proportional to Ω and is given by[3]

$$\Delta\phi = \left(\frac{4\pi LR}{\lambda_o c}\right) \Omega \qquad (21.22)$$

where L is the length of fiber in the coil, λ_o is the free-space wavelength of the light, and c is the speed of light. As can be seen, for a coil of fixed radius R, the phase change and therefore the sensitivity may be increased by increasing L (i.e., the number of turns in the coil). For $R = 10$ cm, $L = 1$ km, $\lambda_o = 1.3$ µm, and $\Delta\phi_{min} = 10^{-6}$ rad (path lengths identical and therefore no phase noise), $\Omega_{min} = 3.1 \times 10^{-7}$ rad/s, or 0.064°/h. State-of-the-art gyroscopes exhibit Ω_{min} on the order of 0.01°/h.

Sources of noise include Rayleigh scattering and polarization phenomena. In the Rayleigh scattering case, if a long coherence length optical source is employed, a beam propagating in one direction will have a portion of its light back-scattered. This back-scattered light interferes with the light propagating in the opposite direction. This leads to noise at the output which can be many orders of magnitude greater than the state-of-the-art signals measured. To eliminate this noise source, super luminescent LEDs[32] having short coherence lengths are employed as the optical source. With respect to polarization noise, the two orthogonal polarization modes in a single-mode fiber propagate with slightly different phase velocities. Mechanical perturbations cause the two modes to couple, thereby transferring energy between each other. Because of the different velocities of the modes, phase shifts proportional to the amplitude and position of the perturbations result at the output. This noise source is greatly reduced by the use of polarization-maintaining fiber.

21.8 MAGNETIC/ELECTRIC FIELD SENSORS

Communication-grade optical fibers are essentially insensitive to electric and magnetic fields. This is an important advantage of their use for communication and nonelectromagnetic sensors. Thus in order to fabricate a fiber-optic magnetic/ electric field sensor, one must employ a fiber or bulk material having a large Verdet constant or use a magnetostrictive/piezoelectric material. The latter materials undergo a dimensional change when subjected to a magnetic or electric field, respectively. Other techniques such as microbending or reflection are not commonly used.

21.8.1 Magnetic Field Sensors

Interferometric magnetometers presently under development employ transducers based on metallic glass[63] (Metglas) having high magnetostrictive constants.[64] In order to maximize the magnetostrictive coefficient of Metglas, a magnetic anneal is required.[65] The optical fiber in one arm of an interferometer is attached to either a strip of Metglas or is circumferentially wound around a Metglas cylinder.[66,67] When a magnetic field is applied, the Metglas strains the fiber, thereby resulting in a phase shift. If the net magnetic field applied to the Metglas is near 0, the relation[68] between the strain and the magnetic field is approximately $S_{11} = CH^2$, where C is an effective magnetomechanical coupling coefficient. This relationship permits the detection of dc or low-frequency magnetic signals, despite the presence of low-frequency noise (e.g., thermal and laser amplitude noise). The detection is accomplished by the use of a high-frequency (several kilohertz) magnetic dither of amplitude H_d, applied by means of a solenoid. Substituting CH^2 for the strain in Eq. (21.11) and letting $H = H_o + H_d \cos(\omega t)$ yields

$$\Delta\phi = 0.79knLC[H_{dc} + H_\omega \cos(\omega t) + H_{2\omega} \cos(2\omega t)] \qquad (21.23)$$

where the amplitudes $H_{dc} = H_o^2 + H_d^2/2$, $H_\omega = 2H_oH_d$, $H_{2\omega} = H_d^2/2$ and trigonometric substitutions have been made. It is seen that H_ω is proportional to H_o. H_d serves as a gain factor. $H_{2\omega}$ may be used to determine C, since it is independent of H_o. A portable magnetometer[69] employing this approach is under development at Optical Technologies Inc.

The interferometric demodulation schemes employed for these devices are similar to those described in Sec. 21.3.3. Synchronous sampling techniques also may be used.[70] The advantage of this approach is that it allows the coefficient of the fundamental dither frequency to be measured in the presence of a large second-harmonic term. Thus, H_d (the gain factor) may be increased without affecting the dynamic range of the system. Alternative approaches to the demodulation of the magnetic signal use open- and closed-loop configurations.[71] Finally, polarimetric techniques have been demonstrated.[72]

21.8.2 Current Transformers

The development of conventional current transformers used in electrical power transmission occurred over the last century. While this has led to a very mature design, they do not easily lend themselves to new demands. These include higher system voltages, floating tank breakers, insulated cable systems, and a growing dependence on digital information technology. Furthermore, the cost of these systems in higher voltage applications becomes significant if insulators, foundations, engineering, and installation are included in the price. The dielectric and lightweight characteristics as well as the digital transmission capability of optical fibers makes fiber-optic sensors ideal for use as an alternative sensing approach. The primary techniques presently under development utilize the Faraday effect and optical phase modulation.

Faraday magneto-optic–effect current sensors[73] employing optical-fiber or bulk materials are being developed. Each has its own advantages. The former leads to a rather simple, easy-to-fabricate design. The latter permits a far wider choice of optical materials. Light from an LED is injected into a multimode fiber and transmitted up to the power line where the measurement is to be made. A bulk optical material having a large Verdet constant is fabricated into a rectangular frame surrounding the transmission line. A lens located on the end of the fiber is used to collimate the light and direct it into a polarizer attached to the bulk material. This polarized beam makes one or more passes around the transmission line. A portion of the light has its direction of polarization rotated into the orthogonal direction by the magnetic field surrounding the transmission line. The degree of rotation is proportional to the magnetic field (and hence the current) as given in Eq. (21.14). A polarization analyzer serving as the optical demodulator is attached at the output of the device. A second lens couples the intensity-modulated output into a return fiber, where it is transmitted to a photodetector.

Some advantages of this approach relative to conventional current transformers are its explosionproof nature, absence of ferromagnetic saturation or residual magnetism effects, linearity of ±1 percent up to 100 kA, high frequency response, and elimination of coupling of fast transients through interwinding capacitance. A disadvantage is the somewhat complicated nature of the optical geometry in the sensor.

Another approach to the measurement of electric current employs the Verdet constant in the fiber. In this case, the optical fiber may be configured around the transmission line; alternatively, a magnetic yoke may be employed about the transmission line to concentrate the flux in a gap. An optical fiber coil is located in the gap. In either case, a polarizer is located at the input to the fiber coil and an analyzer at the output.

The optical phase modulation approach to the measurement of current employs a pickup coil whose output is a voltage proportional to the current. In the simplest case, a linear coil is oriented with its axis tangent to the magnetic field about the transmission line. To eliminate the effect of crosstalk, a Rogowski coil may be used (see Fig. 21.14). The voltage induced in the coil is

$$V = -N \frac{d(\Phi)}{dt} \qquad (21.24)$$

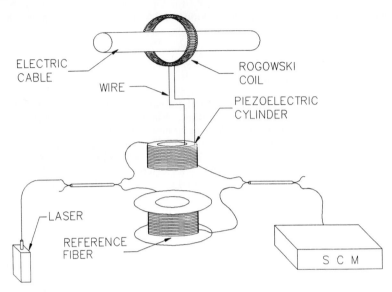

FIGURE 21.14   Fiber-optic current transformer (FOCT).

where N is the number of turns in the coil and Φ is the magnetic flux intersected by the cross-sectional area of the coil. The voltage from the coil is applied to a phase-shifting element[74] to which a portion of the fiber in one arm of an interferometer is attached. The time-varying voltage associated with the current results in a time-varying phase shift in the interferometer. The demodulation techniques described in Sec. 21.3.3 may be used. Using this approach, the authors have demonstrated measurement spans greater than 90 dB. This permits the simultaneous measurement of nominal and fault currents. The latter often exceed the former by a factor of 20 or more. Sensor bandwidths greater than several kilohertz have been shown. This permits the waveform to be analyzed. This approach permits in-situ calibration to be performed by applying a known voltage across the phase-shifting element. The phase modulation approach has all the advantages listed above for the bulk Faraday approach. Finally, in an alternative interferometric technique,[75] a magnetostrictive cylinder, toroidally wound with optical fiber, surrounds the transmission line. A slot in the cylinder allows it to be installed without interruption of the power.

21.8.3 Electric Field/Voltage Sensors

A similar phase-modulation technique may be used to measure voltage. In this case, a pickup coil is not required. Instead, a portion of the voltage to be measured is applied directly to the phase-shifting element (see Fig. 21.15). In the example shown, a voltage divider is employed. In a simple case, the voltage divider is formed by the capacitance of a PZT element and that of the air path to ground. Due to the variations in dielectric constant of the air, the distance of the wire from the ground, and the presence of disturbances such as nearby vehicles, errors of at least

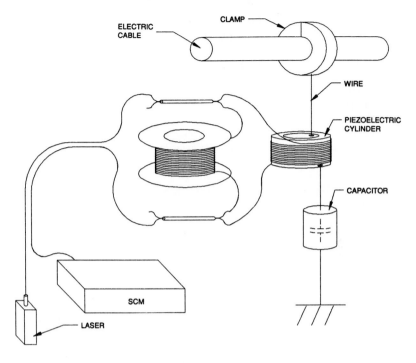

FIGURE 21.15 Fiber-optic voltage transformer (FOVT).

5 percent may be expected. For those measurements requiring greater accuracy, a standard divider must be used. It should be noted that the fiber-optic element in both the phase-modulated current and voltage sensors may be identical.

21.9 REFERENCES

1. Snyder, A., and J. Love, *Optical Waveguide Theory*, Chapman and Hall, London, 1983.
2. Esposito, J., Chap. 5, *Handbook of Fiber Optics: Theory and Applications*, H. Garland Wolf, ed., STPM Press, New York, 1979.
3. C. Davis, et al., *Fiberoptic Sensor Technology Handbook*, Optical Technologies, Herndon, Va., 1982, 1986.
4. Krohn, D., *Fiber Optic Sensors: Fundamentals and Applications*, Instrument Society of America, Research Triangle Park, N.C., 1988.
5. Conley, M., C. Zarobila, and J. Freal, "Reflection Type Fiber-Optic Sensor," *Proc. SPIE*, vol. 718, p. 237, 1986.
6. Krohn, D., "Fiber-Optic Displacement Sensors," *Proc. ISA*, vol. 39, p. 331, 1984.
7. Gloge, D., *Bell Syst. Tech. J.*, vol. 51, p. 1767, 1972.
8. Marcuse, D., *Theory of Dielectric Optical Waveguides*, Academic Press, New York, 1974.

9. Gloge, D., and E. Marcatili, *Bell Syst. Tech. J.*, vol. 52, p. 1563, 1973.

10. Fields, J., C. Asawa, O. Ramer, and M. Barnoski, *J. Acoust. Soc. Am.*, vol. 67, p. 816, 1980. See also Lagakos, N., and J. Bucaro, "Pressure Desensitization of Optical Fibers," *Applied Optics*, vol. 20, p. 2716, 1981.

11. Fields, J., and J. Cole, "Fiber Microbend Acoustic Sensor," *Applied Optics*, vol. 19, p. 3265, 1980.

12. Freal, J., C. Zarobila, and C. Davis, "A Microbend Horizontal Accelerometer for Borehole Deployment," *J. Lightwave Tech.*, vol. LT-5, p. 993, 1987.

13. Bucaro, J., N. Lagakos, J. Cole, and T. Giallorenzi, "Fiber Optic Acoustic Transduction," in *Physical Acoustics*, vol. XVI, Academic Press, New York, 1982.

14. Diemeer, M., and E. Trommel, "Fiber-Optic Microbend Sensors: Sensitivity as a Function of Distortion Wavelength," *Opt. Lett.*, vol. 9, p. 260, 1984.

15. Spillman, W., "Multimode Fiber-Optic Hydrophone Based on a Schlieren Technique," *Applied Optics*, vol. 20, p. 465, 1981.

16. Lagakos, N., J. Bucaro, and J. Jarzynski, "Temperature-Induced Optical Phase Shifts in Fiber," *Applied Optics*, vol. 20, p. 2305, 1981.

17. Yariv, A., *Introduction to Optical Electronics*, Holt, Reinhart, and Winston, New York, 1976.

18. Lagakos, N., et al., "Desensitization of the Ultrasonic Response of Single-Mode Fibers," *J. Lightwave Tech.*, vol. LT-3, p. 1036, 1985.

19. Davis, C., C. Zarobila, J. Rand, and R. Lampman, "Fiber-Optic Sensors for Geophysical Applications," *Proc. SPIE*, vol. 985, p. 26, 1988. See also Dandridge, A., and A. Kersey, "Overview of Mach-Zehnder Sensor Technology and Applications" vol. 985, p. 34.

20. Birch, R., D. Payne, and M. Varnham, "Fabrication of Polarization-Maintaining Fibers Using Gas-Phase Etching," *Electron. Lett.*, vol. 18, p. 1035, 1982. Also see Stolen, R., V. Ramaswamy, P. Kaiser, and Pleibel, W. "Linear Polarization in Birefringent Single-Mode Fibers," *Appl. Phys. Lett.*, vol. 33, p. 699, 1978.

21. Ezekiel, S., and H. Arditty, eds., *Fiber-Optic Rotation Related Technologies*, vol. 32 of Springer Series in Optical Science, Springer-Verlag, Berlin, 1982.

22. Villarruell, C., and R. Moeller, "Fused Single-Mode Fiber Access Couplers," *Electron. Lett.*, vol. 17, p. 243, 1981. See also Lamont, R., D. Johnson, and K. Hill, "Power Transfer in Fused Biconical Taper Single-Mode Fiber Couplers: Dependence on External Refractive Index," *Applied Optics*, vol. 24, p. 327, 1985.

23. Bergh, R., G. Kotler, and H. Shaw, "Single-Mode Fiber-Optic Directional Coupler," *Electron. Lett.*, vol. 16, p. 260, 1980. See also Parriaux, O., S. Gidon, and A. Kuznetsor, "Distributed Coupling On Polished Single-Mode Optical Fibers," *Applied Optics*, vol. 20, p. 2420, 1981.

24. Giallorenzi, T., et al., "Optical Fiber Sensor Technology," *IEEE J. Quantum Electron.*, vol. QE-18, p. 626, 1982.

25. Dandridge, A., "Zero Path-Length Difference in Fiber-Optic Interferometers," *J. Lightwave Tech.*, vol. LT-1, p. 514, 1983.

26. Dandridge, A., A. Tveten, and T. Giallorenzi, "Homodyne Demodulation Scheme for Fiber Optic Sensors Using Phase-Generated Carrier," *IEEE J. Quantum Electron.*, vol. QE-18, p. 1647, 1982.

27. Kersey, A., K. Williams, A. Dandridge, and J. Weller, "Characterization of a Diode Laser-Pumped Nd:Yag Ring Laser for Fiber Sensor Applications," *Proc. OFS '89*, p. 172, Paris, 1989.

28. Tkach, R., and A. Chraplyvy, "Phase Noise and Line Width in an InGaAsP DFB Laser," *J. Lightwave Tech.*, vol. LT-4, p. 1712, 1986.

29. Miles, R., et al., "Feedback-Induced Line Broadening in CW Channel-Substrate Planar Laser Diodes," *Appl. Phys. Lett.*, vol. 37, p. 990, 1980.

30. Agrawal, G. and N. Dutta, *Long-Wavelength Semiconductor Lasers*, Van Nostrand Reinhold, New York, 1986.

31. Byer, R., "Diode Laser-Pumped Solid-State Lasers," *Science*, vol. 239, p. 742, 1988. See also Williams, K., et al., "Interferometric Measurement of Low-Frequency Phase Noise Characteristics of Diode Laser-Pumped Nd:Yag Ring Laser," *Electron. Lett.*, vol. 25, p. 774, 1989.

32. Wang, C., "Broadband Super Radiant Diode," *Proc. SPIE*, vol. 412, p. 54, 1983.

33. Sheem, S., T. Giallorenzi, and K. Koo, "Optical Techniques to Solve the Signal Fading Problem in Fiber Interferometers," *Applied Optics*, February 1982.

34. Jackson, D., R. Priest, A. Dandridge, and A. Tveten, "Elimination of Drift in a Single-Mode Optical Fiber Interferometer Using a Piezoelectrically Stretched Coiled Fiber," *Applied Optics*, vol. 19, p. 2926, 1980.

35. Dandridge, A., A. B. Tveten, and T. G. Giallorenzi, "Homodyne Demodulation Scheme for Fiber Optic Sensors Using Phase Generated Carrier," *IEEE J. Quantum. Electron.*, vol. QE-18, p. 1647, 1982. See also Berkoff, T., A. Kersey, and A. Dandridge, "Noise Aliasing in Interferometric Sensors Utilizing Phase-Generated Carrier Demodulation," *Proc. SPIE*, vol. 1169, p. 80, 1989.

36. Cole, J., B. Danver, and J. Bucaro, "Synthetic-Heterodyne Interferometric Demodulation," *IEEE J. Quantum Electron.*, vol. QE-18, p. 694, 1982.

37. Nightingale, J., and B. Carlson, "Designing Integrated Optics for Foundry Fabrication," *Photonics Spectra*, July 1988, p. 71. See also Alferness, R., "Guided-Wave Devices for Optical Communication," *IEEE J. Quantum Electron.*, vol. QE-17, p. 946, 1981.

38. Brooks, J., et al., "Coherence Multiplexing of Fiber-Optic Interferometric Sensors," *J. Lightwave Tech.*, vol. LT-3, p. 1062, 1985. See also Wagoner, R., and T. Clark, "Overview of Multiplexing Techniques for All-Fiber Interferometric Sensor Arrays," *Proc. SPIE*, vol. 718, p. 80, 1986. Also Kersey, A., A. Dandridge, and A. Tveten, "Overview of Multiplexing Techniques for Interferometric Fiber Sensors," *Proc. SPIE*, vol. 838, p. 184, 1987.

39. Lee, C., R. Atkins, and H. Taylor, "Performance of a Fiber-Optic Temperature Sensor From −200 to 1050°C," *Opt. Lett.*, vol. 13, p. 1038, 1988.

40. Lee, C., H. Taylor, A. Markus, and E. Udd, "Optical-Fiber Fabry-Perot Embedded Sensor," *Opt. Lett.*, vol. 14, 1225, 1989.

41. Davis, C., "Fiber-Optic Interferometric Thermometer," U.S. Patent 4,868,381, Sept. 19, 1989.

42. Zarobila, C., J. Rand, R. Lampman, and C. Davis, "Fiber-Optic Interferometric Temperature Sensor," *Sensors Expo Proc.*, p. 353, Optical Technologies Inc., Herndon, Va., 1987.

43. Lee, C., and H. Taylor, "Fiber-Optic Fabry-Perot Temperature Sensor Using a Low-Coherence Light Source," *J. Lightwave Tech.*, vol. 9, p. 129, 1991.

44. Davis, C., "Fiber-Optic Interferometric Thermometer With Serially Positioned Fiber-Optic Sensors," U.S. Patent 4,755,668, July 5, 1988.

45. Zarobila, C., "Divided Interferometer Employing a Single 3 × 3 Coupler/Splitter," U.S. Patent 4,728,191, March 1, 1988.

46. Saaski, E., J. Hartl, and G., Mitchell, "A Fiber Optic Sensing System Based on Spectral Modulation," *Adv. Instrument.*, vol. 41, p. 1177, 1986. See Also "Fiber-Optic Sensing of Physical Parameters," product feature in *Sensors: The Journal of Machine Perception*, vol. 5, p. 21, 1988.

47. Wickersheim, K., and M. Sun, "Fluoroptic Thermometry," *Medical Electronics*, p. 84, February 1987.

48. Dils, R., "High-Temperature Optical Fiber Thermometer," *J. Appl. Phys.*, vol. 54, p. 1198, 1983. See also Cooper, J., "Optical Fiber Thermometry: A New Advance in Process Temperature Sensing," Accufiber, Beaverton, Or., 1986.

49. Berthold, J., W. Ghering, and D. Varshneya, "Calibration of High-Temperature, Fiber-Optic, Microbend, Pressure Transducers," *Proc. SPIE*, vol. 718, p. 153, 1986.

50. Bucaro, J., B. Houston, and E. Williams, "Fiber-Optic Air-Backed Hydrophone Transduction Mechanisms," *J. Acoust. Soc. Am.*, vol. 89, p. 451, 1991.

51. McDearmon, G., "Theoretical Analysis of a Push-Pull Fiber-Optic Hydrophone," *J. Lightwave Tech.*, vol. LT-5, p. 647, 1987.

52. Lagakos, N., et al., "Planar Flexible Fiber-Optic Acoustic Sensors," *J. Lightwave Tech.*, vol. 8, p. 1298, 1990.

53. Shajenko, P., "Fiber-Optic Acoustic Array," *J. Acoust. Soc. Am.*, vol. 59, p. S27, 1976.

54. Rashleigh, S., "Acoustic Sensing with a Single Coiled Monomode Fiber," *Opt. Lett.*, vol. 5, p. 392, 1980.

55. DePaula, R., L. Flax, J. Cole, and J. Bucaro, "Single-Mode Fiber Ultrasonic Sensor," *IEEE J. Quantum Electron.*, vol. QE-18, p. 680, 1982.

56. Spillman, W., and D. McMahon, "Multimode Fiber-Optic Hydrophone," *Appl. Phys. Lett.*, vol. 37, 145, 1980.

57. Freal, J., C. Zarobila, and C. Davis, "Optical Fiber Microbend Horizontal Accelerometer," U.S. Patent 4,800,267, Jan. 24, 1989.

58. Leary, P., D. Manov, and Y. Li, "A Fiber Optical Borehole Seismometer," *Bull. Seismological Soc. Am.*, vol. 80, p. 218, 1990.

59. Davis, C., "Fiber-Optic Force Measuring Device," U.S. Patent 4,613,752, Sept. 23, 1986.

60. Lagakos, N., et al., "Acoustic Desensitization of Single-Mode Fibers Utilizing Nickel Coatings," Opt. Lett., vol. 7, pp. 460–462, 1982.

61. Gardner, D., et al., "A Fiber-Optic Interferometric Seismometer," *J. Lightwave Tech.*, vol. LT-5, p. 953, 1987.

62. Davis, C., C. Zarobila, and R. Lampman, "Micro-Miniature Fiber-Optic Accelerometer," *Proc. SPIE*, vol. 840, p. 55, 1987.

63. Hasegawa, R., *Glassy Metals: Magnetic, Chemical, and Structural Properties*, CRC, Boca Raton, Fla., 1983.

64. Livingston, J., "Magnetomechanical Properties of Amorphous Metals," *Phys. Stat. Sol.*, vol. 70, p. 591, 1982.

65. Bucholtz, F., et al., "Preparation of Amorphous Metallic Glass Transducers for Use in Fiber-Optic Magnetic Sensors," *J. Appl. Phys.*, vol. 61, p. 3790, 1987.

66. Dandridge, A., et al. "Optical Fibre Magnetic Field Sensors," *Electron. Lett.*, vol. 16, p. 408, 1980.

67. Koo, K., and G. Sigel, Jr., "Characteristics of Fiber Optic Magnetic Field Sensors Employing Metallic Glasses," *Opt. Lett.*, vol. 7, p. 334, 1982.

68. Koo, K., D. Dagenais, F. Bucholtz, and A. Dandridge, "Modified Coherent Rotation Model for Fiber-Optic Magnetometers Using Magnetostrictive Transducers," *Proc. SPIE*, vol. 1169, p. 190, 1989.

69. Freal, J., C. Zarobila, and C. Davis, "Fiber-Optic Magnetometer for Explosive Ordnance Detection and Degaussing Ranges," *Proc. Fiber Optics Conference '90*, AFCEA, p. 419, 1990.

70. Koo, K., F. Bucholtz, and A. Dandridge, "Synchronous Sampling Demodulation Scheme for Nonlinear Fiber-Optic Sensors," *Opt. Lett.*, vol. 11, p. 683, 1986.

71. Koo, K., F. Bucholtz, A. Dandridge, and A. Tveten, "Stability of a Fiber-Optic Magnetometer," *IEEE Trans. Mag.*, vol. Mag-22, p. 141, 1986.

72. Mermelstein, M., "Fiber-Optic Polarimetric DC Magnetometer Utilizing a Composite Metallic Glass Resonator," *J. Lightwave Tech.*, vol. LT-4, p. 1376, 1986.

73. Day, G., and A. Rose, "Faraday Effect Sensors: The State of the Art," *Proc. SPIE*, vol. 985, p. 138, 1988.

74. Carome, E. and C. Davis, "Fiber Optic Electric Field Sensor/Phase Modulator," U.S. Patent 4,477,723, Oct. 16, 1984.

75. Einzig, R., C. Davis, and C. Zarobila, "Fiber-Optic Current Transformer," U.S. Patent 4,868,495, Sept. 19, 1989.

CHAPTER 22
HIGH-RESOLUTION LITHOGRAPHY FOR OPTOELECTRONICS

Martin Peckerar, P.-T. Ho, and Y. J. Chen

22.1 INTRODUCTION

For many years, miniature light sources such as light-emitting diodes (LEDs) and solid-state lasers have been combined with diode sensors through a variety of optical channels and modulators on a single integrated substrate.[1] Prismatic couplers allowed off-chip communication.[2] Standard integrated circuit patterning technology (optical lithography and wet chemical etching) was used to make these structures. Although demands on resolution and process robustness were not great in the past, new generations of optoelectronic devices place stringent demand on patterning technology.

Semiconductor layered structures can be fabricated with predetermined bandgaps through the emerging field of bandgap engineering.[3,4] Artificial superlattices created by multilayer epitaxy enable the fabrication of laser sources and sensors tuned to a wide range of optical wavelengths. To improve laser efficiency, grating structures and confining cavities are created as integral parts of these devices. Minimum feature sizes of these gratings are on the order of the wavelength of light propagating through the device.

State-of-the-art devices may even make use of quantum mechanical phenomena to achieve low threshold currents and narrow tunable linewidths, etc. Such quantum-effect (QE) devices are possible as a result of the ability to create materials structures whose critical dimensions are on the order of the diameter of an electron-wave packet in the transport medium. Initially, QE device fabrication was accomplished by using the precise thickness control possible in thin-film layered deposition. This vertical thickness control, combined with the ultraclean crystal-growing capability of molecular beam epitaxy systems allowed the formation of the first artificial potential wells of wave packet dimensions. This gave rise to quantum wells, to multiple quantum wells, and (in the case in which electrons in different confining layers interact) to artificial superlattices.

All of these synthetic microstructures create interesting electron-hole transport and optical properties. These properties arise from the creation of localized energy states and from miniband formation (in the case of the artificial superlattice). Optical and electrical properties can be tuned by adjusting layer thickness. An added degree of "tunability" is afforded by creating confinement in horizontal as well as in vertical dimensions. In addition, horizontal confinement structures can give rise to a new class of device based on quantum phase interference principles such as the Bohm-Aharanov effect. It is this *three-dimensional* ultrasubmicron capability which has spawned the new field of nanoelectronics.

It might be thought that horizontal confinement can easily be achieved by conventional lithographic techniques. This is far from the case. The confining feature size must be on the order of the de Broglie wavelength for electrons, which is usually smaller than 0.1 μm. This dimension is certainly beyond the capabilities of existing optical tools. Even "advanced" systems, such as electron-beam (e-beam) and x-ray lithographic tools are hard-pressed to meet the needs of QE devices. E-beams are limited by beam-matter interaction, frequently referred to as *the proximity effect*. X-ray lithography does exhibit diffraction effects. As a result, many other approaches have been attempted to create lateral confinement using the self-organizing properties of certain classes of materials. But none of these techniques afford the simplicity and success of lithographic patterning.

The fundamentals of lithography as they apply to advanced microstructures are presented below. Included is a brief discussion of pattern transfer methods (etching). First, conventional optical lithography is described (primarily as a vehicle to introduce the concept of process latitude). This is followed by a discussion of e-beam lithography and of the techniques needed to achieve sub-tenth-micrometer structures with this tool. Next, x-ray lithography is presented as a relatively inexpensive, simple way to achieve sub-tenth-micrometer patterns. It is shown that e-beam pattern generation of x-ray masks surmounts some of the difficulties associated with the proximity effect. Diffraction effects in x-ray pattern replication can be eliminated by using either contact printing or microgap proximity printing. In highly ordered structures, such as gratings, holographic techniques can be exploited to producing high resolution over large areas.

In the discussions that follow, it is assumed that the reader has some familiarity with basic optical principles such as resolution, numerical aperture, and depth of focus. Excellent reviews exist as introductions to these concepts.[5,6,7] The following text concentrates on advanced lithographic procedures as they relate to the fabrication of quantum-effect and electro-optic device structures.

22.2 FUNDAMENTALS OF LITHOGRAPHY

The basic idea of practically all lithographic approaches is the same: to create a pattern in an etch-resistant material called *photoresist*. In the subsequent etch process, the resist pattern is transferred to the underlying material. Where resist was absent, underlying material is removed. Underlying material remains in the region protected by photoresist. The latent image is created in the resist by local exposure to ionizing radiation. Radiation-induced chemical changes make the resist either more or less soluble when it is exposed to solvent developers. Resists which become more soluble on exposure are termed *positive acting*. Those which become less soluble are termed *negative acting*.

Resists are applied to a variety of different planar material surfaces usually by a spinning process. The photosensitive material is dissolved in a spinning solvent and poured onto the surface to be patterned. The workpiece is then spun rapidly, creating a thin film of resist over the workpiece surface. The material is prebaked (i.e., preexposure-baked) to remove remaining spinning solvent. The resist is then exposed and developed. Development is usually accomplished by spraying a solvent over the exposed surface. The material is then postexposure-baked (postbaked) and sent on for etching. This process is illustrated in Fig. 22.1. Note that exposure sources can be of either a flood or a beam variety. Flood exposure is done either by using a shadow mask or by using a lens system to project the high-resolution image onto the wafer surface.

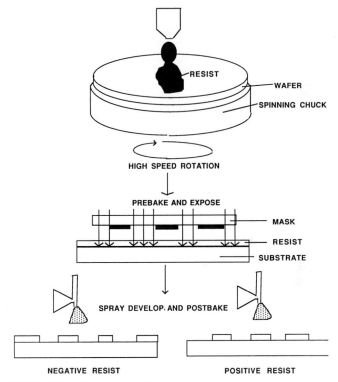

FIGURE 22.1   Photoresist application, exposure, and development.

A number of resists, working on different chemical principles, are available. The simplest type of resist action is exhibited by polymethyl methacrylate (PMMA). This material is widely used (particularly in e-beam lithography, described below). On exposure, the long polymeric chains which make up the material are broken by a process called *bond scission*. The mean molecular weight of the polymer is thus reduced, making it more soluble. Certain materials, such as cyclic rubbers, will actually increase molecular weight on exposure to ionizing radiation. This process is called *cross-linking*.

Most modern resists are referred to as *multicomponent* resists.[8] A resin, such as Novalac, serves as the base on which such materials are built. Other organic groups are grafted onto the novalac base polymeric chain. In the most common multicomponent process, these groups [frequently referred to as *photoactive compounds* (PACs)], protect the polymer base from the action of the solvent. Exposure to ionizing radiation breaks down the protecting group and leaves the chain susceptible to solvation. This leads to a positive-acting resist.

Multicomponent negative-acting resists are possible. An example is a recently developed chemically amplified resist.[8,9] Here, an acid-forming chemical group is grafted onto the polymer base chain. This is simply a molecule which releases hydrogen atoms on exposure to ionizing radiation. The hydrogen catalyzes cross-linking elsewhere in the chain. The result is a negative-acting multicomponent system. One photon may release one hydrogen ion, but that hydrogen ion may create many cross-links. Hence, the chemistry "amplifies" the exposure process.

Resists are usually characterized by their development properties as expressed by plots of the logarithm of dose versus the normalized film thickness remaining (after a fixed development time). Examples are shown in Fig. 22.2 for both positive and negative resists. As expected, for negative resists, dissolution rate lowers for increased dose. For positive resists, increasing dose increases dissolution rate. Resist contrast (sometimes termed *gamma*) is the slope of the dissolution rate plot taken at the midpoint of its log-linear region (the D_g^x point). Two other points are also specified: D_g^0 and D_g^i. These roughly correspond, respectively, to the true exposure dose and to the point at which exposure is initiated.

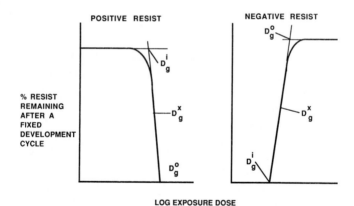

POSITIVE RESIST NEGATIVE RESIST

% RESIST REMAINING AFTER A FIXED DEVELOPMENT CYCLE

LOG EXPOSURE DOSE

FIGURE 22.2 Resist development curves.

High-contrast materials can give developed resist sidewall profiles which are very steep (approaching vertical). Such an exposure is shown in Fig. 22.3. The resist is highly sensitive, jumping from zero exposure to maximum exposure over a very small range of incident doses. The steep sidewall profile such resists create is of particular importance in high-frequency optical modulator fabrication, for reasons described below. While high contrast is useful in obtaining steep sidewall profiles, linewidth control depends critically on the quality of the optical image cast on the resist, as shown in Fig. 22.4. Techniques for creating high-resolution exposure images are discussed in the next section. The mechanism for achieving very steep sidewalls in thick photoresist is also described.

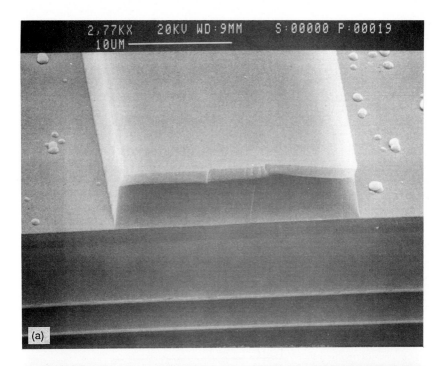

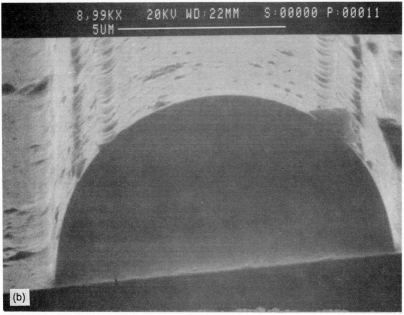

FIGURE 22.3 High-contrast resist exhibiting steep sidewall. (*a*) Optimized result showing high resolution and steep sidewall. (*b*) Unoptimized result.

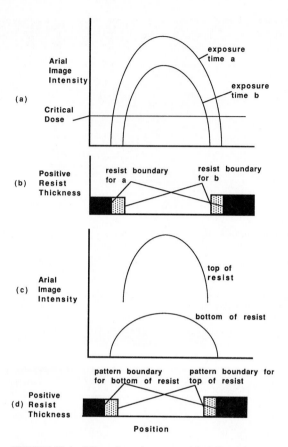

FIGURE 22.4 Effect of area image on critical dimension control in a high-contrast resist. (*a*) Changes in the intensity distribution caused by changing the total incident exposure dose; (*b*) Result of dose changes on resist boundary position; (*c*) Spreading of diffracted intensity distribution as the light propagates to the bottom of the resist; (*d*) Effect of diffraction-induced spreading on the resist boundary.

22.3 LITHOGRAPHIC TECHNIQUES USEFUL IN OPTOELECTRONIC DEVICE FABRICATION

22.3.1 Optical Lithography

Optical lithography is the workhorse[9] for patterning of integrated microelectronics. In the integrated circuit industry, exposure tools are generally projection cameras which either step or scan the image onto the workpiece one exposure field at a time. The workpiece is usually composed of many exposure fields. Projection systems contain highly corrected and complex optical-lens systems.

In the electro-optics community, contact and proximity printers which use shadow masks are still frequently in use. During contact printing, the mask actually makes physical contact with the surface of the workpiece. In proximity printing, the mask is held a small distance above the workpiece (usually about 25 μm). This latter technique is particularly common at universities. Such tools were used extensively in the early days of the integrated circuit. Even though they afforded high resolution, they were severe limiters of functional circuits (yield) as the size and complexity of the circuits increased. However, proximity printing is currently being resurrected for use in x-ray lithography. X-ray lithography is treated in Sec. 22.3.3. A discussion of its fundamental operating principles is deferred until that time.

The resolution of a lens optical system is determined by the wavelength of the light imaged and the numerical aperture. The shortest exposure wavelength contemplated is 193 nm (an ArF excimer laser source). A common exposure source is the mercury lamp. Production systems commonly use filters to pass only the g line of the mercury emission spectrum (436 nm). Advanced systems use the i line (365 nm) for exposure. Numerical apertures of approximately 0.5 are also envisioned. Table 22.1 summarizes estimates of future lithographic system parameters. As a rule of thumb, the minimum resolved feature size R_{min} for such a system is

$$R_{min} = k_1\left(\frac{\lambda}{NA}\right) \tag{22.1}$$

where λ is the wavelength of the exposure light, NA is the numerical aperture of the system, and k_1 is a factor between 0.5 and 1. It is generally conceded that such systems may approach 0.25 μm in R_{min}, which is well above the sub-tenth-micrometer resolution regime needed for nanoelectronics.

TABLE 22.1 Map of Resolution Trends in Optical Lithography

Source wavelength, nm	Limiting NA	Line/Space, μm														
		0.56	0.50	0.45	0.40	0.35	0.31	0.28	0.25	0.22	0.20	0.175	0.156	0.140	0.125	0.110
g 436	0.65		○			□		△								
i 365	0.60			○			□		△							
KrF 248	0.50				○			□		△						
ArF 193	0.50							○		□			△			
F$_2$ 157	0.50								○			□				△
		1989	90	91	92	93	94	95	96	97	98	99	2000	01	02	03	04

Year

○ $k_1 = 0.7$ □ $k_1 = 0.5$ △ $k_1 = 0.35$

Source: Courtesy of Fabian Pease, Stanford University.

Resolution limit is only one parameter defining optical system performance. Another key parameter is depth of focus (DOF). This parameter defines the maximum tolerable displacement of the image plane from its ideal position. It, too, is determined by numerical aperture and wavelength:

$$DOF = k_2 \left(\frac{\lambda}{NA^2} \right)$$
(22.2)

Here, k_2 is a process-related constant similar to k_1 whose range is also between 0.5 and 1. In the submicrometer regime, the DOF is roughly equal to the minimum resolved feature size. In many optoelectronic applications, DOF limitations are less restrictive than in conventional microelectronics. Many surfaces (such as lithium niobate slabs used in Bragg cell work) are optically polished and the surface flatness is controlled to a fraction of an optical wavelength over a lithographic exposure field. In other cases (such as one encounters in compound semiconductor superlattices), surface flatness is poorer than the 1 μm/cm frequently quoted for ultraflat silicon wafers.

Optical lithography does serve to point out one important concept which will be of relevance to any high-resolution patterning technique: the concept of process latitude.[10] To understand this concept, consider the following proposition. It may be possible to create a set of exposure and development conditions for achieving a given lithographic goal. But if tiny variations in these conditions cause major changes in the exposed image, the process is worthless. The change of some critical process parameter (such as linewidth, edge acuity, etc.) with respect to processing conditions (such as exposure time, development time, pre- and/or postbake condition, etc.) is called *process latitude*. Ideally, a process whose critical parameters change as little as possible for a given change in process condition is desired.

As an example of process latitude, consider the change in linewidth caused by over- or underdevelopment. A development profile for a conventional mercury arc lamp g-line (436 nm) exposure of a standard novalac-based resist is given in Fig. 22.5. A number of resist profiles are shown, corresponding to different times in the development process. This figure represents the results of a model which cal-

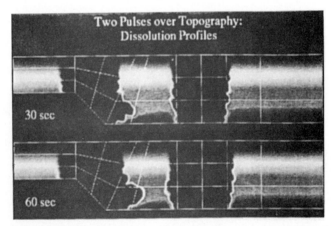

FIGURE 22.5 Image in resist exhibiting standing-wave and notching effects as a function of development time.

culates the exposure energy versus deposition in the resist using a numerically accurate solution to the vector Maxwell equations combined with a dissolution model which predicts the resist dissolution front by solving an Eikonal equation. A number of scallops appear in the sidewall profile of the resist as a result of standing-wave formation. A notch appears in the centerline of the resist due to reflections off the wafer topography.

Figure 22.6 shows the width of the bottom cleared region as a function of development time. If the resist was not affected by the etch process, the width of the bottom cleared region would correspond to the width of the etched line. Note that the top of this curve never flattens out. That is, continued development continuously increases the width of the resist aperture. The *development latitude* can be defined in terms of the slope of the bottom cleared versus development time plot. This slope for the case at hand is 200 Å/s. Thus, over- or underdeveloping by 1 s causes a resist linewidth change of 200 Å. Certainly for the 1-μm process targeted by the exposure system employed, this is not serious. However, as linewidths shrink, this effect becomes even more severe. Furthermore, development latitude is only one relevant latitude: exposure latitude and latitudes relating to resist baking condition are also important.

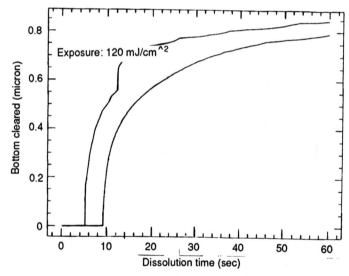

FIGURE 22.6   Width of bottom cleared region as a function of development time for notch slot in Fig. 22.5. The right-most curve refers to the profile in the planar region shown in Fig. 22.5. The left-most curve refers to the feature over topography.

Figure 22.6 also indicates one further problem associated with optical systems— the effect of underlying topography. On going over a step, standing-wave effects may be evident in a direction parallel to the surface. This gives rise to the notch evident on the right side of the exposure. The thin layer of resist remaining at the bottom of the notch will not hold up to subsequent etch processes, leading to further degradation of process latitude.

In Sec. 22.2, the use of high-contrast resists in the formation of steep sidewalls in thick resist was discussed. The results described in this section would lead one to believe that images such as those shown in Fig. 22.3a represent a kind of "super resolution" [resolution beyond the diffraction limit, expressed in Eq. (22.1)]. In fact, such images do appear to exceed diffraction limits.[11] The reason for this is shown in Fig. 22.4. A number of competing effects are evident which can be controlled to our advantage here. As the image propagates through the thick resist, diffraction spreads the beam and lowers the intensity of the central maximum of the transmitted pulse. The total energy in the pulse would remain constant were it not for absorption in the resist, which further lowers the overall image intensity. The width of the pulse broadens as it propagates through the resist. This would tend to broaden the exposed line. But the overall line intensity lowers, moving the boundary at which the critical exposure intensity occurs closer to the center of the pulse. This narrows the pulse, as shown in Fig. 22.4b. Thus, the exposure boundaries at the top and bottom of the resist can be manipulated and made equal (by changing resist thickness and exposure times). If exposure times and thicknesses are not optimized, results such as those shown in Fig. 22.3b are obtained.

To summarize, usable optical system resolution will clearly extend below 0.50 μm, possibly extending to 0.25 μm. The ultraflat imaging surfaces used in many optoelectronic devices tend to make lithography easier by eliminating depth-of-focus problems. But even advanced optical systems will not meet the needs of QE devices. These systems fail on three accounts. First, optical resolving power is too poor. Second, DOF limitations place limits on out-of-plane surface topography. Finally, process latitude is insufficient to support requisite accuracies.

22.3.2 E-Beam Lithography

E-beam lithography provides sufficient resolution to fabricate QE devices. *Single-pass* lines (isolated lines created by a single pass of the electron beam) as small as 80 Å have been demonstrated. Line and dot patterns smaller than 300 Å have routinely been produced (see Fig. 22.7). The small diameter of the beam (80 Å for many nanolithography tools) leads to the conclusion that sub-100-Å lithography is possible with existing e-beam machines. This is not the case. The reason for this lies in the nature of the beam-matter interaction which occurs on exposure. An in-depth understanding of beam-matter interaction is required to achieve ultra-high-resolution with this technique. The fundamentals of beam-matter interaction are presented below. Furthermore, it is shown that even with this understanding, ultra-high-resolution with e-beams is possible only on a restricted class of substrates.

There are two dominant components to this interaction: forward and backward scattering.[12] When the beam strikes the resist, it splays apart as a result of forward-scattering processes. In addition, some fraction of the incident electrons can be turned around due to large-angle scattering processes in the exposure substrate. These backscattered electrons can also contribute to resist exposure. In addition, low-energy secondary electrons produced by the primary beam also contribute to the exposure.

Typically, the exposure field is broken into *pixels* (individual beam probes) placed on a square grid (or *address structure*). The energy dose d, absorbed as a result of a single pixel exposure is usually written as

$$d(r) = \frac{K}{\pi(1 + \eta)} \left[\frac{1}{\beta_f^2} \exp\left(-\frac{r^2}{\beta_f^2}\right) + \frac{\eta}{\beta_r^2} \exp\left(-\frac{r^2}{\beta_r^2}\right) \right] \qquad (22.3)$$

FIGURE 22.7 High-resolution lines in resist exposed with a nanometric e-beam tool.

where the bracket prefactor provides dose normalization, η is the back-to-forward scattering ratio, r is the distance from the center of the pixel, and β_f and β_r reflect the beam-broadening effects of the forward and backscatter process. This double-Gaussian approach can be changed to a triple-Gaussian,[13] or even to an n-term Gaussian expansion approach to account for secondary processes or other processes not effectively modeled by the double-Gaussian. Relevant parameters are included as Table 22.2. The $d(r)$ profile is a function of beam energy and depth into the resist. For most practical modeling projects, though, one chooses a single characterizing depth.

Inspection of Eq. (22.3) and Table 22.2 indicates that, even for single pixels, the dose profile extends well beyond the incident probe diameter. The situation is more complicated in the case of single-pass lines. Consider a dose versus position profile taken in a direction normal to the single-pass line. The Gaussian tails from

TABLE 22.2 Some Typical Values for Constants in Eq. (22.3)

Beam energy	β_f, μm	β_r, μm	η	K
20 kV	0.1317	1.3783	2.3060	0.3308
40 kV	0.11	2.500	2.4958	0.2583

pixels in front of and behind the cross-sectioned pixel sum to form the total dose profile. The result, shown graphically in Fig. 22.8, shows that this profile can be modeled by a double-Gaussian with effective β's which are much larger than the true β_r and β_f. This is referred to as *intraproximity* effect. Of course, the results in the table and the figure scale as the β's scale.

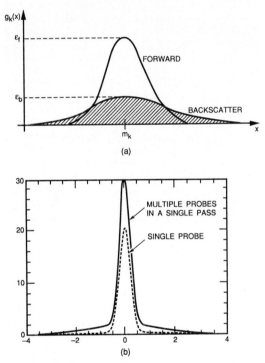

FIGURE 22.8   Intraproximity effect in a single-pass line. (*a*) Forward and backscatter effects in a single pixel exposure; (*b*) intraproximity effect caused by multiple pixel exposures in a single-pass line.

When individual features are written close together, the Gaussian tails sum laterally to create an enhanced exposure in the region between the exposed pixel boundaries. This enhancement can be so severe as to destroy the boundary between the two regions. Even if this does not occur, the slope of the dose-position curve is reduced and process latitude is degraded. This is referred to as *interproximity effect*.

Of course, for near-micrometer applications, a variety of approaches to limit proximity effects are possible. For example, the size of the written feature can be scaled to allow the developed image to equal the desired dimension. More aggressive processes are currently under development. For example, in the GHOST approach,[14] the normally unwritten field is written with a broadened (defocused) Gaussian beam. The summation of the field and feature doses yields a very steeply varying dose versus position curve at a feature boundary, enhancing image contrast. Feature sizes down to 0.35 μm have been written this way.[15] The mechanism of GHOST contrast enhancement is illustrated in Fig. 22.9

Computational approaches are also possible. Here, the dose is adjusted on a pixel-by-pixel basis to optimize the energy deposition profiles.[16,17] It should also be pointed out that Gerber[18] has shown that closed-form explicit solutions of the proximity effect equations are possible for highly restricted geometries. One of the restricted cases Gerber presents is the line and space problem. This is of particular importance when dealing with grating structures used on optoelectronics. All of these techniques become marginally useful as we enter the nanometrics arena. This is because the precise form of the dose versus position profile is necessary for good computational results. The precise form of this relationship is tool-parameter dependent and is not well-known.

There are three ways to get around proximity problems for fabricating nanometric device structures. One way is to reduce the energy of the exposing beam to the kilovolt (or lower) range. Scattering processes cannot carry energy far from the point of initial energy deposition. Such low-energy e-beams cannot provide full thickness resist exposures if the resist has any measurable thickness. The beams are not energetic enough to reach the bottom. To overcome this drawback, top-layer imaging resists (some based on Langmuir-Blodgett film approaches[19]) are being investigated.

In conventional top-layer imaging, the latent image is contained in a very thin surface layer of resist which is not etchable in oxygen plasma. Oxygen plasmas are then used to anisotropically etch a thick, passivating resist layer which is used to transfer the latent image into underlying material. As expected, these thin top layers have relatively high defect counts, and extraneous resist is frequently left in the field. Top-layer imaging is still in its infancy and may become a method of choice in the future.[20,21]

A second method to overcome the proximity effect is to use high-energy e-beams (greater than 50-keV accelerating potential). Such beams are relatively "stiff," and forward scattering is not as much a problem. The total number of backscattered electrons reaching the surface is roughly the same, but the radius of the backscatter Gaussian β_b is much larger. This means that the backscatter-Gaussian tails contributing to the dose summations will be smaller in individual pixels other than those of primary incidence. This mitigates intra- and interproximity effects.

Finally, substrate backscatter effects can be eliminated by removing the substrate. This is really not a facetious remark. The shadow mask can be printed on a membrane whose thickness is smaller than the penetration depth of the e-beam, essentially removing the backscatter component. As a rule of thumb, materials whose density is about that of silicon exhibit penetration depths of 1000 Å/keV. Thus, a 20-keV incident beam will penetrate about 2 μm. Most suitable membrane materials are fairly opaque in the ultraviolet (uv) and deep uv and cannot be used as optical masks. Thin, durable (less than 2 μm thick) membranes can be made out of silicon, silicon nitride, or silicon carbide, which are essentially transparent to x-rays useful for lithography (x-rays in the 0.8- to 2-keV energy range). A typical x-ray mask structure is shown in Fig. 22.10.

In fact, the highest-resolution e-beam lithography has been done on membranes such as those shown in this figure. Line and space patterns with minimum feature sizes less than 100 Å have been made in this way.[22] Features on such membranes can be replicated by x-rays. E-beams alone could not resolve such fine lines on typical substrates of interest in microdevice work. High-Z compound-semiconduc-

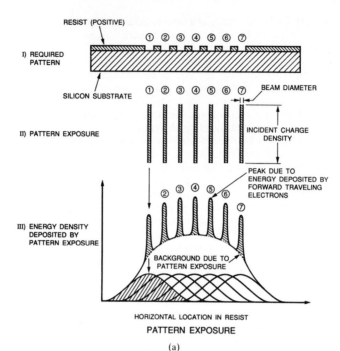

FIGURE 22.9 Contrast enhancement with GHOST. *(After Ref. 14.)* *(a)* Exposure of seven closely spaced lines and the resulting resist energy deposition contour. Note the main peaks superimposed over the backscatter and secondary distribution. This gives rise to spikes on a more gently varying background dose. *(b)* Dose equalization using a GHOST corrector. If the gently varying background dose in Fig. 9a was flat, there would be no linewidth variation. To accomplish this flattening, the normally unexposed field region is written with a defocused beam.

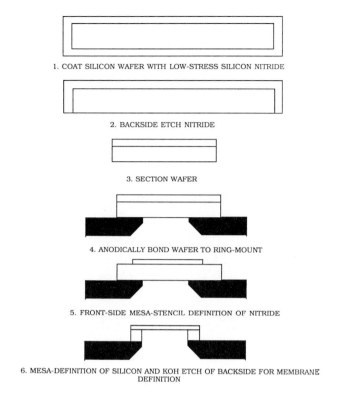

1. COAT SILICON WAFER WITH LOW-STRESS SILICON NITRIDE

2. BACKSIDE ETCH NITRIDE

3. SECTION WAFER

4. ANODICALLY BOND WAFER TO RING-MOUNT

5. FRONT-SIDE MESA-STENCIL DEFINITION OF NITRIDE

6. MESA-DEFINITION OF SILICON AND KOH ETCH OF BACKSIDE FOR MEMBRANE
DEFINITION

FIGURE 22.10   Typical x-ray mask structure.

tor substrates are frequently used in this work. These dense substrates exhibit enhanced backscatter and enhanced proximity effect.

It should be pointed out that ion beams as well as e-beams do exhibit very high resolution.[9] In addition, there is a very small ion backscatter and secondary electron exposure component resulting from application of this technique. This produces a relatively small proximity effect. Ion-beam lithographic systems are not currently commercially available, largely because ion gun reliability and stability have been problems. In addition, exposing ions may implant into the substrate, creating a need for blocking layers.

To summarize, electron beams do provide the highest resolution capability of any widely available lithographic tool. While sub-100-Å resolution is possible with this technique, proximity effects and process latitude make such a goal possible only for certain classes of substrates. Ultrahigh-resolution patterns can be fabricated on membranes useful as x-ray masks. Actual device replication can be accomplished on a wide variety of substrates by x-ray exposure systems.

22.3.3   X-Ray Lithography

X-ray lithography is currently the focus of a worldwide development effort involving government, industry, and university facilities. The goal of this program—insertion of x-ray technology into the manufacturing lines of integrated circuit (IC) manu-facturers—differs significantly from the goals of those interested in electro-optics and nanometric devices. But both programs can be viewed as symbiotic. In IC work, there is a heavy emphasis on yield and throughput. Level-to-level alignment is also a critical issue. In the nanometrics area, emphasis is on resolution. Low yields can be tolerated, since device demands are not high, as yet. Outside the regions in which quantum effects are generated, alignment and resolution can be fairly coarse. In both the IC and nanodevice areas, there is emphasis on critical dimension control, process latitude, and resolution. In this section the near and long-term resolution goals of x-ray lithography are outlined.

X-ray lithography, as it is currently practiced, is an extension of optical proximity printing utilizing very short wavelength radiation. In the past it was felt that pe-numbral blur (source-size effect in shadow printing, illustrated in Fig.22.11) would limit resolution. This was based on the assumption that low-power extended-source

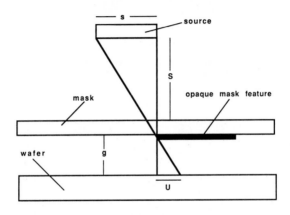

Source induced feature undercut = U = (s/S)g

FIGURE 22.11 Source size effect in x-ray lithography.

x-ray tubes would be used as exposure sources. These sources would have to be set close to the workpiece, aggravating the penumbra problem. Currently, system development calls for use of synchrotron storage rings or high-brightness laser point sources. The characteristics of these sources are given in Table 22.3.[6] The syn-chrotron is a line source of exceptionally small divergence. Synchrotron radiation arises as a result of creating a high-energy relativistic beam of electrons which is caused to run a circular modified elliptical path. Whenever the beam changes direction, radiation is emitted. In a laser-plasma source, a high-energy laser im-pinges on a solid target. The solid material is ablated, ionized, and heated to the point of x-ray emission by the laser beam. The laser-plasma source extent is less than 100 μm. Penumbral blur is not an issue affecting resolution in either of these cases. The two factors which do create resolution limits are diffraction effects and photoelectron generation. These are discussed in turn below.

The discussion of diffraction is begun by considering the replication of a grating structure (an equal line and space pattern). Assume that a grating pattern on a mask is completely resolved when the main beam (zero-order diffraction maxima)

TABLE 22.3 A Comparison of Current X-Ray Sources for Lithographic Application

	X-ray tube	Synchrotron	Laser plasma
Total power on target	1 mW/cm^2 at 10 cm	100 mW/cm^2	10 mW/cm^2 at 10 cm
Source-broadening parameter	Source size ~5 mm	Beam divergence 50 mrad	Source size 100 μm
Source spectrum	Line + continuum	Band of wavelengths usually set to peak at 10 Å	Lines (usually in the 12–14-Å range)

just coincides with the first-order peaks. It can be shown that this condition leads to a minimum resolved period p_{min} for a given mask-to-wafer spacing s and exposure wavelength λ:

$$p_{min} = k'\sqrt{\lambda s} \qquad (22.4)$$

where k' is a constant (usually between 1 and 3) reflecting the contrast factor of the resist as well as pre- and post-development resist temperature cycles.

Even if the mask were to come in contact with the wafer, the photoresist layer would create an effective separation between the mask and the imaging plane. But bringing the mask in contact with the wafer is really not a viable approach, since the mask is made on a delicate membrane which could shatter on contact with a substrate. "Microgap" approaches are possible.[23] Here, the mask-to-wafer separation is about 5 μm, and sub-thousand-Ångstrom minimum feature sizes are possible.

More recently, optical coherence effects associated with the x-ray source have been debated in the literature. In the case of a completely coherent source, all the "bumps and wiggles" normally associated with diffracted image formation manifest themselves fully. The shadow is not clean-cast. Synchrotron sources can be built that exhibit minimal divergence and a fair degree of coherence. In point-source work, in order to achieve effective distance collimation of the illuminating radiation, source-to-mask separation must be large. As the bumpy, spatially incoherent wavefront spreads out from the source, the wave front smooths. The coherence of the illumination increases and the nature of the image on the plane of illumination changes. Wavefront interference effects become much more pronounced. These effects influence process latitude.[24] More recent work by Guo[25] indicates that in practical sources, such effects are minimal.

The final resolution inhibitor to be considered is photoelectron spreading. The incident x-ray photon usually deposits all its energy in the incident solid by creating a single high-energy photoelectron. This primary photoelectron creates a secondary electron shower and creates the chemical alterations in resist necessary for exposure. The range of this photoelectron was thought to create a resolution limit. As a rule of thumb, photoelectron ranges in low-atomic-weight materials are 1000 Å per kilo-electron-volt of starting energy. Thus, for a 1-keV exposure source, resolution is degraded by at least 1000 Å. Recent work by Early[26] has shown that these photoelectron effects are not observed. Detailed study of the energy distribution of the secondary ensemble reveals that most of the secondary photoelectrons are much lower in energy and the mean photoelectron range of this distribution is far less than previously thought. Features smaller than 500 Å have been resolved with microgap printing.[23]

To summarize, x-ray lithography exercised in the microgap mode (with mask-to-wafer separations less than 5 μm) appears to be a viable technique for mask production of quantum-effect devices.

22.3.4 Holographic Image Definition

From the above discussion, it may be thought that e-beam or x-ray lithography can provide a solution to the optoelectronic grating fabrication problem. This is true only for small-area gratings. The reason for this lies in the way e-beams work. These tools expose a field whose dimensions are determined by the maximum deflection of the beam and by the digital-to-analog pattern-generation electronics. Once the field is exposed, the table on which the workpiece sits must be mechanically moved to a new location. The maximum field size is tool-dependent and ranges from 1 mm to 3 mm on a side. Random jogs appear at the field boundary. Dense gratings take a long time to write and machine drifts occur within a field. These factors destroy phase coherence of the grating. Holographic techniques can provide the solution to the large-area grating problem.

It is a fundamental theorem of optics that an object placed at the right-hand focus of a lens system and the image formed at the left-hand focus exist as Fourier pairs. The Fourier transform of a grating pattern (a square-wave intensity pattern of constant period) is another grating pattern. An illustration of this is shown in Fig. 22.12. If we consider the spherical wave emanating from one slit to be a reference and the spherical wave emanating from another slit to be a signal wave,

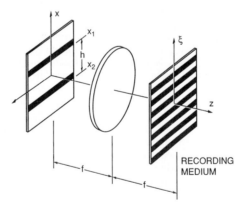

FIGURE 22.12   Fourier transformation relationship of a grating placed at the focal plane of a lens.

the imaging process can be viewed as a form of holography. Some workers have even suggested techniques for holographically imaging complex patterns. This involves development of advanced computational approaches. But for many optoelectronic devices, simple, high-resolution grating structures are sufficient. If the image plane is sufficiently flat, holography is a relatively easy, viable fabrication method.

In an elementary application of this technique,[27] a mask is made on which is written the image to be projected. The lens system demagnifies the image. The demagnification is a function of the focal length of the lens system, f, slit separation on the mask, h, and the wavelength of the exposing light, λ. The actual intensity distribution measured along the y axis (normal to the grating line) is

$$I_d = \frac{I_0}{2} (1 + \cos 2\beta_h) \tag{22.5}$$

where I_0 is the light intensity incident on the resist, and

$$\beta_h = \frac{\pi h y}{\lambda f} \tag{22.6}$$

One of the advantages of holographic imaging is the fact that the exposure intensity of a given point in the image plane is the sum of excitations arising from many points along the transmissive regions of the mask. This creates a type of spatial "averaging" which makes the hologram relatively insensitive to local mask defects.

The ultimate resolution of the hologram is half the wavelength of the exposure beam. However, as we are attempting to create a highly modulated diffraction pattern, the transverse optical coherence length of the exposure source, l_t, is an important contributor to resolution limit. The image blur of a point source, δx, is given as

$$\delta x = \frac{\lambda}{2l_t} Z_s \tag{22.7}$$

where Z_s is the mask-to-image plane separation.

Thus, it is essential to use laser systems with a high degree of coherence to form clean, high-contrast images. Furthermore, it should be pointed out that the image intensity distribution is cosinusoidal. When working near the resolution limits of the optical system, sloped or rounded sidewalls will be obtained. In many applications, this may actually be advantageous. When the pattern is transferred into underlying material in the etch process, a triangle wave, rather than a square wave grating will develop. Such "blaze" gratings have high diffraction efficiency. This is described in greater detail in the next section.

Further consideration must be given to exclude spurious modulations due to the optical system aperture as well as spurious modulations due to the finite size of the mask slit aperture. With reference to Fig. 22.12, the angles of arrival of the wavelets from positions x_1 and x_2 are

$$\theta_1 = -\frac{x_1}{f} \quad \text{and} \quad \theta_2 = \frac{x_2}{f} \tag{22.8}$$

where f is the focal length of the lens system employed. In order to exclude spurious modulations, the difference between θ_1 and θ_2 ($\delta\theta$) should be

$$\delta\theta = \sin^{-1}\left(\frac{3k_\omega\lambda}{2\pi}\right) \tag{22.9}$$

where k_ω is the wave number corresponding to the minimum resolved period.

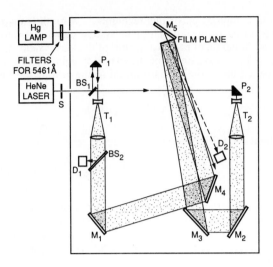

FIGURE 22.13 Jenney's apparatus for holographic grating fabrication. *(After Ref. 30.)*

In more recent applications, two laser beam sources are used to provide[29,30] the reference and the signal created by the mask in the above discussion. Using beam splitters, a single laser source can be used, as shown in Fig. 22.13.[30] The single-source illumination is broken into two flood sources, each incident on the surface with a slightly different angle with respect to the surface normal. If α is the difference in incidence angles, the resulting line spacing is

$$d = \frac{\lambda}{2 \sin \alpha} \tag{22.10}$$

where λ is the exposure wavelength. This is illustrated in Fig. 22.14.

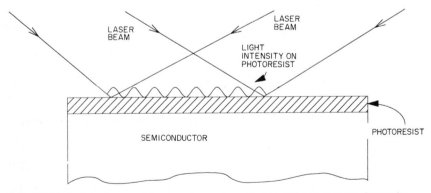

FIGURE 22.14 Interference of two laser beams to produce a grating pattern on photoresist.

The mercury source shown in this figure is used to cure the polymer resist. Such curing may not be necessary for other resist systems. This approach eliminates the need for a mask. But the discussion above relating to resolution is still valid. Newer techniques make use of conventional photoresists.[31]

23.3.5 Pattern Transfer Techniques

Lines drawn in resist are seldom used in finished devices. The resist is used as a guide which defines material boundaries created by subsequent etch processes. In this section we go over some etch techniques commonly used in optoelectronics: wet etching, ion milling, plasma etching, chemically assisted ion beam etching and lift-off.

Wet etching of materials is the easiest and most commonly used pattern-definition technique.[6,32] Here, the material to be etched is soaked in an acid or a base which dissolves where it is not protected by resist. Selective etches are employed which will not attack the underlying material. Highly selective etches are possible in wet etching. The main problem associated with this process (as it is applied to high-resolution patterning) is referred to as *undercut*. The material etches isotropically (i.e., there is no direction dependence to the etch). As shown in Fig. 22.15, this means that the material attacked will etch a considerable distance under the protecting resist. In a completely isotropic process, at the material-resist boundary, material will etch as far in a direction parallel to the surface as it etches down. Anisotropic etches will etch only in a direction normal to the surface.

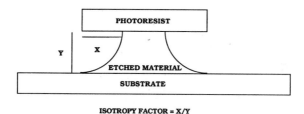

FIGURE 22.15   Definition of etch anisotropy.

One example of an anisotropic etch (direction-dependent etch) is known as *ion milling*.[32] Here, a noble gas (such as argon) ion beam impinges normally to a surface. Unwanted material sputters away. Resist has a relatively low sputtering yield, so it will protect underlying material. Etching is normal to the surface, yielding very steep sidewall profiles in the etched material. But the process is not very selective. Sputter rates for metals and semiconductors are usually not very different. As discussed above, if the resist sidewall has some structure to it, if the resist is consumed in the etch process, the sidewall structure can be transferred to underlying material. Sputter etching meets these criteria since both resist and the underlying material are consumed in the etch process. An example of this transfer is shown in Fig. 22.16.[33] As the sputtering process is angle-of-incidence dependent, interesting faceting effects occur, as shown in the figure.

A synthesis of wet-chemical and sputter etching is known as *anisotropic plasma etching*.[34,35,37] There are two common approaches to anisotropic plasma etching:

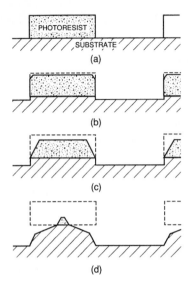

FIGURE 22.16 Faceting effects in sputter etch definition. *(After Ref. 33.)* (*a*) Initial conditions: resist over substrate; (*b*) development of facets in resist due to angle-dependent sputtering process; (*c*) substrate etch commences; (*d*) transfer of facets to substrate.

parallel-plate and reactive-ion etching (RIE). These are illustrated in Fig. 22.17. In parallel-plate etching, an etch gas is bled into an etch chamber. A radio-frequency discharge is struck between two parallel conducting plates. The resulting plasma forms an ion sheath around all the surfaces in the etch chamber, including the surface of the workpiece. Positive ions are accelerated through the sheath space charge. This acceleration enhances the etch rate normal to the surface. In one typical etch process, a normally unreacting gas such as CF_4 is bled into the etch chamber. The discharge releases highly reactive flourine ions, which are accelerated to the surface of the workpiece, etching it. As the etching is still a chemical process, the selectivity of wet etching is possible. Acceleration through the plasma sheath creates anisotropy.

The aim of RIE processing is to increase etching species' acceleration through the plasma sheath in order to increase anisotropy. This is accomplished in three ways. First, the pressure of the chamber is lowered below 1 torr. This prevents ion scattering in the sheath. Next the counterelectrode opposite the workpiece is removed. This removes the capacitive voltage drop across the counterelectrode sheath, doubling the electrostatic drop across the sheath surrounding the workpiece. Finally, the radio-frequency (RF) field is coupled to the workpiece through a coupling capacitor. During the positive swing of the RF cycle, the capacitor plate in contact with the workpiece electrode charges negatively. During the negative swing, the plate must discharge through the positive ion current. Ions are much less mobile than electrons and, at the 13.6-MHz RF frequencies usually used, this discharge is incomplete. Thus, the workpiece assumes a negative bias, enhancing ion acceleration through the sheath. In many RIE processes, acceleration yielding ion incident energies in excess of 500 eV is possible. This creates a sputtering component to the material removal process which can destroy selectivity.

Reactive ion etching and ion-milling are combined in a process called *chemically assisted ion-beam etching* (CAIBE). Here, a broad area ion beam of reactive ions is incident on the workpiece. The degree of sputtering can be controlled by changing the ion-beam energy.

In the lift-off process,[9,36] the material to be defined is not etched. A negative image of the pattern to be defined is created in photoresist. The workpiece is transferred to an evaporator and the resist is covered with metal. The metal film breaks continuity over the resist sidewall. The resist is soaked in its solvent, lifting off the undesired metal. Sidewalls must be very steep to ensure a clean break of the metal. Furthermore, the top of the resist is sometimes chemically hardened by 30-s soaks in hexamethyl disilosane. This creates a reentrant profile which further ensures breakage of the metal over the resist step. This is a very popular process

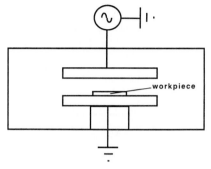

(a) conventional parallel plate system

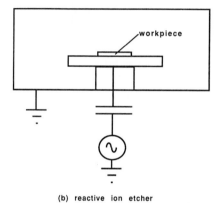

(b) reactive ion etcher

FIGURE 22.17 A comparison of RIE and parallel-plate etch systems.

for patterning materials which are difficult to etch by wet chemistry or by plasma
processes, such as gold and gold alloys. The lift-off process is illustrated in Fig.
22.18.

This completes our discussion of lithographic techniques and pattern transfer
operations useful for fabrication of optoelectronic devices. In the following sections,
fabrication examples are given.

22.4 EXAMPLES

22.4.1 Lasers

Distributed Feedback Lasers. Probably the most prominent example of submicrometer lithography in optoelectronics is the semiconductor laser with distributed
feedback. Semiconductor lasers with distributed feedback are important stable
single-frequency sources for optical communication. A light wave traveling in a
periodic medium can be deflected by the coherent superposition of reflections from

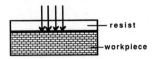

(a) resist application and exposure

(b) chlorobenzene soak and develop - note "re-entrant" lip

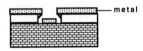

(c) metal deposition and acetone soak

(d) resulting metal definition

FIGURE 22.18   Lift-off patterning.

the nonuniformity of the medium. This fact has been used to provide feedback in a laser.[38]

In semiconductor laser technology, there are two common variations of this scheme. The first, called *distributed feedback* (DFB), uses a periodic variation of the refractive index of the material above or below the gain region of the laser (Fig. 22.19). The second, called the *distributed Bragg reflector* (DBR), places the

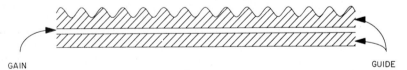

GAIN   GUIDE

FIGURE 22.19   Distributed feedback laser with grating on top of gain.

index gratings at one or both ends of the gain region (Fig. 22.20). The gratings are routinely fabricated with an ultraviolet laser by interfering two laser beams on the semiconductor surface where the gratings are designed to be placed (Fig. 22.14). Fabrication techniques for accomplishing this placement were discussed in depth in Sec. 22.3.4. The developed resist is then used as an etch mask for a subsequent plasma etch process. Sometimes additional material is grown on the top of the

FIGURE 22.20   Distributed Bragg reflector laser with gratings at ends of gain.

etched surface. Regrowth is possible through a chemical vapor deposition process known as *metallo-organic chemical vapor deposition* (MOCVD). More conventional molecular-beam epitaxy processes are of limited use here.

The index of refraction of GaAs-based materials is between 3 and 4; the most common wavelength regions are 0.8 to 0.9 μm (AlGaAs compounds) and near 1.5 μm (InGaAs compounds), which translates into 0.22 μm to 0.4 μm in the semiconductors. A first-order grating has its period equal to half the wavelength in the material, and the deflected wave travels in exactly the opposite direction of the incident wave. Because of the small dimensions involved in first-order gratings, especially in AlGaAs compounds, second-order gratings are frequently used instead, with the grating period equal to the wavelength. Second-order gratings rely on the second harmonic of their period to deflect an incident wave by 180°. A purely sinusoidal second-order grating therefore cannot be used for feedback; however, it can be used to deflect an incident wave into two waves perpendicular to the incident direction. This property of second-order gratings has been used for surface-emitting lasers, where the gain is highest along the surface of the semiconductor chip. In plasma etching, a rounded resist pattern can be converted into a rounded pattern in underlying material as described above.

The index coupling used in DFB lasers introduces a problem of frequency degeneracy: the two counterpropagating waves allowed are coupled through the grating and split into two different frequencies.[39] Two possible independent oscillation frequencies are undesirable in many applications, and a solution was made by introducing a quarter-wavelength shift in the grating.[40] The introduction of a quarter-wavelength shift in the grating can be accomplished by laying positive and negative photoresists side by side (Fig. 22.21).[41]

The second-order grating has been used to great success in fabricating high-power surface-emitting laser arrays.[42,43] The gratings are used for three functions: for surface emitting, reflection, and coupling between different lasers in the array. Figure 22.22 shows an example made by Sarnoff Laboratories. By slightly detuning a second-order grating, reflection is effectively suppressed, and the surface-emitted beam travels slightly off perpendicular. This property has been used to fabricate oscillator-amplifiers on the same chip,[44] where the detuned grating is placed between different amplifier stages to eliminate feedback.

It should be noted that the holographic method used to make submicrometer period gratings are only convenient for simple periodic structures. Even so, the subsequent processing steps are pushed to the limit. The problem arises from the high refractive indices of the semiconductors used. If other materials of substantially lower refractive indices are used, then the wavelengths in these materials will be longer. For example, dielectrics commonly used in GaAs-based technology like SiO_2 or SiN_x have refractive indices from 1.4 to 2, and can be integrated with GaAs. GaAs lasers with monolithically integrated dielectric waveguides and gratings have been demonstrated.[45]

Vertical Cavity Surface-Emitting Lasers.   Vertical cavity surface-emitting lasers (VCSEL), which have optical cavities normal to the substrate, are excellent examples of the application of high-resolution lithography to optoelectronics. Since

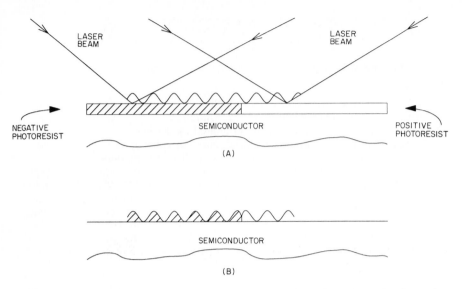

SEMICONDUCTOR POSITIVE
PHOTORESIST

(A)

SEMICONDUCTOR

(B)

FIGURE 22.21 Generation of quarter-wave shift in grating by using positive and negative photoresists.

VCSELs usually have a high-Q cavity (i.e., the reflectivities of both laser facets are near unity), they require (and prefer) very small active volumes. The smallest electrically pumped VCSELs to date have active volumes less than 0.05 μm^3, and optically pumped VCSELs (in which a higher carrier density can be injected into the laser cavity) are as small as 0.002 μm^3.[46]

VCSEL was first demonstrated by Soda et al. in 1979.[47] For the past ten-plus years, enormous progress has been made in developing the VCSEL technology: it can operate continuously at room temperature,[48] two-dimensional diode laser arrays were made from VCSELs,[49] microcavity VCSELs exhibit one of the lowest threshold currents (less than 100 μA).

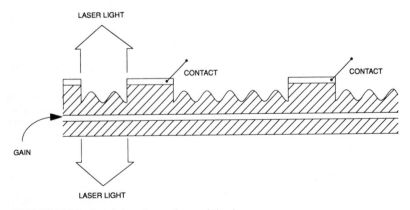

FIGURE 22.22 Coupled-grating surface-emitting laser array.

FIGURE 22.23   Scanning electron microscope image of a semiconductor microdisk laser is shown in a side view. The disk diameter is 2 μm and its thickness is 0.15 μm. The InP pedestal is a rhombus in cross-section and tapers to smaller dimensions as it approaches the substrate. There are six 100Å InGaAs quantum wells with InGaAsP barriers in the plane of the disk that provide optical gain for a microlaser operating at a wavelength of 1.5 μm. The laser emits into a whispering-gallery mode that propagates around the edge of the disk.

The benefits of a microcavity VCSEL structure are more than the small active column alone (which corresponds to a low current injection requirement): when the laser resonator dimensions are comparable to a wavelength of emitted light, the density of radiation states can be drastically different from that in a large space. As a consequence, the total spontaneous emission rate can be drastically reduced, leading to very low threshold current densities.[50] However, for a VCSEL with submicrometer diameter (a necessity for a microcavity VCSEL), strong optical confinement is needed. This can be achieved by etching vertically through the optical resonator. Electron-beam lithography is typically used to define the micro-cavity pattern, and reactive ion etching, in particular chemically assisted ion-beam etching, is used to produce the columnlike structure.[51] Figure 22.23 shows a scanning electron microscope image of ultrasmall microlasers, with widths down to 0.25 μm and aspect ratios as high as 15:1.

22.4.2   Nanofabrication of Integrated Optoelectronic Devices

As we have mentioned earlier in this chapter, many optoelectronic devices require a fine control of their size and/or shape on the order of the wavelength. Since compound semiconductors have a large index of refraction (about 3.5), the typical feature size is around 0.025 μm or smaller. Furthermore, in order to control the coherence of the optical beam, the minimal feature size requirement can be much smaller than that. In this section we shall describe the nanofabrication process of an integrated wavelength division multiplexer (WDM) device as an example.

A WDM device serves the function of separating input optical signals of various wavelengths to different channels, according to their corresponding wavelength. A monolithically integrated WDM device is essentially a spectrometer on a chip! One effective optical design of a spectrometer is to use a concave grating in a Rowland circle configuration.[52] The concave grating used in this design serves a dual purpose as a collimating mirror and a diffraction grating. Therefore, the spacing of the ruled grating is not constant around the grating surface, but it is constant along the tangent axis (of the incident light) of the circle. Considering a circle, with a radius half of that of the concave grating, as shown in Fig. 22.24, one can readily show that for a source of light at point P on the circle, its diffracted beam is focused at spot Q, which is also on the circle. From a light source of many wavelengths (channels) incident to the Rowland circle spectrometer at the entrance slit P, each wavelength (channel) will be diffracted to a different spot (exit slit) on the circle. The implementation of a Rowland circle spectrometer in an integrated optics fashion is quite straightforward, as optical waveguides can be readily employed as the entrance/exit slits.[53,54,55] To achieve high throughput, it is desirable to operate at a low diffraction order and employ a grating device of a small period.

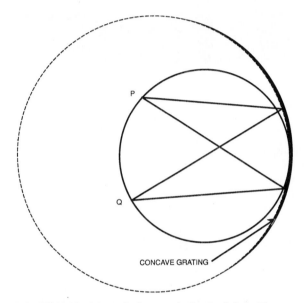

FIGURE 22.24   Schematic diagram of a Rowland circle. The concave grating lies on a large circle of radius R and the small circle, which is in contact with the grating, has a radius of $R/2$. For an incident light from point P on the small circle, the diffracted beam is focused on a point Q, which is also on the small circle.

For a second-order grating, operating at 8500 Å, a grating period of less than 1 µm is needed for a 1-cm Rowland circle spectrometer. Since the shape and blaze angle of a grating is important for its diffraction efficiency, the lithographic resolution requirement for this low-diffraction-order grating is subnanometric!

Figure 22.25 shows an engineering drawing of an integrated Rowland circle spectrometer. The devices consist of a ridge waveguide input channel with a tapered expansion leading to a curved blazed reflection grating. Different wavelengths in the input light are focused by the curved grating into separate output ridge waveguide channels. The incident waveguide and output waveguides are normal to each other so that they can be cleaved at the appropriate crystal axis (110 and 110, respectively).

The starting material for device fabrication was planar AlGaAs/GaAs/AlGaAs double heterojunction waveguide grown by molecular beam epitaxy. First, the input and output waveguide channels and WDM sections were defined by projection photolithography followed by CAIBE using photoresist as the etching mask (Fig. 22.26). The sample was then coated with polymethyl methacrylate (PMMA) positive e-beam resist, and the grating pattern was exposed using a JEOL nanowriter. Since PMMA does not have the process latitude needed to etch a 4-μm-deep grating, a chromium etch mask was used. After development, the grating pattern was transferred into a chromium pattern by a lift-off process. A further photolithographic step left the sample coated with photoresist, except for small windows

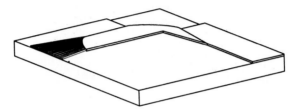

FIGURE 22.25 A schematic diagram of an integrated Rowland circle spectrometer on a chip. The incident light enters the spectrometer via a waveguide, while the incident slit is defined by the termination point of the waveguide. The light expands in the propagation section and is diffracted and refocused by the concave grating. Light beams with different wavelengths are focused to different waveguides at the exit slit.

exposing the grating edge of the Cr pattern and adjacent substrate. The sample was then etched by CAIBE to create wells of grating walls deep enough to cover the expected extent of the optical mode (Fig. 22.27). The final step in creating the WDM device is to thin the sample and cleave end facets to allow coupling to and from the input and output waveguides.

Figure 22.28 is a close-up micrograph of an etched grating. The high fidelity of the blazed grating structure demonstrates the validity of applying high-resolution lithographic technique to the fabrication of optoelectronic devices. Using the process described above, gratings of several periods were fabricated, ranging from 0.3 to 1.2 μm. The grating patterns show smooth and well-defined blazed facets except at the very shortest periods. The excessive roughness exhibited on the 0.3-μm-period grating was attributed mainly to the unoptimized CAIBE process.

22.4.3 Binary Optics

Another example of the application of microfabrication technology to optics is *binary optics*. The advent of high-resolution lithographic techniques has allowed control of the optical path lengths through the transmission region by etching these regions to different depths. In the words of a prominent advocate,[56] "Binary optics is a diffractive-optics technology that . . . utilizes high-resolution lithography and ion beam etching to transfer a binary surface-relief pattern to a dielectric or metallic

FIGURE 22.26 Etched input and output waveguides of the Rowland circle spectrometer. The pattern was defined by a 10 to 1 reduction imaging projection aligner and etched by a chemically assisted ion-beam etcher. The etch mask was photoresist.

substrate. . . . A single etching step produces a two-level surface relief, giving rise to the name 'binary optics.' . . . To be efficient, the relief structure must be smaller than or comparable in size to a wavelength of light and it must typically have a half-wave phase depth." Binary optics can be cheaper or easier to make. Since its fabrication is by the same planar technology of integrated electronics and optoelectronics, it is a natural candidate for monolithic integration with other electronic or optoelectronic components.

Probably the simplest example is the binary phase grating (Fig. 22.29), which has been used to suppress the sidelobes of a semiconductor laser array by phase correction.[56] Coherent combination of laser beams has been achieved with more sophisticated phase gratings.[57] Many functions of refractive optics can be performed with binary optics. For example, collimation and focusing by a conventional lens (Fig. 22.30*a*) can be done with a Fresnel lens (Fig. 22.30*b*), which can be approximated to different degrees by different levels of binary optics (Fig. 22.30*c* and *d*). A two-dimensional Fresnel lens array has been made for optical interconnections.[58] Binary optics can also be used to correct the aberrations of conventional spherical lenses.[59] Applications to "hemispherical vision and amacronic sensors" (layered structures of electronics and microoptics with massive parallel short range interactions between optical layers and elements of the detector array) have been proposed.[60]

The fabrication of a multilevel phase grating is discussed in some detail by Walker and Jahns.[61] In the lens fabrication process, the first pattern is transferred to photoresist on the lens substrate. Next, a thin metal film is deposited and defined by using the lift-off process discussed above. The remaining metal is used as a plasma-etch mask, and is removed after the substrate is etched. The process is

FIGURE 22.27 Etched submicrometer concave grating structure for the Rowland circle spectrometer on a chip. The grating pattern was generated by direct e-beam written by a JEOL nanowriter. The pattern was transferred to a chrome etch mask via a lift-off process. The grating was etched with a chemically assisted ion-beam etcher.

repeated as many times as is necessary to achieve the right number of phase levels. The success of this process depends critically on the ability of the lithographic tool to align the phase levels. The tool employed will most likely be an e-beam, because of its inherently large depth of focus. Lens surfaces need not be planar. E-beams currently can exhibit point repositioning accuracies of roughly 0.1 μm, accounting for all field stitching errors and for machine drift.

22.4.4 Modulators

Optical waveguides may be made in electro-optic materials (such as gallium arsenide or lithium niobate) by diffusing (or by ion implanting) dopants which alter the substrate dielectric constant.[62,63] Optical confinement is achieved as in a standard stand-alone fiber optical element. In addition, the optical index of the medium through which the light propagates can be altered by electro-optic effects.[64] Consider the branched channel waveguide shown in Fig. 22.31. The light path in each arm is the same physical path length. By applying bias in the gold electrodes as shown, the optical path length in the top arm changes. Thus, under bias, the light

FIGURE 22.28 Scanning electron microscope close-up picture of the etched concave grating for the spectrometer on a chip.

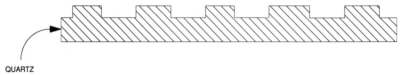

QUARTZ

FIGURE 22.29 A binary-optics phase grating.

waves recombining at the opposite ends of the bifurcation would be out of phase. The split-channel waveguide acts as a Mach-Zehnder interferometer. The intensity of the emerging beam is modulated by the bias on the electrodes.

For high-speed operation, the resistance of the modulator electrodes can create a high-frequency cutoff. This is especially true in densely packed couplers requiring narrow lines. To reduce resistance while keeping the lines narrow, the electrode films must be thicker. Metal thickness ranging from 5 to 10 μm may be required. Interelectrode spacings are small (less than 10 μm) and the metal sidewall slopes must be close to 90°. Most etch processes will not allow for such thick films with 90° sidewalls. As discussed above, steep sidewalls in thick resist layers are possible with high-contrast resist, but process latitude is compromised.

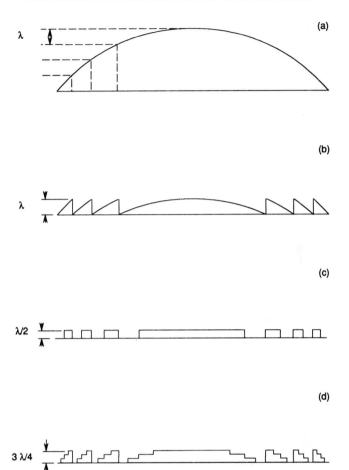

FIGURE 22.30 Implementation of refractive lens by binary optics. *(Adapted from Refs. 46 and 58.)* (*a*) Conventional refractive lens. (*b*) Fresnel lens. Binary-optics lens—two levels. Binary-optics lens—four levels.

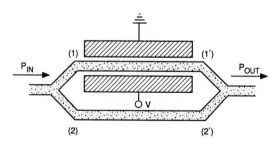

FIGURE 22.31 Mach-Zehnder interferometric modulator.

A process sequence which aims to circumvent these problems is shown in Fig. 22.32. Here, the image is formed in a very thick (greater than 50 μm) high-contrast resist. The basis of this process is discussed in Sec. 22.2, and the results of the lithography are shown in Fig. 22.3. Since the processed surfaces are highly planar and of controlled reflectivity, feature biasing is possible. That is, the size of the feature on the lithographic mask can be adjusted to create the proper size feature

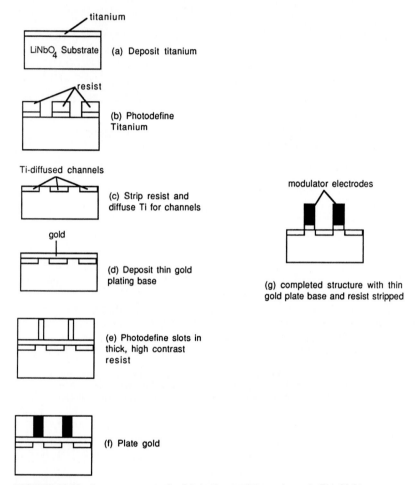

FIGURE 22.32   Process sequence for fabricating modulator shown in Fig. 22.31.

on the workpiece. Electroplating is used to build up a thick metal film where the resist was removed. Note the thin gold plating base required to form electrode contact to the plating solution. This thin layer can be removed by a brief dip etch after plating. The completed structure is shown in Fig. 22.33.

It is not always possible to control surface topography and reflectivity to the levels required by the process summarized in Fig. 22.32. An alternative process

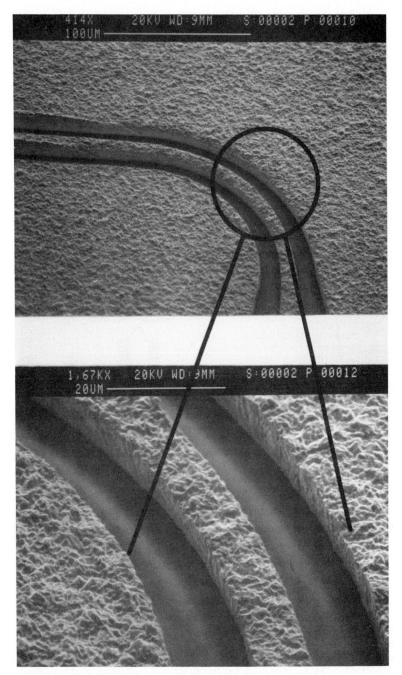

FIGURE 22.33 Results of the process sequence shown in Fig. 22.32.

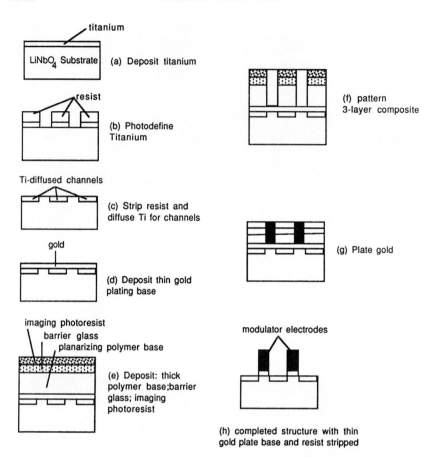

FIGURE 22.34   An alternative processing sequence for the Mach-Zehnder interferometer.

scheme is shown in Fig. 22.34. Here, the surface is "planarized" by a thick polymer undercoat. The primary pattern is created in a thin imaging layer of photoresist separated from the polymer by a transparent layer of sputtered SiO_2. The primary image is transferred into the thick polymer by oxygen reactive ion etching. Once the deep slots are created, the up-plating process commences, as in Fig. 22.32.

22.5 CONCLUDING REMARKS

The recent rapid advances in optoelectronics have been stimulated mostly by the technological progress in materials and fabrication. While there is controversy concerning the feasibility or competitiveness of certain areas, such as optical computers, there is more consensus on the practicality of integrated optoelectronics performing specialized functions better than electronics. The advantages of optics over electronics include higher speed and higher device packing density. To exploit these advantages fully, the physical dimensions of certain optoelectronic devices have to shrink. Thus, the fabrication of small structures will become more important. In this chapter, we have reviewed current methods of high-resolution lithography applicable to optoelectronics and have given a few examples of its applications. The choice of examples is arbitrary, in part dictated by the authors' familiarity with the subject matter. Interested readers can no doubt find other applications in current literature. We anticipate that new lithographic and other methods will develop, and the resolution limit will be advanced, very rapidly.

22.6 ACKNOWLEDGMENTS

The authors would like to thank M. L. Rebbert, the staff of the Naval Research Laboratory Nanoelectronic Processing Facility, and Dr. G. M. Borsuk for many useful inputs to this work.

22.7 REFERENCES

1. Sze, S. M., *The Physics of Semiconductor Devices*, 2d ed., part V, Wiley-Interscience, New York, 1981.
2. Lizuka, K., *Engineering Optics*, 2d ed., Springer-Verlag, New York, 1987.
3. Jaros, M., *Physics and Applications of Semiconductor Microstructures*, Oxford Science Publications, Oxford, U.K. 1990.
4. Bauer, G., F. Kuchar, and H. Heinrich, *Two Dimensional Systems, Heterostructures and Superlattices*, Springer Series in Solid State Science, vol. 53, Springer-Verlag, Berlin, 1984.
5. Born, M., and E. Wolf, *Principles of Optics*, 6th ed., Pergamon Press, Oxford, U.K., 1980.
6. Murarka, S. P., and M. C. Peckerar, *Electronic Materials: Science and Technology*, Academic Press, Cambridge, U.K., 1989.
7. Thompson, L. F., C. G. Willson, and M. J. Bowden, eds., *Introduction to Microlithography*, ACS Symposium series 219, American Chemical Society, Washington, D.C., 1983.
8. Reiser, A., *Photoreactive Polymers: The Science and Technology of Resists*, Wiley Interscience, New York, 1989.
9. Moreau, W., *Semiconductor Lithography: Principles, Practices and Materials*, Plenum, New York, 1988.
10. Barouch, E., et al., "Modeling Process Latitude In UV Projection Lithography," *IEEE Elect. Dev. Lett.*, vol. 12, no. 10, pp. 513–514, 1991.
11. Peckerar, M., M. Rebbert, and G. Gopalakrishan, "Apparent Super-Resolution by High Contrast Photoresist," *Appl. Phys. Lett.*, vol. 61, no. 17, p. 2036, 1992.
12. Rishton, S. A., and D. P. Kern, "Point Exposure Distribution Measurements for Proximity Correction in E-Beam Lithography on a Sub-100nm Scale," *J. Vac. Sci. Technol.*, vol. B5, no. 1, pp. 135–141, 1987.
13. Wind, S. J., et al., "Proximity Correction for Electron Beam Lithography Using a Three-Gaussian Model of the Electron Energy Distribution," *J. Vac. Sci. Technol.*, vol. B7, no. 6, pp. 1507–1512, 1989.

14. Owen, G., and P. Rissman, "Proximity Effect Correction for E-Beam Lithography by Equalization of Background Dose," *J. Appl. Phys.*, vol. 54, pp. 3573–3581, 1983.

15. Muray, A., R. Lozes, K. Milner, and G. Hughes, "Point Proximity Effect Correction at 10 keV using GHOST and Sizing for 0.4 µm Mask Lithography," *J. Vac. Sci. Technol.*, vol. B8, no. 6, pp. 1775–1779, 1990.

16. Parikh, M., and D. F. Kyser, "Correction to Proximity Effects in E-Beam Lithography," *J. Appl. Phys.*, vol. 50, no. 6, pp. 4371–4377, 1979.

17. Pati, Y. C., et al., "An Error Measure for Dose Correction in E-Beam Lithography," *J. Vac. Sci. Technol.*, vol. B8, no. 6, pp. 1882–1888, 1990.

18. Gerber, P. D., "Exact Solution of the Proximity Effect by a Splitting Method," *J. Vac. Sci. Technol.*, vol. B6, no. 1, pp. 432–436, 1987.

19. Stroeve, P., and E. Franses, eds., *Molecular Engineering of Ultrathin Polymeric Films*, Elsevier, London, 1987.

20. Calvert, J. M., et al., "Deep UV Patterning of Monolayer Films for High Resolution Lithography," *J. Vac. Sci. Technol. B*, vol. B9, no. 6, p. 3447, 1991.

21. Ito, H., ed., *Advances in Resist Technology and Processing VIII*, SPIE Press, vol. 1466, 1991.

22. Umbach, C. P., and A. N. Broers, "Experimental Determination of the Proximity Effect from 25–100 KeV in Electron Beam Patterned X-Ray Masks," *J. Vac. Sci. Tech.*, vol. B63, no. 8, 1990.

23. Schattenburg, M. L., and H. I. Smith, "X-Ray Nanolithography—The Clearest Path to 0.1 and Sub 0.1 µm ULSI," *Proc. 1991 International Microelectronics Processing Conf.*, Kanazawa, Japan, July 15–18, 1991.

24. Lin, B. J., "A Comparison of Projection and Proximity Printing from UV to X-Ray," *Proc. SPIE*, vol. 1263, *Electron Beam, X-Ray and Ion Beam Lithography: Submicrometer Lithographies, IX*, p. 80, 1990.

25. Guo, J. Z. Y., et al., "Aerial Image Formation In Synchrotron Radiation Based Lithography: The Whole Picture," *J. Vac. Sci. Tech.*, vol. B8, no. 6, p. 1551, 1990.

26. Early, K., M. L. Schattenburg, and H. I. Smith, "Photoelectron Range Effects in X-Ray Lithography," *Microelectronics Engineering*, vol. 11, pp. 317–321, 1990.

27. Guenther, R., *Modern Optics*, Wiley, New York, 1990.

28. Kogelnik, H., and C. V. Shank, "Stimulated Emission in a Periodic Structure," *Appl. Phys. Lett.*, vol. 18, pp. 152–154, 1971.

29. Labeyrie, A., and J. Flamand, "Spectroscopic Performance of Holographically Made Diffraction Gratings," *Optics Communications*, vol. 1, no. 1, pp. 5–8, 1969.

30. Jenney, J. A., "Holographic Recording With Polymers," *J. Opt. Soc. Am.*, vol. 60, no. 9, pp. 1155–1161, 1970.

31. Shank, C. V., and R. V. Schmidt, "Optical Techniques for Producing 0.1 µm Periodic Surface Structures," *Appl. Phys. Lett.*, vol. 23, pp. 154–155, 1973.

32. Vossen, J. L., and W. Kern, eds., *Thin Film Processes*, Academic Press, New York, 1978.

33. Smith, H. I., "Fabrication Techniques for Surface-Acoustic-Wave and Thin-Film Optical Devices," *Proc. IEEE*, vol. 62, no. 10, pp. 1361–1387, 1974.

34. Vossen, J. L., and W. Kern, eds., *Thin Film Processes II*, Academic Press, New York, 1991.

35. Manos, D. M., and D. L. Flamm, eds., *Plasma Etching: An Introduction*, Academic Press, New York, 1989.

36. Hatzakis, M., "Electron Resists for Microcircuit and Mask Production," *J. Electrochem. Soc.*, vol. 116, p. 1033, 1969.

37. Chapman, B., *Glow Discharge Processes: Sputtering and Plasma Deposition*, Wiley, New York, 1980.

38. Kogelnik, H., and C. V. Shank, "Stimulated Emission in a Periodic Structure," *Appl. Phys. Lett.*, vol. 18, pp. 152–154, 1971.

39. Kogelnik, H., and C. V. Shank, "Coupled-Wave Theory of Distributed Feedback Lasers," *J. Appl. Phys.*, vol. 43, pp. 2328–2335, 1973.

40. Haus, H. A., and H. Kogelnik, "Antisymmetric Taper of Distributed Feedback Lasers," *IEEE J. Quant. Electron.*, vol. QE-12, pp. 532–539, 1976.

41. Okai, M., S. Tsuji, H. Hirao, and M. Matsumura, "New High Resolution Positive and Negative Photoresist Methods for $\lambda/4\pi$ DFB Lasers," *Electron. Lett.*, vol. 23, pp. 370–371, 1987.

42. Evans, G. A., et al., "Characteristics of a Coherent Two-Dimensional Grating Surface Emitting Diode Laser Array during CW Operation," *IEEE J. Quantum Electron.*, vol. 27, pp. 1594–1605, 1991.

43. Menhuys, D., et al., "Characteristics of Multistage Monolithically Integrated Master Oscillator Power Amplifiers," *J. Appl. Phys.*, vol. 27, pp. 1574–1581, 1991.

44. Carlson, N. W., et al., "Demonstration of a Monolithic Grating–Surface-Emitting Laser-Maser-Oscillator Cascaded Power Amplifier Array," *IEEE Photon. Tech. Lett.*, vol. 2, pp. 708–710, 1990.

45. Alferov, Zh. I., "Monolithically-Integrated Hybrid Heterostructure Diode Laser, with Dielectric-Film Waveguide DBR," *IEEE J. Quantum Electron.*, vol. QE-23, pp. 869–881, 1987.

46. Jewell, J. K., et al., "Vertical-Cavity Surface-Emitting Lasers, Design, Growth, Fabrication, Characterization," *IEEE J. Quantum Electron.*, vol. 27, p. 1332, 1991.

47. Soda, H., K. Iga, C. Kitahara, and Y. Suematsu, "GaInAsP/InP Surface Emitting Injection Lasers," *Japan J. Appl. Phys.*, vol. 18, pp. 2329–2330, 1979.

48. Ibariki, A., et al., *Japan J. Appl. Phys.*, vol. 28, L667–L668, 1989.

49. Von Lehmen, A., et al., "Independently Addressable InGaAs/GaAs VCSEL Array," *Electron. Lett.*, vol. 27, p. 583, 1991.

50. Yablonovitch, E., "Inhibited Spontaneous Emission in Solid State Physics and Electronics," *Phys. Rev. Lett.*, vol. 58, pp. 2059–2062, 1987.

51. Van der Gaag, B. P., and A. Scherer, "Microfabrication below 10 nm," *Appl. Phys. Lett.*, vol. 56, p. 481, 1990.

52. Rowland, H. A., *Philos. Mag.*, vol. 13, p. 467, 1982. For a good introduction to the subject see, for example, M. C. Huley, *Diffraction Gratings*, Academic Press, New York, 1982.

53. Soole, J. B. D., et al., "Monolithic InP/InGaAsP grating spectrometer for the 1.48–1.56 mm wavelength range," *Appl. Phys. Lett.*, vol. 58, pp. 1949–1951.

54. Cremer, C., et al., "Grating spectrograph in InGaAsP/InP for dense wavelength division multiplexing," *Appl. Phys. Lett.*, vol. 59, pp. 627–629, 1991.

55. Hryniewicz, J., Y. J. Chen, and R. Tiberio, "Spectrometer on a chip by nanofabrication," *Summary Digest of LEOS Summer Topical Meeting on Microfabrication for Photonics and Optoelectronics*, pp. 16–17, 1991.

56. Leger, J. R., M. Holz, and G. J. Swanson, "Coherent Laser Beam Addition: An Application of Binary-Optics Technology," *The Lincoln Laboratory Journal*, vol. 1, pp. 225–245, 1988.

57. Veldkamp, W. B., J. R. Leger, and G. J. Swanson, "Coherent Summation of Laser Beam Pulses Using Binary Phase Gratings," *Opt. Lett.*, vol. 11, pp. 303–305, 1986.

58. Rastani, K., S. F. Habiby, A. Marrakchi, and W. M. Hubbard, "Fabrication and Analysis of Binary Fresnel Lenses as a Practical Means of Generating Two-Dimensional Source Arrays for Optical Interconnections," *Conference on Lasers and Electro-optics*, paper CMG-7, 1990.

59. Swanson, G. J., and W. B. Veldkamp, "Diffractive Optical Elements for Use in Infrared Systems," *Optical Engineering*, vol. 28, pp. 605–608, 1989.

60. Veldkamp, W. B., "Binary Optics: The Optics Technology of the 90's," *Conference on Lasers and Electro-optics*, paper CMG-6, 1990.

61. Walker, S. J., and J. Jahns, "Array Generation with Multilevel Phase Gratings," *J. Opt. Soc. Am. A*, vol. 7, pp. 1509–1513, 1990.

62. Garmire, E., H. Stoll, A. Yariv, and R. G. Hunspurger, "Optical Waveguiding in Proton Implanted GaAs," *Appl. Phys. Lett.*, vol. 21, pp. 87–88, 1972.

63. Ranganath, T. R., and S. Wang, "Ti-Diffused LiNbO$_3$ Branched-Waveguide Modulators: Performance and Design," *IEEE J. Quantum Electron.*, vol. QE-13, pp. 290–295, 1977.

64. Leonberger, F. J., "High Speed Operation of LiNbO$_3$ Electrooptic Waveguide Modulators," *Opt. Lett.*, vol. 5, pp. 312–314, 1980.

CHAPTER 23
LASER SAFETY IN THE RESEARCH AND DEVELOPMENT ENVIRONMENT

David H. Sliney

23.1 INTRODUCTION

The greatest experience in the safe use of lasers has been derived from over 30 years of working with lasers in the research laboratory. Today, most potentially hazardous laser exposures still occur in the research and engineering laboratory, where engineering safety enclosures are impractical to use.

Only shortly after the development of the first laser were the physiological implications of lasers considered.[1] After several research programs in the 1960s aimed at studying the adverse biological effects of lasers and other optical radiation sources, laser occupational exposure limits were set and general safety standards were developed.[2-4] Today, the experience from laser accidents and the development of new lasers and new applications have altered the format of the exposure limits and the safety procedures.[4-9]

Laser safety exposure limits (ELs) and safety procedures vary considerably with the wavelength and type of laser. It is critically important to distinguish between different biological injury mechanisms. For example, the biological effects of ultraviolet radiation on the skin and eye are additive over a period of at least one workday, and require different safety procedures. The scattered ultraviolet (uv) irradiance from excimer lasers may be quite hazardous, depending on wavelength and action spectra. Since laser technology is young, the exposure of an individual in natural sunlight must be studied to evaluate the potential for chronic effects. The safety measures necessary in the use of lasers depend on a hazard evaluation. The appropriate control measures and alternative means of enclosure, baffling, and operational control measures vary with the type of laser.

Note: The opinions or assertions herein are those of the author and should not be construed as reflecting official positions of the Department of the U.S. Army or Department of Defense.

Lasers are now found in many applications, and safety measures will vary to some degree by application. The first widespread use of lasers outside the research laboratory was for optical alignment, and the helium-neon (He-Ne) alignment laser remains the commonly used laser for this purpose. In terms of numbers of lasers, only the prolific gallium arsenide (GaAs) family of laser diodes—used in communication—outnumber the He-Ne lasers. Today neodymium-doped yttrium aluminum garnet (Nd:YAG) and carbon-dioxide (CO_2) lasers are finding increasing use in material processing throughout industry. The earliest medical lasers were employed for eye surgery, and the argon laser is still the most widespread for retinal photocoagulation. The introduction of the Q-switched and mode-locked pulsed Nd:YAG laser photodisrupters into ophthalmic surgery has greatly increased the use of medical lasers. The clinical use of CO_2 and continuous-wave (cw) Nd:YAG surgical lasers is also on the increase.

23.2 BIOLOGICAL EFFECTS

A solid understanding of the potential hazards resulting from laser use requires a general knowledge of laser biological effects. The different effects of laser exposure of the skin and eye must be understood by the laser user to appreciate the hazards to both the user and bystanders. Indeed, one should have a general understanding of the hazards from exposure to ultraviolet, visible, and infrared radiation from conventional light sources in order to place laser hazards in perspective.

The critical organ of interest is the eye, and the absorption properties of different structures of the eye vary with wavelength, as shown schematically in Fig. 23.1. In discussing photobiological effects, it is customary to divide the optical spectrum into seven spectral bands designated by the CIE (Commission International de l'Eclairage, or International Commission on Illumination). As shown by spectral band in Fig. 23.2, there are at least five separate types of hazards to the eye and skin from lasers and other optical sources:

1. Ultraviolet photochemical injury to the skin (erythema and carcinogenic effects), and to the cornea (photokeratitis, i.e, "welder's flash"), and lens opacities (cataracts) of the eye; from the edge of the vacuum ultraviolet region to the short-wavelength edge of the visible spectrum (180 nm to 400 nm).[4,12–14]

2. Thermal injury to the retina of the eye; occurs in both the visible and IR-A spectral bands (400 nm to 1400 nm).

3. Blue-light photochemical injury to the retina of the eye (photic maculopathy); occurs principally from 400 nm to 550 nm.

4. Near-infrared thermal hazards to the lens (industrial heat cataract) of the eye; occurs from approximately 800 nm to 3000 nm.

5. Thermal injury (burns) of the skin (approximately 400 nm to 1 mm) and of the cornea of the eye (approximately 1400 nm to 1 mm).

Not only will the biological damage mechanism (i.e., thermal, thermomechanical, photochemical, etc.) play a role in the type of possible injury, but the pattern of absorption of energy is important, as shown in Figs. 23.1 and 23.2. Finally, some biological effects produce transient injuries which are repaired and the tissue is normal within a number of days or weeks (e.g., photokeratitis is a painful inflammatory response of the cornea, but because of rapid repair of the corneal epithe-

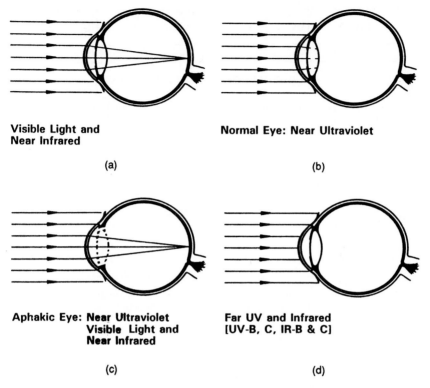

Visible Light and
Near Infrared

(a)

Normal Eye: Near Ultraviolet

(b)

Aphakic Eye: Near Ultraviolet
Visible Light and
Near Infrared

(c)

Far UV and Infrared
[UV-B, C, IR-B & C]

(d)

FIGURE 23.1 The potential for injury to structures of the eye depends on the primary location of energy absorption. The absorption is critically dependent on the spectral band of incident optical radiation.

lium, the signs and symptoms disappear within a day or two after exposure). By contrast, retinal injury normally is permanent, although some degree of tissue repair (and resolution of a hemorrhage, should that occur) will reduce the degree of initial visual loss.

Lasers that operate in the visible and near-infrared (IR-A) "retinal-hazard" region (400 to 1400 nm) pose a serious threat to the retina, and it is these lasers that have caused serious loss of vision from accidental exposure. Hazards differ with each type of laser operating in this spectral region, depending on whether the laser is pulsed or continuous wave (cw), and whether the wavelength is sufficiently short to produce a photochemical lesion. Unlike the cw argon laser which can only coagulate the retina upon accidental exposure, Q-switched or other short-pulse lasers can cause an explosive lesion in the retina if focused there—resulting in a hemorrhage into the vitreous humor matter. The worst-case exposure condition will occur when a collimated beam is focused by the relaxed normal eye or when the diverging beam is imaged to a point on the retina when the eye is focused at the focal spot from whence the beam comes. The minimal retinal image size is

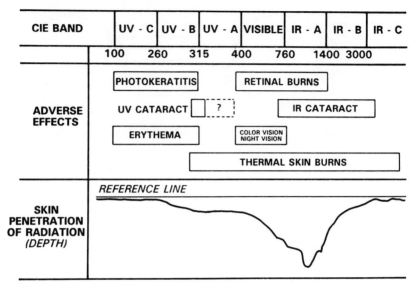

FIGURE 23.2 Optical radiation hazards as a function of CIE Spectral Band. The primary adverse effects for each spectral band are shown; however, with sufficient energy, some effects may occur outside of the boxed spectral regions (e.g., UV-A injury of the retina).

about 10 to 20 μm in diameter, and the optical gain factor from cornea to retina is approximately 100,000 times; hence Q-switched energies of the order of 1 μJ entering the eye can produce a minimal retinal injury.[4]

23.3 SAFETY STANDARDS

The first widespread use of lasers raised a number of laser safety questions, and in 1968 a committee of experts was formed to draft a laser safety standard. Under the umbrella of the American National Standards Institute (ANSI), Committee Z-136 was formed and issued its first consensus standard in 1973.[5] It has since issued three revisions and has drafted specialized standards on fiber-optic and medical applications.[6,7] The 1973 ANSI standard served to a large extent as the basis for a U.S. governmental regulation from the Food and Drug Administration (FDA) which was first issued in 1975, became effective in 1976, and has now been revised.[8] The federal standard (referred to as 21CFR1040) applies only to the manufacturer and does not apply to the user, since it is a product performance (or system safety) standard. The ANSI standard and the federal governmental regulation were the basis of the international standard WS-825 issued in 1985 by the International Electrotechnical Commission (IEC).[9]

The initial safety guidelines were commonsense procedures based on an understanding of optics and a basic assessment of the probability of accidental exposure.[10] From these early guidelines a system of hazard classification[4,10] was developed to simplify hazard analysis. The hazard classification has since been modified and the recommended hazard controls have been updated to keep up with new applications and the greater experience from accidental[11] and near-accidental injuries.

23.4 RISK OF EXPOSURE

Our understanding of laser safety risks from studying accidents points all too often to carelessness with *invisible* laser beams. Indeed, most of the severe injuries have been produced by the neodymium laser. Since the Nd:YAG laser wavelength is infrared, and invisible, potentially hazardous secondary beams are often unnoticed or forgotten. The pulsed Nd:YAG laser has been considered one of the most dangerous lasers in industrial and research applications, because there have been a number of serious retinal injuries caused by improper attention to safety in each incident.[3,11] Stringent safety measures are needed with this type of laser in a laboratory.

The probability of accidental exposure is a critical aspect of any laser safety analysis. A major factor which influences risk is the high collimation characteristic of nonfocused lasers. The collimated beam can present a concentrated, hazardous laser beam at quite some distance from the device. This is particularly true of military laser rangefinders and some light detection and ranging systems (lidars) used outdoors.[4]

Among other factors, beam focusing influences the probability of hazardous exposure. In most surgical and industrial material processing applications, the beam is focused at a point 100 to 250 mm from the laser aperture. With the beam sharply focused, the extent of the specular reflection hazard is minimized, and the risk of eye injury to bystanders and the laser operator is limited. Nevertheless, certain precautions are still necessary even with a focused-beam infrared laser.

Hazards from high-power industrial CO_2 welding and cutting lasers (in excess of a kilowatt) would be expected to be severe because of the high power. However, because of the plasma created at the focal point of the beam, reflections at a distance can be less severe than from a 100-W beam on a flat metal surface which is not melted and reflects the beam in a specular fashion.[12]

Beam visibility influences the risk of exposure. For example, the cw argon laser used in scientific and ophthalmic applications has such a visible beam that it is highly unusual for an individual to place an eye near the beam. Hence its potential hazards are less often realized, even though the beam is normally collimated. Even though the argon laser is continuous wave, or nearly so, the biological effects and potential hazards from this type of laser are quite different from those of a cw CO_2 or the Q-switched Nd:YAG laser. Although the severity of retinal injuries from a cw laser is normally considered to be far less than from a Q-switched or mode-locked laser, eye injuries are still possible.

The potential optical radiation hazard to both the laser operator and to onlookers during laser operation normally results from specular reflections. Figure 23.3 shows the types of reflected beams from unfocused and focused beams incident on a flat or curved specular surface. Flat optical elements such as prism surfaces, beam splitters, and filters can cause hazardous specular reflections, even when an antireflection coating has been applied to reduce the power of a reflected beam from 4 to 1 percent, or even less. For example, the emergent beam of many lasers may exceed the applicable exposure limit (EL) by a factor of 10^5 to 10^6. Thus, a reflected beam with only 1 percent of the initial beam's power can still exceed the EL by 10^3 to 10^4! Overlooking this fact has often led to lack of precautions and serious

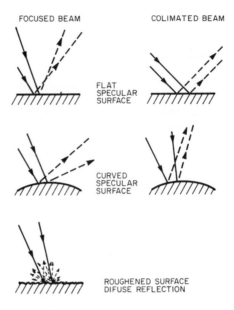

FIGURE 23.3   Potentially hazardous specular reflections. The degree of collimation of the incident laser beam and the surface curvature greatly influence the hazardous viewing distance for the person viewing the laser reflection.

eye injuries, where a reflection was discounted by a scientist or engineer because the surface was "not reflecting." Although a 1-W argon laser is not thought of as very powerful in engineering terms, only 1 percent of 1 W is still 10 mW and exceeds the EL for momentary viewing tenfold. The EL here, based on an aversion response time (the blink reflex, etc., of 0.25 s), is 2.5 mW/cm^2, corresponding to a beam power of 1 mW entering the eye. Fortunately, for a reflected beam to be hazardous, it must be concentrated, and therefore it does not occupy a great deal of space. To produce an accidental retinal injury to the eye, the beam must enter the 3- to 7-mm pupil of the eye, and this is highly unlikely in most environments and laser operations. It is this low probability of exposure that accounts for the limited number of severe accidental ocular injuries from lasers. The laser represents a severe potential hazard but normally does not represent a high probability of injury. Fortunately, the accident rate is still lower because most persons observing laser operations have the good sense to wear eye protectors. Figure 23.4 shows the zones where hazardous reflections from a high-power Nd:YAG laser exist due to dangerous diffuse reflections.[13]

Hazardous specular reflections from the industrial material-processing Nd:YAG or CO_2 lasers are limited in extent because of the spreading of reflected focused beams, as shown in Fig. 23.3, producing areas where an eye hazard exists [referred to as the *nominal hazard zone* (NHZ), Fig. 23.4]. However, the permissible occupational ELs for short-pulsed Nd:YAG lasers are far less in terms of radiant

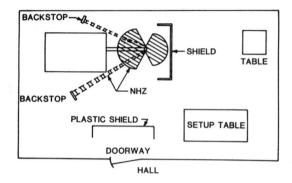

FIGURE 23.4 Reflected beam paths within a laser installation (dashes). Potentially hazardous areas are shaded and referred to as the *nominal hazard zone* (NHZ) in ANSI Standard Z-136.1.

exposure (J/cm²) than for momentary (0.25-s) cw Nd:YAG or even cw argon or He-Ne laser exposures. Relevant ELs for lasers of interest here are given in Table 23.1.

The ELs in Table 23.1 are calculated or measured at the cornea. If a visible or near-infrared (IR-A) laser beam is less than 7 mm in diameter, it is assumed that the entire beam could enter the dark-adapted pupil and one can express the maximal safe power or energy in the beam; it is the above EL multiplied by the area of a 7-mm pupil, i.e., 0.4 cm². For example, for visible cw lasers at 0.25 s, the limiting power is 1.0 mW. This 1-mW value has a special significance in laser safety, since it is a dividing line between two laser safety hazard classifications: Class 2 and Class 3.[4,5] Any cw visible laser (400 to 700 nm) that has an output power less than 1.0 mW is termed a Class 2 (low-risk) laser, and could be considered more or less equivalent in risk to staring at the sun, a tungsten-halogen spotlight, or other bright light which could cause a retinal injury (photic maculopathy) only if one forced oneself to overcome the natural aversion response to bright light. An aiming beam or alignment laser operating at a total power above 1.0 mW would fall into hazard

TABLE 23.1 Selected Occupational Exposure Limits for Some Commonly Encountered Lasers

Laser	Wavelength	Exposure limit
Argon ion laser	488, 514.5 nm	3.2 mW/cm² for 0.1 s; 2.5 mW/cm² for 0.25 s
Helium-neon laser	632.8 nm	1.8 mW/cm² for 1.0 s; 1.0 mW/cm² for 10 s
Krypton ion laser	568, 647 nm	
Nd:YAG laser	1064 nm	5.0 μJ/cm² for 1 ns to 100 μs; no EL for $t < 1$ ns;
	1334 nm	5 mW/cm² for 10 s; 1.6 mW/cm² for $t > 1000$ s
Carbon dioxide laser	10.6 μm	100 mW/cm² for 10 s to 8 h, but limited to 10 mW/cm² for > 10 s for whole-body exposure

Source: ANSI Standard Z-136.1-1993.

Note: To convert ELs in mW/cm² to mJ/cm², multiply by exposure time t in seconds; e.g., the He-Ne or argon EL at 0.1 s is 0.32 mJ/cm².

Class 3 and would be hazardous even if viewed momentarily within the aversion response time. Any cw laser with an output power above 0.5 W (500 mW) would fall into Class 4. The purpose of assigning hazard classes to laser products is to simplify the determination of adequate safety measures; i.e., Class 3 measures are more stringent than Class 2 measures, and Class 4 measures are more stringent than Class 3 measures.

23.5 LASER HAZARD CLASSIFICATION

It was recognized in the early development of laser safety standards that by performing an on-site safety evaluation, one cannot readily predict all possible laser beam exposures to the eye and skin. Measurement of laser power, energy, radiant exposure, and irradiance often pose a severe problem. One would virtually need to be an optical physicist with specialized training in radiometry to make reliable measurements of some laser environments. It was obvious that it was unrealistic to expect general safety personnel, health physicists, or industrial hygienists to obtain the extensive training necessary to perform a battery of comprehensive measurements.

To help health and safety personnel to assess the potential hazards of a laser system, the ANSI Z-136 committee developed a simplified method. This simplified hazard assessment, which has become standardized throughout the world, is laser hazard classification. The basic hazard classification concept of ANSI Z-136.1 is shown in Table 23.2 for some primary laser categories. Subcategories exist within some classes (i.e., Classes 2a and 3a). Representative hazard classifications for a few common laser systems are listed in Table 23.3.

The Food and Drug Administration regulation (21CFR1040) requires manufacturers to classify their laser products and indicate the hazard classification on a label. Since the manufacturer has the greatest knowledge about the laser's capability, it was thought appropriate to have the manufacturer perform the classification task. The technical staff of the manufacturer can perform careful measurements and detailed calculations of the laser's output, and they can assess the uncertainties in measured output parameters and the likelihood of the laser to exceed those parameters under unusual conditions.

The basis of the hazard classification is a set of accessible emission limits (AELs) which in turn are based on the ELs for occupational exposure with certain inherent assumptions as to reasonable exposure durations that might be anticipated through the use of the laser. The underlying assumption for exposure to a Class 1 laser normally is a worst-case, lengthy exposure to the eye of the total output of the laser if it were collimated or recollimated and directed at the eye (or, as in the case of visible and IR-A wavelengths, into the pupil of the eye).

23.5.1 Class 1 Laser Products

A Class 1 laser product is a laser device which is incapable of emitting laser radiation in excess of the AEL for Class 1. This obviously applies to a laser which emits very low power, such as some semiconductor diode lasers. However, most Class 1 laser products are in that category by virtue of an enclosure which limits the accessible emitted laser radiation to Class 1 AELs. A laser video disk player is an example of a Class 1 product. Most laser printers, optical-fiber communication

TABLE 23.2 Intrabeam Laser Ocular Exposure Limits

Wavelength λ, nm	Expos. duration t, s	Exposure limit
	Ultraviolet*	
180 to 302	1 ns to 30 ks	3 mJ/cm^2
303	1 ns to 30 ks	4 mJ/cm^2
304	1 ns to 30 ks	6 mJ/cm^2
305	1 ns to 30 ks	10 mJ/cm^2
306	1 ns to 30 ks	16 mJ/cm^2
307	1 ns to 30 ks	25 mJ/cm^2
308	1 ns to 30 ks	40 mJ/cm^2
309	1 ns to 30 ks	63 mJ/cm^2
310	1 ns to 30 ks	0.1 J/cm^2
311	1 ns to 30 ks	0.16 J/cm^2
312	1 ns to 30 ks	0.25 J/cm^2
313	1 ns to 30 ks	0.40 J/cm^2
314	1 ns to 30 ks	0.63 J/cm^2
315 to 400	1 ns to 10 s	$0.56t^{3/4}$ J/cm^2
315 to 400	10 s to 30 ks	1.0 J/cm^2
	Visible and IR-A†	
400 to 700	1 ns to 18 μs	0.5 μJ/cm^2
400 to 700	18 μs to 10 s	$1.8t^{3/4}$ mJ/cm^2
400 to 550	10 s to 10 ks	10 mJ/cm^2
550 to 700	10 s to T_1 s	$1.8t^{3/4}$ mJ/cm^2
550 to 700	T_1 s to 10 ks	10 C_B mJ/cm^2
400 to 700	10 ks to 30 ks	C_B μW/cm^2
700 to 1050	1 ns to 18 μs	0.5 C_A μJ/cm^2
700 to 1050	18 μs to 1 ks	$1.8C_At^{3/4}$ mJ/cm^2
1051 to 1400	1 ns to 50 μs	$5C_C$ μJ/cm^2
1051 to 1400	50 μs to 1 ks	$9.0C_Ct^{3/4}$ mJ/cm^2
1051 to 1400	1 ks to 30 ks	320 $C_A \cdot C_C$ μW/cm^2
	Far infrared‡	
1400 to 1500 nm	1 ns to 1.0 ms	0.1 J/cm^2
1400 to 1500 nm	1.0 ms to 10 s	$0.56t^{1/4}$ J/cm^2
1500 to 1800 nm	1 ns to 10 s	1.0 J/cm^2
1801 to 2600 nm	1 ns to 1.0 ms	0.1 J/cm^2
1801 to 2600 nm	1.0 ms to 10 s	$0.56t^{1/4}$ J/cm^2
2601 nm to 1 mm	1 ns to 100 ns	10 mJ/cm^2
2601 nm to 1 mm	100 ns to 10 s	$0.56t^{1/4}$ J/cm^2
1400 nm to 1 mm	10 s to 30 ks	100 mW/cm^2

Notes: 1 ks = 1000 s; 30 ks = 8 h.
C_A = 1 for λ = 400 to 700 nm; C_A = $10^{[0.02(\lambda - 700)]}$ if λ = 700 to 1050 nm
C_B = 1 for λ < 550 nm; C_B = $10^{[0.015(\lambda - 550\,nm)]}$ for λ = 550 to 700 nm
T_1 = 10 × $10^{[0.02(\lambda - 550\,nm)]}$ for λ = 550 nm to 700 nm
*All ELs for less than 315 nm must be $0.56t^{3/4}$ J/cm^2.
†7 mm limiting aperture.   ‡1 mm limiting aperture.
For all EL tables, exposure limit is used as it is by International Radiation Protection Association (IRPA); The same values are termed MPEs (maxium permissible exposure limits) by ANSI and TLVs (threshold limit values) by American Conference of Governmental Industrial Hygienists (ACGIH). Essentially all have the same limit values.
C_C is a new factor which permits increased TLVs for ocular exposure because of pre-retinal absorption in the ocular media of radiant energy in the spectral region between 1150 and 1400 nm. C_C = $10^{[0.0181(\lambda - 1150\,nm)]}$ for wavelengths greater than 1150 and less than 1200 nm and is 8.0 from 1200 to 1400 nm, thus raising earlier TLVs by 8 fold.

TABLE 23.3 Laser Exposure Limits for the Skin

Wavelength λ, nm	Exposure duration t, s	Exposure limit
	Ultraviolet	
200 to 400	1 ns to 30 ks	Same as eye EL
	Visible and IR-A*	
400 nm to 1 mm	1 ns to 100 ns	20 C_A mJ/cm^2
400 nm to 1 mm	100 ns to 10 s	1.1$C_A t^{1/4}$ J/cm^2
400 nm to 1 mm	10 s to 30 ks	0.2 C_A W/cm^2
	Far infrared	
1400 nm to 1 mm	1 ns to 30 ks	Same as eye EL

Notes: 1 ks = 1000 s; 30 ks = 8 h.
C_A = 1 for λ = 400 to 700 nm; C_A = 10[0.02(λ − 700)] if λ = 700 − 1050 nm
*1 mm limiting aperture for $t < 0.5$ s
3.5 mm limiting aperture for $t > 0.5$ s

systems, and certain laboratory chemical assay instruments (such as some commercial Raman spectrophotometers) are other examples. If a laser product is Class 1 by virtue of enclosure, the more hazardous laser inside the enclosure is referred to as an *embedded* laser.

Any removable portion of the protective housing of such a laser product would have to be secured or interlocked, if one could gain access to potentially hazardous laser radiation. No hazardous light levels could be emitted if removal of the panel by the final user was intended. If the panel required tools to remove, then it would not have to be interlocked, but service personnel could gain access to potentially harmful laser energy. In this case a warning label should be placed on the innermost panel, so that, prior to panel removal, there would be clear indications of the presence of potentially hazardous laser beams. If the panel were intended for removal by the user, labeling on the outside of the panel should indicate that accessible hazardous emissions exist if the panel cover were removed and the interlock defeated.

The AEL for a Class 1 laser product is the maximum permissible exposure (MPE) for the eye multiplied by the area of the limiting aperture. The limiting aperture is defined for the spectral region in which the laser emits. For wavelengths in the retinal hazard region from 400 to 1400 nm, the limiting aperture is 7 mm in diameter (0.4 cm^2), simulating the dark-adapted pupil. For wavelengths outside this spectral region, the limiting aperture is 1 mm, corresponding to the smallest reasonable area of direct beam exposure which could be measured. AELs for representative lasers are given in Table 23.5.

TABLE 23.4 Extended Source Ocular Exposure Limits

For extended-source laser radiation (e.g., diffuse reflection viewing) at wavelengths between 400 and 1400 nm, the intrabeam viewing ELs can be increased by the following correction factor C_E provided that the angular subtense of the source (measured at the viewer's eye) is greater than α_{min} (e.g., greater than 1.5 mrad for $t < 0.7$ s; $\alpha_{min} = 2t^{3/4}$ mrad for 0.7 s $< t < 10$ s, and $\alpha_{min} = 11$ mrad for $t > 10$ s).

$CE = \alpha/\alpha_{min}$ for $\alpha_{min} < \alpha < 100$ mrad.

$CE = \alpha^2/(\alpha_{min} \cdot \alpha_{max})$ for $\alpha > 100$ mrad.

The angle of 100 mrad may also be referred to as α_{max} at which point the extended source limits can be expressed as a constant radiance using the last equation written in terms of α_{max}.

$L_{EL} = (8.5 \times 10^3) \, (EL_{pt\ source}) \, \text{J}/(\text{cm}^2 \cdot \text{sr})$ for $t < 0.7$ s

$L_{EL} = (6.4 \times 10^3 \, t^{1/3}) \, (EL_{pt\ source}) \, \text{J}/(\text{cm}^2 \cdot \text{sr})$ for 0.7 s $< t < 10$ s

$L_{EL} = (1.2 \times 10^3) \, (EL_{pt\ source}) \, \text{J}/(\text{cm}^2 \cdot \text{sr})^*$ for $t > 10$ s

*[or $W/(\text{cm}^2 \cdot \text{sr})$ for point source limits expressed in W/cm^2]

TABLE 23.5 Some Examples of Laser Hazard Classifications

Some CW visible and IR-A lasers (U.S. system: FDA and ANSI):

Hazard classification	Power output Φ @ $\lambda = 400–700$ nm	Power output Φ^c @ $\lambda = 1050–1150$ nm	Emission duration
Class 1	$\Phi \leq 0.4 \ \mu W$[a]	$\Phi \leq 0.64$ mW	$t > 10$ s
Class 2	$0.4 \ \mu W < \Phi \leq 1$ mW	Does not exist	$t > 10$ s
Class 3A	1 mW $< \Phi \leq 5$ mW[b]	0.64 mW $< \Phi \leq 3.2$ mW	$t > 10$ s
Class 3B	5 mW $< \Phi \leq 500$ mW	3.2 mW $< \Phi \leq 500$ mW	$t > 10$ s
Class 4	$\Phi > 500$ mW	$\Phi > 500$ mW	$t > 10$ s

Some CW IR-B and IR-C infrared lasers (U.S. System: FDA and ANSI):

Hazard classification	Power output Φ @ $\lambda = 1.5–1.8 \ \mu m$	Power output Φ^c @ $\lambda = 2.8 \ \mu m$	Emission duration
Class 1 (pre-1993)	$\Phi \leq 0.79 \ W$[d]	$\Phi \leq 0.79$ mW	$t > 10$ s
Class 1 (post-1993)	$\Phi \leq 10 \ mW$[e]	$\Phi \leq 10$ mW	$t > 10$ s
Class 2	Does not exist	Does not exist	$t > 10$ s
Class 3A(1992)	0.79 mW $< \Phi \leq 4$ mW[b]	0.79 mW $< \Phi \leq 4$ mW	$t > 10$ s
Class 3A(1993)	10 mW $< \Phi \leq 50$ mW[b]	10 mW $< \Phi \leq 50$ mW	$t > 10$ s
Class 3B	50 mW $< \Phi \leq 500$ mW	50 mW $< \Phi \leq 500$ mW	$t > 10$ s
Class 4	$\Phi > 500$ mW	$\Phi > 500$ mW	$t > 10$ s

[a]AEL for class 1 is greater for $\lambda > 550$ nm in ANSI Z136.1 and IEC.
[b]Additional restrictions on irradiance apply in IEC system.
[c]All AELs for class 1 and class 3A raised 12-fold in 1992/1993 ANSI/IEC revised standards.
[d]Class 3A exists only in the visible in the FDA system.
[e]AEL established in 1992–1993 by ANSI/IEC.

23.5.2 Class 2 Laser Products

Class 2 laser products are visible lasers which are normally not considered hazardous unless one were to force oneself to stare directly into the visible laser beam. For example, a 0.5-mW He-Ne laser used to align a high-power infrared beam would be a Class 2 laser product. Since Class 2 lasers are of very low risk because of the aversion response to bright light, this class is limited to lasers emitting visible light between 400 and 700 nm (ANSI) or 710 nm [Center for Devices and Radiological Health (CDRH)].

Class 2 lasers are typically used in industry and medicine for alignment or to mark the path of an invisible laser beam. These guide beams are of critical safety importance when used to indicate the location of a beam of higher power such as a Nd:YAG laser beam.

The accessible emission level for a Class 2 laser is 1.0 mW. This power corresponds to the MPE for the eye for a 0.25-s (aversion response) exposure where the entire beam could enter the eye. The MPE is 2.5 mW/cm^2 at 0.25 s, leading to 1 mW passing through a 7-mm (0.4 cm^2) limiting aperture. The 7-mm limiting aperture has been standardized for all laser measurements and calculations in laser safety standards in the retinal hazard wavelength region of 400 to 1400 nm.

The potential hazard of such low-power lasers should be compared to that of a movie projector or a slide projector. If one were to force oneself to stare into a movie projector, or into a helium-neon alignment laser beam of less than 1 mW, a permanent retinal injury could result from lengthy exposures of many seconds, minutes, or hours, depending on power level. However, such exposures are considered highly unrealistic. Therefore, one should consider a Class 2 laser to pose a theoretical hazard, but not a realistic hazard for most situations.

This absence of a realistic hazard from the low power of a Class 2 laser applies to the awake, task-oriented individual. One must recognize that exposure of the retina of an anesthetized patient to Class 2 levels for some length of time may be potentially harmful.

Class 2 lasers must have a CAUTION label telling the user not to stare into the beam, but other precautions are generally not needed. Again, for emphasis, the AEL for a Class 2 laser is currently based on the aversion response to bright visible light. Hence, this type of laser must emit only a visible beam having a wavelength between 400 and 710 nm (FDA, 1988) or between 400 and 700 nm (ANSI and IEC standards).

There has been some consideration in the standards community to extend Class 2 into the IR-A band and include lasers that do not exceed the AEL of Class 1 for 10 s. The rationale is that Class 2 lasers currently are those where behavioral considerations place them into a category of posing a theoretical hazard, but not a realistic hazard. The ANSI standard recommends a maximal viewing duration of 10 s for infrared lasers when calculating MPEs for applications where viewing is not intended.

23.5.3 Class 3 Laser Products

Class 3 laser products are potentially hazardous on direct, "instantaneous" exposure of the eye. The collimated beam, if directed into the eye, could result in injury within a time less than the aversion response (0.25 s) for visible lasers. The Class 3 category is divided into two subcategories: Class 3A and Class 3B.

Class 3A lasers emitting in the visible spectrum must have an output power between 1.0 and 5.0 mW. Originally this subclass 3A was limited to those lasers where the beam diameter has been expanded to greater than 7 mm, such that less than 1.0 mW could enter a 7-mm pupillary aperture. This requirement was later dropped in U.S. standards, but not in European, IEC, and Australian standards. Any laser now sold in the United States with an output power between 1 and 5 mW and wavelength between 400 and 700 nm (710 in the FDA standard) is defined as a Class 3A laser product. A Class 3A laser product must meet fewer FDA performance standards and requires less stringent precautions than does a Class 3B laser system. Despite the differences between U.S. and other standards, the performance requirements are the same for the same beam power and beam diameter, and the use of CAUTION labels and DANGER labels, depending on whether the laser exceeds 2.5 mW/cm^2, remains the same.

The ANSI Z-136 and IEC 825 standards also include in Class 3A those lasers emitting in the ultraviolet and infrared regions of the spectrum which have output power or energy within 5 times the AEL of Class 1, but with an expanded beam such that the AEL of the emitted laser beam is not exceeded if measured by the limiting aperture for that laser wavelength. The limiting aperture varies with laser wavelength. It is 7 mm (corresponding to the dark-adapted pupil diameter) for wavelengths in the retinal hazard region (400 to 1400 nm), and is 1.00 mm in the ultraviolet, IR-B, and most of the IR-C spectral bands. [Laser safety standards may be revised to use an aperture larger than 1 mm (i.e., 3.5 or 3 mm) for cw IR-B and IR-C lasers.]

Additional performance requirements and safety measures apply to Class 3B lasers, which are capable of injuring the eye upon direct intrabeam exposure. Any cw laser above 5 mW output (in the visible) and less than or equal to 500 mW falls into Class 3B. Virtually all cw lasers used in material processing, surgery, and laser chemistry have an output power exceeding 0.5 W (500 mW) and therefore fall into the next class, Class 4.

23.5.4 Class 4 Laser Products

Class 4 lasers are those devices which may present a serious fire hazard, skin hazard, or diffuse reflection hazard. Initially the Class 4 laser category was based on hazard assessments of very high powered Q-switched crystal lasers such as the ruby and Nd:YAG lasers, which were often capable of producing a hazardous diffused reflection in a research laboratory. If such a hazardous diffuse reflection existed, it was likely to pose a serious risk of eye injury to anyone in the research laboratory room who could see the laser target. Hence the probability of exposure was very high by comparison to exposure from a small collimated pencil beam of laser light which would have to be aligned to the pupil of an individual in the laboratory for an injury to result from direct exposure (as from a Class 3B laser).

Any laser with an average power exceeding 0.5 W will be in Class 4. As noted previously, most material processing and surgical lasers are Class 4 laser products; most Q-switched lasers are generally Class 4.

23.6 LASER HAZARD ASSESSMENT

The concept of a hazard classification is always based on a worst-case assessment of laser exposure to persons in or near the laser. For example, a 3-mm diameter, collimated, far-infrared laser beam with an average power of 1 W could cause a burn to the skin as well to the eye. However, if the same power were expanded to a 10-cm-diameter beam, the irradiance would fall below the exposure limit for even lengthy exposures of several hours or a full day.

Hence there are situations where, by virtue of the laser beam characteristics, the classification can grossly overstate the real risk of one particular laser. Nevertheless, even the expanded 10-cm-diameter laser beam could be recollected by a concave mirror or by some optical element which could focus the infrared energy onto tissue and then pose a significant hazard. Hence the present classification scheme is defended on the basis of worst-case conditions.

The above situation illustrates that laser classification is a valuable tool that will have exceptions. It provides only a first step in the hazard evaluation. Actual use conditions may permit relaxation of the precautions normally followed with that type of laser. Other delivery systems may also alter the hazard and even the class of the laser system.

23.7 LASER SYSTEM SAFETY

System safety pertains to the design and manufacture of a laser product. Examples of system safety features are warning lights, built-in eye-protective filters in viewing optics, protective covers over a firing switch, and electrical grounding. In the United States, certain laser system safety features are mandated by the FDA federal governmental regulations[8] under the Radiation Control for Health and Safety Act of 1968. Under these regulations (specifically 21CFR1040), certain performance standards apply to all laser products marketed in the United States, whereas others apply only to specific laser hazard classifications or to specialized laser uses. The manufacturer must certify that its laser product meets these requirements and must file documents which detail this certification with the Center for Devices and Radiological Health of the Food and Drug Administration, Rockville, Md.

The details of the FDA laser product performance standards can be quite complex, and shall not be dealt with here, but the typical laser user can be assured that reasonable system safeguards have been designed into all commercially available lasers. This may not be true of prototype or experimental laser devices, but final product lines should be in compliance. The FDA performance standards vary with hazard classification (and FDA denotes classes by Roman numerals I through IV).

The laser system safety features required by the CDRH performance standards which are applicable to most commercially constructed research and industrial lasers are

1. An interlocked or secured protective housing
2. A remote connector which can be used to interlock an entrance door
3. A key-operated switch
4. An emission indicator, such as a pilot light
5. A beam attenuator, e.g., a mechanical shutter
6. Specified warning labels
7. Protective viewing optics (i.e., a filter or shutter system)
8. Operator controls located to limit the chance for exposure.

In addition to these general FDA requirements, requirements for specific-purpose laser products are found in 21CFR1040.11 of the Federal Performance Standard. All medical laser products must also comply with three other requirements: (1) a means to measure the output within ±20 percent, (2) a measurement calibration schedule, and (3) a laser aperture label. In some instances, a self-monitoring fixed laser output is considered to fulfill the first medical requirement. Each manufacturer can obtain variances from the above standards if alternative and effective controls are provided. Surveying, alignment, and leveling laser products, and demonstration laser products, are limited to Class 3A or below. The term *demonstration* here refers to lasers used for demonstration in schools or art displays. However, laser light shows obviously employ much more powerful lasers and these must be operated within a variance issued by CDRH.[4,8]

Of key interest to the operator of a Nd:YAG microwelder/drilling instrument is the protective filter in the viewing optics. The most common type of filter is 3 to 4 mm of Schott KG-3 filter glass which has an optical density of 4.5 to 6.0 (i.e., an attenuation factor of 31,000 to 1,000,000) at 1064 nm. This filter is absolutely essential, since reflections of the laser beam from the material target can enter the operator's eye. Obviously, any servicing on the viewing optics, or replacement of a part of the viewing optics, must leave the user with the assurance that the protective filter is in place. Since the 1064-nm wavelength is well into the infrared spectrum, a safety filter does not need to filter in the visible and can therefore be nearly transparent; thus the filter should be fixed in position.

The key switch is helpful, if used, to prevent unauthorized use of the device by untutored persons. Since the output of many lasers is sufficient to seriously injure a bystander if a hazardous reflection exists, removal of the key when not in use is recommended.

Since hazardous reflections and high voltages are accessible within the protective housing, the interlocked or screw-fastened enclosure should not be tampered with except by a well-trained service person.

Other system safety features that are not mandated by the FDA-CDRH regulations are possible and can be very useful. For example, beam baffles in the work area can reduce the likelihood of personnel exposure. If the remote control connector on Class 3B and 4 laser products is not used to activate a door interlock, it may be valuable to connect it to a warning light that illuminates only when the laser is operational, or to a beam delivery interconnect switch (required by ANSI Z-136.3 for surgical lasers) which would disable the laser when the focusing objective is removed.

23.8   THE SAFE INDUSTRIAL LASER LABORATORY

Most industrial lasers are Class 4 unless totally enclosed to fulfill Class 1 requirements. High-power CO_2 laser systems are often not as hazardous as one would expect, since the focused beam creates an optically absorbing and diffuse plasma and reflected levels are low unless the beam power fails and the beam does not create a plasma—resulting in specular reflections. Such a laser operating as an unfocused heat-treating beam could be specularly reflected.

By contrast, a much lower powered Nd:YAG laser, which operates in the retinal hazard region, can create dangerous reflections over a far wider area. It should be situated in a closed room with a controlled entrance. Windows should be nonexistent or covered with a lighttight, opaque screen. In this regard, many plastic "opaque" curtains actually transmit the 1064-nm wavelength. Where feasible, the beam path should be terminated so that there is no real likelihood of an individual standing within the direct beam. A strict interpretation of the 1980 edition of the ANSI Z-136.1 standard[8] indicates that a door interlock should be installed to preclude operation when the door is ajar. However, it has been argued that interlocks are probably unwarranted for many Nd:YAG (and CO_2) laser devices, if the beam is focused, and therefore the potentially hazardous area (NHZ) is of limited extent (e.g., about 1 m around the focal spot). A warning sign and/or light above the door should indicate when the laser is in use.

During the 1980s, the clamor for a more reasonable policy on door interlocks for laser facilities forced the ANSI Z-136 Committee on Safe Use of Lasers to reconsider the requirement. In the earlier issues of ANSI Standard Z-136.1, "Safe Use of Lasers" (1973, 1976, 1980), the use of either safety latches or door interlocks at the entrance to Class 4 laser facilities had been mandatory.[8] It has been an advisory requirement (i.e., recommended by *should* rather than by *shall*) for Class 3 laser facilities since the 1980 edition of the standard. Clearly, a door interlock is necessary if during laser operation the unexpected entry of an unprotected person would probably result in eye injury to the visitor. However, in many laboratory and industrial settings, the laser beam path may be largely shielded or enclosed and the probability of accidental exposure upon entry into the facility would be extremely remote. Indeed, if the control measures within the facility are sufficient that eye protectors are not mandated, then surely door interlocks are unwarranted.

For these reasons, revisions were made in the 1986 edition of the standard which introduced the concept of a nominal hazard zone to define a space wherein certain control measures are essential. If the entryway of a laser installation were within the NHZ, then interlocks would be required. This concept is an extension of what appears in the new international standard from the International Electrotechnical Commission (IEC), Publication 825, which defined the concept of nominal ocular hazard area (NOHA), which is essentially the same idea.[9]

23.8.1 Why Entryway Interlocks?

It is useful to review the original concept of a Class 4 laser system, and why door interlocks were initially mandated for such facilities in safety proposals of the 1960s.[2-4,10] The probability of accidental eye exposure to a small, collimated laser beam—either the direct beam or a specular reflection—is quite small in most operations[4,10] unless intentional viewing is anticipated. However, if a laser is capable of producing a hazardous diffuse reflection, then the probability of individuals within a nearby viewing area being exposed to hazardous levels is very high. Since most raw-beam (i.e., an undiverged, unexpanded beam) Q-switched ruby and neodymium lasers were known to be capable of causing a hazardous diffuse reflection within the confines of a laboratory, and the high probability of dangerous exposure was recognized, the greatly increased risk of injury from this type of laser led to the initial definition of the high-risk Class 4 laser group. Since then, lasers which had beam irradiances likely to cause either severe skin injuries or a fire hazard were included in Class 4. It is virtually impossible to create a hazardous diffuse reflection with a cw laser since the beam irradiance would burn most targets, so the latter refinement was both realistic and useful. Lasers which posed a potential hazard to the eye only on intrabeam viewing were then placed in the less dangerous

Class 3 category where control measures were designed to prevent (or at least greatly reduce the probability of) direct intrabeam exposure of the eye.

23.8.2 The Nominal Hazard Zone

The new approach of defining an NHZ, within which Class 3 and 4 control measures would apply, is more realistic, but it requires a careful analysis by the locally designated laser safety expert (or responsible staff member), the laser safety officer (LSO). If we calculate the distance from a diffuse reflector beyond which the exposure limit (EL) or maximum permissible exposure for accidental (momentary) viewing is not exceeded, we find that this is a relatively short distance in most practical applications. For example, an argon laser which emits a power Φ of 10 W will produce a hazardous irradiance E of 2.5 mW/cm² (0.25-s aversion response time) by diffuse reflection from a target of reflectance ρ to a distance r_{Haz} of

$$r_{\text{Haz}} = \frac{\rho\ \Phi\ \cos\theta}{?\ E}$$

$$= \frac{(1.0)(10\ \text{W})(1.0)}{(3.14)(0.0025\ \text{W/cm}^2)}$$

$$= 35.7\ \text{cm} \tag{23.1}$$

where θ is taken at the worst-case angle of 0° and ρ is taken to be 1.0 (100 percent) for a worst-case condition. Even if the person entering the room were to overcome the aversion response and stare at the bright reflection for 10 s, r_{Haz} would only be increased to 56 cm. By comparison, a Nd:YAG Q-switched laser with an output energy Q of 0.1 J would have a distance r_{Haz} of 80 cm for a single-pulse exposure at 1064 nm. A more energetic Q-switched ruby laser with an output energy of 1.0 J would have a distance r_{Haz} of 8 m. The last laser would likely require door interlocks, since the 800-cm hazard distance would reach the entryway of most laboratories.

Not only hazardous diffuse reflections can lead to likely ocular exposures that would exceed the appropriate EL or MPE. In some instances, exposure to the direct or specularly reflected beam is likely. In these situations, the NHZ would also include the hazard distance along such a beam. For example, a laser robotic welding system could have the freedom to point in the direction of the entrance or other occupied areas, and the NHZ should extend into those areas where beam pointing is likely. Fortunately, many laser applications in material processing and in surgery employ a focused beam which limits the on-axis hazard distance. In this instance, a relatively high-power CO_2 surgical laser with an output power of 100 W, a 2-cm-diameter exit beam, and 10-cm focal distance (i.e., a divergence of 2/10 = 0.2 rad) could have a hazard distance to 100 mW/cm² of only 178 cm from the focal spot. Hence the most conservative NHZ would be only in the immediate vicinity of the operating table.

Although the introduction of more calculations and a more involved hazard analysis for the LSO may be an unwelcome development to some, it appears to be a necessary evolution in laser safety as more laser applications develop. For those who are unconcerned about any adverse impacts from the installation of interlocks and other hazard control procedures, then these added assessments become unnecessary. Manufacturers of industrial and surgical laser systems could provide a very valuable service to their customers by including NHZ data in user manuals. Of course, this service would only be practical for laser systems with fixed beams or fixed-focus beams where the NHZ would remain fixed.

23.9 LASER EYE PROTECTION

Probably no single hazard control measure is more important in the laboratory setting than the proper use of laser protective eyewear. In an industrial laser application, it is always desirable to enclose the potentially hazardous laser beam so that workers are not exposed. Preventing injuries to the human eye is, of course, the primary objective of most laser safety programs. However, for some laser environments, such as in surgical suites and research and development laboratories, it is not always possible to totally enclose the beam, and eye protectors may be required if exposure to hazardous laser radiation is possible.

If exposure of the body to concentrated, very high power laser beams (e.g., above 100 W in a 1-cm-diameter beam) is likely, then eye protection alone cannot be relied on. Skin protection becomes a concern, and physical barriers or beam enclosures must be employed.[4,5,12]

Laser eye protectors are available in the form of spectacles, dual lens goggles, coverall goggles or even a face mask, as shown in Fig. 23.5. The objective of laser eye protection is to filter out the specific laser wavelength while transmitting as much visible light as possible so that the wearer can perform a task.[4,14]

Commercially available laser eye protectors are normally designed and specified for just a few specific laser wavelengths. For example, most manufacturers of laser eye protectors supply protection against the most common laser wavelengths of the argon laser (488 and 514.5 nm), excimer ultraviolet lasers (193, 224, 248, 308,

EYE PROTECTORS

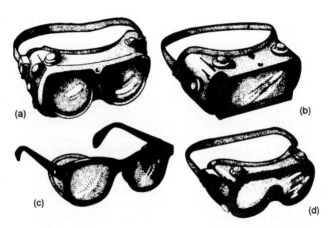

(a)

(b)

(c)

(d)

FIGURE 23.5 Laser eye protectors. Laser eye protection is available in several general designs: two-lens goggles, coverall goggles, spectacles with side shields, and wraparound spectacles.

and 350 nm), the Nd:YAG laser (1064 nm), the frequency-doubled Nd:YAG (532 nm), the GaAs lasers (840 to 910 nm), and the CO_2 laser (10.6 μm). With the use of new laser wavelengths, special tests may be required to determine whether eye protection designed for another type of laser could be used for the new wavelength.

Obviously, laser eye protection is generally considered necessary if an open beam exists at levels above the maximum permissible exposure limit. If there is a potential for a hazardous beam to be directed at an individual's eyes, either directly or indirectly by reflection, eyewear becomes mandatory. Even in instances where the risk of a hazardous reflection or direct exposure is extremely small, most safety authorities feel that one should still insist on the wearing of eye protectors regardless of how unlikely may be the hazardous exposure condition.[15]

At the CO_2 laser wavelength of 10.6 μm, virtually all clear visors and glasses are nearly opaque and will substantially eliminate the risk of hazardous exposure. Even contact lenses afford a degree of protection. Nevertheless, most experts caution users to at least label street eyewear or clear safety glasses/goggles for use at the 10.6 μm wavelength. Any currently available transparent lens should have an optical density (OD) of at least 4. Polycarbonate lenses have superior resistance to burn-through of any transparent plastics. Because contact lenses do not reliably cover the cornea, authorities warn against their use as laser eye protection.

23.9.1 Specifying Laser Eye Protectors

To determine the adequacy of any given eye protector, one should read the manufacturer's data sheet, and one may need to make a simple calculation based on the laser's output. ANSI standard Z-136.1 (1986) requires the eyewear manufacturer to label laser protective goggles with certain technical specifications such as wavelength and optical density (for example "Nd:YAG, OD 6 at 1064 nm"). In Germany, DIN Standard 58 215 (1985)[16] requires a label encoded on the eye protector which also indicates the protected wavelength and a density based on the damage threshold of the filter material and frame (e.g., "D/RI 1 000 − 1 000 L6A X DIN," which means little to those not familiar with the special nomenclature of that standard).

Obviously, it is very important that goggles designed for one laser wavelength are not mistakenly used for a different wavelength where filtration is not afforded. For this reason, some safety specialists favor the current ANSI approach where clear wording appropriate to the knowledge and training of the user is provided. On the other hand, if one understands the details of the DIN standard, more technical information is available.

The DIN standard also requires testing of filter damage. While the ANSI standard recommends that the user consider the filter damage threshold when selecting eye protectors, there is no requirement for a filter to withstand a given irradiance (W/cm^2) or radiant exposure (J/cm^2) over a particular exposure duration. When one thinks of the wide variety of wavelengths, pulse durations (from femtoseconds to seconds), and beam spot sizes that could apply for the same goggle, it becomes clearly a major challenge to design a general testing program to rate a filter's vulnerability. The approach of the DIN standard is to choose a few common wavelengths and pulse durations and to test for relatively worst case conditions (e.g., a small laser spot size on the filter). By such an approach, the DIN standard virtually rules out the use of lightweight plastic filters for use with most Class 4 lasers, even though in cases where the beam would never be directly viewed as a collimated beam, the plastic filter would be perfectly adequate.

The color of the filter provides a clue as to the wavelength range of protection, but only a clue. It would be dangerous to attempt to judge the appropriateness of a protector by its color. While it is true that green filters transmit green light and attenuate red light, and blue filters transmit blue and absorb red, one cannot ascertain the degree of filtration for a laser wavelength. Of course, one can be sure that a filter is inappropriate if the filter appears visually to transmit the same color as the laser wavelength. With an ever-increasing number of near-infrared and middle-infrared lasers (e.g., alexandrite at 780 to 850 nm, Er:YAG at 2.94 μm, Ho:Cr:Th:YAG at 2.09 μm, Ho:Th:YAG at 2.08 μm, Er:glass at 1.54 μm, Th:Cr:YAG at 2.01 μm, Th:YLF at 1.95 μm), the problem of distinguishing different filters and labeling them with varying optical densities becomes serious. In the IR-B (1.4 to 3 μm) spectral region, the one group of filter materials that covers the spectrum of the aforementioned lasers is Schott KG-3 or KG-5 filter glass.[17,18]

Sunglasses would normally be inappropriate as laser protection. The protection factor is normally quite small because the visual transmittance is normally greater than 15 percent, i.e., a protection factor of less than one optical density unit.

23.9.2 Optical Density of Protective Filters

The amount of energy or power that can enter the eye will determine the degree of protection needed at the laser wavelength. The protective level of a filter is normally specified by optical density. OD is a logarithmic phenomenon, and it is a valuable parameter because of the enormous filtration factors that are generally necessary for laser eye protection. For example, instead of specifying a goggle filter with a transmission of 0.0001 percent, one can specify it with an optical density of 6.0. The transmittance τ is expressed as

$$\tau = \frac{H_t \text{ (transmitted)}}{H_i \text{ (incident)}} \tag{23.2}$$

and

$$OD = \log_{10}\left(\frac{1}{\tau}\right) = \log_{10}\left(\frac{H_i}{H_t}\right) \tag{23.3}$$

where the expression $(1/\tau)$ is frequently termed the *attenuation factor*. The attenuation factor is the ratio of the laser beam input power or energy divided by the transmitted power or energy. The OD can also be expressed as the logarithm of the corresponding ratios of either irradiances or radiant exposures. It is a logarithmic ratio and therefore is unitless.

A filter with an attenuation factor of 10 will reduce the incident laser beam power to 10 percent of its initial value and has an optical density of 1.0. If a filter reduces the output power by a factor of 10,000 (i.e., 0.01 percent transmission), the optical density is 4. Of course, in realistic situations, the measured optical density may not be a whole number and a decimal number may be used, such as 5.3 (which corresponds to an attenuation factor of 2×10^5 or 200,000 times).

23.9.3 Visual Transmittance of Protective Filters

The objective of a protective filter is not only to filter out the hazardous wavelengths but to transmit as much of the rest of the visible spectrum as possible. Thus, the visual transmittance of the goggle is often specified. Although it does not indicate the degree of any color distortion, this parameter is useful to indicate how significantly normal vision will be affected.

Technically, the luminous (visual) transmittance is the ratio of the photopic incidence energy to photopic transmitted energy. *Photopic* refers to the daylight vision of the cone receptor cells in the retina. This visual transmittance may not be the same for dark-adapted vision, i.e., for *scotopic*, or rod vision.[19] Hence, if one is working in a darkened laboratory, then one should also check the scotopic transmittance of the eye protection to be worn. For example, argon laser eye protectors, which are orange in color, and yellow or greenish-yellow goggles, will each attenuate more in the blue end of the spectrum than in the green and yellow regions of the spectrum where photopic vision dominates. This will result in severely reduced scotopic or rod vision if these filters are worn in a darkened environment. A blue-green filter specified to protect against the Nd:YAG infrared wavelength should have a good scotopic transmittance and could be worn in a darkened room. By contrast, another filter which is green-yellow in appearance and designed to protect to the same degree at the same infrared wavelength, would transmit very little blue light and would therefore have poor scotopic transmission.

Some users may find it puzzling that not all manufacturers will report the same visual transmittance for the same goggle filter, but this can occur when two different light sources are used in the measurement. As a technical aside, the differing values in both photopic and scotopic luminous transmittance can occur if the laboratories use different *source functions*. For example, one may use a theoretical white-light source, noontime sunlight, a so-called CIE C source, and when two sources spectrally differ from one another, the measured or calculated transmittance through a colored filter will differ. Most laboratories that measure the spectral transmittance in a spectrophotometer will use a neutral (white-light) spectrum.

From a realistic standpoint, the type of light sources viewed through color filters may have unusual characteristics that will greatly influence the visibility of different objects. For example, viewing a light-emitting diode (LED) display, an oscilloscope or computer screen through colored filters may pose special problems. If the display contains phosphors (e.g., P43) which emit only in a narrow waveband, the display may become invisible if the phosphor emission is not transmitted by the colored protective filter.

23.9.4 Protective Filter Damage

One often asks at what power level the filter material will degrade or be burned or shattered when hit by the direct beam of a laser. Most U.S. eye-protector manufacturers specify that the goggles are "not intended for direct (intrabeam) viewing" because of concerns about filter damage. Indeed, in most realistic situations, one would never intentionally aim the focused beam directly on any eye protector. Nevertheless, the required optical density is frequently based on the assumption that the entire beam could be incident on the eye.[16] Hence, it would be desirable to filter the incident beam energy or power to a level which is below the exposure limit (MPE).

The full beam power of large CO_2 industrial lasers and some surgical lasers may indeed be capable of burning through most protective filter materials.[15,20,21] As worrisome as this may appear, in reality it would not be expected that someone would stabilize his or her head in a fixed position while a laser beam turns a glass filter red hot or burns through a plastic goggle. With a flame (or at least bright light or luminescence) emanating from the impact site and smoke engulfing the area, a person would, realistically, rapidly move away from the beam, thereby preventing the beam from penetrating the protective filter. Even a 100-W CO_2 laser beam takes a few seconds to burn through a typical clear transparent plastic industrial safety goggle.[22]

One laser accident victim has reported that he continued to look through a goggle as a beam burned through it in 30 s and did not realize the bright light he experienced from the burn-through was from his goggle.[23] This report was greeted with some skepticism[21] by those knowledgeable in physiological optics and those who have witnessed burn-through experiments with plastic goggles.

Polycarbonate lenses were shown to be clearly superior when compared to other plastics in burn-through time. The added protection afforded by polycarbonates exists because a graphite char grows from the burning surface and this protects the underlying plastic surface. Figure 23.6 shows examples of burns made in polycarbonate plastic lens blanks with high-power industrial CO_2 lasers. The lens blanks were exposed to irradiances from 20 W/cm² to more than 2 kW/cm². The irradiances are what might be encountered from realistic collimated-beam reflections in an industrial laser operation, and the tests clearly show that any normal degree of lens-versus-beam movement—as would certainly be expected on exposure—would protect the wearer. This type of goggle filter material could serve as both laser eye protection for far-infrared CO_2 lasers as well as for protection against flying objects; ANSI Standard Z87 relates to impact-resistant eye and face protection. Normally,

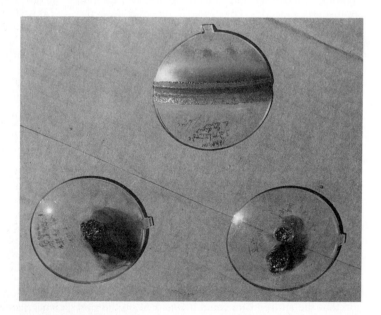

FIGURE 23.6 Examples of polycarbonate lens blanks exposed to high-power CO_2 laser radiation. Upper left: An 8-mm-diameter 50-W beam did not penetrate a 1-mm thickness in 100 s at a 45° angle, but did penetrate in 40 s at normal incidence. Upper right: a 20- by 20-mm-square, 450-W beam at normal incidence did not penetrate 2.5-mm-thick filter in 10-s exposure; note the graphite char which protected the underlying plastic from burnthrough. Lower left: An 8-mm-diameter 1600-W beam did not even penetrate halfway when moving at 5 cm/s. Lower right: Polycarbonate plate tested at 62 W/cm²; note soot from extensive flame.

several watts of cw power are required before a beam can even start to penetrate the protective lens material.[21,22] If needed, eyewear with quartz lenses can be used for CO_2 laser exposures to irradiances exceeding 1 kW/cm².

The exact thresholds for filter damage are determined by a complicated combination of such factors as beam irradiance, beam diameter, total beam power, exposure duration, wavelength, and material. Although glass filters will withstand higher irradiances before damage, the damage mode is frequently to shatter at a time after exposure when the glass surface cools and creates a tension relative to the central zone of the glass plate.

As noted previously, the DIN standard[16] for laser eye protectors strongly emphasizes the factor of filter vulnerability upon direct exposure. Some German laser eye protector manufacturers have produced glass goggles with coated lenses capable of withstanding 1 kW/cm² for 10 to 100 s. Of course, the skin would be severely burned at such intensities, and beam barriers should normally be used instead of relying on eye protectors.[20,21,27] Yet some argue for eye protectors for the "added degree of protection."[23]

In any case, other tests in the DIN standard are not controversial and are generally considered very useful, e.g., abrasion resistance, visual transmittance, ultraviolet solarization, thermal stability, and optical quality tests, as well as the measurement of optical density.

If a committee were charged with developing a standard for laser eye protector specification today, the challenge would be immense. The choice of reasonable densities and OD labeling would be the least of the problem. The primary difficulty would be to achieve a consensus on laser damage. What constitutes damage that would be hazardous? It would be difficult for many practical safety specialists to accept the rigorous, worst-case approach of the DIN standard. For example, the use of a 1-mm-diameter beam size and a burn-through time of 100 s which had been proposed in one European draft standard was not very realistic, as noted above. A realistic specified time for burn-through testing would be 3 s, and certainly no more than 10 s.

23.9.5 Eyewear Comfort and Fit

Other factors of importance in selecting eye protectors are related to comfort of fit and visual performance. Compliance with a rule to wear eye protectors will be very poor if a goggle is not comfortable to wear and if it is difficult to see through. Problems with vision may arise from the absorption properties of the protector or from fogged or degraded surfaces. It is, therefore, important that the laser safety officer assure that eye protection is both comfortable and protective.

It is not necessary from a laser safety standpoint that the goggle be so tightly fitting that it is uncomfortable and thereby limit air circulation, leading to fogging. Comfortable fitting is appropriate and does not significantly increase the risk to the wearer.

Any laser user who wears prescription spectacles with a frequent need for protective eyewear may find it worth the expense of having a pair of protective spectacles made containing a refractive prescription. Large coverall goggles are available which can be worn over the user's normal spectacles. However, some goggle designs lead to tunnel vision which may impair visual performance necessary for certain operations.

A variety of commercial vendors provide laser eye protection, and a partial list of primary sources is provided in Table 23.6.

TABLE 23.6   Principal Manufacturing Sources of Laser Eye Protectors*

---

1. American Optical Co, Inc.
   Safety Products Division
   Sturbridge, MA 01550, U.S.A.

2. Bollé
   rue Tacon/B.P. 139
   F-01104 Oyannex, France

3. Glendale Optical Co. Inc.
   130 Crossways Park Drive
   Woodbury, Long Island, NY 11797, U.S.A.

4. Fish-Shurman Corp.
   PO Box 319
   New Rochelle, NY 10802, U.S.A.

4. LaserVision, GmbH
   Berliner Strasse 9b
   D-8550 Forchheim, Germany
   (in U.S.A.: Winter Optical
                Smithfield, RI 02197)

5. Fred Reed Optical Co.
   P.O. Box 27010
   Albuquerque, NM 87125, U.S.A

6. Laser-R Shield, Inc.
   P.O. Box 91957
   Albuquerque, NM 87199, U.S.A.

7. Yamamoto Kogaku Co.
   Safety and Healthcare Div.
   1-2, Chodo-3, Higashiosaka City
   Osaka 577, Japan

---

*Many vendors use one or several of these principal sources or may fabricate custom laser eye protectors.

23.9.6   Viewing the Laser Beam

Many who work with lasers in the research and development laboratory complain that they cannot wear laser eye protectors because they cannot see the beam. They seem to ignore the fact that those who work with invisible Nd:YAG, CO_2, and other infrared lasers manage to align beams and perform laser experiments without using their normal vision. The tricks used to visualize the invisible beam are useful for those who work with visible laser beams that are not visible through their eye protective filters. Visible and ultraviolet laser beams are normally visible by fluorescence emitted by ordinary phosphor cards, and even by white paper (which actually contains phosphors). The beam is often easier to align because the beam center is pronounced and the visible light is not so intense that it is uncomfortable to view. Pulsed laser beams may be visible by using laser burn paper. Infrared beams can be made visible through the use of special infrared phosphor cards. Finally, electronic image converters or vidicon cameras may be used to render the beam visible. The LSO should never accept the statement, "I cannot see the beam," as an excuse for not wearing laser eye protectors within the NHZ.

23.9.7 Eyewear Filter Testing

It is not unreasonable to question whether eye protective filters should be period-ically tested to assure that they remain in conformity with the protective require-ments. This, however, is not a simple matter.[17,20,25] Measurement of optical densities above 3 is beyond the capability of most common laboratory spectro-photometers and similar testing instruments.

Unless protective filter materials are placed in an unusual environment such as in the bright sun or at extreme temperatures for an extended period of time, there has been no indication that commercially available filter materials degrade seriously with time. Some plastic filters have become darker with aging but not less protective. Therefore, in the past it has not appeared necessary to require routine testing of the filtration factor of the filters within the protective eyewear. However, eye protection devices should be periodically inspected for general degradation such as loss of fit, missing ventilation caps, severely and obviously visually degraded filters, and missing straps. If the eye protectors are faulty, they must be repaired or replaced.

Filters used to protect against short-pulse lasers must not saturate or undergo reversible bleaching to the point that they do not meet the specified OD. Some filters have shown this effect.[4,25] From a technical standpoint, the saturation effect—particularly characteristic of certain types of organic dyes—is dependent largely on the radiant exposure and not the irradiance.[26]

23.9.8 Protection against Nonlaser Optical Radiation

The bright, visible, white light created during laser welding and cutting may be annoying, and neutral filtration may be desired to bring the bright flash down to a comfortable viewing level. If one experiences afterimages, viewing this light can be a serious potential hazard to the retina (i.e., it is similar to staring at the sun) and dark filters should be mandatory. Retinal injuries have occurred to researchers who thought that the laser eye protection they were wearing would protect them when they viewed the optical plasma produced by high-power cw laser material cutting.

23.9.9 Methods of Eyewear Construction

Laser eye protectors are constructed of either filter glass or molded plastic. Plastic filter materials may have an organic dye impregnated in the plastic either during the plastic mixing process or by surface diffusion. Plastics generally degrade more readily by abrasion than glass, but are of lighter weight than glass filter materials. There are advantages to both types of filter materials.

Glass is generally more resistant to abrasion. Glass filter materials degrade less with time and use and frequently have been noted to have better optical quality.[24] It is easier to place antireflection coatings on glass and to stack filters of different absorbance to achieve a higher resistance to damage by intense laser exposure. However, glass filter lenses are heavier and more fragile upon impact than their plastic counterparts. These factors should be borne in mind in selecting eye pro-tection.

The frame can be separated from the filter material as in most goggle designs using glass filters. The frame design may be either an open design similar to a conventional spectacle frame or a tightly fitting design similar to protective goggles used by aviators, motorcyclists, and welders.

As previously noted, the degree of ventilation is affected by the closeness of the fit. Interference coatings may be used to protect the surface and increase resistance to damage. This damage resistant factor is emphasized far more in German than in American laser safety standards.

23.10 LASER ACCIDENTS

Most severe laser-induced eye injuries result from inadequate attention to laser safety procedures. Most often injured are research workers and service technicians. Service technicians frequently must gain access to locations within the protective housing and may have access to the laser beam prior to the focusing objective lens. This beam is normally collimated, and a reflection from a flat surface, as shown in Fig. 23.3, can be exceedingly dangerous. The collimated beam can travel a great distance before the irradiance falls below safe levels. In many instances, service technicians must wear laser eye protectors and onlookers should be barred from the closed room during servicing. As noted previously, lasers that operate in the visible and near-infrared, retinal-hazard region (400 to 1400 nm) pose a serious threat to the retina, and it is these lasers that have caused serious loss of vision, and therefore deserve special attention. The risk of accidental exposure differs with each type of laser operating in this spectral region, depending on whether the laser is pulsed or cw, and whether the beam is visible. Unlike the cw argon laser, which can only coagulate the retina upon accidental exposure, Q-switched or other short-pulse lasers can cause an explosive lesion in the retina if focused there, resulting in a hemorrhage into the vitreous matter. The minimal-image-size, worst-case exposure condition will occur when the collimated beam is focused by the relaxed normal eye or when the diverging beam is imaged to a point on the retina when the eye is focused at the origin of the diverging laser beam, e.g. at a point behind a curved lens surface or at the focal point from which some diverging beams originate. These two instances are most likely to occur during laser beam alignment, if at all. Each accident to a research worker or service technician occurred when the individual was not convinced of the need for eye protectors. When the beam is invisible, it is hard to remember its high irradiance, and secondary beam reflections can readily be overlooked.

23.11 ELECTRICAL HAZARDS

At least eight, and perhaps as many as a dozen laser servicemen, technicians, and researchers have been electrocuted by high voltage laser power supplies over the past 25 years. The voltages in many Class 4 laser power supplies can be lethal, and should be dealt with accordingly. Only experienced technicians should attempt servicing, and capacitors should be discharged prior to repair.[4]

23.12 VISITORS AND OBSERVERS

A visiting friend, a new student, an apprentice to the industrial laser operator, a salesperson, or another observer may be present during laser operation. Because of the presence of dangerous reflected beams, laser eye protectors should be made available.[28] Although it is true that the zone of reflections is normally limited to the immediate vicinity of the laser as shown in Fig. 23.4, and an observer standing 1 m to the side of a laser beam may not be at risk, it may be wise legal advice to have these observers wear eye protectors with an optical density sufficient to safely attenuate the main beam.

Sometimes the nontechnical visitor will find high-technology laser devices hard to comprehend. The event will remain in memory, and if any visual change is noted some time later (perhaps years), the individual may associate it with the visit to the laser laboratory. No matter how unreasonable this assumption may be scientifically, it has happened often in the past in laser research laboratories, and has sometimes led to lawsuits.

The use of a video presentation to permit observation of the laser operation is not only safe, but probably superior to most other forms of viewing. If secondary observation ports are used by aides or visiting staff, care must be taken that protective filters are present in all optical viewing paths of any observer.

23.13 DELAYED EFFECTS AND FUTURE CONSIDERATIONS

At the present time there is no indication for long-term delayed effects on the skin from laser exposure at wavelengths greater than approximately 350 nm in the ultraviolet.[29] For ocular exposure, the potential hazards from chronic exposure to ultraviolet and blue-violet light are still under study, but levels of cumulative exposure from sunlight are more likely to overwhelm any contributions from laser radiation exposure that may lead to effects such as cataract or macular degeneration.[30]

Extensive medical surveillance of laser research workers has failed to reveal even the presence of retinal lesions unknown to the worker.[31,32] It could be argued that since a Q-switched or mode-locked laser exposure can produce acute retinal injury that could disrupt Bruch's membrane in the retina, delayed retinal effects are conceivable. However, monitoring of accident victims has not revealed any such delayed effects.[11,31,32] In the past few years, excimer lasers operating at carcinogenic ultraviolet wavelengths have found use in industry. The potential exposure levels to the skin from reflected beams from this type of laser application could approach levels which could be of serious concern. The 308-nm line of the Xe-Cl excimer laser, which falls near the peak of the action spectrum for cataract and accidental ocular exposure, could be very dangerous to the lens. The Xe-Cl laser has even been referred to as the "cataract machine."[33]

The two more recent standards, on laser safety with fiber-optic communications systems[6] and safe use of lasers in health-care facilities,[7] are most likely to be revised by the ANSI Z-136 Committee after these have been in use for several years. The ANSI Z-136 Committee expected to issue a new standard on laser measurement in 1992. Exposure limits for extended sources were also revised in 1992 when the ANSI Z-136.1 standard was reissued.

23.14 CONCLUSIONS AND GENERAL GUIDELINES

Experience has shown that most laser-induced injuries occur in the laboratory setting, and that in almost all cases the cause of the accident was a refusal to wear laser eye protectors that were supplied to the user. In most laboratory environments, laser beams must remain unenclosed, and the principal protective measure is the use of laser eye protectors.

Laser eye protectors are available for the most commonly encountered laser systems. However, caution must be exercised in selecting suitable eye protectors for a specific work environment. Comfort and visual performance while wearing the protectors are of great importance. Uncomfortable goggles with poor visual performance may not be worn by the user without constant supervision, and accidental eye injuries may result. Although one must consider the possibility of filter damage by direct-beam exposure, unrealistic safety filter standards requiring damage resistance to even lengthy (10-s) exposures may result in costly and heavier, less comfortable eye protectors.

Several key steps should be followed to achieve a safe laser laboratory operation (Fig. 23.7). The laboratory laser user should always first attempt to minimize beam access through the use of baffles and enclosures; then wear suitable eye protectors if the beam is at all accessible. The laser should be disabled with the key switch master control if untrained (unauthorized) persons may have access to the laboratory. The laser beam attenuator and low-power alignment laser should be used during initial setup and beam alignment. A warning sign should be posted during laser operation, and most importantly, all those working in the laser environment (Class 3B and Class 4) should be adequately trained in the safe use of lasers. Finally, doorway interlocks should be employed with Class 4 laser installations if the NHZ extends to the area near the entryway to the laboratory. Such interlocks are required (denoted by the verb "shall") in ANSI standard Z-136.1, whereas the interlocks are advisory (as denoted by the verb "should" used in the standard) for Class 3B. It is important that the laser laboratory researcher be aware of one "tailoring" phrase in the ANSI standard if there is a legitimate problem in meeting the exact requirements of the standard. Paragraph 4.1.2 in the 1986 edition of the standard is entitled "Substitution of Alternate Control Measures (Classes 3B and 4)." Recognizing the constantly changing laboratory environment, the paragraph states that with LSO approval one can

Laser Control Measures:

– Beam enclosures and baffles
– Laser eye protectors
– Key switch master control
– Beam attenuator
– Low–power alignment
– Warning signs and labels
– Training (most important)
– Doorway interlocks if NHZ...

FIGURE 23.7 Summary of laboratory control measures.

substitute alternative control measures " . . . for example, in medical or research and development environments."

23.15   REFERENCES

1. Solon, L. R., G. Gould, and R. Aaronson, "The physiological implications of laser beams," *Science*, vol. 134, pp. 1506–1508, 1961.

2. Sliney, D. H., and W. A. Palmisano, "The Evaluation of Laser Hazards," *Amer. Industr. Hyg. Assoc. J.*, vol. 29, no. 5, pp. 325–431, 1968.

3. Wolbarsht, M. L., and D. H. Sliney, "Historical development of the ANSI laser safety standard," *J. Las. Appl.*, vol. 3, no. 1, pp. 5–11, 1991.

4. Sliney, D. H., and M. L. Wolbarsht, *Safety with Lasers and Other Optical Sources, A Comprehensive Handbook*, Plenum, New York, 1980.

5. American National Standards Institute, "Safe Use of Lasers," Standard Z-136.1, ANSI, New York, 1973, 1976, 1980, 1986.

6. American National Standards Institute, "Safe Use of Fiber Optic Lasers," Standard Z-136.2, Laser Institute of America, ANSI, New York, 1988.

7. American National Standards Institute, "Laser Safety in Health Care Facilities," Standard Z-136.3, Laser Institute of America, ANSI, New York, 1988.

8. Center for Devices and Radiological Health, "Federal Performance Standards for Laser Products, Food and Drugs," Title 21, Code of Federal Regulations, Sec. 1040, CDRH, Rockville, N.Y., 1985.

9. International Electrotechnical Commission (IEC), "Radiation Safety of Laser Products, Equipment Classification, and User's Guide," Publication WS-802, Geneva, 1984.

10. Sliney, D. H., "Evaluating hazards—and controlling them," *Laser Focus*, pp. 39–42, August 1969.

11. Boldrey, E. E., H. L. Little, M. Flocks, and A. Vassiliadis, "Retinal Injury Due to Industrial Laser Burns," *Ophthalmology*, vol. 88, no. 2, pp. 101–107, 1981.

12. Sliney, D. H., K. W. Vorpahl, and D. C. Winburn, "Environmental Health Hazards of High Powered Infrared Laser Devices," *Arch. Environ. Health*, vol. 30, no. 4, pp. 174–179, 1975.

13. Sliney, D. H., and H. Lebodo, "Laser eye protectors," *J. Laser Applications*, vol. 2, no. 3, pp. 9–13, 1990.

14. Chisum, G. T., "Concepts in Laser Eye Protection," pp. 350–355, *Proc. First Int. Symp. Laser Biological Effects and Exposure Limits*, Lasers et Normes de Protection, Paris, Nov. 24–26, 1986, Commissariat à l'Energie Atomique, Fontenay-aux-Roses, France, 1988.

15. Rockwell, R. J., "Selecting Laser Eyeware," *Medical Laser Buyers Guide*, PennWell, Westford, Mass., 1989.

16. Deutsche Institüt für Normung (DIN), DIN Standard 58 215, "Laserschutzfilter und Lasersuchtzbrillen; Sicherheitstechnische Anforderungen und Prufung" (Filters and Eye Protectors against Laser Radiation; Safety Requirements and Testing) DIN, Berlin, 1985.

17. Eriksen, P., and P. K. Galoff, "Measurements of laser eye protective filters," *Health Physics*, vol. 56, no. 3, pp. 741–742, 1989.

18. Galoff, P. K., and D. H. Sliney, "Evaluation of laser eye protectors in the ultraviolet and infrared," pp. 367–386, *Proc. 1st Int. Symp. Laser Biological Effects and Exposure Limits*, Lasers et Normes de Protection, Paris, Nov. 24–26, 1986, Commissariat à l'Energie Atomique, Fontenay-aux-Roses, France, 1988.

19. Holst, G. C., "Proper selection of and testing of laser protective materials," *Am. J. Optometry*, vol. 50, pp. 477–483, 1973.

20. Sliney, D. H., "Laser protective eyewear," pp. 163–238, in M. L.Wolbarsht, ed., *Laser Applications in Medicine and Biology*, vol. 2, Plenum, New York, 1971.

21. Sliney, D. H., "Laser eye protection II," *Optics Laser Tech.*, vol. 21, no. 4, p. 258, 1989.

22. Swearengen, P. M., W. F. Vance, and D. L. Counts, "A study of burn-through times for laser protective eyewear," *Am. Ind. Hyg. Assoc. J.*, vol. 49, no. 12, pp. 608–612, 1988.

23. Yeo, R., "Laser Eye Protection I," *Optics Laser Tech.*, vol. 21, no. 4, p. 257, 1989.

24. Swope, C. H., "Design considerations for laser eye protection," *Arch. Environ. Health*, vol. 20, pp. 184–187, 1970.

25. Lyon, T. L., and W. J. Marshall, "Nonlinear properties of optical filters—implications for laser safety," *Health Physics*, vol. 51, no. 1, pp. 95–96, 1986.

26. Robinson, A., "A study of saturation in commercial laser goggles," *Proc. SPIE*, Laser Safety, Eyesafe Laser Systems, and Laser Eye Protection, vol. 1207, pp. 202–213, 1990.

27. Rockwell, R. J., "On the Surface of It All . . . ," *J. Las. Appl.*, vol. 3, no. 1, pp. 55–56, 1991.

28. Zwick, H., M. Belkin, and E. S. Beatrice, "Effects of broadbanded eye protection on dark adaptation," pp. 356–366, *Proc. 1st Int. Symp. Laser Biological Effects and Exposure Limits*, Lasers et Normes de Protection, Paris, Nov. 24–26, 1986, Commissariat à l'Energie Atomique, Fontenay-aux-Roses, 1988.

29. van der Leun, J. C., "UV Carcinogenisis," *Photochem. Photobiol.*, vol. 39(6), pp. 861–868, 1984.

30. Young, R. W., "A Theory of Central Retinal Disease," in M. L. Sears, ed., *New Directions in Ophthalmic Research*, Chap. 14, pp. 237–370, Yale University Press, New Haven, 1981.

31. Hathaway, J. A., N. Stern, E. M. Soles, and E. Leighton, "Ocular Medical Surveillance on Microwave and Laser Workers," *J. Occup. Med.*, vol. 19, no. 10, pp. 683–688, 1977.

32. Pitts, W. G., and D. H. Sliney, eds., *Proc. Symp. Medical (Ophthalmic) Surveillance of Personnel Potentially Exposed to Laser Radiation*, 8–9 Sept. 1982, U.S. Army Environmental Hygiene Agency, Aberdeen Proving Ground, Md., undated.

33. Sliney, D. H., and S. Trokel, *Medical Lasers and Their Safe Use*, Springer-Verlag, New York, 1992.

CHAPTER 24
LASERS IN MEDICINE

Ashley J. Welch and M. J. C. van Gemert

24.1 INTRODUCTION

The ability of the surgeon to focus a laser beam to a small spot and precisely coagulate or vaporize tissue led to the immediate acceptance of lasers as a photothermal device for medical applications. Retinal photocoagulation with an argon laser (λ = 488 nm, 514.5 nm) became the method of choice for treatment of detached retina. Also, because of the excellent absorption of the argon wavelength by blood, the laser was initially selected for coagulation of enlarged blood vessels in the treatment of port wine stain. For procedures that required coagulation of tissue to depths of a centimeter or more, the neodymium-doped yttrium aluminum garnet (Nd:YAG) laser (1.06 μm) was selected because of the deep penetration of near-infrared wavelengths in tissue. In contrast, the CO_2 laser (10.6 μm) became the standard for ablation of tissue because of its shallow penetration depth of approximately 20 μm.

The rapid development of continuous-wave (cw) and pulsed lasers has produced a cornucopia of wavelengths, powers, and exposure durations available for medical applications. Laser-tissue interactions now include photochemical, photomechanical, and photodissociation in addition to classical photothermal interactions. The public demand for the use of lasers in medicine has been so great that *Consumer Reports* tried to evaluate the effectiveness of medical applications in an article "Laser Surgery: Too Much, Too Soon?"[1]

The article reports that lasers are the preferred treatment for:

Diabetic retinopathy	Photothermal
Advanced glaucoma	Photothermal
Cataract surgery follow-up	Photomechanical
Port wine stain birthmarks	Photothermal
Facial vascular conditions (spider veins, rosacea, hemangenomas)	Photothermal
Obstructive cancers in the windpipe, esophagus, or colon	Photothermal
Multiple warts that resist other treatments.	Photothermal

Other "legitimate, established" treatments are for:

Growths inside mouth and nasal passages	Photothermal
Abnormal cervical tissue	Photothermal
Tubal ligation	Photothermal
Endometriosis	Photothermal
Brain surgery	Photothermal
Rhinophyma (enlarged nose)	Photothermal
Mastectomy/lumpectomy	Photothermal
Laparoscopic gall bladder removal	Photothermal
Gastrointestinal bleeding	Photothermal
Hemorrhoid removal	Photothermal
Tattoo removal	Photothermal

Questionable uses, according to *Consumer Reports*, that are "experimental or without demonstrated advantage" are

Angioplasty	Photothermal, photodissociation
Facial plastic surgery	Photothermal
Corneal sculpting for nearsightedness	Photothermal, photodissociation
Spinal disk decomposition	Photothermal
Arthroscopic joint surgery	Photothermal
Treatment of dental decay or gum disease	Photothermal

Consumer Reports further states that it is inappropriate or unproven treatment to use lasers for:

Leg spider veins
Cellulite removal
Smoking cessation
Weight loss
Wrinkle removal
Pain control
Biostimulation

Obviously, the list of accepted procedures is incomplete; for example, the approved fragmentation of kidney stones (photomechanical) and experimental photodynamic therapy (photochemical) are not included. Although some experimental procedures will become accepted, the limited number of medical successes illustrates that the laser is not a magic wand and medical applications must be based on an understanding of the interaction of laser light with tissue.

The type of interaction is a function of laser wavelength, pulse duration, and irradiance. Chromophores that abosrb ultraviolet (uv) wavelengths are different than the chromophores that absorb infrared (ir) wavelengths. In the uv, protein and amino acids are the primary absorbers of the 193-nm wavelength of the ArF excimer laser. At this wavelength, there is sufficient photon energy (6.4 eV) to directly break molecular bonds. This form of interaction, called *photodissociation*, is the basis for experimental corneal shaping devices[2] that change the curvature of the cornea by removing successive layers of cornea that are less than 1 μm thick. Developers hope this technology will eliminate the need for glasses. Also photo-dissociation may be a factor in laser angioplasty systems that use the 308-nm XeCl excimer laser to remove plaque.[3]

Srinivasan et al. note that 30-ns uv pulse ablation is an explosive event occurring within less than 1 μs because of the decomposition of a significant fraction of proteins and amino acids; excess energy over the amount needed for bond breaking causes a local increase in pressure[4] that removes the irradiated tissue. However, the resulting explosive ejection of debris at supersonic speed causes stress waves which are undesirable in angioplasty.[4,5]

At longer wavelengths, the absorbed light energy is converted to heat by molecular vibrational modes. This is the classical *photothermal* mode of laser tissue interaction. Absorption of visible and ir radiation is by chromophores such as blood, melanin pigment, and water. As indicated in the above list of medical procedures, photothermal coagulation and ablation are the principle accepted laser-tissue interactions.

When irradiance is increased to levels of 10^8 W/cm^2, plasma formation can be achieved. The resulting acoustic transients can produce pressure waves that can fragment kidney stones or mechanically stress tissue. This form of interaction is termed *photomechanical*.

The above modes of operation are typically associated with high irradiances and exposure durations from picoseconds to a few seconds. They are used to either coagulate, ablate, or fragment tissue during medical procedures. In contrast, long-duration (minutes to hours), low-level irradiances are used to promote photo-chemical reactions. Typically, reciprocity holds; reactions are a function of total energy. For example, extensive exposure of the retina to low-level blue light (400 to 500 nm) can cause actinic insult from photochemical effects. Irradiances of less than 0.5 W/cm^2 for 100 s have produced damage in rhesus retina. The expected temperature rise at this irradiance is less than 1°C.[6]

An excellent review of laser surgery that describes various laser-tissue interactions and medical applications for these interactions has been published by M. Berns in *Scientific American*.[7]

Not all medical laser-tissue interactions are destructive. An important aspect is the rapidly growing diagnostic applications of lasers. Various forms of spectroscopy are being considered for the detection of cardiovascular plaque,[8] tumors,[9] and oxygen content of blood.[10] Both diagnostic and treatment applications of lasers are governed by the optical-thermal response of tissue to laser irradiation.

24.2 OPTICAL-THERMAL INTERACTIONS

The optical response of tissue to laser irradiation is depicted in Fig. 24.1. A portion of the laser beam is reflected at the surface according to Fresnel's relation:

$$R(\theta_i) = \frac{1}{2} \left[\frac{\sin^2 (\theta_i - \theta_t)}{\sin^2 (\theta_i + \theta_t)} + \frac{\tan^2 (\theta_i - \theta_t)}{\tan^2 (\theta_i + \theta_t)} \right) \tag{24.1}$$

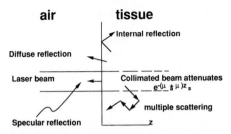

FIGURE 24.1 Optical interaction of collimated laser beam with tissue. Specular reflection for air-tissue interface is approximately 2.5 percent when laser beam is normal to tissue. Scattering diffuses light in tissue and produces diffuse reflection.

where θ_i is the angle of incidence on the boundary and the angle of transmission is given by Snell's law using the indices of refraction n_i and n_t of the respective layers:

$$\sin \theta_t = \frac{n_i}{n_t} \sin \theta_i \tag{24.2}$$

For an air-tissue interface ($n_a : n_t \approx 1 : 1.4$), specular reflection is about 2.5 percent light. A uniform laser beam is attenuated as it propagates through tissue by absorption and scattering according to the relation

$$\phi_c(z) = E_0 e^{-\mu_t z} (1 - R) \tag{24.3}$$

where E_0 is the surface irradiance (W/m²), μ_t is the attenuation coefficient (1/m), and R is the Fresnel reflection coefficient. Light scattered from the beam is re-scattered until it is either absorbed, remitted (diffuse reflectance), or transmitted (diffuse transmission). Because of the volume of tissue involved, propagation of scattered light is described by the transport equation rather than Maxwell equations. The radiance L(W/m² · sr) in direction **s** at position **r** is given by

$$\nabla \cdot L(\mathbf{r}, \mathbf{s}) = -\mu_t L(\mathbf{r}, \mathbf{s}) + \mu_s \int_{4\pi} p(\mathbf{s}, \mathbf{s}') L(\mathbf{r}, \mathbf{s}') \, d\omega' \tag{24.4}$$

where $\mu_t = \mu_a + \mu_s$, μ_a is the absorption coefficient, μ_s is the scattering coefficient, and $p(\mathbf{s}, \mathbf{s}')$ is the phase function which describes the probability of scattering from direction $\mathbf{s}'$ to $\mathbf{s}$.[11] We assume that tissue is isotropic in the sense that orientation of the tissue with respect to light direction $\mathbf{s}'$ does not affect the scattering angle. That is,

$$p(\mathbf{s}, \mathbf{s}') = p(\mathbf{s} \cdot \mathbf{s}') = p(\cos \theta) \tag{24.5}$$

where θ is the scattering angle. The expected cosine of the phase function is called the anisotropy factor g. Light is forwardscattered in tissue, and typical values of g

are in the range 0.7 to 0.99. Experimental measurements of $p(\theta)$ suggest it has the form of a Henyey-Greenstein function.[12] The fluence rate at **r** is

$$\phi(\mathbf{r}) = \int_{4\pi} L(\mathbf{r}, \mathbf{s}') \, d\omega' \qquad (24.6)$$

The measurement of *in vitro* optical properties typically involves the measurement of diffuse reflection, diffuse transmission, and collimated transmission for a tissue sample with the aid of integrating spheres. By assuming the form of the phase function and employing an iterative solution which varies the optical properties which are parameters in an approximation to Eq. (24.4), unique values for the absorption coefficient μ_a, scattering coefficient μ_s, and anisotropy factor g can be determined.[13] A summary of values extracted by Cheong et al.[13] from published material is reproduced in part in Table 24.1. Some in vivo measurements of optical properties have been obtained and are summarized in Table 24.2.

Solution of Eq. (24.4) is required when scattering dominates absorption (i.e., $\mu_s \gg \mu_a$). This condition occurs most often in the spectrum from 600 nm to 1.2 μm. In this range the absorption coefficient is typically small relative to the scattering coefficient. Approximate values of μ_a for protein, amino acids, blood (HbO), pigment, and water are presented in Fig. 24.2. The 600-nm to 1.2-μm spectrum provides a window for rather deep penetration of light in tissue. Within this window and even lower visible wavelengths, when tissue does not contain significant pigment or blood, the fluence rate associated with diffuse light is much larger than the fluence rate of the attenuated collimated laser beam. As a result of scattering, a rather large portion of the irradiance may be remitted from tissue. Measurements of diffuse reflection as a function of wavelength just below the surface of irradiated tissue may be greater than the irradiance. For example, consider irradiance of a human artery at 476 nm with a uniform beam that is 2.0 mm in radius. The fluence rate for a 1.0 W/cm² irradiance calculated with a Monte Carlo model[14] for $\mu_a = 6\ \text{cm}^{-1}$, $\mu_s = 414\ \text{cm}^{-1}$, and $g = 0.91$ is shown in Fig. 24.3*b*. Note that the fluence rate just below the surface of the tissue is 2.4 times larger than the irradiance. As the radius of the uniform irradiance increases, the fluence rate of the central cylinder has a one-dimensional distribution (Fig. 24.3).

The rate of generation of heat Q (W/m³) associated with the absorption of light in tissue is theoretically equal to

$$Q(\mathbf{r}) = \mu_a(\mathbf{r})\phi(\mathbf{r}) \qquad (24.7)$$

where the fluence rate ϕ includes collimated and diffuse light.[11] When scattering is important, fluence rate must be determined from Eqs. (24.4) and (24.6) to estimate the rate of heat generation with Eq. (24.7). However, if absorption is dominant, then Eq. (24.3) is sufficient. Determination of the rate of heat generation is central in the prediction of photothermal interactions.

The relation of fluence rate, absorption, and rate of heat generation is illustrated in Fig. 24.4 for two layers of tissue. Each layer is homogeneous and absorption dominates scattering in both layers. The absorption of the posterior layer is 4 times the absorption of the first layer. The fluence rate is obtained using Beer's law of attenuation $(e^{-\mu_a z})$. Even though the light reaching the second layer is less than the fluence rate of the first layer, the rate of heat generation for the first few millimeters of the second layer is larger. Thus, by selection of wavelengths, it may be possible to target interior layers of tissue.

TABLE 24.1 Optical Properties: Scattering, Absorption, and Anisotropy Parameters (*in Vitro*), Coefficients in cm^{-1}

Tissue	λ, nm	μ_t	μ_a	μ_s	μ_s'	g	μ_{eff}
Adipose:							
Bovine	632.8						3.4
Porcine	630	376 ($\pm$69)^a				0.77	
Aorta:							
Human	632.8	316.00	0.52	315.5	41.02	0.87	
Human intima	476	251.8	14.8	237	45.0	0.81	
	580	191.3	8.9	183	34.8	0.81	
	600	182.0	4.0	178	33.8	0.81	
	633	174.6	3.6	171	25.7	0.85	
Human media	476	251.8	7.3	410	45.1	0.89	
	580	191.3	4.8	331	33.1	0.90	
	600	182.0	2.5	323	35.5	0.89	
	633	312.3	2.3	310	31.0	0.90	
Human adventitia	476	251.8	18.1	267	69.4	0.74	
	580	191.3	11.3	217	49.9	0.77	
	600	182.0	6.1	211	46.4	0.78	
	633	200.8	5.8	195	37.1	0.81	
Human	1060		2.0				
Biliary caculi (gallstones):							
Porcinement	351		102 ($\pm$16)				
Stones	488		179 ($\pm$28)				
	580		125 ($\pm$29)				
	630		85 ($\pm$11)				
	1060		121 ($\pm$12)				
Cholesterol	351		88 ($\pm$7)				
Stones	488		62 ($\pm$15)				
	580		36 ($\pm$7)				
	630		44 ($\pm$10)				
	1060		60 ($\pm$9)				
Bladder:							
Canine	630	59.6	0.6	59.0	8.85	0.85	
Canine	633	52.0	1.25	50.8	2.54	0.95	
Canine	632.8	45.10	1.10	44.0	3.52	0.92	

Tissue preparation	Sample geometry	Experimental parameters	Theory	Reference
	Thick slabs	Total T measurement with interstitial fiber detectors	Diffusion theory	Preuss, 1982
Ground, frozen, and sliced	Very thin slabs	Direct T measurement, μ_r; goniophotometry	Direct method for μ_r; Mie theory	Flock, 1987
Postmortem, kept in saline	Plane sections	Diffuse T measurement, phase function with goniophotometry	Asymptotic diffusion, Henyey-Greenstein (H-G) phase function	12
Postmortem, frozen and sliced	Plane sections	Total T and R, axial (unscattered) T measurements	∂-Eddington phase function in diffusion theory	14
Postmortem, frozen and sliced	Plane sections	Total T and R, axial (unscattered) T measurements	∂-Eddington phase function in diffusion theory	14
Postmortem, frozen and sliced	Plane sections	Total T and R, axial (unscattered) T measurements	∂-Eddington phase function in diffusion theory	14
	Thick slabs	Magnitude of acoustic signal, neglect scattering	Photoacoustic spectroscopy	MacLeod, 1988
Dehydrated stones, embedded in plastic and sliced	~1 mm slabs	Time response of PPTR signal	Pulsed photothermal radiometry (PPTR)	Long, 1987
Dehydrated stones, embedded in plastic and sliced	~1 mm slabs	Time response of PPTR signal	PPTR	Long, 1987
Intact bladder	Whole bladder	μ_t, μ_{eff}, and radiance pattern with isotropic detectors	Numerical transport solution by van de Hulst	Star, 1987
Postmortem, intact in saline	Slabs	Diffuse R and T; axial transmission to get μ_t	Three-flux model, transform Kubelka-Munk (KM) to transport coefficient	Splinter, 1989
~1 day postmortem, in saline	Slabs	Diffuse R and T; axial transmission to get μ_t	Three-flux model, transform KM to transport coefficient	13

TABLE 24.1 *(Continued)*

Tissue	λ, nm	μ_t	μ_a	μ_s	$\mu_s{}'$	g	μ_{eff}
Bladder:							
Human	632.8	89.40	1.40	88.0	3.52	0.96	
Human	633	30.7	1.40	29.3	2.64	0.91	
Whole blood:							
Human (HbO$_2$) ([Hb] = 0.41)	685	1415.65	2.65	1413.0		0.99	
Human (HbO$_2$) ([Hb] = 0.41)	665	1247.30	1.30	1246.0	6.11	0.995	
	960	507.84	2.84	505.0	3.84	0.992	
Human (HB) (Hem = 0.41)	665	513.87	4.87	509.0	2.49	0.995	
	960	669.68	1.68	668.0	5.08	0.992	
Human	633	29.0				0.974	
Canine	632.8					0.9845	
	660					0.9840	
	800					0.980	
Brain:							
Calf	633		0.19		6.6		3.4[b]
	1064		0.36		6.7		2.5[b]
	1320		0.84		5.4		4.0
Porcine	633	1036.6[c]	0.26	1036.4[c]	57.0	0.945[d]	6.7
	633						4.3–14.2
	630		0.64		52.0		
	630	687.0				0.945	
Human, adult	488						14.0–25.0
	514						14.0–16.7
	630						7.0–12.5
	1060						2.3–3.4
	630		0.3–1.0		30.0–40.0		8.3

Tissue preparation	Sample geometry	Experimental parameters	Theory	Reference
~1 day postmortem, in saline	Slabs	Diffuse R and T; axial transmission to get μ_t	Three-flux model, transform KM to transport coefficient	13
Postmortem, intact, in saline	Slabs	Diffuse R and T; axial transmission to get μ_t	Three-flux model, transform KM to transport coefficient	Splinter, 1989
Diluted		Radial distribution of reflectance by Chandrasekhar	Curve-fit experimental data to H and S functions H-G phase function	Pedersen, 1976
Nonhemolyzed, heparinized blood	In curvettes	Absorbance as function of sample thickness, angular light distribution	Mie scattering theory	Reynolds, 1976
As above	In curvettes	As above	Mie scattering theory	Reynolds, 1976
As above	In curvettes	As above	Mie scattering theory	Reynolds, 1976
As above	In curvettes	As above	Mie scattering theory	Reynolds, 1976
Diluted in 1% PBS (phospate buffered solution), nonhemolyzed	In curvettes	Direct T measurement, goniophotometry	Direct method for μ_t; Mie scattering theory	Flock, 1987
Heparinized	In curvettes	Goniophotometry	Two-parameter phase function by Reynolds and McCormick	10
Frozen sections, postmortem	Mounted on slides	Total T and diffuse R	Numerical iterations, two-parameter phase function, similarity transform	Karagiannes, 1989
Postmortem	Thick slabs in situ	Total T at different depths; added absorber approach	Total attenuation, μ_{eff}, diffusion theory	Wilson, 1986
Postmortem	In situ, thick (~40–50 mm)	Total T at different distance from irradiation surface, two interstitial fiber-optic detectors	Diffusion theory	Wilson, 1985
	Thick slabs	Total T measurement with	Diffusion theory interstitial fiber detectors	Preuss, 1982
Frozen, then thawed	Thin slabs	Direct T measurement; phase function with goniophotometry	Direct method for μ_t, Mie theory	Flock, 1987
1–2 days postmortem, no fix, no irrigation of blood vessel	Bulk tissue (250 cm^3), in situ	Total attenuation using interstitial source and fiber-optic detectors	One-dimensional diffusion theory	Svaasand and Ellingsen, 1984, 1983
Cadaver (postmortem)	Slabs	Diffuse R and T, on-axis T	From KM into transport (P_1) theory	Sterenborg, 1988

24.9

TABLE 24.1 *(Continued)*

Tissue	λ, nm	μ_t	μ_a	μ_s	μ_s'	g	μ_{eff}
Brain:							
Human, neonate	488						5.9–7.9
	514						5.8–9.0
	630						2.5–3.3
	1060						1.1–1.4
Human:							
White matter	633	52.6	1.58	51.0	2.04	0.96	
Gray matter	633	62.8	2.63	60.2	7.22	0.88	
Canine:							
White matter	633	92.2	2.02	90.2	6.31	0.93	
Gray matter	633	58.0	1.65	56.3	1.97	0.97	
Brain tumors:							
Tumors	630						3.8–8.3
Glioma	630		5.0		7.0		
Melanoma	630				8.0		
Bovine	630						2.5
Feline	630						5.3–8.9
	514.5						13.3
	488						10.9
Breast tissue:							
Human, fibrous	514	202.0					
	633	188.7					
	1060	165.0					
Human, fatty	514	775.0					
	633	676.0					
	1060	524.0					
Human	635		$\leq$0.2	395 ($\pm$35)			
Skin—dermis:							
Human	630	243.0	1.8				
Human (caucasian)	633	189.7	2.7	187.0	35.5	0.81	

Tissue preparation	Sample geometry	Experimental parameters	Theory	Reference
1–2 days postmortem, no fix, no irrigation of blood vessel	Bulk tissue, (250 cm³), in situ	Total attenuation using interstitial source and fiber-optic detectors	Diffusion theory	Svaasand and Ellingsen, 1984, 1983
Postmortem, intact, in saline	Plane sections	Diffuse R and T; axial transmission	Three-flux model, transform KM to transport coefficient	Splinter, 1989
Postmortem, intact, in saline	Plane sections	Diffuse R and T; axial transmission	Three-flux model, transform KM to transport coefficient	Splinter, 1989
Postmortem, intact	In situ	Total T with interstitial fiber-optic detectors	Diffusion theory	Svaasand and Ellingsen, 1985
Cadaver, postmortem	Plane sections	Diffuse R and T, axial T	Transform KM into transport coefficients	Sterenborg, 1988
Postmortem	In situ	Total T with interstitial fiber-optic detectors	Diffusion theory	Doiron, 1983
Postmortem	In situ	As above	As above	Doiron, 1983
Freshly resected	~20-µm slices enclosed in glass cells	Direct total attenuation, include scattered light at angle < 0.8°	Exponential (direct) method	Key, 1988
Freshly resected	~20-µm slices enclosed in glass cells	As above	As above	Key, 1988
Frozen sections	Plane slices, enclosed between glass slides	Absorbance with integrating sphere, axial T from goniophotometry	Total attenuation	Marchesini, 1989
Postmortem	0.05–0.2 mm slabs	Direct transmission with detecting angle of 2 × 10⁻⁵ sr; absorbance with integrating sphere, goniophotometry.	Exponential attenuation, phase function	Andreola, 1988
Bloodless dermis, 85% hydrated in normal saline, fresh and frozen slices	Plane sections	goniophotometry, total R and T	Iterative diffusion approximation, HG phase function	Jacques, 1987

TABLE 24.1 *(Continued)*

Tissue	λ, nm	μ_t	μ_a	μ_s	μ_s'	g	μ_{eff}
Skin—dermis:							
Human	635		1.8	244			
Murine (albino)	488	241.8	2.82	39.0	62.14	0.74	
Skin—Epidermis:							
Human stratum corneum	193		6000				
Heart:							
Endocardium	1060		0.07	136		0.973	
Epicardium	1060		0.35	167		0.983	
Kidney:							
Human	630						4.0
Bovine	630						7.9
Porcine (cortex)	630						4.8
Liver:							
Bovine	630						8.1
Bovine	633		3.21		5.23		6.8[b]
	1064		0.53		1.76		3.2[b]
	1320		0.70		1.2		2.0
Human	630						11.0
	630		3.2	414		0.95[e]	
Human	635	315	2.3	313		0.68	26.6
	515	304	18.9	285			

Tissue preparation	Sample geometry	Experimental parameters	Theory	Reference
Frozen sections	Plane slices, enclosed between glass slides	Absorbance with integrating sphere, axial T from goniophotometry	Total attenuation	Marchesini, 1989
Fresh whole dermis	Slabs, on one slide	Total R and T, axial T measurements	Iterative diffusion approximation	Jacques, 1987
Frozen sections	Plane sections	Direct T measurement as function of thickness	Exponential (direct) method	Watanabe, 1988
Postmortem, intact, in saline	Plane sections	Simultaneous diffuse R and T; axial T	Three-flux model, transform KM to transport coefficient	Splinter, 1989
Postmortem, intact, in saline	Plane sections	Simultaneous diffuse R and T; axial T	Three-flux model, transform KM to transport coefficient	Splinter, 1989
Postmortem	Plane sections	Direct transmission	Exponential attenuation	Eichler, 1977
Postmortem	*In situ*	Total T measurement with interstitial fiber detectors	Diffusion theory	Preuss, 1982
Postmortem	*In situ*	Direct T using interstitial fiber-optic detectors	Diffusion theory	Doiron, 1983
Postmortem	*In situ*	Direct T measurement with interstitial fiber detectors	Diffusion theory	Preuss, 1982
Frozen sections, postmortem	Mounted on slides	Total T and diffuse R	Numerical iterations, two-parameter Groenhius' method	Karagiannes, 1989
Postmortem	Plane sections	Direct transmission	Exponential attenuation	Eichler, 1977
Postmortem	0.05–0.2 mm slabs	Direct transmission with detecting angle of 2×10^{-5} sr; absorbance with integrating sphere; goniophotometry.	Exponential attenuation, phase function	Andreola, 1988
Frozen sections	Plane slices, enclosed between glass slides	Absorbance with integrating sphere, axial T from goniophotometry	Total attenuation	Marchesini, 1989

TABLE 24.1 *(Continued)*

Tissue	λ, nm	μ_t	μ_a	μ_s	μ_s'	g	μ_{eff}
Liver:							
Murine	488		12.2	173.5		0.93	29.9
(albino)	633		6.5	143.7		0.95	16.3
	800		5.7	97.0		0.94	14.0
	1064		5.9	60.9		0.92	13.8
	1320		6.6	44.2		0.91	14.5
	2100		27.2	24.5		0.80	51.2
Porcine	630						13.0
	630		2.7	17.0			
Rabbit	630						12.5
	1060		10.0				
Lung:							
Human lung substance, deflated	633						11.0
Squamous cell carcinoma	633	57.0					6.3
Bronchial mucosa	633						9.1
Human, normal	630		8.4	35.9		0.95^e	
	635	332	8.1	324		0.75	
	515	380	25.5	356			
Muscle:							
Bovine	633	8.30	0.40	7.9	5.53	0.30	2.7
	633	120.1^c	1.50	118.6^c	7.0	0.941^d	6.2
	630						5.6
	630						6.9
	630	328 (±37)				0.941	

Tissue preparation	Sample geometry	Experimental parameters	Theory	Reference
Fresh sections for total T and R measurements; frozen sections for axial T	Sandwiched between glass slides	Total T and R measurements; axial T	Iterative ∂-Eddington phase function in diffusion model	Parwane, 1989
Postmortem	*In situ*	Direct T using interstitial fiber-optic detectors	Diffusion theory	Doiron, 1983
Postmortem	*In situ*	Direct transmission	Diffusion theory	Wilson, 1986
Postmortem, surface moist	*In situ*, thick ($\sim$15 mm)	Direct transmission, interstitial fiber-optic detectors	Diffusion theory	Wilson, 1985
	Thick slabs	Neglected scattering	Photoacoustic spectroscopy	MacLeod, 1988
Postmortem	*In situ*, bulk	Direct transmission using interstitial fiber-optic detectors	Diffusion theory	Doiron, 1983; Profio, 1981
Postmortem	*In situ*	As above	As above	Doiron, 1983
Postmortem	*In situ*	As above	As above	Doiron, 1983
Frozen, rehydrated	0.05–0.2 mm slabs	Direct transmission with detecting angle of 2×10^{-5} sr; absorbance with integrating sphere; goniophotometry.	Exponential attenuation, phase function	Andreola, 1988
Frozen sections	Plane slices, enclosed between glass slides	Absorbance with integrating sphere, axial T from goniophotometry	Total attenuation	Marchesini, 1989
Chopped	Bulk	μ_t, μ_{eff}, and radiance pattern with isotropic detectors	Numerical transport solution by van de Hulst	Marijnissen, 1987
Postmortem	Thick slabs	Direct T; added absorber technique	Diffusion theory	Wilson, 1986
Postmortem	*In situ*	Direct transmission using interstitial fiber-optic detectors	Diffusion theory	Doiron, 1983
Postmortem	*In situ*	Direct T measurement with interstitial fiber detectors	Diffusion theory	Preuss, 1982
Ground, frozen and then thawed	Thin slabs	Direct T measurement; phase function with goniophotometry	Direct method for μ_t, Mie theory	Flock, 1987

TABLE 24.1 *(Continued)*

Tissue	λ, nm	μ_t	μ_a	μ_s	$\mu_s{}'$	g	μ_{eff}
Muscle:							
	630		3.5	45.0			5.9
	630						4.3–5.6
	633		1.7		4.4		3.9[b]
	1064		1.2		2.8		2.3[b]
	1320		2.3		2.4		5.6
Chicken	633	4.30	0.17	4.1	3.3	0.20	1.34
	633	230.0[c]	0.12	228.6[c]	8.0	0.965[d]	1.7
	630	345 ($\pm$42)				0.965	
Human	515	541	11.2	530			
Porcine	633	41.00	1.0	40.0	1.2	0.97	
	1060		2.0				
Rabbit	630						1.1–1.5
	514.5						2.0–2.5
	630						2.7–12.5
	514						3.7–10.0
Stomach:							
Canine	1060	10.0	0.11	9.89			
Tumors:							
R3327-AT rat prostate solid tumor	633	270.5	0.49	270.0	8.1–5.4	0.97–0.98	3.6–2.9
Rhabdomyo-sarcoma (rat)	630		1.1		7.0		
	514		2.3		11.1		
	405		42.9		24.8		

Tissue preparation	Sample geometry	Experimental parameters	Theory	Reference
	Bulk tissue	Isodoses recorded on photographic film, diffusion theory	Diffusion theory contours yield μ_{eff}	McKenzie, 1988 Bolin, 1987
Frozen sections, postmortem	Mounted on slides	Total T and diffuse R	Numerical iterations, two-parameter phase function, similarity transform	Karagiannes, 1989
Chopped	Bulk, *in situ*	μ_t, μ_{eff}, and radiance pattern with isotropic detectors	Numerical transport solution by van de Hulst	Marijnissen and Star, 1984
Postmortem, coarsely ground	Thick slabs	Direct T; added absorber technique	Diffusion theory	Wilson, 1986
Ground, frozen, and then thawed	Thin slabs	Direct T measurement; phase function with goniophotometry	Direct method for μ_t, Mie theory	Flock, 1987
Frozen	Plane slices, enclosed between glass slides	Absorbance with integrating sphere, axial T from goniophotometry	Total attenuation	Marchesini, 1989
Fresh and frozen sections	Thin sections	Total T and diffuse R	Monte Carlo	Wilksch, 1984
	Thick slabs	Neglected scattering	Photoacoustic spectroscopy	MacLeod, 1988
Postmortem	*In situ*	Direct transmission using interstitial fiber-optic detectors	Diffusion (spherical) theory	Doiron, 1983
Postmortem	*In situ*, bulk	As above	As above	Doiron, 1983
Postmortem, surface moist	*In situ*, thick (~30–40 mm)	Direct transmission, interstitial fiber-optic detectors	Diffusion theory	Wilson, 1985
Postmortem	As above	As above	As above	Wilson, 1985
Postmortem	Whole intact stomach	Surface temperatures by thermal imaging	Multiple scattering theory with solution of heat diffusion equation	Halldorsson, 1978
Postmortem, frozen sections	120-μm sections	Goniophotometry, absorbance with integrating sphere	Diffusion approximation, anisotropic phase function	Arnfield, 1988
Postmortem	Thin slabs	Diffuse R and T	KM converted to transport coefficients using equations	van Gemert, 1985

TABLE 24.1   *(Continued)*

Tissue	λ, nm	μ_t	μ_a	μ_s	μ_s'	g	μ_{eff}
Tumors:							
Human intracranial tumors (meningiomas, astrocytomas, glioblastomas)	488 514 635 1060						7.1–20.0 7.1–20.0 5.9–3.9 3.3–1.9
VX2 rabbit tumor	630	628 (±106)				0.639	
Murine sarcoma	630 514.5						2.3 4.8
Murine fibrosarcoma	630						4.4–9.8
Uterus:							
Human	635	394.4	0.35	394		0.69	

[a]Numbers in parentheses are the standard deviation.
[b]Experimental measurement using interstitial fiber-optic detectors.
[c]Calculated from g (ref. 43, Flock) and μ_s'.
[d]From Ref. 43, Flock 1987.
[e]Averaged value

The optical and thermal processes of laser-tissue interaction are summarized in Fig. 24.5. For example, development of a mathematical model for thermal injury must include determination of the light distribution, rate of heat generation, and heat transfer. At each step in the process, it is necessary to know the corresponding physical properties of the tissue under consideration.

24.3 MEDICAL APPLICATIONS

After the introduction of lasers in medicine in the '60s and '70s, little thought was given to the development of a laser for a specific medical application. The goal was simply to find as many uses as possible for available cw argon (488, 514.5 nm), Nd:YAG (1.06 μm), and CO_2 (10.6 μm) lasers. The first widespread applications in the early 1970s were in ophthalmology. All lines of the argon laser (approximately 100 mW) were focused to a spot of about 200 to 400 μm on the retina. An irradiation time of approximately 100 ms coagulated a region of the retina and attached it to the underlying structures. This procedure was successful for welding detached retinas in place. At the same time laser retinal photocoagulation surgery was es-

Tissue preparation	Sample geometry	Experimental parameters	Theory	Reference
Postmortem	Tissue volume ~5–10 cm^3, in situ	In situ transmission with embedded fiber-optic detectors	Diffusion theory	Svaasand, 1985
Ground, frozen, and then thawed	Thin slabs	Direct T measurement; phase function with goniophotometry	Direct method for μ_t, Mie theory	Flock, 1987
Postmortem	In situ	Transmission using interstitial fiber-optic detectors oriented in three directions	Diffusion theory	Doiron, 1982
		Direct transmission	Exponential attenuation	Driver, 1988
Frozen sections	Plane slices, enclosed between glass slides	Absorbance with integrating sphere, axial T from goniophotometry	Total attenuation	Marchesini, 1989

tablished, the military funded a number of laboratories to investigate retinal hazards to laser irradiation so safety standards could be established for cw and pulsed devices. Several laboratories noted that as little as 20 mW of argon radiation on the cornea produced a threshold lesion at the retina during the eye blink reaction time of 100 ms.

24.3.1 Retinal Photocoagulation

The process of producing retinal lesions with laser irradiation provides an excellent example for describing photocoagulation which is employed in a majority of surgical techniques.

A cross section of the eye is illustrated in Fig. 24.6. Laser light is focused by the cornea and lens, and as the light passes through the pigment epithelium and choroid, heat is generated as photons are absorbed by local chromophores. The light is attenuated by Beer's law [see Eq. (24.3)], since absorption dominates scattering. Prior to heat conduction, the temperature at any location is proportional to the local rate of heat generation. The temperature of any differential volume is

TABLE 24.2 Optical Properties: Scattering, Absorption, and Anisotropy Parameters, *in Vivo*

Tissue	λ, nm	μ_t	μ_a	μ_s	$\mu_s{}'$	g	μ_{eff}
Brain:							
Human	630						2.2–3.7
	630						4.8–10.0
Porcine	630						3.7–4.5
Brain tumors	630						2.4
	630						2.2–6.6
Feline	630						5.0–9.8
	577						25.9
	545						34.4
	405–410						44.1
Muscle:							
Rabbit	630						2.6–4.8
	514						4.5–6.3
	630						1.6–2.3
	514.5						4.8–7.7
Liver:							
Rabbit	630						9.0–25.0

$\mu_t = \mu_s + \mu_a$
$\mu_s{}' = (1 - g)\mu_s$
$\mu_{eff} = 1$

given by the rate of heat storage which is equal to the local rate of heat generation plus the difference of heat conducted into and out of a differential volume. The temperature response can be described by the heat-conduction equation

$$\rho c \frac{\partial T}{\partial t} = Q + \nabla k \nabla T \qquad (24.8)$$

where T = temperature, °C
ρc = volumetric specific heat, J/cm³·°C

Tissue preparation	Sample geometry	Experimental parameters	Theory	Reference
In situ	Intact, spherical field	Direct *T* measured during PDT, interstitially; irradiated with embedded inflated balloon light source	Diffusion theory— spherical solution, added absorber technique	Wilson, 1986
In situ	Intact	As above	As above	Muller, 1986
Postmortem, in situ	Intact, spherical field	Direct *T* with distance from irradiation surface, interstitial fiber-optic detectors	Diffusion theory	Wilson, 1985
In situ	Intact	Direct *T* at different distances from interstitial spherical source, post-PDT	Diffusion theory— spherical solution, added absorber	Wilson, 1986
In situ	Intact	As above	As above	Muller, 1986
Postmortem	In situ	Direct transmission using interstitial fiber-optic detectors	Diffusion theory	Doiron, 1982
In situ	Intact bulk, ~30–40 mm	Direct transmission, interstitial fiber-optic detectors	Diffusion theory	Wilson, 1985
In situ	Intact	Direct transmission using interstitial fiber-optic detectors	Diffusion theory	Doiron, 1983
Postmortem	In situ, thick	Direct *T* with distance from irradiation surface, interstitial fiber-optic detectors	Diffusion theory	Wilson, 1985

Q = rate of heat generation, W/cm^3
k = thermal conductivity, W/cm·°C

The rate of heat generation, Q, at any point **r** is given by Eq. (24.7).

Since the retina is transparent for visible and near-infrared wavelengths, absorption of these wavelengths takes place in the melanin and hemoglobin chromophores of the pigment epithelium (PE) and choroid (Ch) respectively. Melanin contains dark pigments about 1.0 μm in diameter that are densely clustered in a 4-μm layer of the PE. Hemoglobin is concentrated in a 25-μm blood plexus layer

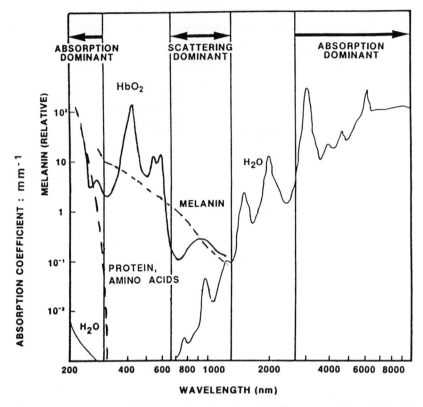

FIGURE 24.2   Absorption for (*a*) proteins and amino acids, (*b*) oxygenated hemoglobin, (*c*) melanin, and (*d*) water, in units of cm^{-1}.

of the choroid. During argon irradiation of the eye, about 50 percent of the light is absorbed in the PE and most of the remaining light is absorbed in the Ch. Retinal tissue is not directly heated by laser light; it is indirectly heated by conduction from the PE.

The effect of heat conduction from a thin layer on surrounding protein was examined in a series of experiments in our laboratory using a phantom for tissue. The model consisted of clear gel of dehydrated egg white that was covered by a thin (40 μm) layer of black paint. The top layer was clear egg white. The phantom was irradiated with all lines from an argon laser. The beam had a Gaussian profile, $1/e^2$ diameter of 2.3 mm, and power of 2.4 W. The zone of coagulation after a 5-s exposure is graphically indicated in the egg white with white-light illumination. The coagulated tissue appears white because of scattering in the "cooked" egg white as shown in Fig. 24.6*b*.

The absorption coefficient is a function of wavelength, and, for the pigment melanin, μ_a decreases monatonically with increasing wavelength. At a wavelength of 800 nm only 15 to 20 percent of the laser light is absorbed in the PE, whereas 50 to 60 percent is absorbed at 500 nm.[15]

In contrast to melanin, hemoglobin has a rather complex absorption spectrum as indicated in Fig. 24.2. The excellent absorption by blood of the argon wavelengths

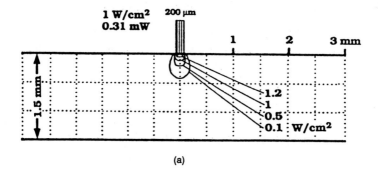

(a)

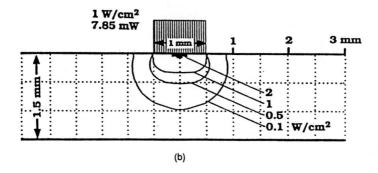

(b)

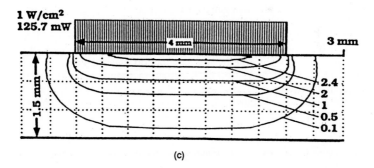

(c)

FIGURE 24.3 Distributions of 476-nm light in human aorta for flat collimated incident beams with radii of (a) 100 μm, (b) 0.5 mm, and (c) 2 mm. The optical properties for the 1.5-mm-thick sample were μ_a = 6 cm^{-1}, μ_s = 414 cm^{-1}, g = 0.91, index of refraction = 1.37. *(From Ref. 13, with permission.)*

allows little of the 488-, 514.5-nm wavelength light to be transmitted beyond the Ch. In contrast, 800-nm radiation is not significantly absorbed by blood, so when this wavelength, or light in the 750- to 850-nm spectrum, is used for retinal photocoagulation, a rather large percent of the light is transmitted beyond the posterior boundaries of the eye. Geeraets et al. measured a transmission of the entire globe of 45 percent at 800 nm for two human eyes.[16]

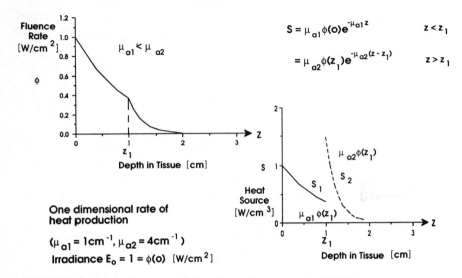

$$S = \mu_{a1}\phi(o)e^{-\mu_{a1}z} \qquad z < z_1$$

$$= \mu_{a2}\phi(z_1)e^{-\mu_{a2}(z-z_1)} \qquad z > z_1$$

Fluence Rate ϕ [W/cm^2]

$\mu_{a1} < \mu_{a2}$

Depth in Tissue [cm]

One dimensional rate of heat production

$(\mu_{a1} = 1\,cm^{-1}, \mu_{a2} = 4\,cm^{-1})$

Irradiance $E_o = 1 = \phi(o)$ [W/cm^2]

Heat Source [W/cm^3]

$\mu_{a2}\phi(z_1)$

S_1 S_2

$\mu_{a1}\phi(z_1)$

Depth in Tissue [cm]

FIGURE 24.4 Relative fluence rate (left) and rate of heat generation (right) for two homogeneous layers of tissue. The absorption coefficient for the anterior layer is 1 cm^{-1} and for the second layer, 4 cm^{-1}. There is no scattering.

Thus, efforts to replace the argon ion laser with a near-ir diode laser have encountered some problems. More irradiance is required at the cornea to produce an equivalent rate of heat production at the PE. There is increased heat production in the Ch, and radiation that is not absorbed in the PE and Ch is transmitted through the globe of the eye. Heat generation in the Ch by 800-nm radiation appears to be conducted to Ch nerve fibers, causing the pain that has been reported during retinal treatment with diode lasers. Because of the wavelength-dependent nature of the optical properties of tissue, care must be taken if the wavelength of the laser source is changed. Also, the role of heat conduction must be carefully considered in any laser-tissue interaction. Retinal coagulation requires the transfer of heat from an adjunct structure (PE) to the target tissue (retina). However, in other photothermal medical procedures, the requirement is the coagulation of a target tissue with minimal damage to surrounding tissue.

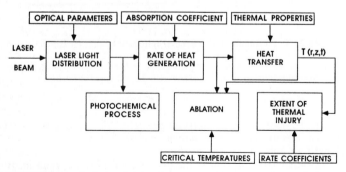

FIGURE 24.5 Optical and thermal processes of laser-tissue interaction.

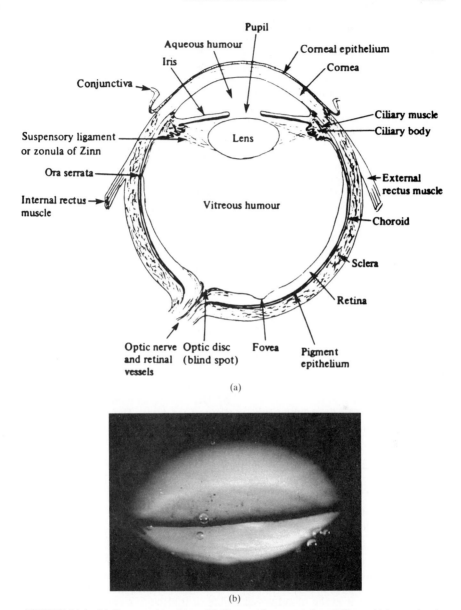

(a)

(b)

FIGURE 24.6 (*a*) Cross section of eye. (*b*) Pigmented center region of 40 μm thickness absorbs laser radiation. White region is coagulated egg white due to heat conduction from center region. Sample was irradiated with a Gaussian argon laser with $1/e^2$ diameter of 2.3 mm; the delivered power was 2.4 W for 5 s.

24.3.2 Treatment of Port Wine Stain

Port wine stain (PWS) is a congenital vascular malformation due to ectatic venules (30 to 300 μm) scattered throughout the dermis.[17] PWS appears as a light-colored pink or reddish flat lesion during infancy. The birthmark may darken with age as the vessels enlarge and the skin becomes rough. A cartoon depicting the layers of the skin and the enlarged vessels is illustrated in Fig. 24.7. Laser treatment consists of coagulating the enlarged vessels; the destroyed vessels are replaced by normal sized vessels and over time the appearance of the skin becomes more normal.[18] One complication is scarring that may take place when an excessive volume of normal tissue is damaged and/or when complications occur during the healing process.

The target for treatment of PWS is the enlarged blood vessels. Berns[7] notes:

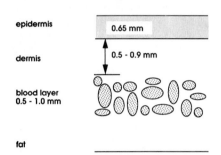

epidermis   0.65 mm

dermis   0.5 - 0.9 mm

blood layer
0.5 - 1.0 mm

fat

FIGURE 24.7 Cartoon of skin cross section associated with port wine stain. The ectatic blood vessels have a diameter between 0.03 and 0.1 mm. Treatment consists of photocoagulation of enlarged vessels. Coagulated vessels are replaced with smaller, normal vessels.

> Lasers make good scalpels. Rather than having to slice through everything they encounter, these instruments can be highly selective. That specificity allows lasers to penetrate to the interior of a cell or organ, while leaving the exterior intact—something no surgeon's knife can do.

To achieve the specificity suggested by Dr. Berns, care must be taken to match the characteristics of the target to those of the laser. The requirement for treatment of PWS is the coagulation of subsurface blood vessels without damage to other tissue. The most successful strategy for treatment of PWS has been

1. Selection of a wavelength that maximizes absorption of laser light in blood
2. Selection of a wavelength that minimizes absorption in the epidermis (the pigmented layer of skin that contains melanin)
3. Selection of a wavelength that coagulates the full thickness of enlarged vessels
4. Irradiation duration that minimizes temperature rise in tissue anterior to the blood layer[19]

The peaks of the absorption curve for blood shown in Fig. 24.2 are all candidates for the treatment wavelength. However, since absorption in melanin decreases with wavelength, the selection of 577 nm minimizes heat production in the epidermis while maximizing it in the blood vessels. Clinical trials at 577 nm have been disappointing.[20] The depth of effective coagulation using 577 nm has not been sufficient for clearing the enlarged vessels. In view of the large value of the absorption coefficient at 577 nm [350 cm^{-1}], this should have been anticipated, assuming a blood vessel diameter of 100 μm and exponential attenuation of the laser beam in the blood vessel, the decrease in light through the vessel is $e^{-3.5}$ (0.03). The blood so strongly absorbs 577-nm irradiation that anterior vessels shadow posterior vessels. Tan et al. have demonstrated effective treatment by increasing the treatment wavelength to 585 nm.[20] Absorption in the blood is reduced by a factor of 2 [μ_a

(585 nm) = 175 cm^{-1}], and the attenuation through a 100-μm vessel is $e^{-1.75}$ (0.17). Thus transmission through blood is increased by a factor of 5.8 and the effective treatment depth is increased. More details are given in Ref. 21. It is interesting to note that 586 nm is the isosbestic wavelength for oxygenated and deoxygenated hemoglobin, that is, the wavelength where they have equal absorption coefficients.

The selection of exposure duration of the small vessels (50 to 100 μm) is based on the assumption that we want to minimize the temperature increase in tissue surrounding the target vessels. If laser irradiation is maintained for a long time, the temperature within a vessel quickly increases until the rate of heat conduction from the vessel approaches the rate of heat production due to absorption of laser radiation. Continued irradiation raises the temperature of the vessel but at a much slower rate. Heat flow to cooler regions increases the overall temperature of surrounding tissue. The rate of temperature increase or decrease is associated with the volume of tissue heated; values can be estimated by computing the diffusion time τ:

$$\tau = \frac{l^2}{4\kappa} \tag{24.9}$$

where l is a characteristic length and κ is the thermal diffusivity ($k/\rho c$). During the heating cycle, the characteristic length for a 100-μm-diameter vessel is 50 μm and the diffusion time is approximately 2.5 ms. If the irradiation time is much less than the diffusion time, then there is insufficient time for significant heat conduction to surrounding tissue. Once irradiation ends, the temperature returns to its pre-irradiated value. The cooling diffusion time once again is a function of the volume of tissue heated. At the end of the irradiation, if the exposure duration t_0 is much less than 2.5 ms, then the decay diffusion time is somewhat larger than 2.5 ms. Thus any pulse duration less than a few milliseconds should be suitable for treatment of PWS. However, there are limits. If a laser pulse contains sufficient energy for coagulation and its pulse duration is less than a few microseconds, microexplosions occur in the blood due to the high rate of heat deposition. Garden et al. have demonstrated that pulse durations of 100 μs and less tear normal vessels.[22] Suggested pulse durations for treatment of PWS are 100 μs to 1.0 ms. For more details on laser treatment of PWS we refer to the recent book edited by O. T. Tan.[23]

24.3.3 Vessel Welding

A new application for photocoagulation may be vessel welding. Laser-assisted vascular anastomosis (LAVA) uses laser irradiation to rejoin the cut edges of severed blood vessels by thermally fusing the tissues of the vascular wall to form a bloodtight bond of acceptable tensile strength. The procedure requires placing the ends of two vessels together using a minimal number of stay sutures to hold the vessels in place. The junction then is irradiated with laser light to coagulate the tissue. Typically the laser beam is scanned across the junction until a visual dehydration of the tissue is noted. The advantages of LAVA over conventional suture surgical techniques include shorter surgical times, decreased suture (foreign body) scar tissue formation, and no postoperative vascular stenosis in the growing artery.[24] Aneurysm formation at the anastomotic site is a significant complication of LAVA (30 to 40 percent) when applied to small arteries (less than 2 mm in diameter).[24] We have noted acceptable levels of aneurysm formation (around 4 to 6 percent) when minimal thermal damage, determined by the gross and histologic appearance of the anastomosis, is produced. The gross appearance of the ideal LAVA microsurgical vascular bond shows slight "drying and crinkling" of the adventitial surfaces (anterior surface) with no carbonization or anastomotic contraction.

Surface temperatures associated with successful end-to-end anastomoses have been in the range from 60 to 80°C (Ref. 24), which coincides with our measurements for laser anastomoses of rat femoral and carotid arteries. Surface temperatures as low as 45°C for argon LAVA have been reported; however, that surface was being cooled with a saline drip.[25] Reports of high incidence of aneurysms routinely describe slight vascular discolorization (brown), carbonization and contraction of the anastomotic site, severe shrinkage, vacuolization (water vapor pockets), condensation and dense drying artifact of all tissue components histologically.[26]

Histologically, the LAVA bond of small arteries is formed by thermal coagulation of both cellular and extracellular components of the vessel wall as shown by light microscopy (LM) and transmission electron microscopy (TEM) (unpublished data, S. Thomsen). The adventitial portion of the anastomosis, a thermally produced coagulum of collagen and other extracellular proteins, may be the most important component contributing to the bloodtight integrity of the bond. A coagulum of thermally damaged, vascular-smooth muscle cells and collagen are the major components of the medial (middle section of vessel wall) portion of the anastomosis. The more heat-resistant, histologically intact elastin membranes and fibers of the media and adventitia are embedded in the thermal coagulum. However, the heated elastic membranes and fibers do not contribute to the bond. Thermally coagulated intimal (posterior section) cells are found in the anastomotic bond, but this thin arterial lining contributes little to the bond. The relative contributions of the adventitia and media to the integrity of LAVA bonds of larger arteries have not been studied.

Typically investigators have selected commercially available lasers for anastomoses. Because of differences in the wavelength of these lasers, it is not surprising there have been inconsistent results. The CO_2 laser at 10,600 nm has a penetration depth ($1/e$ attenuation of collimated beam) of about 20 μm, whereas the Nd:YAG at 1060 nm has an attenuation depth on the order of several millimeters. Although the blue-green light of the argon laser has an intermediate penetration depth in vessel walls, it is highly absorbed by blood (absorption coefficient for oxygenated hemoglobin is about 100 cm^{-1}). An advantage of the Nd:YAG 1320-nm laser is that, at this wavelength, hemoglobin absorption is much less than water absorption, which is about 7 cm^{-1}; however, the penetration depth is still rather deep (about 1.4 mm). Between 1200 nm and 2000 nm, water absorption varies from approximately 1.0 cm^{-1} to 10^4 cm^{-1}. Thus we anticipate that the desired wavelengths for LAVA will be in this portion of the ir spectrum.

Recent advances in our laboratories and elsewhere have provided procedures for measuring the optical and thermal properties of tissues such as vessel walls.[12,13] By incorporating these parameters in optical-thermal models, it is possible to predict the optical-thermal behavior of the vessel under laser irradiation. These models include light scattering, rate of heat generation, heat conduction, and rate process denaturation.

So far, during LAVA the visual appearance of the tissue has been used as an end point for laser irradiation. This is a subjective and unreliable assessment of the gross appearance of the anastomosis by the surgeon. Once the bond begins to form, observations show that exposure times are critical, and unwanted thermal damage can occur very rapidly as the tissue dehydrates. Tissue temperatures necessary to form a thermal bond of sufficient strength without producing unwanted thermal damage associated with aneurysm formation in the range of 65 to 85°C are necessary for the successful application of LAVA in vascular surgery.

The difficulty of LAVA suggests that some form of feedback control is needed to produce acceptable welds. The first attempt to develop such a device is to measure temperature at the site of irradiation and cut off the laser beam if tissue temperature exceeds a preselected limit. Preliminary *in vitro* tests using human vessels and argon laser irradiation have demonstrated that acceptable welds can be formed using this technique.[27]

24.4 ABLATION

A striking feature of an intense laser beam is its ability to cut or vaporize tissue. Laser ablation is highly wavelength-dependent. For wavelengths such as argon (488 nm, 514.5 nm) and Nd:YAG (1.06 μm) that penetrate several millimeters in tissue, the ablation process consists of a number of distinct events; however, for pulsed irradiation these events appear to occur simultaneously.

24.4.1 Continuous-Wave Ablation

The temperature response at the surface of an aorta irradiated with an argon laser beam is illustrated in Fig. 24.8. When temperature increases beyond 100°C, there is an increase in subsurface pressure. Just prior to ablation, surface temperature exceeds 100°C, and subsurface temperature exceeds the surface temperature at the onset of ablation.[28] A subsurface pressure of perhaps 10 atm explodes through the

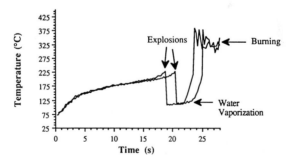

FIGURE 24.8 CW argon laser ablation of pig aorta. Temperatures were measured with a thermal camera at a rate of 60 fields per second.

surface with a distinct sound which gives rise to the term *popcorn effect*. The supersonic ejection of tissue leaves an underlying surface at 100°C. Further irradiation dehydrates this layer, which reduces thermal conductivity. Temperature of the dehydrated surface rapidly increases to several hundred degrees, and charring occurs as the tissue burns at nucleation sites. The explosive onset of ablation produces dissections that extend beyond the crater that is formed.[29]

Ideally, ablation should remove tissue with minimal thermal damage and dissections. This ideal is best achieved by a pulsed laser with wavelengths that penetrate tissue only a few micrometers. Lasers in this class are either uv devices below 360 nm or ir lasers at the water absorption peaks of 1.94 μm and 2.96 μm. The CO_2 laser at 10.6 μm is also included as a device whose irradiation is absorbed near the surface. The penetration depths of several lasers used for medical applications are given in Fig. 24.9. Notice that the excimer ArF laser at 193 nm and the Er:YAG at 2.94 μm both have penetration depths of less than 1.0 μm. Demonstration that the 193-nm wavelength can remove submicrometer layers of tissue without dissections or thermal damage is the key feature of an experimental system for reshaping the cornea to correct refractive errors.[30]

CO_2 laser radiation at 10.6 μm is absorbed by water in tissue, and the penetration depth at this wavelength is about 20 μm in soft tissue. The virtual surface absorption of CO_2 radiation is ideal for the removal of tissue. The rate of heat generation at the surface is equal to the irradiance (W/cm^2) times the absorption coefficient μ_a. For soft tissue $\mu_a \approx 500$ cm^{-1} at 10.6 μm. Thus irradiances on the order of 1000 W/cm^2 produce a rate of heat generation of 5×10^5 W/cm^3.

At a high irradiance, the CO_2 beam can act like a scalpel. By placing the focal point of the CO_2 beam on the surface of tissue, sufficient energy is deposited to rapidly vaporize the tissue. Beyond the focal point, a lower irradiance can be achieved to coagulate bleeding vessels without further vaporization of tissue.

The major limitation of CO_2 surgery has been the limitation of line-of-sight delivery. To date, a practical optical fiber for delivery of 10.6 μm radiation is not available. Standard silica fibers will not transmit wavelengths beyond 2.5 μm. Thus endoscopic ablation applications such as laser angioplasty have been limited to wavelengths that can be transmitted by silica fibers. However, novel plastic waveguides are available for transmission of CO_2 radiation. Kaplan has delivered 20 W

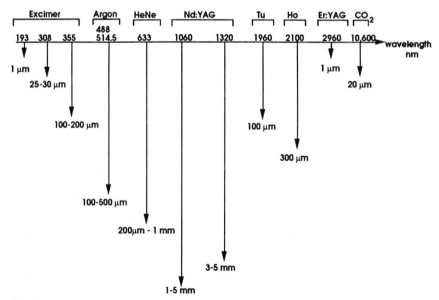

FIGURE 24.9   Optical penetration depth in soft tissue.

of output power for an input of 30 W with a waveguide that was 70 cm long and had an internal diameter of 1.9 mm.[31]

Laser Angioplasty. The goal of laser angioplasty is the ablation of plaque in partially or fully clogged arteries. In concept, a fiber is inserted through a catheter to the site of blockage and laser light vaporizes the plaque. Ideally, there would be a wavelength that is preferentially absorbed by plaque so accidental irradiation of the normal vessel wall would not injure the vessel. Unfortunately, such a wavelength does not exist, so considerable effort has been extended to use fluorescence to differentiate between plaque and the normal vessel wall. In the future, angioplasty may consist of a relatively smart system that optically interrogates the target tissue to determine if it is plaque.[32]

Modified Fiber Tips. A critical feature of an endoscopic ablation or coagulation system is the fiber-optic delivery system. An interface at the distal end of the fiber must (1) protect the fiber, (2) project a desired radiation pattern, and (3) prevent the mechanical tearing of tissue by the sharp edge of a fiber. Modified tips consisting of fused silica or sapphire with rounded or tapered shapes have been developed. The sapphire tapered tips concentrate the laser light at the tip of the probe because of the high index of refraction of the sapphire and angle of the taper.[33] These probes are used to concentrate Nd:YAG cw irradiation for cutting tissue. Light transmitted beyond the point of concentration, light along the sides of the probe, and the temperature of the probe wall are sufficient to coagulate bleeding vessels to provide hemostasis. One disadvantage of the tapered tip is that it is not pressed against the tissue. This lack of contact eliminates the tactile feedback a surgeon has with a scalpel.

In contrast, rounded-tip probes are placed firmly against tissue for the purposes of coagulation or ablation. In air the indexes of refraction and the curvature of these probes focuses the laser beam just in front of the probe. The ability to focus disappears when probes are placed against tissue since there is little difference in the indices of refraction of quartz and tissue, as shown in Fig. 24.10. A ray-tracing

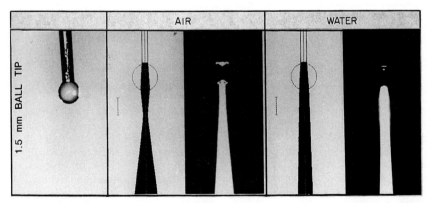

FIGURE 24.10 Beam profiles of ball-shaped fiber tips photographed in air and water and their corresponding profiles calculated by ray tracing. From left to right for 1.5-mm Advanced Cardiovascular Systems (ACS) ball: probe shape, calculated beam in air, measured beam in air, calculated beam in water, measured beam in water. *(From Ref. 24, with permission.)*

analysis and experimental verification of modified tips and fiber optics has been published by Verdaasdonk.[34]

Many catheter systems for endoscopic ablation are based on multiple fibers that are oriented to produce an irradiation pattern that covers a target area. Laser angioplasty systems may consist of eight or more fibers with cores 80 to 400 μm in diameter. Often a shield is placed in front of the fiber bundle to provide a smoother, rounded surface which protects fibers from debris associated with ablation and prevents tearing of the vessel by a sharp edge.

A series of lasers have been tested as possible sources for endoscopic systems. Although none have been totally successful, desirable features for ablation are (1) penetration depth of laser light less than 100 μm and (2) pulse duration 200 ns to a few milliseconds. The lower limit for uv lasers is imposed by the fiber optics and not the tissue. Below 200 ns, the uv irradiance required for ablation damages the surface of fiber optics.[35]

Current candidates for laser angioplasty are the XeCl and XeF excimer lasers at 308 or 351 nm respectively, tripled Nd:YAG at 355 nm, and Tu:YAG at 2.01 μm. Both the CO_2 (10.6 μm) and Er:YAG (2.94 μm) are waiting for the availability of flexible delivery systems that can transport these wavelengths. Descriptions of laser angioplasty and diagnostic systems for differentiating between normal vessel wall and plaque may be found in Ref. 36.

24.5 PHOTOCHEMICAL INTERACTIONS

Laser irradiation can react with either natural light-sensitive agents or photosensitizing chemicals that are injected into the body. Photochemical reactions can produce products that lead to irreversible damage of tissues that have accumulated the photosensitizer.

Because of the ability to destroy tissue at specific locations, considerable interest has been generated in the use of photosensitizers for the treatment of cancer. These photosensitizers must have the following properties:

1. Selective accumulation in targeted malignant tissue
2. Nontoxic to normal tissue
3. Efficient activation with light between 600 and 900 nm
4. Fluorescence

Hematoporphyrin derivative (HpD) which can be injected intravenously has an affinity for malignant tumors. When an HpD molecule absorbs a photon, one of two possible photochemical reactions take place. In one reaction the excited HpD returns to the ground state and emits fluorescent light in the spectrum between 600 and 700 nm. This light provides information for localization of tumors.

The second reaction path is that energy transferred from the excited porphyrin molecule to oxygen creates an excited singlet oxygen. This aggressive excited state of oxygen causes oxidation of the host cell which leads to destruction of cells. This photochemical mechanism for treatment of cancer is called photodynamic therapy (PDT).

Typically, diagnostic fluorescence in the 600- to 700-nm spectrum is excited with blue-violet light, and treatment is accomplished with red light (625 to 633 nm). As shown in Fig. 24.11, the red wavelength is the smallest of the excitation peaks for

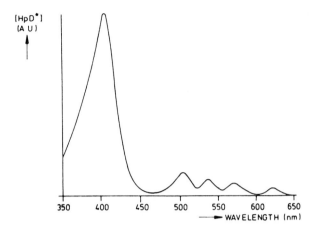

FIGURE 24.11 Excitation probability for HpD (in arbitrary units) as a function of wavelength.

HpD. Nevertheless, the advantage of the red wavelength for PDT is its tissue penetration depth of several millimeters. Usually treatment involves application of a low level of irradiance (about 25 mW/cm^2) for rather long periods of time (30 min to several hours). For small embedded tumors, fiber optics are inserted for localized treatment.

Although HpD has received considerable attention as a photosensitizer, it does not fully meet the requirements of an ideal agent. HpD can be toxic to normal tissue, and the low efficiency and penetration depth at its red excitation peak are not ideal. The former will ultimately affect the efficiency of treatment. The second affects the cost of treatment. Currently the 628-nm treatment wavelength is obtained from an argon (20 to 25 W) pumped dye laser. If a photosensitizer with a strong excitation peak from 750 to 900 nm can efficiently attack cancerous tissue, then it will have several practical advantages. The penetration depth of the laser light will be deeper and treatment of a larger tissue volume will be possible, and most important, the laser source can be an inexpensive laser diode.

Once a photosensitizer is approved for PDT, it will be necessary to have an accurate dosimetry for treatment. Key questions are the concentration of photosensitizer, fluence rate of activating light, specific absorption of the light by the photosensitizer, and the quantum yield for singlet oxygen.

The effective dose D (J/m^3) at position $\mathbf{r}$ for a constant monochromatic irradiance over time t is

$$D(\mathbf{r}) = \int \phi(\mathbf{r})\mu_c(\mathbf{r}, t)F(\mathbf{r}, t) \, dt \qquad (24.10)$$

where ϕ = fluence rate, W/m^2
μ_c = absorption coefficient of photosensitizer, m^{-1}
F = quantum yield for singlet oxygen

If μ_c and F remain constant over t, then Eq. (24.10) reduces to

$$D(\mathbf{r}) = \phi(\mathbf{r})\mu_c(\mathbf{r})F(\mathbf{r})t \qquad (24.11)$$

The absorption coefficient is

$$\mu_c(\mathbf{r}) = C(\mathbf{r})\epsilon \tag{24.12}$$

where $c(\mathbf{r})$ is concentration of photosensitizer in tissue, M, and ϵ is molar extinction coefficient, $m^{-1}M^{-1}$.

Any effective dosimetry requires determination of fluence rate by using light transport Eqs. (24.4) and (24.6). The absorption coefficient for tissue μ_a must include any significant effect of photosensitizer concentration.

In addition to computing dosimetry, the governing equations for light propagation can be used to model fluorescence. A description of physics, chemistry, and clinical treatment associated with PDT can be found in Ref. 37.

24.6 PHOTOACOUSTIC MECHANISMS

The absorption of laser light by tissue produces an increase in local temperature and temperature gradients. Expansion causes a strain and produces an acoustic wave that propagates away from the heated zone. If the absorbed energy due to a short pulse exceeds the heat of vaporization, an explosion occurs which is accompanied with rapidly expanding gas bubbles. Bubble formation and its dynamics in fluid has been noted by several investigators. Cavitation associated with CO_2 and excimer (351 nm) radiation in blood and in water was first reported by Isner et al.[38] They coined the term *Moses effect* to describe the effects of the pressure wave in removing or reducing a highly absorbing fluid field between the irradiance source and target. High-speed photography graphically illustrated the interaction of the laser light with absorbing fluid. Bubble expansion and implosion in water at 15 to 450 μs after the start of a holmium laser (2.1 μm) pulse has been described by van Leeuwen et al.[39] The greatest effect of bubble formation is not damage to surrounding tissue by acoustic waves; it is the abrupt change of optical properties in the path of the laser beam. Also, the formation of vapor in tissue just below the laser impact site may allow the laser radiation to penetrate farther than expected and cause excessive damage.

At fluence rates above 10^8 W/cm^2, the large electric field causes dielectric breakdown. The resulting plasma produces an acoustic shockwave which is used in laser lithotripsy to fragment urinary calculi. Typically laser pulses are directed through a quartz fiber which is in contact with the calculi. Fair[40] notes that 0.2- to 35-kbar shocks have been associated with the destruction of urinary calculi by laser-induced stress waves. A description of plasma-mediated laser ablation with a microsecond-pulse dye laser has been established by Teng et al.[41]

Another application is in the field of ophthalmology. After removal of the lens for treatment of a cataract, clouding of the posterial capsule may occur, causing a secondary cataract. The treatment of choice is disruption of the membrane by creating a plasma at its surface. Typically treatment is with a Q-switched Nd:YAG laser. The resulting shock wave tears the membrane. By the opening of the clouded membrane, sight is restored in the affected eye.

24.7 FUTURE DIRECTIONS

Extraordinary advances in laser medicine have occurred during the past 10 years. The old "burn and learn" philosophy has been replaced by careful selection of laser characteristics and a delivery system to match a medical task. All modes of laser-tissue interaction are employed in treatment procedures: photochemical, photothermal (coagulation and ablation), photodisruptive (bond breaking for ablation), and photomechanical (acoustic).

Extraordinary delivery systems include the use of diagnostic information to set irradiation parameters. The excimer ArF (193 nm) corneal ablation system reshapes the cornea by ablation of 1.0 μm or less thicknesses. The system predetermines the amount of cornea to be removed and the excimer laser is programmed to proceed with a series of irradiations selected to achieve the desired curvature. The technology is designed to correct a 20/200 myopic eye to 20/20 vision.

As diagnostics are added, "smart" systems will be available for all medical procedures. Although speculation beyond a few years is tenuous at best, advances in the near future will include:

1. Advances in endoscopic delivery system that (*a*) provide optical (reflectance, radiation, and fluorescence) and ultrasonic diagnostic signals and (*b*) incorporate fibers and/or waveguides for delivery and/or detection of ir wavelengths beyond 2.5 μm.

2. Development of feedback control systems for control of irradiation parameters and for placement of the laser beam. These systems will undoubtedly use optical, thermal, ultrasonic, and/or MRI signals for control decisions. MRI has the potential for monitoring temperature and/or zones of denatured tissue during surgery involving laser radiation. Three-dimensional temperature distributions can provide feedback control for hyperthermia, or zones of necrosis can be monitored during photodynamic therapy (PDT).

3. Most exciting are advances that have been made in the field of laser diodes. By creating high-power (greater than 10 W) cw devices and high-pulse-energy (greater than 200 mJ/pulse) devices, small, low-cost systems can replace today's expensive (to buy and to maintain) excimer and gas laser systems.

CW laser diodes are now being investigated for retinal surgery, cyclophotocoagulation, and photodynamic therapy. However, their greatest potential may be as pumps for infrared crystals such as Er:YAG which can be frequency doubled to obtain uv and visible wavelengths.

Undoubtedly advances in diagnostic systems, delivery systems, and lasers will occur rapidly given the technological base that has been established and the expanding medical applications for these systems.

24.8 REFERENCES

1. "Laser Surgery: Too Much, Too Soon?" *Consumer Reports*, vol. 56, no. 8, pp. 536–540, 1991.

2. Cotliar, A. M., H. D. Schubert, E. R. Mandel, and S. L. Trokel, "Excimer laser radial keratotomy," *Ophthalmology*, vol. 92, pp. 206–208, 1985.

3. Singleton, D. L., G. Papaskevopollos, R. S. Taylor, and L. A. J. Higginson, "Excimer laser angioplasty: tissue ablation, arterial response and fiber optic delivery," *IEEE J. Quan. Electron.*, vol. QE-23, pp. 1772–1777, 1987.

4. Srinivasan, R., K. G. Casey, and J. D. Haller, "Subnanosecond probing of the ablation of the soft plaque from arterial wall by 308 nm laser pulses delivered through a fiber," *IEEE J. Quant. Electron.*, vol. 26, pp. 2279–2283, 1990.

5. Haller, J. D., et al., "A sober view of laser angioplasty," *Cardiol.*, vol. 2, pp. 31–33, 1985.

6. Ham, W. T., and H. A. Mueller, "The photopathology and nature of blue light and near uv retinal lesions produced by lasers and other optical sources," *Laser Applications in Medicine and Biology*, M. L. Wolbarsht, ed., Plenum, New York, pp. 191–246, 1989.

7. M. W. Berns, "Laser Surgery," *Scientific American*, vol. 264, no. 6, pp. 84–90, 1991.

8. Richards-Kortum, R. R., et al., "A one layer model of laser induced fluorescence for diagnosis of disease in human tissue: Applications to atherosclerosis," *IEEE Trans. Biomed. Eng.*, vol. 36, pp. 1222–1232, 1989.

9. Andersson-Engels, S., et al., "Malignant tumor and atherosclerotic plaque diagnosis using laser-induced fluorescence," *IEEE J. Quant. Electron.*, vol. 36, no. 12, pp. 2207–2217, 1990.

10. Steinke, J. M., and A. P. Shepherd, "Role of light scattering in whole blood oximetry," *IEEE Trans. Biomed. Eng.*, vol. 33, no. 3, pp. 294–301, 1986.

11. Ishimaru, A., *Wave Propagation and Scattering in Random Media*, vol. 1, Academic, New York, 1978.

12. Yoon, G., A. J. Welch, M. Motamedi, and M. J. C. van Gemert, "Development and application of a three-dimensional light distribution model for laser irradiated tissue," *IEEE J. Quant. Electron.*, vol. 23, no. 10, pp. 1721–1733, 1987.

13. Cheong, W. F., S. A. Prahl, and A. J. Welch, "A review of the optical properties of tissues," *IEEE J. Quantum Electron.*, vol. 26, pp. 2166–2186, 1990.

14. Kaijzer, M., S. L. Jacques, S. A. Prahl, and A. J. Welch, "Light distribution in artery tissue: Monte Carlo simulations for finite laser beams," *Lasers in Surgery and Medicine*, vol. 9, pp. 148–154, 1989.

15. Geeraets, W. J., et al., "The relative absorption of thermal energy in retina and choroid," *Investigative Ophthalmology*, vol. 1, pp. 340–347, 1962.

16. Geeraets, W. J., et al., "The loss of light energy in retina and choroid," *Archives of Ophthalmology*, vol. 64, pp. 606–615, 1960.

17. Borsky, S. H., S. Rosen, D. E. Geern, and J. M. Noe, "The nature and evolution of port wine stain: a computer assisted study," *J. Invest. Dermatol.*, vol. 74, pp. 154–157, 1980.

18. Tan, O. T., D. Whitaker, J. M. Garden, and G. Murphy, "Pulsed dye laser (577 nm) treatment of portwine stains: ultrastructural evidence of neovascularization and most cell degranulation in heated lesions," *J. Invest. Dermatol.*, vol. 90, pp. 395–398, 1988.

19. van Gemert, M. J. C., A. J. Welch, I. D. Miller, and O. T. Tan, "Can physical modeling lead to an optimal laser treatment strategy for port wine stains?" *Laser Applications in Medicine and Biology*, vol. 5, M. L. Wolbarsht, ed., Plenum, New York, pp. 199–275, 1991.

20. Tan, O. T., P. Morrison, and A. K. Kurban, "585 nm for treatment of port wine stains," *Plast. Reconstr. Surgery*, vol. 86, pp. 1112–1117, 1990.

21. Pickering, J. W., and M. J. C. van Gemert, "585 nm for the laser treatment of port wine stains: a possible mechanism," *Lasers in Surgery and Medicine*, vol. 11 and 6, pp. 616–618, 1991.

22. Garden, J. M., et al., "Effect of dye laser pulse duration on selective cutaneous vascular injury," *J. Invest. Dermatol.*, vol. 87, pp. 653–657, 1986.

23. O. T. Tan, ed., *Management and Treatment of Benign Cutaneous Vascular Lesions*, Philadelphia, 1991, Lea and Febiger, Philadelphia, 1992.

24. Neblet, C. R., J. R. Morris, and S. Thomsen, "Laser-assisted microsurgical anastomosis," *Neurosurgery*, vol. 19, pp. 14–34, 1986.

25. Kopchak, G. E., et al., "CO_2 and argon vascular welding: acute histologic and thermodynamic comparison," *Lasers in Surgery and Medicine*, vol. 8, pp. 584–588, 1988.

26. Quigley, M. D., J. E. Bailes, and H. C. Kwann, "Aneurysm formation after low level carbon dioxide laser-assisted vascular anastomosis," *Neurosurgery*, vol. 18, no. 3, pp. 92–99, 1986.

27. Springer, T., "An automated system for laser anastomosis," Ph.D. dissertation, The University of Texas at Austin, 1991.

28. LeCarpentier, G. L., M. Motamedi, L. P. McMath, and A. J. Welch, "The effect of wavelength on ablation mechanisms during CW laser irradiation: argon versus Nd:YAG (1.32 μm)," *Proc. IEEE EMBS '89*, Seattle.

29. LeCarpentier, G. L., et al., "Continuous wave laser ablation of tissue: analysis of thermal and mechanical events," *IEEE Trans. Biomed. Eng.*, (in press).

30. Marshall, J., S. Trokel, S. Rothery, and R. R. Krueger, "Photoablative reprofiling of the cornea using an excimer laser: photorefractive keratectomy," *Lasers Ophthalmol.*, vol. 1, pp. 21–48, 1986.

31. Kaplan, I., et al., "Preliminary experiments of possible uses in medicine of novel plastic hollow fibers for transmission of CO_2 radiation," *Lasers in Surgery and Medicine*, vol. 10, pp. 291–294, 1990.

32. Richards-Kortum, R. R., et al., "A one-layer model of laser induced fluorescence for diagnosis of disease in human tissue: applications to atherosclerosis," *IEEE Trans. Biomed. Eng.*, vol. 36, pp. 1222–1232, 1989.

33. Verdaasdonk, R., and C. Borst, "Ray tracing of optically modified fiber tips II: laser scalpels," *Applied Optics*, vol. 30, pp. 2172–2178, 1991.

34. Verdaasdonk, R., and C. Borst, "Ray tracing of optically modified fiber tips I: Spherical probes," *Applied Optics*, vol. 30, pp. 2159–2171, 1991.

35. Singleton, D. L., G. Papaskevopoulos, R. S. Taylor, and L. A. J. Higginson, "Excimer laser angioplasty: tissue ablation, arterial response and fiber optic delivery," *IEEE J. Quant. Electron.*, vol. QE-23, pp. 1772–1782, 1987.

36. Abela, G. S., ed. *Lasers in Cardiovascular Medicine and Surgery: Fundamentals and Techniques*, Kluwer Academic, Boston, 1990.

37. Doiron, D. R., and C. J. Gomer, eds., *Porphyrin Localization and Treatment of Tumors*, Liss, New York, 1984.

38. Isner, J. M., et al., "Mechanism of laser ablation in an absorbing fluid field," *Lasers in Surgery and Medicine*, vol. 8, pp. 543–554, 1988.

39. van Leeuwen, T. G., M. J. van der Veen, R. M. Verdaasdonk, and C. Borst, "Noncontact tissue ablation by Holmium:YSGG laser pulses in blood," *Lasers in Surgery and Medicine*, vol. 11, pp. 26–34, 1991.

40. Fair, H. D., "*In vitro* destruction of urinary calculi by laser induced stress waves," *Medical Instrumentation*, vol. 12, pp. 100–105, 1978.

41. Teng, P., N. S. Nishioka, R. R. Anderson, and T. R. Deutsch, "Acoustic studies of the role of immersion in plasma-mediated laser ablation," *IEEE J. Quant. Electron.*, vol. QE-23, pp. 1845–1852, 1987.

42. Preuss, L. E., F. P. Bolin, and B. Cain, "Tissue as a medium for laser light transport implications for photoradiation therapy," *Proc. SPIE 357 Lasers in Surg. & Med*, pp. 77–84, 1982.

43. Flock, S. T., B. C. Wilson, and M. S. Patterson, "Total attenuation coefficients and scattering phase functions of tissue and phantom materials at 633 nm," *Med. Phys.*, vol. 14, pp. 835–841, 1987.

44. MacLeod, J. S., D. Blanc, and M. J. Cottes, "Measurement of the optical absorption coefficients at 1.06 μm of various tissues using photoacoustic effect," *Lasers Surg. & Med.*, vol. 8, pp. 143, 1988 (Abstr.).

45. Long, F. H., N. S. Nishioka, and T. F. Deutsch, "Measurement of the optical and thermal properties of biliary calculi using pulsed photothermal radiometry," *Lasers Surg. & Med.*, vol. 7, pp. 461–466, 1987.

46. Star, W. M., et al., "Light dosimetry for photodynamic therapy by whole bladder wall irradiation," *Photochem. Photobiol.*, vol. 46, pp. 619–624, 1987.

47. Splinter, R., et al., "In vitro optical properties of human and canine brain and urinary bladder tissues at 633 nm," *Lasers Surg. & Med.*, vol. 9, pp. 37–41, 1989.

48. Pedersen, G. D., N. J. McCormick, and L. O. Reynolds, "Transport calculations for light scattering in blood," *Biophys. J.*, vol. 16, pp. 199–207, 1976.

49. Reynolds, L. O., C. C. Johnson, and A. Ishimaru, "Diffuse reflectance from a finite blood medium: application to the modeling of fiber optic catheters," *Appl. Optics*, vol. 15, pp. 2059–2067, 1976.

50. Karagiannes, J. L., et al., "Applications of the 1-dimensional diffusion approximation to the optics of tissues and tissue phantoms," *Appl. Optics*, vol. 28, pp. 2311–2317, 1989.

51. Wilson, B. C., M. S. Patterson, and D. M. Burns, "The Effect of Photosensitizer Concentration in Tissue on the Penetration Depth of Photoactivating Light," *Laser Med. Sci.*, vol. 1, pp. 235–244, 1986.

52. Wilson, B. C., W. P. Jeeves, and D. M. Lowe, "In vivo and post mortem measurements of the attenuation spectra of light in mammalian tissues," *Photochem. Photobiol.*, vol. 42, pp. 153–162, 1985.

53. Svaasand, L. O. and R. Ellingsen, "Optical penetration of human intracranial tumors," *Photochem. Photobiol.*, vol. 41, pp. 73–76, 1985.

54. Svaasand, L. O. and R. Ellingsen, "Optical properties of human brain," *Photochem. Photobiol.*, vol. 38, pp. 283–299, 1983.

55. Sterenborg, H. J., et al., "The spectral dependence of the optical properties of the human brain," *Lasers Med. Sci.*, vol. 4, pp. 221–227, 1989.

56. Doiron, D. R., L. O. Svaasand, and A. E. Profio, "Light dosimetry in tissue applications to photoradiation therapy," in *Porphyrin Photosensitization* ed. D. Kessel, T. J. Dougherty, Plenum, New York, pp. 63–75, 1983.

57. Key, H., P. C. Jackson, and P. N. T. Wells, "Light scattering and propagation in tissue," *Poster Presentation*, World Congress on Medical Physics and Bioengineering, San Antonio, Texas, August 1988.

58. Marchesini, R., et al., "Extinction and absorption coefficients and scattering phase functions of human tissues in vitro," *Appl. Optics*, vol. 28, pp. 2318–2324, 1989.

59. Andreola, S., et al., "Evaluation of optical characteristics of different human tissues in vitro," *Lasers in Surg. & Med.*, vol. 8, p. 142 (Abstr.), 1988.

60. Jacques, S. L., C. A. Alter, and S. A. Prahl, "Angular dependence of He-Ne light light scattering by human dermis," *Lasers in Life Sc.*, vol. 1, pp. 309–333, 1987.

61. Watanabe, S., et al., "Putative photoacoustic damage in skin induced by pulsed ArF excimer laser," *J. Invest. Derm*, vol 90, pp. 761–766, 1988.

62. Eichler, J., J. Knof, and H. Lenz, "Measurement of the depth of penetration of light (0.35 μm-1.0 μm) in tissue," *Rad. Environ. Biophys.*, vol. 14, pp. 239–242, 1977.

63. Parsa, P., S. L. Jacques, and N. S. Nishioka, "Optical properties of rat liver between 350 and 2200 nm," *Appl. Optics*, vol. 28, pp. 2325–2330, 1989.

64. Marijnissen, J. P. A. and Star, W. M., "Quantitative light dosimetry in vitro and in vivo," *Lasers in Med. Sc.*, vol. 2, pp. 235–242, 1987.

65. McKenzie, A. L., "Can photography be used to measure isodose distribution of space irradiance for laser photoradiation therapy?," *Phys. Med. Biol.*, vol. 33, pp. 113–131, 1988.

66. Bolin, F. P., et al., "A study of the 3-dimensional distribution of light (632.8 nm) in tissue," *IEEE J. Quant. Elec.*, vol. 23, 1734–1738, 1987.

67. Marijnissen, J. P. A. and W. M. Star, "Phantom measurements for light dosimetry using isotropic and small aperture detectors," in *Porphyrin Localization and Treatment of Tumors* ed. D. R. Doiron, C. J. Gomer, and Alan R. Liss, New York, pp. 133–148, 1984.

68. Wilksch, P. A., F. Jacka, and A. J. Blake, "Studies of light propagation in tissue," in *Porphyrin Localization and Treatment of Tumors* ed. D. R. Doiron, C. J. Gomer, and Alan R. Liss, New York, pp. 149–161, 1984.

69. Halldorsson, T., et al., "Theoretical and experimental investigations prove Nd:YAG laser treatment to be safe," *Lasers. Surg. Med.*, vol. 1, pp. 253–262, 1981.

70. Arnfield, M. R., J. Tulip, and M. S. McPhee, "Optical propagation in tissue with anisotropic scattering," *IEEE Trans. Biomed. Eng.*, vol. 35, pp. 372–381, 1988.

71. van Gemert, M. J. C., et al., "Optical properties of human blood vessel wall and plaque," *Lasers in Surg. & Med.*, vol. 5, pp. 235–237, 1985.

72. Driver I., et al., "In vivo light dosimetry in interstitial photoradiation therapy (PRT)," *Proc. of SPIE Int. Soc. of Opt. Engr. OE 'Lase 88*, Los Angeles, Mike Berns (ed.), pp. 98–102, 1988.

73. Muller, P. J. and B. C. Wilson, "An update on the penetration depth of 630 nm light in normal and malignant human brain tissues in vivo," *Phys. Med. Biol.*, vol. 31, pp. 1295–1297, 1986.

CHAPTER 25
MATERIAL PROCESSING APPLICATIONS OF LASERS

James T. Luxon

In this chapter we deal with the types of lasers used in material processing applications and the characteristics of these lasers which are pertinent to the applications. Appropriate optical materials and devices will be described. The general advantages and disadvantages of lasers in materials processing are discussed. Brief discussions of the major processing applications are presented. The applications of lasers in microelectronic manufacturing are too numerous to discuss in detail; hence, a brief overview is presented.

25.1 MATERIAL PROCESSING LASERS

This discussion will be limited to three types of lasers, all of which have been discussed to some extent in previous chapters. These are the neodymium-doped yttrium aluminum garnet (Nd:YAG) laser, the CO_2 laser, and the excimer laser.

25.1.1 Nd:YAG Lasers

The Nd:YAG laser is a solid-state laser which emits radiation at a wavelength of 1.06 μm. Power levels are available from a few watts to over 1000 W average power. Applications are typically hole piercing, cutting, welding, and marking. Nearly all these applications are to metals. Absorption by metals at this wavelength is fairly high. Polymeric materials and glass are too transparent unless coated or an absorptive filler material is present.

In most applications, this laser is pulsed to increase peak power and to decrease thermal damage to the workpiece. In marking applications, Q switching is used to create rapidly repeating short pulses with peak powers that reach 1200 times the average power. For cutting, drilling, and welding applications, the laser is electronically pulsed. Pulse shaping is used to enhance the performance of the laser. Intercavity techniques are used to decrease divergence and to improve the spiking characteristics of the laser pulse.

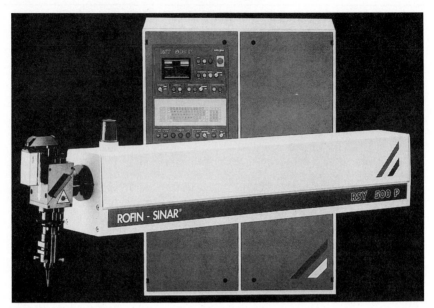

FIGURE 25.1 A modern Nd:YAG laser. (Courtesy of Rotin-Sinar, Inc.)

For marking or engraving applications, the Q-switched pulse length might be 150 ns with a pulse repetition rate of 10 kHz. An average power of 50 W is typical. Pulse lengths for cutting and welding are generally in the 0.1- to 2-ms range. Most welding, even seam welding, is done in the pulse mode with pulse lengths in the 1- to 10-ms range. With the advent of Nd:YAG lasers capable of power levels over 1 kW, the possibility of continuous seam welding has been developed.

A serious drawback to Nd:YAG lasers in the past has been the poor beam quality produced at high power levels. The cause was optical distortion of the beam in the YAG rod due to thermal lensing and other thermally induced inhomogeneities. This tends to produce a beam with poor mode quality and, consequently, high divergence. High divergence translates to low brightness and low power density in the focused spot. Recent developments in Nd:YAG laser resonator design have overcome this problem to a large extent. It is now possible to drill 1-in-thick steel in a few seconds, compared to over one minute previously. Holes can be drilled through steel with the beam making an angle of less than 10° to the surface (>80° to the normal) and steel over 1 in thick can be cut by a modern Nd:YAG laser. Figure 25.1 is a photograph of a modern Nd:YAG laser.

25.1.2 CO$_2$ Lasers

The CO$_2$ laser has long been considered the workhorse of the material processing industry because of the wide power range available and the large variety of materials that can be processed with it. All commercially important lasers can be classified in one of several basic designs. Some low-power lasers, generally under 100 W, are sealed-off designs. There are conventional glass-tube types and ceramic-tube waveguide designs. In the latter, the bore of the tube is so narrow that it literally acts as a waveguide. The output of these low-power lasers is very near Gaussian. Periodically the gas fill in these lasers must be replaced. They typically operate at 110 V and require air or water cooling.

Medium- to high-power CO_2 lasers are all flowing-gas systems which fall into three categories: (1) slow axial flow with axial discharge (SAFAD), (2) fast axial flow with axial discharge (FAFAD), and (3) fast transverse flow with transverse discharge (FTFTD). Average power ranges from about 50 W to over 25 kW. SAFAD lasers are limited to about 1.5 kW because of their large size. FAFAD lasers are available from around 600 W up to 6 kW. FTFTD lasers are available in the 1.5- to over 25-kW range.

SAFAD lasers typically operate in near-Gaussian mode or low-order mode such as the TEM_{01} doughnut mode or a mixture of TEM_{01} and TEM_{00}. FAFAD lasers are capable of operating with mode quality similar to SAFAD lasers when operated in the 600- to 1500-W range. At higher power levels, the mode tends to be somewhat higher, such as a TEM_{20} mode. FTFTD lasers operate near Gaussian (1.5 kW or less) to higher-order modes at multikilowatt levels.

It is common to use an unstable resonator configuration in lasers designed to operate at an average power over 5 kW. The purpose of this is to avoid the problem of thermal lensing in the output coupler.

AC, radio-frequency (rf), and dc discharges are utilized in axial discharge lasers. RF and dc discharges are used in transverse discharge lasers. Enhanced pulsing capability is also available in axial discharge lasers. Peak powers around 5 times the average power can be achieved. RF discharges produce a very uniform discharge region for high-quality beam output, but the electronics and required rf shielding are quite bulky. Figure 25.2 shows modern FAFAD and FTFTD lasers.

25.1.3 Excimer Lasers

The term *excimer* is literally a contraction of the phrase *excited dimer*. A dimer is a molecule containing two identical atoms. The molecules in excimer lasers are actually complexes because they are composed of a halogen and a noble gas atom. The term *exciplex* would be more accurate, but the term *excimer* is accepted.

As discussed in Chap. 3 there are several gas combinations that lead to excimer laser operation. All emit radiation in the ultraviolet region of the electromagnetic spectrum. The combinations and the wavelengths at which they emit are ArF, 193 nm; KrF, 248 nm; XeCl, 308 nm; and XeF, 351 nm. These molecules are formed only when the noble gas atom is in an excited state and the lifetime of this state is very short. Consequently, the output of excimer lasers is in the form of pulses on the order of 100 ns in length. Power levels of 200 W or more can be achieved with pulse repetition rates of 1000 pps or more.

One machine can operate at all of the excimer wavelengths with different gas mixes. Changeover from one mix to another does require careful flushing of the system. These same lasers could also be used as pulsed CO_2 lasers.

There are numerous applications for excimer lasers in micromachining of thin metals and polymers, marking, and microelectronics. A few of these are discussed later in this chapter. One of the most interesting applications is not strictly a materials processing operation, at least not for production. This involves the use of the excimer laser to polymerize a layer of polymeric material to form a plastic prototype model of a part layer by layer in a process called Stereo Lithography developed by 3D Systems, Inc. Other lasers and nonlaser light sources are used

(a)

(b)

FIGURE 25.2   (*a*) Modern FAFAD and (*b*) FTFTD lasers. (Courtesy of PRC Corp.)

for similar processes. Essentially the surface of the liquid is irradiated by the laser under computer control directly from computer-aided design (CAD) data. As each layer of the part is formed, the part is lowered to expose additional liquid until the entire three-dimensional part is formed.

25.2 LASER CHARACTERISTICS FOR MATERIAL PROCESSING: ADVANTAGES AND DISADVANTAGES

The primary advantages that lasers have in material processing will be discussed first. Clearly, these advantages do not make lasers the best choice in all applications. What one has to look for are those situations in which the laser gives a cost or quality advantage not achievable by more conventional processing methods. The laser is rapidly becoming a conventional tool itself. More and more designers are considering the unique capabilities of lasers when designing parts or products. The advent of synchronous engineering is accelerating the pace of designing for manufacturability and the laser is playing an important role in this philosophical revolution in manufacturing. Some of the characteristics of lasers that may provide an advantage in material processing are listed below.

Advantages
1. *Versatility.* A given laser may be capable of performing more than one type of application, such as cutting and welding. These operations could be performed on the same or different parts at the same or different workstations.

2. *Deliverability.* Because a laser beam has low divergence, it can be propagated over large distances before reaching the point of application. Because it is a beam of light, it can be split into two or more beams for delivery to multiple workstations or to do simultaneous multiple tasks on a single part. Alternatively, the same beam can be shared by multiple workstations. In the case of Nd:YAG lasers the beam can be delivered by fiber optics to the workpiece.

3. *Atmospheric effects.* Industrial laser beams propagate through most atmospheres without significant attenuation. Thermal blooming can occur if the air or cover gas is heavily contaminated with oil or dust. Strong plasma effects can alter the focus location and focused spot size.

4. *Field effects.* Magnetic and electric fields do not affect the behavior of laser beams in any way. In fact, it is common to use magnetic fixturing to hold steel parts while laser-processing them.

5. *Noncontact processing.* Since the energy delivered to the part is from a beam of light, the force exerted on the part, though nonzero, is negligible. This is very important in processing certain delicate parts and materials. There is, of course, no tool wear.

6. *Focusability.* Low beam divergence (high radiance) makes it possible to focus the beam to a very small spot size. This in turn produces very high power density (irradiance) in cw applications and extremely high energy density (fluence) in pulsed applications. High irradiance or fluence allows the laser to perform tasks in a very short time, thereby minimizing the energy input, which, in turn, minimizes the heat-affected zone (HAZ). Reducing the HAZ minimizes chem-

ical and metallurgical effects that can be detrimental to the performance and longevity of the part.

7. *Adaptability.* The propagation characteristics of laser beams make this device easily adapted to automated manufacturing systems.

8. *Controllability.* Different lasers range from a few watts to tens of kilowatts average power or a few millijoules to tens of joules per pulse. The average power or pulse energy of a given laser can generally be varied over a wide range. A factor of 10 variation in power or pulse energy is not unreasonable to expect. Pulse characteristics are also variable. The pulse shape, pulse width, and repetition rate can be varied to best suit the application.

Disadvantages. There are aspects and characteristics of lasers and laser systems that may represent serious disadvantages in some applications. Some of these are listed below:

1. *High capital cost.* Lasers are expensive machines. You can expect to pay $100 to $200 per watt of average power for a CO_2 laser and up to $1000 per watt for a Nd:YAG or excimer laser. This is only part of the cost. To make the laser useful, beam-delivery optics and some sort of part handling or beam delivery system are required. Depending on the sophistication needed, the total cost can run well over $1,000,000 for a complete system.

2. *Low efficiency.* The overall "wall plug" power efficiency for lasers is low. This refers to the ratio of useful light power output to the total electrical power input. Nd:YAG and excimer lasers have a "wall plug" efficiency around 3 to 5 percent, whereas CO_2 lasers have a wall plug efficiency in the 10 to 25 percent range. When total process efficiency is taken into account, the laser is frequently better than conventional methods. This refers to the ratio of power that actually does useful work to the total power that goes into the system.

3. *High technology.* Actually the laser or laser system is not the most complicated technology used in modern industry. Electron-beam systems are much more difficult to understand, operate, and maintain. Nevertheless, operators and maintenance personnel must be well-trained and must deal with situations that are unique to lasers. This is particularly true with respect to the optical components.

4. *Safety.* Lasers and laser systems, when properly operated and maintained, are safe machine tools. One concern, however, regards exposure to the beam. Wherever lasers or laser systems are installed, someone should have laser safety training and be designated the laser safety officer (LSO). This person should have the responsibility to see that operators and others working with lasers or laser systems receive the proper instructions to carry out their jobs in a manner that is safe for them and their coworkers. The LSO should have the responsibility and authority to shut down a laser operation if proper safety practices are not being followed.

5. *Small focused spot size.* Although this is an advantage from a power or energy density standpoint, it presents problems in some applications. Part fit-up and seam tracking require careful attention in welding applications. Additional, a small focused spot size results in a relatively short depth of focus (working depth), which must be taken into consideration in welding and material-removal applications.

25.3 LASER SURFACE MODIFICATION

The primary laser surface modification applications fall into two categories: transformation hardening and material addition. The latter breaks down into two approaches, alloying and cladding. A brief description of these processes is presented here.

25.3.1 Transformation Hardening

In this process, a defocused CO_2 laser beam impinges on a ferrous metal part to cause case hardening. The material must have a minimum of 0.02 percent carbon to be hardenable, and the optimum hardening occurs when the carbon is finely dispersed, as in a pearlitic structure in steel or iron. Typically the spot is 2 to 10 mm across, depending on the power level and the coverage rate needed.

Because of the high reflectance of metals for 10.6-μm radiation and low power density, an absorptive coating is needed to couple the energy into the part. Materials such as manganese phosphate (Lubrite) or black paint are used. Coating thickness is important. Too thin a layer will burn off too soon, allowing the beam to be reflected from the metal surface. Too thick a coating will actually insulate the metal from the laser energy.

Laser surface hardening is a purely thermal process. The energy of the laser beam is converted into heat at the surface of the metal. Because of the large temperature rise at the surface, heat is rapidly conducted into the material. The maximum temperature reached decreases with distance into the material. The metal will harden to the depth that the maximum temperature exceeds the transition temperature (the austenitic temperature in metallurgical terms). Depending on laser power and coverage rate, the hardened depth ranges from about 0.25 to 1.5 mm. Unless the parts are very small or thin, the temperature will drop rapidly (thousands of degrees per second) after processing so that external quenching is not necessary; this is referred to as *self-quenching*. Figure 25.3 contains a photo-

FIGURE 25.3 Laser-hardened track in medium-carbon steel.

graph of a laser-hardened track in medium carbon steel. Figure 25.4 is a plot of hardness versus distance into the material for a typical case. This plot shows the abrupt transition from near ultimate hardness to base metal hardness at the point where the maximum temperature did not reach the transition temperature.

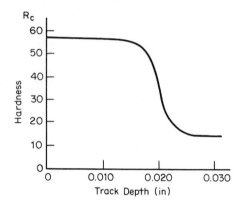

FIGURE 25.4   Hardness versus distance into material.

The main advantage of laser hardening is that the heat input is minimal. This frequently results in elimination of postprocessing to remove deformations induced by heat treating. Since the heat-treated tracks are relatively small, selective hardening is possible and complex geometries can be hardened with simple fixturing. One of the most common applications is hardening of gear teeth. For small gears, hardening can be accomplished all the way to the roots of the teeth by simply rotating the gear under a properly defocused beam. For larger gears, two beams or two passes may be required to accomplish the job.

The major disadvantage of laser hardening is that the coverage rate is low. Rates of ~1300 cm^2/min (200 in^2/min) are typical. Also, CO_2 laser beams frequently do not have the proper power distribution for heat treating. A Gaussian or near-Gaussian beam tends to cause excessive heating at the center of the track, leading to undesirable melting and a track with a semicircular cross section instead of the more desirable near-rectangular cross section. Higher-order or multi mode beams may have hot spots that cause melting. Sometimes the mode is not stable with time. In all these cases, it is necessary to use a beam-integrating device. This device can be a segmented mirror, a transmissive device which converts the incident beam into a near top-hat configuration, or a scanning device which rotates or dithers the beam to average out the power distribution. A passive beam-integration device is generally preferable.

25.3.2 Alloying and Cladding

These processes are similar in that an alloy material, such as Stellite powder, is added to the surface of a metal. The powder can be applied as a coating before being heated by the laser, or by spraying during the laser heating process. In both cases a defocused laser beam is used to melt the alloying material. In alloying, the process is controlled in such a way that the substrate is melted to a given depth, usually a fraction of a millimeter, and the alloying material is mixed into the substrate material by the violent flowing action of the molten metal. Depth and concentration of the alloying material can be precisely controlled.

Alloying is used to increase resistance to wear and corrosion. The main advantage of laser alloying is the ability to selectively apply the alloying material. The distribution of the material is extremely uniform and the resulting grain structure is very fine as a result of the rapid cooling process. This procedure adds to the ability of the material to resist wear and corrosion.

25.4 WELDING

Two things determine how a laser weld is formed: thermal conduction and keyholing. When laser radiation is incident on a clean metal surface, it is absorbed by interaction with the free electrons in the metal. This occurs within about 1 μm of the surface. If the average power is under 1 kW or the power density is below 0.5 mW, the welding process will generally be dominated by energy transported into the material by thermal conduction. Depth of penetration (melt depth) is determined by the depth to which the temperature exceeds the melting temperature. The depth of penetration (melt depth) in thermal-conduction-limited welding is usually 1 to 2 mm, and weld nuggets or beads are generally twice as wide as deep due to lateral heat conduction.

Keyholing is the name given to the phenomenon depicted in Fig. 25.5. In this process the average power and power density are sufficiently high that the surface temperature reaches such a high level that the vapor pressure exceeds the surface tension. When this happens, the molten metal is literally blown out of the way, producing a hole that allows the laser beam to proceed directly into the metal relatively unimpeded. The molten metal flows up and around the hole toward the backside (opposite the direction of beam travel) and flows back into the hole as the beam passes. The term *keyhole* was coined because of the analogy with looking through a keyhole.

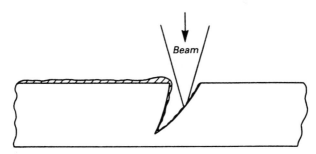

FIGURE 25.5   The keyholing phenomenon.

Keyholing was first observed in high-power electron beam welding. It is this phenomenon which enables deep penetration in laser and electron-beam welding. Absorption of a laser beam in the keyhole is a controversial process. It probably involves a combination of multiple reflections and absorptions by the walls of the keyhole and absorption by the plasma present in the keyhole. When keyholing predominates, nearly 100 percent of the incident light is absorbed. When keyholing is not dominant, a substantial portion of the incident light is reflected.

Figure 25.6 is a photograph of a laser weld cross section. The upper part of the nugget is flared, partially due to thermal conduction and partially due to the fact that the laser beam was focused below the surface to enhance penetration. The lower part of the weld has relatively straight sides because it behaves as a light guide for the beam.

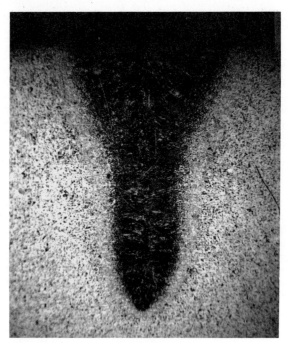

FIGURE 25.6   Laser weld cross section.

Solid-state pulsed lasers generally produce welds by thermal conduction, whereas multikilowatt CO_2 lasers produce welds chiefly by the keyhole process. As the average power of solid-state lasers increases, the keyhole phenomenon will become more predominant.

Laser welding is applied to a wide variety of materials and parts. Space does not allow a thorough survey of these diverse applications. Nd:YAG lasers are used for spot and seam welding. Seam welding is accomplished by overlapping spot welds. Excellent hermetic welds can be made this way, since the amount of overlap can be controlled to ensure that pores produced by one laser pulse are sealed by the next. This process is used quite extensively in the electronics industry for electronic packaging of items like heart pacemakers and aerospace electronics. Nd:YAG lasers are also used for lead wire welding and soldering. As the average power of Nd:YAG lasers increases and fiber-optic delivery systems become more prevalent, they will compete directly with CO_2 lasers in more robust applications.

CO_2 lasers have a wide range of available power levels, from a few watts to tens of kilowatts, and therefore, find wider applicability. CO_2 lasers are used to weld such items as fuel filter packages, turbine engine heat recuperator plates, and automotive transmission gears. Generally, pulsed operation of the laser produces deeper penetration and better weld quality by producing a more stable keyhole. Problems occur in keyhole welding if the weld parameters are not properly optimized. Attempting to weld too fast will cause the keyhole to fluctuate and cause porosity and humping. Welding too slow may reduce the penetration because too much metal is melted and it obscures the keyhole. Figure 25.7 is a photograph of a CO_2 laser seam weld made on a straight pipe which was then bent into the shape shown.

Virtually every type of joint configuration can be welded by a laser if the beam has unrestricted access to the joint. One should never underestimate the effect of clipping the beam by a fixture or a portion of the part being welded. What seems to be a negligible interference can cause a drastic change in weld quality. Butt welds require careful fixturing and beam positioning, and the edge quality and fit-up must be good. A higher-mode beam which does not produce such a sharp focus allows a wider fit-up tolerance than a near-Gaussian beam. In lap welding, the fit-up between the sheets of metal must be reasonable. Since the laser melts a relatively small amount of metal, it is not possible to provide fusion of the sheets if the gap between them is too large.

FIGURE 25.7 CO_2 laser seam weld on straight pipe. (Courtesy of Rotin-Sinar, Inc.)

In butt or seam welding of dissimilar metals with widely different thermal properties, the beam can be adjusted to impinge more on the side of the metal, requiring more energy to melt it. Also, when welding high-carbon steel to low-carbon steel, the beam can be focused more on the low-carbon steel side to dilute the carbon in the fusion zone. This helps to minimize embrittlement of the weld zone by transformation hardening.

Nearly any material that can be welded by conventional techniques can be welded by a laser. Because laser welding is a rapid process, problems can arise with some materials, such as high-carbon steel. Sometimes dissimilar materials can be welded more successfully with a laser than by other methods. This is particularly true if the materials are thin or the parts are small and lightweight.

The cover gas plays an important role in the welding process. The most commonly used gases are helium and argon. Helium has a higher thermal conductivity and higher ionization potential than argon, but argon is about 10 times heavier than helium. Consequently, argon does a better job of displacing air, but tends to produce a more intense plasma, which can absorb a substantial portion of the beam's power. Plasma absorption by argon is a severe problem for multikilowatt CO_2 lasers. Helium, on the other hand, may actually provide cooling that is too rapid, causing brittleness and cracking. Some materials, like titanium and aluminum, must be thoroughly covered top and bottom, if full penetration is achieved, to prevent serious oxidation. The best solution may be a mixture of helium and argon, to take advantage of the best features of each.

25.5 CUTTING AND DRILLING

Most laser cutting is done with either CO_2 or Nd:YAG lasers. CO_2 lasers can be used in the continuous-wave (cw) or pulsed mode. When operated in pulsed mode, the pulse rate is typically 1 kilohertz or higher. Many modern CO_2 lasers are rf-excited and naturally produce a pulsed output. The majority of CO_2 lasers employed in cutting have power outputs in the range of 500 to 1500 W with near-Gaussian or low-order mode outputs.

The list of materials that can be cut with CO_2 lasers is extensive. It includes most metals, ceramics, plastics, wood, rubber, fabrics, and paper. Steel up to 35 mm thick has been cut, but the maximum thickness cut in common practice is about 5 to 6 mm. Wood and plastic over 25 mm thick are easily cut, though it is impossible to maintain perfectly straight sides due the focusing of the beam.

Laser cutting is a gas-assisted process, and either air or oxygen is frequently used to take advantage of the exothermic oxidation to enhance the process. Although much research has been done on high-pressure and supersonic or near-supersonic cutting nozzles, the typical nozzle is fairly simple. The top opening of the nozzle must be large enough for the focusing beam to enter, and the taper angle must be sufficient to allow the beam to pass through without clipping. The outlet hole is 1 to 1.5 mm in diameter, or just large enough to let the beam through. Standoff from the part is about 1 to 2 mm. The gas pressure in the nozzle is typically 3 to 10 lb/in². With this arrangement, sufficient pressure is exerted on the part to blow away molten metal; thus, it is not necessary to raise the material to the vaporization temperature to cut it. High-pressure, large-opening nozzles result in excessive consumption of the cutting gas, besides being extremely noisy. Metals that oxidize violently (including titanium and aluminum) must be cut with an inert gas. The gas most commonly used for this is argon.

Cutting with Nd:YAG lasers is almost exclusively limited to metals. Nd:YAG cutting is generally a process of overlapping hole drilling, since these lasers usually have an average power output of under 200 W. Pulse lengths are in the 0.1- to 2.0-ms range with pulse repetition rates from 20 to 200 pulses per second. With modern Nd:YAG lasers, the pulse shape can actually be programmed to produce the optimum cutting conditions. This typically involves a leading edge spike of high peak power, followed by a period of lower power to complete the drilling process. The pulse is broken up into several channels and the amount of energy in each channel is programmed in by the operator. Figure 25.8 is a photograph of a CO_2-laser-cut saw blade body.

Drilling (hole piercing is a more accurate description, but drilling is in common use) can be accomplished with either CO_2 or Nd:YAG lasers. CO_2 lasers are used to put holes in filter paper, plastic pipe, and ceramic circuit boards, to name a few of the dozens of applications. When large numbers of holes are required in thin material, it is often done on the fly with very rapid pulsing to maintain near roundness of the holes. When very little energy is required to drill each hole, a masking technique may be used to define the holes and their locations.

FIGURE 25.8 CO_2-laser-cut saw blade body. (Courtesy of Rotin-Sinar, Inc.)

Nd:YAG lasers are used to drill diamonds and other jewels and turbine blades and vanes, to name just a few of the many applications. Deep holes with 20 to 1 aspect ratios are commonly drilled in metal. Recent innovations in laser design have made it possible to drill metal over 25 mm thick in just a few seconds. Entrance angles of 15° to the surface are routine, but angles as low as 5° can be attained. Hole drilling with Nd:YAG lasers is generally applied to materials which are difficult to drill by conventional techniques and where chemical machining or electrical discharge machining is impractical or too slow. It is an excellent technique for producing a large number of precisely placed holes for airflow cooling purposes, such as in jet engine parts.

25.6 MARKING

The lasers used for marking are CO_2, Nd:YAG, and excimer. Occasionally, CO_2 lasers of fairly high average power are used to put identifying numbers or letters on heavy-section metal parts, such as engine blocks, by scanning the beam. Most CO_2 marking applications utilize mask imaging to produce a pattern. The laser used in these applications must have a very high energy pulsed output, such as a transversely excited atmospheric (TEA) laser. Most metals, wood, glass, and plastic can be marked this way. This technique is most appropriate when the same pattern must be applied to a large number of parts.

The most commonly used marking laser is the Nd:YAG laser. It is excellent for marking nearly all metals and many colored plastics. The typical laser in a marking system is capable of 50 W average power output. They are Q-switched to produce pulses on the order of 150 ns in length at repetition rates up to 10 kHz. The peak power attainable is around 60 kW. Marking or engraving is accomplished by drilling tiny holes in the material or a coating on a substrate, such as anodizing on aluminum. The holes may or may not be overlapping, depending on the quality of mark required.

Nd:YAG marking systems employ galvanometer mirrors (electronically controlled angular positioning) to direct the beam to the part being marked. In this respect it is like writing with a beam of light. Since this is done under computer control, it is very flexible, and serialization can be programmed into the system. Patterns can be changed almost instantly by calling up a different program.

Excimer lasers, because of their short wavelength and extremely short pulse length, interact differently with materials than traditional material processing lasers. When an excimer laser is used to drill or cut polymeric materials, there is virtually no evidence of thermal interaction. The short wavelength allows direct interaction with molecular bonds, instead of building up enough heat to break them, which takes time and allows thermal conduction away from the direct interaction zone. Even if there is some thermal interaction taking place, the pulses are so short and the hot material is so quickly expelled that little, if any, heat-affected zone is produced. Excimer lasers can be used in marking applications in either the mask imaging or beam deflection mode. Figure 25.9 is a photograph of excimer-marked wire insulation. The laser induces a photochemical change which does not affect the insulating ability of the material.

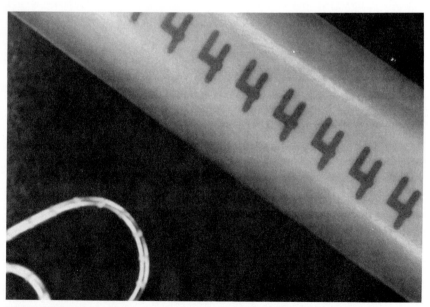

FIGURE 25.9   Excimer-marked wire insulation. (Courtesy of Lumonics, Inc.)

25.7 MICROELECTRONICS APPLICATIONS

There are hundreds of actual and potential applications of lasers in microelectronics manufacturing. Only a brief overview of these applications will be presented here.
Lasers have long been used in interferometry-based step-and-repeat systems to achieve the high accuracy and repeatability required for modern very large scale integration (VLSI) circuit manufacturing. Such systems are capable of maintaining better than 0.05-μm alignment accuracy between the wafer and reticle (mask with step-and-repeat pattern).
Krypton fluoride (KrF) excimer lasers operate at a wavelength of 248 nm. These lasers are capable of producing 0.3- to 0.35-μm details in integrated circuit chips when the wavelength spread is reduced by a technique called *line narrowing*. In this application, the excimer laser replaces a mercury lamp system for exposing the photoresist.
It is also possible to use a laser to write directly in the resist, rather than expose the resist through a mask. Helium-cadmium (HeCd) lasers are used for this purpose to achieve 2.0- to 3.0-μm accuracy and ±0.3-μm repeatability.
Frequency-doubled and -quadrupled Nd:YAG lasers may be used to generate x-rays which will be used in mask exposure to achieve even smaller linewidths in future integrated circuits.
A significant application of lasers is in mask repair. IC masks are made of chromium deposited on glass. Opaque defects are extraneous deposits of chromium. These are removed with a Q-switched, frequency-doubled Nd:YAG laser without damaging the mask. Clear defects are the absence of chromium where it should be. CW argon lasers are used to vapor-deposit chromium to repair this type of defect. Lasers are also used in microelectronics to cut conductive links, deposit conductive links, drill via holes, mark ICs with identifying marks, trim resistors, detect alignment marks, and to inspect the surface, to name a few of the myriad of applications.

25.8 BIBLIOGRAPHY

Luxon, J. T., and D. E. Parker, *Industrial Lasers and Their Applications* 2nd Edition, Prentice-Hall, Englewood Cliffs, N.J., 1992.
Bass, M., volume editor, *Laser Material Processing*, New York, North Holland, 1983.
Schick, L., "Laser Micromachining in Circuit Production," *Photonic Spectra*, pp. 90–94, November 1989.
Carts, Y. A., "IC Processing," *Laser Focus World*, pp. 105–118, May 1989.

CHAPTER 26
OPTICAL INTEGRATED CIRCUITS

Hiroshi Nishihara, Masamitsu Haruna, and Toshiaki Suhara

26.1 FEATURES OF OPTICAL INTEGRATED CIRCUTIS

An *optical integrated circuit* (OIC) is a thin-film-type optical circuit designed to perform a function by integrating a laser diode light source, functional components such as switches/modulators, interconnecting waveguides, and photodiode detectors, all on a single substrate. Through integration, a more compact, stable, and functional optical system can be produced. The key components are slab [two-dimensional (2-D)] or channel [three dimensional (3-D)] waveguides. Therefore, the important point is how to design and fabricate good waveguides using the right materials and processes. Some theories and technologies have been investigated by many researchers, and published in several technical books.[1-4]

The features of OICs are[4]

1. Single-mode structure: waveguide widths are on the order of micrometers and are such that a single-mode optical wave propagates.
2. Stable alignment by integration: the device can withstand vibration and temperature change; that is the greatest advantage of OICs.
3. Easy control of the guided wave.
4. Low operating voltage and short interaction length.
5. Faster operation due to shorter electrodes and less capacitance.
6. Larger optical power density.
7. Compactness and light weight.

26.2 WAVEGUIDE THEORY, DESIGN, AND FABRICATION

26.2.1 2-D Waveguides

The basic structure of a 2-D (or slab) waveguide is shown in Fig. 26.1 with the index profiles along the depth, where the indices of the cladding layer, guiding

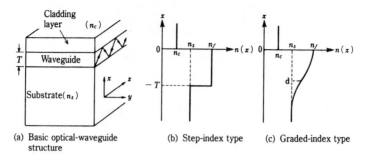

(a) Basic optical-waveguide structure   (b) Step-index type   (c) Graded-index type

FIGURE 26.1 2-D optical waveguides. (*a*) The basic optical-waveguide structure; (*b*) the step-index type; and (*c*) the graded-index type. *(From Ref. 4.)*

layer, and substrate are n_c, n_f, and n_s, respectively. In the case that $n_f > n_s > n_c$, the light is confined in the guiding layer by the total internal reflections at two interfaces and propagates along a zigzag path, as shown in Fig. 26.1*a*. Such a confined lightwave is called a *guided mode* whose propagation constant β along the z direction exists in the range of $k_0 n_s < β < k_0 n_f$, where $k_0 = 2π/λ$. Usually, the guided mode is characterized by the effective index N, where $β = k_0 N$ and $n_s < N < n_f$. N must have discrete values in this range because only zigzag rays with certain incident angles can propagate as the guided modes along the guiding layer.

The dispersion characteristics of the guided modes in the 2-D waveguide with a step-index distribution are straightforward, being derived from Maxwell's equations (See Fig. 26.1*b*). The 2-D wave analysis indicates that pure TE and TM modes can propagate in the waveguide. The TE mode consists of field components, E_y, H_x, and H_z, while the TM mode has E_x, H_y, and E_z. A unified treatment of the TE modes is made possible by introducing the normalized frequency V and the normalized guide index b_E, defined as

$$V = k_0 T \sqrt{n_f^2 - n_s^2}$$

$$b_E = \frac{N^2 - n_s^2}{n_f^2 - n_s^2} \tag{26.1}$$

The asymmetric measure of the waveguide is also defined as

$$a_E = \frac{(n_s^2 - n_c^2)}{(n_f^2 - n_s^2)} \tag{26.2}$$

When $n_s = n_c$, $a_E = 0$. This implies symmetric waveguides. However, the 2-D waveguides are generally asymmetric ($n_s \neq n_c$). By using the above definitions, the dispersion equation of the TE$_m$ modes can be expressed in the normalized form

$$V\sqrt{1 - b_E} = (m + 1)π - \tan^{-1}\sqrt{\frac{1 - b_E}{b_E}} - \tan^{-1}\sqrt{\frac{1 - b_E}{b_E + a_E}} \tag{26.3}$$

The normalized dispersion curve is shown in Fig. 26.2, where $m = 0, 1, 2, \ldots$, which is the mode number corresponding to the number of nodes of the electric field distribution $E_y(x)$. When the waveguide parameters, such as the material

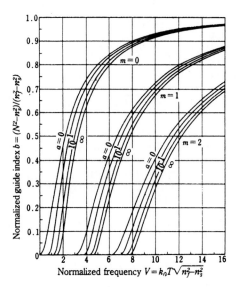

FIGURE 26.2 Dispersion curves of step-index 2-D waveguides.

indices and the guide thickness, are given, the effective index N of the TE mode is obtained graphically. The waveguide parameters are usually defined on the basis of cutoff of the guided mode, in which $N = n_s$ ($b_E = 0$). From Eq. (26.3), the value of V_m at the cutoff is given by

$$V_m = V_0 + m\pi \qquad V_0 = \tan^{-1} \sqrt{a_E} \qquad (26.4)$$

V_0 is the cutoff value of the fundamental mode. If V ranges over $V_m < V < V_{m+1}$, the number of TE modes supported in the waveguide is $m + 1$. In symmetric waveguides ($n_s = n_c$, $a_E = 0$), the fundamental mode is not cut off. On the other hand, the dispersion equation of the TM mode is rather complex. In an actual waveguide, however, the index difference between the guiding layer and the substrate is small enough that the condition ($n_f - n_s$) $\ll n_s$ is satisfied. Under this condition, all dispersion curves in Fig. 26.2 are made applicable to the TM modes simply by replacing the asymmetric measure a_E with a_M, defined as

$$a_M = \left(\frac{n_f}{n_c}\right)^4 \frac{n_s^2 - n_c^2}{n_f^2 - n_s^2} \qquad (26.5)$$

Low-loss optical waveguides are usually fabricated by metal diffusion and ion-exchange techniques that provide a graded-index profile along the depth, as shown in Fig. 26.1c. Two analytical methods, the ray approximation[5] and Wentzel-Kramers-Brillouin (WKB) methods, are often used to obtain the mode dispersion of such graded-index slab waveguides. The index distribution is generally given by

$$n(x) = n_s + \Delta n f\left(\frac{x}{d}\right) \qquad \Delta n = n_f - n_s \qquad (26.6)$$

where n_f is the maximum index of the waveguide and d is the diffusion depth. The distribution function $f(x/d)$ is assumed to be a function that decreases monotonically with x, and $f(x/d)$ takes on values between 0 and 1. Using the normalized diffusion depth, defined as

$$V_d = k_0 d\sqrt{n_f^2 - n_s^2} \tag{26.7}$$

the dispersion equation is expressed in the normalized form

$$2V_d \int_0^{\zeta_t} \sqrt{f(\zeta) - b}\, d\zeta = \left(2m + \frac{3}{2}\right)\pi \tag{26.8}$$

where $\zeta = x/d$, $\zeta_t = x_t/d$, and $b = f(\zeta_t)$. b has already been defined as Eq. (26.1). x_t denotes the turning point, and is regarded as the effective waveguide depth. Equation (26.8) is also usable as long as the condition $(n_f - N) \ll (N - n_c)$ is satisfied. The mode dispersion is calculated from Eq. (26.8) if the index distribution $f(\zeta)$ is specified. Titanium-diffused LiNbO$_3$ waveguides, for example, have the Gaussian index distribution, that is $f(\zeta) = \exp(-\zeta^2)$. The $V_d - b$ diagram for the Gaussian index distribution is shown in Fig. 26.3 where n_s is n_0 or n_e for the ordinary or extraordinary wave used as the guided mode. In addition, the values V_d for the guided-mode cutoff are found by putting $b = 0$ and $x_t \to \infty$, resulting in

$$V_{dm} = \sqrt{2\pi}\left(m + \frac{3}{4}\right) \tag{26.9}$$

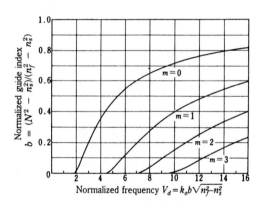

FIGURE 26.3 Dispersion curves of graded-index 2-D waveguides with a Gaussian index profile.

26.2.2 3-D Waveguides

Optical waveguide devices having functions of light modulation/switching require 3-D (or channel) waveguides in which the light is transversely confined in the y direction in addition to confinement along the depth. In 3-D waveguides, a guided mode is effectively controlled without light spreading due to diffraction on the guide surface. The 3-D waveguides are divided into four different types, as shown in Fig. 26.4. Among them, the buried type of 3-D waveguides, including Ti-diffused LiNbO$_3$ and ion-exchanged waveguides, are more suitable for optical waveguide devices. The reasons why this type of waveguide has advantages are that the propagation loss is usually lower than 1 dB/cm even for visible light and that planar electrodes are easily placed on the guide surface to achieve light modulation/switching. On the contrary, ridge waveguides are formed by removing undesired

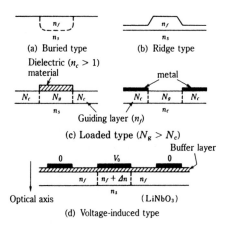

FIGURE 26.4 Basic structures of 3-D optical waveguides.

higher-index film with dry etching and lift-off of deposited film. These waveguides tend to suffer a significant scattering loss due to waveguide wall roughness. This shortcoming, however, is overcome by deposition of rather thick lower-index material as a cladding layer on the waveguides.

In the 3-D waveguides consisting of dielectric materials, pure TE and TM modes are not supported, and two families of hybrid modes exist. The hybrid modes are classified according to whether the main electric field component lies in the x or y direction (see Fig. 26.5). The mode having the main electric field E_x is called the E^x_{pq} mode. This mode resembles the TM mode in a slab waveguide; hence the E^x_{pq} mode is sometimes called the TM-like mode. The subscripts p and q denote the number of nodes of the electric field E_x in the x and y directions, respectively. Similarly, the E^y_{pq} mode (that is the TE-like mode) has the main electric field E_y. To obtain the mode dispersion of 3-D waveguides, two approximate analyses are often used: (1) Marcatili's method[6] and (2), the effective index method.[2] Both are available if the guided mode is far from the cutoff and the aspect ratio W/T is larger than unity. In the analytical model for the effective index method, as shown in Fig. 26.5, a buried 3-D waveguide is divided into two 2-D waveguides, I and II. Consider here the E^x_{pq} mode having main field components E_x and H_y in a 3-D waveguide with step-index distribution. In a 2-D waveguide I, the dispersion equation (26.3) yields the effective index N_I of the TM mode. In the symmetric 2-D waveguide II, the guided mode of interest is regarded as the TE mode which sees the effective index N_I as the index of the guiding layer because it is mainly polarized along the x direction. The dispersion equation of the TE mode in the symmetric 2-D waveguide is easily derived by putting $a_E = 0$ in Eq. (26.3), resulting in

$$V_{\mathrm{II}}\sqrt{1 - b_{\mathrm{II}}} = (q + 1)\pi - 2 \tan^{-1}\sqrt{\frac{1 - b_{\mathrm{II}}}{b_{\mathrm{II}}}} \qquad (26.10)$$

where

$$V_{\mathrm{II}} = k_o W\sqrt{N_I^2 - n_s^2} \quad \text{and} \quad b_{\mathrm{II}} = \frac{N^2 - n_s^2}{N_I^2 - n_s^2} \qquad (26.11)$$

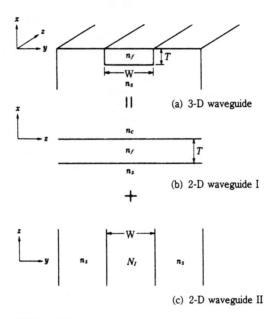

FIGURE 26.5   Analytical model for the effective index
method.

The effective index method discussed here is also adopted even for graded-index
3-D waveguides if the dispersion equation (26.8) is used. This method has thus an
advantage over Marcatili's method in that the mode dispersion is easily obtained
by a short calculation. If the field distributions are required as well as the mode
dispersion, however, Marcatili's method must be chosen.

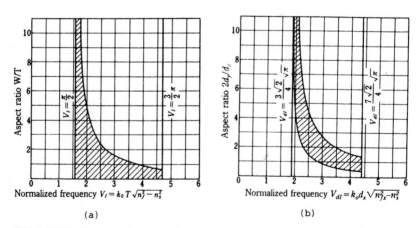

FIGURE 26.6   Single-mode propagation range (indicated by the hatched area) in (a) step-
index 3-D waveguides and (b) graded-index 3-D waveguides with Gaussian index profiles in
both x and y directions.

Design of Single-Mode 3-D Waveguides.   Usually, most waveguide devices consist of single-mode 3-D waveguides to attain highly efficient control of the guided mode. It, therefore, is important to provide design consideration of single-mode waveguides. Once all waveguide parameters are specified, it is found on the basis of the effective index method described above that, in a buried, step-index 3-D waveguide with air cladding, single-mode propagation is restricted in the hatched area of Fig. 26.6*a*, showing the relation of the aspect ratio W/T and the normalized frequency V_I in the 2-D waveguide I. If the 3-D waveguide has Gaussian index profiles in both x and y directions, the diagram for single-mode propagation range is shifted as shown in Fig. 26.6*b*. In practical use, waveguide devices require single-mode waveguides in which light is as strongly confined as possible to minimize scattering loss due to bending and branching. To meet this requirement, the aspect ratio and the normalized frequency should be close to the upper boundary of the shaded areas of Fig. 26.6*a* and *b*.

26.2.3 Waveguide Materials and Fabrication

Higher-index guiding layers are formed on substrates by deposition, thermal indiffusion, ion exchange, epitaxial growth, and so on. The relatively popular materials and the relevant fabrication techniques for 2-D waveguides are summarized in Table 26.1. Furthermore, microfabrication techniques, including photolithography, dry or chemical etching, and lift-off techniques, are required for fabrication of 3-D waveguides, as shown in Fig. 26.4. Our attention is focused on two representative waveguide materials, $LiNbO_3$ and glass, and their fabrication process will be described as well as the waveguide characteristics.

TABLE 26.1   Optical Waveguide Materials and Fabrication Techniques

	Waveguide materials							
Fabrication techniques	Polymer	Glass	Chalcogenide	$LiNbO_3$, $LiTaO_3$	ZnO	Nb_2O_5, Ta_2O_5	Si_3N_4	YIG
Deposition:								
Spin-coating	○							
Vacuum evaporation			◎					
RF or dc sputtering		◎	◎	○		◎	○	
CVD		○				○		○
Polymerization	○							
Thermal diffusion				◎				
Ion exchange		◎		◎				
Ion implantation		○						
Epitaxial growth:								
LPE				○				○
VPE					○			

◎ = Often-used fabrication techniques
CVD = chemical vapor deposition
LPE = liquid-phase epitaxy
VPE = vapor-phase epitaxy

LiNbO₃ Waveguides. Low-loss 3-D waveguides can be formed near the surface of LiNbO₃ by the lift-off of Ti stripes, followed by thermal indiffusion, as illustrated in Fig. 26.7. In most cases, Z-cut LiNbO₃ is used as the substrate, and Ti is indiffused into the $-Z$ surface to prevent domain inversion. Both the thickness and width of Ti stripes depend on the wavelength of interest. Ti stripes, for instance, are 4 μm wide and 400 Å thick at the 0.8-μm wavelength; in this case, the thermal in-diffusion is performed in flowing oxygen gas or synthetic air at 1025°C for nearly 6 h.

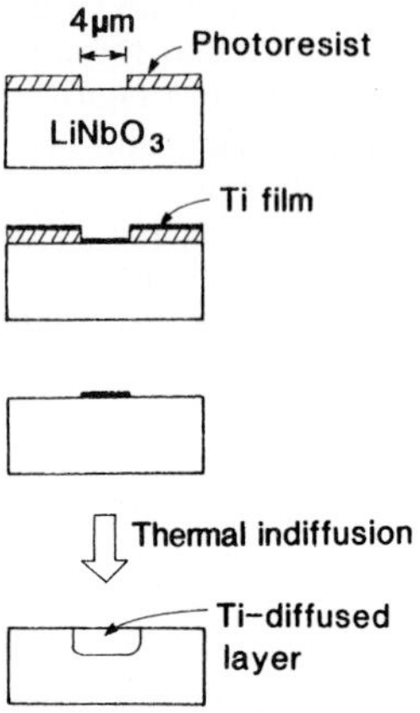

FIGURE 26.7 Fabrication procedure for Ti-diffused LiNbO₃ waveguides.

The use of moistened flowing gas is effective to suppress outdiffusion of Li₂O, which leads to weak light confinement in Ti-diffused channel waveguides. Surface roughness of the diffused waveguides should be remarkably less if the LiNbO₃ is loosely closed within a platinum foil or crucible. The resulting Ti-diffused waveguides provide single-mode propagation for both TE- and TM-like modes with propagation loss of 0.5 dB/cm or less.

Single-mode waveguides for the use of the 1.3- or 1.5-μm wavelength are also fabricated under the conditions that Ti stripes are 6 to 8 μm wide and more than 700 Å thick, and the diffusion time is above 8 h. In such Ti-diffused waveguides, the input power level should be limited to a few tens of microwatts by the optical damage threshold of the waveguide itself, especially for visible light and the 0.8-μm wavelength. To avoid this problem, MgO-doped LiNbO₃ is used as the substrate, resulting in a hundredfold increase in the damage threshold. Another way is to use Z-propagating LiNbO₃, where both TE and TM modes are ordinary waves that are much less influenced by optical damage. Besides Ti indiffusion, the other important fabrication technique of LiNbO₃ waveguides is proton exchange, which provides an extremely high index increment ($\Delta n_e = 0.13$) only for the extraordinary wave; on the contrary, the index change (Δn_o) for the ordinary wave is nearly -0.04. It is noted, however, that the electro-optic and acousto-optic effects of LiNbO₃ itself are drastically reduced by the proton exchange, and therefore the proton-exchanged waveguides have a lower susceptibility to optical damage by one-tenth or less compared to Ti-diffused waveguides. The proton exchange is usually performed by immersing the LiNbO₃ in molten benzoic acid (C₆H₅COOH) or pyrophosphoric acid (H₄P₂O₇). The waveguide depth is determined by the exchange time and temperature. The proton-exchanged waveguides exhibit significant scattering loss due to a large amount of H⁺ ions localized very close to the crystal surface. Therefore, an annealing is necessary after the proton exchange to obtain low-loss waveguides. The electro-optic effect is also recovered by the annealing. Typical fabrication conditions for proton-exchanged/annealed single-mode waveguides are as follows: a shallow high-index layer is formed on a X-cut LiNbO₃ surface by exchanging in pure benzoic acid at 200°C for 10 min through a 3.5-μm window of a Ta mask, followed by annealing the LiNbO₃ at 350°C for 2 h. The resulting waveguide exhibits the propagation loss of 0.15 dB/cm at the 0.8-μm wavelength.[7]

Glass Waveguides. The most popular glass waveguide fabrication technique is ion exchange in which, for instance, soda-lime glass is immersed in molten salt ($AgNO_3$, KNO_3, or $TlNO_3$) to exchange Na^+ ions with univalent ions such as Ag^+, K^+, or Tl^+. The index change Δn is greatly dependent on the electronic polarizability of metal ions; typically, $\Delta n > 0.1$ for Tl^+ ions, $\Delta n = 2$ to 8×10^{-2} for Ag ions, and $\Delta n = 8$ to 20×10^{-3} for K^+ ions. Three-dimensional waveguides are easily fabricated by waveguide patterning of a suitable metal mask deposited on the glass substrate before the ion exchange. The Tl^+ and Ag^+ ion exchanges provide multimode waveguides because the index change is quite large. On the other hand, the K^+ ion exchange is suitable for fabricating single-mode waveguides. A microscope slide, for example, is immersed in molten KNO_3 at 370°C to be selectively exchanged through aluminum-film windows. The K^+ ion exchange takes nearly 1 h to form 4-μm-wide single-mode waveguides. The resulting waveguide has a propagation loss of less than 1 dB/cm, even for visible light. The ion exchange is sometimes performed under application of an electric field E; in this case, the exchanged ion density becomes nearly constant within the depth $E\mu t$ where t is the exchange time and μ is the ion mobility, which depends on temperature. The electric-field-assisted ion exchange thus provides a rigid step-index waveguide. Sputtering is another popular technique for depositing waveguide films on a glass substrate such as Corning 7059 and Pyrex glass. Silicon is also used as a substrate instead of glass. In this case, thermal oxidation of Si is necessary before deposition of a waveguide film to form a SiO_2 buffer layer nearly 2 μm thick. Recently, a research group at NTT developed a promising fabrication technique for low-loss 3-D silica waveguides using the SiO_2/Si substrate. Their procedure is shown in Fig. 26.8. The propagation loss is as low as 0.1 dB/cm at 1.3 μm.[8]

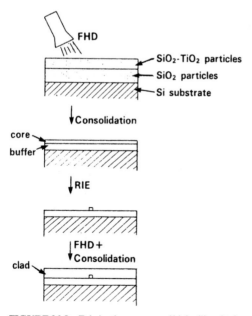

FIGURE 26.8   Fabrication process of high-silica single-mode waveguides using flame hydrolysis deposition (FHD).

26.3 GRATING COMPONENTS FOR OPTICAL INTEGRATED CIRCUITS

Periodic structures or gratings in waveguide are one of the most important elements for OICs, since they can perform various passive functions and provide effective means of guided-wave control.[9,10]

26.3.1 Coupling of Optical Waves by Gratings

Classification of Gratings. Figure 26.9 illustrates examples of passive grating components for OIC. They include input/output couplers, interwaveguide couplers, deflectors, guided-beam splitters, reflectors, mode converters, wavelength filters and dividers, and guided-wavefront converters such as waveguide lenses and focusing grating couplers. Periodic modulation of the refractive index can be induced

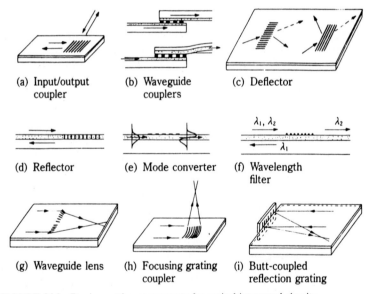

(a) Input/output   (b) Waveguide   (c) Deflector
    coupler   couplers

(d) Reflector   (e) Mode converter   (f) Wavelength
                                              filter

(g) Waveguide lens   (h) Focusing grating   (i) Butt-coupled
                           coupler   reflection grating

FIGURE 26.9 Passive grating components for optical integrated circuits.

through acousto-optic (AO) and electro-optic (EO) effects. They can be considered a controllable grating, and have many applications to functional devices. Optical coupling by a grating is classified as either guided-mode to guided-mode coupling or guided-mode to radiation-mode coupling, the former subdivided into collinear coupling and coplanar coupling. Gratings are also classified by structure into index-modulation and relief types, as shown in Fig. 26.10.

Phase Matching Condition. Various grating structures can be described by the change in distribution of relative dielectric permittivity $\Delta\epsilon$ caused by attaching a

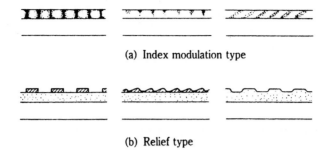

(a) Index modulation type

(b) Relief type

FIGURE 26.10   Various cross sections of gratings.

grating to a waveguide. Since the grating is periodic, $\Delta\epsilon$ can be written by Fourier expansion as

$$\Delta\epsilon(x, y, z) = \sum_q \Delta\epsilon_q(x) \exp(-jq\boldsymbol{K}\cdot\boldsymbol{r}) \qquad (26.12)$$

using a grating vector $\boldsymbol{K}$ ($|\boldsymbol{K}| = K = 2\pi/\Lambda$, Λ = period). When an optical wave with propagation vector $\boldsymbol{\beta}$ is incident in the grating region, space harmonics of propagation vectors $\boldsymbol{\beta} + q\boldsymbol{K}$ are produced. The harmonics can propagate as a guided mode, if a coupling condition

$$\boldsymbol{\beta}_b = \boldsymbol{\beta}_a + q\boldsymbol{K} \qquad q = \pm1, \pm2, \ldots \qquad (26.13)$$

is satisfied between two waves, a and b, with propagation vectors $\boldsymbol{\beta}_a$, $\boldsymbol{\beta}_b$. In many cases, $\Delta\epsilon$ is nonzero only in the vicinity of the waveguide (y-z) plane, and Eq. (26.13) need not be satisfied for the x component. Each part of Eq. (26.13) is called a *phase matching* condition, while the three-dimensional relation is called the *Bragg condition*. The relation can be depicted in a wave vector diagram, which is used to determine the waves involved in the coupling.

26.3.2 Collinear Coupling

Two guided modes propagating along the z axis couple with each other in a grating of vector $\boldsymbol{K}$ parallel to the z axis, as shown in Fig. 26.11, if a and b satisfy approximately the phase matching condition $\boldsymbol{\beta}_b = \boldsymbol{\beta}_a + q\boldsymbol{K}$. The interaction is de-

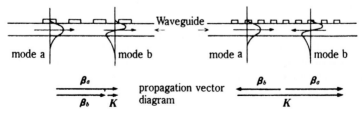

(a) Codirectional coupling   (b) Contradirectional coupling

FIGURE 26.11   Collinear coupling of guided modes by a grating.

scribed by coupled mode equations for the amplitude $A(z)$ and $B(z)$ of modes a and b:

$$\pm \frac{d}{dz} A(z) = -j\kappa^* B(z) \exp (-j2\Delta z) \qquad (\beta_a \gtrless 0) \qquad (26.14a)$$

$$\pm \frac{d}{dz} B(z) = -j\kappa A(z) \exp (+j2\Delta z) \qquad (\beta_b \gtrless 0) \qquad (26.14b)$$

where κ is the coupling coefficient and the parameter 2Δ denotes the deviation from the exact phase matching.

Codirectional Coupling.   For coupling between two different modes propagating in the same direction ($\beta_a > 0$, $\beta_b > 0$), Eq. (26.14), with boundary conditions $A(0) = 1$, $B(0) = 0$ gives a solution which indicates periodic transfer of the guided mode power. The efficiency for a grating of length L is given by

$$\eta = \left| \frac{B(L)}{A(0)} \right|^2 = \frac{\sin^2 \{ \sqrt{|\kappa|^2 + \Delta^2} L}{1 + \Delta^2/|\kappa|^2} \qquad (26.15)$$

When the phase matching is exactly satisfied ($\Delta = 0$), the efficiency is given by the $\sin^2$ function. Complete power transfer takes place when L equals the coupling length $L_c = \pi/2\kappa$.

Contradirectional Coupling.   For the coupling of modes propagating in the opposite directions ($\beta_a > 0$, $\beta_b < 0$), Eq. (26.14) with $A(0) = 1$, $B(L) = 0$ gives a solution which shows a monotonous power transfer. The efficiency is given by

$$\eta = \left| \frac{B(0)}{A(0)} \right|^2 = \left[1 + \frac{1 - \Delta^2/|\kappa|^2}{\sinh^2 (\sqrt{|\kappa|^2 - \Delta^2} L)} \right]^{-1} \qquad (26.16)$$

Complete power transfer takes place for $L \to \infty$, provided that $|\Delta| < |\kappa|$. When $\Delta = 0$, the efficiency is given by the $\tanh^2$ function; most of the power is transferred ($\eta > 0.84$) when $L > L_c = \pi/2\kappa$. A grating reflector, called a *distributed Bragg reflector* (DBR), exhibits a sharp wavelength selectivity.

Coupling Coefficient.   Coupling coefficient κ can be evaluated by integrating the multiple of index modulation profile $\Delta\epsilon$ and the profiles of modes a and b. The mathematical expressions of κ depend on the polarizations of the coupling modes. For an index-modulation grating, κ can be written as a multiple of $\kappa_b = \pi \Delta n/\lambda$. κ_b is the value for coupling in a bulk medium, and a factor describing the effect of confinement in waveguide. From mode orthogonality, κ for uniform index modulation and well-guided modes, can be written as $\kappa = \kappa_b \delta_{ab}$, which implies that coupling with mode conversion hardly takes place and substantial coupling is limited to contradirectional coupling (reflection) of the same mode. For relief gratings with groove depth much smaller than guiding layer thickness, a simple analytical expression of κ is given by approximating the mode profiles by the values at the guide surface. Coupling with mode conversion ($\text{TE}_m \leftrightarrow \text{TE}_n$, $\text{TM}_m \leftrightarrow \text{TM}_n$, $m \neq n$) may take place.

Brillouin Diagram.   The dispersion of a waveguide grating can be illustrated by the Brillouin diagram, i.e., an $\omega/c(=k)$-β diagram, as shown in Fig. 26.12. Curve

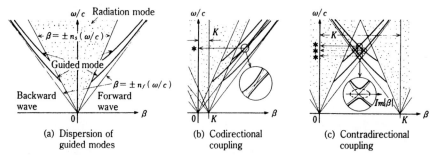

FIGURE 26.12 Brillouin diagrams for guided-wave coupling by a grating. The coupling occurs at a frequency indicated by *.

a shows the dispersion of a waveguide without grating. The curves for the qth-order space harmonics are obtained by shifting the curves in *a* by qK along the β axis. Coupling occurs in the vicinity of the intersection with the original curve where phase matching holds. Curves *b* and *c* show the diagrams for co- and contradirectional couplings, respectively. Coupling occurs only at or in the vicinity of the wavelength corresponding to $\omega/c = k = 2\pi/\lambda$ indicated by *, and, therefore, gratings can be used as wavelength filters and dividers.

26.3.3 Coplanar Coupling

In a planar waveguide (in the y-z plane), coupling is made to take place between guided waves propagating in different directions by using a grating (with length L in the z direction) of appropriate orientation. The qth-order Bragg condition for two waves of vector β_a and β_b can be written as Eq. (26.13). For the coupling, Eq. (26.13) must be satisfied exactly for the y component, but the z component need not be satisfied exactly; the allowance depends on K and L. Since the coupling exhibits different behavior for different values of K and L, a parameter Q defined by $Q = K^2 L/\beta$ is used for classification.

Raman-Nath Diffraction. When $Q \ll 1$, many diffraction orders appear, since the relatively small value of L allows coupling without exact matching for the z component. The solution of the coupled-mode equation can be written by using Bessel functions, and the diffraction efficiency for the qth order is given by $\eta_q = J_q^2(2\kappa L)$. The fundamental efficiency $\eta_{\pm 1}$ takes the maximum value 0.339 at $2\kappa L = 1.84$. The incident-angle dependence of the efficiency is small, and accordingly, gratings barely exhibit angular and wavelength selectivities.

Bragg Diffraction. When $Q \gg 1$, the coupling takes place only between waves at the Bragg condition because of the relatively large length L. As shown in Fig. 26.13, a diffracted wave of a specific order appears only when the incident angle satisfies the Bragg condition. The wave vector diagram to determine the diffraction angle is shown in Fig. 26.14, where the wave vectors of the incident and diffracted waves are denoted by ρ and σ. The phase mismatch 2Δ can be correlated with the deviation of the incident angle from the Bragg angle. When the incident angle is fixed at the Bragg angle, changing the wavelength results in a deviation from the

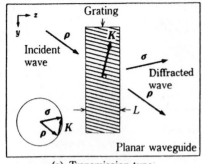
(a) Transmission type

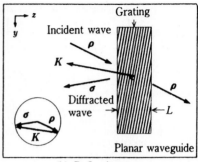
(b) Reflection type

FIGURE 26.13 Bragg diffraction of a guided wave.

Bragg condition. The phase mismatch 2Δ also can be correlated to such a wavelength change. The coupled-mode equations are written as

$$C_R \frac{d}{dz} R(z) = -j\kappa^* S(z) \exp(-j2\,\Delta z) \qquad (26.17a)$$

$$C_S \frac{d}{dz} S(z) = -j\kappa R(z) \exp(+j2\,\Delta z) \qquad (26.17b)$$

where $c_R = \cos\theta_i$, $C_S = \cos\theta_d$, and κ = coupling coefficient.

Transmission Grating. Equation (26.17) is solved with the boundary conditions $R(0) = 1$, $S(0) = 0$, and $c_R > 0$, $c_S > 0$. The diffraction efficiency η can be written as

$$\eta = \frac{\sin^2(\nu^2 + \xi^2)^{1/2}}{(1 + \xi^2/\nu^2)} \qquad \nu = \kappa L/\sqrt{c_R c_S}, \qquad \xi = \Delta L \qquad (26.18)$$

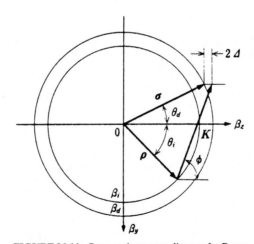

FIGURE 26.14 Propagation vector diagram for Bragg diffraction.

Under Bragg condition $\xi = 0$, the efficiency takes a maximum value of 100 percent at $\nu = \pi/2$. Efficiency decreases with deviation from the Bragg condition. Since, for $\nu = \pi/2$, $\eta/\eta_0 = 0.5$ at $\xi \approx 1.25$, the angular and wavelength selectivity can be evaluated by combining $\xi = \Delta L \approx 1.25$ with relations between 2Δ and angular/wavelength deviations.

Reflection Grating. Equation (26.14) is solved with $R(0) = 1$, $S(L) = 0$ and $c_R > 0$, $c_S < 0$. The diffraction efficiency can be written as

$$\eta = \left\{ 1 + \frac{(1 - \xi^2/\nu^2)}{\sinh^2 (\nu^2 - \xi^2)^{1/2}} \right\}^{-1} \qquad \nu = \kappa L / \sqrt{c_R |c_S|} \qquad \xi = \Delta L \qquad (26.19)$$

Under Bragg conditions, the efficiency increases monotonously with ν. The efficiency is 84.1 percent at $\nu = \pi/2$ and larger than 99.3% for ν/π. The angular and wavelength selectivities depend on ν or η_0, since the ξ value giving $\eta/\eta_0 = 0.5$ depends on ν. For $\nu = \pi/2$, for example, $\eta/\eta_0 = 0.5$ at $\xi = \Delta L \approx 2.5$.

Coupling Coefficient. Coupling coefficient $\kappa(\theta_d, \theta_i)$ for coplanar coupling can be written as $\kappa_{\text{TE-TE}} \cos \theta_{di}$, $\kappa_{\text{TM-TM}}$, and $\kappa_{\text{TM-TE}} \sin \theta_{di}$, for TE-TE, TM-TM, and TE-TM coupling, respectively, where κ is the coupling coefficient for collinear coupling and $\theta_{di} = \theta_d - \theta_i$ denotes the diffraction angle. The coefficient for TE-TE depends on θ_{di}, whereas that for TM-TM has very little dependence. When $\theta_{di} = \pi/2$, the former coupling does not occur, since the electric vectors are perpendicular to each other. A grating of $\theta_{di} = \pi/2$ serves as a TE-TM mode divider. It should also be noted that TE-TM mode conversion, which does not take place in collinear coupling, may occur when $\theta_{di} \neq 0$, although $\kappa_{\text{TE-TE}}$ is considerably smaller than $\kappa_{\text{TE-TE}}$ or $\kappa_{\text{TM-TM}}$.

26.3.4 Guide-Mode to Radiation-Mode Coupling

Output Coupling. Figure 26.15 illustrates the coupling between a guided mode and radiation modes. Coupling takes place between waves satisfying phase matching

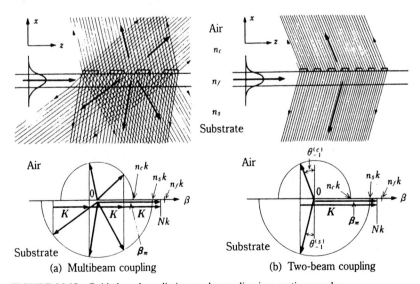

(a) Multibeam coupling   (b) Two-beam coupling

FIGURE 26.15 Guided-mode–radiation-mode coupling in a grating coupler.

for z components. When a guided wave of propagation constant β_0 is incident, the qth harmonics radiate into air and/or substrate at angles determined by

$$n_c k \sin \theta_q^{(c)} = n_s k \sin \theta_q^{(s)} = \beta_q = Nk + qK \tag{26.20}$$

The number of radiation beams is determined by the number of real values of $\theta_q^{(s)}$ and $\theta_q^{(c)}$ satisfying Eq. (26.20). An order results in radiation into either the substrate alone or both air and substrate. Figure 26.15a shows multibeam coupling where more than three beams are yielded and Fig. 26.15b shows two-beam coupling where only a single beam for the fundamental order ($q = -1$) is yielded in both air and substrate. Another possibility is one-beam coupling where a beam radiates only into the substrate. The amplitude of the guided and radiation wave decays as $g(z) = \exp(-\alpha_r z)$ due to the power leakage by radiation. Since the guided-wave attenuation corresponds to the power transferred to radiation modes, the output coupling efficiency for a grating of length L can be written as

$$\eta_{\text{out}} = P_q^i \{1 - \exp(-2\alpha_r L)\} \tag{26.21}$$

for the qth-order (i) radiation, where i(equal to c or s) distinguishes air and substrate. Here α_r denotes the radiation decay factor and P_q^i is the fractional power to q-i radiation.

Input Coupling. A guided wave can be excited through reverse input coupling of an external beam incident on a grating. When the incident angle coincides with one of the angles satisfying Eq. (26.20), one of the produced space harmonics synchronizes with a guided mode and the guided mode is excited. Figure 26.16 correlates output and input couplings. A reciprocity theorem analysis shows that the input coupling efficiency can be written as

$$\eta_{\text{in}} = P_q^i \cdot I(g, h) \qquad I(g, h) = \frac{[\int gh\, dz]^2}{\int g^2\, dz \int h^2\, dz} \tag{26.22}$$

where $h(z)$ is the input beam profile. The overlap integral $I(g, h)$ takes the maximum value 1 when the beam profiles are similar $[h(z) \approx g(z)]$. Practically, high efficiency can be achieved by (1) making a grating of $\alpha_r L \gg 1$, (2) making $P_q^i \approx 1$ for one beam q, i, and (3) feeding an input beam satisfying $h(z) \approx g(z)$. For an input beam with Gaussian profile, the maximum value of $I(g, h)$ is 0.801.

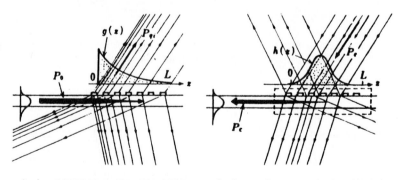

(a) OUTPUT COUPLING   **(B) INPUT COUPLING**

FIGURE 26.16 Input and output coupling by a grating.

Radiation Decay Factor. The radiation decay factor α_r can be calculated by various methods, e.g., a coupled-mode analysis, a rigorous numerical analysis to calculate the complex propagation constant of normal modes by space harmonics expansion based on Floquet's theorem, and approximate perturbation analyses based on a Green's function approach or a transmission-line approach. Figure 26.17 illustrates typical dependence of the decay factor α_r of couplers of the relief type on the grating groove depth h. For small h, α_r increases monotonously with h and is approximately proportional to h^2. For larger h, the coupling saturates because of the limited penetration of the guided-mode evanescent tail into the grating layer. In the saturation region, interference of the reflection at upper and lower interfaces of the grating gives rise to a weak periodic fluctuation.

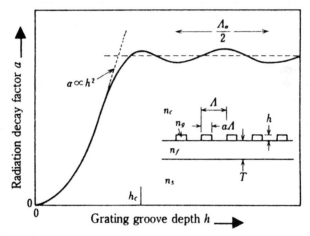

FIGURE 26.17 Dependence of radiation decay factor on the grating groove depth for grating coupler of the relief type.

High-Efficiency Grating Couplers. One-beam coupling is desirable to achieve high efficiency. Such coupling can be realized by using backward coupling by a grating of short period. Two-beam couplers, as shown in Fig. 26.15b, are more widely used, but they have the drawback that the power is halved for air and substrate. The drawback can be eliminated by inserting a reflection layer on the substrate side. Methods for confining the power into single q and i include use of the Bragg effect in a thick index-modulation grating, and use of the blazing effect in a relief grating having an asymmetrical triangular cross section.

26.3.5 Fabrication of Gratings

Grating Patterning

Two-Beam Interference. The most effective optical method for obtaining fine periodic patterns is holographic interference lithography, which utilizes the interference fringe resulting from interference of two coherent optical waves. The fringe is recorded in a photoresist layer. Grating patterns of the desired period are obtained by appropriate choice of recording wavelength and incidence angles. The recording optics are arranged on a vibrationfree optical bench. An Ar laser or He-

Cd laser is used for the light source. The laser output is divided into two beams by a beam splitter, and the beam angles are adjusted for the required grating period. To maximize fringe visibility, the two beams must have equal intensities and path lengths. A spatial filter (pinhole) is used in the beam expander lens to remove spatial noise and obtain a uniform pattern. Inclined periodic structures, e.g., Bragg-effect index-modulation gratings, can be formed in a thick recording layer. Recording of inclined fringes in a thin resist and development result in a sawtooth cross section useful for blazed grating fabrication. The minimum feasible grating period is half the recording wavelength. For shorter periods, a prism or liquid immersion is used. Fabrication of chirped/curved gratings is also possible by using an appropriate combination of spherical and cylindrical lenses, but the flexibility is limited. The advantages of the two-beam interference are fabrication of small period gratings with simple apparatus, good period uniformity, and easy fabrication of large-area gratings.

Electron-Beam Writing. Since many gratings for OIC require very small periods, but have rather small areas, computer-controlled electron-beam (EB) writing can be used, it is more convenient and effective to use a system with a specialized scanning controller, in which digital control and analog signal processing are incorporated to enable writing of very smooth straight and curved lines. The EB writing area with submicrometer resolution is typically 3×3 mm^2. Grating patterns can be written by (1) a painting-out method which writes a half period by many scanning lines, (2) a line-drawing method which writes one period by a scanning line, and (3) a gradient-dose method which involves continuous changes in EB dose. Methods 1 and 2 are suitable for gratings with large and small periods, respectively. Method 3 allows fabrication of blazed gratings (gradient-thickness cross sections after development) and gradient-index gratings. Resolution of EB writing is limited by EB diameter and EB scattering in the resist. If the substrate is an insulator, a very thin (about 100-Å) conductive (Au, Al) layer must be deposited to avoid the charging-up problem. The EB writing technique has features almost complementary to those of the interference technique. The advantages are extremely high resolution, large flexibility in fabrication of modulated gratings, and easy parameter change by computer control.

Grating Processing. If a resist is used as a grating material, the patterning is the final process to obtain a relief grating. If gratings are fabricated in a waveguide material whose refractive index can be changed by light or EB irradiation, index-modulation gratings are obtained by the patterning. Usually the resist pattern is transferred to waveguide, cladding, or hard-mask layers. Gratings of the relief type are fabricated by etching the waveguide surface of a cladding layer, using the resist pattern as a mask. Although gratings can be produced by chemical etching, better results are obtained with dry etching, e.g., sputter etching, plasma etching, reactive ion etching, and (reactive) ion-beam etching. Another method to obtain relief gratings is deposition and lift-off patterning of a thin cladding layer on the waveguide. Techniques to obtain index-modulation gratings include ion (proton) exchange using a hard mask and in-diffusion of a patterned metal layer.

26.4 PASSIVE WAVEGUIDE DEVICES

OIC elements which exhibit static characteristics, i.e., those without optical-wave control by an external signal, are called passive devices. Although direct modification into two-dimensional versions from classic bulk components can be used in OIC, there are many cases where such implementation is difficult or results in poor performance. Implementation of waveguide components may require different structures and working principles, but novel functions and improved performance can possibly be obtained by effective use of waveguides.

26.4.1 Optical Path-Bending Components

Implementation of OIC by integration of several components often requires changing optical-path direction or translating paths.

Elements for Planar Waveguide.   A prism can be implemented by loading a thin film on a triangular region of a waveguide. The wavefront is refracted according to Snell's law. The deflection angle, however, cannot be large, since the available mode-index difference is small. Large changes of path are realized with geodesic components, in which the ray travels along the geodesic on a concave part produced by deformation of the waveguide plane. Another element is a waveguide end-face mirror, prepared by polishing at a right angle with respect to the guide plane. The path can be bent by the total internal reflection (TIR) at the end face. For deflection larger than the critical angle, the end face should be coated with a reflective (metal) film. Similar TIR can be accomplished by a tapered termination of a guiding layer. Mirrors and beam splitters can be obtained by making a ridge in the waveguide, which produces a quasi-abrupt change of the mode index. Reflection- and transmission-type Bragg grating components can also be used for path bending.

Bent Waveguides.   Path bending for connecting channel-guide components can be accomplished simply by bending the channel. Although the simplest method is to use corner-bent waveguides, the guided wave suffers a large scattering loss. The loss can be reduced by using a carefully designed multisection corner-bent waveguide. Another method often adopted for connecting two parallel channels with an offset is to use smoothly curved (S-shaped) waveguides.

26.4.2 Power Dividers

Power dividers are an important component to divide an optical signal into many branches in optical-fiber subscriber networks.

Single-Mode Power Dividers.   Figure 26.18a and b shows two-branch waveguides. The waveguide should have a small branching angle and a tapered part for maintaining the fundamental-mode propagation of the incident wave. Although a multibranch waveguide (Fig. 26.18c) can perform multidividing, control of the branching ratio is easier in a tandem two-branch structure (Fig. 26.18d). Directional couplers, shown in Fig. 26.19, are used as dividers of low insertion loss. The couplers are wavelength-sensitive because their operation is based on phase matching. Greater bandwidth and larger fabrication-error tolerances can be obtained in couplers modified to have variable spacing.

Multimode Power Dividers.   Multimode branching waveguides suffer from the problem that the dividing ratio is influenced by guided-mode excitation conditions, and, therefore, a mode-mixing region is required to stabilize the ratio. Figure 26.20 illustrates $N \times N$ power dividers (star couplers) using ion-exchanged glass waveguides.[11,12]

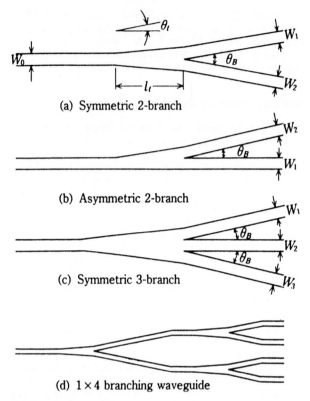

(a) Symmetric 2-branch

(b) Asymmetric 2-branch

(c) Symmetric 3-branch

(d) 1×4 branching waveguide

FIGURE 26.18   Basic structures of single-mode branching waveguides.

26.4.3 Polarizers and Mode Splitters

Since optical waves transmitted by a fiber are usually elliptically polarized, and many waveguide devices are polarization-dependent, polarizers/mode splitters are required for filtering single-polarization/mode waves to avoid performance degradation.

Polarizers.   A metal-clad waveguide is used as a polarizer, since TE and TM modes are transmitted and absorbed, respectively, in the cladded part. The most suitable cladding material is Al, which has a large value for the imaginary part of the dielectric constant. A typical extinction ratio for a 5-mm-long cladding is 30 dB. Polarizers of smaller insertion loss are obtained by using an anisotropic crystal (calcite, etc.) for cladding as shown in Fig. 26.21.[13] In such structures, one of the TE and TM waves is transmitted and the other leaks into the crystal. A high extinction ratio is feasible, although high-grade crystal polishing and complete contact with the guide surface is required.

Mode Splitters.   Mode splitters for separating two modes (order or polarization) can be realized using directional couplers consisting of two different waveguides

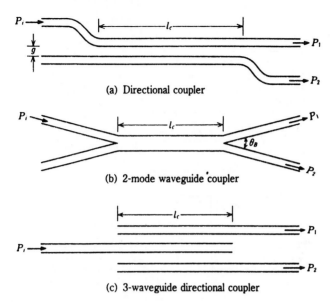

(a) Directional coupler

(b) 2-mode waveguide coupler

(c) 3-waveguide directional coupler

FIGURE 26.19 Waveguide directional couplers.

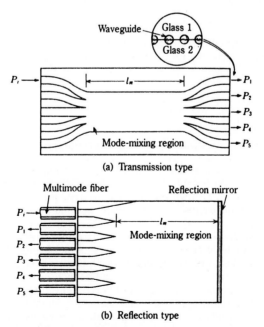

(a) Transmission type

(b) Reflection type

FIGURE 26.20 Multimode star couplers using ion-exchanged glass waveguides.[11,12]

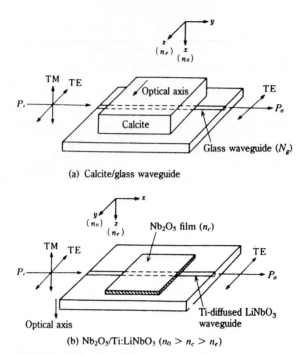

(a) Calcite/glass waveguide

(b) Nb_2O_5/Ti:$LiNbO_3$ ($n_0 > n_c > n_e$)

FIGURE 26.21   Waveguide polarizers using anisotropic crystals.[13]

(e.g., a multilayer structure). Since the two guides have different mode dispersions, the coupling takes place only for a specific mode of the incident wave. The desired mode is transferred to another guide, whereas other modes are transmitted through the input guide. A TE/TM mode splitter also can be realized by using a branching waveguide structure. A branching waveguide of small branching angle acts as a mode splitter, in which lower and higher modes are divided into different branches.

26.4.4   Wavelength Multiplexers and Demultiplexers

Wavelength multiplexers and demultiplexers are important devices for constructing transmitter and receiver terminals for a wavelength-division-multiplexing (WDM) communication system. Diffraction gratings are used for dispersion elements because of their large dispersion, high efficiency, and integration compatibility.

Single-Mode Waveguide Type.   Collinear and coplanar Bragg diffraction gratings are used. An example of the collinear device is a band-stop filter based on contradirectional coupling, with which very narrow bandwidth can be obtained. Coplanar Bragg gratings, both transmission and reflection types, are suitable for multiplexers. These gratings diffract guided waves only when the Bragg condition is satisfied with the wavelength and incidence angle. Demultiplexers can be constructed by (1) a cascade array of gratings with different periods/orientations or (2) a chirped grating with periodic gradient; type (1) has the advantages of flexibility in wavelength layout and high efficiency, whereas type (2) can demultiplex many (even continuous) wavelengths with a single component. Use of a Si substrate allows monolithic integration of photodetectors with a demultiplexer. Figure 26.22 shows demultiplexers consisting of a grating array and Schottky diodes,[14] and Figure 26.23 illustrates a WDM receiver terminal in which a chirped grating, collimating and focusing lenses, and *pin* photodiodes are integrated.[15]

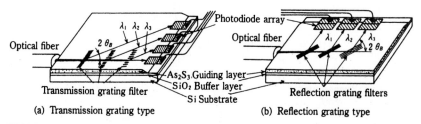

(a) Transmission grating type   (b) Reflection grating type

FIGURE 26.22   Wavelength demultiplexers with integrated micrograting array and Schottky photodiode array.[14]

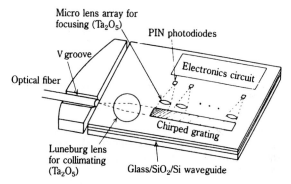

FIGURE 26.23   Wavelength demultiplexer with integrated chirped grating, lenses, and photodiodes.[15]

Multimode Waveguide Type.   Bragg gratings are not suitable, since the Bragg condition can not be satisfied simultaneously for all modes. An effective device construction is to make a miniaturized two-dimensional version of a grating monochromator using a planar waveguide. Thin reflection gratings butt-coupled to the waveguide are used with various configurations to perform the lens function. A blazed grating is required to attain high efficiencies. A device of Rowland construction shown in Figure 26.24 uses a concave grating bonded on a circular-polished edge of a sandwich-glass waveguide.[16]

Echelette gratings fabricated by anisotropic etching of Si are also used. Currently demonstrated multiplexers have 5 to 10 channels, 100- to 300-Å wavelength separation, and a few decibels insertion loss lenses. Figure 26.25 shows a device using a chirped grating designed to incorporate a lens function with dispersion.[17] Since the diffraction angle and focal length are wavelength-dependent, the demultiplexer can be achieved by connecting the output channels at focal points for each channel wavelength. The chirped grating was fabricated by EB lithography and bonded to the edge of a patterned ion-exchanged glass waveguide. Littrow construction of

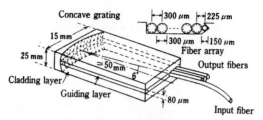

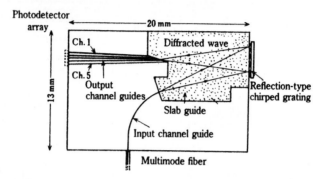

FIGURE 26.24 Wavelength demultiplexer using a concave grating and a glass-plate waveguide.[16]

FIGURE 26.25 Wavelength demultiplexer using a reflection-type chirped grating and an ion-exchanged glass waveguide.[17]

the demultiplexer is also possible by using a geodesic lens, which is a unique waveguide lens that exhibits no chromatic/mode aberrations, and which can be fabricated by thermal casting of the glass substrate.

26.4.5 Waveguide Lenses

Waveguide lenses, which perform focusing, imaging, and Fourier transformation of guided waves in a planar waveguide, are a very important component, especially for constructing IOCs for signal processing. An important lens characteristic is the focus spot size. The theoretical diffraction-limited 3-dB spot width $2w$ of a waveguide lens having mode index n_e, focal length f, and aperture D (F number $F = f/D$) is given by $2w = 0.88F\lambda/n_e$.

Mode-Index Lenses. The effective index of a waveguide, i.e., mode index, can be changed by changing the thickness of the guiding layer, cladding, impurity diffusion, etc. The lens function based on ray refraction can be obtained by making a lens-shaped area with an index increment. Whereas lenses of circular boundary exhibit large aberrations, spherical aberration can be removed by using hyperbolic or elliptic arcs for the lens boundaries. A class of aberration-free mode-index lens is the Luneburg lens. It has a rotation-symmetrical graded mode-index distribution, as shown in Fig. 26.26. The aberrations, except for field curvature, can be eliminated by designing an appropriate distribution of mode index $n(r)$, which is given as a solution of an integral equation deduced from Fermat's principle. The usual method

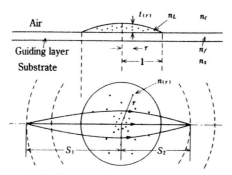

FIGURE 26.26 Luneburg lens.

to realize $n(r)$ is to deposit a high-index lens layer of gradient thickness $t(r)$ on the guiding layer by sputtering or evaporation using shadow masks.

Geodesic Lenses. When a planar waveguide is partly deformed into a curved surface, the guided ray changes direction and travels along the geodesic according to Fermat's principle. A lens function can be realized by forming an appropriate curved surface as shown in Fig. 26.27. One method to determine the shape of an aberration-free geodesic lens surface is to convert a Luneburg lens into an equivalent geodesic lens. A more practical design procedure is to express the profile by a function including parameters and perform ray tracing to determine the parameters for minimum aberration. A geodesic lens is inherently free of chromatic aberration and mode-independent. The simplest method for making the lens surface applicable to a glass substrate is thermal casting, which results in spherical depression. Fabrication of aberration-free lenses requires aspheric surface machining with submicrometer accuracy, which can be accomplished by ultrasonic machining, diamond honing, or diamond grinding and polishing.

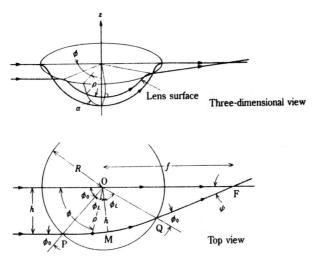

FIGURE 26.27 Geodesic lens.

Diffraction Lenses.   Lenses based on light diffraction in a periodic structure are called *diffraction lenses*. An element which imposes a phase modulation corresponding to phase difference between parallel and converging waves serves as a lens. *Fresnel lenses* give a modulation which results from modulus-2π segmentation of such modulation, by gradient distribution of thickness or index, as shown in Fig. 26.28. An efficiency of 100 percent can be obtained in thin lenses. The focusing properties are determined primarily by the zone arrangement and are not sensitive to deviation from the ideal distribution, but the efficiency is reduced by the deviations. *Grating lenses* use the coplanar guided-wave diffraction provided by a transmission grating. Since the diffraction angle depends on the grating period, lens function can be realized by making a chirped grating which has a continuous period variation as shown in Fig. 26.9*g*. The structure has the same periodicity as the zone arrangement of Fresnel lens. To obtain high efficiency, a Bragg grating of $Q > 10$ is required, and the grating lines should be gradiently inclined to satisfy the Bragg condition over the whole aperture. The condition necessary for nearly 100 percent efficiency is $\kappa L = \pi/2$. Diffraction elements with lens function include focusing grating couplers and butt-coupled gratings as shown in Fig. 26.9*h* and *f*.

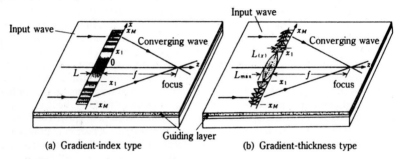

(a) Gradient-index type   (b) Gradient-thickness type

FIGURE 26.28   Fresnel lenses.

26.5 FUNCTIONAL WAVEGUIDE DEVICES

In various kinds of functional waveguide devices developed so far, guided modes are controlled via physical phenomena such as electro-optic (EO), acousto-optic (AO), magneto-optic (MO), nonlinear-optic (NO), and thermo-optic (TO) effects. In these waveguide devices, light can be more effectively controlled than in bulk-optic devices, because the interaction between light and an externally applied signal is restricted to the region surrounding the waveguide. A variety of waveguide structures are also utilized to attain a desired function (i.e., greater freedom in device design). This section describes key points of design and characteristics of the representative functional waveguide devices for each physical phenomenon used to control guided modes.

26.5.1 Electro-Optic Devices

High-speed light modulation/switching is attained in EO devices consisting of Ti-diffused single-mode waveguides in $LiNbO_3$. The most popular one is a Mach-Zehnder interferometeric modulator, as shown in Fig. 26.29*a*, where push-pull

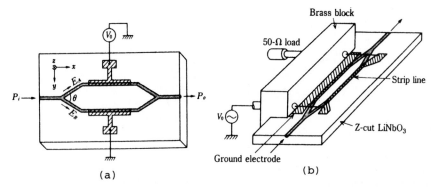

FIGURE 26.29   Waveguide interferometric modulators: (*a*) the lumped-circuit type; (*b*) the traveling-wave type.

operation is possible because Z-cut LiNbO$_3$ is used as the substrate. This modulator has the advantage that high extinction ratios are easily obtained because of its large fabrication tolerance; for instance, the extinction ratio becomes more than 15 dB if the power dividing ratio $(E_A/E_B)^2$ is below 2 at the input Y-junction waveguide. The modulation bandwidth Δf is determined by the electrode length l because the modulator of Fig. 26.29*a* is the lumped-circuit type. When $l = 5$ mm, Δf is nearly 4 GHz. On the other hand, a higher-speed modulation is attained by using the traveling-wave type of modulator, as shown in Fig. 26.29*b*, where Δf is determined by the degree of velocity matching between a modulating microwave and a guided wave. A modulation bandwidth of up to 20 GHz has already been achieved, and therefore, this type will be used as an external modulator for large-capacity optical communication.[18] The resonant type of modulator can also provide a frequency modulation of more than 30 GHz.[19] Furthermore, the Y-junction waveguides can be replaced by 3-dB couplers in the interferometric modulator. This is called a *balanced bridge modulator*, and can be used as a 2 × 2 switch with a low drive voltage.

Efficient spatial switching is possible by placing planar electrodes on directional couplers consisting of two identical single-mode waveguides close to each other, as shown in Fig. 26.30, where Z-cut LiNbO$_3$ is again used as the substrate and the TM-like mode is excited in a Ti-diffused waveguide. It is here noted that the propagation constant difference $\Delta\beta$ between two waveguides is variable via the EO effect with an applied voltage V, while the coupling coefficient κ is insensitive to V under weak coupling condition. The coupling length L for complete power transfer from one waveguide to the other is also defined as $\pi/2\kappa$ when $\Delta\beta = 0$. In the uniform-$\Delta\beta$ switch of Fig. 26.30*a*, the coupler length l must be adjusted to be an odd multiple of L so that the incident light on waveguide A is totally transferred to waveguide B at the output in the absence of the applied voltage (the crossover state). When V is tuned so that $\Delta\beta/\kappa = 2\sqrt{3}$, the power transfer of waveguide A to A is then obtained (the through state). The crossover state is thus not obtained unless the condition $l = (2m + 1)L$ is satisfied. This requires high accuracy in fabricating the directional coupler. This shortcoming is overcome by dividing planar electrodes into two equal-length sections, as shown in Fig. 26.30*b*, where phase mismatch with opposite signs is induced via the EO effect. This is called a *reversed-*$\Delta\beta$ (or *stepped-*$\Delta\beta$) switch in which both crossover and through states can be obtained by voltage tuning as long as the coupler length l ranges from L to $3L$.

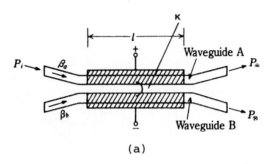

(a)

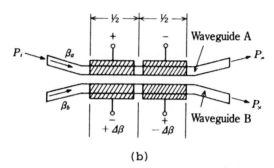

(b)

FIGURE 26.30 Directional-waveguide-coupler switches; (a) uniform-$\Delta\beta$ configuration; (b) stepped (reversed)-$\Delta\beta$ configuration.

The guided mode is deflected by an electro-optically induced periodic index change with an interdigital electrode placed on a Ti:LiNbO$_3$ slab waveguide, as shown in Fig. 26.31a, which is a Bragg deflector. The periodic index change is also used for mode conversion via the electro-optic coefficient r_{51} in X-cut, Y-propagating Ti:LiNbO$_3$, as shown in Fig. 26.31b, where the period $\Lambda = \lambda/(n_o - n_e)$. The mode converter acts as a wavelength filter with a narrow bandwidth, nearly 1 nm in near-infrared light, whose center wavelength is tunable via r_{33}.[20]

Electro-optic control of the index distribution inside a waveguide leads to compact waveguide switches such as total internal reflection, branching waveguide, and cutoff switches. In all these switches, the driving voltage is relatively high.

26.5.2 Acousto-Optic Devices

Surface acoustic waves (SAWs) excited in waveguides produce index-modulation gratings via the acousto-optic effect, in which the grating period is variable with the frequency of radio-frequency (rf) power applied to an interdigital transducer (IDT). By utilizing such unique gratings, interesting functions including optical beam scanning, tunable wavelength filtering, and spatial modulation corresponding to input rf time signals become possible. From a viewpoint of the interaction scheme of SAWs and guided modes, AO waveguide devices are classified into collinear and coplanar devices, as will be described below.

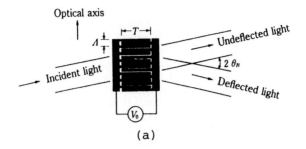

(a)

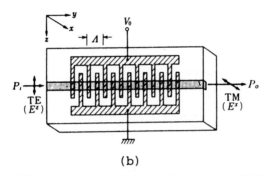

(b)

FIGURE 26.31   EO grating devices: (a) Bragg deflector; (b) TE-TM mode converter.

The collinear AO interaction is utilized for TE-TM mode conversion in an anisotropic waveguide like the Ti-diffused LiNbO₃ (see Fig. 26.32), in which the frequency f of the rf power applied to an IDT is adjusted so that $|\beta_{TE} - \beta_{TM}| = 2\pi f/v$ where v is the SAW velocity and β_{TE} and β_{TM} are propagation constants of the TE and TM modes, respectively. The conversion efficiency depends on the SAW power P_s. The response time is also determined by the SAW transit time over the interaction length L. When $L = 8$ mm, $f = 250$ MHz, and the rf power is 0.55 W at $\lambda = 1.15$ μm, the mode conversion efficiency is nearly 70 percent. This type of mode converter can operate as a tunable wavelength filter by tuning the rf frequency f in response to variation of the light wavelength. The filtering bandwidth $\Delta\lambda$ becomes narrower as the interaction length L increases; for instance, $\Delta\lambda$ is only 1 nm when L is 10 mm at $\lambda = 1$ μm.

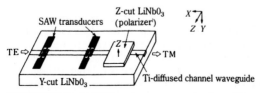

FIGURE 26.32   TE-TM mode converter using the collinear AO interaction.

There are two types of diffraction, Raman-Nath and Bragg, which are based on coplanar AO interaction. The latter type is widely used for AO waveguide devices because of its high diffraction efficiency. In AO Bragg cells, an IDT must be formed on piezoelectric materials such as Y-cut quartz, $LiNbO_3$, and ZnO film; even nonpiezoelectric films can be used as the waveguide. A typical example of AO Bragg cells is shown in Fig. 26.33; in a Ta_2O_5-film waveguide formed on Y-cut quartz, a diffraction efficiency of 93 percent was obtained when $P_s = 0.175$ W and $f = 290$ MHz at $\lambda = 0.633$ μm. Such Bragg cells are used as modulators/switches; in this case, the important characteristic is the response time, which can be as short as 10 ns. Further high-speed operation is performed in a specific deflector where an IDT is fabricated on a waveguide in order to excite a periodic strain propagating along the waveguide depth.[21]

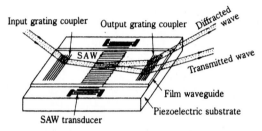

FIGURE 26.33 AO Bragg cell using a nonpiezoelectric film waveguide on a piezoelectric substrate.

The Bragg cells described above are mainly used as light deflectors whose angles are variable with the rf frequency f. For wideband deflection, the waveguide material should have a small propagation loss of SAWs at certain high frequencies as well as a large electromechanical coefficient. Titanium-diffused $LiNbO_3$ is the best material to meet these requirements. In $LiNbO_3$ Bragg cells, the wideband operation is actually accomplished by modification of the IDT configuration. Some different types of IDT configurations were reported: multiple tilted array, curved-finger, phased-array, and tilted-finger chirped transducers. Among them, the best result was obtained in a Bragg cell with a two-stage array of tilted-finger chirped transducers which exhibited light deflection as wide as 1 GHz, as shown in Fig. 26.34.[22] The so-called optimum anisotropic Bragg deflection is also important for wideband deflection.

26.5.3 Magneto-Optic Devices

An optical isolator is required to maintain stable lasing of a laser diode without influence of the reflected light returning to the light source. Such a nonreciprocal optical device is formed by utilizing the Faraday effect in $Y_3Fe_5O_{12}$ (YIG) and paramagnetic glass whose dielectric tensor is asymmetric under an application of a magnetic field. In MO thin-film waveguides, however, the Faraday rotation is not found because circularly polarized waves are not supported as guided modes. Therefore, TE-TM mode conversion is utilized, instead of Faraday rotation, to obtain the polarization-direction rotation required for optical isolation. The mode

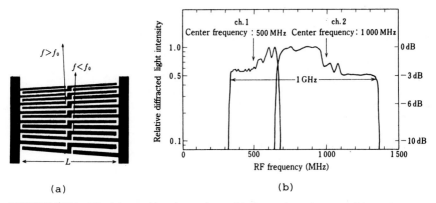

FIGURE 26.34 Tilted-finger chirped transducer: (a) the configuration, and (b) the frequency response of a wideband Bragg cell using the transducer.[22]

converter consists of YIG film epitaxially grown on a Gadolinium Gallium Garnet $(Gd_3Ga_5O_{12})$ (GGG) substrate, in which the film waveguide exhibits an optical anisotropy induced by the lattice constant mismatch. The undesirable anisotropy is canceled out for phase matching between TE and TM modes by the following methods: application of a periodic magnetic field, double epitaxial growth of YIG films with different substitution ratios (Ga/Sc), and loading of an anisotropic dielectric film like $LiIO_3$ on the YIG film. By the method of double epitaxial growth, conversion efficiencies as high as 96 percent were obtained at $\lambda = 1.5$ μm.[23]

Possible configurations of waveguide isolators reported so far are shown in Fig. 26.35. When polarization of the input light is tilted by 45°, the isolator function is attained without a polarizer by using a nonreciprocal 45° mode converter (Fig. 26.35a). On the other hand, the input light is generally polarized parallel to the waveguide surface; in this case, polarizers are placed at both the input and output of the device. The YIG-film waveguide is covered with an anisotropic dielectric crystal $(LiIO_3)$ to provide nonreciprocal and reciprocal 45° mode conversions simultaneously (Fig. 26.35b). In the configuration of Fig. 26.35c, Faraday and Cotton-Mouton effects are incorporated.[24]

26.5.4 Thermo-Optic Devices

The thermo-optic effect that originates from the temperature dependency of the refractive index can be used for modulation/switching of guided modes in times on the order of milli- to microseconds. Many kinds of transparent materials are available for TO waveguide devices if the temperature for the waveguide formation is much higher than the operating temperature. A TO interferometric modulator is shown in Fig. 26.36, where Ti-film heaters are placed on K^+ ion-exchanged waveguides in soda-lime glass.[25] A voltage is applied to one film heater, while the other film heater acts as a heat absorber. The output light is intensity-modulated with respect to the applied electric power P_0. The half-wave electric power P_π is 135 mW with a response time of 0.25 ms. Switching on the order of microseconds is also possible by applying a pulse voltage to the film heater.

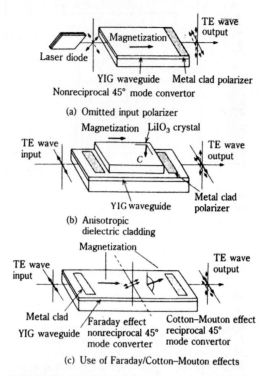

(a) Omitted input polarizer

(b) Anisotropic dielectric cladding

(c) Use of Faraday/Cotton–Mouton effects

FIGURE 26.35 Waveguide isolator configurations using nonreciprocal mode converters: (*a*) without polarizers; (*b*) anisotropic dielectric cladding; (*c*) use of Faraday/Cotton-Mouton effects.[24]

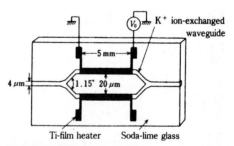

FIGURE 26.36 TO waveguide interferometric modulator/switch in glass.[25]

26.5.5 Nonlinear-Optic Devices

The nonlinear-optic effect, widely found in semiconductors, dielectric crystals, organic materials, and polymers, gives rise to numerous interesting phenomena based on light-to-light interaction. In particular, the second-order NO effect is utilized for second-harmonic generation (SHG), which enables us to realize compact coherent-light sources in the short wavelength region. The $LiNbO_3$ waveguide is a promising material for practical SHG devices. In the design of SHG waveguide devices, a key point is phase matching between the fundamental wave and the second harmonic wave. The E^y_{00} mode (ordinary wave) and the E^z_{00} mode (extraordinary wave) are chosen as the fundamental and the second harmonic waves, respectively; in this case, the phase matching of two modes is attained by temperature control of the birefringence and the waveguide dispersion. An experimental result on this type of SHG device is reported in which the conversion efficiency was 0.77 percent for the incident power of 65 mW at λ = 1.09 μm. On the other hand, phase matching is achieved automatically by using the Cherenkov radiation scheme in proton-exchanged $LiNbO_3$ (see Fig. 26.37), where the fundamental wave is the guided mode, while the second harmonic wave is the radiation mode.[26] In

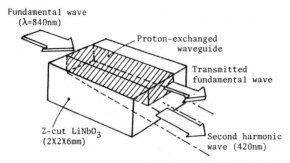

FIGURE 26.37 SHG device using the Cherenkov radiation scheme.[26]

this scheme, an efficient frequency doubling is possible because the largest nonlinear coefficient d_{33} is used; in contrast, the shortcoming is divergence of the SHG output with an angle of nearly 16°. A compact blue light source of a few tens of milliwatts was already developed by assembling a laser diode and a $LiNbO_3$ waveguide. The other interesting scheme is the so-called quasi-phase-matching SHG, in which a ferroelectric domain-inverted grating appearing on the $+Z$ surface of $LiNbO_3$ is used. Both fundamental and second-harmonic waves are guided in a channel waveguide, and therefore, exact phase matching is required as well as a strong confinement of optical fields. To meet these requirements, a SHG device with a fan-out domain-inverted grating is reported, as shown in Fig. 26.38.[27]

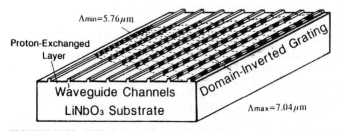

FIGURE 26.38 SHG device phase-matched with fan-out ferroelectric domain-inverted grating.[27]

In addition to the $LiNbO_3$ SHG devices described above, KTP ($KTiOPO_4$) is a promising material for SHG because the optical damage resistance is much higher than for $LiNbO_3$. The waveguide is also formed by Rb^+ ion exchange. A balanced phase matching in Rb^+ ion-exchanged KTP was reported.[28]

26.6 EXAMPLES OF OPTICAL INTEGRATED CIRCUITS

It is very difficult to integrate several passive and functional discrete devices on a single substrate, because the number of fabrication processes increases with the number of discrete components. The processes are even more complicated when different materials are used for different components. Currently, research on OICs is in progress in such fields as fiber communications, information processing, sensing, metrology, and laser arrays.

26.6.1 Optical-Fiber Communications

Directional-Coupler Type Switches.   The directional-coupler EO switch has a high extinction ratio with a relatively low drive voltage, although it requires a highly accurate fabrication technique for phase matching. A 4 × 4 optical switch on Z-cut $LiNbO_3$ is shown in Fig. 26.39,[29] where stepped directional-coupler switches are integrated. The Ti-diffused waveguide is 10 μm wide with a 4-μm gap, and the coupler length chosen is 8 mm. The total length of the OIC is 40 mm. When a 1.3-μm laser diode was used as a light source, the required voltages for the crossover and through states were 12 V and 28 V, respectively, with crosstalk of −18 dB in each directional-coupler switch. The insertion loss of the OIC was also 6.25 dB.

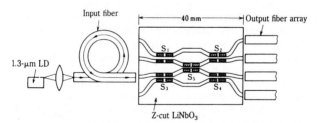

FIGURE 26.39   A 4 × 4 optical switch fabricated by integrating five stepped directional-coupler switches in Z-cut $LiNbO_3$.[29]

Other Switches.   Switch research has been extended to 8 × 8 (Ref. 30) and 16 × 16 (Ref. 31) switches on $LiNbO_3$; a 4 × 4 carrier-injection type switch[32] on InGaAsP/InP; and an 8 × 8 space-division TO switch,[33] and a 128 × 128 frequency-division multiplex TO switch,[34] both on silica/silicon.

26.6.2 Optical Information Processing

RF Spectrum Analyzers.   An integrated-optic rf spectrum analyzer (IOSA) is a representative integrated circuit for signal processing. Research and development of IOSAs has been aimed at the immediate application to radar signal processing. Future applications are anticipated in various types of signal processing such as radio astronomy and remote sensing, especially where compactness and light weight are strictly required.

An IOSA is constructed by integrating a wideband acousto-optic Bragg cell and a pair of geodesic waveguide lenses for guided-wave collimating and Fourier transforming, as shown in Fig. 26.40.[35] In the Bragg cell, the guided wave is deflected at an angle approximately proportional to the frequency of the rf signal fed into the SAW transducer, and the diffraction efficiency is approximately proportional to the rf power in the small-signal range. After the Fourier transformation of the guided wave by the second lens, the power frequency spectrum of the input rf signal is obtained on the focal plane in the form of light-intensity distribution. The spectrum signal is converted into an electric signal and read out by a photodetector array (linear image sensor).

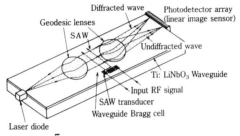

FIGURE 26.40 Integrated-optic rf spectrum analyzer using geodesic lenses.[35]

The frequency resolution of an IOSA is given by the inverse of the SAW transit time. The frequency bandwidth equals that of the Bragg cell, so that the number of resolvable points equals the time-bandwidth product. The IOSA response speed is determined by the speed of the image sensor. The dynamic range is limited by the guided-wave scattering and photodetector noise level. 1-GHz bandwidth and 4-MHz resolution have been obtained in a folded-type IOSA using reflection-type chirped grating lenses.[22]

Signal-processing devices for the convolution and correlation[36] of rf signals can be implemented by modifying the IOSA configuration, and have been investigated.

26.6.3 Optical Sensing and Metrology

Integrated-optic sensors may be divided into the following two types: (1) the waveguide sensors in which the waveguide itself is used as a sensor for temperature, humidity, gas, position, displacement, and so on, and (2) integrated devices of optical components required for sensor-signal processing. Such integrated-optic sensors are more compact and rugged than fiber-optic sensors assembled with micro-optic bulk components.

Fiber Gyroscopes. In a fiber-optic gyroscope, the laser light is coupled to two fiber ends of a multiturn single-mode fiber coil with a typical diameter of 10 cm.

The phase difference between two waves propagating clockwise and counterclockwise along the fiber coil is then measured by the interference fringe of two output lights from the fiber coil. Rotation of the fiber coil produces the Sagnac effect, by which the angular velocity is measured with high accuracy (for instance, 10^{-3}/h). One of the problems in practical fiber gyroscopes is that, because the optical system is constructed by combining bulk optical components, including beam splitters and phase shifters on an optical bench, the system is too bulky and suffers from vibration. A good deal of effort has been directed toward integration of this optical system on a $LiNbO_3$ substrate to create a compact and vibration-free rotation sensor. An example of the OICs for the fiber gyroscope is shown in Fig. 26.41.[37]

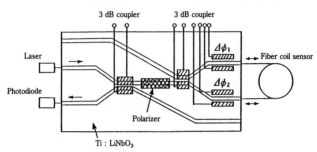

FIGURE 26.41   Integrated-optic device for fiber gyroscope.[37]

The authors proposed and demonstrated an OIC for the fiber laser Doppler velocimeter (fiber LDV),[38] as shown in Fig. 24.42. The fiber LDV has the advantages of high spatial and temporal resolution, excellent accessibility to a moving object, and minimum electronic induction noises. The heterodyne optics of the prototype system consisted of bulk optical components on a 30- $\times$ 30-cm^2 optical bench. On the other hand, in the OIC in Fig. 26.42, the heterodyne optics can be used on the $LiNbO_3$ substrate of only 32 $\times$ 7 mm^2 by integrating a waveguide interferometer. A piece of polarization-maintaining fiber is also pigtailed with a

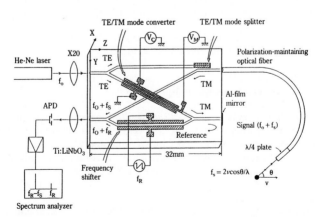

FIGURE 26.42   Integrated-optic device for fiber LDV.[38]

3.5-μm-wide Ti-diffused waveguide to pick up the Doppler-shifted frequency corresponding to the velocity of a moving object. The OIC presented here has a wide variety of applications including the measurement of displacement and position in addition to velocity.

Disk Pickup. Optical-disk pickup heads are currently constructed with bulk microoptics and need complex and time-consuming fabrication processes. If an integrated-optic disk pickup (IODPU) is put in practical use, however, there would be a great improvement in producibility, reduction of size, and application flexibility.

The schematic view of a proposed IODPU is shown in Fig. 26.43.[39] The waveguide is formed with a glass guiding layer and a SiO$_2$ buffer layer on a Si substrate.

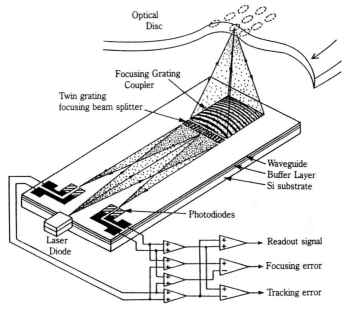

FIGURE 26.43   Integrated-optic disk pickup device.[39]

A focusing grating coupler (FGC), a twin-grating focusing beam splitter (TGFBS), and a photodiode array are integrated in this film waveguide. The TGFBS has a tilted and chirped pattern with a 4-μm period. The guided wave diverging from the butt-coupled laser diode is focused by the FGC into a point on a disk. The wave reflected by the disk is collected and coupled into the waveguide by the same FGC. The TGFBS then divides the reflected wavefront into halves, deflects it (by beam splitting), and simultaneously focuses it on four photodiodes. The complex functions of the TGFBS minimize the number of components. In this way, the disk signal is read out. The focusing and tracking error signals are also obtained by using the photocurrents based on the Foucault and push-pull methods, respectively. This device is an example of an OIC that might be used in consumer electronics in the future.

26.6.4 Laser Diode Arrays

Waveguide integration techniques will be extended to fabricate surface-emitting-type laser diode arrays with gratings[41] and edge-emitting-type laser diode arrays with more than 100 waveguides[40] to increase total power and beam cross section.

26.7 REFERENCES

1. Miller, S. E., "Integrated optics: an introduction," *Bell Syst. Tech. J.*, vol. 48, no. 7, pp. 2059–2068, 1969.
2. Tamir, T., *Integrated Optics*, Springer-Verlag, New York, 1975.
3. Hunsperger, R. G., *Integrated Optics: Theory and Technology*, Springer-Verlag, New York, 1982.
4. Nishihara, H., M. Haruna, and T. Suhara, *Optical Integrated Circuits*, McGraw-Hill, New York, 1989.
5. Hocker, G. B., and W. K. Burns, "Modes in diffused optical waveguides of arbitrary index profiles," *IEEE J. Quantum Electron.*, vol. QE-11, no. 6, pp. 270–276, 1975.
6. Marcatilli, E. A. J., "Dielectric rectangular waveguide and directional coupler for integrated optics," *Bell Syst. Tech. J.*, vol. 48, no. 9, pp. 2071–2102, 1969.
7. Suchoski, P. G., T. K. Findakly, and F. J. Leonberger, "Stable low-loss proton-exchanged $LiNbO_3$ waveguide devices with no electro-optic degradation," *Opt. Lett.*, vol. 13, no. 11, pp. 1050–1052, 1988.
8. Takato, N., et al., "Silica-based single-mode waveguides on silicon and their application to guided-wave optical interferometers," *J. Lightwave Tech.*, vol. 6, no. 6, pp. 1003–1010, 1988.
9. Yariv, A., and M. Nakamura, "Periodic structures for integrated optics," *IEEE J. Quantum Electron.*, vol. QE-13, no. 4, p. 233, 1977.
10. Suhara, T., and H. Nishihara, "Integrated optics components and devices using periodic structures," *IEEE J. Quantum Electron.*, vol. QE-22, no. 6, pp. 845–867, 1986.
11. Kaede, K., and R. Ishikawa, "A ten-port graded-index waveguide star coupler fabricated by dry ion diffusion process," *9th European Conf. Opt. Commun.*, *Tech. Dig.*, pp. 209–212, Geneva, October 1983.
12. Tangonan, G. L., et al., "Planar coupler devices of multimode fiber optics," *Topical Meeting Opt. Fiber Commun.*, WG2, Washington, D.C., March 1979.
13. Uehara, S., T. Izawa, and H. Nakagome: "Optical waveguide polarizer," *Applied Optics*, vol. 13, no. 8, pp. 1753–1754, 1974.
14. Suhara, T., Y. Handa, H. Nishihara, and J. Koyama, "Monolithic integrated microgratings and photodiodes for wavelength demultiplexing," *Appl. Phys. Lett.*, vol. 40, no. 2, p. 120, 1982.
15. Rice, R. R., et al., "Multiwavelength monolithic integrated fiber-optic terminal," *Proc. SPIE*, vol. 176, p. 133, 1979.
16. Watanabe, R., and K. Nosu, "Slab waveguide demultiplexer for multimode optical transmission in the 1.0–1.4 micron wavelength region," *Applied Optics*, vol. 19, no. 21, p. 3588, 1980.
17. Suhara, T., J. Viljanen, and M. Leppihalme, "Integrated-optic wavelength multi- and demultiplexers using a chirped grating and an ion-exchanged waveguide," *Applied Optics*, vol. 21, no. 12, p. 2159, 1982.
18. Nakajima, H., "High-speed $LiNbO_3$ modulator and application," *Optoelectronics Conf. (OEC '88) Proc.*, pp. 162–163, Tokyo, 1988.
19. Izutsu, M., and T. Sueta, "Millimeter-wave light modulation using $LiNbO_3$ waveguide with resonant electrode," *Conf. Lasers & Electro-Optics (CLEO '88) Proc. PD*, pp. 485–486, Anaheim, Calif., 1988.

20. Warzanski, W., F. Heismann, and R. C. Alferness, "Polarization-independent electro-optically tunable narrow-band wavelength filter," *Appl Phys. Lett.*, vol. 53, no. 1, pp. 13–15, 1988.

21. Shah, M. L., "Fast acoustic diffraction-type optical waveguide modulator," *Appl. Phys. Lett.*, vol. 23, no. 22, pp. 556–558, 1973.

22. Suhara, T., H. Nishihara, and J. Koyama: "One-gigahertz-bandwidth demonstration in integrated-optic spectrum analyzer," *Int. Conf. Integrated Opt. & Opt. Fiber Commun. (IOOC '83)*, 29C5-5, Tokyo, 1983.

23. Shibukawa, A., and M. Kobayashi, "Optical TE-TM mode conversion in double epitaxial garnet waveguides," *Applied Optics*, vol. 20, no. 14, pp. 2444–2447, 1981.

24. Castera, J. P., and G. Hapner, "Isolator in integrated optics using Farady and Cotton-Mouton effects," *Applied Optics*, vol. 16, no. 8, pp. 2031–2034, 1977.

25. Haruna, M., and J. Koyama, "Thermo-optic waveguide interferometric modulator/switch in glass," *IEEE Proc.*, vol. 131, pt. H, no. 5, pp. 322–324, 1984.

26. Taniuchi, T., and K. Yamamoto, "Miniaturized light source of coherent blue radiation," *Conf. Lasers & Electro-Optics (CLEO '87) Proc.*, WP6, Baltimore, 1987.

27. Ishigame, I., T. Suhara, and H. Nishihara, "LiNbO₃ waveguide second-harmonic-generation device phase matched with a fan-out domain-inverted grating," *Opt. Lett.*, vol. 16, no. 6, pp. 375–377, 1991.

28. Bierlein, J. D., D. B. Laubacher, and J. B. Brown, "Balanced phase matching in segment KTiPO₄ waveguides," *Appl. Phys. Lett.*, vol. 56, no. 18, pp. 1725–1727, 1990.

29. Schmidt, R. V., and L. L. Buhl, "Experimental 4 × 4 optical switching network," *Electron. Lett.*, vol. 12, no. 22, pp. 575–577, 1976.

30. Thylen, L., "Integrated optics in LiNbO₃: recent developments in devices for telecommunications," *J. Lightwave Tech.*, vol. 6, no. 6, pp. 847–861, 1988.

31. Duthie, P. J., M. J. Wale, and I. Bennoin, "Size, transparency and control in optical space switch fabrics: a 16 × 16 single chip array in Lithium Niobate and its applications," *Photonic Switching (PS '90)*, 13A-3, Kobe, 1990.

32. Inoue, H., et al., "An 8 mm length nonblocking 4 × 4 optical switch array," *IEEE J. Select. Areas Commun.*, vol. SAC-6, no. 7, pp. 1262–1266, 1988.

33. Sugita, A., M. Okuno, T. Matsunaga, and M. Kawachi, "Strictly nonblocking 8 × 8 integrated optical matrix switch with silica-based waveguides on silicon substrate," *16th European Conf. Opt. Commun. (ECOC '90) Proc.*, WeG4.1, pp. 545–548, Amsterdam, 1990.

34. Nakato, N., et al., "128-channel polarization-insensitive frequency-selection-switch using high-silica waveguides on Si," *IEEE Photon. Technol. Lett.*, vol. 2, no. 6, pp. 441–443, 1990.

35. Mergerian, D., et al., "Operational integrated optical RF spectrum analyzer," *Applied Optics*, vol. 19, no. 18, pp. 3033–3034, 1980.

36. Verber, C. M., R. P. Kenan, and J. R. Busch, "Correlator based on an integrated optical spatial light modulator." *Applied Optics*, vol. 20, no. 9, pp. 1626–1629, 1981.

37. Ezekiel, S., and H. J. Arditty, *Fiber Optic Rotation Sensors*, Springer-Verlag, New York, 1982.

38. Toda, H., M. Haruna, and H. Nishihara, "Optical integrated circuit for a fiber laser Doppler velocimeter," *IEEE J. Lightwave Tech.*, vol. LT-5, no. 7, pp. 901–905, 1987.

39. Ura, T., T. Suhara, H. Nishihara, and J. Koyama, "An integrated-optic disk pickup device," *IEEE J. Lightwave Tech.*, vol. LT-4, no. 7, pp. 913–918, 1986.

40. Harnagel, G. L., P. S. Cross, D. R. Scifres, and D. P. Worland, "11 W quasi-cw monolithic laser diode array," *Electron. Lett.*, vol. 22, no. 5, pp. 231–233, 1986.

41. Evans, G. A., et al., "Grating-surface emitting laser array with 1.2 cm output aperture," *Int. Conf. Integrated Opt. & Opt. Fiber Commun. (IOOC '89)*, 18B2-3, Kobe, 1989.

CHAPTER 27
OPTOELECTRONIC INTEGRATED CIRCUITS

Osamu Wada

27.1 INTRODUCTION

Since the successful continuous-wave operation of semiconductor lasers in 1970, optoelectronic devices based on III-V semiconductors have received increasing attention. Discrete optoelectronic devices have thus become the key to optical telecommunications, data processing, and sensing systems. To meet increases in information transmission and processing capacity, optoelectronic devices must be enhanced to provide better performance, a broader range of functions, improved reliability, and lower cost. The bit rate in optical transmission systems, for example, is already beyond gigabits per second, but is expected to advance to the terabits per second level by the early 2000s.

Optoelectronic integrated circuits (OEICs) help provide the above enhancements by integrating multiple optical and electronic functions. The first experiment demonstrating OEIC was done in 1978–79 by integrating GaAs-based lasers and driver diodes and transistors.[1] During this decade, many advances were made in different areas of integration technology. Not only have OEIC transmitters and receivers been developed for telecommunications,[2,3] but novel optical functional circuits have been investigated for next-generation data processing.[4,5]

This chapter reviews current technology in semiconductor-based optoelectronic integration which incorporates optical and electronic devices. Categories and features of optoelectronic integration are first explained. Basic integration techniques, including materials, basic devices, and fabrication processes, are then described. The current status of development and possible applications are reviewed, and projected trends are discussed.

27.2 CATEGORIES AND FEATURES

The term *optoelectronic integrated circuit* has often been used to indicate the integration of optical and electronic devices.[1,6] Here, we use a somewhat broader definition that covers all integrated devices and circuits that incorporate optoelectronic elements. Figure 27.1 shows the three major categories of optoelectronic

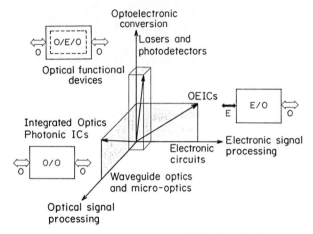

FIGURE 27.1 Three major categories (shaded areas) of optoelectronic integration with respect to three basic functions and devices (axes). The construction of the integrated circuit and the signal form are indicated by box and arrows. (O: optical; E: electronic)

integration, referring to three basic functions: optoelectronic conversion, electronic signal processing, and optical signal processing. Lasers and photodetectors perform optoelectronic conversion. Standard OEICs combine these with electronic circuits that process signals. OEIC input and output differ in that one is optical and the other is electronic. The integration of lasers and photodetectors with other optical signal processing circuits using waveguide devices and micro-optic elements produces useful components providing advanced optical signal processing functions. The integration of waveguide devices using semiconductors and dielectric materials such as $LiNbO_3$ has been called *integrated optics* or *optical integrated circuits*[7] and is described in detail in Chap. 26. More recent integration of various waveguide-based devices and optoelectronic conversion devices on a common semiconductor substrate is referred to as *photonic integrated circuits*.[8] Optical signals are provided at both interfaces and a variety of functions are performed by using the wave properties of light. Such integration schemes fully encompass existing and developing device technology based on III-V semiconductors such as GaAs and InP alloy systems. As the third category, novel optical functions such as optical switching and storage are provided by incorporating optoelectronic conversion and optical and electronic processing elements within an integrated device structure.[5] This is done by efficient interaction between photon and electron systems within the integrated structure. Devices with optical interfaces are expected to be the key to future optical systems, including photonic switching and optical computing.

Integration affects three areas: performance, function, and manufacturability. The integration of optoelectronic devices with electronic circuits on a chip reduces parasitic reactances which have inevitably been introduced by electrical interconnections between discrete devices. This improves compactness as well as speed and noise characteristics of optoelectronic devices and is useful in very high-speed telecommunication and coherent optical telecommunication and sensing systems for which large-bandwidth light sources and receivers are required. Optical signal processing functions provided by integration simplify the overall design of optical components and systems. Integration of multiple optical devices on a common substrate helps eliminate problems in the delicate alignment of elements and is prerequisite for making complicated optical circuits stable enough for practical application. Novel optical functions including logic operation, wavelength control, and light beam steering are provided by integrated device structures, opening the way to photonic switching and optical computing architectures. Integration enhances manufacturability by reducing the overall number of components needed, simplifying assembly, and improving reliability, all of which help lower cost. Such improvements will extend optical technology applications into wide-area network services and optical interconnections in high-speed processors and switching systems.

27.3 MATERIALS, BASIC DEVICES, AND FABRICATION TECHNIQUES

27.3.1 Materials

The materials most often used in optoelectronic integration are GaAs - and InP-based III-V semiconductor alloy systems, as shown in Table 27.1. The high electron mobility and drift velocity in these systems, together with their high-quantum-efficiency heterostructures, provides a variety of advantages in optical and electronic device application. The semi-insulating (SI) substrates made possible with these material systems greatly simplify electrical isolation and aid in high-speed operation. Heterostructures made using these material systems are applied according to the wavelength. The InP-based system emits long-wavelength light useful in optical transmission over distances exceeding tens of kilometers due to low optical fiber loss. The AlGaAs/GaAs system is, in contrast, more useful in short-distance applications, and its mature electronic device technology provides a strong base for optoelectronic integration involving electronic circuits. There is also an emerging technology built on non-lattice-matched heteroepitaxy such as GaAs on Si.[9] Epitaxy techniques allowing low-defect-density heterointerfaces are essential to this. Recent research has achieved a defect density on the order of 10^5 cm^{-2} or less, and improvements are expected to reach 10^3 cm^{-2} in the near future, sufficient for fabricating reliable lasers. Once this is possible, optical devices can be built on Si very large scale integrated circuits (VLSIs), allowing powerful signal processing functions.

TABLE 27.1   Representative Materials for Optoelectronic Integrated Circuits

Substrate	Heterostructure	Wavelength	Features
GaAs	GaAs/AlGaAs	~0.8 μm	Short distance; advanced electronics
InP	InP/InGaAsP, InP/InGaAlAs	1.3 to ~1.55 μm	Long distance; potentially very fast electronics
InP, GaAs, Si heteroepitaxy	GaAs, InP	0.8 to ~1.55 μm	Low-defect density growth needed; possible use of Si very large scale integration

27.3.2 Basic Devices

Light Sources. Heterostructure light-emitting diodes (LEDs) and lasers are used as light sources in optoelectronic integration. Lasers are more advantageous because of their high output power, response speed, beam directionality, and narrow spectrum. Quantum-well (QW) laser structures are very promising because of their low threshold current, high-temperature stability, and fast modulation speed.[10] Figure 27.2 shows a cross section of an AlGaAs/GaAs QW laser with a graded-index waveguide separate confinement heterostructure (GRIN-SCH) with a very thin (6-nm) GaAs well layer.[10,11] The AlGaAs/GaAs superlattice buffer (SLB) layer improves the quality of molecular beam epitaxy (MBE) layers grown on the substrate. Lateral mode stabilization is achieved by using the index-guided structure with simple, narrow (1-μm) ridge waveguide. Such lasers having a 300-μm cavity length have shown a threshold current of 5 mA, a differential quantum efficiency of 80 percent, and a characteristic temperature T_0 of 160 K.[11] Recent studies have shown that a submilliampere threshold current[82] and a modulation frequency up to 30 GHz are possible.[12]

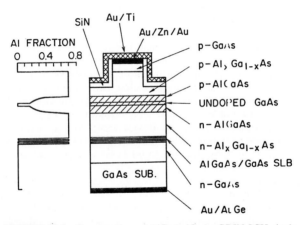

FIGURE 27.2 Cross section of AlGaAs/GaAs GRIN-SCH single-quantum-well laser with ridge waveguide structure.[11]

To stabilize the lateral modes of lasing, a buried heterostructure (BH) having an active layer buried in lower-refractive-index materials is often used. This requires at least two separate crystal-growth steps, however, and thus complicates fabrication. A structure called a *lateral current injection laser*, as shown in Fig. 27.3, can simplify this process while maintaining the planar surface and index-guided waveguide.[13] In this structure, a multiple quantum well (MQW) structure is sandwiched between high-resistivity (HR) AlGaAs cladding layers, and the MQW disordering occurring during the Zn and Si impurity diffusion processes is used for obtaining the lower-index burying regions. The impact of planar lasers on integration is particularly great, since a common MQW structure can be used for other optical and electrical devices on the same wafer as indicated in Fig. 27.3. Contin-

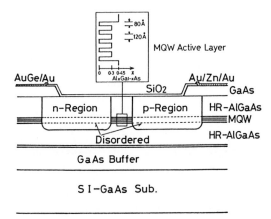

FIGURE 27.3   Cross section of AlGaAs/GaAs MQW lateral current injection laser. An index-guided structure and planar contacts are realized by impurity-induced disordering of MQW.[13]

uous-wave (cw) operation of lasers and field-effect transistors (FETs) has already been demonstrated.[14]

Figure 27.4 diagrams a three-section distributed Bragg reflector (DBR) laser made of a GaInAsP/InP heterostructure.[15,16] It is an example of the optoelectronic integrated structure and also serves as a basic device extremely useful in photonic integrated circuits because its wavelength is tunable and its structure does not require cleaved facets. It consists of the serial coupling of an electronically controlled gain medium, variable phase shifter, and tunable Bragg reflection filter, allowing continuous wavelength tunability. These lasers have shown output powers of 20 to 30 mW, minimum linewidths of 1 to 2 MHz, and wavelength tuning exceeding 10 nm, while maintaining linewidths below 16 MHz around 1.53 μm.[15] A three-section λ/4-shift distributed-feedback (DFB) laser, in which both the active layer and the grating extend over the whole cavity length, has shown a linewidth narrower than 1 MHz over an entire tuning range of 2.2 nm.[15]

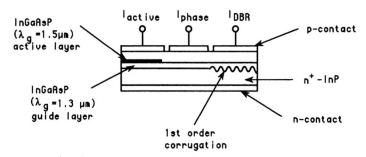

FIGURE 27.4   Schematic diagram of three-section DFB laser.

Photodetectors.   The photodetectors most widely used in integration are PIN photodiodes. Figure 27.5 illustrates a cross section of a planar, embedded GaInAs/InP *pin* photodiode.[17] A small-diameter (less than 30-µm) junction and an interconnection lead formed on SI-InP reduce the capacitance and enable a cutoff frequency well beyond 10 GHz. The PIN photodiode has a vertical junction structure but planarization, as shown here, simplifies the integration process.

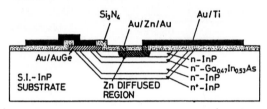

FIGURE 27.5   Cross section of GaInAs/InP *pin* photodiode with planar embedded structure formed on SI-InP substrate.[17]

Figure 27.6 shows a metal-semiconductor-metal (MSM) photodiode structure with a pair of interdigitated Schottky barrier contacts on an undoped GaAs photoabsorption layer.[18,19] The optical coupling efficiency is reduced by half due to contact shadowing, but such a lateral junction structure is extremely useful for integration because it is planar and process-compatible in fabricating metal-semiconductor (MES) FET circuits on GaAs substrates. The capacitance is considerably lower than that of vertical junction structures. A very fast response exceeding 100 GHz has been demonstrated by using MSM photodiodes with fine contact patterns.[20]

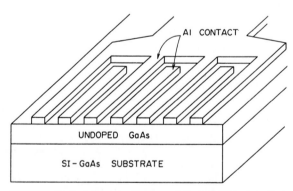

FIGURE 27.6   Schematic diagram of GaAs MSM photodiode with interdigital Schottky contacts.[19]

In InP-based circuits, the use of lateral MSM photodiodes has been inhibited by the low Schottky barrier height in this material system. However, some solutions to this have been suggested recently, and Fig. 27.7 shows an example. An AlInAs barrier-enhancement layer and a graded bandgap superlattice layer are introduced

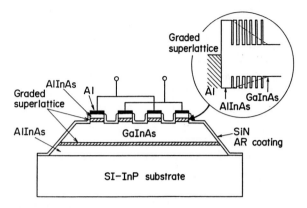

FIGURE 27.7 Cross section of GaInAs MSM photodiode with AlInAs/GaInAs barrier enhancement layer for dark current suppression.[21]

on the GaInAs photoabsorption layer to achieve a dark current of less than 100 nA and a response speed of 15 ps full width at half maximum (FWHM).[21] Other barrier-enhancement techniques include non-lattice-matched GaAs[22] and SI-InP[23] layers. Back illumination using the transparency of InP substrates to long wavelengths is effective to increase the optical coupling efficiency in lateral photodiodes, and 80 percent efficiency has been shown in lateral *pin* photodiodes.[24]

The phototransistor is another element useful in integration. Its greatest advantage is the optical gain produced by the transistor action. The optical gain and response time, however, usually depend on the optical input power, and care must be taken in heterostructure design.[25] Both photoconductors and FETs are used as photodetectors in integration because of their simple planar structures. Optical gain is achieved at the sacrifice of response speed; despite this, very fast responses have been shown in AlGaAs/GaAs HEMTs (FWHM: 22 ps).[26] AlInAs/GaInAs modulation-doped photoconductors have been used for receivers operating at 2.7 Gb/s.[27]

Waveguide Devices. Details on waveguides and optical integration are given in Chap. 26. The application of semiconductor materials and processing technology can expand their flexibility in functioned design and device structure. One of the main issues in waveguides is optical loss. High-purity GaAs and InP materials have made possible very low propagation losses—as low as 0.2 dB/cm. The rather large fiber-coupling loss previously a problem in implementing waveguide devices in practical systems can be overcome by applying semiconductor processes. Figure 27.8 shows the cross section of a ridge waveguide with a thick-core diluted-MQW (D-MQW) structure. An extremely small refractive index difference between the core and the cladding, Δ_n, as small as 3×10^{-3}, is achieved by inserting 15 extremely thin (3.5-nm) layers of high-index material (GaInAsP with $\lambda_g = 1.13$ μm) in the primary InP core layer (2 μm) by using metal-organic chemical vapor deposition (MOCVD) techniques, allowing a large core size. A core size over 8 μm has been observed in a deep-rib guide structure and a coupling loss as low as 0.2 dB has been achieved for single-mode fiber.[28]

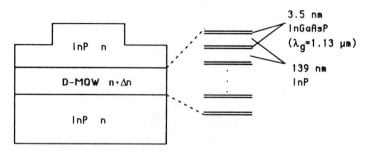

FIGURE 27.8   Structure of InP/GaInAsP diluted-MQW waveguide with large-core size.

The coupling loss between devices needs to be considered. Figure 27.9 shows three representative structures for device coupling. Butt-joint coupling provides low-loss coupling, although the fabrication becomes complicated. This has been used often in laser integration. Graduated coupling using a device on top of the waveguide is a simpler structure for fabrication. Efficient optical transfer is achieved by either inserting an impedance-matching layer with an intermediate refractive index[65] or incorporating a built-in second-order grating. Coupling with a 45° angle mirror can be applied to transparent substrates such as InP and is particularly useful for integrating detectors.

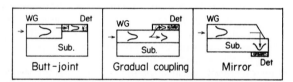

FIGURE 27.9   Three representative techniques of coupling optical devices and waveguides for monolithic integration.[5]

A variety of optical signal processing functions can be generated by waveguide devices. Figure 27.10 shows just one example of a total-reflection optical switch/ modulator in which the carrier-induced decrease of the refractive index in the crosspoint area is used to reflect the optical beam from one port to the other.[29] Directional couplers can also be used for switches and modulators. Many other functions for controlling the wavelength, phase, and mode can be provided by waveguide devices.[30]

Electronic Devices.   Transistors and diodes are essential to electronic circuits. GaAs MESFET IC technology, as the most mature in this area, has accelerated the development of GaAs-based OEICs. Figure 27.11a and b shows two GaAs MESFET structures with channel layers formed by epitaxial growth and ion implantation. To fabricate circuits operating at gigabit rates, the gate length should be reduced down to the submicrometer level.

In AlGaAs/GaAs modulation-doped structures, electrons in the doped, large-bandgap AlGaAs layer are transferred to the undoped, small-bandgap GaAs layer

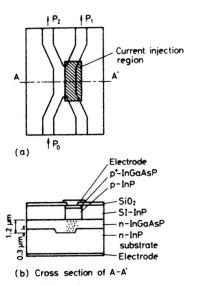

(a)

(b) Cross section of A-A'

FIGURE 27.10 (*a*) Schematic diagram of InP-based current-injection waveguide switch with planar SI-InP current blocking layer. (*b*) Structure of current injection section. *(After Wakao et al.[29])*

to form a two-dimensional electron gas (2DEG) at the interface. Such a layer can serve as a high-electron-mobility channel for FETs. The cross section of the device and band diagram are shown in Fig. 27.11c and *d*. Such devices are called high-electron-mobility transistors (HEMTs).[31] The enhanced performance of HEMTs has been used in both microwave and digital circuits. AlInAs/GaInAs heterostructures are used in the InP-based system, and recent discrete AlInAs/GaInAs HEMTs with submicrometer gates have shown excellent performance, including a cutoff frequency exceeding 250 GHz.[32]

Heterojunction bipolar transistors (HBTs) also have been intensively studied to improve the circuit performance at ultrahigh speeds.[33] Heterostructure systems being applied to HBTs include AlGaAs/GaAs and AlInAs/GaInAs. Ultra-high-speed digital circuits such as optical receiver front ends and laser drivers have been demonstrated in the AlGaAs/GaAs system.

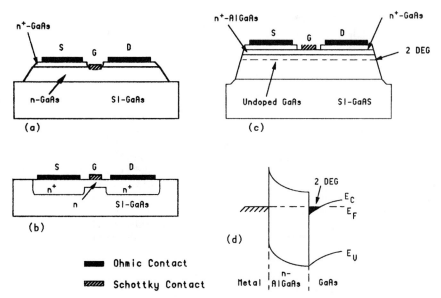

FIGURE 27.11 Cross sections of GaAs MESFETs with (*a*) epitaxial and (*b*) ion-implanted channel layers. (*c*) Cross section and (*d*) band diagram of AlGaAs/GaAs HEMT.

27.3.3 Fabrication Techniques

Techniques for fabricating optoelectronic integrated circuits include epitaxial growth, device processing, and packaging. In epitaxial growth, liquid-phase epitaxy (LPE) has most often been used for optoelectronic devices. Recent growth trends have, however, increasingly used MBE[34] and MOCVD or metal-organic vapor-phase epitaxy (MOVPE)[35] to produce large, high-quality epitaxial wafers with improved uniformity and reproducibility. In an MBE growth system, solid metal sources such as Ga, Al, and As and dopant sources such as Si and Be are evaporated in an ultra-high-vacuum chamber onto a rotating, heated substrate. The growth of very thin layers with abrupt interfaces can be done at relatively low temperatures (450 to 800°C for GaAs) favorable to heterostructure growth. For growing GaInAsP compounds by low-pressure MOCVD, the substrate InP is charged in a low-pressure (10- to 100-torr) quartz tube and rf-heated. Source gases, including metal organics, such as triethyl indium and triethyl gallium, and hydrides, such as AsH_3 and PH_3, are introduced into the reactor tube, thermally decomposed and then deposited onto the substrate. Both MBE and MOCVD have been shown to produce sufficient quality in both GaAs- and InP-based heterostructures. More recent proposals include chemical-beam epitaxy (CBE) (vapor III's, vapor V's)[36] and gas-source MBE (GSMBE) (solid III's, vapor V's)[37] used in intermediate vacuum in pursuit of higher crystal quality and better layer uniformity.

In device processing for optoelectronic integration, the generation of fine device geometries is a basic requirement, as is recognized in Si VLSI technology. Laser waveguides, photodetector junctions, and transistor electrodes all must have micrometer to submicrometer dimensions. High-resolution photolithography and electron-beam writing techniques are used for submicrometer patterning. An optical interference patterning technique is used for forming submicrometer corrugations. Fine pattern etching is also important, and is being rapidly improved with dry etching techniques. Figure 27.12 diagrams a reactive-ion-beam etching (RIBE) system, in which reactive ions, for example, Cl_2 ions for AlGaAs/GaAs etching, are accelerated in a high vacuum toward the target material to be etched.[38] Smooth, anisotropic geometries can be generated without the problem of oxidation. This is extremely useful in forming laser facets, while avoiding conventional substrate

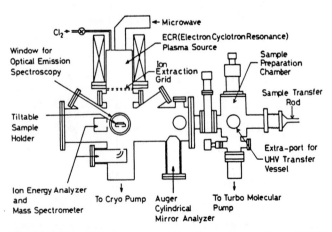

FIGURE 27.12   Schematic diagram of reactive ion-beam etching (RIBE) system.   (*After Asakawa et al.*[38])

cleavage. Etching in InP-based material can also be done by using reactive-ion etching (RIE) in a simpler parallel-plate plasma chamber because of their resistance to oxidation. Various gas systems, including Cl_2/Ar and CH_3/H_2, are used for waveguide and facet etching. C_2H_6 has been applied to forming highly uniform DFB laser gratings.[89]

Another important challenge in the fabrication area is how to integrate growth and other processes necessary for device fabrication. Figure 27.13 shows the diagram of integrated in-situ process equipment consisting of an MBE chamber, a focused-ion-beam implanter, and other chambers for etching, cleaning, and annealing, integrated through a high-vacuum wafer-transfer chamber.[39] A wafer is grown and processed sequentially without exposure to the ambient atmosphere so that complicated three-dimensional device structures with highly pure crystals can be prepared. Process integration should improve the fabrication quality and reproducibility.

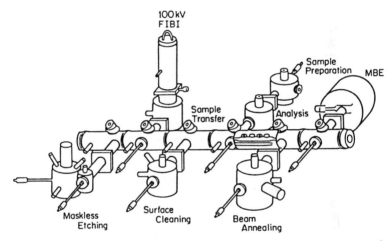

FIGURE 27.13   Schematic diagram of in-situ process equipment incorporating growth, patterning, and various other processing chambers combined through an ultra-high-vacuum sample transfer chamber.   *(After Takamori et al.[39])*

Requirements in the packaging of optoelectronic integrated devices and circuits include compactness, low-loss optical coupling, high-speed electrical interfacing, and efficient heat removal. Low crosstalk is a critical issue for multichannel circuits. Efforts have been made in a variety of areas to develop manufacturable optoelectronic packaging techniques. The use of optical self-alignment techniques is one of the important issues in this area. Waveguide-to-fiber self-alignment using solder-reflow bonding techniques has been proposed.[40]

27.4   OPTOELECTRONIC INTEGRATED CIRCUITS

27.4.1   OEIC Transmitters and Receivers

Transmitters and receivers consist of optical and electronic devices having very different structures and fabrication processes. Most work in this regard has gone into attaining process compatibility and developing reproducible fabrication techniques. The sections that follow review state-of-the-art development of OEIC transmitters and receivers.

Integrated Structures. Figure 27.14 shows representative monolithic structures for a GaAs-based receiver. The vertical structure can be formed by a single growth procedure, but it needs a tight electrical isolation layer and suffers from performance degradation at high frequencies due to interdevice capacitive coupling. Recent fabrication has involved horizontal integration on a semi-insulating substrate. The most important requirements here are surface planarity and process compatibility.

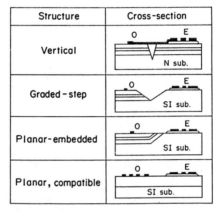

Structure	Cross-section
Vertical	O E N sub.
Graded-step	O E SI sub.
Planar-embedded	O E SI sub.
Planar, compatible	O E SI sub.

FIGURE 27.14 Four representative structures of OEIC receivers.[82]

In the graded-slope and planar-embedded structures, engineering of a dry etching technique has provided surfaces sufficiently planar for device processing. Process compatibility has been achieved by introducing MSM photodiodes with a planar, lateral structure. The same basic structures can be used in integrating lasers, although the fabrication is more complicated because of the need to form built-in waveguides with vertical facets.

The InP-based material system has not yet had a standard IC technology set up like that for the GaAs MESFET. Developments are in progress for high-performance circuits using InP-based HEMTs and HBTs. Other challenges being examined include heteroepitaxy and flip-chip integration techniques, and they are discussed in the following sections.

Transmitters. In transmitter fabrication, low-threshold current lasers are important in minimizing heating. Quantum-well lasers grown by either MBE or MOVPE are extensively used to lower the threshold current. Transverse mode-stabilized laser structures such as BH and ridge-waveguide structures are used to ensure stable lasing.

Figure 27.15 shows the structure and the circuit diagram of a multichannel OEIC transmitter array fabricated on a SI-GaAs substrate.[41] MBE-grown AlGaAs/GaAs GRIN-SCH single-quantum-well (SQW) lasers having ridge waveguide structures

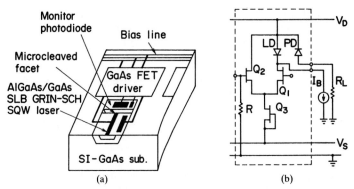

(a)   (b)

FIGURE 27.15   (a) Structure and (b) circuit diagram of multichannel GaAs-based OEIC transmitter array containing QW lasers, monitor photodiodes, and GaAs MESFET driver circuits.[41]

are planar-embedded in a SI-GaAs substrate, and the laser's inner facets are formed by microcleavage techniques.[42] Laser power monitoring photodiodes are formed by etching the same heterostructure. The driver circuit consists of three MBE-grown GaAs MESFETs with 2-μm gates.

Figure 27.16 shows the laser and monitoring performance for four channels. The laser threshold current is 15 mA to 21 mA, the differential quantum efficiency is 50 to 60 percent, and the monitor efficiency is 1.8 to 3.0 μA/mW, indicating fair uniformity for the four channels. The overall conversion ratio of this transmitter is 6 mW/V, sufficiently large for an emitter-coupled logic (ECL) interface. Each channel shows high-speed modulation up to 2 Gb/s. The crosstalk between channels is less than −20 dB in the useful frequency range below 0.6 GHz.

More complicated circuits such as a 1:4 serializing circuit composed of more than 200 MESFETs[43] and a ring oscillator circuit[44] have been integrated with lasers. Fabrication techniques such as RIBE for facet formation and ion implantation for FET channel formation have been demonstrated for GaAs-based OEICs.[45]

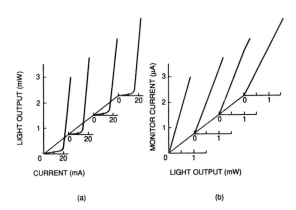

(a)   (b)

FIGURE 27.16   (a) Laser light-current characteristics. (b) Monitor characteristics of four-channel GaAs-based OEIC transmitter array.[41]

In InP-based materials systems, simple transmitters involving a cleaved-facet laser and a single to a few driver transistors have been fabricated. A GaInAsP/InP transmitter consisting of a BH laser and three HBTs fabricated by vertical integration with LPE showed 0.4-Gb/s operation.[46] A BH laser/InP MESFET transmitter with a horizontal structure has demonstrated high-speed operation at 6.6 GHz.[47] Recent fabrication of a transmitter incorporating a $\lambda/4$-shift DFB laser and an InAlAs/GaInAs HEMT with a horizontal structure has shown ultra-high-speed operation of up to 10 Gb/s.[48]

Great advances in non-lattice-matched heteroepitaxy have opened up the possibility of combining materials for enhancing the advantages of integration. The combination of high-reliability, lattice-matched InP-based optoelectronic devices and standard GaAs-based MESFETs is a good example of this. The interface quality problem is less severe in MESFETs, because it is a majority carrier device with a current direction parallel to the interface. A transmitter consisting of a GaInAsP/InP BH laser and GaAs MESFETs has shown excellent transmission characteristics at 1.2 Gb/s.[3] Laser transmitter fabrication on Si substrates has not yet been achieved, but AlGaAs/GaAs LED/Si MOSFET integration[9] and single GaAs-based laser fabrication on Si[49] have already been demonstrated.

Receivers.   Simple receivers consisting of *pin* photodiodes and preamplifiers have been fabricated using vertical and horizontal integration structures. Planar, compatible structures incorporating an MSM photodiode and MESFETs on SI-GaAs have been attained and could increase the integration scale without compromising manufacturability. Figure 27.17 shows the cross section and the circuit diagram for a four-channel MSM/amplifier receiver array involving 52 devices.[19] An MSM photodiode 100 μm^2 in area with 3-μm lines and spaces for interdigital electrodes has been integrated with 2-μm gate MESFETs to construct a transimpedance amplifier. Overall receiver sensitivity is uniform, within 105 V/W $\pm$ 10 V/W over an even eight-channel array, confirming the advantage of a planar, compatible process. MSM photodiodes also help achieve low-capacitance, high-speed, and low-noise performance. The output response, input equivalent noise current, and interchannel crosstalk are plotted versus the frequency in Fig. 27.18.[88] The cutoff frequency of 1.1 GHz and the noise floor of 5 pA/$Hz^{1/2}$ are consistent with the circuit parameters used. The crosstalk is extremely low, less than -37 dB up to 0.6 GHz. Essentially the same structures as used for MSM photodiodes have been used in many applications for large-scale OEIC receivers. A large-scale-integration- (LSI-) level OEIC receiver with deserializing and clock recovery circuits and 8000 MESFETs has been demonstrated.[50] A receiver incorporating an MSM photodiode with 0.5-μm contacts

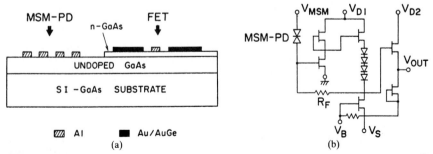

FIGURE 27.17   (*a*) Structure and (*b*) circuit diagram of GaAs MSM/amplifier OEIC receiver.[19]

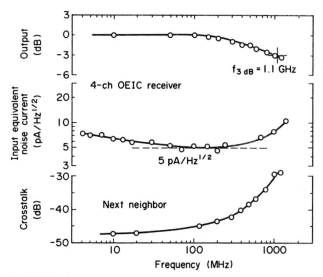

FIGURE 27.18   Frequency response characteristics of output, noise current, and crosstalk in four-channel GaAs OEIC receiver array.[88]

has been shown to operate as fast as 5.2 GHz.[51] Other GaAs-based OEIC receivers with planar structures include photoconductor/FET receivers. An AlGaAs/GaAs modulation-doped photoconductor has been integrated with a simple HEMT amplifier.[52]

In InP-based OEIC receivers, most fabrication up to now has used GaInAs/InP PIN photodiodes. Four-channel PIN/amplifier receivers with junction FETs (JFETs) have been fabricated by using quasiplanar, embedded structures, and gigabit rates have been demonstrated.[53] Sensitivity at high speed has been improved by introducing high-cut-off frequency transistors such as AlInAs/GaInAs HEMTs.[54,55] Integration of a PIN photodiode with an HBT amplifier has shown a very good receiver sensitivity of −21 dB at 4 Gb/s.[56] The development of planar, compatible structures has been slow in this material due to the low Schottky barrier height, but recent integration of AlInAs-capped MSM photodiodes with HEMTs is encouraging in simplifying fabrication.[57]

Non-lattice-matched heteroepitaxy has been used to combine a GaInAs/InP *pin* photodiode and a GaAs MESFET amplifier on an InP substrate.[58] As a totally different way of combining different materials on a single substrate, flip-chip bonding has been applied for receiver integration, since no problem of device heating arises.[59] Figure 27.19 shows the cross section of a flip-chip integrated receiver consisting of a GaInAs/InP *pin* photodiode and a GaAs MESFET amplifier.[60] In this example, the *pin* photodiode has a back-illuminated structure with a monolithic microlens for maximizing the fiber alignment tolerance (greater than 50 μm). Recent application of this technique to integrating a GaInAs/InP avalanche photodiode (APD) with a Si bipolar transistor preamplifier has shown excellent receiver sensitivity at 10 Gb/s.[61] The flip-chip bonding technique can thus provide simple optoelectronic packaging.

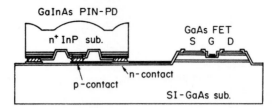

FIGURE 27.19 Structure of flip-chip integrated long-wavelength receiver combining a GaInAs/InP *pin* photodiode and a GaAs MESFET amplifier.[60]

27.4.2 Photonic Integrated Circuits

Very useful functions and performance are produced by integrating optical devices with other waveguide devices. Figure 27.20 shows the structure of a butt-joint integrated light source consisting of a DFB laser and a waveguide absorption

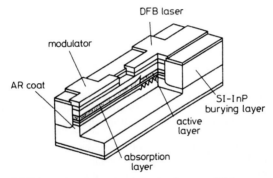

FIGURE 27.20 Structure of ultrafast integrated light source consisting of a DFB laser and an absorption waveguide modulator butt-joint-coupled on an InP substrate. *(After Soda et al.[62])*

modulator.[62] This device has very high performance, exhibiting very low chirp (less than 0.01 nm at 1.55 μm) under an ultrahigh bit rate of 10 Gb/s with a fiber-coupled output power of several milliwatts and -10 dB extinction with only 3- to 5-V bias. Figure 27.21 diagrams a wavelength-division multiplexing (WDM) source containing four tunable MQW distributed Bragg reflector (DBR) lasers combined through an MQW amplifier (approximately 7 dB) to a single waveguide fiber-coupling output port. The center frequency of each DBR laser is shifted by changing the guide thickness and thus the effective index.[63] WDM transmission at the wavelengths near 1.3 μm with a 2.5-nm interchannel separation has been demonstrated at 2 Gb/s for four channels.

Figure 27.22 shows a monolithic coherent optical receiver chip in which a multiple-electrode tunable DFB laser as a local oscillator, a 3-dB directional coupler, and a pair of *pin* photodiodes are butt-joint-coupled on an *n*-InP substrate.[64] Heterodyne receiver operation has been shown for similar circuits by using either a

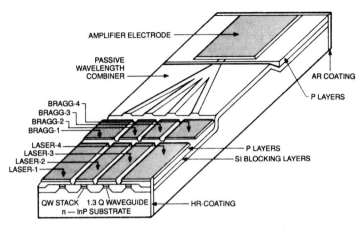

FIGURE 27.21 Schematic diagram of a four-wavelength-multiplexing photonic integrated circuit containing four tunable MQW-DBR lasers combining through an MQW amplifier to a single waveguide fiber-coupling port. *(After Koch et al.[63])*

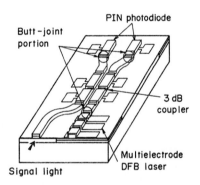

FIGURE 27.22 Schematic diagram of heterodyne coherent optical receiver photonic integrated circuit containing a tunable local oscillator laser, a 3-dB directional coupler, and two *pin* photodiodes, butt-joint-coupled on an InP substrate. *(After Takeuchi et al.[64])*

single-detector[64] or a balanced dual-detector[63] configuration.

Graduated coupling structures are used in detector integration. In InP-based detector/waveguide integration, an impedance-matching layer with an intermediate alloy (GaInAsP) composition can be inserted between the Ga-InAs detector absorption layer and the InP waveguide layer to transfer the optical power efficiently. An internal quantum efficiency of 70 percent has been achieved by a photodiode as short as 100 μm; the reduced capacitance has allowed high speeds with a cutoff frequency of 11 GHz.[65] For smoothly transferring the optical power from one device to another, a passive-mode conversion taper structure has also been proposed in which the thickness of the waveguide core is varied longitudinally.[63]

To provide efficient coupling between the fiber and waveguide device, the mode size in between must match. One approach to this uses a diluted-MQW structure as described in Sec. 27.3.2 under "Waveguide Devices." Figure 27.23 shows the structure of a low-loss coherent receiver consisting of a waveguide directional coupler and a balanced detector pair.[66] Diluted-MQW waveguides are used for low-loss fiber coupling, and 45° mirrors provide the coupling between waveguides and photodiodes through a transparent SI-InP substrate. Heterodyne operation at 1.8 GHz has been demonstrated.[67]

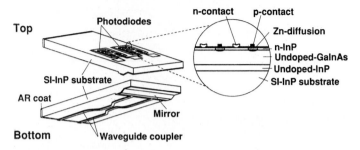

FIGURE 27.23 Schematic diagram showing top and bottom surfaces of coherent receiver incorporating a directional coupler and twin *pin* photodiodes coupled through a transparent InP substrate.[67]

27.4.3 Integrated Optical Functional Devices

Novel optical functions not possible under conventional conditions with discrete optoelectronic devices or their combinations can be assembled by using the interaction between the photon and electron systems efficiently within integrated structures. Figure 27.24 shows a monolithic colliding-pulse mode-locked (CPM) quantum-well laser composed of a pair of GRIN-SCH MQW laser gain sections connected through a saturable absorber section within a common waveguide.[68] The integration eliminates tedious optical alignment and simplifies the electrical bias control for optimizing operating conditions. This integrated laser has generated transform-limited optical pulses near 1.56 μm with a duration of 0.64 ps at a repetition rate of 350 GHz.

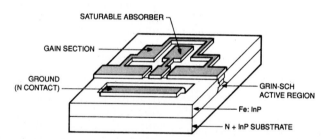

FIGURE 27.24 Schematic diagram of colliding pulse mode-locked laser integrated on an InP substrate. *(After Chen et al.[68])*

Optical bistability is required for optical logic functions such as switching and memory. One approach to optical bistable devices assumes that a tandem laser structure having a low-bias section acting as the saturable absorber enables optical set operation.[69] Optical reset operation can also be accomplished by using the gain quenching effect which occurs at optical powers exceeding the critical level.[70] Figure 27.25 shows the structure of a wavelength conversion laser consisting of a two-section bistable laser and a phase shifter/DBR wavelength tuner.[71] This device exhibits a wavelength tunability in excess of 4.5 nm. Adjusting the active laser

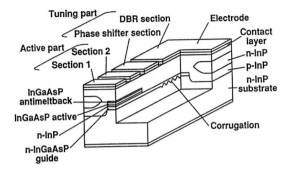

FIGURE 27.25 Structure of wavelength conversion laser integrating a bistable gain section, a phase shifter section, and a DBR tuning section. This device exhibits various optical logic functions in both intensity and wavelength.[72]

current for the light-current hysteresis loop enables operation mode selection, e.g., OR/NOT, XOR, flip-flop, or NOR/NAND.[72] Logic operation at 1 Gb/s with 10- to 100-fJ switching energy has been achieved. The major drawback of bistable laser-based devices is the rather large dc power consumption, but reducing the laser threshold current is expected to solve this problem in the future.

The highly directional propagation of light beams in free space is extremely important in systems transmitting and processing large-capacity signals. Figure 27.26 shows an example of a beam-scanning laser which incorporates two mutually coupled AlGaAs/GaAs gain-guided waveguides. The beam waist location is controlled by adjusting bias currents.[73,74] A beam steering over ±8° around the normal direction and various space division switching operations have been demonstrated with such devices. Beam deflection has also been shown in an AlGaAs/GaAs

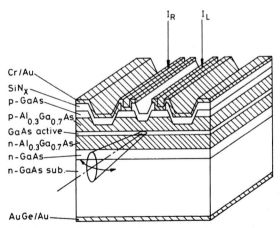

FIGURE 27.26 Structure of beam-scanning laser consisting of two parallel gain-guided waveguides. The beam angle is varied by the ratio of currents in two waveguides. *(After Itoh et al.[74])*

surface-emitting DFB laser in which the current-induced refractive index change is used in a second-order grating for vertical emission.[75]

Two-dimensional integration of optoelectronic devices is another important technique for realizing highly parallel data transmission and processing. Many demonstrations so far reported include vertical cavity surface-emitting lasers and various bistable devices such as self-electro-optic effect devices (SEEDs),[76] heterojunction phototransistor (HPT)/light-emitting diode integrated devices,[77] and *pnn* double heterostructure optoelectronic switches (DOES).[78] Figure 27.27 shows the structure of a vertical-to-surface transmission electrophotonic (VSTEP) device used for 1-kbit two-dimensional integration.[79] Built-in light-emitting structures provide an optical gain offering an important advantage in cascadability and a large fan-out for optical processor applications. Dynamic memory operation with 2.5-ns turnoff time and 20-μW holding power has been demonstrated. More recently, a VSTEP device having a built-in vertical cavity laser has been demonstrated.[80] Two-dimensional array integration has been reported for the integration level of 1 to 3 kbit for the SEED,[81] HPT/LED, and VSTEP.[80] Supported by the development in optoelectronic packaging for two-dimensional array devices, these devices will be important keys in future optical switching and computing systems.

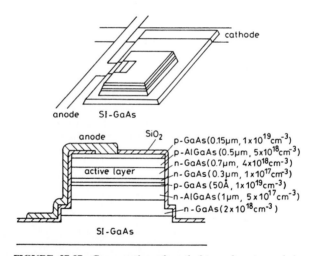

FIGURE 27.27 Cross section of vertical-to-surface transmission electrophotonic (VSTEP) device with AlGaAs/GaAs *pnpn* structure. *(After Kurihara et al.[79])*

27.5 SYSTEM APPLICATIONS

Because of the advantages of optoelectronic integrated circuits in performance, manufacturability, and functions, they are potentially applicable in many areas.[82] Optical telecommunication is the most straightforward application for various optoelectronic integrated circuits, particularly OEIC transmitters and receivers. Significant advances in both performance and circuit integration scale have recently been achieved. To represent the current status of OEICs, OEIC receiver sensitivity versus bit rate is shown in Fig. 27.28 for recent InP-based OEIC receivers. Early OEICs suffered from degraded performance, but this problem has nearly been solved by the recent introduction of low-capacitance photodetectors and high-speed transistors such as HEMTs, HBTs, and heteroepitaxial GaAs MESFETs as shown in Fig. 27.28. Flip-chip integrated receivers are already comparable to the best hybrid circuits. Further improving low-noise receiver circuit design and high-yield device processing techniques will produce practical OEICs for applications in local-area networks (LANs) and subscriber loops.[83] In this regard, the high-performance advantages of OEICs have already been demonstrated in transmission experiments in both short-[2] and long-wavelength regions, including over-50-km transmission at a 1.3-μm wavelength using GaAs/InP heteroepitaxial OEIC transmitter/receiver pairs operating at 1.2 Gb/s.[3]

The enhancement of signal processing functions by integration is significant in advanced telecommunication and switching system applications. Optical processing functions implemented by waveguide-based photonic integrated circuits are extremely important to ultra-high-speed coherent optical WDM systems. Laser/modulator light sources and coherent receivers as described in the foregoing sections have clearly shown their importance. The WDM scheme is particularly significant for handling large-capacity signals in switching and processing.[63] Electronic signal processing functions implemented by standard OEICs are extremely useful in many

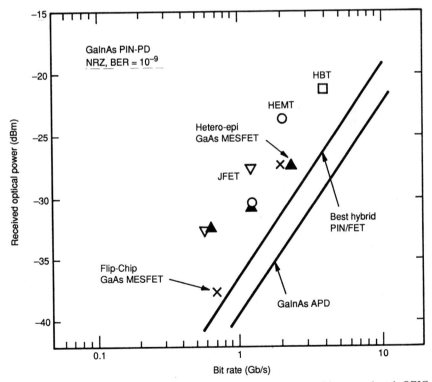

FIGURE 27.28 Plot of receiver sensitivity versus bit rate for reported long-wavelength OEIC receivers.[5]

existing systems because of easy electrical interfacing. Figure 27.29 diagrams a 4 × 4 optical switch incorporating GaAs-based, four-channel OEIC transmitter and receiver arrays coupled with a GaAs electronic switch circuit.[84] Optoelectronic packaging techniques play an important role in producing practical components. Low-loss, highly uniform optical coupling and high-frequency microstripline connections have been used in this fabrication, and full optical switching operation has been demonstrated at 560 Mb/s with a crosstalk less than −20 dB.

Optical interconnection will become increasingly important in future data processing systems, since it can eliminate electromagnetic noise, crosstalk, grounding problems, and capacitance-resistance product (CR)-limited delay time in existing electronic systems. OEIC transmitters and receivers can be basic components for optical interconnections within electronic switching and computing systems. GaAs-based four-channel array transmitters and receivers involving time-division serializer/deserializer and timing recovery circuits have been fabricated for application to a computer network operating at 1 Gb/s.[50] Optical interconnections will be important also at various levels of system implementation. Intrachip interconnections using waveguides formed over the chip surface will be useful, for example, in distributing clock signals over the chip.[85] Another possibility is in interwafer communication using parallel optical signal transfer in free space. A three-dimensional optically coupled common memory as shown in Fig. 27.30 has been proposed recently; it features fast optical data transfer in the vertical direction combined with conventional lateral electrical interconnections providing intelligent memory for real-time parallel processor systems.[86] Such intra- and interchip optical interconnections will be made possible by extensive development of non-lattice-matched heteroepitaxy, particularly III-V on Si technology.

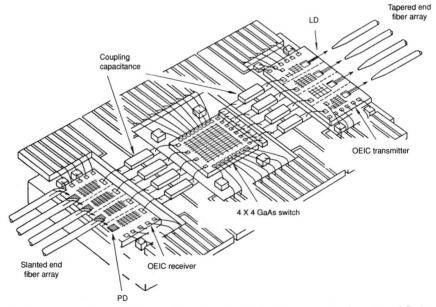

FIGURE 27.29   Schematic diagram of four-channel optical switch composed of four-channel GaAs-based OEIC receiver and transmitter arrays and a 4 × 4 GaAs-IC switch.[84]

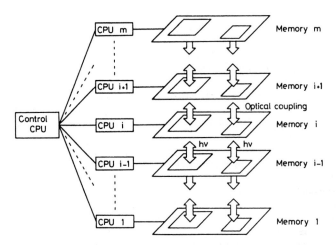

FIGURE 27.30   Schematic diagram showing multilayer structure for three-dimensional optically coupled common memory. Vertical interlayer data transfer is carried out through optical interconnections.   *(After Koyanagi et al.*[86]*)*

Integrated optical functional devices such as optical logic gates will play an important role in constructing optical switching and computing systems. Two-dimensional integration of these devices will be a prerequisite for using the parallelism of light. Several basic switching and computing experiments have already indicated the importance of such devices.[87] System design for fully utilizing such novel devices and packaging techniques for free-space optical coupling will be keys for making two-dimensionally integrated devices that can be implemented in practical systems. In this view, future device development should continue with a strong link to the system architecture.

27.6 SUMMARY

This chapter has described trends in the development of optoelectronic integrated circuits based on III-V semiconductor materials, and reviewed recent progress in different categories of integration having different target applications. The integration of different kinds of devices requires advances in fabrication techniques. Developments made so far have already demonstrated the advantage of integration in improving performance and functions of optical components. Further development of simple, high-yield fabrication and packaging techniques will provide the advantage of high manufacturability in optoelectronic integrated circuits, enabling widespread system applications. Looking into the future, the needs in system data rate and volume processing will keep increasing, and optical components' capability must expand from just simple data transmission toward optical processing of massive data. Optoelectronic integrated circuits are opening this possibility and continuous progress will lead to the construction of advanced optical processing systems.

27.7 REFERENCES

1. Yariv, A., *IEEE Trans. Electron Devices*, vol. ED-31, p. 1656, 1984.
2. Wada, O., T. Sakurai, and T. Nakagami, *IEEE J. Quantum Electron.*, vol. QE-22, p. 805, 1986.
3. Suzuki, A., K. Kasahara, and M. Shikada, *IEEE J. Lightwave Tech.*, vol. LT-10, p. 1479, 1987.
4. Dagenais, M., R. F. Leheny, H. Temkin, and P. Bhattacharya, *IEEE J. Lightwave Tech.*, vol. LT-8, p. 846, 1990.
5. Wada, O., *Proc. SPIE*, "Physical Concepts of Materials for Novel Optoelectronic Device Applications II: Device Physics and Applications," vol. 1362, p. 598, 1991.
6. Hayashi, I., *Tech. Dig.*, *Int. Conf. Integrated Optics and Optical Fiber Commun. (IOOC '83)*, Tokyo, p. 170. 1983.
7. Hunsperger, R. G., "Integrated Optics: Theory and Technology," Springer-Verlag, Berlin, 1982.
8. Koren, U., T. L. Koch, B. I. Miller, and A. Shahar, *Tech. Dig. Integrated and Guided Wave Optics Conf.*, paper MDD2, Houston, 1989.
9. Choi, H. K., G. W. Turner, T. H. Windhorn, and B. Y. Tsaur, *IEEE Electron Device Lett.*, vol. EDL-7, p. 500, 1986.
10. Tsang, W. T., "Quantum Confinement Heterostructure Semiconductor Lasers," Chap. 7 in *Semiconductors and Semimetals*, vol. 24, R. Dingle, ed., Academic, New York, 1987.
11. Wada, O., T. Sanada, M. Kuno, and T. Fujii, *Electron. Lett.*, vol. 21, p. 1025, 1985.
12. Uomi, K., T. Mishima, and N. Chinone, *Japan J. Appl. Phys.*, vol. 29, p. 88, 1990.
13. Furuya, A., et al., *Japan J. Appl. Phys.*, vol. 26, p. L134, 1987.
14. Wada, O., A. Furuya, and M. Makiuchi, *IEEE Photon. Technol. Lett.*, vol. 1, no. 16, 1989.
15. Kotaki, Y., and H. Ishikawa, *IEEE Proc.—J. Optoelectronics*, vol. 138, p. 171, 1991.
16. Koch, T. L., et al., *Electron. Lett.*, vol. 24, p. 1431, 1988.
17. Miura, S., H. Kuwatsuka, T. Mikawa, and O. Wada, *IEEE J. Lightwave Tech.*, vol. LT-5, p. 1371, 1987.
18. Sugeta, T., T. Urisu, S. Sakata, and Y. Mizushima, *Japan J. Appl. Phys.*, vol. 19, suppl. 19-1, 1980, p. 459.
19. Wada, O., et al., *IEEE J. Lightwave Tech.*, vol. LT-4, p. 1694, 1986.
20. Zeghbroeck, B. J. V., W. Patrick, J-M. Halbout, and P. Vettiger, *IEEE Electron Device Lett.*, vol. 9, p. 527, 1988.
21. Wada, O., et al., *Appl. Phys. Lett.*, vol. 54, p. 16, 1989.
22. Schumacher, H., H. Leblanc, J. Soole, and R. Bhat, *IEEE Electron Device Lett.*, vol. 36, p. 659, 1989.
23. Young, L., et al., *IEEE Photon. Technol. Lett.*, vol. 2, p. 56, 1990.
24. Yasuoka, N., et al., *Extended Abstracts 22d (1990 Int.) Conf. Sol. State Devices and Materials*, paper D-7-4, p. 637, Sendai, Japan, 1990.
25. Campbell, J. C., "Phototransistors for Lightwave Communication," pt. D, chap. 5 in *Semiconductors and Semimetals*, vol. 22, W. T. Tsang, ed., Academic, New York, 1985.
26. Umeda, T., Y. Cho, and A. Shibatomi, *Japan J. Appl. Phys.*, vol. 25, p. L801, 1986.
27. Chen, C. Y., et al., *Appl. Phys. Lett.*, vol. 44, p. 99, 1984.
28. Deri, R. J., et al., *Appl. Phys. Lett.*, vol. 55, p. 1495, 1989.

29. Wakao, K., K. Nakai, M. Kuno, and S. Yamakoshi, *IEEE J. Selected Areas Commun.*, vol. 6, p. 1199, 1988.

30. Leonberger, F. J., and J. F. Donnelly, Chap. 6 in *"Guided-Wave Optoelectronics,"* T. Tamir, ed., Springer-Verlag, Berlin, 1988.

31. Abe, M., et al., Chap. 4 in *Semiconductors and Semimetals*, vol. 24, R. Dingle, ed., Academic, New York, 1987.

32. Mishra, U. K., *IEEE Electron. Device Lett.*, vol. EDL-9, p. 647, 1988.

33. Chen, Y. K., et al., *IEEE Electron Device Lett.*, vol. EDL-10, p. 267, 1989.

34. Tsang, W. T., "Molecular Beam Epitaxy for III-V Compound Semiconductors," Chap. 2 in *Semiconductors and Semimetals*, vol. 22, pt. A, W. T. Tsang, ed., Academic, New York, 1985.

35. Razeghi, M., Chap. 5 in *Semiconductors and Semimetals*, vol. 22, pt. A, W. T. Tsang, ed., Academic, New York, 1985.

36. Tsang, W. T., *J. Crystal Growth*, vol. 105, p. 1, 1990.

37. Panish, M. B., et al., *J. Vac. Sci. Technol.*, vol. B3, p. 657, 1985.

38. Asakawa, K., and S. Sugata, *Japan J. Appl. Phys.*, vol. 22, p. L653, 1983.

39. Takamori, A., et al., *Japan J. Appl. Phys.*, vol. 26, p. L142, 1987.

40. Wale, M. J., C. Edge, F. A. Randle, and D. J. Pedder, *Tech. Dig., European Conf. Optical Commun. (ECOC '89)*, Gothenburg, Sweden, vol. 2, p. 368, ThA19-7, 1989.

41. Wada, O., et al., *IEEE J. Lightwave Tech.*, vol. LT-7, p. 186, 1989.

42. Wada, O., et al., *Electron. Lett.*, vol. 18, p. 189, 1982.

43. Kerney, J. K., et al., *Tech. Dig. GaAs IC Symp.*, New Orleans, p. 38, 1982.

44. Hamada, K., et al. *Extended Abstracts, 18th Conf. Solid State Devices and Materials*, p. 181, Tokyo, 1986.

45. Matsueda, H., et al., *Tech. Dig. Int. Symp. GaAs and Related Compounds*, Karuizawa, Japan, Inst. Phys. Conf. Ser. no. 79, p. 655, 1985.

46. Shibata, J., et al., *Appl. Phys. Lett.*, vol. 45, p. 191, 1984.

47. Suzuki, N., et al., *Electron. Lett.*, vol. 24, p. 467, 1988.

48. Lo, Y. H., et al., *IEEE Photon. Technol. Lett.*, vol. 2, p. 673, 1990.

49. Deppe, D. G., et al., *Appl. Phys. Lett.*, vol. 51, p. 637, 1987.

50. Crow, J. D., *Tech. Dig., Conf. Optical Fiber Commun. (OFC '89)*, p. 83, and *Tech. Dig., Int. Conf. Integrated Optics and Optical Commun. (IOCC '89)*, vol. 4, p. 86, Kobe, Japan, 1989.

51. Harder, C. S., et al., *IEEE Electron Device Lett.*, vol. EDL-9, p. 171, 1988.

52. Chen, C. Y., N. A. Olsson, C. W. Tu, and P. A. Garbinski, *Appl. Phys. Lett.*, vol. 46, p. 681, 1985.

53. Lee, W. S., D. A. H. Spear, P. J. G. Dawe, and S. W. Bland, *Electron. Lett.*, vol. 26, p. 1834, 1990.

54. Nobuhara, H., et al., *Electron. Lett.*, vol. 19, p. 1246, 1988.

55. Hayashi, H., *IEE Proc.—J. Optoelectronics*, vol. 138, p. 164, 1991.

56. Chandrasekhar, S., et al., *Electron. Lett.*, vol. 26, p. 1880, 1990.

57. Hong, W.-P., et al., *Electron. Lett.*, vol. 25, p. 1562, 1989.

58. Suzuki, A., et al., *Electron. Lett.*, vol. 23, p. 954, 1987.

59. Sussmann, R. S., R. M. Ash, A. J. Moseley, and R. C. Goodfellow, *Electron. Lett.*, vol. 21, p. 593, 1985.

60. Hamaguchi, H., et al., *Tech. Dig., Optoelectronics Conf. (OEC '89)*, p. 194, Tokyo, 1989.

61. Hamano, H., et al., *Electron. Lett.*, vol. 27, p. 1602, 1991.
62. Soda, H., et al., *Electron. Lett.*, vol. 26, no. 9, 1990.
63. Koch, T. L., and U. Koren, *IEEE J. Quant. Electron.*, vol. 27, no. 641, 1991.
64. Takeuchi, H., et al., *IEEE Photon. Technol. Lett.*, vol. 1, p. 398, 1989.
65. Deri, R. J., et al., *Appl. Phys. Lett.*, vol. 56, p. 1737, 1990.
66. Deri, R. J., et al., *IEEE Photon. Technol. Lett.*, vol. 2, p. 581, 1990.
67. Yasuoka, N., et al., *Tech. Dig. and Proc. 3d Int. Conf. InP and Related Materials*, IEEE LEOS, paper TH02, p. 580, Cardiff, 1991.
68. Chen, Y. K., et al., *Appl. Phys. Lett.*, vol. 58, p. 1253, 1991.
69. Kawaguchi, H., *Appl. Phys. Lett.*, vol. 45, p. 1264, 1984.
70. Odagawa, T., et al., *IEE Proc. J.*, vol. 138, p. 75, 1991.
71. Kondo, K., H. Nobuhara, S. Yamakoshi, and K. Wakao, *Tech. Dig. Int. Topical Meeting Photonic Switching*, 13D-9, Kobe, Japan, 1990.
72. Wada, O., and S. Yamakoshi, "Digital Optical Computing II," *Proc. SPIE*, vol. 1215, p. 28, 1990.
73. Mukai, S., *Opt. Quant. Electron.*, vol. 17, p. 431, 1985.
74. Itoh, H., et al., *IEE Proc. J.*, vol. 138, p. 113, 1991.
75. Yamashita, K., H. Nagata, Y. Kubota, and K. Tone, *Tech. Dig. Int. Topical Meeting Photonic Switching*, p. 125, Kobe, Japan, 1990.
76. Streibel, N., et al., *Proc. IEEE*, vol. 77, p. 1954, 1989.
77. Matsuda, K., K. Takimoto, D. H. Lee, and J. Shibata, *IEEE Trans. Electron Devices*, vol. 37, p. 1630, 1990.
78. Taylor, G. W., R. S. Mand, J. G. Simmons, and A. Y. Cho, *Appl. Phys. Lett.*, vol. 49, p. 1406, 1986.
79. Kurihara, K., *IEE Proc.—J. Optoelectronics*, vol. 138, p. 161, 1991.
80. Numai, T., *Appl. Phys. Lett.*, vol. 58, p. 1250, 1991.
81. Lentine, A. L., et al., *IEEE Photon. Technol. Lett.*, vol. 2, p. 51, 1990.
82. Wada, O., *Int. J. High Speed Electron.*, vol. 1, p. 47, 1990.
83. Leheny, R. F., *IEEE Circuits and Devices Mag.*, vol. 5, p. 38, 1989.
84. Iwama, T., et al., *IEEE J. Lightwave Tech.*, vol. LT-6, p. 772, 1988.
85. Goodman, J. W., F. J. Leonberger, S. Kung, and R. A. Athale, *Proc. IEEE*, vol. 72, p. 850, 1984.
86. Koyanagi, M., H. Takata, H. Mori, and J. Iba, *IEEE J. Solid-State Circuits*, vol. 25, p. 109, 1990.
87. Athale, R. A., ed., "Digital Optical Computing," *Critical Reviews of Optical Science and Technology*, vol. CR35, SPIE Press, Bellingham, Wash., 1990.
88. Wada, O., *Opt. Quantum Electron.*, vol. 20, p. 441, 1988.
89. Matsuda, M., Y. Kotaki, H. Ishikawa and O. Wada, *Tech. Dig. and Proc. 3d Int. Conf. InP and Related Materials*, IEEE LEOS, paper TuF4, p. 256, Cardiff, 1991.

INDEX